A Problem Solving Approach to Mathematics

for Elementary School Teachers

TENTH EDITION

RICK BILLSTEIN
University of Montana

SHLOMO LIBESKIND
University of Oregon

JOHNNY W. LOTT
University of Mississippi

Addison-Wesley

Boston • New York • San Francisco

London • Toronto • Sydney • Tokyo • Singapore • Madrid

Mexico City • Munich • Paris • Cape Town • Hong Kong • Montreal

Executive Editor   Anne Kelly
Acquisitions Editor   Marnie Greenhut
Executive Project Manager   Christine O'Brien
Senior Project Editor   Joanne Dill
Assistant Editor   Leah Goldberg
Editorial Assistant   Leah Driska
Senior Managing Editor   Karen Wernholm
Senior Production Supervisor   Kathleen A. Manley
Senior Designer   Barbara T. Atkinson
Executive Media Manager   Peter Silvia
Software Development   Eileen Moore (Math XL), Marty Wright (TestGen)
Executive Marketing Manager   Becky Anderson
Marketing Assistant   Katherine Minton
Senior Author Support/Technology Specialist   Joe Vetere
Rights and Permissions Advisor   Shannon Barbe
Manufacturing Manager   Evelyn Beaton
Cover and Text Design   Susan Raymond
Production Coordination, Composition, and Illustrations   Pre-Press PMG

Cover image: From **CRITICAL THINKING ACTIVITIES IN PATTERNS, IMAGERY, LOGIC, GRADES 4–6,** © 1988 by Pearson Education, Inc. or its affiliate(s). Used by permission. All rights reserved.

For permission to use copyrighted material, grateful acknowledgment is made to the copyright holders on page 1008, which is hereby made part of this copyright page.

Many of the designations used by manufacturers and sellers to distinguish their products are claimed as trademarks. Where those designations appear in this book, and Pearson was aware of a trademark claim, the designations have been printed in initial caps or all caps.

Library of Congress Cataloging-in-Publication Data

Billstein, Rick.
   Problem solving approach to mathematics for elementary school teachers /
Rick Billstein, Shlomo Libeskind, [and] Johnny W. Lott. — 10th ed.
         p. cm.
   Includes bibliographical references and index.
   ISBN 0-321-57054-5 (recover : alk. paper) — ISBN 0-321-57055-3 (math for elementary teachers :
alk. paper)
   1. Mathematics—Study and teaching (Elementary)   2. Problem solving—Study and teaching
(Elementary)   I. Libeskind, Shlomo.   II: Lott, Johnny W., 1944–   III. Title.
   QA135.6.B55 2009
   372.7—dc22

                                                                                    2008039186

1 2 3 4 5 6 7 8 9 10–QWT–11 10 09 08

Addison-Wesley
is an imprint of

ISBN 10: 0-321-57055-3
ISBN 13: 978-0-321-57055-0

www.pearsonhighered.com

To all the students and teachers who have used
this book from its inception—RWB, SL, and JWL

To Jane for her patience through 10 editions—RB

To the memory of my beloved grandfather Itzhak Białowąs and my beloved
uncle Marian Białowąs—SL

To the next generation of mathematics students including Hamilton Grey Lott,
William Thomas Falk, and Grant Warren Falk—JWL

Contents

* Optional section

* Optional section

~ Section can be found online at www.pearsonhighered.com/Billstein10einfo

Preface

The tenth edition of *A Problem Solving Approach to Mathematics for Elementary School Teachers* is designed to meet the education needs of prospective elementary teachers who will be the future's high-quality teachers. This edition continues to be heavily concept- and skill-based, with a new emphasis on active and collaborative learning. The content has been revised and updated to better prepare students for when they will be instructors in their own classrooms.

STANDARDS OF THE NCTM
- **Principles and Standards** We focus on the National Council of Teachers of Mathematics (NCTM) publication, *Principles and Standards of School Mathematics* (2000) (hereafter referred to as *Principles and Standards*).

New!
- **Curriculum Focal Points** The National Council of Teachers of Mathematics (2006) published *Curriculum Focal Points for Pre-kindergarten through Grade 8 Mathematics* that describes the essential mathematical concepts and skills to which the mathematics in each chapter relates. *Focal Points* are referred to and set off throughout the text.

The complete text of both the NCTM *Principles and Standards* and *Curriculum Focal Points* can be found online at www.nctm.org.

OUR GOALS
- To present appropriate mathematics in an intellectually honest and mathematically correct manner.
- To use problem solving as an integral part of mathematics.
- To approach mathematics in a sequence that instills confidence AND challenges students.
- To provide opportunities for alternate forms of teaching and learning.
- To provide communication problems to develop writing skills and allow students to practice explanation.
- To encourage the integration of technology tools.
- To provide core mathematics for prospective elementary and middle school teachers in a way so that they are challenged to determine why mathematics is done as it is.
- To provide core mathematics that allows instructors to use methods integrated with content.
- To assist prospective teachers with connecting mathematics, its ideas, and its applications.

The tenth edition allows instructors a variety of approaches to teaching, encourages discussion and collaboration among students and with their instructors, and allows for the integration of projects into the curriculum. Most importantly, it promotes discovery and active learning for the students and the teacher.

NEW TO THIS EDITION
- Because algebraic thinking is so important at all levels, we have created a new separate chapter on the subject, Chapter 4 "Algebraic Thinking," early in the text while continuing to integrate algebra throughout the text.
- A new, separate chapter, Chapter 8 "Proportional Reasoning, Percents, and Applications," has been added to more fully address the needs of future middle school teachers.
- The assessments are now organized in a more logical and easily accessible manner. Assessment A problems have answers in the text so that students can check their work. Assessment B contains problems similar to those in Assessment A but no answers are given in the student text. By creating parallel exercise sets, we increased the number of problems, giving instructors more choices.

Mathematical Connections problems have been separated out because those tend to be more open-ended and allow students and instructors to work either alone or in groups to find possible solutions. They are divided into the following categories: *Communication*, *Open-Ended*, *Cooperative Learning*, *Questions from the Classroom*, and *Review*. The Annotated Instructor's Edition of the text contains answers or solutions to Assessments A, B, and Mathematical Connections. The problem sets also include sample questions from both the TIMMS and NAEP exams so that future teachers can examine the types of questions that students are being asked on national and international exams.
- Updated data analysis and statistical thinking—this material has been expanded to include more content on populations, sampling, and surveys.

CONTENT HIGHLIGHTS

Chapter 1 An Introduction to Problem Solving
The reorganization of this chapter puts mathematics and problem solving first, followed by an expanded section on exploration with patterns. New problems and new student pages are other additions, as is a new section on Fibonacci sequences. The final section on reasoning and logic is included for those who wish to pursue these topics in the course.

Chapter 2 Numeration Systems and Sets
This chapter has been reorganized and shortened. The development of numeration systems is now the first section of this chapter because of the historical development of the systems, which were in place long before more formal set concepts were developed. The chapter includes all of the more traditional set concepts later.

Chapter 3 Whole Numbers and Their Operations
This chapter explores whole numbers and operations on whole numbers. Various algorithms are investigated and explained in detail. Mental mathematics and estimation with whole numbers are prominently featured.

Chapter 4 Algebraic Thinking
In response to greater emphasis on learning and teaching algebra throughout the elementary school curriculum, a new chapter on algebraic thinking has been incorporated. Only whole numbers are used, but in each subsequent chapter, algebraic thinking is reinforced when integers, rational numbers, and, finally, real numbers are introduced. Algebraic thinking is further reinforced in the chapter on probability and statistics as well as in the geometry chapters.

Chapter 5 Integers and Number Theory
This chapter deals with integers and operations on integers. New models for operations and algorithms on integers are introduced with explanations. Divisibility and prime numbers are discussed along with explanations on why the divisibility rules work. Greatest common divisors and least common multiples are introduced. An optional section on clock and modular arithmetic is included for those who would like to examine how a different number system works.

Chapter 6 Rational Numbers as Fractions
New examples in this chapter emphasize algebraic skills by simplifying algebraic expressions and solving equations and word problems. The concept of division is enhanced with more thorough explanations and examples. Functions are revisited with the domain of rational numbers.

Chapter 7 Decimals and Real Numbers
This chapter has been shortened with the addition of the new Chapter 8. Many more student pages have been added and a new optional section, "Using Real Numbers in Equations," adds more algebra emphasis to this reorganized chapter.

Chapter 8 Proportional Reasoning, Percents, and Applications
Because proportional reasoning and percents are so important in the middle school, an entire chapter is devoted to the subject. The chapter includes a discussion of how the relationship between two ratios is multiplicative instead of additive and why this is important. Work with percents has been expanded

to include percent bars and estimations involving percents. An optional section on computing interest is included to show an application of percents.

Chapter 9 Probability

The Preliminary Problem, involving the artwork of François Morellet, gives the signal that probability is used in the real world and in the world that students experience. Student pages are added to show how the concepts appear in the grades; more drawings, cartoons, and diagrams illustrate concepts.

Chapter 10 Data Analysis/Statistics: An Introduction

In both the ninth and tenth editions, emphasis has been paid to *Guidelines for Assessment and Instruction in Statistics Education (GAISE) Report: A Pre-K–12 Curriculum Framework* by the American Statistical Association (2005). One section, Designing Experiments and Collecting Data based on this *Statistics Framework*, has been developed and may be accessed online. Many new problems have been added, and algebraic notions are used in the development of the chapter.

Chapter 11 Introductory Geometry

The various concepts of geometry are more thoroughly explained and there is more detailed treatment of interior and exterior angles in convex polygons. Algebraic thinking is prominent throughout the chapter.

Chapter 12 Constructions, Congruence, and Similarity

The discussion on congruence and noncongruence of triangles has been expanded to include the ambiguous SSA case, and congruence of quadrilaterals has been added. The discussion of systems of linear equations has been expanded to include an algebraic explanation of when a system of two equations with two unknowns has no solution and when it has infinitely many solutions.

Chapter 13 Concepts of Measurement

This chapter includes work with both the English system and the metric system, along with conversion among and between systems. Linear, area, volume, mass, and temperature measurements are included. Formulas for computing measurements are derived to show where they come from. The Pythagorean Theorem and distance formulas are developed along with a new section on the equation of a circle.

Chapter 14 Motion Geometry and Tessellations

Although most of the features of the ninth edition remain in the new edition of this chapter, there are many more drawings and more ties to student pages than in the past. We have tried to build on what prospective teachers need to know, and that is more than their future students may need to know. This chapter takes a look at the fun and interest that can be had from motion geometry.

Calculator Usage

As stated in *Principles and Standards*, coverage of calculators is necessary and timely. The use of the graphing calculator is presented, where relevant, in the Technology Corners. In addition, problems involving the use of both scientific/fraction and graphing calculators appear in the problem sets.

FEATURES

We continue to incorporate various study aids and features that facilitate learning.

Professional Development

New + Improved!

- Updated *School Book Pages* are included to show how the mathematics is actually introduced to the K–8 student and are referenced throughout the text. Students are asked to complete many of the activities on the student pages so they can see what is expected in elementary schools.
- *Research Notes*, in the margin, highlight current research projects in mathematics and mathematics education as it relates to the content.
- *Historical Notes* add context and humanize the mathematics.

- Relevant quotations from the *NCTM Principles and Standards* and *NCTM Focal Points* are incorporated throughout the text.
- *Questions from the Classroom* present questions as they might be posed by K–8 students. A substantial number of new questions have been added. They now appear at the end of each section as part of Mathematical Connections.

Active Learning

New!
- A new *e-Manipulatives CD* is included with every book and has been added to help students investigate, explore, and practice new concepts and is referenced throughout the book.

New!
- References to the *Activity Manual* are found throughout the Annotated Instructor's Edition as a guide to more fully integrate activities into the course.
- *Brain Teasers* provide a different avenue for problem solving. They are solved in the Annotated Instructor's Edition and may be assigned or used by the teacher to challenge students.
- *Laboratory Activities* are integrated throughout the book to provide hands-on learning exercises. Answers are in the Annotated Instructor's Edition. A separate activities book is also available as a supplement.
- *Now Try This* activities appear throughout each chapter, and are intended to help students become actively involved in their learning, to facilitate the development and improvement of their critical thinking and problem-solving skills, and to stimulate both in-class and out-of-class discussion. Answers are in both the Annotated Instructor's Edition and student text.
- *Technology Corners* include use of spreadsheets, both graphing and scientific calculators, The Geometer's Sketchpad, and computer activities. Answers are in the Annotated Instructor's Edition.

Pedagogical Tools

- *Definitions, Properties, and Theorems* are set off in the text for quick review.
- *Problem-Solving Strategies* are highlighted in italics, and Problem Solving boxes help students put these strategies to work.
- *Cartoons* teach or emphasize important material, and add levity.
- *Chapter Outlines* at the end of each chapter help students review the chapter.
- *Chapter Reviews* at the end of each chapter allow students to effectively test themselves in preparation for an in-class exam.
- *Selected Bibliographies* at the end of each chapter have been updated and revised.

Assessment

New + Improved!
- *Problem Sets:* These have been thoroughly revised and reorganized into Assessments A, B, and Mathematical Connections. Assessment A problems have answers in the text so that students can check their work. Assessment B contains problems similar to those in Assessment A, but answers are not given in the student text. Mathematical Connections is divided into the following categories of problems: Communication, Open-Ended, Cooperative Learning, Questions from the Classroom, and Review. Answers to all Mathematical Connections questions are in the Annotated Instructor's Edition while the student text includes just odd answers.
- Relevant and realistic problems are more accessible and appealing to students of diverse backgrounds.

STUDENT AND INSTRUCTOR RESOURCES

For the Student

E-Manipulatives CD (bound in book) *New!*
- ISBN-13: 978-0-321-56931-8; ISBN-10: 0-321-56931-8
- 21 flash-based manipulatives investigate, explore, and practice new concepts and solve specific problems that will help students develop a conceptual understanding of key ideas.
- Helps explore the way elementary students would use manipulatives in the classroom.
- References to these e-manipulatives are in the margin of the main text.

Activities Manual–*Mathematics Activities for Elementary School Teachers: A Problem Solving Approach, Tenth Edition,* by Daniel Dolan, *Wesleyan University*; Jim Williamson, *University of Montana*; and Mari Muri, *Connecticut Department of Education*
- ISBN-13: 978-0-321-57568-5; ISBN-10: 0-321-57568-7
- Activities help develop, reinforce, and/or apply mathematical concepts. They can also be adapted for use with elementary students at a later time.

- Revised to closely match the text by topics and skills within each chapter.
- References to these activities are in the margin of the AIE.

Student's Solutions Manual, by Louis L. Levy
- ISBN-13: 978-0-321-56927-1; ISBN-10: 0-321-56927-X
- Provides detailed, worked-out solutions to all of the newly organized **Assessment A** problems and **Chapter Review** exercises.

Connecting Mathematics for Elementary Teachers, by David Feikes, *Purdue University North Central*, Keith Schwingendorf, *Purdue University North Central*, and Jeff Gregg, *Purdue University Calumet* *New!*
- ISBN-13: 978-0-321-54266-3; ISBN-10: 0-321-54266-5
- Understanding mathematics and how children think about mathematics can help you become a better teacher. Provides general descriptions of children's learning and shows how children approach mathematics differently than adults.

Technology Manual: Using Spreadsheets, Graphing Calculators, and a Geometry Drawing Utility by Rick Billstein, *University of Montana*, Shlomo Libeskind, *University of Oregon*, and Johnny W. Lott, *University of Mississippi*
- ISBN-13: 978-0-321-62929-6; ISBN-10: 0-321-62929-9
- Provides tutorials and exercises on technologies often used in this course
- Materials can be found online at www.pearsonhighered.com/Billstein10einfo.
- (Available online within MyMathLab or from the Instructor Resource Center at www.pearsonhighered.com/irc.)

When Will I Ever Teach This? by Sharon E. Taylor, *Georgia Southern University* and Susie Lanier, *Georgia Southern University*
- ISBN-13: 978-0-321-23717-0; ISBN-10: 0-321-23717-X
- The best way to demonstrate to students the need to learn certain topics is to show pages from a real K–8 textbook. This allows students to see when and where a topic occurs in the curriculum and also to see how it is presented in a text.

Video Lectures on DVD with Optional Subtitles
- ISBN-13: 978-0-321-56928-8; ISBN-10: 0-321-56928-8
- Complete set of affordable and portable digitized videos for student use at home or on campus. Ideal for distance learning and supplemental instruction. They include optional English subtitles.
- Includes examples and exercises from the text and supports visualization and problem solving.

For the Instructor

Annotated Instructor's Edition *New!*
- ISBN-13: 978-0-321-57179-3; ISBN-10: 0-321-57179-7
- This special edition includes answers to the text exercises on the page where they occur and includes answers to the Preliminary Problems, Now Try This activities, Mathematical Connections questions, Laboratory Activities, Technology Corners, Brain Teasers, and Questions from the Classroom.

Instructor's Solutions Manual, by Louis L. Levy
- ISBN-13: 978-0-321-56924-0; ISBN-10: 0-321-56924-5
- Provides detailed, worked-out solutions to all of the newly organized **Assessment A** & **B** problems and **Chapter Review** exercises.

Printed Test Bank, by Mark Oursland, *Central Washington University*
- Comprehensive worksheets contain two forms of chapter assessments with answers for each.
- (Available online within MyMathLab or from the Instructor Resource Center at www.pearsonhighered.com/irc.)

Insider's Guide *New!*
- ISBN-13: 978-0-321-57331-5; ISBN-10: 0-321-57331-5
- Includes resources to assist faculty with course preparation and classroom management.
- Provides helpful teaching tips correlated to the sections of the text, as well as general teaching advice and tips on using manipulatives.

Instructor's Guide to *Mathematics Activities for Elementary School Teachers: A Problem Solving Approach, Tenth Edition*, by Daniel Dolan, *Wesleyan University*; Jim Williamson, *University of Montana*; and Mari Muri, *Connecticut Department of Education*
- Contains answers for all activities, as well as additional teaching suggestions for some activities.
- (Available online within MyMathLab or from the Instructor Resource Center at www.pearsonhighered.com/irc.)

Instructor's Guide to *Connecting Mathematics for* *New!* *Elementary Teachers*, by David Feikes, *Purdue University North Central*; Keith Schwingendorf, *Purdue University North Central*; Jeff Gregg, *Purdue University Calumet*; and Marcela Perlwitz, *Ohio State University, Mansfield*.
- Contains correlations to all of Pearson's math for elementary teachers textbooks, as well as tips and teaching suggestions on how to incorporate the book into your course and syllabus.
- (Available online within MyMathLab or from the Instructor Resource Center at www.pearsonhighered.com/irc.)

PowerPoint® Lecture Slides
- Provides section-by-section coverage of key topics and concepts.
- (Available online within MyMathLab or from the Instructor Resource Center at www.pearsonhighered.com/irc.)

TestGen®
- Enables instructors to build, edit, print, and administer tests using a computerized bank of questions developed to cover all the objectives of the text:
- Algorithmically based, allowing instructors to create multiple but equivalent versions of the same question or test with the click of a button.
- Tests can be printed or administered online.

- Instructors can also modify test bank questions or add new questions.
- (The software is available for download from Pearson Education's online catalog: www.pearsonhighered.com/testgen.)

Instructor Resource Center
All instructor resources can be downloaded from www.pearsonhighered.com/irc. This is a password-protected site that requires instructors to set up an account or, alternatively, instructor resources can be ordered from your Pearson Higher Education sales representative. Instructors may use their instructor log-in from MyMathLab or MyStatLab to access the Instructor Resource Center.

Pearson Math Adjunct Support Center
The Pearson Math Adjunct Support Center (http://www.pearsontutorservices.com/math-adjunct.html) is staffed by qualified instructors with more than 50 years of combined experience at both the community college and university levels. Assistance is provided for faculty in the following areas:
- Suggested syllabus consultation
- Tips on using materials packed with your book
- Book-specific content assistance
- Teaching suggestions, including advice on classroom strategies

Online Learning

MathXL® Online Course (Access code required.)
MathXL® is a powerful online homework, tutorial, and assessment system that accompanies Pearson Education's textbooks in mathematics or statistics.
- With MathXL, instructors can create, edit, and assign online homework and tests using algorithmically generated exercises correlated at the objective level to the textbook. They can also create and assign their own online exercises and import TestGen tests for added flexibility.
- All student work is tracked in MathXL's online gradebook. Students can take chapter tests in MathXL and receive personalized study plans based on their test results.
- The study plan diagnoses weaknesses and links students directly to tutorial exercises for the objectives they need to study and retest. Students can also access supplemental animations and video clips directly from selected exercises.
- MathXL is available to qualified adopters. For more information, visit our website at www.mathxl.com, or contact your sales representative.

MyMathLab® Online Course (Access code required.)
MyMathLab is a text-specific, easily customizable online course that integrates interactive multimedia instruction with textbook content. MyMathLab gives you the tools you need to deliver all or a portion of your course online, whether your students are in a lab setting or working from home.
- **Interactive Tutorial Exercises:** A comprehensive set of exercises—correlated to your textbook at the objective level—are algorithmically generated for unlimited practice and mastery. Most exercises are free-response and provide guided solutions, sample problems, and learning aids for extra help at point-of-use.
- **Personalized Study Plan:** When students complete a test or quiz in MyMathLab, the program generates a personalized study plan for each student that indicates which topics have been mastered and links students directly to tutorial exercises for topics they need to study and retest
- **Multimedia Learning Aids:** Students can use online learning aids, such as video lectures, animations, and a complete multimedia textbook, to help them independently improve their understanding and performance.
- **Assessment Manager:** An easy-to-use assessment manager lets instructors create online homework, quizzes, and tests that are automatically graded and correlated directly to your textbook. Assignments can be created using a mix of questions from the MyMathLab exercise bank, instructor-created custom exercises, and/or TestGen test items.
- **Gradebook:** Designed specifically for mathematics and statistics, the MyMathLab gradebook automatically tracks students' results and gives you control over how to calculate final grades. You can also add offline (paper-and-pencil) grades to the gradebook.
- **Math Exercise Builder:** You can use the MathXL Exercise Builder to create static and algorithmic exercises for your online assignments. A library of sample exercises provides an easy starting point for creating questions, and you can also create questions from scratch.
- **Pearson Tutor Center** (www.pearsontutorservices.com): Access is automatically included with MyMathLab. The Tutor Center is staffed by qualified mathematics instructors who provide textbook-specific tutoring for students via toll-free phone, fax, email, and interactive Web sessions.

MyMathLab is powered by CourseCompass™, Pearson Education's online teaching and learning environment, and by MathXL®, our online homework, tutorial, and assessment system. MyMathLab is available to qualified adopters. **For more information**, visit our website at www.mymathlab.com or contact your Pearson sales representative.

Acknowledgments

For past editions of this book, many noted and illustrious mathematics educators and mathematicians have served as reviewers. To honor the work of the past as well as to honor the reviewers of this edition, we list all but place asterisks by this edition's reviewers. We would like to thank Jerrold Grossman for his contributions on accuracy checking this text.

Leon J. Ablon
Paul Ache
G.L. Alexanderson
Haldon Anderson
Bernadette Antkoviak
Richard Avery
Sue H. Baker
Jane Barnard
Joann Becker
Cindy Bernlohr
James Bierden
Jackie Blagg
Jim Boone
Sue Boren
Barbara Britton
Beverly R. Broomell
*Anne Brown
Jane Buerger
Maurice Burke
David Bush
Laura Cameron
Louis J. Chatterley
Phyllis Chinn
Donald J. Dessart
Ronald Dettmers
Jackie Dewar
*Nicole Duvernoy
Amy Edwards
Lauri Edwards
Margaret Ehringer
*Rita Eisele
Albert Filano
Marjorie Fitting
Michael Flom
Martha Gady
*Edward A. Gallo
Dwight Galster
Sandy Geiger
Glenadine Gibb
Don Gilmore

Diane Ginsbach
Elizabeth Gray
*Jerrold Grossman
Alice Guckin
Jennifer Hegeman
Joan Henn
Boyd Henry
Linda Hintzman
Alan Hoffer
E. John Hornsby, Jr.
*Patricia A. Jaberg
Judith E. Jacobs
Donald James
Thomas R. Jay
*Jeff Johannes
Jerry Johnson
Wilburn C. Jones
Robert Kalin
Sarah Kennedy
Steven D. Kerr
Leland Knauf
Margret F. Kothmann
Kathryn E. Lenz
Hester Lewellen
Ralph A. Liguori
*Richard Little
*Susan B. Lloyd
Don Loftsgaarden
Sharon Louvier
Stanley Lukawecki
*Lou Ann Martin
Judith Merlau
Barbara Moses
Cynthia Naples
Charles Nelson
Glenn Nelson
Kathy Nickell
*Bethany Noblitt
Dale Oliver
Mark Oursland

Linda Padilla
Dennis Parker
Clyde Paul
Keith Peck
Barbara Pence
Glen L. Pfeifer
Debra Pharo
Jack Porter
Edward Rathnell
Sandra Rucker
Jennifer Rutherford
Helen R. Santiz
Sherry Scarborough
Jane Schielack
Barbara Shabell
M. Geralda Shaefer
Nancy Shell
Wade H. Sherard
Gwen Shufelt
Julie Sliva
Ron Smit
Joe K. Smith
William Sparks
Virginia Strawderman
Mary M. Sullivan
Viji Sundar
Sharon Taylor
Jo Temple
C. Ralph Verno
Hubert Voltz
John Wagner
Edward Wallace
Virginia Warfield
Lettie Watford
Mark F. Weiner
Grayson Wheatley
Jim Williamson
Ken Yoder
Jerry L. Young
Deborah Zopf

CHAPTER 1

An Introduction to Problem Solving

Preliminary Problem

There are three bowls of fruit sitting on a shelf so high that you can't see into any of them. One bowl contains all apples, one bowl contains all oranges, and one bowl contains apples and oranges. Each bowl is visibly labeled with one of the labels: APPLES, ORANGES, or APPLES AND ORANGES. However, each bowl is incorrectly labeled. Your task is to select one bowl and reach in and grab one piece of fruit. Having done this and using the information above can you label each bowl correctly? Explain your answer.

Problem solving has long been recognized as one of the hallmarks of mathematics. What does *problem solving* mean? George Pólya (1887–1985), one of the great mathematicians and teachers of the twentieth century, pointed out that "solving a problem means finding a way out of difficulty, a way around an obstacle, attaining an aim which was not immediately attainable." (Pólya 1981, p. ix)

In *Principles and Standards of School Mathematics PSSM*, (NCTM 2000), we find the following:

> Problem solving means engaging in a task for which the solution method is not known in advance. In order to find a solution, students must draw on their knowledge, and through this process, they will often develop new mathematical understandings. Solving problems is not only a goal of learning mathematics but also a major means of doing so. Students should have frequent opportunities to formulate, grapple with, and solve complex problems that require a significant amount of effort and should then be encouraged to reflect on their thinking. (p. 52)

Further, we find that

> Instructional programs from pre-kindergarten through grade 12 should enable all students to
> • build new mathematical knowledge through problem solving;
> • solve problems that arise in mathematics and in other contexts;
> • apply and adapt a variety of appropriate strategies to solve problems;
> • monitor and reflect on the process of mathematical problem solving. (p. 52)

Students learn mathematics as a result of solving problems. Exercises that are routine practice for skill building serve a purpose in learning mathematics, but problem solving must be a focus of school mathematics. As pointed out in the Research Note, a reasonable amount of tension and discomfort improves problem-solving performance. Mathematical experience often determines whether situations are *problems* or *exercises*.

 **Research Note**

A reasonable amount of tension and discomfort improves students' problem-solving performance. The motivation is the release of tension after the problem has been solved. If the tension is not present, the problem is either an *exercise* or the students are "generally unwilling to attack the problem in a serious way" (Bloom and Broder 1950; McLeod 1985). ◆

Worthwhile, interesting problems, not just routine word problems, must be a part of elementary students' mathematical experience. To engage students in worthwhile tasks, problems should be introduced in a familiar context, as seen in the cartoon below.

Good mathematical problem solving occurs when all of the following are present:

1. Students are presented with a situation that they understand but do not know how to proceed directly to a solution.
2. Students are interested in finding the solution and attempt to do so.
3. Students are required to use mathematical ideas to solve the problem.

In this text, you will have many opportunities to solve problems. Each chapter opens with a problem that can be solved by using the concepts developed in the chapter. We give a hint for the solution to the problem at the end of each chapter. Throughout the text, numerous problems are solved using a four-step process and others solved using other formats.

Research Note

Students who are involved in justifying their solutions to other students, especially if there is a disagreement, will gain better mathematical understanding. Discussions of differing points of view are a valuable part of the learning experience. Mathematical language is learned in this way, as is the appreciation of the need for precision in the language (Hatano and Ingaki 1991).

As the Research Note indicates, working with other students to solve problems can enhance problem-solving ability and communication skills. We encourage *cooperative learning* and working in groups whenever possible. To encourage group work and help identify when cooperative learning could be useful, we identify activities that involve tasks where it might be helpful to have several people gathering data, or problems where group discussions might lead to strategies for solving the problem.

1-1 Mathematics and Problem Solving

If problems are approached in only one way, a mind-set may be formed. For example, consider the following: Spell the word *spot* three times out loud. "S-P-O-T! S-P-O-T! S-P-O-T!" Now answer the question "What do you do when you come to a green light?" Write an answer. If you answered "Stop," you may be guilty of having formed a mind-set. You do not stop at a *green* light.

Consider the following problem: "A shepherd had 36 sheep. All but 10 died. How many lived?" Did you answer "10"? If you did, you are catching on and are ready to try some problems. If you did not answer "10," then you did not understand the question. *Understanding the problem* is the first step in the four-step problem-solving process developed by George Pólya. Using the four-step process does not guarantee a solution to a problem, but it does provide a systematic means of attacking problems.

Historical Note

George Pólya (1887–1985) was born in Hungary and received his Ph.D. from the University of Budapest. He moved to the United States in 1940, and after a brief stay at Brown University he joined the faculty at Stanford University. In addition to being a preeminent mathematician, he focused on the vital importance of mathematics education. At Stanford, he published 10 books, including *How To Solve It* (1945), which has been translated into 23 languages.

Four-Step Problem-Solving Process

1. **Understanding the problem**
   a. Can you state the problem in your own words?
   b. What are you trying to find or do?
   c. What are the unknowns?
   d. What information do you obtain from the problem?
   e. What information, if any, is missing or not needed?

2. **Devising a plan**
   The following list of strategies, although not exhaustive, is very useful:
   a. Look for a pattern.
   b. Examine related problems and determine if the same technique applied to them can be applied to the current problem.
   c. Examine a simpler or special case of the problem to gain insight into the solution of the original problem.
   d. Make a table or list.
   e. Make a diagram.
   f. Write an equation.
   g. Use guess and check.
   h. Work backward.
   i. Identify a subgoal.
   j. Use indirect reasoning.
   k. Use direct reasoning.

3. **Carrying out the plan**
   a. Implement the strategy or strategies in step 2 and perform any necessary actions or computations.
   b. Check each step of the plan as you proceed. This may be intuitive checking or a formal proof of each step.
   c. Keep an accurate record of your work.

4. **Looking back**
   a. Check the results in the original problem. (In some cases, this will require a proof.)
   b. Interpret the solution in terms of the original problem. Does your answer make sense? Is it reasonable? Does it answer the question that was asked?
   c. Determine whether there is another method of finding the solution.
   d. If possible, determine other related or more general problems for which the techniques will work.

What role should Pólya's problem-solving process play in the teaching of elementary mathematics? This question is answered in *Principles and Standards* in the following way:

An obvious question is, How should these strategies be taught? Should they receive explicit attention, and how should they be integrated with the mathematics curriculum? As with any other component of the mathematical tool kit, strategies must receive instructional attention if students are expected to learn them. In the lower grades, teachers can help children express, categorize, and compare their strategies. Opportunities to use strategies must be embedded naturally in the curriculum across the content areas. By the time students reach the middle grades, they should be skilled at recognizing when various strategies are appropriate to use and should be capable of deciding when and how to use them. (p. 54)

♦ *Research Note*

Problem-solving ability develops slowly over time, perhaps because the many understandings and skills needed for problem solving develop at different rates. A key element in developing problem-solving skills is multiple, continuous experience in solving problems with different contexts and at different levels of ability (Kantowski 1981).   ♦

Strategies for Problem Solving

We next provide a variety of problems with different contexts so that you may gain experience in problem solving, as mentioned in the Research Note. Frequently, a variety of problem-solving strategies is necessary to solve these and other problems.

Strategies are tools that might be used to discover or construct the means to achieve a goal. For each strategy described next, we give an example that can be solved with that strategy. Often, problems can be solved in more than one way, as seen in the cartoon below. You may devise a different strategy to solve the sample problems. There is no one best strategy to use.

BLAIR

NONTRADITIONAL SOLUTION

♦ *Historical Note*

Carl Gauss (1777–1855) is regarded as the greatest mathematician of the nineteenth century and one of the greatest mathematicians of all time. Born to humble parents in Brunswick, Germany, he was an infant prodigy who, it is said, at age 3 corrected an arithmetic error in his father's bookkeeping. Gauss made contributions in the areas of astronomy, geodesy, and electricity. After Gauss's death, the King of Hanover ordered a commemorative medal prepared in his honor. On the medal was an inscription referring to Gauss as the "Prince of Mathematics," a title that stayed with his name.   ♦

Strategy: Look for a Pattern

When Carl Gauss was a child, his teacher required the students to find the sum of the first 100 natural numbers. The teacher expected this problem to keep the class occupied for some time. Gauss gave the answer almost immediately. Can you?

Understanding the Problem   The **natural numbers** are 1, 2, 3, 4, ... Thus, the problem is to find the sum $1 + 2 + 3 + 4 + \ldots + 100$.

Devising a Plan   The strategy *look for a pattern* is useful here. One version of the story about young Gauss reports that he listed the numbers as shown in Figure 1-1.

Let $S = 1 + 2 + 3 + 4 + 5 + \ldots + 98 + 99 + 100$. Then,

$$
\begin{array}{rcccccccccc}
S = & 1 + & 2 + & 3 + & 4 + & 5 + \ldots + & 98 + & 99 + & 100 \\
S = & 100 + & 99 + & 98 + & 97 + & 96 + \ldots + & 3 + & 2 + & 1 \\
\hline
2S = & 101 + & 101 + & 101 + & 101 + & 101 + \ldots + & 101 + & 101 + & 101
\end{array}
$$

Figure 1-1

To discover the original sum, Gauss then divided the sum, $2S$, in Figure 1-1 by 2.

Carrying Out the Plan   There are 100 sums of 101. Thus, $2S = 100 \cdot 101$ and $S = \dfrac{100 \cdot 101}{2}$, or 5050.

Looking Back   The method is mathematically correct because addition can be performed in any order, and multiplication is repeated addition. Also, the sum in each pair is always 101 because when we move from any pair to the next, we add 1 to the top and subtract 1 from the bottom, which does not change the sum; for example, $2 + 99 = (1 + 1) + (100 - 1) = 1 + 100$, $3 + 98 = (2 + 1) + (99 - 1) = 2 + 99 = 101$, and so on.

A more general problem is to find the sum of the first n natural numbers $1 + 2 + 3 + 4 + 5 + 6 + \ldots + n$. We use the same plan as before and notice the relationship in Figure 1-2. There are n sums of $n + 1$ for a total of $n(n + 1)$. Therefore, $2S = n(n + 1)$ and $S = \dfrac{n(n + 1)}{2}$.

$$
\begin{array}{rcccccc}
S = & 1 + & 2 + & 3 + & 4 + \ldots + & n \\
S = & n + & (n - 1) + & (n - 2) + & (n - 3) + \ldots + & 1 \\
\hline
2S = & (n + 1) + & (n + 1) + & (n + 1) + & (n + 1) + \ldots + & (n + 1)
\end{array}
$$

Figure 1-2

A different strategy for finding the sum $1 + 2 + 3 + \ldots + n$ involves the strategy of *making a diagram* and thinking of the sum geometrically as a stack of blocks. To find the sum, consider the stack in Figure 1-3(a) and a stack of the same size placed differently next to the original stack, as in Figure 1-3(b). The total number of blocks in the stack in Figure 1-3(b) is $n(n + 1)$, which is twice the desired sum. Thus, the desired sum is $n(n + 1)/2$.

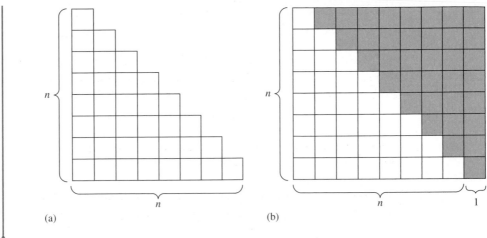

Figure 1-3

REMARK We will examine the sum $1 + 2 + 3 + 4 + 5 + \ldots + n = \dfrac{n(n + 1)}{2}$ further in the next section when we discuss arithmetic sequences.

 NOW TRY THIS 1-1 One cut on a log produces two pieces, two cuts produce three pieces, and three cuts produce four pieces. How many pieces are produced by ten cuts? Assume the cuts are made in the same manner as the first three cuts. How many pieces are produced by n cuts?

Strategy: Examine a Related Problem

Problem Solving   Sums of Even Natural Numbers

Find the sum of the even natural numbers less than or equal to 100. Devise a strategy for finding that sum and generalize the result.

Understanding the Problem   Even natural numbers are 2, 4, 6, 8, 10, ... The problem is to find the sum of the even natural numbers $2 + 4 + 6 + 8 + \ldots + 100$.

Devising a Plan   Recognizing that the sum can be separated into two simpler parts related to Gauss's original problem helps us devise a plan. Consider the following:

$$2 + 4 + 6 + 8 + \ldots + 100 = 2 \cdot 1 + 2 \cdot 2 + 2 \cdot 3 + 2 \cdot 4 + \ldots + 2 \cdot 50$$
$$= 2(1 + 2 + 3 + 4 + \ldots + 50)$$

Thus, we can use Gauss's method to find the sum of the first 50 natural numbers and then double that.

Carrying Out the Plan   We carry out the plan as follows:

$$2 + 4 + 6 + 8 + \ldots + 100 = 2(1 + 2 + 3 + 4 + \ldots + 50)$$
$$= 2 \cdot [50(50 + 1)/2]$$
$$= 2550$$

Thus, the sum is 2550.

Looking Back   A different way to approach this problem is to realize that there are 25 sums of 102, as shown in Figure 1-4.

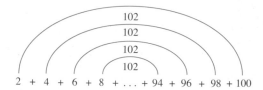

Figure 1-4

Thus, the sum is $25 \cdot 102$, or 2550.

NOW TRY THIS 1-2

a. Find the sum of the odd natural numbers less than 100.
b. Let $a_1, a_2, a_3, a_4, \ldots, a_n$ be any sequence of n terms where $a_2 - a_1 = a_3 - a_2 = a_4 - a_3 = \ldots = a_n - a_{n-1} = d$, where d is a fixed number. Write an expression for the sum of the terms in this sequence in terms of a_1, a_n, and n.

Strategy: Examine a Simpler Case

One strategy for solving a complex problem is to *examine a simpler case* of the problem and then consider other parts of the complex problem. An example is shown on the student page on page 9.

NOW TRY THIS 1-3   Sixteen people in a round-robin handball tournament played every person once. How many games were played?

Strategy: Make a Table

An often-used strategy in elementary school mathematics is *making a table*. A table can be used to look for patterns that emerge in the problem, which in turn can lead to a solution. An example of this strategy is shown on page 10. Did Plan II really pay $128?

NOW TRY THIS 1-4   Molly and Karly started a new job the same day. After they start work, Molly is to visit the home office every 15 days and Karly is to visit the home office every 18 days. How many days will it be before they both visit the home office the same day?

School Book Page SOLVING A SIMPLER PROBLEM

Lesson 11-8

Problem-Solving Strategy

Reading Helps!

Activating prior knowledge **can help you with...** the problem-solving strategy, *Solve a Simpler Problem.*

Key Idea
Learning how and when to solve a simpler problem can help you solve problems.

Solve a Simpler Problem

LEARN

How do you solve a simpler problem?

Triangle Trains Each side of each triangle in the figure at the right is one inch. If there are 12 triangles in a row, what is the perimeter of the figure?

Read and Understand

What do you know?

Triangles are being connected. Each side of each triangle is one inch.

What are you trying to find?

Find the perimeter of the figure with 12 triangles.

Plan and Solve

What strategy will you use?

Strategy: Solve a Simpler Problem.

I can look at 1 triangle, then 2 triangles, then 3 triangles.

Step 1 Break apart or change the problem into problems that are simpler to solve.

Step 2 Solve the simpler problems.

Step 3 Use the answers to the simpler problems to solve the original problem.

perimeter = 3 inches

perimeter = 4 inches

perimeter = 5 inches

Answer: The perimeter is 2 more than the number of triangles. For 12 triangles, the perimeter is 14 inches.

Look Back and Check

Is your work correct?   Yes, I saw a correct pattern.

☑ **Talk About It**

1. How was the problem broken apart into simpler problems?

2. Describe the pattern in the simpler problems.

648

Source: Scott Foresman-Addison Wesley, Grade 4, 2008 (p. 648).

School Book Page MAKING A TABLE

Problem-Solving Strategy

Reading Helps!

Understanding graphic sources such as tables and charts

can help you with...

the problem-solving strategy, *Make a Table.*

Key Idea
Learning how and when to make a table can help you solve problems.

Make a Table

LEARN

How can you make and use a table to solve a problem?

Babysitting Carrie is offered an afternoon babysitting job that will last 10 days. The parents who want to hire her offer two plans for payment. Which payment should Carrie accept?

Plan I: A single $100 payment for the 10 days worked

Plan II: Pay for the first day would be $0.25. Then each day thereafter, the total amount of pay would double.

Read and Understand

What do you know? There are two different plans.
What are you trying to find? Find the total pay for 10 days under Plan II.

Plan and Solve

What strategy will you use? **Strategy:** Make a Table

Days			
Amount			

Days	1	2	3
Amount	$0.25	$0.50	$1

Days	1	2	3	4	5	6	7	8	9	10
Amount	$0.25	$0.50	$1	$2	$4	$8	$16	$32	$64	$128

Days	1	2	3	4	5	6	7	8	9	10
Amount	$0.25	$0.50	$1	$2	$4	$8	$16	$32	$64	$128

Answer: Carrie should accept Plan II which pays $128.

How to Make a Table

Step 1 Set up the table with the correct labels.
Step 2 Enter known data into the table.
Step 3 Look for a pattern. Extend the table.
Step 4 Find the answer in the table.

Look Back and Check

Is your answer reasonable?

Yes, the answer should be an even number because the amounts in the table were doubled.

156

Source: Scott Foresman-Addison Wesley, Grade 6, 2008 (p. 156).

Strategy: Identify a Subgoal

In attempting to devise a plan for solving some problems, we may realize that the problem could be solved if the solution to a somewhat easier or more familiar related problem could be found. In such a case, finding the solution to the easier problem may become a *subgoal* of the primary goal of solving the original problem. The following Magic Square problem shows an example of this.

Figure 1-5

> **Problem Solving** A Magic Square
>
> Arrange the numbers 1 through 9 into a square subdivided into nine smaller squares like the one shown in Figure 1-5 so that the sum of every row, column, and main diagonal is the same. (The result is a *magic square*.)

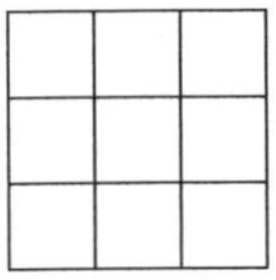

Understanding the Problem We need to put each of the nine numbers $1, 2, 3, \ldots, 9$ in the small squares, a different number in each square, so that the sums of the numbers in each row, in each column, and in each of the two major diagonals are the same.

Devising a Plan If we knew the fixed sum of the numbers in each row, column, and diagonal, we would have a better idea of which numbers can appear together in a single row, column, or diagonal. Thus our *subgoal* is to find that fixed sum. The sum of the nine numbers, $1 + 2 + 3 + \ldots + 9$, equals 3 times the sum in one row (why?). Consequently, the fixed sum is obtained by dividing $1 + 2 + 3 + \ldots + 9$, by 3. Using the process developed by Gauss, we have $(1 + 2 + 3 + \ldots + 9) \div 3 = \left(\dfrac{9 \cdot 10}{2}\right) \div 3$, or $45 \div 3 = 15$, so the sum in each row, column, and diagonal must be 15. Next, we need to decide what numbers could occupy the various squares. The number in the center space will appear in four sums, each adding to 15 (two diagonals, the second row, and the second column). Each number in the corners will appear in three sums of 15. (Do you see why?) If we write 15 as a sum of three different numbers 1 through 9 in all possible ways, we could then count how many sums contain each of the numbers 1 through 9. The numbers that appear in at least four sums are candidates for placement in the center square, whereas the numbers that appear in at least three sums are candidates for the corner squares. Thus our new *subgoal* is to write 15 in as many ways as possible as a sum of three different numbers from the set $\{1, 2, 3, \ldots, 9\}$.

Carrying Out the Plan The sums of 15 can be written systematically as follows:

$$9 + 5 + 1$$
$$9 + 4 + 2$$
$$8 + 6 + 1$$
$$8 + 5 + 2$$
$$8 + 4 + 3$$
$$7 + 6 + 2$$
$$7 + 5 + 3$$
$$6 + 5 + 4$$

Note that $1 + 5 + 9$ and $5 + 1 + 9$, for example, are counted as the same. Notice that 1 appears in only two sums, 2 in three sums, 3 in two sums, and so on. Table 1-1 summarizes this pattern.

Table 1-1

Number	1	2	3	4	5	6	7	8	9
Number of sums containing the number	2	3	2	3	4	3	2	3	2

The only number that appears in four sums is 5; hence, 5 must be in the center of the square. (Why?) Because 2, 4, 6, and 8 appear 3 times each, they must go in the corners. Suppose we choose 2 for the upper left corner. Then 8 must be in the lower right corner. (Why?) This is shown in Figure 1-6(a). Now we could place 6 in the lower left corner or upper right corner. If we choose the upper right corner, we obtain the result in Figure 1-6(b). The magic square can now be completed, as shown in Figure 1-6(c).

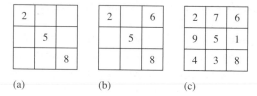

(a)   (b)   (c)

Figure 1-6

Looking Back   We have seen that 5 was the only number among the given numbers that could appear in the center. However, we had various choices for a corner, and so it seems that the magic square we found is not the only one possible. Can you find all the others?

Another way to see that 5 could be in the center square is to consider the sums $1 + 9, 2 + 8, 3 + 7, 4 + 6$, as shown in Figure 1-7. We could add 5 to each to obtain 15.

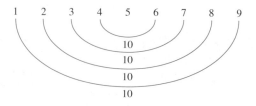

Figure 1-7

NOW TRY THIS 1-5 Five friends decided to give a party and split the costs equally. Al spent $4.75 on invitations, Betty spent $12 for drinks and $5.25 on vegetables, Carl spent $24 for pizza, Dani spent $6 on paper plates and napkins, and Ellen spent $13 on decorations. Determine who owes money to whom and how the money can be paid.

Strategy: Make a Diagram

It has often been said that a picture is worth a thousand words. This is particularly true in problem solving. In the following problem, *making a diagram* helps us to understand the problem and work towards a solution.

Problem Solving   50-m Race Problem

Bill and Jim ran a 50-m race 3 times. The speed of the runners does not vary. In the first race, Jim was at the 45-m mark when Bill crossed the finish line.

a. In the second race, to make the race closer Jim started 5 m ahead of Bill, who lined up at the starting line. Who will win the race?
b. In the third race, Jim starts at the starting line and Bill starts 5 m behind. Who will win the race?

Understanding the Problem   When Bill and Jim run a 50-m race, Bill wins by 5 m; that is, whenever Bill covers 50 m, at the same time Jim covers only 45 m. If Bill starts at the starting line and Jim is given a 5-m head start, we are to determine who will win the race. If Jim starts at the starting line and Bill starts 5 m behind, we are to determine who will win.

Devising a Plan   A strategy to determine the winner under each condition is to *make a diagram*. A diagram for the first 50-m race is given in Figure 1-8(a). In this case, Bill wins by 5 m. In the second race, Jim is given a 5-m head start and hence when Bill runs 50 m to the finish line, Jim runs only 45 m. Because Jim is 45 m from the finish line, he reaches the finish line at the same time as Bill does. This is shown in Figure 1-8(b). In the third race, because Bill starts 5 m behind, we use Figure 1-8(a) and move Bill back 5 m, as shown in Figure 1-8(c). From the diagram we can determine the results in each case.

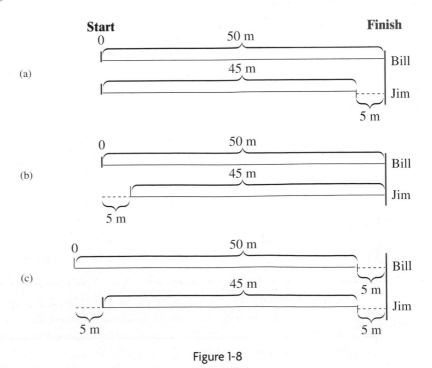

Figure 1-8

Carrying Out the Plan   From Figure 1-8(b) we see that if Jim is given a 5-m head start, then the race will end in a tie. If Bill starts 5 m behind Jim, then at 45 m they will be tied. Because Bill is faster than Jim, Bill will cover the last 5 m faster than Jim and win the race.

Looking Back   The diagrams show the solution makes sense and is appropriate. Other problems can be investigated involving racing and handicaps. For example, if Bill and Jim run on a 50-m oval track, how many laps will it take for Bill to lead Jim by one full lap. (Assume the same speeds as earlier.)

> **REMARK** In many cases, students' solutions may involve processes that occur simultaneously: thinking through the problem and supporting that thinking by diagram making.

 NOW TRY THIS 1-6 An elevator stopped at the middle floor of a building. It then moved up 4 floors, stopped, moved down 6 floors, stopped, and then moved up 10 floors and stopped. The elevator was now 3 floors from the top floor. How many floors does the building have?

Strategy: Guess and Check

 In the strategy of *guess and check*, we first guess at a solution using as reasonable a guess as possible. Then we check to see whether the guess is correct. If not, the next step is to learn as much as possible about the solution based on the guess before making a next guess. This strategy can be regarded as a form of trial and error, where the information about the error helps us choose what trial to make next. The guess-and-check strategy is often used when a student does not know how to solve the problem more efficiently or if the student does not yet have the tools to solve the problem in a faster way. Notice on the student page on page 15, students benefit from the observed "errors," as mentioned in the Research Note.

Research Note   Students in grades 1–3 rely primarily on the *guess-and-check* strategy when faced with a mathematical problem. As students enter grades 6–12 this tendency decreases. Older students benefit more from the observed "errors" after a guess when formulating a new "trial" (Lester 1975). ◆

 NOW TRY THIS 1-7 A cryptarithm is a collection of words in which each unique letter represents a unique number. Find the digits that can be substituted in the following:

$$
\begin{array}{r}
S\,U\,N \\
+\,F\,U\,N \\
\hline
S\,W\,I\,M
\end{array}
$$

School Book page GUESS AND CHECK

Lesson 5-7

Problem-Solving Strategy

Reading Helps!

Predicting and verifying
can help you with...

the problem-solving strategy,
Try, Check, and Revise.

Key Idea
The strategy Try,
Check, and Revise
can help you solve
problems.

Try, Check, and Revise

LEARN

How do you try, check, and revise?

Sale Suzanne spent $27, not including tax,
on dog supplies. She bought two of one
item and one other item. What did she buy?

Dog Supplies Sale!

Leash	$8
Collar	$6
Bowls	$7
Medium Beds	$15
Toys	$12

Read and Understand

What do you know? She bought three items.
Two of the items were the same.
The prices are in the sign.
She paid $27 for all three.

What are you trying to find? What three items did she buy?

Plan and Solve

What strategy will you use? Strategy: **Try, Check, and Revise**

How to Try, Check, and Revise

Step 1 Think to make a reasonable first try.

Step 2 Check using the information given in the problem.

Step 3 Revise. Use your first try to make a reasonable second try. Check.

Step 4 Use previous tries to continue trying and checking until you get the answer.

Two beds are too much. I'll try one. Then I'll
try 2 of the smaller items. I'll try leashes first.

$8 + $8 + $15 = $31
That's too high, but very close.

If I keep the bed, I need to come down $4
total or $2 for each item. I'll try the collars.

$6 + $6 + $15 = $27   That's it!

Answer: She bought two collars and one medium bed.

Look Back and Check

Is your work correct? Yes, the sum is $27 and
there are two of one item
and one of another item.

278

Source: Scott Foresman-Addison Wesley, Grade 4, 2005 (p. 278).

Strategy: Work Backward

In some problems, it is easier to start with the result and to work backward. This is demonstrated on the student page on page 17. Notice also that the *make a diagram* strategy is used.

NOW TRY THIS 1-8 Linda has an 80 average (mean) on her 11 math tests. Her teacher tells her she can drop her single low score of 50. What is her new average?

Strategy: Use Indirect Reasoning

To show that a statement is true, it is sometimes easier to show that it is impossible for the statement to be false. We can do this by showing that if the statement were false, something contradictory or impossible would follow. This approach is useful when it is difficult to start a direct argument and when negating the given statement gives us something tangible with which to work. An example follows.

Problem Solving Checkerboard Problem

In Figure 1-9, we are given a checkerboard with the two squares on opposite corners removed and dominoes such that each domino can cover two adjacent squares on the board. Can the dominoes be arranged in such a way that all the remaining squares on the board can be covered with no dominoes overlapping or hanging off the board? If not, why not?

Figure 1-9

Understanding the Problem Two red spaces on opposite corners were removed from the checkerboard in Figure 1-9. We are asked whether it is possible to cover the remaining 62 squares with dominoes the size of 2 squares.

Devising a Plan If we try to cover the board in Figure 1-9 with dominoes, we will find that the dominoes do not fit and some squares will remain uncovered. To show that there is no way to cover the board with dominoes, we use *indirect reasoning*. If the 62 squares in Figure 1-9 could be covered with no dominoes overlapping or hanging, it would take 31 dominoes to accomplish the task. We want to show that this implies something impossible.

School Book page   WORK BACKWARD

Lesson 8-9
Problem-Solving Strategy

Reading Helps!
Identifying steps in the process
can help you with...
the problem-solving strategy,
Work Backward.

Key Idea
Learning how and when to work backward can help you solve problems.

Work Backward

LEARN

How do you work backward to solve a problem?

Tunnel Vision It took workers 5 weeks to dig a 10-mile tunnel. How much had the workers completed after the first 3 weeks of digging?

During the fourth week, the workers dug $2\frac{1}{2}$ miles. The next week, the workers dug $1\frac{1}{4}$ miles to complete the tunnel.

Read and Understand

What do you know?   The workers completed a 10-mile tunnel in 5 weeks. During week 4, they dug $2\frac{1}{2}$ miles. During week 5, they dug $1\frac{1}{4}$ miles.

What are you trying to find?   How many miles of tunnel did the workers dig in the first 3 weeks?

Plan and Solve

What strategy will you use?   **Strategy: Work Backward**

The number of miles they dug in the first 3 weeks is not known.

distance dug in the first 3 weeks = n miles

How to Work Backward
Step 1   Identify the unknown initial amount.
Step 2   Draw a picture to show each change, starting with the initial amount.
Step 3   Start at the end result. Work backward, using the inverse of each change.

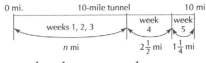

0 mi.   10-mile tunnel   10 mi

weeks 1, 2, 3 | week 4 | week 5

n mi $2\frac{1}{2}$ mi $1\frac{1}{4}$ mi

$10 - 1\frac{1}{4} - 2\frac{1}{2} = n;$ $n = 6\frac{1}{4}$

Answer: The workers dug $6\frac{1}{4}$ miles of tunnel during the first 3 weeks.

Look Back and Check

Is your answer reasonable?   Yes, because when I work forward from the initial amount, I get the end result.

$6\frac{1}{4}$ mi + $2\frac{1}{2}$ mi + $1\frac{1}{4}$ mi = 10 mi

484

Source: Scott Foresman–Addison Wesley, Grade 5, 2008 (p. 484).

Carrying Out the Plan   Each domino must cover 1 black and 1 red square. Hence, 31 dominoes would cover 31 red and 31 black squares. This is impossible, however, because the board in Figure 1-9 has 30 red and 32 black squares. Consequently, our assumption that the board in Figure 1-9 can be covered with dominoes is wrong.

Looking Back   The counting of black and red squares implies that if we remove any number of squares from a checkerboard so that the number of remaining red squares differs from the number of remaining black squares, the board cannot be covered with dominoes. (Do you see why?) We could also investigate what happens when two squares of the same color are removed from an 8-by-7 board and other-sized boards. Also, is it always possible to cover the remaining board if two squares of opposite colors are removed?

NOW TRY THIS 1-9   Al, Bob, Carl, and Dan each participate in exactly one sport: swimming, baseball, basketball, or tennis. Bob plays baseball. Al can't swim. Carl plays basketball. In what sports does each person participate?

Strategy: Use Direct Reasoning

Problem Solving   Checker Games

Two people each won three games of checkers. Is it possible that only five games were played?

Solution   We know that each person won three games. *Reasoning directly*, we see that if they each won three games and they played each other, they would have had to play six games. Thus, they could not play each other in all games and have three wins each. Could they in fact each win three games while playing only five total games and not have played each other? The answer is no and the situation is impossible.

Strategy: Write an Equation

A problem-solving strategy used in algebraic thinking is *writing an equation*. This strategy is very important, and it is covered in Chapter 4, "Algebraic Thinking."

Assessment 1-1A

1. Use the approach in Gauss's Problem to find the following sums (do not use formulas):
   a. $1 + 2 + 3 + 4 + \ldots + 99$
   b. $1 + 3 + 5 + 7 + \ldots + 1001$
2. Find the sum of $36 + 37 + 38 + 39 + \ldots + 146 + 147$.
3. Cookies are sold singly or in packages of two or six. How many ways can you buy a dozen cookies?
4. The sign says you are leaving Missoula, Butte is 120 mi away, and Bozeman is 200 mi away. There is a rest stop halfway between Butte and Bozeman. How far is the rest stop from Missoula if both Butte and Bozeman are in the same direction?
5. Alababa, Bubba, Cory, and Dandy are in a horse race. Bubba is the slowest, Cory is faster than Alababa but slower than Dandy. Name the finishing order of the horses.
6. Frankie and Johnny began reading a novel on the same day. Frankie reads eight pages a day and Johnny reads five pages a day. If Frankie is on page 72, what page is Johnny on?

7. What is the largest sum of money—all in coins and no silver dollars—that you could have in your pocket without being able to give change for a dollar, a half-dollar, a quarter, a dime, or a nickel?

8. **a.** Place the digits 1, 2, 4, 5, and 7 in the following boxes so that in (i) the greatest product is obtained and in (ii) the greatest quotient is obtained:

(i) □□□
× □□

(ii) □□)‾□□□‾

b. Use the same digits as in (a) to obtain (i) the least product and (ii) the least quotient.

9. Suppose you could spend $10 every minute, night and day. How much could you spend in a year? (Assume there are 365 days in a year.)

10. How many different four-digit numbers have the same digits as 1993?

11. A compass and a ruler together cost $4. The compass costs 90¢ more than the ruler. How much does the compass cost?

12. Kathy stood on the middle rung of a ladder. She climbed up three rungs, moved down five rungs, and then climbed up seven rungs. Then she climbed up the remaining six rungs to the top of the ladder. How many rungs are there in the whole ladder?

★ 13. Same-sized cubes are glued together to form a staircase-like sequence of solids as shown:

All of the unglued faces of the cubes need to be painted. How many squares will need to be painted in **(a)** the 100th solid? **(b)** the *n*th solid?

14. A farmer needs to fence a rectangular piece of land. She wants the length of the field to be 80 ft longer than the width. If she has 1080 ft of fencing material, what should the length and the width of the field be?

15. One winter night the temperature fell 15°F between midnight and 5 A.M. By 9 A.M., the temperature had doubled from what it was at 5 A.M. By noon, it had risen another 10° to 32°F. What was the temperature at midnight?

16. Al, Betty, Carl, and Dan were each born in a different season. Al was born in February. Betty was not born in the fall. Carl was born in the spring. Determine which season each child was born in.

17. The 14 digits of a credit card are written in the boxes shown. If the sum of any three consecutive digits is 20, what is the value of A?

A	7											7	4

Assessment 1-1B

1. Use the approach in Gauss's Problem to find the following sums (do not use formulas):
   a. $1 + 2 + 3 + 4 + \ldots + 49$
   b. $1 + 3 + 5 + 7 + \ldots + 2009$

2. Find the sum of $58 + 59 + 60 + 61 + \ldots + 203$.

3. How many ways can you make change for a $50 bill using $5, $10, and $20 bills?

4. How many squares are in the following figure?

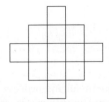

5. **a.** Without computing each sum, find which is greater, *O* or *E*, and by how much.

$$O = 1 + 3 + 5 + 7 + \ldots + 97$$
$$E = 2 + 4 + 6 + 8 + \ldots + 98$$

b. If $P = 1 + 3 + 5 + 7 + \ldots + 99$, which is greater, *E* or *P*, and by how much?

6. **a.** Place the digits 4, 5, 6, 7, and 9 in the following boxes so that in (i) the greatest product is obtained and in (ii) the greatest quotient is obtained:

(i) □□□
× □□

(ii) □□)‾□□□‾

b. Use the same digits as in (a) to obtain (i) the least product and (ii) the least quotient.

7. Marc goes to the store with exactly $1.00 in change. He has at least one of each coin less than a half-dollar coin, but he does not have a half-dollar coin.
   a. What is the least number of coins he could have?
   b. What is the greatest number of coins he could have?

8. Find a 3-by-3 magic square using the numbers 3, 5, 7, 9, 11, 13, 15, 17, and 19.

9. Eight marbles look alike, but one is slightly heavier than the others. Using a balance scale, explain how you can determine the heavier one in exactly
   a. three weighings
   b. two weighings

10. **a.** Find the sum of all the numbers in the following array:

1	2	3	4	5	6	...	100
2	4	6	8	10	12	...	200
3	6	9	12	15	18	...	300
⋮	⋮	⋮	⋮	⋮	⋮	⋮	
100	200	300	400	500	600	...	$100 \cdot 100$

   ★**b.** Generalize part (a) to a similar array in which each row has n numbers and there are n rows.

11. Recall the song "The Twelve Days of Christmas":

    On the first day of Christmas my true love gave to me a partridge in a pear tree.
    On the second day of Christmas my true love gave to me two turtle doves and a partridge in a pear tree.
    On the third day of Christmas my true love gave to me three French hens, two turtle doves, and a partridge in a pear tree.

    This pattern continues for 9 more days. After 12 days,
    a. which gifts did my true love give the most? (Yes, you will have to remember the song.)
    b. how many total gifts did my true love give to me?

12. **a.** Using the existing lines on the checkerboard shown, how many squares are there?

    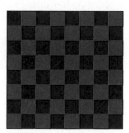

    b. If the number of rows and columns of the checkerboard is doubled, is the number of squares doubled? Justify your answer.

13. Suppose you throw three darts at the target pictured below. All of the darts hit the target for a score. What are all the different possible scores?

14. The following is a magic square (all rows, columns, and diagonals sum to the same number). Find the value of each variable.

17	a	7
12	22	b
c	d	27

15. Two cards are on a table. A 12 is written on one of the cards and 9 on the other. Each card has a number written on the flip side. By turning over one card, both cards, or neither card and adding the two numbers, you can get the sums of 15, 16, 20, and 21. What number is written on the flip side of each card?

16. Suppose you buy lunch for the math club. You have enough money to buy 20 salads or 15 sandwiches. The group wants 12 sandwiches. How many salads can you buy?

17. **a.** Suppose you have quarters, dimes, and pennies with a total value of $1.19. How many of each coin can you have without being able to make change for a dollar?
    b. Tell why the combination of coins you have in part (a) is the greatest amount of money that you can have without being able to make change for a dollar.

18. You have two containers. One holds 7 cups and one holds 4 cups. How can you measure exactly 5 cups of water if you have an unlimited amount of water to start with?

Mathematical Connections 1-1

Communication

1. Why is teaching problem solving an important part of mathematics?

2. Look at the NCTM Process standard for problem solving inside the cover of this book, Discuss how Pólya's four-step problem-solving process addresses the last two expectations listed in this process standard.

3. Explain when you might use the *guess-and-check* strategy.

Open-Ended

4. Use exactly four 4s and any mathematical symbols to create the natural numbers 1 to 20 inclusive; for example, $4/4 + 4/4 = 2$ and $4 \times 4 + 4 - \sqrt{4} = 18$.

5. Choose a problem-solving strategy and make up a problem that would use this strategy. Write the solution using Pólya's four-step approach.

Cooperative Learning

6. Have each person in your group work the following problem: If eight people shake hands with one another, how many handshakes take place?
   a. Compare your strategies for working the problem. How are they the same? How are they different?
   b. Find as many ways as possible to do the problem.
   c. Generalize the solution for n people.

7. The distance around the world is approximately 40,000 km. Approximately how many people of average size in your group would it take to stretch around the world if they were holding hands?

8. Work in pairs on the following version of a game called NIM. A calculator is needed for each pair.
   a. Player 1 presses $\boxed{1}$ and $\boxed{+}$ or $\boxed{2}$ and $\boxed{+}$. Player 2 does the same. The players take turns until the target number of 21 is reached. The first player to make the display read 21 is the winner. Determine a strategy for deciding who always wins.
   b. Try a game of NIM using the digits 1, 2, 3, and 4, with a target number of 104. The first player to reach 104 wins. What is the winning strategy?
   c. Try a game of NIM using the digits 3, 5, and 7, with a target number of 73. The first player to exceed 73 loses. What is the winning strategy?
   d. Now play Reverse NIM with the keys $\boxed{1}$ and $\boxed{2}$. Instead of $\boxed{+}$, use $\boxed{-}$. Put 21 on the display. Let the target number be 0. Determine a strategy for winning Reverse NIM.
   e. Try Reverse NIM using the digits 1, 2, and 3 and starting with 24 on the display. The target number is 0. What is the winning strategy?
   f. Try Reverse NIM using the digits 3, 5, and 7 and starting with 73 on the display. The first player to display a negative number loses. What is the winning strategy?

9. When a book is printed, pages are passed through a printing press and then folded to form a book. To see how this works, start out with a simple book made from one $8\frac{1}{2} \times 11$ in. sheet of paper. Fold the page in half lengthwise to form a book and number the pages 1 through 4. When you open your sheet of paper, the numbers 2 and 3 are on one side of the paper and the numbers 1 and 4 are on the other side. The sum of the numbers on each side of the sheet equals 5, and the sum of all the page numbers is 10. If two sheets of paper are used similarly to build an eight-page book and the pages are numbered, predict the sum of the numbers on each page and the sum of all of the page numbers. Build your book to see if you were correct. Try the same thing with three pages.
   a. Suppose you are building a 100-page book. How many sheets will you need?
   b. What is the sum of the two page numbers that occur on the same side of a sheet?
   c. What is the sum of all the page numbers in this book?
   d. Suppose you have n sheets of paper. Generalize to find the number of book pages, the sum of the numbers on the same side of the sheet, and the sum of all the page numbers.

Questions from the Classroom

10. Amy asks what "problem solving" is and whether 3×8 is a problem? What do you tell her?

11. John asks why the last step of Pólya's four-step problem-solving process, *looking back*, is necessary since he has already given the answer. What could you tell him?

12. A student asks why he just can't make "random guesses" rather than "intelligent guesses" when using the guess-and-check problem-solving strategy. How do you respond?

13. Rob says that it is possible to create a magic square with the numbers 1, 3, 4, 5, 6, 7, 8, 9, and 10. How do you respond?

Third International Mathematics and Science Study (TIMSS) Question

4	11	6
9		5
8	3	10

The rule for the table is that numbers in each row and column must add up to the same number. What number goes in the center of the table?
   a. 1   b. 2
   c. 7   d. 12

TIMSS 2003, Grade 4

National Assessment of Educational Progress (NAEP) Question

There will be 58 people at a breakfast and each person will eat 2 eggs. There are 12 eggs in each carton. How many cartons of eggs will be needed for the breakfast?
   a. 9   b. 10
   c. 72   d. 116

NAEP 2007, Grade 4

BRAIN TEASER Ten women are fishing all in a row in a boat. One seat in the center of the boat is empty. The five women in the front of the boat want to change seats with the five women in the back of the boat. A person can move from her seat to the next empty seat or she can step over one person without capsizing the boat. What is the minimum number of moves needed for the five women in front to change places with the five in back?

LABORATORY ACTIVITY Place a half-dollar, a quarter, and a nickel in position *A* as shown in Figure 1-10. Try to move these coins, one at a time, to position *C*. At no time may a larger coin be placed on a smaller coin. Coins may be placed in position *B*. How many moves does it take to get them to position *C*? Now add a penny to the pile and see how many moves are required. This is a simple case of the famous Tower of Hanoi problem, in which ancient Brahman priests were required to move a pile of 64 disks of decreasing size, after which the world would end. How long would it take at a rate of one move per second?

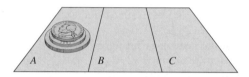

A   B   C

Figure 1-10

1-2   Explorations with Patterns

E-Manipulative Activity

Additional practice with patterns not involving numbers can be found in the *Patterns* activity on the E-Manipulatives disk. The activity involves completing simple and complex patterns of symbols and colored blocks.

Mathematics has been described as the study of patterns. Patterns are everywhere—in wallpaper, tiles, traffic, and even television schedules. Police investigators study case files to find the *modus operandi*, or pattern of operation, when a series of crimes is committed. Scientists look for patterns in order to isolate variables so that they can reach valid conclusions in their research. In *Principles and Standards*, we find the following:

> . . . students should investigate numerical and geometric patterns and express them mathematically in words or in symbols. They should analyze the structure of the pattern and how it grows or changes, organize this information systematically, and use their analysis to develop generalizations about the mathematical relationships in the pattern. (p. 159)

Patterns do not always have to be numerical, as shown in Now Try This 1-10.

NOW TRY THIS 1-10

a. Find three more terms to continue a pattern:

o, △, △, o, △, △, o ___, ___, ___

b. Describe in words the pattern found in part (a).

Patterns can be surprising. Consider Example 1-1.

Example 1-1

a. Describe any patterns seen in the following:

$$1 + 0 \cdot 9 = 1.$$
$$2 + 1 \cdot 9 = 11$$
$$3 + 12 \cdot 9 = 111$$
$$4 + 123 \cdot 9 = 1111$$
$$5 + 1234 \cdot 9 = 11111$$

b. Do the patterns continue? Why or why not?

Solution a. There are several possible patterns. For example, the numbers on the far left are natural numbers, that is, numbers from the set {1, 2, 3, 4, 5, . . . }. The pattern starts with 1 and continues to the next greater natural number in each successive line. The numbers "in the middle" are products of two numbers, the second of which is 9. The first number in the first product is 0; after that the first number is formed using natural numbers and including one more in each successive line. The resulting numbers on the right are formed using 1s and include an additional 1 in each successive line.

b. The pattern in the complete equation appears to continue for a number of cases, but it does not continue in general; for example,

$$13 + 123456789101112 \cdot 9 = 1,111,111,101,910,021$$

This pattern breaks down when the pattern of digits in the number being multiplied by 9 contains previously used digits.

As seen in Example 1-1, determining a pattern on the basis of a few cases is not reliable. For all patterns found, we should either find a counterexample to show the pattern does not hold in general or justify that the pattern always works.

In *Principles and Standards* we find the following:

When students make a discovery or determine a fact, rather than tell them whether it holds for all numbers or if it is correct, the teacher should help the students make that determination themselves. Teachers should ask such questions as "How do you know it is true?" and should also model ways that students can verify or disprove their conjectures. In this way, students gradually develop the abilities to determine whether an assertion is true, a generalization valid, or an answer correct and to do it on their own instead of depending on the authority of the teacher or the book. (p. 126)

Inductive Reasoning

Scientists make observations and propose general laws based on patterns. Statisticians use patterns when they form conclusions based on collected data. This process, **inductive reasoning**, is the method of making generalizations based on observations and patterns. Although inductive reasoning may lead to new discoveries, its weakness is that conclusions are drawn only from the collected evidence. If not all cases have been checked, another case may prove the conclusion false. In mathematics, inductive reasoning may lead to a **conjecture**, a statement thought to be true but not yet proved true or false. For example, considering only that $0^2 = 0$ and that $1^2 = 1$, we might conjecture that *every number squared is equal to itself*. When we find an example that contradicts the conjecture, we provide a **counterexample** and prove the conjecture false in general. Students have trouble with the concept of a counterexample, as pointed out in the Research Note. To show that the preceding conjecture is false, we need exhibit only one counterexample, for example, $2^2 = 4$. Sometimes finding a counterexample is difficult, and not finding one right away does not make a conjecture true.

Next, consider a pattern that does work and helps solve a problem. How can you find the sum of three consecutive natural numbers without performing the addition? Several examples are given here. Look for a pattern in these examples.

$$14 + 15 + 16 \qquad (\mathbf{45})$$
$$19 + 20 + 21 \qquad (\mathbf{60})$$
$$99 + 100 + 101 \qquad (\mathbf{300})$$

After studying the sums, you see a pattern of multiplying the middle number by 3. You could try other numbers to see if a *counterexample* can be found. The pattern suggests other mathematical questions you could consider. For example,

1. Does this work for any three consecutive natural numbers?
2. How can you find the sum of any odd number of consecutive natural numbers?
3. What happens if there is an even number of consecutive natural numbers?

To answer question (1), we give a proof showing that the sum of three consecutive natural numbers is equal to 3 times the middle number.

Proof

Let n be the first of three consecutive natural numbers. Then the three numbers are n, $n + 1$, and $n + 2$. The sum of these three numbers is $n + (n + 1) + (n + 2) = 3n + 3 = 3(n + 1)$. Therefore, the sum of the three consecutive natural numbers is 3 times the middle number.

The Danger of Making Conjectures Based on a Few Cases

In *Principle and Standards* we find the following:

> During grades 3–5, students should move toward reasoning that depends on relationships and properties. Students need to be challenged with questions such as, What if I gave you twenty more problems like this to do—would they all work the same way? How do you know? (p. 190)

This concept is further emphasized in the Research Note.

The following discussion illustrates the danger of making a conjecture based on a few cases. In Figure 1-11, we choose points on a circle and connect them to form distinct, nonoverlapping regions. In this figure, 2 points determine 2 regions, 3 points determine 4 regions, and 4 points determine 8 regions. What is the maximum number of regions that would be determined by 10 points?

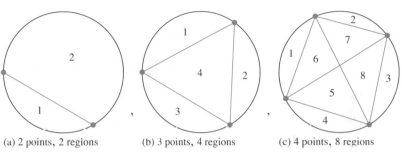

| (a) 2 points, 2 regions | (b) 3 points, 4 regions | (c) 4 points, 8 regions |

Figure 1-11

The data from Figure 1-11 are recorded in Table 1-2. It appears that each time we increase the number of points by 1, we double the number of regions. If this were true, then for 5 points we would have 2 times the number of regions with 4 points, or $2 \cdot 8 = 16 = 2^4$, and so on. If we base our conjecture on this pattern, we might believe that for 10 points, we would have 2^9, or 512 regions. (Why?)

Table 1-2

Number of Points	2	3	4	5	6	...	10
Maximum Number of Regions	2	4	8				?

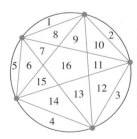

Figure 1-12

An initial check for this conjecture is to see whether we obtain 16 regions for 5 points. We obtain a diagram similar to that in Figure 1-12, and our guess of 16 regions is confirmed. For 6 points, the pattern predicts that the number of regions will be 32. Choose the points so that they are neither symmetrically arranged nor equally spaced and count the regions carefully. You should obtain 31 regions and not 32 regions as predicted. No matter how the points are located on the circle, the guess of 32 regions is not correct. The counterexample tells us that the doubling pattern is not correct; note that it does not tell us whether or not there are 512 regions with 10 points, but only that the pattern is not what we conjectured.

A natural-looking pattern of 2, 4, 8, 16, ... is suggested in this example, but the pattern does not continue, as shown when actual pictures are drawn. If we look only at the first four terms of the sequence 2, 4, 8, 16 without context, the doubling pattern is logical. In the context of counting the number of regions of a circle, however, the pattern is incorrect.

NOW TRY THIS 1-11   A *prime number* is a natural number with exactly two distinct positive numbers that divide it; 1 and the number itself; for example, 2, 3, 5, 7, 11, 13 are primes. One day Amy makes a *conjecture* that the formula, $y = x^2 + x + 11$ will produce only prime numbers if she substitutes the natural numbers, 1, 2, 3, 4, 5. . . . for *x*. She shows her work so far in Table 1-3 for $x = 1, 2, 3, 4$.

Table 1-3

x	1	2	3	4
y	13	17	23	31

a. What type of reasoning is Amy using?
b. Try the next several natural numbers and see if they seem to work.
c. Can you find a counterexample to show that Amy's conjecture is false?

Arithmetic Sequences

A **sequence** is an ordered arrangement of numbers, figures, or objects. A sequence has items or terms identified as *1st*, *2nd*, *3rd*, and so on. Often, sequences can be classified by their properties. For example, what property do the following first three sequences have that the fourth does not?

a. 1, 2, 3, 4, 5, 6, . . .
b. 0, 5, 10, 15, 20, 25, . . .
c. 2, 6, 10, 14, 18, 22, . . .
d. 1, 11, 111, 1111, 11111, 111111, . . .

In each of the first three sequences, each term—starting from the second—is obtained from the preceding one by adding a fixed number called the **common difference** or **difference**. In part (a) the difference is 1, in part (b) the difference is 5, and in part (c) the difference is 4. Sequences such as the first three are arithmetic sequences. An **arithmetic sequence** is one in which each successive term from the second term on is obtained from the previous term by the addition or subtraction of a fixed number. The sequence in part (d) is not arithmetic because there is no single fixed number you can add to or subtract from the previous term to obtain the next term.

Arithmetic sequences can be generated from objects, as shown in Example 1-2.

Example 1-2

Find a pattern in the number of matchsticks required to continue the pattern shown in Figure 1-13.

Figure 1-13

Solution   Assume the matchsticks are arranged so that each figure has one more square on the right than the preceding figure. Note that the addition of a square to an arrangement requires the addition of three matchsticks each time. Thus, the numerical pattern obtained is 4, 7, 10, 13, 16, 19, . . . , an arithmetic sequence with a difference of 3.

An informal description of an arithmetic sequence is one that can be described as an "add d" pattern, where d is the common difference. In Example 1-2, $d = 3$. In the language of children, the pattern in Example 1-2 is "add 3." This is an example of a **recursive pattern**. In a recursive pattern, after one or more consecutive terms are given to start, each successive term of the sequence is obtained from the previous term(s). For example, 3, 6, 9, ... is another "add 3" sequence starting with 3, and 1, 2, 3, 5, 8, 13, ... is a recursive pattern in which the next term starting with the third is obtained by adding the two previous terms.

A recursive pattern is typically used in a spreadsheet, as seen in Table 1-4 where column A tracks the order of the terms; the headers for the columns are A, B, etc. The first entry in the B column (in the B1 cell) is 4; and to find the term in the B2 cell, we use the number in the B1 cell and add 3. Once we find the B2 cell entry, the pattern is continued using the *Fill Down* command. In spreadsheet language, the formula $= B1 + 3$ finds any term after the first by adding 3 to the previous term. The formula, based on a recursive pattern, is a **recursive formula**. (For more explicit directions on using a spreadsheet, see the Technology Manual.)

Table 1-4

	A	B
1	1	4
2	2	7
3	3	10
4	4	13
5	5	16
6	6	19
7	7	22
8	8	25
9	9	28
10		
11		
12		
13		

If you want to find the number of matchsticks in the 100th figure in Example 1-2, you can use the spreadsheet or you can find an explicit formula or a general rule for finding the number of matchsticks when given the number of the term. The problem-solving strategy of *making a table* is again helpful here.

The spreadsheet in Table 1-4 provides an easy way to *make a table*. Column A gives the numbers of the terms and column B gives the terms of the sequence. If you are building such a table without a spreadsheet, it might look like Table 1-5. An **ellipsis**, denoted by three dots, indicates that the sequence continues in the same manner. Notice that each term is a sum of 4 and a certain number of 3s. We see that the number of 3s is 1 less than the number of the term. This pattern should continue, since the first term is $4 + 0 \cdot 3$ and each time we increase the number of the term by 1, we add one *more* 3. Thus, it seems that the 100th term is $4 + (100 - 1)3$, and, in general, the **nth term**, a_n, is $4 + (n - 1)3$. We write this as $a_n = 4 + (n - 1)3$. Note that $4 + (n - 1)3$ could be written as $3n + 1$.

Table 1-5

Number of Term	Term
1	4
2	$7 = 4 + 3 = 4 + 1 \cdot 3$
3	$10 = (4 + 1 \cdot 3) + 3 = 4 + 2 \cdot 3$
4	$13 = (4 + 2 \cdot 3) + 3 = 4 + 3 \cdot 3$
.	.
.	.
.	.
n	$4 + (n - 1)3$

Still a different approach to finding the number of matchsticks in the 100th term of Figure 1-13 might be as follows: If the matchstick figure has 100 squares, we could find the total number of matchsticks by adding the number of horizontal and vertical sticks. There are $2 \cdot 100$ placed horizontally. (Why?) Notice that in the first figure, there are 2 matchsticks placed vertically; in the second, 3; and in the third, 4. In the 100th figure, there should be $100 + 1$ vertical matchsticks. Altogether there will be $2 \cdot 100 + (100 + 1)$, or 301, matchsticks in the 100th figure. Similarly, in the nth figure, there would be $2n$ horizontal and $(n + 1)$ vertical matchsticks, for a total of $3n + 1$. This discussion is summarized in Table 1-6.

Table 1-6

Number of Term	Number of Matchsticks Horizontally	Number of Matchsticks Vertically	Total
1	2	2	4
2	4	3	7
3	6	4	10
4	8	5	13
.	.	.	.
.	.	.	.
.	.	.	.
100	200	101	301
.	.	.	.
.	.	.	.
.	.	.	.
n	$2n$	$n + 1$	$2n + (n + 1) = 3n + 1$

If we are given the value of the term, we can use the formula for the nth term in Table 1-6 to *work backward* to find the number of the term. For example, given the term 1798, we know that $3n + 1 = 1798$. Therefore, $3n = 1797$ and $n = 599$. Consequently, 1798 is the 599th term. We could obtain the same answer by solving $4 + (n - 1)3 = 1798$ for n.

In the matchstick problem, we found the nth term of a sequence. If the nth term of a sequence is given, we can find any term of the sequence, as shown in Example 1-3.

Example 1-3

Find the first four terms of a sequence the nth term of which is given by the following and determine whether the sequence seems to be arithmetic:

a. $a_n = 4n + 3$ **b.** $a_n = n^2 - 1$

Solution a.

Number of Term	Term
1	$4 \cdot 1 + 3 = 7$
2	$4 \cdot 2 + 3 = 11$
3	$4 \cdot 3 + 3 = 15$
4	$4 \cdot 4 + 3 = 19$

Hence, the first four terms of the sequence are 7, 11, 15, 19. This sequence seems arithmetic with difference 4.

b.

Number of Term	Term
1	$1^2 - 1 = 0$
2	$2^2 - 1 = 3$
3	$3^2 - 1 = 8$
4	$4^2 - 1 = 15$

Thus, the first four terms of the sequence are 0, 3, 8, 15. This sequence is not arithmetic because it has no common difference.

Generalizing Arithmetic Sequences

To generalize our work with arithmetic sequences, suppose the first term in an arithmetic sequence is a_1 and the difference is d. The strategy of *making a table* can be used to investigate the general term for the sequence $a_1, a_1 + d, a_1 + 2d, a_1 + 3d,$ as shown in Table 1-7. *The nth term of any sequence with first term a_1 and difference d is given by $a_n = a_1 + (n - 1)d$.* For example, in the arithmetic sequence 5, 9, 13, 17, 21, 25,..., the first term is 5 and the difference is 4. Thus, the nth term is given by $a_1 + (n - 1)d = 5 + (n - 1)4$. Simplifying algebraically, we obtain $5 + (n - 1)4 = 5 + 4n - 4 = 4n + 1$. Check to see if $4n + 1$ generates the sequence 5, 9, 13, 17, 21, ...

Table 1-7

Number of Term	Term
1	a_1
2	$a_1 + d$
3	$a_1 + 2d$
4	$a_1 + 3d$
5	$a_1 + 4d$
·	·
·	·
·	·
n	$a_1 + (n - 1)d$

REMARK The nth term of any arithmetic sequence with first term a_1 and difference d is $a_n = a_1 + (n - 1)d$, where n is a natural number but there are no restrictions on d.

NOW TRY THIS 1-12 In an arithmetic sequence with the 2nd term 11 and the 5th term 23, find the 100th term.

Example 1-4

The diagrams in Figure 1-14 show the molecular structure of alkanes, a class of hydro-carbons. C represents a carbon atom and H a hydrogen atom. A connecting segment shows a chemical bond. (Remark: CH_4 stands for C_1H_4.)

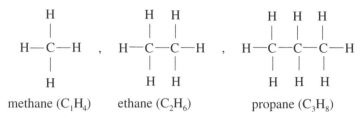

Figure 1-14

a. Hectane is an alkane with 100 carbon atoms. How many hydrogen atoms does it have?

b. Write a general rule for alkanes C_nH_m showing the relationship between m and n.

Solution   **a.** To determine the relationship between the number of carbon and hydrogen atoms, we first study the drawing of the alkanes and disregard the extreme left and right hydrogen atoms in each. With this restriction, we see that for every carbon atom, there are two hydrogen atoms. Therefore, there are twice as many hydrogen atoms as carbon atoms plus the two hydrogen atoms at the extremes. For example, when there are 3 carbon atoms, there are $(2 \cdot 3) + 2$, or 8, hydrogen atoms. This notion is summarized in Table 1-8. If we extend the table for 4 carbon atoms, we get $(2 \cdot 4) + 2$, or 10, hydrogen atoms. For 100 carbon atoms, there are $(2 \cdot 100) + 2$, or 202, hydrogen atoms.

b. In general, for n carbon atoms there would be n hydrogen atoms attached above, n attached below, and 2 attached on the sides. Hence, the total number of hydrogen atoms would be $2n + 2$. Because the number of hydrogen atoms was designated by m, it follows that $m = 2n + 2$.

Table 1-8

No. of Carbon Atoms	No. of Hydrogen Atoms
1	4
2	6
3	8
.	.
.	.
.	.
100	?
.	.
.	.
.	.
n	m

Example 1-5

A theater is set up in such a way that there are 20 seats in the first row and 4 additional seats in each consecutive row. The last row has 144 seats. How many rows are there in the theater?

Solution   Because there are 4 additional seats in each consecutive row, the numbers of seats in the rows forms an arithmetic sequence. The first term, a_1, of the sequence is 20 and the difference, d, is 4. The last term in the sequence is 144. A computerized spreadsheet could easily be used to count the number of terms in the sequence, 20, 24, 28, ..., 144. However, without technology, we could find the number of the terms as follows: In an arithmetic sequence, $a_n = a_1 + (n-1)d$, where a_1 is the first term, d is the difference, and n is the number of the term. In this case, $a_1 = 20$ and $d = 4$. Therefore,

$$a_n = a_1 + (n-1)d = 20 + (n-1)4$$

We now want to find the number of the term when $a_n = 20 + (n-1)4$ is equal to 144. Therefore,

$$20 + (n-1)4 = 144$$
$$(n-1)4 = 124$$
$$n-1 = 31$$
$$n = 32$$

This shows that there are 32 rows in the theater.

Fibonacci Sequence

The popular book *The Da Vinci Code* brought renewed interest to one of the most famous sequences of all time, the ***Fibonacci sequence***. The Fibonacci sequence is hinted at in the following cartoon. Can you figure out a rule for the Fibonacci Sequence?

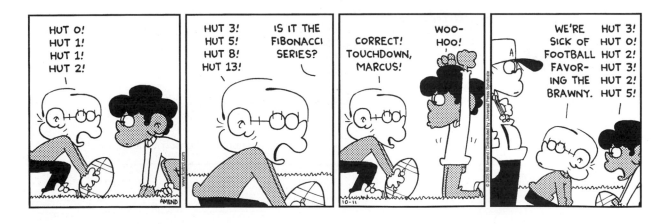

Leonardo de Pisa was born in Italy around 1170. His real family name was Bonaccio but he preferred the nickname Fibonacci, derived from the Latin for *filius Bonacci*, meaning "son of Bonacci." In his travels, Leonardo learned the Hindu-Arabic number system from the Moors. In his book *Liber Abaci* (1202) he described the workings of the Hindu-Arabic system. One of the problems in his book was the now-famous rabbit problem, whose solution is the sequence 1, 1, 2, 3, 5, 8, 13, 21, ..., which became known as the *Fibonacci sequence.* ◆

The Fibonacci sequence in the cartoon had 0 as a starting term. More typically, the sequence is seen as follows:

$$1, 1, 2, 3, 5, 8, 13, 21, 34, 55, 89, 144, \ldots$$

The sequence is named after the Italian Leonardo de Pisa, better known by the nickname Fibonacci. This sequence is not *arithmetic* as there is no fixed difference, d.

The standard mathematical way to represent a Fibonacci number has F_1 representing the first term, F_2 representing the second term, F_3 representing the third term, and in general F_n representing the nth term. If we want to indicate the Fibonacci numbers that come after F_n, we write them as F_{n+1}, F_{n+2}, and so on. The number that comes before F_n is F_{n-1}. With this notation, the rule for generating the Fibonacci sequence can be written as

$$F_n = F_{n-1} + F_{n-2}, \quad \text{for } n = 3, 4, 5, \ldots$$

Notice that this rule cannot be applied to the first two Fibonacci numbers. Because $F_1 = 1$ and $F_2 = 1$, then $F_3 = 1 + 1 = 2$. The *seeds* $F_1 = 1$ and $F_2 = 1$ and the rule $F_n = F_{n-1} + F_{n-2}$ give another example of a *recursive* definition because the rule in the sequence defines a number using previous numbers in the same sequence. Using the seeds and the rule, we can find any Fibonacci number. To find F_{100} with what we know right now we would have to know F_{98} and F_{99}. A spreadsheet can easily generate this sequence.

NOW TRY THIS 1-13

a. Add the first three Fibonacci numbers.
b. Add the first four Fibonacci numbers.
c. Add the first five Fibonacci numbers.
d. Add the first six Fibonacci numbers.
e. Add the first seven Fibonacci numbers.
f. What pattern is there in the sums in parts (a)–(e) and any of the remaining numbers in the Fibonacci sequence?
g. Write a rule for your pattern in part (f) using the notation for Fibonacci numbers.

Geometric Sequences

A child has 2 biological parents, 4 grandparents, 8 great grandparents, 16 great-great grandparents, and so on. The number of generational ancestors form the **geometric sequence** 2, 4, 8, 16, 32, ... Each successive term of a geometric sequence is obtained from its predecessor by multiplying by a fixed nonzero number, the **ratio**. In this example, both the first term and the ratio are 2. (The ratio is 2 because each person has two parents.) To find the nth term, a_n, examine the pattern in Table 1-9.

In Table 1-9, when the given term is written as a power of 2, the number of the term is the **exponent**. Following this pattern, the 10th term, a_{10}, is 2^{10}, or 1024, the 100th term,

Table 1-9

Number of Term	Term
1	$2 = 2^1$
2	$4 = 2 \cdot 2 = 2^2$
3	$8 = (2 \cdot 2) \cdot 2 = 2^3$
4	$16 = (2 \cdot 2 \cdot 2) \cdot 2 = 2^4$
5	$32 = (2 \cdot 2 \cdot 2 \cdot 2) \cdot 2 = 2^5$
.	.
.	.
.	.

a_{100}, is 2^{100}, and the nth term, a_n, is 2^n. Thus, the number of ancestors in the nth previous generation is 2^n. The notation used in Table 1-9 can be generalized as follows.

Definition

$$\overbrace{a^n = a \cdot a \cdot a \cdot \ldots \cdot a}^{n \text{ factors}}$$

If n is a natural number, then $a^n = \overbrace{a \cdot a \cdot a \cdot \ldots \cdot a}^{n \text{ factors}}$. If $n = 0$ and $a \neq 0$, then $a^0 = 1$.

Geometric sequences play an important role in everyday life. For example, suppose you have $1000 in a bank that pays 5% interest annually. If no money is added or taken out, then at the end of the first year you have all of the money you started with plus 5% more, that is,

Year 1: $\$1000 + 0.05(\$1000) = \$1000(1 + 0.05) = \$1000(1.05) = \$1050$

If no money is added or taken out, then at the end of the second year you would have 5% more money than the previous year.

Year 2: $\$1050 + 0.05(\$1050) = \$1050(1 + 0.05) = \$1050(1.05) = \$1102.50$

The amount of money in the account after any number of years can be found by noting that every dollar invested for one year becomes $1 + 0.05 \cdot 1$, or 1.05 dollars. Therefore, the amount in each year is obtained by multiplying the amount from the previous year by 1.05. The amounts in the bank after each year form a geometric sequence because the amount in each year (starting from year 2) is obtained by multiplying the amount in the previous year by the same number, 1.05. This is summarized in Table 1-10.

Table 1-10

Number of Term (Year)	Term (Amount at the beginning of each year)
1	$1000
2	$\$1000(1.05)^1 = \$1050.00$
3	$\$1000(1.05)^2 = \$1102.50$
4	$\$1000(1.05)^3 = \$1157.63$
.	.
.	.
.	.
n	$\$1000(1.05)^{n-1}$

Table 1-11

Number of Term	Term
1	a_1
2	$a_1 r$
3	$a_1 r^2$
4	$a_1 r^3$
5	$a_1 r^4$
.	.
.	.
.	.
n	$a_1 r^{n-1}$

Finding the *n*th Term for a Geometric Sequence

It is possible to find the *n*th term, a_n, of any geometric sequence when given the first term and the ratio. If the first term is a_1 and the ratio is r, then the terms are as listed in Table 1-11. Notice that the second term is $a_1 r$, the third term is $a_1 r^2$, and the fourth term is $a_1 r^3$. The power of r in each term is 1 less than the number of the term. This pattern continues since we multiply by r to get the next term. Thus, the *n*th term, a_n, is $a_1 r^{n-1}$. For $n = 1$, we have $a_1 r^{1-1} = a_1 r^0$. Because the first term is a_1, $a_1 r^0 = a_1$. For all numbers $r \neq 0$, we define $r^0 = 1$. For the geometric sequence $3, 12, 48, 192, \ldots$, the first term is 3 and the ratio is 4, and so the *n*th term, a_n, is given by $a_n = a_1 r^{n-1} = 3 \cdot 4^{n-1}$.

> **REMARK** The *n*th term of any geometric sequence with first term a_1 and a ratio r is $a_n = a_1 \cdot r^{n-1}$, when n is a natural number and $r \neq 0$.

NOW TRY THIS 1-14

a. Two bacteria are in a dish. The number of bacteria triples every hour. Following this pattern, find the number of bacteria in the dish after 10 hours and after n hours.

b. Suppose that instead of increasing geometrically as in part (a), the number of bacteria increases arithmetically by 3 each hour. Compare the growth after 10 hours and after n hours. Comment on the difference in growth of a geometric sequence versus an arithmetic sequence.

Other Sequences

Figurate numbers provide examples of sequences that are neither arithmetic nor geometric. Such numbers can be represented by dots arranged in the shape of certain geometric figures. The number 1 is the beginning of most patterns involving figurate numbers. The array in Figure 1-15 represents the first four terms of the sequence of **triangular numbers**.

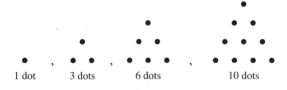

Figure 1-15

The triangular numbers can be written numerically as $1, 3, 6, 10, 15, \ldots$. The sequence $1, 3, 6, 10, 15, \ldots$ is not an arithmetic sequence because there is no common difference, as Figure 1-16 shows. It is not a geometric sequence because there is no common ratio. It is not a Fibonacci sequence.

$$\begin{array}{ccccccccc} 1 & & 3 & & 6 & & 10 & & 15 \\ & \vee & & \vee & & \vee & & \vee & \\ \text{(First difference)} & 2 & & 3 & & 4 & & 5 & \end{array}$$

Figure 1-16

However, the sequence of differences, 2, 3, 4, 5, ..., is an arithmetic sequence with difference 1, as Figure 1-17 shows. The next successive terms for the original sequence are shown in color in Figure 1-17.

$$
\begin{array}{ccccccc}
1 & 3 & 6 & 10 & 15 & 21 & 28 \\
\end{array}
$$

(First difference)
$$
\begin{array}{cccccc}
2 & 3 & 4 & 5 & 6 & 7
\end{array}
$$

(Second difference)
$$
\begin{array}{ccccc}
1 & 1 & 1 & 1 & 1
\end{array}
$$

Figure 1-17

Table 1-12 suggests a pattern for finding the next terms and the nth term for the triangular numbers. The second term is obtained from the first term by adding 2; the third term is obtained from the second term by adding 3; and so on.

Table 1-12

Number of Term	Term
1	1
2	$3 = 1 + 2$
3	$6 = 1 + 2 + 3$
4	$10 = 1 + 2 + 3 + 4$
5	$15 = 1 + 2 + 3 + 4 + 5$
.	.
.	.
.	.
10	$55 = 1 + 2 + 3 + 4 + 5 + 6 + 7 + 8 + 9 + 10$

In general, because the nth triangular number has n dots in the nth row, it is equal to the sum of the dots in the previous triangular number (the $(n - 1)$st one) plus the n dots in the nth row. Following this pattern, the 10th term is $1 + 2 + 3 + 4 + 5 + 6 + 7 + 8 + 9 + 10$, or 55, and the nth term, a_n, is $1 + 2 + 3 + 4 + 5 + \ldots + (n - 1) + n$. This problem is similar to Gauss's Problem in Section 1-1. Because of the work done in Section 1-1, we know that

$$
a_n = \frac{n(n + 1)}{2}
$$

Next consider the first four *square numbers* in Figure 1-18. These square numbers, 1, 4, 9, 16, ..., can be written as $1^2, 2^2, 3^2, 4^2$, and so on. The number of dots in the 10th array is 10^2, the number of dots in the 100th array is 100^2, and the number of dots in the nth array is n^2. The sequence of square numbers is neither arithmetic nor geometric. Investigate whether the sequence of first differences is an arithmetic sequence and tell why.

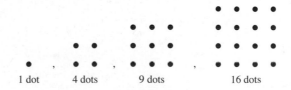

| 1 dot | 4 dots | 9 dots | 16 dots |

Figure 1-18

Example 1-6

Use differences to find a pattern. Then assuming that the pattern discovered continues, find the seventh term in each of the following sequences:

a. 5, 6, 14, 29, 51, 80, ...
b. 2, 3, 9, 23, 48, 87, ...

Solution   **a.** Following is the sequence of first differences:

$$
\begin{array}{ccccccc}
5 & 6 & 14 & 29 & 51 & 80 \\
\end{array}
$$
(First difference)   1   8   15   22   29

To discover a pattern for the original sequence, we try to find a pattern for the sequence of differences 1, 8, 15, 22, 29, ... This sequence is an arithmetic sequence with fixed difference 7:

$$
\begin{array}{ccccccc}
5 & 6 & 14 & 29 & 51 & 80 \\
\end{array}
$$
(First difference)   1   8   15   22   29
(Second difference)   7   7   7   7

Thus, the sixth term in the first difference row is $29 + 7$, or 36, and the seventh term in the original sequence is $80 + 36$, or 116. What number follows 116?

b. Because the second difference is not a fixed number, we go on to the third difference:

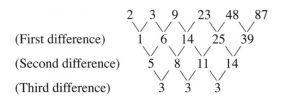

(First difference)   1   6   14   25   39
(Second difference)   5   8   11   14
(Third difference)   3   3   3

The third difference is a fixed number; therefore, the second difference is an arithmetic sequence. The fifth term in the second-difference sequence is $14 + 3$, or 17; the sixth term in the first-difference sequence is $39 + 17$, or 56; and the seventh term in the original sequence is $87 + 56$, or 143.

NOW TRY THIS 1-15 Figure 1-19 shows the first three figures of arrays of sticks.

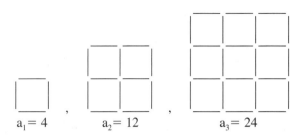

$$a_1 = 4 \qquad a_2 = 12 \qquad a_3 = 24$$

Figure 1-19

a. Draw the next array of sticks.
b. Build a table showing the term number and the number of sticks for $n = 1, 2, 3, 4$.
c. Use differences to predict the number of sticks for $n = 5, 6, 7$.
d. Is finding differences the best way to determine how many sticks there are for a_{100}? Tell how you would find a_{100} and a_n.

When asked to find a pattern for a given sequence, you first look for some easily recognizable pattern and determine whether the sequence is arithmetic or geometric. If a pattern is unclear, taking successive differences may help. *It is possible that none of the methods described reveal a pattern.*

Assessment 1-2 A

1. For each of the following sequences of figures, determine a possible pattern and draw the next figure according to that pattern:
   a.
   b.
   c.

2. In each of the following, list terms that continue a possible pattern. Which of the sequences are arithmetic, which are geometric, and which are neither?
   a. 1, 3, 5, 7, 9
   b. 0, 50, 100, 150, 200
   c. 3, 6, 12, 24, 48
   d. 10, 100, 1,000, 10,000, 100,000
   e. 9, 13, 17, 21, 25, 29
   f. 1, 8, 27, 64, 125

3. Find the 100th term and the nth term for each of the sequences in exercise 2.

4. Use a traditional clock face to determine the next three terms in the following sequence:

$$1, 6, 11, 4, 9, \ldots$$

5. In the pattern, 8, 16, 14, 10, ..., the sum of digits can be used to create the next number. In this case, each succeeding number is double the sum of the digits in the previous number.
   a. Find the next three numbers in the sequence described.
   b. Find the next three numbers in the sequence 4, 16, 49, 169, 256, _____, _____, _____. Describe the rule you used.
   ★c. Find the next three numbers in the sequence 4, 16, 37, 58, 89, 145, 42, 20, _____, _____, _____. Describe the rule you used.
   d. What will happen if the sequence in part (c) is continued indefinitely?

6. The following geometric arrays suggest a sequence of numbers:

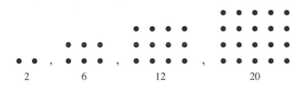

2   6   12   20

  a. Find the next three terms.
  b. Find the 100th term.
  c. Find the nth term.

7. The first windmill takes 5 matchstick squares to build, the second takes 9 to build, and the third takes 13 to build, as shown. How many matchstick squares will it take to build **(a)** the 10th windmill? **(b)** the nth windmill? **(c)** How many matchsticks will it take to build the nth windmill?

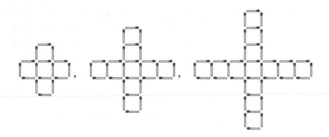

8. In the following sequence, the figures are made of cubes that are glued together. If the exposed surface needs to be painted, how many squares will be painted in **(a)** the 10th figure? **(b)** the nth figure?

9. The school population for a certain school is predicted to increase by 50 students per year for the next 10 years. If the current enrollment is 700 students, what will the enrollment be after 10 years?

10. Joe's annual income has been increasing each year by the same amount. The first year his income was $24,000, and the ninth year his income was $31,680. In which year was his income $45,120?

11. The first difference of a sequence is 2, 4, 6, 8, 10, ... Find the first six terms of the original sequence in each of the following cases:
  a. The first term of the original sequence is 3.
  b The sum of the first two terms of the original sequence is 10.
  c. The fifth term of the original sequence is 35.

12. List the next three terms to continue a pattern in each of the following. (Finding differences may be helpful.)
  a. 5, 6, 14, 32, 64, 115, 191
  b. 0, 2, 6, 12, 20, 30, 42

13. How many terms are there in each of the following sequences?
  a. 51, 52, 53, 54, ..., 151
  b. 1, 2, 2^2, 2^3, ..., 2^{60}
  c. 10, 20, 30, 40, ..., 2000
  d. 1, 2, 4, 8, 16, 32, ..., 1024

14. Find the first five terms in each of the following:
  a. $a_n = n^2 + 2$
  b. $a_n = 5n - 1$
  c. $a_n = 10^n - 1$
  d. $a_n = 3n + 2$

15. Find a counterexample for each of the following:
  a. If x is a natural number, then $(x + 5)/5 = x + 1$.
  b. If x is a natural number, then $(x + 4)^2 = x + 16$.

16. Assume that the following pattern of square tile figures, ($\square$), continues and answer the questions that follow.

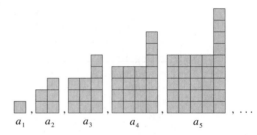

a_1 a_2 a_3 a_4 a_5

  a. How many square tiles are there in the sixth figure, a_6?
  b. How many square tiles are in the nth shape, a_n?
  c. Is there a figure that has exactly 1259 square tiles? If so, which one?

17. Find the third, fourth, and fifth terms in the sequence if $a_1 = 2, a_2 = 5$, and $a_n = 2a_{n-1} - a_{n-2}$.

18. Consider the following sequences:

$$300, 500, 700, 900, 1100, 1300, \ldots$$
$$2, 4, 8, 16, 32, 64, \ldots$$

Find the number of the term in which the geometric sequence becomes greater than the arithmetic sequence.

19. Start out with a piece of paper. Next, cut the piece of paper into five pieces. Take any one of the pieces and cut it into five pieces, and so on.
  a. What number of pieces can be obtained in this way?
  b. What is the total number of pieces obtained after nth cuts?

20. The sequence 32, a, b, c, 512, ... is a geometric sequence. Find a, b, c.

21. Assume the following pattern of dots continues:
  a. How many dots are there in a_6?
  b. How many dots are in the nth shape, a_n?

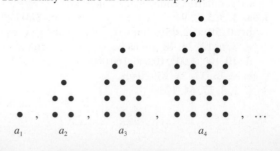

a_1 a_2 a_3 a_4

Assessment 1-2 B

1. For each of the following sequence of figures, determine a possible pattern and draw the next figure according to that pattern:

a.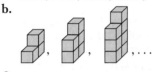

b.

c.

2. In each of the following, list terms that continue a possible pattern. Which of the sequences are arithmetic, which are geometric, and which are neither?

   a. 8, 11, 14, 17, 20, ...
   b. 1, 16, 81, 256, 625, ...
   c. 5, 15, 45, 135, 405, ...
   d. 2, 7, 12, 17, 22, ...
   e. $1, \frac{1}{2}, \frac{1}{4}, \frac{1}{8}, \frac{1}{16}, \ldots$

3. Find the 100th term and the nth term for each of the sequences in exercise 2.

4. Observe the following pattern:

$$1 + 3 = 2^2,$$
$$1 + 3 + 5 = 3^2,$$
$$1 + 3 + 5 + 7 = 4^2$$

   a. State a generalization based on this pattern.
   b. Based on the generalization in (a), find

$$1 + 3 + 5 + 7 + \ldots + 35$$

5. In the following pattern, one hexagon takes 6 toothpicks to build, two hexagons take 11 toothpicks to build, and so on. How many toothpicks would it take to build **(a)** 10 hexagons? **(b)** n hexagons?

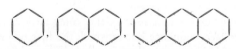

6. Each of the following figures is made of small triangles like the first one in the sequence. (The second figure is made of four small triangles.) Make a conjecture concerning the number of small triangles needed to make **(a)** the 100th figure and **(b)** the nth figure.

7. At the end of the day, a tank contains 15,360 L of water. At the end of each subsequent day, half of the water is removed and not replaced. How much water is left in the tank after 10 days?

8. Each side of each pentagon below is 1 unit long.

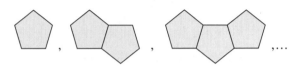

   a. Draw a next figure in the sequence.
   b. What is the perimeter (distance around) of each of the first four figures?
   c. What is the perimeter of the 100th figure?
   d. What is the perimeter of the nth figure?

9. Washington Middle School has a schedule that forms an arithmetic sequence. Each period is the same length and includes a 4th period lunch. The first three periods begin at 8:10 A.M., 9:00 A.M., and 9:50 A.M., respectively. At what time does the eighth period begin?

10. The first difference of a sequence is 3, 6, 9, 12, 15, ... Find the first six terms of the original sequence in each of the following cases:

    a. The first term of the original sequence is 3.
    b. The sum of the first two terms of the original sequence is 7.
    c. The fifth term of the original sequence is 34.

11. List the next three terms to continue a pattern in each of the following. (Finding differences may be helpful.)

    a. 3, 8, 15, 24, 35, 48, ...
    b. 1, 7, 18, 37, 67, 111, ...

12. How many terms are there in each of the following sequences?

    a. $1, 2, 2^2, 2^3, \ldots, 2^{60}$
    b. $9, 13, 17, 21, 25, \ldots, 353$
    c. $38, 39, 40, 41, \ldots, 198$

13. Find the first five terms in each of the following:

    a. $a_n = 5n + 6$
    b. $a_n = 6n - 2$
    c. $a_n = 5^n + 1$
    d. $a_n = 3n - 3$

14. Find a counterexample for each of the following:

    a. If x is a natural number, then $(3 + x)/3 = x$.
    b. If x is a natural number, then $(x - 2)^2 = x^2 - 2^2$.

15. Assume the following pattern of square tile figures, ($\square$), continues and answer the questions that follow.

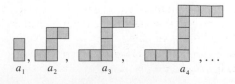

a. How many square tiles are there in the sixth figure, a_6?

b. How many square tiles are in the nth shape, a_n?

c. Is there a figure that has exactly 449 square tiles? If so, which one?

16. Write consecutive odd numbers in triangular form as shown.

$$1$$
$$3 \quad 5$$
$$7 \quad 9 \quad 11$$
$$13 \quad 15 \quad 17 \quad 19$$
and so on

a. Find the sum of each of the first five horizontal rows.

b. What patterns do you notice?

17. Find the third, fourth, and fifth terms in the sequence if $a_1 = 3$, $a_2 = 6$, and $a_n = 3a_{n-1} - 2a_{n-2}$.

18. Consider the following sequences:

$$200, 500, 800, 1100, 1400, 1700, \ldots$$
$$1, 3, 9, 27, 81, 243, \ldots$$

Find the number of the term after which the geometric sequence becomes greater than the arithmetic sequence.

19. The sequence $17, a, b, c, 1377, \ldots$ is a geometric sequence. Find a, b, c.

20. Find the sum of the first 43 terms of an arithmetic sequence in which the 11th term is 83 and the 62nd term is 440.

21. Female bees are born from fertilized eggs, and male bees are born from unfertilized eggs. This means that a male bee has only a mother, whereas a female bee has a mother and a father. If the ancestry of a male bee is traced 10 generations, how many bees are there in all 10 generations? (*Hint:* The Fibonacci sequence might be helpful.) Explain how you arrived at your answer.

Mathematical Connections 1-2

Communication

1. Explain how the two sequences in each part are the same and how they are different.

a. $2, 4, 6, 8, 10, \ldots$ and $2, 4, 8, 16, 32, \ldots$

b. $2, 4, 6, 8, 10, \ldots$ and $3, 5, 7, 9, 11, \ldots$

c. $5, 10, 15, 20, 25, \ldots$ and $50, 100, 150, 200, 250, \ldots$

2. Give two examples of how inductive reasoning might be used in everyday life. Is a conclusion based on inductive reasoning certain?

3. a. If a fixed number is added to each term of an arithmetic sequence, is the resulting sequence an arithmetic sequence? Justify the answer.

b. If each term of an arithmetic sequence is multiplied by a fixed number, will the resulting sequence always be an arithmetic sequence? Justify the answer.

c. If the corresponding terms of two arithmetic sequences are added, is the resulting sequence arithmetic?

4. A student says she read that Thomas Robert Malthus (1766–1834), a renowned British economist and demographer, claimed that the increase of population will take place, if unchecked, in a geometric sequence, whereas the supply of food will increase in only an arithmetic sequence. This theory implies that population increases faster than food production. The student is wondering why. How do you respond?

Open-Ended

5. Patterns can be used to count the number of dots on the Chinese checkerboard; two patterns are shown here. Determine several other patterns to count the dots.

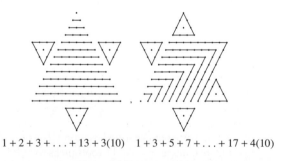

$$1 + 2 + 3 + \ldots + 13 + 3(10) \qquad 1 + 3 + 5 + 7 + \ldots + 17 + 4(10)$$

6. Make up a pattern involving figurate numbers and find a formula for the 100th term. Describe the pattern and how to find the 100th term.

7. A sequence that follows the same pattern as the Fibonacci sequence but in which the first two terms are not 1s but any numbers is called a *Fibonacci type sequence*. Choose a few such sequences and answer the questions in Now Try This 1-13. Do these sequences behave in the same way?

Cooperative Learning

8. The following pattern is called *Pascal's triangle*. It was named for the mathematician Blaise Pascal (1623–1662).

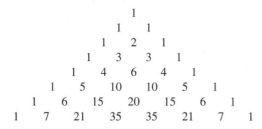

```
                    1
                 1     1
              1     2     1
           1     3     3     1
        1     4     6     4     1
     1     5    10    10     5     1
  1     6    15    20    15     6     1
1     7    21    35    35    21     7     1
```

    a. Have each person in the group find four different patterns in the triangle and then share them with the rest of the group.

    b. Add the numbers in each horizontal row. Discuss the pattern that occurs.

    c. Use what you have learned in part (b) to find the sum in the 16th row.

    d. What is the sum of the numbers in the nth row?

9. If the following pattern continued indefinitely, the resulting figure would be called the *Sierpinski triangle*, or *Sierpinski gasket*.

In a group, determine each of the following. Discuss different counting strategies.

    a. How many black triangles would be in the fifth figure?

    b. How many white triangles would be in the fifth figure?

    c. If the pattern is continued for n figures, how many black triangles will there be?

    d. If the pattern is continued for n figures, how many white triangles will there be?

10. Create your own sequence of numbers that follow a pattern. Share your sequence with the other members of your group. If they can't determine a correct rule for your pattern, explain it to them.

Questions from the Classroom

11. Joey said that 4, 24, 44, and 64 all have remainder 0 when divided by 4, so all numbers that end in 4 must have 0 remainder when divided by 4. How do you respond?

12. Al and Betty were asked to extend the sequence 2, 4, 8, ... Al said his answer of 2, 4, 8, 16, 32, 64, . . . was the correct one. Betty said Al was wrong and it should be 2, 4, 8, 14, 22, 32, 44, ... What do you tell these students?

13. A student claims that if both the numerator and denominator of a fraction are greater than the numerator and denominator, respectively, of another fraction, then the first fraction must be the greater of the two. How do you respond?

14. A student claims the sequence 6, 6, 6, 6, 6, . . . never changes, so it is neither arithmetic nor geometric. How do you respond?

15. A student claims that two terms are enough to determine any sequence. For example, 3, 6, . . . means the sequence is 3, 6, 9, 12, 15, ... How do you respond?

16. Lisa claims subtracting the first term from the last term and dividing by the common difference can find the number of terms in any finite arithmetic sequence. How do you respond?

Review Problems

17. In a baseball league consisting of 10 teams, each team plays each of the other teams twice. How may games will be played?

18. How many ways can you make change for 40¢ using only nickels, dimes, and quarters?

19. Tents hold 2, 3, 5, 6, or 12 people. What combinations of tents are possible to sleep 26 people if only one 12-person tent is used?

Third International Mathematics and Science Study (TIMSS) Questions

The numbers in the sequence 7, 11, 15, 19, 23, . . . increase by four. The numbers in the sequence 1, 10, 19, 28, 37, . . . increase by nine. The number 19 is in both sequences. If the two sequences are continued, what is the next number that is in BOTH the first and the second sequences?

TIMSS 2003, Grade 8

The three figures below are divided into small congruent triangles.

Figure 1   Figure 2   Figure 3

a. Complete the table below. First, fill in how many small triangles make up Figure 3. Then, find the number of small triangles that would be needed for the fourth figure if the sequence of figures is extended.

Figure	Numbers of Small Triangles
1	2
2	8
3	
4	

b. The sequence of figures is extended to the 7th figure. How many small triangles would be needed for Figure 7?

c. The sequence of figures is extended to the 50th figure. Explain a way to find the number of small triangles in the 50th figure that does not involve drawing it and counting the number of triangles.

TIMSS, Grade 8

BRAIN TEASER   Find the next row in the following pattern and explain your pattern:

```
            1
          1   1
          2   1
      1   2   1   1
  1   1   1   2   2   1
```

*1-3   Reasoning and Logic: An Introduction

Logic is a tool used in mathematical thinking and problem solving. It is essential for reasoning and, as pointed out in the Research Note, cannot be taught in a single unit on logic. However, we present a very brief view of some "basics" in this section. In logic, a **statement** *is a sentence that is either true or false, but not both.* The following expressions are not statements because their truth values cannot be determined without more information:

Research Note

Reasoning and proof cannot simply be taught in a single unit on logic, for example, or by "doing proofs" in geometry. Proof is a very difficult area for undergraduate mathematics students. Perhaps students at the post-secondary level find proof so difficult because their only experience in writing proofs has been in a high school geometry course, so they have a limited perspective (Moore 1994). ◆

1. She has blue eyes.
2. $x + 7 = 18$.
3. $2y + 7 > 1$.
4. $2 + 3$

5. How are you?
6. Look out!
7. Reagan was the best president.

Expressions (**1**), (**2**), and (**3**) become statements if, for (**1**), "she" is identified, and for (**2**) and (**3**), values are assigned to x and y, respectively. However, an expression involving he or she or x or y may already be a statement. For example, "If he is over 210 cm tall, then he is over 2 m tall" and "$2(x + y) = 2x + 2y$" are both statements because they are true no matter who *he* is or what the numerical values of x and y are.

Negation and Quantifiers

From a given statement, it is possible to create a new statement by forming a **negation**. The negation of a statement is *a statement with the opposite truth value of the given statement.* If a statement is true, its negation is false, and if a statement is false, its negation is true. Consider the statement "It is snowing now." The negation of this statement may be stated simply as "It is not snowing now."

Example 1-7

Negate each of the following statements:

a. $2 + 3 = 5$.
b. A hexagon has six sides.

Solution **a.** $2 + 3 \neq 5$.
   b. A hexagon does not have six sides.

Sentences like "The shirt is blue" and "The shirt is green" are statements if put in context. However, they are not negations of each other. A statement and its negation must have opposite truth values in all possible cases. If the shirt is actually red, then both of the statements are false and, hence, cannot be negations of each other. However, the statements "The shirt is blue" and "The shirt is not blue" are negations of each other because they have opposite truth values no matter what color the shirt really is.

Some statements involve **quantifiers** and are more complicated to negate. Quantifiers include words such as *all*, *some*, *every*, and *there exists*.

- The quantifiers *all*, *every*, and *no* refer to each and every element in a set and are called *universal quantifiers*.
- The quantifiers *some* and *there exists at least one* refer to one or more, or possibly all, of the elements in a set and are called *existential quantifiers*.
- *All*, *every*, and *each* have the same mathematical meaning. Similarly, *some* and *there exists at least one* both have the same meaning.

Consider the following statement involving the existential quantifier *some* and known to be true: "Some professors at Paxson University have blue eyes." This means that at least one professor at Paxson University has blue eyes. It does not rule out the possibilities that *all* the Paxson professors have blue eyes or that some of the Paxson professors do not have blue eyes. Because the negation of a true statement is false, neither "Some professors at Paxson University do not have blue eyes" nor "All professors at Paxson have blue eyes" are negations of the original statement. One possible negation of the original statement is "No professors at Paxson University have blue eyes."

To discover if one statement is a negation of another, we use arguments similar to the preceding one to determine if they have opposite truth values in all possible cases.

General forms of quantified statements with their negations follow:

Statement	**Negation**
Some *a* are *b*.	No *a* is *b*.
Some *a* are not *b*.	All *a* are *b*.
All *a* are *b*.	Some *a* are not *b*.
No *a* is *b*.	Some *a* are *b*.

Example 1-8

Negate each of the following regardless of its truth value:

a. All students like hamburgers.
b. Some people like mathematics.
c. There exists a natural number x such that $3x = 6$.
d. For all natural numbers, $3x = 3x$.

> **Solution** **a.** Some students do not like hamburgers.
> **b.** No people like mathematics.
> **c.** For all natural numbers x, $3x \neq 6$.
> **d.** There exists a natural number x such that $3x \neq 3x$.

Truth Tables and Compound Statements

To investigate the truth of statements, consider the following puzzle by one of today's foremost writers of logic puzzles, Raymond Smullyan. He has written several books on logic, including *The Lady or the Tiger?* This title is taken from the Frank Stockton short story about a prisoner who must make a choice between two doors: Behind one is a hungry tiger and behind the other is a beautiful lady.

Smullyan proposes that each door has a sign and the prisoner knows that only *one* sign is true. The sign on Door 1 reads:

IN THIS ROOM THERE IS A LADY AND
IN THE OTHER ROOM THERE IS A TIGER.

The sign on Door 2 reads:

IN ONE OF THESE ROOMS THERE IS A LADY AND
IN ONE OF THESE ROOMS THERE IS A TIGER.

With this information, the man can choose the correct door. Discuss this problem and try to find a solution before reading on.

Solution *If the sign on Door 1 is true, then the sign on Door 2 must be true. Since this cannot happen, the sign on Door 2 must be true, making the sign on Door 1 false. Because the sign on Door 1 is false, the lady can't be in Room 1 and must be in Room 2.*

There is a symbolic system defined to help in the study of logic. If p represents a statement, the negation of the statement p is denoted by $\sim p$ and is read "not p." **Truth tables** are often used to show all possible true-false patterns for statements. Table 1-13 summarizes the truth tables for p and $\sim p$.

From two given statements, it is possible to create a new, **compound statement** by using a connective such as *and*. A compound statement may be formed by combining two or more statements. For example, "It is snowing" and "The ski run is open" together with *and* give "It is snowing and the ski run is open." Other compound statements can be obtained by using the connective *or*. For example, "It is snowing or the ski run is open." The symbols $\wedge$ and $\vee$ are used to represent the connectives *and* and *or*, respectively. For example, if p represents "It is snowing" and q represents "The ski run is open," then "It is snowing and the ski run is open" is denoted by $p \wedge q$. Similarly, "It is snowing or the ski run is open" is denoted by $p \vee q$.

The truth value of any compound statement, such as $p \wedge q$, is defined using the truth value of each of the simple statements. Because each of the statements p and q may be either true or false, there are four distinct possibilities for the truth value of $p \wedge q$, as shown

Table 1-13

Statement p	Negation $\sim p$
T	F
F	T

in Table 1-14. The compound statement is the **conjunction** of p and q and is defined to be true if, and only if, both p and q are true. Otherwise, it is false.

Table 1-14

p	q	Conjunction $p \wedge q$
T	T	T
T	F	F
F	T	F
F	F	F

Table 1-15

p	q	Disjunction $p \vee q$
T	T	T
T	F	T
F	T	T
F	F	F

The compound statement $p \vee q$—that is, p *or* q—is a **disjunction**. In everyday language, *or* is not always interpreted in the same way. In logic, we use an *inclusive or.* The statement "I will go to a movie or I will read a book" means I will either go to a movie, or read a book, or do both. Hence, in logic, p or q, symbolized $p \vee q$, is defined to be false if both p and q are false and true in all other cases. This is summarized in Table 1-15.

Example 1-9

Classify each of the following as true or false:

$$p: 2 + 3 = 5 \qquad q: 2 \cdot 3 = 6 \qquad r: 5 + 3 = 9$$

a. $p \wedge q$ **c.** $\sim p \vee r$ **e.** $\sim(p \wedge q)$
b. $q \vee r$ **d.** $\sim p \wedge \sim q$ **f.** $(p \wedge q) \vee \sim r$

Solution **a.** p is true and q is true, so $p \wedge q$ is true.
  b. q is true and r is false, so $q \vee r$ is true.
  c. $\sim p$ is false and r is false, so $\sim p \vee r$ is false.
  d. $\sim p$ is false and $\sim q$ is false, so $\sim p \wedge \sim q$ is false.
  e. $p \wedge q$ is true so $\sim(p \wedge q)$ is false.
  f. $p \wedge q$ is true and $\sim r$ is true, so $(p \wedge q) \vee \sim r$ is true.

Truth tables are used not only to summarize the truth values of compound statements; they also are used to determine if two statements are logically equivalent. Two statements are **logically equivalent** if, and only if, they have the same truth values in every possible situation. If p and q are logically equivalent, we write $p \equiv q$.

Historical Note

George Boole (1815–1864), born in Lincoln, England, is called "the father of logic." At age 15, he began a teaching career. In 1849, he was appointed professor at Queens College in Cork, Ireland. In his work he employed symbols to represent concepts and developed a system of algebraic manipulations to accompany the symbols. His work was a marriage of logic and mathematics. Many of Boole's ideas, such as Boolean algebra, have applications in computer science and in the design of telephone switching devices. ◆

Example 1-10

Show that $\sim(p \wedge q) \equiv \sim p \vee \sim q$.

Solution   Two statements are logically equivalent if they have the same truth values. Truth tables for these statements are given in Table 1-16 and Table 1-17.

Table 1-16

p	q	$p \wedge q$	$\sim(p \wedge q)$
T	T	T	F
T	F	F	T
F	T	F	T
F	F	F	T

Table 1-17

p	q	$\sim p$	$\sim q$	$\sim p \vee \sim q$
T	T	F	F	F
T	F	F	T	T
F	T	T	F	T
F	F	T	T	T

Because the two statements have the same truth values, we know that $\sim(p \wedge q) \equiv \sim p \vee \sim q$.

NOW TRY THIS 1-16 Example 1-10 shows that $\sim(p \wedge q) \equiv \sim p \vee \sim q$. In the same way we can show that $\sim(p \vee q) \equiv \sim p \wedge \sim q$. We call these equivalencies **De Morgan's Laws.** Use truth tables to confirm the second De Morgan Law.

Conditionals and Biconditionals

Statements expressed in the form "if p, then q" are **conditionals**, or **implications**, and are denoted by $p \rightarrow q$. Such statements also can be read "p implies q." The "if" part of a conditional is the **hypothesis** of the implication and the "then" part is the **conclusion**. Many types of statements can be put in "if-then" form. An example follows:

Statement: All equilateral triangles have acute angles.

If-then form: If a triangle is equilateral, then it has acute angles.

$\underbrace{\hspace{3.5cm}}_{\text{Hypothesis}}$ $\underbrace{\hspace{3cm}}_{\text{Conclusion}}$

An implication may also be thought of as a promise. Suppose Betty makes the promise "If I get a raise, then I will take you to dinner." If Betty keeps her promise, the implication is true; if Betty breaks her promise, the implication is false. Consider the following four possibilities:

	p	q	
(1)	T	T	Betty gets the raise; she takes you to dinner.
(2)	T	F	Betty gets the raise; she does not take you to dinner.
(3)	F	T	Betty does not get the raise; she takes you to dinner.
(4)	F	F	Betty does not get the raise; she does not take you to dinner.

The only case in which Betty breaks her promise is when she gets her raise and fails to take you to dinner, case (2). If she does not get the raise, she can either take you to dinner or not without breaking her promise. The definition of implication is summarized in

Table 1-18

		Implication
p	**q**	**p → q**
T	T	T
T	F	F
F	T	T
F	F	T

Table 1-18. Observe that the only case for which the implication is false is when p is true and q is false.

An implication can be worded in several equivalent ways, as follows:

1. If the sun shines, then the swimming pool is open. (If p, then q.)
2. If the sun shines, the swimming pool is open. (If p, q.)
3. The swimming pool is open if the sun shines. (q if p.)
4. The sun is shining implies the swimming pool is open. (p implies q.)
5. The sun is shining only if the pool is open. (p only if q.)
6. The sun's shining is a sufficient condition for the swimming pool to be open. (p is a sufficient condition for q.)
7. The swimming pool's being open is a necessary condition for the sun to be shining. (q is a necessary condition for p.)

Any implication $p \to q$ has three related implication statements, as follows:

Statement:	If p, then q.	$p \to q$
Converse:	If q, then p.	$q \to p$
Inverse:	If not p, then not q.	$\sim p \to \sim q$
Contrapositive:	If not q, then not p.	$\sim q \to \sim p$

Example 1-11

Write the converse, the inverse, and the contrapositive for the following statement:

If I am in San Francisco, then I am in California.

Solution
Converse:	If I am in California, then I am in San Francisco.
Inverse:	If I am not in San Francisco, then I am not in California.
Contrapositive:	If I am not in California, then I am not in San Francisco.

Example 1-11 can be used to show that if an implication is true, its converse and inverse are not necessarily true. However, the contrapositive is true. Let's check these observations on the following true statement: *If a number is a natural number, the number is not 0.* The set of natural numbers are the numbers $N = \{1, 2, 3, 4, 5, 6, \ldots\}$. We check the truth of the converse, inverse, and contrapositive.

Inverse: *If a number is not a natural number, then it is 0.* This is false, since -6 is not a natural number but it also is not 0.

Converse: *If a number is not 0, then it is a natural number.* This is false, since -6 is not 0 but neither is it a natural number.

Contrapositive: *If a number is 0, then it is not a natural number.* This is true because $N = \{1, 2, 3, 4, 5, 6, \ldots\}$.

The contrapositive of the last statement is the original statement. Hence, the preceding discussion suggests that if $p \to q$ is true, its contrapositive $\sim q \to \sim p$ is also true, and if the contrapositive is true, the original statement must be true. It follows that a statement and its contrapositive cannot have opposite truth values. We summarize this in the following theorem.

> **Theorem 1–1:   Equivalence of a statement and its contrapositive**
>
> The implication $p \rightarrow q$ and its contrapositive $\sim q \rightarrow \sim p$ are logically equivalent.

Example 1-12

Use truth tables to show that $p \rightarrow q \equiv \sim q \rightarrow \sim p$.

Solution   Truth tables for these statements are given in Table 1-19 and Table 1-20.

Table 1-19

p	q	$p \rightarrow q$
T	T	T
T	F	F
F	T	T
F	F	T

Table 1-20

p	q	$\sim q$	$\sim p$	$\sim q \rightarrow \sim p$
T	T	F	F	T
T	F	T	F	F
F	T	F	T	T
F	F	T	T	T

Because the two statements have the same truth values, we know that $p \rightarrow q \equiv \sim q \rightarrow \sim p$.

NOW TRY THIS 1-17 Previously we used the implication "If Betty gets a raise (p), then she will take you to dinner (q)" to motivate the truth table for $p \rightarrow q$. We said that the only case where Betty breaks her promise is when she gets her raise and then fails to take you to dinner, that is, $p \wedge \sim q$. Therefore, $p \wedge \sim q$ is a candidate for the negation of $p \rightarrow q$. To investigate whether $p \wedge \sim q$ is the negation of $p \rightarrow q$, use truth tables to determine if $\sim(p \rightarrow q) \equiv p \wedge \sim q$.

Connecting a statement and its converse with the connective *and* gives $(p \rightarrow q) \wedge (q \rightarrow p)$. This compound statement can be written as $p \leftrightarrow q$ and usually is read "**p if, and only if, q.**" The statement "p if, and only if, q" is a **biconditional**.

NOW TRY THIS 1-18 Build a truth table to determine when a biconditional statement is true.

Valid Reasoning

In problem solving, the reasoning is said to be **valid** if the conclusion follows unavoidably from true hypotheses. Thus, in all arguments in this section we assume the hypotheses are true. Consider the following examples:

Hypotheses:	All dogs are animals.
	Goofy is a dog.
Conclusion:	Therefore, Goofy is an animal.

The statement "All dogs are animals" can be pictured with the **Euler diagram** in Figure 1-20(a).

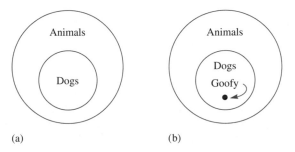

Figure 1-20

The information "Goofy is a dog" implies that Goofy now also belongs to the circle containing dogs, as pictured in Figure 1-20(b). Goofy must also belong to the circle containing animals. Thus, the reasoning is valid because it is impossible to draw a picture that satisfies the hypotheses and contradicts the conclusion.

Consider the following argument.

Hypotheses:	All elementary schoolteachers are mathematically literate.
	Some mathematically literate people are not children.
Conclusion:	Therefore, no elementary schoolteacher is a child.

Let E be the set of elementary schoolteachers, M be the set of mathematically literate people, and C be the set of children. Then the statement "All elementary schoolteachers are mathematically literate" can be pictured as in Figure 1-21(a). The statement "Some mathematically literate people are not children" can be pictured in several ways. Three of these are illustrated in Figure 1-21(b) through (d). According to Figure 1-21(d), it is possible that some elementary schoolteachers are children, and yet the given statements are satisfied. Therefore, the conclusion that "No elementary schoolteacher is a child" does not follow from the given hypotheses. Hence, the reasoning is not valid.

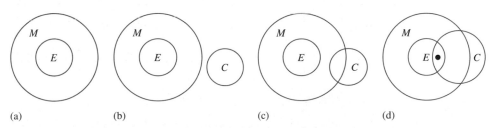

Figure 1-21

If even one picture can be drawn to satisfy the hypotheses of an argument and contradict the conclusion, the argument is not valid. However, to show that an argument is valid, *all* possible pictures must show that there are no contradictions. There must be no way to satisfy the hypotheses and contradict the conclusion if the argument is valid.

Example 1-13

Determine if the following argument is valid:

Hypotheses: In Washington, D.C., all lobbyists wear suits.
No one in Washington, D.C., over 6 ft tall wears a suit.

Conclusion: Persons over 6 ft tall are not lobbyists in Washington, D.C.

Solution   If L represents the lobbyists in Washington, D.C., and S the people who wear suits, the first hypothesis is pictured as shown in Figure 1-22(a). If W represents the people in Washington, D.C., over 6 ft tall, the second hypothesis is pictured in Figure 1-22(b). Because people over 6 ft tall are outside the circle representing suit wearers and lobbyists are in the circle S, the conclusion is valid and no person over 6 ft tall is a lobbyist in Washington, D.C.

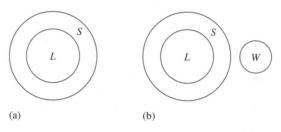

(a)   (b)

Figure 1-22

A different method for determining whether an argument is valid uses **direct reasoning** and a form of argument called the **law of detachment** (or *modus ponens*). For example, assume the following are true statements:

If the sun is shining, then we shall go to the beach.
The sun is shining.

Using these two statements, we can conclude that we shall go to the beach. In general, the law of detachment (or *modus ponens*) is stated as follows:

| If the statement "if p, then q" is true, and p is true, then q must be true.

Example 1-14

Determine whether the following argument is valid if the hypotheses are true and x is a natural number:

Hypotheses: If $x > 2$, then $x^2 > 4$.
$x > 2$.

Conclusion: Therefore, $x^2 > 4$.

Solution   Using the law of detachment, *modus ponens*, we see that the conclusion is valid.

Example 1-15

Show that $[(p \rightarrow q) \wedge p] \rightarrow q$ is always true.

Solution   A truth table for this implication is given in Table 1-21.

Table 1-21

p	q	$p \rightarrow q$	$(p \rightarrow q) \wedge p$	$[(p \rightarrow q) \wedge p] \rightarrow q$
T	T	T	T	T
T	F	F	F	T
F	T	T	F	T
F	F	T	F	T

> **REMARK** The statement $[(p \rightarrow q) \wedge\ ] \rightarrow q$ is a **tautology;** that is, a statement that is true all the time.

A different type of reasoning, **indirect reasoning**, uses a form of argument called *modus tollens*. For example, consider the following true statements:

> If a figure is a square, then it is a rectangle.
> The figure is not a rectangle.

The conclusion is that the figure cannot be a square. *Modus tollens* can be interpreted as follows:

> If we have a conditional accepted as true, and we know the conclusion is false, then the hypothesis must be false.

Example 1-16

Determine conclusions for each of the following pairs of true statements:

a. If a person lives in Boston, then the person lives in Massachusetts. Jessica does not live in Massachusetts.
b. If $x = 3$, then $2x \neq 7$. We know that $2x = 7$.

Solution   **a.** Jessica does not live in Boston (*modus tollens*).
            b. $x \neq 3$ (*modus tollens*).

The final reasoning argument we consider here involves the **chain rule (transitivity).** Consider the following statements:

> If I save, I will retire early.
> If I retire early, I will play golf.

What is the conclusion? The conclusion is that if I save, I will play golf. In general, the chain rule can be stated as follows:

> If "if p, then q" and "if q, then r" are true, then "if p, then r" is true.

People often make invalid conclusions based on advertising or other information. Assume, for example, the statement "Healthy people eat Super-Bran cereal" is true. Are the following conclusions valid?

If a person eats Super-Bran cereal, then the person is healthy.

If a person is not healthy, the person does not eat Super-Bran cereal.

If the original statement is denoted by $p \rightarrow q$, where p is "a person is healthy" and q is "a person eats Super-Bran cereal," then the first conclusion is the converse of $p \rightarrow q$, that is, $q \rightarrow p$, and the second conclusion is the inverse of $p \rightarrow q$, that is, $\sim p \rightarrow \sim q$. Hence, neither is valid.

Example 1-17

Determine valid conclusions for the following true statements:

a. If a triangle is equilateral, then it is isosceles. If a triangle is isosceles, it has at least two congruent sides.

b. If a number is a whole number, then the number is an integer. If a number is an integer, then the number is a rational number. If a number is a rational number, then the number is a real number.

Solution a. If a triangle is equilateral, then it has at least two congruent sides.

b. If a number is a whole number, then it is a real number.

Assessment 1-3A

1. Determine which of the following are statements and then classify each statement as true or false:
   a. $2 + 4 = 8$.
   b. Los Angeles is a state.
   c. What time is it?
   d. $3 \cdot 2 = 6$.
   e. This statement is false.

2. Use quantifiers to make each of the following true, where x is a natural number:
   a. $x + 8 = 11$.   b. $x^2 = 4$.
   c. $x + 3 = 3 + x$.   d. $5x + 4x = 9x$.

3. Use quantifiers to make each equation in problem 2 false.

4. Write the negation of each of the following statements:
   a. This book has 500 pages.
   b. $3 \cdot 5 = 15$.
   c. All dogs have four legs.
   d. Some rectangles are squares.
   e. Not all rectangles are squares.
   f. No dogs have fleas.

5. Identify the following as true or false:
   a. For some natural numbers x, $x < 6$ and $x > 3$.
   b. For all natural numbers x, $x > 0$ or $x < 5$.

6. Complete each of the following truth tables:
   a.

p	$\sim p$	$\sim(\sim p)$
T		
F		

   b.

p	$\sim p$	$p \vee \sim p$	$p \wedge \sim p$
T			
F			

   c. Based on part (a), is p logically equivalent to $\sim(\sim p)$?
   d. Based on part (b), is $p \vee \sim p$ logically equivalent to $p \wedge \sim p$?

7. If q stands for "This course is easy" and r stands for "Lazy students do not study," write each of the following in symbolic form:
   a. This course is easy, and lazy students do not study.
   b. Lazy students do not study, or this course is not easy.
   c. It is false that both this course is easy and lazy students do not study.
   d. This course is not easy.

8. If p is false and q is true, find the truth values for each of the following:
   a. $p \wedge q$
   b. $\sim p$
   c. $\sim(\sim p)$
   d. $p \wedge \sim q$
   e. $\sim(\sim p \wedge q)$
9. Find the truth value for each statement in problem 8 if p is false and q is false.
10. For each of the following, is the pair of statements logically equivalent?
    a. $\sim(p \vee q)$ and $\sim p \vee \sim q$
    b. $\sim(p \wedge q)$ and $\sim p \wedge \sim q$
11. Complete the following truth table:

p	q	$\sim p$	$\sim p \wedge q$
T	T		
T	F		
F	T		
F	F		

12. Write each of the following in symbolic form if p is the statement "It is raining" and q is the statement "The grass is wet."
    a. If it is raining, then the grass is wet.
    b. If it is not raining, then the grass is wet.
    c. If it is raining, then the grass is not wet.
    d. The grass is wet if it is raining.
    e. The grass is not wet implies that it is not raining.
    f. The grass is wet if, and only if, it is raining.
13. For each of the following implications, state the converse, inverse, and contrapositive:
    a. If $x = 5$, then $2x = 10$.
    b. If you do not like this book, then you do not like mathematics.
    c. If you do not use Ultra Brush toothpaste, then you have cavities.
    d. If you are good at logic, then your grades are high.
14. Consider the statement "If every digit of a number is 6, then the number is divisible by 3." Which of the following is logically equivalent to the statement?

a. If every digit of a number is not 6, then the number is not divisible by 3.
b. If a number is not divisible by 3, then some digit of the number is not 6.
c. If a number is divisible by 3, then every digit of the number is 6.
15. Write a statement logically equivalent to the statement "If a number is a multiple of 8, then it is a multiple of 4."
16. Investigate the validity of each of the following arguments:
    a. All squares are quadrilaterals.
       All quadrilaterals are polygons.
       Therefore, all squares are polygons.
    b. All teachers are intelligent.
       Some teachers are rich.
       Therefore, some intelligent people are rich.
    c. If a student is a freshman, then the student takes mathematics.
       Jane is a sophomore.
       Therefore, Jane does not take mathematics.
17. For each of the following, form a conclusion that follows logically from the given statements:
    a. Some freshmen like mathematics.
       All people who like mathematics are intelligent.
    b. If I study for the final, then I will pass the final.
       If I pass the final, then I will pass the course.
       If I pass the course, then I will look for a teaching job.
    c. Every equilateral triangle is isosceles.
       There exist triangles that are equilateral.
18. Write the following in if-then form:
    a. Every figure that is a square is a rectangle.
    b. All integers are rational numbers.
    c. Polygons with exactly three sides are triangles.
19. Use De Morgan's Laws from Now Try This 1-16 to write a negation of each of the following:
    a. $3 \cdot 2 = 6$ and $1 + 1 \neq 3$.
    b. You can pay me now or you can pay me later.

Assessment 1-3B

1. Determine which of the following are statements and then classify each statement as true or false:
   a. Shut the window.
   b. He is in town.
   c. $2 \cdot 2 = 2 + 2$.
   d. $2 + 3 = 8$.
   e. Stay put!
2. Use quantifiers to make each of the following true, where x is a natural number:
   a. $x + 0 = x$
   b. $x + 1 = x + 2$
   c. $3(x + 2) = 12$
   d. $x^3 = 8$

3. Use quantifiers to make each equation in problem 2 false.

4. Write the negation of each of the following statements:
   a. Six is less than 8.
   b. Some cats do not have nine lives.
   c. All squares are rectangles.
   d. Not all numbers are positive.
   e. Some people have blond hair.

5. Identify the following as true or false:
   a. For some natural numbers x, $x > 5$ and $x > 2$.
   b. For all natural numbers x, $x > 5$ or $x < 5$.

6. If you know that p is true, what can you conclude about the truth value of $p \lor q$, even if you don't know the truth value of q?

7. If you know that p is false, what can you conclude about the truth value of $p \rightarrow q$, even if you don't know the truth value of q?

8. If q stands for "You said goodbye" and r stands for "I said hello," write each of the following in symbolic form:
   a. You said goodbye and I said hello.
   b. You said goodbye and I did not say hello.
   c. I did not say hello or you did not say goodbye.
   d. It is false that both you said goodbye and I said hello.

9. If p is false and q is true, find the truth values for each of the following:
   a. $p \lor q$
   b. $\sim q$
   c. $\sim p \lor q$
   d. $\sim(p \lor q)$
   e. $\sim q \land \sim p$

10. Find the truth value for each statement in problem 9 if p is false and q is false.

11. For each of the following, is the pair of statements logically equivalent?
    a. $\sim(p \lor q)$ and $\sim p \land \sim q$
    b. $\sim(p \land q)$ and $\sim p \lor \sim q$

12. Complete the following truth table:

p	q	$\sim q$	$p \lor \sim q$
T	T		
T	F		
F	T		
F	F		

13. Write each of the following in symbolic form if p is the statement "You build it" and q is the statement "They will come":
    a. If you build it, they will come.
    b. If you do not build it, then they will come.
    c. If you build it, they will not come.
    d. They will come if you build it.
    e. If you do not build it, then they will not come.
    f. If they will not come, then you do not build it.

14. For each of the following implications, state the converse, inverse, and contrapositive:
    a. If $x = 3$, then $x^2 = 9$.
    b. If it snows, then classes are canceled.

15. Iris makes the true statement "If it rains, then I am going to the movies." Does it follow logically that if it does not rain, then Iris does not go to the movies?

16. Investigate the validity of each of the following arguments:
    a. All women are mortal.
       Hypatia was a woman.
       Therefore, Hypatia was mortal.
    b. All rainy days are cloudy.
       Today is not cloudy.
       Therefore, today is not rainy.
    c. Some students like skiing.
       Al is a student.
       Therefore, Al likes skiing.

17. For each of the following, form a conclusion that follows logically from the given statements:
    a. All college students are poor.
       Helen is a college student.
    b. All engineers need mathematics.
       Ron does not need mathematics.
    c. All bicycles have tires.
       All tires use rubber.

18. Write each of the following in if-then form:
    a. All natural numbers are real numbers.
    b. Every circle is a closed figure.

19. Use De Morgan's Law from Now Try This 1-16 to write a negation of each of the following:
    a. $3 + 5 \neq 9$ and $3 \cdot 5 = 15$.
    b. I am going or she is going.

Mathematical Connections 1-3

Communication

1. Explain why commands, questions, and opinions are not statements.

2. Explain how to write the negation of a quantified statement in the form "Some A are B." Give an example.

3. Describe what is meant by a compound statement.

4. a. Describe under what conditions a disjunction is true.
   b. Describe under what conditions an implication is true.

5. What does the use of an "inclusive" *or* mean?

6. Explain how to determine if two statements are logically equivalent.
7. Describe Dr. No as completely as possible.

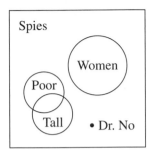

8. Consider the poem:

   For want of a nail, the shoe was lost.
   For want of a shoe, the horse was lost.
   For want of a horse, the rider was lost.
   For want of a rider, the battle was lost.
   For want of a battle, the war was lost.
   Therefore, for want of a nail, the war was lost.

   a. Write each line as an *if-then* statement.
   b. Does the conclusion follow logically. Why?
9. Most students today use Internet search engines such as Yahoo or Google. Efficient use of a search engine requires some knowledge of the connectives AND, OR, and NOT. One common type of advanced search is called a *Boolean search* (see Historical Note). With a Boolean search you can increase the accuracy of your search by specifying relationships among the keywords and phrases. The operator AND tells the search engine to search for all documents that contain both words, for example, "sports AND baseball." Go online and explore the connectives AND, OR, and NOT. Explain your findings.

Open-Ended

10. Give two examples from mathematics for each of the following:
    a. A statement and its converse are true.
    b. A statement is true, but its converse is false.
    c. An "if, and only if," true statement.
    d. An "if, and only if," false statement.

Cooperative Learning

11. Each person in a group makes five sequences of statements similar to the ones in Example 1-13 but concerning mathematical objects, each with a valid or invalid conclusion. The statements should be as varied as possible. Each group member exchanges statements with another person—not revealing which conclusions are valid and which are not—and determines which of the other person's conclusions are valid and which are not. The two group members compare their answers and discuss any discrepancies.
12. Discuss the paradox arising from the following:
    a. This textbook is 1000 pages long.
    b. The author of this textbook is Dante.
    c. The statements (a), (b), and (c) are all false.

Questions from the Classroom

13. A student says he does not understand the difference between $\sim(p \wedge q)$ and $\sim p \wedge q$. How do you explain this?
14. A student says that she does not see how a compound statement consisting of two simple sentences that are false can be true. How do you respond?
15. A student says that if the hypothesis is false, an argument cannot be valid. How do you respond?

Hint for Solving the Preliminary Problem

The strategy of *guess and check* might be useful here. For example, determine what would happen if you choose each particular bowl with its incorrect label. What fruit could you draw from that bowl, and based on that information, what labeling changes should be made? Do any of the particular choices along with logical reasoning lead to the correct labeling? Is there only one possible starting point that leads to the correct labeling?

Chapter Outline

I. Problem solving
  A. Problem solving can be guided by the following four-step process:
   1. Understanding the problem
   2. Devising a plan
   3. Carrying out the plan
   4. Looking back
  B. Important problem-solving strategies include the following:
   1. Look for a pattern.
   2. Make a table.
   3. Examine a simpler or special case of the problem to gain insight into the solution of the original problem.
   4. Identify a subgoal.
   5. Examine related problems and determine if the same technique can be applied.
   6. Work backward.
   7. Write an equation.
   8. Draw a diagram.
   9. Guess and check.
   10. Use indirect reasoning.
   11. Use direct reasoning.
  C. Beware of mind-sets!
II. Mathematical patterns
  A. Patterns are an important part of problem solving.
  B. Patterns are used in **inductive reasoning** to form conjectures. Inductive reasoning is the method of making generalizations based on observations and patterns. A **conjecture** is a statement that is thought to be true but that has not yet been proved to be true or false. One way to prove a conjecture false is to produce a **counterexample**.
  C. A **sequence** is a group of terms in a definite order.
   1. Arithmetic sequence: Each successive term is obtained from the previous one by the addition of a fixed number called the **difference**. The nth term, a_n, is given by $a_n = a_1 + (n - 1)d$, where a_1 is the first term and d is the difference.
   2. Geometric sequence: Each successive term is obtained from its predecessor by multiplying it by a fixed, nonzero number called the **ratio**. The nth term, a_n, is given by $a_1 r^{n-1}$, where a_1 is the first term and r is the ratio.

   3. In a **recursive pattern**, after one or more terms are given to start the sequence, each successive term of the sequence is obtained from previous term(s). The **Fibonacci sequence**, $1, 1, 2, 3, 5, 8, 13, 21, \ldots$, is an example of a recursive sequence where $F_1 = 1, F_2 = 1, F_{n+2} = F_n + F_{n+1}$.
   4. $a^n = \underbrace{a \cdot a \cdot a \cdot a \cdot a \cdot \ldots \cdot a}_{n \text{ factors}}$, where $n \neq 0$.
   5. $a^0 = 1$, where $a \neq 0$.
   6. Finding differences for a sequence is one technique for finding the next terms.
* III. Reasoning and logic
  A. A **statement** is a sentence that is either true or false but not both.
  B. The **negation** of a statement is a statement with the opposite truth value of the given statement. The negation of p is denoted by $\sim p$.
  C. The **compound statement** $p \wedge q$ is the **conjunction** of p and q and is defined to be true if, and only if, both p and q are true.
  D. The compound statement $p \vee q$ is the **disjunction** of p and q and is true if either p or q or both are true.
  E. Statements of the form "if p, then q" are **conditionals** or **implications** and are false only if p is true and q is false.
  F. Given the conditional $p \rightarrow q$, the following can be found:
   1. Converse: $q \rightarrow p$
   2. Inverse: $\sim p \rightarrow \sim q$
   3. Contrapositive: $\sim q \rightarrow \sim p$
  G. If $p \rightarrow q$ is true, the converse and the inverse are not necessarily true, but the contrapositive is true.
  H. Two statements are **logically equivalent** if, and only if, they have the same truth value. An implication and its contrapositive are logically equivalent.
  I. The statement "$p \rightarrow q$ and $q \rightarrow p$" is written $p \leftrightarrow q$, a **biconditional**, and referred to as "p if, and only if, q."
  J. Laws to determine the validity of arguments include the **law of detachment** (*modus ponens*), *modus tollens*, and the **chain rule**.

Chapter Review

1. List three more terms that complete a pattern in each of the following:
   a. 0, 1, 3, 6, 10, ____, ____, ____
   b. 52, 47, 42, 37, ____, ____, ____
   c. 6400, 3200, 1600, 800, ____, ____, ____
   d. 1, 2, 3, 5, 8, 13, ____, ____, ____
   e. 2, 5, 8, 11, 14, ____, ____, ____
   f. 1, 4, 16, 64, ____, ____, ____
   g. 0, 4, 8, 12, ____, ____, ____
   h. 1, 8, 27, 64, ____, ____, ____

2. Classify each sequence in problem 1 as arithmetic, geometric, or neither.

3. Find a possible nth term in each of the following:
   a. 5, 8, 11, 14, …
   b. 0, 7, 26, 63, …
   c. 3, 9, 27, 81, 243, …

4. Find the first five terms of the sequences whose nth term is given as follows:
   a. $3n - 2$
   b. $n^2 + n$
   c. $4n - 1$

5. Find the following sums:
   a. $2 + 4 + 6 + 8 + 10 + … + 200$
   b. $51 + 52 + 53 + 54 + … + 151$

6. Produce a counterexample, if possible, to disprove each of the following:
   a. If two odd numbers are added, then the sum is odd.
   b. If a number is odd, then it ends in a 1 or a 3.
   c. If two even numbers are added, then the sum is even.

7. Complete the following magic square; that is, complete the square so that the sum in each row, column, and diagonal is the same.

16	3	2	13
	10		
9		7	12
4		14	

8. How many people can be seated at 12 square tables lined up end to end if each table individually holds four persons?

9. A shirt and a tie sell for $9.50. The shirt costs $5.50 more than the tie. What is the cost of the tie?

10. If fence posts are to be placed in a row 5 m apart, how many posts are needed for 100 m of fence?

11. A total of 129 players entered a single-elimination handball tournament. In the first round of play, the top-seeded player received a bye and the remaining 128 players played in 64 matches. Thus, 65 players entered the second round of play. How many matches must be played to determine the tournament champion?

12. a. Use patterns to predict the next two lines.

$$3 = \frac{3 \cdot 2}{2}$$

$$3 + 6 = \frac{6 \cdot 3}{2}$$

$$3 + 6 + 9 = \frac{9 \cdot 4}{2}$$

$$3 + 6 + 9 + 12 = \frac{12 \cdot 5}{2}$$

   b. Show that this pattern works in general for adding consecutive multiples of 3.

13. If a complete turn of a car tire moves a car forward 6 ft, how many turns of the tire occur before the tire goes off its 50,000 mi warranty?

14. The members of Mrs. Grant's class are standing in a circle, they are evenly spaced and are numbered in order. The student with number 7 is standing directly across from the student with number 17. How many students are in the class?

15. A carpenter has three large boxes. Inside each large box are two medium-sized boxes. Inside each medium-sized box are five small boxes. How many boxes are there altogether?

16. How many triangles are there in the following figure? Explain your reasoning.

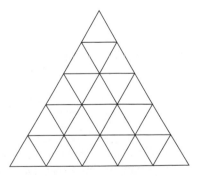

17. Mary left her home and averaged 16 km/hr riding her bicycle on an uphill trip to Larry's house. On the return trip over the same route, she averaged 20 km/hr. If it took 4 hr to make the return trip, how much cycling time did the entire trip take?

18. Use differences to find the next term in the following pattern:

5, 15, 37, 77, 141, ____

19. An ant farm can hold 100,000 ants. If the farm held 1500 ants on the first day, 3000 ants on the second day, 6000 ants on the third day, and so on forming a geometric sequence, in how many days will the farm be full?

20. Toma's team entered a mathematics contest where teams of students compete by answering questions that are worth either 3 points or 5 points. No partial credit was given. Toma's team scored 44 points on 12 questions. How many 5-point questions did the team answer correctly?

21. Three pieces of wood are needed for a project. They are to be cut from a 90-cm-long piece of wood. The longest piece is to be 3 times as long as the middle-sized piece and the shortest piece is to be 10 cm shorter than the middle-sized piece. Can this be done with two cuts? If so, tell how.

22. I am thinking of a number. If I double it, square the result, then divide by 2 and add 8, I get 40. What is my number?

* 23. Explain the difference between the following two statements: (i) All students passed the final. (ii) Some students passed the final.

* 24. Which of the following are statements?
  a. The moon is inhabited.
  b. $3 + 5 = 8$.
  c. $x + 7 = 15$.
  d. Some women have Ph.D.'s in mathematics.

* 25. Negate each of the following:
  a. Some women smoke.
  b. $3 + 5 = 8$.
  c. All heavy-metal rock is loud.
  d. Beethoven wrote only classical music.

* 26. Write the converse, inverse, and contrapositive of the following: If we have a rock concert, someone will faint.

* 27. Use truth tables to show that $p \rightarrow \sim q \equiv q \rightarrow \sim p$.

* 28. Construct truth tables for each of the following:
  a. $(p \wedge \sim q) \vee (p \wedge q)$
  b. $[(p \vee q) \wedge \sim p] \rightarrow q$

* 29. Find valid conclusions for the following hypotheses:
  a. All Americans love Mom and apple pie.
     Joe Czernyu is an American.
  b. Steel eventually rusts.
     The Statue of Liberty has a steel structure.
  c. Albertina passed Math 100 or Albertina dropped out.
     Albertina did not drop out.

* 30. Write the following argument symbolically and then determine its validity:
  If you are fair-skinned, you will sunburn.
  If you sunburn, you will not go to the dance.
  If you do not go to the dance, your parents will want to know why you didn't go to the dance.
  Your parents do not want to know why you didn't go to the dance.
  Therefore, you are not fair-skinned.

* 31. State whether the conclusion is valid in each case and tell why.
  a. If Bob scores at least 80 on the final, he will pass the course.
     Bob did not pass the course.
     Therefore, Bob did not score at least 80 on the final.
  b. If you build it, they will come.
     You build it.
     Therefore, they will come.

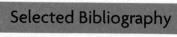

Selected Bibliography

Ameis, J. "Stories Invite Children to Solve Mathematical Problems." *Teaching Children Mathematics* 8 (January 2002): 260–264.

Artz, S., and E. Armour-Thomas. "Development of a Cognitive-Metacognitive Framework for Protocol Analysis of Mathematical Problem Solving in Small Groups." *Cognition and Instruction* 9 (1992): 137–175.

Bloom, B., and L. Broder. *Problem Solving Processes of College Students.* Chicago, IL: University of Chicago Press, 1950.

Buschman, L. "Becoming a Problem-Solver." *Teaching Children Mathematics* 9 (October 2002): 98–103.

_____. "Children Who Enjoy Problem-Solving." *Teaching Children Mathematics* 9 (May 2003): 539–544.

———"Teaching Problem Solving in Mathematics." *Teaching Children Mathematics* 10 (February 2004): 302–309.

Buyea, R. "Problem Solving in a Structured Mathematics Program." *Teaching Children Mathematics* 13 (February 2007): 300–307.

Clement, L., and J. Bernhard. "A Problem-Solving Alternative to Using Key Words." *Mathematics Teaching in the Middle School* 10 (March 2005): 360–365.

Crespo, C., and A. Kyriakides. "To Draw or Not to Draw: Exploring Children's Drawings for Solving Mathematics Problems." *Teaching Children Mathematics* 14 (September 2007): 118–125.

Dugdale, S., J. Matthews, and S. Guerro. "The Art of Posing Problems and Guiding Investigations." *Mathematics Teaching in the Middle School* 10 (October 2004): 140–147.

Ferrucci, B., B. Yeap, and J. Carter. "A Modeling Approach for Enhancing Problem-Solving in the Middle Grades." *Mathematics Teaching in the Middle School* 8 (May 2003): 470–475.

Hatano, G., and K. Ingaki. "Sharing Cognition through Collective Comprehension Activity." In *Perspectives on Socially Shared Cognition*, edited by L. Resnick, J. Levine, and S. Teasley. Washington, D.C.: American Psychological Association, 1991, pp. 331–348.

Hylton-Lindsay, A. "Problem-Solving, Patterns, Probability, Pascal, and Palindromes." *Mathematics Teaching in the Middle School* 8 (February 2003): 288–293.

Hoosain, E., and R. Chance. "Problem-Solving Strategies of First Graders." *Teaching Children Mathematics* 10 (May 2004): 474–479.

Kantowski, M. "Problem Solving." In *Mathematics Education Research: Implications for the 80s*, edited by E. Fennema. Alexandria, VA: ASCD, 1981.

Krebs, A. "Studying Students' Reasoning in Writing Generalizations." *Mathematics Teaching in the Middle School* 10 (February 2005): 284–287.

Lee, L., and V. Freiman. "Developing Algebraic Thinking through Pattern Exploration." *Mathematics Teaching in the Middle School* 11 (May 2006): 428–433.

Lester, F. "Developmental Aspects of Children's Ability to Understand Mathematical Proof." *Journal for Research in Mathematics Education* 6 (1975): 14–25.

Maher, C., and A. Martino. "The Development of the Idea of Mathematical Proof: A Five-Year Case Study." *Journal for Research in Mathematics Education* 27 (March 1996): 194–214.

McLeod, D. "Affective Issues in Research on Teaching Mathematical Problem Solving." In *Teaching and Learning Mathematical Problem Solving: Multiple Research Perspectives*, edited by E. Silver. Hillsdale, NJ: LEA, 1985.

Martinez-Cruz, A., and E. Barger. "Adding a la Gauss." *Mathematics Teaching in the Middle School* 10 (October 2004): 152–155.

Moore, R. C. "Making the Transition to Formal Proof." *Educational Studies in Mathematics* 27, no. 3 (1994): 249–266.

Moran, G. "X-tending the Fibonacci Sequence." *Mathematics Teaching in the Middle School* 7 (April 2002): 452–454.

O'Donnell, B. "On Becoming a Better Problem-Solving Teacher." *Teaching Children Mathematics* 12 (March 2006): 346–351.

Pólya, G. *How to Solve It.* Princeton, NJ: Princeton University Press, 1945.

———. *Mathematical Discovery, Combined Edition.* New York: John Wiley & Sons, Inc., 1981.

Reeves, A., and R. Gleichowski. "Engaging Contexts for the Game of Nim." *Mathematics Teaching in the Middle School* 12 (December 2006/January 2007): 251–255.

Reid, D. "Describing Reasoning in Early Elementary School Mathematics." *Teaching Children Mathematics* 9 (December 2002): 234–237.

Rigelman, N. "Fostering Mathematical Thinking and Problem Solving: The Teacher's Role." *Teaching Children Mathematics* 13 (February 2007): 308–314.

Rivera, F., and J. Becker. "Figural and Numerical Modes of Generalizing in Algebra." *Mathematics Teaching in the Middle School* 11 (November 2005): 198–203.

Rubenstein, R. "Building Explicit and Recursive Forms of Patterns with the Function Game." *Mathematics Teaching in the Middle School* 7 (April 2002): 426–431.

Siegel, M. "The Sum of Cubes: An Activity Review and Conjecture." *Mathematics Teaching in the Middle School* 10 (March 2005): 356–359.

Smith, M., A. Hillen, and C. Catania. "Using Pattern Tasks to Develop Mathematical Understandings and Set Classroom Norms." *Mathematics Teaching in the Middle School* 13 (August 2007): 38–44.

Steele, D. "Understanding Students' Problem-Solving Knowledge Through Their Writing." *Mathematics Teaching in the Middle School* 13 (September 2007): 102–109.

Strutchens, M. "Multicultural Literature as a Context for Problem Solving: Children and Parents Learning Together." *Teaching Children Mathematics* 8 (April 2002): 448–454.

Thomas, K. "Students THINK: A Framework for Improving Problem Solving." *Teaching Children Mathematics* 13 (September 2006): 86–95.

Turner, E., and B. Strawhun. "Posing Problems That Matter: Investigating School Overcrowding." *Teaching Children Mathematics* 13 (May 2007): 457–463.

Van de Walle, J. *Elementary and Middle School Mathematics: Teaching Developmentally.* New York: Addison Wesley Longman, 2001.

Van Reeuwijk, M., and M. Wijers. "Students' Construction of Formulas in Context." *Mathematics Teaching in the Middle School* 2 (February 1997): 230–236.

Verzoni, K. "Turning Students into Problem Solvers." *Mathematics Teaching in the Middle School* 3 (October 1997): 102–107.

Wallace, A. "Anticipating Student Responses to Improve Problem Solving." *Mathematics Teaching in the Middle School* 12 (May 2007): 504–511.

Wells, P., and D. Coffey. "Are They Wrong? Or Did They Just Answer a Different Question?" *Teaching Children Mathematics* 12 (November 2005): 202–207.

Whitin, D. "Problem Posing in the Elementary Classroom." *Teaching Children Mathematics* 13 (August 2006): 14–18.

Wilburne, J. "Preparing Preservice Elementary Teachers to Teach Problem Solving." *Teaching Children Mathematics* 12 (May 2006): 454–463.

Williams, E. "An Investigation of Senior High School Students' Understanding of Mathematical Proof." *Journal for Research in Mathematics Education* 11 (May 1980): 165–166.

Yolles, A. "Using Friday Puzzles to Discover Arithmetic Sequences." *Mathematics Teaching in the Middle School* 9 (November 2003): 180–185.

Numeration Systems and Sets

Preliminary Problem

On a tour, several Egyptologists/guides were talking about the people on the latest mathematics education conference tour, all of them from either Mississippi or Tennessee. The guides could not remember the total number in the group; however, they compiled the following statistics about the group. It contained 26 Mississippi females, 17 Tennessee women, 17 Tennessee males, 29 girls, 44 Mississippi residents, 29 women, and 24 Mississippi adults. Find the total number of people in the group.

The National Council of Teachers of Mathematics (NCTM) in 2006 recognized the need for more coherence in the elementary grades mathematics curriculum. In its document *Curriculum Focal Points for Prekindergarten through Grade 8 Mathematics: A Quest for Coherence*, the Council suggested specific topics that must be taught in grades pre–k through 8. In that document as early as pre–k, we find the following:

> Children develop an understanding of the meanings of whole numbers [0, 1, 2, 3, . . .] and recognize the number of objects in small groups without counting—the first and most basic mathematical algorithm. They understand that number words refer to quantity. They use one-to-one correspondence to solve problems by matching sets and comparing number amounts and in counting objects to 10 and beyond. (p. 11)

In this chapter, we introduce different and early historical counting systems. Next, we present a mathematical historical development by Georg Cantor that added structure to the number system and provided methods for treating the system theoretically.

2-1 Numeration Systems

In this section, we introduce various number systems and compare them to the system of numbers that we use today in the United States. By comparing our current system with ancient systems that used other bases, we may develop a clearer appreciation of numbers. Our system now relies on 10 digits—0, 1, 2, 3, 4, 5, 6, 7, 8, and 9. The written symbols for the digits, such as 2 or 5, are **numerals**. Different cultures developed different numerals over the years to represent numbers. Table 2-1 shows other representations along with how they relate to the digits 0 through 9 and the number 10.

Table 2-1

Babylonian		▼	▼▼	▼▼▼	▼▼▼▼	▼▼▼ ▼▼	▼▼▼ ▼▼▼	▼▼▼▼ ▼▼▼	▼▼▼▼ ▼▼▼▼	▼▼▼▼▼ ▼▼▼▼	<
Egyptian		ǀ	ǀǀ	ǀǀǀ	ǀǀǀǀ	ǀǀǀ ǀǀ	ǀǀǀ ǀǀǀ	ǀǀǀǀ ǀǀǀ	ǀǀǀǀ ǀǀǀǀ	ǀǀǀ ǀǀǀ ǀǀǀ	∩
Mayan	𝋠	•	••	•••	••••	—	• —	•• —	••• —	•••• —	— —
Greek		α	β	γ	δ	∊	ϕ	ζ	η	υ	ι
Roman		I	II	III	IV	V	VI	VII	VIII	IX	X
Hindu	०	१	२	३	४	५	६	७	८	९	
Arabic	٠	١	٢	٣	٤	٥	٦	٧	٨	٩	
Hindu-Arabic	0	1	2	3	4	5	6	7	8	9	10

Table 2-1 shows rudiments of different sets of numbers. A **numeration system** is a collection of properties and symbols agreed upon to represent numbers systematically. Through the study of various numeration systems, we explore the evolution of our current system, the Hindu-Arabic system.

Hindu-Arabic Numeration System

The Hindu-Arabic numeration system that we use today was developed by the Hindus and transported to Europe by the Arabs—hence, the name *Hindu-Arabic*. The Hindu-Arabic system relies on the following properties:

1. All numerals are constructed from the 10 digits—0, 1, 2, 3, 4, 5, 6, 7, 8, and 9.
2. Place value is based on powers of 10, the number base of the system.

Because the Hindu-Arabic system is based on powers of 10, the system is a base-ten, or a decimal, system. **Place value** assigns a value to a digit depending on its placement in a numeral. To find the value of a digit in a whole number, we multiply the place value of the digit by its **face value**, where the face value is a digit. For example, in the numeral 5984, the 5 has place value "thousands," the 9 has place value "hundreds," the 8 has place value "tens," and the 4 has place value "units," as seen in Figure 2-1.

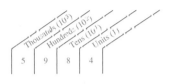

Figure 2-1

We could write 5984 in **expanded form** as $5 \cdot 10^3 + 9 \cdot 10^2 + 8 \cdot 10 + 4 \cdot 1$. In the expanded form of 5984, exponents are used. For example, 1000, or $10 \cdot 10 \cdot 10$, is written as 10^3. In this case, 10 is a **factor** of the product. In general, we have the following:

Definition of a^n

If a is any number and n is any natural number, then
$$a^n = \underbrace{a \cdot a \cdot a \cdot \ldots \cdot a}_{n \text{ factors}}.$$

A set of base-ten blocks, shown in Figure 2-2, consists of *units*, *longs*, *flats*, and *blocks*, representing 1, 10, 100, and 1000, respectively. Such base-ten blocks, a subset of multibase blocks, can be used to teach place value.

◆ *Research Note*

Base-ten blocks improve students' understanding of place value, accuracy in computing multidigit addition and subtraction problems, and explanations of the trading/regrouping involved. (Fuson 1992). ◆

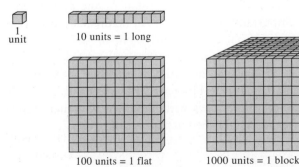

1 unit   10 units = 1 long

100 units = 1 flat   1000 units = 1 block

Figure 2-2

$$1 \, \text{long} \rightarrow 10^1 = 1 \text{ row of 10 units}$$
$$1 \, \text{flat} \rightarrow 10^2 = 1 \text{ row of 10 longs, or 100 units}$$
$$1 \, \text{block} \rightarrow 10^3 = 1 \text{ row of 10 flats, or 100 longs, or 1000 units}$$

Students trade blocks by regrouping. That is, they take a set of base-ten blocks representing a number and trade them until they have the fewest possible pieces representing the same number. For example, suppose you have 58 units and want to trade them for other base-ten blocks. You start trading the units for as many longs as possible. Five sets of 10 units each can be traded for 5 longs. Thus, 58 units can be traded so that you now have 5 longs and 8 units. In terms of numbers, this is analogous to rewriting 58 as $5 \cdot 10 + 8$. The use of manipulatives has been shown to improve student understanding, as seen in the Research Note on page 63.

Example 2-1

What is the fewest number of pieces you can receive in a fair exchange for 11 flats, 17 longs, and 16 units?

Solution

11 flats	17 longs	16 units	(16 units = 1 long and 6 units)
	1 long	6 units	(Trade)
11 flats	18 longs	6 units	(After the first trade)

11 flats	18 longs	6 units	(18 longs = 1 flat and 8 longs)
1 flat	8 longs		(Trade)
12 flats	8 longs	6 units	(After the second trade)

	12 flats	8 longs	6 units	(12 flats = 1 block and 2 flats)
1 block	2 flats			(Trade)
1 block	2 flats	8 longs	6 units	(After the third trade)

Therefore, the fewest number of pieces is $1 + 2 + 8 + 6 = 17$. This trading is analogous to rewriting $11 \cdot 100 + 17 \cdot 10 + 16$ as $1 \cdot 10^3 + 2 \cdot 10^2 + 8 \cdot 10 + 6$, which implies that there are 1286 units.

Historical Note

The invention of the Hindu-Arabic numeration system is considered one of the most important developments in mathematics. The system was introduced in India and then transmitted by the Arabs to North Africa and Spain and then to the rest of Europe. Historians trace the use of zero as a placeholder to the fourth century BCE (Before the Common Era). Arab mathematicians extended the decimal system to include fractions. The Italian mathematician Fibonacci, also known as Leonardo of Pisa (1170–1250), studied in Algeria and brought back with him the new numeration system, which he described and used in a book he published in 1202. ◆

NOW TRY THIS 2-1 Use trading with base-ten blocks (shown in Figure 2-2) to write 3 blocks, 12 flats, 11 longs, and 17 units as a Hindu-Arabic number.

Next, we discuss other numeration systems. The study of such systems will give us a historical perspective on the development of numeration systems and will help us better understand our own system.

Tally Numeration System

The **tally numeration system** used single strokes, or tally marks, to represent each object that was counted; for example, the first 10 counting numbers are

$$I, II, III, IIII, IIIII, IIIIII, IIIIIII, IIIIIIII, IIIIIIIII, IIIIIIIIII$$

In a tally system there is a correspondence between the marks and the items being counted. The system is simple, but it requires many symbols, especially when numbers become greater. Also as numbers become greater, they are harder to read.

As we see in the "Barney Google and Snuffy Smith" cartoon, the tally system can be improved by *grouping*. We see that the tallies are grouped into fives by placing a diagonal across four tallies to make a group of five. Grouping makes it easier to read the numeral.

Egyptian Numeration System

The Egyptian numeration system, which dates back to about 3400 BCE, used tally marks. The first nine numerals in the Egyptian system in Table 2-1 show the use of tally marks. The Egyptians improved on the system based only on tally marks by developing a *grouping system* to represent certain sets of numbers. This makes the numbers easier to record. For example, the Egyptians used a heel bone symbol, $\cap$, to stand for a grouping of 10 tally marks.

$$IIIIIIIIII \rightarrow \cap$$

Table 2-2 shows other numerals that the Egyptians used in their system, and some of the symbols from the Karnak temple in Luxor are depicted in Figure 2-3.

Table 2-2

Egyptian Numeral	Description	Hindu-Arabic Equivalent
I	Vertical staff	1
∩	Heel bone	10
9	Scroll	100
₰	Lotus flower	1,000
⧸	Pointing finger	10,000
◔	Polliwog or burbot	100,000
⚲	Astonished man	1,000,000

Figure 2-3

Note that in Figure 2-3, the symbol for 100 is carved in a different direction from that of Table 2-2.

In its simplest form, the Egyptian system involved an **additive property**; that is, the value of a number was the sum of the face values of the numerals. The Egyptians customarily wrote the numerals in decreasing order from left to right, as in ◔999∩∩II. The number can be converted to base ten as shown here:

◔	represents	100,000	
999	represents	300	(100 + 100 + 100)
∩∩	represents	20	(10 + 10)
II	represents	2	(1 + 1)
◔999∩∩II	represents	100,322	

NOW TRY THIS 2-2

a. Use the Egyptian system to represent 1,312,322.

b. Use the Hindu-Arabic system to represent ⌒⌒𓏤𓏤𓏤∩∩∩||||.

c. What disadvantages do you see in the Egyptian system compared to the Hindu-Arabic system?

Babylonian Numeration System

The Babylonian numeration system was developed at about the same time as the Egyptian system. The symbols in Table 2-3 were made using a stylus either vertically or horizontally on clay tablets.

Table 2-3

Babylonian Numeral	Hindu-Arabic Equivalent
▼	1
<	10

The Babylonian numerals 1 through 59 were similar to the Egyptian numerals, but the vertical staff and the heel bone were replaced by the symbols shown in Table 2-3. For example, **<<▼▼** represented 22.

The Babylonian numeration system used a *place value system*. Numbers greater than 59 were represented by repeated groupings of 60, much as we use groupings of 10 today. For example, **▼▼ <<** might represent $2 \cdot 60 + 20$, or 140. The space indicates that **▼▼** represents $2 \cdot 60$ rather than 2. Numerals immediately to the left of a second space have a value $60 \cdot 60$ times their face value, and so on.

<< ▼	represents	$20 \cdot 60 + 1$, or 1201
<▼ <▼ ▼	represents	$11 \cdot 60 \cdot 60 + 11 \cdot 60 + 1$, or $11 \cdot 60^2 + 11 \cdot 60 + 1$, or 40,261
▼ <▼ <▼ ▼	represents	$1 \cdot 60 \cdot 60 \cdot 60 + 11 \cdot 60 \cdot 60 + 11 \cdot 60 + 1$, or $1 \cdot 60^3 + 11 \cdot 60^2 + 11 \cdot 60 + 1$, or 256,261

The initial Babylonian system was inadequate by today's standards. For example, the symbol **▼▼** could have represented 2 or $2 \cdot 60$. Later, the Babylonians introduced the symbol **⸶** as a placeholder for missing position values. Using this symbol, **< <<▼** represents $10 \cdot 60 + 21$ and **< ⸶ <<▼** represents $10 \cdot 60^2 + 0 \cdot 60 + 21$. In this sense, **⸶** represented 0.

NOW TRY THIS 2-3

a. Use the Babylonian system to represent 12,321.

b. Use the Hindu-Arabic system to represent **▼▼ ▼▼ <▼ ▼**.

c. What advantages does the Hindu-Arabic system have over the Babylonian system?

Mayan Numeration System

In the early development of numeration systems, people frequently used parts of their bodies to count. Fingers could be matched to objects to stand for one, two, three, four, or five objects. Two hands could then stand for a set of ten objects. In warmer climates where people went barefoot, people may have used their toes as well as their fingers for counting. The Mayans introduced an attribute that was not present in the Egyptian or early Babylonian systems, namely, a symbol for zero. The Mayan system used only three symbols, which Table 2-4 shows, and based their system on 20 with vertical groupings.

Table 2-4

Mayan Numeral	Hindu-Arabic Equivalent
•	1
——	5
 ⬭	0

The symbols for the first numerals in the Mayan system are shown in Table 2-1. Notice the groupings of five, where each horizontal bar represents a group of five. Thus, the symbol for 19 was ≣, or three 5s and four 1s. The symbol for 20 was ⬭, which represents one group of 20 plus zero 1s. In Figure 2-4(a), we have $2 \cdot 5 + 3 \cdot 1$, or thirteen groups of 20 plus $2 \cdot 5 + 1 \cdot 1$, or eleven 1s, for a total of 271. In Figure 2-4(b), we have $3 \cdot 5 + 1 \cdot 1$, or 16, groups of 20 and zero 1s, for a total of 320.

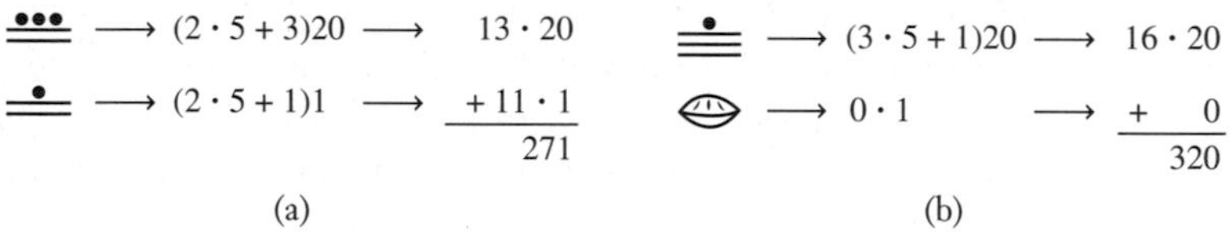

$$\overset{\bullet\bullet\bullet}{\equiv} \longrightarrow (2 \cdot 5 + 3)20 \longrightarrow 13 \cdot 20$$
$$\overset{\bullet}{=} \longrightarrow (2 \cdot 5 + 1)1 \longrightarrow \underline{+ 11 \cdot 1}$$
$$271$$
(a)

$$\overset{\bullet}{\equiv} \longrightarrow (3 \cdot 5 + 1)20 \longrightarrow 16 \cdot 20$$
$$⬭ \longrightarrow 0 \cdot 1 \longrightarrow \underline{+ \quad 0}$$
$$320$$
(b)

Figure 2-4

In a true base-twenty system, the place value of the symbols in the third position vertically from the bottom should be 20^2, or 400. However, the Mayans used $20 \cdot 18$, or 360, instead of 400. (The number 360 is an approximation of the length of a calendar year, which consisted of 18 months of 20 days each, plus 5 "unlucky" days.) Thus, instead of place values of $1, 20, 20^2, 20^3, 20^4$, and so on, the Mayans used $1, 20, 20 \cdot 18, 20^2 \cdot 18, 20^3 \cdot 18$, and so on. For example, in Figure 2-5(a), we have $5 + 1$ (or 6) groups of 360, plus $2 \cdot 5 + 2$ (or 12) groups of 20, plus $5 + 4$ (or 9) groups of 1, for a total of 2409. In Figure 2-5(b), we have $2 \cdot 5$ (or 10) groups of 360, plus 0 groups of 20, plus two 1s, for a total of 3602. Spacing is important in the Mayan system. For example, if two horizontal bars are placed close together, as in ≡, the symbols represent $5 + 5 = 10$. If the bars are spaced apart, as in ⹀, then the value is $5 \cdot 20 + 5 \cdot 1 = 105$.

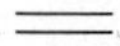

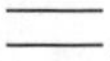

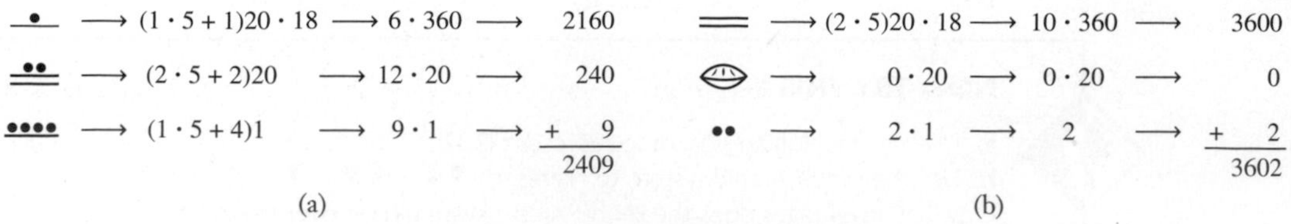

$$\overset{\bullet}{=} \longrightarrow (1 \cdot 5 + 1)20 \cdot 18 \longrightarrow 6 \cdot 360 \longrightarrow 2160$$
$$\overset{\bullet\bullet}{=} \longrightarrow (2 \cdot 5 + 2)20 \longrightarrow 12 \cdot 20 \longrightarrow 240$$
$$\overset{\bullet\bullet\bullet\bullet}{=} \longrightarrow (1 \cdot 5 + 4)1 \longrightarrow 9 \cdot 1 \longrightarrow \underline{+ \quad 9}$$
$$2409$$
(a)

$$\equiv \longrightarrow (2 \cdot 5)20 \cdot 18 \longrightarrow 10 \cdot 360 \longrightarrow 3600$$
$$⬭ \longrightarrow 0 \cdot 20 \longrightarrow 0 \cdot 20 \longrightarrow 0$$
$$\bullet\bullet \longrightarrow 2 \cdot 1 \longrightarrow 2 \longrightarrow \underline{+ \quad 2}$$
$$3602$$
(b)

Figure 2-5

Roman Numeration System

The Roman numeration system was used in Europe in its early form from the third century BCE. It remains in use today, as seen on cornerstones, on the opening pages of books, and on the faces of some clocks. The Roman system uses only the symbols shown in Table 2-5.

Table 2-5

Roman Numeral	Hindu-Arabic Equivalent
I	1
V	5
X	10
L	50
C	100
D	500
M	1000

Roman numerals can be combined by using an additive property. For example, MDCLXVI represents $1000 + 500 + 100 + 50 + 10 + 5 + 1 = 1666$; CCCXXVIII represents 328, and VI represents 6.

To avoid repeating a symbol more than three times, as in IIII, a **subtractive property** was introduced in the Middle Ages. For example, I is less than V, so if it is to the left of V, it is subtracted. Thus, IV has a value of 5 − 1, or 4, and XC represents 100 − 10, or 90. Some extensions of the subtractive property could lead to ambiguous results. For example, IXC could be 91 or 89. By custom, 91 is written XCI and 89 is written LXXXIX. In general, only one smaller number symbol can be to the left of a larger number symbol and the pair must be one of those listed in Table 2-6.

Table 2-6

Roman Numeral	Hindu-Arabic Equivalent
IV	5 – 1, or 4
IX	10 – 1, or 9
XL	50 – 10, or 40
XC	100 – 10, or 90
CD	500 – 100, or 400
CM	1000 – 100, or 900

In the Middle Ages, a bar was placed over a Roman number to multiply it by 1000. The use of bars is based on a **multiplicative property**. For example, $\overline{V}$ represents $5 \cdot 1000$, or 5000, and $\overline{CDX}$ represents $410 \cdot 1000$, or 410,000. To indicate even greater numbers, more bars appear. For example, $\overline{\overline{V}}$ represents $(5 \cdot 1000)1000$, or 5,000,000; $\overline{\overline{\overline{CXI}}}$ represents $111 \cdot 1000^3$, or 111,000,000,000; and $\overline{CXI}$ represents $110 \cdot 1000 + 1$, or 110,001.

Several properties might be used to represent some numbers, for example:

$$\overline{D}CLIX = \underbrace{\underbrace{(500 \cdot 1000)}_{\text{Multiplicative}} + \underbrace{(100 + 50)}_{\text{Additive}} + \underbrace{(10 - 1)}_{\text{Subtractive}}}_{\text{Additive}} = 500{,}159$$

Also, the Roman system evolved over time, so there exist examples where not all rules are followed.

Other Number Base Systems

To better understand our base-ten system and to investigate some of the problems that students might have when learning the Hindu-Arabic system, we investigate similar systems that have different number bases.

Base Five

The Luo peoples of Kenya used a *quinary*, or base-five, system. A system of this type can be modeled by counting with only one hand. The digits available for counting are 0, 1, 2, 3, and 4. In the "one-hand system," or base-five system, you count 1, 2, 3, 4, 10, where *10 represents one hand and no fingers*. Counting in base five proceeds as shown in Figure 2-6. We write the small "five" below the numeral as a reminder that the number is written in base five. If no base is written, a number is assumed to be in base ten. Also note that 1, 2, 3, 4 are the same, and have the same meaning, in both base five and base ten.

One-Hand System	Base-Five Symbol	Base-Five Blocks
0 fingers	0_{five}	
1 finger	1_{five}	
2 fingers	2_{five}	
3 fingers	3_{five}	
4 fingers	4_{five}	
1 hand and 0 fingers	10_{five}	
1 hand and 1 finger	11_{five}	
1 hand and 2 fingers	12_{five}	
1 hand and 3 fingers	13_{five}	
1 hand and 4 fingers	14_{five}	
2 hands and 0 fingers	20_{five}	
2 hands and 1 finger	21_{five}	

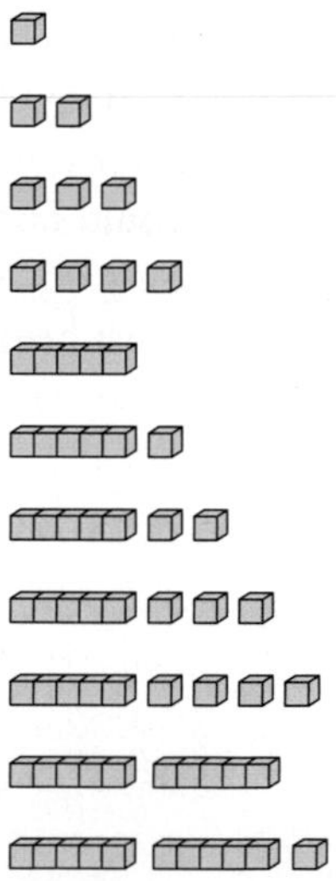

Figure 2-6

Counting in base five is similar to counting in base ten. Because we have only five digits (0_{five}, 1_{five}, 2_{five}, 3_{five}, and 4_{five}), 4_{five} plays the role of 9 in base ten. Figure 2-7 shows how we can find the number that comes after 34_{five} by using base-five blocks.

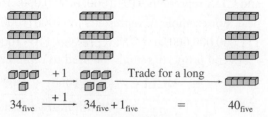

$$34_{\text{five}} \xrightarrow{+1} 34_{\text{five}} + 1_{\text{five}} = 40_{\text{five}}$$

Figure 2-7

What number follows 44_{five}? There are no more two-digit numbers in the system after 44_{five}. In base ten, the same situation occurs at 99. We use 100 to represent ten 10s, or one 100. In the base-five system, we need a symbol to represent five 5s. To continue the analogy with base ten, we use 100_{five} to represent one group of five 5s, zero groups of five, and zero units. To distinguish from "one hundred" in base ten, the name for 100_{five}, is read "one-zero-zero base five." The number 100 means $1 \cdot 10^2 + 0 \cdot 10^1 + 0$, whereas the number 100_{five} means $(1 \cdot 10^2 + 0 \cdot 10^1 + 0)_{\text{five}}$, or $1 \cdot 5^2 + 0 \cdot 5^1 + 0$, or 25.

Examples of base-five numerals along with their base-five block representations and conversions to base ten are given in Figure 2-8. Multibase blocks will be used throughout the text to illustrate various concepts.

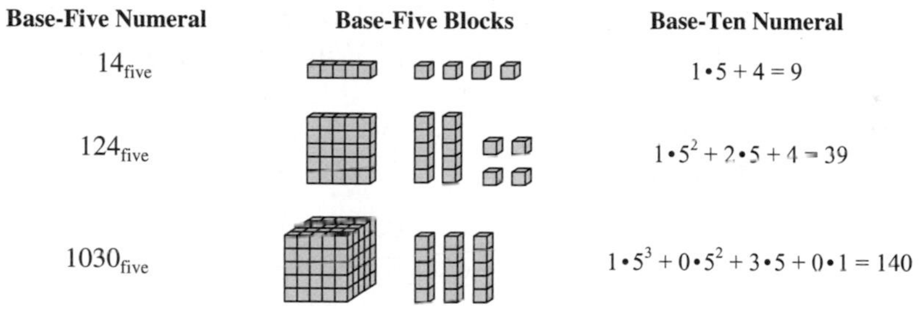

Base-Five Numeral	Base-Five Blocks	Base-Ten Numeral
14_{five}		$1 \cdot 5 + 4 = 9$
124_{five}		$1 \cdot 5^2 + 2 \cdot 5 + 4 = 39$
1030_{five}		$1 \cdot 5^3 + 0 \cdot 5^2 + 3 \cdot 5 + 0 \cdot 1 = 140$

Figure 2-8

Example 2-2

Convert 11244_{five} to base ten.

Solution

$$\begin{aligned}
11244_{\text{five}} &= 1 \cdot 5^4 + 1 \cdot 5^3 + 2 \cdot 5^2 + 4 \cdot 5^1 + 4 \cdot 1 \\
&= 1 \cdot 625 + 1 \cdot 125 + 2 \cdot 25 + 4 \cdot 5 + 4 \cdot 1 \\
&= 625 + 125 + 50 + 20 + 4 \\
&= 824
\end{aligned}$$

Example 2-2 suggests a method for changing a base-ten numeral to a base-five numeral using powers of 5. To convert 824 to base five, we divide by successive powers of 5 starting with the greatest power of 5 less than or equal to 824. A shorthand method for illustrating this conversion is the following:

$5^4 = 625 \rightarrow 625\overline{)824}\ \underline{1}$ How many groups of 625 in 824?
$\qquad\qquad\qquad -625$

$5^3 = 125 \rightarrow 125\overline{)199}\ \underline{1}$ How many groups of 125 in 199?
$\qquad\qquad\qquad -125$

$5^2 = 25 \rightarrow 25\overline{)74}\ \underline{2}$ How many groups of 25 in 74?
$\qquad\qquad\qquad -50$

$5^1 = 5 \rightarrow 5\overline{)24}\ \underline{4}$ How many groups of 5 in 24?
$\qquad\qquad\qquad -20$

$5^0 = 1 \rightarrow 1\overline{)4}\ \underline{4}$ How many 1s in 4?
$\qquad\qquad\qquad -4$
$\qquad\qquad\qquad\ \ 0$

Thus, $824 = 11244_{\text{five}}$.

NOW TRY THIS 2-4 A different method of converting 824 to base five is shown using successive divisions by 5. The quotient in each case is placed below the dividend and the remainder is placed on the right, on the same line with the quotient. The answer is read from bottom to top, that is, as 11244_{five}.

```
5|824
5|164   4
5|32    4
5|6     2
  1     1
```

a. Why does this method work?
b. Use this method to convert 728 to base five.

Calculators with the integer division feature—$\boxed{\text{INT}\div}$ on a Texas Instruments calculator or $\boxed{\div R}$ on a Casio—can be used to change base-ten numbers to different number bases. For example, to convert 8 to base five, we enter $\boxed{8}$ $\boxed{\text{INT}\div}$ $\boxed{5}$ $\boxed{=}$ and obtain $\underset{Q}{\rightharpoondown 1 \rightharpoondown}$ $\underset{R}{\rightharpoondown 3 \rightharpoondown}$. This implies that $8 = 13_{\text{five}}$. Will this technique work to convert 34 to base five? Why or why not?

Base Two

Historians tell of early tribes that used base two. Some aboriginal tribes still count "one, two, two and one, two twos, two twos and one, ..." Because base two has only two digits, it is called the **binary system**. Base two is especially important because of its use in computers. One of the two digits is represented by the presence of an electrical signal and the other by the absence of an electrical signal. Although base two works well for some purposes, it is inefficient for everyday use because multidigit numbers are reached very rapidly in counting in this system. In the following cartoon, we see an infant working with the binary system.

Conversions from base two to base ten, and vice versa, can be accomplished in a manner similar to that used for base-five conversions.

Example 2-3

a. Convert 10111_{two} to base ten.
b. Convert 27 to base two.

Solution a. $10111_{\text{two}} = 1 \cdot 2^4 + 0 \cdot 2^3 + 1 \cdot 2^2 + 1 \cdot 2^1 + 1$
$= 16 + 0 + 4 + 2 + 1$
$= 23$

b. 16 | 27 | 1   How many groups of 16 in 27?
-16

8 | 11 | 1   How many groups of 8 in 11?
-8

4 | 3 | 0   How many groups of 4 in 3?
-0

2 | 3 | 1   How many groups of 2 in 3?
-2

1 | 1 | 1   How many 1s in 1?
-1
0

Alternative Solution:

2 | 27
2 | 13 | 1
2 | 6 | 1
2 | 3 | 0
1 | 1

Thus, 27 is equivalent to 11011_{two}.

Base Twelve

Another commonly used number-base system is the base-twelve, or the duodecimal ("dozens"), system. Eggs are bought by the dozen, and pencils are bought by the *gross* (a dozen dozen). In base twelve, there are 12 digits, just as there are 10 digits in base ten, 5 digits in base five, and 2 digits in base two. In base twelve, new symbols are needed to represent the following groups of x's:

$$10\,x\text{'s} \qquad\qquad 11\,x\text{'s}$$

$$\overbrace{xxxxxxxxxx} \quad \text{and} \quad \overbrace{xxxxxxxxxxx}$$

The new symbols chosen are T and E, respectively, so that the base-twelve digits are 0, 1, 2, 3, 4, 5, 6, 7, 8, 9, T, and E. Thus, in base twelve we count "1, 2, 3, 4, 5, 6, 7, 8, 9, T, E, 10, 11, 12, …, 17, 18, 19, $1T$, $1E$, 20, 21, 22, …, 28, 29, $2T$, $2E$, 30, …."

Example 2-4

▲▲▲▲▲▲▲▲▲▲

a. Convert $E2T_{twelve}$ to base ten.
b. Convert 1277 to base twelve.

Solution **a.** $E2T_{twelve} = 11 \cdot 12^2 + 2 \cdot 12^1 + 10 \cdot 1$
$= 11 \cdot 144 + 24 + 10$
$= 1584 + 24 + 10$
$= 1618$

b. 144 | 1277 | 8   How many groups of 144 in 1277?
-1152

12 | 125 | T How many groups of 12 in 125?
-120

1 | 5 | 5   How many 1s in 5?
-5
0

Thus, $1277 = 8T5_{twelve}$.

Example 2-5

Rob used base twelve to write the following:

$$g36_{twelve} = 1050_{ten}$$

What is the value of g?

Solution   Using expanded form, we could write the following equations:

$$g \cdot 12^2 + 3 \cdot 12 + 6 \cdot 1 = 1050$$
$$144g + 36 + 6 = 1050$$
$$144g + 42 = 1050$$
$$144g = 1008$$
$$g = 7$$

Assessment 2-1A

1. For each of the following, tell which numeral represents the greater number and why:
   a. $\overline{\text{MCDXXIV}}$ and $\overline{\overline{\text{MCDXXIV}}}$
   b. 4632 and 46,032
   c. and
   d. 999∩∩II and
   e. and

2. For each of the following, name both the succeeding and preceding numbers (one more and one less):
   a. MCMXLIX
   b.
   c.
   d.

3. If the cornerstone represents when a building was built and it reads MCMXXII, when was this building built?

4. Write each of the following in Roman symbols:
   a. 121   b. 42

5. Complete the following table, which compares symbols for numbers in different numeration systems:

	Hindu-Arabic	Babylonian	Egyptian	Roman	Mayan
a.	72				
b.		⟨ ▼▼			
c.			⸸99∩∩III		

6. For each of the following decimal numerals, give the place value of the underlined digit:
   a. 827,367
   b. 8,421,000

7. Rewrite each of the following as a base-ten numeral:
   a. $3 \cdot 10^6 + 4 \cdot 10^3 + 5$
   b. $2 \cdot 10^4 + 1$

8. A certain three-digit whole number has the following properties: The hundreds digit is greater than 7; the tens digit is an odd number; and the sum of the digits is 10. What could the number be?

9. Study the following counting frame. In the frame, the value of each dot is represented by the number in the box below the dot. For example, the following figure represents the number 154:

••	•••	••
64	8	1

What numbers are represented in the frames in (a) and (b)?

a.

•••	••	•
25	5	1

b.

•		•	•
8	4	2	1

10. Write the base-four numeral for the base-four representation shown.

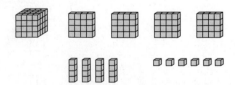

11. Write the first 15 counting numbers for each of the following bases:
    a. base two   b. base four

12. How many different digits are needed for base twenty?

13. Write 2032_{four} in expanded notation.

14. Determine the greatest three-digit number in each of the following bases:
    a. base two   **b.** base twelve

15. Find the number preceding and succeeding each of the following:
    a. $EE0_{twelve}$
    b. 100000_{two}
    c. 555_{six}

16. What, if anything, is wrong with the following numerals?
    a. 204_{four}
    b. 607_{five}

17. The smallest number of base-four blocks needed to represent 214 is _____ block(s) _____ flat(s) _____ long(s) _____ unit(s).

18. Draw multibase blocks to represent 231_{five}.

19. An introduction to base five is especially suitable for early learning in elementary school, as children can think of making change using quarters, nickels, and pennies. Use only these coins to answer the following:
    a. What is the fewest number of quarters, nickels, and pennies you can receive in a fair exchange for two quarters, nine nickels, and eight pennies?
    b. How could you use the approach in (a) to write 73 in base five?

20. Recall that with base-ten blocks, 1 long = 10 units, 1 flat = 10 longs, and 1 block = 10 flats (see Figure 2-2). In a set of multibase pieces make all possible exchanges to obtain the smallest number of pieces and write the corresponding numeral in the given base.
    a. Ten flats in base ten
    b. Twenty flats in base twelve

21. Change 42_{eight} to base two without first changing to base ten.

22. Write each of the following numbers in base ten:
    a. 432_{five} **b.** 101101_{two}
    c. $92E_{twelve}$

23. You are asked to distribute $900 in prize money. The dollar amounts for the prizes are $625, $125, $25, $5, and $1. How should this $900 be distributed in order to give the fewest number of prizes?

24. Convert each of the following:
    a. 58 days to weeks and days
    b. 29 hours to days and hours

25. For each of the following, find b if possible. If not possible, tell why.
    a. $b2_{seven} = 44_{ten}$ **b.** $5b2_{twelve} = 734_{ten}$

26. The Chinese abacus, depicted as follows, shows the number 5857. (*Hint:* The beads above the bar represent 5s, 50s, 500s, and 5000s.)

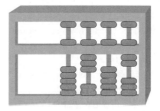

    Discuss how the number 5857 is depicted and show how the number 4869 could be depicted.

27. On a calculator, using only the non-zero number keys, fill the calculator's display to show the greatest number possible if each key may be used only once.

28. In a game called WIPEOUT, we are to "wipe out" digits from a calculator's display without changing any of the other digits. "Wipeout" in this case means to replace the chosen digit(s) with a 0. For example, if the initial number is 54,321 and we are to wipe out the 4, we could subtract 4000 to obtain 50,321. Complete the following two problems and then try other numbers or challenge another person to wipe out a number from the number you have placed on the screen:
    a. Wipe out the 2s from 32,420.
    b. Wipe out the 5 from 67,357.

Assessment 2-1B

1. For each of the following, tell which numeral represents the greater number and why:
   a. $\overline{\text{MDCXXIV}}$ and $\overline{\text{MCDXXIV}}$
   b. 3456 and 30,456
   c. $<\blacktriangledown$ and $< \; \blacktriangledown\blacktriangledown$
   d. 99∩I and 999
   e. ⚏ and ☉

2. For each of the following, name both the preceding and succeeding numbers (one more and one less):

   a. $\overline{\text{MI}}$
   b. CMXCIX
   c. $< \; <\blacktriangledown$
   d. ⚟9
   e. ⚏

3. Write each of the following in Roman symbols:
   a. 89
   b. 5202

4. Complete the following table, which compares symbols for numbers in different numeration systems:

Hindu-Arabic	Babylonian	Egyptian	Roman	Mayan
a. 78				
b.	< ▼			
c.				

5. For each of the following decimal numbers, give the place value of the underlined digit:
   a. 97,9̲98
   b. 8̲10,485

6. Rewrite each of the following as a base-ten numeral:
   a. $3 \cdot 10^3 + 5 \cdot 10^2 + 6 \cdot 10$
   b. $9 \cdot 10^6 + 9 \cdot 10 + 9$

7. A two-digit number has the property that the units digit is 4 less than the tens digit and the tens digit is twice the units digit. What is the number?

8. On a counting frame, the following number is represented. What might the number be? Explain your reasoning.

•	••	••
27	9	1

9. Write the base-three numeral for the base-three representation shown.

10. Write the first 10 counting numbers for each of the following bases:
   a. base three
   b. base eight

11. How many different digits are needed for base eighteen?

12. Write 2022_{three} in expanded form.

13. Determine the greatest three-digit number in each of the following bases:
   a. base three   **b.** base twelve

14. Find the number preceding and succeeding each of the following:
   a. 100_{seven}
   b. 10000_{two}
   c. 101_{two}

15. What, if anything, is wrong with the following numerals?
   a. 306_{four}
   b. 1023_{two}

16. The smallest number of base-three blocks needed to represent 79 is _____ block(s) _____ flat(s) _____ long(s) _____ unit(s).

17. Draw multibase blocks to represent 1001_{two}.

18. Using a number system based on dozen and gross, how would you describe the representation for 277?

19. Without converting to base ten, tell which is the lesser for each of the following pairs:
   a. $EET9E_{\text{twelve}}$ or $E0T9E_{\text{twelve}}$
   b. 1011011_{two} or 101011_{two}
   c. 50555_{six} or 51000_{six}

20. What is the smallest number of pieces of multibase blocks that can be used to write the corresponding numeral in the given base?
   a. 10 longs in base four
   b. 10 longs in base three

21. Convert each of the following base-ten numbers to a numeral in the indicated bases:
   a. 234 in base four
   b. 1876 in base twelve
   c. 303 in base three
   d. 22 in base two

22. Write each of the following numbers in base ten:
   a. 432_{six}
   b. 11011_{two}
   c. $E29_{\text{twelve}}$

23. *Who Wants the Money*, a game show, distributes prizes that are powers of 2. What is the minimum number of prizes that could be distributed from $900?

24. A coffee shop sold 1 cup, 1 pint, and 1 quart of coffee. Express the number of cups sold in base two.

25. For each of the following, find b, if possible. If not possible, tell why.
   a. $b3_{\text{eight}} - 31_{\text{ten}}$
   b. $1b2_{\text{twelve}} = 1534_{\text{six}}$

26. Using only the number keys on a calculator, fill the display to show the greatest four-digit number if each key can be used only once.

Mathematical Connections 2-1

Communication

1. Ben claims that zero is the same as nothing. Explain how you as a teacher would respond to Ben's statement.

2. What are the major drawbacks to each of the following systems?
   a. Egyptian   **b.** Babylonian   **c.** Roman

3. a. Why are large numbers in the United States written with commas separating groups of three digits?

b. Find examples from other countries that do not use commas to separate groups of three digits.

4. Marcy bets that if you do a series of mathematical computations and activities, she can guess the color you chose. First you must find your special number by completing the following:

Take the number of your birth month.

Add 24.

Add the difference you obtain when you subtract the number of your birth month from 12.

Divide by 3.

Add 13.

The result is your special number.

To each letter in the alphabet, assign a number that is the letter's order when alphabetical, that is, $a = 1, b = 2, c = 3, d = 4$, and so on. Find the letter that corresponds to your special number. Next write the name of a color that starts with this letter. What color will Marcy predict is your color? Explain why this works.

Open-Ended

5. An inspector of weights and measures uses a special set of weights to check the accuracy of scales. Various weights are placed on a scale to check accuracy of any amount from 1 oz through 15 oz. What is the least number of weights the inspector needs? What weights are needed to check the accuracy of scales from 1 oz through 15 oz? From 1 oz through 31 oz?

Cooperative Learning

6. a. Create a numeration system with unique symbols and write a paragraph explaining the properties of the system.

b. Complete the following table using the system:

Hindu-Arabic Numeral	Your System Numeral
1	
5	
10	
50	
100	
5,000	
10,000	
115,280	

Questions from the Classroom

7. While studying different number bases, a student asks if it is possible to have a negative number for a base. What do you tell this student?

8. A student claims that the Roman system is a base-ten system since it has symbols for 10, 100, and 1000. How do you respond?

9. When using Roman numerals, a student asks whether it is correct to write $\overline{\text{II}}$, as well as MI for 1001. How do you reply?

10. A parent complains about the use of manipulatives in the classroom and likens it to the use of fingers to count. What do you say?

Third International Mathematics and Science Study (TIMSS) Questions

Which digit is in the hundreds place in 2345?

   a. 2   **b.** 3   **c.** 4   **d.** 5

   TIMSS, 2003, Grade 4

Which of these is a name for 9740?

   a. Nine thousand seventy-four

   b. Nine thousand seven hundred forty

   c. Nine thousand seventy-four hundred

   d. Nine hundred seventy-four thousand

   TIMSS, 2003, Grade 4

National Assessment of Educational Progress (NAEP) Question

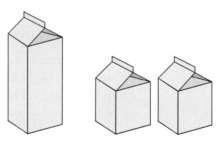

1 quart = 2 pints

Mr. Harper bought 6 pints of milk. How many quarts of milk is this equal to?

a. 3   **b.** 4   **c.** 6   **d.** 12

NAEP, 2007, Grade 4

2-2 Describing Sets

Having had a look at different numeration systems and recalling some things about the Hindu-Arabic system that is used in the United States today, it is time to consider one of the major developments around the turn of the twentieth century that gave a theoretical basis for the number system we have become so accustomed to using. In the years from 1871 through 1884, Georg Cantor created *set theory* and had a profound effect on research and mathematics teaching.

Sets, and relations between sets, are a basis to teach children the *concept* of whole numbers and the concept of "less than" as well as addition, subtraction, and multiplication of whole numbers. We introduce set notation, relations between sets, set operations, and their properties. The concept of a set helps define relations and functions (see Chapter 4).

In NCTM's *Principles and Standards for School Mathematics* (2000), we find:

> Instructional programs from pre-kindergarten through grade 12 should enable all students to –
>
> • understand numbers, ways of representing numbers, relationships among numbers, and number systems;
>
> • understand meanings of operations and how they relate to one another. . . . (p. 32)

Teacher understanding of number and operation can be enhanced by a deep understanding of the mathematics behind the number system. A part of that understanding includes the notions of sets.

The Language of Sets

A **set** is understood to be any collection of objects. Individual objects in a set are **elements**, or **members**, of the set. For example, each letter is an element of the set of letters in the English language. The set A of lowercase letters of the English alphabet can be written in set notation as follows:

$$A = \{a, b, c, d, e, f, g, h, i, j, k, l, m, n, o, p, q, r, s, t, u, v, w, x, y, z\}$$

The order in which the elements are written makes no difference, and *each element is listed only once*. Thus, $\{b, o, k\}$ and $\{k, o, b\}$ are considered to be the same set.

We symbolize an element belonging to a set by using the symbol $\in$. For example, $b \in A$. If an element does not belong to a set, we use the symbol $\notin$. For example, $3 \notin A$.

Historical Note

Georg Cantor (1845–1918) was born in St. Petersburg, Russia. His family moved to Frankfurt when he was 11. Against his father's advice to become an engineer, Cantor pursued a career in mathematics and obtained his doctorate in Berlin at age 22. Most of his academic career was spent at the University of Halle. His hope of becoming a professor at the University of Berlin did not materialize because his work gained little recognition during his lifetime.

However, after his death Cantor's work was praised as an "astonishing product of mathematical thought, one of the most beautiful realizations of human activity." ◆

> **REMARK** In mathematics, the same letter, one lowercase and the other uppercase, cannot be freely interchanged. For example, in the set A mentioned earlier, we have $b \in A$ but $B \notin A$.

For a given set to be useful in mathematics, it must be **well defined**; that is, if we are given a set and some particular object, then we must be able to tell whether the object does or does not belong to the set. For example, the set of all citizens of Pasadena, California, who ate rice on January 1, 2009, is well defined. We personally may not know if a particular resident of Pasadena ate rice or not, but that resident either belongs or does not belong to the set. On the other hand, the set of all tall people is not well defined because there is no clear meaning of "tall."

We may use sets to define mathematical terms. For example, the set N of *natural numbers* is defined by the following:

$$N = \{1, 2, 3, 4, \dots\}$$

An *ellipsis* (three dots) indicates that the sequence continues in the same manner.

Two common methods of describing sets are the **listing method** and **set-builder notation** as seen in the examples:

Listing method: $C = \{1, 2, 3, 4\}$

Set-builder notation: $C = \{x \mid x \in N \text{ where } x < 5\}$

The latter notation is read as follows:

C	$=$	$\{$	x	$\mid$	$x \in N$	where	$x < 5\}$
Set C	is	the	all	such	x is a	where	x is less
	equal	set	elements	that	natural		than 5
	to	of	x		number		

Set-builder notation is useful when the individual elements of a set are not known or they are too numerous to list. For example, the set of decimals between 0 and 1 can be written as

$$D = \{x \mid x \text{ is a decimal between 0 and 1}\}$$

This is read "D is the set of all elements x such that x is a decimal between 0 and 1." It would be impossible to list all the elements of D. Hence the set-builder notation is indispensable here.

Example 2-6

Write the following sets using set-builder notation:

a. $\{2, 4, 6, 8, 10, \dots\}$ **b.** $\{1, 3, 5, 7, \dots\}$

Solution **a.** $\{x \mid x \text{ is an even natural number}\}$. Or because every even natural number can be written as 2 times some natural number, this set can be written as $\{x \mid x = 2n, \text{ where } n \in N\}$ or, in a somewhat simpler form, as $\{2n \mid n \in N\}$.

b. $\{x \mid x \text{ is an odd natural number}\}$. Or because every odd natural number can be written as some even number minus 1, this set can be written as $\{x \mid x = 2n - 1, \text{ where } n \in N\}$ or $\{2n - 1 \mid n \in N\}$.

Example 2-7   Each of the following sets is described in set-builder notation. Write each of the sets by listing its elements.

a. $A = \{2k + 1 \mid k = 3, 4, 5\}$
b. $B = \{a^2 + b^2 \mid a = 2 \text{ or } 3, \text{ and } b = 2, 3, \text{ or } 4\}$

Solution   **a.** We substitute $k = 3, 4, 5$ in $2k + 1$ and obtain the corresponding values shown in Table 2-7. Thus, $A = \{7, 9, 11\}$.

Table 2-7

k	$2k + 1$
3	$2 \cdot 3 + 1 = 7$
4	$2 \cdot 4 + 1 = 9$
5	$2 \cdot 5 + 1 = 11$

b. Here $a = 2$ or 3 and $b = 2, 3,$ or 4. Table 2-8 shows all possible combinations of a and b and the corresponding values of $a^2 + b^2$. Thus, $B = \{8, 13, 20, 18, 25\}$. Notice that 13 appears twice in the table but only once in the set. Why?

Table 2-8

b / a	2	3	4
2	$2^2 + 2^2 = 8$	$2^2 + 3^2 = 13$	$2^2 + 4^2 = 20$
3	$3^2 + 2^2 = 13$	$3^2 + 3^2 = 18$	$3^2 + 4^2 = 25$

As noted earlier, the order in which the elements are listed does not matter. If sets A and B are equal, written $A = B$, then every element of A is an element of B, and every element of B is an element of A. If A does not equal B, we write $A \neq B$.

Definition of Equal Sets

Two sets are **equal** if, and only if, they contain exactly the same elements.

One-to-One Correspondence

One of the most useful tools in mathematics is a **one-to-one correspondence** between two sets. Here the sets may be equal or not. For example, consider the set of people $P = \{$Tomas, Dick, Mari$\}$ and the set of swimming lanes $S = \{1, 2, 3\}$. Suppose each person in P is to swim in a lane numbered 1, 2, or 3 so that no two people swim in the same lane. Such a person-lane pairing is a one-to-one correspondence. One way to exhibit a one-to-one correspondence is shown in Figure 2-9 with arrows connecting corresponding elements.

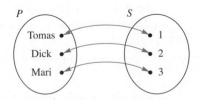

Figure 2-9

Other possible one-to-one correspondences exist between the sets P and S. All six possible one-to-one correspondences between sets P and S can be listed as follows:

1. Tomas ↔ 1   **2.** Tomas ↔ 1   **3.** Tomas ↔ 2
   Dick ↔ 2   Dick ↔ 3   Dick ↔ 1
   Mari ↔ 3   Mari ↔ 2   Mari ↔ 3

4. Tomas ↔ 2   **5.** Tomas ↔ 3   **6.** Tomas ↔ 3
   Dick ↔ 3   Dick ↔ 1   Dick ↔ 2
   Mari ↔ 1   Mari ↔ 2   Mari ↔ 1

Notice that the listing in (**1**) as well as Figure 2-9 represent a single one-to-one correspondence between the sets P and S. The correspondence Tomas ↔ 1 can also be a one-to-one correspondence but between two different sets, namely, the sets {Tomas} and {1}. The complete set of one-to-one correspondences between sets P and S is seen in Table 2-9.

Table 2-9

Pairings \ Lanes	1	2	3
1.	Tomas	Dick	Mari
2.	Tomas	Mari	Dick
3.	Dick	Tomas	Mari
4.	Dick	Mari	Tomas
5.	Mari	Tomas	Dick
6.	Mari	Dick	Tomas

The general definition of one-to-one correspondence follows:

Definition of One-to-One Correspondence

If the elements of sets P and S can be paired so that for each element of P there is exactly one element of S and for each element of S there is exactly one element of P, then the two sets P and S are said to be in **one-to-one correspondence**.

NOW TRY THIS 2-5 Consider a set of four people {A, B, C, D} and a set of four swimming lanes {1, 2, 3, 4}.

a. Exhibit all the one-to-one correspondences between the two sets.
b. How many such one-to-one correspondences are there?
c. Find the number of one-to-one correspondences between two sets with five elements each and explain your reasoning.

A tree diagram also lists the possible one-to-one correspondences in Figure 2-10. To read the tree diagram and see the one-to-one correspondence, we follow each branch. The person occupying a specific lane in a correspondence is listed below the lane number. For example, the top branch gives the pairing (Tomas, 1), (Dick, 2), and (Mari, 3).

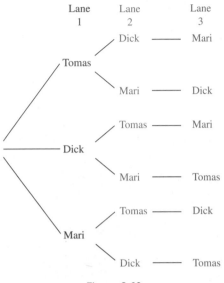

Figure 2-10

Observe in Figure 2-10 when assigning a swimmer to lane 1 we have a choice of three people: Tomas, Dick, or Mari. If we put Tomas in lane 1, then he cannot be in lane 2, and hence the second lane must be occupied by either Dick or Mari. In the same way, we see that if Dick is in lane 1, then there are two choices for lane 2: Tomas or Mari. Similarly, if Mari is in lane 1, then again there are two choices for the second lane: Tomas or Dick. Thus, for each of the three ways we can fill the first lane, there are two subsequent ways to fill the second lane, and hence there are $2 + 2 + 2$, or $3 \cdot 2$, or 6 ways to arrange the swimmers in the first two lanes. Notice that for each arrangement of the swimmers in the first two lanes, there remains only one possible swimmer to fill the third lane. For example, if Mari fills the first lane and Dick fills the second, then Tomas must be in the third. Thus, the total number of arrangements for the three swimmers is equal to $3 \cdot 2$, or 6.

Similar reasoning can be used to find how many ice-cream arrangements are possible on a two-scoop cone if 10 flavors are offered. If we count chocolate and vanilla (chocolate on bottom and vanilla on top) different from vanilla and chocolate (vanilla on bottom and chocolate on top) and allow two scoops to be of the same flavor, we can proceed as follows. There are 10 choices for the first scoop, and for each of these 10 choices there are 10 subsequent choices for the second scoop. Thus, the total number of arrangements is $10 \cdot 10$, or 100.

The counting argument used to find the number of possible one-to-one correspondences between the set of swimmers and the set of lanes and the previous problem about ice-cream-scoop arrangements are examples of the Fundamental Counting Principle.

Theorem 2–1:   Fundamental Counting Principle

If event M can occur in m ways and, after it has occurred, event N can occur in n ways, then event M followed by event N can occur in mn ways.

NOW TRY THIS 2-6 Explain how the Fundamental Counting Principle can be extended to any number of events.

Equivalent Sets

Closely associated with one-to-one correspondences is the concept of **equivalent sets**. For example, suppose a room contains 20 chairs and one student is sitting in each chair with no one standing. There is a one-to-one correspondence between the set of chairs and the set of students in the room. In this case, the set of chairs and the set of students are equivalent sets.

Definition of Equivalent Sets

Two sets A and B are **equivalent**, written $A \sim B$, if and only if there exists a one-to-one correspondence between the sets.

The term *equivalent* should not be confused with *equal*. The difference should be made clear by Example 2-8.

Example 2-8

Let

$$A = \{p, q, r, s\}, \quad B = \{a, b, c\}, \quad C = \{x, y, z\}, \quad \text{and} \quad D = \{b, a, c\}.$$

Compare the sets, using the terms *equal* and *equivalent*.

Solution   Each set is both equivalent to and equal to itself.

Sets A and B are not equivalent ($A \not\sim B$) and not equal ($A \neq B$).

Sets A and C are not equivalent ($A \not\sim C$) and not equal ($A \neq C$).

Sets A and D are not equivalent ($A \not\sim D$) and not equal ($A \neq D$).

Sets B and C are equivalent ($B \sim C$) but not equal ($B \neq C$).

Sets B and D are equivalent ($B \sim D$) and equal ($B = D$).

Sets C and D are equivalent ($C \sim D$) but not equal ($C \neq D$).

NOW TRY THIS 2-7

a. If two sets are equivalent, are they necessarily equal? Explain why or why not.
b. If two sets are equal, are they necessarily equivalent? Explain why or why not.

Cardinal Numbers

The concept of one-to-one correspondence can be used to consider the notion of two sets having the same number of elements. Without knowing how to count, a child might tell that there are as many fingers on the left hand as on the right hand by matching the fingers on one hand with the fingers on the other hand, as in Figure 2-11. Naturally placing the fingers so that the left thumb touches the right thumb, the left index finger touches the right index finger, and so on, exhibits a one-to-one correspondence between the fingers of the two hands. Similarly, without counting, children realize that if every student in a class sits in a chair and no chairs are empty, there are as many chairs as students, and vice versa.

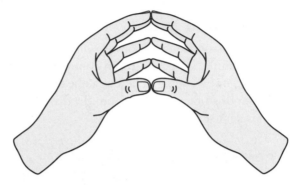

Figure 2-11

One-to-one correspondence between sets helps explain the concept of a number. Consider the five sets $\{a, b\}$, $\{p, q\}$, $\{x, y\}$, $\{b, a\}$, and $\{*, \#\}$; the sets are equivalent to one another and share the property of "twoness"; that is, these sets have the same cardinal number, namely, 2. The **cardinal number** of a set S, denoted $n(S)$, indicates the number of elements in the set S. If $S = \{a, b\}$, the cardinal number of S is 2, and we write $n(S) = 2$. If two sets, A and B, are equivalent, then A and B have the same cardinal number; that is, $n(A) = n(B)$.

A set that contains no elements has cardinal number 0 and is an **empty**, or **null, set**. The empty set is designated by the symbol $\varnothing$ or $\{\ \}$. Two examples of sets with no elements are the following:

$$C = \{x \mid x \text{ was a state of the United States before 1200 CE}\}$$
$$D = \{x \mid x \text{ is a natural number less than 1}\}$$

> **REMARK** The empty set is often incorrectly recorded as $\{\varnothing\}$. This set is not empty but contains one element. Likewise, $\{0\}$ does not represent the empty set. Why?

A set is a **finite set** if the cardinal number of the set is zero or a natural number. The set of natural numbers N is an **infinite set**; it is not finite. The set W, containing all the natural numbers and 0, is the set of **whole numbers**: $W = \{0, 1, 2, 3, \dots\}$. W is an infinite set.

The following "Peanuts" cartoon demonstrates some set theory concepts related to addition, though a child would not be expected to know all these concepts to add 2 and 2.

NOW TRY THIS 2-8 Use *reasoning* to explain why there can be no greatest natural number. That is, explain why the set of natural numbers is not a finite set as Dolly implies in the cartoon below.

THE FAMILY CIRCUS®   By Bil Keane

"The alphabet ends at 'Z,' but numbers go on forever."

More About Sets

The **universal set**, or the **universe**, denoted U, is the set that contains all elements being considered in a given discussion. Suppose $U = \{x \mid x$ is a person living in California$\}$ and $F = \{x \mid x$ is a female living in California$\}$. The universal set, U, and set F can be represented by a diagram, as in Figure 2-12(a). The universal set is represented by a large rectangle, and F is indicated by the circle inside the rectangle, as shown in Figure 2-12(a). This figure is an example of a **Venn diagram**, named after the Englishman John Venn (1834–1923), who used such diagrams to illustrate ideas in logic. The set of elements in the universe that are not in F, denoted by $\overline{F}$, is the set of males living in California and is the **complement** of F. It is represented by the shaded region in Figure 2-12(b).

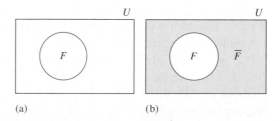

(a)   (b)

Figure 2-12

Definition of Set Complement

The **complement** of a set F, written $\overline{F}$, is the set of all elements in the universal set U that are not in F; that is, $\overline{F} = \{x \mid x \in U \text{ and } x \notin F\}$.

Example 2-9

a. If $U = \{a, b, c, d\}$ and $B = \{c, d\}$, find (i) $\overline{B}$; (ii) $\overline{U}$; (iii) $\overline{\varnothing}$.
b. If $U = \{x \mid x \text{ is an animal in the zoo}\}$ and $S = \{x \mid x \text{ is a snake in the zoo}\}$, describe $\overline{S}$.
c. If $U = N$, $E = \{2, 4, 6, 8, \dots\}$, and $O = \{1, 3, 5, 7, \dots\}$, find (i) $\overline{E}$; (ii) $\overline{O}$.

Solution   **a.** (i) $\overline{B} = \{a, b\}$; (ii) $\overline{U} = \varnothing$; (iii) $\overline{\varnothing} = U$

b. Because the individual animals in the zoo are not known, $\overline{S}$ must be described using set-builder notation:

$$\overline{S} = \{x \mid x \text{ is a zoo animal that is not a snake}\}$$

c. (i) $\overline{E} = O$; (ii) $\overline{O} = E$

Subsets

Consider the sets $A = \{1, 2, 3, 4, 5, 6\}$ and $B = \{2, 4, 6\}$. All the elements of B are contained in A and we say that B is a **subset** of A. We write $B \subseteq A$. In general, we have the following:

> **Definition of Subset**
>
> B is a **subset** of A, written $B \subseteq A$, if, and only if, every element of B is an element of A.

This definition allows B to be equal to A. The definition is written with the phrase "if, and only if," which means "if B is a subset of A, then every element of B is an element of A, and if every element of B is an element of A, then B is a subset of A." *If both $A \subseteq B$ and $B \subseteq A$, then $A = B$.*

When a set A is not a subset of another set B, we write $A \nsubseteq B$. To show that $A \nsubseteq B$, we must find at least one element of A that is not in B. If $A = \{1, 3, 5\}$ and $B = \{1, 2, 3\}$, then A is not a subset of B because 5 is an element of A but not of B. Likewise, $B \nsubseteq A$ because 2 belongs to B but not to A.

It is not obvious how the empty set fits the definition of a subset because no elements in the empty set are elements of another set. To investigate this problem, we use the strategies of *indirect reasoning* and *looking at a special case*.

For the set $\{1, 2\}$, either $\varnothing \subseteq \{1, 2\}$ or $\varnothing \nsubseteq \{1, 2\}$. Suppose $\varnothing \nsubseteq \{1, 2\}$. Then there must be some element in $\varnothing$ that is not in $\{1, 2\}$. Because the empty set has no elements, there cannot be an element in the empty set that is not in $\{1, 2\}$. Consequently, $\varnothing \nsubseteq \{1, 2\}$ is false. Therefore, the only other possibility, $\varnothing \subseteq \{1, 2\}$, is true. The same reasoning can be applied in the case of the empty set and any other set.

If B is a subset of A and B is not equal to A, then B is a **proper subset** of A, written $B \subset A$. This means that every element of B is contained in A and there is at least one element of A that is not in B. To indicate a proper subset, sometimes a Venn diagram like the one shown in Figure 2-13 is used, showing a dot (an element) in A that is not in B.

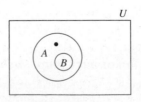

Figure 2-13

Example 2-10

Given $A = \{1, 2, 3, 4, 5\}, B = \{1, 3\}, P = \{x \mid x = 2^n - 1, \text{ where } n \in N\}$:

a. Which sets are subsets of each other?
b. Which sets are proper subsets of each other?
c. If $C = \{2k \mid k \in N\}$ and $D = \{4k \mid k \in N\}$, show that one of the sets is a subset of the other.

Solution **a.** Because $2^1 - 1 = 1, 2^2 - 1 = 3, 2^3 - 1 = 7, 2^4 - 1 = 15, 2^5 - 1 = 31$, and so on, $P = \{1, 3, 7, 15, 31, \dots\}$. Thus, $B \subseteq P$. Also $B \subseteq A, A \subseteq A, B \subseteq B$ and $P \subseteq P$.

b. $B \subset A$ and $B \subset P$

c. Because $4k = 2(2k)$, every element of D is an element of C. Thus, $D \subseteq C$.

NOW TRY THIS 2-9

a. Suppose $A \subset B$. Can we always conclude that $A \subseteq B$?
b. If $A \subseteq B$, does it follow that $A \subset B$?

Subsets and elements of sets are often confused. We say that $2 \in \{1, 2, 3\}$. But because 2 is not a set, we cannot substitute the symbol $\subseteq$ for $\in$. However, $\{2\} \subseteq \{1, 2, 3\}$ and $\{2\} \subset \{1, 2, 3\}$. Notice that the symbol $\in$ cannot be used between $\{2\}$ and $\{1, 2, 3\}$ in a true sentence.

NOW TRY THIS 2-10 Convince a classmate that the following are true:

a. The empty set is a subset of itself.
b. The empty set is not a proper subset of itself.

Inequalities: An Application of Set Concepts

The notion of a proper subset and the concept of one-to-one correspondence can be used to define the concept of "less than" among natural numbers. The set $\{a, b, c\}$ has fewer elements than the set $\{w, x, y, z\}$ because when we try to pair the elements of the two sets, as in

$$\{a, b, c\}$$
$$\big\uparrow\big\downarrow \ \big\uparrow\big\downarrow \ \big\uparrow\big\downarrow$$
$$\{x, y, z, w\}$$

we see that there is an element of the second set that is not paired with an element of the first set. The set $\{a, b, c\}$ is equivalent to a proper subset of the set $\{x, y, z, w\}$.

In general, *if A and B are finite sets, A has fewer elements than B if A is equivalent to a proper subset of B.* We say that $n(A)$ is **less than** $n(B)$ and write $n(A) < n(B)$. We say that a is **greater than** b, written $a > b$, if, and only if, $b < a$. Defining the concept of "less than or equal to" in a similar way is explored in Assessment 2-2A and 2-2B.

We have just seen that if A and B are finite sets and $A \subset B$, then A has fewer elements than B and it is not possible to find a one-to-one correspondence between the sets. Consequently, A and B are not equivalent. However, when both sets are infinite and $A \subset B$, the sets could be equivalent. For example, consider the set N of counting numbers and the set

E of even counting numbers. We have $E \subset N$, but it is still possible to find a one-to-one correspondence between the sets. To do so, we correspond each number in set N to a number in set E that is twice as great. That is, $n \in N$ corresponds to $2n \in E$, as shown next.

$$N = \{1, 2, 3, 4, \ 5, \ldots, n, \ldots\}$$

$$E = \{2, 4, 6, 8, 10, \ldots, 2n, \ldots\}$$

Notice that in the correspondence, every element of N corresponds to a unique element in E and, conversely, every element of E corresponds to a unique element in N. For example, 11 in N corresponds to $2 \cdot 11$, or 22, in E. And 100 in E corresponds to $100 \div 2$, or 50, in N. Thus, $N \sim E$; that is, N and E are equivalent.

Students sometimes have difficulty with infinite sets and especially with their cardinal numbers, called **transfinite numbers**. As shown above, E is a proper subset of N, but because they can be placed in a one-to-one correspondence, they are equivalent and have the same cardinal number. Tsamir and Triosh found that representations of infinite sets caused problems as seen in the Research Note.

Problem Solving   Passing a Senate Measure

A committee of senators consists of Abel, Baro, Carni, and Davis. Suppose each member of the committee has one vote and a simple majority is needed to either pass or reject any measure. A measure that is neither passed nor rejected is considered to be blocked and will be voted on again. Determine the number of ways a measure could be passed or rejected and the number of ways a measure could be blocked.

Understanding the Problem   We are asked to determine how many ways the committee of four could pass or reject a proposal and how many ways the committee of four could block a proposal. To pass or reject a proposal requires a winning coalition, that is, a group of senators who can pass or reject the proposal, regardless of what the others do. To block a proposal, there must be a blocking coalition, that is, a group who can prevent any proposal from passing but who cannot reject the measure.

Devising a Plan   To solve the problem, we can *make a list* of subsets of the set of senators. Any subset of the set of senators with three or four members will form a winning coalition. Any subset of the set of senators with exactly two members will form a blocking coalition.

Carrying Out the Plan   We list all subsets of the set $S = \{$Abel, Baro, Carni, Davis$\}$ that have at least three elements and all subsets that have exactly two elements. For ease, we identify the members as follows: A—Abel, B—Baro, C—Carni, D—Davis. All the subsets are given next:

$\varnothing$	$\{A\}$	$\{A, B\}$	$\{A, B, C\}$	$\{A, B, C, D\}$
	$\{B\}$	$\{A, C\}$	$\{A, B, D\}$	
	$\{C\}$	$\{A, D\}$	$\{A, C, D\}$	
	$\{D\}$	$\{B, C\}$	$\{B, C, D\}$	
		$\{B, D\}$		
		$\{C, D\}$		

There are five subsets with at least three members that can form a winning coalition to pass or reject a measure and six subsets with exactly two members that can block a measure.

Looking Back   Other questions that might be considered include the following:

1. How many minimal winning coalitions are there? In other words, how many subsets are there of which no proper subset could pass a measure?
2. Devise a method to solve this problem without listing all subsets.
3. In "Carrying Out the Plan," 16 subsets of $\{A, B, C, D\}$ are listed. Use that result to systematically list all the subsets of a committee of five senators. Can you find the number of subsets of the 5-member committee without actually counting the subsets?

NOW TRY THIS 2-11 Suppose a committee of U.S. senators consists of five members.

a. Compare the number of winning coalitions having exactly four members with the number of senators on the committee. What is the reason for the result?

b. Compare the number of winning coalitions having exactly three members with the number of subsets of the committee having exactly two members. What is the reason for the result?

Number of Subsets of a Set

How many subsets can be made from a set containing n elements? To obtain a general formula, we use the strategy of *trying simpler cases* first.

1. If $P = \{a\}$, then P has two subsets, $\varnothing$ and $\{a\}$.
2. If $Q = \{a, b\}$, then Q has four subsets, $\varnothing$, $\{a\}$, $\{b\}$, and $\{a, b\}$.
3. If $R = \{a, b, c\}$, then R has eight subsets, $\varnothing$, $\{a\}$, $\{b\}$, $\{c\}$, $\{a, b\}$, $\{a, c\}$, $\{b, c\}$, and $\{a, b, c\}$.

An alternative strategy for listing the number of subsets of a given set is to use a tree diagram. For example, tree diagrams for the subsets of $Q = \{a, b\}$ and $R = \{a, b, c\}$ are given in Figure 2-14.

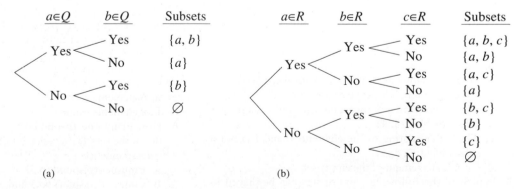

Figure 2-14

Using the information from these cases, we *make a table and search for a pattern*, as in Table 2-10.

Table 2-10

Number of Elements	Number of Subsets
1	2, or 2^1
2	4, or 2^2
3	8, or 2^3
.	.
.	.
.	.

Table 2-10 suggests that for four elements, there might be 2^4, or 16, subsets. Is this correct? If $S = \{a, b, c, d\}$, then all the subsets of $R = \{a, b, c\}$ are also subsets of S. Eight new subsets are also formed by including the element d in each of the eight subsets of R. The eight new subsets are $\{d\}, \{a, d\}, \{b, d\}, \{c, d\}, \{a, b, d\}, \{a, c, d\}, \{b, c, d\}$, and $\{a, b, c, d\}$. Thus, there are twice as many subsets of set S (with four elements) as there are of set R (with three elements). Consequently, there are $2 \cdot 8$, or 2^4, subsets of a set with four elements. Because including one more element in a finite set doubles the number of possible subsets of the new set, a set with five elements will have $2 \cdot 2^4$, or 2^5, subsets and so on. In each case, the number of elements and the power of 2 used to obtain the number of subsets are equal. *Therefore, if there are n elements in a set, 2^n subsets can be formed.* If we apply this result to the empty set—that is when $n = 0$—then we have $2^0 = 1$. The pattern is meaningful because the empty set has only one subset—itself.

NOW TRY THIS 2-12

a. How many proper subsets does a set with four elements have?
b. How many proper subsets does a set with n elements have?

Assessment 2-2A

1. Write the following sets using the listing method or using set-builder notation:
   a. The set of letters in the word *mathematics*
   b. The set of natural numbers greater than 20
2. Rewrite the following using mathematical symbols:
   a. P is equal to the set containing a, b, c, and d.
   b. The set consisting of the elements 1 and 2 is a proper subset of $\{1, 2, 3, 4\}$.
   c. The set consisting of the elements 0 and 1 is not a subset of $\{1, 2, 3, 4\}$.
   d. 0 is not an element of the empty set.
3. Which of the following pairs of sets can be placed in one-to-one correspondence?
   a. $\{1, 2, 3, 4, 5\}$ and $\{m, n, o, p, q\}$

   b. $\{a, b, c, d, e, f, \ldots, m\}$ and $\{1, 2, 3, 4, 5, 6, \ldots, 13\}$
   c. $\{x \mid x$ is a letter in the word *mathematics*$\}$ and $\{1, 2, 3, 4, \ldots, 11\}$
4. How many one-to-one correspondences are there between two sets with
   a. 6 elements each?
   b. n elements each?
5. How many one-to-one correspondences are there between the sets $\{x, y, z, u, v\}$ and $\{1, 2, 3, 4, 5\}$ if in each correspondence
   a. x must correspond to 5?
   b. x must correspond to 5 and y to 1?
   c. x, y, and z must correspond to odd numbers?

6. Which of the following represent equal sets?
$A = \{a, b, c, d\}$ $B = \{x, y, z, w\}$
$C = \{c, d, a, b\}$ $D = \{x \mid 1 \le x \le 4 \text{ where } x \in N\}$
$E = \varnothing$ $F = \{\varnothing\}$
$G = \{0\}$ $H = \{ \ \}$
$I = \{x \mid x = 2n+1 \text{ where } n \in W\}$, and
$W = \{0, 1, 2, 3, \ldots\}$
$J = \{x \mid x = 2n - 1 \text{ where } n \in N\}$

7. Find the cardinal number of each of the following sets:
   a. $\{101, 102, 103, \ldots, 1100\}$
   b. $\{1, 3, 5, \ldots, 1001\}$
   c. $\{1, 2, 4, 8, 16, \ldots, 1024\}$
   d. $\{x \mid x = k^2 \text{ where } k = 1, 2, 3, \ldots, \text{ or } 100\}$
   e. $\{i + j \mid i \in \{1, 2, 3\} \text{ and } j \in \{1, 2, 3\}\}$

8. If U is the set of all college students and A is the set of all college students with a straight-A average, describe $\bar{A}$.

9. Suppose B is a proper subset of C.
   a. If $n(C) = 8$, what is the maximum number of elements in B?
   b. What is the least possible number of elements in B?

10. Suppose C is a subset of D and D is a subset of C.
    a. If $n(C) = 5$, find $n(D)$.
    b. What other relationship exists between sets C and D?

11. Indicate which symbol, $\in$ or $\notin$, makes each of the following statements true:
    a. 0 _____ $\varnothing$
    b. $\{1\}$ _____ $\{1, 2\}$
    c. 1024 _____ $\{x \mid x = 2^n \text{ where } n \in N\}$
    d. 3002 _____ $\{x \mid x = 3n - 1 \text{ where } n \in N\}$

12. Indicate which symbol, $\subseteq$ or $\nsubseteq$, makes each part of problem 11 true.

13. Answer each of the following. If your answer is *no*, tell why.
    a. If $A = B$, can we always conclude that $A \subseteq B$?
    b. If $A \subseteq B$, can we always conclude that $A \subset B$?
    c. If $A \subset B$, can we always conclude that $A \subseteq B$?
    d. If $A \subseteq B$, can we always conclude that $A = B$?

14. Use the definition of *less than* to show each of the following:
    a. $3 < 100$
    b. $0 < 3$

15. On a certain senate committee there are seven senators: Abel, Brooke, Cox, Dean, Eggers, Funk, and Gage. Three of these members are to be appointed to a subcommittee. How many possible subcommittees are there?

16. How many two-digit numbers in base ten can be formed if the tens digit cannot be 0 and no digit can be repeated?

Assessment 2-2B

1. Write the following sets using the listing method or set-builder notation:
   a. the set of letters in the word *geometry*
   b. the set of natural numbers greater than 7

2. Rewrite the following using mathematical symbols:
   a. Q is equal to the set whose elements are a, b, and c.
   b. The set containing 1 and 3 only is a proper subset of the set of natural numbers.
   c. The set containing 1 and 3 only is not a subset of $\{1, 4, 6, 7\}$.
   d. The empty set does not contain 0 as an element.

3. Which of the following pairs of sets can be placed in a one-to-one correspondence?
   a. $\{1, 2, 3, 4\}$ and $\{w, c, y, z\}$
   b. $\{1, 2, 3, \ldots, 25\}$ and $\{a, b, c, d, \ldots, x, y\}$
   c. $\{x \mid x \text{ is a letter in the word } geometry\}$ and $\{1, 2, 3, 4, 5, 6, 7, 8\}$

4. How many one-to-one correspondences exist between two sets with
   a. 8 elements each?
   b. $n - 1$ elements each?

5. How many one-to-one correspondences are there between the sets $\{a, b, c, d\}$ and $\{1, 2, 3, 4\}$ if in each correspondence
   a. b must correspond to 3?
   b. b must correspond to 3 and d to 4?
   c. a and c must correspond to even numbers?

6. Which of the following represent unequal sets?
$A = \{a, b, c, d\}$ $B = \{x, y, z, w\}$
$C = \{c, d, a, b\}$ $D = \{x \mid 1 \le x \le 4 \text{ where } x \in N\}$
$E = \varnothing$ $F = \{\varnothing\}$
$G = \{0\}$ $H = \{ \ \}$
$I = \{x \mid x = 2n + 1 \text{ where } n \in W\}$, and
$W = \{0, 1, 2, 3, \ldots\}$
$J = \{x \mid x = 2n - 1 \text{ where } n \in N\}$

7. Find the cardinal number of each of the following sets:
   a. $\{9, 10, 11, \ldots, 99\}$
   b. $\{2, 4, 6, 8, \ldots, 2002\}$
   c. $\{0, 1, 3, 7, 15, \ldots, 1023\}$
   d. $\{x^2 \mid x = 1, 3, 5, 7, \ldots, \text{ or } 99\}$
   e. $\{i \cdot j \mid i \in \{1, 2, 3\} \text{ and } j \in \{1, 2, 3\}\}$

8. If U is the set of all women and G is the set of alumnae of Georgia State University, describe $\overline{G}$.

9. Suppose $A \subseteq B$.
   a. What is the minimum number of elements in set A?
   b. Is it possible for set B to be the empty set? If so, give an example of sets A and B satisfying this. If not, explain why not.

10. If two sets are subsets of each other, what other relationships must they have?

11. Indicate which symbol, $\in$ or $\notin$, makes each of the following statements true:
    a. $\varnothing$ ____ $\varnothing$
    b. $\{2\}$ ____ $\{3, 2, 1\}$
    c. 1022 ____ $\{s \mid s = 2^n - 2 \text{ where } n \text{ is an element of } N\}$
    d. 3004 ____ $\{x \mid x = 3n + 1 \text{ where } n \text{ is a natural number}\}$

12. Indicate which symbol, $\subseteq$ or $\not\subseteq$, makes each part of problem 11 true.

13. Answer each of the following. If your answer is *no*, tell why.
    a. If $A \subseteq B$, can we always conclude that $A = B$?
    b. If $A \subset B$, can we conclude that $A = B$?
    c. If A and B can be placed in a one-to-one correspondence, must $A = B$?
    d. If A and B can be placed in a one-to-one correspondence, must $A \subseteq B$?

14. Use the definition of *less than* to show each of the following:
    a. $0 < 2$
    b. $99 < 100$

15. How many ways are there to stack an ice-cream cone with 4 scoops if the choices are
    a. vanilla, chocolate, rhubarb, and strawberry and each scoop must be different?
    b. vanilla, chocolate, rhubarb, and strawberry and there are no restrictions on different scoops?

16. How many seven-digit phone numbers are there when 0 and 1 cannot be the leading number?

Mathematical Connections 2-2

Communication

1. Explain the difference between a well-defined set and one that is not. Give examples.

2. Which of the following sets are not well defined? Explain.
   a. the set of wealthy schoolteachers
   b. the set of great books
   c. the set of natural numbers greater than 100
   d. the set of subsets of $\{1, 2, 3, 4, 5, 6\}$
   e. the set $\{x \mid x \neq x \text{ and } x \in N\}$

3. Is $\varnothing$ a proper subset of every nonempty set? Explain your reasoning.

4. Explain why $\{\varnothing\}$ has $\varnothing$ as an element and also as a subset.

5. Tell how you would show that $A \not\subseteq B$.

6. Explain why every set is a subset of itself.

7. Define *less than or equal to* in a way similar to the definition of *less than*.

Open-Ended

8. a. Give three examples of sets A and B and a universal set U such that $A \subset B$; find $\overline{A}$ and $\overline{B}$.
   b. Based on your observations, conjecture a relationship between $\overline{B}$ and $\overline{A}$.
   c. Justify your conjecture in (b) using a Venn diagram.

9. Find an infinite set A such that
   a. $\overline{A}$ is finite.
   b. $\overline{A}$ is infinite.

10. Describe two sets from real-life situations such that it is clear from using one-to-one correspondence, and not from counting, that one set has fewer elements than the other.

Cooperative Learning

11. a. Use a calculator if necessary to estimate the time in years it would take a computer to list all the subsets of $\{1, 2, 3, \ldots, 64\}$. Assume the fastest computer can list one subset in approximately 1 microsecond (one-millionth of a second).
    b. Estimate the time in years it would take the computer to exhibit all the one-to-one correspondences between the sets $\{1, 2, 3, \ldots, 64\}$ and $\{65, 66, 67, \ldots, 128\}$.

12. Using people in your class standing in a line, determine the number of possible arrangements of 1, 2, 3, 4, and 5 people that exist. Use your model to validate the Fundamental Counting Principle.

Questions from the Classroom

13. A student argues that $\{\varnothing\}$ is the proper notation for the empty set. What is your response?

14. A student claims that a finite set is any set that has a greatest element. Do you agree?

15. A student argues that $A = \{1, \{1\}\}$ has only one element. How do you respond?

16. A student states that either $A \subseteq B$ or $B \subseteq A$. Is the student correct?

Review Problems

17. Investigate the measuring of lengths in the metric system. Develop a plan for using place value with lengths to convert among different metric units.

18. Write 5280 in expanded form.

19. What is the value of MCDX in Hindu-Arabic numerals?

20. Convert each of the following to base ten:
    a. $E0T_{twelve}$
    b. 1011_{two}
    c. 43_{five}

21. If 1 month is approximately 4 weeks and 1 year is approximately 365 days, or 52 weeks, answer the following:
    a. Lewis and Clark spent approximately 2 years, 4 months, and 9 days exploring the territory in the Northwest. What is this time in weeks?

    b. It took Magellan 1126 days to circle the world. How many years is this?
    c. How many seconds old are you?
    d. Approximately how many times does your heart beat in 1 year?

National Assessment of Educational Progress (NAEP) Question

Four people—A, X, Y, and Z—go to a movie and sit in adjacent seats. If A sits in the aisle seat, list all possible arrangements of the other three people. One of the arrangements is shown below.

NAEP, 1996, Grade 12

BRAIN TEASER Mr. Gonzales's and Ms. Chan's seventh-grade classes in Paxson Middle School have 24 and 25 students, respectively. Linda, a student in Mr. Gonzales's class, claims that the number of school committees that could be formed to contain at least one student from each class is greater than the number of people in the world. Assuming that a committee can have up to 49 students, find the number of committees and determine if Linda is right.

2-3   Other Set Operations and Their Properties

Finding the complement of a set is an operation that acts on only one set at a time. In this section, we consider operations that act on two sets at a time.

Set Intersection

$A \cap B$

Figure 2-15

Suppose that during the fall quarter, a college wants to mail a survey to all its students who are enrolled in both art and biology classes. To do this, the school officials must identify those students who are taking both classes. If A and B are the set of students taking art courses and the set of students taking biology courses, respectively, during the fall quarter, then the desired set of students includes those common to A and B, or the **intersection** of A and B. The intersection of sets A and B is the shaded region in Figure 2-15.

REMARK Figure 2-15 depicts the possibility of A and B containing common elements. The intersection might contain no elements.

> **Definition of Set Intersection**
>
> The **intersection** of two sets A and B, written $A \cap B$, is the set of all elements common to both A and B. $A \cap B = \{x \mid x \in A \text{ and } x \in B\}$.

The key word in the definition of *intersection* is *and* (see Chapter 1). In everyday language, as in mathematics, *and* implies that both conditions must be met. In the example, the desired set is the set of those students enrolled in both art and biology.

If sets such as A and B have no elements in common, they are **disjoint sets**. In other words, two sets A and B are disjoint if, and only if, $A \cap B = \varnothing$. For example, the set of males taking biology and the set of females taking biology are disjoint.

Example 2-11

Find $A \cap B$ in each of the following:

a. $A = \{1, 2, 3, 4\}$, $B = \{3, 4, 5, 6\}$
b. $A = \{0, 2, 4, 6, \ldots\}$, $B = \{1, 3, 5, 7, \ldots\}$
c. $A = \{2, 4, 6, 8, \ldots\}$, $B = \{1, 2, 3, 4, \ldots\}$

Solution
a. $A \cap B = \{3, 4\}$.
b. $A \cap B = \varnothing$; therefore A and B are disjoint.
c. $A \cap B = A$ because all the elements of A are also in B.

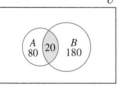

Figure 2-16

If A represents all students enrolled in art classes and B all students enrolled in biology classes, we may use a Venn diagram, taking into account that some students are enrolled in both subjects. If we know that 100 students are enrolled in art and 200 in biology and that 20 of these students are enrolled in both art and biology, then $100 - 20$, or 80, students are enrolled in art but not in biology and $200 - 20$, or 180, are enrolled in biology but not art. We can record this information as in Figure 2-16. Notice that the total number of students in set A is 100 and the total in set B is 200.

Set Union

If A is the set of students taking art courses during the fall quarter and B is the set of students taking biology courses during the fall quarter, then the set of students taking art or biology or both during the fall quarter is the **union** of sets A and B. The union of sets A and B is pictured in Figure 2-17.

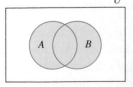

$A \cup B$

Figure 2-17

> **Definition of Set Union**
>
> The **union** of two sets A and B, written $A \cup B$, is the set of all elements in A or in B, $A \cup B = \{x \mid x \in A \text{ or } x \in B\}$.

The key word in the definition of *union* is *or* (see Chapter 1). In mathematics, *or* usually means "one or the other or both." This is known as the *inclusive or*.

Example 2-12

Find $A \cup B$ for each of the following:

a. $A = \{1,2,3,4\}$, $B = \{3,4,5,6\}$
b. $A = \{0,2,4,6,\dots\}$, $B = \{1,3,5,7,\dots\}$
c. $A = \{2,4,6,8,\dots\}$, $B = \{1,2,3,4,\dots\}$

Solution   **a.** $A \cup B = \{1,2,3,4,5,6\}$.
   b. $A \cup B = \{0,1,2,3,4,\dots\}$.
   c. Because every element of A is already in B, we have $A \cup B = B$.

NOW TRY THIS 2-13 Notice that in Figure 2-16, $n(A \cup B) = 80 + 20 + 180 = 280$, but $n(A) + n(B) = 100 + 200 = 300$; hence in general, $n(A \cup B) \neq n(A) + n(B)$. Use the concept of intersection of sets to write a formula for $n(A \cup B)$.

Set Difference

If A is the set of students taking art classes during the fall quarter and B is the set of students taking biology classes, then the set of all students taking biology but not art is called the **complement of A relative to B**, or the **set difference** of B and A.

Definition of Relative Complement

The **complement of A relative to B**, written $B - A$, is the set of all elements in B that are not in A; $B - A = \{x \mid x \in B \text{ and } x \notin A\}$.

REMARK Note that $B - A$ is not read as "B minus A." *Minus* is an operation on numbers and *set difference* is an operation on sets.

A Venn diagram representing $B - A$ is shown in Figure 2-18(a). The shaded region represents all the elements that are in B but not in A. A Venn diagram for $B \cap \overline{A}$ is given in Figure 2-18(b). The shaded region represents all the elements that are in B and in $\overline{A}$. Notice that $B \cap \overline{A} = B - A$ because $B \cap \overline{A}$ is, by definition of intersection and complement, the set of all elements in B and not in A.

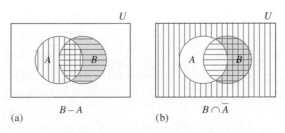

Figure 2-18

Example 2-13

If $A = \{d, e, f\}, B = \{a, b, c, d, e, f\}$, and $C = \{a, b, c\}$, find each of the following:

a. $A - B$
b. $B - A$
c. $B - C$
d. $C - B$
e. To answer parts (a)–(d), does it matter what the universal set is?

Solution a. $A - B = \emptyset$
  b. $B - A = \{a, b, c\}$
  c. $B - C = \{d, e, f\}$
  d. $C - B = \emptyset$
  e. Parts (a)–(d) can be answered independently of the universal set. The definition of set difference relates one set to another, independent of the universal set.

Properties of Set Operations

Because the order of elements in a set is not important, $A \cup B$ is equal to $B \cup A$. This is the **commutative property of set union**. It does not matter in which order we write the sets when the union of two sets is involved. Similarly, $A \cap B = B \cap A$. This is the **commutative property of set intersection**.

NOW TRY THIS 2-14 Use Venn diagrams and other means to find whether grouping is important when the same operation is involved. For example, is it always true that $A \cap (B \cap C) = (A \cap B) \cap C$? Similar questions should be investigated involving union and set difference.

In answering Now Try This 2-14, you may have discovered the following properties:

> **Theorem 2–2: Associative Property of Set Intersection and Associative Property of Set Union**
>
> The property $A \cap (B \cap C) = (A \cap B) \cap C$ is the **associative property of set intersection**.
> Similarly, $A \cup (B \cup C) = (A \cup B) \cup C$ is the **associative property of set union**.

Example 2-14

Is grouping important when two different set operations are involved? For example, is it true that $A \cap (B \cup C) = (A \cap B) \cup C$?

Solution To investigate this, we let $A = \{a, b, c, d\}, B = \{c, d, e\}$, and $C = \{d, e, f, g\}$. Then

$$A \cap (B \cup C) = \{a, b, c, d\} \cap (\{c, d, e\} \cup \{d, e, f, g\})$$
$$= \{a, b, c, d\} \cap \{c, d, e, f, g\}$$
$$= \{c, d\}$$
$$(A \cap B) \cup C = (\{a, b, c, d\} \cap \{c, d, e\}) \cup \{d, e, f, g\}$$
$$= \{c, d\} \cup \{d, e, f, g\}$$
$$= \{c, d, e, f, g\}$$

In this case, $A \cap (B \cup C) \neq (A \cap B) \cup C$. So we have found a counterexample, that is, an example illustrating that the general statement is not always true.

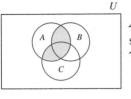

U

To discover an expression that is equal to $A \cap (B \cup C)$, consider the Venn diagram for $A \cap (B \cup C)$ shown by the shaded region in Figure 2-19. In the figure, $A \cap C$ and $A \cap B$ are subsets of the shaded region. The union of $A \cap C$ and $A \cap B$ is the entire shaded region. Thus, $A \cap (B \cup C) = (A \cap B) \cup (A \cap C)$. This property is stated formally next.

Figure 2-19

> **Theorem 2–3:   Distributive Property of Set Intersection over Union**
>
> For all sets A, B, and C,
> $$A \cap (B \cup C) = (A \cap B) \cup (A \cap C)$$

NOW TRY THIS 2-15 If on both sides of the equation in the distributive property of set intersection over union the symbol $\cap$ is replaced by $\cup$ and the symbol $\cup$ is replaced by $\cap$, is the new property true? Explain why. What should this property be called?

Example 2-15

Use set notation to describe the shaded portions of the Venn diagrams in Figure 2-20.

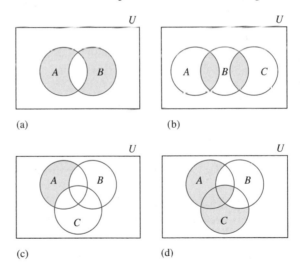

(a)   (b)

(c)   (d)

Figure 2-20

Solution   The solutions can be described in many different, but equivalent, forms. Some possible answers follow:

a. $(A \cup B) - (A \cap B), (A \cup B) \cap \overline{(A \cap B)}$, or $(A - B) \cup (B - A)$
b. $(A \cap B) \cup (B \cap C)$ or $B \cap (A \cup C)$
c. $(A - B) - C, A - (B \cup C)$, or $(A - (A \cap B)) - (A \cap C)$
d. $((A \cup C) - B) \cup (A \cap B \cap C)$ or $(A - (B \cup C)) \cup (C - (A \cup B)) \cup (A \cap C)$

Using Venn Diagrams as a Problem-Solving Tool

Venn diagrams can be used as a problem-solving tool for modeling information, as shown in the following examples.

Example 2-16

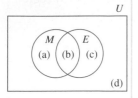

Figure 2-21

Suppose M is the set of all students taking mathematics and E is the set of all students taking English. Identify the students described by each region in Figure 2-21.

Solution

Region (a) contains all students taking mathematics but not English.

Region (b) contains all students taking both mathematics and English.

Region (c) contains all students taking English but not mathematics.

Region (d) contains all students taking neither mathematics nor English.

NOW TRY THIS 2-16 The following student page is another example of the use of Venn diagrams in modeling information. Answer question 27 on the student page.

Example 2-17

In a survey of 110 college freshmen that investigated their high school backgrounds, the following information was gathered:

25 took physics.

45 took biology.

48 took mathematics.

10 took physics and mathematics.

8 took biology and mathematics.

6 took physics and biology.

5 took all three subjects.

a. How many students took biology but neither physics nor mathematics?

b. How many took physics, biology, or mathematics?

c. How many did not take any of the three subjects?

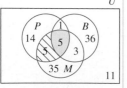

Figure 2-22

Solution   To solve this problem, we *build a model* using sets. Because there are three distinct subjects, we should use three circles. In Figure 2-22, P is the set of students taking physics, B is the set taking biology, and M is the set taking mathematics. The shaded region represents the 5 students who took all three subjects. The lined region represents the students who took physics and mathematics, but who did not take biology.

In part (a) we are asked for the number of students in the subset of B that has no element in common with either P or M. That is, $B - (P \cup M)$. In part (b) we are asked for the number of elements in $P \cup B \cup M$. Finally, in part (c) we are asked for the number of students in $\overline{P \cup B \cup M}$, or $U - (P \cup B \cup M)$. Our strategy is to find the number of students in each of the eight nonoverlapping regions.

One mindset to beware of in this problem is thinking, for example, that the 25 students who took physics, took only physics. That is not necessarily the case. If those students had been taking only physics, then we should have been told so.

a. Because a total of 10 students took physics and mathematics and 5 of those also took biology, $10 - 5$, or 5, students took physics and mathematics but not biology. Similarly,

School Book Page VENN DIAGRAMS

For a class project, students collect data about the number of boys or girls in the families of their classmates. Use the table below to answer Exercises 27–30.

Name	Number of Boys in the Family	Number of Girls in the Family
Anya	0	2
Brian	8	0
Charlie	1	2
Diane	0	1
Elisha	1	1
Felix	2	0
Gloria	0	2
Han	1	2
Ivan	1	1
Jorge	4	1

27. Anya wants to make a Venn diagram with the groups "Has Boys in the Family" and "Has Girls in the Family." She begins by placing herself on the diagram. Copy and complete her diagram below.

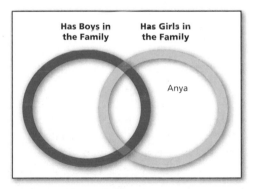

28. Make a bar graph that shows that the usual number of children in the family is two and that Brian's family is unusual.

29. Charlie wants to make a circle graph. What could the parts of the circle be labeled?

30. Make a graph to see if there is any relationship between the number of boys in a family and the number of girls in a family.

Investigation 4 Relating Two Variables   **77**

Source: Connected Mathematics 2, 2006 (p. 77).

because 8 students took biology and mathematics and 5 took all three subjects, $8 - 5$, or 3, took biology and mathematics but not physics. Also $6 - 5$, or 1, student took physics and biology but not mathematics. To find the number of students who took biology but neither physics nor mathematics, we subtract from 45 (the total number that took biology) the number of those that are in the distinct regions that include biology and other subjects, that is, $1 + 5 + 3$, or 9. Because $45 - 9 = 36$, we know that 36 students took biology but neither physics nor mathematics.

b. To find the number of students in all the distinct regions in P, M, or B, we proceed as follows. The number of students who took physics but neither mathematics nor biology is $25 - (1 + 5 + 5)$, or 14. The number of students who took mathematics but neither physics nor biology is $48 - (5 + 5 + 3)$, or 35. Hence the number of students who took mathematics, physics, or biology is $35 + 14 + 36 + 3 + 5 + 5 + 1$, or 99.

c. Because the total number of students is 110, the number that did not take any of the three subjects is $110 - 99$, or 11.

E-Manipulative Activity

Additional work with Venn diagrams can be found in the *Venn Diagrams* activity on the E-Manipulatives disk.

Cartesian Products

Another way to produce a set from two given sets is by forming the **Cartesian product**. This formation pairs the elements of one set with the elements of another set in a specific way to create elements in a new set. Suppose a person has three pairs of pants, $P = \{$blue, white, green$\}$, and two shirts, $S = \{$blue, red$\}$. According to the Fundamental Counting Principle, there are $3 \cdot 2$, or 6, possible different pant-and-shirt pairs, as shown in Figure 2-23.

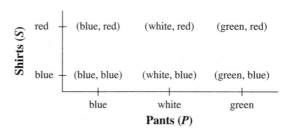

Figure 2-23

The pants-shirt combinations are elements of the set of all possible pairs in which the first member of the pair is an element of set P and the second member is an element of set S. The set of all possible pairs is given in Figure 2-23. Because the first component in each pair represents pants and the second component in each pair represents shirts, the order in which the components are written is important. Thus (green, blue) represents green pants and a blue shirt, whereas (blue, green) would represent blue pants and a green shirt. Therefore, the two pairs represent different outfits. Because the order in each pair is important, the pairs are **ordered pairs**. The positions that the ordered pairs occupy within the set of outfits is immaterial. Only the order of the **components** within each pair is significant.

The pants-and-shirt pairs suggest the following definition of **equality for ordered pairs**: $(x, y) = (m, n)$ *if, and only if, the first components are equal and the second components are equal.* A set consisting of ordered pairs is an example of a Cartesian product. A formal definition follows.

Definition of Cartesian Product

For any sets A and B, the **Cartesian product** of A and B, written $A \times B$, is the set of all ordered pairs such that the first component of each pair is an element of A and the second component of each pair is an element of B.

$$A \times B = \{(x,y) \mid x \in A \text{ and } y \in B\}$$

REMARK $A \times B$ is commonly read as "A cross B" and should never be read "A times B."

Example 2-18

If $A = \{a,b,c\}$ and $B = \{1,2,3\}$, find each of the following:

a. $A \times B$ **b.** $B \times A$ **c.** $A \times A$

Solution **a.** $A \times B = \{(a,1),(a,2),(a,3),(b,1),(b,2),(b,3),(c,1),(c,2),(c,3)\}$
   b. $B \times A = \{(1,a),(1,b),(1,c),(2,a),(2,b),(2,c),(3,a),(3,b),(3,c)\}$
   c. $A \times A = \{(a,a),(a,b),(a,c),(b,a),(b,b),(b,c),(c,a),(c,b),(c,c)\}$

It is possible to form a Cartesian product involving the null set. Suppose $A = \{1,2\}$. Because there are no elements in $\varnothing$, no ordered pairs (x,y) with $x \in A$ and $y \in \varnothing$ are possible, so $A \times \varnothing = \varnothing$. This is true for all sets A. Similarly, $\varnothing \times A = \varnothing$ for all sets A. There is an analogy between the last equation and the multiplication fact that $0 \cdot a = 0$, where a is a natural number. In Chapter 3 we use the concept of Cartesian product to define multiplication of natural numbers.

Assessment 2-3A

1. If $N = \{1,2,3,4,\dots\}$, $A = \{x \mid x = 2n - 1$ where $n \in N\}$, $B = \{x \mid x = 2n$ where $n \in N\}$, and $C = \{x \mid x = 2n + 1$ where $n = 0$ or $n \in N\}$, find the simplest possible expression for each of the following:
   a. $A \cup C$ **b.** $A \cup B$ **c.** $A \cap B$
2. Decide whether the following pairs of sets are always equal:
   a. $A \cap B$ and $B \cap A$
   b. $A \cup B$ and $B \cup A$
   c. $A \cup (B \cup C)$ and $(A \cup B) \cup C$
   d. $A \cup A$ and $A \cup \varnothing$
3. Tell whether each of the following is true for all sets A and B. If false, give a counterexample.
   a. $A \cup \varnothing = A$
   b. $A - B = B - A$
   c. $\overline{A \cap B} = \overline{A} \cap \overline{B}$
   d. $(A \cup B) - A = B$
   e. $(A - B) \cup A = (A - B) \cup (B - A)$
4. If $B \subseteq A$, find a simpler expression for each of the following:
   a. $A \cap B$ **b.** $A \cup B$

5. For each of the following, shade the portion of the Venn diagram that illustrates the set:
   a. $A \cup B$ **b.** $\overline{A \cap B}$
   c. $(A \cap B) \cup (A \cap C)$ **d.** $(A \cup B) \cap \overline{C}$
   e. $(A \cap B) \cup C$

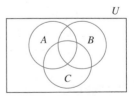

6. If S is a subset of universe U, find each of the following:
   a. $S \cup \overline{S}$ **b.** $\overline{U}$
   c. $S \cap \overline{S}$ **d.** $\varnothing \cap S$
7. For each of the following conditions, find $A - B$:
   a. $A \cap B = \varnothing$
   b. $B = U$
8. If for sets A and B we know that $A - B = \varnothing$, is it necessarily true that $A \subseteq B$? Justify your answer.

9. Use set notation to identify each of the following shaded regions:

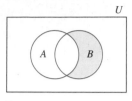

a.

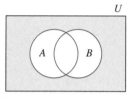

b.

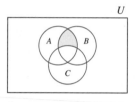

c.

10. In the following, shade the portion of the Venn diagram that represents the given set:

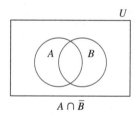

$A \cap \bar{B}$

11. Use Venn diagrams to determine if each of the following is true:
    a. $A \cup (B \cap C) = (A \cup B) \cap C$
    b. $A - (B - C) = (A - B) - C$

12. For each of the following pairs of sets, explain which is a subset of the other. If neither is a subset of the other, explain why.
    a. $A \cap B$ and $A \cap B \cap C$
    b. $A \cup B$ and $A \cup B \cup C$

13. a. If A has three elements and B has two elements, what is the greatest number of elements possible in (i) $A \cup B$? (ii) $A \cap B$? (iii) $B - A$? (iv) $A - B$?
    b. If A has n elements and B has m elements, what is the greatest number of elements possible in (i) $A \cup B$? (ii) $A \cap B$? (iii) $B - A$? (iv) $A - B$?

14. If $n(A) = 4, n(B) = 5$, and $n(C) = 6$, what is the greatest and least number of elements possible in
    a. $A \cup B \cup C$?
    b. $A \cap B \cap C$?

15. Given that the universe is the set of all humans, $B = \{x \mid x$ is a college basketball player$\}$, and $S = \{x \mid x$ is a college student more than 200 cm tall$\}$, describe each of the following in words:
    a. $B \cap S$ b. $\bar{S}$ c. $B \cup S$
    d. $\overline{B \cup S}$ e. $\bar{B} \cap S$ f. $B \cap \bar{S}$

16. Of the eighth graders at the Paxson School, 7 played basketball, 9 played volleyball, 10 played soccer, 1 played basketball and volleyball only, 1 played basketball and soccer only, 2 played volleyball and soccer only, and 2 played volleyball, basketball, and soccer. How many played one or more of the three sports?

17. In a fraternity with 30 members, 18 take mathematics, 5 take both mathematics and biology, and 8 take neither mathematics nor biology. How many take biology but not mathematics?

18. In Paul's bicycle shop, 40 bicycles are inspected. If 20 needed new tires and 30 needed gear repairs, answer the following:
    a. What is the greatest number of bikes that could have needed both?
    b. What is the least number of bikes that could have needed both?
    c. What is the greatest number of bikes that could have needed neither?

19. The Red Cross looks for three types of antigens in blood tests: A, B, and Rh. When the antigen A or B is present, it is listed, but if both these antigens are absent, the blood is type O. If the Rh antigen is present, the blood is positive; otherwise, it is negative. If a laboratory technician reports the following results after testing the blood samples of 100 people, how many were classified as O negative? Explain your reasoning.

Number of Samples	Antigen in Blood
40	A
18	B
82	Rh
5	A and B
31	A and Rh
11	B and Rh
4	A, B, and Rh

20. Classify the following as true or false. If false, give a counterexample. Assume that A and B are finite sets.
    a. If $n(A) = n(B)$, then $A = B$.
    b. If $A - B = \emptyset$, then $A = B$.
    c. If $A \subset B$, where A and B are finite, then $n(A) < n(B)$.

21. Three announcers each tried to predict the winners of Sunday's professional football games. The only team not picked that is playing Sunday was the Giants. The choices for each person were as follows:

Phyllis: Cowboys, Steelers, Vikings, Bills
Paula: Steelers, Packers, Cowboys, Redskins
Rashid: Redskins, Vikings, Jets, Cowboys

If the only teams playing Sunday are those just mentioned, which teams will play which other teams?

22. Let $A = \{x, y\}$ and $B = \{a, b, c\}$. Find each of the following:
  a. $A \times B$
  b. $B \times A$

23. For each of the following, the Cartesian product $C \times D$ is given by the sets listed. Find C and D.
  a. $\{(a, b), (a, c), (a, d), (a, e)\}$
  b. $\{(1, 1), (1, 2), (1, 3), (2, 1), (2, 2), (2, 3)\}$
  c. $\{(0, 1), (0, 0), (1, 1), (1, 0)\}$

Assessment 2-3B

1. If $W = \{0, 1, 2, 3, \dots\}$, $A = \{x \mid x = 2n + 1 \text{ where } n \in W\}$, $B = \{x \mid x = 2n \text{ where } n \in W\}$, and $N = \{1, 2, 3, \dots\}$, find the simplest possible expression for each of the following:
  a. $W - A$ **b.** $A \cap B$ **c.** $W \cap N$

2. Decide whether the following pairs of sets are always equal.
  a. $X \cap Y$ and $Y \cap X$
  b. $X \cup Y$ and $Y \cup X$
  c. $A \cap (B \cap C)$ and $(A \cap B) \cap C$
  d. $B \cup \emptyset$ and $B \cap B$

3. Tell whether each of the following is true for all sets A, B, or C. If false, give a counterexample.
  a. $A - B = A - \emptyset$
  b. $\overline{A \cup B} = A \cup \overline{B}$
  c. $A \cap (B \cup C) = (A \cap B) \cup C$
  d. $(A - B) \cap A = A$
  e. $A - (B \cap C) = (A - B) \cap (A - C)$

4. If $X \subseteq Y$, find a simpler expression for each of the following:
  a. $X - Y$ **b.** $X \cap \overline{Y}$

5. For each of the following, shade the portion of the Venn diagram that illustrates the set:
  a. $A \cap \overline{C}$ **b.** $\overline{A \cup B}$
  c. $(A \cap B) \cup (B \cap C)$ **d.** $A \cup (B \cap C)$
  e. $A \cup \overline{(B \cap C)}$

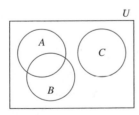

6. If A is a subset of universe U, find each of the following:
  a. $A \cup U$ **b.** $U - A$
  c. $A - \emptyset$ **d.** $\overline{\emptyset} \cap A$

7. For each of the following conditions, find $B - A$.
  a. $A = B$ **b.** $B \subseteq A$

8. Give two examples of sets A and B for which $B - A = \emptyset$. Show that in each example $B \subseteq A$.

9. Use set notation to identify each of the following shaded regions:

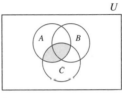
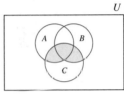

a.   **b.**

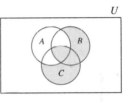

c.

10. In the following, shade the portion of the Venn diagram that represents the given set:

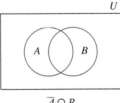

$\overline{A} \cap B$

11. Use Venn diagrams to determine if each of the following is true:
  a. $A - (B \cap C) = (A - B) \cap (A - C)$
  b. $A - (B \cup C) = (A - B) \cup (A - C)$

12. For each of the following pairs of sets, explain which is a subset of the other. If neither is a subset of the other, explain why.
  a. $A - B$ and $A - (B - C)$
  b. $A \cup B$ and $(A \cup B) - \emptyset$

13. a. If $n(A \cup B) = 22, n(A \cap B) = 8$, and $n(B) = 12$, find $n(A)$.
  b. If $n(A) = 8$, $n(B) = 14$, and $n(A \cap B) = 5$, find $n(A \cup B)$.

14. The equation $\overline{A \cup B} = \overline{A} \cap \overline{B}$ and a similar equation for $\overline{A \cap B}$ are referred to as *DeMorgan's Laws* in honor of the famous British mathematician who first discovered them.
  a. Use Venn diagrams to show that $\overline{A \cup B} = \overline{A} \cap \overline{B}$.

b. Discover an equation similar to the one in part (a) involving $\overline{A \cap B}$, $\overline{A}$, and $\overline{B}$. Use Venn diagrams to show that the equation holds.

c. Verify the equations in (a) and (b) for specific sets.

15. Suppose P is the set of all eighth-grade students at the Paxson School, with B the set of all students in the band and C the set of all students in the choir. Identify in words the students described by each region of the following figure:

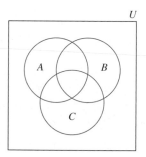

16. Fill in the Venn diagram with the appropriate numbers based on the following information:

$n(A) = 26$ $n(B \cap C) = 12$
$n(B) = 32$ $n(A \cap C) = 8$
$n(C) = 23$ $n(A \cap B \cap C) = 3$
$n(A \cap B) = 10$ $n(U) = 65$

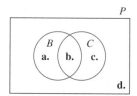

17. Write the letters in the appropriate sections of the following Venn diagram using the following information:

Set A contains the letters in the word *Iowa*.
Set B contains the letters in the word *Hawaii*.
Set C contains the letters in the word *Ohio*.

The universal set U contains the letters in the word *Washington*.

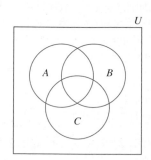

18. When three sets A, B, and C intersect as in the diagram of problem 17, eight nonoverlapping regions are created. Describe each of the regions using set notation.

19. A pollster interviewed 500 university seniors who owned credit cards. She reported that 240 owned Goldcard, 290 had Supercard, and 270 had Thriftcard. Of those seniors, the report said that 80 owned only a Goldcard and a Supercard, 70 owned only a Goldcard and a Thriftcard, 60 owned only a Supercard and a Thriftcard, and 50 owned all three cards. When the report was submitted for publication in the local campus newspaper, the editor refused to publish it, claiming the poll was not accurate. Was the editor right? Why or why not?

20. A baseball manager examined his roster and noticed the following:

- Every outfielder was a switch hitter.
- A third of the infielders were switch hitters.
- Half of all the switch hitters were outfielders.
- There are 12 infielders and 8 outfielders and no person played both positions.

How many switch hitters are neither infielders nor outfielders?

21. On the first day of tryouts for Little League, 128 boys of ages 10 (T), 11 (E), and 12 (W) showed up. They were asked what positions besides pitcher they wanted to play: infield (I), outfield (O), or catcher (C). The results are shown in the following table:

	I	O	C	Totals
10 (T)	28	14	12	54
11 (E)	18	20	8	46
12 (W)	10	12	6	28
Totals	56	46	26	128

Tell what each of the following means in words along with the number of boys indicated in each part:

a. $I \cap W$
b. $C \cap (T \cup E)$
c. $(I \cup O) \cap T$
d. $(T \cup E) \cap O$

22. Tell whether each of the following is true or false and tell why:

a. $(2, 5) = (5, 2)$ **b.** $(2, 5) = \{2, 5\}$

23. Answer each of the following:

a. If A has five elements and B has four elements, how many elements are in $A \times B$?

b. If A has m elements and B has n elements, how many elements are in $A \times B$?

c. If A has m elements, B has n elements, and C has p elements, how many elements are in $(A \times B) \times C$?

Mathematical Connections 2-3

Communication

1. Answer each of the following and justify your answer:
   a. If $a \in A \cap B$, is it true that $a \in A \cup B$?
   b. If $a \in A \cup B$, is it true that $a \in A \cap B$?
2. Explain how $\overline{A}$ is related to $U - A$.
3. Is the operation of forming Cartesian products commutative? Explain why or why not.
4. If A and B are sets, is it always true that $n(A - B) = n(A) - n(B)$? Explain.

Open-Ended

5. Make up and solve a story problem concerning specific sets A, B, and C for which $n(A \cup B \cup C)$ is known and it is required to find $n(A), n(B)$, and $n(C)$.
6. Describe a real-life situation that can be represented by each of the following:
   a. $A \cap \overline{B}$
   b. $A \cap B \cap C$
   c. $A - (B \cup C)$

Cooperative Learning

7. Use set operations of union, intersection, complement, and set difference to describe the shaded region in the following figure in as many ways as possible. Compare your expressions with those of other groups to see which has the most. What is the total number of different expressions found by all the groups? Which expressions appeared in all the groups?

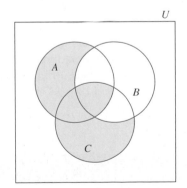

Questions from the Classroom

8. A student asks, "If $A = \{a,b,c\}$ and $B = \{b,c,d\}$, why isn't it true that $A \cup B = \{a,b,c,b,c,d\}$?" What is your response?
9. A student says that she can show that if $A \cap B = A \cap C$, then it is not necessarily true that $B = C$; but she thinks that whenever $A \cap B = A \cap C$ and $A \cup B = A \cup C$, then $B = C$. What is your response?
10. A student claims that the complement bar can be "broken" over the operation of intersections; that is, $\overline{A \cap B} = \overline{A} \cap \overline{B}$. What is your response?
11. A student is asked to find all one-to-one correspondences between two given sets. He finds the Cartesian product of the sets and claims that his answer is correct because it includes all possible pairings between the elements of the sets. How do you respond?
12. A student argues that adding two sets A and B, or $A + B$, and taking the union of two sets, $A \cup B$, is the same thing. How do you respond?

Review Problems

13. In base two, does the number "two" exist? Explain your reasoning.
14. How would you write 81 in base three? Any power of 3 in base ten?
15. **a.** Write $\{4,5,6,7,8,9\}$ using set-builder notation.
    b. Write $\{x \mid x = 5n, \text{where } n = 3,6, \text{or } 9\}$ using the listing method.
16. Find the number of elements in the following sets:
    a. $\{x \mid x \text{ is a letter in } commonsense\}$
    b. The set of letters appearing in the word *committee*
17. If $A = \{1,2,3,4\}$ and $B = \{1,2,3,4,5\}$, answer the following questions:
    a. How many subsets of A do not contain the element 1?
    b. How many subsets of A contain the element 1?
    c. How many subsets of A contain either the element 1 or 2?
    d. How many subsets of A contain neither the element 1 nor 2?
    e. How many subsets of B contain the element 5 and how many do not?
    f. If all the subsets of A are known, how can all the subsets of B be listed systematically? How many subsets of B are there?
18. **a.** Which of the following sets are equal?
    b. Which sets are proper subsets of the other sets?

    $A = \{2,4,6,8,10, \ldots\}$
    $B = \{x \mid x = 2n + 2 \text{ where } n = 0,1,2,3,4, \ldots\}$
    $C = \{x \mid x = 4n \text{ where } n \in N\}$

19. Give examples from real life for each of the following:
    a. A one-to-one correspondence between two sets
    b. A correspondence between two sets that is not one-to-one

20. If there are six teams in the Alpha league and five teams in the Beta league and if each team from one league plays each team from the other league exactly once, how many games are played?

21. José has four pairs of slacks, five shirts, and three sweaters. From how many combinations can he choose if he chooses a pair of slacks, a shirt, and a sweater each day?

National Assessment of Educational Progress (NAEP) Question

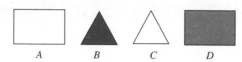

A   B   C   D

Melissa chose one of the figures above.
- The figure she chose was shaded.
- The figure she chose was *not* a triangle.

Which figure did she choose?
a. *A*   **b.** *B*   **c.** *C*   **d.** *D*

NAEP, 2007, Grade 4

LABORATORY ACTIVITY   A set of attribute blocks consists of 32 blocks. Each block is identified by its own shape, size, and color. The four shapes in a set are square, triangle, rhombus, and circle; the four colors are red, yellow, blue, and green; the two sizes are large and small. In addition to the blocks, each set contains a group of 20 cards. Ten of the cards specify one of the attributes of the blocks (for example, red, large, square). The other 10 cards are negation cards and specify the lack of an attribute (for example, not green, not circle). Many set-type problems can be studied with these blocks. For example, let *A* be the set of all green blocks and *B* be the set of all large blocks. Using the set of all blocks as the universal set, describe elements in each set listed here to determine which are equal:

1. $A \cup B$; $B \cup A$
2. $\overline{A \cap B}$; $\overline{A} \cap \overline{B}$
3. $\overline{A \cap B}$; $\overline{A} \cup \overline{B}$
4. $A - B$; $A \cap \overline{B}$

Hint for Solving the Preliminary Problem

The Venn diagram of this chapter is a good tool for sorting the data. Try distinguishing among disjoint sets of people. For example, consider one circle with adults, one with Mississippi citizens, and one with females. Sorting the information with these circles should provide guidance in finding a solution. Remember to determine what types of people are in the intersections and what the complements of the sets are.

Chapter Outline

I. Numeration systems
   A. A **numeration** system consists of a set of symbols with operations and properties to represent numbers systematically.
   B. Properties of numeration systems give basic structure to the systems.
   1. Additive property
   2. Place value property
   3. Subtractive property
   4. Multiplicative property
   5. **Place value** assigns a value to a digit depending on its placement in a numeral. The value of a digit is the product of its place value and its **face value**.
   C. The **Hindu-Arabic numeration system** is a base-ten system that uses the digits 0, 1, 2, 3, 4, 5, 6, 7, 8, and 9.

II. Exponents
   A. For any whole number a and any natural number n,

   $$a^n = \underbrace{a \cdot a \cdot a \cdot \ldots \cdot a,}_{n \text{ factors}}$$

   where a is the **base** and n is the **exponent**.
   B. $a^0 = 1$ where $a \in N$
III. Set definitions and notation
   A. A **set** can be described as any collection of objects.
   B. Sets should be **well defined** so that an object either does or does not belong to the set.
   C. An **element** is any **member** of a set.
   D. Sets can be specified by either **listing** all the elements or using **set-builder notation**.
   E. The **empty set**, written ∅, contains no elements.
   F. The **universal set** contains all the elements being discussed.
IV. Relationships and operations on sets
   A. Two sets are **equal** if, and only if, they have exactly the same elements.
   B. Two sets A and B are in **one-to-one correspondence** if, and only if, each element of A can be paired with exactly one element of B and each element of B can be paired with exactly one element of A.
   C. Two sets A and B are **equivalent** if, and only if, their elements can be placed into one-to-one correspondence (written $A \sim B$).
   D. Set A is a **subset** of set B if, and only if, every element of A is an element of B (written $A \subseteq B$).
   E. Set A is a **proper subset** of set B if, and only if, every element of A is an element of B and there is at least one element of B that is not in A (written $A \subset B$).

F. A set containing n elements has 2^n subsets.
G. The **union** of two sets A and B is the set of all elements in A, in B, or in both A and B (written $A \cup B$).
H. The **intersection** of two sets A and B is the set of all elements belonging to both A and B (written $A \cap B$).
I. The **cardinal number** of a finite set S, $n(S)$, indicates the number of elements in the set.
J. A set is **finite** if the number of elements in the set is zero or a natural number. Otherwise, the set is **infinite**.
K. Two sets A and B are **disjoint** if they have no elements in common.
L. The **complement** of a set A is the set consisting of the elements of the universal set not in A (written $\overline{A}$).
M. The **complement of set A relative to set B** (set difference) is the set of all elements in B that are not in A (written $B - A$).
N. The **Cartesian product** of sets A and B, written $A \times B$, is the set of all ordered pairs such that the first element in each pair is from A and the second element of each pair is from B.
O. Properties of set operations
   1. Commutative properties of set union and intersection
   2. Associative properties of set union and intersection
   3. Distributive property of set intersection over union and of set union over intersection
P. **Fundamental Counting Principle:** If event M can occur in m ways and, after it has occurred, event N can occur in n ways, then event M followed by event N can occur in mn ways.

Chapter Review

1. For each of the following base-ten numerals, tell the place value for each of the circled digits:
   a. 4③2   **b.** ③432
   c. 19③24
2. Convert each of the following to base ten:
   a. $\overline{CDXLIV}$
   b. 432_{five}
   c. $ET0_{\text{twelve}}$
   d. 1011_{two}
   e. 4136_{seven}
3. Convert each of the following numbers to numerals in the indicated system:
   a. 999 to Roman
   b. 86 to Egyptian
   c. 123 to Mayan

   d. 346_{ten} to base five
   e. 27_{ten} to base two
4. Simplify each of the following, if possible. Write your answers in exponential form, a^b.
   a. $3^4 \cdot 3^7 \cdot 3^6$ **b.** $2^{10} \cdot 2^{11}$
5. Write the base-three numeral for the base-three blocks shown.

6. The smallest number of base-three blocks needed to represent 51 is _____ block(s) _____ flat(s) _____ long(s) _____ unit(s).
7. Draw multibase blocks to represent
   a. 123_{four}.   **b.** 24_{five}.

8. **a.** The first digit from the left (the lead digit) of a base-ten numeral is 4 followed by 10 zeros. What is the place value of 4?

   b. A number in base five has 10 digits. What is the place value of the second digit from the left?

   c. A number in base two has lead digit 1 followed by 30 zeros and units digit 1. What is the place value of the lead digit?

9. Write the following base-ten numerals in the indicated base without performing any multiplications:

   a. $10^{10} + 23$ in base ten

   b. $2^{10} + 1$ in base two

   c. $5^{10} + 1$ in base five

   d. $10^{10} - 1$ in base ten

   e. $2^{10} - 1$ in base two

   f. $12^5 - 1$ in base twelve

10. Write an example of a base other than ten used in a real-life situation. How is it used?

11. Describe the important characteristics of each of the following systems:

   a. Egyptian   **b.** Babylonian

   c. Roman   **d.** Hindu-Arabic

12. Write 128 in each of the following bases:

   a. five   **b.** two

   c. twelve

13. Write each of the following in the indicated bases without multiplying out the various powers:

   a. $4 \cdot 5^6 + 11 \cdot 5^4 + 9$ in base five

   b. $2^{10} + 2^3$ in base two

   c. $11 \cdot 12^5 + 10 \cdot 12^3 + 20$ in base twelve

   d. $9 \cdot 8^5 + 8$ in base eight

14. List all the subsets of $\{m, a, t, h\}$.

15. Let

   $U = \{u, n, i, v, e, r, s, a, l\}$,

   $A = \{r, a, v, e\}$, $C = \{l, i, n, e\}$,

   $B = \{a, r, e\}$, $D = \{s, a, l, e\}$.

   Find each of the following:

   a. $A \cup B$ **b.** $C \cap D$

   c. $\overline{D}$ **d.** $A \cap \overline{D}$

   e. $\overline{B \cup C}$

   f. $(B \cup C) \cap D$

   g. $(\overline{A} \cup B) \cap (C \cap \overline{D})$

   h. $(C \cap D) \cap A$

   i. $n(\overline{C})$ **j.** $n(C \times D)$

16. Indicate the following sets by shading the figure:

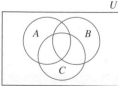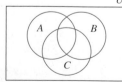

   a. $A \cap (B \cup C)$ **b.** $(\overline{A \cup B}) \cap C$

17. Suppose you are playing a word game with seven distinct letters. How many seven-letter words can there be?

18. **a.** Show one possible one-to-one correspondence between sets D and E if $D = \{t, b, e\}$ and $E = \{e, n, d\}$.

   b. How many one-to-one correspondences between sets D and E are possible?

19. Use a Venn diagram to determine whether $A \cap (B \cup C) = (A \cap B) \cup C$ for all sets A, B, and C.

20. According to a student survey, 16 students liked history, 19 liked English, 18 liked mathematics, 8 liked mathematics and English, 5 liked history and English, 7 liked history and mathematics, 3 liked all three subjects, and every student liked at least one of the subjects. Draw a Venn diagram describing this information and answer the following questions:

   a. How many students were in the survey?

   b. How many students liked only mathematics?

   c. How many students liked English and mathematics but not history?

21. Describe, using symbols, the shaded portion in each of the following figures:

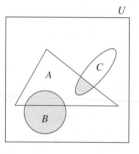

   a.   **b.**

22. Classify each of the following as true or false. If false, tell why.

   a. For all sets A and B, either $A \subseteq B$ or $B \subseteq A$.

   b. The empty set is a proper subset of every set.

   c. For all sets A and B, if $A \sim B$, then $A = B$.

   d. The set $\{5, 10, 15, 20, \ldots\}$ is a finite set.

   e. No set is equivalent to a proper subset of itself.

   f. If A is an infinite set and $B \subseteq A$, then B also is an infinite set.

   g. For all finite sets A and B, if $A \cap B \neq \varnothing$, then $n(A \cup B) \neq n(A) + n(B)$.

   h. If A and B are sets such that $A \cap B = \varnothing$, then $A = \varnothing$ or $B = \varnothing$.

23. Use Venn diagrams to decide whether each of the following is always true for finite sets A and B.

   a. $n(A \cup B) = n(A - B) + n(B - A) + n(A \cap B)$

   b. $n(A \cup B) = n(A - B) + n(B) = n(B - A) + n(A)$

24. Suppose P and Q are equivalent sets and $n(P) = 17$.

   a. What is the minimum number of elements in $P \cup Q$?

   b. What is the maximum number of elements in $P \cup Q$?

c. What is the minimum number of elements in $P \cap Q$?

d. What is the maximum number of elements in $P \cap Q$?

25. Case Eastern Junior College awarded 26 varsity letters in crew, 15 in swimming, and 16 in soccer. If awards went to 46 students and only 2 lettered in all sports, how many students lettered in just two of the three sports?

26. Consider the set of northwestern states or provinces {Montana, Washington, Idaho, Oregon, Alaska, British Columbia, Alberta}. If a person chooses one element, show that in three *yes* or *no* questions, we can determine the element.

27. Using the definitions of less than or greater than, prove that each of the following inequalities is true:
a. $3 < 13$ **b.** $12 > 9$

28. Heidi has a brown pair and a gray pair of slacks; a brown blouse, a yellow blouse, and a white blouse; and a blue sweater and a white sweater. How many different outfits does she have if each outfit she wears consists of slacks, a blouse, and a sweater?

Selected Bibliography

Barkley, C. "Other Ways to Count." *Student Math Notes* (November 2003).

Framer, J., and R. Powers. "Exploring Mayan Numerals." *Teaching Children Mathematics* 12 (September 2005): 69.

Fuson, K. "Research on Learning and Teaching Addition and Subtraction of Whole Numbers." In *Handbook of Research on Mathematics Teaching and Learning,* edited by D. Grouws. New York: MacMillan, 1992.

Ginsburg, H., A. Klein, and P. Starkey. "The Development of Children's Mathematical Thinking: Connecting Research with Practice." In *Child Psychology in Practice,* edited by Irving E. Sigel and K. Ann Renninger, pp. 401–476, vol. 4 of *Handbook of Child Psychology,* edited by William Damon. New York: John Wiley & Sons, 1998.

Klein, A., M. Beishuizen, and A. Treffers. "The Empty Number Line in Dutch Second Grades: Realistic Versus Gradual Program Design." *Journal of Research in Mathematics Education* 29 (July 1998): 443–464.

Moldovan, C. "Culture in the Curriculum: Enriching Numeration and Number Operations." *Teaching Children Mathematics* 8 (December 2001): 238–243.

Overbay, S., and M. Brod. "Magic with Mayan Mathematics." *Mathematics Teaching in the Middle School* 12 (February 2007): 340.

Pickreign, J. "Alternative Base Arithmetic Activities." *ON-Math* 5 (2006–7).

Resnick, L. "From Protoquantities to Operators: Building Mathematical Competence on a Foundation of Everyday Knowledge." In *Analysis of Arithmetic for Mathematics Teaching,* edited by D. Leinhardt, R. Putnam, and R. Hattrup. Hillsdale, NJ: LEA, 1992.

Siegler, R. *Emerging Minds: The Process of Change in Children's Thinking.* New York: Oxford University Press, 1996.

Tsamir, P., and D. Triosh. "Consistency and Representations: The Case of Actual Infinity." *Journal for Research in Mathematics Education* 30 (March 1999): 213–219.

Uy, F. "The Chinese Numeration System and Place Value." *Teaching Children Mathematics* 9 (January 2003): 243.

Walmsley, A. "Math Roots: Understanding Aztec and Mayan Numeration Systems." *Mathematics Teaching in the Middle School* 12 (August 2006): 55.

Zaslavsky, C. "Developing Number Sense: What Can Other Cultures Tell Us?" *Teaching Children Mathematics* 7 (February 2001): 312–319.

Zaslavsky, C. "The Influence of Ancient Egypt on Greek and Other Numeration Systems." *Mathematics Teaching in the Middle School* 9 (November 2003): 174.

CHAPTER 3

Whole Numbers and Their Operations

Preliminary Problem

Using exactly five 5s and only addition, subtraction, multiplication, and division, write an expression that equals each of the numbers from 1 to 10. You do not have to use all operations. Numbers such as 55 are permitted; for example, 5 could be written as $5 + [(5 - 5) \cdot 55]$.

I n Section 2-1 we saw that the concept of one-to-one correspondence between sets can be used to introduce children to the concept of a number. In NCTM's *Curriculum Focal Points for Prekindergarten through Grade 8 Mathematics*, we find the following:

> Children develop an understanding of the meanings of whole numbers and recognize the number of objects in small groups without counting and by counting—the first and most basic mathematical algorithm. They understand that number words refer to quantity. They use one-to-one correspondence to solve problems by matching sets and comparing number amounts and in counting objects to 10 and beyond. They understand that the last word that they state in counting tells "how many," they count to determine number amounts and compare quantities (using language such as "more than" and "less than"), and they order sets by the number of objects in them. (p. 11)

In the following *Peanuts* cartoon, it seems that Lucy's little brother has not yet learned to associate number words with a collection of objects. He will soon learn that this set of fingers can be put into one-to-one correspondence with many sets of objects that can be counted. He will associate the word *three* not only with Lucy's three upheld fingers but with other sets of objects with this same cardinal number.

In this chapter, we will investigate operations involving whole numbers. As stated in the *Principles and Standards (PSSM)* in "Number and Operation" for grades pre-K–2, all students on this level should understand meanings of operations and how they relate to one another. In particular, *PSSM* states that in pre-K–2 all students should:

- understand various meanings of addition and subtraction of whole numbers and the relationship between the two operations;

- understand the effects of adding and subtracting whole numbers;

- understand situations that entail multiplication and division, such as equal groupings of objects and sharing equally. (p. 78)

3-1   Addition and Subtraction of Whole Numbers

When zero is included with the set of natural numbers, $N = \{1, 2, 3, 4, 5, \dots\}$, we have the set of whole numbers, denoted $W = \{0, 1, 2, 3, 4, 5, \dots\}$. In this section, we provide a variety of models for teaching computational skills involving whole numbers and allow you to revisit mathematics for the deeper understanding that teachers need.

Addition of Whole Numbers

Children encounter addition in preschool years by combining objects and wanting to know how many objects there are in the combined set. They may "count on" as suggested by Carpenter and Moser in the Research Note, or they may count the objects to find the cardinal number of the combined set.

Research Note

Simple addition and subtraction problems can be better understood when students solve "joining" and "take-away" problems by directly modeling the situation or by using counting strategies such as counting on or counting back (Carpenter and Moser 1984).   ◆

Set Model

A set model is one way to represent addition of whole numbers. Suppose Jane has 4 blocks in one pile and 3 in another. If she combines the two groups, how many objects are there in the combined group? Figure 3-1 shows the solution as it might appear in an elementary school text. The combined set of blocks is the union of the disjoint sets of 4 blocks and 3 blocks. After the sets have been combined, some children count the objects to determine that there are 7 in all. *Note the importance of the sets being disjoint or having no elements in common.* If the sets have common elements, then an incorrect conclusion can be drawn.

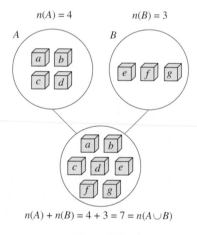

$$n(A) + n(B) = 4 + 3 = 7 = n(A \cup B)$$

Figure 3-1

Using set terminology, we define addition formally.

Definition of Addition of Whole Numbers

Let A and B be two disjoint finite sets. If $n(A) = a$ and $n(B) = b$, then $a + b = n(A \cup B)$.

The numbers a and b in $a + b$ are the **addends** and $a + b$ is the **sum**.

NOW TRY THIS 3-1 If the sets in the preceding definition of addition of whole numbers are not disjoint, explain why the definition is incorrect.

Historical Note

Historians think that the word *zero* originated from the Hindu word *sūnya*, which means "void." Then *sūnya* was translated into the Arabic *sifr*, which when translated to Latin became *zephirum*, from which the word *zero* was derived.   ◆

Number-Line (Measurement) Model

The set model for addition may not be the best model for some additions. For example, consider the following questions:

1. Josh has 4 feet of red ribbon and 3 feet of white ribbon. How many feet of ribbon does he have altogether?
2. One day, Gail drank 4 ounces of orange juice in the morning and 3 ounces at lunchtime. If she drank no other orange juice that day, how many ounces of orange juice did she drink for the entire day?

A *number line* can be used to model whole-number addition and answer questions 1 and 2. Any line marked with two fundamental points, one representing 0 and the other representing 1, can be turned into a number line. The points representing 0 and 1 mark the ends of a *unit segment*. Other points can be marked and labeled, as shown in Figure 3-2. Any two consecutive whole numbers on the number line in Figure 3-2 mark the ends of a segment that has the same length as the unit segment.

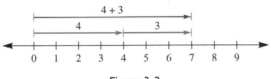

Figure 3-2

Addition problems may be modeled using directed arrows (vectors) on the number line. For example, the sum of 4 + 3 is shown in Figure 3-2. Arrows representing the addends, 4 and 3, are combined into one arrow representing the sum 4 + 3. Figure 3-2 poses an inherent problem for students. If an arrow starting at 0 and ending at 3 represents 3, why should an arrow starting at 4 and ending at 7 represent 3? Students need to understand that the sum represented by any two directed arrows can be found by placing the endpoint of the first directed arrow at 0 and then joining to it the directed arrow for the second number with no gaps or overlaps. The sum of the numbers can then be read. We have depicted the addends as arrows (or vectors) above the number line, but students typically concatenate (connect) the arrows directly on the line.

NOW TRY THIS 3-2 A common error is that students represent 3 as an arrow on the number line sometimes starting at 1, as shown in Figure 3-3. Explain why this is not appropriate.

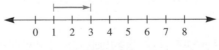

Figure 3-3

The symbol "+" first appeared in a 1417 manuscript and was a short way of writing the Latin word *et*, which means "and." The word *minus* means "less" in Latin. First written as an *m*, it was later shortened to a horizontal bar. Johannes Widman wrote a book in 1489 that made use of the + and − symbols for addition and subtraction. ◆

Ordering Whole Numbers

In the NCTM grade 1 *Curriculum Focal Points*, we find the following:

> Children compare and order whole numbers (at least to 100) to develop an understanding of and solve problems involving the relative sizes of these numbers. They think of whole numbers between 10 and 100 in terms of groups of tens and ones (especially recognizing the numbers 11 to 19 as 1 group of ten and particular numbers of ones). They understand the sequential order of the counting numbers and their relative magnitudes and represent numbers on a number line. (p. 13)

In Chapter 2, we used the concept of a set and the concept of a one-to-one correspondence to define *greater-than* relations. A horizontal number line can also be used to describe **greater-than** and **less-than** relations in the set of whole numbers. For example, in Figure 3-2, notice that 4 is to the left of 7 on the number line. We say, "four is less than seven," and we write $4 < 7$. We can also say "seven is greater than four" and write $7 > 4$. Since 4 is to the left of 7, there is a natural number that can be added to 4 to get 7, namely, 3. Thus, $4 < 7$ because $4 + 3 = 7$. We can generalize this discussion to form the following definition of *less than*.

Definition of Less Than

For any whole numbers a and b, a is **less than** b, written $a < b$, if, and only if, there exists a natural number k such that $a + k = b$.

Sometimes equality is combined with the inequalities, greater than and less than, to give the relations **greater than or equal to** and **less than or equal to**, denoted $\geq$ and $\leq$. Thus, $a \leq b$ means $a < b$ or $a = b$. The emphasis with respect to these symbols is on the *or*, so $3 \leq 5, 5 \geq 3$, and $3 \geq 3$ are all true statements.

Whole-Number Addition Properties

Any time two whole numbers are added, we are guaranteed that a unique whole number will be obtained. This property is sometimes referred to as the *closure property of addition of whole numbers*. We say that "the set of whole numbers is closed under addition."

Theorem 3–1:   Closure Property of Addition of Whole Numbers

If a and b are whole numbers, then $a + b$ is a whole number.

REMARK The closure property implies that the sum of two whole numbers *exists* and that the sum is a *unique* whole number; for example, $5 + 2$ is a unique whole number and we identify that number as 7.

NOW TRY THIS 3-3 Determine whether each of the following sets is closed under addition:

a. $E = \{2, 4, 6, 8, 10, \ldots\}$
b. $F = \{1, 3, 5, 7, 9\}$

In *Principles and Standards* we find the following:

> In developing the meaning of addition and subtraction with whole numbers, students should also encounter the properties of operations, such as the commutativity and associativity of addition. Although some students discover and use properties of operations naturally, teachers can bring these properties to the forefront through class discussions. (p. 83)

Figure 3-4(a) shows two additions. Pictured above the number line is 3 + 5 and below the number line is 5 + 3. The sums are the same. Figure 3-4(b) shows the same sums obtained with colored rods and the result being the same. Both illustrations in Figure 3-4 demonstrate the idea that two whole numbers can be added in either order. This property is true in general and is the *commutative property of addition of whole numbers*. We say that "addition of whole numbers is commutative." The word *commutative* is derived from *commute*, which means "to interchange."

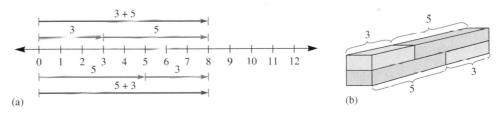

(a)   (b)

Figure 3-4

Theorem 3–2:   Commutative Property of Addition of Whole Numbers

If a and b are any whole numbers, then $a + b = b + a$.

The commutative property of addition of whole numbers is not obvious to many young children. They may be able to find the sum 9 + 2 and not be able to find the sum 2 + 9. Using *counting on*, 9 + 2 can be computed by starting at 9 and then counting on two more as "ten" and "eleven." To compute 2 + 9 without the commutative property, the *counting on* is more involved. Students need to understand that 2 + 9 is another name for 9 + 2.

NOW TRY THIS 3-4 Use the set model to show the commutative property for 3 + 5 = 5 + 3.

Another property of addition is demonstrated when we select the order in which to add three or more numbers. For example, we could compute 24 + 8 + 2 by grouping the 24 and the 8 together: (24 + 8) + 2 = 32 + 2 = 34. (The parentheses indicate that the first two numbers are grouped together.) We might also recognize that it is easy to add any number to 10 and compute it as 24 + (8 + 2) = 24 + 10 = 34. This example illustrates the *associative property of addition of whole numbers*. The word *associative* is derived from the word *associate*, which means "to unite."

> **Theorem 3–3:   Associative Property of Addition of Whole Numbers**
>
> If a, b, and c are whole numbers, then $(a + b) + c = a + (b + c)$.

When several numbers are being added, the parentheses are usually omitted since the grouping does not alter the result.

Another property of addition of whole numbers is seen when one addend is 0. In Figure 3-5, set A has 5 blocks and set B has 0 blocks. The union of sets A and B has only 5 blocks.

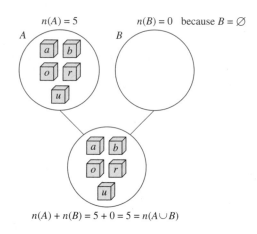

$$n(A) + n(B) = 5 + 0 = 5 = n(A \cup B)$$

Figure 3-5

This example uses the set model to illustrate the following property of whole numbers:

> **Theorem 3–4:   Identity Property of Addition of Whole Numbers**
>
> There is a unique whole number 0, the **additive identity**, such that for any whole number a, $a + 0 = a = 0 + a$.

 Notice how the associative and identity properties are introduced on the grade 3 student page on p. 117. Work through parts 4–7.

Example 3-1

Which properties justify each of the following?

a. $5 + 7 = 7 + 5$
b. $1001 + 733$ is a unique whole number.
c. $(3 + 5) + 7 = (5 + 3) + 7$
d. $(8 + 5) + 2 = 2 + (8 + 5) = (2 + 8) + 5$

Solution   **a.** Commutative property of addition
                b. Closure property of addition
                c. Commutative property of addition
                d. Commutative and associative properties of addition

School Book Page   PROPERTIES OF ADDITION

What is the Associative Property?

The **Associative (grouping) Property of Addition** says that you can group addends in any way and the sum will be the same.

$(3 + 1) + 2 = 6$

> Grouping symbols, like parentheses, ( ), show which numbers to add first.

$3 + (1 + 2) = 6$

So $(3 + 1) + 2 = 3 + (1 + 2)$.

☑ Talk About It

4. Evan says, "You can rewrite $(5 + 3) + 2$ as $8 + 2$." Do you agree? Explain.

What is the Identity Property?

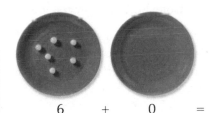

The **Identity (zero) Property of Addition** says that the sum of any number and zero is that same number.

$6 + 0 = 6$

☑ Talk About It

5. How could you use the Identity Property of Addition to find $536 + 0$?

CHECK ✓

For another example, see Set 2-1 on p. 116.

Find each sum.

1. $2 + (5 + 3)$ **2.** $3 + (1 + 4)$ **3.** $3 + 2 + 6$

Write each missing number.

4. $5 + 2 = 2 + \blacksquare$ **5.** $\blacksquare + 4 = 4$ **6.** $(1 + 4) + 3 = \blacksquare + (4 + 3)$

7. Number Sense What property of addition is shown in the following number sentence? Explain. $4 + (5 + 2) = (5 + 2) + 4$

Source: Scott Foresman-Addison Wesley Math, Grade 3, 2008 (p. 67).

Mastering Basic Addition Facts

Certain mathematical facts are *basic addition facts*. Basic addition facts are those involving a single digit plus a single digit. In the *Dennis The Menace* cartoon, it seems that Dennis has not mastered the basic addition facts.

DENNIS THE MENACE

"MAKE UP YOUR MIND. FIRST YOU TELL ME 3 PLUS 3 IS SIX, AND NOW YOU SAY 4 PLUS 2 IS SIX!"

One method of learning the basic facts is to organize them according to different derived fact strategies, as suggested in the Research Note. Several strategies are listed below.

1. **Counting on.** The strategy of counting on from the greater of the addends can be used any time we need to add whole numbers, but it is inefficient. It is usually used when one addend is 1, 2, or 3. For example, in the cartoon Dennis could have computed 4 + 2 by starting at 4 and then counting on 5, 6. Likewise, 3 + 3 would be computed by starting at 3 and then counting on 4, 5, 6.

2. **Doubles.** The next strategy involves the use of *doubles*. Doubles such as 3 + 3 in the cartoon receive special attention. After students master doubles, *doubles* + 1 and *doubles* + 2 can be learned easily. For example, if a student knows 6 + 6 = 12, then 6 + 7 is (6 + 6) + 1, or 1 more than the double of 6, or 13. Likewise, 7 + 9 is (7 + 7) + 2, or 2 more than the double of 7, or 16.

3. **Making 10.** Another strategy is that of *making 10* and then adding any leftover. For example, we could think of 8 + 5 as shown in Figure 3-6. Notice that we are really using the associative property of addition.

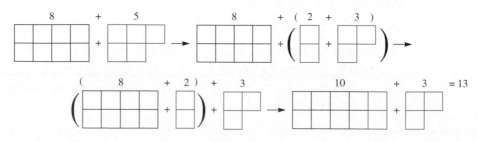

Figure 3-6

4. **Counting back.** The strategy of *counting back* is usually used when one number is 1 or 2 less than 10. For example, because 9 is 1 less than 10, then $9 + 7$ is 1 less than $10 + 7$, or 16. In symbols, this is $9 + 7 = (10 + 7) - 1 = 17 - 1 = 16$. Also, $8 + 7 = (10 + 7) - 2 = 17 - 2 = 15$.

Many basic facts might be classified under more than one strategy. For example, we could find $9 + 8$ using *making 10* as $9 + (1 + 7) = (9 + 1) + 7 = 10 + 7 = 17$ or using a *double + 1* as $(8 + 8) + 1$.

Subtraction of Whole Numbers

In the grade 1 *Focal Points* we find the following:

> By comparing a variety of solution strategies, children relate addition and subtraction as inverse operations. (p. 13)

In elementary school, operations that "undo" each other are **inverse operations**. Subtraction is the inverse operation for addition. It is sometimes hard for students to understand this inverse relationship between the two operations, as the cartoon demonstrates.

In *Principles and Standards* we find the following:

> Students develop further understandings of addition when they solve missing-addend problems that arise from stories or real situations. Further understandings of subtraction are conveyed by situations in which two collections need to be made equal or one collection needs to be made a desired size. Some problems, such as "Carlos had three cookies. María gave him some more, and now he has eight. How many did she give him?" can help students see the relationship between addition and subtraction. (p. 83)

We can model subtraction of whole numbers using several solution strategies, including the *take-away* model, the *missing-addend* model, the *comparison* model, and the *number-line (measurement)* model.

Take-Away Model

In addition, we imagine a second set of objects as being joined to a first set, but in subtraction, we imagine a second set as being *taken away* from a first set. For example, suppose we have 8 blocks and take away 3 of them. We illustrate this in Figure 3-7 and record this process as $8 - 3 = 5$.

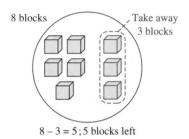

$$8 - 3 = 5 \text{ ; } 5 \text{ blocks left}$$

Figure 3-7

NOW TRY THIS 3-5 Recall how addition of whole numbers was defined using the concept of union of two disjoint sets. Similarly, write a definition of subtraction of whole numbers using the concepts of subsets and set difference.

Missing-Addend Model

A second model for subtraction, the *missing-addend* model, relates subtraction and addition. In Figure 3-7, $8 - 3$ is pictured as 8 blocks "take away" 3 blocks. The number of blocks left is the number $8 - 3$, or 5. This can also be thought of as the number of blocks that could be added to 3 blocks in order to get 8 blocks, that is,

$$3 + \boxed{8 - 3} = 8$$

The number $8 - 3$, or 5, is the **missing addend** in the equation

$$3 + \square = 8.$$

We can also relate the *missing-addend* approach to sets or to a number line. The subtraction $8 - 3$ is illustrated in Figure 3-8(a) using sets and in Figure 3-8(b) using the number line.

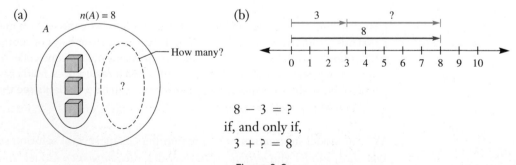

$$8 - 3 = ?$$
if, and only if,
$$3 + ? = 8$$

Figure 3-8

The missing-addend model gives elementary school students an opportunity to begin algebraic thinking. An unknown is a major part of the problem of trying to decide the difference of $8 - 3$.

Refer to the student page on p. 122 to see how grade 3 students are shown how addition and subtraction are related using a **fact family**. Answer the *Talk About It* questions on the bottom of the student page.

Cashiers often use the missing-addend model. For example, if the bill for a movie is $8 and you pay $10, the cashier might calculate the change by saying "8 and 2 is 10." This idea can be generalized: For any whole numbers a and b such that $a \geq b$, $a - b$ is the unique whole number such that $b + (a - b) = a$. That is, $a - b$ is the unique solution of the equation $b + \square = a$. The definition can be written as follows:

Definition of Subtraction of Whole Numbers

For any whole numbers a and b such that $a \geq b$, $a - b$ is the unique whole number c such that $b + c = a$.

Comparison Model

Another way to consider subtraction is by using a *comparison* model. Suppose Juan has 8 blocks and Susan has 3 blocks and we want to know how many more blocks Juan has than Susan. We can pair Susan's blocks with some of Juan's blocks, as shown in Figure 3-9, and determine that Juan has 5 more blocks than Susan. We also write this as $8 - 3 = 5$.

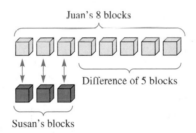

Figure 3-9

Number-Line (Measurement) Model

Subtraction of whole numbers can also be modeled on a number line using directed arrows, as suggested in Figure 3-10, which shows that $5 - 3 = 2$.

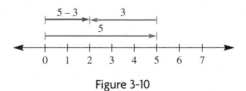

Figure 3-10

The following four problems illustrate each of the four models for subtraction. In all four problems the answer is 5 but each can be thought of using a different model.

1. **Take-away model.** Al had $9 and spent $4. How much did he have left?
2. **Missing-addend model.** Al has read 4 chapters of a 9-chapter book. How many chapters does he have left to read?
3. **Comparison model.** Al has 9 books and Betty has 4 books. How many more books does Al have than Betty?
4. **Number-line model.** Al completed a 9-mile hike in two days. He hiked 4 miles on the second day. How far did he hike on the first day?

School Book Page

RELATING ADDITION AND SUBTRACTION

Lesson 2-2

Algebra

Key Idea
Fact families show how addition and subtraction are related.

Vocabulary
• fact family
• difference

Think It Through
I can **draw a parts-whole picture** to show addition and subtraction.

Relating Addition and Subtraction

LEARN

How are addition and subtraction related?

You can think about parts and the whole to show how addition and subtraction are related.

Whole

	13	
5		8

Part   Part

You can write a **fact family** when you know the parts and the whole.

Fact family:

$5 + 8 = 13$ $13 - 8 = 5$

$8 + 5 = 13$ $13 - 5 = 8$

difference

Example

Find $12 - 7$.

What You **Think**	What You **Write**
$7 + \,? = 12$	
$7 + \mathbf{5} = 12$	$12 - 7 = 5$

✔ **Talk About It**

1. What are the other three number sentences in the fact family with $6 + 3 = 9$?

2. What addition fact can help you find $11 - 3$?

70

Source: Scott Foresman-Addison Wesley Math, Grade 3, 2008 (p. 70).

Properties of Subtraction

In an attempt to find $3 - 5$, we use the definition of subtraction: $3 - 5 = c$ if, and only if, $c + 5 = 3$. Since there is no whole number c that satisfies the equation, $3 - 5$ is not meaningful in the set of whole numbers. In general, it can be shown that if $a < b$, then $a - b$ is not meaningful in the set of whole numbers. Therefore, subtraction is not closed on the set of whole numbers.

NOW TRY THIS 3-6 Which of the following properties hold for subtraction of whole numbers? Explain.

a. Closure property
c. Commutative property
b. Associative property
d. Identity property

Introductory Algebra Using Whole-Number Addition and Subtraction

Sentences such as $9 + 5 = x$ and $12 - y = 4$ can be true or false depending on the values of x and y. For example, if $x = 10$, then $9 + 5 = x$ is false. If $y = 8$, then $12 - y = 4$ is true. If the value that is used makes the equation true, it is a **solution** to the equation.

NOW TRY THIS 3-7 Find the solution for each of the following where x is a whole number:

a. $x + 8 = 13$ b. $15 - x = 8$ c. $x > 9$ and $x < 11$

Assessment 3-1A

1. Give an example to show why, in the definition of addition, sets A and B must be disjoint.
2. For which of the following is it true that $n(A) + n(B) = n(A \cup B)$?
   a. $A = \{a, b, c\}, B = \{d, e\}$
   b. $A = \{a, b, c\}, B = \{b, c\}$
   c. $A = \{a, b, c\}, B = \varnothing$
3. If $n(A) = 3$, $n(B) = 5$, and $n(A \cup B) = 6$, what do you know about $n(A \cap B)$?
4. If $n(A) = 3$ and $n(A \cup B) = 6$,
   a. what are the possible values for $n(B)$?
   b. If $A \cap B = \varnothing$, what are the possible values of $n(B)$?
5. Explain whether the following given sets are closed under addition:
   a. $B = \{0\}$
   b. $T = \{0, 3, 6, 9, 12, \dots\}$
   c. $N = \{1, 2, 3, 4, 5, \dots\}$
   d. $V = \{3, 5, 7\}$
   e. $\{x \mid x \in W \text{ and } x > 10\}$

6. Each of the following is an example of one of the properties for addition of whole numbers. Identify the property illustrated.
   a. $6 + 3 = 3 + 6$
   b. $(6 + 3) + 5 = 6 + (3 + 5)$
   c. $(6 + 3) + 5 = (3 + 6) + 5$
   d. $5 + 0 = 5 = 0 + 5$
   e. $5 + 0 = 0 + 5$
   f. $(a + c) + d = a + (c + d)$
7. In the definition of *less than*, can the natural number k be replaced by the whole number k? Why or why not?
8. a. Recall how we have defined *less-than* and *greater-than* relations, and give a similar definition using the concept of subtraction for each of the following:
      i. $a < b$
      ii. $a > b$
   b. Use subtraction to define $a \geq b$.

9. Find the next three terms in each of the following arithmetic sequences:
   a. 8, 13, 18, 23, 28, ____, ____, ____
   b. 98, 91, 84, 77, 70, 63, ____, ____, ____

10. If A, B, and C each stand for a different single digit from 1 to 9, answer the following if

$$A + B = C$$

   a. What is the greatest digit that C could be? Why?
   b. What is the greatest digit that A could be? Why?
   c. What is the smallest digit that C could be? Why?
   d. If A, B, and C are even, what number(s) could C be? Why?
   e. If C is 5 more than A, what number(s) could B be? Why?
   f. If A is 3 times B, what number(s) could C be? Why?
   g. If A is odd and A is 5 more than B, what number(s) could C be? Why?

11. Assuming the same pattern continues, find the total of the terms in the 50th row in the following figure:

1	1st row
1 − 1	2nd row
1 − 1 + 1	3rd row
1 − 1 + 1 − 1	4th row
1 − 1 + 1 − 1 + 1	5th row

12. Make each of the following a magic square (a magic square was defined in Chapter 1):

a.

	1	6
	5	7
4		2

b.

17	10	
	14	
13	18	

13. a. At a volleyball game, the players stood in a row ordered by height. If Kent is shorter than Mischa, Sally is taller than Mischa, and Vera is taller than Sally, who is the tallest and who is the shortest?
    b. Write possible heights for the players in part (a).

14. Rewrite each of the following subtraction problems as an equivalent addition problem:
    a. $9 − 7 = x$
    b. $x − 6 = 3$
    c. $9 − x = 2$

15. Refer to the student page on page 122 to recall the description of a *fact family*.
    a. Write the fact family for $8 + 3 = 11$.
    b. Write the fact family for $13 − 8 = 5$.

16. What conditions, if any, must be placed on a, b, and c in each of the following to make sure that the result is a whole number?
    a. $a − b$ b. $a − (b − c)$

17. Illustrate $8 − 5 = 3$ using each of the following models:
    a. Take-away   b. Missing addend
    c. Comparison   d. Number line

18. Find the solution for each of the following:
    a. $3 + (4 + 7) = (3 + x) + 7$
    b. $8 + 0 = x$
    c. $5 + 8 = 8 + x$
    d. $x + 8 = 12 + 5$

Assessment 3-1B

1. For which of the following is it true that $n(A) + n(B) = n(A \cup B)$?
   a. $A = \{a, b\}$, $B = \{d, e\}$
   b. $A = \{a, b, c\}$, $B = \{b, c, d\}$
   c. $A = \{a\}$, $B = \varnothing$

2. If $n(A) = 3$, $n(B) = 5$, and $n(A \cap B) = 1$, what do you know about $n(A \cup B)$?

3. Explain whether the following given sets are closed under addition:
   a. $B = \{0, 1\}$
   b. $T = \{0, 4, 8, 12, 16, \dots\}$
   c. $F = \{5, 6, 7, 8, 9, 10, \dots\}$
   d. $\{x \mid x \in W \text{ and } x > 100\}$

4. Set A contains the element 1. What other whole numbers must be in set A for it to be closed under addition?

5. Set A is closed under addition and contains the numbers 2, 5, and 8. List six other elements that must be in A.

6. Each of the following is an example of one of the properties of whole-number addition. Fill in the blank to make a true statement and identify the property.
   a. $3 + 4 = $ ____ $+ 3$
   b. $5 + (4 + 3) = (4 + 3) + $ ____
   c. $8 + $ ____ $= 8$
   d. $3 + (4 + 5) = (3 + $ ____ $) + 5$
   e. $3 + 4$ is a unique ____ number.

7. Each of the following is an example of one of the properties for addition of whole numbers. Identify the property illustrated.
   a. $6 + 8 = 8 + 6$
   b. $(6 + 3) + 0 = 6 + 3$

c. $(6 + 8) + 2 = (8 + 6) + 2$
d. $(5 + 3) + 2 = 5 + (3 + 2)$

8. Find the next three terms in each of the following arithmetic sequences:
   a. 5, 12, 19, 26, 33, ____, ____, ____
   b. 63, 59, 55, 51, 47, ____, ____, ____

9. If A, B, C, and D each stand for a different single digit from 1 to 9, answer each of the following if

$$\begin{array}{r} A \\ + B \\ \hline CD \end{array}$$

   a. What is the value of C? Why?
   b. Can D be 1? Why?
   c. If D is 7, what values can A be?
   d. If A is 6 greater than B, what is the value of D?

10. **a.** A domino set contains all number pairs from double-0 to double-6, with each number pair occurring only once; for example, the following domino counts as 2-4 and 4-2. How many dominoes are in the set?

   b. When considering the sum of all dots on a single domino in an ordinary set of dominoes, explain how the commutative property might be important.

11. Rewrite each of the following subtraction problems as an equivalent addition problem:
   a. $9 - 3 = x$

b. $x - 5 = 8$
c. $11 - x = 2$

12. Refer to the student page on page 122 to recall the description of a *fact family*.
   a. Write the fact family for $9 + 8 = 17$.
   b. Write the fact family for $15 - 7 = 8$.

13. Show that each of the following is true. Give a property of addition to justify each step.
   a. $a + (b + c) = c + (a + b)$
   b. $a + (b + c) = (c + b) + a$

14. Illustrate $7 - 3 = 4$ using each of the following models:
   a. Take-away
   b. Missing addend
   c. Comparison
   d. Number line

15. Find the solution for each of the following:
   a. $12 - x = x + 6$
   b. $(9 - x) - 6 = 1$
   c. $3 + x = x + 3$
   d. $15 - x = x - 7$
   e. $14 - x = 7 - x$

16. Rob has 11 pencils. Kelly has 5 pencils. Which number sentence shows how many more pencils Rob has than Kelly?
   (i) $11 + 5 = 16$
   (ii) $16 - 5 = 11$
   (iii) $11 - 5 = 6$
   (iv) $11 - 6 = 5$

Mathematical Connections 3-1

Communication

1. In a survey of 52 students, 22 said they were taking algebra and 30 said they were taking biology. Are there necessarily 52 students taking algebra or biology? Why?

2. To find $9 + 7$ a student says she thinks of $9 + 7$ as $9 + (1 + 6) = (9 + 1) + 6 = 10 + 6 = 16$. What property or properties is she using?

3. In Figure 3-2, arrows were used to represent numbers in completing an addition. Explain whether you think an arrow starting at 0 and ending at 3 represents the same number as an arrow starting at 4 and ending at 7. How would you explain this to students?

4. When subtraction and addition appear in an expression without parentheses, it is agreed that the operations are performed in order of their appearance from left to right. Taking this into account, answer the following:
   a. Use an appropriate model for subtraction to explain why

$$a - b - c = a - c - b$$

   assuming that all expressions are meaningful.

   b. Use an appropriate model for subtraction to explain why

$$a - b \quad c = a - (b + c)$$

5. Explain whether it is important for elementary students to learn more than one model for performing the operations of addition and subtraction.

6. Do elementary students still have to learn their basic facts when the calculator is part of the curriculum? Why or why not?

7. Explain how the model shown can be used to illustrate each of the following addition and subtraction facts:
   a. $9 + 4 = 13$ **b.** $4 + 9 = 13$
   c. $4 = 13 - 9$ **d.** $9 = 13 - 4$

13

9	4

8. How are addition and subtraction related? Explain.
9. Why is 0 not an identity for subtraction? Explain.

Open-Ended

10. Describe any model not in this text that you might use to teach addition to students.
11. Suppose $A \subseteq B$. If $n(A) = a$ and $n(B) = b$, then $b - a$ could be defined as $n(B - A)$. Choose two sets A and B and illustrate this definition.
12. **a.** Create a word problem in which the set model would be more appropriate to show $25 + 8 = 33$.
    b. Create a word problem in which the number-line (measurement) model would be more appropriate to show $25 + 8 = 33$.

Cooperative Learning

13. Discuss with your group each of the following. Use the basic addition fact table for whole numbers shown.

+	0	1	2	3	4	5	6	7	8	9
0	0	1	2	3	4	5	6	7	8	9
1	1	2	3	4	5	6	7	8	9	10
2	2	3	4	5	6	7	8	9	10	11
3	3	4	5	6	7	8	9	10	11	12
4	4	5	6	7	8	9	10	11	12	13
5	5	6	7	8	9	10	11	12	13	14
6	6	7	8	9	10	11	12	13	14	15
7	7	8	9	10	11	12	13	14	15	16
8	8	9	10	11	12	13	14	15	16	17
9	9	10	11	12	13	14	15	16	17	18

    a. How does the table show the closure property?
    b. How does the table show the commutative property?
    c. How does the table show the identity property?
    d. How do the addition properties help students learn their basic facts?
14. Suppose that a number system used only four symbols, a, b, c, and d, and the operation Δ; and the system operated as shown in the table. Discuss with your group each of the following:

Δ	a	b	c	d
a	a	b	c	d
b	b	c	d	a
c	c	d	a	b
d	d	a	b	c

    a. Is the system closed? Why?
    b. Is the operation commutative? Why?
    c. Does the operation have an identity? If so, what is it?
    d. Try several examples to investigate the associative property of this operation.

15. Have each person in your group choose a different grade textbook and report on when and how subtraction of whole numbers is introduced. Compare with different ways subtraction is introduced in this section.

Questions from the Classroom

16. A student says that 0 is the identity for subtraction. How do you respond?
17. A student claims that on the following number line, the arrow doesn't really represent 3 because the end of the arrow does not start at 0. How do you respond?

18. A student asks why we use subtraction to determine how many more pencils Sam has than Karly if nothing is being taken away. How do you respond?
19. A student claims that subtraction is closed with respect to the whole numbers. To show this is true, she shows that $8 - 5 = 3, 5 - 2 = 3, 6 - 1 = 5$, and $12 - 7 = 5$ and says she can show examples like this all day that yield whole numbers when the subtraction is performed. How do you respond?
20. John claims that he can get the same answer to the problem below by adding up (begin with $4 + 7$) or by adding down (begin with $8 + 7$). He wants to know why and if this works all the time. How do you respond?

$$\begin{array}{r} 8 \\ 7 \\ + 4 \\ \hline \end{array}$$

Third International Mathematics and Science Study (TIMSS) Questions

Ali had 50 apples. He sold some and then had 20 left. Which of these is a number sentence that shows this?
    a. $\Box - 20 = 50$
    b. $20 - \Box = 50$
    c. $\Box - 50 = 20$
    d. $50 - \Box = 20$

TIMSS 2003, Grade 4

The rule for the table is that numbers in each row and column must add up to the same number. What number goes in the center of the table?
    a. 1
    b. 2
    c. 7
    d. 12

TIMSS 2003, Grade 4

4	11	6
9		5
8	3	10

BRAIN TEASER Use Figure 3-11 to design an *unmagic square*. That is, use each of the digits 1, 2, 3, 4, 5, 6, 7, 8, and 9 exactly once so that every column, row, and diagonal adds to a different sum.

Figure 3-11

3-2   Algorithms for Whole-Number Addition and Subtraction

In the grade 2 *Curriculum Focal Points*, we find the following regarding fluency with multi-digit addition and subtraction:

> Children use their understanding of addition to develop quick recall of basic addition facts and related subtraction facts. They solve arithmetic problems by applying their understanding of models of addition and subtraction (such as combining or separating sets or using number lines), relationships and properties of number (such as place value), and properties of addition (commutativity and associativity). Children develop, discuss, and use efficient, accurate, and generalizable methods to add and subtract multidigit whole numbers. They select and apply appropriate methods to estimate sums and differences or calculate them mentally, depending on the context and numbers involved. They develop fluency with efficient procedures, including standard algorithms, for adding and subtracting whole numbers, understand why the procedures work (on the basis of place value and properties of operations), and use them to solve problems. (p. 14)

In *Principles and Standards* we also find a discussion of *computational fluency* and *standard algorithms*.

> By the end of grade 2, students should know the basic addition and subtraction combinations, should be fluent in adding two-digit numbers, and should have methods for subtracting two-digit numbers. At the grades 3–5 level, as students develop the basic number combinations for multiplication and division, they should also develop reliable algorithms to solve arithmetic problems efficiently and accurately. These methods should be applied to larger numbers and practiced for fluency . . . students must become fluent in arithmetic computation—they must have efficient and accurate methods that are supported by an understanding of numbers and operations. "Standard" algorithms for arithmetic computation are one means of achieving this fluency. (p. 35)

The previous section introduced the operations of addition and subtraction of whole numbers, and now, as pointed out in the *Focal Points* and the *Principles and Standards*, it is time to focus on *computational fluency*—having and using efficient and accurate methods for computing. *PSSM* suggest that "standard algorithms" are one means to achieve this fluency. An **algorithm** (named for the ninth-century Arabian mathematician Abu al Khwarizmi) is a systematic procedure used to accomplish an operation. In the 1998 National Council of Teachers of Mathematics (NCTM) Yearbook, *Teaching and Learning Algorithms in School Mathematics*, Usiskin stated that "algorithms are generalizations that embody one of the main reasons for studying mathematics—to find ways of solving *classes* of problems. When we know an algorithm, we complete not just one task but all tasks of a particular kind and we are guaranteed an answer or answers. The power of an algorithm derives from the breadth of its applicability." (p. 10)

This section focuses on developing and understanding algorithms involving addition and subtraction. We develop alternative algorithms as well as standard algorithms.

Addition Algorithms

In teaching mathematics to young children, it is important that we support them in the transition from concrete to abstract thinking by using techniques that parallel their developmental processes. To help children understand the use of paper-and-pencil algorithms, they should explore addition by first using manipulatives. If children can touch and move around items such as chips, bean sticks, and an abacus or use base-ten blocks, they can be led (and often will proceed naturally on their own) to the creation of algorithms for addition. In what follows, we use base-ten blocks to illustrate the development of an algorithm for whole-number addition.

Suppose we want to add 14 + 23. We start with a concrete model in Figure 3-12(a), move to the expanded algorithm in Figure 3-12(b), and then to the standard algorithm in Figure 3-12(c).

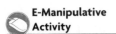

E-Manipulative Activity

Practice using base-ten blocks to perform additions can be found in the *Adding Blocks* activity.

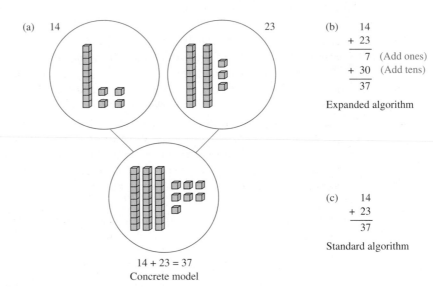

(b)
```
    14
  + 23
  ────
     7   (Add ones)
  + 30   (Add tens)
  ────
    37
```
Expanded algorithm

(c)
```
    14
  + 23
  ────
    37
```
Standard algorithm

14 + 23 = 37
Concrete model

Figure 3-12

A more formal justification for this addition not usually presented at the elementary level is the following:

$$
\begin{aligned}
14 + 23 &= (1 \cdot 10 + 4) + (2 \cdot 10 + 3) && \text{Place value} \\
&= (1 \cdot 10 + 2 \cdot 10) + (4 + 3) && \text{Commutative and associative properties of} \\
&&& \text{addition} \\
&= (1 + 2)10 + (4 + 3) && \text{Distributive property of multiplication over} \\
&&& \text{addition} \\
&= 3 \cdot 10 + 7 && \text{Single-digit addition facts} \\
&= 37 && \text{Place value}
\end{aligned}
$$

On the student page (page 129), we see an example of adding two-digit numbers with regrouping using base-ten blocks. Kim's way leads to the *expanded algorithm* and Henry's way leads to the *standard algorithm*. Each of these algorithms is discussed in more detail on page 130. Notice that on the student page, students are asked to estimate their answers before performing an algorithm. This is good practice and leads to the development of number sense and also makes students consider if their answers are reasonable. Study the student page and answer the *Talk About It* questions.

School Book Page   ADDING TWO-DIGIT NUMBERS

Lesson 3-1

Key Idea
You can break apart numbers, using place value, to add.

Vocabulary
• regroup

TEST TALK

Think It Through
• I should **estimate** so I will know if my answer is reasonable.
• I can **use place-value blocks** to show addition.

Adding Two-Digit Numbers

LEARN

How do you add two-digit numbers?

Example

Cal counted 46 ladybugs on a log and 78 more on some bushes. How many ladybugs did he count all together?

Find 46 + 78.

Estimate: 46 rounds to 50. 78 rounds to 80.

50 + 80 = 130, so the answer should be about 130.

What You **Think**		What You **Write**

Kim's Way
• Add the ones.
  6 + 8 = **14** ones
• Add the tens.
  4 tens + 7 tens = 11 tens = 110
• Find the sum.

11 tens   **14** ones

```
   46
 + 78
 ----
   14
  110
 ----
  124
```

Henry's Way
• Add the ones.
  6 + 8 = **14** ones
• **Regroup** 14 ones into **1** ten **4** ones.
• Add the tens.
  1 ten + 4 tens + 7 tens = 12 tens
• Find the sum.

14 ones = **1** ten **4** ones

```
    1
   46
 + 78
 ----
  124
```

Cal counted 124 ladybugs all together.

✎ **Talk About It**

1. Why did Henry write a small 1 above the 4 in the tens place?

2. Why should you estimate when adding two-digit numbers?

126

Source: Scott Foresman-Addison Wesley Mathematics, Grade 3, 2008 (p. 126)

Once children have mastered the use of concrete models with regrouping, they should be ready to use the expanded and standard algorithms. Figure 3-13 shows the computation 37 + 28 using both algorithms. In Figure 3-13(b), you will notice that when there were more than 10 ones, we regrouped 10 ones as a ten and then added the tens. Notice that the words *regroup* or *trade* are now commonly used in the elementary classroom to describe what we used to call *carrying*.

(a)

$$
\begin{array}{r}
37 \\
+\ 28 \\
\hline
15 \quad \text{(Add ones)} \\
+\ 50 \quad \text{(Add tens)} \\
\hline
65
\end{array}
$$

Expanded algorithm

(b)

$$
\begin{array}{r}
\overset{1}{3}7 \\
+\ 28 \\
\hline
65 \quad \text{(Add the ones, regroup, and add the tens)}
\end{array}
$$

Standard algorithm

Figure 3-13

Next we add two three-digit numbers involving two regroupings. Figure 3-14 shows how to add 186 + 127 using base-ten blocks and how this concrete model carries over to the standard algorithm.

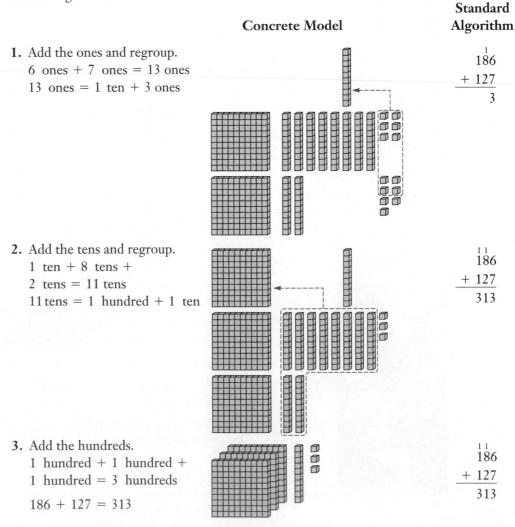

| | **Concrete Model** | **Standard Algorithm** |

1. Add the ones and regroup.
   6 ones + 7 ones = 13 ones
   13 ones = 1 ten + 3 ones

$$
\begin{array}{r}
\overset{1}{1}86 \\
+\ 127 \\
\hline
3
\end{array}
$$

2. Add the tens and regroup.
   1 ten + 8 tens +
   2 tens = 11 tens
   11 tens = 1 hundred + 1 ten

$$
\begin{array}{r}
\overset{1\,1}{1}86 \\
+\ 127 \\
\hline
313
\end{array}
$$

3. Add the hundreds.
   1 hundred + 1 hundred +
   1 hundred = 3 hundreds

   186 + 127 = 313

$$
\begin{array}{r}
\overset{1\,1}{1}86 \\
+\ 127 \\
\hline
313
\end{array}
$$

Figure 3-14

Students often develop algorithms on their own, and learning can occur by investigating how and if various algorithms work. Addition of whole numbers using blocks has a natural carryover to the expanded form and trading used earlier. For example, consider the following addition:

$$
\begin{array}{r}
376 \\
459 \\
+\ 8716 \\
\end{array}
\quad \text{or} \quad
\begin{array}{r}
3\cdot 10^2 +\ 7\cdot 10 +\ 6 \\
4\cdot 10^2 +\ 5\cdot 10 +\ 9 \\
8\cdot 10^3 +\ 7\cdot 10^2 +\ 1\cdot 10 +\ 6 \\
\hline
8\cdot 10^3 + 14\cdot 10^2 + 13\cdot 10 + 21 \\
\end{array}
$$

To complete the addition, trading is used. However, consider an analogous algebra problem of adding polynomials:

$$(3x^2 + 7x + 6) + (4x^2 + 5x + 9) + (8x^3 + 7x^2 + x + 6) \quad \text{or}$$

$$
\begin{array}{r}
3x^2 +\ 7x +\ 6 \\
4x^2 +\ 5x +\ 9 \\
+\ 8x^3 +\ 7x^2 +\ x +\ 6 \\
\hline
8x^3 + 14x^2 + 13x + 21 \\
\end{array}
$$

Note that if $x = 10$, the addition is the same as given earlier. Also note that knowledge of place value in addition problems aids in algebraic thinking. Next we explore several algorithms that have been used throughout history.

Left-to-Right Algorithm for Addition

Since children learn to read from left to right, it might seem natural that they may try to add from left to right. When working with base-ten blocks, many children in fact do combine the larger pieces first and then move to combining the smaller pieces. This method has the advantage of emphasizing place value. A left-to-right algorithm is as follows:

$$
\begin{array}{ccc}
& \begin{array}{r} 568 \\ +\ 757 \\ \hline \end{array} & \begin{array}{r} 568 \\ +\ 757 \\ \hline \end{array} \\
(500 + 700) \rightarrow & 1200 & 1215 \rightarrow 1325 \\
(60 + 50) \rightarrow & 110 & {\scriptstyle 3\ 2} \\
(8 + 7) \rightarrow & 15 & \\
& \hline \\
& 1325 & \\
\end{array}
$$

Explain why this technique works and try it with $9076 + 4689$.

Lattice Algorithm for Addition

We introduce this algorithm by working through an addition involving two four-digit numbers. For example,

$$
\begin{array}{r}
3\ 5\ 6\ 7 \\
+\ 5\ 6\ 7\ 8 \\
\end{array}
$$

$$
\begin{array}{c}
\boxed{0\!\diagup 1\!\diagup 1\!\diagup 1\!\diagup} \\
\boxed{\diagup 8\ \diagup 1\ \diagup 3\ \diagup 5} \\
9\ \ 2\ \ 4\ \ 5 \\
\end{array}
$$

To use this algorithm, add the single-digit numbers by place value on top to the single digit numbers on the bottom and record the results in a lattice. Then add the sums along the diagonals, as shown. This is very similar to the expanded algorithm introduced earlier. Try this technique with $4578 + 2691$.

Scratch Algorithm for Addition

The scratch algorithm for addition is often referred to as a *low-stress algorithm* because it allows students to perform complicated additions by doing a series of additions that involve only two single digits. An example follows:

1.

$$\begin{array}{r} 87 \\ 6\not{8}_2 \\ + \ 49 \\ \end{array}$$

Add the numbers in the units place starting at the top. When the sum is 10 or more, record this sum by scratching a line through the last digit added and writing the number of units next to the scratched digit. For example, since $7 + 5 = 12$, the "scratch" represents 10 and the 2 represents the units.

2.

$$\begin{array}{r} 87 \\ 6\not{8}_2 \\ + \ 4\not{9}_1 \\ \end{array}$$

Continue adding the units, including any new digits written down. When the addition again results in a sum of 10 or more, as with $2 + 9 = 11$, repeat the process described in (1).

3.

$$\begin{array}{r} {}^{2}87 \\ 6\not{8}_2 \\ + \ 4\not{9}_1 \\ \hline 1 \end{array}$$

When the first column of additions is completed, write the number of units, 1, below the addition line in the proper place value position. Count the number of scratches, 2, and add this number to the second column.

4.

$$\begin{array}{r} {}^{2}\not{8}_07 \\ 6 \ \not{8}_2 \\ \not{4}_0\not{9}_1 \\ \hline 2 \ 0 \ 1 \end{array}$$

Repeat the procedure for each successive column until the last column with non-zero values. At this stage, sum the scratches and place the number to the left of the current value.

Try this technique with $56 + 23 + 34 + 67$.

Subtraction Algorithms

As with addition, we can use base-ten blocks to provide a concrete model for subtraction. Consider how the base-ten blocks are used to perform the subtraction $243 - 61$: First we represent 243 with 2 flats, 4 longs, and 3 units, as shown in Figure 3-15.

Figure 3-15

To subtract 61 from 243, we try to remove 6 longs and 1 unit from the blocks in Figure 3-15. We can remove 1 unit, as in Figure 3-16.

Figure 3-16

To remove 6 longs from Figure 3-16, we have to trade 1 flat for 10 longs, as shown in Figure 3-17.

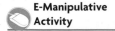

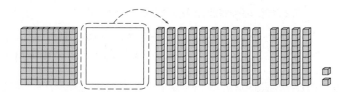

Figure 3-17

Now we can remove, or "take away," 6 longs, leaving 1 flat, 8 longs, and 2 units, or 182, as shown in Figure 3-18.

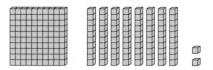

Figure 3-18

 Student work with base-ten blocks along with discussions and recorded work lead to the development of the standard algorithm as seen on the student page on page 134. Work through the student page (a)–(f).

 NOW TRY THIS 3-8 Use base-ten blocks and addition to check that $243 - 61 = 182$.

Equal-Additions Algorithm

The equal-additions algorithm for subtraction is based on the fact that the difference between two numbers does not change if we add the same amount to both numbers. For example, $93 - 27 = (93 + 3) - (27 + 3)$. Thus, the difference can be computed as $96 - 30 = 66$. Using this approach, the subtraction on the student page could be performed as follows:

$$
\begin{array}{r} 255 \\ -\,163 \end{array}
\;\rightarrow\;
\begin{array}{r} 255 + 7 \\ -\,(163 + 7) \end{array}
\;\rightarrow\;
\begin{array}{r} 262 \\ -\,170 \end{array}
\;\rightarrow\;
\begin{array}{r} 262 + 30 \\ -\,(170 + 30) \end{array}
\;\rightarrow\;
\begin{array}{r} 292 \\ -\,200 \\ \hline 92 \end{array}
$$

Subtraction of whole numbers using blocks has a natural carryover to the expanded form and trading. For example, consider the following subtraction problem done earlier with blocks:

$$
\begin{array}{r} 243 \\ -\,61 \end{array}
\;\text{or}\;
\begin{array}{r} 2\cdot10^2 + 4\cdot10 + 3 \\ -\,(6\cdot10 + 1) \end{array}
\;\text{or}\;
\begin{array}{r} 1\cdot10^2 + 14\cdot10 + 3 \\ -\,(6\cdot10 + 1) \\ \hline 1\cdot10^2 + (14 - 6)10 + (3 - 1) \end{array}
$$

which results in 182. Note that to complete the subtraction, trading was used. As with addition, we see that understanding place value aids in subtraction computations.

School Book Page

MODELS FOR SUBTRACTING THREE-DIGIT NUMBERS

Lesson 3-8

Key Idea
You can use blocks to show regrouping for subtraction.

Materials
• place-value blocks
  or tools

Think It Through
I can **use objects** to show a subtraction problem with regrouping.

Models for Subtracting Three-Digit Numbers

WARM UP
1. 96 − 42   2. 77 − 39
3. 43 − 8   4. 50 − 27

LEARN

Activity

How can you subtract with place-value blocks?

Find 255 − 163.

	What You **Show**	What You **Write**

a. Show 255 with place-value blocks.

```
  2 5 5
−1 6 3
```

b. Subtract the ones. Regroup if needed.

5 > 3. No regrouping is needed.   **5 ones − 3 ones = 2 ones**

```
  2 5 5
−1 6 3
      2
```

c. Subtract the tens. Regroup if needed.

5 tens < 6 tens.

So, regroup 1 hundred for 10 tens.

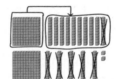

15 tens − 6 tens = 9 tens

```
  1 15
  2 5 5
−1 6 3
    9 2
```

d. Subtract the hundreds.

1 hundred − 1 hundred = 0 hundreds

```
  1 15
  2 5 5
−1 6 3
    9 2
```

e. Find the value of the remaining blocks in Step d: 9 tens 2 ones = 92, so 255 − 163 = 92.

f. In Step b, did you have to regroup to subtract the ones? Explain.

g. In Step c, did you have to regroup to subtract the tens? Explain.

h. Use place-value blocks to subtract.
243 − 72   145 − 126   223 − 156

150

Source: Scott Foresman-Addison Wesley Mathematics, Grade 3, 2008 (p. 150).

NOW TRY THIS 3-9 Jessica claims that a method similar to *equal additions* for subtraction also works for addition. She says that in an addition problem, "you may add the same amount to one number as you subtract from the other." For example, $68 + 29 = (68 - 1) + (29 + 1)$. Thus, the sum can be computed as $67 + 30 = 97$ or as $(68 + 2) + (29 - 2) = 70 + 27 = 97$. (i) Explain why this method is valid and (ii) use it to compute $97 + 69$.

Understanding Addition and Subtraction in Bases Other Than Ten

A look at computation in other bases may provide insight into computation in base ten. Use of multibase blocks may be helpful in building an addition table for different bases and is highly recommended. Table 3-1 is a base-five addition table.

Table 3-1   Base-Five Addition Table

+	0	1	2	3	4
0	0	1	2	3	4
1	1	2	3	4	10
2	2	3	4	10	11
3	3	4	10	11	12
4	4	10	11	12	13

NOW TRY THIS 3-10 Write each of the following as numerals in base five·

a. $444_{\text{five}} + 1_{\text{five}}$ **b.** $13_{\text{five}} + 13_{\text{five}}$

Using the addition facts in Table 3-1, we can develop algorithms for base-five addition similar to those we used for base-ten addition. We show the computation using a concrete model in Figure 3-19(a); in Figure 3-19(b), we use an expanded algorithm; in Figure 3-19(c), we use the standard algorithm.

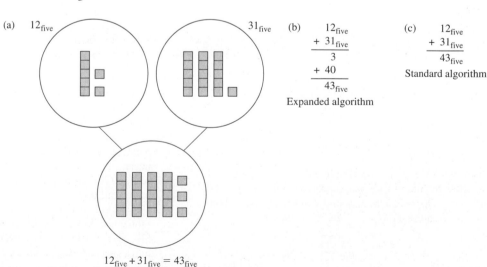

(a) 12_{five} 31_{five}

(b)
$$
\begin{array}{r}
12_{\text{five}} \\
+ \ 31_{\text{five}} \\
\hline
3 \\
+ \ 40 \\
\hline
43_{\text{five}}
\end{array}
$$
Expanded algorithm

(c)
$$
\begin{array}{r}
12_{\text{five}} \\
+ \ 31_{\text{five}} \\
\hline
43_{\text{five}}
\end{array}
$$
Standard algorithm

$12_{\text{five}} + 31_{\text{five}} = 43_{\text{five}}$
Concrete model

Figure 3-19

The subtraction facts for base five can also be derived from the addition-facts table by using the definition of subtraction. For example, to find $12_{\text{five}} - 4_{\text{five}}$, recall that $12_{\text{five}} - 4_{\text{five}} = c_{\text{five}}$ if, and only if, $c_{\text{five}} + 4_{\text{five}} = 12_{\text{five}}$. From Table 3-1, we see that $c = 3_{\text{five}}$. An example of subtraction involving regrouping, $32_{\text{five}} - 14_{\text{five}}$, is developed in Figure 3-20.

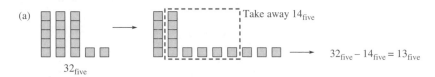

(a)   Take away 14_{five}

32_{five}

$32_{\text{five}} - 14_{\text{five}} = 13_{\text{five}}$

(b)

Fives	Ones
3	2
− 1	4

→

Fives	Ones
2	12
− 1	4
1	3

(c)
$$\overset{2\,1}{\cancel{3}2}_{\text{five}}$$
$$-\ 14_{\text{five}}$$
$$\overline{13_{\text{five}}}$$

Figure 3-20

NOW TRY THIS 3-11

a. Build an addition table for base two.

b. Use the addition table from part (a) to perform (i) $1101_{\text{two}} - 111_{\text{two}}$, (ii) $1111_{\text{two}} + 111_{\text{two}}$.

BRAIN TEASER   The number on a license plate consists of five digits. When the license plate is looked at upside down, you can still read it, but the value of the upside-down number is 78,633 greater than the real license number. What is the license number?

Assessment 3-2A

1. Find the missing digits in each of the following:

   a.
```
    _ _ 1
  + 4 2 _
  _ 4 0 2
```
   b.
```
    _ 0 2 5
    1 1 _ 6
  + 3 1 4 8
    6 _ 6 _
```

2. Make an appropriate drawing like the one in Figure 3-14 to show the use of base-ten blocks to compute $29 + 37$.

3. Place the digits 7, 6, 8, 3, 5, and 2 in the boxes to obtain
   a. the greatest sum.   **b.** the least sum.

```
   □ □ □
 + □ □ □
```

4. Tom's diet allows only 1500 calories per day. For breakfast, Tom had skim milk (90 calories), a waffle with no syrup (120 calories), and a banana (119 calories). For lunch, he had $\frac{1}{2}$ cup of salad (185 calories) with mayonnaise (110 calories) and tea (0 calories). Then he had pecan pie (570 calories). Can he have dinner consisting of fish (250 calories), a $\frac{1}{2}$ cup of salad with no mayonnaise, and tea?

5. Wally kept track of last week's money transactions. His salary was $150 plus $54 in overtime and $260 in tips. His transportation expenses were $22, his food expenses were $60, his laundry costs were $15, his entertainment expenditures were $58, and his rent was $185.

After expenses, did he have any money left? If so, how much?

6. In the following problem, the sum is correct but the order of the digits in each addend has been scrambled. Correct the addends to obtain the correct sum.

$$\begin{array}{r} 2834 \\ + 6315 \\ \hline 9059 \end{array} \qquad \begin{array}{r} \square\square\square\square \\ + \square\square\square\square \\ \hline 9059 \end{array}$$

7. Use the equal-additions approach to compute each of the following:

   a.
   $$\begin{array}{r} 93 \\ - 37 \\ \hline \end{array}$$

   b.
   $$\begin{array}{r} 321 \\ - 38 \\ \hline \end{array}$$

8. Janet worked her addition problems by placing the partial sums as shown here:

   $$\begin{array}{r} 569 \\ + 645 \\ \hline 14 \\ 10 \\ 11 \\ \hline 1214 \end{array}$$

   a. Use this method to work the following:

   (i)
   $$\begin{array}{r} 687 \\ + 549 \\ \hline \end{array}$$
   (ii)
   $$\begin{array}{r} 359 \\ + 673 \\ \hline \end{array}$$

   b. Explain why this algorithm works.

9. Analyze the following computations. Explain what is wrong in each case.

   a.
   $$\begin{array}{r} 28 \\ + 75 \\ \hline 913 \end{array}$$
   b.
   $$\begin{array}{r} 28 \\ + 75 \\ \hline 121 \end{array}$$
   c.
   $$\begin{array}{r} 305 \\ - 259 \\ \hline 154 \end{array}$$
   d.
   $$\begin{array}{r} \overset{210}{3\cancel{0}5} \\ - 259 \\ \hline 56 \end{array}$$

10. Give reasons for each of the following steps:

$$\begin{aligned} 16 + 31 &= (1 \cdot 10 + 6) + (3 \cdot 10 + 1) \\ &= (1 \cdot 10 + 3 \cdot 10) + (6 + 1) \\ &= (1 + 3)10 + (6 + 1) \\ &= 4 \cdot 10 + 7 \\ &= 47 \end{aligned}$$

11. In each of the following, justify the standard addition algorithm using place value of the numbers, the commutative and associative properties of addition, and the distributive property of multiplication over addition:

   a. 68 + 23
   b. 174 + 285
   c. 2458 + 793

12. Use the lattice algorithm to perform each of the following:

   a. 4358 + 3864
   b. 4923 + 9897

13. Perform each of the following operations using the bases shown:

   a. $43_{\text{five}} + 23_{\text{five}}$
   b. $43_{\text{five}} - 23_{\text{five}}$
   c. $432_{\text{five}} + 23_{\text{five}}$
   d. $42_{\text{five}} - 23_{\text{five}}$
   e. $110_{\text{two}} + 11_{\text{two}}$
   f. $10001_{\text{two}} - 111_{\text{two}}$

14. Construct an addition table for base eight.

15. Perform each of the following operations:

   a.
   $$\begin{array}{r} 3 \text{ hr } 36 \text{ min } 58 \text{ sec} \\ + 5 \text{ hr } 56 \text{ min } 27 \text{ sec} \\ \hline \end{array}$$
   b.
   $$\begin{array}{r} 5 \text{ hr } 36 \text{ min } 38 \text{ sec} \\ - 3 \text{ hr } 56 \text{ min } 58 \text{ sec} \\ \hline \end{array}$$

16. Andrew's calculator was not functioning properly. When he pressed $\boxed{8}\ \boxed{+}\ \boxed{6}\ \boxed{=}$, the numeral 20 appeared on the display. When he pressed $\boxed{5}\ \boxed{+}\ \boxed{4}\ \boxed{=}$, 13 was displayed. When he pressed, $\boxed{1}\ \boxed{5}\ \boxed{-}\ \boxed{3}\ \boxed{=}$, 9 was displayed. What do you think Andrew's calculator was doing?

17. Use scratch addition to perform the following:

   a.
   $$\begin{array}{r} 432 \\ 976 \\ + 1418 \\ \hline \end{array}$$

   b.
   $$\begin{array}{r} 32_{\text{five}} \\ 13_{\text{five}} \\ 22_{\text{five}} \\ 43_{\text{five}} \\ 23_{\text{five}} \\ + 12_{\text{five}} \\ \hline \end{array}$$

18. Perform each of the following operations:

   a.
   $$\begin{array}{r} 4 \text{ gross } 4 \text{ doz } 6 \text{ ones} \\ - \qquad 5 \text{ doz } 9 \text{ ones} \\ \hline \end{array}$$
   b.
   $$\begin{array}{r} 2 \text{ gross } 9 \text{ doz } 7 \text{ ones} \\ + 3 \text{ gross } 5 \text{ doz } 9 \text{ ones} \\ \hline \end{array}$$

19. Determine what is wrong with the following:

$$\begin{array}{r} 22_{\text{five}} \\ + 33_{\text{five}} \\ \hline 55_{\text{five}} \end{array}$$

20. Fill in the missing numbers in each of the following:

   a.
   $$\begin{array}{r} 2\_\_{\text{five}} \\ - \quad 2\ 2_{\text{five}} \\ \hline \_\ 0\ 3_{\text{five}} \end{array}$$

b. $2\ 0\ 0\ 1\ 0_{\text{three}}$
$\underline{-\quad 2\ \_\ 2\ \_\text{three}}$
$1\ \_\ 2\ \_\ 1_{\text{three}}$

21. Find the numeral to put in the blank to make each equation true. Do not convert to base ten.
a. $3423_{\text{five}} -$ _____ $= 2132_{\text{five}}$
b. $11011_{\text{two}} +$ _____ $= 100000_{\text{two}}$
c. $TEE_{\text{twelve}} -$ _____ $= 1$
d. $1000_{\text{five}} +$ _____ $= 10000_{\text{five}}$

22. A palindrome is any number that reads the same backward as forward, for example, 121 and 2332. Try the following. Begin with any multi-digit number. Is it a palindrome? If not, reverse the digits and add this reversed number to the original number. Is the result a palindrome? If not, repeat the procedure until a palindrome is obtained. For example, start with 78. Because 78 is not a palindrome, we add: $78 + 87 = 165$. Because 165 is not a palindrome, we add: $165 + 561 = 726$. Again, 726 is not a palindrome, so we add $726 + 627$ to obtain 1353. Finally, $1353 + 3531$ yields 4884, which is a palindrome.
a. Try this method with the following numbers:
(i) 93   **(ii)** 588   **(iii)** 2003
b. Find a number for which the procedure described takes more than five steps to form a palindrome.

Assessment 3-2B

1. Find the missing digits in each of the following:
a. $3\ \_\ \_$
$\underline{-\ 1\ 5\ 9}$
$\_\ 2\ 4$
b. $1\ \_\ \_\ \_\ 6$
$\underline{-\qquad 8\ 3\ 0\ 9}$
$4\ 9\ 8\ 7$

2. Make an appropriate drawing like the one in Figure 3-12 to show the use of base-ten blocks to compute $46 + 38$.

3. Place the digits 7, 6, 8, 3, 5, and 2 in the boxes to obtain
a. the greatest difference.
b. the least difference.

$$\square\square\square$$
$$-\square\square\square$$

4. In the following problem, the sum is correct but the order of the digits in each addend has been scrambled. Correct the addends to obtain the correct sum.

$$\begin{array}{r} 8354 \\ + 3456 \\ \hline 11729 \end{array} \qquad \begin{array}{r} \square\square\square\square \\ + \square\square\square\square \\ \hline 1\ 1\ 7\ 2\ 9 \end{array}$$

5. Use the equal-additions approach to compute each of the following:
a. $\begin{array}{r} 86 \\ -38 \end{array}$ **b.** $\begin{array}{r} 582 \\ -44 \end{array}$

6. Janet worked her addition problems by placing the partial sums as shown here:

$$\begin{array}{r} 569 \\ + 645 \\ \hline 14 \\ 10 \\ 11 \\ \hline 1214 \end{array}$$

Use this method to work the following:
a. $\begin{array}{r} 985 \\ + 356 \end{array}$
b. $\begin{array}{r} 413 \\ +89 \end{array}$

7. Analyze the following computations. Explain what is wrong in each case.
a. $\begin{array}{r} 135 \\ + 47 \\ \hline 172 \end{array}$
b. $\begin{array}{r} 87 \\ + 25 \\ \hline 1012 \end{array}$
c. $\begin{array}{r} 57 \\ - 38 \\ \hline 21 \end{array}$
d. $\begin{array}{r} 56 \\ - 18 \\ \hline 48 \end{array}$

8. George is cooking an elaborate meal for Thanksgiving. He can cook only one thing at a time in his microwave oven. His turkey takes 75 min; the pumpkin pie takes 18 min; rolls take 45 sec; and a cup of coffee takes 30 sec to heat. How much time does he need to cook the meal?

9. Give reasons for each of the following steps:

$$\begin{aligned} 123 + 45 &= (1 \cdot 10^2 + 2 \cdot 10 + 3) + (4 \cdot 10 + 5) \\ &= 1 \cdot 10^2 + (2 \cdot 10 + 4 \cdot 10) + (3 + 5) \\ &= 1 \cdot 10^2 + (2 + 4)10 + (3 + 5) \\ &= 1 \cdot 10^2 + 6 \cdot 10 + 8 \\ &= 168 \end{aligned}$$

10. In each of the following justify the standard addition algorithm using place value of the numbers, the

commutative and associative properties of addition, and the distributive property of multiplication over addition:
a. $46 + 32$
b. $3214 + 783$

11. Use the lattice algorithm to perform each of the following:
a. $2345 + 8888$
b. $8713 + 4214$

12. Perform each of the following operations using the bases shown:
a. $43_{five} - 24_{five}$
b. $143_{five} + 23_{five}$
c. $32_{five} - 23_{five}$
d. $232_{five} + 43_{five}$
e. $110_{two} + 111_{two}$
f. $10001_{two} - 101_{two}$

13. Construct an addition table for base six.

14. Perform each of the following operations ($2\,c = 1\,pt$, $2\,pt = 1\,qt$, $4\,qt = 1\,gal$):
a. 1 qt 1 pt 1 c
    $+$ 1 pt 1 c
b. 1 qt   1 c
    $-$ 1 pt 1 c
c. 1 gal 3 qt 1 c
    $-$ 4 qt 2 c

15. The following is a supermagic square taken from an engraving called *Melancholia* by Dürer. Notice 1514 in the bottom row, the year the engraving was made.

16	3	2	13
5	10	11	8
9	6	7	12
4	15	14	1

a. Find the sum of each row, the sum of each column, and the sum of each diagonal.
b. Find the sum of the four numbers in the center.
c. Find the sum of the four numbers in each corner.
d. Add 11 to each number in the square. Is the square still a magic square? Explain your answer.
e. Subtract 11 from each number in the square. Is the square still a magic square?

16. Use scratch addition to perform the following:
a.   537
       318
    $+ 2345$
b. 41_{six}
      32_{six}
      22_{six}
      43_{six}
      22_{six}
    $+ 54_{six}$

17. Determine what is wrong with the following:

$$\begin{array}{r} 23_{six} \\ + \ 43_{six} \\ \hline 66_{six} \end{array}$$

18. Find the numeral to put in the blank to make each equation true. Do not convert to base ten.
a. $342_{five} - \underline{\quad} = 213_{five}$
b. $1101_{two} - \underline{\quad} = 1011_{two}$
c. $E08_{twelve} - \underline{\quad} = 9_{twelve}$
d. $100_{two} + \underline{\quad} = 10000_{two}$

19. The Hawks played the Elks in a basketball game. Based on the following information, complete the scoreboard showing the number of points scored by each team during each quarter and the final score of the game.

Teams	Quarters				Final Score
	1	2	3	4	
Hawks					
Elks					

a. The Hawks scored 15 points in the first quarter.
b. The Hawks were behind by 5 points at the end of the first quarter.
c. The Elks scored 5 more points in the second quarter than they did in the first quarter.
d. The Hawks scored 7 more points than the Elks in the second quarter.
e. The Elks outscored the Hawks by 6 points in the fourth quarter.
f. The Hawks scored a total of 120 points in the game.
g. The Hawks scored twice as many points in the third quarter as the Elks did in the first quarter.
h. The Elks scored as many points in the third quarter as the Hawks did in the first two quarters combined.

20. a. Place the numbers 24 through 32 in the following circles so that the sums are the same in each direction:

b. How many different numbers can be placed in the middle to obtain a solution?

Mathematical Connections 3-2

Communication

1. Discuss the merit of the following algorithm for addition where we first add the ones, then the tens, then the hundreds, and then the total:

$$\begin{array}{r} 479 \\ + 385 \\ \hline 14 \\ 150 \\ + 700 \\ \hline 864 \end{array}$$

2. The following example uses a regrouping approach to subtraction. Discuss the merit of this approach in teaching subtraction.

$$\begin{array}{r} 843 \\ -568 \end{array} \rightarrow \begin{array}{r} 800 + 40 + 3 \\ -(500 + 60 + 8) \end{array} \rightarrow$$

$$\begin{array}{r} 800 + 30 + 13 \\ -(500 + 60 + 8) \end{array} \rightarrow \begin{array}{r} 700 + 130 + 13 \\ -(500 + 60 + 8) \\ \hline 200 + 70 + 5 \end{array} = 275$$

3. Tira, a fourth grader, performs addition by adding and subtracting the same number. She added as follows:

$$\begin{array}{r} 39 \\ + 84 \end{array} \rightarrow \begin{array}{r} 39 + 1 \\ + 84 - 1 \end{array} \rightarrow \begin{array}{r} 40 \\ + 83 \\ \hline 123 \end{array}$$

How would you respond if you were her teacher?
4. Explain why the scratch addition algorithm works.
5. The *equal-additions* algorithm was introduced in this section. The following shows how this algorithm works for 1464 − 687:

$$\begin{array}{r} 1\,4\,6^1 4 \\ -\ 6^9 8\,7 \\ \hline 7 \end{array}$$ (Add 10 to the 4 ones to get 14 ones.)
(Add 1 ten to the 8 tens to get 9 tens.)
(Subtract the ones.)

Now we move to the next column.

$$\begin{array}{r} 1\ 4^1\ 6^1\ 4 \\ -\ ^7 6^9\ 8\ 7 \\ \hline 7\ \ 7\ 7 \end{array}$$ (Add 10 tens to the 6 tens to get 16 tens.)
(Add 1 hundred to the 6 hundreds to get 7 hundreds.)
(Subtract the 9 tens from the 16 tens and then the 7 hundreds from the 14 hundreds.)

a. Try the technique on three more subtractions.
b. Explain why the equal-additions algorithm works.

6. Cathy found her own algorithm for subtraction. She subtracted as follows:

$$\begin{array}{r} 97 \\ -\ 28 \\ \hline -\ 1 \\ +\ 70 \\ \hline 69 \end{array}$$

How would you respond if you were her teacher?
7. Discuss why the words *regroup* and *trade* are used rather than *carry* and *borrow* for whole-number addition and subtraction algorithms.
8. Consider the following subtraction algorithm.
   a. Explain how it works.
   b. Use this algorithm to find 787 − 398.

$$\begin{array}{r} 585 \\ -\ 277 \\ \hline 385 \\ 285 \\ +\ 23 \\ \hline 308 \end{array}$$

Open-Ended

9. Search for or develop an algorithm for whole-number addition or subtraction and write a description of your algorithm so that others can understand and use it.

Cooperative Learning

10. In this section you have been exposed to many different algorithms. Discuss in your group whether children should be encouraged to develop and use their own algorithms for whole-number addition and subtraction or whether they should be taught only one algorithm per operation and all students should use only one algorithm.

Questions from the Classroom

11. To find 68 − 19, Joe began by finding 9 − 8. How do you help?
12. Jill subtracted 415 − 212 and got 303. She asked if this was correct. How would you respond?
13. Betsy found 518 − 49 = 469. She was not sure she was correct so she tried to check her answer by adding 518 + 49. How could you help her?

14. A child is asked to compute $7 + 2 + 3 + 8 + 11$ and writes $7 + 2 = 9 + 3 = 12 + 8 = 20 + 11 = 31$. Noticing that the answer is correct, if you were the teacher how would you react?

Review Problems

15. Is the set $\{1, 2, 3\}$ closed under addition? Why?

16. Give an example of the associative property of addition of whole numbers.

National Assessment of Educational Progress (NAEP) Questions

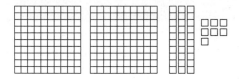

The figure above represents 237. Which number is

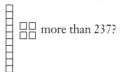 more than 237?

a. 244 **b.** 249 **c.** 251 **d.** 377

NAEP 2007, Grade 4

The Ben Franklin Bridge was 75 years old in 2001. In what year was the bridge 50 years old?

a. 1951 **b.** 1976 **c.** 1984 **d.** 1986

NAEP 2007, Grade 4

LABORATORY ACTIVITY

1. One type of Japanese abacus, *soroban*, is shown in Figure 3-21(a). In this abacus, a bar separates two sets of bead counters. Each counter above the bar represents 5 times the counter below the bar. Numbers are illustrated by moving the counter toward the bar. The number 7632 is pictured. Practice demonstrating and adding numbers on this abacus.

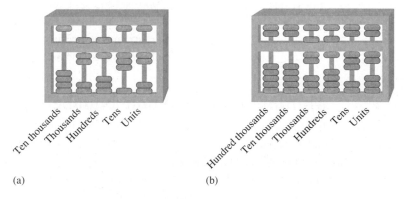

(a)   (b)

Figure 3-21

2. The Chinese abacus, *suan pan* (see Figure 3-21(b)), is still in use today. This abacus is similar to the Japanese abacus but has two counters above the bar and 5 counters below the bar. The number 7632 is also pictured on it. Practice demonstrating and adding numbers on this abacus. Compare the ease of using the two versions.

3-3   Multiplication and Division of Whole Numbers

In the grade 3 *Focal Points* we find the following concerning multiplication and division of whole numbers:

> Students understand the meanings of multiplication and division of whole numbers through the use of representations (e.g., equal-sized groups, arrays, area models, and equal "jumps" on number lines for multiplication, and successive subtraction, partitioning, and sharing for division). They use properties of addition and multiplication (e.g., commutativity, associativity, and the distributive property) to multiply whole numbers and apply increasingly sophisticated strategies based on these properties to solve multiplication and division problems involving basic facts. By comparing a variety of solution strategies, students relate multiplication and division as inverse operations. (p. 15)

Further in the grade 3 *Focal Points*, we see the connection between studying multiplication and division of whole numbers and the study of algebra.

> Understanding properties of multiplication and the relationship between multiplication and division is a part of algebra readiness that develops at grade 3. The creation and analysis of patterns and relationships involving multiplication and division should occur at this grade level. Students build a foundation for later understanding of functional relationships by describing relationships in context with such statements as, "The number of legs is 4 times the number of chairs." (p. 15)

These quotes from the *Focal Points* set the tone and agenda for this section. We discuss representations that can be used to help students understand the meanings of multiplication and division. We develop the distributive property of multiplication over addition along with the relationship of multiplication and division as inverse operations.

Multiplication of Whole Numbers

In this section, we explore the kind of problems that Grampa is having in the *Peanuts* cartoon. Why do you think he would have more troubles with "9 times 8" rather than "3 times 4"? If multiplication facts are only memorized, they may be forgotten. If students have a conceptual understanding of the basic facts, then all of the basic facts can be determined even if not automatically recalled.

Repeated-Addition Model

On the student page on page 144 we see that if we have 4 groups of 3 brushes, we can use addition to put the groups together. When we put equal-sized groups together we can use multiplication. We can think of this as combining 4 sets of 3 objects into a single set. The 4 sets of 3 suggest the following addition:

$$\underbrace{3 + 3 + 3 + 3}_{\text{four 3s}} = 12$$

We write $3 + 3 + 3 + 3$ as $4 \cdot 3$ and say "four times three" or "three multiplied by four." The advantage of the multiplication notation over repeated addition is evident when the number of addends is great, for example, if we have 25 groups of 3 brushes, we could find the total number of brushes by adding 25 threes or $25 \cdot 3$.

The *repeated-addition* model can be illustrated in several ways, including number lines and arrays. For example, using colored rods of length 4, we could show that the combined length of five of the rods can be found by joining the rods end-to-end, as in Figure 3-22(a). Figure 3-22(b) shows the process using arrows on a number line.

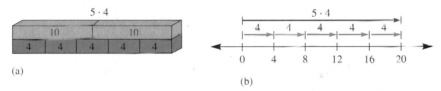

Figure 3-22

The constant feature on a calculator can help relate multiplication to addition. Students can find products on the calculator without using the $\boxed{\times}$ key. For example, if a calculator has the *constant feature*, then 5×3 can be found by starting with and pressing $\boxed{+}\,\boxed{3}\,\boxed{=}\,\boxed{=}\,\boxed{=}\,\boxed{=}\,\boxed{=}$. Each press of the equal sign will add 3 to the display. (Some calculators will work differently.)

As pointed out in the Research Note, access to only the "repeated-addition" model for multiplication can lead to misunderstanding. In this section, we introduce three other multiplication models: the *array* and *area* models and the *Cartesian-product* model.

Research Note

Students learning multiplication as a conceptual operation need exposure to a variety of models (for example, array and area). Access only to the "multiplication as repeated addition" and the term *times* leads to basic misunderstandings of multiplication that complicate future extensions to decimals and fractions (Bell et al. 1989; English and Halford 1995). ♦

Historical Note

William Oughtred (1574–1660), an English mathematician, was interested in mathematical symbols. He was the first to introduce the "St. Andrew's cross" (×) as the symbol for multiplication. This symbol was not readily adopted because, as Gottfried Wilhelm von Leibniz (1646–1716) objected, it was too easily confused with the letter *x*. Leibniz used the dot (·) for multiplication, which has become common. ♦

School Book Page

MULTIPLICATION AS REPEATED ADDITION

Lesson 5-1

Algebra

Key Idea
Multiplying is a quick way of adding equal groups.

Vocabulary
• multiplication
• factor
• product

Materials
• counters
or ⚙ tools

TEST TALK

Think It Through
I can **use objects** to show equal groups.

Multiplication as Repeated Addition

LEARN

Activity

How can you find the total?

There are **4 groups of 3** paintbrushes.

You can use addition to put together groups.

$$3 + 3 + 3 + 3 = 12$$ **Addition sentence**

When you put together **equal groups,** you can also use **multiplication.**

What You **Say:**   4 times 3 equals 12

What You **Write:**   4 × 3 = 12   **Multiplication sentence**
                       ↑   ↑   ↑
                    factor factor product

a. Write an addition sentence and a multiplication sentence to show the total number of counters below.

b. Use counters and draw a picture to show the groups described below. For each picture, write an addition sentence and a multiplication sentence to show how many counters in all.

5 groups of 2
4 groups of 5
3 groups of 3

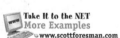
Take It to the NET
More Examples
www.scottforesman.com

260

Source: Scott Foresman-Addison Wesley Math, Grade 3, 2008 (p. 260).

The Array and Area Models

Another representation useful in exploring multiplication of whole numbers is an *array*. An array is suggested when we have objects in equal-sized rows, as in Figure 3-23.

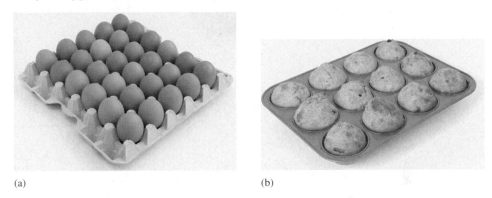

(a)   (b)

Figure 3-23

In Figure 3-24(a), we cross sticks to create intersection points, thus forming an array of points. The number of intersection points on a single vertical stick is 4 and there are 5 sticks, forming a total of $5 \cdot 4$ points in the array. In Figure 3-24(b), the area model is shown as a 4-by-5 grid. The number of unit squares required to fill in the grid is 20. These models motivate the following definition of multiplication of whole numbers.

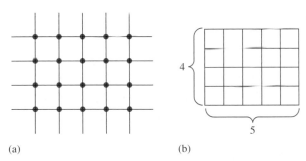

(a)   (b)

Figure 3-24

Definition of Multiplication of Whole Numbers

For any whole numbers a and $n \neq 0$,
$$n \cdot a = \underbrace{a + a + a + \ldots + a}_{n \text{ terms}}.$$

If $n = 0$, then $0 \cdot a = 0$.

REMARK We typically write $n \cdot a$ as na, where a is not a number but a variable.

Cartesian-Product Model

The *Cartesian-product* model offers another way to discuss multiplication. Suppose you can order a soyburger on light or dark bread with one condiment: mustard, mayonnaise, or horseradish. To show the number of different soyburger orders that a waiter could write for the cook, we use a *tree diagram*. The ways of writing the order are listed in Figure 3-25, where the bread is chosen from the set $B = \{\text{light, dark}\}$ and the condiment is chosen from the set $C = \{\text{mustard, mayonnaise, horseradish}\}$.

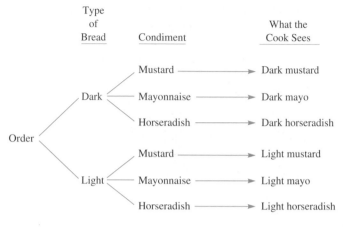

Figure 3-25

Each order can be written as an ordered pair, for example, (dark, mustard). The set of ordered pairs forms the Cartesian product $B \times C$. The Fundamental Counting Principle tells us that the number of ordered pairs in $B \times C$ is $2 \cdot 3$.

The preceding discussion demonstrates how multiplication of whole numbers can be defined in terms of Cartesian products. Thus, an alternative definition of multiplication of whole numbers is as follows:

> **Alternative Definition of Multiplication of Whole Numbers**
> For finite sets A and B, if $n(A) = a$ and $n(B) = b$, then $a \cdot b = n(A \times B)$.

In this alternative definition, sets A and B do not have to be disjoint. The expression $a \cdot b$, or simply ab, is the **product** of a and b, and a and b are **factors**. Note that $A \times B$ indicates the Cartesian product, not multiplication. We multiply numbers, not sets.

NOW TRY THIS 3-12 How would you use the repeated-addition definition of multiplication to explain to a child unfamiliar with the Fundamental Counting Principle that the number of possible outfits consisting of a shirt and pants combination—given 6 shirts and 5 pairs of pants—is $6 \cdot 5$?

The following problems illustrate each of the models shown for multiplication. In all five problems, the answer can be thought of using a different model. Work through each problem using the suggested model.

1. *Repeated-addition model.* One piece of gum costs 5¢; how much do three pieces cost?
2. *Number-line model.* If Al walks 5 mph for 3 hr, how far has he walked?
3. *Array model.* A panel of stamps has 4 rows of 5 stamps. How many stamps are there in a panel?
4. *Area model.* If a carpet is 5 ft by 3 ft, what is the area of the carpet?
5. *Cartesian-product model.* Al has 5 shirts and 3 pairs of pants; how many different shirt-pants combinations are possible?

Properties of Whole-Number Multiplication

The set of whole numbers is *closed* under multiplication. That is, if we multiply any two whole numbers, the result is a unique whole number. This property is referred to as the *closure property of multiplication of whole numbers*. Multiplication on the set of whole numbers, like addition, has the commutative, associative, and identity properties.

> **Theorem 3–5: Properties of Multiplication of Whole Numbers**
>
> **Closure property of multiplication of whole numbers**   For whole numbers a and b, $a \cdot b$ is a unique whole number.
>
> **Commutative property of multiplication of whole numbers**   For whole numbers a and b, $a \cdot b = b \cdot a$.
>
> **Associative property of multiplication of whole numbers**   For whole numbers a, b, and c, $(a \cdot b) \cdot c = a \cdot (b \cdot c)$.
>
> **Identity property of multiplication of whole numbers**   There is a unique whole number 1 such that for any whole number a, $a \cdot 1 = a = 1 \cdot a$
>
> **Zero multiplication property of whole numbers**   For any whole number a, $a \cdot 0 = 0 = 0 \cdot a$.

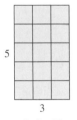

3

5

$3 \cdot 5 = 15$

5

3

$5 \cdot 3 = 15$

Figure 3-26

The *commutative property of multiplication of whole numbers* is illustrated easily by building a 3-by-5 grid and then turning it sideways, as shown in Figure 3-26. We see that the number of 1×1 squares present in either case is 15; that is, $3 \cdot 5 = 15 = 5 \cdot 3$. The commutative property can be verified by recalling that $n(A \times B) = n(B \times A)$.

The *associative property of multiplication of whole numbers* can be illustrated as follows. Suppose $a = 3, b = 5$, and $c = 4$. In Figure 3-27(a), we see a picture of $3(5 \cdot 4)$ blocks. In Figure 3-27(b), we see the same blocks, this time arranged as $4(3 \cdot 5)$. By the commutative property this can be written as $(3 \cdot 5)4$. Because both sets of blocks in Figure 3-27(a) and (b) compress to the set shown in Figure 3-27(c), we see that $3(5 \cdot 4) = (3 \cdot 5)4$. The associative property is useful in computations such as the following:

$$3 \cdot 40 = 3(4 \cdot 10) = (3 \cdot 4)10 = 12 \cdot 10 = 120$$

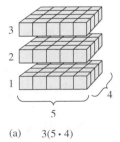
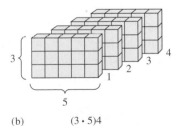
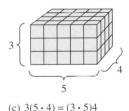

(a) $3(5 \cdot 4)$ (b) $(3 \cdot 5)4$ (c) $3(5 \cdot 4) = (3 \cdot 5)4$

Figure 3-27

The *multiplicative identity for whole numbers* is 1. For example, $3 \cdot 1 = 1 + 1 + 1 = 3$. In general, for any whole number a,

$$a \cdot 1 = \underbrace{1 + 1 + 1 + \ldots + 1}_{a \text{ terms}} = a$$

Thus, $a \cdot 1 = a$, which, along with the commutative property for multiplication implies that $a \cdot 1 = a = 1 \cdot a$. Cartesian products can also be used to show that $a \cdot 1 = a = 1 \cdot a$.

Next, consider multiplication involving 0. For example, $0 \cdot 6$ by definition means we have 0 6's or 0. Also $6 \cdot 0 = 0 + 0 + 0 + 0 + 0 + 0 = 0$. Thus we see that multiplying 0 by 6 or 6 by 0 yields a product of 0. This is an example of the *zero multiplication property*. This property can also be verified by using the definition of multiplication in terms of Cartesian products. In algebra, $3x$ means 3 x's or $x + x + x$. Therefore, $0 \cdot x$ means 0 sets of x, or 0.

The Distributive Property of Multiplication over Addition and Subtraction

We now investigate the basis for understanding multiplication algorithms for whole numbers. The area of the large rectangle in Figure 3-28 equals the sum of the areas of the two smaller rectangles and hence $5(3 + 4) = 5 \cdot 3 + 5 \cdot 4$.

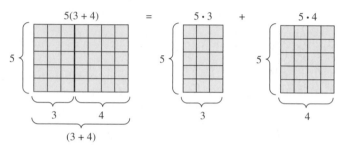

Figure 3-28

The properties of addition and multiplication of whole numbers also can be used to justify this result:

$$5(3 + 4) = \underbrace{(3 + 4) + (3 + 4) + (3 + 4) + (3 + 4) + (3 + 4)}$$

Five terms

Definition of multiplication
$$= (3 + 3 + 3 + 3 + 3) + (4 + 4 + 4 + 4 + 4)$$

Commutative and associative properties of addition
$$= 5 \cdot 3 + 5 \cdot 4 \quad \text{Definition of multiplication}$$

Note that $5(3 + 4)$ can be thought of as 5 $(3 + 4)$'s.

This example illustrates the *distributive property of multiplication over addition* for whole numbers. A similar property of subtraction is also true. Because in algebra it is customary to write $a \cdot b$ as ab, we state the distributive property of multiplication over addition and the distributive property of multiplication over subtraction as follows:

> **Theorem 3–6:   Distributive Property of Multiplication over Addition for Whole Numbers**
>
> For any whole numbers a, b, and c,
> $$a(b + c) = ab + ac$$

> **Theorem 3–7:   Distributive Property of Multiplication over Subtraction for Whole Numbers**
>
> For any whole numbers a, b, and c with $b > c$,
> $$a(b - c) = ab - ac$$

> **REMARK** Because the commutative property of multiplication of whole numbers holds, the distributive property of multiplication over addition can be rewritten as $(b + c)a = ba + ca$.
>   The distributive property can be generalized to any finite number of terms. For example, $a(b + c + d) = ab + ac + ad$.

The distributive property can be written as

$$ab + ac = a(b + c)$$

This is commonly referred to as *factoring*. Thus, the factors of $ab + ac$ are a and $(b + c)$.

Students find the distributive property of multiplication over addition useful when doing mental mathematics. For example, $13 \cdot 7 = (10 + 3)7 = 10 \cdot 7 + 3 \cdot 7 = 70 + 21 = 91$. The distributive property of multiplication over addition is important in the study of algebra and in developing algorithms for arithmetic operations. For example, it is used to combine like terms when we work with variables, as in $3x + 5x = (3 + 5)x = 8x$ or $3ab + 2b = (3a + 2)b$.

Example 3-2

a. Use an area model to show that $(x + y)(z + w) = xz + xw + yz + yw$.

b. Use the distributive property of multiplication over addition to justify the result in part (a).

Solution a. Consider the rectangle in Figure 3-29, whose height is $x + y$ and whose length is $z + w$. The area of the entire rectangle is $(x + y)(z + w)$. If we divide the rectangle into smaller rectangles as shown, we notice that the sum of the areas of the four smaller rectangles is $xz + xw + yz + yw$. Because the area of the original rectangle equals the sum of the areas of the smaller rectangles, the result follows.

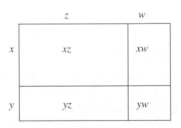

Figure 3-29

b. To apply the distributive property of multiplication over addition, we think about $x + y$ as one number and proceed as follows:

$$(x + y)(z + w) = (x + y)z + (x + y)w \qquad \text{The distributive property of multiplication over addition}$$

$$= xz + yz + xw + yw \qquad \text{The distributive property of multiplication over addition}$$

$$= xz + xw + yz + yw \qquad \text{The commutative and associative properties of addition}$$

The properties of whole-number multiplication reduce the 100 basic multiplication facts involving numbers 0–9 that students have to learn. For example, 19 facts involve multiplication by 0, and 17 more have a factor of 1. Therefore, knowing the zero multiplication property and the identity multiplication property allows students to know 36 facts. Next, 8 facts are *squares*, such as $5 \cdot 5$, that students seem to know, and that leaves 56 facts. The commutative property cuts this number in half, because if students know $7 \cdot 9$, then they know $9 \cdot 7$. This leaves 28 facts that students can learn or use the associative and distributive properties to figure out. For example, $6 \cdot 5$ can be thought of as $(5 + 1)5 = 5 \cdot 5 + 1 \cdot 5$, or 30.

Division of Whole Numbers

We discuss division using three models: the *set (partition)* model, the *missing-factor* model, and the *repeated-subtraction* model.

Set (Partition) Model

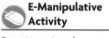

E-Manipulative Activity

Practice using the set (partition) model to perform divisions can be found in the *Division* activity.

Suppose we have 18 cookies and want to give an equal number of cookies to each of three friends: Bob, Dean, and Charlie. How many should each person receive? If we draw a picture, we can see that we can divide (or partition) the 18 cookies into three sets, with an equal number of cookies in each set. Figure 3-30 shows that each friend receives 6 cookies.

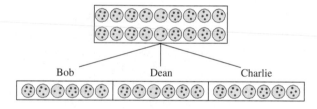

Figure 3-30

The answer may be symbolized as $18 \div 3 = 6$. Thus, $18 \div 3$ is the number of cookies in each of three disjoint sets whose union has 18 cookies. In this approach to division, we partition a set into a number of equivalent subsets.

Missing-Factor Model

Another strategy for dividing 18 cookies among three friends is to use the *missing-factor* model. If each friend receives c cookies, then the three friends receive $3c$, or 18, cookies. Hence, $3c = 18$. Because $3 \cdot 6 = 18$, we have $c = 6$. We have answered the division computation by using multiplication. This leads us to the following definition of division of whole numbers:

Definition of Division of Whole Numbers

For any whole numbers a and b, with $b \neq 0$, $a \div b = c$ if, and only if, c is the unique whole number such that $b \cdot c = a$.

The number a is the **dividend**, b is the **divisor**, and c is the **quotient**. Note that $a \div b$ can also be written as $\dfrac{a}{b}$ or $b\overline{)a}$.

Repeated-Subtraction Model

Suppose we have 18 cookies and want to package them in cookie boxes that hold 6 cookies each. How many boxes are needed? We could reason that if one box is filled, then we would have $18 - 6$ (or 12) cookies left. If one more box is filled, then there are $12 - 6$ (or 6) cookies left. Finally, we could place the last 6 cookies in a third box. This discussion can be summarized by writing $18 - 6 - 6 - 6 = 0$. We have found by repeated subtraction that $18 \div 6 = 3$. Treating division as repeated subtraction works well if there are no cookies left over. If there are cookies left over a nonzero remainder, will arise.

Calculators can illustrate the repeated subtraction operation. For example, consider $135 \div 15$. If the calculator has a constant key, $\boxed{K}$, press $\boxed{1}\boxed{5}\boxed{-}\boxed{K}\boxed{1}\boxed{3}\boxed{5}\boxed{=}\ldots$ and then count how many times you must press the $\boxed{=}$ key in order to make the display read 0. Calculators with a different constant feature may require a different sequence of entries. For example, on some calculators, we can press $\boxed{1}\boxed{3}\boxed{5}\boxed{-}\boxed{1}\boxed{5}\boxed{=}$ and then count the number of times we press the $\boxed{=}$ key to make the display read $\boxed{0}$.

The Division Algorithm

Just as subtraction of whole numbers is not closed, division of whole numbers is not closed. For example, to find $27 \div 5$, we look for a whole number c such that $5c = 27$.

Table 3-2 shows several products of whole numbers times 5. Since 27 is between 25 and 30, there is no whole number c such that $5c = 27$. Because no whole number c satisfies this equation, we see that $27 \div 5$ has no meaning in the set of whole numbers, and the set of whole numbers is not closed under division.

Table 3-2

$5 \cdot 1$	$5 \cdot 2$	$5 \cdot 3$	$5 \cdot 4$	$5 \cdot 5$	$5 \cdot 6$
5	10	15	20	25	30

Even though the set of whole numbers is not closed under division, practical applications with whole number divisions are common. For example, if 27 apples were to be divided among 5 students, each student would receive 5 apples and 2 apples would remain. The number 2 is the **remainder**. Thus, 27 contains five 5s with a remainder of 2. Observe that the remainder is a whole number less than 5. This operation is illustrated in Figure 3-31. The concept illustrated is the **division algorithm**.

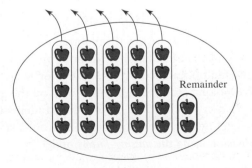

$27 = 5 \cdot 5 + 2$ with $0 \le 2 < 5$

Figure 3-31

Division Algorithm

Given any whole numbers a and b with $b \neq 0$, there exist unique whole numbers q (quotient) and r (remainder) such that

$$a = bq + r \quad \text{with } 0 \leq r < b$$

When a is "divided" by b and the remainder is 0, we say that a is *divisible* by b or that b is a *divisor* of a or that b *divides* a. By the division algorithm, a is divisible by b if $a = bq$ for a unique whole number q. Thus, 63 is divisible by 9 because $63 = 9 \cdot 7$. Notice that 63 is also divisible by 7 and that the remainder is 0.

Example 3-3

If 123 is divided by a number and the remainder is 13, what are the possible divisors?

Solution   If 123 is divided by b, then from the division algorithm we have

$$123 = bq + 13 \quad \text{and} \quad b > 13$$

Using the definition of subtraction, we have $bq = 123 - 13$, and hence $110 = bq$. Now we are looking for two numbers whose product is 110, where one number is greater than 13. Table 3-3 shows the pairs of whole numbers whose product is 110.

Table 3-3

1	110
2	55
5	22
10	11

We see that 110, 55, and 22 are the only possible values for b because each is greater than 13.

NOW TRY THIS 3-13 When the marching band was placed in rows of 5, one member was left over. When the members were placed in rows of 6, there was still one member left over. However, when they were placed in rows of 7, nobody was left over. What is the smallest number of members that could have been in the band?

Relating Multiplication and Division as Inverse Operations

In Section 3-1, we saw that subtraction and addition were related as inverse operations and we looked at fact families for both operations. In a similar way, division with remainder 0 and multiplication are related. Division is the inverse of multiplication. We can again see this by looking at fact families as shown on the grade 3 student page on page 153. Notice that question 1 in the *Talk About It* has students think of division using the *repeated-subtraction* model by skip counting backward from a starting point. Question 2 has students think of division using the *missing-factor* model.

School Book Page

RELATING MULTIPLICATION AND DIVISION

Lesson 7-5

Algebra

Key Idea
Fact families show how multiplication and division are connected.

Vocabulary
- array (p. 262)
- fact family (p. 70)
- factor (p. 260)
- product (p. 260)
- dividend
- divisor
- quotient

Think It Through
I can use **what I know** about multiplication to understand division.

384

Relating Multiplication and Division

LEARN

How does an array show division?

In 1818, there were only 20 stars on the United States flag.

There were 4 equal rows of stars.

How many stars were in each row?

The **array** shows:

Multiplication

4 rows of **5** stars = 20 stars

$4 \times 5 = 20$

Division

20 stars in 4 equal rows = **5** stars in each row

$20 \div 4 = 5$

So, there were 5 stars in each row.

How can a fact family help you divide?

A **fact family** shows how multiplication and division are related.

Fact family for 4, 5, and 20:

$$4 \times 5 = 20 \qquad 20 \div 4 = 5$$
$$5 \times 4 = 20 \qquad 20 \div 5 = 4$$

factor × factor = product   dividend ÷ divisor = quotient

✔ **Talk About It**

1. Skip count by 5s to find 4×5. Then start at 20 and skip count by 5s backward to 0. The number of times you count back is the quotient for $20 \div 5$.

2. How can you use the fact $3 \times 6 = 18$ to find $18 \div 3$?

3. **Number Sense** Is $3 \times 5 = 15$ part of the fact family for 3, 4, and 12? Explain.

Source: Scott Foresman-Addison Wesley Math, Grade 3, 2008 (p. 384).

Next consider how the four operations of addition, subtraction, multiplication, and division are related for the set of whole numbers. This is shown in Figure 3-32. Note that addition and subtraction are inverses of each other, as are multiplication and division with remainder 0. Also note that multiplication can be viewed as repeated addition, and division can be accomplished using repeated subtraction.

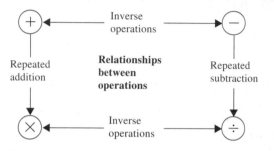

Figure 3-32

In Section 3-1, we have seen that the set of whole numbers is closed under addition and that addition is commutative and associative and has an identity. On the other hand, subtraction did not have these properties. In this section, we have seen that multiplication has some of the same properties that hold for addition. Does it follow that division behaves like subtraction? Investigate this in Now Try This 3-14.

NOW TRY THIS 3-14

a. Provide counterexamples to show that the set of whole numbers is not closed under division and that division is neither commutative nor associative.
b. Why is 1 not the identity for division?

Division by 0 and 1

Division by 0 and by 1 are frequently misunderstood by students. Before reading on, try to find the values of the following three expressions:

$$\textbf{1. } 3 \div 0 \quad \textbf{2. } 0 \div 3 \quad \textbf{3. } 0 \div 0$$

Consider the following explanations:

1. By definition, $3 \div 0 = c$ if there is a unique whole number c such that $0 \cdot c = 3$. Since the zero property of multiplication states that $0 \cdot c = 0$ for any whole number c, there is no whole number c such that $0 \cdot c = 3$. Thus, $3 \div 0$ is undefined because there is no answer to the equivalent multiplication problem.
2. By definition, $0 \div 3 = c$ if there exists a unique whole number such that $3 \cdot c = 0$. Any number c times 0 is 0, and in particular $3 \cdot 0 = 0$. Therefore, $c = 0$ and $0 \div 3 = 0$. Note that $c = 0$ is the only number that satisfies $3 \cdot c = 0$.
3. By definition, $0 \div 0 = c$ if there is a unique whole number c such that $0 \cdot c = 0$. Notice that for *any* c, $0 \cdot c = 0$. According to the definition of division, c must be unique. Since there is no *unique* number c such that $0 \cdot c = 0$, it follows that $0 \div 0$ is undefined.

School Book Page DIVISION BY 0 AND 1

Lesson 7-10

Key Idea
Thinking of related multiplication facts can help you understand division rules for 0 and 1.

Dividing with 0 and 1

LEARN

What are the division rules for 0 and 1?

Example A

	What You **Think**	What You **Write**
Divide a number by 1. 4 ÷ 1 =	1 times what number = 4? 1 × 4 = 4 So, 4 ÷ 1 = 4.	$4 \div 1 = 4$ or $1\overline{)4}$ with 4 above

Rule: When any number is divided by 1, the quotient is that number.

Example B

Divide a number by itself. 7 ÷ 7 =	7 times what number = 7? 7 × 1 = 7 So, 7 ÷ 7 = 1.

$7 \div 7 = 1$ or $7\overline{)7}$ with 1 above

Rule: When any number (except 0) is divided by itself, the quotient is 1.

Example C

Divide zero by a number. 0 ÷ 2 =	2 times what number = 0? 2 × 0 = 0 So, 0 ÷ 2 = 0.

$0 \div 2 = 0$ or $2\overline{)0}$ with 0 above

Rule: When zero is divided by a number (except 0) the quotient is 0.

Example D

Divide a number by zero. 3 ÷ 0 =	0 times what number = 3? There is no number that works, so, 3 ÷ 0 cannot be done.

3 ÷ 0 cannot be done.

Rule: You cannot divide a number by 0.

✔ **Talk About It**

1. How can you tell without dividing that 427 ÷ 1 = 427?

396

4 ÷ 1 = 4 or $1\overline{)4}$

Source: Scott Foresman–Addison Wesley Math, Grade 3, 2008 (p. 396).

Division involving 0 may be summarized as follows. Let n be any nonzero whole number. Then,

1. $n \div 0$ is undefined;   **2.** $0 \div n = 0$;   **3.** $0 \div 0$ is undefined.

Recall that $n \cdot 1 = n$ for any whole number n. Thus, by the definition of division, $n \div 1 = n$. For example, $3 \div 1 = 3, 1 \div 1 = 1$, and $0 \div 1 = 0$. A grade 3 discussion of division by 0 and 1 can be found on the student page on page 155.

Order of Operations

Difficulties involving the order of arithmetic operations sometimes arise. For example, many students will treat $2 + 3 \cdot 6$ as $(2 + 3)6$, while others will treat it as $2 + (3 \cdot 6)$. In the first case, the value is 30; in the second case, the value is 20. To avoid confusion, mathematicians agree that when no parentheses are present, multiplications and divisions are performed *before* additions and subtractions. The multiplications and divisions are performed in the order they occur, and then the additions and subtractions are performed in the order they occur. Thus, $2 + 3 \cdot 6 = 2 + 18 = 20$. This order of operations is not built into some calculators that display an incorrect answer of 30. The computation $8 - 9 \div 3 \cdot 2 + 3$ is performed as

$$8 - 9 \div 3 \cdot 2 + 3 = 8 - 3 \cdot 2 + 3$$
$$= 8 - 6 + 3$$
$$= 2 + 3$$
$$= 5$$

Assessment 3-3A

1. For each of the following, find, if possible, the whole numbers that make the equations true:
  a. $3 \cdot \square = 15$ **b.** $18 = 6 + 3 \cdot \square$
  c. $\square \cdot (5 + 6) = \square \cdot 5 + \square \cdot 6$

2. In terms of set theory, the product na could be thought of as the number of elements in the union of n sets with a elements in each. If this were the case, what must be true about the sets?

3. Determine if the following sets are closed under multiplication:
  a. $\{0, 1\}$ **b.** $\{2, 4, 6, 8, 10, \ldots\}$
  c. $\{1, 4, 7, 10, 13, \ldots\}$

4. a. If 5 is removed from the set of whole numbers, is the set closed with respect to addition? Explain.
  b. If 5 is removed from the set of whole numbers, is the set closed with respect to multiplication? Explain.
  c. Answer the same questions as (a) and (b) if 6 is removed from the set of whole numbers.

5. Rename each of the following using the distributive property of multiplication over addition so that there are no parentheses in the final answer:
  a. $(a + b)(c + d)$

  b. $\square(\Delta + \bigcirc)$
  c. $a(b + c) - ac$

6. Place parentheses, if needed, to make each of the following equations true:
  a. $5 + 6 \cdot 3 = 33$
  b. $8 + 7 - 3 = 12$
  c. $6 + 8 - 2 \div 2 = 13$
  d. $9 + 6 \div 3 = 5$

7. Using the distributive property of multiplication over addition, we can factor as in $x^2 + xy = x(x + y)$. Use the distributive property and other multiplication properties to factor each of the following:
  a. $xy + y^2$ **b.** $xy + x$
  c. $a^2b + ab^2$

8. For each of the following, find whole numbers to make the statement true, if possible:
  a. $18 \div 3 = \square$ **b.** $\square \div 76 = 0$
  c. $28 \div \square = 7$

9. A sporting goods store has designs for six shirts, four pairs of pants, and three vests. How many different shirt-pants-vest outfits are possible?

10. Which property is illustrated in each of the following:
  a. $6(5 \cdot 4) = (6 \cdot 5)4$

b. $6(5 \cdot 4) = 6(4 \cdot 5)$

c. $6(5 \cdot 4) = (5 \cdot 4)6$

d. $1(5 \cdot 4) = 5 \cdot 4$

e. $(3 + 4) \cdot 0 = 0$

f. $(3 + 4)(5 + 6) = (3 + 4)5 + (3 + 4)6$

11. Students are overheard making the following statements. What properties justify their statements?

   a. I know that $9 \cdot 7$ is either 63 or 69 and I know they can't both be right.

   b. I know that $9 \cdot 0$ is 0 because I know that any number times 0 is 0.

   c. Any number times 1 is the same as the number we started with, so $9 \cdot 1$ is 9.

12. The product $6 \cdot 14$ can be found by thinking of the problem as $6(10 + 4) = (6 \cdot 10) + (6 \cdot 4) = 60 + 24 = 84$.

   a. What properties are being used?

   b. Use this technique to mentally compute $32 \cdot 12$.

13. The distributive property of multiplication over subtraction is

$$a(b - c) = ab - ac$$

Use this property to find each of the following:

   a. $9(10 - 2)$

   b. $20(8 - 3)$

14. Show that $(a + b)^2 = a^2 + 2ab + b^2$ using

   a. the distributive property of multiplication over addition and other properties.

   b. an area model.

15. If a and b are whole numbers with $a > b$, use the rectangles in the figure to explain why $(a + b)^2 - (a - b)^2 = 4ab$.

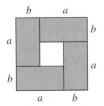

16. In each of the following, show that the left side of the equation is equal to the right side and give a reason for every step:

   a. $(ab)c = (ca)b$ **b.** $(a + b)c = c(b + a)$

17. Factor each of the following:

   a. $xy - y^2$

   b. $47 \cdot 101 - 47$

   c. $ab^2 - ba^2$

18. Rewrite each of the following division problems as a multiplication problem:

   a. $40 \div 8 = 5$ **b.** $326 \div 2 = x$

19. Show that, in general, each of the following is false if a, b, and c are whole numbers:

   a. $(a \div b) \div c = a \div (b \div c)$

   b. $a \div (b + c) = (a \div b) + (a \div c)$

20. Suppose c is a divisor of a and of b. Show that $(a + b) \div c = (a \div c) + (b \div c)$ using

   a. a model.

   b. the definition of division in terms of multiplication and the distributive property of multiplication over addition.

21. Find the solution for each of the following:

   a. $5x + 2 = 22$ **b.** $3x + 7 = x + 13$

   c. $3(x + 4) = 18$

22. Millie and Samantha began saving money at the same time. Millie plans to save \$3 a month, and Samantha plans to save \$5 a month. After how many months will Samantha have saved exactly \$10 more than Millie?

23. There were 17 sandwiches for 7 people on a picnic. How many whole sandwiches were there for each person if they were divided equally? How many were left over?

24. **a.** Find all pairs of whole numbers whose product is 36.

   b. Plot the points found in (a) on a grid.

   c. Compare the pattern shape formed by the points to the graph of the pattern shape that could be found using all pairs of whole numbers whose sum is 36.

25. A new model of car is available in 4 exterior colors and 3 interior colors. Use a tree diagram and specific colors to show how many color schemes are possible for the car.

26. To find $7 \div 5$ on the calculator, press $\boxed{7}\ \boxed{\div}\ \boxed{5}\ \boxed{=}$, which yields 1.4. To find the whole-number remainder, ignore the decimal portion of 1.4, multiply $5 \cdot 1$, and subtract this product from 7. The result is the remainder. Use a calculator to find the whole-number remainder for each of the following divisions:

   a. $28 \div 5$ **b.** $32 \div 10$

   c. $29 \div 3$ **d.** $41 \div 7$

   e. $49{,}382 \div 14$

27. Is it possible to find a whole number less than 100 that when divided by 10 leaves remainder 4 and when divided by 47 leaves remainder 17?

28. In each of the following, tell what computation must be done last:

   a. $5(16 - 7) - 18$

   b. $54/(10 - 5 + 4)$

   c. $(14 - 3) + (24 \cdot 2)$

   d. $21{,}045/345 + 8$

29. Write an algebraic expression for each of the following:

   a. Width of a rectangle whose area is A and length is l

   b. f feet in yards

   c. h hours in minutes

   d. d days in weeks

Assessment 3-3B

1. For each of the following, find, if possible, the whole numbers that make the equations true:
   a. $8 \cdot \square = 24$ b. $28 = 4 + 6 \cdot \square$
   c. $\square \cdot (8 + 6) = \square \cdot 8 + \square \cdot 6$

2. Determine if the following sets are closed under multiplication:
   a. $\{0\}$ b. $\{1, 3, 5, 7, 9, \dots\}$
   c. $\{0, 1, 2\}$

3. Rename each of the following using the distributive property of multiplication over addition so that there are no parentheses in the final answer. Simplify when possible.
   a. $3(x + y + 5)$
   b. $(x + y)(x + y + z)$
   c. $x(y + 1) - x$

4. Place parentheses, if needed, to make each of the following equations true:
   a. $4 + 3 \cdot 2 = 14$
   b. $9 \div 3 + 1 = 4$
   c. $5 + 4 + 9 \div 3 = 6$
   d. $3 + 6 - 2 \div 1 = 7$

5. The generalized distributive property for three terms states that for any whole numbers a, b, c, and d, $a(b + c + d) = ab + ac + ad$. Justify this property using the distributive property for two terms.

6. Using the distributive property of multiplication over addition, we can factor as in $x^2 + xy = x(x + y)$. Use the distributive property and other multiplication properties to factor each of the following:
   a. $47 \cdot 99 + 47$
   b. $(x + 1)y + (x + 1)$
   c. $x^2 y + zx^3$

7. For each of the following, find whole numbers to make the statement true, if possible:
   a. $27 \div 9 = \square$ b. $\square \div 52 = 1$
   c. $13 \div \square = 13$

8. A new car comes in 5 exterior colors and 3 interior colors. How many different-looking cars are possible?

9. What multiplication is suggested by the following models?
   a.

   b. ▦

10. Which property of whole numbers is illustrated in each of the following:
    a. $(5 \cdot 4)0 = 0$
    b. $7(3 \cdot 4) = 7(4 \cdot 3)$
    c. $7(3 \cdot 4) = (3 \cdot 4)7$
    d. $(3 + 4)1 = 3 + 4$
    e. $(3 + 4)5 = 3 \cdot 5 + 4 \cdot 5$
    f. $(1 + 2)(3 + 4) = (1 + 2)3 + (1 + 2)4$

11. Students are overheard making the following statements. What properties justify their statements?
    a. I know if I remember what $7 \cdot 9$ is, then I also know what $9 \cdot 7$ is.
    b. To find $9 \cdot 6$, I just remember that $9 \cdot 5$ is 45 and so $9 \cdot 6$ is just 9 more than 45, or 54.

12. The product $5 \cdot 24$ can be found by thinking of the problem as $5(20 + 4) = 5 \cdot 20 + 5 \cdot 4 = 100 + 20 = 120$.
    a. What property is being used?
    b. Use this technique to mentally compute $8 \cdot 34$.

13. The distributive property of multiplication over subtraction is
    $$a(b - c) = ab - ac$$
    Use this property to find each of the following:
    a. $15(10 - 2)$ b. $30(9 - 2)$

14. Show that if $b > c$, then $a(b - c) = ab - ac$ using:
    a. an area model suggested by the given figure (express the shaded area in two different ways).

    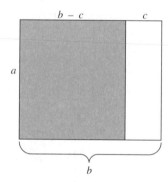

    b. the definition of subtraction in terms of addition and the distributive property of multiplication over addition.

15. Show that the left-hand side of the equation is equal to the right-hand side and give a reason for every step.
    a. $(ab)c = b(ac)$ b. $a(b + c) = ac + ab$

16. Factor each of the following:
    a. $xy - y$
    b. $(x + 1)y - (x + 1)$
    c. $a^2 b^3 - ab^2$

17. Rewrite each of the following division problems as a multiplication problem:
    a. $48 \div x = 16$ b. $x \div 5 = 17$

18. Think of a number. Multiply it by 2. Add 2. Divide by 2. Subtract 1. How does the result compare with your original number? Will this work all the time? Explain your answer.

19. Show that, in general, each of the following is false if a, b, and c are whole numbers:
    a. $a \div b = b \div a$
    b. $a - b = b - a$

20. Find the solution for each of the following:
    a. $5x + 8 = 28$ **b.** $5x + 6 = x + 14$
    c. $5(x + 3) = 35$
21. String art is formed by connecting evenly spaced nails on the vertical and horizontal axes by segments of string. Connect the nail farthest from the origin on the vertical axis with the nail closest to the origin on the horizontal axis. Continue until all nails are connected, as shown in the figure that follows. How many intersection points are created with 10 nails on each axis?

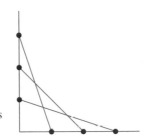

3 nails per axis
3 intersections

22. Students were divided into eight teams with nine on each team. Later, the same students were divided into teams with six on each team. How many teams were there then?
23. Jonah has a large collection of marbles. He notices that if he borrows 5 marbles from a friend, he can arrange the marbles in rows of 13 each. What is the remainder when he divides his original number of marbles by 13?
24. In the following problems, use only the designated number keys on the calculator. You may use any function keys.

a. Use the keys ①, ⑨, and ⑦ exactly once each in any order and use any operations available to write as many of the whole numbers as possible from 1 to 20. For example, $9 - 7 - 1 = 1$ and $1 \cdot 9 - 7 = 2$.
b. Use the ④ key as many times as desired with any operations to display 13.
c. Use the ② key three times with any operations to display 24.
d. Use the ① key five times with any operations to display 100.
25. In each of the following, tell what computation must be done last:
    a. $5 \cdot 6 - 3 \cdot 4 + 2$
    b. $19 - 3 \cdot 4 + 9 \div 3$
    c. $15 - 6 \div 2 \cdot 4$
    d. $5 + (8 - 2)3$
26. Find infinitely many whole numbers that leave remainder 3 upon division by 5.
27. The operation $\odot$ is defined on the set $S = \{a, b, c\}$, as shown in the following table. For example, $a \odot b = b$ and $b \odot a = b$.

$\odot$	a	b	c
a	a	b	c
b	b	a	c
c	c	c	c

a. Is S closed with respect to $\odot$?
b. Is $\odot$ commutative on S?
c. Is there an identity for $\odot$ on S? If yes, what is it?
d. Try several examples to investigate the associative property for $\odot$ on S.

Mathematical Connections 3-3

Communication

1. A number leaves remainder 6 when divided by 10. What is the remainder when the number is divided by 5? Justify your reasoning.
2. Can 0 be the identity for multiplication? Explain why or why not.
3. Suppose you forgot the product of $9 \cdot 7$. Give several ways that you could find the product using different multiplication facts and properties.
4. Is $x \div x$ always equal to 1? Explain your answer.
5. Is $x \cdot x$ ever equal to x? Explain your answer.
6. Describe all pairs of whole numbers whose sum and product are the same.

Open-Ended

7. Describe a real-life situation that could be represented by the expression $3 + 2 \cdot 6$.
8. How would you explain to a child that an even number has the form $2q$ and an odd number has the form $2q + 1$, where q is a whole number?

Cooperative Learning

9. Multiplication facts that most children have memorized can be stated in the table that is partially filled:

×	1	2	3	4	5	6	7	8	9
1									
2									
3									
4				16					
5						35			
6									
7									
8									72
9									81

a. Fill out the table of multiplication facts. Find as many patterns as you can. List all the patterns that your group discovered and explain why some of those patterns occur in the table.

b. How can the multiplication table be used to solve division problems?

c. Consider the odd number 35 shown in the multiplication table. Consider all the numbers that surround it. Note that they are all even. Does this happen for all odd numbers in the table? Explain why or why not.

Questions from the Classroom

10. Suppose a student argued that $0 \div 0 = 0$ because "nothing divided by nothing" is "nothing." How would you help that person?

11. Sue claims the following is true by the distributive law, where a and b are whole numbers:

$$3(ab) = (3a)(3b)$$

How might you help her?

12. a. A student claims that for all whole numbers $(ab) \div b = a$. How do you respond?

   b. The student in part (a) claims that $0 \div 0 = 0$. The student's reasoning is, "If $a = 0$ and $b = 0$ are substituted in the equation in part (a), the result is $0 \cdot 0 \div 0 = 0$. But because $0 \cdot 0 = 0$, it follows that $0 \div 0 = 0$." How do you respond?

13. A student asks if division on the set of whole numbers is distributive over subtraction. How do you respond?

14. A student says that 1 is the identity for division. How do you respond?

Review Problems

15. Give a set that is not closed under addition.

16. Is the operation of subtraction for whole numbers commutative? If not, give a counterexample.

17. What is wrong in each of the following?

a.	**b.**	**c.**	**d.**
137	35	56	46
+ 56	+ 47	− 29	− 17
183	712	33	39

Third International Mathematics and Science Study (TIMSS) Questions

In Toshi's class there are twice as many girls as boys. There are 8 boys in the class. What is the total number of boys and girls in the class?

a. 12
b. 16
c. 20
d. 24

TIMSS 2003, Grade 4

A piece of rope 204 cm long is cut into 4 equal pieces. Which of these gives the length of each piece in centimeters?

a. $204 + 4$
b. 204×4
c. $204 - 4$
d. $204 \div 4$

TIMSS 2003, Grade 4

National Assessment of Educational Progress (NAEP) Question

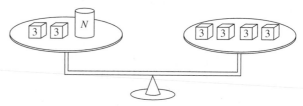

The weights on the scale above are balanced. Each cube weighs 3 pounds. The cylinder weighs N pounds. Which number sentence best describes this situation?

a. $6 + N = 12$
b. $6 + N = 4$
c. $2 + N = 12$
d. $2 + N = 4$

NAEP, 2007, Grade 4

LABORATORY ACTIVITY   Enter a natural number less than 20 on the calculator. If the number is even, divide it by 2; if it is odd, multiply it by 3 and add 1. Next, use the number on the display. Follow the given directions. Repeat the process.

1. Will the display eventually reach 1?

2. Which number less than 20 takes the most steps before reaching 1?

3. Do even or odd numbers reach 1 more quickly?

4. Investigate what happens with numbers greater than 20.

3-4   Algorithms for Whole-Number Multiplication and Division

In the grade 4 *Focal Points*, we find the following with respect to multiplication and division and the use of algorithms for doing computations:

> Students use understandings of multiplication to develop quick recall of the basic multiplication facts and related division facts. They apply their understanding of models for multiplication (i.e., equal-sized groups, arrays, area models, equal intervals on the number line), place value, and properties of operations (in particular, the distributive property) as they develop, discuss, and use efficient, accurate, and generalizable methods to multiply multidigit whole numbers. They select appropriate methods and apply them accurately to estimate products or calculate them mentally, depending on the context and numbers involved. They develop fluency with efficient procedures, including the standard algorithm, for multiplying whole numbers, understand why the procedures work (on the basis of place value and properties of operations), and use them to solve problems. (p. 16)

In this section, multiplication and division algorithms will be developed using various models.

Multiplication Algorithms

To develop algorithms for multiplying multidigit whole numbers, we use the strategy of *examining simpler computations first.* Consider $4 \cdot 12$. This computation could be pictured as in Figure 3-33(a) with 4 rows of 12 blocks, or 48 blocks. These blocks in Figure 3-33(a) can also be partitioned to show that $4 \cdot 12 = 4(10 + 2) = 4 \cdot 10 + 4 \cdot 2$. The numbers $4 \cdot 10$ and $4 \cdot 2$ are *partial products*.

(a)

(b)

Tens	Ones
1	2
$\times$	4

$$10 + 2$$
$$\times \quad 4$$
$$\overline{40 + 8}$$

$$12$$
$$\times 4$$
$$\overline{8}$$
$$40$$
$$\overline{48}$$

$$12$$
$$\times 4$$
$$\overline{48}$$

Figure 3-33

Figure 3-33(a) illustrates the distributive property of multiplication over addition on the set of whole numbers. The process leading to an algorithm for multiplying $4 \cdot 12$ is seen in Figure 3-33(b). Notice the similarity between the multiplication in Figure 3-33 and the following algebra multiplication:

$$4(x + 2) = 4x + 4 \cdot 2$$
$$= 4x + 8$$

Similarly, notice the analogy between the product

$$23 \cdot 14 = (2 \cdot 10 + 3)(1 \cdot 10 + 4) \quad \text{and} \quad (2x + 3)(1x + 4)$$

The analogy is continued as shown below:

$$
\begin{array}{r}
2 \cdot 10 + 3 \\
\times\ (1 \cdot 10 + 4) \\
\hline
12 \\
8 \cdot 10 \\
3 \cdot 10 \\
2 \cdot 10^2 \\
\hline
2 \cdot 10^2 + 11 \cdot 10 + 12
\end{array}
\qquad
\begin{array}{r}
2x + 3 \\
\times\ (1x + 4) \\
\hline
8x + 12 \\
2x^2 +\ 3x \\
\hline
2x^2 + 11x + 12
\end{array}
$$

Multiplication of a three digit number by a one-digit factor will be explored after we discuss multiplication by a power of 10.

Multiplication by 10^n

Next we consider multiplication by powers of 10. First consider what happens when we multiply a given number by 10, such as $10 \cdot 23$. If we start out with the base-ten block representation of 23, we have 2 longs and 3 units. To multiply by 10, we must replace each piece with a base-ten piece that represents the next higher power of 10. This is shown in Figure 3-34. Notice that the 3 units in 23 when multiplied by 10 become 3 longs or 3 tens. Therefore, after multiplication by 10 there are no units and hence we have 0 in the units place. In general, if we multiply any natural number by 10, we append a 0 at the end of the number.

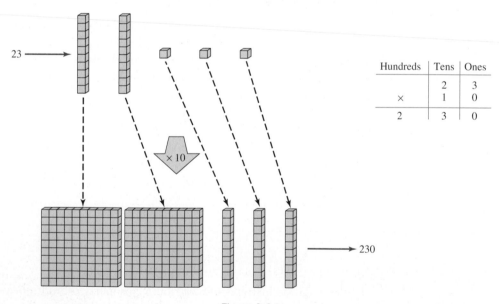

Hundreds	Tens	Ones
	2	3
×	1	0
2	3	0

Figure 3-34

The computation $23 \cdot 10$ from Figure 3-34 can be justified as follows:

$$
\begin{aligned}
23 \cdot 10 &= (2 \cdot 10 + 3)10 \\
&= (2 \cdot 10)10 + 3 \cdot 10 \\
&= 2(10 \cdot 10) + 3 \cdot 10 \\
&= 2 \cdot 10^2 + 3 \cdot 10 \\
&= 2 \cdot 10^2 + 3 \cdot 10 + 0 \cdot 1 \\
&= 230
\end{aligned}
$$

To compute products such as $3 \cdot 200$, we proceed as follows:

$$\begin{aligned}
3 \cdot 200 &= 3(2 \cdot 10^2) \\
&= (3 \cdot 2)10^2 \\
&= 6 \cdot 10^2 \\
&= 6 \cdot 10^2 + 0 \cdot 10 + 0 \cdot 1 \\
&= 600
\end{aligned}$$

We see that multiplying 6 by 10^2 results in appending two zeros to the right of 6. This idea can be generalized to the statement that *multiplication of any natural number by 10^n, where n is a natural number, results in appending n zeros to the right of the number*.

REMARK The appending of n zeros to a natural number when multiplying by 10^n can also be explained as follows. We first multiply by 10, resulting in appending of one zero (as in $23 \cdot 10 = 230$). When we multiply by another 10, another zero is appended (as in $230 \cdot 10 = 2300$). Since we multiply n times by 10, n zeros are appended to the natural number factor.

When multiplying powers of 10, the definition of exponents is used. For example, $10^2 \cdot 10^1 = (10 \cdot 10)10 = 10^3$, or 10^{2+1}. In general, where a is a natural number and m and n are whole numbers, $a^m \cdot a^n$ is given by the following:

$$a^m \cdot a^n = \underbrace{(a \cdot a \cdot a \cdot \ldots \cdot a)}_{m \text{ factors}} \cdot \underbrace{(a \cdot a \cdot a \cdot \ldots \cdot a)}_{n \text{ factors}}$$

$$= \underbrace{a \cdot a \cdot a \cdot \ldots \cdot a}_{m + n \text{ factors}} = a^{m+n}$$

Consequently, $a^m \cdot a^n = a^{m+n}$.

NOW TRY THIS 3-15 Use the fact that $a^m \cdot a^n = a^{m+n}$ along other multiplication properties to explain why the computations in the cartoon are both true.

Multiplication by a power of 10 is helpful in calculating the product of a one-digit number and a three digit number. In the following example, we assume the previously developed algorithm for multiplying a one-digit numeral times a two-digit numeral:

$$4 \cdot 367 = 4(3 \cdot 10^2 + 6 \cdot 10 + 7)$$
$$= 4(3 \cdot 10^2) + 4(6 \cdot 10) + 4 \cdot 7 \quad \rightarrow$$
$$= (4 \cdot 3)10^2 + (4 \cdot 6)10 + 4 \cdot 7$$
$$= 1200 + 240 + 28$$
$$= 1468$$

$$
\begin{array}{r}
367 \\
\times\ 4 \\
\hline
28 \\
240 \\
1200 \\
\hline
1468
\end{array}
$$

 NOW TRY THIS 3-16 Use expanded notation and an approach similar to the preceding to calculate $7 \cdot 4589$.

Multiplication with Two-Digit Factors

Consider $14 \cdot 23$. Model this computation by first using base-ten blocks, as shown in Figure 3-35(a), and then showing all the *partial products* and adding, as shown in Figure 3-35(b).

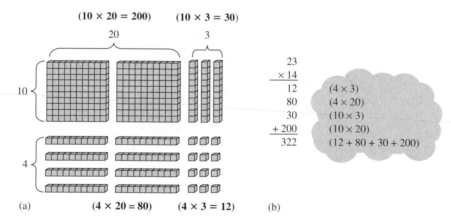

Figure 3-35

This last approach leads to an algorithm for multiplication:

$$
\begin{array}{r}
23 \\
\times\ 14 \\
\hline
92 \\
230 \\
\hline
322
\end{array}
\quad
\begin{array}{l}
(10 + 4) \\
(4 \cdot 23) \\
(10 \cdot 23)
\end{array}
\quad \text{or} \quad
\begin{array}{r}
23 \\
\times\ 14 \\
\hline
92 \\
23 \\
\hline
322
\end{array}
$$

We are accustomed to seeing the partial product 230 written without the zero, as 23. The placement of 23 with 3 in the tens column obviates having to write the 0 in the units column. But when children are first learning multiplication algorithms, we should encourage them to include the zero. This promotes better understanding and helps to avoid errors.

The distributive property of multiplication over addition can be used to explain why the algorithm for multiplication works. Again, consider $14 \cdot 23$.

$$14 \cdot 23 = (10 + 4)23$$
$$= 10 \cdot 23 + 4 \cdot 23$$
$$= 230 + 92$$
$$= 322$$

Because algorithms are powerful, there is sometimes a tendency to overapply them or to use paper and pencil for a task that should be done mentally. For example, consider

$$
\begin{array}{r}
213 \\
\times\ 1000 \\
\hline
000 \\
000 \\
000 \\
213 \\
\hline
213000
\end{array}
$$

This application is not wrong but is inefficient. Mental math and estimation are important skills in learning mathematics and should be practiced in addition to paper-and-pencil computations. Children should be encouraged to *estimate* whether their answers are reasonable. In the computation $14 \cdot 23$, we know that the answer must be between $10 \cdot 20 = 200$ and $20 \cdot 30 = 600$ because $10 < 14 < 20$ and $20 < 23 < 30$.

Lattice Multiplication

Lattice multiplication has the advantage of delaying all additions until the single-digit multiplications are complete. Because of this, it is sometimes referred to as a "low-stress algorithm." Low-achieving students especially seem to like this algorithm, perhaps because of the structure provided by the lattice. The **lattice multiplication** algorithm for multiplying 14 and 23 is shown in Figure 3-36. (Determining the reasons why lattice multiplication works is left as an exercise.)

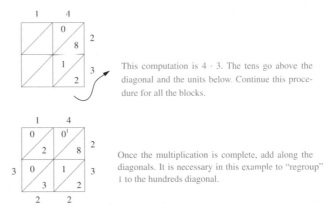

This computation is $4 \cdot 3$. The tens go above the diagonal and the units below. Continue this procedure for all the blocks.

Once the multiplication is complete, add along the diagonals. It is necessary in this example to "regroup" 1 to the hundreds diagonal.

Figure 3-36

♦ Historical Note

Lattice multiplication dates back to tenth-century India. This algorithm was imported to Europe and was popular in the fourteenth and fifteenth centuries. Napier's rods (or bones), developed by John Napier in the early 1600s, were modeled on lattice multiplication. Napier's rods can be used in a multiplication procedure.   ♦

Division Algorithms

Using Repeated Subtraction to Develop the Standard Division Algorithm

One algorithm for division of whole numbers was developed in an earlier section using repeated subtraction. However, it could have been done more efficiently. Consider the following:

A shopkeeper is packaging juice in cartons that hold 6 bottles each. She has 726 bottles. How many cartons does she need?

We might reason that if 1 carton holds 6 bottles, then 10 cartons hold 60 bottles and 100 cartons hold 600 bottles. If 100 cartons are filled, there are $726 - 100 \cdot 6$, or 126, bottles remaining. If 10 more cartons are filled, then $126 - 10 \cdot 6$, or 66, bottles remain. Similarly, if 10 more cartons are filled, $66 - 10 \cdot 6$, or 6, bottles remain. Finally, 1 carton will hold the remaining 6 bottles. The total number of cartons necessary is $100 + 10 + 10 + 1$, or 121. This procedure is summarized in Figure 3-37(a). A more efficient method is shown in Figure 3-37(b).

$$
\begin{array}{lll}
\text{(a)} & 6\overline{)726} & \\
& \underline{-\ 600} & \text{100 sixes} \\
& 126 & \\
& \underline{-\ 60} & \text{10 sixes} \\
& 66 & \\
& \underline{-60} & \text{10 sixes} \\
& 6 & \\
& \underline{-\ 6} & \underline{\text{1 six}} \\
& 0 & \text{121 sixes}
\end{array}
\qquad
\begin{array}{lll}
\text{(b)} & 6\overline{)726} & \\
& \underline{-\ 600} & \text{100 sixes} \\
& 126 & \\
& \underline{-\ 120} & \text{20 sixes} \\
& 6 & \\
& \underline{-\ 6} & \underline{\text{1 six}} \\
& 0 & \text{121 sixes}
\end{array}
$$

Figure 3-37

Divisions such as the one in Figure 3-38 are usually shown in elementary school texts in the most efficient form, as in Figure 3-38(b), in which the numbers in color in Figure 3-38(a) are omitted. The technique used in Figure 3-38(a) is often called

$$
\begin{array}{lr}
\text{(a)} & 121 \\
& 1 \\
& 20 \\
& 100 \\
& 6\overline{)726} \\
& \underline{-\ 600} \\
& 126 \\
& \underline{-\ 120} \\
& 6 \\
& \underline{-\ 6} \\
& 0
\end{array}
\qquad
\begin{array}{lr}
\text{(b)} & 121 \\
& 6\overline{)726} \\
& \underline{-\ 6} \\
& 12 \\
& \underline{-\ 12} \\
& 6 \\
& \underline{-\ 6} \\
& 0
\end{array}
$$

Figure 3-38

"scaffolding" and may be used as a preliminary step to achieving the standard algorithm, as in Figure 3-38(b). Notice that scaffolding takes the numbers on the right in Figure 3-37(b), and places them on the top as in Figure 3-38(a). Note that the scaffolding shows place value and, as indicated in the Research Note, place value is important to understanding the standard algorithms.

Using Base-Ten Blocks to Develop the Standard Division Algorithm

As pointed out in the Research Note, students need to see why each move in an algorithm is appropriate rather than just what sequence of moves to make. Next we use base-ten blocks to justify why each move in the standard algorithm is appropriate. In Table 3-4, the base-ten model is on the left with the corresponding steps in the standard algorithm on the right.

 *Research Note*   Students constructing meanings underlying an operation such as long division need to focus on understanding why each move in an algorithm is appropriate rather than on which moves to make and in which sequence. Also, teachers should encourage students to invent their own personal procedures for the operations but expect them to explain why their inventions are legitimate (Lampert 1992). ◆

Table 3-4

Base-Ten Blocks	Algorithm
1. First we represent 726 with base-ten blocks.	$6\overline{)726}$
2. We next determine how many sets of 6 flats (hundreds) there are in the representation. There is 1 set of 6 flats with 1 flat, 2 longs (tens), and 6 units (ones) left over.	1 set of 6 flats $\begin{array}{r} 1 \\ 6\overline{)726} \\ -6 \\ \hline 1 \end{array}$ 1 flat 2 longs 6 units left over
3. Next, we convert the one leftover flat to 10 longs (tens). Now we have 12 longs (tens) and 6 units (ones). 1 flat = 10 longs	1 set of 6 flats $\begin{array}{r} 1 \\ 6\overline{)726} \\ -6 \\ \hline 12 \end{array}$ 12 longs 6 units left over

(continued)

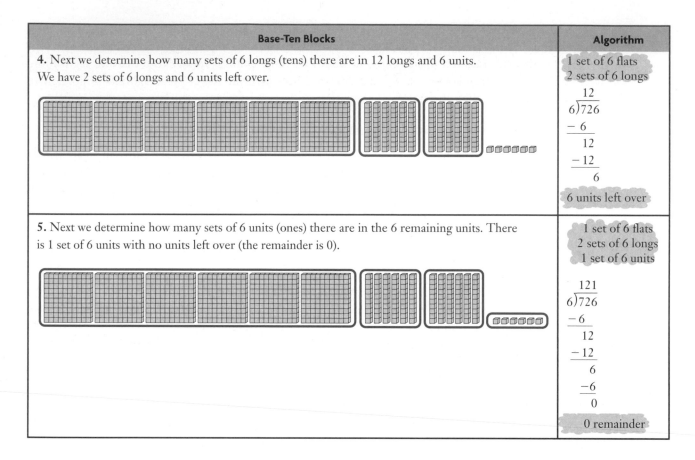

Base-Ten Blocks	Algorithm
4. Next we determine how many sets of 6 longs (tens) there are in 12 longs and 6 units. We have 2 sets of 6 longs and 6 units left over.	1 set of 6 flats 2 sets of 6 longs $\begin{array}{r} 12 \\ 6\overline{)726} \\ -6 \\ \hline 12 \\ -12 \\ \hline 6 \end{array}$ 6 units left over
5. Next we determine how many sets of 6 units (ones) there are in the 6 remaining units. There is 1 set of 6 units with no units left over (the remainder is 0).	1 set of 6 flats 2 sets of 6 longs 1 set of 6 units $\begin{array}{r} 121 \\ 6\overline{)726} \\ -6 \\ \hline 12 \\ -12 \\ \hline 6 \\ -6 \\ \hline 0 \end{array}$ 0 remainder

Therefore, we see that in the base-ten block representation of 726, there is 1 group of 6 flats (hundreds), 2 groups of 6 longs (tens), and 1 group of 6 units (ones) with none left over. Hence, the quotient is 121 with a remainder of 0. The steps in the algorithm are shown alongside the work with the base-ten blocks.

Short Division

The process used in Figure 3-38(b) is usually referred to as "long" division. Another technique, called "short" division, can be used when the divisor is a one-digit number and most of the work is done mentally. An example of the short division algorithm is given in Figure 3-39.

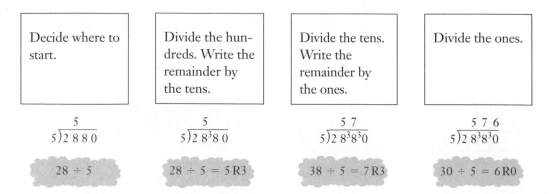

Decide where to start.	Divide the hundreds. Write the remainder by the tens.	Divide the tens. Write the remainder by the ones.	Divide the ones.
$\begin{array}{r} 5 \\ 5\overline{)2\,8\,8\,0} \end{array}$	$\begin{array}{r} 5 \\ 5\overline{)2\,8^3 8\,0} \end{array}$	$\begin{array}{r} 5\,7 \\ 5\overline{)2\,8^3 8^3 0} \end{array}$	$\begin{array}{r} 5\,7\,6 \\ 5\overline{)2\,8^3 8^3 0} \end{array}$
$28 \div 5$	$28 \div 5 = 5\,\text{R}3$	$38 \div 5 = 7\,\text{R}3$	$30 \div 5 = 6\,\text{R}0$

Figure 3-39

Division in many elementary texts is taught using a four-step algorithm: *estimate, multiply, subtract,* and *compare.* This is demonstrated in the student page on page 170. Notice that students check the division by using the inverse operation of multiplication. Study the student page and answer the question at the bottom of the page.

Division by a Two-Digit Divisor

An example of division by a divisor of more than one digit is given next. Consider $32\overline{)2618}$.

1. Estimate the quotient in $32\overline{)2618}$. Because $1 \cdot 32 = 32, 10 \cdot 32 = 320, 100 \cdot 32 = 3200$, we see that the quotient is between 10 and 100.

2. Find the number of tens in the quotient. Because $26 \div 3$ is approximately 8, 26 hundreds divided by 3 tens is approximately 8 tens. We then write the 8 in the tens place, as shown:

$$
\begin{array}{r}
80 \\
32\overline{)2618} \\
-\ 2560 \quad (32\cdot 80) \\
\hline
58
\end{array}
$$

3. Find the number of units in the quotient. Because $5 \div 3$ is approximately 1, 5 tens divided by 3 tens is approximately 1. This is shown on the left, with the standard algorithm shown on the right.

$$
\begin{array}{rcr}
81 & & \\
1 & & \\
80 & & 81\ \text{R}26 \\
32\overline{)2618} & & 32\overline{)2618} \\
-\ 2560 & \rightarrow & -\ 256 \\
\hline
58 & & 58 \\
-\ 32 \quad (32\cdot 1) & & -\ 32 \\
\hline
26 & & 26
\end{array}
$$

4. Check: $32 \cdot 81 + 26 = 2618$.

   Normally in grade-school books, we see the format shown on the right, which places the remainder beside the quotient.

Multiplication and Division in Different Bases

In multiplication, as with addition and subtraction, we need to identify the basic facts of single-digit multiplication before we can develop any algorithms. The multiplication facts for base five are given in Table 3-5. These facts can be derived by using repeated addition.

Table 3-5   Base-Five Multiplication Table

x	0	1	2	3	4
0	0	0	0	0	0
1	0	1	2	3	4
2	0	2	4	11	13
3	0	3	11	14	22
4	0	4	13	22	31

School Book Page DIVIDING THREE-DIGIT NUMBERS

Lesson 7-7

Key Idea
You can divide
larger numbers
the same way
you divide
smaller numbers.

Dividing Three-Digit Numbers

LEARN

How do you divide larger numbers?

A school bus company has 273 buses and five parking lots. If the company wants to park the same number of buses in each lot, how many buses should be parked in each lot?

Since the company wants to park the same number of buses in each lot, you can divide.

273

WARM UP
Estimate each quotient.
1. 75 ÷ 4   **2.** 62 ÷ 3
3. 485 ÷ 5   **4.** 823 ÷ 4

Example A

Find 273 ÷ 5.

Estimate: 273 is close to 250 and 250 ÷ 5 is 50, so the quotient is a little more than 50.

STEP 1	STEP 2	CHECK
Divide the tens.	Bring down the ones and divide.	Multiply the quotient by the divisor and add the remainder.

STEP 1

Divide the tens.

```
     5
 5)273   Multiply.
 − 25    Subtract.
    2    Compare. 2 < 5
```

STEP 2

Bring down the ones and divide.

```
    54 R3
 5)273
 − 25↓
    23   Multiply.
  − 20   Subtract.
     3   Compare.
         3 < 5
```

CHECK

Multiply the quotient by the divisor and add the remainder.

```
   2
  54      270
× 5     + 3
 270      273
```

The answer checks.
The dividend is 273.

So, 273 ÷ 5 = 54 R3.

The company can park 54 buses in each lot. They will have 3 buses left over.

✔ Talk About It

1. In Example A, why do you start dividing with the tens?

Take It to the NET
More Examples
www.scottforesman.com

386

Source: Scott Foresman-Addison Wesley Math, Grade 4, 2008 (p. 386).

There are various ways to do the multiplication $21_{\text{five}} \cdot 3_{\text{five}}$:

Fives	Ones
2	1
$\times$	3

$\rightarrow$
$$\begin{array}{r} (20 + 1)_{\text{five}} \\ \times \qquad 3_{\text{five}} \\ \hline (110 + 3)_{\text{five}} \end{array}$$
$\rightarrow$
$$\begin{array}{r} 21_{\text{five}} \\ \times\, 3_{\text{five}} \\ \hline 3 \\ 110 \\ \hline 113_{\text{five}} \end{array}$$
$\rightarrow$
$$\begin{array}{r} 21_{\text{five}} \\ \times\, 3_{\text{five}} \\ \hline 113_{\text{five}} \end{array}$$

The multiplication of a two-digit number by a two-digit number is developed next:

$$\begin{array}{r} 23_{\text{five}} \\ \times\, 14_{\text{five}} \\ \hline 22 \\ 130 \\ 30 \\ 200 \\ \hline 432_{\text{five}} \end{array}$$

$$\begin{array}{l} (10 + 4)_{\text{five}} \\ (4 \cdot 3)_{\text{five}} \\ (4 \cdot 20)_{\text{five}} \\ (10 \cdot 3)_{\text{five}} \\ (10 \cdot 20)_{\text{five}} \end{array}$$

$$\begin{array}{r} 23_{\text{five}} \\ \times 14_{\text{five}} \\ \hline 202 \\ 230 \\ \hline 432_{\text{five}} \end{array}$$

Lattice multiplication can also be used to multiply numbers in various number bases. This is explored in Assessment 3-4.

Division in different bases can be performed using the multiplication facts and the definition of division. For example, $22_{\text{five}} \div 3_{\text{five}} = c$ if, and only if, $c \cdot 3_{\text{five}} = 22_{\text{five}}$. From Table 3-5, we see that $c = 4_{\text{five}}$. As in base ten, computing multidigit divisions efficiently in different bases requires practice. The ideas behind the algorithms for division can be developed by using repeated subtraction. For example, $3241_{\text{five}} \div 43_{\text{five}}$ is computed by means of the repeated-subtraction technique in Figure 3-40(a) and by means of the conventional algorithm in Figure 3-40(b). Thus, $3241_{\text{five}} \div 43_{\text{five}} = 34_{\text{five}}$ with remainder 14_{five}.

(a)
$$\begin{array}{r} 43_{\text{five}} \overline{)3241_{\text{five}}} \\ -\,430 \\ \hline 2311 \\ -\,430 \\ \hline 1331 \\ -\,430 \\ \hline 401 \\ -\,141 \\ -\,210 \\ -\,141 \\ \hline 14 \end{array}$$

$$\begin{array}{l} (10 \cdot 43)_{\text{five}} \\[1.2em] (10 \cdot 43)_{\text{five}} \\[1.2em] (10 \cdot 43)_{\text{five}} \\[1.2em] (2 \cdot 43)_{\text{five}} \\[0.6em] (2 \cdot 43)_{\text{five}} \\ (34 \cdot 43)_{\text{five}} \end{array}$$

(b)
$$\begin{array}{r} 34_{\text{five}} \text{ R}14_{\text{five}} \\ 43_{\text{five}} \overline{)3241_{\text{five}}} \\ -\,234 \\ \hline 401 \\ -\,332 \\ \hline 14_{\text{five}} \end{array}$$

Figure 3-40

Computations involving base two are demonstrated in Example 3-4.

Example 3-4

a. Multiply:

$$\begin{array}{r} 101_{\text{two}} \\ \times 11_{\text{two}} \end{array}$$

b. Divide:

$$101_{\text{two}} \overline{)110110_{\text{two}}}$$

Solution **a.** 101_{two}
$\underline{\times\ 11_{\text{two}}}$
101
$\underline{101}$
1111_{two}

b.

$$101_{\text{two}}\overline{)110110_{\text{two}}}\quad 1010_{\text{two}}\ \text{R}100_{\text{two}}$$
$$\underline{-\ 101}$$
$$111$$
$$\underline{-\ 101}$$
$$100_{\text{two}}$$

Assessment 3-4A

1. Fill in the missing numbers in each of the following:

a.
```
    4_6
×  783
  1_78
  3408
 _982
3335_8
```

b.
```
    327
×  9_1
    327
  1_08
 _9_3
30__07
```

2. Perform the following multiplications using the lattice multiplication algorithm:

a. 728
$\underline{\times\ \ 94}$

b. 306
$\underline{\times\ \ 24}$

3. Explain why the lattice multiplication algorithm works.

4. Simplify each of the following using properties of exponents. Leave answers as powers.

a. $5^7 \cdot 5^{12}$ **b.** $6^{10} \cdot 6^2 \cdot 6^3$
c. $10^{296} \cdot 10^{17}$ **d.** $2^7 \cdot 10^5 \cdot 5^7$

5. a. Which is greater, $2^{80} + 2^{80}$ or 2^{100}? Why?
b. Which is greatest, $2^{101}, 3 \cdot 2^{100}$, or 2^{102}? Why?

6. The following model illustrates $22 \cdot 13$:

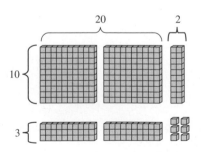

a. Explain how the partial products are shown in the figure.
b. Draw a similar model for $15 \cdot 21$.
c. Draw a similar base-five model for the product $43_{\text{five}} \cdot 23_{\text{five}}$. Explain how the model can be used to find the answer in base five.

7. Consider the following:

$$
\begin{array}{r}
476 \\
\times\ 293 \\
\hline
952 \\
4284 \\
1428 \\
\hline
139468
\end{array}
\quad
\begin{array}{l}
(2 \cdot 476) \\
(9 \cdot 476) \\
(3 \cdot 476)
\end{array}
$$

a. Use the conventional algorithm to show that the answer is correct.
b. Explain why the algorithm works.
c. Try the method to multiply 84×363.

8. The Russian peasant algorithm for multiplying 27×68 follows. (Disregard remainders when halving.)

Halves				Doubles	
	→	27 ×		⃝68	
Halve 27	→	13		⃝136	Double 68
Halve 13	→	6		272	Double 136
Halve 6	→	3		⃝544	Double 272
Halve 3	→	1		⃝1088	Double 544

In the "Halves" column, choose the odd numbers. In the "Doubles" column, circle the numbers paired with the odds from the "Halves" column. Add the circled numbers.

```
   68
  136
  544
 1088
 1836   This is the product of 27 · 68.
```

Try this algorithm for $17 \cdot 63$ and other numbers.

9. Answer the following questions based on the activity chart given next:

Activity	Calories Burned per Hour
Playing tennis	462
Snowshoeing	708
Cross-country skiing	444
Playing volleyball	198

a. How many calories are burned during 3 hr of cross-country skiing?

b. Jane played tennis for 2 hr while Carolyn played volleyball for 3 hr. Who burned more calories, and how many more?

c. Lyle went snowshoeing for 3 hr and Maurice went cross-country skiing for 5 hr. Who burned more calories, and how many more?

10. On a 14-day vacation, Glenn increased his caloric intake by 1500 calories per day. He also worked out more than usual by swimming 2 hr a day. Swimming burns 666 calories per hour, and a net gain of 3500 calories adds 1 lb of weight. Did Glenn gain at least 1 lb during his vacation?

11. Perform each of the following divisions using both the repeated-subtraction and standard algorithms:

a. $8\overline{)623}$ **b.** $36\overline{)298}$
c. $391\overline{)4001}$

12. Using a calculator, Ralph multiplied by 10 when he should have divided by 10. The display read 300. What should the correct answer be?

13. The following figure shows four operation machines. The output from one machine becomes the input for the one below it. Complete the accompanying chart.

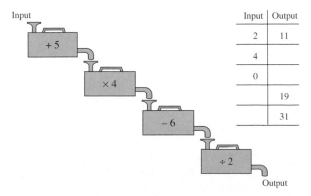

Input	Output
2	11
4	
0	
	19
	31

14. Consider the following multiplications. Notice that when the digits in the factors are reversed, the products are the same.

$$\begin{array}{r} 36 \\ \times 42 \\ \hline 1512 \end{array} \qquad \begin{array}{r} 63 \\ \times 24 \\ \hline 1512 \end{array}$$

a. Find other multiplications where this procedure works.

b. Find a pattern for the numbers that work in this way.

15. Molly read 160 pages in her book in 4 hr. Her sister Karly took 4 hr to read 100 pages in the same book. If the book is 200 pages long and if the two girls continued to read at these rates, how much longer would it take Karly to read the book than Molly?

16. Dan has 4520 pennies in three boxes. He says that there are 3 times as many pennies in the first box as in the third and twice as many in the second box as in the first. How much does he have in each box?

17. Gina buys apples from an orchard and then sells them at a country fair in bags of 3 for $1 a bag. She bought 50 boxes of apples, 36 apples in a box, and paid $452. If she sold all but 18 apples, what was her total profit?

18. Discuss possible error patterns in each of the following:

a. $\begin{array}{r} 35 \\ \times 26 \\ \hline 90 \end{array}$ **b.** $\begin{array}{r} 5\,3 \\ 5\overline{)2515} \\ -25 \\ \hline 15 \\ -15 \\ \hline 0 \end{array}$

19. a. Give reasons for each of the following steps.

$$\begin{aligned} 56 \cdot 10 &= (5 \cdot 10 + 6) \cdot 10 \\ &= (5 \cdot 10) \cdot 10 + 6 \cdot 10 \\ &= 5 \cdot (10 \cdot 10) + 6 \cdot 10 \\ &= 5 \cdot 10^2 + 6 \cdot 10 \\ &= 5 \cdot 10^2 + 6 \cdot 10 + 0 \cdot 1 \\ &= 560 \end{aligned}$$

b. Give reasons for each step in computing $34 \cdot 10^2$.

20. To transport the complete student body of 1672 students to a talk given by the governor, the school plans to rent buses that can hold 29 students each. How many buses are needed? Will all the buses be full?

21. Place the digits 7, 6, 8, and 3 in the boxes to obtain

$$\begin{array}{c} \square\square\square \\ \times \quad \square \end{array}$$

a. the greatest product.
b. the least product.

22. For what possible bases are each of the following computations correct?

a. $\begin{array}{r} 213 \\ +308 \\ \hline 522 \end{array}$ **b.** $\begin{array}{r} 213 \\ \times 32 \\ \hline 430 \\ 1043 \\ \hline 11300 \end{array}$

23. a. Use lattice multiplication to compute $323_{\text{five}} \cdot 42_{\text{five}}$.

b. Find the smallest values of a and b such that $32_a = 23_b$.

24. Perform each of these operations using the bases shown:

a. $32_{\text{five}} \cdot 4_{\text{five}}$
b. $32_{\text{five}} \div 4_{\text{five}}$
c. $43_{\text{six}} \cdot 23_{\text{six}}$
d. $143_{\text{five}} \div 3_{\text{five}}$
e. $10010_{\text{two}} \div 11_{\text{two}}$
f. $10110_{\text{two}} \cdot 101_{\text{two}}$

Assessment 3-4B

1. Fill in the missing numbers in the following:

$$
\begin{array}{r}
4\_4 \\
\times\, 327 \\
\hline
3\_88 \\
968 \\
\underline{\_452} \\
1\,5\,8\,2\_8
\end{array}
$$

2. Perform the following multiplications using the lattice multiplication algorithm:

   a. 327
         × 43

   b. 2618
         × 137

3. The following chart gives average water usage for one person for one day:

Use	Average Amount
Taking bath	110 L (liters)
Taking shower	75 L
Flushing toilet	22 L
Washing hands, face	7 L
Getting a drink	1 L
Brushing teeth	1 L
Doing dishes (one meal)	30 L
Cooking (one meal)	18 L

   a. Use the chart to calculate how much water you use each day.
   b. The average American uses approximately 200 L of water per day. Are you average?
   c. If there are 310,000,000 people in the United States, on average approximately how much water is used in the United States per day?

4. Simplify each of the following using properties of exponents. Leave answers as powers.

   a. $3^8 \cdot 3^4$ **b.** $5^2 \cdot 5^4 \cdot 5^2$
   c. $6^2 \cdot 2^2 \cdot 3^2$

5. **a.** Which is greater, $2^{20} + 2^{20}$ or 2^{21}? Why?
   b. Which is greatest, $3^{31}, 9 \cdot 3^{30}$, or 3^{33}? Why?

6. The following model illustrates $13 \cdot 12$:

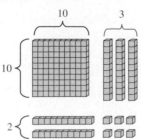

   a. Explain how the partial products are shown in the figure.
   b. Draw a similar model for $12 \cdot 22$.

7. **a.** Use base-five blocks to compute $14_{\text{five}} \cdot 23_{\text{five}}$.
   b. Use the distributive property of multiplication over addition to explain why multiplication of a natural number in base five by 10_{five} results in annexation of 0 to the number.
   c. Explain why multiplication of a natural number in base five by 100_{five} results in annexation of two 0s to the number.
   d. Use the distributive property of multiplication over addition and part (b) to compute $14_{\text{five}} \cdot 23_{\text{five}}$.

8. Complete the following table:

a	b	$a \cdot b$	$a + b$
	56	3752	
32			110
		270	33

9. Sue purchased a $30,000 life-insurance policy at the price of $24 for each $1000 of coverage. If she pays the premium in 12 monthly installments, how much is each installment?

10. Perform each of the following divisions using both the repeated-subtraction and the standard algorithms:
   a. $7\overline{)392}$
   b. $37\overline{)925}$
   c. $423\overline{)5002}$

11. Place the digits 7, 6, 8, and 3 in the boxes $\square\overline{)\square\square\square}$ to obtain
   a. the greatest quotient.
   b. the least quotient.

12. Twenty members of the band plan to attend a festival. The band members washed 245 cars at $2 per car to help cover expenses. The school will match every dollar the band raises with a dollar from the school budget. The cost of renting the bus to take the band is 72¢ per mile and the round-trip is 350 mi. The band members can stay in the dorm for two nights at $5 per person per night. Meals for the trip will cost $28 per person. Has the band raised enough money yet? If not, how many more cars do they have to wash?

13. The following figure shows three operation machines. The output from one machine becomes the input for the one below it. Complete the accompanying chart.

Input

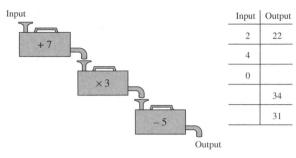

Output

Input	Output
2	22
4	
0	
	34
	31

14. Choose three different digits.
    a. Create all possible two-digit numbers from the numbers you chose. Each number can be used only once.
    b. Add the six numbers.
    c. Add the three digits you chose.
    d. Divide the answer in (b) by the answer in (c).
    e. Repeat (a) through (d) with three different digits.
    f. Is the final result always the same? Why?
15. Xuan saved $5340 in 3 years. If he saved $95 per month in the first year and a fixed amount per month for the next 2 years, how much did he save per month during the last 2 years?
16. A group of fourth-grade children had to cut four pieces of ribbon each 4 ft long from a ribbon of 44 yd. What is the length of the remaining ribbon?
17. Discuss possible error patterns in each of the following:

    a. $\begin{array}{r} 34 \\ \times\ 8 \\ \hline 2432 \end{array}$
    b. $\begin{array}{r} 34 \\ \times 6 \\ \hline 114 \end{array}$

18. Give reasons for each of the following steps:

$$\begin{aligned}
35 \cdot 100 &= (3 \cdot 10 + 5)100 \\
&= (3 \cdot 10 + 5)10^2 \\
&= (3 \cdot 10)10^2 + 5 \cdot 10^2 \\
&= 3(10 \cdot 10^2) + 5 \cdot 10^2 \\
&= 3 \cdot 10^3 + 5 \cdot 10^2 \\
&= 3 \cdot 10^3 + 5 \cdot 10^2 + 0 \cdot 10 + 0 \cdot 1 \\
&= 3500
\end{aligned}$$

19. a. Find all the whole numbers that leave remainder 1 upon division by 4. Write your answer using set builder notation.
    b. Write the numbers from part (a) in a sequence starting from the smallest.
    c. What kind of sequence is the one in part (b)?
20. Perform each of these operations using the bases shown:
    a. $42_{\text{five}} \cdot 3_{\text{five}}$
    b. $22_{\text{five}} \div 4_{\text{five}}$
    c. $32_{\text{five}} \cdot 42_{\text{five}}$
    d. $1313_{\text{five}} \div 23_{\text{five}}$
    e. $101_{\text{two}} \cdot 101_{\text{two}}$
    f. $1001_{\text{two}} \div 11_{\text{two}}$
21. For what possible bases are each of the following computations correct?

    a. $\begin{array}{r} 322 \\ -\ 233 \\ \hline 23 \end{array}$
    b. $\begin{array}{r} 101 \\ 11\overline{)1111} \\ -\ 11 \\ \hline 11 \\ -\ 11 \\ \hline 0 \end{array}$

22. a. Use lattice multiplication to compute $423_{\text{five}} \cdot 23_{\text{five}}$.
    b. Find the smallest values of a and b such that $41_a = 14_b$.
23. Place the digits 7, 6, 8, 3, and 2 in the boxes to obtain

    a. the greatest product.   b. the least product.
24. Find the products of the following and describe the pattern that emerges:
    a. $\begin{array}{l} 1 \times 1 \\ 11 \times 11 \\ 111 \times 111 \\ 1111 \times 1111 \end{array}$
    b. $\begin{array}{l} 99 \times 99 \\ 999 \times 999 \\ 9999 \times 9999 \end{array}$
    c. Test the patterns discovered. If the patterns do not continue as expected, determine when the patterns stop.

Mathematical Connections 3-4

Communication

1. How would you explain to children how to multiply $345 \cdot 678$, assuming that they know and understand multiplication by a single digit and multiplication by a power of 10?
2. What happens when you multiply any two-digit number by 101? Explain why this happens.
3. Pick a number. Double it. Multiply the result by 3. Add 24. Divide by 6. Subtract your original number. Is the result always the same? Write a convincing argument for your answer.

4. Do you think it is valuable for students to see more than one method of doing computation problems? Why or why not?
5. Choose what you consider the "best" algorithm studied in this section. Explain the reasoning behind your choice.
6. Tom claims that long division should receive reduced attention in elementary classrooms. Do you agree or disagree? Defend your answer.
7. Prove that all numbers of the form $abba$ (a and b are digits in base ten) leave remainder 0 upon division by 11.

Is the same true for all the numbers of the form *abccba*? Why or why not?

Open-Ended

8. If a student presented a new "algorithm" for computing with whole numbers, describe the process you would recommend to the student to determine if the algorithm would always work.

Cooperative Learning

9. The traditional sequence for teaching operations in the elementary school is first addition, then subtraction, followed by multiplication, and finally division. Some educators advocate teaching addition followed by multiplication, then subtraction followed by division. Within your group prepare arguments for teaching the operations in either order listed.

Questions from the Classroom

10. A student divides as follows. How would you help?

$$
\begin{array}{r}
4\ 5 \\
3\overline{)1215} \\
-12 \\
\hline
15 \\
-15 \\
\hline
0
\end{array}
$$

11. A student divides as follows. How do you help?

$$
\begin{array}{r}
15 \\
6\overline{)36} \\
-6 \\
\hline
30 \\
\hline
30
\end{array}
$$

12. A student asks how you can find the quotient and the remainder in a division problem like $592 \div 36$ using a calculator without an integer division button.

13. A student claims that to divide a number with the units digit 0 by 10, she just crosses out the 0 to get the answer. She wants to know if this is always true and why and if the 0 has to be the units digit. How do you respond?

14. A student claims that if the remainder when m is divided by n is 0, then the dividend (m) and the divisor (n) can each be multiplied by the same nonzero whole number c and the answer to the division stays the same. That is, $m \div n = (mc) \div (nc)$. She wants to know why. How would you respond, assuming the student does not know anything about fractions?

15. **a.** A student asks if $39 + 41 = 40 + 40$, is it true that $39 \cdot 41 = 40 \cdot 40$. How do you reply?

b. Another student says that he knows that $39 \cdot 41 \neq 40 \cdot 40$ but he found that $39 \cdot 41 = 40 \cdot 40 - 1$. He also found that $49 \cdot 51 = 50 \cdot 50 - 1$. He wants to know if this pattern continues. How would you respond?

Review Problems

16. Illustrate the identity property of addition for whole numbers.

17. Rename each of the following using the distributive property of multiplication over addition:
    a. $ax + bx + 2x$
    b. $3(a + b) + x(a + b)$

18. At the beginning of a trip, the odometer registered 52,281. At the end of the trip, the odometer registered 59,260. How many miles were traveled on this trip?

19. Write each of the following division problems as a multiplication problem:
    a. $36 \div 4 = 9$
    b. $112 \div 2 = x$
    c. $48 \div x = 6$
    d. $x \div 7 = 17$

Third International Mathematics and Science Study (TIMSS) Question

Each student needs 8 notebooks for school. How many notebooks are needed for 115 students?

Use the tiles $\boxed{1}$, $\boxed{4}$, and $\boxed{5}$. Write the numbers on the tiles in the boxes below to make the largest answer when you multiply.

$$
\begin{array}{r}
\square\square \\
\times\ \ \square \\
\hline
\end{array}
$$

$37 \times \blacksquare = 703$.
What is the value of $37 \times \blacksquare + 6$?

TIMSS, 2003, Grade 4

National Assessment of Educational Progress (NAEP) Question

There will be 58 people at a breakfast and each person will eat 2 eggs. There are 12 eggs in each carton. How many cartons of eggs will be needed for the breakfast?
a. 9
b. 10
c. 72
d. 116

NAEP 2007, Grade 4

BRAIN TEASER For each of the following, replace the letters with digits in such a way that the computation is correct. Each letter may represent only one digit.

a. LYNDON
      × B
    ‾‾‾‾‾‾‾‾
    JOHNSON

b.   MA
      MA
    + MA
    ‾‾‾‾‾
     EEL

LABORATORY ACTIVITY

1. Messages can be coded on paper tape in base two. A hole in the tape represents 1, whereas the absence of a space represents 0. The value of each hole depends on its position; from left to right, 16, 8, 4, 2, 1 (all powers of 2). Letters of the alphabet may be coded in base two according to their position in the alphabet. For example, G is the seventh letter. Since $7 = 1 \cdot 4 + 1 \cdot 2 + 1$, the holes appear as they do in Figure 3-41:

    ◌ ◠ ◠
16 8 4 2 1

Figure 3-41

   a. Decode the message in Figure 3-42.

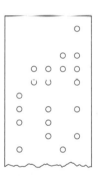

Figure 3-42

   b. Write your name on a tape using base two.

2. Consider the cards in Figure 3-43 that are modeled on base-two arithmetic.

Card E		Card D		Card C		Card B		Card A	
16	24	8	24	4	20	2	18	1	17
17	25	9	25	5	21	3	19	3	19
18	26	10	26	6	22	6	22	5	21
19	27	11	27	7	23	7	23	7	23
20	28	12	28	12	28	10	26	9	25
21	29	13	29	13	29	11	27	11	27
22	30	14	30	14	30	14	30	13	29
23	31	15	31	15	31	15	31	15	31

Figure 3-43

   a. Suppose a person's age appears on cards E, C, and B, and the person is 22. Can you discover how this works and why?
   b. Design card F and redesign cards A to E so that the numbers 1 through 63 can be used.

3-5   Mental Mathematics and Estimation
for Whole-Number Operations

In the *Principles and Standards* we find the following:

Part of being able to compute fluently means making smart choices about which tools to use and when. Students should have experiences that help them learn to choose among mental computation, paper-and-pencil strategies, estimation, and calculator use. The particular context, the question, and the numbers involved all play roles in those choices. Do the numbers allow a mental strategy? Does the context call for an estimate? Does the problem require repeated and tedious computations? Students should evaluate problem situations to determine whether an estimate or an exact answer is needed, using their number sense to advantage, and be able to give a rationale for their decision. (p. 36)

In addition, we find in the *Focal Points* the following statements with respect to estimation at the various grade levels. Notice that as the grade level advances, additional operations are included until all four operations are covered.

In the grade 2 *Focal Points*:

They (students) select and apply appropriate methods to estimate sums and differences or calculate them mentally, depending on the context and numbers involved. (p. 14)

In the grade 4 *Focal Points*:

They [students] select appropriate methods and apply them accurately to estimate products or calculate them mentally, depending on the context and numbers involved. (p. 16)

In the grade 5 *Focal Points*:

They [students] select appropriate methods and apply them accurately to estimate quotients or calculate them mentally, depending on the context and numbers involved. (p. 17)

In the previous sections in this chapter, we focused mainly on paper-and-pencil strategies. Next we focus on mental mathematics and computational estimation. **Mental mathematics** is the process of producing an answer to a computation without using computational aids. **Computational estimation** is the process of forming an *approximate* answer to a numerical problem. Facility with estimation strategies helps to determine whether or not an answer is

◆ *Research Note*

Good estimators tend to have strong self-concepts relative to mathematics, attribute their success in estimation to their ability rather than mere effort, and believe that estimation is an important tool. In contrast, poor estimators tend to have a weak self-concept relative to mathematics, attribute the success of others to effort, and believe that estimation is neither important nor useful (J. Sowder 1989). ◆

Calvin and Hobbes   by Bill Watterson

◆ *Research Note*

Mental computation becomes efficient when it involves algorithms different from the standard algorithms done using pencil and paper. Also, mental computational strategies are quite personal, being dependent on a student's creativity, flexibility, and understanding of number concepts and properties. For example, consider the skills and thinking involved in computing the sum $74 + 29$ by mentally representing the problem as $70 + (29 + 1) + 3 = 103$ (J. Sowder 1989). ◆

reasonable. In the *Calvin and Hobbes* cartoon, we see that Calvin is a poor estimator and may very well believe, as mentioned in the Research Note on page 178, that estimation is neither important nor useful.

Proficiency in mental mathematics can help in your everyday estimation skills. It is essential that you have these skills even in a time when calculators are readily available. You must be able to judge the reasonableness of answers obtained on a calculator. Mental mathematics makes use of a variety of strategies and properties. As mentioned in the Research Note on the left, mental computation becomes efficient when it involves algorithms different from standard paper-and-pencil algorithms. We consider next several of the most common algorithms for performing operations mentally on whole numbers. Note that the *trading off* algorithm is just the *equal additions* algorithm discussed earlier.

Mental Mathematics: Addition

1. *Adding from the left*

   a.   67
       + 36

$60 + 30 = 90$ (Add the tens.)
$7 + 6 = 13$ (Add the units.)
$90 + 13 = 103$ (Add the two sums.)

   b.   36
       + 36

$30 + 30 = 60$ (Double 30.)
$6 + 6 = 12$ (Double 6.)
$60 + 12 = 72$ (Add the doubles.)

2. *Breaking up and bridging*

    67
  + 36

$67 + 30 = 97$ (Add the first number to the tens in the second number.)
$97 + 6 = 103$ (Add this sum to the units in the second number.)

3. *Trading off*

   a.   67
       + 36

$67 + 3 = 70$ (Add 3 to make a multiple of 10.)
$36 - 3 = 33$ (Subtract 3 to compensate for the 3 that was added.)
$70 + 33 = 103$ (Add the two numbers.)

   b.   67
       + 29

$67 + 30 = 97$ (Add 30 (next multiple of 10 greater than 29).)
$97 - 1 = 96$ (Subtract 1 to compensate for the extra 1 that was added.)

4. *Using compatible numbers*
   Compatible numbers are numbers whose sums are easy to calculate mentally.

$130 + 70 = 200$
$50 + 50 = 100$
$100 + 200 = 300$
$300 + 20 = 320$

5. *Making compatible numbers*

    25
  + 79

$25 + 75 = 100$ (25 + 75 adds to 100.)
$100 + 4 = 104$ (Add 4 more units.)

Mental Mathematics: Subtraction

1. *Breaking up and bridging*

$\begin{array}{r} 67 \\ -36 \\ \hline \end{array}$ $67 - 30 = 37$ (Subtract the tens in the second number from the first number.)

$37 - 6 = 31$ (Subtract the units in the second number from the difference.)

2. *Trading off*

$\begin{array}{r} 71 \\ -39 \\ \hline \end{array}$ $71 + 1 = 72; 39 + 1 = 40$ (Add 1 to both numbers. Perform the

$72 - 40 = 32$ subtraction, which is easier than the original problem.)

Notice that adding 1 to both numbers does not change the answer. Why?

3. *Drop the zeros*

$\begin{array}{r} 8700 \\ -\ 500 \\ \hline \end{array}$ $87 - 5 = 82$ (Notice that there are two zeros in each number. Drop

$82 \rightarrow 8200$ these zeros and perform the computation. Then replace the two zeros to obtain proper place value.)

Another mental-mathematics technique for subtraction is called "adding up." This method is based on the *missing addend* approach and is sometimes referred to as the "cashier's algorithm." An example of *adding up* or the *cashier's algorithm* follows.

Example 3-5

Noah owed $11 for his groceries. He used a $50 check to pay the bill. While handing Noah the change, the cashier said, "11, 12, 13, 14, 15, 20, 30, 50." How much change did Noah receive?

Solution   Table 3-6 shows what the cashier said and how much money Noah received each time. Since $11 plus $1 is $12, Noah must have received $1 when the cashier said $12. The same reasoning follows for $13, $14, and so on. Thus, the total amount of change that Noah received is given by $1 + $1 + $1 + $1 + $5 + $10 + $20 = $39. In other words, $50 − $11 = $39 because $39 + $11 = $50.

Table 3-6

What the Cashier Said	$11	$12	$13	$14	$15	$20	$30	$50
Amount of Money Noah Received Each Time	0	$1	$1	$1	$1	$5	$10	$20

NOW TRY THIS 3-17 Perform each of the following computations mentally and explain what technique you used to find the answer:

a. $40 + 160 + 29 + 31$
b. $3679 - 474$
c. $75 + 28$
d. $2500 - 700$

Mental Mathematics: Multiplication

As with addition and subtraction, mental mathematics is useful for multiplication. For example, consider 8×26. Students may think of this computation in a variety of ways, as shown here.

$26 = 20 + 6$	$26 = 25 + 1$	$26 = 30 - 4$
8×20 is 160 and	8×25 is 200, then	8×30 is 240, then
8×6 is 48, so	8×1 is 8 more, so	take off $8 \times 4 = 32$,
8×26 is $160 + 48$,	8×26 is $200 + 8$,	so 8×26 is
or 208.	or 208.	$240 - 32 = 208$.

Next we consider several of the most common strategies for performing mental mathematics using multiplication.

1. *Front-end multiplying*

$$\begin{array}{r} 64 \\ \times\, 5 \\ \hline \end{array}$$

$60 \times 5 = 300$ (Multiply the number of tens in the first number by 5.)
$4 \times 5 = 20$ (Multiply the number of units in the first number by 5.)
$300 + 20 = 320$ (Add the two products.)

2. *Using compatible numbers*

$2 \times 9 \times 5 \times 20 \times 5$ Rearrange as $9 \times (2 \times 5) \times (20 \times 5) = 9 \times 10 \times 100 = 9000$.

3. *Thinking money*

a. $\begin{array}{r} 64 \\ \times\, 5 \\ \hline \end{array}$ Think of the product as 64 nickels, which can be thought of as 32 dimes, which is $32 \times 10 = 320$ cents.

b. $\begin{array}{r} 64 \\ \times 50 \\ \hline \end{array}$ Think of the product as 64 half-dollars, which is 32 dollars, or 3200 cents.

c. $\begin{array}{r} 64 \\ \times 25 \\ \hline \end{array}$ Think of the product as 64 quarters, which is 32 half-dollars, or 16 dollars. Thus we have 1600 cents.

Mental Mathematics: Division

1. *Breaking up the dividend*

$7)\overline{4256}$ $7)\overline{42|56}$ (Break up the dividend into parts.)

$\begin{array}{r} 600 + 8 \\ \hline 7)\overline{4200 + 56} \end{array}$ (Divide both parts by 7.)

$600 + 8 = 608$ (Add the answers together.)

2. *Using compatible numbers*

a. $3)\overline{105}$ $105 = 90 + 15$ (Look for numbers that you recognize as divisible by 3 and having a sum of 105.)

$\dfrac{30 + 5}{3)\overline{90 + 15}} = 35$ (Divide both parts and add the answers.)

b. $8)\overline{232}$ $232 = 240 - 8$ (Look for numbers that are easily divisible by 8 and whose difference is 232.)

$\dfrac{30 - 1}{8)\overline{240 - 8}} = 29$ (Divide both parts and take the difference.)

NOW TRY THIS 3-18 Perform each of the following computations mentally and explain what technique you used to find the answer:

a. $25 \cdot 32 \cdot 4$ b. $123 \cdot 3$
c. $25 \cdot 35$ d. $5075 \div 25$

Computational Estimation

Computational estimation may help determine whether an answer is reasonable or not. This is especially useful when the computation is done on a calculator. Some of the common estimation strategies for addition are given next.

1. *Front-end with adjustment*
   Front-end with adjustment estimation begins by focusing on the lead, or front, digits of the addition. These front, or lead, digits are added and assigned an appropriate place value. At this point we may have an underestimate that needs to be adjusted. The adjustment is made by focusing on the next group of digits. The following example shows how front-end estimation works:

$$4 + 3 + 5$$
$$12 \text{ hundred}$$

$$\begin{array}{r} 423 \\ 338 \\ + 561 \end{array} \begin{array}{l} \underline{\quad} 20 \\ \\ 100 \end{array} \Big\rangle 120$$

Steps: **(1.) Add front-end digits**
$4 + 3 + 5 = 12$.
(2.) Place value $= 1200$.
(3.) Adjust $61 + 38 \approx 100$
and $20 + 100$ is 120.
(4.) Adjusted estimate is
$1200 + 120 = 1320$.

2. *Grouping to nice numbers*
   The strategy used to obtain the adjustment in the preceding example is the *grouping to nice numbers* strategy, which means that numbers that "nicely" fit together are grouped. Another example is given here.

$$\begin{array}{r} 23 \\ 39 \\ 32 \\ 64 \\ + 49 \end{array}$$

About 100   32 → About 100   Therefore, the sum is about $100 + 100$, or 200.

3. *Clustering*
   Clustering is used when a group of numbers cluster around a common value. This strategy is limited to certain kinds of computations. In the next example, the numbers seem to cluster around 6000.

$$\begin{array}{r} 6200 \\ 5842 \\ 6512 \\ 5521 \\ + 6319 \end{array}$$

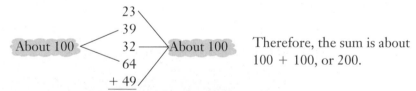

Estimate the "average"—about 6000

Multiply the average by the number of values to obtain $5 \cdot 6000 = 30{,}000$.

4. *Rounding*

Rounding is a way of cleaning up numbers so that they are easier to handle. Rounding enables us to find approximate answers to calculations, as follows:

4724	5000	(Round 4724 to 5000.)
+ 3192	+ 3000	(Round 3192 to 3000.)
	8000	(Add the rounded numbers.)

1267	1300	(Round 1267 to 1300.)
− 510	− 500	(Round 510 to 500.)
	800	(Subtract the rounded numbers.)

Performing estimations requires a knowledge of place value and rounding techniques. We illustrate a rounding procedure that can be generalized to all rounding situations. For example, suppose we wish to round 4724 to the nearest thousand. We may proceed in four steps (see also Figure 3-44).

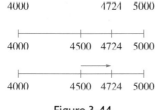

a. Determine between which two consecutive thousands the number lies.
b. Determine the midpoint between the thousands.
c. Determine which thousand the number is closer to by observing whether it is greater than or less than the midpoint. (*Not all texts use the same rule for rounding when a number falls at a midpoint.*)
d. If the number to be rounded is greater than or equal to the midpoint, round the given number to the greater thousand; otherwise, round to the lesser thousand. In this case, we round 4724 to 5000.

Figure 3-44

5. *Using the range*

It is often useful to know into what *range* an answer falls. The range is determined by finding a low estimate and a high estimate and reporting that the answer falls in this interval. An example follows:

Problem	Low Estimate	High Estimate
378	300	400
+ 524	+ 500	+ 600
	800	1000

Thus, a range for this problem is from 800 to 1000.

 The student page on page 184 shows both the *rounding* and *front-end* estimation strategies applied to a problem.

Estimation: Multiplication and Division

Examples of estimation strategies for multiplication and division are given next.

1. *Front-end*

524	$500 \times 8 = 4000$	(Start multiplying at the front to obtain a first estimate.)
× 8		
	$20 \times 8 = 160$	(Multiply the next important digit by 8.)
	$4000 + 160 = 4160$	(Adjust the first estimate by adding the two numbers.)

School Book Page

ESTIMATING SUMS AND DIFFERENCES

Lesson 1-9

Key Idea
There is more than one way to estimate sums and differences.

Vocabulary
• front-end estimation
• rounding (p. 26)

TEST TALK

Think It Through
I only need an **estimate** because it asks about how many pounds.

Estimating Sums and Differences

LEARN

How can you estimate sums?

Students at Skyline Elementary collected aluminum cans for recycling. About how many pounds of cans did they collect in all?

✓ **WARM UP**

Round each number to the place of the underlined digit.

1. 1<u>7</u>.333

2. 5<u>6</u>7,642

3. 38.<u>0</u>45

Recycling Cans				
Grade	3rd	4th	5th	6th
Pounds Collected	398	257	285	318

You can estimate 398 + 257 + 285 + 318 two ways.

Jon used **rounding**.

I'll round each number to the nearest hundred.

$$398 \rightarrow 400$$
$$257 \rightarrow 300$$
$$285 \rightarrow 300$$
$$+ 318 \rightarrow + 300$$
$$\overline{ 1{,}300}$$

About 1,300 pounds

Kylie used **front-end estimation** and adjusted the estimate.

I'll first add the front-end digits.

$$398 \rightarrow 300$$
$$257 \rightarrow 200$$
$$285 \rightarrow 200$$
$$+ 318 \rightarrow + 300$$
$$\overline{ 1{,}000}$$

Then I'll adjust to include the remaining numbers.
98 → 100.
85 → 100.
57 + 18 → 100.
Less than 1,300 pounds

Source: Scott Foresman-Addison Wesley Math, Grade 5, 2008 (p. 28).

28

2. *Compatible numbers*

$$5\overline{)4163}$$ $$5\overline{)4000}$$ (Change 4163 to a number close to it that you know is divisible by 5.)

$$\begin{array}{r} 800 \\ 5\overline{)4000} \end{array}$$ (Carry out the division and obtain the first estimate of 800. Various techniques can be used to adjust the first estimate.)

NOW TRY THIS 3-19 Estimate each of the following mentally and explain what technique you used to find the answer:

a. A sold-out concert was held in a theater with a capacity of 4525 people. Tickets were sold for $9 each. How much money was collected?
b. Fliers are to be delivered to 3625 houses and there are 42 people who will be doing the distribution. If distributed equally, how many houses will each person visit?

Assessment 3-5A

1. Compute each of the following mentally:
   a. $180 + 97 - 23 + 20 - 140 + 26$
   b. $07 \quad 12 \mid 70 \quad 30 \mid 13$
2. Use compatible numbers to compute each of the following mentally:
   a. $2 \cdot 9 \cdot 5 \cdot 6$ **b.** $8 \cdot 25 \cdot 7 \cdot 4$
3. Use breaking up and bridging or front-end multiplying to compute each of the following mentally:
   a. $567 + 38$ **b.** $321 \cdot 3$
4. Use trading off to compute each of the following mentally:
   a. $85 - 49$ **b.** $87 + 33$
   c. $143 - 97$ **d.** $58 + 39$
5. A car trip took 8 hr of driving at an average of 62 mph. Mentally compute the total number of miles traveled. Describe your method.
6. Compute each of the following using the *adding up* (cashier's) algorithm:
   a. $53 - 28$ **b.** $63 - 47$
7. Compute each of the following mentally. In each case, briefly explain your method.
   a. $86 + 37$
   b. $97 + 54$
   c. $230 + 60 + 70 + 44 + 40 + 6$
8. Round each number to the place value indicated by the digit in bold.
   a. 5**2**80   **b.** **1**15,234
   c. 1**1**5,234   **d.** 2,3**2**5
9. Estimate each answer by rounding.
   a. $878 \div 29$
   b. $25,201 - 19,987$

   c. $32 \cdot 28$
   d. $2215 + 3023 + 5967 + 975$
10. Use front-end estimation with adjustment to estimate each of the following:
   a. $2215 + 3023 + 5987 + 975$
   b. $234 + 478 + 987 + 319 + 469$
11. a. Would the clustering strategy of estimation be a good one to use in each of the following cases? Why or why not?

   (i) $\begin{array}{r} 474 \\ 1467 \\ 64 \\ + 2445 \end{array}$ (ii) $\begin{array}{r} 483 \\ 475 \\ 530 \\ 503 \\ + 528 \end{array}$

   b. Estimate each part of (a) using the following strategies:
     (i) Front-end
     (ii) Grouping to nice numbers
     (iii) Rounding
12. Use the range strategy to estimate each of the following. Explain how you arrived at your estimates.
   a. $22 \cdot 38$
   b. $145 + 678$
   c. $278 + 36$
13. Suppose you had a balance of $3287 in your checking account and you wrote checks for $85, $297, $403, and $523. Estimate your balance and tell what you did and whether you think your estimate is too high or too low.

14. A theater has 38 rows with 23 seats in each row. Estimate the number of seats in the theater and tell how you arrived at your estimate.

15. Without computing, tell which of the following have the same answer. Describe your reasoning.
  a. $44 \cdot 22$ and $22 \cdot 11$
  b. $22 \cdot 32$ and $11 \cdot 64$
  c. $13 \cdot 33$ and $39 \cdot 11$

16. The following is a list of the areas in square miles of Europe's largest countries. Mentally use this information to decide if each of the given statements is true.

France	211,207
Spain	194,896
Sweden	173,731
Finland	130,119
Norway	125,181

  a. Sweden is less than 40,000 mi² larger than Finland.
  b. France is more than twice the size of Norway.
  c. France is more than 100,000 mi² larger than Norway.
  d. Spain is about 21,000 mi² larger than Sweden

17. The attendance at a World's Fair for 1 week follows:

Monday	72,250
Tuesday	63,891
Wednesday	67,490
Thursday	73,180
Friday	74,918
Saturday	68,480

Estimate the week's attendance and tell what strategy you used and why you used it.

18. In each of the following, determine if the estimate given in parentheses is high (higher than the actual answer) or low (lower than the actual answer). Justify your answers without computing the exact values.
  a. $299 \cdot 300$ (90,000)
  b. $6001 \div 299$ (20)
  c. $6000 \div 299$ (20)
  d. $999 \div 99$ (10)

19. Use your calculator to calculate 25^2, 35^2, 45^2, and 55^2 and then see if you can find a pattern that will let you find 65^2 and 75^2 mentally.

Assessment 3-5B

1. Compute each of the following mentally:
  a. $160 + 92 - 32 + 40 - 18$
  b. $36 + 97 - 80 + 44$

2. Use compatible numbers to compute each of the following mentally:
  a. $5 \cdot 11 \cdot 3 \cdot 20$
  b. $82 + 37 + 18 + 13$

3. Supply reasons for each of the first four steps given here.

$$(525 + 37) + 75 = 525 + (37 + 75)$$
$$= 525 + (75 + 37)$$
$$= (525 + 75) + 37$$
$$= 600 + 37$$
$$= 637$$

4. Use breaking and bridging or front-end multiplying to compute each of the following mentally:
  a. $997 - 32$ **b.** $56 \cdot 30$

5. Use trading off to compute each of the following mentally:
  a. $75 - 38$ **b.** $57 + 35$
  c. $137 - 29$ **d.** $78 + 49$

6. Compute each of the following using the *adding up* (cashier's) algorithm:
  a. $74 - 63$ **b.** $73 - 57$

7. Compute each of the following mentally. In each case, briefly explain your method.
  a. $81 - 46$ **b.** $98 - 19$
  c. $9700 - 600$

8. Round each number to the place value indicated by the digit in bold.
  a. $3\underline{5}87$
  b. $\underline{1}48,213$
  c. $23,\underline{7}85$
  d. $2,3\underline{5}7$

9. Estimate each answer by rounding.
  a. $937 \div 28$ **b.** $32,285 - 18,988$
  c. $52 \cdot 48$ **d.** $3215 + 3789 + 5987$

10. Use front-end estimation with adjustment to estimate each of the following:
  a. $2345 + 5250 + 4210 + 910$
  b. $345 + 518 + 655 + 270$

11. a. Would the clustering strategy of estimation be a good one to use in each of the following cases? Why or why not?

(i)	318	(ii)	2350
	2314		1987
	57		2036
	+ 3489		2103
			+ 1890

  b. Estimate each part of (a) using the following strategies:
    (i) Front-end with adjustment
    (ii) Grouping to nice numbers
    (iii) Rounding

12. Use the range strategy to estimate each of the following. Explain how you arrived at your estimates.
    a. $32 \cdot 47$
    b. $123 + 780$
    c. $482 + 246$
13. Tom estimated $31 \cdot 179$ in the three ways shown.
    (i) $30 \cdot 200 = 6000$
    (ii) $30 \cdot 180 = 5400$
    (iii) $31 \cdot 200 = 6200$
    Without finding the actual product, which estimate do you think is closer to the actual product? Why?
14. About 3540 calories must be burned to lose 1 lb of body weight. Estimate how many calories must be burned to lose 6 lb.
15. Without computing, tell which of the following have the same answer. Describe your reasoning.
    a. $88 \cdot 44$ and $44 \cdot 22$
    b. $93 \cdot 15$ and $31 \cdot 45$
    c. $12 \cdot 18$ and $20 \cdot 17$
16. In each of the following, answer the question using estimation methods if possible. If estimation is not appropriate, explain why not.
    a. Josh has $380 in his checking account. He wants to write checks for $39, $28, $59, and $250. Will he have enough money in his account to cover these checks?
    b. Gila deposited two checks into her account, one for $981 and the other for $1140. Does she have enough money in her account to cover a check for $2000 if we know she has a positive balance to start with?

c. Alberto and Juan are running for city council. They receive votes from two districts. Alberto receives 3473 votes from one district and 5615 votes from the other district. Juan receives 3463 votes from the first district and 5616 from the second. Who gets elected?
    d. Two rectangular parcels have dimensions 101 ft by 120 ft and 103 ft by 129 ft. Which parcel has greater area? (Recall that the area of a rectangle is length times width.)
17. In each of the following, determine if the estimate given in parentheses is high (higher than the actual answer) or low (lower than the actual answer). Justify your answers without computing the exact values.
    a. $398 \cdot 500$ (200,000)
    b. $8001 \div 398$ (20)
    c. $10,000 \div 999$ (10)
    d. $1999 \div 201$ (10)
18. Use your calculator to multiply several two-digit numbers times 99. Then see if you can find a pattern that will let you find the product of any two-digit number and 99 mentally.

Mathematical Connections 3-5

Communication

1. What is the difference between mental mathematics and computational estimation?
2. Is the front-end estimate for addition before adjustment always less than the exact sum? Explain why or why not.
3. In the new textbooks, there is an emphasis on mental mathematics and estimation. Do you think these topics are important for today's students? Why?
4. Suppose x and y are positive (greater than 0) whole numbers. If x is greater than y and you estimate $x - y$ by rounding x up and y down, will your estimate always be too high or too low or could it be either? Explain.

Open-Ended

5. Give several examples from real-world situations where an estimate, rather than an exact answer, is sufficient.
6. a. Give a numerical example of when front-end estimation and rounding can produce the same estimate.
   b. Give an example of when they can produce a different estimate.

Cooperative Learning

7. Have each person in a group choose a different grade (3–6) textbook and make a list of mental math or estimation strategies done for each grade level. How do the lists compare?
8. As a group, and without actually finding the answers, find whether $19,876 \cdot 43$ or $19,875 \cdot 44$ is greater. Prepare a group response to present to the rest of the class.

Questions from the Classroom

9. Molly computed $261 - 48$ by first subtracting 50 from 261 to obtain 211; then, to make up for adding 2 to 48, she subtracted 2 from 211 to obtain an answer of 209. Is her thinking correct? If not, how could you help her?
10. A student asks why he has to learn about any estimation strategy other than rounding. What is your response?
11. In order to finish her homework quickly, an elementary student does her estimation problems by using a calculator to find the exact answers and then rounds them to get her estimate. What do you tell her?

Review Problems

12. Explain why when a number is multiplied by 10 we append a zero to the number.

13. Perform each of the following divisions using both the repeated-subtraction and the standard algorithm.

   a. $18\overline{)623}$

   b. $21\overline{)493}$

   c. $97\overline{)1000}$

14. Write each of the answers in problem 13 in the form $a = b \cdot q + r$, where $0 \le r < b$.

National Assessment of Educational Progress (NAEP) Questions

Which of these would be easiest to solve by using mental math?

   a. $\$65.12 - \$28.19$

   b. 358×2

   c. $1{,}625 \div 3$

   d. $\$100.00 + \$10.00$

NAEP 2007, Grade 4

Rockville 128 miles

Mika and her mother noticed the road sign shown above while in their car on their way to Rockville. If their speed is about 65 miles per hour, approximately how many more hours are needed to finish the trip?

   a. 1   **b.** 2   **c.** 3   **d.** 4   **e.** 5

NAEP 2007, Grade 8

BRAIN TEASER The Washington School PTA set up a phone tree in order to reach all of its members. Each person's responsibility, after receiving a call, is to call two other assigned members until all members have been called. Assume that everyone is home and answers the phone and that each phone call takes 30 seconds. If one of the 85 members, the PTA president, makes the first phone call and starts the clock, what is the least amount of time necessary to reach all 85 members of the group?

Hint for Solving the Preliminary Problem

The use of multiple 5s, such as 55 and 555, does not often come into play and the problem can be solved using multiple 5s only twice. A major factor in solving this problem, as was shown in the example, is the use of grouping symbols.

Chapter Outline

I. Whole numbers
  A. The set of **whole numbers** W is $\{0, 1, 2, 3, \dots\}$.
  B. The basic operations for whole numbers are addition, subtraction, multiplication, and division.
    1. *Addition:* If $n(A) = a$ and $n(B) = b$, where $A \cap B = \varnothing$, then $a + b = n(A \cup B)$. The numbers a and b are **addends** and $a + b$ is the **sum**.
    2. *Subtraction:* If a and b are any whole numbers, then $a - b$ is the unique whole number c such that $a = b + c$.
    3. *Multiplication:* If a and b are any whole numbers, and $a \neq 0$, then
$$ab = \underbrace{b + b + b + \dots + b}_{a \text{ terms}}$$
      where a and b are **factors** and ab is the **product**.
    4. *Multiplication:* If A and B are sets such that $n(A) = a$ and $n(B) = b$, then $ab = n(A \times B)$.
    5. *Division:* If a and b are any whole numbers with $b \neq 0$, $a \div b$ is the unique whole number c such that $bc = a$. The number a is the **dividend**, b is the **divisor**, and c is the **quotient**.
    6. **Division algorithm:** Given any whole numbers a and b, with $b \neq 0$, there exist unique whole numbers q and r such that $a = bq + r$, with $0 \leq r < b$.
  C. Properties of addition and multiplication of whole numbers
    1. *Closure:* If $a, b \in W$, then $a + b \in W$ and $ab \in W$.
    2. *Commutative:* If $a, b \in W$, then $a + b = b + a$ and $ab = ba$.
    3. *Associative:* If $a, b, c \in W$, then $(a + b) + c = a + (b + c)$ and $(ab)c = a(bc)$.
    4. *Identity:* 0 is the unique identity element for addition of whole numbers; 1 is the unique identity element for multiplication.
    5. *Distributive property of multiplication over addition:* If $a, b, c \in W$, then $a(b + c) = ab + ac$.
    6. *Distributive property of multiplication over subtraction:* If $a, b, c \in W$, with $b \geq c$, $a(b - c) = ab - ac$.
    7. *Zero multiplication property:* For any whole number a, $a \cdot 0 = 0 = 0 \cdot a$.
  D. Relations on whole numbers
    1. $a < b$ if, and only if, there is a natural number c such that $a + c = b$.
    2. $a > b$ if, and only if, $b < a$.
II. Algorithms for whole-number operations
  A. Addition and subtraction algorithms
    1. Concrete models
    2. Expanded algorithms
    3. Standard algorithms
    4. Addition and subtraction with regrouping
    5. Scratch addition
    6. Equal additions
    7. Addition and subtraction in different number bases
  B. Multiplication and division algorithms
    1. Concrete models
    2. Expanded algorithms
    3. Standard algorithms
    4. Lattice multiplication
    5. Scaffolding with division
    6. Short division
    7. Multiplication and division in different number bases
III. Mental mathematics and computational estimation strategies
  A. Mental mathematics
    1. Adding from the left
    2. Breaking up and bridging
    3. Trading off
    4. Using compatible numbers
    5. Making compatible numbers
    6. Dropping the zeros
    7. Using the cashier's algorithm (adding up)
    8. Front-end multiplying
    9. Thinking money
    10. Breaking up the dividend
  B. Computational estimation strategies
    1. Front-end
    2. Grouping to nice numbers
    3. Clustering
    4. Rounding
    5. Range
    6. Compatible numbers

Chapter Review

1. For each of the following, identify the properties of the operation(s) for whole numbers illustrated:
   a. $3(a + b) = 3a + 3b$
   b. $2 + a = a + 2$
   c. $16 \cdot 1 = 1 \cdot 16 = 16$
   d. $6(12 + 3) = 6 \cdot 12 + 6 \cdot 3$
   e. $3(a \cdot 2) = 3(2a)$
   f. $3(2a) = (3 \cdot 2)a$

2. Using the definitions of less than or greater than given in this chapter, prove that each of the following inequalities is true:
   a. $3 < 13$
   b. $12 > 9$

3. For each of the following, find all possible replacements to make the statements true for whole numbers:
   a. $4 \cdot \square - 37 < 27$
   b. $398 = \square \cdot 37 + 28$
   c. $\square \cdot (3 + 4) = \square \cdot 3 + \square \cdot 4$
   d. $42 - \square \geq 16$

4. Use the distributive property of multiplication over addition, other multiplication properties, and addition facts, if possible, to rename each of the following:
   a. $3a + 7a + 5a$
   b. $3x^2 + 7x^2 - 5x^2$
   c. $x(a + b + y)$
   d. $(x + 5)3 + (x + 5)y$

5. How many 12-oz cans of juice would it take to give 60 people one 8-oz serving each?

6. Heidi has a brown pair and a gray pair of slacks; a brown blouse, a yellow blouse, and a white blouse; and a blue sweater and a white sweater. How many different outfits does she have if each outfit she wears consists of slacks, a blouse, and a sweater?

7. I am thinking of a whole number. If I divide it by 13, then multiply the answer by 12, then subtract 20, and then add 89, I end up with 93. What was my original number?

8. A ski resort offers a weekend ski package for $80 per person or $6000 for a group of 80 people. Which would be the less expensive option for a group of 80?

9. Josi has a job in which she works 30 hr/wk and gets paid $5/hr. If she works more than 30 hr in a week, she receives $8/hr for each hour over 30 hr. If she worked 38 hr this week, how much did she earn?

10. In a television game show, there are five questions to answer. Each question is worth twice as much as the previous question. If the last question was worth $6400, what was the first question worth?

11. a. Think of a number.
       Add 17.
       Double the result.
       Subtract 4.
       Double the result.
       Add 20.
       Divide by 4.
       Subtract 20.
    Your answer will be your original number. Explain how this trick works.
    b. Fill in two more steps that will take you back to your original number.
       Think of a number.
       Add 18.
       Multiply by 4.
       Subtract 7.
       .
       .
       .
    c. Make up a series of instructions such that you will always get back to your original number.

12. Use both the scratch and the traditional algorithms to perform the following:

$$
\begin{array}{r}
316 \\
712 \\
+ \ 91 \\
\end{array}
$$

13. Use both the traditional and the lattice multiplication algorithms to perform the following:

$$
\begin{array}{r}
613 \\
\times \ 98 \\
\end{array}
$$

14. Use both the repeated-subtraction and the conventional algorithms to perform the following:
    a. $912\overline{)4803}$
    b. $11\overline{)1011}$
    c. $23_{\text{five}}\overline{)3312_{\text{five}}}$
    d. $11_{\text{two}}\overline{)1011_{\text{two}}}$

15. Use the division algorithm to check your answers in problem 14.

16. In some calculations a combination of mental math and a calculator is most appropriate. For example, because

$$
\begin{aligned}
200 \cdot 97 \cdot 146 \cdot 5 &= 97 \cdot 146(200 \cdot 5) \\
&= 97 \cdot 146 \cdot 1000
\end{aligned}
$$

we can calculate $97 \cdot 146$ on a calculator and then mentally multiply by 1000. Show how to calculate each of the following using a combination of mental math and a calculator:
    a. $19 \cdot 5 \cdot 194 \cdot 2$
    b. $379 \cdot 4 \cdot 193 \cdot 25$

c. $8 \cdot 481 \cdot 73 \cdot 125$
d. $374 \cdot 200 \cdot 893 \cdot 50$

17. You had a balance in your checking account of $720 before writing checks for $162, $158, and $33 and making a deposit of $28. What is your new balance?

18. Jim was paid $320 a month for 6 mo and $410 a month for 6 mo. What were his total earnings for the year?

19. A soft-drink manufacturer produces 15,600 cans of his product each hour. Cans are packed 24 to a case. How many cases could be filled with the cans produced in 4 hr?

20. A limited partnership of 120 investors sold a piece of land for $461,040. If divided equally, how much did each investor receive?

21. Apples normally sell for 32¢ each. They go on sale for 3 for 69¢. How much money is saved if you purchase 2 doz apples while they are on sale?

22. The owner of a bicycle shop reported his inventory of bicycles and tricycles in an unusual way. He said he counted 126 wheels and 108 pedals. How many bikes and how many trikes did he have?

23. Perform each of the following computations:

a. 123_{five}
$+\ 34_{\text{five}}$

b. 1010_{two}
$-\ 101_{\text{two}}$

c. 23_{five}
$\times\ 34_{\text{five}}$

d. 1001_{two}
$\times\ 101_{\text{two}}$

24. Tell how to use compatible numbers mentally to perform each of the following:

a. $26 + 37 + 24 - 7$ **b.** $4 \cdot 7 \cdot 9 \cdot 25$

25. Compute each of the following mentally. Name the strategy you used to perform your mental math (strategies vary).

a. $63 \cdot 7$ **b.** $85 - 49$
c. $(18 \cdot 5)2$ **d.** $2436 \div 6$

26. Estimate the following addition using **(a)** front-end estimation with adjustment and **(b)** rounding.

$$543$$
$$398$$
$$255$$
$$408$$
$$+\ 998$$

27. Using clustering, estimate the sum $2345 + 2854 + 2234 + 2203$.

28. Explain how the standard division algorithm works for the following division:

$$\begin{array}{r} 23 \\ 14\overline{)322} \\ -28 \\ \hline 42 \\ -42 \\ \hline 0 \end{array}$$

29. In some cases, the distributive property of multiplication over addition or distributive property of multiplication over subtraction can be used to obtain an answer quickly. Use one of the distributive properties to calculate each of the following in as simple a way as possible:

a. $999 \cdot 47 + 47$
b. $43 \cdot 59 + 41 \cdot 43$
c. $1003 \cdot 79 - 3 \cdot 79$
d. $1001 \cdot 113 - 113$
e. $101 \cdot 35$
f. $98 \cdot 35$

30. Recall that addition problems like $3478 + 521$ can be written and computed using expanded notation as shown here, and answer the questions that follow.

$$\begin{array}{r} 3 \cdot 10^3 + 4 \cdot 10^2 + 7 \cdot 10 + 8 \\ +\ \ 5 \cdot 10^2 + 2 \cdot 10 + 1 \\ \hline 3 \cdot 10^3 + 9 \cdot 10^2 + 9 \cdot 10 + 9 \end{array}$$

a. Write a corresponding addition algebra problem (use x for 10) and find the answer.

b. Write a subtraction problem and the corresponding algebra problem and find the answer.

c. Write a multiplication problem and the corresponding algebra problem and compute the answer.

Selected Bibliography

Baek, J. "Children's Mathematical Understanding and Invented Strategies for Multidigit Multiplication." *Teaching Children Mathematics* 12 (December 2005): 242–247.

Baroody, A. "Why Children Have Difficulties Mastering the Basic Number Combinations and How to Help Them." *Teaching Children Mathematics* 13 (August 2006): 22–31.

Bass, H. "Computational Fluency, Algorithms, and Mathematical Proficiency: One Mathematician's Perspective." *Teaching Children Mathematics* 9 (February 2003): 322–329.

Bell, A., B. Greer, C. Mangan, and L. Grimison. "Children's Performance on Multiplicative Word Problems: Elements of a Descriptive Theory." *Journal for Research in Mathematics Education* 1989, 20(5): 434–449.

Bobis, J. "The Empty Number Line: A Useful Tool or Just Another Procedure?" *Teaching Children Mathematics* 13 (April 2007): 410–413.

Broadent, F. "Lattice Multiplication and Division." *Arithmetic Teacher* 34 (January 1987): 28–31.

Brownell, W. "From NCTM's Archives: Meaning and Skill—Maintaining the Balance." *Teaching Children Mathematics* 9 (February 2003): 310–316.

Carpenter, T., and J. Moser. "The Acquisition of Addition and Subtraction Concepts in Grades One Through Three." *Journal for Research in Mathematics Education* 15 (May 1984): 179–202.

Crespo, S., A. Kyriakides, and S. McGee. "Nothing Basic about Basic Facts: Exploring Addition Facts with Fourth Graders." *Teaching Children Mathematics* 12 (September 2005): 60–67.

deGroot, C., and T. Whalen, "Longing for Division." *Teaching Children Mathematics* 12 (April 2006): 410–418.

Ebdon, S., M. Coakley, and D. Legnard. "Mathematical Mind Journeys: Awakening Minds to Computational Fluency." *Teaching Children Mathematics* 9 (April 2003): 486–493.

English, L., and G. Halford. *Mathematics Education Models and Processes*. Mahwah, New Jersey: Laurence Erlbaum, 1995.

Flowers, J., K. Kline, and R. Rubenstein. "Developing Teachers' Computational Fluency: Examples in Subtraction." *Teaching Children Mathematics* 9 (February 2003): 330–346.

Fuson, K. "Research on Learning and Teaching Addition and Subtraction of Whole Numbers." In *Handbook of Research on Mathematics Teaching and Learning*, edited by D. Grouws. New York: MacMillan, 1992.

———. "Toward Computational Fluency." *Teaching Children Mathematics* 9 (February 2003): 300–309.

Ginsburg, H., A. Klein, and P. Starkey. "The Development of Children's Mathematical Thinking: Connecting Research with Practice." In *Child Psychology in Practice*, edited by Irving E. Sigel and K. Ann Renninger, pp. 401–476, vol. 4 of *Handbook of Child Psychology*, edited by William Damon. New York: John Wiley & Sons, 1998.

Grant, T., J. Lo, and J. Flowers. "Shaping Prospective Teachers' Justifications for Computation: Challenges and Opportunities." *Teaching Children Mathematics* 14 (September 2007): 112–116.

Gregg, J. "Interpreting the Standard Division Algorithm in a 'Candy Factory' Context." *Teaching Children Mathematics* 14 (August 2007): 25–31.

Hedges, M., D. Huinker, and M. Steinmeyer. "Unpacking Division to Build Teachers' Mathematical Knowledge." *Teaching Children Mathematics* 11 (May 2005): 478–483.

Heuser, D. "Teaching without Telling: Computational Fluency and Understanding through Invention." *Teaching Children Mathematics* 11 (April 2005): 404–412.

Huinker, D., J. Freckman, and M. Steinmeyer. "Subtraction Strategies from Children's Thinking: Moving Toward Fluency with Greater Numbers." *Teaching Children Mathematics* 9 (February 2003): 347–353.

Lampert, M. "Teaching and Learning Long Division for Understanding in School." In *Analysis of Arithmetic for Mathematics Teaching*, edited by G. Leinhardt, R. Putnam, and R. Hattrup. Hillsdale, NJ: LEA, 1992.

Postlewait, K., M. Adams, and J. Shih. "Promoting Meaningful Mastery of Addition and Subtraction." *Teaching Children Mathematics* 9 (February 2003): 354–357.

Randolph, T., and H. Sherman. "Alternative Algorithms: Increasing Options, Reducing Errors." *Teaching Children Mathematics* 7 (April 2001): 480–484.

Resnick, L. "From Protoquantities to Operators: Building Mathematical Competence on a Foundation of Everyday Knowledge." In *Analysis of Arithmetic for Mathematics Teaching*, edited by D. Leinhardt, R. Putnam, and R. Hattrup. Hillsdale, NJ: LEA, 1992.

Reys, R. "Mental Computation and Estimation: Past, Present, and Future." *Elementary School Journal* 84 (1984): 547–557.

———. "Computation Versus Number Sense." *Mathematics Teaching in the Middle School* 4 (October 1998): 110–112.

Reys, B., and R. Reys. "Computation in the Elementary Curriculum: Shifting the Emphasis." *Teaching Children Mathematics* 5 (December 1998): 236–241.

Russell, S. "Developing Computational Fluency with Whole Numbers." *Teaching Children Mathematics* 7 (November 2000): 154–158.

Scharton, S. "I Did It My Way: Providing Opportunities for Students to Create, Explain, and Analyze Computation Procedures." *Teaching Children Mathematics* 10 (January 2004): 278–283.

Siegler, R. *Emerging Minds: The Process of Change in Children's Thinking*. New York: Oxford University Press, 1996.

Silver, E., L. Shapiro, and A. Deutsch. "Sense Making and the Solution of Division Problems Involving Remainders: An Examination of Middle School Students' Solution Processes and Their Interpretations of Solutions." *Journal for Research in Mathematics Education* 24 (March 1993): 117–135.

Sisul, J. "Fostering Flexibility with Numbers in the Primary Grades." *Teaching Children Mathematics* 9 (December 2002): 202–204.

Sowder, J. "Affective Factors and Computational Estimation Abilities." In *Affect and Problem Solving: A New Perspective*, edited by D. McLeod and V. Adams. New York: Springer-Verlag, 1989.

———. "Mental Computation and Number Sense." *Arithmetic Teacher* 37 (March 1990): 18–20.

Wallace, A., and S. Gurganus. "Teaching for Mastery of Multiplication." *Teaching Children Mathematics* 12 (August 2005): 26–33.

Whitenack, J., N. Knipping, S. Novinger, and G. Underwood. "Second Graders Circumvent Addition and Subtraction Difficulties." *Teaching Children Mathematics* 8 (December 2001): 228–233.

Wickett, M. "Discussion as a Vehicle for Demonstrating Computational Fluency in Multiplication." *Teaching Children Mathematics* 9 (February 2003): 318–321.

Algebraic Thinking

Preliminary Problem

Annie, a fifth-grade teacher, asks every student in her class to think of a number, multiply the number by 6, add 4, then divide the result by 2, add 5, multiply the new result by 2, and then subtract 18. She then asks each student for the final result, and as they answer, tells each one the initial number the student chose. How was Annie able to find each student's number so quickly?

Because algebraic thinking is so important in mathematics at all levels—from the early grades on—we include a separate chapter on the subject. In this chapter, we will focus not only on patterns (introduced in Chapter 1) but on other features of algebraic thinking as well, including solving equations, word problems, functions, and graphing.

In years past, schoolchildren were not introduced to algebra until at least late middle school. Today, however, we realize the importance of integrating algebraic thinking and problem solving at all levels, beginning with kindergarten. In fact, as the research note points out, algebraic thinking must be taught to *all* students.

Principles and Standards recommends that students in pre-K–2 be able to:

- use concrete, pictorial, and verbal representations to develop an understanding of invented and conventional symbolic notations;
- model situations that involve the addition and subtraction of whole numbers, using objects, pictures, and symbols. (p. 90)

And that in grades 3–5 students be able to:

- represent and analyze patterns and functions, using words, tables, and graphs;
- represent the idea of a variable as an unknown quantity using a letter or a symbol;
- express mathematical relationships using equations;
- investigate how a change in one variable relates to a change in a second variable;
- identify and describe situations with constant or varying rates of change and compare them. (p. 158)

And that students in grades 6–8 be able to:

- identify functions as linear or nonlinear and contrast their properties from tables, graphs, or equations;
- develop an initial conceptual understanding of different uses of variables;
- explore relationships between symbolic expressions and graphs of lines, paying particular attention to the meaning of intercept and slope;
- use symbolic algebra to represent situations and to solve problems, especially those that involve linear relationships;
- recognize and generate equivalent forms for simple algebraic expressions and solve linear equations. (p. 222)

Focal Points states that students in grade 6 should be taught to:

- write mathematical expressions and equations that correspond to given situations
- evaluate expressions
- use expressions and formulas to solve problems
- understand that variables represent numbers whose exact values are not yet specified
- use variables appropriately

Research Note

We have found that elementary children can learn to engage in algebraic reasoning. Furthermore, learning the big ideas and practices of mathematics is not just for a few mathematically gifted students. In fact, a strong case can be made that it is most critical for students at risk of failing in mathematics to engage with these ideas and practices (Carpenter et al. 2003).

- understand that expressions in different forms can be equivalent

- rewrite an expression to represent a quantity in a different way

- know that the solutions of an equation are the values of the variables that make the equation true

- solve simple one-step equations by using number sense, properties of operations, and the idea of maintaining equality on both sides of an equation

- construct and analyze tables

- use equations to describe simple relationships (such as $3x = y$) shown in a table (p. 18)

In this chapter, we use the basic knowledge of operations to build tenets of algebraic thinking. Subsequent chapters take a closer look at the mathematics assumed here as well as at new mathematics and will delve more deeply into algebraic thinking.

Algebra is a branch of mathematics in which symbols—usually letters—represent numbers or members of a given set. Elementary algebra is used to generalize arithmetic. For example, the fact that $7 + (3 + 5) = (7 + 3) + 5$, or that $9 + (3 + 8) = (9 + 3) + 8$, are special cases of $a + (b + c) = (a + b) + c$, where a, b, and c are numbers from a given set, for example, whole numbers, integers, rational numbers, or real numbers. Similarly, $2 + 3 = 3 + 2$ and $2 \cdot 3 = 3 \cdot 2$ are special cases of $a + b = b + a$ and $a \cdot b = b \cdot a$ for all whole numbers a and b.

◆ Historical Note

al-Khowarizmi

Fibonacci

The word *algebra*—the Latinized version of the Arabic word *al-jabr*—comes from the book *Hidab al-jabr wa'l muqabalah*, written by Mohammed ibn Musa al-Khowarizmi (ca. 825 CE). Al-Khowarizmi (from whose name we get the word *algorithm*) was part of the House of Wisdom (Bayt al-Hikma), an institution for education and research founded by the caliph al-Ma'mun. In his book he synthesized previous Hindu work on the notions of algebra and used the words *jabr* and *muqubalah* to designate two basic operations in solving equations: *jabr* meant to transpose subtracted terms to the other side of the equation; *muqubalah* meant to cancel like terms on opposite sides of the equation. The title of his book translates as *The Science of Restoring What Is Missing and Equating Like With Like*.

Another major contributor to the development of algebra was Diophantus (ca. 200–284 CE). The *Arithmetica* is the major work of Diophantus and the most prominent work on algebra in Greek mathematics. Of the original thirteen books of which *Arithmetica* consisted, only six have survived.

About 900 years later, Leonardo di Pisa (ca. 1170–1250) introduced algebra to Europe. He was also known as Fibonacci, which means the son of Bonacci. Fibonacci was the greatest mathematician of his age; he made mathematics more accessible because he brought the Hindu-Arabic numeration system, including zero, to Western Europe. Algebra was referred to at that time as *Ars Magna*—The Great Art.

A third major contributor to algebra was Francois Viete (1540–1603), known as "the father of modern algebra," who introduced the first systematic algebraic notation in his book *In Artem Analyticam*. A prominent lawyer, he also served as privy councillor to Henry IV, for whom he decoded wartime messages. ◆

4-1　Variables

A major aspect of algebraic thinking is the concept of a **variable**, an understanding of which is fundamental to algebra. Whereas in basic arithmetic we have only fixed numbers, or **constants**, as in $4 + 3 = 7$, in algebra we also have values that vary—hence the term *variable*. However, *variable* can mean several different things in mathematics.

A variable may stand for a missing element or for an unknown, as in $x + 2 = 5$. In this situation, while we could replace the variable in the sentence with any number, there is exactly one number that makes the sentence true. Here, when we replace the unknown x with 3, we make the statement true.

In a different situation, a variable can represent more than one thing. For example, in a group of children, you could say that their heights vary with their ages. If h represents height and a represents age, then both h and a can have different values for different children in the group. Here a variable represents a changing quantity.

Variables can also be used in generalizations of patterns, as we saw in Section 1-2. If we were to use actual values instead of variables, the instructions would only apply in a limited set of situations.

A variable can also be an element of a set, or a set itself; for example, in the definition of the intersection of two sets $A \cap B = \{x \mid x \in A \text{ and } x \in B\}$, x is any element that belongs to both sets.

To apply algebra in solving problems, we frequently need to translate given information into a mathematical expression involving variables designated by letters or words. In all such examples, we may name the variables as we choose.

Variables are useful because they allow instructions to be specified in a general way. For example, if we ask each student to think of a number, double it, and add 1 to the result, these instructions may be written as $2x + 1$. In mathematics, the most common letters for variables are x, y, and z, but any other letter from the Latin or even Greek alphabet may be used. If a variable is named x, then each occurrence of x in a given problem, equation, or proof refers to the same quantity.

In Examples 4-1 and 4-2, as well as in the student page that follows it, simple word statements are translated into **algebraic expressions** (see the student page for a definition). Notice in the student page the algebraic expressions for division. In this chapter, we will use the fact that division is the inverse of multiplication and vice versa, that is, that

♦ Historical Note

Mary Everest Boole

Mary Everest (1832–1916), born in England and raised in France, was a self-taught mathematician and is most well known for her works on mathematics and science education. In 1855, Everest married her friend and fellow mathematician George Boole. (Mt. Everest was named after her uncle Sir George Everest.)

In *Philosophy and Fun of Algebra* (London: C. W. Daniel, LTD, 1909), a book for children, she writes:

But when we come to the end of our arithmetic we do not content ourselves with guesses; we proceed to algebra—that is to say, to dealing logically with the fact of our own ignorance. . . .

Instead of guessing whether we are to call it nine, or seven, or a hundred and twenty, or a thousand and fifty, let us agree to call it *x*, and let us always remember that *x* stands for the Unknown. . . . This method of solving problems by honest confession of one's ignorance is called Algebra. ♦

$(a \div b) \cdot b = a$ and $(a \cdot b) \div b = a$ (where $(b \neq 0)$. However, we will follow the common practice and write $\dfrac{a}{b}$ for $a \div b$; therefore $\dfrac{a}{b} \cdot b = a$ and $\dfrac{a \cdot b}{b} = a$ $(b \neq 0)$.

Example 4-1

Write each of the following statements in algebraic form:

a. 2 more than a number
b. 2 greater than a number
c. 2 less than a number
d. 2 times a number
e. A number times itself
f. The cost of renting a car for any number of days if the charge per day is $40
g. The distance a car traveled at a constant speed of 65 mph for any number of hours

Solution **a.** $n + 2$
 b. $n + 2$
 c. $n - 2$
 d. $2 \cdot n$ or $2n$
 e. $n \cdot n$ or n^2
 f. If n is the number of days, the cost of renting the car for n days at $40 per day is $40 \cdot n$ or $40n$ dollars.
 g. If h is the number of hours traveled at 60 mph, the total distance traveled in h hours is $60 \cdot h$ or $60h$ miles.

To apply algebra in solving problems we frequently need to translate information into a mathematical expression involving variables designated by letters. In such problems we may name the variables as we choose.

Example 4-2

In each of the following, translate the given information into a symbolic expression involving quantities designated by letters:

a. One weekend, a store sold twice as many CDs as full size DVDs and 25 fewer mini DVDs than CDs. If the store sold d full size DVDs, how many mini DVDs and CDs did it sell?
b. French fries have about 12 calories apiece. A hamburger has about 600 calories. Akiva is on a diet of 2000 calories per day. If he ate f french fries and one hamburger, how many more calories can he consume that day?

Solution **a.** Because d full size DVDs were sold, twice as many CDs as full size DVDs implies $2d$ CDs. Thus, 25 fewer mini DVDs than CDs implies $2d - 25$ mini DVDs.
 b. First, find how many calories Akiva consumed eating f french fries and one hamburger. Then, to find how many more calories he can consume, subtract this expression from 2000.

| 1 french fry | 12 calories |
| f french fries | 12f calories |

Therefore, the number of calories in f french fries and one hamburger is

$$600 + 12f$$

The number of calories left for the day is $2000 - (600 + 12f)$, or $2000 - 600 - 12f$, or $1400 - 12f$.

School Book Page **VARIABLE AND EXPRESSIONS**

Lesson 1-13

Algebra

Key Idea
Relationships among quantities can be written using algebra.

Vocabulary
• variable
• evaluate

Variables and Expressions

LEARN

How can you write an algebraic expression?

Example A

Nita bought some candles costing $4 each. How can you represent their total cost?

Make a table to show the cost for different quantities of candles. Use a letter such as *n* to represent the number of candles. Because *n* represents a quantity whose value can vary, it is called a **variable.**

The total cost of the candles is represented by $4 \times n$ or $4n$.

Number of Candles	Total Cost ($)
1	4 × 1
2	4 × 2
3	4 × 3
4	4 × 4
⋮	⋮
n	4 × *n*

An **algebraic expression** is a mathematical expression containing variables, numbers, and operation symbols. Before you write an algebraic expression, identify the operation. The table below shows how two or more word phrases can refer to an operation.

Word Phrase	Operation	Algebraic Expression
the **sum** of 9 and a number *n* a number *m* **increased** by 8 six **more than** a number *t* **add** eighteen to a number *h* seventy-seven **plus** a number *r*	Addition	$9 + n$ $m + 8$ $t + 6$ $h + 18$ $77 + r$
the **difference** of 12 and a number *n* seven **less than** a number *y* ten **decreased** by a number *p*	Subtraction	$12 - n$ $y - 7$ $10 - p$
the **product** of 4 and a number *k* fifteen **times** a number *t* two **multiplied** by a number *m*	Multiplication	$4k$ $15t$ $2m$
the **quotient** of a number divided by five twenty-five **divided** by a number *m*	Division	$\frac{a}{5}$ $\frac{25}{m}$

40

Source: Mathematics, Diamond Edition, Grade Six, Scott Foresman-Addison Wesley 2008 (p. 40).

Example 4-3   A teacher instructed her class as follows:

▶▶▶▶▶▶▶▶▶▶

> Take any number and add 15 to it. Now multiply that sum by 4. Next subtract 8 and divide the difference by 4. Now subtract 12 from the quotient and tell me the answer. I will tell you the original number.

Analyze the instructions to see how the teacher was able to determine the original number.

Solution   Translate the information into an algebraic form.

Instructions	Discussion	Symbols
Take any number.	Since any number is used, we need a variable to represent the number. Let n be that variable.	n
Add 15 to it.	We are told to add 15 to "it." "It" refers to the variable n.	$n + 15$
Multiply that sum by 4.	We are told to multiply "that sum" by 4. "That sum" is $n + 15$.	$4(n + 15)$
Subtract 8.	We are told to subtract 8 from the product.	$4(n + 15) - 8$
Divide the difference by 4.	The difference is $4(n + 15) - 8$. Divide it by 4.	$\dfrac{4(n + 15) - 8}{4}$
Subtract 12 from the quotient and tell me the answer.	We are told to subtract 12 from the quotient.	$\dfrac{4(n + 15) - 8}{4} - 12$

Translating what the teacher told the class to do results in the algebraic expression $\dfrac{4(n + 15) - 8}{4} - 12$. We are also told that we have to tell the teacher the answer obtained and she then produces the original number. Let's use the strategy of *working backward* to see if we can determine what happens. Suppose we tell the teacher that our final result is r. Think about how r was obtained. Just before we told the teacher "r," we had subtracted 12. To reverse that operation, we could add 12 to obtain $r + 12$. Prior to that we had divided by 4. To reverse that, we could multiply by 4 to obtain $4r + 48$. To get that result, we had subtracted 8, so that now we add 8 to obtain $4r + 56$. Just previous to that we had multiplied by 4, so now we divide $4r + 56$ by 4 to obtain $r + 14$. The first operation had been to add 15, so now we subtract 15 from $r + 14$ to get $r - 1$. Thus, the teacher knows when we tell her that our final result is r, it is 1 more than the number with which we started, or the number with which we started, n, is the result minus 1.

This can be shown as follows:

$$\frac{4(n + 15) - 8}{4} - 12 = \frac{4(n + 15 - 2)}{4} - 12$$
$$= (n + 13) - 12$$
$$= n + 1$$

◆

Example 4-4

Figure 4-1 shows a sequence of figures containing small square tiles. Some of the tiles are shaded. Notice that the first figure has one shaded tile. The second figure has $2 \cdot 2$, or 2^2, shaded tiles. The third figure has $3 \cdot 3$, or 3^2, shaded tiles. Answer the following:

a. How many shaded tiles are there in the nth figure?
b. How many white tiles are there in the nth figure?

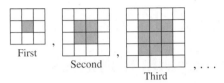

Figure 4-1

Solution **a.** The squares of shaded tiles have sides with increasing lengths 1, 2, 3, and so on. In the nth figure, the length of a side of the shaded region would be n. Hence, the nth figure has n^2 shaded tiles.

b. One way to think about the number of white tiles is to recognize that the number of white tiles on a side is 2 more than n, or $n + 2$. The number of white tiles could be 4 times $(n + 2)$, less any overlapping counting. In this case, each corner tile would be duplicated so 4 white tiles are overcounted giving us $4(n + 2) - 4$, or $4n + 4$, white tiles.

Another way to count the white tiles in the nth figure is to count the total number of tiles and then subtract from this total the number of shaded tiles. We have seen that the number of white tiles on the bottom side of the nth square is $n + 2$, and the number of the shaded tiles on a side is n. Thus, the number of white tiles is $(n + 2)^2 - n^2$. It can be shown that this answer is the same as $4n + 4$ obtained earlier.

NOW TRY THIS 4-1

a. There is another way to count the white tiles in Example 4-4. First, remove the four white corner tiles and then count the number of remaining white tiles. Complete this approach.
b. Noah has some white square tiles and some blue square tiles. They are all the same size. He first makes a row of white tiles and then surrounds the white tiles with a single layer of blue tiles, as shown in Figure 4-2.

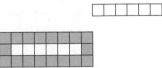

Figure 4-2

How many gray tiles does he need:

i. to surround a row of 100 white tiles?
ii. to surround a row of n white tiles?

Variables are commonly used in spreadsheets. To compute the 50th term of the Fibonacci sequence, 1, 1, 2, 3, 5, 8, 13, ..., in which the first two terms are 1, 1 and each subsequent term is the sum of the two preceding terms, could take a very long time by hand or even by calculator. However, using a spreadsheet, any desired term of the Fibonacci sequence and the previous term appear instantaneously. The student page above shows how to create the Fibonacci sequence on a spreadsheet using two variables, $A1$ and $A2$.

$\mathcal{S}$chool Book Page LEARNING WITH TECHNOLOGY

Learning with Technology

Spreadsheet/Data/Grapher eTool: Generating a Sequence

Almost 800 years ago, an Italian mathematician named Leonardo Fibonacci discovered this sequence of numbers: 1, 1, 2, 3, 5, 8, 13, 21, 34, 55, 89, 144, 233, 377, 610,…

Beginning with the number 1, each number in the sequence is the sum of the previous two numbers.
$1 + 1 = \mathbf{2}$, $1 + 2 = \mathbf{3}$, $2 + 3 = \mathbf{5}$, $3 + 5 = \mathbf{8}$, $5 + 8 = \mathbf{13}$, $8 + 13 = \mathbf{21}$, $13 + 21 = \mathbf{34}$, and so on.

Create a spreadsheet that will generate the first 32 terms of the Fibonacci sequence. Copy the formula in A5 to cells A6–A35, and B6 to cells B7–B35.

Leonardo Fibonacci

1. Do you think there is an infinite number of terms in the sequence? Explain.

2. Change the number in cell C2 to 3. What happens to the numbers in column B? How does the sequence depend on the numbers in cells B2 and C2?

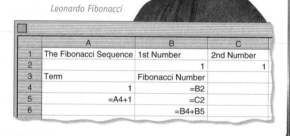

	A	B	C
1	The Fibonacci Sequence	1st Number	2nd Number
2		1	1
3	Term	Fibonacci Number	
4	1	=B2	
5	=A4+1	=C2	
6		=B4+B5	

Source: Mathematics, Diamond Edition, Grade Six, Scott Foresman-Addison Wesley 2008 (p. 163).

Algebraic thinking can occur in different ways. One example that uses pictures is seen in Example 4-5.

Example 4-5

At a local farmer's market, three purchases were made for the prices shown in Figure 4-3. What is the cost of each object?

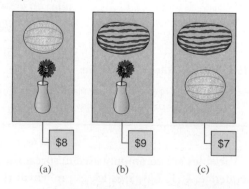

(a)   (b)   (c)

Figure 4-3

Solution Approaches to this problem may vary. For example, if the objects in the first two purchases are put together, the total cost would be $8 + $9, or $17. That cost would be for two vases and one each of the cantaloupe and watermelon, as in Figure 4-4.

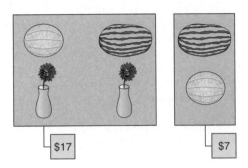

$17 $7

Figure 4-4

Now if the cantaloupe and watermelon are taken away from that total, then according to the cost of those two objects from the tag on the right, the cost should be reduced to $10 for two vases. That means each of the two vases costs $5. This in turn tells us that the cantaloupe costs $8 − $5, or $3, and the watermelon costs $9 − $5, or $4.

The solution in Example 4-5 could involve the strategy of *writing an equation*. But first we need a basic knowledge of solving equations.

Assessment 4-1A

1. Write each of the following statements in algebraic form:
   a. The third term of an arithmetic sequence whose first term is 10 and whose difference is d
   b. 10 less than twice a number
   c. 10 times the square of a number
   d. The difference between the square of a number and twice the number
2. a. Translate the following information into an algebraic form: Take any number, add 3 to it, multiply the sum by 7, subtract 14, and divide the difference by 7. Finally, subtract the original number.
   b. Simplify your answer in part (a).
3. In the tile pattern in the sequence of figures shown each figure starting from the second has two more blue squares than the preceding one. Answer the following:

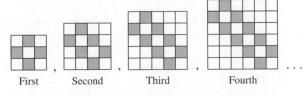

First , Second , Third , Fourth . . .

   a. How many shaded tiles are there in the nth figure?
   b. How many white tiles are there in the nth figure?

4. In the following, write an expression in terms of the given variable that represents the indicated quantity. For example, the distance traveled at a constant speed of 60 mph during t hr could be written as $60t$ miles.
   a. The cost of having a plumber spend h hr at your house if the plumber charges $20 for coming to the house and $25 per hour for labor
   b. The amount of money in cents in a jar containing d dimes and some nickels and quarters if there are 3 times as many nickels as dimes and twice as many quarters as nickels
   c. The sum of three consecutive integers if the least integer is x
   d. The amount of bacteria after n min if the initial amount of bacteria is q and the amount of bacteria doubles every minute. (*Hint:* The answer should contain q as well as n.)
   e. The temperature after t hr if the initial temperature is 40°F and each hour it drops by 3°F
   f. Pawel's total earning after 3 yr if the first year his salary was s dollars, the second year it was $5000 higher, and the third year it was twice as much as the second year

g. The sum of three consecutive odd natural numbers if the least is x

h. The sum of three consecutive natural numbers if the middle is m

5. If the number of professors in a college is P and the number of students S, and there are 20 times as many students as professors, write an algebraic equation that shows this relationship.

6. If g is the number of girls in a class and b the number of boys and if there are five more girls (g) than boys (b) in a class, write an algebraic equation that shows this relationship.

7. Ryan is building matchstick square sequences as shown. How many matchsticks will he use for the nth figure?

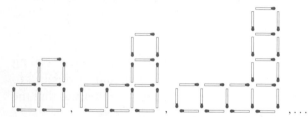

8. Write an algebraic equation relating the variables described in each of the following situations:

   a. The pay, P, for t hr if you are paid \$8 an hour

   b. The pay, P, for t hr if you are paid \$15 for the first hour and \$10 for each additional hour

9. For a particular event, a student pays \$5 per ticket and a nonstudent pays \$13 per ticket. If x students and 100 nonstudents buy tickets, find the total revenue from the sale of the tickets in terms of x.

10. Suppose a will decreed that three siblings will each receive a cash inheritance according to the following: The eldest receives 3 times as much as the youngest, and twice as much as the middle sibling. Answer the following:

    a. If the youngest sibling receives \$$x$, how much do the other two receive in terms of x?

    b. If the middle sibling receives \$$y$, how much do the other two receive in terms of y?

    c. If the oldest sibling receives \$$z$, how much do the other two receive in terms of z?

Assessment 4-1B

1. Write each of the following statements in algebraic form:

   a. 10 more than a number

   b. 10 less than a number

   c. 10 times a number

   d. The sum of a number and 10

   e. The difference between the square of a number and the number

2. Translate the following into algebraic form:

   a. Take any number, add 25 to it, multiply the sum by 3, subtract 60, and divide the difference by 3. Finally, add 5.

   b. Simplify your answer in part (a).

3. Discover a possible tile pattern in the following sequence and answer the following:

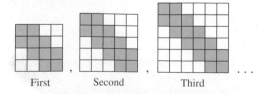

| First | Second | Third |

   a. How many shaded tiles are there in the nth figure of your pattern?

   b. How many white tiles are there in the nth figure of your pattern?

4. If a school has w women and m men and you know that there are 100 more men than women, write an algebraic equation relating w and m.

5. Suppose there are 15 more chairs (c) and tables (t) in a classroom and there are 15 more chairs than tables. Write an algebraic equation relating c and t.

6. In the following, write an expression in terms of the given variables that represents the indicated quantity:

   a. The cost of having a plumber spend h hr at your house if the plumber charges \$30 for coming to the house and \$$x$ per hour for labor

   b. The amount of money in cents in a jar containing some nickels and d dimes and some quarters if there are 4 times as many nickels as dimes and twice as many quarters as nickels

   c. The sum of three consecutive integers if the greatest integer is x

   d. The amount of bacteria after n min if the initial amount of bacteria is q and the amount of bacteria triples every 30 sec. (*Hint:* The answer should contain q as well as n.)

   e. The temperature t hr ago if the present temperature is 40°F and each hour it drops by 3°F

   f. Pawel's total earnings after 3 yr if the first year his salary was s dollars, the second year it was \$5000 higher, and the third year it was twice as much as the first year

   g. The sum of three consecutive even whole numbers if the greatest is x

7. Ryan is building matchstick square sequences so that one square is added to the right each time, as shown. How many matchsticks will he use for the *n*th figure and for the figure one before the *n*th?

8. Write an algebraic equation relating the variables described in each of the following situations:
   a. The pay, *P*, for *t* hr if you are paid $*d* an hour
   b. The pay, *P*, for *t* hr if you are paid $15 for the first hour and $*k* for each additional hour
   c. The total pay, *P*, for a visit and *t* hr of gardening if you are paid $20 for the visit and $10 for each hour of gardening

d. The total cost, *C*, of membership in a health club that charges a $300 initiation fee and $4 for each of *n* days attended
e. The cost, *C*, of renting a midsized car for 1 day of driving *m* mi if the rent is $30 per day plus 35¢ per mile.

9. A teacher instructed her class as follows:

   Take any odd number, multiply it by 4, add 16, and divide the result by 2. Subtract 7 from the quotient and tell me your answer. I will tell you the original number.

   Explain how the teacher was able to tell each student's original number.

10. Matt has twice as many stickers as David. If David has *d* stickers and Matt *m* stickers, and Matt gives David 10 stickers, how many stickers does each have in terms of *d* ?

Mathematical Connections 4-1

Communication

1. Students were asked to write an algebraic expression for the sum of three consecutive natural numbers. One student wrote $x + (x + 1) + (x + 2) = 3x + 3$. Another wrote $(x - 1) + x + (x + 1) = 3x$. Explain who is correct and why.

Open-Ended

2. A teacher instructed her class to take any number and perform a series of computations using that number. The teacher was able to tell each student's original number by subtracting 1 from the student's answer. Create similar instructions for students so that the teacher needs to do only the following to obtain the student's original number:
   a. Add 1 to the answer.
   b. Multiply the answer by 2.
   c. Multiply the answer by 1.
3. Create instructions for students similar to those in problem 2, involving addition, subtraction, multiplication, and division so that following the instructions, each student will get her original number.

Cooperative Learning

4. Examine several elementary school textbooks for grades 1 through 5 and report on which algebraic concepts involving variables are introduced in each.

Questions from the Classroom

5. A student claims that the sum of five consecutive integers is equal to 5 times the middle integer and would like to know if this is always true, and if so, why. He would like to know if the statement generalizes to the sum of five consecutive terms in any arithmetic sequence. How do you respond?

6. A student writes $a \cdot (b \cdot c) - (a \cdot b) \cdot (a \cdot c)$. How do you respond?
7. A student wonders if sets can ever be considered as variables. What do you tell her?
8. A student thinks that if *A* and *B* are sets, then the statements $A \cup B = B \cup A$ and $A \cap B = B \cap A$ are algebraic generalizations of set properties in a way similar to the statements $a + b = b + a$ and $ab = ba$ are generalizations of arithmetic properties of numbers. How do you respond?

Third International Mathematics and Science Study (TIMSS) Question

☐ represents the number of magazines that Lina reads each week. Which of these represents the total number of magazines that Lina reads in 6 weeks?
   a. 6 + ☐
   b. 6 × ☐
   c. ☐ + 6
   d. (☐ + ☐) × 6

TIMSS 2003, Grade 4

National Assessment of Educational Progress (NAEP) Question

N stands for the number of hours of sleep Ken gets each night. Which of the following represents the number of hours of sleep Ken gets in 1 week?
   a. *N* + 7
   b. *N* − 7
   c. *N* × 7
   d. *N* ÷ 7

NAEP, Grade 4, 2005

4-2   Equations

Variables are frequently associated with equations. When variables are thought of as un-knowns, we can consider expressions such as $w + c = 7$. The equal sign indicates that the values on both sides of the equation are the same even though they do not look the same. As pointed out in the Research Note, students sometimes mistakenly think of the equal sign only as a symbol for separation.

To solve equations, we need several properties of equality. Children discover many of these by using a balance scale. For example, consider two weights of amounts a and b on the balances, as in Figure 4-5(a). If the balance is level, then $a = b$. When we add an equal amount of weight, c, to both sides, the balance is still level, as in Figure 4-5(b).

♦ *Research Note*

Van de Walle writes that students tend to see an equation such as $3x + 7 = 5 + 9$ as having two sepa-rate sides with things to do, rather than two names for the same thing. The equal sign is often viewed as a symbol used to separate a problem from its answer (Van de Walle 2007).   ♦

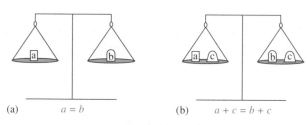

(a) $a = b$ (b) $a + c = b + c$

Figure 4-5

This demonstrates that *if $a = b$, then $a + c = b + c$*, which is *the addition property of equality*.

Similarly, if the scale is balanced with amounts a and b, as in Figure 4-6(a), and we put ad-ditional a's on one side and an equal number of b's on the other side, the scale remains level, as in Figure 4-6(b).

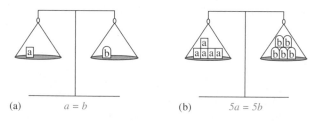

(a) $a = b$ (b) $5a = 5b$

Figure 4-6

Figure 4-6 suggests that *if c is any natural number and $a = b$, then $ac = bc$*, which is the *multiplication property of equality*. These properties are summarized next. They are true for all numbers, but in this chapter we use them just for whole numbers.

REMARK
a. In algebra it is common to omit the multiplication sign in a product when letters are involved. Thus, we write ac instead of $a \cdot c$ and $3x$ instead of $3 \cdot x$
b. Under certain conditions the properties listed here can be proved and therefore are called theorems.

Theorem 4–1:   The Addition Property of Equality

For any numbers a, b, and c, if $a = b$, then $a + c = b + c$.

The Multiplication Property of Equality   For any numbers a, b, and c, if $a = b$, then $ac = bc$.

The properties imply that we may add the same number to both sides of an equation or multiply both sides of the equation by the same number without affecting the equality. A new statement results from reversing the order of the *if* and *then* parts of the addition property of equality. The new statement is the *converse* of the original statement. In the case of the addition property, the converse is a true statement. The converse of the multiplication property of equality is also true when $c \neq 0$. These properties are summarized next.

Theorem 4–2:   Cancellation Properties of Equality

1. For any numbers a, b, and c, if $a + c = b + c$, then $a = b$.
2. For any numbers a, b, and c, with $c \neq 0$, if $ac = bc$, then $a = b$.

REMARK Notice that if we divide each side of the equation $ac = bc$ by c we get $a = b$.

Equality is not affected if we substitute a number for its equal. This property is referred to as the **substitution property**. Examples of substitution follow:

1. If $a + b = c + d$ and $d = 5$, then $a + b = c + 5$.
2. If $a + b = c + d$, $b = e$, and $d = f$, then $a + e = c + f$.

Using the substitution property we can see that equations can be added or subtracted "side by side"; that is, we have the following:

Theorem 4–3:   Addition and Subtraction Property of Equations

If $a = b$ and $c = d$, then $a + c = b + d$ and $a - c = b - d$.

This property can be justified as follows: Using the addition property of equality, if $a = b$, then $a + c = b + c$. Substituting d for c on the right side, we get $a + c = b + d$. Similarly, $a - c = b - d$.

Widely used properties in early grades are the commutative and associative properties of addition and multiplication that can be performed in any order, for example, $2 + 8 = 8 + 2$ and $2 \cdot 8 = 8 \cdot 2$. Also, $2 + (8 + 5) = (2 + 8) + 5$, and in general, we have the following:

Theorem 4–4:   Commutative Properties of Addition and Multiplication

$$a + b = b + a, \quad ab = ba$$

Theorem 4–5:   Associative Properties of Addition and Multiplication

$$(a + b) + c = a + (b + c), \quad (ab)c = a(bc)$$

 On the student page that follows you will see two of the properties just discussed.

School Book Page MULTIPLICATION PATTERNS

Lesson 2-1

Key Idea
You can use basic facts, patterns, and properties to multiply mentally.

Vocabulary
• factor
• product
• Commutative Property of Multiplication
• Associative Property of Multiplication

Materials
• calculator

TEST TALK

Think It Through
I can **look for a pattern** to find a rule.

Multiplication Patterns

LEARN

Activity

What's the pattern?

a. Use a calculator to find each product.

3 × 5	3 × 50	3 × 500
30 × 5	30 × 50	30 × 500
300 × 5	300 × 50	300 × 500

Factors are numbers that are multiplied to get a *product*.

b. Find the following products without a calculator. Then check your answers with a calculator.

5 × 8, 50 × 8, 50 × 80, 500 × 8, 500 × 80, 500 × 800

c. Describe a rule that tells how to find each product.

WARM UP

1. 4 × 8   2. 6 × 8
3. 7 × 7   4. 7 × 8
5. 6 × 9   6. 7 × 9

How can properties help you multiply more easily?

Commutative Property of Multiplication

You can change the order of the factors.

34 × 8 = 8 × 34

Associative Property of Multiplication

You can change the grouping of factors.

(7 × 25) × 4 = 7 × (25 × 4)

Example A

Find 20 × 5 × 6.

Using the Associative Property, you can think:

(20 × 5) × 6 = 100 × 6 = 600 OR
20 × (5 × 6) = 20 × 30 = 600.

Example B

Find 2 × 70 × 50.

Use the properties to change the order and the groupings.

2 × 70 × 50 = 2 × (70 × 50)
= 2 × (50 × 70)
= (2 × 50) × 70
= 100 × 70
= 7,000

✔ **Talk About It**

1. How is the Associative Property used in Example B?

2. How is the Commutative Property used in Example B?

3. Can you use the Associative Property for 2 × (5 + 6)? Explain.

66

Source: Mathematics, Diamond Edition, Grade Five, Scott Foresman-Addison Wesley 2008 (p. 66).

NOW TRY THIS 4-2 Answer the three questions at the bottom of the student page.

A vital property used in solving equations—and in algebraic thinking in general—is the *distributive property of multiplication over addition*. This property is used in the early grades in multiplication problems such as finding the product $12 \cdot 65$. If we think about the product as the number of plants in 12 rows of 65 plants each, that number equals the number of plants in $10 + 2$ rows, or the number in 10 rows plus the number in 2 rows, or $10 \cdot 65 + 2 \cdot 65$. Thus, $12 \cdot 65 = (10 + 2)65 = 10 \cdot 65 + 2 \cdot 65$.

> **Theorem 4–6:   The Distributive Property of Multiplication over Addition**
>
> For all numbers a, b, and c, $a(b + c) = ab + ac$.

Notice that we used this property in the solution of Example 4-3, where we wrote $4(n + 15)$ as $4n + 4 \cdot 15$. Also notice the analogous property among sets: $A \cap (B \cup C) = (A \cap B) \cup (A \cap C)$. Similarly, we have

> **Theorem 4–7:   The Distributive Property of Multiplication over Subtraction**
>
> For all numbers a, b, and c, $a(b - c) = ab - ac$.

Notice that using the commutative property of multiplication, each of the distributive properties above can be written in the equivalent forms

$$(b + c)a = ba + ca, \text{ and}$$
$$(b - c)a = ba - ca$$

When the distributive properties are written from right to left, we refer to them as *factoring*. Thus, $ab + ac = a(b + c)$ and $ab - ac = a(b - c)$. We say that a has been "factored out."

Solving Equations

Part of algebraic thinking involves operations on numbers and other elements represented by symbols. Finding solutions to equations is a major part of algebra. As the Research Note points out, use of tangible objects can increase students engagement and comprehension when they work with equations. The balance-scale model is an excellent way to foster understanding of the basic concepts used in solving equations and inequalities, and equations can be explored using a balance scale. Inequalities tilt the scale.

Research Note

Use of objects and scale models resulted in significant gains and improved attitude in solving equations (Quinlan 1992). ◆

For example, consider Figure 4-7. If we release the pan on the left, what will happen? Upon release, the scale will tilt down on the right side and we have an *inequality*, $2 \cdot 3 < 3 + (2 \cdot 2)$.

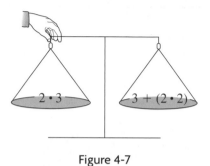

Figure 4-7

Next consider Figure 4-8. If we release the pan, then the sides will balance and we have the *equality* $2 \cdot 3 = (1 + 1) + 4$.

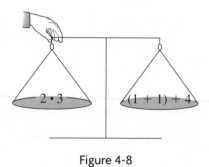

Figure 4-8

A balance scale can also be used to reinforce the idea of a replacement set for a variable. Name some solutions in Figure 4-9 that will keep the scale balanced. For example, $3 \cdot 2$ balances $2 \cdot 3$, $3 \cdot 6$ balances $2 \cdot 9$, and so on. Do you see any patterns in the numbers that will balance the scale?

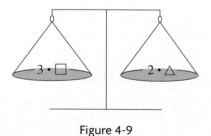

Figure 4-9

Other types of balance-scale problems can also help students get ready for algebra. Work through Now Try This 4-3 before proceeding.

NOW TRY THIS 4-3 What are the values of □ and △ in Figure 4-10?

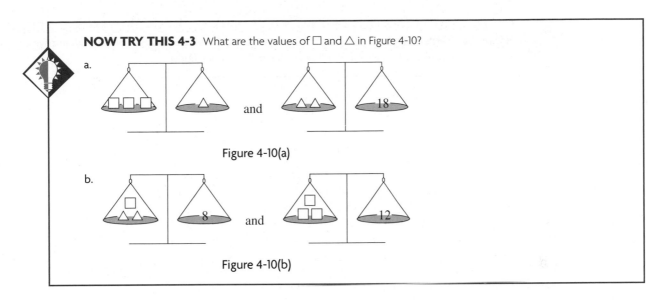

a.

and

Figure 4-10(a)

b.

and

Figure 4-10(b)

To solve equations, we may use the properties of equality developed earlier. Consider $3x - 14 = 1$. Put the equal expressions on the opposite pans of the balance scale. Because the expressions are equal, the pans should be level, as in Figure 4-11.

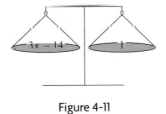

Figure 4-11

To solve for x, we use the properties of equality to manipulate the expressions on the scale so that after each step, the scale remains level and, at the final step, only an x remains on one side of the scale. The number on the other side of the scale represents the solution to the original equation. To find x in the equation of Figure 4-11, consider the scales pictured in successive steps in Figure 4-12. In Figure 4-12 each successive scale represents an equation that is equivalent to the original equation; that is, each has the same solution as the original. The last scale shows $x = 5$. To check that 5 is the correct solution, we substitute 5 for x in the original equation. Because $3 \cdot 5 - 14 = 1$ is a true statement, 5 is the solution to the original equation.

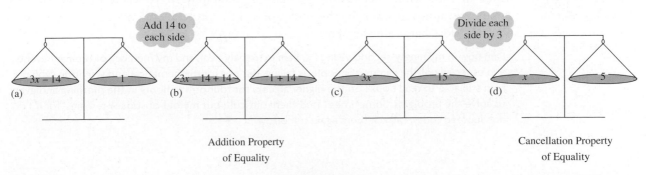

Figure 4-12

NOW TRY THIS 4-4 Notice the use of concrete objects in solving equations on the student page and answer the "Talk About It" question that follows the balance-scale model.

Example 4-6

Solve each of the following for x:

a. $x + 4 = 20$
b. $3x = x + 10$
c. $4x + 5x = 99$
d. $4(x + 3) + 5(x + 3) = 99$

Solution **a.** $x + 4 = 20$ implies $(x + 4) - 4 = 20 - 4$; $x = 16$.

b. $3x = x + 10$ implies $3x - x = 10 + x - x$; $3 \cdot x - 1 \cdot x = 10$; $(3 - 1)x = 10$; $2x = 10$; $x = 5$.

c. $4x + 5x = 99$; $(4 + 5)x = 99$; $9x = 99$; $x = 11$

d. We could multiply out and get

$$4(x + 3) + 5(x + 3) = 99; 4x + 12 + 5x + 15 = 99$$
$$9x + 27 = 99; 9x + 27 - 27 = 99 - 27; 9x = 72, x = 8$$

Or we could have thought of $x + 3$ as a new unknown $\square$. So if $x + 3 = \square$, we get $4 \cdot \square + 5 \cdot \square = 99$, so $9 \cdot \square = 99$, $\square = 11$, which implies $x + 3 = 11$ and hence $x = 8$.

Mary Fairfax Somerville

Historical Note

Born into a well-to-do family in Scotland, Mary Fairfax (1780–1872) first studied simple arithmetic only briefly at the age of 13. Also at about this time she came upon some mysterious symbols in a women's fashion magazine and, after persuading her brother's tutor to purchase some elementary literature for her on the subject, began her study of algebra. Later as a young mother and widow, she obtained a small library of works to provide her with a sound background in mathematics. Throughout the rest of her life, Somerville distinguished herself as a skilled scientific writer respected among her colleagues, publishing a number of works. Her last scientific book, *Molecular and Microscopic Science*, was published in 1869 when she was 89. In her autobiography Mary wrote of how she "was sometimes annoyed when in the midst of a difficult problem" a visitor would enter. Shortly before her death she wrote

I am now in my ninety-second year, . . . , I am extremely deaf, and my memory of ordinary events, and especially of the names of people, is failing, but not for mathematical and scientific subjects. I am still able to read books on the higher algebra for four or five hours in the morning and even to solve the problems. Sometimes I find them difficult, but my old obstinacy remains, for if I do not succeed today, I attack them again tomorrow.   ◆

School Book Page

SOLVING EQUATIONS WITH WHOLE NUMBERS

Lesson 1-15

Algebra

Key Idea
You can use inverse operations and the properties of equality to solve equations.

Vocabulary
• equation (p. 44)
• inverse operations (p. 45)
• properties of equality (p. 44)

Think It Through
I can **think of a pan balance** to help solve the problem.

Solving Equations with Whole Numbers

WARM UP
Explain how to get each variable alone.
1. $12x = 60$
2. $d - 10 = 10$
3. $32 = 8 + a$

LEARN

How can you solve an equation?

When you **solve** an equation, you find the value of the variable that makes the equation true.

Example A

Wynn sold 6 sketches, each for the same amount, and made $180 in sales. How much did he charge for each sketch?

Let s equal the amount for each sketch.

Then the equation is $6s = 180$.

What You Write	Balancing the Pans	
$6s = 180$		The pans are balanced.
$6s \div 6 = 180 \div 6$		180 has been separated into 6 equal parts.
$s = 30$		Each ⬚ equals ⬚.

Wynn charged $30 for each sketch.

Talk About It

1. Why was each side of the equation in Example A divided by 6?

How can you check your answer?

To check your answer, substitute it for the variable in the original equation. In Example A, substitute 30 for s in $6s = 180$.

Check: $6s = 180$
      $6(30) = 180$
        $180 = 180$ When both sides of the equation can be simplified to the same number, the value of the variable is correct.

48

Source: Mathematics, Diamond Edition, Grade 6, Scott Foresman-Addison Wesley 2008 (p. 48).

Application Problems

The simple model in Figure 4-13 demonstrates a method for solving application problems. Formulate the problem with a mathematical model, solve that mathematical model, and then interpret the solution in terms of the original problem.

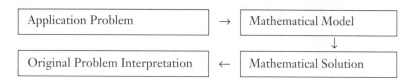

| Application Problem | → | Mathematical Model |
| Original Problem Interpretation | ← | Mathematical Solution |

Figure 4-13

An example of this model at the third-grade level appears in Figure 4-14.

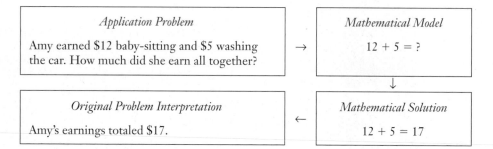

| *Application Problem*
 Amy earned \$12 baby-sitting and \$5 washing the car. How much did she earn all together? | → | *Mathematical Model*
 $12 + 5 = ?$ |
| *Original Problem Interpretation*
 Amy's earnings totaled \$17. | ← | *Mathematical Solution*
 $12 + 5 = 17$ |

Figure 4-14

NOW TRY THIS 4-5 Read the following student page from a fourth-grade textbook and answer the "Talk About It" questions.

We can apply Pólya's four-step problem-solving process to solving word problems in which the use of algebraic thinking is appropriate. In Understanding the Problem, we identify what is given and what is to be found. In Devising a Plan, we assign letters to the unknown quantities and try to translate the information in the problem into a model involving equations. In Carrying Out the Plan, we solve the equations or inequalities. In Looking Back, we interpret and check the solution in terms of the original problem.

In the following problems, we demonstrate Pólya's four-step problem-solving process.

School Book Page

TRANSLATING WORDS TO EQUATIONS

Lesson 12-2

Algebra

Key Idea
Equations can help you solve problems.

Vocabulary
• equation (p. 100)

TEST TALK

Think It Through
• I can **draw a picture** and **write an equation** to represent the situation.
• I can **try, check, and revise** to solve an equation.

Translating Words to Equations

LEARN

How do you write an equation?

A Great Dane puppy weighed 4 pounds when it was born. After 3 weeks, it weighed 6 pounds.

Example

Write an equation to show how much weight the puppy gained in 3 weeks.

| 4 lb | p | 6 lb |

| Weight at birth | + | Pounds gained | = | Weight at 3 weeks |
| 4 | + | p | = | 6 |

The **equation** $4 + p = 6$ shows how much weight the puppy gained in 3 weeks.

WARM UP

Evaluate each expression.

1. $n + 7$ for $n = 14$

2. $n - 8$ for $n = 28$

3. $5n$ for $n = 7$

✔ Talk About It

1. What does p stand for in the Example?

2. After 5 weeks, the puppy weighed 8 pounds. Write an equation to show how much weight the puppy gained in 5 weeks.

CHECK ✔

For another example, see Set 12-2 on p. 728.

Write an equation for each sentence.

1. p pages plus 7 pages equals 17 pages.

2. 8 less than k is 15.

3. 9 times n is 27.

4. 36 divided by y is 12.

5. **Reasoning** Kate wanted to find how many centimeters equal 80 millimeters. She used the equation $10x = 80$. Is Kate's equation correct? Explain.

690

Source: Mathematics, Diamond Edition, Grade Four, Scott Foresman-Addison Wesley 2008 (p.690).

Problem Solving   Overdue Books

Bruno has five books overdue at the library. The fine for overdue books is 10¢ a day per book. He remembers that he checked out an astronomy book a week before he checked out four novels. If his total fine was $8.70, how long was each book overdue?

Understanding the Problem   Bruno has five books overdue. He checked out an astronomy book seven days earlier than the four novels, so the astronomy book is overdue seven days more than the novels. The fine per day for each book is 10¢, and the total fine was $8.70. We need to find out how many days each book is overdue.

Devising a Plan   Let d be the number of days that each of the four novels is overdue. The astronomy book is overdue seven days longer, that is, $d + 7$ days. To *write an equation* for d, we express the total fine in two ways. The total fine is $8.70. This fine in cents equals the fine for the astronomy book plus the fine for the four novels.

$$\text{Fine for each of the novels} = \underbrace{\text{Fine per day}}_{10} \underbrace{\text{times}}_{\cdot} \underbrace{\text{number of overdue days}}_{d}$$

$$\text{Fine for the four novels} = \underbrace{\text{1 day's fine for novels}}_{4\cdot10} \underbrace{\text{times}}_{\cdot} \underbrace{\text{number of overdue days}}_{d}$$

$$= (4 \cdot 10)d$$
$$= 40d$$

$$\text{Fine for the astronomy book} = \underbrace{\text{Fine per day}}_{10} \underbrace{\text{times}}_{\cdot} \underbrace{\text{number of overdue days}}_{(d + 7)}$$

$$= 10(d + 7)$$

Because each of the expressions is in cents, we need to write the total fine of $8.70 as 870¢ to produce the following:

$$\text{Fine for the four novels} + \text{Fine for the astronomy book} = \text{Total fine}$$
$$40d \qquad + \qquad 10(d + 7) \qquad = 870$$

Carrying Out the Plan   Solve the equation for d.

$$40d + 10(d + 7) = 870$$
$$40d + 10d + 70 = 870$$
$$50d + 70 = 870$$
$$50d = 870 - 70$$
$$50d = 800$$
$$d = 16$$

Thus, each of the four novels was 16 days overdue, and the astronomy book was overdue $d + 7$, or 23, days.

Looking Back   To check the answer, follow the original information. Each of the four novels was 16 days overdue, and the astronomy book was 23 days overdue. Because the fine was 10¢ per day per book, the fine for each of the novels was $16 \cdot 10$¢, or 160¢. Hence, the fine for all four novels was $4 \cdot 160$¢, or 640¢. The fine for the astronomy book was $23 \cdot 10$¢, or

230¢. Consequently, the total fine was 640¢ + 230¢, or 870¢, which agrees with the given information of $8.70 as the total fine.

The problem can also be solved without algebra. One way is to notice that the astronomy book was overdue for 7 days for a fine of 70¢ before the other four books were overdue. Thus, 870¢ − 70¢, or 800¢, is the fine for the remaining five books. Therefore, the fine for one book is 800¢/5, or 160¢. Because the fine is 10¢ per day, each book was overdue 160/10, or 16, days. The astronomy book was checked out a week earlier and hence was overdue 7 days longer for 23 days.

Problem Solving   Newspaper Delivery

In a small town, three children deliver all the newspapers. Abby delivers 3 times as many papers as Bob, and Connie delivers 13 more than Abby. If the three children delivered a total of 496 papers, how many papers does each deliver?

Understanding the Problem   The problem asks for the number of papers that each child delivers. It gives information that compares the number of papers that each child delivers as well as the total number of papers delivered in the town.

Devising a Plan   Let a, b, and c be the number of papers delivered by Abby, Bob, and Connie, respectively. We translate the given information into algebraic *equations* as follows:

$$\text{Abby delivers 3 times as many papers as Bob: } a = 3b$$
$$\text{Connie delivers 13 more papers than Abby: } c = a + 13$$
$$\text{Total delivery is 496: } a + b + c = 496$$

To reduce the number of variables, substitute $3b$ for a in the second and third equations:

$$c = a + 13 \qquad \text{becomes} \quad c = 3b + 13$$
$$a + b + c = 496 \quad \text{becomes} \quad 3b + b + c = 496$$

Next, make an equation in one variable, b, by substituting $3b + 13$ for c in the equation $3b + b + c = 496$, solve for b, and then find a and c.

Carrying Out the Plan

$$3b + b + 3b + 13 = 496$$
$$7b + 13 = 496$$
$$7b = 483$$
$$b = 69$$

Thus, $a = 3b = 3 \cdot 69 = 207$. Also, $c = a + 13 = 207 + 13 = 220$. So, Abby delivers 207 papers, Bob delivers 69 papers, and Connie delivers 220 papers.

Looking Back   To check the answers, follow the original information, using $a = 207$, $b = 69$, and $c = 220$. The information in the first sentence, "Abby delivers 3 times as many papers as Bob," checks, since $207 = 3 \cdot 69$. The second sentence, "Connie delivers 13 more papers than Abby," is true because $220 = 207 + 13$. The information on the total delivery checks, since $207 + 69 + 220 = 496$.

NOW TRY THIS 4-6

1. Solve the *Newspaper Delivery* problem above by introducing only one unknown for the number of newspapers Bob delivered.
2. Recall that Example 4-5 was solved without using equations, and following the solution we mentioned that the problem can be solved using the strategy of *writing an equation*. You are ready now to approach the problem in this way. Suppose the price of the cantaloupe in dollars is *x*, the price of the flower and the vase *y*, and the watermelon *z*. Check that from the information in Figure 4-3(a) we get $x + y = 8$, and write the equations corresponding to parts (b) and (c) of the figure. Solve the equations by reducing them to two equations with two unknowns and then one equation with one unknown.

Assessment 4-2A

1.
If 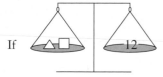 12

and 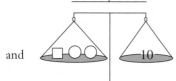 18

and 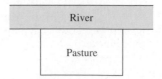 10

what is the value of each shape? Tell why.

2. Solve each of the following, if possible:
   a. $x - 3 = 21$
   b. $2x + 5 = x + 25$
   c. $2x + 5 = 3x - 4$
   d. $5(2x + 1) + 7(2x + 1) = 84$
   e. $3(2x - 6) = 4(2x - 6)$

   Solve problems 3 through 10 by setting up and solving an equation.

3. Ryan is building matchstick square sequences so that one square is added to the right each time a new figure is formed, as shown. He used 67 matchsticks to form the last figure in his sequence. How many squares are in his last figure?

4. For a particular event, 812 tickets were sold for a total of $1912. If students paid $2 per ticket and nonstudents paid $3 per ticket, how many student tickets were sold?

5. An estate of $486,000 is left to three siblings. The eldest receives 3 times as much as the youngest. The middle sibling receives $14,000 more than the youngest. How much did each receive?

6. A 10 ft board is to be cut into three pieces, two equal-length ones and the third 3 in. shorter than each of the other two. If the cutting does not result in any length being lost, how long are the pieces?

7. A box contains 67 coins, only dimes and nickels. The amount of money in the box is $4.20. How many dimes and how many nickels are in the box?

8. Miriam is 10 years older than Ricardo. Two years ago, Miriam was 3 times as old as Ricardo is now. How old are they now?

9. In a college, 15 times as many undergraduate students than graduate students are enrolled. If the total student enrollment at the college is 10,000, how many graduate students are there?

10. A farmer has 700 yd of fencing to enclose a rectangular pasture for her goats. Since one side of the pasture borders a river, that side does not need to be fenced. The side parallel to the river must be twice as long as the side perpendicular to the river. Find the dimensions of the rectangular pasture.

River
Pasture

Assessment 4-2B

1.

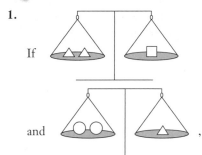

If

and

,

a. Which shape weighs the most? Tell why.
b. Which shape weighs the least? Tell why.

2. Solve each of the following, if possible:
 a. $3x + 13 = 2x + 100$
 b. $2x + 5 = 2(x + 5)$
 c. $7(3x + 6) + 5(3x + 6) = 144$
 d. $22 - x = 3x + 6$
 e. $22 - (2x - 6) = 3(2x - 6) + 6$
 f. $5(2x - 10) = 4(2x - 10)$

Solve problems 3 through 11 by setting up and solving an equation.

3. Ryan is building matchstick square sequences, as shown. He used 599 matchsticks to form the last two figures in his sequence. How many matchsticks did he use in each of the last two figures?

4. The sum of two consecutive terms in the arithmetic sequence 1, 4, 7, 10, . . . is 299; find these two terms.

5. The sum of the first two terms of a geometric sequence is 100 times the first term. What is the common ratio?

6. An estate of $1,000,000 is left to four siblings. The eldest is to receive twice as much as the youngest. The other two siblings are each to receive $16,000 more than the youngest. How much will each receive?

7. Miriam is four years older than Ricardo. Ten years ago Miriam was 3 times as old as Ricardo was then.
 a. How old are they now?
 b. Determine whether your answer is correct by checking that it satisfies the conditions of the problem.

8. In a college there are 13 times as many students as professors. If together the students and professors number 28,000, how many students are there in the college?

9. A farmer has 800 yd of fencing to enclose a rectangular pasture. Since one side of the pasture borders a river, that side does not need to be fenced. If the side perpendicular to the river must be twice as long as the side parallel to the river, what are the dimensions of the rectangular pasture?

10. Ten years from now Alex's age will be 3 times her present age. Find Alex's age now.

11. Matt has twice as many stickers as David. How many stickers must Matt give David so that they will each have 120 stickers? Check that your answer is correct.

Mathematical Connections 4-2

Communication

1. Students were asked to find three consecutive whole numbers whose sum is 393. One student wrote the equation $x + (x + 1) + (x + 2) = 393$. Another wrote $(x - 1) + x + (x + 1) = 393$. Can either approach work to find the answer to the question? Explain why or why not.

2. Explain how to solve the equation $3x + 5 = 5x - 3$ using a balance scale.

Open-Ended

3. Create an equation with x on both sides of the equation for each of the following:
 a. Every whole number is a solution.
 b. No whole number is a solution.
 c. 0 is a solution.

Cooperative Learning

4. Examine several elementary school textbooks for grades 1 through 5 and report on how algebraic concepts involving equations are introduced in each.

Questions from the Classroom

5. A student claims that the equation $3x = 5x$ has no solution because $3 \neq 5$. How do you respond?

6. A student claims that because in the following problem we need to find three unknown quantities, he must set up equations with three unknowns. How do you respond?

 Abby delivers twice as many papers as Jillian, and Brandy delivers 50 more papers than Abby. How many papers does each deliver if the total number of papers delivered is 550?

7. A student was told that in order to check a solution to a word problem like the one in problem 6, it is not enough to check that the solution she found satisfies the equation she set up, but rather that it is necessary to check the answer against the original problem. She would like to know why. How do you respond?

8. On a test, a student was asked to solve the equation $4x + 5 = 3(x + 15)$. She proceeded as follows:

$$4x + 5 = 3x + 45 = x + 5 = 45 = x = 40$$

Hence, $x = 40$. She checked that $x = 40$ satisfies the original equation; however, she did not get full credit for the problem. She wants to know why. How do you respond?

Review Problems

9. If the number of sophomores, juniors, and seniors combined is denoted by x and it is 3 times the number of freshmen, denoted by y, write an algebraic equation that shows the relationship.

10. Write the sum of five consecutive even numbers if the middle one is n. Simplify your answer.

11. If Julie has twice as many CDs as Jack and Tira has 3 times as many as Julie, write an algebraic expression for the number of CDs each has in terms of one variable.

12. Write an algebraic equation relating the variables described in each of the following:
    a. The pay P for t hr if you are paid $30 for the first hour and $5 more than the preceding hour for each hour thereafter.

b. Jimmy's total pay after 4 years if the first year his salary was d dollars and then each year thereafter his salary is twice as much as in the preceding year.

Third International Mathematics and Science Study (TIMSS) Questions

Ali had 50 apples. He sold some and then had 20 left. Which of these is a number sentence that shows this?
a. □ − 20 = 50
b. 20 − □ = 50
c. □ − 50 = 20
d. 50 − □ = 20

TIMSS 2003, Grade 4

The objects on the scale make it balance exactly. On the left pan there is a 1 kg weight (mass) and half a brick. On the right pan there is one brick.

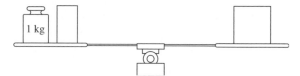

What is the weight (mass) of one brick?
a. 0.5 kg
b. 1 kg
c. 2 kg
d. 3 kg

TIMSS 2003, Grade 8

4-3   Functions

The concept of a function is central to all of mathematics and particularly to algebra, as elaborated in the following excerpt from the *Principles and Standards*:

Research Note

Students tend to view a function as either a collection of points or of ordered pairs. This limited view may hinder students' development of the concept of a function (Adams 1997). ◆

By viewing algebra as a strand in the curriculum from prekindergarten on, teachers can help students build a solid foundation of understanding and experience as a preparation for more sophisticated work in algebra in the middle grades and high school. For example, systematic experience with patterns can build up to an understanding of the idea of function (Erick Smith forthcoming), and experience with numbers and their properties lays a foundation for later work with symbols and algebraic expressions. By learning that situations often can be described using mathematics, students can begin to form elementary notions of mathematical modeling. (p. 37)

Functions can model many real-world phenomena, as we shall see in this section and in later chapters. In this section, we will explore different ways to represent functions—as *rules, machines, equations, arrow diagrams, tables, ordered pairs,* and *graphs.* It is important that students see a variety of ways of representing functions, as indicated in the research note.

Functions as Rules

The following is an example of a game called "guess my rule," often used to introduce the concept of a function.

> When Tom said 2, Noah said 5. When Dick said 4, Noah said 7. When Mary said 10, Noah said 13. When Liz said 6, what did Noah say? What is Noah's rule?

The answer to the first question may be 9, and the rule could be "Take the original number and add 3"; that is, for any number n, Noah's answer is $n + 3$.

Example 4-7

Guess the teacher's rule for the following responses:

a.

You	Teacher
1	3
0	0
4	12
10	30

b.

You	Teacher
2	5
3	7
5	11
10	21

c.

You	Teacher
2	0
4	0
7	1
21	1

Solution   a. The teacher's rule could be "Multiply the given number n by 3," that is, $3n$.

b. The teacher's rule could be "Double the original number n and add 1,"; that is, $2n + 1$.

c. The teacher's rule could be "If the number n is even, answer 0; if the number is odd, answer 1." Another possible rule is "If the number is less than 5, answer 0; if greater than or equal to 5, answer 1."

Functions as Machines

Another way to prepare students for the concept of a function is by using a "function machine." The following student page shows an example of a function machine. What goes in the machine is referred to as *input* and what comes out as *output*. Thus, on the student page, if the input to the function is 2, the output is 110. Note that the output here is denoted by d. In later grades, a special notation for the output is used. For any input element x, the output is denoted $f(x)$*, read "f of x". For the function $d = 55t$ in the example on the student page, if f is the function then when the input is 2, the output could be written as $d(2)$. Because the output is 110, we have $d(2) = 110$. Since the function works according to the rule $d(t) = 55t$, we have $d(2) = 2 \cdot 55 = 110$.

Historical Note

Leonhard Euler

The Babylonians of Mesopotamia (ca. 2000 BCE) developed a precursor to what we today call a function. To them, a function was a table or a correspondence. Two tablets found at Senkerah on the Euphrates in 1854 give squares of the numbers up to 59 and cubes of the numbers up to 32.

In the seventeenth century the idea of a function underwent further development. In his book *Geometry* (1637), René Descartes (1596–1650) used the concept to describe many mathematical relationships. Almost 50 years after the publication of Descartes' book, Gottfried Wilhelm Leibniz (1646–1716) introduced the term *function*. The idea of a function was further formalized by Leonhard Euler (pronounced "oiler," 1707–1783), who introduced the notation for a function, $y = f(x)$. Further contributions to the concept came from the mathematicians Joseph-Louis Lagrange (1736–1813) and Jean Joseph Fourier (1768–1830). ◆

* When a function designates a specific quantity, such as d (distance) on the student page, $d(x)$ can be used instead of $f(x)$.

School Book Page FUNCTION RULES

Algebra

9-4 Function Rules

Source: Mathematics, Course 2, Pearson Prentice Hall 2008 (p. 452).

✔ Check Skills You'll Need

1. Vocabulary Review
Why is 5 + 2 not an *algebraic expression*?

Evaluate $-4x + 1$ for each value of x.

2. -2 **3.** 0

4. $\frac{1}{4}$ **5.** $-\frac{1}{4}$

GO for Help
Lesson 4-1

What You'll Learn

To write and evaluate functions

◀» **New Vocabulary** function

Why Learn This?

The distance you travel in a car depends on the driving time. When one quantity depends on another, you say that one is a *function* of the other. So distance is a function of time. You can use functions to help you make predictions.

In the diagram at the right, an input goes through the "function machine" to produce an output.

Input (time)

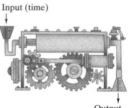

Output (distance)

A **function** is a relationship that assigns exactly one output value for each input value.

EXAMPLE   Writing a Function Rule

1 **Cars** You are traveling in a car at an average speed of 55 mi/h. Write a function rule that describes the relationship between the time and the distance you travel.

You can *make a table* to solve this problem.

Input: time (h)	1	2	3	4
Output: distance (mi)	55	110	165	220

distance in miles $= 55 \cdot$ time in hours   ← **Write the function rule in words.**

$d = 55t$ ← **Use variables d and t for distance and time.**

✔ Quick Check

1. Write a function rule for the relationship between the time and the distance you travel at an average speed of 62 mi/h.

> **REMARK** On most graphing calculators, the function notation used is Y1, Y2, Y3, ..., and so on. Here, Y1 acts like d as described on the student page or like $d(x)$ if the function rule is written in terms of x.

Example 4-8
▶▶▶▶▶▶▶▶▶▶

Consider the function machine in Figure 4-15. For the function named f, what will happen if the numbers 0, 1, 3, and 6 are entered?

Solution If the numbers output are denoted by $f(x)$, the corresponding values can be described using Table 4-1. Notice that $f(4)$ in Figure 4-15 is the output from the "f" function when the input is 4.

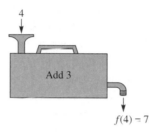

Table 4-1

x	$f(x)$
0	3
1	4
3	6
4	7
6	9

Figure 4-15

Functions as Equations

We can write an equation to depict the rule in Example 4-8 as follows. If the input is x, the output is $x + 3$; that is, $f(x) = x + 3$. The output values can be obtained by substituting the values 0, 1, 3, 4, and 6 for x in $f(x) = x + 3$, as shown:

$$f(0) = 0 + 3 = 3$$
$$f(1) = 1 + 3 = 4$$
$$f(3) = 3 + 3 = 6$$
$$f(4) = 4 + 3 = 7$$
$$f(6) = 6 + 3 = 9$$

Notice the input and output terminology in the function machine representation on the last student page.

In many applications, both the inputs and the outputs of a function machine are numbers. However, inputs and outputs can be any objects. For example, consider a particular candy machine that accepts only 25¢, 50¢, and 75¢ and outputs one of three types of candy with costs of 25¢, 50¢, and 75¢, respectively. A function machine associates *exactly one output with each input*. If you enter some element x as input and obtain $f(x)$ as output, then every time you enter the same x as input, you will obtain the same $f(x)$ as output. The idea of a function machine associating exactly one output with each input according to some rule leads to the following definition.

E-Manipulative Activity

Additional practice with function machines can be found in *Function Machine* on the E-Manipulative Disk.

> **Definition of Function**
>
> A **function** from set A to set B is a correspondence from A to B in which each element of A is paired with one, and only one, element of B.

The set A in the previous definition is the set of all allowable inputs and is the **domain** of the function. The set B, the **codomain**, is any set that includes all the possible outputs. The set of all outputs is the **range** of the function. Set B in the definition is any set that includes the range and can be the range itself. The distinction is made for convenience sake, since sometimes the range is not easy to find. For example, consider corresponding to each student at a university the student's identification number. This is a function from the set of all students to the set W of whole numbers. Thus the codomain is a subset of the range. The range in this case is all the identification numbers of students who are enrolled at the university. The range is a proper subset of the set W. Normally, *if no domain is given to describe a function, then the domain is assumed to contain all elements for which the rule is meaningful.* As the Research Note points out, these concepts can be difficult for students to grasp.

A calculator contains many functions. Suppose a student enters $\boxed{9}\boxed{\times}\boxed{K}$ on the calculator that has a constant key, $\boxed{K}$. The student then presses $\boxed{0}$ and hands the calculator to another student. The other student is to determine the rule by entering various numbers followed by the $\boxed{=}$ key. Machines with an automatic constant feature can also be used.

Other buttons on a calculator are function buttons. For example, the $\boxed{\pi}$ button always displays an approximation for π, such as 3.1415927; the $\boxed{+/-}$ button either displays a negative sign in front of a number or removes an existing negative sign; and the $\boxed{x^2}$ and $\boxed{\sqrt{}}$ buttons square numbers and take the square root of numbers, respectively.

Are all input-output machines function machines? Consider the machine in Figure 4-16. For any natural-number input x, the machine outputs a number that is less than x. If, for example, you input the number 10, the machine may output 9, since 9 is less than 10. If you input 10 again, the machine may output 3, since 3 is less than 10. Such a machine is not a function machine because the same input may give different outputs.

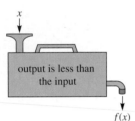

x

output is less than the input

$f(x)$

Figure 4-16

Example 4-9

A high-end bicycle manufacturer incurs a daily fixed cost of $1400 for overhead expenses and a cost of $500 per bike manufactured. Answer the following:

a. Find the cost $C(x)$ of manufacturing x bikes in a day.
b. If the manufacturer sells each bike for $700, and the profit (or loss) in producing and selling x bikes in a day is $P(x)$, find $P(x)$ in terms of x.
c. Find the break-even point, that is, the number of bikes, x, produced and sold at which break-even occurs (to break even means to make neither a profit nor a loss).

Solution   **a.** Since the cost of producing a single bike is $500, the cost of producing x bikes is $500x$ dollars. Because of the fixed cost of $1400 per day, the total cost, $C(x)$ in dollars, of producing x bikes in a given day is $C(x) = 500x + 1400$.

b. $P(x) = 700x - (500x + 1400)$
$\qquad = 200x - 1400$

c. We need to find the number of bikes x to be produced so that $P(x) = 0$; that is, we need to solve $200x - 1400 = 0$. This equation is equivalent to

$$200x = 1400 \quad \text{or} \quad x = \frac{1400}{200} \text{ or } 7$$

Thus, the manufacturer needs to produce and sell 7 bikes to break even.

Functions as Arrow Diagrams

Arrow diagrams can be used to examine whether a correspondence represents a function. This representation is normally used when sets A and B are finite sets with few elements. The following example shows how arrow diagrams can be used to examine both functions and nonfunctions.

Example 4-10

Which, if any, of the parts of Figure 4-17 exhibit a function from A to B? If a correspondence is a function from A to B, find the range of the function.

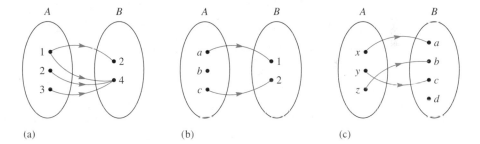

(a)   (b)   (c)

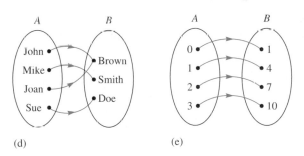

(d)   (e)

Figure 4-17

Solution a. Figure 4-17(a) does not define a function from A to B, since the element 1 is paired with both 2 and 4.
  b. Figure 4-17(b) does not define a function from A to B, since the element b is not paired with any element of B. (It is a function from a subset of A to B.)
  c. Figure 4-17(c) does define a function from A to B, since there is one, and only one, arrow leaving each element of A. The fact that d, an element of B, is not paired with any element in the domain does not violate the definition. The range is $\{a, b, c\}$ and does not include d because d is not an output of this function, as no element of A is paired with d.
  d. Figure 4-17(d) illustrates a function, since there is one and only one arrow leaving each element in A. It does not matter that an element of set B, Brown, has two arrows pointing to it. The range is {Brown, Smith, Doe}.
  e. Figure 4-17(e) illustrates a function whose range is $\{1, 4, 7, 10\}$.

Figure 4-17(e) also illustrates a one-to-one correspondence between A and B. In fact, any one-to-one correspondence between A and B defines a function from A to B as well as a function from B to A.

NOW TRY THIS 4-7   Determine which of the following are functions from the set of natural numbers to {0, 1}. Justify your reasoning.

a. For every natural-number input, the output is 0.
b. For every natural-number input, the output is 0 if the input is an even number, and the output is 1 if the input is an odd number.

Functions as Tables and Ordered Pairs

Another useful way to describe a function is with a table. Consider the information in Table 4-2 relating the amount spent on advertising and the resulting sales in a given month for a small business. Note that for *Amount of Advertising* and *Amount of Sales*, the information is actually given in thousands of dollars. We could talk about a function between the amount of dollars in *Advertising* and the amount of dollars in *Sales*, or we could simply define the function as follows: If $A = \{0, 1, 2, 3, 4\}$ and $S = \{1, 3, 5, 7, 9\}$, the table describes a function from A to S, where A represents thousands of dollars in advertising and S represents thousands of dollars in sales.

Table 4-2

Amount of Advertising (in $1000s)	Amount of Sales (in $1000s)
0	1
1	3
2	5
3	7
4	9

The function could be given using ordered pairs. When 0 is the input and 1 is the output, that is recorded as the ordered pair (0, 1). Similarly, the information in the second row is recorded as (1, 3) and the rest of the information as (2, 5), (3, 7), and (4, 9). The first component in the ordered pair is always an input element in the domain and the second is the corresponding output.

Example 4-11

Which of the following sets of ordered pairs represent functions? If a set represents a function, give its domain and range. If it does not, explain why.

a. $\{(1,2), (1,3), (2,3), (3,4)\}$ b. $\{(1,2), (2,3), (3,4), (4,5)\}$
c. $\{(1,0), (2,0), (3,0), (4,4)\}$ d. $\{(a,b) \mid a \in N \text{ and } b = 2a\}$

Solution   a. This is not a function because the input 1 has two different outputs.
b. This is a function with domain $\{1, 2, 3, 4\}$. Because the range is the possible set of outputs, the range is $\{2, 3, 4, 5\}$.

c. This is a function with domain $\{1,2,3,4\}$ and range $\{0,4\}$. The output 0 appears more than once, but this does not contradict the definition of a function in that each input corresponds to only one output.

d. This is a function with domain N and range E, the set of all even natural numbers.

Functions as Graphs

Perhaps one of the most widely recognized representations of a function is as a graph. Graphs, as visual representations of functions, appear in newspapers and books and on television. To graph the function created from Table 4-2, consider the set of ordered pairs $\{(0,1),(1,3),(2,5),(3,7),(4,9)\}$ and match each ordered pair to a point on the grid in Figure 4-18. We use the horizontal scale for the inputs and the vertical scale for the outputs and mark the point corresponding to $(0,1)$ by starting at 0 on the horizontal scale and going up 1 unit on the vertical scale. To mark the point that corresponds to $(1,3)$, we start at 0 and move 1 unit horizontally and then 3 units vertically. Marking the point that corresponds to an ordered pair is referred to as **graphing** the ordered pair. The set of all points that correspond to all the ordered pairs is the graph of the function. Notice that the graph consists of five points. The points are connected by a dashed line to emphasize that they lie on a straight line, but not every point on the line belongs to the graph.

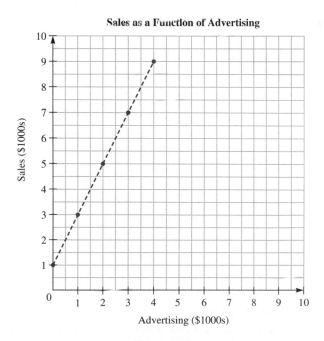

Figure 4-18

Notice Examples A and C on the two partial student pages that follow. The graph in Example C consists of points that lie on a straight line. Why may we connect the points by a dashed line rather than a solid line?

School Book Page RULES, TABLES, AND GRAPHS

Lesson 3-15

Algebra

Key Idea
Rules, tables, and graphs can be used to show how one quantity is related to another.

Vocabulary
• variable (p. 100)
• table of values

Materials
• grid paper or
 tools

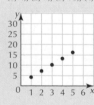

Think It Through
I know that a variable is any symbol or letter that is used to represent a number.

Rules, Tables, and Graphs

LEARN

How do you use a rule to make a table?

Tickets to the county fair are $3 per person plus $1 for parking. A rule can be written to show that the total cost is $3 times the number of people, plus $1.

Rule in words: **Multiply by 3, then add 1.**

Rule using a variable: $3x + 1$.

Example A

Make a **table of values** for the rule.

Multiply by 3, then add 1: $3x + 1$

Evaluate the expression $3x + 1$ using 1, 2, 3, 4, and 5 for x.

x	$3x + 1$
1	4
2	7
3	10
4	13
5	16

For $x = 1$, $3x + 1 = 3 \times 1 + 1 = 4$.
For $x = 2$, $3x + 1 = 3 \times 2 + 1 = 7$.
For $x = 3$, $3x + 1 = 3 \times 3 + 1 = 10$.
For $x = 4$, $3x + 1 = 3 \times 4 + 1 = 13$.
For $x = 5$, $3x + 1 = 3 \times 5 + 1 = 16$.

✓ **WARM UP**

Evaluate each expression for $x = 3$.

1. $x + 8$ 2. $7x$
3. $17 - x$ 4. $\frac{6}{x}$
5. $7x + 8$ 6. $\frac{6}{x} - 2$

How do you make a graph for a table of values?

Example C

Make a graph for the table of values in Example A.

Graph each of the ordered pairs in the table:
(1, 4), (2, 7), (3, 10), (4, 13), (5, 16).

Think It Through
I can use multiples along the x-axis or y-axis so that the graph is a reasonable size.

✓ **Talk About It**

4. What do you notice about the points in the graph?

5. Reasoning How could you use the graph to find another ordered pair that fits the rule in Example A?

6. Do you think (6, 10) fits the rule? How can you tell without doing any computation?

Source: Mathematics, Diamond Edition, Grade Five, Scott Foresman-Addison Wesley 2008 (pp. 176, 177).

NOW TRY THIS 4-8   Answer questions 4–6 on the last student page.

Example 4-12

Explain why a telephone company would not set rates for telephone calls as depicted on the graph in Figure 4-19.

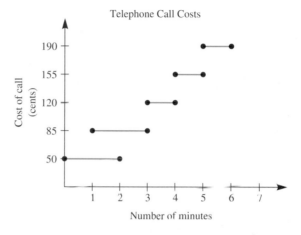

Figure 4-19

Solution   The graph does not depict a function. For example, a customer could be charged either $0.50 or $0.85 for a 2-min call; hence, not every input has a unique output. Also notice that the endpoints of each segment are shaded on both sides of each segment. This implies that for 3-, 4-, or 5-min calls there is no unique charge—another reason that the graph does not depict a function.

Suppose you join a DVD rental club where your cost per rental is $5. We have seen that one way to describe a function is by writing an equation. Based on the information in Table 4-3, the equation relating the number of videos rented to cost is $C = n \cdot 5$, or $C = 5n$, where n is the number of DVDs rented.

Table 4-3

Number of DVDs Rented	Cost in Dollars
1	$1 \cdot 5 = 5$
2	$2 \cdot 5 = 10$
3	$3 \cdot 5 = 15$
4	$4 \cdot 5 = 20$
5	$5 \cdot 5 = 25$
.	.
.	.
.	.
n	$n \cdot 5$ or $5n$

This could also be written as $f(n) = 5n$, where $f(n)$ is the cost of the rental in dollars. If we restrict the number of rentals to the first five natural numbers, the function can be described as the set of ordered pairs $\{(1,5)(2,10),(3,15),(4,20),(5,25)\}$. Figure 4-20 shows the graph of the function, which consists of five points that are not connected by a solid line. In graphing the function in Figure 4-20, we assume the domain to be the set $\{1,2,3,4,5\}$. It does not make any sense to connect the points because one cannot, for example, rent 1.5 DVDs. However, to show that the points lie on a straight line, we connected them by a dashed line.

Figure 4-20

NOW TRY THIS 4-9 Find the range of the function in Figure 4-20.

Sequences as Functions

Arithmetic, geometric, and other sequences introduced in Chapter 1 can be thought of as functions whose inputs are natural numbers and whose outputs are the terms of a particular sequence. For example, the arithmetic sequence 2, 4, 6, 8, ..., whose nth term is $2n$ can be described as a function from the set N (natural numbers) to the set E (even natural numbers) using the rule $f(n) = 2n$, where n is a natural number and $f(n)$ stands for the value of the nth term.

Example 4-13

If $f(n)$ denotes the nth term of a sequence, find $f(n)$ in terms of n for each of the following:

a. An arithmetic sequence whose first term is 3 and whose difference is 3
b. A geometric sequence whose first term is 3 and whose ratio is 3
c. The sequence $1, 1 + 2, 1 + 2 + 3, 1 + 2 + 3 + 4, \ldots$

Solution **a.** The first term is 3, the second term is $3 + 3$, or $2 \cdot 3$, the third is $2 \cdot 3 + 3$, or $3 \cdot 3$, and the fourth term is $3 \cdot 3 + 3$, or $4 \cdot 3$, the nth term is $n \cdot 3$, and hence $f(n) = n \cdot 3 = 3n$, where n is a natural number.

b. The first term is 3, the second $3 \cdot 3$, or 3^2, the third $3 \cdot 3^2$, or 3^3, and so on. Hence, the nth term is 3^n and therefore $f(n) = 3^n$, where n is a natural number.

c. The nth term is $1 + 2 + 3 + 4 + \ldots + n$. In Chapter 1 we saw that this sum equals $\dfrac{n(n+1)}{2}$, so the function is $f(n) = \dfrac{n(n+1)}{2}$, where n is a natural number.

Composition of Functions

Consider the function machines in Figure 4-21. If 2 is entered in the top machine, then $f(2) = 2 + 4 = 6$. The number 6 is then entered in the second machine and $g(6) = 2 \cdot 6 = 12$. The functions in Figure 4-21 illustrate the **composition of two functions**. In the composition of two functions, the range of the first function becomes the domain of the second function.

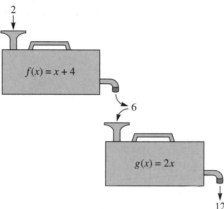

Figure 4-21

If the first function f is followed by a second function g, as in Figure 4-21, we symbolize the composition of the functions as $g \circ f$. If we input 3 in the function machines of Figure 4-21, then the output is symbolized by $(g \circ f)(3)$. Because f acts first on 3, to compute $(g \circ f)(3)$ we find $f(3) = 3 + 4 = 7$ and then $g(7) = 2 \cdot 7 = 14$. Hence, $(g \circ f)(3) = 14$. Notice that $(g \circ f)(3) = g(f(3))$. Also note that $(g \circ f)(x) = g(f(x)) = 2 \cdot f(x) = 2(x + 4)$ and hence $g(f(3)) = 2(3 + 4) = 14$.

Example 4-14

If $f(x) = 2x + 3$ and $g(x) = x - 3$, find the following:

a. $(f \circ g)(3)$ **b.** $(g \circ f)(3)$ **c.** $(f \circ g)(x)$ **d.** $(g \circ f)(x)$

Solution **a.** $(f \circ g)(3) = f(g(3)) = f(3 - 3) = f(0) = 2 \cdot 0 + 3 = 3$

  b. $(g \circ f)(3) = g(f(3)) = g(2 \cdot 3 + 3) = g(9) = 9 - 3 = 6$

  c. $(f \circ g)(x) = f(g(x)) = 2 \cdot g(x) + 3 = 2(x - 3) + 3 = 2x - 6 + 3 = 2x - 3$

  d. $(g \circ f)(x) = g(f(x)) = f(x) - 3 = (2x + 3) - 3 = 2x$

REMARK Example 4-14 shows that composition of functions is not commutative, since $(f \circ g)(3) \neq (g \circ f)(3)$.

We have seen that a function can be represented in a variety of ways. Pictures of sets with arrows and function machines are used mostly as pedagogical devices in learning the concept of a function. The most common representations are a table, an equation, or a graph. Depending on the situation, one representation may be more useful than another. For example, if the domain of a function has many elements, a table is not a convenient representation. Graphing calculators can be used to display a graph of most functions given by equations with specified domains.

TECHNOLOGY CORNER A sketch of the function $y = 2x + 1$ for x between 0 and 5 is shown in Figure 4-22. Use a graphing calculator to sketch the graphs of $y = 2x + b$ for three choices of b. What do the graphs seem to have in common? Why?

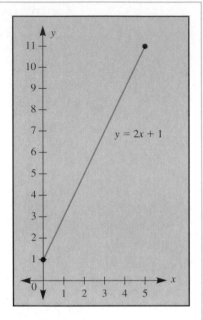

Figure 4-22

Relations

As in the definition of a function from set A to set B, a *relation from set A to set B* is a correspondence between elements of A and elements of B, but unlike functions, we do not require that each element of A be paired with one, and only one, element of B. Consequently, any set of ordered pairs is a relation. Notice that every function is a relation, but not every relation is a function. Examples of relations include the following:

"is a daughter of"   "is the same color as"

"is less than"   "is greater than or equal to"

Consider the relation "is a sister of." Figure 4-23 illustrates this relation among children on a playground, with letters A through J representing the childrens' names. An arrow from I to J indicates that I "is a sister of" J. Notice the arrows from F to G and from G to F, which indicate that F is a sister of G and G is a sister of F. This implies that F and G are girls. On the other hand, the absence of an arrow from J to I implies that J is not a sister of I. Thus, I is a girl and J is a boy.

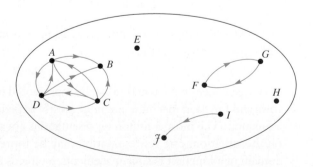

Figure 4-23

NOW TRY THIS 4-10 If all sister relationships are indicated in Figure 4-23, determine

a. which children are boys and which are girls.
b. for which children is there not enough information to determine gender.

Another way to show the relation "is a sister of" is to write the relation "A is a sister of B" as an ordered pair (A, B). Notice that (B, A) means that B is a sister of A. Using this notation, the relation "is a sister of" can be described for the children on the playground as the set

$$\{(A,B),(A,C),(A,D),(C,A),(C,B),(C,D),(D,A),(D,B),(D,C),(F,G),(G,F),(I,\mathcal{J})\}$$

Observe that this set is a subset of $\{A,B,C,D,E,F,G,H,I,\mathcal{J}\} \times \{A,B,C,D,E,F,G,H,I,\mathcal{J}\}$.
This observation motivates the following definition of a relation.

> **Definition**
>
> **A relation from set A to set B** Given any two sets A and B, a **relation** from A to B is a subset of $A \times B$; that is, R is a relation from set A to set B if, and only if, $R \subseteq A \times B$.

In the definition, the phrase "from A to B" means that the first components in the ordered pairs are elements of A and the second components are elements of B. If $A = B$, we say that the **relation is on A**.

Properties of Relations

Figure 4-24 represents a set of children in a small group. The children have drawn all possible arrows representing the relation "has the same first letter in his or her name as." Notice that the children were very careful to observe that each child in the group has the same first initial as himself or herself. Three properties of relations are illustrated in Figure 4-24.

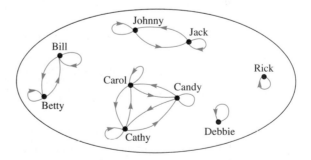

Figure 4-24

> **Definition of the Reflexive Property**
>
> A relation R on a set X is reflexive if, and only if, for every element $a \in X$, a is related to a. That is, for every $a \in X$, $(a, a) \in R$.

In the diagram, there is a loop at every point. For example, Rick has the same first initial as himself, namely R. A relation such as "is taller than" is not reflexive because people cannot be taller than themselves.

Definition of the Symmetric Property

A relation R on a set X is symmetric if, and only if, for all elements a and b in X, whenever a is related to b, then b also is related to a. That is, if $(a, b) \in R$, then $(b, a) \in R$.

In terms of the diagram, every pair of points that has an arrow headed in one direction also has a return arrow. For example, if Bill has the same first initial as Betty, then Betty has the same first initial as Bill. A relation such as "is a brother of" is not symmetric since Dick can be a brother of Jane, but Jane cannot be a brother of Dick.

Definition of the Transitive Property

A relation R on a set X is transitive if, and only if, for all elements a, b, and c of X, whenever a is related to b and b is related to c, then a is related to c. That is, if $(a, b) \in R$ and $(b, c) \in R$, then $(a, c) \in R$.

REMARK a, b, and c do not have to be different. Three symbols are used to allow for difference.

Notice that the relation in Figure 4-24 is transitive. For example, if Carol has the same first initial as Candy, and Candy has the same first initial as Cathy, then Carol has the same first initial as Cathy. A relation such as "is the father of" is not transitive since, if Tom Jones is the father of Tom Jones, Jr. and Tom Jones, Jr. is the father of Joe Jones, then Tom Jones is not the father of Joe Jones. He is, instead, the grandfather of Joe Jones.

The relation "is the same color as" is reflexive, symmetric, and transitive. The common relation "is equal to" also satisfies all three properties. In general, relations that satisfy all three properties are called **equivalence relations**.

Definition

An **equivalence relation** is any relation R that satisfies the reflexive, symmetric, and transitive properties.

The most natural equivalence relation encountered in elementary school is "is equal to" on the set of all numbers. In subsequent chapters, we will see more examples of equivalence relations.

The symmetric property of a relation is particularly useful in determining a symmetric nature of functions.* Consider the relation $x + y = 10$, where x and y are whole numbers. The

---

* Using the terminology of Chapter 14, a relation is symmetric if the line $y = x$ is the line of symmetry of its graph. However, a graph of a relation can have other lines of symmetry.

relation consists of the 11 ordered pairs graphed in Figure 4-25; the points lie on the dotted straight line. Notice that if the pair (a, b) is in the relation—that is, is on the graph—so is the pair (b, a). For example, $(1, 9)$ is on the graph because $1 + 9 = 10$, but then $(9, 1)$ is also on the graph because $9 + 1 = 10$. Also notice that the relation is a function. This can be seen from the graph, where for each input x, $x = 0, 1, 2, 3, \ldots, 10$, there is exactly one output y. We could also see that the relation is a function because $x + y = 10$ gives $y = 10 - x$; that is, for each x in the domain, there is a unique y. The domain and range of the function are the same set $\{0, 1, 2, 3, \ldots, 10\}$.

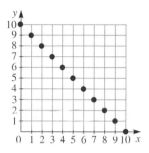

Figure 4-25

NOW TRY THIS 4-11

a. Explain why the domain and the range of the function $y = 10 - x$, where both x and y are whole numbers, is the set $\{0, 1, 2, 3, \ldots, 10\}$.

b. Show why the function $y = x + 10$, where x and y are whole numbers, is not a symmetric relation.

Assessment 4-3A

1. The following sets of ordered pairs are functions. Give a rule that could describe each function.
    a. $\{(2, 4), (3, 6), (9, 18), (12, 24)\}$
    b. $\{(2, 8), (5, 11), (7, 13), (4, 10)\}$
2. Which of the following are functions from the set $\{1, 2, 3\}$ to the set $\{a, b, c, d\}$? If the set of ordered pairs is not a function, explain why not.
    a. $\{(1, a), (2, b), (3, c), (1, d)\}$
    b. $\{(1, a), (2, b), (3, a)\}$

3. a. Draw an arrow diagram of a function with domain $\{1, 2, 3, 4, 5\}$ and range $\{a, b\}$.
    b. How many possible functions are there in part (a)?
4. Suppose $f(x) = 2x + 1$ and the domain is $\{0, 1, 2, 3, 4\}$. Describe the function in the following ways:
    a. Draw an arrow diagram involving two sets.
    b. Use ordered pairs.
    c. Make a table.
    d. Draw a graph to depict the function.

5. Determine which of the following are functions from $W = \{0, 1, 2, 3, \ldots\}$ or a subset of W to W. If your answer is that it is not a function, explain why not.
   a. $f(x) = 2$ for all $x \in W$
   b. $f(x) = x$
   c. $f(x)$ is the sum of the digits in x for all $x \in W$.

6. a. Make an arrow diagram for each of the following:
   (i) Rule: "when doubled is"

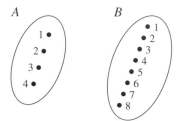

   (ii) Rule: "is greater than"

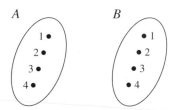

   b. Which, if any, of the parts in (a) exhibits a function from A to B? If it is a function, tell why and find the range of the function.

7. The dosage of a certain drug is related to the weight of a child as follows: 50 mg of the drug and an additional 15 mg for each 2 lb or fraction of 2 lb of body weight above 30 lb. Sketch the graph of the dosage as a function of the weight of a child for children who weigh between 20 and 40 lb.

8. If taxi fares are $3.50 for the first half mile and $0.75 for each additional quarter mile, answer the following:
   a. What is the fare for a 2-mi trip?
   b. Write a rule for computing the fare for an n-mile trip by taxi if n is a natural number.

9. For each of the following, guess what might be Latifah's rule. In each case, if n is your input and $L(n)$ is Latifah's answer, express $L(n)$ in terms of n.

a.
You	Latifah
3	8
4	11
5	14
10	29

b.
You	Latifah
0	1
3	10
5	26
8	65

10. In the *Principles and Standards* for grades 6–8, the "Algebra" section (p. 229) poses the following problem. Quick-Talk advertises monthly cellular phone service for $0.50 a minute for the first 60 min but only $0.10 a minute for each minute thereafter. Quick-Talk charges for the exact amount of time used. Answer the following:
   a. Make one graph showing the cost per minute as a function of number of minutes and the other showing the total cost for calls as a function of the number of minutes up to 100 min.
   b. If you connect the points in the second graph in part (a), what kind of assumption needs to be made about the way the telephone company charges phone calls?
   c. Why does the total cost for calls consist of two line segments? Why is one part steeper than the other?
   d. The function representing the total cost for calls as a function of number of minutes talked can be represented by two equations. Write these equations.

11. For each of the following sequences, find a possible function $f(n)$ whose domain is the set of natural numbers and whose outputs are the terms of the sequence.
    a. $3, 8, 13, 18, 23, \ldots$
    b. $3, 9, 27, 81, 243, \ldots$

12. Consider the following two function machines. Find the final output for each of the following inputs:
    a. 5
    b. 10

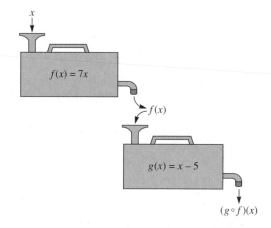

13. Let $t(n)$ represent the nth term of a sequence for $n \in N$. Answer the following:
    a. If $t(n) = 4n - 3$, determine which of the following are output values of the function:
       (i) 1   (ii) 385   (iii) 389   (iv) 392
    b. If $t(n) = n^2$, determine which of the following are output values of the function:
       (i) 0   (ii) 25   (iii) 625   (iv) 1000   (v) 90
    c. If $t(n) = n(n - 1)$, determine which of the following are in the range of the function:
       (i) 0   (ii) 2   (iii) 20   (iv) 999

14. Consider a function machine that accepts inputs as ordered pairs. Suppose the components of the ordered pairs are natural numbers and the first component is the length of a rectangle and the second is its width. The following machine computes the perimeter (the distance around a figure) of the rectangle. Thus, for a

rectangle whose length, l, is 3 and whose width, w, is 2, the input is (3, 2) and the output is $2 \cdot 3 + 2 \cdot 2$, or 10. Answer each of the following:

a. For each of the following inputs, find the corresponding output: (1, 7), (2, 6), (6, 2), (5, 5).

b. Find the set of all the inputs for which the output is 20.

c. What is the domain and the range of the function?

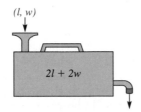

15. The following graph shows the relationship between the number of cars on a certain road and the time of day for times between 5:00 A.M. and 9:00 A.M.:

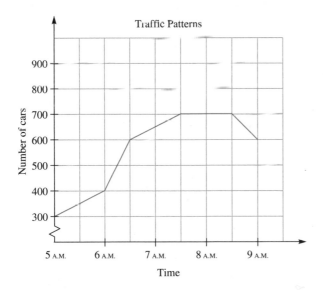

a. What was the increase in the number of cars on the road between 6:30 A.M. and 7:00 A.M.?

b. During which half hour was the increase in the number of cars the greatest?

c. What was the increase in the number of cars between 8:00 A.M. and 8:30 A.M.?

d. During which half hour(s) did the number of cars decrease? By how much?

e. The graph for this problem is composed of segments rather than just points as in Figure 4-20. Why do you think segments are used here instead of just points?

16. A ball is shot out of a cannon at ground level. We know that its height H in feet after t sec is given by the function $H(t) = 128t - 16t^2$.

a. Find $H(2), H(6), H(3)$, and $H(5)$. Why are some of the outputs equal?

b. Graph the function and from the graph find at what instant the ball is at its highest point. What is its height at that instant?

c. How long will it take the ball to hit the ground?

d. What is the domain of H?

e. What is the range of H?

17. For each of the following sequences of matchstick figures, let $S(n)$ be the function giving the total number of matchsticks in the nth figure.

a. For each of the following, find the total number of matchsticks in the fourth figure.

b. For each of the following, find as simple a formula as possible for $S(n)$ in terms of n.

(i)

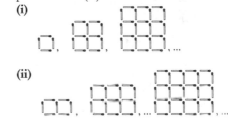

(ii)

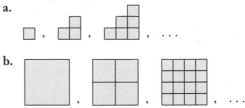

18. Assume the pattern continues for each of the following sequences of square tile figures. Let $S(n)$ be the function giving the total number of tiles in the nth figure. For each of the following, find a formula for $S(n)$ in terms of n. In part (b), each square is divided into four squares in the subsequent figure.

a.

b.

19. A function can be represented as a set of ordered pairs where the set of all the first components is the domain and where the set of all the second components is the range. Is the converse also true? That is, is every set of ordered pairs a function whose domain is the set of first components and whose range is the set of second components? Justify your answer.

20. Which of the following equations or inequalities represent functions and which do not? In each case x and y are whole numbers. Justify your answers.

a. $x + y = 2$

b. $x - y < 2$

c. $y = x^3 + x$

d. $xy = 2$

21. Which of the following are graphs of functions and which are not? Justify your answers.

a.

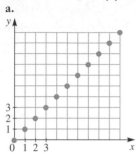

b.

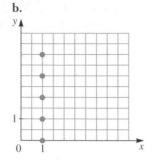

c.

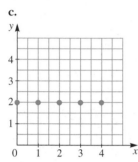

22. Suppose each point in the figure represents a child on a playground, the letters represent their names, and an arrow going from I to J means that I "is the sister of" J.

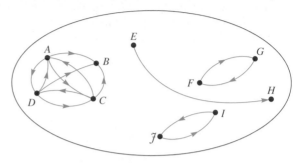

a. Based on the information in the figure, who are definitely girls and who are definitely boys?
b. Suppose we write "A is the sister of B" as an ordered pair (A, B). Based on the information in the diagram, write the set of all such ordered pairs.
c. Is the set of all ordered pairs in (b) a function with the domain equal to the set of all first components of the ordered pairs and with the range equal to the set of all second components?

Assessment 4-3B

1. The following sets of ordered pairs are functions. Give a rule that could describe each function.
   a. $\{(5,3),(7,5),(11,9),(14,12)\}$
   b. $\{(2,5),(3,10),(4,17),(5,26)\}$

2. Which of the following are functions from the set $\{1,2,3\}$ to the set $\{a,b,c,d\}$? If the set of ordered pairs is not a function, explain why not.
   a. $\{(1,c),(3,d)\}$
   b. $\{(1,a),(1,b),(1,c)\}$

3. a. Draw an arrow diagram of a function with domain $\{1,2,3\}$ and range $\{a,b\}$.
   b. How many possible functions are there in part (a)?

4. Suppose $f(x) = 2(x + 1)$ and the domain is $\{0,1,2,3,4\}$. Describe the function in the following ways:
   a. Draw an arrow diagram involving two sets.
   b. Use ordered pairs.   c. Make a table.
   d. Draw a graph to depict the function.

5. Determine which of the following are functions from $W = \{0,1,2,3,\dots\}$ or a subset of W to W. If your answer is that it is not a function, explain why not.
   a. $f(x) = 0$ if $x \in \{0,1,2,3\}$, and $f(x) = 3$ if $x \notin \{0,1,2,3\}$
   b. $f(x) = 0$ for all $x \in W$ and $f(x) = 1$ if $x \in \{3,4,5,6,\dots\}$
   c. $f(x)$ is the unit digit in x for all $x \in W$.

6. Given the following arrow diagrams for functions from A to B, give a possible rule for the function:

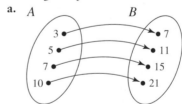

7. According to wildlife experts, the rate at which crickets chirp is a function of the temperature; specifically, $C = T - 40$, where C is the number of chirps every 15 sec and T is the temperature in degrees Fahrenheit.
   a. How many chirps does the cricket make per second if the temperature is 70°F?
   b. What is the temperature if the cricket chirps 40 times in 1 min?

8. For each of the following, guess what might be Latifah's rule. In each case, if n is your input and $L(n)$ is Latifah's answer, express $L(n)$ in terms of n.

a.

You	Latifah
6	42
0	0
8	72
2	6

b.

You	Latifah
0	1
1	2
5	32
6	64
10	1024

9. The *Principles and Standards* for grades 6–8 points out that "in their study of algebra, middle-grades students should encounter questions that focus on quantities that change" (p. 229). It poses the following problem.

ChitChat charges $0.45 a minute for cellular phone calls. The cost per minute does not change, but the total cost changes as the telephone is used.

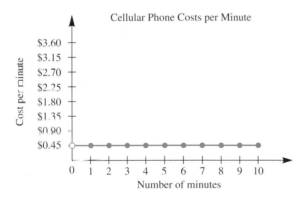

Cellular Phone Costs per Minute

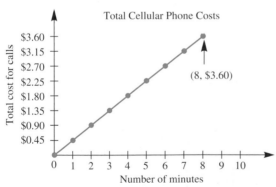

Total Cellular Phone Costs

(8, $3.60)

a. When the number of minutes is 6, what do the values of the corresponding point on each graph represent?

b. What kind of assumption about the charges needs to be made to allow the connection of the points on each graph? Explain.

c. If time in minutes is t and the total cost in dollars for calls is c, write c as a function of t for each graph.

10. For each of the following sequences, discover a possible pattern and find a function whose domain is the set of natural numbers and whose outputs are the terms of the sequence:

a. 2, 4, 6, 8, 10, ...
b. 1, 3, 9, 27, 81, ...
c. 2, 2 + 4, 2 + 4 + 6, 2 + 4 + 6 + 8, ...

11. Consider two function machines that are placed as shown. Find the final output for each of the following inputs:

a. 5   **b.** 3   **c.** 10   **d.** a

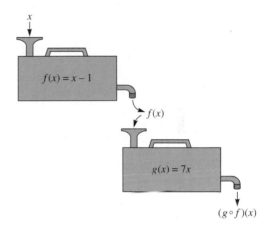

12. Let $t(n)$ represent the nth term of a sequence for $n \in N$. Answer the following:

a. If $t(n) = n^2$, determine which of the following are output values of the function:
   (i) 1   **(ii)** 4   **(iii)** 9   **(iv)** 10   **(v)** 900

b. If $t(n) = n(n + 1)$, determine which of the following are in the range of the function:
   (i) 2   **(ii)** 12   **(iii)** 2550   **(iv)** 2600

13. Consider a function machine that accepts inputs as ordered pairs. Suppose the components of the ordered pairs are natural numbers and the first component is the length of a rectangle and the second is its width. The following machine computes the perimeter (the distance around a figure) of the rectangle. Thus, for a rectangle whose length, l, is 3 and whose width, w, is 1, the input is $(3, 1)$ and the output is $2 \cdot 3 + 2 \cdot 1$, or 8. Answer each of the following:

a. For each of the following inputs, find the corresponding output: $(1, 4), (2, 1), (1, 2), (2, 2), (x, y)$.

b. Find the set of all the inputs for which the output is 20.

c. Is $(2, 2)$ a possible output? Explain.

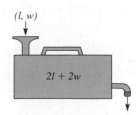

14. A health club charges a one-time initiation fee of $100 plus a membership fee of $40 per month.

  a. Write an expression for the cost function $C(x)$ that gives the total cost for membership at the health club for x months.
  b. Draw the graph of the function in (a).
  c. The health club decided to give its members an option of a higher initiation fee but a lower monthly membership charge. If the initiation fee is $300 and the monthly membership fee is $30, use a different color and draw on the same set of axes the cost graph under this plan.
  d. Determine after how many months the second plan is less expensive for the member.

15. A ball is shot straight up at ground level. We know that its height H in feet after t sec is given by the function $H(t) = 128t - 16t^2$.
  a. Graph the function and from the graph find at what instant the ball is at its highest point. What is its height at that instant?
  b. Find from the graph all t such that $H(t) = H(1)$.
  c. How long will it take the ball to hit the ground? Check your answer.
  d. What is the domain of H?
  e. What is the range of H?

16. A rectangular plot is bounded on one side by a straight river and on the other sides by a fence. Suppose 900 yd of fence are available and the length of the side of the rectangle parallel to the river is denoted by x.
  a. Find an expression for the area $A(x)$ in terms of x.
  b. Graph $A(x)$.
  c. Use the graph in (b) or your calculator to estimate the length and width of the rectangle for which the area will be the largest.

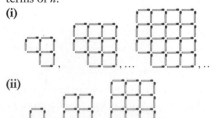

17. For each of the following sequences of matchstick figures, assume that your discovered pattern continues and let $S(n)$ be the function giving the total number of matchsticks in the nth figure.
  a. For each of the following figures assume that your pattern continues and find the total number of matchsticks in the fourth figure.
  b. For each of the following, find a formula for $S(n)$ in terms of n.
  (i)

  (ii)

18. You are 20 mi away from your home and start driving away from home at a constant speed of 60 mph. Write the distance S from home as a function of t, the time in hours of driving.

19. Assume the pattern continues for each of the following sequences of square tile figures. Let $S(n)$ be the function giving the total number of tiles in the nth figure. For each of the following, find a formula for $S(n)$ in terms of n.
  a.
  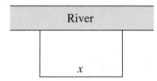
  b.
  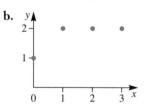

20. A function can be represented as a set of ordered pairs where the set of all the first components is the domain and the set of all the second components is the range. If each ordered pair (a,b) is replaced by (b,a), is the new set still a function?

21. Which of the following equations or inequalities represent functions and which do not? In each case x and y are whole numbers. Justify your answers.
  a. $x - y = 2$ **b.** $x + y < 20$
  c. $y = 2x^2$ **d.** $y = x^3 - 1$

22. Which of the following are graphs of functions and which are not? Justify your answers.
  a.
  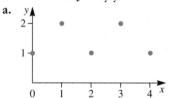
  b.
  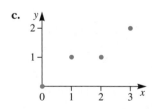
  c.
  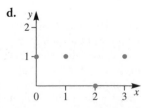
  d.

23. a. Which of the following relations from the set W of whole numbers to W have the symmetric property? Justify your answers.

    (i) $x + y = 10$ **(ii)** $x - y = 100$

    (iii) $xy = 100$ **(iv)** $y = x$

    (v) $y = x^2$

  b. Which of the relations in part (a) are functions? Justify your answer.

24. Suppose each point in the figure represents a child on a playground, the letters represent their names, and an arrow going from I to J means that I "is the sister of" J.

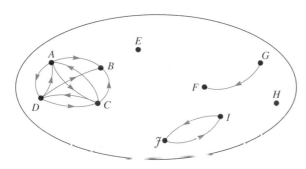

  a. Based on the information in the figure, who are definitely girls and who are definitely boys?

  b. Suppose we write "A is the sister of B" as an ordered pair (A, B). Based on the information in the diagram, write the set of all such ordered pairs.

  c. Is the set of all ordered pairs in (b) a function with the domain equal to the set of all first components of the ordered pairs and with the range equal to the set of all second components?

25. Which of the following are functions and which are relations but not functions from the set of first components of the ordered pairs to the set of second components?

  a. {(Montana, Helena), (Oregon, Salem), (Illinois, Springfield), (Arkansas, Little Rock)}

  b. {(Pennsylvania, Philadelphia), (New York, Albany), (New York, Niagara Falls), (Florida, Ft. Lauderdale)}

  c. $\{(x,y) \mid x$ resides in Birmingham, Alabama, and x is the mother of y, where y is a U.S. resident$\}$

  d. $\{(1,1), (2,4), (3,9), (4,16)\}$

  e. $\{(x,y) \mid x$ and y are natural numbers and $x + y$ is an even number$\}$

26. a. Consider the relation consisting of ordered pairs (x, y) such that y is the biological mother of x. Is this a function whose domain is the set of all people?

  b. Like in part (a) but now y is a brother of x. Is the relation a function from the set of all boys to the set of all boys?

27. Tell whether each of the following is reflexive, symmetric, or transitive on the set of all people. Which are equivalence relations?

  a. "Is a parent of"

  b. "Is the same age as"

  c. "Has the same last name as"

  d. "Is the same height as"

  e. "Is married to"

  f. "Lives within 10 mi of"

  g. "Is older than"

28. Tell whether each of the following is reflexive, symmetric, or transitive on the set of subsets of a nonempty set. Which are equivalence relations?

  a. "Is equal to"

  b. "Is a proper subset of"

  c. "Is not equal to"

  d. "Has the same cardinal number as"

Mathematical Connections 4-3

Communication

1. Does the diagram define a function from A to B? Why or why not?

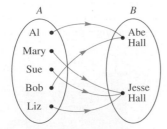

2. Is a one-to-one correspondence a function? Explain your answer and give an example.

3. Which of the following are functions from A to B? If your answer is "not a function," explain why not.

  a. A is the set of mathematics faculty at the university. B is the set of all mathematics classes. To each mathematics faculty member, we associate the class that person is teaching during a given term.

  b. A is the set of mathematics classes at the university. B is the set of mathematics faculty. To each mathematics class, we associate the teacher who is teaching the class.

  c. A is the set of all U.S. senators and B is the set of all senate committees. We associate each senator to the committee of which the senator is chairperson.

4. If S is the set of students in Ms. Carmel's class, and A is any subset of S, we define: $f(A) = \overline{A}$ (where $\overline{A}$ is the complement of A). Notice that the input in this function is a subset of S and the output is a subset of S. Answer the following:

a. Explain why f is a function and describe the domain and the range of f.

b. If there are 20 children in the class, what are the number of elements in the domain and the number in the range? Explain.

c. Is the function in this question a one-to-one correspondence? Justify your answer.

Open-Ended

5. Examine several newspapers and magazines and describe at least three examples of functions that you find. What is the domain and range of each function?

6. Give at least three examples of functions from A to B where neither A nor B is a set of numbers.

7. Draw a sequence of matchstick figures and describe the pattern in words. Find as simple an expression as possible for $S(n)$, the total number of matchsticks in the nth figure.

8. A function whose output is always the same regardless of the input is a *constant function*. Give several examples of constant functions from real life.

9. A function whose output is the same as its input is an *identity function*. Give several concrete examples of identity functions.

Cooperative Learning

10. Each person in a group picks a natural number and uses it as an input in the following function machine:

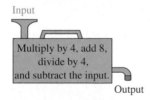

a. Compare your answers. Based on the answers, make a conjecture about the range of the function.

b. Based on your answer in (a), graph the function.

c. Write the function in the simplest possible way using $f(x)$ notation.

d. Justify your conjecture in (a).

e. Make up similar function machines and try different inputs in your group.

f. Devise a function machine in which the machine performs several operations but the output is always the same as the input. Exchange your answer with someone in the group and check that the other person's function machine performs as required.

11. In a group of four, work through the following. You will need a metric tape or meterstick.

a. Place your math book on a desk and measure the distance (to the nearest centimeter) from the floor to the top of the book. Record the distance.

b. Place a second math book on top of the first and measure the distance (to the nearest centimeter) from the floor to the top of the second book. Record the distance.

c. Continue this procedure for all four of your math books and complete the following table and graph:

Number of books	Distance from floor
1	
2	
3	
4	

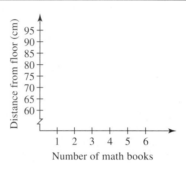

d. Without measuring, what is the distance from the floor with 0 books? 5 books?

e. Write a rule or function for $d(x)$, where $d(x)$ is the distance above the floor to the top of the stack of books and x is the number of books.

f. Suppose the distance from the floor to the ceiling is 2.5 m. If you stack the books as described above, how many books would be needed to reach the ceiling?

g. The function $h(x) = 34x + 70$ represents the height of another stack of x math books (in centimeters) on a cabinet. What does the function tell you about the height of the cabinet? What does it tell you about the thickness of each book?

h. Suppose that a table with a stack of similar math books (more than 10) is 200 cm high. If the top math book is removed, the height is 197 cm. If a second book is removed, the height is 194 cm. What is the height if 5 books are removed?

i. Write a function $h(x)$ for the height of the stack in part (h) after x books are removed.

Questions from the Classroom

12. A student claims that the following machine does not represent a function machine because it accepts two inputs at once rather than a single input. How do you respond?

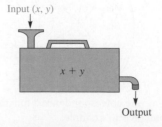

13. A student asks, "If every sequence is a function, is it also true that every function is a sequence?" How do you respond?

14. A student claims that the following does not represent a function, since all the values of x correspond to the same number.

x	0	1	2	3	4	5
y	1	1	1	1	1	1

How do you respond?

15. A student thinks that the function $f(x) = 3x + 5$ with domain of all whole numbers is a one-to-one correspondence and he would like to know why. How do you respond?

16. A student wants to know why sometimes it is incorrect to connect points on the graph of a function. How do you respond?

Review Problems

17. Solve the following equations, if possible:
    a. $3x - 1 = x + 99$
    b. $2(5x + 1) - 11 = x + 9$
    c. $3(x - 1) = 2(x - 1) + 99$
    d. $5(2x - 6) = 3(2x - 6)$

18. Solve the following problem by setting up an appropriate equation:

    Two cars, each traveling at a constant speed—one 60 mph and the other 70 mph—start at the same time from the same point traveling in the same direction. After how many hours will the distance between them be 40 mi?

National Assessment of Educational Progress (NAEP) Questions

In	Out
2	5
3	7
4	9
5	11
15	31
38	

The table shows how the "In" numbers are related to the "Out" numbers. When 38 goes in, what number comes out?

a. 41   b. 51   c. 54   d. 77

NAEP, Grade 4, 2007

Each figure in the pattern below is made of hexagons that measure 1 centimeter on each side.

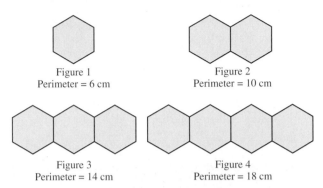

Figure 1
Perimeter = 6 cm

Figure 2
Perimeter = 10 cm

Figure 3
Perimeter = 14 cm

Figure 4
Perimeter = 18 cm

If the pattern of adding one hexagon to each figure is continued, what will be the perimeter of the 25th figure in the pattern? Show how you got your answer.

NAEP, Grade 8, 2007

In the equation $y = 4x$, if the value of x is increased by 2, what is the effect on the value of y?

a. It is 8 more than the original amount.
b. It is 6 more than the original amount.
c. It is 2 more than the original amount.
d. It is 16 times the original amount.
e. It is 8 times the original amount.

NAEP, Grade 8, 2007

Third International Mathematics and Science Study (TIMSS) Questions

A number machine takes a number and operates on it. When the Input Number is 5, the Output Number is 9, as shown below.

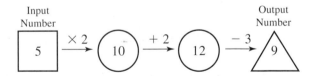

When the Input Number is 7, which of these is the Output Number?

a. 11
b. 13
c. 14
d. 25

TIMSS 2003, Grade 4

Ali had 50 apples. He sold some and then had 20 left. Which of these is a number sentence that shows this?

a. □ − 20 = 50
b. 20 − □ = 50
c. □ − 50 = 20
d. 50 − □ = 20

TIMSS 2003, Grade 4

The objects on the scale make it balance exactly. On the left pan there is a 1 kg weight (mass) and half a brick. On the right pan there is one brick.

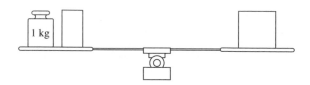

What is the weight (mass) of one brick?

a. 0.5 kg   **b.** 1 kg
c. 2 kg   **d.** 3 kg

TIMSS 2003, Grade 8

Hint for Solving the Preliminary Problem

Call each student number x and the final answer a. Write an equation involving x and a and solve for x in terms of a.

Chapter Outline

I. Variables
  A. Unknown in an equation
  B. Changing quantity
  C. To apply algebra in solving problems
  D. In a spreadsheet

II. Equations
  A. Addition property: For numbers a, b, and c, if $a = b$, then $a + c = b + c$.
  B. Multiplication property: For numbers a, b, and c, if $a = b$, then $ac = bc$.
  C. Cancellation properties: For numbers a, b, and c,
    1. if $a + c = b + c$, then $a = b$.
    2. if $c \neq 0$, and $ac = bc$, then $a = b$.
  D. Equality is not affected if we substitute a number for its equal.
  E. Addition and subtraction property of equations
    1. If $a = b$ and $c = d$, then $a + c = b + d$ and $a - c = b - d$.
  F. The distributive properties
    1. $a(b + c) = ab + ac$
    2. $a(b - c) = ab - ac$
  G. Solving equations
  H. Application problems

III. Functions and Relations
  A. A function from set A to B is a correspondence in which each element $a \in A$ is paired with one, and only one, element $b \in B$. If the function is denoted by f, we write $f(a) = b$. The element $a \in A$ is the input and $f(a)$ is the output. A is the **domain** of the function. B is any set containing all the outputs. The set of all the outputs is the **range** of the function.
  B. A function can be represented by a rule, a table, an equation, an arrow diagram, a function machine, a set of ordered pairs, or a graph.
  C. A sequence is a function whose domain is N, the set of natural numbers.
  D. Any set of ordered pairs is a relation.
    1. A relation from A to B is a subset of $A \times B$.
    2. A relation may have one or more of the following properties:
      a. Reflexive
      b. Symmetric
      c. Transitive

Chapter Review

1. There are 13 times as many students as professors at a college. Use S for the number of students and P for the number of professors to represent the given information.

2. Write a sentence that gives the same information as the following equation: $A = 103 \cdot B$, where A is the number of girls in a neighborhood and B is the number of boys.

3. Write an equation to find the number of feet given the number of yards (let f be the number of feet and y be the number of yards).

4. The sum of a set of n whole numbers is S. If each number is multiplied by 10 and then decreased by 10, what is the sum of the new set in terms of n and S?

5. I am thinking of a whole number. If I divide it by 13, then multiply the answer by 12, then subtract 20, and then add 89, I end up with 93. What was my original number?

6. a. Think of a number.
    Add 17.
    Double the result.
    Subtract 4.
    Double the result.
    Add 20.
    Divide by 4.
    Subtract 20.
    Your answer will be your original number. Explain how this trick works.

   b. Fill in three more steps that will take you back to your original number.
    Think of a number.
    Add 18.
    Multiply by 4.
    Subtract 7.
    .
    .
    .

   c. Make up a series of instructions such that you will always get back to your original number.

7. Find all the values of x that satisfy the following equations:
   a. $4x - 2 = 3x + 10$
   b. $4(x - 12) = 2x + 10$
   c. $4(7x - 21) = 14(7x - 21)$
   d. $2(3x + 5) = 6x + 11$
   e. $3(x + 1) + 1 = 3x + 4$

8. Mike has 3 times as many baseball cards as Jordan, who has twice as many cards as Paige. Together, the three children have 999 cards. Set up an equation in one variable and find how many cards each child has.

9. Jeannie has 10 books overdue at the library. She remembers she checked out 2 science books two weeks before she checked out 8 children's books. The daily fine per book is $0.20. If her total fine was $11.60, how long was each book overdue?

10. Three children deliver all the newspapers in a small town. Jacobo delivers twice as many papers as Dahlia, who delivers 100 more papers than Rashid. If altogether 500 papers are delivered, how many papers does each child deliver?

11. Which of the following sets of ordered pairs are functions from the set of first components to the set of second components?
    a. $\{(a,b), (c,d), (e,a), (f,g)\}$
    b. $\{(a,b), (a,c), (b,b), (b,c)\}$
    c. $\{(a,b), (b,a)\}$

12. Given the following function rules and the domains, find the associated ranges:
    a. $f(x) = x + 3$; domain $= \{0, 1, 2, 3\}$
    b. $f(x) = 3x - 1$; domain $= \{5, 10, 15, 20\}$
    c. $f(x) = x^2$; domain $= \{0, 1, 2, 3, 4\}$
    d. $f(x) = x^2 + 3x + 5$; domain $= \{0, 1, 2\}$

13. Which of the following correspondences from A to B describe a function? If a correspondence is a function, find its range. Justify your answers.
    a. A is the set of college students, and B is the set of majors. To each college student corresponds his or her major.
    b. A is the set of books in the library, and B is the set N of natural numbers. To each book corresponds the number of pages in the book.
    c. $A = \{(a,b) \mid a \in N$ and $b \in N\}$, and $B = N$. To each element of A corresponds the number $4a + 2b$.
    d. $A = N$ and $B = N$. If x is even, then $f(x) = 0$, and if x is odd, then $f(x) = 1$.
    e. $A = N$ and $B = N$. To each natural number corresponds the sum of its digits.

14. A health club charges an initiation fee of $200, which gives 1 month of free membership, and then charges $55 per month.
    a. If $C(x)$ is the total cost of membership in the club for x months, express $C(x)$ in terms of x.
    b. Graph $C(x)$ for the first 12 months.
    c. Use the graph in (b) to find when the total cost of membership in the club will exceed $600.
    d. When will the total cost of membership exceed $6000?

15. If the rule for the function is $f(x) = 4x - 5$ and $f(x) = 15$ is the output, what is the input?

16. Which of the following graphs represent functions? Tell why.

a.

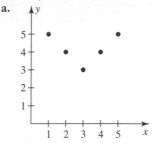

b.

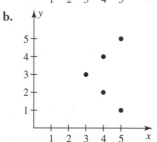

c.

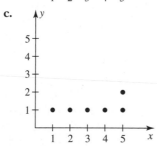

17. a. Jilly is building towers with cubes, placing one cube on top of another and painting the tower (including the top and the bottom, but not the faces touching each other). Find the number of square faces that Jilly needs to paint for towers made of 1, 2, 3, 4, 5, and 6 cubes by filling in the following table:

# of Cubes	# of Squares to Paint
1	6
2	10
3	
4	
5	
6	

b. Graph the information you found in part (a) where the number of cubes in a tower is on the horizontal x-axis and the number of squares to be painted is on the vertical axis.

c. If x is the number of cubes in a tower and y is the corresponding number of squares to be painted, write an equation that gives y as a function of x.

d. Is the graph describing the number of squares as a function of the number of cubes used a straight line?

Selected Bibliography

Adams, T.L. "Addressing Students' Difficulties with the Concept of Function: Applying Graphing Calculators and a Model of Conceptual Change." *Focus on Learning Problems in Mathematics* (1997) 19(2): 43–57.

Billings, E., T.L. Tiedt, and L.H. Slater. "Algebraic Thinking and Pictorial Growth Patterns." *Teaching Children Mathematics* 14 (December 2007–January 2008): 302–308.

Bishop, J., A. Otto, and C. Lubinski. "Promoting Algebraic Reasoning Using Student Thinking." *Mathematics Teaching in the Middle School* 6 (May 2001): 508–514.

Booth, L. "Children's Difficulties in Beginning Algebra." In *The Ideas of Algebra, K-12*, edited by A. Coxford and A. Shulte. Reston, VA: NCTM, 1988.

Bradley. E. "Is Algebra in the Cards?" *Mathematics Teaching in the Middle School* 2 (May 1997): 398–403.

Carpenter, T.P., M.L. Franke, and L. Levi. *Thinking Mathematically: Integrating Arithmetic and Algebra in Elementary School.* Portsmouth, NH: Heinemann (2003).

Chappell, M., and M. Strutchens. "Creating Connections: Promoting Algebraic Thinking with Concrete Models." *Mathematics Teaching in the Middle School* 7 (September 2001): 20–25.

Ferrucci, B., B. Yeap, and J. Carter. "A Modeling Approach for Enhancing Problem-Solving in the Middle Grades." *Mathematics Teaching in the Middle School* 8 (May 2003): 470–475.

Fouche, K. "Algebra for Everyone: Start Early." *Mathematics Teaching in the Middle School* 2 (February 1997): 226–229.

Joram, E., and V. Oleson. "How Fast Do Trees Grow? Using Tables and Graphs to Explore Slope."

Mathematics Teaching in the Middle School 13 (January 2008): 260–265.

Joram, E., V. Oleson, and K. Sabey. "Is It a Good Deal? Developing Number Sense in Algebra by Comparing Housing Prices." *Journal of the Iowa Council of Teachers of Mathematics* 32 (Winter 2005): 15–19.

Kalchman, M.S. "Walking Through Space: A New Approach for Teaching Functions." *Mathematics Teaching in the Middle School* 11 (August 2005): 12–17.

Koirala, H., and P. Goodwin. "Teaching Algebra in the Middle Grades Using Mathmagic." *Mathematics Teaching in the Middle School* 5 (May 2000): 562–566.

Krebs, A. "Studying Students' Reasoning in Writing Generalizations." *Mathematics Teaching in the Middle School* 10 (February 2005): 284–287.

Lamdin, D., R. Lynch, and H. McDaniel. "Algebra in the Middle Grades." *Mathematics Teaching in the Middle School* 6 (November 2000): 195–198.

Lannin, J. "Developing Algebraic Reasoning Through Generalization." *Mathematics Teaching in the Middle School* 8 (March 2003): 342–348.

Lubinski, C., and A. Otto. "Meaningful Mathematical Representation and Early Algebraic Reasoning." *Teaching Children Mathematics* 9 (October 2002): 76–80.

MacGregor, M., and Quinlan, C. "Research in Teaching and Learning Algebra." In *Research in Mathematics Education in Australasia: 1992–1995*, edited by W. Atweh, K. Owens, and P. Sullivan. 365–381. (1996)

Markovits, Z., B.-S. Eylon, and M. Bruckheimer. "Functions—Linearity unconstrained." In *Proceedings of the Seventh International Conference for the Psychology of Mathematics Education*, edited by R. Hershkowitz. Rehovot, Israel: Weizmann Institute of Science, 1983: 271–277.

Markovits, Z., B.-S. Eylon, and M. Bruckheimer. "Functions Today and Yesterday?" *For the Learning of Mathematics* (1986), 6(2): 18–24.

Markovits, Z., B.-S. Eylon, and M. Bruckheimer. "Difficulties Students have with the Function Concept." In *The Ideas of Algebra, K–12*, edited by A. Coxford and A. Shulte, Reston, VA: NCTM, 1988.

Martinez-Cruz, A., and E. Barger. "Adding a la Gauss." *Mathematics Teaching in the Middle School* 10 (October 2004): 152–155.

Ploger, D. "Spreadsheets, Patterns, and Algebraic Thinking." *Teaching Children Mathematics* 3 (February 1997): 330–334.

Pólya, G. *How to Solve It*. Princeton, NJ: Princeton University Press, 1957.

———— *Mathematical Discovery, Combined Edition*. New York: John Wiley & Sons, Inc., 1981.

Quinlan, Cyril R. E. (1992) *Developing an Understanding of Algebraic Symbols*. Ph.D. thesis, University of Tasmania.

Rubenstein, R. "Building Explicit and Recursive Forms of Patterns with the Function Game." *Mathematics Teaching in the Middle School* 7 (April 2002): 426–431.

————. "The Function Game." *Mathematics Teaching in the Middle School* 2 (November–December 1996): 74–78.

Sakshaug, L., and K. Wohlhuter. Responses to the Which Graph Is Which Problem." *Teaching Children Mathematics* 7 (February 2001): 352–353.

Sand, M. "A Function Is a Mail Carrier." *Mathematics Teacher* 89 (September 1996): 468–469.

Schneider, S., and C. Thompson. "Incredible Equations Develop Incredible Number Sense." *Teaching Children Mathematics* 7 (November 2000): 146–148, 165–168.

Shealy, B. "Becoming Flexible with Functions: Investigating United States Population Growth." *Mathematics Teacher* 89 (May 1996): 414–418.

Siegel, M. "The Sum of Cubes: An Activity Review and Conjecture." *Mathematics Teaching in the Middle School* 10 (March 2005): 356–359.

Smith, E. "Patterns, Functions, and Algebra." In *A Research Companion to NCTM's Standards*, edited by J. Kirkpatrick, W. G. Martin, and D. S. Schifter. Reston, VA: NCTM, 2000.

Smith, J., and E. Phillips. "Listening to Middle School Students' Algebraic Thinking." *Mathematics Teaching in the Middle School* 6 (November 2000): 156–161.

Steinberg, R., D. Sleeman, and D. Ktorza. "Algebra Students Knowledge of Equivalence of Equations." *Journal for Research in Mathematics Education* (1990), 22(2): 112–121.

Suh, J. M. "Developing "Algebra-'Rithmetic" in the Elementary Grades." *Teaching Children Mathematics* 14 (November 2007): 246–252.

Thompson, F. M. "Algebraic Instruction for the Younger Child." In *The Ideas of Algebra K–12*, NCTM Yearbook, 1988: 69–77.

Thornton, S. "New Approaches to Algebra: Have We Missed the Point?" *Mathematics Teaching in the Middle School* 6 (March 2001): 388–392.

Usiskin, Z. "Doing Algebra in Grades K–4." *Teaching Children Mathematics* 3 (February 1997): 346–356.

Van de Walle, J. *Elementary and Middle School Mathematics: Teaching Developmentally*. New York: Addison Wesley Longman, 2007.

Van Dyke, F., and J. Tomback. "Collaborating to Introduce Algebra," *Mathematics Teaching in the Middle School* 10 (December/January 2005): 236–242.

Van Reeuwijk, M., and M. Wijers. "Students' Construction of Formulas in Context." *Mathematics Teaching in the Middle School* 2 (February 1997): 230–236.

CHAPTER 5

Integers and Number Theory

Preliminary Problem

Find a quick way to compute the sum without using a calculator.

$$50^2 - 49^2 + 48^2 - 47^2 + \ldots + 2^2 - 1^2$$

T he *Principles and Standards* expectations for students in grades 3–5 include the following:

- explore numbers less than 0 by extending the number line and through familiar applications;
- describe classes of numbers according to characteristics such as the nature of their factors. (p. 148)

The expectations for students in grades 6–8 include:

- use factors, multiples, prime factorization, and relatively prime numbers to solve problems;
- develop meaning for integers and represent and compare quantities with them. (p. 214)

In addition, the *Principles and Standards* points out that in grades 6–8:

Students can also work with whole numbers in their study of number theory. Tasks, such as the following, involving factors, multiples, prime numbers, and divisibility can afford opportunities for problem solving and reasoning.

1. Explain why the sum of the digits of any multiple of 3 is itself divisible by 3.

2. A number of the form *abcabc* always has several prime-number factors. Which prime numbers are always factors of a number of this form? Why?

Middle-grades students should also work with integers. In lower grades, students may have connected negative integers in appropriate ways to informal knowledge derived from everyday experiences, such as below-zero winter temperatures or lost yards on football plays. In the middle grades, students should extend these initial understandings of integers. Positive and negative integers should be seen as useful for noting relative changes or values. Students can also appreciate the utility of negative integers when they work with equations whose solution requires them, such as $2x + 7 = 1$. (pp. 217–18)

In this chapter, we start with the system of integers and then develop an understanding of number theory.

Negative numbers are useful in everyday life. For example, Mount Everest is 29,028 ft above sea level, and the Dead Sea is 1293 ft below sea level. We may symbolize these elevations as 29,028 and ⁻1293.

In mathematics, the need for negative integers arises because subtractions cannot always be performed in the set of whole numbers. To compute $4 - 6$ using the definition of subtraction for whole numbers, we must find a whole number n such that $6 + n = 4$. There is no such whole number n. To perform the computation, we must invent a new number, a

♦ *Historical Note*

The Hindu mathematician Brahmagupta (ca. 598–665 CE) provided the first systematic treatment of negative numbers and of zero. Only about 1000 years later did the Italian mathematician Gerolamo Cardano (1501–1576) consider negative solutions to certain equations. Still uncomfortable with the concept of negative numbers, he called them "fictitious" numbers. ♦

negative integer. If we attempt to calculate $4 - 6$ on a number line, then we must draw intervals to the left of 0. In Figure 5-1, $4 - 6$ is pictured as an arrow that starts at 0 and ends 2 units to the left of 0. The new number that corresponds to a point 2 units to the left of 0 is *negative two*, symbolized by $^-2$.

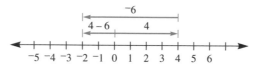

Figure 5-1

Other numbers to the left of 0 are created similarly. The new set of numbers $\{^-1, ^-2, ^-3, ^-4, \ldots\}$ is the set of **negative integers**. The set $\{1, 2, 3, 4, \ldots\}$ is the set of **positive integers**. The integer 0 is neither positive nor negative. The union of the set of negative integers, the set of positive integers, and $\{0\}$ is the set of **integers**, denoted by I.

$$I = \{\ldots, ^-4, ^-3, ^-2, ^-1, 0, 1, 2, 3, 4, \ldots\}$$

As a field of study, number theory started to flourish in the seventeenth century with the work of Pierre de Fermat (1601–1665). Topics in number theory that occur in the elementary school curriculum include factors, multiples, divisibility tests, prime numbers, prime factorizations, greatest common divisors, and least common multiples. The topic of congruences, introduced by Carl Gauss (1777–1855), is also incorporated into the elementary curriculum through clock arithmetic and modular arithmetic. Clock and modular arithmetic give students a look at a mathematical system.

5-1   Integers and the Operations of Addition and Subtraction

Representations of Integers

It is unfortunate that we use the symbol "$-$" to indicate both a subtraction and a negative sign. To reduce confusion between the uses of this symbol in this text, a raised "$^-$" sign is used for negative numbers, as in $^-2$, and for the opposite of a number, as in ^-x, in contrast to the lower sign for subtraction. To emphasize that an integer is positive, sometimes a raised plus sign is used, as in $^+3$. In this text, we use the plus sign for addition only and write $^+3$ simply as 3.

Historical Note

The dash has not always been used for both the subtraction operation and the negative sign. Other notations were developed but never adopted universally. One such notation was used by Abu al-Khwârizmî (ca. 825), who indicated a negative number by placing a small circle over it. For example, $^-4$ was recorded as $\overset{\circ}{4}$. The Hindus denoted a negative number by enclosing it in a circle; for example, $^-4$ was recorded as ④. The symbols $+$ and $-$ first appeared in print in European mathematics in the late fifteenth century, at which time the symbols referred not to addition or subtraction nor positive or negative numbers, but to surpluses and deficits in business problems. ◆

The negative integers are **opposites** of the positive integers. For example, the opposite of 5 is ⁻5. Similarly, the positive integers are the opposites of the negative integers. Because the opposite of 4 is denoted ⁻4, the opposite of ⁻4 can be denoted ⁻(⁻4), which equals 4. The opposite of 0 is 0. In the set of integers I, every element has an opposite that is also in I.

> **REMARK** Using addition of integers, we shall soon see that when an opposite of an integer is added to the integer the sum is 0. In fact, ⁻a can be defined as the solution of $x + a = 0$.

Example 5-1

For each of the following, find the opposite of x:

a. $x = 3$
b. $x = {}^-5$
c. $x = 0$

Solution **a.** ⁻$x = {}^-3$
      b. ⁻$x = {}^-({}^-5) = 5$
      c. ⁻$x = {}^-0 = 0$

The value of ⁻x in Example 5-1(b) is 5. Note that ⁻x is the opposite of x and might *not represent a negative number*. In other words, x is a variable that can be replaced by some number either positive, zero, or negative. *Note:* ⁻x *is read "the opposite of x" not "minus x" or "negative x."*

In the grade 7 *Focal Points* we find the following:

> By applying properties of arithmetic and considering negative numbers in everyday contexts (e.g., situations of owing money or measuring elevations above and below sea level), students explain why the rules for adding, subtracting, multiplying, and dividing with negative numbers make sense. (p. 19)

We next investigate many informal ways to introduce operations on integers; we first begin by looking at addition of integers.

♦ Research Note

It is important to use manipulatives when working with negative numbers (Thompson 1988). ♦

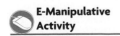

E-Manipulative Activity

Use the *Token Addition* activity for additional practice using the chip or charged-field models.

Integer Addition

As mentioned in the Research Note, it is important to use hands-on materials when working with integers. Several models are presented next to motivate integer addition. Teachers can actually build a number line for students to walk when using the number-line model.

Chip Model for Addition

In the chip model, positive integers are represented by black chips and negative integers by red chips. One red chip neutralizes one black chip. Hence, the integer ⁻1 can be represented by 1 red chip, or 2 red and 1 black, or 3 red and 2 black, and so on. Similarly, every integer can be represented in many ways using chips, as shown in Figure 5-2.

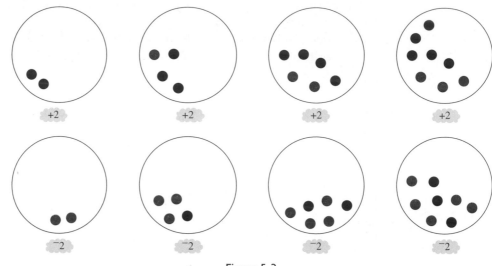

Figure 5-2

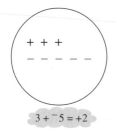

$$^-4 + 3 = ^-1$$

Figure 5-3

Figure 5-3 shows a chip model for the addition $^-4 + 3$. We put 4 red chips together with 3 black chips. Because 3 red chips neutralize 3 black ones, Figure 5-3 represents the equivalent of 1 red chip or $^-1$.

$$3 + ^-5 = +2$$

Figure 5-4

Charged-Field Model for Addition

A model similar to the chip model uses positive and negative charges. A field has 0 charge if it has the same number of positive $(+)$ and negative $(-)$ charges. As in the chip model, a given integer can be represented in many ways using the charged-field model. Figure 5-4 uses the model for $3 + ^-5$. Because 3 positive charges "neutralize" 3 negative charges, the net result is 2 negative ones. Hence, $3 + ^-5 = ^-2$.

Number-Line Model

Another model for addition of integers involves a number line, and it can be introduced with the idea of a hiker walking the number line, as seen on the student page on page 253. Study the student page to see how the model for the hiker works and then how the hiker's moves might be recorded on a number line. Without the hiker, $^-3 + ^-5$ can be pictured as in Figure 5-5.

E-Manipulative Activity

Use the *Number Line* activity to illustrate the number-line model.

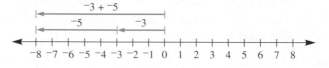

Figure 5-5

School Book Page ADDING INTEGERS

Lesson 8-5

Algebra

Key Idea
You can use a number line to add integers.

Vocabulary
• absolute value (p. 408)

Think It Through
When adding integers on the number line, I need to **start at zero**.

Adding Integers

LEARN

✓ WARM UP	
1. 5 + 8	2. 12 + 9
3. 15 − 6	4. 23 − 18

How can you add integers using a number line?

Think of walking along a number line. Walk forward for positive integers and walk backward for negative integers.

Example A

On the first down after getting the football, a football team gained 5 yards. On the next down, it lost 7 yards. Has the team gained or lost yardage after two downs?

Find 5 + (−7).

   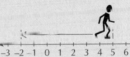

Start at zero, facing the positive integers. Walk forward 5 steps for 5.

Then walk backward 7 steps for −7. You stop at −2.

So, 5 + (−7) = −2.

The team has lost 2 yards.

Example B

Find −4 + (−2).

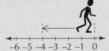

Start at zero, facing the positive integers. Walk backward 4 steps for −4.

Then walk backward 2 steps for −2. You stop at −6.

So, −4 + (−2) = −6.

Example C

Find −6 + 11.

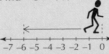   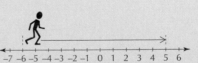

Start at zero, facing the positive integers. Walk backward 6 steps for −6.

Then walk forward 11 steps for 11. You stop at 5.

So, −6 + 11 = 5.

418

Source: Scott Foresman-Addison Wesley, Grade 6, 2008 (p. 418).

Figure 5-6 similarly depicts integer addition of 3 + ⁻5.

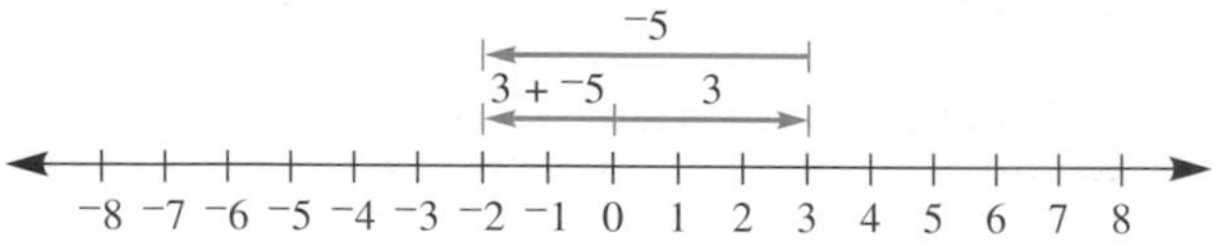

$$
\begin{array}{c}
\xleftarrow{\hspace{2cm}} \; {}^{-}5 \\
\boxed{3 + {}^{-}5 \quad \quad 3}
\end{array}
$$

-8 -7 -6 -5 -4 -3 -2 -1 0 1 2 3 4 5 6 7 8

Figure 5-6

NOW TRY THIS 5-1

a. Refer to Example B on the student page. Is the sum of two negative integers always negative?

b. Refer to Examples A and C on the student page. Is the sum of a positive and a negative integer positive or negative? Explain.

c. Use a number line to add 6 + (⁻8) + (⁻2).

Example 5-2 involves a thermometer with a scale in the form of a vertical number line.

Example 5-2

The temperature was ⁻4°C. In an hour, it rose 10°C. What is the new temperature?

Solution   Figure 5-7 shows that the new temperature is 6°C and that ⁻4 + 10 = 6.

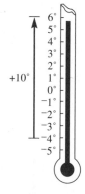

Figure 5-7

Pattern Model for Addition

Addition of whole numbers was discussed in Chapter 3. Addition of integers can also be motivated by using patterns of addition of whole numbers. Notice that in the left column of the following list, the first four facts are known from whole-number addition. Also notice that the 4 stays fixed and as the numbers added to 4 decrease by 1, the sum decreases by 1. Following this pattern, 4 + ⁻1 = 3 and we can complete the remainder of the first column. Similar reasoning can be used to complete the computations in the right column, where ⁻2 stays fixed and the other numbers decrease by 1 each time.

4 + 3 = 7	⁻2 + 4 = 2
4 + 2 = 6	⁻2 + 3 = 1
4 + 1 = 5	⁻2 + 2 = 0
4 + 0 = 4	⁻2 + 1 = ⁻1
4 + ⁻1 = 3	⁻2 + 0 = ⁻2
4 + ⁻2 = 2	⁻2 + ⁻1 = ⁻3
4 + ⁻3 = 1	⁻2 + ⁻2 = ⁻4
4 + ⁻4 = 0	⁻2 + ⁻3 = ⁻5
4 + ⁻5 = ⁻1	⁻2 + ⁻4 = ⁻6
4 + ⁻6 = ⁻2	⁻2 + ⁻5 = ⁻7

Note that the reasoning with patterns is inductive reasoning and therefore does not constitute a proof.

TECHNOLOGY CORNER On a spreadsheet, in column A enter 4 and fill down 20 rows. (For help on spreadsheets, see the Technology Manual.) In column B, enter 3 as the first entry and then write a formula to add $^-1$ to 3 for the second entry, add $^-1$ to the second entry to get the third entry, and fill down continuing the pattern. In column C, find the sum of the respective entries in columns A and B. What patterns do you observe? Repeat the problem by changing the entries in column A to $^-4$ and repeating the process.

Absolute Value

Because 4 and $^-4$ are opposites, they are on opposite sides of 0 on the number line and are the same distance (4 units) from 0, as shown in Figure 5-8.

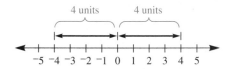

Figure 5-8

Distance is always a positive number or zero. The distance between the point corresponding to an integer and 0 is the **absolute value** of the integer. Thus, the absolute value of both 4 and $^-4$ is 4, written $|4| = 4$ and $|^-4| = 4$, respectively. Notice that if $x \geq 0$, then $|x| = x$, and if $x < 0$, then ^-x is positive. Therefore, we have the following:

Definition of Absolute Value

$$|x| = x \quad \text{if } x \geq 0$$
$$|x| = {}^-x \quad \text{if } x < 0$$

REMARK Some students want to shorten the above definition to $|x| = \pm x$. This is not true, as $|x|$ has only one value.

Example 5-3

Evaluate each of the following:

a. $|20|$
b. $|^-5|$
c. $|0|$
d. $^-|^-3|$
e. $|2 + {}^-5|$

Solution **a.** $|20| = 20$
b. $|^-5| = 5$
c. $|0| = 0$
d. $^-|^-3| = {}^-3$
e. $|2 + {}^-5| = |^-3| = 3$

NOW TRY THIS 5-2 Write each of the following in simplest form without the absolute value notation in the final answer. Show your work.

a. $|x| + x$ if $x \leq 0$
b. $^-|x| + x$ if $x \leq 0$
c. $^-|x| + x$ if $x \geq 0$

REMARK It is possible to describe addition of integers as the process of finding the difference or the sum of the absolute values of the integers and attaching an appropriate sign.

Properties of Integer Addition

Integer addition has all the properties of whole-number addition. These properties can be proved if addition of integers is defined in terms of whole number addition and subtraction.

Theorem 5–1:   Properties

Given integers a, b, and c:

Closure property of addition of integers $a + b$ is a unique integer.

Commutative property of addition of integers $a + b = b + a$.

Associative property of addition of integers $(a + b) + c = a + (b + c)$.

Identity element of addition of integers   0 is the unique integer such that, for all integers a, $0 + a = a = a + 0$.

Notice the name *identity element* in Theorem 5–1. Zero is the identity element of addition because when it is added to any integer it does not change the result; it leaves the integer unchanged.

We have seen that every integer has an opposite. This opposite is the **additive inverse** of the integer. The fact that each integer has a unique (one and only one) additive inverse is stated in Theorem 5–2.

Theorem 5–2:   Uniqueness of the Additive Inverse

For every integer a, there exists a unique integer ^-a, the additive inverse of a, such that $a + {}^-a = 0 = {}^-a + a$.

REMARK Notice that by definition the additive inverse, ^-a, is the solution of the equation $x + a = 0$. The fact that the additive inverse is unique is equivalent to saying that the preceding equation has only one solution. In fact, for any integers a and b, the equation $x + a = b$ has a unique solution, $b + {}^-a$.

The uniqueness of additive inverses can be used to justify other theorems. For example, the opposite, or the additive inverse, of ^-a can be written $^-(^-a)$. However, because

$a + {}^-a = 0$, the additive inverse of ${}^-a$ is also a. Because the additive inverse of ${}^-a$ must be unique, we have ${}^-({}^-a) = a$. Other theorems of addition of integers can be investigated by considering previously developed notions. For example, we saw that ${}^-2 + {}^-4 = {}^-6$, and we know that ${}^-6$ is the additive inverse of 6, or $2 + 4$. This leads us to the following:

$${}^-2 + {}^-4 = {}^-(2 + 4)$$

This relationship is true in general and is stated along with its proof.

Theorem 5–3

For any integers, a and b:

1. ${}^-({}^-a) = a$
2. ${}^-a + {}^-b = {}^-(a + b)$

We prove the second part of Theorem 5–3 as follows: By definition ${}^-(a + b)$ is the additive inverse of $(a + b)$, that is, $(a + b) + {}^-(a + b) = 0$. If we could show that ${}^-a + {}^-b$ is also the additive inverse of $a + b$, the uniqueness of the additive inverse implies that ${}^-(a + b)$ and ${}^-a + {}^-b$ are equal. To show that ${}^-a + {}^-b$ is also the additive inverse of $a + b$, we need only to show that $(a + b) + ({}^-a + {}^-b) = 0$. This can be shown using the associative and commutative properties of integer addition and the definition of the additive inverse as follows:

$$(a + b) + ({}^-a + {}^-b) = (a + {}^-a) + (b + {}^-b)$$
$$= 0 + 0$$
$$= 0$$

Now we have

$$(a + b) + ({}^-a + {}^-b) = 0$$
$$(a + b) + {}^-(a + b) = 0$$

Hence, ${}^-(a + b) = {}^-a + {}^-b$.

REMARK Notice that in the proof some of the steps involving the commutative and associative properties of integer addition were omitted. A more detailed proof showing all the steps follows:

$$(a + b) + ({}^-a + {}^-b) = (a + b) + ({}^-b + {}^-a)$$
$$= [(a + b) + {}^-b] + {}^-a$$
$$= [a + (b + {}^-b)] + {}^-a$$
$$= (a + 0) + {}^-a$$
$$= a + {}^-a$$
$$= 0$$

Example 5-4

Find the additive inverse of each of the following:

a. ${}^-(3 + x)$
b. $a + {}^-4$
c. ${}^-3 + {}^-x$

Solution   **a.** $3 + x$

 b. $^-(a + {}^-4)$, which can be written as $^-a + {}^-({}^-4)$, or $^-a + 4$

 c. $^-({}^-3 + {}^-x)$, which can be written $^-({}^-3) + {}^-({}^-x)$, or $3 + x$

Integer Subtraction

As with integer addition, we explore several models for integer subtraction.

Chip Model for Subtraction

To find $3 - {}^-2$, we want to subtract $^-2$ (take away 2 red chips) from 3 black chips. As seen in Figure 5-9(a), if we just have 3 black chips, we can't take 2 red ones away. Therefore, we need to represent 3 so that at least 2 red chips are present. Recall that 1 red chip neutralizes 1 black chip and so adding a black chip and a red chip (or 2 black chips and 2 red chips) is the same as adding 0, and the problem does not change. Because we need 2 red chips, we can add 2 black chips and 2 red chips without changing the problem. In Figure 5-9(b), we now see 3 represented using 5 black chips and 2 red chips. Now when the 2 red chips are "taken away," in Figure 5-9(c), 5 black chips are left and hence, $3 - {}^-2 = 5$.

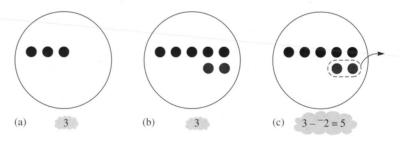

(a)   3   (b)   3   (c) $3 - {}^-2 = 5$

Figure 5-9

Charged-Field Model for Subtraction

Integer subtraction can be modeled with a charged field. For example, consider $^-3 - {}^-5$. To subtract $^-5$ from $^-3$, we first represent $^-3$ so that at least 5 negative charges are present. An example is shown in Figure 5-10(a). To subtract $^-5$, remove the 5 negative charges, leaving 2 positive charges, as in Figure 5-10(b). Hence, $^-3 - {}^-5 = 2$.

E-Manipulative Activity

Use *Token Subtraction* for additional practice using the chip and charged-field models.

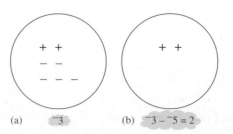

(a) $^-3$ (b) $^-3 - {}^-5 = 2$

Figure 5-10

Notice that the chip model and the charged-field model are combined on the partial student page shown next.

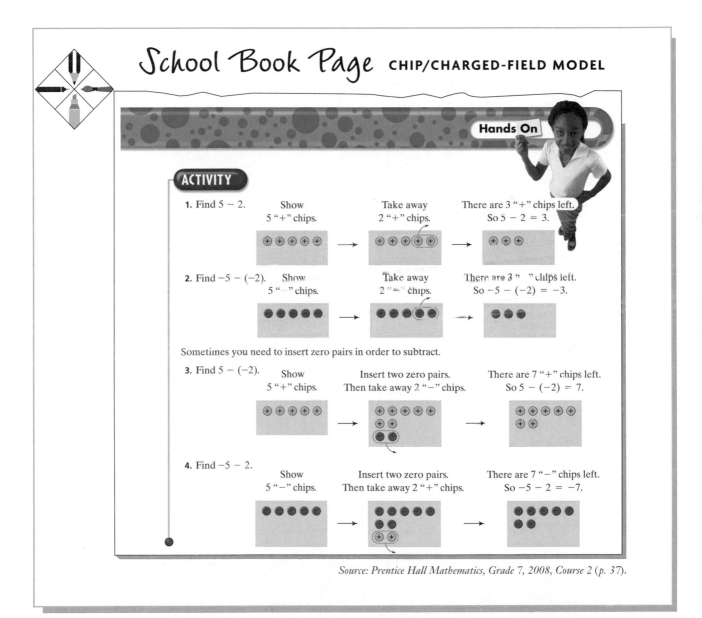

Number-Line Model for Subtraction

The number-line model used for integer addition can also be used to model integer subtraction. While integer addition is modeled by maintaining the same direction and moving forward or backward depending on whether a positive or negative integer is added, subtraction is modeled by turning around. To see how this works, examine the student page on page 260. Study the model to make sure you are comfortable with it and use it to find $5 - {}^-3 - {}^-2$.

School Book Page SUBTRACTING INTEGERS

Lesson 12-7

Algebra

Key Idea
You can use a number line to show the subtraction of integers.

Vocabulary
• positive integers (p. 712)
• negative integers (p. 712)

Materials
• number lines

TEST TALK

Think It Through
I can **draw a picture of a number line** to show the main idea.

Subtracting Integers

LEARN

How do you subtract integers?

Example A

Find ⁺4 − ⁻3.

To subtract integers, you can think of walking along a number line. Start at 0. Face the positive integers.

| Walk forward 4 steps for ⁺4. | The subtraction sign (−) means *turn around.* | Then walk backward 3 steps for ⁻3. |

You stop at ⁺7. So ⁺4 − ⁻3 = ⁺7

The problem above shows how to subtract a negative integer from a positive integer. Other cases that exist are shown below.

Example B

Find ⁺2 − ⁺4.

Start at 0. Face the positive integers.

| Walk forward 2 steps for ⁺2. | The subtraction sign (−) means *turn around.* | Then walk forward 4 steps for ⁺4. |

You stop at ⁻2. So ⁺2 − ⁺4 = ⁻2

Example C

Find ⁻6 − ⁻3.

Start at 0. Face the positive integers.

| Walk backward 6 steps for ⁻6. | The subtraction sign (−) means *turn around.* | Then walk backward 3 steps for ⁻3. |

You stop at ⁻3. So ⁻6 − ⁻3 = ⁻3

718

Source: Scott Foresman-Addison Wesley, Grade 5, 2008 (p. 718).

NOW TRY THIS 5-3 Suppose a mail carrier brings you three letters, one with a check for $25 and the other two with bills for $15 and $20, respectively. You record this as $25 + {}^-15 + {}^-20$, or ${}^-10$; that is, you are $10 poorer. Suppose that the next day you find out that the bill for $20 was actually intended for someone else and therefore you give it back to the delivery person. You record your new balance as

$$^-10 - {}^-20$$

or as

$$25 + {}^-15 + {}^-20 - {}^-20$$

which equals $25 + {}^-15$, or 10.

   For each of the following, make up a mail delivery story and explain how your story can help to find the answer.

a. $23 + {}^-13 + {}^-12$
b. $18 - {}^-37$

Pattern Model for Subtraction

By using inductive reasoning, we can find the difference of two integers by considering the following patterns, where we start with subtractions that we already know how to do. Both the following pattern on the left and the pattern on the right start with $3 - 2 = 1$.

$$
\begin{array}{ll}
3 - 2 = 1 & \quad 3 - 2 = 1 \\
3 - 3 = 0 & \quad 3 - 1 = 2 \\
3 - 4 = ? & \quad 3 - 0 = 3 \\
3 - 5 = ? & \quad 3 - {}^-1 = ?
\end{array}
$$

In the pattern on the left, the difference decreases by 1. If we continue the pattern, we have $3 - 4 = {}^-1$ and $3 - 5 = {}^-2$. In the pattern on the right, the difference increases by 1. If we continue the pattern, we have $3 - {}^-1 = 4$ and $3 - {}^-2 = 5$.

Subtraction Using the Missing-Addend Approach

Subtraction of integers, like subtraction of whole numbers, can be defined in terms of addition. Using the missing-addend approach, $5 - 3$ can be computed by finding a whole number n as follows:

$$5 - 3 = n \quad \text{if, and only if,} \quad 5 = 3 + n$$

Because $3 + 2 = 5, n = 2$.

   Similarly, we compute $3 - 5$ as follows:

$$3 - 5 = n \quad \text{if, and only if,} \quad 3 = 5 + n$$

Because $5 + {}^-2 = 3, n = {}^-2$. In general, for integers a and b, we have the following definition of *subtraction*.

Definition of Subtraction

For integers a and b, $a - b$ is the unique integer n such that $a = b + n$.

> **REMARK** Addition "undoes" subtraction; that is, $(a - b) + b = a$. Also, subtraction "undoes" addition; that is, $(a + b) - b = a$.

Example 5-5

Use the definition of subtraction to compute the following:

a. $3 - 10$
b. $^-2 - 10$

Solution **a.** Let $3 - 10 = n$. Then $10 + n = 3$, so $n = {}^-7$. Therefore, $3 - 10 = {}^-7$.
b. Let $^-2 - 10 = n$. Then $10 + n = {}^-2$, so $n = {}^-12$. Therefore, $^-2 - 10 = {}^-12$.

Subtraction Using Adding the Opposite Approach

Next, study the partial student page from *Scott Foresman-Addison Wesley Mathematics, Grade 6, 2008* (p. 423). Consider the subtractions and additions in parts A–D and then the rule that the students discovered. This technique is very useful in performing integer subtractions.

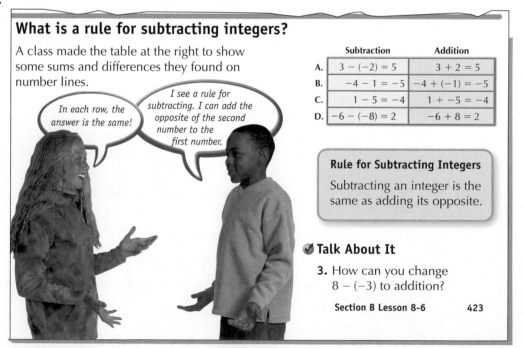

School Book Page SUBTRACTING INTEGERS

What is a rule for subtracting integers?

A class made the table at the right to show some sums and differences they found on number lines.

	Subtraction	Addition
A.	$3 - (-2) = 5$	$3 + 2 = 5$
B.	$-4 - 1 = -5$	$-4 + (-1) = -5$
C.	$1 - 5 = -4$	$1 + -5 = -4$
D.	$-6 - (-8) = 2$	$-6 + 8 = 2$

In each row, the answer is the same!

I see a rule for subtracting. I can add the opposite of the second number to the first number.

Rule for Subtracting Integers

Subtracting an integer is the same as adding its opposite.

☑ **Talk About It**

3. How can you change $8 - (-3)$ to addition?

Section B Lesson 8-6   423

Source: Scott Foresman-Addison Wesley, Mathematics 2008, Grade 6 (p. 423).

From our previous work with addition of integers, we know that $3 - 5 = {}^-2$ and $3 + {}^-5 = {}^-2$. Hence, $3 - 5 = 3 + {}^-5$. In general, the following is true.

Theore 5–4

For all integers a and b, $a - b = a + {}^-b$.

The preceding theorem can be justified using the fact that the equation $b + x = a$ has a unique solution for x. From the definition of subtraction, the solution of the equation is $a - b$. To show that $a - b = a + {}^-b$, we only need to show that $a + {}^-b$ is also a solution. For that purpose, we substitute $a + {}^-b$ for x and check if $b + (a + {}^-b) = a$:

$$b + (a + {}^-b) = b + ({}^-b + a)$$
$$= (b + {}^-b) + a$$
$$= 0 + a$$
$$= a$$

Consequently, $a - b = a + {}^-b$.

> **REMARK** Sometimes the preceding theorem is used as the definition of subtraction.

NOW TRY THIS 5-4

a. Is the set of integers closed under subtraction? Why?
b. Do the commutative, associative, or identity properties hold for subtraction of integers? Why?

 Many calculators have a change-of-sign key, either $\boxed{\text{CHS}}$ or $\boxed{+/-}$. Other calculators use $\boxed{(-)}$, a key that allows computation with integers. For example, to compute $8 - ({}^-3)$, we would press $\boxed{8}\ \boxed{-}\ \boxed{3}\ \boxed{+/-}\ \boxed{=}$. Investigate what happens if you press $\boxed{8}\ \boxed{-}\ \boxed{-}\ \boxed{3}\ \boxed{=}$.

Example 5-6

Using the fact that $a - b = a + {}^-b$, compute each of the following:

a. $2 - 8$ **b.** $2 - {}^-8$ **c.** ${}^-12 - {}^-5$ **d.** ${}^-12 - 5$

Solution   **a.** $2 - 8 = 2 + {}^-8 = {}^-6$
         b. $2 - {}^-8 = 2 + {}^-({}^-8) = 2 + 8 = 10$
         c. ${}^-12 - {}^-5 = {}^-12 + {}^-({}^-5) = {}^-12 + 5 = {}^-7$
         d. ${}^-12 - 5 = {}^-12 + {}^-5 = {}^-17$

Example 5-7

Use the fact that $a - b = a + ({}^-b)$ and the theorems involving the additive inverse to write expressions equal to each of the following without parentheses.

a. ${}^-(b - c)$ **b.** $a - (b + c)$

Solution   **a.** ${}^-(b - c) = {}^-(b + {}^-c) = {}^-b + {}^-({}^-c) = {}^-b + c$
         b. $a - (b + c) = a + {}^-(b + c) = a + ({}^-b + {}^-c) = (a + {}^-b) + {}^-c = a + {}^-b + {}^-c$

REMARK It is possible to simplify the answers in Example 5-7(a) and (b) further, as follows: $^-b + c = c + ^-b = c - b$; and $a + ^-b + ^-c = (a - b) - c$.

Example 5-8

Simplify each of the following:

a. $2 - (5 - x)$ **b.** $5 - (x - 3)$ **c.** $^-(x - y) - y$

Solution

a. $2 - (5 - x) = 2 + ^-(5 + ^-x)$
$= 2 + ^-5 + ^-(^-x)$
$= 2 + ^-5 + x$
$= ^-3 + x$ or $x - 3$

b. $5 - (x - 3) = 5 + ^-(x + ^-3)$
$= 5 + ^-x + ^-(^-3)$
$= 5 + ^-x + 3$
$= 8 + ^-x$
$= 8 - x$

c. $^-(x - y) - y = ^-(x + ^-y) + ^-y$
$= [^-x + ^-(^-y)] + ^-y$
$= (^-x + y) + ^-y$
$= ^-x + (y + ^-y)$
$= ^-x + 0$
$= ^-x$

Order of Operations

Subtraction in the set of integers is neither commutative nor associative, as illustrated in these counterexamples:

$$5 - 3 \neq 3 - 5 \quad \text{because} \quad 2 \neq ^-2$$
$$(3 - 15) - 8 \neq 3 - (15 - 8) \quad \text{because} \quad ^-20 \neq ^-4$$

An expression such as $3 - 15 - 8$ is ambiguous unless we know in which order to perform the subtractions. Mathematicians agree that $3 - 15 - 8$ means $(3 - 15) - 8$; that is, the subtractions in $3 - 15 - 8$ are performed in order from left to right. Similarly, $3 - 4 + 5$ means $(3 - 4) + 5$ and not $3 - (4 + 5)$. Thus, $(a - b) - c$ may be written without parentheses as $a - b - c$. Order of operations for integers will be revisited after multiplication and division are discussed.

Example 5-9

Compute each of the following:

a. $2 - 5 - 5$ **b.** $3 - 7 + 3$ **c.** $3 - (7 - 3)$

Solution

a. $2 - 5 - 5 = ^-3 - 5 = ^-8$
b. $3 - 7 + 3 = ^-4 + 3 = ^-1$
c. $3 - (7 - 3) = 3 - 4 = ^-1$

TECHNOLOGY CORNER

a. On a graphing calculator or a spreadsheet, graph the function with equation $y = x - ^-4$.
b. Using the graph in (a), describe what happens as x takes on values that are less than $^-4$, equal to $^-4$, and greater than $^-4$.

Assessment 5-1A

1. Find the additive inverse of each of the following integers. Write your answer in the simplest possible form.
   a. 2   **b.** $^-5$
   c. m **d.** 0
   e. ^-m **f.** $a + b$

2. Simplify each of the following:
   a. $^-(^-2)$ **b.** $^-(^-m)$ **c.** $^-0$

3. Evaluate each of the following:
   a. $|^-5|$ **b.** $|10|$
   c. $^-|^-5|$ **d.** $^-|5|$

4. Demonstrate each of the following additions using the charged-field or chip model:
   a. $5 + {}^-3$ **b.** $^-2 + 3$
   c. $^-3 + 2$ **d.** $^-3 + {}^-2$

5. Demonstrate each of the additions in problem 4 using a number-line model.

6. Compute each of the following using $a - b = a + {}^-b$:
   a. $3 - {}^-2$
   b. $^-3 - 2$
   c. $^-3 - {}^-2$

7. Answer each part of problem 6 using the definition of subtraction with the missing-addend approach.

8. Write an addition fact that corresponds to each of the following sentences and then answer the question:
   a. A certain stock dropped 17 points and the following day gained 10 points. What was the net change in the stock's worth?
   b. The temperature was $^-10°C$ and then it rose by 8°C. What is the new temperature?
   c. The plane was at 5000 ft and dropped 100 ft. What is the new altitude of the plane?

9. On January 1, Jane's bank balance was $300. During the month, she wrote checks for $45, $55, $165, $35, and $100 and made deposits of $75, $25, and $400.
   a. If a check is represented by a negative integer and a deposit by a positive integer, express Jane's transactions as a sum of positive and negative integers.
   b. What was the balance in Jane's account at the end of the month?

10. Use a number-line model to find the following:
    a. $^-4 - {}^-1$ **b.** $^-4 - {}^-3$

11. Use patterns to show the following:
    a. $^-4 - {}^-1 = {}^-3$ **b.** $^-2 - 1 = {}^-3$

12. Perform each of the following:
    a. $^-2 + (3 - 10)$
    b. $[8 - (^-5)] - 10$
    c. $(^-2 - 7) + 10$

13. In each of the following, write a subtraction problem that corresponds to the question and an addition problem that corresponds to the question and then answer the questions:

a. The temperature is 55°F and is supposed to drop 60°F by midnight. What is the expected midnight temperature?
b. Moses has overdraft privileges at his bank. If he had $200 in his checking account and he wrote a $220 check, what is his balance?

14. Motor oils protect car engines over a range of temperatures. These oils have names like 10W–40 or 5W–30. The following graph shows the temperatures, in degrees Fahrenheit, at which the engine is protected by a particular oil. Using the graph, find which oils can be used for the following temperatures:
    a. Between $^-5°$ and 90°
    b. Below $^-20°$
    c. Between $^-10°$ and 50°
    d. From $^-20°$ to over 100°
    e. From $^-8°$ to 90°

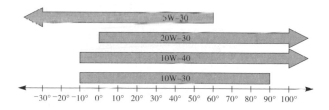

15. Simplify each of the following as much as possible. Show all work.
    a. $3 - (2 - 4x)$
    b. $x - (^-x - y)$

16. For which integers a, b, and c does $a - b - c = a - (b - c)$? Justify your answer.

17. Let W stand for the set of whole numbers, I the set of integers, I^+ the set of positive integers, and I^- the set of negative integers. Find each of the following:
    a. $W \cup I$
    b. $W \cap I$
    c. $I^+ \cup I^-$
    d. $I^+ \cap I^-$
    e. $W - I$
    f. $I - W$

18. Let $f(x) = {}^-x - 1$ with domain I. Find the following:
    a. $f(^-1)$ **b.** $f(100)$
    c. $f(^-2)$ **d.** $f(^-a)$ in terms of a
    e. For which values of x will the output be 3?

19. Find all integers x, if there are any, such that the following are true:
    a. ^-x is positive.
    b. ^-x is negative.
    c. $^-x - 1$ is positive.
    d. $|x| = 2$.

20. Let $f(x) = |1 - x|$ with domain I. Find the following:
    a. $f(10)$ **b.** $f(^-1)$
    c. All the inputs for which the output is 1
    d. The range

21. In each of the following, find all integers x satisfying the given equation:
    a. $|x - 6| = 6$
    b. $|x| + 2 = 10$
    c. $|^-x| = |x|$

22. Determine how many integers there are between the following given integers (not including the given integers):
    a. 10 and 100   **b.** $^-30$ and $^-10$

23. Suppose $a = 6, b = 5, c = 4$, and $d = ^-3$. Insert parentheses in the expression $a - b - c - d$ to obtain the greatest possible and the least possible values. What are these values?

24. An arithmetic sequence may have a positive or negative difference. In each of the following arithmetic sequences, find the difference and write the next two terms:
    a. $0, ^-3, ^-6, ^-9$
    b. $x + y, x, x - y$

25. In an arithmetic sequence, the eighth term minus the first term equals 21. The sum of the first and the eighth term is $^-5$. Find the fifth term of the sequence.

26. Classify each of the following as true or false. If false, show a counterexample that makes it false.
    a. $|^-x| = |x|$
    b. $|x - y| = |y - x|$
    c. $|^-x + ^-y| = |x + y|$

27. Solve the following equations:
    a. $x + 7 = 3$ **b.** $^-10 + x = ^-7$
    c. $^-x = 5$

28. Complete each of the following integer arithmetic problems on your calculator, making use of the change-of-sign key. For example, on some calculators to find $^-5 + ^-4$, press $\boxed{5}\ \boxed{+/-}\ \boxed{+}\ \boxed{4}\ \boxed{+/-}\ \boxed{=}$.
    a. $^-12 + ^-6$ **b.** $^-12 + 6$
    c. $27 + ^-5$ **d.** $^-12 - 6$
    e. $16 - ^-7$

Assessment 5-1B

1. Find the additive inverse of each of the following integers. Write your answer in the simplest possible form.
    a. 3   **b.** $^-4$
    c. q **d.** 6
    e. ^-n **f.** $3 + x$

2. Simplify each of the following:
    a. $^-(^-5)$
    b. $^-(^-x)$

3. Evaluate each of the following:
    a. $|^-3|$ **b.** $|15|$ **c.** $^-|^-3|$

4. Demonstrate each of the following additions using the charged-field or chip model:
    a. $^-2 + 5$ **b.** $^-5 + 2$ **c.** $^-3 + ^-3$

5. Demonstrate each of the additions in problem 4 using a number-line model.

6. Compute each of the following using $a - b = a + ^-b$.
    a. $^-3 - 5$
    b. $5 - (^-3)$

7. Answer each part of problem 6 using the definition of subtraction with the missing-addend approach.

8. Write an addition fact that corresponds to each of the following sentences and then answer the question:
    a. A visitor in a Las Vegas casino lost $200, won $100, and then lost $50. What is the change in the gambler's net worth?

    b. In four downs, the football team lost 2 yd, gained 7 yd, gained 0 yd, and lost 8 yd. What is the total gain or loss?

9. Use a number-line model to find the following:
    a. $^-3 - ^-2$ **b.** $^-4 - 3$

10. Use patterns to show the following:
    a. $^-2 - ^-3 = 1$ **b.** $^-3 - 2 = ^-5$

11. Perform each of the following:
    a. $^-2 - (7 + 10)$
    b. $8 - 11 - 10$
    c. $^-2 - 7 + 3$

12. Answer each of the following:
    a. In a game of Triominoes, Jack's scores in five successive turns are $17, ^-8, ^-9, 14$, and 45. What is his total at the end of five turns?
    b. The largest bubble chamber in the world is 15 ft in diameter and contains 7259 gal of liquid hydrogen at a temperature of $^-247°C$. If the temperature is dropped by $11°C$ per hour for 2 consecutive hours, what is the new temperature?
    c. The greatest recorded temperature ranges in the world are around the "cold pole" in Siberia. Temperatures in Verkhoyansk have varied from $^-94°F$ to $98°F$. What is the difference between the high and low temperatures in Verkhoyansk?

13. Simplify each of the following as much as possible. Show all work.
    a. $4x - 2 - 3x$ **b.** $4x - (2 - 3x)$
14. Let W stand for the set of whole numbers, I the set of integers, I^+ the set of positive integers, and I^- the set of negative integers. Find each of the following:
    a. $W - I^+$ **b.** $W - I^-$ **c.** $I \cap I$
15. **a.** Prove that $^-x - y = {}^-y - x$; for all integers x and y.
    b. Does part (a) imply that subtraction is commutative? Explain.
16. Complete the magic square using the following integers: $^-13, \ ^-10, \ ^-7, \ ^-4, 2, 5, 8, 11$.

17. Let $f(x) = {}^-3x - 2$ with domain I. Find the following:
    a. $f(^-1)$ **b.** $f(100)$
    c. $f(^-2)$ **d.** $f(^-a)$ in terms of a
    e. For which values of x will the output be $^-11$?
18. Find all integers x, if there are any, such that the following are true:
    a. $^-|x| = 2$.
    b. $^-|x|$ is negative.
    c. $^-|x|$ is positive.
    d. $^-x - 1$ is positive.
    e. $^-x - 1$ is negative.
19. Let $f(x) = |x - 5|$ with domain I. Find the following:
    a. $f(10)$ **b.** $f(^-1)$
    c. All the inputs for which the output is 7
    d. The range
20. **a.** For each of the following functions, find $f(f(x))$:
       i. $f(x) = x$
       ii. $f(x) = {}^-x$
       iii. $f(x) = {}^-x + 2$
    b. Interpret your answers in part (a) using the function machine model.
    c. Find other functions for which $f(f(x)) = x$. Justify your answer.
21. By the definition of absolute value, the function $f(x) = |x|$ can be written as follows:

    $$f(x) = \begin{cases} x, & \text{if } x \geq 0 \\ -x, & \text{if } x < 0 \end{cases}$$

    Write the function $f(x) = |x - 6|$ in a similar way without absolute value.

22. Determine how many integers there are between the following given integers (not including the given integers):
    a. $^-10$ and 10
    b. x and y (if $x < y$)
23. From midnight to 1:00 A.M in January, the temperature dropped 15°F. After it dropped, the outside temperature was $^-12$°F. What was the temperature at midnight?
24. An arithmetic sequence may have a positive or negative difference. In each of the following arithmetic sequences, find the difference and write the next two terms:
    a. $7, 3, ^-1, ^-5$
    b. $1 - 3x, 1 - x, 1 + x$
25. Find the sums of the following arithmetic sequences:
    a. $^-20 + {}^-19 + {}^-18 + \ldots + 18 + 19 + 20$
    b. $100 + 99 + 98 + \ldots + {}^-50$
    c. $100 + 98 + 96 + \ldots + {}^-6$
26. Classify each of the following as true or false. If false, show a counterexample that makes it false.
    a. $|x^2| = x^2$ **b.** $|x^3| = x^3$
    c. $|x^3| = x^2|x|$
27. Solve the following equations:
    a. $^-x + 5 = 7$ **b.** $1 - x = {}^-13$
    c. $^-x - 8 = {}^-9$
28. Assume that gear A has 56 teeth and gear B has 14 teeth. Suppose that the number of counterclockwise rotations is designated by a positive number and the number of clockwise rotations by a negative number. If gear A rotates 7 times per minute, how many times per minute does gear B rotate? Explain your reasoning.

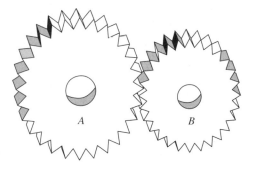

29. Estimate each of the following and then use a calculator to find the actual answer:
    a. $343 + {}^-42 - 402$
    b. $^-1992 + 3005 - 497$
    c. $992 - {}^-10,003 - 101$
    d. $^-301 - {}^-1303 + 4993$
30. Find a quick way to calculate the following:
    $1 - 2 + 3 - 4 + 5 - 6 + \ldots - 2004 + 2005 - 2006 + 2007?$

Mathematical Connections 5-1

Communication

1. A turnpike driver had car trouble. He knew that he had driven 12 mi from milepost 68 before the trouble started. Assuming he is confused and disoriented when he calls on his cellular phone for help, how can he determine his possible location? Explain.

2. Dolores claims that the best way to understand that $a - b = a + {}^-b$, for all integers a and b, is to show that when you add b to each expression you get equal answers.
   a. Explain why you think Dolores is making this claim.
   b. Do you agree with Dolores that her approach is the "best way"? If not, what is a better approach?

3. Addition of integers with like signs can be described using absolute values as follows:

   To add integers with like signs, add the absolute values of the integers. The sum has the same sign as the integers.

   Describe in a similar way how to add integers with unlike signs.

4. Explain why $b - a$ and $a - b$ are additive inverses of each other.

5. a. The absolute value of an integer is never negative. Does this contradict the fact that the absolute value of x could be equal to ${}^-x$? Explain why or why not.
   b. Explain how to write the additive inverse of $a - b - c$ using the least number of symbols.

6. If an integer a is pictured on the number line, then the distance from the point on the number line that represents the integer to the origin is $|a|$. Using this idea, answer the following:
   a. Explain why $|a - b|$ is the distance between the points that represent the integers a and b.
   b. One way to define "less than" for integers is as follows: $a < b$ if, and only if, a is to the left of b on the number line. Consequently, $b > a$ if, and only if, b is to the right of a. Use these ideas to mark on a number line all integers x such that
      i. $|x| < 5$.   ii. $|x| < 1$.
      iii. $|x| \geq 5$.   iv. $|x| > {}^-1$.

7. Recall the definition of less than for whole numbers using addition and define $a < b$ when a and b are any integers. Use your definition to show that ${}^-8 < {}^-7$.

Open-Ended

8. Describe a realistic word problem that models ${}^-50 + ({}^-85) - ({}^-30)$.

9. In a library some floors are below ground level and others are above ground level. If the ground-level floor is designated the zero floor, design a system to number the floors.

10. a. I am choosing an integer. I then subtract 10 from the integer, take the opposite of the result, add ${}^-3$, and find the opposite of the new result. My result is ${}^-3$. What is the original number?
    b. Judy wants to do the activity in part (a) with her classmates. Each classmate probably chooses a different number and Judy wants to tell each classmate quickly what number was chosen. Judy figures out that the only thing she needs to do is to add 7 to each answer she gets. Does this always work? Explain why or why not.
    c. Come up with your own "trick" similar to the one in part (b) that works for each answer you get from your classmates.

11. a. Write a function $f(x)$ such that for all integer inputs the outputs will be negative.
    b. Write a function $f(x)$ such that the sequence $f({}^-1), f({}^-2), f({}^-3), \ldots$ is an arithmetic sequence.

Cooperative Learning

12. Examine several elementary mathematics textbooks. Report on how addition and subtraction of integers is treated, and on how various properties are justified. Discuss in your group how the treatment of addition and subtraction of integers presented in this section compares to the treatment in elementary textbooks.

13. Look at several history of mathematics books and the Internet and report in your group on when and how negative integers were introduced first.

14. Number each card in a set of 21 3-by-5 cards with an integer from ${}^-10$ to 10. Lay the cards on the floor to form a number line. Choose someone from your group to act like the hiker on the student pages for number-line addition and subtraction. Give the hiker directions to walk the number line to solve problems 5 and 9 in Problem Set 5-1A. Try other addition and subtraction problems to make sure that the number-line model is understood by everyone in your group and could be used in an elementary classroom.

Questions from the Classroom

15. A fourth-grade student devised the following subtraction algorithm for subtracting 84 − 27:
   4 minus 7 equals negative 3.

$$\begin{array}{r} 84 \\ -\ 27 \\ \hline ^-3 \end{array}$$

   80 minus 20 equals 60.

$$\begin{array}{r} 84 \\ -\ 27 \\ \hline ^-3 \\ 60 \end{array}$$

   60 plus negative 3 equals 57.

$$\begin{array}{r} 84 \\ -\ 27 \\ \hline ^-3 \\ +\ 60 \\ \hline 57 \end{array}$$

   Thus the answer is 57. What is your response as a teacher?

16. An eighth-grade student claims she can prove that subtraction of integers is commutative. She points out that if a and b are integers, then $a - b = a + {}^-b$. Since addition is commutative, so is subtraction. What is your response?

17. A student had the following picture of an integer and its opposite. Other students in the class objected, saying that ^-a should be to the left of 0. How do you respond?

18. A student found that addition of integers can be performed by finding the sum or the difference of the absolute values of these integers and then attaching the "−" sign if necessary. She would like to know if this is always true. How do you respond?

Third International Mathematics and Science Study (TIMSS) Question

When Tracy left for school, the temperature was minus 3 degrees.

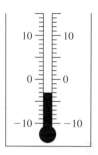

At recess, the temperature was 5 degrees.

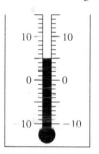

How many degrees did the temperature rise?
a. 2 degrees **b.** 3 degrees **c.** 5 degrees **d.** 8 degrees

TIMSS, Grade 4, 2003

National Assessment of Educational Progress (NAEP) Question

Paco had 32 trading cards. He gave N trading cards to his friend. Which expression tells how many trading cards Paco has now?
a. $32 + N$ **b.** $32 - N$ **c.** $N - 32$ **d.** $32 \div N$

NAEP, Grade 4, 2007

BRAIN TEASER If the digits 1 through 9 are written in order, it is possible to place plus and minus signs between the numbers or to use no operation symbol at all to obtain a total of 100. For example,

$$1 + 2 + 3 - 4 + 5 + 6 + 78 + 9 = 100$$

Can you obtain a total of 100 using fewer plus or minus signs than in the given example? Note that digits, such as 7 and 8 in the example, may be combined.

5-2   Multiplication and Division of Integers

We approach multiplication of integers through a variety of models: *patterns*, *charged-field*, *chip*, and *number-line*. Note that the reasoning with these models is inductive reasoning and therefore does not constitute a proof.

Patterns Model for Multiplication of Integers

We may approach multiplication of integers by using repeated addition. For example, if a running back lost 2 yd on each of three carries in a football game, then he had a net loss of $^-2 + {}^-2 + {}^-2$, or $^-6$, yards. Since $^-2 + {}^-2 + {}^-2$ can be written as $3(^-2)$, using repeated addition, we have $3(^-2) = {}^-6$.

Consider $(^-2)3$. It is meaningless to say that there are $^-2$ threes in a sum. But if the commutative property of multiplication is to hold for all integers, we must have $(^-2)3 = 3(^-2) = {}^-6$.

Next, consider $(^-3)(^-2)$. We can develop the following pattern:

$$3(^-2) = {}^-6$$
$$2(^-2) = {}^-4$$
$$1(^-2) = {}^-2$$
$$0(^-2) = 0$$
$$^-1(^-2) = ?$$
$$^-2(^-2) = ?$$
$$^-3(^-2) = ?$$

The first four products, $^-6, {}^-4, {}^-2$, and 0, are terms in an arithmetic sequence with a fixed difference of 2. If the pattern continues, the next three terms in the sequence are 2, 4, and 6. Thus it appears that $(^-3)(^-2) = 6$. Likewise, $(^-2)(^-3) = 6$.

REMARK Notice the phrase "it appears that $(^-3)(^-2) = 6$." Later in this section, we explore why $^-3(^-2) = 6$.

TECHNOLOGY CORNER On a spreadsheet, in column A enter 5 as the first entry and then write a formula to add $^-1$ to 5 for the second entry. Then add $^-1$ to the second entry and fill down, continuing the pattern. In column B, repeat the process. In column C, find the product of the respective entries in columns A and B. What patterns do you observe?

Next we approach multiplication of integers using the chip model, the charged-field model, and the number-line model. In all of these models we start with 0, represented possibly in various ways.

Chip Model and the Charged-Field Model for Multiplication

The *chip model* and the *charged-field model* can both be used to illustrate multiplication of integers. Consider Figure 5-11(a), where $3(^-2)$ is pictured using a chip model. The product $3(^-2)$ is interpreted as putting in 3 groups of 2 red chips each. In Figure 5-11(b), $3(^-2)$ is pictured as 3 groups of 2 negative charges.

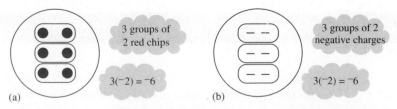

Figure 5-11

To find $(^-3)(^-2)$ using the chip model, we interpret the signs as follows: $^-3$ is taken to mean "*remove 3 groups of*"; $^-2$ is taken to mean "*2 red chips.*" To do this, we start with a value of 0 that includes at least 6 red chips, as shown in Figure 5-12(a). When we remove 6 red chips, we are left with 6 black chips. The result is a positive 6, so $(^-3)(^-2) = 6$. Similar reasoning can be used in Figure 5-12(b) with the charged-field model.

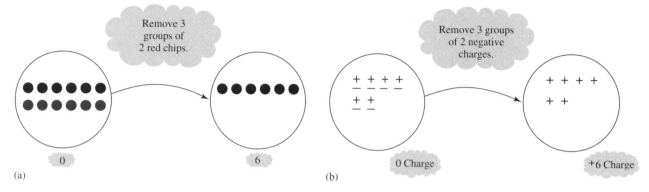

Figure 5-12

Number-Line Model

As with addition and subtraction, we demonstrate multiplication by using a hiker moving along a number line, according to the following rules:

1. Traveling to the left (west) means moving in the negative direction, and traveling to the right (east) means moving in the positive direction.
2. Time in the future is denoted by a positive value, and time in the past is denoted by a negative value.

Consider the number line shown in Figure 5-13. Various cases using this number line are given next.

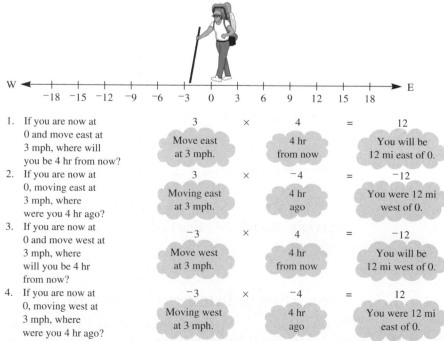

Figure 5-13

An alternative use of the number line to show multiplication of integers is given in Figure 5-14.

a. $3 \cdot 2$ means three groups of 2 each: $3 \cdot 2 = 6$.

b. $3(^-2)$ means three groups of $^-2$ each: $3(^-2) = ^-6$.

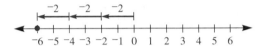

c. The integers 3 and $^-3$ are opposites. You can think of $(^-3)2$ as the opposite of three groups of 2 each. So $(^-3)2 = ^-6$.

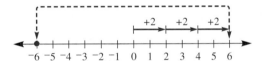

d. You can think of $(^-3)(^-2)$ as the opposite of three groups of $^-2$ each. Since $3(^-2) = ^-6, ^-3(^-2) = 6$.

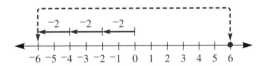

Figure 5-14

These models illustrate the following:

Theorem 5–5

For any whole numbers a and b, the following hold:

1. $(^-a)(^-b) = ab$
2. $(^-a)b = b(^-a) = ^-(ab)$

REMARK We will show later in this section that this theorem is true for all integers a and b.

Properties of Integer Multiplication

The set of integers has properties under multiplication analogous to those of the set of whole numbers under multiplication. These properties are summarized next.

Theorem 5–6:   Properties of Integer Multiplication

The set of integers I satisfies the following properties of multiplication for all integers $a, b, c \in I$:

Closure property of multiplication of integers ab is a unique integer.

Commutative property of multiplication of integers $ab = ba$.

Associative property of multiplication of integers $(ab)c = a(bc)$.

Multiplicative identity property   1 is the unique integer such that for all integers a, $1 \cdot a = a = a \cdot 1$.

Distributive properties of multiplication over addition for integers $a(b + c) = ab + ac$ and $(b + c)a = ba + ca$.

Zero multiplication property of integers   0 is the unique integer such that for all integers a, $a \cdot 0 = 0 = 0 \cdot a$.

An approach to showing that $(^-2)3 = \,^-(2 \cdot 3)$ uses the uniqueness theorem of additive inverses. If we can show that $(^-2)3$ and $^-(2 \cdot 3)$ are additive inverses of the same number, then they must be equal. By definition, the additive inverse of $(2 \cdot 3)$ is $^-(2 \cdot 3)$. That $(^-2)3$ is also the additive inverse of $2 \cdot 3$ can be proved by showing that $(^-2)3 + 2 \cdot 3 = 0$. The proof follows:

$$(^-2)3 + 2 \cdot 3 = (^-2 + 2)3 \qquad \text{Distributive property of multiplication over addition}$$
$$= 0 \cdot 3 \qquad \text{Additive inverse}$$
$$= 0 \qquad \text{Zero multiplication}$$

Now we have

$$(^-2)3 + 2 \cdot 3 = 0$$
$$-(2 \cdot 3) + 2 \cdot 3 = 0$$

Because $(^-2)3$ and $^-(2 \cdot 3)$ are both additive inverses of $(2 \cdot 3)$ and because the additive inverse must be unique, $(^-2)3 = \,^-(2 \cdot 3)$. Using this approach, we could prove the following theorem (the proof is explored in Assessment 5-2A):

Theorem 5–7

For every integer a, $(^-1)a = \,^-a$.

It is important to keep in mind that $(^-1)a = \,^-a$ is true for all integers a. Thus if we substitute $^-1$ for a, we get $(^-1)(^-1) = \,^-(^-1)$. Because $^-(^-1) = 1$, we have another justification for the fact that $(^-1)(^-1) = 1$. Using this result, the preceding property, and

Historical Note

Emmy Noether (1882–1935) made lasting contributions to the study of *rings*, algebraic systems among which is the set of integers. When she entered the University of Erlangen (Germany) in 1900, Emmy Noether was one of only two women enrolled. After completing her doctorate in 1907, she could not find a suitable job because she was a woman. In 1919, she got a university appointment without pay and only later a very modest salary. In 1933, along with many other scholars, she was dismissed from the University at Göttingen because she was Jewish. She emigrated to the United States and taught at Bryn Mawr College until her untimely death only 18 months after arriving in the United States.   ◆

the properties of integers listed earlier, we can show that $(^-a)b = ^-(ab)$ and that $(^-a)(^-b) = ab$ for all integers a and b as follows:

$$(^-a)b = [(^-1)a]b$$
$$= (^-1)(ab)$$
$$= ^-(ab)$$

Also:

$$(^-a)(^-b) = [(^-1)a][(^-1)b]$$
$$= [(^-1)(^-1)](ab)$$
$$= 1(ab)$$
$$= ab$$

We have established the following theorems:

Theorem 5–8

For all integers a and b,
$$(^-a)b = ^-(ab)$$
$$(^-a)(^-b) = ab$$

REMARK It is important to note that in Theorem 5–8, ^-a and ^-b are not necessarily negative and a and b are not necessarily positive.

The distributive property of multiplication over subtraction follows from the distributive property of multiplication over addition:

$$a(b - c) = a(b + ^-c)$$
$$= ab + a(^-c)$$
$$= ab + ^-(ac)$$
$$= ab - ac$$

Consequently, $a(b - c) = ab - ac$. From this and the commutative property of multiplication we see that $(b - c)a = ba - ca$.

Theorem 5–9: Distributive Property of Multiplication over Subtraction for Integers

For any integers a, b, and c,
$$a(b - c) = ab - ac \quad \text{and} \quad (b - c)a = ba - ca$$

Example 5-10

Simplify each of the following so that there are no parentheses in the final answer:

a. $(^-3)(x - 2)$ **b.** $(a + b)(a - b)$

Solution   **a.** $(^-3)(x - 2) = (^-3)x - (^-3)(2) = ^-3x - (^-6) = ^-3x + ^-(^-6) = ^-3x + 6$

b. $(a + b)(a - b) = (a + b)a - (a + b)b$
$$= (a^2 + ba) - (ab + b^2)$$
$$= a^2 + ba + ^-(ab + b^2)$$
$$= a^2 + ab + ^-(ab) + ^-b^2 \quad \text{(Note: } ^-b^2 \text{ means } ^-(b^2).)$$
$$= a^2 + 0 + ^-b^2$$
$$= a^2 - b^2$$

Thus, $(a + b)(a - b) = a^2 - b^2$.

The result $(a + b)(a - b) = a^2 - b^2$ in Example 5-10(b) is the **difference-of-squares** formula.

Example 5-11

Use the difference-of-squares formula to simplify the following:

a. $(4 + b)(4 - b)$ **b.** $(^-4 + b)(^-4 - b)$ **c.** $(x + 3)^2 - (x - 3)^2$

Solution   **a.** $(4 + b)(4 - b) = 4^2 - b^2 = 16 - b^2$
b. $(^-4 + b)(^-4 - b) = (^-4)^2 - b^2 = 16 - b^2$
c. $(x + 3)^2 - (x - 3)^2 = [(x + 3) + (x - 3)][(x + 3) - (x - 3)]$
$= 2x(x + 3 - x + 3)$
$= 2x \cdot 6$
$= 12x$

NOW TRY THIS 5-5 Determine how to use the difference-of-squares formula to compute the following mentally:

a. $101 \cdot 99$ **b.** $22 \cdot 18$ **c.** $24 \cdot 36$ **d.** $998 \cdot 1002$

When the distributive property of multiplication over subtraction is written in reverse order as

$$ab - ac = a(b - c) \quad \text{and} \quad ba - ca = (b - c)a$$

and similarly for addition, the expressions on the right of each equation are in *factored* form. We say that the common factor a has been *factored out*. Both the difference-of-squares formula and the distributive properties of multiplication over addition and subtraction can be used for factoring.

Example 5-12

Factor each of the following completely:

a. $x^2 - 9$ **b.** $(x + y)^2 - z^2$ **c.** $^-3x + 5xy$ **d.** $3x - 6$ **e.** $5x^2 - 2x^2$

Solution   **a.** $x^2 - 9 = x^2 - 3^2 = (x + 3)(x - 3)$
b. $(x + y)^2 - z^2 = (x + y + z)(x + y - z)$
c. $^-3x + 5xy = x(^-3 + 5y)$
d. $3x - 6 = 3(x - 2)$
e. $5x^2 - 2x^2 = (5 - 2)x^2 = 3x^2$

Integer Division

In the set of whole numbers, $a \div b$, where $b \neq 0$, is the unique whole number c such that $a = bc$. If such a whole number c does not exist, then $a \div b$ is undefined. Division on the set of integers is defined analogously.

Definition of Integer Division

If a and b are any integers, then $a \div b$ is the unique integer c, if it exists, such that $a = bc$.

REMARK Notice that $a \div b$, if it exists, is the solution of $a = bx$.

Example 5-13

Use the definition of integer division, if possible, to evaluate each of the following:

a. $12 \div (^{-}4)$ **b.** $^{-}12 \div 4$ **c.** $^{-}12 \div (^{-}4)$ **d.** $^{-}12 \div 5$
e. $(ab) \div b, b \neq 0$ **f.** $(ab) \div a, a \neq 0$

Solution   **a.** Let $12 \div (^{-}4) = c$. Then $12 = ^{-}4c$ and consequently $c = ^{-}3$. Thus, $12 \div (^{-}4) = ^{-}3$.

b. Let $^{-}12 \div 4 = c$. Then $^{-}12 = 4c$ and therefore $c = ^{-}3$. Thus, $^{-}12 \div 4 = ^{-}3$.

c. Let $^{-}12 \div (^{-}4) = c$. Then $^{-}12 = ^{-}4c$ and consequently $c = 3$. Thus, $^{-}12 \div (^{-}4) = 3$.

d. Let $^{-}12 \div 5 = c$. Then $^{-}12 = 5c$. Because no integer c exists to satisfy this equation (why?), we say that $^{-}12 \div 5$ is undefined over the set of integers.

e. Let $(ab) \div b = x$. Then $ab = bx$ and consequently $x = a$.

f. Let $(ab) \div a = x$. Then $ab = ax$ and hence $x = b$.

Example 5-13 suggests that *the quotient of two negative integers, if it exists, is a positive integer and the quotient of a positive and a negative integer, if it exists, or of a negative and a positive integer, if it exists, is negative.*

NOW TRY THIS 5-6   Use the definition of division for integers to show that dividing by 0 is not possible.

Order of Operations on Integers

When addition, subtraction, multiplication, division, and exponentiation appear without parentheses, exponentiation is done first in order from right to left, then multiplications and divisions in the order of their appearance from left to right, and then additions and subtractions in the order of their appearance from left to right. Arithmetic operations that appear inside parentheses must be done first.

Example 5-14

Evaluate each of the following:

a. $2 - 5 \cdot 4 + 1$ **b.** $(2 - 5)4 + 1$ **c.** $2 - 3 \cdot 4 + 5 \cdot 2 - 1 + 5$
d. $2 + 16 \div 4 \cdot 2 + 8$ **e.** $(^{-}3)^4$ **f.** $^{-}3^4$

Solution   **a.** $2 - 5 \cdot 4 + 1 = 2 - 20 + 1 = ^{-}18 + 1 = ^{-}17$

b. $(2 - 5)4 + 1 = (^{-}3)4 + 1 = ^{-}12 + 1 = ^{-}11$

c. $2 - 3 \cdot 4 + 5 \cdot 2 - 1 + 5 = 2 - 12 + 10 - 1 + 5 = 4$

d. $2 + 16 \div 4 \cdot 2 + 8 = 2 + 4 \cdot 2 + 8 = 2 + 8 + 8 = 10 + 8 = 18$

e. $(^{-}3)^4 = (^{-}3)(^{-}3)(^{-}3)(^{-}3) = 81$

f. $^{-}3^4 = ^{-}(3^4) = ^{-}(81) = ^{-}81$

REMARK   Notice that from Example 5-14(e) and (f), we have $(^{-}3)^4 \neq ^{-}3^4$. By convention, $(^{-}x)^4$ means $(^{-}x)(^{-}x)(^{-}x)(^{-}x)$ and $^{-}x^4$ means $^{-}(x^4)$ and therefore equals $^{-}(x \cdot x \cdot x \cdot x)$.

Ordering Integers

As with whole numbers, a number line as shown in Figure 5-15 can be used to describe greater-than and less-than relations for the set of integers. Because $^-5$ is to the left of $^-3$ on the number line, we say that "$^-5$ is less than $^-3$," and we write $^-5 < {}^-3$. We can also say that "$^-3$ is greater than $^-5$," and we can write $^-3 > {}^-5$.

Figure 5-15

Notice that since $^-5$ is to the left of $^-3$, there is a positive integer that can be added to $^-5$ to get $^-3$, namely, 2. Thus, $^-5 < {}^-3$ because $^-5 + 2 = {}^-3$. The definition of *less than* for integers is similar to that for whole numbers.

Definition of Less Than for Integers

For any integers a and b, a is less than b, written $a < b$, if, and only if, there exists a positive integer k such that $a + k = b$.

The last equation implies that $k = b - a$. Thus we have proved the following theorem:

Theorem 5–10

$a < b$ (or equivalently, $b > a$) if, and only if, $b - a$ is equal to a positive integer; that is, $b - a$ is greater than 0.

◆ *Research Note*

Students do not have a good understanding of the concepts of equivalent equations. For example, though able to use transformations to solve simple equations ($x + 2 = 5$ becomes $x + 2 - 2 = 5 - 2$), students seem unaware that each transformation produces an equivalent equation (Steinberg et al. 1990).   ◆

Using this theorem, $^-5 < {}^-3$ because $^-3 - ({}^-5) = {}^-3 + {}^-({}^-5) = {}^-3 + 5 = 2 > 0$. (Also $a \le b$ means that $a < b$ or $a = b$. Notice that $b > a$ if, and only if, $a < b$. Also $b \ge a$ if, and only if, $a \le b$.)

The preceding theorem can be used to justify each of the following for integers x, y, and n:

Theorem 5–11

a. If $x < y$ and n is any integer, then $x + n < y + n$.
b. If $x < y$, then $^-x > {}^-y$.
c. If $x < y$ and $n > 0$, then $nx < ny$.
d. If $x < y$ and $n < 0$, then $nx > ny$.

The justifications are given next.

a. Because $x < y$, $y - x > 0$. We need to show that $(y + n) - (x + n) > 0$. We have $y + n - (x + n) = y + n - x - n = y - x$. Because $y - x > 0$, we have $y + n - (x + n) > 0$, and hence $x + n < y + n$.

b. Because $x < y$, $y - x > 0$. We need to show that $^-x - ({}^-y) > 0$. We have $^-x - ({}^-y) = {}^-x + {}^-({}^-y) = {}^-x + y = y + {}^-x = y - x$. Because $y - x > 0$, we have $^-x - ({}^-y) > 0$, and hence $^-x > {}^-y$.

c. Because $x < y, y - x > 0$. We need to show that $ny - nx > 0$. We have $ny - nx = n(y - x)$. Because n is a positive integer and $y - x$ is positive, $n(y - x)$ must also be positive. Because $ny - nx > 0$, we have $nx < ny$.

d. To show that $nx > ny$, we need only show that $nx - ny > 0$. We have $nx - ny = n(x - y)$. Since $y - x > 0, x - y < 0$ (why?). Because $n < 0$ and $x - y < 0, n(x - y)$ is positive. Thus, $nx - ny > 0$ and hence, $nx > ny$.

From the Research Note on page 277 we see that students do not have good understanding of the concept of equivalent equations. Practice on this concept is given in Example 5-15.

Example 5-15

Use the theorems developed above to find all integers x that satisfy the following:

a. $x + 3 < {}^-2$
b. ${}^-x - 3 < 5$
c. If $x \leq {}^-2$, find the values of $5 - 3x$.

Solution   **a.** If $x + 3 < {}^-2$ then by Theorem 5–11 (a),

$$x + 3 + {}^-3 < {}^-2 + {}^-3$$
$$x < {}^-5, \quad x \text{ is an integer.}$$

We can also write the solution set (the set of all solutions) as

$$\{{}^-6, {}^-7, {}^-8, {}^-9, \ldots\}$$

Notice that, strictly speaking, we have only shown that every x that satisfies the first inequality also satisfies $x < {}^-5$. To be sure that $x < {}^-5$ represents all the solutions, we need to show the converse; that is, if $x < {}^-5$ then $x + 3 < {}^-2$. This can be easily done by adding 3 to both sides of $x < {}^-5$.

b. If ${}^-x - 3 < 5$, then

$$\begin{aligned}
{}^-x - 3 + 3 &< 5 + 3 \\
-x &< 8 \\
{}^-({}^-x) &> {}^-8 \quad \text{by Theorem 5–11 (b)} \\
x &> {}^-8, \quad x \text{ is an integer}
\end{aligned}$$

c. If $x \leq {}^-2$, then

$$\begin{aligned}
{}^-3x &\geq {}^-3({}^-2) \\
{}^-3x &\geq 6 \\
5 + {}^-3x &\geq 5 + 6 \\
5 - 3x &\geq 5 + 6 \\
5 - 3x &\geq 11; \text{ that is, all integers in the set } \{11, 12, 13, 14, \ldots\}.
\end{aligned}$$

Extending the Coordinate System

We extended a number line to include all integers. This new extended number line can become the x- and y-axes of a coordinate system. In Chapter 14 we investigate the coordinate system for all real numbers. Meanwhile, look at the student page where the coordinate system is extended to include negative integers. Answer the questions on the student page. Notice that when $x = 0$, the descriptions left or right of the y-axis and similarly for $y = 0$ are not true.

School Book Page THE COORDINATE SYSTEM

Lesson 8-11

Algebra

Key Idea
The location of a point on the coordinate plane can be described by an ordered pair of numbers.

Vocabulary
• coordinate plane
• quadrant
• ordered pair
• origin
• x-coordinate
• y-coordinate
• x-axis
• y-axis

Materials
• grid paper or
🛠 **tools**

Graphing Ordered Pairs

◄ LEARN ►

What is a coordinate plane?

The term *pixel* is a contraction of *picture elements* and is used to describe the dots on a computer display. Computer programmers use ordered pairs to position pixels on a display.

A **coordinate plane** is a grid containing two number lines that intersect in a right angle at zero. The number lines, which are called the **x- and y-axes,** divide the plane into four **quadrants.**

An **ordered pair** (x, y) of numbers gives the coordinates and location of a point.

✓ **WARM UP**
1. $\frac{-36}{-4}$
2. $(-9) \bullet |-9|$
3. $\frac{-105}{5}$
4. $(-13)(-11)$

The **x-coordinate** shows the position left or right of the y-axis.

The **y-coordinate** shows the position above or below of the x-axis.

If a point lies on an axis, it does not lie in a quadrant. Point S lies on the y-axis.

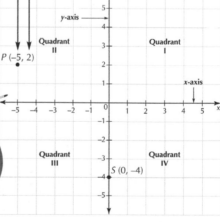

To locate any point, I need to know its horizontal and vertical distance and direction from (0, 0) or the **origin.**

✓ **Talk About It**

1. Where would the point Q lie if both its x- and y-coordinate are negative?

2. Where would the point R lie if its y-coordinate is 0?

Take It to the NET
More Examples
🌐 www.scottforesman.com

440

Source: Scott Foresman-Addison Wesley, Mathematics 2008, Grade 6 (p. 440).

NOW TRY THIS 5-7 In each of the following find and graph all points (x, y) satisfying the given condition, where x and y are integers.

a. $y = x$ **b.** $y = {}^-x$ **c.** $y = |x|$ **d.** $|x| + |y| = 5$

BRAIN TEASER   Express each of the numbers from 1 through 10 using four 4s and any operations. For example,

$$1 = 44 \div 44, \text{or}$$
$$1 = (4 \div 4)^{44}, \text{or}$$
$$1 = {}^-4 + 4 + (4 \div 4)$$

Assessment 5-2A

1. Use patterns to show that $({}^-1)({}^-1) = 1$.
2. Use the charged-field model to show that $({}^-4)({}^-2) = 8$.
3. Use the number-line model to show that $2({}^-4) = {}^-8$.
4. In each of the following charged-field models, the encircled charges are removed. Write the corresponding integer multiplication problem with its solution based on the model.

   a. 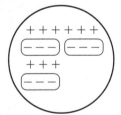   **b.**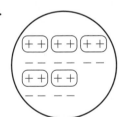

5. The number of students eating in the school cafeteria has been decreasing at the rate of 20 per year for many years. Assuming this trend continues, write a multiplication problem that describes the change in the number of students eating in the school cafeteria for each of the following:
   a. The change over the next 4 years
   b. The situation 4 years ago
   c. The change over the next n years
   d. The situation n years ago
6. Use the definition of division to find each quotient, if possible. If a quotient is not defined, explain why.
   a. ${}^-40 \div {}^-8$ **b.** ${}^-143 \div 13$
   c. ${}^-5 \div 0$
7. Evaluate each of the following, if possible:
   a. $({}^-10 \div {}^-2)({}^-2)$
   b. $({}^-10 \cdot 5) \div 5$
   c. ${}^-8 \div ({}^-8 + 8)$

   d. $({}^-6 + 6) \div ({}^-2 + 2)$
   e. $|{}^-24| \div [4(9 - 15)]$
8. Evaluate each of the following products and then, if possible, write two division statements that are equivalent to the given multiplication statements. If two division statements are not possible, explain why.
   a. $({}^-6)5$
   b. $({}^-5)({}^-4)$
   c. $({}^-3)0$
   d. $0 \cdot 0$
9. In each of the following, x and y are integers $y \neq 0$. Use the definition of division in terms of multiplication to perform the indicated operations. Write your answers in simplest form.
   a. $(4x) \div 4$
   b. $({}^-xy) \div y$
10. In a lab, the temperature of various chemical reactions is changing by a fixed number of degrees per minute. Write a numeric expression that describes each of the following:
   a. The temperature at 8:00 P.M. is 32°C. If it dropped 3°C per minute, what is the temperature at 8:30 P.M.?
   b. The temperature at 8:20 P.M. is 0°C. If it dropped 4°C per minute, what was the temperature at 7:55 P.M.?
   c. The temperature at 8:00 P.M. is ${}^-20$°C. If it is dropping 4°C per minute, what was the temperature at 7:30 P.M.?
   d. The temperature at 8:00 P.M. is 25°C. If it increased every minute by 3°C, what was the temperature at 7:40 P.M.?
11. If it was predicted that the farmland acreage lost to family dwellings over the next 9 years would be 12,000 acres per year, how much acreage would be lost to homes during this time period?

12. Show that the distributive property of multiplication over addition, $a(b + c) = ab + ac$, is true for each of the following values of a, b, and c:
    a. $a = {}^-1, b = {}^-5, c = {}^-2$
    b. $a = {}^-3, b = {}^-3, c = 2$

13. Compute each of the following:
    a. $({}^-2)^3$ b. $({}^-2)^4$
    c. $({}^-10)^5 \div ({}^-10)^2$
    d. $({}^-3)^5 \div ({}^-3)$
    e. $({}^-1)^{50}$ f. $({}^-1)^{151}$
    g. $^-2 + 3 \cdot 5 - 1$ h. $10 - 3 \cdot 7 - 4({}^-2) + 3$

14. Compute the following without using a calculator:
    a. $({}^-2)^{64} - 2^{64}$ b. $^-2^8 + 2^8$
    c. $^-({}^-2)^5 + 0 \cdot 9 - |7 - 15| - 15$

15. If x is an integer and $x \neq 0$, which of the following are always positive and which are always negative?
    a. $^-x^2$ b. x^2 c. $({}^-x)^2$
    d. $^-x^3$ e. $({}^-x)^3$

16. Which of the expressions in problem 15 are equal to each other for all values of x?

17. Identify the property of integers being illustrated in each of the following:
    a. $({}^-3)(4 + 5) = (4 + 5)({}^-3)$
    b. $({}^-4)({}^-7) \in I$
    c. $5[4({}^-3)] = (5 \cdot 4)({}^-3)$
    d. $({}^-9)[5 + ({}^-8)] = ({}^-9) \cdot 5 + ({}^-9)({}^-8)$

18. Simplify each of the following:
    a. $({}^-x)({}^-y)$ b. $^-2x({}^-y)$
    c. $^-2({}^-x + y) + x + y$ d. $^-1 \cdot x$

19. Multiply each of the following and combine terms where possible:
    a. $^-2(x - y)$ b. $x(x - y)$
    c. $^-x(x - y)$
    d. $^-2(x + y - z)$

20. Find all integers x (if possible) that make each of the following true:
    a. $^-3x = 6$ b. $^-3x = {}^-6$
    c. $^-2x = 0$ d. $5x = {}^-30$
    e. $x \div 3 = {}^-12$ f. $x \div ({}^-3) = {}^-2$
    g. $x \div ({}^-x) = {}^-1$
    h. $0 \div x = 0$

21. Solve each of the following for x.
    a. $^-3x - 8 = 7$
    b. $^-2(5x - 3) = 26$
    c. $3x - x - 2x = 3$
    d. $^-2(5x - 6) - 30 = {}^-x$
    e. $x^2 = 4$ f. $(x - 1)^2 = 9$
    g. $(x - 1)^2 = (x + 3)^2$
    h. $(x - 1)(x + 3) = 0$

22. Use the difference-of-squares formula to simplify each of the following, if possible:
    a. $52 \cdot 48$
    b. $(5 - 100)(5 + 100)$
    c. $({}^-x - y)({}^-x + y)$

23. Factor each of the following expressions completely.
    a. $3x + 5x$
    b. $xy + x$
    c. $x^2 + xy$
    d. $3xy + 2x - xz$
    e. $abc + ab - a$
    f. $16 - a^2$
    g. $4x^2 - 25y^2$

24. a. Use the distributive property of multiplication over addition or over subtraction and other properties to show that

$$(a - b)^2 = a^2 - 2ab + b^2$$

    b. Use your results from (a) to compute each of the following mentally:
    (i) 98^2 (*Hint:* Write $98 = 100 - 2$.)
    (ii) 99^2
    (iii) 997^2

25. In each of the following, find the next two terms. Assume the sequence is arithmetic or geometric, and find its difference or ratio and the nth term.
    a. $^-10, ^-7, ^-4, ^-1, 2, 5, \_, \_$
    b. $^-2, ^-4, ^-8, ^-16, ^-32, ^-64, \_, \_$
    c. $2, ^-2^2, 2^3, ^-2^4, 2^5, ^-2^6, \_, \_$

26. Find the sum of the first 100 terms in the arithmetic sequence $^-10, ^-7, ^-4, ^-1, 2, 5, \ldots$.

27. Find the first five terms of the sequences whose nth term is
    a. $n^2 - 10$.
    b. $^-5n + 3$.
    c. $({}^-2)^n - 1$.

28. Find the first two terms of an arithmetic sequence in which the fourth term is $^-8$ and the 101st term is $^-493$.

29. Tira noticed that every 30 sec, the temperature of a chemical reaction in her lab was decreasing by the same number of degrees. Initially, the temperature was 28°C and 5 min later, $^-12$°C. In a second experiment, Tira noticed that the temperature of the chemical reaction was initially $^-57$°C and was decreasing by 3°C every minute. If she started the two experiments at the same time, when were the temperatures of the reactions the same? What was that temperature?

30. Find all integer values (if any) of x and y for which the following are true.
    a. $xy = {}^-|x||y|$
    b. $^-x^2 = x^2$
    c. $x^2 > y^2$.

Assessment 5-2B

1. Use patterns to show that $(^-2)(^-2) = 4$.
2. Use the charged-field model to show that $(^-2)(^-2) = 4$.
3. Use the number-line model to show that $2(^-2) = ^-4$.
4. In each of the following charged-field models, the encircled charges are removed. Write the corresponding integer multiplication problem with its solution based on the model.

   a.   **b.**

   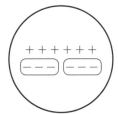   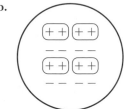

5. Use the definition of division to find each quotient, if possible. If a quotient is not defined, explain why.
   a. $143 \div (^-11)$ **b.** $0 \div (^-5)$
   c. $0 \div 0$

6. Evaluate each of the following, if possible:
   a. $(a \div b)b$ **b.** $(ab) \div b$
   c. $(^-8 + 8) \div 8$ **d.** $(^-23 - ^-7) \div 4$

7. Evaluate each of the following products and then, if possible, write two division statements that are equivalent to your multiplication statements. If two division statements are not possible, explain why.
   a. $(^-5)4$
   b. $(^-4)(^-3)$

8. Suppose a and b are integers and $a \div b$ is an integer.
   a. Use the definition of integer division to prove that if $c \neq 0$, then $(ac) \div (bc) = a \div b$.
   b. Why is the statement in (a) not true if $c = 0$? Is it true if $a \div b = 0$ and $c \neq 0$? Justify your answers.

9. In each of the following, x and y are integers. Use the definition of division in terms of multiplication to perform the indicated operations. Write your answers in simplest form.
   a. $(^-4x) \div x$
   b. $(^-10x + 5) \div 5$

10. In a lab, the temperature of various chemical reactions was changing by a fixed number of degrees per minute. Write a numeric expression that describes each of the following:
    a. The temperature at 8:00 A.M. is $^-5°C$. If it increases by d degrees per minute, what will the temperature be m minutes later?
    b. The temperature at 8:00 P.M. is $0°C$. If it dropped d degrees per minute, what was the temperature m minutes before?

c. The temperature at 8:00 P.M. is 20°C. If it increased every minute by d degrees, what was the temperature m minutes before?

11. **a.** On each of four consecutive plays in a football game, a team lost 11 yd. If lost yardage is interpreted as a negative integer, write the information as a product of integers and determine the total number of yards lost.
    b. If Jack Jones lost a total of 66 yd in 11 plays, how many yards, on the average, did he lose on each play?

12. Show that the distributive property of multiplication over addition, $a(b + c) = ab + ac$, is true for each of the following values of a, b, and c:
    a. $a = ^-5, b = 2, c = ^-6$
    b. $a = ^-2, b = ^-3, c = 4$

13. Compute each of the following:
    a. $10 - 3 - 12$ **b.** $10 - (3 - 12)$
    c. $(^-3)^2$ **d.** $^-3^2$
    e. $^-5^2 + 3(^-2)^2$ **f.** $^-2^3$
    g. $(^-2)^5$ **h.** $^-2^4$

14. Compute the following without using a calculator:
    a. $^-2^{63} + 2^{64}$
    b. $^-|^-6| - 8^2 + (^-1)^{49} \cdot 48 \div (^-4) \cdot 3 + (^-5)^3$

15. If x is an integer and $x \neq 0$, which of the following are always positive and which are always negative?
    a. $^-x^4$ **b.** $(^-x)^4$ **c.** x^4
    d. x **e.** ^-x

16. Which of the expressions in problem 15 are equal to each other for all values of x?

17. Identify the property of integers being illustrated in each of the following:
    a. $(^-2)(3) \in I$
    b. $(^-4)0 = 0$
    c. $^-2(3 + 4) = ^-2(3) + (^-2)4$
    d. $(^-2)3 = 3(^-2)$

18. Simplify each of the following:
    a. $x - 2(^-y)$
    b. $a - (a - b)(^-1)$
    c. $y - (y - x)(^-2)$
    d. $^-1(x - y) + x$

19. Multiply each of the following and combine terms where possible:
    a. $^-x(x - y - 3)$
    b. $(^-5 - x)(5 + x)$
    c. $(x - y - 1)(x + y + 1)$
    d. $(^-x^2 + 2)(x^2 - 1)$

20. Find all integers x (if possible) that make each of the following true:
    a. $x \div 0 = 1$

b. $x^2 = 9$

c. $x^2 = {}^-9$

d. ${}^-x \div {}^-x = 1$

e. ${}^-x^2$ is negative.

f. ${}^-(1 - x) = x - 1$

g. $x - 3x = {}^-2x$

h. ${}^-3(x + 2) = {}^-3x + 6$

21. Solve the following for x or find the values of the indicated expression:

a. $(2x - 1)^2 = (1 - 2x)^2$

b. $x^3 = {}^-2^9$

c. ${}^-6x > {}^-x + 20$

d. ${}^-5(x - 3) > {}^-5$

e. If $x > {}^-2$, find the values of $3 - 5x$.

f. If $x < 0$, find the values of $2 - 7x$.

22. Use the difference-of-squares formula to simplify each of the following, if possible:

a. $(2 + 3x)(2 - 3x)$

b. $(x - 1)(1 + x)$

c. $213^2 - 13^2$

23. Factor each of the following expressions completely and then simplify, if possible:

a. $ax + 2x$ b. $ax - 2x$

c. $3x - 4x + 7x$ d. $3x^2 + xy - x$

e. $(a + b)(c + 1) - (a + b)$

f. $x^2 - 9y^2$

g. $(x^2 - y^2) + x + y$

24. If x and y are integers, classify each of the following as true or false. If true, explain why. If false, give a counterexample.

a. $|x + y| = |x| + |y|$

b. $|xy| = |x||y|$

c. $|x^2| = x^2$

d. $|x|^2 = x^2$

25. a. Given a calendar for any month of the year, such as the one that follows, pick several 3×3 groups of numbers and find the sum of these numbers. How are the sums obtained related to the middle number?

JULY						
S	M	T	W	T	F	S
		1	2	3	4	5
6	7	8	9	10	11	12
13	14	15	16	17	18	19
20	21	22	23	24	25	26
27	28	29	30	31		

★ b. Prove that the sum of any nine digits in any 3×3 set of numbers selected from a monthly calendar will always be equal to 9 times the middle number.

26. In each of the following, find the next two terms. Assume the sequence is arithmetic or geometric, and find its difference or ratio and the nth term.

a. $10, 7, 4, 1, {}^-2, {}^-5, \_, \_$

b. ${}^-2, 4, {}^-8, 16, {}^-32, 64, \_, \_$

27. Find the sum of the first 100 terms of the arithmetic sequence $10, 7, 4, 1, {}^-2, {}^-5, \ldots$.

28. Find the first five terms of the sequences whose nth term is

a. $({}^-2)^n + 2^n$

b. $n^2({}^-1)^n$

c. $|10 - n^2|$

29. In the geometric sequence $1, {}^-2, 4, {}^-8, \ldots$, determine if there is a term equal to the following numbers:

a. 512   b. 1024

30. If x and y are integers, classify each of the following as true or false. Justify your answers.

a. $({}^-x)^3 = {}^-x^3$ b. $|x| > {}^-1$

c. If $x < y$ then $a - x < a - y$, for all integers a.

31. Jon has two checking accounts. In the first one, he is $120 overdrawn, and in the second, his balance is $300. If he deposits $40 every day in the first account but withdraws $20 daily from the second account, after how many days will the balance in each account be the same? Explain your solution.

Mathematical Connections 5-2

Communication

1. Can $({}^-x - y)(x + y)$ be multiplied by using the difference-of-squares formula? Explain why or why not.

2. Kahlil said that using the equation $(a + b)^2 = a^2 + 2ab + b^2$, he can find a similar equation for $(a - b)^2$. Examine his argument. If it is correct, supply any missing steps or justifications; if it is incorrect, point out why.

$$(a - b)^2 = [a + ({}^-b)]^2$$
$$= a^2 + 2a({}^-b) + ({}^-b)^2$$
$$= a^2 - 2ab + b^2$$

3. a. Use the distributive property of multiplication over addition to show that $({}^-1)a + a = 0$. (*Hint:* Write a as $1 \cdot a$.)

b. Use part (a) to show that $({}^-1)a = {}^-a$.

4. Nancy gave the following argument to show that $({}^-a)b = {}^-(ab)$, for all integers a and b: *I know that* $({}^-1)a = {}^-a$; *hence:*

$$({}^-a)b = [({}^-1)a]b$$
$$= ({}^-1)(ab)$$
$$= {}^-(ab)$$

If the argument is valid, complete its details; if it is not valid, explain why not.

5. Hosni gave the following argument that $^-(a + b) = {}^-a + {}^-b$, for all integers a and b. If the argument is correct, supply the missing reasons. If it is incorrect, explain why.

$$^-(a + b) = (-1)(a + b)$$
$$= (-1)a + (-1)b$$
$$= {}^-a + {}^-b$$

6. The Swiss mathematician Leonhard Euler (1707–1783) argued that $(^-1)(^-1) = 1$ as follows: "The result must be either $^-1$ or 1. If it is $^-1$, then $(^-1)(^-1) = {}^-1$. Because $^-1 = (^-1)1$, we have $(^-1)(^-1) = (^-1)1$. Now dividing both sides of the last equation by $^-1$ we get $^-1 = 1$, which of course cannot be true. Hence $(^-1)(^-1)$ must be equal to 1."
   a. What is your reaction to this argument? Is it logical? Why or why not?
   b. Can Euler's approach be used to justify other properties of integers? Explain.

7. If $5x + 3 < {}^-20$, answer the following:
   a. Find the greatest integer x for which the inequality is true. Explain your thinking.
   b. Is there a least integer x for which the inequality is true? Explain why or why not.

8. Jill asks each of her classmates to choose a number, then multiply the number by $^-3$, add 2 to the product, multiply the result by $^-2$, and then subtract 14. Finally, each student is asked to divide the result by 6 and record the answer. When Jill gets an answer from a classmate, she just adds 3 to it in her head and announces the number that classmate originally chose. How did Jill know to add 3 to each answer?

Open-Ended

9. Make up a problem similar to problem 8 but with all numbers different and solve it.

10. On a national mathematics competition, scoring is accomplished using the formula 4 times the number done correctly minus the number done incorrectly. In this scheme, problems left blank are considered neither correct nor incorrect. Devise a scenario that would allow a student to have a negative score.

11. Select a current middle-school text that introduces multiplication and division of integers and discuss any models that were used and how effective you think they would be with a group of students.

Cooperative Learning

12. Devise a scheme for determining a grade-point average for a college student that allows negative quality points for a failing grade.
   a. Use your scheme to determine possible grades for students with positive, zero, and negative grade-point averages.
   b. Compare your scheme with that of another class group and write a rationale for the best scheme.

13. a. How would you introduce multiplication of integers in a middle-school class and how would you explain

that a product of two negative numbers is positive? Write a rationale for your approach.
   b. Present your answers and compare them to those of another class group and together decide the most appropriate way to introduce the concepts. Write a rationale for your approach.

14. Discuss in your group what each person's favorite approach is to justify the fact that $(^-1)(^-1) = 1$ and why it is the favorite.

Questions from the Classroom

15. A seventh-grade student does not believe that $^-5 < {}^-2$. The student argues that a debt of $5 is greater than a debt of $2. How do you respond?

16. A student computes $^-8 - 2(^-3)$ by writing $^-10(^-3) = 30$. How would you help this student?

17. A student says that his father showed him a very simple method for dealing with expressions like $^-(a - b + 1)$ and $x - (2x - 3)$. The rule is, if there is a negative sign before the parentheses, change the signs of the expressions inside the parentheses. Thus, $^-(a - b + 1) = {}^-a + b - 1$ and $x - (2x - 3) = x - 2x + 3$. What is your response?

18. Al said that $4(^-2)$ and $^-4(^-2)$ can't both be 8. He said he knows that $4(^-2) = {}^-8$ because $4(^-2) = {}^-2 + {}^-2 + {}^-2 + {}^-2 = {}^-8$. Therefore, $^-4(^-2)$ must be 8. What would you tell Al?

19. Betty used the charged-field model to show that $^-2(^-3) = 6$. She said that this proves that any negative integer times a negative integer is a positive integer. How do you respond?

Review Problems

20. Illustrate $^-8 + {}^-5$ on a number line.

21. Find the additive inverse of each of the following:
   a. $^-5$ b. 7   c. 0

22. Compute each of the following:
   a. $|^-14|$
   b. $|^-14| + 7$
   c. $8 - |^-12|$
   d. $|11| + |^-11|$

23. In the 1400s, European merchants used positive and negative numbers to label barrels of flour. For example, a barrel labeled +3 meant the barrel was 3 lb overweight, whereas a barrel labeled $^-5$ meant the barrel was 5 lb underweight. If the following numbers were found on 100 lb barrels, what was the total weight of the barrels?

24. Write the function $f(x) = (x + |x|) \div 2$ without absolute value. (Distinguish two cases: $x \geq 0$ and $x < 0$.)

25. Solve for x if possible:

    a. $|x| = 3$ **b.** $|x| + 1 = 0$

    c. $|x| = x$ **d.** $|x| = {}^-x$

Third International Mathematics and Science Study (TIMSS) Questions

If n is a negative integer, which of these is the largest number?

    a. $3 + n$ **b.** $3 \times n$ **c.** $3 - n$ **d.** $3 \div n$

If $x = {}^-3$, what is the value of ${}^-3x$?

    a. ${}^-9$

    b. ${}^-6$

    c. ${}^-1$

    d. 1

    e. 9

TIMSS, Grade 8, 2003

BRAIN TEASER If $a, \ldots, z$ are consecutive letters of the alphabet representing integers, find the product:

$$(x - a)(x - b)(x - c) \cdot \ldots \cdot (x - z)$$

5-3   Divisibility

The concepts of *even* and *odd* integers are commonly used. For example, during water shortages in the summer in some parts of the country, houses with even-number addresses can water on even-numbered days of the month and houses with odd-number addresses can water on odd-numbered days. An even integer is an integer that is divisible by 2; that is, it has 0 remainder when divided by 2. An odd integer is an integer that is not divisible by 2. The fact that 12 is divisible by 2 can be stated in the following equivalent statements in the left column:

Example	General Statement
12 is divisible by 2.	a is divisible by b.
2 is a divisor of 12.	b is a divisor of a.
12 is a multiple of 2.	a is a multiple of b.
2 is a factor of 12.	b is a factor of a.
2 divides 12.	b divides a.

The statement that "2 divides 12" is written with a vertical segment, as in $2\,|\,12$, where the vertical segment means **divides**. Likewise, "b divides a" can be written $b\,|\,a$. Each statement in the right column above can be written $b\,|\,a$. We write $5 \nmid 12$ to symbolize that 5 does not divide 12 or that 12 is not divisible by 5. The notation $5 \nmid 12$ also implies that 12 is not a multiple of 5 and 5 is not a factor of 12.

In general, if a is a nonnegative integer and b is a positive integer, we say that a is divisible by b or equivalently that b divides a if, and only if, the remainder when a is divided by b is 0. Using the division algorithm, this means that there is a unique q (quotient) such that $a = bq$. We extend this concept to divisibility for all integers in the following definition.

> **Definition of "Divides"**
>
> If a and b are any integers, then b divides a, written $b\,|\,a$, if, and only if, there is a unique integer q such that $a = bq$.

If $b\,|\,a$, then b is a **factor**, or a **divisor**, of a, and a is a **multiple** of b.

> **REMARK**   In this chapter whenever the "divides" symbol is used as in $b|a$, we assume that a and b are integers and $b \neq 0$.

Do not confuse $b|a$ with b/a, which is interpreted as $b \div a$. The former, a relation, is either true or false. The latter, an operation, has a numerical value if $a \neq 0$. Note that if b/a is an integer, then $a|b$. Also note that for positive integers $a \nmid b$ is equivalent to saying that the remainder when b is divided by a is not 0.

Example 5-16

Classify each of the following as true or false. Explain your answers.

a. $^{-}3|12$ b. $0|2$ c. 0 is even.
d. $8 \nmid 2$ e. For all integers a, $1|a$.   f. For all integers a, $^{-}1|a$.
g. $3|6n$ for all integers n.   h. $(a - b)|(a^2 - b^2)$ if a and b are integers and $a \neq b$.
i. $0|0$

Solution
a. $^{-}3|12$ is true because $12 = ^{-}4(^{-}3)$.
b. $0|2$ is false because there is no integer c such that $2 = c \cdot 0$.
c. $2|0$ is true because $0 = 0 \cdot 2$; therefore, 0 is even.
d. $8 \nmid 2$ is true because there is no integer c such that $2 = c \cdot 8$.
e. $1|a$ is true for all integers a because $a = a \cdot 1$.
f. $^{-}1|a$ is true for all integers a because $a = (^{-}a)(^{-}1)$.
g. $3|6n$ is true. Because $6n = 3 \cdot 2n$, $6n$ is a multiple of 3 and hence $3|6n$.
h. $(a - b)|(a^2 - b^2)$ is true because $a^2 - b^2 = (a - b)(a + b)$ and $a \neq b$.
i. $0|0$ is false because $0 = 0 \cdot q$ for all integers q, so q is not unique.

Suppose we have one pack of gum and we know that the number of pieces, a, is divisible by 5. Then if we had two packs, the total number of pieces of gum is still divisible by 5. The same is true if we had 10 packs, or 100 packs or in general n packs where n is any positive integer. We could record this observation as follows:

If $5|a$, then $5|na$, where a and n are integers and $n > 0$.

Even more generally, if $d|a$ then d divides any multiple of a. We state this fact in the following theorem.

> **Theorem 5–12**
>
> For any integers a and d, if $d|a$ and n is any integer, then $d|na$.

Historical Note

Pierre de Fermat (1601–1665) was a lawyer and a magistrate who served in the provincial parliament in Toulouse, France. He devoted his leisure time to mathematics—a subject in which he had no formal training. After his death, his son decided to publish a new edition of Diophantus's *Arithmetica* with Fermat's notes. One of the notes in the margin of Fermat's copy asserted that the equation $x^n + y^n = z^n$ has no positive-integer solutions if n is an integer greater than 2 and commented, "I have found an admirable proof of this, but the margin is too narrow to contain it." Many great mathematicians spent years trying to prove Fermat's assertion, now called "Fermat's last theorem." In 1995, Andrew Wiles, a Princeton University mathematician, proved Fermat's last theorem. ◆

The theorem can be stated in an equivalent form:

If d is a factor of a (that is, a equals some integer times d), then d is a factor of any multiple of a.

Figure 5-16

Next consider two different brands of chewing gum each having five pieces, as in Figure 5-16. We can divide each package of gum evenly among five students. In addition, if we opened both packages and put all of the pieces in a bag, we could still divide the number of pieces of gum evenly among the five students. To generalize this notion, if we buy gum in larger packages with a pieces in one package and b pieces in a second package with both a and b divisible by 5, we can record the preceding discussion as follows:

$$\text{If } 5\,|\,a \text{ and } 5\,|\,b, \text{ then } 5\,|\,(a + b).$$

If the number, a, of pieces of gum in one package is divisible by 5, but the number, b, of pieces in the other package is not, then the total, $a + b$, cannot be divided evenly among the five students. This can be recorded as follows:

$$\text{If } 5\,|\,a \text{ and } 5\nmid b, \text{ then } 5\nmid(a + b).$$

NOW TRY THIS 5-8 If $a, b \in I$, what, if anything, is wrong with the statement: if $5\nmid a$ and $5\nmid b$, then $5\nmid(a + b)$?

Since subtraction can be defined in terms of addition, results similar to addition hold for subtraction. These ideas are generalized in Theorem 5–13.

Theorem 5–13

For any integers a, b, and d, $d \neq 0$,

a. If $d\,|\,a$ and $d\,|\,b$, then $d\,|\,(a + b)$.
b. If $d\,|\,a$ and $d\nmid b$, then $d\nmid(a + b)$.
c. If $d\,|\,a$ and $d\,|\,b$, then $d\,|\,(a - b)$.
d. If $d\,|\,a$ and $d\nmid b$, then $d\nmid(a - b)$.

REMARK Theorem 5–13 can be extended. For example, if a, b, c, and d are integers, $d \neq 0$, then,

$$\text{If } d\,|\,a, d\,|\,b \text{ and } d\,|\,c, \text{ then } d\,|\,(a + b + c).$$

The proofs of most theorems in this section are left as exercises, but the proof of Theorem 5–13(a) is given as an illustration.

Proof

Theorem 5–13(a) is equivalent to the following:

If a is a multiple of d and b is a multiple of d, then $a + b$ is a multiple of d.

Notice that "a is a multiple of d" means $a = md$, for some integer m. Similarly "b is a multiple of d" means $b = nd$, for some integer n. To show that $a + b$ is a multiple of d, we add these equations as follows:

$$a + b = md + nd$$

Is $md + nd$ a multiple of d? Notice that $md + nd = (m + n)d$, so $a + b = (m + n)d$. By the closure property of integer addition, $m + n$ is an integer. Therefore, $a + b$ is a multiple of d and therefore $d|(a + b)$.

Example 5-17

Classify each of the following as true or false, where x, y, and z are integers. If a statement is true, prove it. If a statement is false, provide a counterexample.

a. If $3|x$ and $3|y$, then $3|xy$.   **b.** If $3|(x + y)$, then $3|x$ and $3|y$.
c. If $9 \nmid a$, then $3 \nmid a$.

Solution   **a.** True. By Theorem 5–12, if $3|x$, then, for any integer y, $3|yx$ or $3|xy$.
              b. False; for example, $3|(7 + 2)$, but $3 \nmid 7$ and $3 \nmid 2$.
              c. False; for example, $9 \nmid 21$, but $3|21$.

 NOW TRY THIS 5-9 If $x, y \in I$, and if $3|x$, is it true that $3|xy$ regardless of whether $3|y$ or $3 \nmid y$? Why?

Example 5-18

Five students found a padlocked money box that had a deposit slip attached to it. The deposit slip was water-spotted, so the currency total appeared as shown in Figure 5-17. One student remarked that if the money listed on the deposit slip was in the box, it could easily be divided equally among the five students without using coins. How did the student know this?

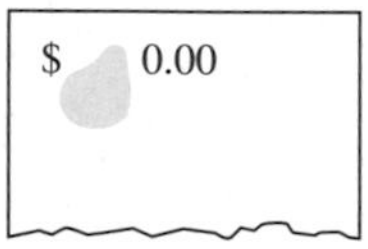

Solution   Because the units digit of the amount of the currency is 0, the solution to the problem is to determine whether all natural numbers whose units digit is 0 are divisible by 5. To solve this problem, *look for a pattern*. Natural numbers whose units digit is 0 form a pattern, that is, 10, 20, 30, 40, 50, ... These numbers are multiples of 10. We are to determine whether 5 divides all multiples of 10. Since $5|10$, by Theorem 5–12, 5 divides any multiple of 10. Hence, 5 divides the amount of money in the box, and the student is correct.

$ 0.00

Figure 5-17

Divisibility Rules

As shown in Example 5-18, sometimes it is handy to know if one number is divisible by another just by looking at it or by performing a simple test. We discovered that if a number ends in 0, then the number is divisible by 5. The same argument can be used to show that if a number ends in 5, it is divisible by 5. This is an example of a divisibility rule. Moreover, if the last digit of a number is neither 0 nor 5, then the number is not divisible by 5.

Elementary texts frequently state divisibility rules. However, such rules have limited use except for mental arithmetic. It is possible to determine whether 1734 is divisible by 17, either by using pencil and paper or a calculator. To check divisibility and avoid decimals, we can use a calculator with an integer division button, $\boxed{\text{INT} \div}$. On such a calculator, integer division can be performed using the following sequence of buttons:

 $\boxed{1}\ \boxed{7}\ \boxed{3}\ \boxed{4}\ \boxed{\text{INT} \div}\ \boxed{1}\ \boxed{7}\ \boxed{=}$

to obtain the display $\underset{Q}{\underline{102}}$ $\underset{R}{\underline{0.}}$

This implies $1734/17 = 102$ with a remainder of 0, which, in turn, implies $17|1734$.

We could have determined this same result mentally by considering the following:

$$1734 = 1700 + 34$$

Because $17 \mid 1700$ and $17 \mid 34$, by Theorem 5–13(a), we have $17 \mid (1700 + 34)$, or $17 \mid 1734$. Similarly, we could determine mentally that $17 \nmid 1735$.

REMARK Notice that $17 \mid 1734$ implies $17 \mid (^-1)1734$, that is, $17 \mid {}^-1734$. In general, $d \mid {}^-a$ if, and only if, $d \mid a$.

Divisibility Tests for 2, 5, and 10

To determine mentally whether a given integer n is divisible by another integer d, we think of n as the sum or difference of integers, where d divides at least one of the numbers. We use a concrete example to get an idea of how this concept works. Consider the number 1362. This number can be represented as in Figure 5-18. Notice that 2 divides each part of the figure (see dashed lines). Since 2 divides each part of the figure, by the extension of Theorem 5–13, 2 divides the sum of all of the parts and hence $2 \mid 1362$.

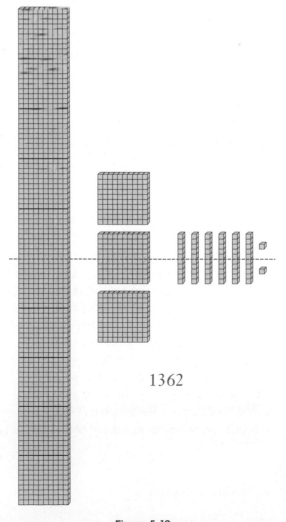

1362

Figure 5-18

Note that if the original number was 1363, then $2|1362$ and $2\nmid1$, so $2\nmid1363$. We see that all we have to do is to determine whether the units digit is divisible by 2 in order to determine if the number is divisible by 2. We can develop a similar test for divisibility by 5 and 10.

> ### Theorem 5–14:   Divisibility Test for 2
> An integer is divisible by 2 if, and only if, its units digit is divisible by 2.

> ### Theorem 5–15:   Divisibility Test for 5
> An integer is divisible by 5 if, and only if, its units digit is divisible by 5; that is, if, and only if, the units digit is 0 or 5.

> ### Theorem 5–16:   Divisibility Test for 10
> An integer is divisible by 10 if, and only if, its units digit is divisible by 10; that is, if, and only if, the units digit is 0.

Divisibility Tests for 4 and 8

When we consider divisibility rules for 4 and 8, we see that $4\nmid10$ and $8\nmid10$, so it is not a matter of checking the units digit for divisibility by 4 and 8. However, 4 (which is 2^2) divides 10^2, and 8 (which is 2^3) divides 10^3.

We first develop a divisibility rule for 4. Consider any four-digit number n such that $n = a10^3 + b10^2 + c10 + d$. Our *subgoal* is to *write the given number as a sum of two numbers*, one of which is as great as possible and divisible by 4. We know that $4|10^2$ because $10^2 = 4 \cdot 25$ and, consequently, $4|10^3$. Because $4|10^2$, then $4|b10^2$; and because $4|10^3$, then $4|a10^3$. Finally, $4|a10^3$ and $4|b10^2$ imply $4|(a10^3 + b10^2)$. Now the divisibility of $a10^3 + b10^2 + c10 + d$ by 4 depends on the divisibility of $(c10 + d)$ by 4. Notice that $c10 + d$ is the number represented by the last two digits in the given number n. We summarize this in the following test.

> ### Theorem 5–17:   Divisibility Test for 4
> An integer is divisible by 4 if, and only if, the last two digits of the integer represent a number divisible by 4.

To investigate divisibility by 8, we note that the least positive power of 10 divisible by 8 is 10^3 since $10^3 = 8 \cdot 125$. Consequently, all integral powers of 10 greater than 10^3 also are divisible by 8. Hence, the following is a divisibility test for 8.

Theorem 5–18:   Divisibility Test for 8

An integer is divisible by 8 if, and only if, the last three digits of the integer represent a number divisible by 8.

Example 5-19

a. Determine whether 97,128 is divisible by 2, 4, and 8.
b. Determine whether 83,026 is divisible by 2, 4, and 8.

Solution   **a.** $2 | 97{,}128$ because $2 | 8$.
$4 | 97{,}128$ because $4 | 28$.
$8 | 97{,}128$ because $8 | 128$.

b. $2 | 83{,}026$ because $2 | 6$.
$4 \nmid 83{,}026$ because $4 \nmid 26$.
$8 \nmid 83{,}026$ because $8 \nmid 026$.

Example 5-20

Use Theorems 5–12 and 5–13 to show why the divisibility test for 8 works in Example 5-19(b).

Solution   We can write 97,128 as 97,000 + 128. Because $8 | 1000$, then $8 | 97 \cdot 1000$ or $8 | 97{,}000$ (Theorem 5–12). Next we need to check if $8 | 128$. It does, so $8 | (97{,}000 + 128)$ or $8 | 97{,}128$ (Theorem 5–13).

REMARK In Example 5–19(a), it would have been sufficient to check that the given number is divisible by 8 because if $8 | a$, then $2 | a$ and $4 | a$. Why? This relationship is shown in Figure 5-19.

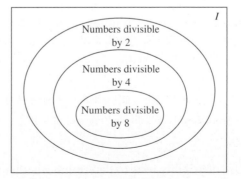

Figure 5-19

Notice that if $8 \nmid a$, we cannot conclude that $4 \nmid a$ or $2 \nmid a$. Why?

Divisibility Tests for 3 and 9

Next, we consider a divisibility test for 3. No power of 10 is divisible by 3, but the numbers 9, and 99, and 999, and others of this type are divisible by 3. For example, to determine whether 5721 is divisible by 3, we rewrite the number using 999, 99, and 9, as follows:

$$
\begin{aligned}
5721 &= 5 \cdot 10^3 + 7 \cdot 10^2 + 2 \cdot 10 + 1 \\
&= 5(999 + 1) + 7(99 + 1) + 2(9 + 1) + 1 \\
&= 5 \cdot 999 + 5 \cdot 1 + 7 \cdot 99 + 7 \cdot 1 + 2 \cdot 9 + 2 + 1 \\
&= (5 \cdot 999 + 7 \cdot 99 + 2 \cdot 9) + (5 + 7 + 2 + 1)
\end{aligned}
$$

The sum in the first set of parentheses in the last line is divisible by 3, so the divisibility of 5721 by 3 depends on the sum in the second set of parentheses. In this case, $5 + 7 + 2 + 1 = 15$ and $3 \mid 15$, so $3 \mid 5721$. Hence, to test 5721 for divisibility by 3, we test $5 + 7 + 2 + 1$ for divisibility by 3. Notice that $5 + 7 + 2 + 1$ is the sum of the digits of 5721. The example suggests the following test for divisibility by 3.

> **Theorem 5–19:   Divisibility Test for 3**
>
> An integer is divisible by 3 if, and only if, the sum of its digits is divisible by 3.

We can use an argument similar to the one used to demonstrate that $3 \mid 5721$ to prove the test for divisibility by 3 on any integer and in particular for any four-digit number $n = a10^3 + b10^2 + c10 + d$. Even though $a10^3 + b10^2 + c10 + d$ is not necessarily divisible by 3, the number $a999 + b99 + c9$ *is* divisible by 3. We have the following:

$$
\begin{aligned}
a10^3 + b10^2 + c10 + d &= a1000 + b100 + c10 + d \\
&= a(999 + 1) + b(99 + 1) + c(9 + 1) + d \\
&= (a999 + b99 + c9) + (a1 + b1 + c1 + d) \\
&= (a999 + b99 + c9) + (a + b + c + d)
\end{aligned}
$$

Because $3 \mid 9$, $3 \mid 99$, and $3 \mid 999$, it follows that $3 \mid (a999 + b99 + c9)$. If $3 \mid (a + b + c + d)$, then $3 \mid [(a999 + b99 + c9) + (a + b + c + d)]$; that is, $3 \mid n$. If, on the other hand, $3 \nmid (a + b + c + d)$, it follows from Theorem 5–13(b) that $3 \nmid n$.

Since $9 \mid 9$, $9 \mid 99$, $9 \mid 999$, and so on, a test similar to that for divisibility by 3 applies to divisibility by 9. Why?

> **Theorem 5–20:   Divisibility Test for 9**
>
> An integer is divisible by 9 if, and only if, the sum of the digits of the integer is divisible by 9.

Example 5-21

Use divisibility tests to determine whether each of the following numbers is divisible by 3 and divisible by 9:

a. 1002   **b.** 14,238

Solution **a.** Because $1 + 0 + 0 + 2 = 3$ and $3 \mid 3$, it follows that $3 \mid 1002$. Because $9 \nmid 3$, it follows that $9 \nmid 1002$.

b. Because $1 + 4 + 2 + 3 + 8 = 18$ and $3|18$, it follows that $3|14{,}238$. Because $9|18$, it follows that $9|14{,}238$.

Example 5-22

The store manager has an invoice for 72 four-function calculators. The first and last digits on the receipt are illegible. The manager can read

$$\$\blacksquare 67.9\blacksquare$$

What are the missing digits, and what is the cost of each calculator?

Solution   Let the missing digits be x and y so that the number is $x67.9y$ dollars, or $x679y$ cents. Because there were 72 calculators sold, the number on the invoice must be divisible by 72. Because the number is divisible by 72 and $72 = 8 \cdot 9$, it must be divisible by 8 and 9, which are factors of 72. For the number on the invoice to be divisible by 8, the three-digit number $79y$ must be divisible by 8. Because $79y$ must be divisible by 8, it is an even number. Therefore, $79y$ must be either 790, 792, 794, 796, or 798. Only the number 792 is divisible by 8, so we know the last digit, y, on the invoice must be 2.

Because the number on the invoice must be divisible by 9, we know that 9 must divide $x + 6 + 7 + 9 + 2$, or $(x + 24)$. Since 3 is the only single digit that will make $(x + 24)$ divisible by 9, then x must be 3. Therefore, the number on the invoice must be $\$367.92$. The calculators must cost $\$367.92/72$, or $\$5.11$, each.

Divisibility Tests for 11 and 6

The divisibility test for 7 is usually harder to use than actually performing the division, so we omit the test. We state the divisibility test for 11 but omit the proof. Interested readers might try to find a proof.

> **Theorem 5–21:   Divisibility Test for 11**
>
> An integer is divisible by 11 if, and only if, the sum of the digits in the places that are even powers of 10 minus the sum of the digits in the places that are odd powers of 10 is divisible by 11.

For example, to test whether 8,471,986 is divisible by 11, we check whether 11 divides the difference $(6 + 9 + 7 + 8) - (8 + 1 + 4)$, or 17. Because $11 \nmid 17$, it follows from the divisibility test for 11 that $11 \nmid 8{,}471{,}986$. A number like 2772 is divisible by 11 because $(2 + 7) - (7 + 2) = 9 - 9 = 0$ and 0 is divisible by 11.

The divisibility test for 6 is related to the divisibility tests for 2 and 3. In Section 5-4, we will ask you to show that if $2|n$ and $3|n$, then $(2 \cdot 3)|n$, and in general: if a and c have no factors in common, then if $a|b$ and $c|b$, we can conclude that $ac|b$. Consequently, the following divisibility test is true.

> **Theorem 5–22:   Divisibility Test for 6**
>
> An integer is divisible by 6 if, and only if, the integer is divisible by both 2 and 3.

Divisibility tests for other numbers are explored in Assessments 5-3A and 5-3B.

Example 5-23

The number 57,729,364,583 has too many digits for most calculator displays. Determine whether it is divisible by each of the following:

a. 2   **b.** 3   **c.** 5   **d.** 6   **e.** 8   **f.** 9   **g.** 10   **h.** 11

Solution **a.** No, the units digit, 3, is not divisible by 2.
 b. No, the sum of the digits is 59, which is not divisible by 3.
 c. No, the units digit is neither 0 nor 5.
 d. No, the number is not divisible by 2.
 e. No, the number formed by the last three digits, 583, is not divisible by 8.
 f. No, the sum of the digits is 59, which is not divisible by 9.
 g. No, the units digit is not 0.
 h. Yes, $(3 + 5 + 6 + 9 + 7 + 5) - (8 + 4 + 3 + 2 + 7) = 35 - 24 = 11$ and 11 is divisible by 11.

NOW TRY THIS 5-10 Fill in the following blanks so that the number is divisible by 9. List all possibilities.

$$12,506,5\_\_.$$

Problem Solving   A Mistake in the Inventory

A class from Washington School visited a neighborhood cannery warehouse. The warehouse manager told the class that there were 11,368 cans of juice in the inventory and that the cans were packed in boxes of 6 or 24, depending on the size of the can. One of the students, Sam, thought for a moment and announced that there was a mistake in the inventory. Is Sam's statement correct? Why or why not?

Understanding the Problem   The problem is to determine if the manager's inventory of 11,368 cans was correct. To solve the problem, we must assume there are no partial boxes of cans; that is, a box must contain exactly 6 or exactly 24 cans of juice.

Devising a Plan   We know that the boxes contain either 6 cans or 24 cans, but we do not know how many boxes of each type there are. One strategy for solving this problem is to *find an equation* that involves the total number of cans in all the boxes.

◆ Historical Note

A 20th century mathematician who worked in the area of number theory was American Julia Robinson (1919–1985). Robinson was the first woman mathematician to be elected to the National Academy of Sciences and the first woman president of the American Mathematical Society. She died of leukemia at the age of 65.   ◆

The total number of cans, 11,368, equals the number of cans in all the 6-can boxes plus the number of cans in all the 24-can boxes. If there are n boxes containing 6 cans each, there are $6n$ cans altogether in those boxes. Similarly, if there are m boxes with 24 cans each, those boxes contain a total of $24m$ cans. Because the total was reported to be 11,368 cans, we have the equation $6n + 24m = 11,368$. Sam claimed that $6n + 24m \neq 11,368$.

One way to show that $6n + 24m \neq 11,368$ is to show that $6n + 24m$ and 11,368 do not have the same divisors. Both $6n$ and $24m$ are divisible by 6. This implies that $6n + 24m$ must be divisible by 6. If 11,368 is not divisible by 6, then Sam is correct.

Carrying Out the Plan   The divisibility test for 6 states that a number is divisible by 6 if, and only if, the number is divisible by both 2 and 3. Because 11,368 ends in 8, it is divisible by 2. Is it divisible by 3?

The divisibility test for 3 states that a number is divisible by 3 if, and only if, the sum of the digits in the number is divisible by 3. We see that $1 + 1 + 3 + 6 + 8 = 19$, which is not divisible by 3, so 11,368 is not divisible by 3. Hence, Sam is correct.

Looking Back   Suppose 11,368 had been divisible by 6. Would that have implied that the manager was correct? The answer is no; it would have implied only that we would have to change our approach to the problem.

As a further Looking Back activity, suppose that, given different data, the manager is correct. Can we determine values for m and n? In fact, this can be done. If a computer is available, a program can be written to determine all possible natural-number values of m and n.

BRAIN TEASER   The following is an argument to show that an ant weighs as much as an elephant. What is wrong?

Let e be the weight of the elephant and a the weight of the ant. Let $e - a = d$. Consequently, $e = a + d$. Multiply each side of $e = a + d$ by $e - a$. Then simplify.

$$e(e - a) = (a + d)(e - a)$$
$$e^2 - ea = ae + de - a^2 - da$$
$$e^2 - ea - de = ae - a^2 - da$$
$$e(e - a - d) = a(e - a - d)$$
$$e = a$$

Thus, the weight of the elephant equals the weight of the ant.

Assessment 5-3A

1. Classify each of the following as true or false. If false, tell why.
   - **a.** 6 is a factor of 30.
   - **b.** 6 is a divisor of 30.
   - **c.** 6|30.
   - **d.** 30 is divisible by 6.
   - **e.** 30 is a multiple of 6.
   - **f.** 6 is a multiple of 30.

2. Using divisibility tests, answer each of the following:
   - **a.** There are 1379 children signed up to play in a baseball league. If exactly 9 players are to be placed on each team, will any team be short of players?
   - **b.** A forester has 43,682 seedlings to be planted. Can these be planted in an equal number of rows with 11 seedlings in each row?

c. There are 261 students to be assigned to 9 teachers so that each teacher has the same number of students. Is this possible?

3. Without using a calculator, test each of the following numbers for divisibility by 2, 3, 4, 5, 6, 8, 9, 10, and 11:
   a. 746,988
   b. 81,342
   c. 15,810

4. Determine each of the following without actually performing the division. Explain how you did it in each case.
   a. Is 34,015 divisible by 17?
   b. Is 34,051 divisible by 17?
   c. Is 19,031 divisible by 19?
   d. Is $2 \cdot 3 \cdot 5 \cdot 7$ divisible by 5?
   e. Is $(2 \cdot 3 \cdot 5 \cdot 7) + 1$ divisible by 5?

5. Justify each of the given statements, assuming that a, b, and c are integers. If a statement cannot be justified by one of the theorems in this section, answer "none."
   a. $4 \mid 20$ implies $4 \mid 113 \cdot 20$.
   b. $4 \mid 100$ and $4 \nmid 13$ imply $4 \nmid (100 + 13)$.
   c. $4 \mid 100$ and $4 \nmid 13$ imply $4 \nmid 1300$.
   d. $3 \mid (a + b)$ and $3 \nmid c$ imply $3 \nmid (a + b + c)$.
   e. $3 \mid a$ implies $3 \mid a^2$.

6. Classify each of the following as true or false. Justify your answers.
   a. If $b \mid a$, then $(b + c) \mid (a + c)$.
   b. If $b \mid a$, then $b^2 \mid a^3$.
   c. If $b \mid a$, then $b \mid {}^-a$ and ${}^-b \mid {}^-a$.

7. Justify each of the following:
   a. $7 \mid 210$
   b. $19 \mid (1900 + 38)$
   c. $6 \mid 2^3 \cdot 3^2 \cdot 17^4$
   d. $7 \nmid (4200 + 22)$

8. Classify each of the following as true or false:
   a. If every digit of a number is divisible by 3, the number itself is divisible by 3.
   b. If a number is divisible by 3, then every digit of the number is divisible by 3.

9. Fill each of the following blanks with the greatest digit that makes the statement true:
   a. $3 \mid 74\_$
   b. $9 \mid 83\_45$
   c. $11 \mid 6\_55$

10. Place a digit in the square, if possible, so that the number
$$527{,}4 \,\square\, 2$$
is divisible by
   a. 2   **b.** 3
   c. 4   **d.** 9
   e. 11

11. The bookstore marked some notepads down from $2.00 but still kept the price over $1.00. It sold all of them. The total amount of money from the sale of the pads was $31.45. How many notepads were sold?

12. A group of people ordered pencils. The bill was $2.09. If the original price of each was 12¢ and the price has risen, how much does each cost?

13. Leap years occur in years that are divisible by 4. However, if the year ends in two zeros, in order for the year to be a leap year, it must be divisible by 400. Determine which of the following are leap years:
   a. 1776
   b. 1986
   c. 2000
   d. 2100

14. A test for checking computations is called *casting out nines*. Consider the sum $193 + 24 + 786 = 1003$. The remainders when 193, 24, and 786 are divided by 9 are 4, 6, and 3, respectively. The sum of the remainders, 13, has a remainder of 4 when divided by 9, as does 1003. Checking the remainders in this manner provides a quasi-check for the computation. Find the following sums and use casting out nines to check your sums:
   a. $12{,}343 + 4546 + 56$
   b. $987 + 456 + 8765$
   c. $10{,}034 + 3004 + 400 + 20$
   d. Try the check on the subtraction $1003 - 46$.
   e. Try the check on the multiplication $345 \cdot 56$.
   f. Would it make sense to try the check on division? Why or why not?

15. **a.** If 21 divides n, what other natural numbers divide n? Why?
    b. If 16 divides n, what other natural numbers divide n? Why?

16. The numbers x and y are divisible by 5.
    a. Is the sum of x and y divisible by 5? Why?
    b. Is the difference of x and y divisible by 5? Why?
    c. Is the product of x and y divisible by 5? Why?

17. Using only divisibility tests, explain whether 6,868,395 is divisible by 15.

18. Classify each of the following as true or false, assuming that a, b, c, and d are integers. If a statement is false, give a counterexample.
    a. If $d \mid (a + b)$, then $d \mid a$ and $d \mid b$.
    b. If $d \mid (a + b)$, then $d \mid a$ or $d \mid b$.
    c. If $d \mid ab$, then $d \mid a$ or $d \mid b$.
    d. If $ab \mid c$, then $a \mid c$ and $b \mid c$.
    e. If $a \mid b$ and $b \mid a$, then $a = b$.

19. Prove the test for divisibility by 9 for any five-digit number.

Assessment 5-3B

1. Classify each of the following as true or false. If false, tell why.
   a. 5 is a multiple of 20.
   b. 10 is a divisor of 30.
   c. 8|32.
   d. 10 is divisible by 1.
   e. 30 is a factor of 6.
   f. 6 is a multiple of 20.

2. Using divisibility tests, answer each of the following:
   a. Six friends win with a lottery ticket. The payoff is $242,800. Can the money be divided evenly?
   b. Jack owes $7812 on a new car. Can this amount be paid in 12 equal monthly installments?

3. Without using a calculator, test each of the following numbers for divisibility by 2, 3, 4, 5, 6, 8, 9, 10, and 11:
   a. 4,201,012   b. 1001
   c. 10,001

4. Determine each of the following without actually performing the division. Explain how you did it in each case.
   a. Is 24,013 divisible by 12?
   b. Is 24,036 divisible by 12?
   c. Is 17,034 divisible by 17?
   d. Is $2 \cdot 3 \cdot 5 \cdot 7$ divisible by 3?
   e. Is $(2 \cdot 3 \cdot 5 \cdot 7) + 1$ divisible by 6?

5. Justify each of the following.
   a. $a^3 | a^4$, if $a \neq 0$.
   b. $a^4 | a^{10}$, if $a \neq 0$.
   c. $a^n | a^m$, if $0 \leq n \leq m$, if $a \neq 0$.
   d. If $b|a$ and $c \neq 0$, then $bc|ac$.

6. Justify each of the following:
   a. $26|(13^4 \cdot 100)$
   b. $13 \nmid (2^4 \cdot 5^3 \cdot 26 + 1)$
   c. $2^4 \nmid (2 \cdot 4 \cdot 6 \cdot 8 \cdot 17^{10} + 1)$
   d. $2^4|(10^4 + 6^4)$

7. Classify each of the following as true or false:
   a. If a number is divisible by 6, then it is divisible by 2 and by 3.
   b. If a number is divisible by 2 and 3, then it is divisible by 6.
   c. If a number is divisible by 2 and 4, then it is divisible by 8.
   d. If a number is divisible by 8, then it is divisible by 2 and 4.

8. Devise a test for divisibility by each of the following numbers:
   a. 16
   b. 25

9. When the two missing digits in the following number are replaced, the number is divisible by 99. What is the number?

$$85\_\_1$$

10. Without using a calculator, classify each of the following as true or false. Justify your answers.
    a. $7|280021$
    b. $19 \nmid 3,800,018$
    c. $23|46^{10}$
    d. $23 \nmid 460,046$

11. In a football game, a touchdown with an extra point is worth 7 points and a field goal is worth 3 points. Suppose that in a game the only scoring done by teams are touchdowns with extra points and field goals.
    a. Which of the scores 1 to 25 are impossible for a team to score?
    b. List all possible ways for a team to score 40 points.
    c. A team scored 57 points with 6 touchdowns and 6 extra points. How many field goals did the team score?

12. Complete the following table where n is the given integer.

n	Remainder When n Is Divided by 9	Sum of the Digits of n	Remainder When the Sum of the Digits of n Is Divided by 9
a. 31			
b. 143			
c. 345			
d. 2987			
e. 7652			

   f. Make a conjecture about the remainder and the sum of the digits in an integer when they are divided by 9.

13. A palindrome is a number that reads the same forward as backward.
    a. Check the following four-digit palindromes for divisibility by 11:
       i. 4554   ii. 9339   iii. 2002   iv. 2222
    b. Are all four-digit palindromes divisible by 11? Why or why not?
    c. Are all five-digit palindromes divisible by 11? Why or why not?
    d. Are all six-digit palindromes divisible by 11? Why or why not?

14. The numbers 5872 and 2785 are a palindromic pair of numbers because reversing the order of the digits of one number gives the other number. Explain why in a

palindromic pair, if one number is divisible by 3, then so is the other.

15. Classify each of the following as true or false, assuming that a, b, c, and d are integers. If a statement is false, give a counterexample.
   a. If $d|a$ and $d|b$, then $d|(ax + by)$ for any integers x and y.
   b. If $d \nmid a$ and $d \nmid b$, then $d \nmid (a + b)$.
   c. If $d|a^2$, then $d|a$.
   d. If $d \nmid a$, then $d \nmid a^2$.

16. Prove Theorem 5–13(b).

17. a. Choose a two-digit number such that the number in the tens place is 1 greater than the number in the units place. Reverse the digits in your number, and subtract this number from your original number; for example,

$87 - 78 = 9$. Make a conjecture concerning the result of performing this operation.
   b. Choose any two-digit number such that the number in the tens place is 2 greater than the number in the units place. Reverse the digits in your number, and subtract this number from your original number; for example, $31 - 13 = 18$. Make a conjecture concerning the result of performing this operation.
   c. Prove that for any two-digit number, if the digits are reversed and the numbers subtracted, the difference is a multiple of 9.
   d. Investigate what happens whenever two-digit numbers with equal digit sums are subtracted; for example, $62 - 35 = 27$.

Mathematical Connections 5-3

Communication

1. A customer wants to mail a package. The postal clerk determines the cost of the package to be $18.95, but only 6¢ and 9¢ stamps are available. Can the available stamps be used for the exact amount of postage for the package? Why or why not?

2. a. Jim uses his calculator to see if a number n having eight or fewer digits is divisible by a number d. He finds that $n \div d$ has a display of 32. Does $d|n$? Why?
   b. If $n \div d$ gives a display of 16.8, does $d|n$? Why?

3. Is the area of each of the following rectangles divisible by 4? Explain why or why not.
   a.

   52,832 cm

   324,518 cm

   b.

   52,834 cm

   324,514 cm

4. Can you find three consecutive natural numbers none of which is divisible by 3? Explain your answer.

5. Answer each of the following and justify your answers.
   a. If a number is not divisible by 5, can it be divisible by 10?
   b. If a number is not divisible by 10, can it be divisible by 5?

6. A number in which each digit except 0 appears exactly 3 times is divisible by 3. For example, 777,555,222 and

414,143,313 are divisible by 3. Explain why this statement is true.

7. Enter any three-digit number on the calculator; for example, enter 243. Repeat it: 243,243. Divide by 7. Divide by 11. Divide by 13. What is the answer? Try it again with any other three-digit number. Will this always work? Why?

8. Alexa claims that she can justify the divisibility test by 11. She says: *I noticed that each even power of 10 can be written as a multiple of 11 plus 1 and every odd power of 10 can be written as a multiple of 11 minus 1. In fact:*

$$10 = 11 - 1$$
$$10^2 = 99 + 1 = 9 \cdot 11 + 1$$
$$10^3 = 10 \cdot 10^2 = 10(9 \cdot 11 + 1) = 90 \cdot 11 + 10$$
$$= 90 \cdot 11 + 11 - 1 = 91 \cdot 11 - 1$$
$$10^4 = 10^2 \cdot 10^2 = 100(9 \cdot 11 + 1)$$
$$= 900 \cdot 11 + 9 \cdot 11 + 1 = 909 \cdot 11 + 1$$

and so on.

   Now I look at a four-digit number abcd and proceed as in the divisibility by 3. I collect the parts that are divisible by 11 regardless of what the digits are and put together the rest, which is

$$d - c + b - a$$

Complete the details of Alexa's argument and justify the test for divisibility by 11.

9. Take a number written in base ten with three or more digits and subtract the units digit from the indicated expression. By what numbers can you be sure that the difference is divisible? Justify your answers.
   a. The units digit
   b. The number formed by the last two digits (that is, the tens digit followed by the units digit).

c. The sum of the digits

d. Answer the preceding questions for a three- or more digit number written in base five.

10. **a.** In what bases will divisibility by 2 depend only on the last digit? Justify your answer.

b. In what bases will divisibility by 2 depend only on the sum of the digits being even or odd? Justify your answer.

Open-Ended

11. A breakfast-food company had a contest in which numbers were placed in breakfast-food boxes. A prize of $1000 was awarded to anyone who could collect numbers whose sum was 100. The company had thousands of cards made with the following numbers on them:

3   12   15   18   27   33   45   51   66   75   84   90

a. If the company did not make any more cards, is there a winning combination?

b. If the company is going to add one more number to the list and it wants to make sure the contest has at most 1000 winners, suggest a strategy for it to use.

12. How would you use concrete materials to explain to young children the following:

a. A number being even or odd

b. A number being divisible by 3 or not being divisible by 3

c. That if $4|a$, then $2|a$

Cooperative Learning

13. In your group, discuss the value of teaching various divisibility tests in middle school. If a teacher decides to discuss the various tests, how should they be introduced?

Questions from the Classroom

14. Jane claimed that a number is divisible by 4 if each of the last two digits is divisible by 4. Is this claim accurate? If not, how would you suggest that Jane change it to make it accurate?

15. Jim says that $a|b$ and a/b mean the same thing. How would you respond?

16. Betty noticed that $2|36$, $9|36$, and $18|36$. She noticed that $18 = 2 \cdot 9$. She then noticed that $4|36$ or $6|36$ so she thought $4 \cdot 6$ or 24 should divide 36. What do you tell her?

17. A student claims that $a|a$ and $a|a$ implies $a|(a - a)$, and hence, $a|0$. Is the student correct?

18. A student writes, "If $d \nmid a$ and $d \nmid b$, then $d \nmid (a + b)$." How do you respond?

19. Your seventh-grade class has just completed a unit on divisibility rules. One of the better students asks why divisibility by numbers other than 3 and 9 cannot be tested by dividing the sum of the digits by the tested number. How should you respond?

20. A student says that a number with an even number of digits is divisible by 7 if, and only if, each of the numbers formed by pairing the digits into groups of two is divisible by 7. For example, 49,562,107 is divisible by 7, since each of the numbers 49, 56, 21, and 07 is divisible by 7. Is this true?

21. A student claims that a number is divisible by 21 if, and only if, it is divisible by 3 and by 7, and, in general, a number is divisible by $a \cdot b$ if, and only if, it is divisible by a and by b. What is your response?

22. A student found that all three-digit numbers of the form aba, where $a + b$ is a multiple of 7, are divisible by 7. She would like to know why this is so. How do you respond?

Review Problems

23. Find all integers x (if possible) that make each of the following true:

a. $3(^-x) = 6$

b. $(^-2)|x| = 6$

c. $(^-x) \div 0 = ^-1$

d. $^-(x - 1) = 1 - x$

e. $^-|^-x| = 5$

f. $^-x < 0$

24. Simplify each of the following:

a. $3x - (1 - 2x)$

b. $(^-2x)^2 - 3x^2$

c. $y - x - 2(y - x)$

d. $(x - 1)^2 - x^2 + 2x$

25. Consider the function $y = f(x) = ^-2x - 3$ whose domain is the set of integers, and answer the following questions:

a. Find $f(^-5)$.

b. For what value of x will the value of y be 17?

c. Is 2 a possible output? Explain your answer.

d. Find the range of the function.

National Assessment of Educational Progress (NAEP) Question

Write each of the following numbers in the circle where it belongs.

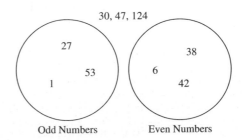

30, 47, 124

Odd Numbers   Even Numbers

NAEP, 2007, Grade 4

BRAIN TEASER Dee finds that she has an extraordinary Social Security number. Its nine digits contain all the numbers from 1 through 9. They also form a number with the following characteristics: When read from left to right, its first two digits form a number divisible by 2, its first three digits form a number divisible by 3, its first four digits form a number divisible by 4, and so on, until the complete number is divisible by 9. What is Dee's Social Security number?

5-4   Prime and Composite Numbers

Principles and Standards states:

> Students should recognize that different types of numbers have particular characteristics; for example, square numbers have an odd number of factors and prime numbers have only two factors. (p. 151)

One method used in elementary schools to determine the positive factors of a natural number is to use squares of paper or cubes and to represent the number as a rectangle. Such a rectangle resembles a candy bar formed with small squares. The dimensions of the rectangle are divisors or factors of the number. For example, Figure 5-20 shows rectangles to represent 12.

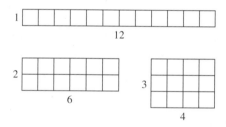

Figure 5-20

As the figure shows, the number 12 has six positive divisors: 1, 2, 3, 4, 6, and 12. If rectangles were used to find the divisors of 7, then we would find only a 1×7 rectangle, as Figure 5-21 shows. Thus, 7 has exactly two divisors: 1 and 7.

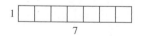

Figure 5-21

To illustrate further the number of positive divisors of a natural number, we construct Table 5-1. Below each number listed across the top, we identify numbers less than or equal to 37 that have that number of positive divisors. For example, 12 is in the 6 column because it has six positive divisors, and 7 is in the 2 column because it has only two positive divisors.

Table 5-1 Number of Positive Divisors

1	2	3	4	5	6	7	8	9
1	2	4	6	16	12		24	36
	3	9	8		18		30	
	5	25	10		20			
	7		14		28			
	11		15		32			
	13		21					
	17		22					
	19		26					
	23		27					
	29		33					
	31		34					
	37		35					

NOW TRY THIS 5-11

a. What patterns do you see forming in Table 5-1?
b. Will there be other entries in the 1 column? Why?
c. What are the next three numbers in the 3 column?
d. Find an entry for the 7 column.
e. What kinds of numbers have an odd number of factors? Why?

The numbers in the 2 column in Table 5-1 are of particular importance. Notice that they have exactly two positive divisors, namely, 1 and themselves. Any positive integer with exactly two distinct, positive divisors is a *prime number*, or a **prime**. Any integer greater than 1 that has a positive factor other than 1 and itself is a *composite number*, or a **composite**. For example, 4, 6, and 16 are composites because they have positive factors other than 1 and themselves. The number 1 has only one positive factor, so it is neither prime nor composite. From the 2 column in Table 5-1, we see that the first 12 primes are 2, 3, 5, 7, 11, 13, 17, 19, 23, 29, 31, and 37. The number 2 is sometimes called the "oddest" prime because it is the only even prime. Other patterns in the table are explored in the problem set.

Example 5-24

Show that the following numbers are composite:

a. 1564
b. 2781
c. 1001
d. $3 \cdot 5 \cdot 7 \cdot 11 \cdot 13 + 1$

Solution a. Since $2 \mid 4$, 1564 is divisible by 2 and is composite.
   b. Since $3 \mid (2 + 7 + 8 + 1)$, 2781 is divisible by 3 and is composite.
   c. Since $11 \mid [(1 + 0) - (0 + 1)]$, 1001 is divisible by 11 and is composite.
   d. Because a product of odd numbers is odd (why?), $3 \cdot 5 \cdot 7 \cdot 11 \cdot 13$ is odd. If we add 1 to an odd number, the sum is even. An even number (other than 2) has a factor of 2 and is therefore composite.

NOW TRY THIS 5-12 The *FoxTrot* cartoon concerns divisibility and prime numbers. Answer the following questions based on the cartoon.

a. Select a number at random from the cartoon and divide it by 13; then divide it by 17; then divide it by 19. Keep doing this until you find a number that when divided by 13, 17, or 19 gives you a whole-number answer. What does it mean when you get a whole-number answer?

b. The cartoonist made one mistake. The number 2261 appears in the left center of the cartoon. Explain why this number can't be included in the cartoon.

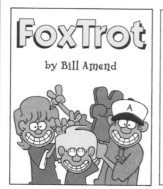
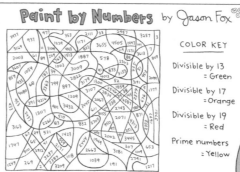

Prime Factorization

In the grade 7 *Focal Points* we find this statement:

Students continue to develop their understanding of multiplication and division and the structure of numbers by determining if a counting number greater than 1 is a prime, and if it is not, by factoring it into a product of primes. (p. 19)

Composite numbers can be expressed as products of two or more whole numbers greater than 1. For example, $18 = 2 \cdot 9, 18 = 3 \cdot 6$, or $18 = 2 \cdot 3 \cdot 3$. Each expression of 18 as a product of factors is a **factorization**.

A factorization containing only prime numbers is a **prime factorization**. To find a prime factorization of a given composite number, first rewrite the number as a product of two smaller numbers greater than 1. Continue the process, factoring the lesser numbers until all factors are primes. For example, consider 260:

$$260 = 26 \cdot 10 = (2 \cdot 13)(2 \cdot 5) = 2 \cdot 2 \cdot 5 \cdot 13 = 2^2 \cdot 5 \cdot 13$$

The procedure for finding a prime factorization of a number can be organized using a **factor tree,** as Figure 5-22(a) demonstrates. The last branches of the tree display the prime factors of 260. A second way to factor 260 is shown in Figure 5-22(b). The two trees produce the same prime factorization, except for the order in which the primes appear in the products.

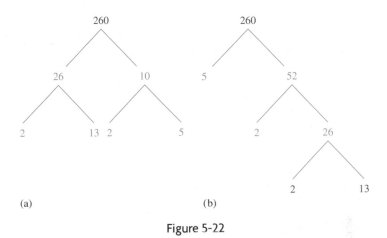

Figure 5-22

The *Fundamental Theorem of Arithmetic*, or the *Unique Factorization Theorem*, states that in general, if order is disregarded, the prime factorization of a number is unique.

Theorem 5–23

Fundamental Theorem of Arithmetic   Each composite number can be written as a product of primes in one, and only one, way except for the order of the prime factors in the product.

The Fundamental Theorem of Arithmetic assures us that once we find a prime factorization of a number, a different prime factorization of the same number cannot be found. For example, consider 260. We start with the least prime, 2, and see whether it divides 260. If not, we try the next greater prime and check for divisibility by this prime. Once we find a prime that divides the number in question, we must find the quotient of the number divided by the prime. This step in the prime factorization of 260 is shown in Figure 5-23(a). Next we check whether the prime divides the quotient. If so, we repeat the process; if not, we try the next greater prime, 3, and check to see if it divides the quotient. We see that 130 divided by 2 yields 65, as shown in Figure 5-23(b). We continue the procedure, using greater primes, until a quotient of 1 is reached. The original number is the product of all the prime divisors used. The complete procedure for 260 is shown in Figure 5-23(c). An alternative form of this procedure is shown in Figure 5-23(d).

$$
\begin{array}{ll}
2\,\big|\,260 \\
130
\end{array}
\qquad
\begin{array}{ll}
2\,\big|\,260 \\
2\,\big|\,130 \\
65
\end{array}
\qquad
\begin{array}{ll}
2\,\big|\,260 \\
2\,\big|\,130 \\
5\,\big|\,65 \\
13\,\big|\,13 \\
1
\end{array}
\qquad
\begin{array}{ll}
\,260 \\
2\,\big|\,130 \\
2\,\big|\,65 \\
5\,\big|\,13 \\
13\,\big|\,1
\end{array}
$$

(a)   (b)   (c)   (d) Alternative form

Figure 5-23

The primes in the prime factorization of a number are typically listed in increasing order from left to right, and if a prime appears in a product more than once, exponential notation is used. Thus, the factorization of 260 is written as $2^2 \cdot 5 \cdot 13$. Prime factorization is demonstrated in the student page on page 304. Notice that the factor tree is developed in two different ways leading to the same result. Work through the divided practice on the student page.

School Book Page

HOW CAN YOU WRITE A NUMBER AS A PRODUCT OF PRIME FACTORS?

Lesson
5-2

Understand It!
Every whole number greater than 1 is either a prime number or a composite number.

Prime Factorization

How can you write the prime factorization of a number?

Whole numbers greater than 1 are either prime or composite numbers.

A prime number has exactly two factors, 1 and itself. The numbers 2, 3, and 5 are prime numbers.

Model	Dimension	Factors
	1×2	1, 2
	1×3	1, 3
	1×4	1, 4
	2×2	2
	1×5	1, 5

Another Example   How can you use a factor tree to find the prime factorization of a number?

One Way

To find the prime factorization of 72, begin with the smallest prime factor. Write factors until all the factors are prime numbers.

```
        72
       /  \
     2 × 36
     / \ / \
   2 × 2 × 18
   / \ / \ / \
 2 × 2 × 2 × 9
 / \ / \ / \ / \
2 × 2 × 2 × 3 × 3
```

$72 = 2 \times 2 \times 2 \times 3 \times 3$
$72 = 2^3 \times 3^2$

Another Way

To find the prime factorization of 72, begin with any two factors of 72. Write factors until all the factors are prime numbers.

Arrange prime factors in order.

```
        72
       /  \
      6 × 12
     / \  / \
   2 × 3 3 × 4
   / \ / \ / \ / \
 2 × 3 × 3 × 2 × 2
```

$72 = 2 \times 2 \times 2 \times 3 \times 3$
$72 = 2^3 \times 3^2$

There is only one prime factorization for any number.

Guided Practice*

Do you know HOW?

In **1** through **8**, write the prime factorization of each number. If it is prime, write *prime*.

1. 18   **2.** 23   **3.** 32   **4.** 45

5. 89   **6.** 169   **7.** 216   **8.** 243

Do you UNDERSTAND?

9. How do the two factor trees above show that there is only one prime factorization for 72?

10. Is 1 prime or composite?

124   *For another example, see Set B on page 140.*

Source: Scott Foresman-Addison Wesley, enVisionMATH, 2008, Grade 6 (p. 124).

NOW TRY THIS 5-13 Colored rods are used in the elementary-school classroom to teach many concepts. The rods vary in length from 1 cm to 10 cm. Various lengths have different colors; for example, the 5 rod is yellow. The rods and their colors are shown in Figure 5-24. A row with all the same color rods is called a *one-color train*.

a. What rods can be used to form a one-color train for 18?
b. What one-color trains are possible for 24?
c. How many one-color trains of two or more rods are possible for each prime number?
d. If a number can be represented by an all-red train, an all-green train, and an all-yellow train, what is the least number of factors it must have? What are they?

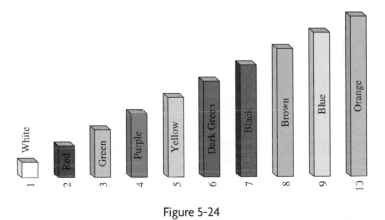

Figure 5-24

Number of Divisors

How many positive divisors does 24 have? Note that the question asks for the number of divisors, not just prime divisors. To aid in the listing, we group divisors as follows:

$$1, 2, 3, 4, 6, 8, 12, 24$$

The positive divisors of 24 occur in pairs, where the product of the divisors in each pair is 24. If 3 is a divisor of 24, then 24/3, or 8, is also a divisor of 24. In general, if a natural number k is a divisor of 24, then $24/k$ is also a divisor of 24.

Another way to think of the number of positive divisors of 24 is to consider the prime factorization $24 = 2^3 \cdot 3$. The positive divisors of 2^3 are $2^0, 2^1, 2^2$, and 2^3. The positive divisors of 3 are 3^0 and 3^1. We know that 2^3 has $(3 + 1)$, or 4, divisors and 3^1 has $(1 + 1)$, or 2, divisors. Because each divisor of 24 is the product of a divisor of 2^3 and a divisor of 3^1, then we use the Fundamental Counting Principle to conclude that 24 has $4 \cdot 2$, or 8, positive divisors. This is summarized in Table 5-2.

Table 5-2

Positive Divisors of 2^3	$2^0 = 1$	$2^1 = 2$	$2^2 = 4$	$2^3 = 8$
Positive Divisors of 3^1	$3^0 = 1$	$3^1 = 3$		
Positive Divisors of $3^1 \cdot$ Positive Divisors of 2^3 (Positive Divisors of 24)	$3^0 \cdot 2^0 = 1$ $3^1 \cdot 2^0 = 3$	$3^0 \cdot 2^1 = 2$ $3^1 \cdot 2^1 = 6$	$3^0 \cdot 2^2 = 4$ $3^1 \cdot 2^2 = 12$	$3^0 \cdot 2^3 = 8$ $3^1 \cdot 2^3 = 24$

This discussion can be generalized as follows: If p is any prime and n is any natural number, then the positive divisors of p^n are p^0, p^1, p^2, p^3, ..., p^n. Therefore, there are positive $(n + 1)$ divisors of p^n. Now, using the Fundamental Counting Principle, we can find the number of divisors of any number whose prime factorization is known.

Theorem 5–24

If p and q are different primes, then $p^n q^m$ has $(n + 1)(m + 1)$ positive divisors. In general, if $p_1, p_2 ..., p_k$ are primes and $n_1, n_2 ..., n_k$ are whole numbers, then $p_1^{n_1} \cdot p_2^{n_2} \cdot ... \cdot p_k^{n_k}$ has $(n_1 + 1)(n_2 + 1) \cdot ... \cdot (n_k + 1)$ positive divisors.

Example 5-25

Find the number of positive divisors of each of the following:

a. 1,000,000   **b.** 210^{10}

Solution   **a.** We first find the prime factorization of 1,000,000.

$$1,000,000 = 10^6 = (2 \cdot 5)^6 = (2 \cdot 5)(2 \cdot 5)(2 \cdot 5)(2 \cdot 5)(2 \cdot 5)(2 \cdot 5)$$
$$= (2 \cdot 2 \cdot 2 \cdot 2 \cdot 2 \cdot 2)(5 \cdot 5 \cdot 5 \cdot 5 \cdot 5 \cdot 5)$$
$$= 2^6 \cdot 5^6$$

Because 2^6 has $6 + 1$ divisors and 5^6 has $6 + 1$ divisors, then by the Fundamental Counting Principle $2^6 \cdot 5^6$ has $(6 + 1)(6 + 1)$, or 49, divisors.

b. The prime factorization of 210 is

$$210 = 21 \cdot 10 = 3 \cdot 7 \cdot 2 \cdot 5 = 2 \cdot 3 \cdot 5 \cdot 7,$$
$$210^{10} = (2 \cdot 3 \cdot 5 \cdot 7)^{10} = 2^{10} \cdot 3^{10} \cdot 5^{10} \cdot 7^{10}$$

By the Fundamental Counting Principle, the number of divisors of 210^{10} is $(10 + 1)(10 + 1)(10 + 1)(10 + 1) = 11^4 = 14,641$.

NOW TRY THIS 5-14 To determine whether it is necessary to divide 97 by 2, 3, 4, 5, 6, ..., 96 to check if it is prime, answer the following (justify your answers):

a. If 2 is not a divisor of 97, could any multiple of 2 be a divisor of 97?
b. If 3 is not a divisor of 97, what other numbers could not be divisors of 97?
c. If 5 is not a divisor of 97, what other numbers could not be divisors of 97?
d. If 7 is not a divisor of 97, what other numbers could not be divisors of 97?
e. Conjecture what numbers we have to check for divisibility in order to determine if 97 is prime.

Determining If a Number Is Prime

As depicted in the following cartoon by Sidney Harris, prime numbers have fascinated people of various backgrounds. In Now Try This 5-14, you might have found that to determine if a number is prime, you must check only divisibility by prime numbers less than the given number. Why? However, do we need to check all the primes less than the number? Suppose we want to check if 97 is prime and we find that 2, 3, 5, and 7 do not divide 97. Could a greater prime divide 97? If p is a prime greater than 7, then $p \geq 11$. If $p|97$, then $97/p$ also divides 97. However, because $p \geq 11$ then $97/p$ must be less than 10 and hence cannot divide 97. Why? So we see that there is no need to check for divisibility by numbers other than 2, 3, 5, and 7. These ideas are generalized in the following theorems.

Theorem 5–25

If d is a divisor of n, then $\dfrac{n}{d}$ is also a divisor of n.

Suppose that p is the *least* divisor of n (greater than 1). Such a divisor must be prime (why?). Then $n = pk, k \neq 1$. Since $k \mid n$ and p was the least divisor of $n, k \geq p$. Therefore, $n = pk \geq pp = p^2$. Since $n \geq p^2$, we have $p^2 \leq n$. This idea is summarized in the following theorem.

Theorem 5–26

If n is composite, then n has a prime factor p such that $p^2 \leq n$.

Theorem 5–26 can be used to help determine whether a given number is prime or composite. For example, consider the number 109. If 109 is composite, it must have a prime divisor p such that $p^2 \leq 109$. The primes whose squares do not exceed 109 are 2, 3, 5, and 7. Mentally, we can see that $2 \nmid 109, 3 \nmid 109, 5 \nmid 109$, and $7 \nmid 109$. Hence, 109 is prime. The argument used leads to the following theorem.

Theorem 5–27

If n is an integer greater than 1 and not divisible by any prime p, such that $p^2 \leq n$, then n is prime.

REMARK Because $p^2 \leq n$ implies that $p \leq \sqrt{n}$, Theorem 5–27 says that to determine if a number n is prime, it is enough to check if any prime less than or equal to $\sqrt{n}$ is a divisor of n.

Example 5-26

a. Is 397 composite or prime?   **b.** Is 91 composite or prime?

Solution **a.** The possible primes p such that $p^2 \leq 397$ are 2, 3, 5, 7, 11, 13, 17, and 19. Because $2 \nmid 397, 3 \nmid 397, 5 \nmid 397, 7 \nmid 397, 11 \nmid 397, 13 \nmid 397, 17 \nmid 397$, and $19 \nmid 397$, the number 397 is prime.

b. The possible primes p such that $p^2 \leq 91$ are 2, 3, 5, and 7. Because 91 is divisible by 7, it is composite.

One way to find all the primes less than a given number is to use the *Sieve of Eratosthenes*, named after the Greek mathematician Eratosthenes (ca. 276–194 BCE). If all the natural numbers greater than 1 are considered (or placed in the sieve), the numbers that are not prime are methodically crossed out (or drop through the holes of the sieve). The remaining numbers are prime. The partial student page on page 309 illustrates this process. Before reading on, build a sieve, work through the process, and answer the questions on the student page.

The Sieve of Eratosthenes is another way to motivate Theorem 5–27. Notice the observations from the sieve in Table 5-3 as we crossed out numbers.

Table 5-3

Prime	Observation
2	First number not crossed out that 2 divides is $4 = 2^2$.
3	First number not crossed out that 3 divides is $9 = 3^2$.
5	First number not crossed out that 5 divides is $25 = 5^2$.
7	First number not crossed out that 7 divides is $49 = 7^2$.

We didn't need to continue the procedure for 11 because the first number not crossed out that 11 divides is 11^2, or 121, and the table only goes to 100. Therefore, to test whether a number such as 137 is a prime, we first test for divisibility by all primes up to but not including the first prime whose square is greater than 137. Because $13^2 = 169$, any prime greater than or equal to 13 would give a quotient less than or equal to 13, and we have already checked these primes. This shows that when testing to see if a number is prime, we need try as divisors only primes whose squares are less than or equal to the number being tested.

More About Primes

There are infinitely many whole numbers, infinitely many odd whole numbers, and infinitely many even whole numbers. Are there infinitely many primes? Because prime numbers do not appear in any known pattern, the answer to this question is not obvious. Euclid was the first to prove that there are infinitely many primes.

Historical Note

Eratosthenes (276–194 BCE), a Greek scholar, was born in Cyrene but spent most of his life in Alexandria as the chief librarian at the museum there. In his work *Geographica*, he gave arguments for the spherical shape of the Earth. Today, Eratosthenes is best known for his "sieve"—a systematic procedure for isolating the prime numbers—and for a simple method for calculating the circumference of the Earth. ◆

School Book Page   SIEVE OF ERATOSTHENES

Enrichment

Sieve of Eratosthenes

About 230 B.C., the Greek mathematician Eratosthenes developed a method for identifying prime numbers. The method is called the **Sieve of Eratosthenes.**

Follow the steps to identify the prime numbers from 1 to 100.

Step 1 Copy the table at the right.

Step 2 Cross out 1 because it is neither prime nor composite.

Step 3 Circle 2. Then, cross out all other multiples of 2.

Step 4 Go to the first number that is not crossed out. Circle it and cross out its other multiples.

Step 5 Repeat Step 4 until all numbers in the table are either crossed out or circled. The circled numbers are prime.

1	2	3	4	5	6	7	8	9	10
11	12	13	14	15	16	17	18	19	20
21	22	23	24	25	26	27	28	29	30
31	32	33	34	35	36	37	38	39	40
41	42	43	44	45	46	47	48	49	50
51	52	53	54	55	56	57	58	59	60
61	62	63	64	65	66	67	68	69	70
71	72	73	74	75	76	77	78	79	80
81	82	83	84	85	86	87	88	89	90
91	92	93	94	95	96	97	98	99	100

1. List the prime numbers from 1 to 100.

2. Explain why some numbers could be crossed out more than once.

All text pages available online and on CD-ROM.   **Section A Lesson 3-2**   149

Source: Scott Foresman-Addison Wesley, Mathematics, 2008, Grade 6 (p. 149).

Mathematicians have long looked for a formula that produces only primes, but no one has found one. One result was the expression $n^2 - n + 41$, where n is a whole number. Substituting $0, 1, 2, 3, \ldots, 40$ for n in the expression always results in a prime number. However, substituting 41 for n gives $41^2 - 41 + 41$, or 41^2, a composite number. In 1998, Roland Clarkson, a 19-year-old student at California State University, showed that $2^{3021377} - 1$ is prime. The number has 909,526 digits. The full decimal expansion of the number would fill several hundred pages. Since then, more large primes have been discovered: $2^{32,582,657} - 1$ (9,808,358 digits) and $2^{30,402,457} - 1$ (9,152,052 digits). These are examples of *Mersenne primes*. A Mersenne prime, named after the French monk Marin Mersenne

♦ *Historical Note*

Sophie Germain (1776–1831) was born in Paris and grew up during the French Revolution. She wanted to study at the prestigious École Polytechnique but women were not allowed as students. Consequently, she studied from lecture notes and from Gauss's monograph on number theory. She made major contributions to the mathematical theory of elasticity, for which she was awarded the prize of the French Academy of Sciences. Germain's work was highly regarded by Gauss, who recommended her for an honorary degree from the University of Göttingen. She died before the degree could be awarded.   ♦

(1588–1648), is a prime of the form $2^n - 1$, where n is prime. On August 23, 2008, a UCLA computer discovered the 45th known Mersenne prime, $2^{43,112,609} - 1$ (12,978,189 digits). On September 6, 2008, the 46th Mersenne prime, $2^{37,156,661} - 1$ (11,185,272 digits) was discovered in Germany.

Searching for large primes has led to advances in *distributed computing*, that is, using the Internet to engage the unused computing power of great numbers of computers. Searching for Mersenne primes has been used as a test for computer hardware.

Another type of interesting prime is a *Sophie Germain prime*, which is an odd prime p for which $2p + 1$ is also a prime. Notice that $p = 3$ is a Sophie Germain prime, since $2 \cdot 3 + 1$, or 7, is also a prime. Check that 5, 11, and 23 are also such primes. The primes were named after the French mathematician Sophie Germain. In 2007, the greatest Sophie Germain prime discovered had 51,910 digits.

Problem Solving   How Many Bears?

A large toy store carries one kind of stuffed bear. On Monday the store sold a certain number of the stuffed bears for a total of $1843 and on Tuesday, without changing the price, the store sold a certain number of the stuffed bears for a total of $1957. How many toy bears were sold each day if the price of each bear is a whole number and greater than $1?

Understanding the Problem   One day a store sold a number of stuffed bears for $1843 and on the next day a number of them for a total of $1957. We need to find the number of bears sold on each day.

Devising a Plan   If x bears were sold the first day and y bears the second day, and if the price of each bear was c dollars, we would have $cx = 1843$ and $cy = 1957$. Thus, 1843 and 1957 should have a common factor—the price c. We could factor each number and find the possible factors. If the problem is to have a unique solution, the two numbers should have only one common factor other than 1. Any common factor of 1957 and 1843 will also be a factor of $1957 - 1843 = 114$ and the factors of 114 are easier to find.

Carrying Out the Plan   We have $114 = 2 \cdot 57 = 2 \cdot 3 \cdot 19$. Thus, if 1957 and 1843 have a common prime factor, it must be 2, 3, or 19. But neither 2 nor 3 divides the numbers, hence the only possible common factor is 19. We divide each number by 19 and find

$$1843 = 19 \cdot 97$$
$$1957 = 19 \cdot 103$$

Historical Note

In the 1970s, determining large prime numbers became extremely useful in coding and decoding secret messages. In all coding and decoding, the letters of an alphabet correspond in some way to nonnegative integers. A "safe" coding system, in which messages are unintelligible to everyone except the intended receiver, was devised by three Massachusetts Institute of Technology scientists (Ronald Rivest, Adi Shamir, and Leonard Adleman) and is referred to as the RSA (their initials) system. The secret deciphering key consists of two large prime numbers chosen by the user. The enciphering key is the product of these two primes. Because it is extremely difficult and time-consuming to factor large numbers, it was practically impossible to recover the deciphering key from a known enciphering key. In 1982, new methods for factoring large numbers were invented, which resulted in the use of even greater primes to prevent the breaking of decoding keys. ◆

Notice that neither 97 nor 103 is divisible by 2, 3, 5, or 7. Hence 97 and 103 are primes (why?) and therefore the only common factor (greater than 1) of 1843 and 1957 is 19. Consequently, the price of each bear was $19. The first day 97 bears were sold and the next day 103 bears were sold.

Looking Back Notice that the problem had a unique solution because the only common factor (greater than 1) of the two numbers was 19. We could create similar problems by having the price of the item be a prime number and the number of items sold each day also be prime numbers. For example, the total sale on the first day could have been $23 \cdot 101$, or $2323 and on the second day $23 \cdot 107$, or $2461 (notice that 23, 101, and 107 are prime numbers).

To find a common factor of 1957 and 1843, we found all the common factors of $1957 - 1843 = 114 = 2 \cdot 3 \cdot 19$ and checked which of the factors of the difference was a common factor of the original numbers. We have used Theorem 5–13: If $d|a$ and $d|b$, then $d|(a - b)$. This theorem assures us that every common factor of a and b will also be a factor of $a - b$.

Assessment 5-4A

1. Find the least positive number that is divisible by three different primes.
2. Determine which of the following numbers are primes:
   a. 109   **b.** 119
   c. 33   **d.** 101
   e. 463   **f.** 97
   g. $2 \cdot 3 \cdot 5 \cdot 7 + 1$ **h.** $2 \cdot 3 \cdot 5 \cdot 7 - 1$
3. Use a factor tree to find the prime factorization for each of the following:
   a. 504   **b.** 2475   **c.** 11,250
4. **a.** Fill in the missing numbers in the following factor tree:

   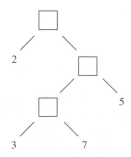

   b. How could you find the top number without finding the other two numbers?
5. What is the greatest prime you must consider to test whether 5669 is prime?
6. Find the prime factorizations of the following:
   a. $1 \cdot 2 \cdot 3 \cdot 4 \cdot 5 \cdot 6 \cdot 7 \cdot 8 \cdot 9 \cdot 10$
   b. $10^2 \cdot 26 \cdot 49^{10}$
   c. 251
   d. 1001

7. **a.** When the U.S. flag had 48 stars, the stars formed a 6 by 8 rectangular array. In what other rectangular arrays could they have been arranged?
   b. How many rectangular arrays of stars could there be if there were only 47 states?
8. **a.** Use the Fundamental Theorem of Arithmetic to justify that if $2|n$ and $3|n$, then $6|n$.
   b. Is it always true that if $a|n$ and $b|n$, then $ab|n^2$? Either prove the statement or give a counterexample.
9. Mr. Arboreta wants to plant fruit trees in a rectangular array. For each of the following numbers of trees, find all possible numbers of rows if each row is to have the same number of trees:
   a. 36
   b. 28
   c. 17
   d. 144
10. Some of the divisors of a locker number are 2, 5, and 9. If there are exactly nine additional positive divisors, what is the locker number?
11. Extend the Sieve of Eratosthenes to find all primes between 100 and 200.
12. The prime numbers 11 and 13 are called *twin primes* because they differ by 2. (The existence of infinitely many twin primes has not been proved.) Find all the twin primes less than 200.
13. If $42|n$, what other positive integers divide n?
14. If 1000 is a factor of n, what other positive integers divide n? How many such integers are there?
15. It is not known whether there are infinitely many primes in the infinite sequence consisting of the numbers whose only digits are ones; 1, 11, 111, 1111, ... Find infinitely many composite numbers in the sequence.

16. Is $3^2 \cdot 2^4$ a factor of $3^4 \cdot 2^7$? Explain why or why not.

17. Explain why each of the following numbers is composite:
   a. $3 \cdot 5 \cdot 7 \cdot 11 \cdot 13$
   b. $(3 \cdot 4 \cdot 5 \cdot 6 \cdot 7 \cdot 8) + 2$
   c. $(3 \cdot 5 \cdot 7 \cdot 11 \cdot 13) + 5$
   d. $10! + 7$ (*Note:* $10! = 10 \cdot 9 \cdot 8 \cdot 7 \cdot 6 \cdot 5 \cdot 4 \cdot 3 \cdot 2 \cdot 1$.)

18. Explain why $2^3 \cdot 3^2 \cdot 25^3$ is not a prime factorization and find the prime factorization of the number.

19. Find the prime factorizations of each of the following:
   a. $36^{10} \cdot 49^{20} \cdot 6^{15}$
   b. $100^{60} \cdot 300^{40}$
   c. $2 \cdot 3^4 \cdot 5^{110} \cdot 7 + 4 \cdot 3^4 \cdot 5^{110}$
   d. $2 \cdot 3 \cdot 5 \cdot 7 \cdot 11 + 1$

20. I am a prime number greater than 40 and less than 90. My units digit and my tens digit are both prime. The difference between my units digit and tens digit is not 2. What number am I?

Assessment 5-4B

1. Determine which of the following numbers are primes:
   a. 89   **b.** 147
   c. 159   **d.** 187
   e. $2 \cdot 3 \cdot 5 \cdot 7 + 5$ **f.** $2 \cdot 3 \cdot 5 \cdot 7 - 5$

2. Use a factor tree to find the prime factorization for each of the following:
   a. 304
   b. 1570
   c. 9550

3. a. Fill in the missing numbers in the following factor tree:

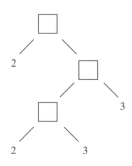

   b. How could you find the top number without finding the other two numbers?

4. What is the greatest prime you must consider to test whether 503 is prime?

5. Find the prime factorizations of the following:
   a. 1001
   b. 1001^2
   c. 999^{10}
   d. $111^{10} - 111^9$

6. Suppose the 435 members of the House of Representatives are placed on committees consisting of more than 2 members but fewer than 30 members. Each committee is to have an equal number of members and each member is to be on only one committee.
   a. What size committees are possible?
   b. How many committees are there of each size?

7. Find the least number divisible by each natural number less than or equal to 12.

8. Find the greatest four-digit number that has exactly three positive factors.

9. Show that if 1 were considered a prime, every number would have more than one prime factorization.

10. Is it possible to find positive integers x, y, and z such that $2^x \cdot 3^y = 5^z$? Why or why not?

11. a. Show that there are infinitely many composite numbers in the arithmetic sequence $1, 5, 9, 13, 17, \ldots$.
   b. Does every arithmetic sequence consisting of integers with difference greater than 0 have infinitely many composite numbers? Justify your answer.

12. If $2N = 2^6 \cdot 3^5 \cdot 5^4 \cdot 7^3 \cdot 11^7$, explain why $2 \cdot 3 \cdot 5 \cdot 7 \cdot 11$ is a factor of N.

13. Is $3^2 \cdot 2^4$ a factor of $3^3 \cdot 2^2$? Explain why or why not.

14. Explain why each of the following numbers is composite:
   a. $7 \cdot 11 \cdot 13 \cdot 17 + 17$
   b. $10! + k$, where $k = 2, 3, 4, 5, 6, 7, 8, 9,$ or 10

15. Explain why $2^2 \cdot 5^3 \cdot 9^2$ is not a prime factorization and find the prime factorization of the number.

16. A prime such as 7331 is a *superprime* because any integers obtained by deleting digits from the right of 7331 are prime; namely, 733, 73, and 7.
   a. For a prime to be a superprime, what digits cannot appear in the number?
   b. Of the digits that can appear in a superprime, what digit cannot be the leftmost digit of a superprime?
   c. Find all of the two-digit superprimes.
   d. Find a three-digit superprime other than 733.

17. Is the following always true? (Justify your answer.) If $m \mid ab$, then $m \mid a$ or $m \mid b$.

18. Find the prime factorizations of each of the following:
   a. $16^4 \cdot 81^4 \cdot 6^6$
   b. $8^4 \cdot 32^5$
   c. $2^2 \cdot 3^5 \cdot 7^{55} + 2^4 \cdot 3^4 \cdot 7^{55}$

19. Use the Fundamental Theorem of Arithmetic to justify the following statement for whole numbers a and b greater than 1: If p is prime and $p|ab$, then $p|a$ or $p|b$.

20. The product of three prime numbers that are less than 30 is 1955. What are the three primes?

Mathematical Connections 5-4

Communication

1. Explain why the product of any three consecutive integers is divisible by 6.

2. Explain why the product of any four consecutive integers is divisible by 24.

3. In order to test for divisibility by 12, one student checked to determine divisibility by 3 and 4; another checked for divisibility by 2 and 6. Are both students using a correct approach to divisibility by 12? Why or why not?

4. In the Sieve of Eratosthenes for numbers less than 100, explain why, after we cross out all the multiples of 2, 3, 5, and 7, the remaining numbers are primes.

5. Let $M = 2 \cdot 3 \cdot 5 \cdot 7 + 11 \cdot 13 \cdot 17 \cdot 19$. Without multiplying, show that none of the primes less than or equal to 19 divides M.

6. A woman with a basket of eggs finds that if she removes the eggs from the basket 3 or 5 at a time, there is always 1 egg left. However, if she removes the eggs 7 at a time, there are no eggs left. If the basket holds up to 100 eggs, how many eggs does she have? Explain your reasoning.

7. Explain why, when a number is composite, its least positive divisor, other than 1, must be prime.

8. Euclid proved that given any finite list of primes, there exists a prime not in the list. Read the following argument and answer the questions that follow.

Let $2, 3, 5, 7, \ldots, p$ be a list of all the primes less than or equal to a certain prime p. We will show that there exists a prime not on the list. Consider the product

$$2 \cdot 3 \cdot 5 \cdot 7 \cdot \ldots \cdot p$$

Notice that every prime in our list divides that product. However, if we add 1 to the product, that is, form the number $N = (2 \cdot 3 \cdot 5 \cdot 7 \cdot \ldots \cdot p) + 1$, then none of the primes in the list will divide N. Notice that whether N is prime or composite, some prime q must divide N. Because no prime in our list divides N, q is not one of the primes in our list. Consequently $q > p$. We have shown that there exists a prime greater than p.

a. Explain why no prime in the list will divide N.

b. Explain why some prime must divide N.

c. Someone discovered a prime that has 65,050 digits. How does the preceding argument assure us that there exists an even larger prime?

d. Does the argument show that there are infinitely many primes? Why or why not?

e. Let $M = 2 \cdot 3 \cdot 5 \cdot 7 \cdot 11 \cdot 13 \cdot 17 \cdot 19 + 1$. Without multiplying, explain why some prime greater than 19 will divide M.

Open-Ended

9. a. In which of the following intervals do you think there are the most primes? Why? Check to see if you were correct.

    i. 0–99   **ii.** 100–199

b. What is the longest string of consecutive composite numbers in the intervals?

c. How many twin primes are there in each interval?

d. What patterns, if any, do you see for any of the preceding questions? Predict what might happen in other intervals.

10. A number is a perfect number if the sum of its factors (other than the number itself) is equal to the number. For example, 6 is a perfect number because its factors sum to 6, that is, $1 + 2 + 3 = 6$. An abundant number has factors whose sum is greater than the number itself. A deficient number is a number with factors whose sum is less than the number itself.

a. Classify each of the following numbers as perfect, abundant, or deficient:

    i. 12   **ii.** 28   **iii.** 35

b. Find at least one more number that is deficient and one that is abundant.

Cooperative Learning

11. A class of 23 students was using square tiles to build rectangular shapes. Each student had more than 1 tile and each had a different number of tiles. Each student was able to build only one shape of rectangle. All tiles had to be used to build a rectangle and the rectangle could not have holes. For example, a 2 by 6 rectangle uses 12 tiles and is considered the same as a 6 by 2 rectangle but is different from a 3 by 4 rectangle. The class did the activity using the least number of tiles. How many tiles did the class use? Explore the various rectangles that could be made.

Questions from the Classroom

12. Mary says that her factor tree for 72 begins with 3 and 24 so her prime factors will be different from Larry's because he is going to start with 8 and 9. What do you tell Mary?

13. Bob says that to check if a number is prime he just uses the divisibility rules he knows for 2, 3, 4, 5, 6, 8, and 10. He says if the number is not divisible by these numbers, then it is prime. How do you respond?

14. Joe says that every odd number greater than 3 can be written as the sum of two primes. To convince the class, he wrote $7 = 2 + 5$, $5 = 2 + 3$, and $9 = 7 + 2$. How do you respond?

15. An eighth grader at the Roosevelt Middle School claims that because there are as many even numbers as odd numbers between 1 and 1000, there must be as many numbers that have an even number of positive divisors as numbers that have an odd number of positive divisors between 1 and 1000. Is the student correct? Why or why not?

16. A sixth-grade student argues that there are infinitely many primes because "there is no end to numbers." How do you respond?

17. A student claims that every prime greater than 3 is a term in the arithmetic sequence whose nth term is $6n + 1$ or in the arithmetic sequence whose nth term is $6n - 1$. Is this true? If so why?

Review Problems

18. Classify the following as true or false:
  a. 11 is a factor of 189.
  b. 1001 is a multiple of 13.
  c. $7|1001$ and $7 \nmid 12$ imply $7 \nmid (1001 - 12)$.

19. Check each of the following for divisibility by 2, 3, 4, 5, 6, 7, 8, 9, 10, and 11:
  a. 438,162
  b. 2,345,678,910

20. Prove that if a number is divisible by 12, then it is divisible by 3.

21. Could $3376 be divided exactly among either seven or eight people?

LABORATORY ACTIVITY In Figure 5-25, a spiral starts with 41 at its center and continues in a counterclockwise direction. Primes are shaded. Check the primes along the shaded diagonal. Can you find each of the primes from the formula $n^2 + n + 41$ by substituting appropriate values for n?

265	264	263	262	261	260	259	258	257	256	255	254	253	252	251
210	209	208	207	206	205	204	203	202	201	200	199	198	197	250
211	162	161	160	159	158	157	156	155	154	153	152	151	196	249
212	163	122	121	120	119	118	117	116	115	114	113	150	195	248
213	164	123	90	89	88	87	86	85	84	83	112	149	194	247
214	165	124	91	66	65	64	63	62	61	82	111	148	193	246
215	166	125	92	67	50	49	48	47	60	81	110	147	192	245
216	167	126	93	68	51	42	⟨41⟩	46	59	80	109	146	191	244
217	168	127	94	69	52	43	44	45	58	79	108	145	190	243
218	169	128	95	70	53	54	55	56	57	78	107	144	189	242
219	170	129	96	71	72	73	74	75	76	77	106	143	188	241
220	171	130	97	98	99	100	101	102	103	104	105	142	187	240
221	172	131	132	133	134	135	136	137	138	139	140	141	186	239
222	173	174	175	176	177	178	179	180	181	182	183	184	185	238
223	224	225	226	227	228	229	230	231	232	233	234	235	236	237

Figure 5-25

BRAIN TEASER One Saturday Jody cut short her visit with her friend Natasha to take three other friends to a movie. "How old are they?" asked Natasha. "The product of their ages is 2450 and the sum is exactly twice your age," replied Jody. Natasha said: "I need more information." To that Jody replied, "I should have mentioned that I am at least one year younger than the oldest of my three friends." With this information Natasha found the ages of the friends. How did Natasha figure the ages of the friends and what were their ages?

5-5 Greatest Common Divisor and Least Common Multiple

Consider the following situation:

Two bands are to be combined to march in a parade. A 24-member band will march behind a 30-member band. The combined bands must have the same number of columns. Each column must be the same size. What is the greatest number of columns in which they can march?

The bands could each march in 2 columns, and we would have the same number of columns, but this does not satisfy the condition of having the greatest number of columns. The number of columns must divide both 24 and 30. Why? Numbers that divide both 24 and 30 are 1, 2, 3, and 6. The greatest of these numbers is 6, so the bands should each march in columns. The first band would have 6 columns with 4 members in each column, and the second band would have 6 columns with 5 members in each column. In this problem, we have found the greatest number that divides both 24 and 30, that is, the **greatest common divisor (GCD)** of 24 and 30.

> ### Research Note
> Possibly because students often confuse factors and multiples, the greatest common factor and the least common multiple are difficult topics for students to grasp (Graviss and Greaver 1992).

Definition
The **greatest common divisor (GCD)** of two natural numbers a and b is the greatest natural number that divides both a and b.

In the Research Note we see that greatest common divisors (GCD) and least common multiples (LCM) are difficult for students. We provide many methods for finding GCDs and LCMs to help clarify these concepts.

Colored Rods Method

We can build a model of two or more integers with colored rods to determine the GCD of two positive integers. For example, consider finding the GCD of 6 and 8 using the 6 rod and the 8 rod, as in Figure 5-26.

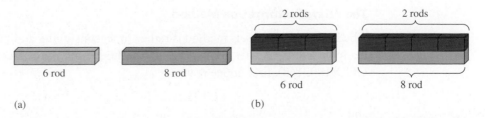

(a)

2 rods   2 rods

6 rod   8 rod   6 rod   8 rod

(b)

Figure 5-26

To find the GCD of 6 and 8, we must find the longest rod such that we can use multiples of that rod to build both the 6 rod and the 8 rod. The 2 rods can be used to build both the 6 and 8 rods, as shown in Figure 5-26(b); the 3 rods can be used to build the 6 rod but not the 8 rod; the 4 rods can be used to build the 8 rod but not the 6 rod; the 5 rods can be used to build neither; and the 6 rods cannot be used to build the 8 rod. Therefore, $GCD(6,8) = 2$.

NOW TRY THIS 5-15 Explain how you could use colored rods to solve the marching bands' problem, stated at the beginning of this section.

The Intersection-of-Sets Method

In the *intersection-of-sets* method, we list all members of the set of positive divisors of the two integers, then find the set of all *common divisors*, and, finally, pick the *greatest* element in that set. For example, to find the GCD of 20 and 32, denote the sets of divisors of 20 and 32 by D_{20} and D_{32}, respectively.

$$D_{20} = \{1, 2, 4, 5, 10, 20\}$$
$$\updownarrow \updownarrow \updownarrow$$
$$D_{32} = \{1, 2, 4, 8, 16, 32\}$$

The set of all common positive divisors of 20 and 32 is

$$D_{20} \cap D_{32} = \{1,2,4\}$$

Because the greatest number in the set of common positive divisors is 4, the GCD of 20 and 32 is 4, written $GCD(20,32) = 4$.

NOW TRY THIS 5-16 The Venn diagram in Figure 5-27 shows the factors of 24 and 40. Answer the following:

a. What is the meaning of each of the shaded regions?
b. Which factor is the GCD?
c. Draw a similar Venn diagram to find the GCD of 36 and 44.

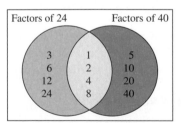

Figure 5-27

The Prime Factorization Method

The intersection-of-sets method is rather time-consuming and tedious if the numbers have many divisors. Another, more efficient, method is the prime factorization method. To find $GCD(180, 168)$, first notice that

$$180 = 2 \cdot 2 \cdot 3 \cdot 3 \cdot 5 = (2^2 \cdot 3)3 \cdot 5$$

and

$$168 = 2 \cdot 2 \cdot 2 \cdot 3 \cdot 7 = (2^2 \cdot 3)2 \cdot 7$$

We see that 180 and 168 have two factors of 2 and one of 3 in common. These common primes divide both 180 and 168. In fact, the only numbers other than 1 that divide both 180 and 168 must have no more than two 2s and one 3 and no other prime factors in their prime factorizations. The possible common divisors are $1, 2, 2^2, 3, 2 \cdot 3$, and $2^2 \cdot 3$. Hence, the greatest common divisor of 180 and 168 is $2^2 \cdot 3$. The procedure for finding the GCD of two or more numbers by using the prime factorization method is summarized as follows:

> To find the GCD of two or more positive integers, first find the prime factorizations of the given numbers and then identify each common prime factor of the given numbers. The GCD is the product of the common factors, each raised to the lowest power of that prime that occurs in any of the prime factorizations.

If we apply the prime factorization technique to finding GCD(4, 9), we see that 4 and 9 have no common prime factors. But that does not mean there is no GCD. We still have 1 as a common divisor, so GCD(4, 9) = 1. Numbers, such as 4 and 9, whose GCD is 1 are **relatively prime**. Both the intersection-of-sets method and the prime factorization method are found on the student page on page 318. Study the page and work through the Talk About It questions on the bottom of the student page. Notice that the GCF in the student page stands for *Greatest Common Factor*, which is the same as GCD.

Example 5-27

Find each of the following:

a. GCD(108, 72)
b. GCD(0, 13)
c. GCD(x, y) if $x = 2^3 \cdot 7^2 \cdot 11 \cdot 13$ and $y = 2 \cdot 7^3 \cdot 13 \cdot 17$
d. GCD(x, y, z) if $z = 2^2 \cdot 7$, using x and y from (c)
e. GCD(x, y), where $x = 5^4 \cdot 13^{10}$ and $y = 3^{10} \cdot 11^{20}$

Solution **a.** Since $108 = 2^2 \cdot 3^3$ and $72 = 2^3 \cdot 3^2$, it follows that GCD(108, 72) = $2^2 \cdot 3^2 = 36$.
   b. Because $13 \mid 0$ and $13 \mid 13$, it follows that GCD(0, 13) = 13.
   c. GCD(x, y) = $2 \cdot 7^2 \cdot 13 = 1274$.
   d. Because $x = 2^3 \cdot 7^2 \cdot 11 \cdot 13$, $y = 2 \cdot 7^3 \cdot 13 \cdot 17$, and $z = 2^2 \cdot 7$, then GCD(x, y, z) = $2 \cdot 7 = 14$. Notice that GCD(x, y, z) can also be obtained by finding the GCD of z and 1274, the answer from (c).
   e. Because x and y have no common prime factors, GCD(x, y) = 1.

Calculator Method

Calculators with a ⬚Simp⬚ key can be used to find the GCD of two numbers. For example, to find the GCD(120, 180), use the following sequence of buttons to start: First, press ⬚1⬚ ⬚2⬚ ⬚0⬚ ⬚/⬚ ⬚1⬚ ⬚8⬚ ⬚0⬚ ⬚Simp⬚ ⬚=⬚ to obtain the display ⬚N/D→n/d 60/90⬚. By pressing the ⬚x⊝y⬚ button, we see ⬚2⬚ on the display as a common divisor of 120 and 180. By pressing the ⬚x⊝y⬚ button again and pressing, ⬚Simp⬚ ⬚=⬚ ⬚x⊝y⬚, we see 2 again as a factor. The process is repeated to reveal 3 and 5 as other common factors. The GCD of 120 and 180 is the product of the common prime factors $2 \cdot 2 \cdot 3 \cdot 5$, or 60.

School Book Page GREATEST COMMON FACTOR

Lesson 3-3

Key Idea
There are different ways to find the factors that are common to two or more numbers.

Vocabulary
• common factor
• greatest common factor (GCF)
• prime factorization (p. 147)

Think It Through
• I can **use factors** to identify equal groups for sharing.
• I can **make an organized list** to find the common factors and GCF.

Greatest Common Factor

LEARN

How can you use factors?

Janelle is making snack packs for a group hike. Each pack should have the same number of bags of trail mix and the same number of bottles of water. What is the greatest number of snack packs that she can make with no refreshments left over?

To solve this problem, you need to find the numbers that are factors of both 60 and 90. These are the **common factors** of 60 and 90. The **greatest common factor (GCF)** is the *greatest* number that is a factor of both 60 and 90.

✔ **WARM UP**

List the factors of each number.

1. 12   2. 31

3. 54   4. 100

Hike Refresments
60 bags of trail mix
90 bottles of water

Example

Find the GCF of 60 and 90.

One Way

List the factors of each number.

60: 1, 2, 3, 4, 5, 6, 10, 12, 15, 20, 30, 60
90: 1, 2, 3, 5, 6, 9, 10, 15, 18, 30, 45, 90

Circle pairs of common factors. Select the greatest one.

60: 1, 2, 3, 4, 5, 6, 10, 12, 15, 20, 30, 60
90: 1, 2, 3, 5, 6, 9, 10, 15, 18, 30, 45, 90

The GCF is 30.

Another Way

Use prime factorization.

$60 = 2 \times 2 \times 3 \times 5$
$90 = 2 \times 3 \times 3 \times 5$

Find the product of the common prime factors. If there are no common prime factors, the GCF is 1.

$60 = 2 \times 2 \times 3 \times 5$
$90 = 2 \times 3 \times 3 \times 5$ $2 \times 3 \times 5 = 30$

The GCF is 30.

The greatest number of snack packs she can make is 30.

✔ **Talk About It**

1. In the second method, why is 2 used only once as a factor of the GCF?

2. In each of the 30 snack packs, how many bags of trail mix and how many bottles of water will there be?

3. **Reasoning** Find the GCF of 48 and 120. Which method did you use? Why?

Take It to the NET
More Examples
www.scottforesman.com

150

Source: Scott Foresman-Addison Wesley Mathematics 2008, Grade 6 (p. 150).

Some calculators have a GCD feature built in, which you will probably have to go to the *MATH* menu to find. With this feature, you select GCD and enter the numbers separated by a comma and closed within a parenthesis; for example, GCD(120, 180). When the $=$ is pressed, the GCD of 60 will be displayed.

Euclidean Algorithm Method

Large numbers may be hard to factor. For these numbers, another method is more efficient than factorization for finding the GCD. For example, suppose we want to find GCD(676, 221). If we could find two smaller numbers whose GCD is the same as GCD(676, 221), our task would be easier. From Theorem 5–13(c), every divisor of 676 and 221 is also a divisor of $676 - 221$ and 221. Conversely, every divisor of $676 - 221$ and 221 is also a divisor of 676 and 221. Thus, the set of all the common divisors of 676 and 221 is the same as the set of all common divisors of $676 - 221$ and 221. Consequently, GCD(676, 221) = GCD($676 - 221$, 221). This process can be continued to subtract three 221s from 676 so that GCD(676, 221) = GCD($676 - 3 \cdot 221$, 221) = GCD(13, 221). To determine how many 221s can be subtracted from 676, we could have divided as follows:

$$\overset{3 \text{ R } 13}{221\overline{)676}} \qquad \overset{17 \text{ R } 0}{13\overline{)221}}$$

When 0 is reached as a remainder, the divisions are complete. Because GCD(0, 13) = 13, GCD(676, 221) = 13. Based on this illustration, we make the generalization outlined in the following theorem.

> ### Theorem 5–28
> If a and b are any whole numbers greater than 0 and $a \geq b$, then GCD(a, b) = GCD(r, b), where r is the remainder when a is divided by b.

> **REMARK** Because GCD(x, y) = GCD(y, x) for all whole numbers x and y not both 0, Theorem 5–28 can be written
>
> $$\text{GCD}(a, b) = \text{GCD}(b, r).$$

Finding the GCD of two numbers by repeatedly using Theorem 5–28 until the remainder 0 is reached is referred to as the **Euclidean algorithm**. This method is found in Book IV of Euclid's *Elements* (300 BC). A flowchart for using the Euclidean algorithm is given in Figure 5-28.

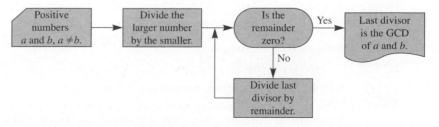

Figure 5-28

Example 5-28

Use the Euclidean algorithm to find GCD(10764, 2300).

Solution

$$\begin{array}{r} 4 \\ 2300\overline{)10764} \\ 9200 \\ \hline 1564 \end{array}$$ Thus, GCD(10764, 2300) = GCD(2300, 1564).

$$\begin{array}{r} 1 \\ 1564\overline{)2300} \\ 1564 \\ \hline 736 \end{array}$$ Thus, GCD(2300, 1564) = GCD(1564, 736).

$$\begin{array}{r} 2 \\ 736\overline{)1564} \\ 1472 \\ \hline 92 \end{array}$$ Thus, GCD(1564, 736) = GCD(736, 92).

$$\begin{array}{r} 8 \\ 92\overline{)736} \\ 736 \\ \hline 0 \end{array}$$ Thus, GCD(736, 92) = GCD(92, 0).

Because GCD(92, 0) = 92, it follows that GCD(10764, 2300) = 92.

REMARK The procedure for finding the GCD by using the Euclidean algorithm can be stopped at any step at which the GCD is obvious.

A calculator with the integer division feature can also be used to perform the Euclidean algorithm. This feature yields the quotient and the remainder when doing a division. For example, if the integer division key looks like $\boxed{\text{INT} \div}$, then to find GCD(10764, 2300) we proceed as follows:

$\boxed{1}\boxed{0}\boxed{7}\boxed{6}\boxed{4}\boxed{\text{INT} \div}\boxed{2}\boxed{3}\boxed{0}\boxed{0}\boxed{=}$ which displays $\underset{Q}{\underline{\lfloor 4 \rfloor}}\ \underset{R}{\underline{\lfloor 1564 \rfloor}}$

$\boxed{2}\boxed{3}\boxed{0}\boxed{0}\boxed{\text{INT} \div}\boxed{1}\boxed{5}\boxed{6}\boxed{4}\boxed{=}$ which displays $\underset{Q}{\underline{\lfloor 1 \rfloor}}\ \underset{R}{\underline{\lfloor 736 \rfloor}}$

$\boxed{1}\boxed{5}\boxed{6}\boxed{4}\boxed{\text{INT} \div}\boxed{7}\boxed{3}\boxed{6}\boxed{=}$ which displays $\underset{Q}{\underline{\lfloor 2 \rfloor}}\ \underset{R}{\underline{\lfloor 92 \rfloor}}$

$\boxed{7}\boxed{3}\boxed{6}\boxed{\text{INT} \div}\boxed{9}\boxed{2}\boxed{=}$ which displays $\underset{Q}{\underline{\lfloor 8 \rfloor}}\ \underset{R}{\underline{\lfloor 0 \rfloor}}$

The last number we divided by when we obtained the 0 remainder is 92, so

$$\text{GCD}(10764, 2300) = 92$$

Sometimes shortcuts can be used to find the GCD of two or more numbers, as in the following example.

Example 5-29

Find each of the following:

a. GCD(134791, 6341, 6339)
b. The GCD of any two consecutive whole numbers.

Solution **a.** Any common divisor of three numbers is also a common divisor of any two of them (why?). Consequently, the GCD of three numbers cannot be greater than the GCD of any two of the numbers. The numbers 6341 and 6339 are close to each other and therefore it is easy to find their GCD:

$$GCD(6341, 6339) = GCD(6341 - 6339, 6339)$$
$$= GCD(2, 6339)$$
$$= 1$$

Because $GCD(134791, 6341, 6339)$ cannot be greater than 1, it follows that it must equal 1.

b. Notice that $GCD(4, 5) = 1$, $GCD(5, 6) = 1$, $GCD(6, 7) = 1$, and $GCD(99, 100) = 1$. It seems that the GCD of any two consecutive whole numbers is 1. To justify this conjecture, we need to show that for all whole numbers n, $GCD(n, n + 1) = 1$. We have

$$GCD(n, n + 1) = GCD(n + 1, n) = GCD(n + 1 - n, n)$$
$$= GCD(1, n)$$
$$= 1$$

Least Common Multiple

Hot dogs are usually sold 10 to a package, while hot dog buns are usually sold 8 to a package. This mismatch causes troubles when one is trying to match hot dogs and buns. What is the least number of packages of each you could order so that there is an equal number of hot dogs and buns? The numbers of hot dogs that we could have are just the multiples of 10, that is, 10, 20, 30, 40, 50, ... Likewise, the possible numbers of buns are 8, 16, 24, 32, 40, 48, ... We can see that the number of hot dogs matches the number of buns whenever 10 and 8 have multiples in common. This occurs at 40, 80, 120, ... In this problem, we are interested in the least of these multiples, 40. Therefore, we could obtain the same number of hot dogs and buns in the least amount by buying four packages of hot dogs and five packages of buns. The answer 40 is the **least common multiple (LCM)** of 8 and 10.

Definition

Suppose that a and b are natural numbers. Then the least common multiple (LCM) of a and b is the least natural number that is simultaneously a multiple of a and a multiple of b.

As with GCDs, there are several methods for finding least common multiples.

Number-Line Method

A number line can be used to find the LCM of two numbers. For example, to find $LCM(3, 4)$, we can show the multiples of 3 and 4 on the number line using intervals of 3 and 4, as shown in Figure 5-29.

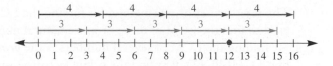

Figure 5-29

Beginning at 0, we see that the arrows do not coincide until the point 12 on the number line. If the line were continued, the arrows would coincide again at 24, 36, 48, and so on. We see that there are an infinite number of common multiples of 3 and 4, but the least common multiple is 12. Note that this number-line approach is instructive and promotes understanding but is not practical for large numbers.

Colored Rods Method

We can use colored rods to determine the LCM of two numbers. For example, consider the 3 rod and the 4 rod in Figure 5-30(a). We build trains of 3 rods and 4 rods until they are the same length, as shown in Figure 5-30(b). The LCM is the common length of the train.

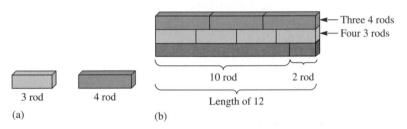

Figure 5-30

The Intersection-of-Sets Method

In the *intersection-of-sets* method, we first find the set of all positive *multiples* of both the first and second numbers, then find the set of all *common multiples* of both numbers, and finally pick the *least* element in that set. For example, to find the LCM of 8 and 12, denote the sets of positive multiples of 8 and 12 by M_8 and M_{12}, respectively.

$$M_8 = \{8, 16, 24, 32, 40, 48, 56, 64, 72, \dots\}$$

$$M_{12} = \{12, 24, 36, 48, 60, 72, 84, 96, 108, \dots\}$$

The set of common multiples is

$$M_8 \cap M_{12} = \{24, 48, 72, \dots\}$$

Because the least number in $M_8 \cap M_{12}$ is 24, the LCM of 8 and 12 is 24, written LCM(8, 12) = 24.

 NOW TRY THIS 5-17 Draw a Venn diagram showing M_8 and M_{12} and show how to find LCM(8, 12) using the diagram.

The Prime Factorization Method

The intersection-of-sets method for finding the LCM is often lengthy, especially when it is used to find the LCM of three or more natural numbers. Another, more efficient, method for finding the LCM of several numbers is the *prime factorization method*. For example, to find LCM(40, 12), first find the prime factorizations of 40 and 12, namely, $2^3 \cdot 5$ and $2^2 \cdot 3$, respectively.

If $m = \text{LCM}(40, 12)$, then m is a multiple of 40 and must contain both 2^3 and 5 as factors. Also, m is a multiple of 12 and must contain 2^2 and 3 as factors. Since 2^3 is a multiple of 2^2, then $m = 2^3 \cdot 5 \cdot 3 = 120$. In general, we have the following:

> To find the LCM of two natural numbers, first find the prime factorization of each number. Then take each of the primes that are factors of either of the given numbers. The LCM is the product of these primes, each raised to the greatest power of the prime that occurs in either of the prime factorizations.

Example 5-30

Find the LCM of 2520 and 10,530.

Solution

$$2520 = 2^3 \cdot 3^2 \cdot 5 \cdot 7.$$
$$10,530 = 2 \cdot 3^4 \cdot 5 \cdot 13.$$
$$\text{LCM}(2520, 10530) = 2^3 \cdot 3^4 \cdot 5 \cdot 7 \cdot 13 = 294,840$$

The prime factorization method can also be used to find the LCM of more than two numbers. For example, to find LCM(12, 108, 120), we can proceed as follows:

$$12 = 2^2 \cdot 3$$
$$108 = 2^2 \cdot 3^3$$
$$120 = 2^3 \cdot 3 \cdot 5$$

Then, $\text{LCM}(12, 108, 120) = 2^3 \cdot 3^3 \cdot 5 = 1080$.

The GCD-LCM Product Method

To see the connection between the GCD and LCM, consider the GCD and LCM of 24 and 30. The prime factorizations of these numbers are

$$24 = 2^3 \cdot 3$$
$$30 = 2 \cdot 3 \cdot 5$$

A diagram showing the prime factorization is given in Figure 5-31.

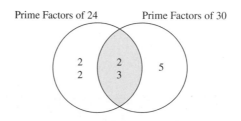

Figure 5-31

Notice that $\text{GCD}(24, 30) = 2 \cdot 3$ is the product of the factors in the shaded region, and $\text{LCM}(24, 30) = 2^3 \cdot 3 \cdot 5$ is the product of the factors in the combined regions. Also notice that

$$\text{GCD}(24, 30) \cdot \text{LCM}(24, 30) = (2 \cdot 3)(2^3 \cdot 3 \cdot 5) = (2^3 \cdot 3)(2 \cdot 3 \cdot 5) = 24 \cdot 30$$

This shows that the product of the GCD and LCM of 24 and 30 is equal to $24 \cdot 30$. In general, the connection between their GCD and LCM of any pair of natural numbers is given by Theorem 5–29.

Theorem 5–29

For any two natural numbers a and b,
$$\text{GCD}(a, b) \cdot \text{LCM}(a, b) = ab$$

Theorem 5–29 can be justified in several ways. Here is a specific example that suggests how the theorem might be proved. Suppose
$$a = 5^{13} \cdot 7^{20} \cdot 11^4 \quad \text{and} \quad b = 5^{10} \cdot 7^{25} \cdot 11^6 \cdot 13$$

Then,
$$\text{LCM}(a, b) = 5^{13} \cdot 7^{25} \cdot 11^6 \cdot 13 \quad \text{and} \quad \text{GCD}(a, b) = 5^{10} \cdot 7^{20} \cdot 11^4$$

Now we have
$$\text{LCM}(a, b) \cdot \text{GCD}(a, b) = 5^{13+10} \cdot 7^{25+20} \cdot 11^{6+4} \cdot 13 \quad \text{and} \quad ab = 5^{13+10} \cdot 7^{20+25} \cdot 11^{4+6} \cdot 13$$

For the preceding values of a and b, Theorem 5–29 is true. Notice, however, that in the product $\text{LCM}(a, b) \cdot \text{GCD}(a, b)$ we have all the powers of the primes appearing in a or in b, because for the LCM we take the greater of the powers of the common primes and for the GCD the lesser. Also in ab we have all the powers. Hence, Theorem 5–29 is true in general.

The Euclidean Algorithm Method

Theorem 5–29 is useful for finding the LCM of two numbers a and b when their prime factorizations are not easy to find. $\text{GCD}(a, b)$ can be found by the Euclidean algorithm, the product ab can be found by simple multiplication, and $\text{LCM}(a, b)$ can be found by division.

Example 5-31

Find $\text{LCM}(731, 952)$.

Solution   By the Euclidean algorithm, $\text{GCD}(731, 952) = 17$. By Theorem 5–29,
$$17 \cdot \text{LCM}(731, 952) = 731 \cdot 952$$

Consequently,
$$\text{LCM}(731, 952) = \frac{731 \cdot 952}{17} = 40{,}936$$

The Division-by-Primes Method

Another procedure for finding the LCM of several natural numbers involves *division by primes*. For example, to find $\text{LCM}(12, 75, 120)$, we start with the least prime that divides at least one of the given numbers and divide as follows:

$$2 \underline{\,|\, 12, \, 75, \, 120}$$
$$6, \, 75, \;\; 60$$

Because 2 does not divide 75, simply bring down the 75. To obtain the LCM using this procedure, continue the division process until the row of answers consists of relatively prime numbers as shown next.

$$
\begin{array}{r|r r r}
2 & 12, & 75, & 120 \\
\hline
2 & 6, & 75, & 60 \\
\hline
2 & 3, & 75, & 30 \\
\hline
3 & 3, & 75, & 15 \\
\hline
5 & 1, & 25, & 5 \\
\hline
& 1, & 5, & 1 \\
\end{array}
$$

Thus, $\mathrm{LCM}(12, 75, 120) = 2 \cdot 2 \cdot 2 \cdot 3 \cdot 5 \cdot 1 \cdot 5 \cdot 1 = 2^3 \cdot 3 \cdot 5^2 = 600$.

Assessment 5-5A

1. Find the GCD and the LCM for each of the following using the intersection-of-sets method:
   a. 18 and 10   b. 24 and 36
   c. 8, 24, and 52   d. 7 and 9
2. Find the GCD and the LCM for each of the following using the prime factorization method:
   a. 132 and 504   b. 65 and 1690
   c. 900, 96, and 630 d. 108 and 360
3. Find the GCD for each of the following using the Euclidean algorithm:
   a. 220 and 2924   b. 14,595 and 10,856
4. Find the LCM for each of the following using any method:
   a. 24 and 36
   b. 72 and 90 and 96
   c. 90 and 105 and 315
   d. 9^{100} and 25^{100}
5. Find the LCM for each of the following pairs of numbers using Theorem 5–29 and the answers from problem 3:
   a. 220 and 2924
   b. 14,595 and 10,856
6. Use colored rods to find the GCD and the LCM of 6 and 10.
7. In Quinn's dormitory room, there are three snooze-alarm clocks, each of which is set at a different time. Clock A goes off every 15 min, clock B goes off every 40 min, and clock C goes off every 60 min. If all three clocks go off at 6:00 A.M., answer the following:
   a. How long will it be before the clocks go off simultaneously again after 6:00 A.M.?
   b. Would the answer to (a) be different if clock B went off every 15 min and clock A went off every 40 min?
8. Midas has 120 gold coins and 144 silver coins. He wants to place his gold coins and his silver coins in stacks so that there are the same number of coins in each stack. What is the greatest number of coins that he can place in each stack?

9. By selling cookies at 24¢ each, José made enough money to buy several cans of pop costing 45¢ per can. If he had no money left over after buying the pop, what is the least number of cookies he could have sold?
10. Two bike riders ride around in a circular path. The first rider completes one round in 12 min and the second rider completes it in 18 min. If they both start at the same place and the same time and go in the same direction, after how many minutes will they meet again at the starting place?
11. Three motorcyclists ride around a circular race course starting at the same place and the same time. The first passes the starting point every 12 min, the second every 18 min, and the third every 16 min. After how many minutes will all three pass the starting point again at the same time? Explain your reasoning.
12. Assume a and b are natural numbers and answer the following:
    a. If $\mathrm{GCD}(a, b) = 1$, find $\mathrm{LCM}(a, b)$.
    b. Find $\mathrm{GCD}(a, a)$ and $\mathrm{LCM}(a, a)$.
    c. Find $\mathrm{GCD}(a^2, a)$ and $\mathrm{LCM}(a^2, a)$.
    d. If $a|b$, find $\mathrm{GCD}(a, b)$ and $\mathrm{LCM}(a, b)$.
13. Classify each of the following as true or false:
    a. If $\mathrm{GCD}(a, b) = 1$, then a and b cannot both be even.
    b. If $\mathrm{GCD}(a, b) = 2$, then both a and b are even.
    c. If a and b are even, then $\mathrm{GCD}(a, b) = 2$.
14. To find $\mathrm{GCD}(24, 20, 12)$, it is possible to find $\mathrm{GCD}(24, 20)$, which is 4, and then find $\mathrm{GCD}(4, 12)$, which is 4. Use this approach and the Euclidean algorithm to find
    a. $\mathrm{GCD}(120, 75, 105)$.
    b. $\mathrm{GCD}(34578, 4618, 4619)$.
15. Show that 97,219,988,751 and 4 are relatively prime.
16. The radio station gave away a discount coupon for every twelfth and thirteenth caller. Every twentieth caller received free concert tickets. Which caller was first to get both a coupon and a concert ticket?

17. Jackie spent the same amount of money on DVDs that she did on compact discs. If DVDs cost $12 and CDs $16, what is the least amount she could have spent on each?

18. At the Party Store, paper plates come in packages of 30, paper cups in packages of 15, and napkins in packages of 20. What is the least number of plates, cups, and napkins that can be purchased so that there is an equal number of each?

19. Diagrams can be used to show factors of two or more numbers. Draw diagrams to show the prime factors for each of the following sets of three numbers:
    a. 10, 15, 60   **b.** 8, 16, 24

20. What are the factors of 4^{10}?

21. In algebra it is often necessary to factor an expression as much as possible. For example, $a^3b^2 + a^2b^3 =$

$a^2b^2(a + b)$ and no further factoring is possible without knowing the values of a and b. Notice that a^2b^2 is the GCD of a^3b^2 and a^2b^3. Factor each of the following as much as possible:
    a. $12x^4y^3 + 18x^3y^4$
    b. $12x^3y^2z^2 + 18x^2y^4z^3 + 24x^4y^3z^4$

22. Label the following statements as "always true," "sometimes true," or "never true." Justify your answers.
    a. $\text{GCD}(a, b) = \text{GCD}(|a|, b) = \text{GCD}(|a|, |b|)$
    b. $\text{GCD}(-a, b) = \text{GCD}(a, -b) = \text{GCD}(-a, -b)$

23. Find the GCD and the LCM of each of the following. (Do not compute the products.)
    a. 10!, 11!
    b. 10!, 10! + 1

24. Factor 1 billion into a product of two numbers, neither of which contains any zeros.

Assessment 5-5B

1. Find the GCD and the LCM for each of the following using the intersection-of-sets method:
    a. 12 and 18   **b.** 18 and 36
    c. 12, 18, and 24   **d.** 6 and 11

2. Find the GCD and the LCM for each of the following using the prime factorization method:
    a. 11 and 19   **b.** 140 and 320
    c. 800, 75, and 450   **d.** 104 and 320

3. Find the GCD for each of the following using the Euclidean algorithm:
    a. 14,560 and 8250   **b.** 8424 and 2520

4. Find the LCM for each of the following using any method:
    a. 25 and 36
    b. 82 and 90 and 50
    c. 80 and 105 and 315
    d. 8^{100} and 50^{100}

5. Find the LCM for each of the following pairs of numbers using Theorem 5–29 and the answers from problem 3:
    a. 14,560 and 8250
    b. 8424 and 2520

6. A movie rental store gives a free popcorn to every fourth customer and a free movie rental to every sixth customer. Use the number-line method to find which customer was the first to win both prizes.

7. Use colored rods to find the GCD and the LCM of 4 and 10.

8. Bill and Sue both work at night. Bill has every sixth night off and Sue has every eighth night off. If they are both off tonight, how many nights will it be before they are both off again?

9. Bijous I and II start their movies at 7:00 P.M. The movie at Bijou I takes 75 min, while the movie at Bijou II takes

90 min. If the shows run continuously, when will they start at the same time again?

10. A rectangular field with dimensions 75 ft by 625 ft is to be divided into same-size square plots. If the sides of the squares need to be whole numbers of feet long, what are
    a. the largest squares possible and how many such squares will fit in the field?
    b. what are the smallest squares possible?
    c. what other size squares are possible?

11. The principal of Valley Elementary School wants to divide each of the three fourth-grade classes into small same-size groups with at least 2 students in each. If the classes have 18, 24, and 36 students, respectively, what size groups are possible?

12. Assume a and b are natural numbers and answer the following:
    a. If a and b are two primes, find $\text{GCD}(a, b)$ and $\text{LCM}(a, b)$.
    b. What is the relationship between a and b if $\text{GCD}(a, b) = a$?
    c. What is the relationship between a and b if $\text{LCM}(a, b) = a$?

13. Classify each of the following as true or false: for all natural numbers a and b.
    a. $\text{LCM}(a, b) | \text{GCD}(a, b)$.
    b. For all natural numbers a and b, $\text{LCM}(a, b) | ab$.
    c. $\text{GCD}(a, b) \leq a$.
    d. $\text{LCM}(a, b) \geq a$.

14. To find $\text{GCD}(24, 20, 12)$, it is possible to find $\text{GCD}(24, 20)$, which is 4, and then find $\text{GCD}(4, 12)$, which is 4. Use this approach and the Euclidean algorithm to find
    a. $\text{GCD}(180, 240, 306)$.
    b. $\text{GCD}(5284, 1250, 1280)$.

15. Show that 181,345,913 and 11 are relatively prime.

16. Larry and Mary bought a special 360-day joint membership to a tennis club. Larry will use the club every other day, and Mary will use the club every third day. They both use the club on the first day. How many days will neither person use the club in the 360 days?

17. Determine how many complete revolutions gear 2 in the following must make before the arrows are lined up again.

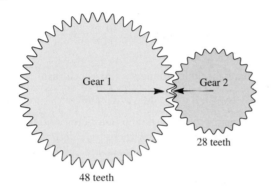

Gear 1

Gear 2

28 teeth

48 teeth

18. Determine how many complete revolutions each gear in the following must make before the arrows are lined up again:

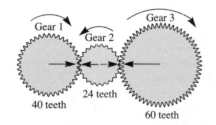

Gear 1

Gear 2

Gear 3

40 teeth

24 teeth

60 teeth

19. Diagrams can be used to show factors of two or more numbers. Draw diagrams to show the prime factors for each of the following sets of three numbers:
   a. 12, 14, 70   b. 6, 8, 18

20. Find all natural numbers x such that $\text{GCD}(25, x) = 1$ and $1 \leq x \leq 25$.

21. In algebra it is often necessary to factor an expression as much as possible. For example, $a^3b^2 + a^2b^3 = a^2b^2(a + b)$ and no further factoring is possible without knowing the values of a and b. Notice that a^2b^2 is the GCD of a^3b^2 and a^2b^3. Factor each of the following as much as possible:
   a. $6(x^2 - y^2) - 3(x - y) + 9(y - x)$
   b. $6(x^2 - y^2) + 12(x^2 - y^2) + 18(y^2 - x^2)$

22. Find all values of a and b for which the following are true.
   a. If $\text{GCD}(a, b) = 1$, then $\text{GCD}(a^2, b^2) = \text{GCD}(a, b^3)$.
   b. If $\text{LCM}(a, b, c) = abc$, then $\text{GCD}(a, b, c) = 1$.

23. Find the GCD and the LCM of each of the following. (Do not compute the products.)
   a. pqr, qrs (where p, q, r, s are prime numbers)
   b. $2^{10}, 2^8$

24. If you find the sum of any two-digit number and the number formed by reversing its digits, the resulting number is always divisible by which three positive integers?

Mathematical Connections 5-5

Communication

1. Can two natural numbers have a greatest common multiple? Explain your answer.

2. Describe to a sixth-grade student the difference between a divisor and a multiple.

3. Is it true that $\text{GCD}(a, b, c) \cdot \text{LCM}(a, b, c) = abc$? Explain your answer.

4. A rectangular plot of land is 558 m by 1212 m. A surveyor needs to divide the plot into the largest possible square plots of the same size, being a whole number of meters long. What is the size of each square and how many square plots can be created? Explain your reasoning.

5. Suppose that $\text{GCD}(a, b, c) = 1$. Is it necessarily true that $\text{GCD}(a, b) = \text{GCD}(b, c) = 1$? Explain your reasoning.

6. Suppose $\text{GCD}(a, b) = \text{GCD}(b, c) = 2$. Does that always imply that $\text{GCD}(a, b, c) = 2$? Justify your answer.

7. How can you tell from the prime factorization of two numbers if their LCM equals the product of the numbers? Explain your reasoning.

8. Can the LCM of two positive numbers ever be greater than the product of the numbers? Explain your reasoning.

9. Let $\text{GCD}(m, n) = g$ and $\text{LCM}(m, n) = l$. Jackie conjectures that $\text{GCD}(m + n, l) = g$ for all integers m and n. Check Jackie's conjecture for three different pairs of integers.

Open-Ended

10. Make up a word problem that can be solved by finding the GCD and another that can be solved by finding the LCM. Solve your problems and explain why you are sure that your approach is correct.

11. Find three pairs of numbers for which the LCM of the numbers in a pair is smaller than the product of the two numbers.

12. Describe infinitely many pairs of numbers whose GCD is
   a. 2
   b. 6
   c. 91

Cooperative Learning

13. Each member of your group should examine a different elementary-school book that covers GCDs and LCMs. Report to the group on what methods were used and how they were used.

14. a. In your group, discuss whether the Euclidean algorithm for finding the GCD of two numbers should be introduced in middle school (to all students? to some?). Why or why not?
   b. If you decide that it should be introduced in middle school, discuss how it should be introduced. Report your group's decision to the class.

Questions from the Classroom

15. Alba asked why we don't talk about the LCD (least common divisor) and GCM (greatest common multiple). How do you respond?

16. A student says that for any two natural numbers a and b, GCD(a, b) divides LCM(a, b) and, hence, GCD(a, b) < LCM(a, b). Is the student correct? Why or why not?

17. A student asks about the relation between least common multiple and least common denominator. How do you respond?

18. A student wants to know how many integers between 1 and 10,000 inclusive are either multiples of 3 or multiples of 5. She wonders if it is correct to find the number of those integers that are multiples of 3 and add the number of those that are multiples of 5. How do you respond?

19. Dolores claims that she has a shortcut for finding the GCD using the Euclidean algorithm. She says when the remainder is large she uses a negative "remainder." For example, to find GCD(2132, 534), she divides 2132 by 534 and gets $2132 = 3 \cdot 534 + 530$, which gives remainder 530. In such case, she writes $2132 = 4 \cdot 534 - 4$ and claims that

$$GCD(2132, 534) = GCD(^-4, 534)$$
$$= GCD(4, 534)$$
$$= 2 \ (\text{because } 4 \nmid 534)$$

Is the approach correct and if so, why?

Review Problems

20. Find two whole numbers x and y such that

$$xy = 1,000,000$$

and neither x nor y contains any zeros as digits.

21. Fill each blank space with a single digit that makes the corresponding statement true. Find all possible answers.
   a. $3|83\_51$
   b. $11|8\_691$
   c. $23|103\_6$

22. Is 3111 a prime? Prove your answer.

23. Find a number that has exactly six prime factors.

24. Produce the least positive number that is divisible by 2, 3, 4, 5, 6, 7, 8, 9, 10, and 11.

25. What is the greatest prime that must be used to determine if 2089 is prime?

National Assessment of Educational Progress (NAEP) Question

The least common multiple of 8, 12, and a third number is 120. Which of the following could be the third number?
   a. 15   b. 16   c. 24   d. 32   e. 48

NAEP, Grade 8, 1990

TECHNOLOGY CORNER

1. Write a spreadsheet to generate the first 50 multiples of 3 and the first 50 multiples of 4. Describe the intersection of the two sets.

2. Use a spreadsheet to find the factors of 2486. How far down do you need to copy the formula to be sure you have found all the divisors?

	A	B
1	1	= 2486/A1
2	2	
3	3	

3. Make a spreadsheet with four columns:
   Column A—the multiples of 6
   Column B—the multiples of 9
   Column C—the multiples of 12
   Column D—the multiples of 15
   a. What is the least number that appears in all four columns?
   b. Explain how to find this number without using a spreadsheet.

BRAIN TEASER For any $n \times m$ rectangle such that $GCD(n, m) = 1$, find a rule for determining the number of unit squares (1×1) that a diagonal passes through. For example, in the drawings in Figure 5-32, the diagonal passes through 8 and 6 unit squares, respectively.

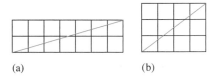

(a)   (b)

Figure 5-32

---

*5-6   Clock and Modular Arithmetic

In this section, we investigate clock arithmetic. Consider the following:

a. A doctor's prescription says to take a pill every 8 hr. If you take the first pill at 7:00 A.M., when should you take the next two pills?

b. Suppose you are following a bean soup recipe that calls for letting the beans soak for 12 hr. If you begin soaking them at 8:00 P.M., when should you take them out?

c. The odometer on a car gives the total miles traveled up to 99,999 mi and then starts counting from 0. If the odometer shows 99,124 mi, what will it show after a trip of 2,116 mi?

Some of these situations involve the ability to solve arithmetic problems using clocks. Most people can solve these problems without thinking much about what they are doing. It is possible to use the clock in Figure 5-33 to determine that 8 hr after 7:00 A.M. is 3:00 P.M. and 8 hr after that is 11:00 P.M. Also, 12 hr after 8:00 P.M. is 8:00 A.M. We could record these additions on the clock as

$$7 \oplus 8 = 3, \qquad 3 \oplus 8 = 11, \qquad 8 \oplus 12 = 8$$

where $\oplus$ indicates addition on a 12-hr clock.

Figure 5-33

You probably noticed the special role of 12 when you found that $8 \oplus 12 = 8$. In the 12-hr clock arithmetic, 12 acts like a 0 if you were adding in the set of whole numbers. An addition table for the finite system based on the clock is shown in Table 5-4.

Table 5-4

+	12	1	2	3	4	5	6	7	8	9	10	11
12	12	1	2	3	4	5	6	7	8	9	10	11
1	1	2	3	4	5	6	7	8	9	10	11	12
2	2	3	4	5	6	7	8	9	10	11	12	1
3	3	4	5	6	7	8	9	10	11	12	1	2
4	4	5	6	7	8	9	10	11	12	1	2	3
5	5	6	7	8	9	10	11	12	1	2	3	4
6	6	7	8	9	10	11	12	1	2	3	4	5
7	7	8	9	10	11	12	1	2	3	4	5	6
8	8	9	10	11	12	1	2	3	4	5	6	7
9	9	10	11	12	1	2	3	4	5	6	7	8
10	10	11	12	1	2	3	4	5	6	7	8	9
11	11	12	1	2	3	4	5	6	7	8	9	10

NOW TRY THIS 5-18 Examine Table 5-4 to determine if the following properties hold for $\oplus$ on the set of numbers in the table:

a. Commutative property of addition
b. Identity property of addition
c. Inverse property of addition

When we allow numbers other than those on the 12-hr clock to be added, such as $8 \oplus 24 = 8$, we find that numbers such as $24, 36, 48, \ldots$ act like 12 (or 0). Likewise, the numbers $13, 25, 37, \ldots$ act like the number 1. Similarly, we can generate classes of numbers that act like each of the numbers on the 12-hr clock. The members of any one class differ by multiples of 12. Consequently, to perform additions on a 12-hr clock, we perform regular addition, divide by 12, and record the remainder as the answer. For example, we can find $11 \oplus 8$ and $8 \oplus 12$ as follows:

$11 + 8 = 19$. Next divide $19 \div 12$. The quotient is 1 with a remainder of 7. Therefore, $11 \oplus 8 = 7$.

$8 + 12 = 20$. Next divide $20 \div 12$. The quotient is 1 with a remainder of 8. Therefore, $8 \oplus 12 = 8$.

Clock multiplication can be defined using repeated addition, as with whole numbers. For example, $2 \otimes 8 = 8 \oplus 8 = 4$, where $\otimes$ denotes clock multiplication. Similarly, $3 \otimes 5 = (5 \oplus 5) \oplus 5 = 10 \oplus 5 = 3$. We could also find $2 \otimes 8$ as $(2 \cdot 8) \div 12$ has remainder 4, so $2 \otimes 8 = 4$. Likewise, we could find $3 \otimes 5$ as $(3 \cdot 5) \div 12$ has remainder 3, so $3 \otimes 5 = 3$. This leads to the following definition.

Definition

12-Hr Clock Sums and Products To compute a sum or product in 12-hr clock arithmetic, perform the computation as with whole numbers, divide by 12, and take the remainder as the answer.

To perform other operations on the clock, such as $2 \ominus 9$, where $\ominus$ denotes clock subtraction, we could interpret it as the time 9 hr before 2 o'clock. Counting backward (counterclockwise) 9 units from 2 reveals that $2 \ominus 9 = 5$. If subtraction on the clock is defined in terms of addition, we have $2 \ominus 9 = x$ if, and only if, $2 = 9 \oplus x$. Consequently, $x = 5$.

Clock division can be defined in terms of multiplication. For example, $8 \oslash 5 = x$, where $\oslash$ denotes clock division, if, and only if, $8 = 5 \otimes x$ for a unique x in the set $\{1, 2, 3, \ldots, 12\}$. Because $5 \otimes 4 = 8$ and 8 is unique, we have then $8 \oslash 5 = 4$.

Example 5-32

Perform each of the following computations on a 12-hr clock:

a. $8 \oplus 8$ b. $4 \ominus 12$

c. $4 \ominus 4$ d. $4 \ominus 8$

Solution
  a. $(8 + 8) \div 12$ has remainder 4. Hence, $8 \oplus 8 = 4$.
  b. $4 \ominus 12 = 4$, since by counting forward or backward 12-hr, you arrive at the original position.
  c. $4 \ominus 4 = 12$. This should be clear from looking at the clock, but it can also be found by using the definition of subtraction in terms of addition.
  d. $4 \ominus 8 = 8$ because $8 \oplus 8 = 4$.

Example 5-33

Perform the following operations on a 12-hr clock, if possible:

a. $3 \otimes 11$ b. $2 \oslash 7$

c. $3 \oslash 2$ d. $5 \oslash 12$

Solution
  a. $3 \otimes 11 = (11 \oplus 11) \oplus 11 = 10 \oplus 11 = 9$ or $(3 \cdot 11) \div 12$ has remainder 9. Hence, $3 \otimes 11 = 9$.
  b. $2 \oslash 7 = x$ if, and only if, $2 = 7 \otimes x$ and x is unique. Consequently, $x = 2$.
  c. $3 \oslash 2 = x$ if, and only if, $3 = 2 \otimes x$ and x is unique. Multiplying each of the numbers $1, 2, 3, 4, \ldots, 12$ by 2 shows that none of the multiplications yields 3. Thus, the equation $3 = 2 \otimes x$ has no solution, and consequently, $3 \oslash 2$ is undefined.
  d. $5 \oslash 12 = x$ if, and only if, $5 = 12 \otimes x$ and x is unique. However, $12 \otimes x = 12$ for every x in the set $\{1, 2, 3, 4, \ldots, 12\}$. Thus, $5 = 12 \otimes x$ has no solution on the clock, and therefore $5 \oslash 12$ is undefined.

Adding or subtracting 12 on a 12-hr clock leaves a number unchanged. Thus 12 behaves as 0 does in integer addition or subtraction and is the additive identity for addition on the 12-hr clock. Similarly, on a 5-hr clock 5 behaves as 0 does.

Addition, subtraction, and multiplication on a 12-hr clock can be performed for any two numbers but, as shown in Example 5-33(d), not all divisions can be performed. Division by 12, the additive identity, on a 12-hr clock either can never be performed or is not meaningful, since it does not yield a unique answer. However, there are clocks on which all divisions can be performed, except by the corresponding additive identities. One such clock is a 5-hr clock, shown in Figure 5-34.

Table 5-5

(a)

$\oplus$	1	2	3	4	5
1	2	3	4	5	1
2	3	4	5	1	2
3	4	5	1	2	3
4	5	1	2	3	4
5	1	2	3	4	5

(b)

$\otimes$	1	2	3	4	5
1	1	2	3	4	5
2	2	4	1	3	5
3	3	1	4	2	5
4	4	3	2	1	5
5	5	5	5	5	5

Figure 5-34

On this clock, $3 \oplus 4 = 2$, $2 \ominus 3 = 4$, $2 \otimes 4 = 3$, and $3 \oslash 4 = 2$. Since adding 5 to any number yields the original number, 5 is the additive identity for this 5 hr clock, as seen in Table 5-5(a). Consequently, you might suspect that division by 5 is not possible on a 5-hr clock. To determine which divisions are possible, consider Table 5-5(b), a multiplication table for 5-hr clock arithmetic. To find $1 \oslash 2$, we write $1 \oslash 2 = x$, which is equivalent to $1 = 2 \otimes x$. The second row of Table 5-5(b) shows that $2 \otimes 1 = 2$, $2 \otimes 2 = 4$, $2 \otimes 3 = 1$, $2 \otimes 4 = 3$, and $2 \otimes 5 = 5$. The unique solution of $1 = 2 \otimes x$ is $x = 3$, so $1 \oslash 2 = 3$. The information given in the second row of the table can be used to determine the following divisions:

$$2 \oslash 2 = 1 \quad \text{because} \quad 2 = 2 \otimes 1$$
$$3 \oslash 2 = 4 \quad \text{because} \quad 3 = 2 \otimes 4$$
$$4 \oslash 2 = 2 \quad \text{because} \quad 4 = 2 \otimes 2$$
$$5 \oslash 2 = 5 \quad \text{because} \quad 5 = 2 \otimes 5$$

Because every element occurs in the second row, division by 2 is always possible. Similarly, division by all other numbers, except 5, is always possible. In the problem set, you are asked to perform arithmetic on different clocks and to investigate for which clocks all computations, except division by the additive identity, can be performed.

Modular Arithmetic

APRIL						
S	M	T	W	T	F	S
1	2	3	4	5	6	7
8	9	10	11	12	13	14
15	16	17	18	19	20	21
22	23	24	25	26	27	28
29	30					

Figure 5-35

Many of the concepts for clock arithmetic can be used to work problems that involve a calendar. On the calendar in Figure 5-35, the five Sundays have dates 1, 8, 15, 22, and 29. Any two of these dates for Sunday differ by a multiple of 7. The same property is true for any other day of the week. For example, the second and thirtieth days fall on the same day, since $30 - 2 = 28$ and 28 is a multiple of 7. We say that 30 is congruent to 2, modulo 7, and we write $30 \equiv 2 \pmod 7$. Similarly, because 18 and 6 differ by a multiple of 12, we write $18 \equiv 6 \pmod{12}$. This is generalized in the following definition.

> **Definition of Modular Congruence**
>
> For integers a and b, **a is congruent to b modulo m**, written $a \equiv b \pmod m$, if, and only if, $a - b$ is a multiple of m, where m is a positive integer greater than 1.

> **REMARK** This definition could be written as $a \equiv b \pmod m$ if, and only if, $m \mid (a - b)$, where m is a positive integer greater than 1.

Notice that 18 and 25 are congruent modulo 7 and each number leaves the same remainder, 4, upon division by 7. Indeed, $18 = 2 \cdot 7 + 4$ and $25 = 3 \cdot 7 + 4$. In general, we have the following property: *Two whole numbers are congruent modulo m if, and only if, their remainders on division by m are the same.*

Example 5-34

Tell why each of the following is true:

a. $23 \equiv 3 \pmod{10}$ **b.** $23 \equiv 3 \pmod 4$ **c.** $23 \not\equiv 3 \pmod 7$
d. $10 \equiv {}^-1 \pmod{11}$ **e.** $m \equiv 0 \pmod m$

Solution **a.** $23 \equiv 3 \pmod{10}$ because $23 - 3$ is a multiple of 10, or because 23 and 3 leave the same remainder, 3, upon division by 10.
  b. $23 \equiv 3 \pmod 4$ because $23 - 3$ is a multiple of 4.
  c. $23 \not\equiv 3 \pmod 7$ because $23 - 3$ is not a multiple of 7.
  d. $10 \equiv {}^-1 \pmod{11}$ because $10 - ({}^-1) = 11$ is a multiple of 11.
  e. $m \equiv 0 \pmod m$ because $m - 0$ is a multiple of m, or because m and 0 have the same remainder, 0, upon division by m.

REMARK Example 5-34(e) shows that m behaves like 0 modulo m. This is also evident for $m = 12$ in Table 5-4 and for $m = 5$ in Table 5-5.

Example 5-35

Find all integers x such that $x \equiv 1 \pmod{10}$.

Solution $x \equiv 1 \pmod{10}$ if, and only if, $x - 1 = 10k$, where k is any integer. Consequently, $x = 10k + 1$. Letting $k = 0, 1, 2, 3, \ldots$ yields the sequence $1, 11, 21, 31, 41, \ldots$. Likewise, letting $k = {}^-1, {}^-2, {}^-3, {}^-4, \ldots$ yields the negative integers ${}^-9, {}^-19, {}^-29, {}^-39, \ldots$. The two sequences can be combined to give the solution set

$$\{ \ldots, {}^-39, {}^-29, {}^-19, {}^-9, 1, 11, 21, 31, 41, 51, \ldots \}$$

  The $\boxed{\text{INT} \div}$ button on a calculator can be used to work with modular arithmetic. If we press the following sequence of buttons, we see that $4325 \equiv 5 \pmod 9$ because the remainder when 4325 is divided by 9 is 5:

$$\boxed{4}\ \boxed{3}\ \boxed{2}\ \boxed{5}\ \boxed{\text{INT} \div}\ \boxed{9}\ \boxed{=}$$

and the display shows a remainder of 5.

Example 5-36

Heidi signed a promissory note that will become due in 90 days. She is worried that it will become due on a weekend. She signed the note on a Monday. On what day of the week will it be due?

Solution Because $90 = 7 \cdot 12 + 6$, we know that $90 \equiv 6 \pmod 7$. On a fraction calculator, you could enter $\boxed{9}\ \boxed{0}\ \boxed{\text{INT} \div}\ \boxed{7}\ \boxed{=}$, and a quotient of 12 with remainder 6 would be displayed. Therefore, the note will come due 12 wk and 6 days after Monday, which is a Sunday.

Example 5-37

a. If it is now Monday, October 14, on what day of the week will October 14 fall next year if next year is not a leap year?

b. If Christmas falls on Thursday this year, on what day of the week will Christmas fall next year if next year is a leap year?

Solution
a. Because next year is not a leap year, we have 365 days in the year. Because $365 = 52 \cdot 7 + 1$, we have $365 \equiv 1 \pmod 7$. Thus, 365 days after October 14 will be 52 wk and 1 day later. Thus, October 14 will be on a Tuesday.

b. There are 366 days in a leap year, and $366 \equiv 2 \pmod 7$. Thus, Christmas will be 2 days after Thursday, on Saturday.

Assessment 5-6A

1. Dr. Harper prescribed some medicine for Camile. She is supposed to take a dose every 6 hr. If she takes her first dose at 8:00 A.M., when should she take her next dose?

2. Perform each of the following operations on a 12-hr clock, if possible:
   a. $7 \oplus 8$ **b.** $4 \oplus 10$ **c.** $3 \ominus 9$
   d. $4 \ominus 8$ **e.** $3 \otimes 9$ **f.** $2 \otimes 2$
   g. $1 \oslash 3$ **h.** $2 \oslash 5$

3. Perform each of the following operations on a 5-hr clock:
   a. $3 \oplus 4$ **b.** $3 \oplus 3$ **c.** $3 \otimes 4$
   d. $1 \otimes 4$ **e.** $4 \otimes 4$ **f.** $2 \otimes 3$
   g. $3 \oslash 4$ **h.** $1 \oslash 4$

4. **a.** Construct an addition table for a 9-hr clock.
   b. Using the addition table in (a), find $5 \ominus 6$ and $2 \ominus 5$.
   c. Using the definition of subtraction in terms of addition, show that subtraction can always be performed on a 9-hr clock.

5. **a.** Construct a multiplication table for a 9-hr clock.
   b. Use the multiplication table in (a) to find $3 \oslash 5$ and $4 \oslash 6$.
   c. Use the multiplication table to find whether division by numbers different from 9 is always possible.

6. On a 5-hr clock, find each of the following:
   a. Additive inverse of 2   **b.** Additive inverse of 3

   c. $(^-2) \oplus (^-2)$ **d.** $^-(2 \oplus 2)$
   e. $(^-2) \ominus (^-3)$ **f.** $(^-2) \otimes (^-2)$

7. **a.** If April 23 falls on Tuesday, what are the dates of the other Tuesdays in April of that year?
   b. If July 2 falls on Tuesday, list the dates of the Wednesdays in July.
   c. If September 3 falls on Monday, on what day of the week will it fall next year if next year is a leap year?

8. Fill in each of the following blanks so that the answer is nonnegative and the least possible number:
   a. $29 \equiv$ _____ $\pmod 5$
   b. $3498 \equiv$ _____ $\pmod 3$
   c. $3498 \equiv$ _____ $\pmod{11}$
   d. $^-23 \equiv$ _____ $\pmod{10}$

9. **a.** Find all x such that $x \equiv 0 \pmod 2$.
   b. Find all x such that $x \equiv 1 \pmod 2$.
   c. Find all x such that $x \equiv 3 \pmod 5$.

10. A new clock is started on Sunday at 10:00 P.M. If the clock continues running nonstop, on what day and hour rounded to the nearest hour will it be when the clock reaches the 100,000th second?

11. If the following pattern continues,

    CLOCK CLOCK CLOCK CLOCK . . .

    what will be the 101st letter in the pattern?

Assessment 5-6B

1. The Smiths left on a car trip at 6:00 A.M. They traveled for exactly 15 hr. When did they arrive?

2. Perform each of the following operations on a 12-hr clock, if possible:
   a. $6 \oplus 6$ **b.** $5 \oplus 11$ **c.** $4 \ominus 6$

   d. $5 \ominus 8$ **e.** $4 \otimes 9$ **f.** $3 \otimes 3$
   g. $2 \oslash 3$ **h.** $4 \oslash 6$

3. Perform each of the following operations on a 5-hr clock:
   a. $4 \oplus 5$ **b.** $2 \oplus 2$ **c.** $4 \otimes 4$
   d. $1 \otimes 3$ **e.** $3 \otimes 3$ **f.** $5 \otimes 3$
   g. $2 \oslash 4$ **h.** $4 \oslash 4$

4. a. Construct an addition table for a 7-hr clock.
   b. Using the addition table in (a), find $5 \ominus 6$ and $2 \ominus 5$.
   c. Using the definition of subtraction in terms of addition, show that subtraction can always be performed on a 7-hr clock.
5. a. Construct a multiplication table for a 7-hr clock.
   b. Use the multiplication table in (a) to find $3 \odot 5$ and $4 \odot 6$.
   c. Use the multiplication table to find whether division by numbers different from 7 is always possible.
6. On a 12-hr clock, find each of the following:
   a. Additive inverse of 2   **b.** Additive inverse of 3
   c. $(^-2) \oplus (^-3)$ **d.** $^-(2 \oplus 3)$
   e. $(^-2) \ominus (^-3)$ **f.** $(^-2) \otimes (^-3)$
7. a. If April 8 falls on Friday, what are the dates of the other Fridays in April?
   b. If July 4 falls on Tuesday, what day will it fall on next year if next year is not a leap year?

c. Do the 125th day and the 256th day of the year fall on the same day of the week? Explain why.
8. Fill in each of the following blanks so that the answer is nonnegative and the least possible number:
   a. $29 \equiv$ _____ (mod 3)
   b. $3498 \equiv$ _____ (mod 5)
   c. $3498 \equiv$ _____ (mod 10)
   d. $^-23 \equiv$ _____ (mod 11)
9. a. Find all x such that $x \equiv 0$ (mod 3).
   b. Find all x such that $x \equiv 1$ (mod 3).
   c. Find all x such that $x \equiv 3$ (mod 7).
10. Continue a possible pattern by listing the next four terms in each sequence in clock arithmetic.
   a. 3, 8, 1, 6, 11, 4, 9, 2, 7, . . .
   b. 3, 8, 13, 4, 9, 14, 5, 10, 1, . . .

Mathematical Connections 5-6

Communication

1. Explain or find each of the following:
   a. A number congruent modulo 10 to the number formed by its last digit
   b. The last digit of $2^{180} - 1$
   c. A number congruent modulo 100 to the number formed by its last two digits
2. a. For all $a, b, c,$ and d, explain why $abcd \equiv a + b + c + d$ (mod 9).
   b. If $abcd_{\text{five}} \equiv a + b + c + d$ (mod m), what is m? Explain your reasoning.

Open-Ended

3. On a clock we define the additive inverse of a in the same way as the additive inverse was defined for integers. Having this definition in mind, list some similarities and some differences between the number system on the clock and the set of integers. Justify your answers.

Cooperative Learning

4. a. Have members of your group construct the multiplication tables for 3 hr, 4 hr, 6 hr, and 11 hr clocks.

b. Compare your results. On which of the clocks in (a) can divisions by numbers other than the additive identity always be performed?
   c. How do the multiplication tables of clocks for which division can always be performed (except by an additive identity) differ from the multiplication tables of clocks for which division is not always meaningful?

Questions from the Classroom

5. Dan was trying to see how fractions like $\frac{1}{4}$ might work in a 5 hr clock system. He said that he showed that $1 \oslash 4$ is greater than 3. He wants to know if this could possibly be correct. How do you respond?
6. Ally wants to know what elements of a 5 hr clock system have a multiplicative inverse. What do you tell her?
7. Zita claims that on the 5 hr clock shown in Figure 5-34 there is no 0 so there can be no additive identity. How do you respond?

BRAIN TEASER How many primes are in the following sequence?

$$9, 98, 987, 9876, \ldots, 987654321, 9876543219, 98765432198, \ldots$$

Hint for Solving the Preliminary Problem

The difference-of-squares formula along with the work done on the Gauss problem in Chapter 1 will help solve this problem.

Chapter Outline

I. Basic concepts of integers
   A. The set of **integers**, I, is $\{\ldots, {}^-3, {}^-2, {}^-1, 0, 1, 2, 3, \ldots\}$.
   B. The distance from any integer to 0 is the **absolute value** of the integer. The absolute value of an integer x is denoted $|x|$. If $x \geq 0$, then $|x| = x$ and if $x < 0$, then $|x| = {}^-x$.
   C. Operations with integers
      1. Addition: For any integers a and b,

$$ {}^-a + {}^-b = {}^-(a + b) $$

      2. Subtraction
         a. For all integers a and b, then $a - b = n$ if, and only if, $a = b + n$.
         b. For all integers a and b, $a - b = a + {}^-b$.
      3. Multiplication: For any integers a and b,
         a. $({}^-a)({}^-b) = ab$.
         b. $({}^-a)b = b({}^-a) = {}^-(ab)$.
      4. Division: If a and b are any integers with $b \neq 0$, then $a \div b$ is the unique integer c, if it exists, such that $a = bc$.
      5. Order of operations: When addition, subtraction, multiplication, and division appear without parentheses, multiplications and divisions are done first in the order of their appearance from left to right and then additions and subtractions are done in the order of their appearance from left to right. Any arithmetic in parentheses is done first.

II. The system of integers
   A. The set of integers, along with the operations of addition and multiplication, satisfy the following properties:

Property	$+$	$\times$
Closure	Yes	Yes
Commutative	Yes	Yes
Associative	Yes	Yes
Identity	Yes, 0	Yes, 1
Inverse	Yes	No
Distributive Property of Multiplication over Addition		

   B. Zero multiplication property of integers For any integer a, $a \cdot 0 = 0 = 0 \cdot a$.
   C. For all integers a, b, and c,
      1. ${}^-({}^-a) = a$.
      2. $a - (b - c) = a - b + c$.
      3. $(a + b)(a - b) = a^2 - b^2$ (**difference-of-squares formula**).

III. Divisibility
   A. If a and b are any integers, then b **divides** a, denoted $b|a$, if, and only if, there is a unique integer c such that $a = cb$.
   B. The following are basic divisibility theorems for integers a, b, and d:
      1. If $d|a$ and k is any integer, then $d|ka$.
      2. If $d|a$ and $d|b$, then $d|(a + b)$ and $d|(a - b)$.
      3. If $d|a$ and $d \nmid b$, then $d \nmid (a + b)$ and $d \nmid (a - b)$.
   C. Divisibility tests
      1. An integer is divisible by 2, 5, or 10 if, and only if, its units digit is divisible by 2, 5, or 10, respectively.
      2. An integer is divisible by 4 if, and only if, the last two digits of the integer represent a number divisible by 4.
      3. An integer is divisible by 8 if, and only if, the last three digits of the integer represent a number divisible by 8.
      4. An integer is divisible by 3 or 9 if, and only if, the sum of its digits is divisible by 3 or 9, respectively.
      5. An integer is divisible by 11 if, and only if, the sum of the digits in the places that are even powers of 10 minus the sum of the digits in the places that are odd powers of 10 is divisible by 11.
      6. An integer is divisible by 6 if, and only if, the integer is divisible by both 2 and 3.

IV. Prime and composite numbers
   A. Positive integers that have exactly two positive divisors are **primes**. Integers greater than 1 that are not primes are **composites**.
   B. Fundamental Theorem of Arithmetic: Every composite number has one, and only one, prime factorization, aside from variation in the order of the prime factors.

C. Criterion for determining if a given number n is prime: *If n is not divisible by any prime p such that $p^2 \leq n$, then n is prime.*

D. If the prime factorization of a number is $p^n q^m$, where p and q are prime, then the number of divisors of n is $(n + 1)(m + 1)$.

V. Greatest common divisor and least common multiple

  A. The **greatest common divisor (GCD)** of two or more natural numbers is the greatest divisor, or factor, that the numbers have in common.

  B. **Euclidean algorithm:** If a and b are positive integers and $a \geq b$, then $\text{GCD}(a, b) = \text{GCD}(b, r)$, where r is the remainder when a is divided by b. The procedure of finding the GCD of two numbers a and b by using this result repeatedly is the *Euclidean algorithm*.

C. The **least common multiple (LCM)** of two or more natural numbers is the least positive multiple that the numbers have in common.

D. $\text{GCD}(a, b) \cdot \text{LCM}(a, b) = ab$.

E. If $\text{GCD}(a, b) = 1$, then a and b are **relatively prime**.

* **VI.** Modular arithmetic

  A. For any integers a and b, a **is congruent to** b **modulo** m if, and only if, $a - b$ is a multiple of m, where m is a positive integer greater than 1.

  B. Two integers are congruent modulo m if, and only if, their remainders upon division by m are the same.

Chapter Review

1. Find the additive inverse of each of the following:
   a. 3   b. ^-a c. $^-2 + 3$
   d. $x + y$ e. $^-x + y$ f. $^-x - y$
   g. $(^-2)^5$ h. $^-2^5$

2. Perform each of the following operations:
   a. $(^-2 + ^-8) + 3$ b. $^-2 - (^-5) + 5$
   c. $^-3(^-2) + 2$ d. $^-3(^-5 + 5)$
   e. $^-40 \div (^-5)$ f. $(^-25 \div 5)(^-3)$

3. For each of the following, find all integer values of x (if there are any) that make the given equation true:
   a. $^-x + 3 = 0$
   b. $^-2x = 10$
   c. $0 \div (^-x) = 0$
   d. $^-x \div 0 = ^-1$
   e. $3x - 1 = ^-124$
   f. $^-2x + 3x = x$

4. Use a pattern approach to explain why $(^-2)(^-3) = 6$.

5. In each of the following chip models, the encircled chips are removed. Write the corresponding integer problem with its solution.

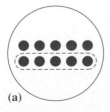

(a)

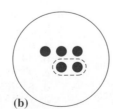

(b)

6. Simplify each of the following expressions:
   a. ^-1x
   b. $(^-1)(x - y)$

   c. $2x - (1 - x)$
   d. $(^-x)^2 + x^2$
   e. $(^-x)^3 + x^3$
   f. $(^-3 - x)(3 + x)$
   g. $(^-2 - x)(^-2 + x)$

7. Factor each of the following expressions and then simplify, if possible:
   a. $x - 3x$
   b. $x^2 + x$
   c. $x^2 - 36$
   d. $81y^4 - 16x^4$
   e. $5 + 5x$
   f. $(x - y)(x + 1) - (x - y)$

8. Classify each of the following as true or false (all letters represent integers). Justify your answers.
   a. $|x|$ always is positive.
   b. For all x and y, $|x + y| = |x| + |y|$.
   c. If $a < ^-b$, then $a < 0$.
   d. For all x and y, $(x - y)^2 = (y - x)^2$.

9. Find a counterexample to disprove each of the following properties on the set of integers:
   a. Commutativity for division
   b. Associativity for subtraction
   c. Closure for division
   d. Distributive property of division over subtraction

10. Solve each of the following for x, where x is an integer:
   a. $x + 3 = ^-x - 17$
   b. $2x = ^-2^{100}$
   c. $2^{10}x = 2^{99}$
   d. $^-x = x$
   e. $|^-x| = 3$
   f. $|x| = ^-x$

g. $|x| > 3$

h. $(x - 1)^2 = 100$

11. Write the first six terms of each of the sequences whose nth term is

    a. $(^-1)^n$

    b. $\dfrac{n}{2}[(^-1)^n + 1]$

    c. $(^-2)^n$

    d. $^-2 - 3n$

12. In each part of problem 11, if a sequence is arithmetic, find its difference, and if it is geometric, find its ratio.

13. Classify each of the following as true or false:

    a. $8|4$

    b. $0|4$

    c. $4|0$

    d. If a number is divisible by 4 and by 6, then it is divisible by 24.

    e. If a number is not divisible by 12, then it is not divisible by 3.

14. Classify each of the following as true or false. If false, show a counterexample.

    a. If $7|x$ and $7 \nmid y$, then $7 \nmid xy$.

    b. If $d \nmid (a + b)$, then $d \nmid a$ and $d \nmid b$.

    c. If $d|(a + b)$ and $d \nmid a$, then $d \nmid b$.

    d. If $d|(x + y)$ and $d|x$, then $d|y$.

    e. If $4 \nmid x$ and $4 \nmid y$, then $4 \nmid xy$.

15. Test each of the following numbers for divisibility by 2, 3, 4, 5, 6, 8, 9, and 11:

    a. 83,160

    b. 83,193

16. Assume that 10,007 is prime. Without actually dividing 10,024 by 17, prove that 10,024 is not divisible by 17.

17. Fill each blank with one digit to make each of the following true (find all the possible answers):

    a. $6|87\_4$

    b. $24|4\_856$

    c. $29|87\_\_4$

18. A student claims that the sum of five consecutive positive integers is always divisible by 5.

    a. Check the student's claim for a few cases.

    b. Prove or disprove the student's claim.

19. Determine whether each of the following numbers is prime or composite:

    a. 143   **b.** 223

20. How can you tell if a number is divisible by 24? Check 4152 for divisibility by 24.

21. Is the LCM of two numbers always greater than the GCD of the numbers? Justify your answer.

22. Explain how to find the LCM of three numbers with the help of the Euclidean algorithm.

23. To find if the number $2 \cdot 3 \cdot 5 \cdot 7 + 11 \cdot 13$ is prime, a student finds that the number equals 353. She checks that $17 \nmid 353$ and $19^2 > 353$ and without

further checking, claims that 353 is prime. Explain why the student is correct.

24. Find the GCD for each of the following:

    a. 24 and 52

    b. 5767 and 4453

25. Find the LCM for each of the following:

    a. $2^3 \cdot 5^2 \cdot 7^3, 2 \cdot 5^3 \cdot 7^2 \cdot 13$ and $2^4 \cdot 5 \cdot 7^4 \cdot 29$

    b. 278 and 279

26. Construct a number that has exactly five positive divisors. Explain your construction.

27. Find all the positive divisors of 144.

28. Find the prime factorization of each of the following:

    a. 172   **b.** 288

    c. 260   **d.** 111

29. Find the least positive number that is divisible by every positive integer less than or equal to 10.

30. Candy bars priced at 50¢ each were not selling, so the price was reduced. Then they all sold in one day for a total of $31.93. What was the reduced price of each candy bar?

31. Two bells ring at 8:00 A.M. For the remainder of the day, one bell rings every half hour and the other bell rings every 45 min. What time will it be when the bells ring together again?

32. If the GCD of two positive whole numbers is 1, what can you say about the LCM of the two numbers? Explain your reasoning.

33. If there were to be 9 boys and 6 girls at a party and the host wanted each to be given exactly the same number of candies that could be bought in packages containing 12 candies, what is the fewest number of packages that could be bought?

34. Jane and Ramon are running laps on a track. If they start at the same time and place and go in the same direction, with Jane running a lap in 5 min and Ramon running a lap in 3 min, how long will it take for them to be at the starting place at the same time if they continue to run at these speeds?

35. June, an owner of a coffee stand, marked down the price of a latte between 7:00 A.M. and 8:00 A.M. from $2.00 a cup. If she grossed $98.69 from the latte sale and we know that she never sells a latte for less than a dollar, how many lattes did she sell between 7:00 A.M. and 8:00 A.M.? Explain your reasoning. (*Note:* 71|9869.)

36. Find the prime factorizations of each of the following.

    a. 6^{10}

    b. 34^n

    c. 97^4

    d. $8^4 \cdot 6^3 \cdot 26^2$

    e. $2^3 \cdot 3^2 + 2^4 \cdot 3^3 \cdot 7$

    f. $2^4 \cdot 3 \cdot 5^7 + 2^4 \cdot 5^6$

37. What are the possible remainders when a prime number greater than 3 is divided by 12? Justify your answer.

38. Prove the test for divisibility by 9 using a three-digit number n such that $n = a \cdot 10^2 + b \cdot 10 + c$.

★**39.** The triplet 3, 5, 7 consists of consecutive odd integers that are all prime. Give a convincing argument that this is the only triplet of consecutive odd integers that are all prime. (*Hint:* use the division algorithm.)

*★**40.** The length of a week was probably inspired by the need for market days and religious holidays. The Romans, for example, once used an 8-day week. Assuming April still had 30 days but was based on an 8-day week, if the first day of the month was on Sunday and the extra day after Saturday was called Venaday, on what day would the last day of the month fall?

*★**41.** To measure angles of rotation (in degrees) that a light on a small island lighthouse sweeps, what mod system would be used and why?

Selected Bibliography

Anthony, G., and M. Walshaw. "Zero A 'None' Number?" *Teaching Children Mathematics* 11 (August 2004): 38–42.

Bay, J. "Developing Number Sense on the Number Line." *Mathematics Teaching in the Middle School* 6 (April 2001): 448–451.

Bennett, A., and L. Nelson. "Divisibility Tests: So Right for Discoveries." *Mathematics Teaching in the Middle School* 7 (April 2002): 460–464.

Bezuszka, S., and M. Kenney. "Even Perfect Numbers: (Update)^2." *Mathematics Teacher* 90 (November 1997): 628–633.

Brown, E., and E. Jones. "Using Clock Arithmetic to Teach Algebra Concepts." *Mathematics Teaching in the Middle School* 11 (September 2005): 104–109.

Graviss, T., and J. Greaver. "Extending the Number Line to Make Connections with Number Theory." *Mathematics Teacher* 85 (September 1992): 418–420.

Gregg, J., and D. Gregg. "A Context for Integer Computation." *Mathematics Teaching in the Middle School* 13 (August 2007): 46–50.

Nurnberger-Haag, J. "Integers Made Easy: Just Walk It Off," *Mathematics Teaching in the Middle School* 13 (September 2007): 118–121.

Peterson, J. "Fourteen Different Strategies for Multiplication of Integers, or Why $(^-1)(^-1) = (^+1)$." *Arithmetic Teacher* 19 (May 1972): 396–403.

Petrella, G. "Subtracting Integers: An Affective Lesson." *Mathematics Teaching in the Middle School* 7 (November 2001): 150–151.

Ponce, G. "It's All in the Cards: Adding and Subtracting Integers." *Mathematics Teaching in the Middle School* 13 (August 2007): 10–17.

Reeves, A., and M. Beasley. "Advanced Paint by Numbers." *Mathematics Teaching in the Middle School* 12 (April 2007): 447.

Robbins, C., and T. Adams. "Get Primed to the Basic Building Blocks of Numbers." *Mathematics Teaching in the Middle School* 13 (September 2007): 122–127.

Schneider, S., and C. Thompson. "Incredible Equations Develop Incredible Number Sense." *Teaching Children Mathematics* 7 (November 2000): 146–148, 165–168.

Shultz, H. "The Postage-Stamp Problem, Number Theory, and the Programmable Calculator." *Mathematics Teacher* 92 (January 1999): 20–22.

Steinberg, R., D. Sleeman, D. Ktorza. "Algebra Students Knowledge of Equivalent Equations." *Journal of Research in Mathematics Education* 22 (February 1990): 112–121.

Rational Numbers
as Fractions

Preliminary Problem

Brandy bought a horse for $270 and immediately started paying for his keep. She sold the horse for $540. Considering the cost of his keep, she found that she had lost an amount equal to half of what she paid for the horse, plus one-fourth of the cost of his keep. How much did Brandy lose on the horse?

Integers such as $^-5$ were invented to solve equations like $x + 5 = 0$. Similarly, a new type of number is needed to solve an equation like $2x = 1$. Because there is no integer that satisfies this equation, a number that does was invented. The new number must be such that it is between 0 and 1. One reason is that $0 < 1 < 2$, which upon substituting $2x$ for 1 becomes $2 \cdot 0 < 2x < 2 \cdot 1$, or $0 < x < 1$. There are no integers between 0 and 1 that meet these conditions. Therefore, similar to the development of new notation for negative integers, we need notation for this new number. If multiplication is to work with this new number as with whole numbers, then $2x = x + x$, so $x + x = 1$. In other words, the number created must be added to itself to get 1. The number invented to solve the equation is *one-half*, denoted $\frac{1}{2}$. It is an element of the set of numbers of the form $\frac{a}{b}$, where $b \neq 0$ and a and b are integers. Moreover, numbers of the form $\frac{a}{b}$ are solutions to equations of the form $bx = a$. This set, denoted Q, is the set of **rational numbers** and is defined as follows:

$$Q = \left\{ \frac{a}{b} \mid a \text{ and } b \text{ are integers and } b \neq 0 \right\}$$

Q is a subset of another set of numbers called *fractions*. Fractions are of the form $\frac{a}{b}$, where $b \neq 0$ but a and b are not necessarily integers. For example, $\frac{1}{\sqrt{2}}$ is a fraction but not a rational number. (In this text, we restrict ourselves to fractions where a and b are real numbers, but that restriction is not necessary.) The fact that $b \neq 0$ is always necessary, because division by 0 is undefined.

As indicated in the *Principles and Standards* excerpt below, children should be introduced to fractions through concrete activities.

Beyond understanding whole numbers, young children can be encouraged to understand and represent commonly used fractions in context, such as 1/2 of a cookie or 1/8 of a pizza, and to see fractions as part of a unit whole or of a collection. Teachers should help students develop an understanding of fractions as division of numbers. And in the middle grades, in part as a basis for their work with proportionality, students need to solidify their understanding of fractions as numbers. (p. 33)

As is evident from the research note, it it important to emphasize that a fraction such as $\frac{3}{4}$ represents a single number.

Rational Numbers as Fractions Expectations

The *Principles and Standards* expectations of students that are covered in this chapter include the following:

In grades K–2, children should understand and represent commonly used fractions such as 1/4, 1/3, and 1/2. (p. 392)

In grades 3–5, children should develop an understanding of fractions to include them as parts of unit wholes, as parts of a collection, as locations on number lines, and as divisions of whole numbers. Children should use models, benchmarks, and equivalent forms to judge the size of fractions. They should be able to develop and use strategies to estimate computations involving fractions. (p. 392)

In grades 6–8, children work flexibly with fractions to solve problems, compare and order fractions, and find their locations on a number line. They understand and use ratios and proportions to represent quantitative relationships. Children understand the meanings and effects of arithmetic operations with fractions. They select appropriate methods and tools for computing with fractions from among mental computation, estimation, calculators, or paper and pencil and apply the selected method. (p. 393)

6-1   The Set of Rational Numbers

In the rational number $\frac{a}{b}$, a is the **numerator** and b is the **denominator**. The rational number $\frac{a}{b}$ may also be represented as a/b or as $a \div b$. The word *fraction* is derived from the Latin word *fractus*, meaning "broken." The word *numerator* comes from a Latin word meaning "numberer," and *denominator* comes from a Latin word meaning "namer." Table 6-1 shows several ways in which we use rational numbers.

Table 6-1 Uses of Rational Numbers

Use	Example
Division problem or solution to a multiplication problem	The solution to $2x = 3$ is $\frac{3}{2}$.
Partition, or part, of a whole	Joe received $\frac{1}{2}$ of Mary's salary each month for alimony.
Ratio	The ratio of Republicans to Democrats on a Senate committee is three to five.
Probability	When you toss a fair coin, the probability of getting heads is $\frac{1}{2}$.

 **Historical Note**

1/3

The early Egyptian numeration system had symbols for fractions with numerators of 1. Most fractions with numerators other than 1 were expressed as a sum of different fractions with numerators of 1, for example, $\frac{7}{12} = \frac{1}{3} + \frac{1}{4}$.

Fractions with denominator 60 or powers of 60 were common in ancient Babylon about 2000 BCE, where 12,35 meant $12 + \frac{35}{60}$. The method was later adopted by the Greek astronomer Ptolemy (approximately BCE 125). The same method was also used in Islamic and European countries and is presently used in the measurements of angles, where $13° 19' 47''$ means $13 + \frac{19}{60} + \frac{47}{60^2}$ degrees.

The modern notation for fractions—a bar between numerator and denominator—is of Hindu origin. It came into general use in Europe in sixteenth-century books.   ◆

Research Note

When students construct an understanding of the concept of a fraction, the area model is preferred over the set model because the total area is a more flexible, visible attribute. The area model allows students to encode almost any fraction, whereas the set model (for example, a group of colored chips) has distinct limitations, especially for a part/whole interpretation. For example, try to represent $\frac{3}{5}$ using either four cookies or a piece of paper (English and Halford 1995; Hope and Owens, 1987). ♦

(a) Area model

(b) Number-line model

(c) Set model

Figure 6-1

Figure 6-1 illustrates the use of rational numbers as part of a whole and as part of a given set. For example, in the area model in Figure 6-1(a), one part out of three congruent parts, or $\frac{1}{3}$ of the largest rectangle, is shaded. In Figure 6-1(b), two parts out of three congruent parts, or $\frac{2}{3}$ of the unit segment, are shaded. In Figure 6-1(c), three circles out of five congruent circles, or $\frac{3}{5}$ of the circles, are shaded. Notice that as indicated in the Research Note, the area model is the preferred model.

Our early exposure to fractions, or rational numbers, usually takes the form of oral description rather than mathematical notation. We hear phrases such as "one-half of a pizza," "one-third of a cake," or "three-fourths of a pie." We encounter such questions as "If three identical fruit bars are distributed equally among four friends, how much does each get?" The answer is that each receives $\frac{3}{4}$ of a bar.

The English words used for rational numbers are the same words we use to tell "order," for example, the *fourth* person in a line and *three-fourths* for $\frac{3}{4}$. This causes confusion for students learning about fractions. In contrast, in Chinese $\frac{3}{4}$ is read "out of four parts, (take) three." The Chinese model enforces the idea of partitioning quantities into equal parts and choosing some number of these parts. The concept of sharing quantities and comparing sizes of shares can provide an entry point to introduce students to rational numbers. For rational numbers, sharing can play the role that counting does for whole numbers.

When rational numbers are introduced as fractions that represent a part of a whole, we must pay attention to the whole from which a rational number is derived. For example, if we talk about $\frac{3}{4}$ of a pizza, then the amount of pizza is determined both by the fractional part, $\frac{3}{4}$, and the size of the pizza. Three-fourths of a large pizza is certainly more than three-fourths of a small pizza. Attention must be paid to the context and the size of the *whole* being considered.

To understand the meaning of any fraction, $\frac{a}{b}$, using the parts-to-whole model, we must consider each of the following:

1. The *whole* being considered.
2. The number b of equal-size parts into which the whole is divided.
3. The number a of parts of the whole that we are selecting.

The cartoon shows that the concept of a fraction is not easy for some children.

THE FAMILY CIRCUS   By Bil Keane

6-2

©2006 Bil Keane, Inc.
Dist. by King Features Synd.
www.familycircus.com

**"How come a quarter is worth
25 cents, and a quarter of an
hour is only 15 minutes?"**

NOW TRY THIS 6-1

a. How would you answer Billy's question in the cartoon?

b. Jim claims that $\frac{1}{3} > \frac{1}{2}$ because in Figure 6-2 the shaded

portion for $\frac{1}{3}$ is larger than the shaded portion

for $\frac{1}{2}$. How would you help him?

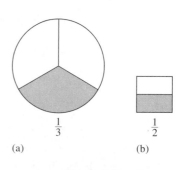

$\frac{1}{3}$ $\frac{1}{2}$

(a)   (b)

Figure 6-2

Rational numbers can be represented on a number line. Once the integers 0 and 1 are assigned to points on a line, the unit segment is defined and every other rational number is assigned to a specific point. For example, to represent $\frac{3}{4}$ on the number line, we divide the segment from 0 to 1 into 4 segments of equal length and mark the line accordingly. Then, starting from 0, we count 3 of these segments and stop at the mark corresponding to the right endpoint of the third segment to obtain the point assigned to the rational number $\frac{3}{4}$. Figure 6-3 shows the points that correspond to $-2, \frac{-5}{4}, -1, \frac{-3}{4}, 0, \frac{3}{4}, 1, \frac{5}{4}$, and 2.

Figure 6-3

In the grade 3 *Focal Points*, we find that students develop an understanding of the meaning and uses of fractions to represent parts of a whole, parts of a set, or points or distances on a number line (p. 15).

NOW TRY THIS 6-2

a. If $\boxed{}$ represents $\dfrac{3}{4}$, draw rectangles for $\dfrac{1}{4}, \dfrac{1}{2}, \dfrac{3}{2}, \dfrac{2}{3}$, and $\dfrac{4}{3}$.

b. In Figure 6-3, locate points that correspond to $\dfrac{1}{2}, \dfrac{-1}{2}, \dfrac{3}{2}$, and $\dfrac{-7}{4}$.

A fraction $\dfrac{a}{b}$, where $0 \le a < b$, is a **proper fraction**. For example, $\dfrac{4}{7}$ is a proper fraction, but $\dfrac{7}{4}, \dfrac{4}{4}$, and $\dfrac{9}{7}$ are not; $\dfrac{7}{4}$ is an **improper fraction**. In general $\dfrac{a}{b}$ is an improper fraction if $a \ge b > 0$.

Typically, in early grades students are introduced only to positive fractions. An example of this is shown on the following student page.

NOW TRY THIS 6-3 Answer the three "Talk About It" questions on the student page (page 346).

REMARK Notice that every integer n can be represented as a rational number because $n = \dfrac{nk}{k}$, where k is any non-zero integer. Thus if $k \ne 0$ then $0 = \dfrac{0 \cdot k}{k} = \dfrac{0}{k}$.

NOW TRY THIS 6-4 Draw a Venn diagram to show the relationship among natural numbers, whole numbers, integers, and rational numbers.

Equivalent or Equal Fractions

Fractions can be introduced in the classroom through a concrete activity such as paper folding. In Figure 6-4(a), one of 3 congruent parts, or $\dfrac{1}{3}$, is shaded. In this case, the whole is the rectangle. In Figure 6-4(b), each of the thirds has been folded in half so that now we have 6 sections, and 2 of 6 congruent parts, or $\dfrac{2}{6}$, are shaded. Thus, both $\dfrac{1}{3}$ and $\dfrac{2}{6}$ represent exactly the same shaded portion. Although the symbols $\dfrac{1}{3}$ and $\dfrac{2}{6}$ do not look alike, they represent the same rational number and are **equivalent fractions** or **equal fractions**.

School Book Page NAMING FRACTIONAL PARTS

Lesson 9-2

Key Idea
You can write a fraction to describe the equal parts of a whole.

Vocabulary
• fraction
• numerator
• denominator

Naming Fractional Parts

LEARN

What is a fraction?

You can use **fractions** to name equal parts of a whole.

☑ **WARM UP**

Tell if each shows equal parts.

1.

2.

Example

The flag of Nigeria has 3 equal parts. What fraction of the flag is green?

Numerator ⟶ $\dfrac{2}{3}$ ← 2 equal parts are green
Denominator ⟶ ← 3 equal parts **in all**

What You **Write**	What You **Say**
$\frac{2}{3}$ of the flag is green.	*Two thirds* of the flag is green.

☑ **Talk About It**

1. What fraction of Nigeria's flag is not green?

2. What fraction of each square is blue?

A   B   C

3. Reggie cut an apple into 4 equal pieces. He ate 1 piece. What fraction tells how much of the apple he ate?

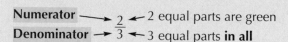

Take It to the NET
More Examples
www.scottforesman.com

502

Source: Mathematics, Diamond Edition, Grade 3, Scott Foresman-Addison Wesley, 2008 (p. 502).

Equivalent fractions are numbers that represent the same point on a number line. Because they represent equal amounts, we write $\frac{1}{3} = \frac{2}{6}$ and say that $\frac{1}{3}$ equals $\frac{2}{6}$.

E-Manipulatives Activity

For practice with the concept of fractions, see the *Naming Fractions* and *Visualizing Fractions* modules on the E-Manipulatives disk.

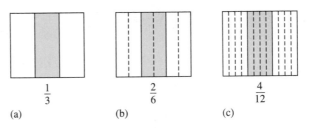

(a)   (b)   (c)

Figure 6-4

Figure 6-4(c) shows the rectangle with each of the original thirds folded into 4 equal parts with 4 of the 12 parts now shaded. Thus, $\frac{1}{3}$ is equal to $\frac{4}{12}$ because the same portion of the model is shaded. Similarly, we could illustrate that $\frac{1}{3}, \frac{2}{6}, \frac{3}{9}, \frac{4}{12}, \frac{5}{15}, \ldots$ are equal.

Fraction strips can also be used for generating equivalent fractions, as seen on the student page. Part (a) on the student page shows that $\frac{1}{2} = \frac{3}{6} = \frac{6}{12}$. What equivalent fractions are modeled in part (b)? Also on the student page on page 348, in Example A we see another way to find equivalent fractions. This technique makes use of the Fundamental Law of Fractions, which can be stated as follows: *The value of a fraction does not change if its numerator and denominator are multiplied by the same nonzero integer.* Under certain assumptions this Law of Fractions can be proved and hence it is stated as a theorem.

Theorem 6–1: Fundamental Law of Fractions

Let $\frac{a}{b}$ be any fraction and n a nonzero integer. Then, $\frac{a}{b} = \frac{an}{bn}$.

REMARK

1. In Theorem 6–1 it can be shown that n can be any nonzero number. This version of the Fundamental Law of Fractions is used later in this book.

2. Theorem 6–1 implies that if d is a common factor of a and b then $\frac{a}{b} = \frac{a \div d}{b \div d}$.

3. At this point, some students justify the Fundamental Law of Fractions as follows:

$$\frac{an}{bn} = \frac{a \cdot n}{b \cdot n} = \frac{a}{b} \cdot \frac{n}{n} = \frac{a}{b} \cdot 1 = \frac{a}{b}$$

This is a correct approach; however in our development of properties of fractions, we have not yet discussed multiplication of fractions.

NOW TRY THIS 6-5 Explain why in the Fundamental Law of Fractions n must be nonzero.

School Book Page EQUIVALENT FRACTIONS

Lesson 3-7

Key Idea
A part of a whole or of a set can be named by equivalent fractions.

Vocabulary
• equivalent fractions
• common factor (p. 150)
• least common denominator (LCD)
• greatest common factor (GCF) (p. 150)
• simplest form

Materials
• fraction strips or ✺ **tools**

Equivalent Fractions

LEARN

Activity

What are equivalent fractions?

Fractions that name the same amount are called **equivalent fractions**.

a. Use fraction strips to identify one or more fractions equivalent to each fraction below.

$\frac{3}{4}$ $\frac{8}{12}$ $\frac{6}{10}$ $\frac{5}{6}$

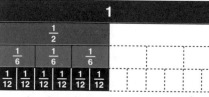

$$\frac{1}{2} = \frac{3}{6} = \frac{6}{12}$$

b. What two equivalent fractions are shown by this model? Explain.

WARM UP

Find the common factors for each pair of numbers.

1. 6, 9   2. 10, 15

3. 4, 12   4. 24, 36

Think It Through
I can **use objects** to help me identify equivalent fractions.

How can you find equivalent fractions?

Example A

Find two fractions that are equivalent to $\frac{18}{24}$.

One Way
Use multiplication.

Multiply both the numerator and denominator by the same nonzero number.

The number 2 is easy to use, so multiply the numerator and denominator by 2.

$$\frac{18}{24} = \frac{18 \times 2}{24 \times 2} = \frac{36}{48}$$

Another Way
Use division.

Divide both the numerator and denominator by the same nonzero number.

The number 3 is a common factor, so divide the numerator and denominator by 3.

$$\frac{18}{24} = \frac{18 \div 3}{24 \div 3} = \frac{6}{8}$$

So, $\frac{18}{24}$, $\frac{36}{48}$, and $\frac{6}{8}$ are all equivalent fractions.

164

Source: Scott Foresman-Addison Wesley Mathematics, Grade 6, 2008 (p. 164).

From the Fundamental Law of Fractions, $\dfrac{7}{-15} = \dfrac{-7}{15}$ because $\dfrac{7}{-15} = \dfrac{7 \cdot (^-1)}{^-15 \cdot (^-1)} = \dfrac{-7}{15}$.

Similarly, $\dfrac{a}{-b} = \dfrac{-a}{b}$. The form $\dfrac{-a}{b}$, where b is a positive number, is usually preferred.

Example 6-1

Find a value for x such that $\dfrac{12}{42} = \dfrac{x}{210}$.

Solution   Because $210 \div 42 = 5$, we use the Fundamental Law of Fractions to obtain $\dfrac{12}{42} = \dfrac{12 \cdot 5}{42 \cdot 5} = \dfrac{60}{210}$. Hence, $\dfrac{x}{210} = \dfrac{60}{210}$, and $x = 60$.

Alternative approach: $\dfrac{12}{42} = \dfrac{2 \cdot 6}{7 \cdot 6} = \dfrac{2}{7} = \dfrac{2 \cdot 30}{7 \cdot 30} = \dfrac{60}{210}$. Therefore $x = 60$.

Simplifying Fractions

The Fundamental Law of Fractions justifies the process of **simplifying fractions**. Consider the following:

$$\frac{60}{210} = \frac{6 \cdot 10}{21 \cdot 10} = \frac{6}{21}$$

Also,

$$\frac{6}{21} = \frac{2 \cdot 3}{7 \cdot 3} = \frac{2}{7}$$

We can simplify $\dfrac{60}{210}$ because the numerator and denominator have a common factor of 10.

We can simplify $\dfrac{6}{21}$ because 6 and 21 have a common factor of 3. However, we cannot simplify $\dfrac{2}{7}$ because 2 and 7 have no positive common factor other than 1. Notice that we could also simplify $\dfrac{60}{120}$ in one step by: $\dfrac{60}{120} = \dfrac{2 \cdot 30}{7 \cdot 30} = \dfrac{2}{7}. \dfrac{2}{7}$ is the **simplest form** of $\dfrac{60}{210}$ because both 60 and 210 have been divided by their greatest common divisor, 30. To write a fraction $\dfrac{a}{b}$ in simplest form; that is, in **lowest terms**, we divide both a and b by the GCD(a, b).

> ### Definition of Simplest Form
>
> A rational number $\dfrac{a}{b}$ is in simplest form if $b > 0$ and GCD$(a,b) = 1$; that is, if a and b have no common factor greater than 1, and $b > 0$.

We can use scientific/fraction calculators to simplify fractions. For example, to simplify $\dfrac{6}{12}$, we enter $\boxed{6}\boxed{/}\boxed{1}\boxed{2}$ and press $\boxed{\text{SIMP}}\boxed{=}$, and $\dfrac{3}{6}$ appears on the screen. At this point, an indicator tells us that this is not in simplest form, so we press $\boxed{\text{SIMP}}\boxed{=}$ again to obtain $\frac{1}{2}$. At any time, we can view the factor that was removed by pressing the $\boxed{\text{x} \circ \text{y}}$ key.

The Fundamental Law of Fractions can be used to simplify algebraic expressions, as seen in the following example.

Example 6-2

Write each of the following fractions in simplest form if they are not already so:

a. $\dfrac{28ab^2}{42a^2b^2}$
b. $\dfrac{(a + b)^2}{3a + 3b}$
c. $\dfrac{x^2 + x}{x + 1}$
d. $\dfrac{3 + x^2}{3x^2}$

e. $\dfrac{3 + 3x^2}{3x^2}$
f. $\dfrac{a^2 - b^2}{a - b}$
g. $\dfrac{a^2 + b^2}{a + b}$

Solution

a. $\dfrac{28\,ab^2}{42a^2b^2} = \dfrac{2(14ab^2)}{3a(14ab^2)} = \dfrac{2}{3a}$

b. $\dfrac{(a + b)^2}{3a + 3b} = \dfrac{(a + b)(a + b)}{3(a + b)} = \dfrac{a + b}{3}$, where $a + b \neq 0$

c. $\dfrac{x^2 + x}{x + 1} = \dfrac{x(x + 1)}{x + 1} = \dfrac{x(x + 1)}{1(x + 1)} = \dfrac{x}{1} = x$, where $x \neq -1$

d. $\dfrac{3 + x^2}{3x^2}$ cannot be simplified because $3 + x^2$ and $3x^2$ have no factors in common except 1.

e. $\dfrac{3 + 3x^2}{3x^2} = \dfrac{3(1 + x^2)}{3x^2} = \dfrac{1 + x^2}{x^2}$

f. Recall that in Chapter 5 we used the distributive properties to show that $(a - b)(a + b) = a^2 - b^2$. Thus,

$$\dfrac{a^2 - b^2}{a - b} = \dfrac{(a - b)(a + b)}{(a - b)1} = \dfrac{a + b}{1} = a + b, \text{ where } a \neq b$$

g. The fraction is already in simplest form because $a^2 + b^2$ does not have $(a + b)$ as a factor. Notice that $a^2 + b^2 \neq (a + b)^2$.

REMARK

1. When we write an algebraic expression that is a fraction, we must assume that the denominator is not 0. Thus, even after the fraction is reduced, this restriction has to be maintained. For example, in part (c) of Example 6-2, $\dfrac{x^2 + x}{x + 1} = x$ if $x \neq -1$, and in part (f) the result holds if $a - b \neq 0$; that is, if $a \neq b$.

2. In part (d) we pointed out that $\dfrac{3 + x^2}{3x^2}$ cannot be simplified. However, for some particular values of x, the corresponding fraction can be simplified. This is the case if x is a multiple of 3; for example, if $x = 3$, then the value of $\dfrac{3 + x^2}{3x^2}$ is $\dfrac{3 + 9}{3 \cdot 9} = \dfrac{3 \cdot 4}{3 \cdot 9} = \dfrac{4}{9}$.

3. Some students, thinking about the Fundamental Law of Fractions as a *cancellation property*, often "simplify" an expression like $\dfrac{6 + a^2}{3a}$ by thinking of it as $\dfrac{2 \cdot 3 + a \cdot a}{3a}$ and "canceling" equal numbers in the products to obtain $2 + a$ as the answer. Emphasizing the factor approach that neither 3 nor a is a factor of $6 + a^2$ may help to avoid such mistakes.

Equality of Fractions

We can use three methods to show that two fractions, such as $\dfrac{12}{42}$ and $\dfrac{10}{35}$, are equal.

1. Simplify both fractions to the same simplest form:

$$\frac{12}{42} = \frac{2^2 \cdot 3}{2 \cdot 3 \cdot 7} = \frac{2}{7} \quad \text{and} \quad \frac{10}{35} = \frac{5 \cdot 2}{5 \cdot 7} = \frac{2}{7}$$

Thus,

$$\frac{12}{42} = \frac{10}{35}$$

2. Rewrite both fractions with the same least common denominator. Since $\text{LCM}(42, 35) = 210$, then

$$\frac{12}{42} = \frac{60}{210} \quad \text{and} \quad \frac{10}{35} = \frac{60}{210}$$

Thus,

$$\frac{12}{42} = \frac{10}{35}$$

3. Rewrite both fractions with a common denominator (not necessarily the least). A common multiple of 42 and 35 may be found by finding the product $42 \cdot 35$, or 1470. Now,

$$\frac{12}{42} = \frac{420}{1470} \quad \text{and} \quad \frac{10}{35} = \frac{420}{1470}$$

Hence,

$$\frac{12}{42} = \frac{10}{35}$$

The third method suggests a general algorithm for determining if two fractions $\dfrac{a}{b}$ and $\dfrac{c}{d}$ are equal. Rewrite both fractions with common denominator bd. That is,

$$\frac{a}{b} = \frac{ad}{bd} \quad \text{and} \quad \frac{c}{d} = \frac{bc}{bd}$$

E-Manipulative Activity

Additional practice with ordering fractions is available in the *Ranking Fractions* module on the E-Manipulative disk.

Because the denominators are the same, $\dfrac{ad}{bd} = \dfrac{bc}{bd}$ if, and only if, $ad = bc$. For example, $\dfrac{24}{36} = \dfrac{6}{9}$ because $24 \cdot 9 = 216 = 36 \cdot 6$. In general, the following theorem results.

Theorem 6–2

Two fractions $\frac{a}{b}$ and $\frac{c}{d}$ are equal if, and only if, $ad = bc$.

Using a calculator, we can determine if two fractions are equal by using the preceding property. Since both $\boxed{2}\boxed{\times}\boxed{2}\boxed{1}\boxed{9}\boxed{6}\boxed{=}$ and $\boxed{4}\boxed{\times}\boxed{1}\boxed{0}\boxed{9}\boxed{8}\boxed{=}$ yield a display of 4392, we see that $\frac{2}{4} = \frac{1098}{2196}$.

Ordering Rational Numbers

As discussed in *Principles and Standards* (p. 392), students in lower grades should experience comparing fractions between 0 and 1 in relation to such benchmarks as $0, \frac{1}{2}, \frac{3}{4}$, and 1. In the middle grades, the comparison of fractions becomes more difficult. We first consider the comparison of fractions with like denominators.

Children know that $\frac{7}{8} > \frac{5}{8}$ because if a pizza is divided into 8 parts of equal size, then 7 parts of a pizza is more than 5 parts. Similarly, $\frac{3}{7} < \frac{4}{7}$. Thus, given two fractions with common positive denominators, the one with the greater numerator is the greater fraction. This can be stated as follows.

Theorem 6–3

If a, b, and c are integers and $b > 0$, then $\frac{a}{b} > \frac{c}{b}$ if, and only if, $a > c$.

Comparing the size of fractions is shown on the following student page.

NOW TRY THIS 6-6 Determine if Theorem 6–3 is true if $b < 0$.

To compare fractions with unlike denominators, some students may incorrectly reason that $\frac{1}{8} > \frac{1}{7}$ because 8 is greater than 7. In other cases, they might falsely believe that $\frac{6}{7}$ is equal to $\frac{7}{8}$ because in both cases the difference between the numerator and the denominator is 1. Comparing fractions with unlike denominators may be aided by using fraction

School Book Page

✓ Talk About It

1. Explain how you know that $\frac{3}{4}$ is greater than $\frac{1}{4}$.

2. Number Sense A cheese pizza is cut into 12 equal slices and a vegetable pizza of the same size is cut into 8 equal slices. Which pizza has the larger slices? Explain.

How do you order fractions?

Example C

Write $\frac{1}{4}$, $\frac{3}{8}$, and $\frac{1}{2}$ in order from greatest to least.

$\frac{1}{2} > \frac{3}{8}$ and $\frac{3}{8} > \frac{1}{4}$.

$\frac{1}{2}$, $\frac{3}{8}$, $\frac{1}{4}$

✓ Talk About It

3. Explain how to order $\frac{1}{5}$, $\frac{1}{2}$, and $\frac{1}{10}$ from least to greatest.

CHECK ✓

For another example, see Set 9-4 on p. 554.

Compare. Write >, <, or =.

1.

$\frac{1}{2}$ ● $\frac{1}{3}$

2.

$\frac{3}{8}$ ● $\frac{4}{8}$

3.

$\frac{3}{12}$ ● $\frac{1}{4}$

4. Order $\frac{2}{4}$, $\frac{1}{3}$, and $\frac{1}{6}$ from least to greatest.

5. Number Sense Can two different unit fractions be equal? Why or why not?

Section A Lesson 9-4   **507**

Source: Mathematics, Diamond Edition, Grade Three, Scott Foresman-Addison Wesley, 2008 (p. 507).

strips to compare the fractions visually. For example, consider the fractions $\frac{4}{5}$ and $\frac{11}{12}$ shown in Figure 6-5.

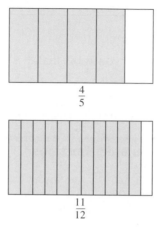

$$\frac{4}{5}$$

$$\frac{11}{12}$$

Figure 6-5

From Figure 6-5, students see that each fraction is one piece less than the same-size whole unit. However, they see that the missing piece for $\frac{11}{12}$ is smaller than the missing piece for $\frac{4}{5}$, so $\frac{11}{12}$ must be greater than $\frac{4}{5}$.

Comparing fractions with unlike denominators can be accomplished by rewriting the fractions with the same positive denominator. Then we can compare them using a previously learned technique. Using the common denominator bd, we can write the fractions $\frac{a}{b}$ and $\frac{c}{d}$ as $\frac{ad}{bd}$ and $\frac{bc}{bd}$. Because $b > 0$ and $d > 0$, then $bd > 0$; and we can apply Theorem 6–3 as follows:

$$\frac{a}{b} > \frac{c}{d} \quad \text{if, and only if,} \quad \frac{ad}{bd} > \frac{bc}{bd} \quad \text{and} \quad \frac{ad}{bd} > \frac{bc}{bd} \quad \text{if, and only if,} \quad ad > bc$$

Therefore, we have the following theorem.

Theorem 6–4

If a, b, c, and d are integers with $b > 0$ and $d > 0$, then $\frac{a}{b} > \frac{c}{d}$ if, and only if, $ad > bc$.

 NOW TRY THIS 6-7 Order the fractions $\frac{3}{4}, \frac{9}{16}, \frac{5}{8},$ and $\frac{2}{3}$ from least to greatest.

Research Note

Students taught the common denominator method for comparing two fractions tend to ignore it and focus on rules associated with ordering whole numbers. Students who correctly compare numerators if the denominators are equal often compare denominators if the numerators are equal (Behr et al. 1984). ◆

As suggested in the Research Note, if students compare denominators of fractions if the numerators are the same, then this can be a good strategy if handled properly. For example, consider $\frac{3}{4}$ and $\frac{3}{10}$. If the whole is the same for both fractions, this means that we have three $\frac{1}{4}$ and three $\frac{1}{10}$. Because $\frac{1}{4}$ is greater than $\frac{1}{10}$, then three of the larger parts is greater than three of the smaller parts, so $\frac{3}{4} > \frac{3}{10}$.

 NOW TRY THIS 6-8 Generalize the approach above for comparing fractions whose numerators and denominators are positive integers.

Denseness of Rational Numbers

The set of rational numbers has a property that the set of whole numbers and the set of integers do not have. Consider $\frac{1}{2}$ and $\frac{2}{3}$. To find a rational number between $\frac{1}{2}$ and $\frac{2}{3}$, we first rewrite the fractions with a common denominator, as $\frac{3}{6}$ and $\frac{4}{6}$. Because there is no whole number between the numerators 3 and 4, we next find two fractions equal, respectively, to $\frac{1}{2}$ and $\frac{2}{3}$ with greater denominators. For example, $\frac{1}{2} = \frac{6}{12}$ and $\frac{2}{3} = \frac{8}{12}$, and $\frac{7}{12}$ is between the two fractions $\frac{6}{12}$ and $\frac{8}{12}$. So $\frac{7}{12}$ is between $\frac{1}{2}$ and $\frac{2}{3}$. This property is generalized as follows and stated as a theorem.

> **Theorem 6–5:   Denseness property for rational numbers**
> Given two different rational numbers $\frac{a}{b}$ and $\frac{c}{d}$, there is another rational number between these two numbers.

 NOW TRY THIS 6-9 Explain why there are infinitely many rational numbers between any two rational numbers.

Example 6-3

a. Find two fractions between $\frac{7}{18}$ and $\frac{1}{2}$.

b. Show that the sequence $\frac{1}{2}, \frac{2}{3}, \frac{3}{4}, \frac{4}{5}, \ldots$ is an *increasing sequence*, that is, that each term starting from the second term is greater than the preceding term.

Solution   **a.** Because $\frac{1}{2} = \frac{1 \cdot 9}{2 \cdot 9} = \frac{9}{18}$, we see that $\frac{8}{18}$, or $\frac{4}{9}$, is between $\frac{7}{18}$ and $\frac{9}{18}$. To find another fraction between the given fractions, we find two fractions equal to

$\frac{7}{18}$ and $\frac{9}{18}$, respectively, but with greater denominators; for example, $\frac{7}{18} = \frac{14}{36}$ and $\frac{9}{18} = \frac{18}{36}$.

We now see that $\frac{15}{36}, \frac{16}{36}$, and $\frac{17}{36}$ are all between $\frac{14}{36}$ and $\frac{18}{36}$ and thus between $\frac{7}{18}$ and $\frac{1}{2}$.

b. Because the nth term of the sequence is $\frac{n}{n + 1}$, the next term is $\frac{n + 1}{(n + 1) + 1}$, or $\frac{n + 1}{n + 2}$. We need to show that for all positive integers n, $\frac{n + 1}{n + 2} > \frac{n}{n + 1}$.

By Theorem 6–2, the inequality will be true if, and only if, $(n + 1)(n + 1) > n(n + 2)$. This inequality is equivalent to $n^2 + 2n + 1 > n^2 + 2n$, which is true for all n since the left side is greater than the right side.

Some students incorrectly add $\frac{a}{b} + \frac{c}{d}$ as $\frac{a + c}{b + d}$. Although the technique does not produce the correct sum, as will be seen in Section 6-2, it provides a way to find a number between any two rational numbers. In Example 6-3, to find a number between $\frac{7}{18}$ and $\frac{1}{2}$ we could add the numerators and add the denominators to produce $\frac{7 + 1}{18 + 2} = \frac{8}{20}$. We see that $\frac{7}{18} < \frac{8}{20}$ because $140 < 144$. Also, $\frac{8}{20} < \frac{1}{2}$ because $16 < 20$. We state the general property in the following theorem, whose proof is explored in Now Try This 6-10.

Theorem 6–6

Let $\frac{a}{b}$ and $\frac{c}{d}$ be any rational numbers with positive denominators, where $\frac{a}{b} < \frac{c}{d}$. Then,

$$\frac{a}{b} < \frac{a + c}{b + d} < \frac{c}{d}.$$

NOW TRY THIS 6-10 Prove Theorem 6–6; that is, if $\frac{a}{b}$ and $\frac{c}{d}$ are any rational numbers with positive denominators, where $\frac{a}{b} < \frac{c}{d}$, then $\frac{a}{b} < \frac{a + c}{b + d} < \frac{c}{d}$. *(Hint:* If $\frac{a}{b} < \frac{c}{d}$, then $ad < bc$. Use this to prove that $\frac{a}{b} < \frac{a + c}{b + d}$ and $\frac{a + c}{b + d} < \frac{c}{d}$.)*

REMARK Notice that the proof of Theorem 6–6 suggested in Now Try This 6-10 also proves Theorem 6–5.

LABORATORY ACTIVITY Obtain tangram pieces or build them and cut them out as shown in Figure 6-6. Answer each of the following.

a. If the area of the entire square is 1 square unit, find the area of each tangram piece.

b. If the area of piece *a* is 1 square unit, find the area of each tangram piece.

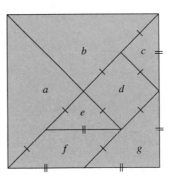

Figure 6-6

Assessment 6-1A

1. Write a sentence that illustrates the use of $\frac{7}{8}$ in each of the following ways:
   a. As a division problem
   b. As part of a whole
   c. As a ratio

2. For each of the following, write a fraction to represent the shaded portion:

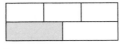

a.   **b.**

        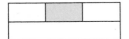

c.   **d.**

3. For each of the following four squares, write a fraction to represent the shaded portion. What property of fractions does the diagram illustrate?

a.   **b.**   **c.**   **d.**

4. Based on your observations could the shaded portions in the following figures represent the indicated fractions? Tell why.

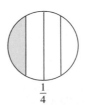   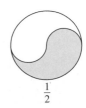

$\frac{1}{4}$ $\frac{3}{4}$ $\frac{1}{2}$

a.   **b.**   **c.**

5. In each case, subdivide the *whole* shown on the right to show the equivalent fraction.

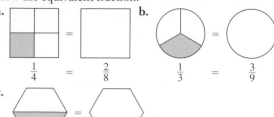

a. $\dfrac{1}{4} = \dfrac{2}{8}$ **b.** $\dfrac{1}{3} = \dfrac{3}{9}$

c.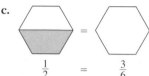

$\dfrac{1}{2} = \dfrac{3}{6}$

6. Referring to the figure, represent each of the following as a fraction:

a. The dots in the interior of the circle as a part of all the dots
b. The dots in the interior of the rectangle as a part of all the dots
c. The dots in the intersection of the interiors of the rectangle and the circle as a part of all the dots
d. The dots outside the circular region but inside the rectangular region as part of all the dots

7. For each of the following, write three fractions equal to the given fraction:

a. $\dfrac{2}{9}$ **b.** $\dfrac{-2}{5}$

c. $\dfrac{0}{3}$ **d.** $\dfrac{a}{2}$

8. Find the simplest form for each of the following fractions:

a. $\dfrac{156}{93}$ **b.** $\dfrac{27}{45}$ **c.** $\dfrac{-65}{91}$

9. For each of the following, choose the expression in parentheses that equals or describes best the given fraction:

a. $\dfrac{0}{0}$ (1, undefined, 0)

b. $\dfrac{5}{0}$ (undefined, 5, 0)

c. $\dfrac{0}{5}$ (undefined, 5, 0)

d. $\dfrac{2+a}{a}$ (2, 3, cannot be simplified)

e. $\dfrac{15+x}{3x}$ $\left(\dfrac{5+x}{x}, 5, \text{cannot be simplified}\right)$

10. Find the simplest form for each of the following:

a. $\dfrac{a^2-b^2}{3a+3b}$ **b.** $\dfrac{14x^2y}{63xy^2}$

11. Determine if the following pairs are equal:

a. $\dfrac{3}{8}$ and $\dfrac{375}{1000}$ **b.** $\dfrac{18}{54}$ and $\dfrac{23}{69}$

12. Determine if the following pairs are equal by changing both to the same denominator:

a. $\dfrac{10}{16}$ and $\dfrac{12}{18}$ **b.** $\dfrac{-21}{86}$ and $\dfrac{-51}{215}$

13. Draw an area model to show that $\dfrac{3}{4} = \dfrac{6}{8}$.

14. If a fraction is equal to $\dfrac{3}{4}$ and the sum of the numerator and denominator is 84, what is the fraction?

15. Mr. Gomez filled his car's 16 gal gas tank. He took a short trip and used 6 gal of gas. Draw an arrow in the following figure to show what his gas gauge looked like after the trip:

16. Solve for x in each of the following:

a. $\dfrac{2}{3} = \dfrac{x}{16}$ **b.** $\dfrac{3}{4} = \dfrac{-27}{x}$

17. For each of the following pairs of fractions, replace the comma with the correct symbol ($<$, $=$, $>$) to make a true statement:

a. $\dfrac{7}{8}, \dfrac{5}{6}$ **b.** $2\dfrac{4}{5}, 2\dfrac{3}{6}$ **c.** $\dfrac{-7}{8}, \dfrac{-4}{5}$

18. Arrange each of the following in decreasing order:

a. $\dfrac{11}{22}, \dfrac{11}{16}, \dfrac{11}{13}$

b. $\dfrac{-1}{5}, \dfrac{-19}{36}, \dfrac{-17}{30}$

19. Show that the sequence $\dfrac{1}{3}, \dfrac{2}{4}, \dfrac{3}{5}, \dfrac{4}{6}, \dfrac{5}{7}, \dfrac{6}{8}, \cdots$ (in which each successive term is obtained from the previous term by adding 1 to the numerator and the denominator), is an increasing sequence; that is, show that each term in the sequence is greater than the preceding one.

20. For each of the following, find two rational numbers between the given fractions:

a. $\dfrac{3}{7}$ and $\dfrac{4}{7}$

b. $\dfrac{-7}{9}$ and $\dfrac{-8}{9}$

21. A scale on a map is 12 mi to the inch. What is the airline mileage between two cities that are 38 in. apart on the map?

22. a. Six oz is what part of a pound? A ton?
b. A dime is what fraction of a dollar?
c. 15 min is what fraction of an hour?
d. 8 hr is what fraction of a day?

Assessment 6-1B

1. Write a sentence that illustrates the use of $\frac{7}{10}$ in each of the following ways:
   a. As a division problem
   b. As part of a whole
   c. As a ratio

2. For each of the following, write a fraction to represent the shaded portion:

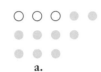

   a.

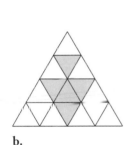

   b.   **c.**

3. Complete each of the following figures so that it shows $\frac{3}{5}$:

   a.   **b.**

   c.

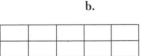

   d.

   e.   **f.**

4. Could the shaded portions in the following figures represent the indicated fractions? Tell why.

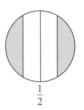 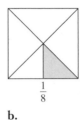 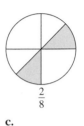

   $\frac{1}{2}$ $\frac{1}{8}$ $\frac{2}{8}$

   a.   **b.**   **c.**

5. If each of the following models represents the given fraction, draw a model that represents the *whole*. Shade your answer.

   a. 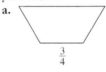   **b.**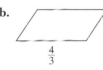

   $\frac{3}{4}$ $\frac{4}{3}$

   c. ◯◯   **d.**

   $\frac{1}{5}$ $\frac{1}{4}$

6. Based on your visual observation write a fraction to represent the shaded portion.

7. Referring to the figure, represent each of the following as a fraction:

   a. The dots outside the circular region as a part of all the dots
   b. The dots outside the rectangular region as a part of all the dots
   c. The dots in the union of the rectangular and the circular regions as a part of all the dots
   d. The dots inside the circular region but outside the rectangular region as part of all the dots

8. Find the simplest form for each of the following fractions:
   a. $\frac{0}{68}$ **b.** $\frac{84^2}{91^2}$ **c.** $\frac{662}{703}$

9. Mr. Gonzales and Ms. Price gave the same test to their fifth-grade classes. In Mr. Gonzales's class, 20 out of 25 students passed the test, and in Ms. Price's class, 24 out

of 30 students passed the test. One of Ms. Price's students heard about the results of the tests and claimed that the classes did equally well. Is the student right? Explain.

10. For each of the following, choose the expression in parentheses that equals or describes best the given fraction:

    a. $\dfrac{6 + x}{3x}$ $\left(\dfrac{2 + x}{x}, 3, \text{cannot be simplified}\right)$

    b. $\dfrac{2^6 + 2^5}{2^4 + 2^7}$ $\left(1, \dfrac{2}{3}, \text{cannot be simplified}\right)$

    c. $\dfrac{2^{100} + 2^{98}}{2^{100} - 2^{98}}$ $\left(2^{196}, \dfrac{5}{3}, \text{too large to simplify}\right)$

11. Find the simplest form for each of the following:

    a. $\dfrac{a^2 + ab}{a + b}$ b. $\dfrac{a}{3a + ab}$

12. Determine if the following pairs are equal:

    a. $\dfrac{3}{8}$ and $\dfrac{3,750}{10,000}$ b. $\dfrac{17}{27}$ and $\dfrac{25}{45}$

13. Determine if the following pairs are equal by changing both to the same denominator:

    a. $\dfrac{3}{-12}$ and $\dfrac{-36}{144}$ b. $\dfrac{-21}{430}$ and $\dfrac{-51}{215}$

14. A board is needed that is exactly $\dfrac{11}{32}$ in. wide to fill a hole.

    Can a board that is $\dfrac{3}{8}$ in. be shaved down to fit the hole? If so, how much must be shaved from the board?

15. The following two parking meters are next to each other with the times left as shown. Which meter has more time left on it? How much more?

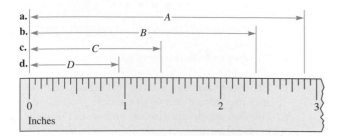

Meter A   Meter B

16. Read each measurement as shown on the following ruler:

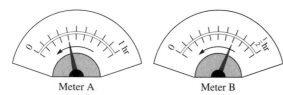

17. Determine if each of the following is always correct. If not find when it is true. Explain.

    a. $\dfrac{ab + c}{b} = a + c$ b. $\dfrac{a + b}{a + c} = \dfrac{b}{c}$

    c. $\dfrac{ab + ac}{ac} = \dfrac{b + c}{c}$ d. $\dfrac{a^2 + a}{a} = a + 1$

18. Solve for x in each of the following:

    a. $\dfrac{2}{3} = \dfrac{x}{18}$

    b. $\dfrac{3}{x} = \dfrac{3x}{x^2}$

19. a. If $\dfrac{a}{c} = \dfrac{b}{c}$, what must be true?

    b. If $\dfrac{a}{b} = \dfrac{a}{c}$, what must be true?

20. In Amy's algebra class, 6 of the 31 students received As on a test. The same test was given to Bren's class and 5 of the 23 students received As. Which class had the higher rate of As?

21. For each of the following pairs of fractions, replace the comma with the correct symbol $(<, =, >)$ to make a true statement:

    a. $\dfrac{1}{-7}, \dfrac{1}{-8}$ b. $\dfrac{2}{5}, \dfrac{4}{10}$ c. $\dfrac{0}{7}, \dfrac{0}{17}$

22. If $0 < a < b$ and $c > 0, d > 0$, compare the size of $\dfrac{c}{d}$ with $\dfrac{ac}{bd}$.

23. Show that the sequence $\dfrac{2}{1}, \dfrac{3}{2}, \dfrac{4}{3}, \dfrac{5}{4}, \dfrac{6}{5}, \dfrac{7}{6}$ is a decreasing sequence; that is, show that each term in the sequence is less than the preceding one.

24. For each of the following, find two rational numbers between the given fractions:

    a. $\dfrac{5}{6}$ and $\dfrac{83}{100}$ b. $\dfrac{-1}{3}$ and $\dfrac{3}{4}$

25. Consider the following number grid. The circled numbers form a rhombus (that is, all sides are the same length).

1	2	3	4	5	6	7	⑧	9	10
11	12	13	14	15	16	17	18	19	⑳
21	22	23	24	25	26	㉗	28	29	30
31	32	33	34	35	36	37	38	㊴	40
41	42	43	44	45	46	47	48	49	50

    a. If A is the sum of the four circled numbers and B is the sum of the four interior numbers, find $\dfrac{A}{B}$.

    b. Form a rhombus by circling the numbers 6, 18, 25, and 37. If A and B are defined as in (a), find $\dfrac{A}{B}$.

    c. How do the answers in (a) and (b) compare? Why does this happen?

26. A scale on a map is 120 mi to the inch. What is the airline mileage between two cities that are $\dfrac{3}{4}$ in. apart on the map?

27. a. 12 oz is what part of a pound?
    b. A nickel is what fraction of a dollar?
    c. 25 min is what fraction of an hour?
    d. 16 hr is what fraction of a 24 hr day?

Mathematical Connections 6-1

Communication

1. Jane has a recipe that calls for 4 c of flour. She wants to make $\frac{3}{4}$ of the recipe. Instead of determining directly how many cups are needed for the new recipe, she fills $\frac{3}{4}$ of a cup 4 times. Explain why Jane's method works.

2. Ann claims that she can't show $\frac{3}{4}$ of the following faces because some are big and some are small. What do you tell her?

3. A student claims that $\frac{2}{3} = \frac{6}{7}$ because if you add 4 to both the top and the bottom of a fraction, the fraction does not change. How do you respond?

4. In each of two different fourth-grade classes, $\frac{1}{3}$ of the members are girls. Does each class have the same number of girls? Explain.

5. Consider the set of all fractions equal to $\frac{1}{2}$. If you take any 10 of those fractions, add their numerators to obtain the numerator of a new fraction and add their denominators to obtain the denominator of a new fraction, how does the new fraction relate to $\frac{1}{2}$? Generalize what you found and explain.

6. Should fractions always be reduced to their simplest form? Why or why not?

7. How would you respond to each of the following students?
   a. Iris claims that if we have two positive rational numbers, the one with the greater numerator is the greater.
   b. Shirley claims that if we have two positive rational numbers, the one with the greater denominator is the lesser.

8. If we take the set of fractions equivalent to $\frac{1}{3}$ and graph them as points on a coordinate system so that the numerator becomes the x-coordinate and the denominator becomes the y-coordinate for each point, explain what type of graph we would get.

9. Write an explanation of how to convert inches to yards and vice versa.

Open-Ended

10. List three types of measure that require rational numbers as the appropriate number of units in the measurements.

11. Some people have argued that the system of integers is more understandable than the system of positive rational numbers. If you could decide which should be taught first in school, which would you choose and why?

12. Make three statements about yourself or your environment and use fractions in each. Explain why your statements are true (for example, your parents have three children, two of whom live at home; hence $\frac{2}{3}$ of their children live at home).

Cooperative Learning

13. Assume the tallest person in your group is 1 unit tall and do the following:
    a. Find rational numbers to approximately represent the heights of other members of the group.
    b. Make a number line and plot the rational number for each person ordered according to height.

Questions from the Classroom

14. A student asks if $\frac{0}{6}$ is in its simplest form. How do you respond?

15. A student writes $\frac{15}{53} < \frac{1}{3}$ because $3 \cdot 15 < 53 \cdot 1$. Another student writes $\frac{15}{53} = \frac{1}{3}$ Where is the fallacy?

16. The numerators and the denominators of the following fractions form arithmetic sequences. A student claims that the fractions form an arithmetic sequence. How would you respond?

$$\frac{1}{2}, \frac{2}{3}, \frac{3}{4}, \frac{4}{5}, \frac{5}{6}, \frac{6}{7}, \frac{7}{8}, \cdots$$

17. A student claims that there are no numbers between $\frac{999}{1000}$ and 1 because they are so close together. What is your response?

18. A student simplified the fraction $\frac{m+n}{p+n}$ to $\frac{m}{p}$. How would you help this student?

19. Without thinking, a student argued that a pizza cut into 12 pieces was more than a pizza cut into 6 pieces. How would you respond?

20. A student asks if adding the same very large number to both the numerator and the denominator of a fraction yields a quotient of 1. How do you respond?

21. Joe reported that if $\frac{a}{b} = \frac{c}{d} \neq 1$, then $\frac{d}{c-d} = \frac{b}{a-b}$. Joan said she didn't believe it. How do you respond?

22. Steve claims that the shaded portions can't represent $\frac{2}{3}$ since there are 5 circles shaded and $\frac{2}{3}$ is less than 1. How do you respond?

23. Daryl says that each piece of the pie shown represents $\frac{1}{3}$ of the pie. How do you respond?

24. **a.** Cassidy noticed that $5 > 4$, but $\frac{1}{5} < \frac{1}{4}$. In general she thinks that if a and b are positive integers, and $a > b$, then $\frac{1}{a} < \frac{1}{b}$. She would like to know if this is always true. How do you respond?

b. Cassidy would like to know if her discovery in part (a) is true when a and b are negative integers, and if so, why or why not. How do you respond?

25. Sven would like to know if the following statement is true, and why or why not. How do you respond?

If $\frac{-1}{2} < x < \frac{1}{2}$ then $\frac{-3}{2} < x < \frac{3}{2}$.

26. Carl says that $\frac{3}{8} > \frac{2}{3}$ because $3 > 2$ and $8 > 3$. How would you help Carl?

Third International Mathematics and Science Study (TIMSS) Questions

Which shows $\frac{2}{3}$ of the square shaded?

TIMSS 2003, Grade 4

In the figure, how many MORE small squares need to be shaded so that $\frac{4}{5}$ of the small squares are shaded?

a. 5   **b.** 4   **c.** 3   **d.** 2   **e.** 1

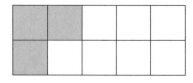

TIMSS 2003, Grade 8

National Assessment of Educational Progress (NAEP) Question

What fraction of the figure is shaded?

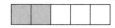

NAEP, Grade 4, 2007

In which of the following are the three fractions arranged from least to greatest?

a. $\frac{2}{7}, \frac{1}{2}, \frac{5}{9}$ **b.** $\frac{1}{2}, \frac{2}{7}, \frac{5}{9}$ **c.** $\frac{1}{2}, \frac{5}{9}, \frac{2}{7}$

d. $\frac{5}{9}, \frac{1}{2}, \frac{2}{7}$ **e.** $\frac{5}{9}, \frac{2}{7}, \frac{1}{2}$

NAEP, Grade 8, 2007

 BRAIN TEASER In an old Sam Loyd puzzle, a watch is described as having stopped when the minute and hour hands formed a straight line and the second hand was not on 12. At what times can this happen?

6-2   Addition, Subtraction, and Estimation with Rational Numbers

Addition and subtraction of rational numbers is very much like addition and subtraction of whole numbers and integers. We first demonstrate the addition of two rational numbers with like denominators, $\frac{2}{5} + \frac{1}{5}$, using an area model in Figure 6-7(a) and a number-line model in Figure 6-7(b).

(a)

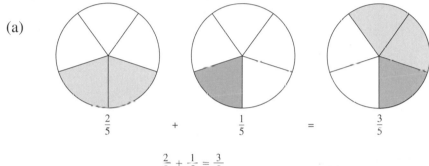

(b)

Figure 6-7

◆ *Research Note*

Students learning computational algorithms involving fractions have difficulty connecting their models involving manipulatives with symbolic procedures. This personal competence with a rote procedure may outstrip their conceptual understanding of fractions; one result can be that students unable to judge the reasonableness of the answer can check their work only by repeating the rote procedure. (Wearne and Hiebert 1988). ◆

Why does the area model in Figure 6-7(a) make sense? Suppose that someone gives you $\frac{2}{5}$ of a pie to start with and then gives you another $\frac{1}{5}$ of the pie. In Figure 6-7(a), $\frac{2}{5}$ is represented by 2 pieces when the pie is cut into 5 equal-size pieces and $\frac{1}{5}$ is represented by 1 piece of the 5 equal-size pieces. So all together you have $2 + 1 = 3$ pieces of the 5 equal-size pieces, or $\frac{3}{5}$ of the total (whole) pie. The number-line model in Figure 6-7(b) works the same as the number-line model for whole numbers.

The ideas illustrated in Figure 6-7 can be applied to the sum of two rational numbers with like denominators and are summarized in the following definition.

Definition of Addition of Rational Numbers with Like Denominators

If $\frac{a}{b}$ and $\frac{c}{b}$ are rational numbers, then $\frac{a}{b} + \frac{c}{b} = \frac{a + c}{b}$.

Next we consider the addition of two rational numbers with unlike denominators, using Polya's four-step, problem-solving process.

Problem Solving   Adding Rational Numbers Problem

Determine how to add the rational numbers $\frac{2}{3}$ and $\frac{1}{4}$.

Understanding the Problem   We can model $\frac{2}{3}$ and $\frac{1}{4}$ as parts of a whole, as seen in Figure 6-8, but we need a way to combine the two drawings to find the sum.

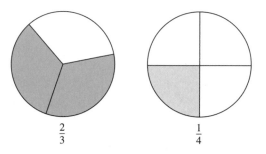

$$\frac{2}{3} \qquad \frac{1}{4}$$

Figure 6-8

Devising a Plan   We use the strategy of *solving a related problem:* adding rational numbers with the same denominators. We can find the sum by writing each fraction with a common denominator and then completing the computation.

Carrying Out the Plan   From earlier work in this chapter, we know that $\frac{2}{3}$ has infinitely many representations, including $\frac{4}{6}, \frac{6}{9}, \frac{8}{12}$, and so on. Also $\frac{1}{4}$ has infinitely many representations, including $\frac{2}{8}, \frac{3}{12}, \frac{4}{16}$, and so on. By comparing the two sets of rational numbers, we see that $\frac{8}{12}$ and $\frac{3}{12}$ have the same denominator. One is 8 parts of 12 equal parts, while the other is 3 parts of 12 equal parts. Consequently, the sum is $\frac{2}{3} + \frac{1}{4} = \frac{8}{12} + \frac{3}{12} = \frac{11}{12}$. Figure 6-9 illustrates the addition.

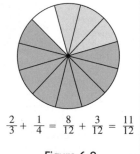

$$\frac{2}{3} + \frac{1}{4} = \frac{8}{12} + \frac{3}{12} = \frac{11}{12}$$

Figure 6-9

Looking Back   To add two rational numbers with unlike denominators, we considered equal rational numbers with like denominators. The common denominator for $\frac{2}{3}$ and $\frac{1}{4}$ is 12. This is also the least common denominator, or the LCM of 3 and 4. To add two fractions with unequal denominators such as $\frac{5}{12}$ and $\frac{7}{18}$, we could find equal fractions with the denominator LCM(12, 18), or 36. However, any common denominator will work as well, for example, 72 or even $12 \cdot 18$.

By considering the sum $\frac{2}{3} + \frac{1}{4} = \frac{2 \cdot 4}{3 \cdot 4} + \frac{1 \cdot 3}{4 \cdot 3} = \frac{8}{12} + \frac{3}{12} = \frac{11}{12}$, we can generalize to finding the sum of two rational numbers with unlike denominators, as in the following theorem.

Theorem 6–7

If $\dfrac{a}{b}$ and $\dfrac{c}{d}$ are any two rational numbers, then $\dfrac{a}{b} + \dfrac{c}{d} = \dfrac{ad + bc}{bd}$.

♦ *Research Note*

The most frequent error students make when adding fractions is adding the numerators and adding the denominators (Bana, Farrell, and McIntosh 1997).   ♦

As pointed out in the Research Note, students don't always add fractions correctly. In grade 5 *Curriculum Focal Points*, students are expected to apply their understanding of fraction models to represent the addition and subtraction of fractions with unlike denominators (p. 17).

REMARK From the definition of addition of rational numbers with like denominators we have $\dfrac{a}{b} + \dfrac{c}{b} = \dfrac{a + c}{b}$. The same result would be obtained if we used Theorem 6–7:

$$\frac{a}{b} + \frac{c}{b} = \frac{ab + cb}{b \cdot b} = \frac{(a + c)b}{b \cdot b} = \frac{a + c}{b}.$$

Example 6-4

Find each of the following sums:

a. $\dfrac{2}{15} + \dfrac{4}{21}$ **b.** $\dfrac{2}{^-3} + \dfrac{1}{5}$ **c.** $\left(\dfrac{3}{4} + \dfrac{1}{5}\right) + \dfrac{1}{6}$

d. $\dfrac{3}{x} + \dfrac{4}{y}$ **e.** $\dfrac{2}{a^2 b} + \dfrac{3}{ab^2}$

Solution   **a.** Because $\text{LCM}(15, 21) = 3 \cdot 5 \cdot 7$, then $\dfrac{2}{15} + \dfrac{4}{21} = \dfrac{2 \cdot 7}{15 \cdot 7} + \dfrac{4 \cdot 5}{21 \cdot 5} = \dfrac{14}{105} + \dfrac{20}{105} = \dfrac{34}{105}.$

b. $\dfrac{2}{^-3} + \dfrac{1}{5} = \dfrac{(2)(5) + (^-3)(1)}{(^-3)(5)} = \dfrac{10 + ^-3}{^-15} = \dfrac{7}{^-15} = \dfrac{7(^-1)}{^-15(^-1)} = \dfrac{^-7}{15}.$

c. $\dfrac{3}{4} + \dfrac{1}{5} = \dfrac{3 \cdot 5 + 4 \cdot 1}{4 \cdot 5} = \dfrac{19}{20}$; hence, $\left(\dfrac{3}{4} + \dfrac{1}{5}\right) + \dfrac{1}{6} = \dfrac{19}{20} + \dfrac{1}{6} = \dfrac{19 \cdot 6 + 20 \cdot 1}{20 \cdot 6} = \dfrac{134}{120}$, or $\dfrac{67}{60}.$

d. $\dfrac{3}{x} + \dfrac{4}{y} = \dfrac{3y}{xy} + \dfrac{4x}{xy} = \dfrac{3y + 4x}{xy}.$

e. $\text{LCM}(a^2 b, ab^2) = a^2 b^2$; $\dfrac{2}{a^2 b} + \dfrac{3}{ab^2} = \dfrac{2b}{a^2 b \cdot b} + \dfrac{3a}{a \cdot ab^2} = \dfrac{2b + 3a}{a^2 b^2}.$

Mixed Numbers

In everyday life, we often use **mixed numbers**; that is, numbers that are made up of an integer and a proper fraction. For example, Figure 6-10 shows that the nail is $2\dfrac{3}{4}$ in. long. The mixed number $2\dfrac{3}{4}$ means $2 + \dfrac{3}{4}$. It is sometimes inferred that $2\dfrac{3}{4}$ means

2 times $\dfrac{3}{4}$, since xy means $x \cdot y$, but this is not correct. Also, the number $^-4\dfrac{3}{4}$ means $-\left(4\dfrac{3}{4}\right)$, or $^-4 - \dfrac{3}{4}$, not $^-4 + \dfrac{3}{4}$.

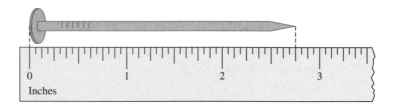

Figure 6-10

REMARK In a National Assessment of Educational Progress (NAEP) test, students were given the following problem:

$$5\dfrac{1}{4} \text{ is the same as:}$$

(a) $5 + \dfrac{1}{4}$ (b) $5 - \dfrac{1}{4}$ (c) $5 \times \dfrac{1}{4}$ (d) $5 \div \dfrac{1}{4}$

Only 47% of the seventh graders chose the correct response, (a), and only 44% of the eleventh graders chose the correct response.

A mixed number is a rational number, and therefore it can always be written in the form $\dfrac{a}{b}$. For example,

$$2\dfrac{3}{4} = 2 + \dfrac{3}{4} = \dfrac{2}{1} + \dfrac{3}{4} = \dfrac{2 \cdot 4 + 1 \cdot 3}{1 \cdot 4} = \dfrac{8 + 3}{4} = \dfrac{11}{4}$$

Example 6-5

Change each of the following mixed numbers to the form $\dfrac{a}{b}$, where a and b are integers:

a. $4\dfrac{1}{3}$ **b.** $^-3\dfrac{2}{5}$

Solution **a.** $4\dfrac{1}{3} = 4 + \dfrac{1}{3} = \dfrac{4}{1} + \dfrac{1}{3} = \dfrac{4 \cdot 3 + 1 \cdot 1}{1 \cdot 3} = \dfrac{12 + 1}{3} = \dfrac{13}{3}$

b. $^-3\dfrac{2}{5} = -\left(3 + \dfrac{2}{5}\right) = -\left(\dfrac{3}{1} + \dfrac{2}{5}\right) = -\left(\dfrac{3 \cdot 5 + 1 \cdot 2}{1 \cdot 5}\right) = -\left(\dfrac{17}{5}\right)$

NOW TRY THIS 6-11 Use the ideas in Example 6-5 to write $2\dfrac{3}{4} + 5\dfrac{3}{8}$ as a mixed number.

Example 6-6

Change $\dfrac{29}{5}$ to a mixed number.

Solution We divide 29 by 5 and use the *division algorithm* as follows:

$$\frac{29}{5} = \frac{5 \cdot 5 + 4}{5} = \frac{5 \cdot 5}{5} + \frac{4}{5} = 5 + \frac{4}{5} = 5\frac{4}{5}$$

REMARK In elementary schools, problems like Example 6-6 are usually computed using division, as follows:

$$
\begin{array}{r}
5 \\
5\overline{)29} \\
25 \\
\hline
4
\end{array}
$$

Hence, $\dfrac{29}{5} = 5 + \dfrac{4}{5} = 5\dfrac{4}{5}$.

 We can use scientific/fraction calculators to change improper fractions to mixed numbers. For example, if we enter ②⑨ / ⑤ and press $\boxed{Ab/c}$, then 5⊔ 4/5 appears, which means $5\dfrac{4}{5}$.

We can also use scientific/fraction calculators to add mixed numbers. For example, to add $2\dfrac{4}{5} + 3\dfrac{5}{6}$, we enter ② $\boxed{\text{Unit}}$ ④ / ⑤ ➕ ③ $\boxed{\text{Unit}}$ ⑤ / ⑥ ⊜, and the display reads 5   49/30. We then press $\boxed{Ab/c}$ to obtain 6 ⊔ 19/30, which means $6\dfrac{19}{30}$.

 Because mixed numbers are rational numbers, the methods of adding rationals can be extended to include mixed numbers. The student page (page 368) shows a method for computing sums of mixed numbers. Work the three problems on the bottom of the student page. Notice that all the fractions on the student page are nonnegative. This is typical in grades 3–6.

Properties of Addition for Rational Numbers

Rational numbers have the following properties for addition: *closure*, *commutative*, *associative*, *additive identity*, and *additive inverse*. To emphasize the additive inverse property of rational numbers, we state it explicitly, as follows.

Theorem 6–8:   Additive Inverse Property of Rational Numbers

For any rational number $\dfrac{a}{b}$, there exists a unique rational number $-\dfrac{a}{b}$, the additive inverse of $\dfrac{a}{b}$, such that

$$\frac{a}{b} + \left(-\frac{a}{b}\right) = 0 = \left(-\frac{a}{b}\right) + \frac{a}{b}$$

School Book Page ADDING MIXED NUMBERS

Source: Mathematics, Diamond Edition, Grade Five, Scott Foresman-Addison Wesley, 2008 (p. 476).

Lesson 8-7

Key Idea
You can add mixed numbers the same way you add whole numbers and fractions.

Think It Through
- I can break the mixed numbers apart to **solve two simpler problems.**
- I can **use estimation to check** if the exact answer is reasonable.

Adding Mixed Numbers

LEARN

How is adding mixed numbers like adding fractions and whole numbers?

Trail mix provides a high-energy food for activities like camping and hiking. Luke used the recipe at the right to make trail mix.

Trail Mix Recipe

$1\frac{3}{4}$ lb dried bananas	$1\frac{1}{2}$ lb dried pineapple
$1\frac{3}{8}$ lb raisins	$1\frac{5}{8}$ lb peanuts
$1\frac{1}{2}$ lb mixed nuts	$\frac{3}{4}$ lb sunflower seeds

Example

How many pounds of dried bananas and raisins did Luke use?

Find $1\frac{3}{4} + 1\frac{3}{8}$.

STEP 1

Write equivalent fractions with the LCD.

$$1\frac{3}{4} = 1\frac{6}{8}$$
$$+ 1\frac{3}{8} = + 1\frac{3}{8}$$

STEP 2

Add the fractions.

$$1\frac{3}{4} = 1\frac{6}{8}$$
$$+ 1\frac{3}{8} = + 1\frac{3}{8}$$
$$\frac{9}{8}$$

STEP 3

Add the whole numbers. Simplify the sum, if necessary.

$$1\frac{3}{4} = 1\frac{6}{8}$$
$$+ 1\frac{3}{8} = + 1\frac{3}{8}$$
$$2\frac{9}{8} = 3\frac{1}{8}$$

So, Luke used $3\frac{1}{8}$ pounds of bananas and raisins.

✓ Talk About It

1. Estimate the sum found in the Example. Is the estimate close to the actual answer? Why is it helpful to compare an exact answer to an estimate?

2. How many pounds of peanuts and mixed nuts did Luke use? Show this sum on a number line.

3. **Reasoning** How is adding mixed numbers like adding fractions and whole numbers?

476

Another form of $-\dfrac{a}{b}$ can be found by considering the sum $\dfrac{a}{b} + \dfrac{-a}{b}$. Because

$$\frac{a}{b} + \frac{-a}{b} = \frac{a + -a}{b} = \frac{0}{b} = 0$$

it follows that $-\dfrac{a}{b}$ and $\dfrac{-a}{b}$ are both additive inverses of $\dfrac{a}{b}$, so $-\dfrac{a}{b} = \dfrac{-a}{b}$.

Example 6-7

Find the additive inverses for each of the following:

a. $\dfrac{3}{5}$ b. $\dfrac{-5}{11}$ c. $4\dfrac{1}{2}$

Solution   a. $-\dfrac{3}{5}$ or $\dfrac{-3}{5}$ b. $-\left(\dfrac{-5}{11}\right) = \dfrac{-(-5)}{11} = \dfrac{5}{11}$ c. $-4\dfrac{1}{2}$, or $\dfrac{-9}{2}$

Properties of the additive inverse for rational numbers are analogous to those of the additive inverse for integers, as shown in Table 6-2. As with the set of integers, the set of rational numbers also has the addition property of equality, which says that you can add the same number to both sides of an equation.

Table 6-2

Integers	Rational Numbers
1. $-(-a) = a$	1. $-\left(\dfrac{a}{b}\right) = \dfrac{a}{b}$
2. $-(a + b) = -a + -b$	2. $-\left(\dfrac{a}{b} + \dfrac{c}{d}\right) = \dfrac{-a}{b} + \dfrac{-c}{d}$

Theorem 6–9:   Addition Property of Equality

If $\dfrac{a}{b}$ and $\dfrac{c}{d}$ are any rational numbers such that $\dfrac{a}{b} = \dfrac{c}{d}$, and if $\dfrac{e}{f}$ is any rational number, then

$$\frac{a}{b} + \frac{e}{f} = \frac{c}{d} + \frac{e}{f}.$$

REMARK Theorem 6–9 can be stated as follows: If $x = y$ then $x + z = y + z$ where x, y, z are any rational numbers.

Subtraction of Rational Numbers

In elementary school, subtraction of rational numbers is usually introduced by using a take-away model. If we have $\dfrac{6}{7}$ of a pizza and $\dfrac{2}{7}$ of the original pizza is taken away, $\dfrac{4}{7}$ of the pizza remains; that is, $\dfrac{6}{7} - \dfrac{2}{7} = \dfrac{6-2}{7} = \dfrac{4}{7}$. In general, subtraction of rational numbers with like denominators is determined as follows:

$$\frac{a}{b} - \frac{c}{b} = \frac{a-c}{b}$$

As with integers, a number line can be used to model subtraction. If a line is marked off in units of length $\dfrac{1}{b}$, then $\dfrac{a}{b} - \dfrac{c}{b}$ is equal to $(a-c)$ units of length $\dfrac{1}{b}$, which implies that $\dfrac{a}{b} - \dfrac{c}{b} = \dfrac{a-c}{b}$. When the denominators are not the same we can perform the subtraction by finding a common denominator. For example,

$$\frac{3}{4} - \frac{2}{3} = \frac{3 \cdot 3}{4 \cdot 3} - \frac{2 \cdot 4}{3 \cdot 4} = \frac{9}{12} - \frac{8}{12} = \frac{9-8}{12} = \frac{1}{12}$$

Subtraction of rational numbers, like subtraction of integers, can be defined in terms of addition as follows.

Definition of Subtraction of Rational Numbers in Terms of Addition

If $\dfrac{a}{b}$ and $\dfrac{c}{d}$ are any rational numbers, then $\dfrac{a}{b} - \dfrac{c}{d}$ is the unique rational number $\dfrac{e}{f}$ such that $\dfrac{a}{b} = \dfrac{c}{d} + \dfrac{e}{f}$.

As with integers, we can see that subtraction of rational numbers can be performed by adding the additive inverses. The following theorem states this.

Theorem 6–10

If $\dfrac{a}{b}$ and $\dfrac{c}{d}$ are any rational numbers, then $\dfrac{a}{b} - \dfrac{c}{d} = \dfrac{a}{b} + \dfrac{{}^{-}c}{d}$.

Now, using Theorem 6–10, we obtain the following:

$$\frac{a}{b} - \frac{c}{d} = \frac{a}{b} + \frac{{}^{-}c}{d} \qquad \text{(Theorem 6–10)}$$

$$= \frac{ad + b({}^{-}c)}{bd} \qquad \text{(Theorem 6–7)}$$

$$= \frac{ad + {}^{-}(bc)}{bd} \qquad \text{(Multiplication with integers)}$$

$$= \frac{ad - bc}{bd}$$

We proved the following theorem:

> **Theorem 6–11**
>
> If $\dfrac{a}{b}$ and $\dfrac{c}{d}$ are any rational numbers, then $\dfrac{a}{b} - \dfrac{c}{d} = \dfrac{ad - bc}{bd}$.

Example 6-8

Find each difference in the following:

a. $\dfrac{5}{8} - \dfrac{1}{4}$ **b.** $5\dfrac{1}{3} - 2\dfrac{3}{4}$

Solution   **a.** One approach is to find the LCM for the fractions. Because LCM$(8, 4) = 8$, we have

$$\frac{5}{8} - \frac{1}{4} = \frac{5}{8} - \frac{2}{8} = \frac{3}{8}$$

An alternative approach is as follows:

$$\frac{5}{8} - \frac{1}{4} = \frac{5 \cdot 4 - 8 \cdot 1}{8 \cdot 4} = \frac{20 - 8}{32} = \frac{12}{32}, \text{ or } \frac{3}{8}$$

b. Two methods of solution are given:

$$
\begin{array}{llll}
5\dfrac{1}{3} = & 5\dfrac{4}{12} = & 4 + 1\dfrac{4}{12} = & 4\dfrac{16}{12} \\[2mm]
-2\dfrac{3}{4} = & -2\dfrac{9}{12} = & -2\dfrac{9}{12} & = -2\dfrac{9}{12} \\[2mm]
& & & 2\dfrac{7}{12}
\end{array}
$$

$$
\begin{aligned}
5\dfrac{1}{3} - 2\dfrac{3}{4} &= \dfrac{16}{3} - \dfrac{11}{4} \\[2mm]
&= \dfrac{16 \cdot 4 - 3 \cdot 11}{3 \cdot 4} \\[2mm]
&= \dfrac{64 - 33}{12} \\[2mm]
&= \dfrac{31}{12}, \text{ or } 2\dfrac{7}{12}
\end{aligned}
$$

The following examples show the use of fractions in algebra.

Example 6-9

Add or subtract each of the following. Write your answer in simplest form.

a. $\dfrac{x}{2} + \dfrac{x}{3}$ **b.** $\dfrac{2 - x}{6 - 3x} + \dfrac{4 - 2x}{3x - 6}$

c. $\dfrac{2}{a + b} - \dfrac{2}{a - b}$ **d.** $\dfrac{1}{x} - \dfrac{1}{2x^2}$

Solution   **a.** $\dfrac{x}{2} + \dfrac{x}{3} = \dfrac{3x}{3 \cdot 2} + \dfrac{2x}{2 \cdot 3}$

$$= \frac{3x + 2x}{6} = \frac{5x}{6}$$

b. We first write each fraction in simplest form:

$$\frac{2-x}{6-3x} = \frac{2-x}{3(2-x)} = \frac{1 \cdot (2-x)}{3 \cdot (2-x)} = \frac{1}{3} \quad \text{if } x \neq 2$$

$$\frac{4-2x}{3x-6} = \frac{^-2(x-2)}{3(x-2)} = \frac{^-2}{3} \quad \text{if } x \neq 2$$

Thus, if $x \neq 2$, the sum is $\dfrac{1}{3} + \dfrac{^-2}{3} = \dfrac{^-1}{3}$.

c. Using Theorem 6–11:

$$\frac{2}{a+b} - \frac{2}{a-b} = \frac{2(a-b) - 2(a+b)}{(a+b)(a-b)}$$

$$= \frac{2a - 2b - 2a - 2b}{(a+b)(a-b)}$$

$$= \frac{^-4b}{(a+b)(a-b)} \quad \text{or} \quad \frac{^-4b}{a^2 - b^2}$$

d. $\dfrac{1}{x} - \dfrac{1}{2x^2} = \dfrac{2x \cdot 1}{2x \cdot x} - \dfrac{1}{2x^2}$

$$= \frac{2x}{2x^2} - \frac{1}{2x^2}$$

$$= \frac{2x - 1}{2x^2}$$

Definition of Greater Than and Less Than in Terms of Subtraction

For integers a and b, $a < b$ (or, equivalently, $b > a$) if, and only if, there exists a positive integer k such that $a + k = b$, from which it followed that $b - a = k$ and hence $b - a > 0$. For rational numbers a definition of greater than and less than follows:

Definition of Less Than and Greater Than for Rational Numbers

$$\frac{a}{b} < \frac{c}{d} \quad \text{if} \quad \frac{c}{d} - \frac{a}{b} > 0,$$

$$\frac{c}{d} > \frac{a}{b} \text{ if, and only if, } \frac{a}{b} < \frac{c}{d}.$$

This definition is useful in justifying certain inequalities. Example 6-10 uses the definition of *greater than* and provides practice with algebra.

Example 6-10

(a) Verify each of the following:

(i) $\dfrac{2}{3} + \dfrac{3}{2} > 2$ **(ii)** $\dfrac{3}{4} + \dfrac{4}{3} > 2$ **(iii)** $\dfrac{5}{7} + \dfrac{7}{5} > 2$

(b) Based on these examples, conjecture a general statement and justify it.

Solution (a) (i) $\frac{2}{3} + \frac{3}{2} - 2 = \frac{4 + 9 - 12}{6} = \frac{1}{6} > 0$, hence $\frac{2}{3} + \frac{3}{2} > 2$

(ii) $\frac{3}{4} + \frac{4}{3} - 2 = \frac{9 + 16 - 24}{12} = \frac{1}{12} > 0$, hence $\frac{3}{4} + \frac{4}{3} > 2$

(iii) $\frac{5}{7} + \frac{7}{5} - 2 = \frac{25 + 49 - 70}{35} = \frac{4}{35} > 0$, hence $\frac{5}{7} + \frac{7}{5} > 2$

(b) *Conjecture:* If a and b are positive integers and $a \neq b$, then $\frac{a}{b} + \frac{b}{a} > 2$.

Justification: $\frac{a}{b} + \frac{b}{a} - 2 = \dfrac{a^2 + b^2 - 2ab}{ab}$

$= \dfrac{(a - b)^2}{ab} > 0$, since $a \neq b$, and $ab > 0$.

Estimation with Rational Numbers

Estimation helps us make practical decisions in our everyday lives. For example, suppose we need to double a recipe that calls for $\frac{3}{8}$ of a cup of flour. Will we need more or less than a cup of flour? Many of the estimation and mental math techniques that we learned to use with whole numbers also work with rational numbers.

The grade 5 *Focal Points* calls for students to make "reasonable estimates of fraction sums and differences." The student page (page 374) from a grade 5 textbook has students estimate fractions of boards for building shelves. Estimation plays an important role in judging the reasonableness of computations. Students do not necessarily have this skill. For example, when asked to estimate $\frac{12}{13} + \frac{7}{8}$, only 24% of 13-year-old students on a national assessment said the answer was close to 2. Most said it was close to 1, 19, or 21. These incorrect estimates suggest common computational errors in adding fractions and a lack of understanding of the operation being carried out. These incorrect estimates also suggest a lack of number sense.

NOW TRY THIS 6-12 A student added $\frac{3}{4} + \frac{1}{2}$ and obtained $\frac{4}{6}$. How would you use estimation to show this student that his answer could not be correct?

Sometimes to obtain an estimate it is desirable to round fractions to a *convenient* fraction, such as $\frac{1}{2}, \frac{1}{3}, \frac{1}{4}, \frac{1}{5}, \frac{2}{3}, \frac{3}{4}$, or 1. For example, if a student had 59 correct answers out of 80 questions, the student answered $\frac{59}{80}$ of the questions correctly, which is approximately $\frac{60}{80}$, or $\frac{3}{4}$. We know $\frac{60}{80}$ is greater than $\frac{59}{80}$. On a number line, the greater fraction is to the right

School Book Page ESTIMATING FRACTIONAL AMOUNTS

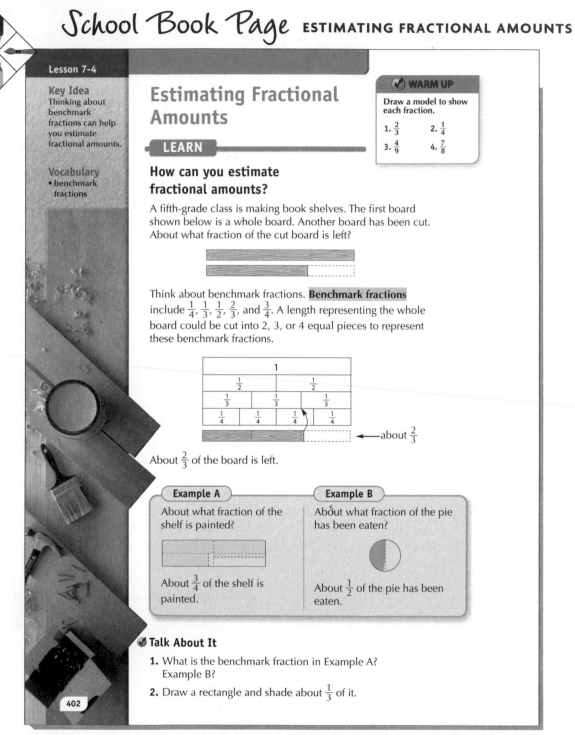

Lesson 7-4

Key Idea
Thinking about benchmark fractions can help you estimate fractional amounts.

Vocabulary
• benchmark fractions

Estimating Fractional Amounts

LEARN

How can you estimate fractional amounts?

A fifth-grade class is making book shelves. The first board shown below is a whole board. Another board has been cut. About what fraction of the cut board is left?

Think about benchmark fractions. **Benchmark fractions** include $\frac{1}{4}, \frac{1}{3}, \frac{1}{2}, \frac{2}{3},$ and $\frac{3}{4}$. A length representing the whole board could be cut into 2, 3, or 4 equal pieces to represent these benchmark fractions.

← about $\frac{2}{3}$

About $\frac{2}{3}$ of the board is left.

WARM UP

Draw a model to show each fraction.

1. $\frac{2}{3}$ 2. $\frac{1}{4}$

3. $\frac{4}{9}$ 4. $\frac{7}{8}$

Example A

About what fraction of the shelf is painted?

About $\frac{3}{4}$ of the shelf is painted.

Example B

About what fraction of the pie has been eaten?

About $\frac{1}{2}$ of the pie has been eaten.

✔ Talk About It

1. What is the benchmark fraction in Example A? Example B?

2. Draw a rectangle and shade about $\frac{1}{3}$ of it.

402

Source: Mathematics, Diamond Edition, Grade Five, Scott Foresman-Addison Wesley 2008 (p. 402).

of the lesser. The estimate $\dfrac{3}{4}$ for $\dfrac{59}{80}$ is a high estimate. In a similar way, we can estimate

$\dfrac{31}{90}$ by $\dfrac{30}{90}$, or $\dfrac{1}{3}$. In this case, the estimate of $\dfrac{1}{3}$ is a low estimate.

Example 6-11

A sixth-grade class is collecting cans to take to the recycling center. Becky's group brought the following amounts (in pounds). About how many pounds does her group have all together?

$$1\dfrac{1}{8},\ 3\dfrac{4}{10},\ 5\dfrac{7}{8},\ \dfrac{6}{10}$$

Solution We can estimate the amount by using front-end estimation and then adjusting by using 0, $\dfrac{1}{2}$, and 1 as reference points. The front-end estimate is $(1 + 3 + 5)$, or 9. The adjustment is $\left(0 + \dfrac{1}{2} + 1 + \dfrac{1}{2}\right)$, or 2. An adjusted estimate would be $9 + 2$ or 11 lb.

Example 6-12

Estimate each of the following:

a. $\dfrac{27}{13} + \dfrac{10}{9}$ **b.** $3\dfrac{9}{10} + 2\dfrac{7}{8} + \dfrac{11}{12}$

Solution **a.** Because $\dfrac{27}{13}$ is slightly more than 2 and $\dfrac{10}{9}$ is slightly more than 1, an estimate is a number close to but more than 3.

b. We first add the whole-number parts to obtain $3 + 2$, or 5. Because each of the fractions, $\dfrac{9}{10}, \dfrac{7}{8}$, and $\dfrac{11}{12}$, is close to but less than 1, their sum is close to but less than 3. The approximate answer is $5 + 3$ or 8.

Assessment 6-2A

1. Perform the following additions or subtractions:

a. $\dfrac{1}{2} + \dfrac{2}{3}$

b. $\dfrac{4}{12} - \dfrac{2}{3}$

c. $\dfrac{5}{x} + \dfrac{^{-}3}{y}$

d. $\dfrac{^{-}3}{2x^2 y} + \dfrac{5}{2xy^2} + \dfrac{7}{x^2}$

e. $\dfrac{5}{6} + 2\dfrac{1}{8}$

f. $^{-}4\dfrac{1}{2} - 3\dfrac{1}{6}$

2. Change each of the following fractions to mixed numbers:

a. $\dfrac{56}{3}$

b. $-\dfrac{293}{100}$

3. Change each of the following mixed numbers to fractions in the form $\frac{a}{b}$, where a and b are integers and $b \neq 0$:

   a. $6\frac{3}{4}$ **b.** $^{-}3\frac{5}{8}$

4. Approximate each of the following situations with a convenient fraction. Explain your reasoning. Tell whether your estimate is high or low.

   a. Giorgio had 15 base hits out of 46 times at bat.

   b. Ruth made 7 goals out of 41 shots.

   c. Laura answered 62 problems correctly out of 80.

   d. Jonathan made 9 baskets out of 19.

5. Use the information in the table to answer each of the following questions:

Team	Games Played	Games Won
Ducks	22	10
Beavers	19	10
Tigers	28	9
Bears	23	8
Lions	27	7
Wildcats	25	6
Badgers	21	5

   a. Which team won more than $\frac{1}{2}$ of its games, but was closest to winning $\frac{1}{2}$ of its games?

   b. Which teams won just under $\frac{1}{2}$ of their games?

   c. Which teams won just over $\frac{1}{3}$ of their games?

6. Sort the following fraction cards into the ovals by estimating in which oval the fraction belongs:

   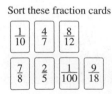 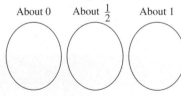

   Sort these fraction cards   About 0   About $\frac{1}{2}$ About 1

7. Approximate each of the following fractions by $0, \frac{1}{4}, \frac{1}{2}, \frac{3}{4}$, or 1. Tell whether your estimate is high or low.

   a. $\frac{19}{39}$ **b.** $\frac{3}{197}$

   c. $\frac{150}{201}$ **d.** $\frac{8}{9}$

8. Without actually finding the exact answer, state which of the numbers given in parentheses in the following is the best approximation for the given sum or difference:

   a. $\frac{6}{13} + \frac{7}{15} + \frac{11}{23} + \frac{17}{35}$ $\left(1, 2, 3, 3\frac{1}{2}\right)$

   b. $\frac{30}{41} + \frac{1}{1000} + \frac{3}{2000}$ $\left(\frac{3}{8}, \frac{3}{4}, 1, 2\right)$

9. Compute each of the following mentally:

   a. $1 - \frac{3}{4}$ **b.** $3\frac{3}{8} + 2\frac{1}{4} - 5\frac{5}{8}$

10. The following ruler has regions marked M, A, T, H:

    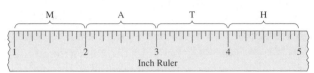

    Use mental mathematics and estimation to determine which region each of the following falls into (for example, $\frac{12}{5}$ in. falls into region A):

    a. $\frac{20}{8}$ in.   **b.** $\frac{36}{8}$ in.

    c. $\frac{60}{16}$ in.   **d.** $\frac{18}{4}$ in.

11. Use *clustering* to estimate the following sum:

    $$3\frac{1}{3} + 3\frac{1}{5} + 2\frac{7}{8} + 2\frac{2}{9}$$

12. A class consists of $\frac{2}{5}$ freshmen, $\frac{1}{4}$ sophomores, and $\frac{1}{10}$ juniors; the rest are seniors. What fraction of the class is seniors?

13. A clerk sold three pieces of one type of ribbon to different customers. One piece was $\frac{1}{3}$ yd long, another was $2\frac{3}{4}$ yd long, and the third was $3\frac{1}{2}$ yd long. What was the total length of that type of ribbon sold?

14. Martine bought $8\frac{3}{4}$ yd of fabric. She wants to make a skirt using $1\frac{7}{8}$ yd, pants using $2\frac{3}{8}$ yd, and a vest using $1\frac{2}{3}$ yd. How much fabric will be left over?

15. Students from Rattlesnake School formed four teams to collect cans for recycling during the months of April and May. The students received 10¢ for each 5 lb of cans. A record of their efforts follows:

    Number of Pounds Collected

	Team 1	Team 2	Team 3	Team 4
April	$28\frac{3}{4}$	$32\frac{7}{8}$	$28\frac{1}{2}$	$35\frac{3}{16}$
May	$33\frac{1}{3}$	$28\frac{5}{12}$	$25\frac{3}{4}$	$41\frac{1}{2}$

    a. Which team collected the most for the 2-month period? How much did they collect?

    b. What was the difference in the total pounds collected in April and the total pounds collected in May?

16. Give an example illustrating each of the following properties of rational numbers addition:
 a. Closure **b.** Commutative **c.** Associative

17. Given that the sequence in part (a) is arithmetic and in part (b) the sequence of numerators is arithmetic and so is the sequence of denominators, answer the following:
 (i) Write three more terms of each sequence.
 (ii) Is the sequence in part (b) arithmetic? Justify your answer.
 a. $\dfrac{1}{4}, \dfrac{1}{2}, \dfrac{3}{4}, 1, \dfrac{5}{4}, \ldots$
 b. $\dfrac{1}{2}, \dfrac{2}{3}, \dfrac{3}{4}, \dfrac{4}{5}, \dfrac{5}{6}, \ldots$

18. Find the nth term in each of the sequences in problem 17.

19. Insert five fractions between the numbers 1 and 2 so that the seven numbers (including 1 and 2) constitute an arithmetic sequence.

20. Let $f(x) = x + \dfrac{3}{4}$, where the domain is the set of rational numbers.
 a. Find the outputs if the inputs are the following:
  (i) 0 **(ii)** $\dfrac{4}{3}$ **(iii)** $\dfrac{^-3}{4}$
 b. For which inputs will the outputs be the following?
  (i) 1 **(ii)** $^-1$ **(iii)** $\dfrac{1}{2}$

21. Let $f(x) = \dfrac{x + 2}{x - 1}$ and let the domain of the function be the set of all integers except 1. Find the following:
 a. $f(0)$ **b.** $f(^-2)$
 c. $f(^-5)$ **d.** $f(5)$

22. a. Check that each of the following is true:
  (i) $\dfrac{1}{3} = \dfrac{1}{4} + \dfrac{1}{3 \cdot 4}$
  (ii) $\dfrac{1}{4} = \dfrac{1}{5} + \dfrac{1}{4 \cdot 5}$
  (iii) $\dfrac{1}{5} = \dfrac{1}{6} + \dfrac{1}{5 \cdot 6}$
 b. Based on the examples in (a), write $\dfrac{1}{n}$ as a sum of two unit fractions, that is, as a sum of fractions with numerator 1.

23. Solve for x in each of the following:
 a. $x + 2\dfrac{1}{2} = 3\dfrac{1}{3}$
 b. $x - 2\dfrac{2}{3} = \dfrac{5}{6}$

24. Find each sum or difference; simplify if possible.
 a. $\dfrac{3x}{xy^2} + \dfrac{y}{x^2}$
 b. $\dfrac{a}{xy^2} - \dfrac{b}{xyz}$
 c. $\dfrac{a^2}{a^2 - b^2} - \dfrac{a - b}{a + b}$

Assessment 6-2B

1. Perform the following additions or subtractions:
 a. $\dfrac{^-1}{2} + \dfrac{2}{3}$ **b.** $\dfrac{5}{12} - \dfrac{2}{3}$
 c. $\dfrac{5}{4x} + \dfrac{^-3}{2y}$
 d. $\dfrac{^-3}{2x^2y^2} + \dfrac{5}{2xy^2} + \dfrac{7}{x^2y}$
 e. $\dfrac{5}{6} - 2\dfrac{1}{8}$ **f.** $^-4\dfrac{1}{2} + 3\dfrac{1}{6}$

2. Change each of the following fractions to mixed numbers:
 a. $\dfrac{14}{5}$ **b.** $-\dfrac{47}{8}$

3. Change each of the following mixed numbers to fractions in the form $\dfrac{a}{b}$, where a and b are integers and $b \neq 0$.
 a. $7\dfrac{1}{2}$
 b. $^-4\dfrac{2}{3}$

4. Place the numbers 2, 5, 6, and 8 in the following boxes to make the equation true:

$$\dfrac{\square}{\square} + \dfrac{\square}{\square} = \dfrac{23}{24}$$

5. Use the information in the table to answer each of the following questions:

Team	Games Played	Games Won
Ducks	22	10
Beavers	19	10
Tigers	28	9
Bears	23	8
Lions	27	7
Wildcats	25	6
Badgers	21	5

  a. Which teams won less than $\frac{1}{3}$ of their games?

  b. Which teams won more than $\frac{1}{4}$ of their games?

  c. Which teams won just under $\frac{1}{4}$ of their games?

6. Sort the following fraction cards into the ovals by estimating in which oval the fraction belongs:

Sort these fraction cards   About 0   About $\frac{1}{2}$ About 1

$\boxed{\frac{1}{100}}$ $\boxed{\frac{36}{70}}$ $\boxed{\frac{19}{36}}$ $\boxed{\frac{1}{30}}$ $\boxed{\frac{7}{800}}$

7. Approximate each of the following fractions by $0, \frac{1}{4}, \frac{1}{2}, \frac{3}{4},$ or 1. Tell whether your estimate is high or low.

  a. $\frac{113}{100}$ **b.** $\frac{3}{1978}$

  c. $\frac{150}{198}$ **d.** $\frac{8}{9}$

8. Without actually finding the exact answer, state which of the numbers given in parentheses in the following is the best approximation for the given sum or difference:

  a. $\frac{6}{13} + \frac{7}{15} + \frac{12}{23} + \frac{18}{35}$ $\left(1, 2, 3, 3\frac{1}{2}\right)$

  b. $\frac{30}{41} + \frac{3}{1000} + \frac{5}{2000}$ $\left(\frac{3}{8}, \frac{3}{4}, 1, 2\right)$

9. Compute each of the following mentally:

  a. $6 - \frac{7}{8}$ **b.** $2\frac{3}{5} + 4\frac{1}{10} + 3\frac{3}{10}$

10. The following ruler has regions marked M, A, T, H:

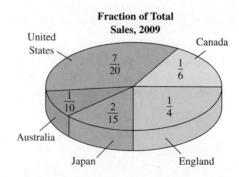

Use mental mathematics and estimation to determine which region each of the following falls into (for example, $\frac{12}{5}$ in. falls in region A):

  a. $\frac{9}{8}$ in.   **b.** $\frac{26}{8}$ in.

  c. $\frac{50}{16}$ in.   **d.** $\frac{14}{4}$ in.

11. A class consists of $\frac{1}{4}$ freshmen, $\frac{1}{5}$ sophomores, and $\frac{1}{10}$ juniors; the rest are seniors. What fraction of the class is seniors?

12. The Naturals Company sells its products in many countries. The following two circle graphs show the fractions of the company's earnings for 2004 and 2009. Based on this information, answer the following questions:

  a. In 2004, how much greater was the fraction of sales for Japan than for Canada?

  b. In 2009, how much less was the fraction of sales for England than for the United States?

  c. How much greater was the fraction of total sales for the United States in 2009 than in 2004?

  d. Is it true that the amount of sales in dollars in Australia was less in 2004 than in 2009? Why?

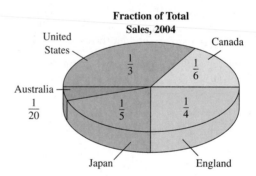

13. A recipe requires $3\frac{1}{2}$ c of milk. Ran put in $1\frac{3}{4}$ c and emptied the container. How much more milk does he need to put in?

14. A $15\frac{3}{4}$ in. board is cut from a $38\frac{1}{4}$ in. board. The saw cut takes $\frac{3}{8}$ in. How much of the $38\frac{1}{4}$ in. board is left after cutting?

15. In each of the following sequences, the numerators form an arithmetic sequence. Write three more terms of the sequence and determine which of the sequences are arithmetic, and which are not? Justify your answers.

    a. $\frac{2}{3}, \frac{5}{3}, \frac{8}{3}, \frac{11}{3}, \frac{14}{3}, \ldots$ **b.** $\frac{5}{4}, \frac{3}{4}, \frac{1}{4}, \frac{^-1}{4}, \frac{^-3}{4}, \ldots$

16. Find the nth term in each of the sequences in problem 15.

17. Insert four fractions between the numbers 1 and 3 so that the six numbers (including 1 and 3) constitute an arithmetic sequence.

18. Let $f(x) = x - \frac{3}{4}$, where the domain is the set of rational numbers.

    a. Find the outputs if the inputs are the following:

        (i) 0   **(ii)** $\frac{4}{3}$ **(iii)** $\frac{^-3}{4}$

    b. For which inputs will the outputs be the following?

        (i) 1   **(ii)** $^-1$ **(iii)** $\frac{1}{2}$

19. Let $f(x) = \frac{x + 2}{x - 1} + \frac{1}{3}$ and let the domain of the function be the set of all integers except 1. Find the following:

    a. $f(0)$ **b.** $f(^-2)$ **c.** $f(^-5)$ **d.** $f(5)$

20. a. Check that each of the following is true:

    (i) $\frac{1}{3} - \frac{1}{4} = \frac{1}{3 \cdot 4}$

    (ii) $\frac{1}{4} - \frac{1}{5} = \frac{1}{4 \cdot 5}$

    (iii) $\frac{1}{5} - \frac{1}{6} = \frac{1}{5 \cdot 6}$

    b. Based on part (a), write $\frac{1}{n(n + 1)}$ as a difference of two fractions with numerators 1. Justify your answer.

21. Solve for x in each of the following:

    a. $x - \frac{5}{6} = \frac{2}{3}$

    b. $x - \frac{7}{2^3 \cdot 3^2} = \frac{5}{2^2 \cdot 3^2}$

22. Find each sum or difference.

    a. $\frac{3x}{xy^2} + \frac{a}{x^2}$

    b. $\frac{a}{xy^2} - \frac{b}{xyz} + \frac{1}{x^2y}$

Mathematical Connections 6-2

Communication

1. Suppose a large pizza is divided into 3 equal-size pieces and a small pizza is divided into 4 equal-size pieces and you get 1 piece from each pizza. Does $\frac{1}{3} + \frac{1}{4}$ represent the amount that you received? Explain why or why not.

2. Explain why we choose a common denominator to add $\frac{1}{3} + \frac{3}{4}$.

3. a. When we add two fractions with unlike denominators and convert them to fractions with the same denominator, must we use the least common denominator? What are the advantages of using the least common denominator?

    b. When the least common denominator is used in adding or subtracting fractions, is the result always a fraction in simplest form?

4. Explain why we can do the following to convert $5\frac{3}{4}$ to a mixed number:

$$\frac{5 \cdot 4 + 3}{4} = \frac{23}{4}$$

5. To show $2\frac{3}{4} = \frac{11}{4}$, the teacher drew the following picture. Ken said this shows a picture of $\frac{11}{12}$, not $\frac{11}{4}$. What is Ken thinking and how should the teacher respond?

6. Kara spent $\frac{1}{2}$ of her allowance on Saturday and $\frac{1}{3}$ of what she had left on Sunday. Can this situation be modeled as $\frac{1}{2} - \frac{1}{3}$? Explain why or why not.

7. Compute $3\frac{3}{4} + 5\frac{1}{3}$ in two different ways and leave your answer as a mixed number. Tell which way you prefer and why.

8. Sally claims that it is easier to add two fractions if she adds the numerators and then adds the denominators. How can you help her?

9. Does each of the following properties hold for subtraction of rational numbers? Justify your answer.
   a. Closure   b. Commutative
   c. Associative   d. Identity
   e. Inverse

10. Explain an error pattern in each of the following:
   a. $\frac{13}{35} = \frac{1}{5}$, $\frac{27}{73} = \frac{2}{3}$, $\frac{16}{64} = \frac{1}{4}$
   b. $\frac{4}{5} + \frac{2}{3} = \frac{6}{8}$, $\frac{2}{5} + \frac{3}{4} = \frac{5}{9}$, $\frac{7}{8} + \frac{1}{3} = \frac{8}{11}$
   c. $8\frac{3}{8} - 6\frac{1}{4} = 2\frac{2}{4}$, $5\frac{3}{8} - 2\frac{2}{3} = 3\frac{1}{5}$, $2\frac{2}{7} - 1\frac{1}{3} = 1\frac{1}{4}$
   d. $\frac{2}{3} \cdot 3 = \frac{6}{9}$, $\frac{1}{4} \cdot 6 = \frac{6}{24}$, $\frac{4}{5} \cdot 2 = \frac{8}{10}$

11. Notice that the sequence $1, \frac{1}{2}, \frac{1}{3}, \frac{1}{4}, \frac{1}{5}, \ldots,$ whose nth term is $\frac{1}{n}$, is a decreasing sequence. Explain how to use this fact to determine if the following sequences are increasing or decreasing:
   a. $2, 1\frac{1}{2}, 1\frac{1}{3}, 1\frac{1}{4}, 1\frac{1}{5}, \ldots, 1\frac{1}{n}$
   b. $\frac{1}{2}, \frac{2}{3}, \frac{3}{4}, \frac{4}{5}, \ldots, \frac{n}{n+1}$ $\left(Hint: \frac{2}{3} = 1 - \frac{1}{3}, \frac{3}{4} = 1 - \frac{1}{4}\right)$
   c. $1, \frac{1}{2^2}, \frac{1}{3^2}, \frac{1}{4^2}, \frac{1}{5^2}, \ldots, \frac{1}{n^2}$

Open-Ended

12. Write a story problem for $\frac{2}{3} - \frac{1}{4}$.

13. a. Write two fractions whose sum is 1. If one of the fractions is $\frac{a}{b}$, what is the other?
    b. Write three fractions whose sum is 1.
    c. Write two fractions whose difference is very close to 1 but not exactly 1.

14. a. With the exception of $\frac{2}{3}$, the Egyptians used only unit fractions (fractions that have numerators of 1).

Every unit fraction can be expressed as the sum of two unit fractions in more than one way, for example, $\frac{1}{2} = \frac{1}{4} + \frac{1}{4}$ and $\frac{1}{2} = \frac{1}{3} + \frac{1}{6}$. Find at least two different unit fraction representations for each of the following:
   (i) $\frac{1}{3}$
   (ii) $\frac{1}{7}$

   b. Compute $\frac{1}{n} - \frac{1}{n+1}$ and simplify your answer.
   c. Re-write part (b) as a sum and then use the sum to answer the question in part (a). Write $\frac{1}{17}$ as a sum of two different unit fractions.

Cooperative Learning

15. Interview 10 people and ask them if and when they add and subtract fractions in their lives. Combine their responses with those of the rest of the class to get a view of how "ordinary" people must use computation of rational numbers in their daily lives.

Questions from the Classroom

16. Joe reported that if $\frac{a}{b} = \frac{c}{d}$, then $\frac{d}{c+d} = \frac{b}{a+b}$. Joan said she didn't believe it. How do you respond?

17. Kendra showed that $\frac{1}{3} + \frac{3}{4} = \frac{4}{7}$ by using the following figure. How would you help her?

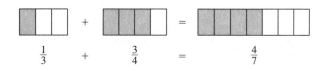

18. Jill claims that for positive fractions, $\frac{a}{b} + \frac{a}{c} = \frac{a}{b+c}$ because the fractions have a common numerator. How do you respond?

Review Problems

19. Simplify if possible.
   a. $\frac{14}{21}$
   b. $\frac{117}{153}$

c. $\dfrac{5^2}{7^2}$

d. $\dfrac{a^2 + a}{1 + a}$

e. $\dfrac{a^2 + 1}{a + 1}$

f. $\dfrac{a^2 - b^2}{a - b}$

20. Determine if the fractions in each of the following pairs are equal:

a. $\dfrac{a^2}{b}$ and $\dfrac{a^2 b}{b^2}$

b. $\dfrac{377}{400}$ and $\dfrac{378}{401}$

c. $\dfrac{0}{10}$ and $\dfrac{0}{-10}$

d. $\dfrac{a}{b}$ and $\dfrac{a + 1}{b + 1}$, where $a \neq b$

21. If we assume 1 yr = 365 days, answer the following:

a. What month of the year has the smallest fraction of days of the year?

b. What fraction of days of the year occur before July 4?

22. a. If the same positive number is added to the numerator and denominator of a positive proper fraction, is the new fraction greater than, less than, or equal to the original fraction? Justify your answer.

b. What if the same positive number, less than the numerator and less than the denominator, is subtracted from the numerator and denominator of a positive proper fraction and the new denominator is positive?

Third International Mathematics and Science Study (TIMSS) Question

Janis, Maija, and their mother were eating a cake. Janis ate $\dfrac{1}{2}$ of the cake. Maija ate $\dfrac{1}{4}$ of the cake. Their mother ate $\dfrac{1}{4}$ of the cake. How much of the cake is left?

a. $\dfrac{3}{4}$ **b.** $\dfrac{1}{2}$ **c.** $\dfrac{1}{4}$ **d.** None

TIMSS 2007, Grade 4

6-3   Multiplication and Division of Rational Numbers

Multiplication of Rational Numbers

To motivate the definition of multiplication of rational numbers, we use the interpretation of multiplication as repeated addition. Using repeated addition, we can interpret $3 \cdot \left(\dfrac{3}{4}\right)$ as follows:

$$3 \cdot \left(\frac{3}{4}\right) = \frac{3}{4} + \frac{3}{4} + \frac{3}{4} = \frac{9}{4} = 2\frac{1}{4}$$

The area model in Figure 6-11 is another way to calculate this product.

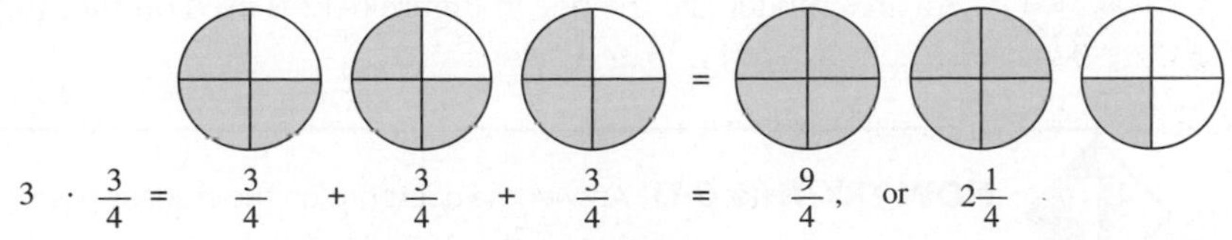

$$3 \quad \cdot \quad \frac{3}{4} = \quad \frac{3}{4} \quad + \quad \frac{3}{4} \quad + \quad \frac{3}{4} \quad = \quad \frac{9}{4}, \quad \text{or} \quad 2\frac{1}{4}$$

Figure 6-11

We next consider $\left(\dfrac{3}{4}\right) \cdot 3$. How should this product be interpreted? If the commutative property of multiplication of rational numbers is to hold, then $\left(\dfrac{3}{4}\right) \cdot 3 = 3 \cdot \left(\dfrac{3}{4}\right) = \dfrac{9}{4}$.

Next, we consider another interpretation of multiplication. What is $\dfrac{3}{4}$ of 3? Recall that $\dfrac{3}{4}$ of a quantity is the amount resulting from dividing the quantity into 4 equal parts and taking 3 of these parts. To see what $\dfrac{1}{4}$ of 3 is consider taking $\dfrac{1}{4}$ of 3 equal size bars. This can be done by taking $\dfrac{1}{4}$ of each of the 3 bars, that is, $\dfrac{1}{4} + \dfrac{1}{4} + \dfrac{1}{4}$ or $\dfrac{3}{4}$. Thus $\dfrac{3}{4}$ of 3 is $\dfrac{3}{4} + \dfrac{3}{4} + \dfrac{3}{4}$, or $3 \cdot \dfrac{3}{4}$, which as we have seen above is equal to $\dfrac{3}{4} \cdot 3$. Thus we can interpret $\dfrac{3}{4} \cdot 3$ as $\dfrac{3}{4}$ of 3 and in general for non-negative rational numbers $\dfrac{a}{b} \cdot \dfrac{c}{d}$ can be represented as $\dfrac{a}{b}$ of $\dfrac{c}{d}$.

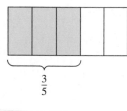

(a)

If forests once covered about $\dfrac{3}{5}$ of Earth's land and only about $\dfrac{1}{2}$ of these forests remain, what fraction of Earth is covered with forests today? We need to find $\dfrac{1}{2}$ of $\dfrac{3}{5}$, and can use an area model to find the answer.

Figure 6-12(a) shows a rectangle representing the *whole* separated into fifths, with $\dfrac{3}{5}$ shaded. To find $\dfrac{1}{2}$ of $\dfrac{3}{5}$, we divide the shaded portion of the rectangle in Figure 6-12(a) into two congruent parts and take one of those parts. The result would be the green portion of Figure 6-12(b). However, the green portion represents 3 parts out of 10, or $\dfrac{3}{10}$, of the whole. Thus,

$$\dfrac{1}{2} \cdot \dfrac{3}{5} = \dfrac{3}{10} = \dfrac{1 \cdot 3}{2 \cdot 5}$$

(b)

Figure 6-12

This discussion leads to the following definition of multiplication for rational numbers.

Definition of Multiplication of Rational Numbers

If $\dfrac{a}{b}$ and $\dfrac{c}{d}$ are any rational numbers, then $\dfrac{a}{b} \cdot \dfrac{c}{d} = \dfrac{a \cdot c}{b \cdot d}$.

REMARK

1. Notice that the definition above follows from the interpretation of $\dfrac{a}{b} \cdot \dfrac{c}{d}$ as $\dfrac{a}{b}$ of $\dfrac{c}{d}$.

2. Also notice that when multiplying fractions it is possible and more efficient to "cancel" common factors before multiplying out the numerators and the denominators.

An area model like the one in Figure 6-12 is used on the student page that follows (page 383).

NOW TRY THIS 6-13 Answer the questions on the student page.

School Book Page — MULTIPLYING FRACTIONS

Lesson 8-12

Key Idea
You can multiply two fractions to find a fraction of a fraction.

Materials
• square sheets of paper
• red and yellow colored pencils
or tools

Think It Through
• I can **make a model** to show the problem.
• I can **look for patterns** in the products to find a multiplication rule.

Multiplying Fractions

LEARN

Activity

How can you use a model to find the product of two fractions?

You can use paper folding to find $\frac{1}{2}$ of $\frac{3}{4}$.

a. Fold a square sheet of paper vertically in half.

b. Fold the paper vertically in half again. Each section is what fraction of the sheet of paper?

c. Now fold the paper in half horizontally. Each section is what fraction of the sheet of paper?

d. Shade $\frac{3}{4}$ of the vertical sections red and $\frac{1}{2}$ of the horizontal sections yellow.

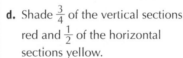
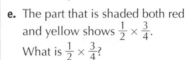

e. The part that is shaded both red and yellow shows $\frac{1}{2} \times \frac{3}{4}$. What is $\frac{1}{2} \times \frac{3}{4}$?

f. Fold paper to find each product.

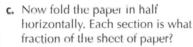

$\frac{1}{2} \times \frac{1}{2} = \frac{\ }{\ }$ $\frac{1}{2} \times \frac{1}{4} = \frac{\ }{\ }$ $\frac{1}{4} \times \frac{1}{4} = \frac{\ }{\ }$ $\frac{3}{4} \times \frac{3}{4} = \frac{\ }{\ }$ $\frac{1}{2} \times \frac{5}{8} = \frac{\ }{\ }$

g. Use the pictures to find the products.

$\frac{3}{4} \times \frac{1}{3} = \frac{\ }{\ }$ $\frac{1}{3} \times \frac{4}{5} = \frac{\ }{\ }$ $\frac{3}{4} \times \frac{5}{6} = \frac{\ }{\ }$

h. Study the numerators and denominators in each problem in parts e and f. What pattern do you see between the fractions multiplied and their products?

WARM UP

Simplify.

1. $\frac{15}{8}$ 2. $\frac{24}{5}$ 3. $\frac{17}{2}$

4. $\frac{21}{6}$ 5. $\frac{24}{7}$ 6. $\frac{45}{10}$

496

Example 6-13

Answer the following:

If $\dfrac{5}{6}$ of the population of a certain city are college graduates and $\dfrac{7}{11}$ of the city's college graduates are female, what fraction of the population of that city is female college graduates?

Solution   The fraction should be $\dfrac{7}{11}$ of $\dfrac{5}{6}$, or $\dfrac{7}{11} \cdot \dfrac{5}{6} = \dfrac{7 \cdot 5}{11 \cdot 6} = \dfrac{35}{66}$.

The fraction of the population that is female college graduates is $\dfrac{35}{66}$.

Properties of Multiplication of Rational Numbers

Multiplication of rational numbers has properties analogous to the properties of addition of rational numbers. These include the following properties for multiplication: closure, commutative, associative, multiplicative identity, and multiplicative inverse. For emphasis, we list the last two properties.

Theorem 6–12:   Multiplicative Identity and Multiplicative Inverse of Rational Numbers

1. The number 1 is the unique number such that for every rational number $\dfrac{a}{b}$,

$$1 \cdot \left(\dfrac{a}{b}\right) = \dfrac{a}{b} = \left(\dfrac{a}{b}\right) \cdot 1$$

2. For any nonzero rational number $\dfrac{a}{b}$, $\dfrac{b}{a}$ is the unique rational number such that $\dfrac{a}{b} \cdot \dfrac{b}{a} = 1 = \dfrac{b}{a} \cdot \dfrac{a}{b}$.

REMARK   The multiplicative inverse of $\dfrac{a}{b}$ is also called the **reciprocal** of $\dfrac{a}{b}$.

Example 6-14

Find the multiplicative inverse, if possible, for each of the following rational numbers:

a. $\dfrac{2}{3}$ **b.** $\dfrac{-2}{5}$ **c.** 4   **d.** 0   **e.** $6\dfrac{1}{2}$

Solution   **a.** $\dfrac{3}{2}$

b. $\dfrac{5}{-2}$, or $\dfrac{-5}{2}$

c. Because $4 = \dfrac{4}{1}$, the multiplicative inverse of 4 is $\dfrac{1}{4}$.

d. Even though $0 = \dfrac{0}{1}$, $\dfrac{1}{0}$ is undefined; there is no multiplicative inverse of 0.

e. Because $6\dfrac{1}{2} = \dfrac{13}{2}$, the multiplicative inverse of $6\dfrac{1}{2}$ is $\dfrac{2}{13}$.

Multiplication and addition are connected through the distributive property of multiplication over addition. Also there is a multiplication property of equality for rational numbers and a multiplication property of zero similar to those for whole numbers and integers.

Theorem 6–13:

1. Distributive Property of Multiplication Over Addition for Rational Numbers

If $\frac{a}{b}, \frac{c}{d}$, and $\frac{e}{f}$ are any rational numbers, then

$$\frac{a}{b}\left(\frac{c}{d} + \frac{e}{f}\right) = \left(\frac{a}{b} \cdot \frac{c}{d}\right) + \left(\frac{a}{b} \cdot \frac{e}{f}\right)$$

2. Multiplication Property of Equality for Rational Numbers

If $\frac{a}{b}$ and $\frac{c}{d}$ are any rational numbers such that $\frac{a}{b} = \frac{c}{d}$, and $\frac{e}{f}$ is any rational number, then $\frac{a}{b} \cdot \frac{e}{f} = \frac{c}{d} \cdot \frac{e}{f}$.

3. Multiplication Property of Inequality for Rational Numbers

(i) If $\frac{a}{b} > \frac{c}{d}$ and $\frac{e}{f} > 0$, then $\frac{a}{b} \cdot \frac{e}{f} > \frac{c}{d} \cdot \frac{e}{f}$.

(ii) If $\frac{a}{b} > \frac{c}{d}$ and $\frac{e}{f} < 0$, then $\frac{a}{b} \cdot \frac{e}{f} < \frac{c}{d} \cdot \frac{e}{f}$.

4. Multiplication Property of Zero for Rational Numbers

If $\frac{a}{b}$ is any rational number, then $\frac{a}{b} \cdot 0 = 0 = 0 \cdot \frac{a}{b}$.

REMARK Theorem 6–13 can be proved using the corresponding properties of integers. For example, $\frac{a}{b} \cdot 0 = \frac{a}{b} \cdot \frac{0}{1} = \frac{a \cdot 0}{b \cdot 1} = \frac{0}{b} = 0$.

Example 6-15

A bicycle is on sale at $\frac{3}{4}$ of its original price. If the sale price is \$330, what was the original price?

Solution Let x be the original price. Then $\frac{3}{4}$ of the original price is $\frac{3}{4}x$. Because the sale price is \$330, we have $\frac{3}{4}x = 330$. Solving for x gives

$$\frac{4}{3} \cdot \frac{3}{4}x = \frac{4}{3} \cdot 330$$
$$1 \cdot x = 440$$
$$x = 440$$

Thus, the original price was \$440.

An alternative approach, which does not use algebra, follows. Because $\frac{3}{4}$ is 3 of the $\frac{1}{4}$ parts, then 3 of the $\frac{1}{4}$ parts is \$330 and 1 of the $\frac{1}{4}$ parts is \$110. If 1 of the $\frac{1}{4}$ parts is \$110, then 4 of these is $4 \cdot \$110$, or \$440.

Multiplication with Mixed Numbers

In the *Peanuts* cartoon, Sally is having trouble multiplying two mixed numbers. If she used an estimate to check whether her answer was reasonable, she would notice that if she multiplies two numbers that are both less than 3, the answer must be less than 9.

One way to multiply $2\frac{1}{2} \cdot 2\frac{1}{2}$ is to change the mixed numbers to improper fractions and use the definition of multiplication, as shown here.

$$2\frac{1}{2} \cdot 2\frac{1}{2} = \frac{5}{2} \cdot \frac{5}{2} = \frac{25}{4}$$

We could then change $\frac{25}{4}$ to the mixed number $6\frac{1}{4}$.

Another way to multiply mixed numbers uses the distributive property of multiplication over addition. For example,

$$\begin{aligned}
2\frac{1}{2} \cdot 2\frac{1}{2} &= \left(2 + \frac{1}{2}\right) \cdot \left(2 + \frac{1}{2}\right) \\
&= \left(2 + \frac{1}{2}\right) \cdot 2 + \left(2 + \frac{1}{2}\right) \cdot \frac{1}{2} \\
&= 2 \cdot 2 + \frac{1}{2} \cdot 2 + 2 \cdot \frac{1}{2} + \frac{1}{2} \cdot \frac{1}{2} \\
&= 4 + 1 + 1 + \frac{1}{4} \\
&= 6 + \frac{1}{4} \\
&= 6\frac{1}{4}
\end{aligned}$$

Multiplication of fractions enables us to obtain equivalent fractions, to perform addition and subtraction of fractions, as well as to solve equations in a different way, as shown in the following example.

Example 6-16

Use the definition of multiplication of fractions and other properties to justify or solve the following:

a. The Fundamental Law of Fractions, $\frac{a}{b} = \frac{an}{bn}$ if $n \neq 0$.

b. Addition of fractions using a common denominator.

c. Enlarging the denominator of a positive fraction by a factor of m is the same as multiplying the fraction by $\frac{1}{m}$.

d. Solve the following equations:

(i) $\dfrac{2}{3x} + \dfrac{5}{4x} = 2$

(ii) $\dfrac{x}{a} + \dfrac{x}{b} = 1$ (find x in terms of a and b)

Solution **a.** $\dfrac{a}{b} = \dfrac{a}{b} \cdot 1 = \dfrac{a}{b} \cdot \dfrac{n}{n} = \dfrac{an}{bn}$ **b.** $\dfrac{a}{b} + \dfrac{c}{d} = \dfrac{a}{b} \cdot \dfrac{d}{d} + \dfrac{c}{d} \cdot \dfrac{b}{b}$

$$= \dfrac{ad}{bd} + \dfrac{bc}{bd}$$

$$= \dfrac{ad + bc}{bd}$$

c. $\dfrac{a}{bm} = \dfrac{a \cdot 1}{b \cdot m} = \dfrac{a}{b} \cdot \dfrac{1}{m}$

d. There are many ways to solve these equations; we show one way that uses multiplication of fractions.

(i) $\dfrac{2}{3x} + \dfrac{5}{4x} = 2$ is equivalent to each of the following:

$$\dfrac{2}{3} \cdot \dfrac{1}{x} + \dfrac{5}{4} \cdot \dfrac{1}{x} = 2, \left(\dfrac{2}{3} + \dfrac{5}{4}\right)\dfrac{1}{x} = 2$$

$$\dfrac{23}{12} \cdot \dfrac{1}{x} = 2, \dfrac{23}{12} \cdot \dfrac{1}{x} \cdot x = 2x$$

$$\dfrac{23}{12} \cdot 1 = 2x, \dfrac{23}{12} \cdot \dfrac{1}{2} = \dfrac{1}{2} \cdot 2x, x = \dfrac{23}{24}.$$

Alternatively we could solve the equation by adding the fractions using the least common denominator of $12x$:

$$\dfrac{2}{3x} + \dfrac{5}{4x} = 2, \dfrac{8 + 15}{12x} = 2$$

$$\dfrac{23}{12x} \cdot 12x = 2 \cdot 12x$$

$$23 = 24x, x = \dfrac{23}{24}$$

(ii) $\dfrac{x}{a} + \dfrac{x}{b} = 1$ is equivalent to each of the following:

$$x \cdot \dfrac{1}{a} + x \cdot \dfrac{1}{b} = 1$$

$$x\left(\dfrac{1}{a} + \dfrac{1}{b}\right) = 1$$

$$x\left(\dfrac{b + a}{ab}\right) = 1$$

$$x\left(\dfrac{b + a}{ab}\right) \cdot \dfrac{ab}{a + b} = 1 \cdot \dfrac{ab}{a + b}$$

$$x \cdot 1 = \dfrac{ab}{a + b}$$

$$x = \dfrac{ab}{a + b}, \text{ and } a \neq {}^{-}b$$

Division of Rational Numbers

In the *Principles and Standards*, we find the following statement concerning division of rational numbers:

> The division of fractions has traditionally been quite vexing for students. Although "invert and multiply" has been a staple of conventional mathematics instruction and although it seems to be a simple way to remember how to divide fractions, students have for a long time had difficulty doing so. Some students forget which number is to be inverted, and others are confused about when it is appropriate to apply the procedure. A common way of formally justifying the "invert and multiply" procedure is to use sophisticated arguments involving the manipulation of algebraic rational expressions—arguments beyond the reach of many middle-grades students. This process can seem very remote and mysterious to many students. Lacking an understanding of the underlying rationale, many students are therefore unable to repair their errors and clear up their confusions about division of fractions on their own. An alternative approach involves helping students understand the division of fractions by building on what they know about the division of whole numbers. (p. 219)

We try to follow that advice in developing the concept of division of rational numbers.

Recall that $6 \div 3$ means "How many 3s are there in 6?" We found that $6 \div 3 = 2$ because $3 \cdot 2 = 6$ and, in general, if $a, b, c \in W$, then $\frac{a}{b} = c$, if, and only if c is the unique whole number such that $bc = a$. Consider $3 \div \left(\frac{1}{2}\right)$, which is equivalent to finding how many halves there are in 3. We see from the area model in Figure 6-13 that there are 6 half pieces in the 3 whole pieces. We record this as $3 \div \left(\frac{1}{2}\right) = 6$. This is true because $\left(\frac{1}{2}\right) \cdot 6 = 3$.

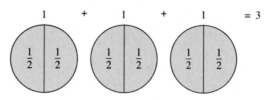

Figure 6-13

With whole numbers one way to think about division was in terms of *repeated subtraction*. We found that $6 \div 2 = 3$ because 2 could be subtracted from 6 three times; that is, $6 - 3 \cdot 2 = 0$. With $3 \div \left(\frac{1}{2}\right)$, we want to know how many halves can be subtracted from 3. Because $3 - 6 \cdot \left(\frac{1}{2}\right) = 0$, we know that $3 \div \left(\frac{1}{2}\right) = 6$.

Next, consider $\left(\frac{3}{4}\right) \div \left(\frac{1}{8}\right)$. This means "How many $\frac{1}{8}$s are in $\frac{3}{4}$?" Figure 6-14 shows that there are six $\frac{1}{8}$s in the shaded portion, which represents $\frac{3}{4}$ of the whole. Therefore, $\left(\frac{3}{4}\right) \div \left(\frac{1}{8}\right) = 6$. This is true because $\left(\frac{1}{8}\right) \cdot 6 = \frac{3}{4}$. Using repeated subtraction, we see that $\frac{3}{4} \div \frac{1}{8} = \frac{6}{8} \div \frac{1}{8}$ and that $\frac{6}{8} - 6\left(\frac{1}{8}\right) = 0$, so $\frac{3}{4} \div \frac{1}{8} = 6$.

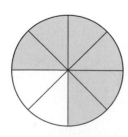

Figure 6-14

Two additional models for division of fractions are shown on the student page on page 390. The measurement or number-line model used in Example A on the student page is also useful. For example, consider $\frac{7}{8} \div \frac{3}{4}$. First we draw a measurement or number line divided into eighths, as shown in Figure 6-15. Next we want to know how many $\frac{3}{4}$s there are in $\frac{7}{8}$. The bar of length $\frac{3}{4}$ is made up of 6 equal-size pieces of length $\frac{1}{8}$. We see that there is at least one length of $\frac{3}{4}$ in $\frac{7}{8}$. If we put another bar of length $\frac{3}{4}$ on the number line, we see there is 1 more of the 6 equal-length segments needed to make $\frac{7}{8}$. Therefore, the answer is $1\frac{1}{6}$, or $\frac{7}{6}$.

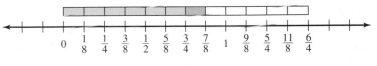

Figure 6-15

In the previous examples, we saw a relationship between division and multiplication of rational numbers. We can define division for rational numbers formally in terms of multiplication in the same way that we defined division for whole numbers.

Definition of Division of Rational Numbers

If $\frac{a}{b}$ and $\frac{c}{d}$ are any rational numbers, then $\frac{a}{b} \div \frac{c}{d} = \frac{e}{f}$ if, and only if, $\frac{e}{f}$ is the unique rational number such that $\frac{c}{d} \cdot \frac{e}{f} = \frac{a}{b}$.

REMARK In the definition of division, $\frac{c}{d} \neq 0$ because if $\frac{c}{d} = 0$, then the equation $\frac{c}{d} x = \frac{a}{b}$ has no solution if $\frac{a}{b} \neq 0$ and has infinitely many solutions if $\frac{a}{b} = 0$.

NOW TRY THIS 6-14 Students often confuse division by 2 and division by $\frac{1}{2}$. Notice that

$a \div 2 = \frac{a}{2} = \frac{1}{2}a$, but $a \div \frac{1}{2} = x$ if, and only if, $\frac{1}{2}x = a$, $2 \cdot \frac{1}{2}x = 2a$, $x = 2a$.

Write a real-life story that will help students see the difference between division by 2 and division by $\frac{1}{2}$.

School Book Page DIVIDING FRACTIONS

Lesson 5-6

Key Idea
You can use multiplication to divide fractions.

Vocabulary
• reciprocal
• multiplicative inverse

Think It Through
• I can **use models** to show division of fractions.
• I can **look for a pattern** to understand how division and multiplication are related.

Dividing Fractions

LEARN

How can you model division of fractions?

Example A

Mr. Wagner makes wooden coasters for glasses. He cuts small round posts into $\frac{1}{2}$-inch thick slices. How many slices can he get from a 5-inch post?

Find $5 \div \frac{1}{2}$.

Think: "How many halves are in 5?"

Divide 5 inches into $\frac{1}{2}$-inch sections.

←There are ten $\frac{1}{2}$-inch slices in 5 inches.

So, $5 \div \frac{1}{2} = 10$.

Mr. Wagner can get 10 slices from a 5-inch post.

Example B

Find $\frac{3}{4} \div 4$.

You can think of $\frac{3}{4} \div 4$ as "What is $\frac{3}{4}$ divided into 4 equal parts?"

So, $\frac{3}{4} \div 4 = \frac{3}{16}$.

Show $\frac{3}{4}$.

Divide $\frac{3}{4}$ into 4 equal parts.

Each of the four equal parts contains 3 sixteenths.

How can you divide fractions?

Activity

a. Study the patterns below. Compare the first and second columns.

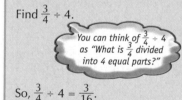

Pattern 1	
$6 \div \frac{3}{1} = 2$	$6 \times \frac{1}{3} = 2$
$3 \div \frac{4}{1} = \frac{3}{4}$	$3 \times \frac{1}{4} = \frac{3}{4}$
$4 \div \frac{1}{2} = 8$	$4 \times \frac{2}{1} = 8$

Pattern 2	
$\frac{3}{4} \div \frac{2}{1} = \frac{3}{8}$	$\frac{3}{4} \times \frac{1}{2} = \frac{3}{8}$
$\frac{1}{4} \div \frac{1}{12} = 3$	$\frac{1}{4} \times \frac{12}{1} = 3$
$\frac{3}{4} \div \frac{3}{8} = 2$	$\frac{3}{4} \times \frac{8}{3} = 2$

b. What do you notice about the divisors in the first column and the factors in the second column of each pattern?

c. How does the quotient compare to the dividend when the divisor is a fraction less than 1?

Source: Mathematics, Diamond Edition, Grade Six, Scott Foresman-Addison Wesley 2008 (p. 266).

Algorithm for Division of Rational Numbers

As we see in the *Peanuts* cartoon, Peppermint Patty doesn't understand why the algorithm for division of fractions works. She is not alone in her confusion, and we will explain why "to divide fractions we use the reciprocal and multiply."

Does the method in the cartoon, often called the *invert-and-multiply* method, make sense based on what we know about rational numbers? We know that one way to interpret $\frac{a}{b}$ is $a \div b$. We also know that $\frac{a}{1} = a$. Therefore, $4 \div 5$ can be written as $\frac{4}{1} \div \frac{5}{1}$. If the invert-and-multiply technique works, then $4 \div 5 = \frac{4}{1} \div \frac{5}{1} = \frac{4}{1} \times \frac{1}{5} = \frac{4 \cdot 1}{1 \cdot 5} = \frac{4}{5}$. Since $4 \div 5 = \frac{4}{5}$, this is consistent with what we have done before.

To develop the general algorithm for division of rational numbers, we consider what such a division might mean. For example,

$$\frac{2}{3} \div \frac{5}{7} = x \quad \text{implies} \quad \frac{2}{3} = \frac{5}{7}x$$

We multiply both sides of the equation by $\frac{7}{5}$, the reciprocal of $\frac{5}{7}$. Thus,

$$\frac{7}{5} \cdot \frac{2}{3} = \frac{7}{5} \cdot \left(\frac{5}{7}x\right) = \left(\frac{7}{5} \cdot \frac{5}{7}\right)x = 1 \cdot x = x$$

Therefore, $\frac{2}{3} \div \frac{5}{7} = \frac{2}{3} \cdot \frac{7}{5}$.

A traditional justification of the division algorithm also follows. The algorithm for division of fractions is usually justified in the middle grades by using the Fundamental Law of Fractions, $\frac{a}{b} = \frac{ac}{bc}$, where a, b, and c are all fractions, or equivalently, the identity property of multiplication. For example,

$$\frac{2}{3} \div \frac{5}{7} = \frac{\frac{2}{3}}{\frac{5}{7}} = \frac{\frac{2}{3}}{\frac{5}{7}} \cdot 1 = \frac{\frac{2}{3} \cdot \frac{7}{5}}{\frac{5}{7} \cdot \frac{7}{5}} = \frac{\frac{2}{3} \cdot \frac{7}{5}}{1} = \frac{2}{3} \cdot \frac{7}{5}$$

Thus,

$$\frac{2}{3} \div \frac{5}{7} = \frac{2}{3} \cdot \frac{7}{5}$$

NOW TRY THIS 6-15 Use an argument similar to the preceding one to show that, in general, if $\frac{a}{b}$ and $\frac{c}{d}$ are rational numbers and $\frac{c}{d} \neq 0$, then $\frac{a}{b} \div \frac{c}{d} = \frac{a}{b} \cdot \frac{d}{c}$.

We summarize the algorithm as follows:

Theorem 6–14:   Algorithm for Division of Fractions

If $\frac{a}{b}$ and $\frac{c}{d}$ are any rational numbers and $\frac{c}{d} \neq 0$, then

$$\frac{a}{b} \div \frac{c}{d} = \frac{a}{b} \cdot \frac{d}{c}$$

Alternate Algorithm for Division of Rational Numbers

An alternative algorithm for division of fractions can be found by first dividing fractions that have equal denominators. For example, $\frac{9}{10} \div \frac{3}{10} = 9 \div 3$ and $\frac{15}{23} \div \frac{5}{23} = 15 \div 5$.

These examples suggest that when two fractions with the same denominator are divided, the result can be obtained by dividing the numerator of the first fraction by the numerator of the second. To divide fractions with different denominators, we rename the fractions so that the denominators are equal. Thus,

$$\frac{a}{b} \div \frac{c}{d} = \frac{ad}{bd} \div \frac{bc}{bd} = ad \div bc, \quad \text{or} \quad \frac{ad}{bc}$$

NOW TRY THIS 6-16 Show that $\frac{a}{b} \div \frac{c}{d} = \frac{a}{b} \cdot \frac{d}{c}$ and $\frac{a}{b} \div \frac{c}{d} = \frac{a \div c}{b \div d}$ are equivalent.

The next three examples illustrate the use of division of rational numbers.

Example 6-17

A radio station provides 36 min for public service announcements for every 24 hr of broadcasting.

a. What part of the broadcasting day is allotted to public service announcements?

b. How many $\frac{3}{4}$-min public service announcements can be allowed in the 36 min?

Solution   a. There are 60 min in an hour and $60 \cdot 24$ min in a day. Thus, $36/(60 \cdot 24)$, or $\frac{1}{40}$, of the day is allotted for the announcements.

b. $36/\left(\frac{3}{4}\right) = 36\left(\frac{4}{3}\right)$, or 48, announcements are allowed.

Example 6-18

We have $35\frac{1}{2}$ yd of material available to make towels. Each towel requires $\frac{3}{8}$ yd of material.

a. How many towels can be made?
b. How much material will be left over?

Solution a. We need to find the integer part of the answer to $35\frac{1}{2} \div \frac{3}{8}$. The division follows:

$$35\frac{1}{2} \div \frac{3}{8} = \frac{71}{2} \cdot \frac{8}{3} = \frac{284}{3} = 94\frac{2}{3}$$

Thus we can make 94 towels.

b. Because the division in (a) was by $\frac{3}{8}$, the amount of material left over is $\frac{2}{3}$ of $\frac{3}{8}$, or $\frac{2}{3} \cdot \frac{3}{8}$, or $\frac{1}{4}$ yd. This can also be answered by noting that the $\frac{2}{3}$ in part (a) is two-thirds of a towel, which requires $\frac{2}{3}$ of $\frac{3}{8}$ yd of material.

Example 6-19

If $\frac{1}{2}$ of a jump rope is $3\frac{3}{4}$ yd long, what is the length of the entire rope?

Solution If the length of the rope is l yd, then $\frac{1}{2}$ of the rope's length is $\frac{1}{2}l$ long. Thus,

$$\frac{1}{2}l = 3\frac{3}{4}$$

$$l = \frac{3\frac{3}{4}}{\frac{1}{2}}$$

$$l = \frac{15}{4} \cdot \frac{2}{1} = \frac{15}{2}, \text{ or } 7\frac{1}{2} \text{ yd}$$

Notice that this problem could be solved without using division, by simply multiplying both sides of the first equation by 2.

Estimation and Mental Math with Rational Numbers

Estimation and mental math strategies that were developed with whole numbers can also be used with rational numbers.

Example 6-20

Use mental math to find

a. $(12 \cdot 25) \cdot \frac{1}{4}$ b. $\left(5\frac{1}{6}\right) \cdot 12$ c. $\frac{4}{5} \cdot 20$

Solution Each of the following is a possible approach:

a. $(12 \cdot 25) \cdot \frac{1}{4} = 25 \cdot \left(12 \cdot \frac{1}{4}\right) = 25 \cdot 3 = 75$

b. $5\frac{1}{6} \cdot 12 = \left(5 + \frac{1}{6}\right)12 = 5 \cdot 12 + \frac{1}{6} \cdot 12 = 60 + 2 = 62$

c. $\frac{4}{5} \cdot 20 = 4\left(\frac{1}{5} \cdot 20\right) = 4 \cdot 4 = 16$

Example 6-21

Estimate each of the following:

a. $3\frac{1}{4} \cdot 7\frac{8}{9}$ **b.** $24\frac{5}{7} \div 4\frac{1}{8}$

Solution   **a.** Using rounding, the product will be close to $3 \cdot 8 = 24$. If we use the range strategy, we can say the product must be between $3 \cdot 7 = 21$ and $4 \cdot 8 = 32$.

b. We can use compatible numbers and think of the estimate as $24 \div 4 = 6$ or $25 \div 5 = 5$.

Extending the Notion of Exponents

Recall that a^m was defined for any integer a and any natural number m as the product of m a's. We define a^m for any rational number in a similar way as follows.

Definition of a to the mth Power

$a^m = \underbrace{a \cdot a \cdot a \cdot \ldots \cdot a}_{m \text{ factors}}$, where a is any rational number and m is any natural number.

From the definition, $a^3 \cdot a^2 = (a \cdot a \cdot a) \cdot (a \cdot a) = a^{3+2}$. In a similar way, it follows that

1. $a^m \cdot a^n = a^{m+n}$

where a is any rational number and m and n are natural numbers. If (1) is to be true for all whole numbers m and n, then because $a^1 \cdot a^0 = a^{1+0} = a^1$, we must have $a^0 = 1$. Hence, it is useful to give meaning to a^0 when $a \neq 0$ as follows.

2. For any nonzero number a, $a^0 = 1$.

The properties above are also true for rational number values of a. For example, consider the following:

$$\left(\frac{2}{3}\right)^2 \cdot \left(\frac{2}{3}\right)^3 = \left(\frac{2}{3} \cdot \frac{2}{3}\right) \cdot \left(\frac{2}{3} \cdot \frac{2}{3} \cdot \frac{2}{3}\right) = \left(\frac{2}{3}\right)^{2+3} = \left(\frac{2}{3}\right)^5$$

If $a^m \cdot a^n = a^{m+n}$ is to be extended to all integer powers of a, then how should a^{-3} be defined? If (1) is to be true for all integers m and n, then $a^{-3} \cdot a^3 = a^{-3+3} = a^0 = 1$. Therefore, $a^{-3} = 1/a^3$. This is true in general and we have the following.

3. For a nonzero number a, $a^{-n} = \frac{1}{a^n}$.

In elementary grades the definition of a^{-n} is typically motivated by looking at patterns. Notice that as the following exponents decrease by 1, the numbers on the right are divided by 10. Thus the pattern might be continued, as shown.

$$10^3 = 10 \cdot 10 \cdot 10$$
$$10^2 = 10 \cdot 10$$
$$10^1 = 10$$
$$10^0 = 1$$
$$10^{-1} = \frac{1}{10} = \frac{1}{10^1}$$
$$10^{-2} = \frac{1}{10} \cdot \frac{1}{10} = \frac{1}{10^2}$$
$$10^{-3} = \frac{1}{10^2} \cdot \frac{1}{10} = \frac{1}{10^3}$$

If the pattern is extended, then we would predict that $10^{-n} = \frac{1}{10^n}$. Notice that this is inductive reasoning and hence is not a full mathematical justification.

Consider whether the property $a^m \cdot a^n = a^{m+n}$ can be extended to include all powers of a, where the exponents are integers. For example, is it true that $2^4 \cdot 2^{-3} = 2^{4+^{-3}} = 2^1$? The definitions of 2^{-3} and the properties of nonnegative exponents ensure this is true, as shown next.

$$2^4 \cdot 2^{-3} = 2^4 \cdot \frac{1}{2^3} = \frac{2^4}{2^3} = \frac{2^1 \cdot 2^3}{2^3} = 2^1$$

Also, $2^{-4} \cdot 2^{-3} = 2^{-4+^{-3}} = 2^{-7}$ is true because

$$2^{-4} \cdot 2^{-3} = \frac{1}{2^4} \cdot \frac{1}{2^3} = \frac{1 \cdot 1}{2^4 \cdot 2^3} = \frac{1}{2^{4+3}} = \frac{1}{2^7} = 2^{-7}$$

In general, with integer exponents, the following theorem holds.

Theorem 6–15

For any nonzero rational number a and any integers m and n, $a^m \cdot a^n = a^{m+n}$.

REMARK   If $a = 0$ then Theorem 6–15 is still valid as long as $m \neq 0, n \neq 0$ and $m + n \neq 0$.

Other properties of exponents can be developed by using the properties of rational numbers. For example,

$$\frac{2^5}{2^3} = \frac{2^3 \cdot 2^2}{2^3} = 2^2 = 2^{5-3} \qquad \frac{2^5}{2^8} = \frac{2^5}{2^5 \cdot 2^3} = \frac{1}{2^3} = 2^{-3} = 2^{5-8}$$

With integer exponents, the following theorem holds.

Theorem 6–16

For any nonzero rational number a and for any integers m and n, $\frac{a^m}{a^n} = a^{m-n}$.

At times, we may encounter an expression like $(2^4)^3$. This expression can be written as a single power of 2 as follows:

$$(2^4)^3 = 2^4 \cdot 2^4 \cdot 2^4 = 2^{4+4+4} = 2^{3 \cdot 4} = 2^{12}$$

In general, if a is any rational number and m and n are positive integers, then

$$(a^m)^n = \underbrace{a^m \cdot a^m \cdot a^m \cdot \ldots \cdot a^m}_{n \text{ factors}} = a^{\overbrace{m+m+\ldots+m}^{n \text{ terms}}} = a^{nm} = a^{mn}$$

Does this theorem hold for negative-integer exponents? For example, does $(2^3)^{-4} = 2^{(3)(-4)} = 2^{-12}$? The answer is yes because $(2^3)^{-4} = \dfrac{1}{(2^3)^4} = \dfrac{1}{2^{12}} = 2^{-12}$. Also, $(2^{-3})^4 = \left(\dfrac{1}{2^3}\right)^4 = \dfrac{1}{2^3} \cdot \dfrac{1}{2^3} \cdot \dfrac{1}{2^3} \cdot \dfrac{1}{2^3} = \dfrac{1^4}{(2^3)^4} = \dfrac{1}{2^{12}} = 2^{-12}$.

Theorem 6–17

For any rational number $a \neq 0$ and any integers m and n,
$$(a^m)^n = a^{mn}$$

Using the definitions and theorems developed, we can derive additional properties. Notice, for example, that

$$\left(\frac{2}{3}\right)^4 = \frac{2}{3} \cdot \frac{2}{3} \cdot \frac{2}{3} \cdot \frac{2}{3} = \frac{2 \cdot 2 \cdot 2 \cdot 2}{3 \cdot 3 \cdot 3 \cdot 3} = \frac{2^4}{3^4}$$

This property can be generalized as follows.

Theorem 6–18

For any nonzero rational number $\dfrac{a}{b}$ and any integer m,
$$\left(\frac{a}{b}\right)^m = \frac{a^m}{b^m}$$

From the definition of negative exponents, the preceding theorem, and division of fractions, we have

$$\left(\frac{a}{b}\right)^{-m} = \frac{1}{\left(\dfrac{a}{b}\right)^m} = \frac{1}{\dfrac{a^m}{b^m}} = \frac{b^m}{a^m} = \left(\frac{b}{a}\right)^m$$

Theorem 6–19

For any nonzero rational number $\dfrac{a}{b}$ and any integer m, $\left(\dfrac{a}{b}\right)^{-m} = \left(\dfrac{b}{a}\right)^m$.

A property similar to the one in Theorem 6–18 holds for multiplication. For example,

$$(2 \cdot 3)^{-3} = \frac{1}{(2 \cdot 3)^3} = \frac{1}{2^3 \cdot 3^3} = \left(\frac{1}{2^3}\right) \cdot \left(\frac{1}{3^3}\right) = 2^{-3} \cdot 3^{-3}$$

and in general, it is true that $(a \cdot b)^m = a^m \cdot b^m$ if a and b are nonzero rational numbers and m is an integer.

The definitions and properties of exponents are summarized in the following definition and theorem. For any nonzero rational numbers a and b and integers m and n (except as noted), we have the following.

Definition of a to an Integer Power

1. $a^m = \underbrace{a \cdot a \cdot a \cdot \ldots \cdot a}_{m \text{ factors}}$, where m is a positive integer and a is any rational number

2. $a^0 = 1$

3. $a^{-m} = \dfrac{1}{a^m}$

Theorem 6–20: Properties of Exponents

1. $a^m \cdot a^n = a^{m+n}$

2. $\dfrac{a^m}{a^n} = a^{m-n}$

3. $(a^m)^n = a^{nm}$

4. $\left(\dfrac{a}{b}\right)^m = \dfrac{a^m}{b^m}$

5. $(ab)^m = a^m b^m$

6. $\left(\dfrac{a}{b}\right)^{-m} = \left(\dfrac{b}{a}\right)^m$

Notice that properties 4 and 5 as well as 7 and 8 are for multiplication and division. Analogous properties do not hold for addition and subtraction. For example, in general, $(a + b)^{-1} \neq a^{-1} + b^{-1}$. To see why, a numerical example is sufficient, but it is instructive to write each side with positive exponents:

$$(a + b)^{-1} = \frac{1}{a + b}$$

$$a^{-1} + b^{-1} = \frac{1}{a} + \frac{1}{b}$$

and we know that in general, $\dfrac{1}{a + b} \neq \dfrac{1}{a} + \dfrac{1}{b}$.

Example 6-22

In each of the following, use properties of exponents to justify the equality or inequality:

a. $(^-x)^{-2} \neq ^-x^{-2}$

b. $(^-x)^{-3} = ^-x^{-3}$

c. $ab^{-1} \neq (ab)^{-1}$

d. $(a^{-2} b^{-2})^{-1} = a^2 b^2$

e. $(a^{-2} + b^{-2})^{-1} \neq a^2 + b^2$

Solution   a. $(^-x)^{-2} = \dfrac{1}{(^-x)^2} = \dfrac{1}{x^2}$

$^-x^{-2} = -(x^{-2}) = {}^-\left(\dfrac{1}{x^2}\right) = \dfrac{^-1}{x^2}$

Hence, $(^-x)^{-2} \neq ^-x^{-2}$.

b. $(-x)^{-3} = \dfrac{1}{(-x)^3} = \dfrac{1}{-x^3} = {}^{-}\!\left(\dfrac{1}{x^3}\right)$

$-x^{-3} = -(x^{-3}) = {}^{-}\!\left(\dfrac{1}{x^3}\right)$

Hence, $(-x)^{-3} = -x^{-3}$.

c. $ab^{-1} = a(b^{-1}) = a \cdot \dfrac{1}{b} = \dfrac{a}{b}$, but $(ab)^{-1} = \dfrac{1}{ab}$. Hence, $ab^{-1} \neq (ab)^{-1}$.

d. $(a^{-2}b^{-2})^{-1} = (a^{-2})^{-1}(b^{-2})^{-1}$

$\qquad = a^{(-2)(-1)}b^{(-2)(-1)}$

$\qquad = a^2 b^2$

e. $(a^{-2} + b^{-2})^{-1} = \left(\dfrac{1}{a^2} + \dfrac{1}{b^2}\right)^{-1} = \left(\dfrac{a^2 + b^2}{a^2 b^2}\right)^{-1} = \dfrac{a^2 b^2}{a^2 + b^2} \neq a^2 + b^2$

Observe that all the properties of exponents refer to powers with either the same base or the same exponent. To evaluate expressions using exponents where different bases and powers are used, perform all the computations or rewrite the expressions in either the same base or the same exponent, if possible. For example, $\dfrac{27^4}{81^3}$ can be rewritten $\dfrac{27^4}{81^3} = \dfrac{(3^3)^4}{(3^4)^3} = \dfrac{3^{12}}{3^{12}} = 1$.

Example 6-23

Write each of the following in simplest form using positive exponents in the final answer:

a. $16^2 \cdot 8^{-3}$ **b.** $20^2 \div 2^4$ **c.** $(10^{-1} + 5 \cdot 10^{-2} + 3 \cdot 10^{-3}) \cdot 10^3$ **d.** $(x^3 y^{-2})^{-4}$

Solution   **a.** $16^2 \cdot 8^{-3} = (2^4)^2 \cdot (2^3)^{-3} = 2^8 \cdot 2^{-9} = 2^{8+-9} = 2^{-1} = \dfrac{1}{2}$

b. $\dfrac{20^2}{2^4} = \dfrac{(2^2 \cdot 5)^2}{2^4} = \dfrac{2^4 \cdot 5^2}{2^4} = 5^2$

c. $(10^{-1} + 5 \cdot 10^{-2} + 3 \cdot 10^{-3}) \cdot 10^3 = 10^{-1} \cdot 10^3 + 5 \cdot 10^{-2} \cdot 10^3 +$

$\qquad\qquad\qquad\qquad\qquad\qquad\qquad\qquad 3 \cdot 10^{-3} \cdot 10^3$

$\qquad\qquad\qquad\qquad = 10^{-1+3} + 5 \cdot 10^{-2+3} + 3 \cdot 10^{-3+3}$

$\qquad\qquad\qquad\qquad = 10^2 + 5 \cdot 10^1 + 3 \cdot 10^0$

$\qquad\qquad\qquad\qquad = 153$

d. $(x^3 y^{-2})^{-4} = x^{-12} y^8 = \dfrac{1}{x^{12}} \cdot y^8 = \dfrac{y^8}{x^{12}}$

BRAIN TEASER   A castle in the faraway land of Aluossim was surrounded by four moats. One day, the castle was attacked and captured by a fierce tribe from the north. Guards were stationed at each bridge. Prince Juan was allowed to take a number of bags of gold from the castle as he went into exile. However, the guard at the first bridge demanded half the bags of gold plus one more bag. Juan met this demand and proceeded to the next bridge. The guards at the second, third, and fourth bridges made identical demands, all of which the prince met. When Juan finally crossed all the bridges, a single bag of gold was left. With how many bags did Juan start?

Assessment 6-3A

1. In the following figures, a unit rectangle is used to illustrate the product of two fractions. Name the fractions and their products.

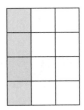 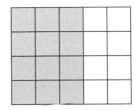

   a.   b.

2. Use a rectangular region to illustrate each of the following products:

   a. $\dfrac{1}{3} \cdot \dfrac{3}{4}$

   b. $\dfrac{2}{3} \cdot \dfrac{1}{5}$

3. Find each of the following products. Write your answers in simplest form.

   a. $\dfrac{19}{65} \cdot \dfrac{26}{98}$ b. $\dfrac{a}{b} \cdot \dfrac{b^2}{a^2}$ c. $\dfrac{xy}{z} \cdot \dfrac{z^2 a}{x^3 y^2}$

4. Use the distributive property to find each product.

   a. $4\dfrac{1}{2} \cdot 2\dfrac{1}{3}$ $\left[Hint: \left(4 + \dfrac{1}{2} \right)\left(2 + \dfrac{1}{3} \right). \right]$

   b. $3\dfrac{1}{3} \cdot 2\dfrac{1}{2}$

5. Find the multiplicative inverse of each of the following:

   a. $\dfrac{^-1}{3}$ b. $3\dfrac{1}{3}$

   c. $\dfrac{x}{y}$, if $x \neq 0$ and $y \neq 0$ d. $^-7$

6. Solve for x in each of the following:

   a. $\dfrac{2}{3}x = \dfrac{7}{6}$ b. $\dfrac{3}{4} \div x = \dfrac{1}{2}$

   c. $\dfrac{5}{6} + \dfrac{2}{3}x = \dfrac{3}{4}$ d. $\dfrac{2x}{3} - \dfrac{1}{4} = \dfrac{x}{6} + \dfrac{1}{2}$

7. Show that the following properties do not hold for the division of rational numbers:

   a. Commutative

   b. Associative

8. Compute the following mentally. Find the exact answers.

   a. $3\dfrac{1}{4} \cdot 8$ b. $7\dfrac{1}{4} \cdot 4$

   c. $9\dfrac{1}{5} \cdot 10$ d. $8 \cdot 2\dfrac{1}{4}$

9. Choose the number that best approximates each of the following from among the numbers in parentheses:

   a. $3\dfrac{11}{12} \cdot 5\dfrac{3}{100}$ $(8, 20, 15, 16)$

   b. $2\dfrac{1}{10} \cdot 7\dfrac{7}{8}$ $(16, 14, 4, 3)$

   c. $\dfrac{1}{101} \div \dfrac{1}{103}$ $\left(0, 1, \dfrac{1}{2}, \dfrac{1}{4} \right)$

10. Estimate the following:

    a. $5\dfrac{4}{5} \cdot 3\dfrac{1}{10}$

    b. $4\dfrac{10}{11} \cdot 5\dfrac{1}{8}$

11. Without actually doing the computations, choose the phrase in parentheses that correctly describes each:

    a. $\dfrac{13}{14} \cdot \dfrac{17}{19}$ (greater than 1, less than 1)

    b. $3\dfrac{2}{7} \div 5\dfrac{1}{9}$ (greater than 1, less than 1)

    c. $4\dfrac{1}{3} \div 2\dfrac{3}{100}$ (greater than 2, less than 2)

12. A sewing project requires $6\dfrac{1}{8}$ yd of material that sells for 62¢ per yard and $3\dfrac{1}{4}$ yd of material that sells for 81¢ per yard. Choose from the best estimate for the cost of the project:

    a. Between $2 and $4   b. Between $4 and $6
    c. Between $6 and $8   d. Between $8 and $10

13. Five-eighths of the students at Salem State College live in dormitories. If 6000 students at the college live in dormitories, how many students are there in the college?

14. Alberto owns $\dfrac{5}{9}$ of the stock in the N.W. Tofu Company. His sister Renatta owns half as much stock as Alberto. What part of the stock is owned by neither Alberto nor Renatta?

15. A suit is on sale for $180. What was the original price of the suit if the discount was $\dfrac{1}{4}$ of the original price?

16. John took all his money out of his savings account. He spent $50 on a radio and $\dfrac{3}{5}$ of what remained on presents. Half of what was left he put back in his checking account, and the remaining $35 he donated to charity. How much money did John originally have in his savings account?

17. Al gives $\frac{1}{2}$ of his marbles to Bev. Bev gives $\frac{1}{2}$ of these to Carl. Carl gives $\frac{1}{2}$ of these to Dani. If Dani was given four marbles, how many did Al have originally?

18. Write each of the following in simplest form using positive exponents in the final answer:

  a. $3^{-7} \cdot 3^{-6}$ **b.** $3^7 \cdot 3^6$

  c. $5^{15} \div 5^4$ **d.** $5^{15} \div 5^{-4}$

  e. $(^-5)^{-2}$ **f.** $\dfrac{a^2}{a^{-3}}$, where $a \neq 0$

19. Write each of the following in simplest form using positive exponents in the final answer:

  a. $\left(\dfrac{1}{2}\right)^3 \cdot \left(\dfrac{1}{2}\right)^7$ **b.** $\left(\dfrac{1}{2}\right)^9 \div \left(\dfrac{1}{2}\right)^6$

  c. $\left(\dfrac{2}{3}\right)^5 \cdot \left(\dfrac{4}{9}\right)^2$ **d.** $\left(\dfrac{3}{5}\right)^7 \div \left(\dfrac{3}{5}\right)^7$

20. If a and b are rational numbers, with $a \neq 0$ and $b \neq 0$, and if m and n are integers, which of the following are true and which are false? Justify your answers.

  a. $a^m \cdot b^n = (ab)^{m+n}$ **b.** $a^m \cdot b^n = (ab)^{mn}$

  c. $a^m \cdot b^m = (ab)^{2m}$ **d.** $(ab)^0 = 1$

  e. $(a + b)^m = a^m + b^m$ **f.** $(a + b)^{-m} = \dfrac{1}{a^m} + \dfrac{1}{b^m}$

21. Solve for the integer n in each of the following:

  a. $2^n = 32$ **b.** $n^2 = 36$

  c. $2^n \cdot 2^7 = 2^5$ **d.** $2^n \cdot 2^7 = 8$

22. Solve each of the following inequalities for x, where x is an integer:

  a. $3^x \leq 81$ **b.** $4^x < 8$

  c. $3^{2x} > 27$ **d.** $2^x > 1$

23. Determine which fraction in each of the following pairs is greater:

  a. $\left(\dfrac{1}{2}\right)^3$ or $\left(\dfrac{1}{2}\right)^4$ **b.** $\left(\dfrac{3}{4}\right)^{10}$ or $\left(\dfrac{3}{4}\right)^8$

  c. $\left(\dfrac{4}{3}\right)^{10}$ or $\left(\dfrac{4}{3}\right)^8$ **d.** $\left(\dfrac{3}{4}\right)^{10}$ or $\left(\dfrac{4}{5}\right)^{10}$

24. Suppose the number of bacteria in a certain culture is given as a function of time by $Q(t) = 10^{10}\left(\dfrac{6}{5}\right)^t$, where t is the time in seconds and $Q(t)$ is the number of bacteria after t sec. Find the following:

  a. The initial number of bacteria (that is, the number of bacteria at $t = 0$)

  b. The number of bacteria after 2 sec

25. Let $S = \dfrac{1}{2} + \dfrac{1}{2^2} + \dfrac{1}{2^3} + \dots + \dfrac{1}{2^{64}}$.

  a. Use the distributive property of multiplication over addition to find an expression for $2S$.

  b. Show that $2S - S = S = 1 - \left(\dfrac{1}{2}\right)^{64}$.

  c. Find a simple expression for the sum

$$\frac{1}{2} + \frac{1}{2^2} + \frac{1}{2^3} + \dots + \frac{1}{2^n}$$

26. If $f(n) = \dfrac{2}{5} \cdot 3^{-n}$, find the following:

  a. $f(0),\ f(^-1),\ f(1)$

  b. n if $f(n) = \dfrac{2}{5} \cdot 9^{-10}$

27. If the nth term of a sequence is given by $a_n = 3 \cdot 2^{-n}$, answer the following:

  a. Find the first five terms.

  b. Show that the first five terms are in a geometric sequence.

  c. Which terms are less than $\dfrac{3}{1000}$?

28. In the following, determine which number is greater:

  a. 32^{50} or 4^{100} **b.** $(^-27)^{-15}$ or $(^-3)^{-75}$

29. Show that the arithmetic mean of two rational numbers is between the two numbers; that is, prove that $0 < \dfrac{a}{b} < \dfrac{1}{2}\left(\dfrac{a}{b} + \dfrac{c}{d}\right) < \dfrac{c}{d}$.

Assessment 6-3B

1. Use a rectangular region to illustrate each of the following products:

  a. $\dfrac{2}{5} \cdot \dfrac{1}{3}$ **b.** $\dfrac{2}{3} \cdot \dfrac{2}{3}$

  c. $\dfrac{^-5}{2} \cdot 2\dfrac{1}{2}$ **d.** $2\dfrac{3}{4} \cdot 2\dfrac{1}{3}$

  e. $\dfrac{a^2}{b^3} \cdot \dfrac{b^2}{a^3}$ **f.** $\dfrac{x^3 y^2}{z} \cdot \dfrac{z}{x^2 y}$

2. Find each of the following products of rational numbers. Write your answers in simplest form.

  a. $2\dfrac{1}{3} \cdot 3\dfrac{3}{4}$ **b.** $\dfrac{22}{7} \cdot 4\dfrac{2}{3}$

3. Use the distributive property to find each product of rational numbers.

  a. $2\dfrac{1}{3} \cdot 4\dfrac{3}{5}$ **b.** $\left(\dfrac{x}{y} + 1\right)\left(\dfrac{y}{x} - 1\right)$

  c. $248\dfrac{2}{5} \cdot 100\dfrac{1}{8}$

4. Solve for x in each of the following:
   a. $\frac{2}{3}x = \frac{11}{6}$ b. $\frac{3}{4} \div x = \frac{1}{3}$
   c. $\frac{5}{6} - \frac{2}{3}x = \frac{3}{4}$ d. $\frac{2x}{3} + \frac{1}{4} = \frac{x}{6} - \frac{1}{2}$

5. Find a fraction such that if you add the denominator to the numerator and place it over the original denominator, the new fraction has triple the value of the original fraction.

6. Compute the following mentally. Find the exact answers.
   a. $3\frac{1}{2} \cdot 8$ b. $7\frac{3}{4} \cdot 4$
   c. $9\frac{1}{5} \cdot 6$ d. $8 \cdot 2\frac{1}{3}$
   e. $3 \div \frac{1}{2}$ f. $3\frac{1}{2} \div \frac{1}{2}$
   g. $3 \div \frac{1}{3}$ h. $4\frac{1}{2} \div 2$

7. Choose the number that best approximates each of the following from among the numbers in parentheses:
   a. $20\frac{2}{3} \div 9\frac{7}{8} \left(2, 180, \frac{1}{2}, 10\right)$
   b. $3\frac{1}{20} \cdot 7\frac{77}{100} \left(21, 24, \frac{1}{20}, 32\right)$
   c. $\frac{1}{10^3} \div \frac{1}{1001} \left(\frac{1}{10^3}, 1, 1001, 0\right)$

8. Without actually doing the computations, choose the phrase in parentheses that correctly describes each:
   a. $4\frac{1}{3} \div 2\frac{13}{100}$ (greater than 2, less than 2)
   b. $16 \div 4\frac{3}{18}$ (greater than 4, less than 4)
   c. $16 \div 3\frac{8}{9}$ (greater than 4, less than 4)

9. When you multiply a certain number by 3 and then subtract $\frac{7}{18}$, you get the same result as when you multiply the number by 2 and add $\frac{5}{12}$. What is the number?

10. Di Paloma University had a faculty reduction and lost $\frac{1}{5}$ of its faculty. If 320 faculty members were left after the reduction, how many members were there originally?

11. A person has $29\frac{1}{2}$ yd of material available to make doll uniforms. Each uniform requires $\frac{3}{4}$ yd of material.
    a. How many uniforms can be made?
    b. How much material will be left over?

12. A suit is on sale for $240. What was the original price of the suit if the discount was $\frac{1}{4}$ of the original price?

13. Every employee's salary at the Sunrise Software Company increases each year by $\frac{1}{10}$ of that person's salary the previous year.
    a. If Martha's present annual salary is $100,000, what will her salary be in 2 yr?
    b. If Aaron's present salary is $99,000, what was his salary 1 yr ago?
    c. If Juanita's present salary is $363,000, what was her salary 2 yr ago?

14. Jasmine is reading a book. She has finished $\frac{3}{4}$ of the book and has 82 pages left to read. How many pages has she read?

15. Peter, Paul, and Mary start at the same time walking around a circular track in the same direction. Peter takes $\frac{1}{2}$ hr to walk around the track. Paul takes $\frac{5}{12}$ hr, and Mary takes $\frac{1}{3}$ hr.
    a. How many minutes does it take each person to walk around the track?
    b. How many times will each person go around the track before all three meet again at the starting line?

16. The formula for converting degrees Celsius (C) to degrees Fahrenheit (F) is $F = \left(\frac{9}{5}\right) \cdot C + 32$.
    a. If Samantha reads that the temperature is 32°C in Spain, what is the Fahrenheit temperature?
    b. If the temperature dropped to ⁻40°F in West Yellowstone, what is the temperature in degrees Celsius?

17. Al gives $\frac{1}{2}$ of his marbles to Bev. Bev gives $\frac{1}{3}$ of these to Carl. Carl gives $\frac{1}{4}$ of these to Dani. If Dani was given four marbles, how many did Al have originally?

18. The normal brain weight for an African bull elephant is approximately 9 1/4 lb. Approximately how much would be the weight of the brains of 13 of these elephants?

19. Write each of the following in simplest form using positive exponents in the final answer:
    a. $\left(\frac{1}{3}\right)^{-1}$ b. $\frac{a^{-3}}{a}$
    c. $\frac{(a^{-4})^3}{a^{-4}}$ d. $\frac{a}{a^{-1}}$
    e. $\frac{a^{-3}}{a^{-2}}$

20. Write each of the following in simplest form using positive exponents in the final answer:
    a. $\left(\frac{1}{2}\right)^{10} \div \left(\frac{1}{2}\right)^2$ b. $\left(\frac{2}{3}\right)^5 \cdot \left(\frac{4}{9}\right)^{-2}$
    c. $\left(\frac{3}{5}\right)^7 \div \left(\frac{5}{3}\right)^4$ d. $\left[\left(\frac{5}{6}\right)^7\right]^3$

21. If a and b are rational numbers, with $a \neq 0$ and $b \neq 0$, and if m and n are integers, which of the following are true and which are false? Justify your answers.

a. $\dfrac{a^m}{b^n} = \left(\dfrac{a}{b}\right)^{m-n}$ b. $(ab)^{-m} = \dfrac{1}{a^m} \cdot \dfrac{1}{b^m}$

c. $\left(\dfrac{2}{a^{-1} + b^{-1}}\right)^{-1} = \dfrac{1}{2} \cdot \dfrac{1}{a+b}$

d. $2(a^{-1} + b^{-1})^{-1} = \dfrac{2ab}{a+b}$

e. $a^{mn} = a^m \cdot a^n$ f. $\left(\dfrac{a}{b}\right)^{-1} = \dfrac{b}{a}$

22. Solve if possible for n in each of the following:

a. $2^n = {}^-32$ b. $n^3 = \dfrac{{}^-1}{27}$

c. $2^n \cdot 2^7 = 1024$ d. $2^n \cdot 2^7 = 64$

e. $(2 + n)^2 = 2^2 + n^2$ f. $3^n = 27^5$

23. Solve each of the following inequalities for x, where x is an integer:

a. $3^x \geq 81$ b. $4^x \geq 8$

c. $3^{2x} \leq 27$ d. $2^x < 1$

24. Determine which fraction in each of the following pairs is greater:

a. $\left(\dfrac{4}{3}\right)^{10}$ or $\left(\dfrac{4}{3}\right)^{8}$ b. $\left(\dfrac{3}{4}\right)^{10}$ or $\left(\dfrac{4}{5}\right)^{10}$

c. $\left(\dfrac{4}{3}\right)^{10}$ or $\left(\dfrac{5}{4}\right)^{10}$ d. $\left(\dfrac{3}{4}\right)^{100}$ or $\left(\dfrac{3}{4} \cdot \dfrac{9}{10}\right)^{100}$

25. Let $S = \dfrac{1}{3} + \dfrac{1}{3^2} + \dfrac{1}{3^3} + \ldots + \dfrac{1}{3^{64}}$.

a. Use the distributive property of multiplication over addition to find an expression for $3S$.

b. Show that $3S - S = 2S = 1 - \left(\dfrac{1}{3}\right)^{64}$.

c. Find a simple expression for the sum

$$\dfrac{1}{3} + \dfrac{1}{3^2} + \dfrac{1}{3^3} + \ldots + \dfrac{1}{3^n}.$$

26. In an arithmetic sequence, the first term is 1 and the hundredth term is 2. Find the following:

a. The 50th term

b. The sum of the first 50 terms

27. If $f(n) = \dfrac{3}{4} \cdot 2^n$, find the following:

a. $f(0)$ b. $f(5)$ c. $f({}^-5)$

d. The greatest integer value of n for which

$$f(n) < \dfrac{3}{1400}$$

28. In the following, determine which number is greater:

a. 32^{100} or 4^{200}

b. $({}^-27)^{-15}$ or $({}^-3)^{-50}$

29. There is a simple method for squaring any number that consists of a whole number and $\dfrac{1}{2}$. For example

$$\left(3\dfrac{1}{2}\right)^2 = 3 \cdot 4 + \left(\dfrac{1}{2}\right)^2 = 12\dfrac{1}{4}; \left(4\dfrac{1}{2}\right)^2 = 4 \cdot 5 +$$

$$\left(\dfrac{1}{2}\right)^2 = 20\dfrac{1}{4}; \left(5\dfrac{1}{2}\right)^2 = 5 \cdot 6 + \left(\dfrac{1}{2}\right)^2 = 30\dfrac{1}{4}.$$

a. Write a statement for $\left(n + \dfrac{1}{2}\right)^2$ that generalizes these examples, where n is a whole number.

★b. Justify this procedure.

30. Consider these products:

First product: $\left(1 + \dfrac{1}{1}\right)\left(1 + \dfrac{1}{2}\right)$

Second product: $\left(1 + \dfrac{1}{1}\right)\left(1 + \dfrac{1}{2}\right)\left(1 + \dfrac{1}{3}\right)$

Third product: $\left(1 + \dfrac{1}{1}\right)\left(1 + \dfrac{1}{2}\right)\left(1 + \dfrac{1}{3}\right)\left(1 + \dfrac{1}{4}\right)$

a. Calculate the value of each product. Based on the pattern in your answers, guess the value of the fourth product. Then check to determine if your guess is correct.

b. Guess the value of the 100th product.

c. Find as simple an expression as possible for the nth product.

31. Show that the arithmetic mean $\dfrac{1}{2}\left(\dfrac{a}{b} + \dfrac{c}{d}\right)$ of two rational numbers $\dfrac{a}{b}$ and $\dfrac{c}{d}$ corresponds to their midpoint on the number line.

Mathematical Connections 6-3

Communication

1. Amy says that dividing a number by $\dfrac{1}{2}$ is the same as taking half of a number. How do you respond?

2. Noah says that dividing a number by 2 is the same as multiplying it by $\dfrac{1}{2}$. He wants to know if he is right, and if so, why. How do you respond?

3. Suppose you divide a natural number, n, by a positive rational number less than 1. Will the answer always be less than n, sometimes less than n, or never less than n? Why?

4. If the fractions represented by points C and D on the following number line are multiplied, what point best represents the product? Explain why.

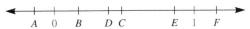

$$A \quad 0 \quad B \quad D\, C \qquad E \quad 1 \quad F$$

5. If the product of two numbers is 1 and one of the numbers is greater than 1, what do you know about the other number? Explain your answer.

6. Bente says to do the problem $12\frac{1}{4} \div 3\frac{3}{4}$ you just find

$12 \div 3 = 4$ and $\frac{1}{4} \div \frac{3}{4} = \frac{1}{3}$ to get $4\frac{1}{3}$. How do you respond?

7. Carl says that every rational number has a multiplicative inverse. How do you respond?

8. Dani says that if we have $\frac{3}{4} \cdot \frac{2}{5}$, we could just do $\frac{3}{5} \cdot \frac{2}{4} = \frac{3}{5} \cdot \frac{1}{2} = \frac{3}{10}$. Is she correct? Explain why.

9. Joel says that $2\frac{2}{5} \cdot 3\frac{4}{5} = 2\frac{4}{5} \cdot 3\frac{2}{5}$ because multiplication is commutative. Is he right? Explain why or why not.

10. a. Amal says that she can finish a job in $\frac{1}{2}$ of a day working by herself. Her son Sharif can finish the same job in $\frac{1}{4}$ of a day working alone. How long will it take to finish the job if they work together?

  b. If Amal can finish a job in a hours and Sharif in b hours, then how long will it take to finish the job if they work together?

11. Ben and Noah started running toward the other's hometown at the same time, each at a constant speed, Ben from Bentown (B) and Noah from Noahtown (N). After 3 hr, Ben reached town N while Noah ran only $\frac{3}{4}$ of the distance between the towns. Answer the following and carefully explain your reasoning so that a fifth grader would understand it.

  a. If Ben's speed was 6 mph, find Noah's speed.

  b. If Ben's speed for the last $\frac{1}{4}$ of the distance between the towns had decreased by 1 mph, from his previous speed, how long would the last $\frac{1}{4}$ of the distance have taken him?

12. Abby walks from town A to town B at a constant speed of 3 mph, and back from town B to town A at a constant speed of 5 mph. Answer the following:

  a. What is Abby's average speed on the round-trip walk? (It is not 4 mph.) Justify your reasoning.

  b. What is Abby's average speed on the round-trip if she walked one way at a mph and returned at b mph?

Open-Ended

13. In the book *Knowing and Teaching Elementary Mathematics*, Liping Ma presents the following scenario:

Imagine that you are teaching division with fractions. To make this situation meaningful for kids, something that many teachers try to do is relate mathematics to other things. Sometimes they try to come up with real-world situations or story problems to show the application of some particular piece of content. What would you say would be a good story or model for $1\frac{3}{4} \div \frac{1}{2}$? (p. 55)

  a. How would you respond to her question?

  b. Use your story to obtain the answers.

  c. If possible, obtain a copy of Ma's book and read how U.S. teachers responded to this task compared to Chinese teachers. Report your findings.

14. Would you use the problem in the following cartoon in your class? Why or why not? Solve the problem.

Cooperative Learning

15. Choose a brick building on your campus. Measure the height of one brick and the thickness of mortar between bricks. Estimate the height of the building and then calculate the height of the building. Were rational numbers used in your computations?

Questions from the Classroom

16. Jim is not sure when to use multiplication by a fraction and when to use division. He has the following list of problems. How would you help him solve these problems in a way that would enable him to solve similar problems on his own?

  a. $\frac{3}{4}$ of a packet of sugar fills $\frac{1}{2}$ c. How many cups of sugar are in a full packet of sugar?

  b. How many packets of sugar will fill 2 c?

  c. If $\frac{1}{3}$ c sugar is required to make two loaves of challah, how many cups of sugar are needed for three loaves?

d. If $\frac{3}{4}$ c sugar is required for 1 gal of punch, how many gallons can be made with 2 c of sugar?

e. If you have $22\frac{3}{8}$ in. of ribbon, and need $1\frac{1}{4}$ in. to decorate one doll, how many dolls can be decorated, and how much ribbon will be left over?

17. When working on the problem of simplifying

$$\frac{3}{4} \cdot \frac{1}{2} \cdot \frac{2}{3}$$

a student did the following:

$$\frac{3}{4} \cdot \frac{1}{2} \cdot \frac{2}{3} = \left(\frac{3 \cdot 1}{4 \cdot 2}\right)\left(\frac{3 \cdot 2}{4 \cdot 3}\right) = \frac{3}{8} \cdot \frac{6}{12} = \frac{18}{96}$$

What was the error, if any?

18. A student claims that division always makes things smaller so $5 \div \left(\frac{1}{2}\right)$ can't be 10 because that is greater than the number she started with. How do you respond?

19. A student simplified the fraction $\frac{m+n}{p+n}$ to $\frac{m}{p}$. How would you help this student?

20. Margie says that $a \div b = \frac{a}{b}$ for any two whole numbers a and $b \neq 0$. She would like to know why. How do you respond?

21. Jillian says she learned that 17 divided by 5 can be written as $17 \div 5 = 3 \text{ R}2$, but she thinks that writing $17 \div 5 = \frac{17}{5} = 3\frac{2}{5}$ is much better. How do you respond?

22. Tyto would like to know how a story problem can help him to figure out the answer to $4\frac{2}{3} \div \frac{1}{3}$ without using the invert-and-multiply algorithm. How do you respond?

23. Fran claimed that $2\frac{1}{2} \cdot \frac{3}{5} = 2\frac{3}{10}$. What did Fran do and how would you help her?

Review Problems

24. Perform each of the following computations. Leave your answers in simplest form.

a. $\frac{^-3}{16} + \frac{7}{4}$ **b.** $\frac{1}{6} + \frac{^-4}{9} + \frac{5}{3}$

c. $\frac{^-5}{2^3 \cdot 3^2} - \frac{^-5}{2 \cdot 3^3}$ **d.** $3\frac{4}{5} + 4\frac{5}{6}$

e. $5\frac{1}{6} - 3\frac{5}{8}$ **f.** $^-4\frac{1}{3} - 5\frac{5}{12}$

25. Each student at Sussex Elementary School takes one foreign language. Two-thirds of the students take Spanish, $\frac{1}{9}$ take French, $\frac{1}{18}$ take German, and the rest take some other foreign language. If there are 720 students in the school, how many do not take Spanish, French, or German?

Third International Mathematics and Science Study (TIMSS) Questions

There are 600 balls in a box, and $\frac{1}{3}$ of the balls are red.

How many red balls are in the box?
    Answer: _____ red balls

TIMSS 2003, Grade 4

A scoop holds $\frac{1}{5}$ kg of flour. How many scoops of flour are needed to fill a bag with 6 kg of flour?
    Answer: _____

TIMSS 2003, Grade 8

National Assessment of Educational Progress (NAEP) Question

Both figures below show the same scale. The marks on the scale have no labels except the zero point.

The weight of the cheese is $\frac{1}{2}$ pound. What is the total weight of the two apples?

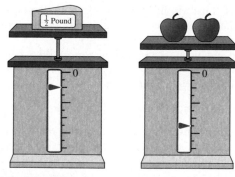

NAEP, Grade 8, 2007

BRAIN TEASER A woman's will decreed that her cats be shared among her three daughters as follows: $\frac{1}{2}$ of the cats to the eldest daughter, $\frac{1}{3}$ of the cats to the middle daughter, and $\frac{1}{9}$ of the cats to the youngest daughter. Since the woman had 17 cats, the daughters decided that they could not carry out their mother's wishes. The judge who held the will agreed to lend the daughters a cat so that they could share the cats as their mother wished. Now, $\frac{1}{2}$ of 18 is 9; $\frac{1}{3}$ of 18 is 6; and $\frac{1}{9}$ of 18 is 2. Since $9 + 6 + 2 = 17$, the daughters were able to divide the 17 cats and return the borrowed cat. They obviously did not need the extra cat to carry out their mother's bequest, but they could not divide 17 into halves, thirds, and ninths. Has the woman's will really been followed?

Hint for Solving the Preliminary Problem

Set up the equation in which the unknown is the cost of the horse's keep. Express the amount of the loss in two different ways in terms of the unknown.

Chapter Outline

I. Fractions and rational numbers

  A. Numbers of the form $\frac{a}{b}$, where a and b are integers and $b \neq 0$, are **rational numbers**. a is the numerator and b is the denominator.

  B. A rational number can be used as follows:

    1. A division problem or the solution to a multiplication problem

    2. A partition, or part, of a whole

    3. A ratio

    4. A probability

  C. **Fundamental Law of Fractions:** For any fraction $\frac{a}{b}$ and any number $c \neq 0$, $\frac{a}{b} = \frac{ac}{bc}$.

  D. Two fractions $\frac{a}{b}$ and $\frac{c}{d}$ are **equal** (**equivalent**) if, and only if, $ad = bc$.

  E. If $\mathrm{GCD}(a, b) = 1$ and $b > 0$, then $\frac{a}{b}$ is said to be in **simplest form**.

  F. If $0 \leq a < b$, then $\frac{a}{b}$ is a **proper fraction**. If $a \geq b > 0$, $\frac{a}{b}$ is an **improper fraction**.

  G. If $b > 0$ and $d > 0$ then: $\frac{a}{b} > \frac{c}{d}$ if, and only if, $ad > bc$.

  H. If $\frac{a}{b} < \frac{c}{d}$, then $\frac{a}{b} < \frac{a + c}{b + d} < \frac{c}{d}$.

  I. Definition of *less than* and *greater than*

$$\frac{a}{b} < \frac{c}{d} \quad \text{if, and only if,} \quad \frac{c}{d} - \frac{a}{b} > 0$$

$$\frac{c}{d} > \frac{a}{b} \quad \text{if, and only if,} \quad \frac{a}{b} < \frac{c}{d}$$

II. Operations on rational numbers

  A. $\dfrac{a}{b} + \dfrac{c}{b} = \dfrac{a + c}{b}$

  B. $\dfrac{a}{b} + \dfrac{c}{d} = \dfrac{ad + bc}{bd}$

  C. $\dfrac{a}{b} - \dfrac{c}{d} = \dfrac{ad - bc}{bd}$

  D. $\dfrac{a}{b} \cdot \dfrac{c}{d} = \dfrac{ac}{bd}$

  E. $\dfrac{a}{b} \div \dfrac{c}{d} = \dfrac{a}{b} \cdot \dfrac{d}{c} = \dfrac{ad}{bc}$, where $c \neq 0$

III. Properties of rational numbers

A.

	Addition	Subtraction	Multiplication	Division
Closure	Yes	Yes	Yes	Yes, except for division by 0
Commutative	Yes	No	Yes	No
Associative	Yes	No	Yes	No
Identity	Yes	No	Yes	No
Inverse	Yes	No	Yes, except 0	No

B. Distributive property of multiplication over addition for rational numbers x, y, and z:

$$x(y + z) = xy + xz$$

C. Denseness property: Between any two rational numbers, there is another rational number.

D. Multiplication property of equality: If $\frac{a}{b}$, $\frac{c}{d}$, and $\frac{e}{f}$ are any rational numbers such that $\frac{a}{b} = \frac{c}{d}$, then $\frac{a}{b} \cdot \frac{e}{f} = \frac{c}{d} \cdot \frac{e}{f}$.

E. Multiplication property of zero for rational numbers: If $\frac{a}{b}$ is any rational number, then $\frac{a}{b} \cdot 0 = 0 = 0 \cdot \frac{a}{b}$.

IV. Exponents

A. $a^m = \underbrace{a \cdot a \cdot a \cdot \ldots \cdot a}_{m \text{ factors}}$, where m is a positive integer and a is a rational number (definition)

B. Definitions and Theorems about exponents involving rational numbers

1. $a^0 = 1$, where $a \neq 0$ (definition)

2. $a^{-m} = \frac{1}{a^m}$, where $a \neq 0$ (definition)

3. $a^m \cdot a^n = a^{m+n}$

4. $\frac{a^m}{a^n} = a^{m-n}$, where $a \neq 0$

5. $(a^m)^n = a^{mn}$

6. $\left(\frac{a}{b}\right)^m = \frac{a^m}{b^m}$, where $b \neq 0$

7. $(ab)^m = a^m \cdot b^m$

8. $\left(\frac{a}{b}\right)^{-m} = \left(\frac{b}{a}\right)^m$

Chapter Review

1. For each of the following, draw a diagram illustrating the fraction:

   a. $\frac{3}{4}$ b. $\frac{2}{3}$

   c. $\frac{3}{4} \cdot \frac{2}{3}$

2. Write three rational numbers equal to $\frac{5}{6}$.

3. Write each of the following rational numbers in simplest form:

   a. $\frac{24}{28}$ b. $\frac{ax^2}{bx}$ c. $\frac{0}{17}$

   d. $\frac{45}{81}$ e. $\frac{b^2 + bx}{b + x}$ f. $\frac{16}{216}$

   g. $\frac{x + a}{x - a}$ h. $\frac{xa}{x + a}$

4. Replace the comma with $>$, $<$, or $=$ in each of the following pairs to make a true statement:

   a. $\frac{6}{10}, \frac{120}{200}$ b. $\frac{-3}{4}, \frac{-5}{6}$

   c. $\left(\frac{4}{5}\right)^{10}, \left(\frac{4}{5}\right)^{20}$ d. $\left(1 + \frac{1}{3}\right)^2, \left(1 + \frac{1}{3}\right)^3$

5. Find the additive and multiplicative inverses for each of the following:

   a. 3 b. $3\frac{1}{7}$

   c. $\frac{5}{6}$ d. $-\frac{3}{4}$

6. Order the following numbers from least to greatest:

   $$-1\frac{7}{8}, 0, -2\frac{1}{3}, \frac{69}{140}, \frac{71}{140}, \left(\frac{71}{140}\right)^{300}, \frac{1}{2}, \left(\frac{74}{73}\right)^{300}$$

7. Can $\frac{4}{5} \cdot \frac{7}{8} \cdot \frac{5}{14}$ be written as $\frac{4}{8} \cdot \frac{7}{14} \cdot \frac{5}{5}$ to obtain the same answer? Why or why not?

8. Use mental math to compute the following. Explain your method.

   a. $\frac{1}{3} \cdot (8 \cdot 9)$ b. $36 \cdot 1\frac{5}{6}$

9. John has $54\frac{1}{4}$ yd of material. If he needs to cut the cloth into pieces that are $3\frac{1}{12}$ yd long, how many pieces can he cut? How much material will be left over?

10. Without actually performing the given operations, choose the most appropriate estimate (among the numbers in parentheses) for the following expressions:

    a. $\dfrac{30\frac{3}{8}}{4\frac{1}{9}} \cdot \dfrac{8\frac{1}{3}}{3\frac{8}{9}}$ (15, 20, 8)

    b. $\left(\dfrac{3}{800} + \dfrac{4}{5000} + \dfrac{15}{6}\right) \cdot 6$ (15, 0, 132)

    c. $\dfrac{1}{407} \div \dfrac{1}{1609}$ $\left(\dfrac{1}{4}, 4, 0\right)$

11. Justify the invert-and-multiply algorithm for division of rational numbers in two different ways.

12. Write a story problem that models $4\frac{5}{8} \div \frac{1}{2}$. Answer the problem by drawing appropriate diagrams.

13. Find two rational numbers between $\frac{3}{4}$ and $\frac{4}{5}$.

14. Suppose the $\boxed{\div}$ button on your calculator is broken, but the $\boxed{1/x}$ button works. Explain how you could compute 504792/23.

15. Jim is starting a diet. When he arrived home, he ate $\frac{1}{3}$ of the half of a pizza that was left from the previous night. The whole pizza contains approximately 2000 calories. How many calories did Jim consume?

16. If a person got heads on a flip of a fair coin one-half the time and obtained 376 heads, how many times was the coin flipped?

17. If a person obtained 240 heads when flipping a coin 1000 times, what fraction of the time did the person obtain heads? Put the answer in simplest form.

18. If the University of New Mexico won $\frac{3}{4}$ of its women's basketball games and $\frac{5}{8}$ of its men's basketball games, explain whether it is reasonable to say that the university won $\frac{3}{4} + \frac{5}{8}$ of its basketball games.

19. Explain why a negative rational number times a negative rational number is a positive rational number.

20. A student argues that the following fraction is not a rational number because it is not the quotient of two integers:

    $$\dfrac{\frac{2}{3}}{\frac{3}{4}}$$

    How would you respond?

21. Molly wants to fertilize 12 acres of park land. If it takes $9\frac{1}{3}$ bags for each acre, how many bags does she need?

22. If $\frac{2}{3}$ of all students in the academy are female and $\frac{2}{5}$ of those are blondes, what fraction describes the number of blond females in the academy?

23. Explain which is greater: $\dfrac{^-11}{9}$ or $\dfrac{^-12}{10}$.

24. Solve for x in each of the following:

    a. $7^x = 343$

    b. $2^{-3x} = \dfrac{1}{512}$

25. Solve for x in each of the following:

    a. $2x - \dfrac{5}{3} = \dfrac{5}{6}$ b. $x + 2\frac{1}{2} = 5\frac{2}{3}$

    c. $\dfrac{20 + x}{x} = \dfrac{4}{5}$ d. $2x + 4 = 3x - \dfrac{1}{3}$

26. Write each of the following in simplest form. Leave all answers with positive exponents.

    a. $\dfrac{(x^3 a^{-1})^{-2}}{xa^{-1}}$

    b. $\left(\dfrac{x^2 y^{-2}}{x^{-3} y^2}\right)^{-2}$

27. Find each sum or difference.

    a. $\dfrac{3a}{xy^2} + \dfrac{b}{x^2 y^2}$ b. $\dfrac{5}{xy^2} - \dfrac{2}{3x}$

    c. $\dfrac{a}{x^3 y^2 z} - \dfrac{b}{xyz}$ d. $\dfrac{7}{2^3 3^2} + \dfrac{5}{2^2 3^3}$

28. Mike drew the following picture to find out how many pieces of ribbon $\frac{1}{2}$ yd long could be cut from a strip of ribbon $1\frac{3}{4}$ yd long.

(1)	(2)	(3)	left over	

0   $\frac{1}{4}$ $\frac{1}{2}$ $\frac{3}{4}$ 1 $1\frac{1}{4}$ $1\frac{1}{2}$ $1\frac{3}{4}$ 2

From the picture he concluded that $1\frac{3}{4} \div \frac{1}{2}$ is 3 pieces with $\frac{1}{4}$ yd left over, so the answer is $3\frac{1}{4}$ pieces. He checked this using the algorithm $\dfrac{7}{4} \cdot \dfrac{2}{1} = \dfrac{14}{4} = 3\frac{1}{2}$ and is confused why he has two different answers. How would you help him?

29. If a, b, and c are nonzero integers, express the following as a rational number $\frac{x}{y}$, where x and y are integers:

    $$\left(\dfrac{a^{-1} + b^{-1} + c^{-1}}{2}\right)^{-1}$$

Selected Bibliography

Alcoro, P., A. Alston, and N. Katjims. "Fractions Attack! Children Thinking and Talking Mathematically." *Teaching Children Mathematics* 6 (May 2000): 562–567.

Anderson, C., E. Anderson, and E. Wenzel. "Oil and Water Don't Mix, But They Teach Fractions." *Teaching Children Mathematics* 7 (November 2000): 174–178.

Bana, J., B. Farrell, and A. McIntosh. "Student Error Patterns in Fraction and Decimal Concepts." In *Proceedings of the 20th Annual Conference of the Mathematics Education Research Group of Australasia* (pp. 81–87), edited by F. Biddulph and K. Carr. Aotearoa, New Zealand: MERGA (1997).

Bay-Williams, J., and S. Martine. "Thinking Rationally about Number and Operations in the Middle School." *Mathematics Teaching in the Middle School* 8 (February 2003): 282–287.

Behr, M., I. Wachsmuth, T. Post, and R. Lesh. "Order and Equivalence of Rational Numbers; A Clinical Teaching Experiment." *Journal for Research in Mathematics Education* 1984, 15 (5): 323–341.

Boston, M., M. Smith, and A. Hillen. "Building on Students' Intuitive Strategies to Make Sense of Cross Multiplication." *Mathematics Teaching in the Middle School* 9 (November 2003): 150–155.

Burrill, G., et al. *Figure This!* Washington, DC: Widmeyer Corporation, 2002. (www.figurethis.org)

Conference Board of the Mathematical Sciences (CBMS). *The Mathematical Education of Teachers*, vol. 11 of *CBMS Issues in Mathematics Education* Washington, DC: The American Mathematical Society and the Mathematical Association of America, 2001. See also http://www.cbmsweb.org/MET_Document/index.htm

Empson, S. "Equal Sharing and the Roots of Fraction Equivalence." *Teaching Children Mathematics* 7 (March 2001): 421–425.

English, L., and G. Halford. *Mathematics Education: Models and Processes*. Mahwah, N.J.: LEA, 1995.

Fennel, F. "Focal Points—What's Your Focus and Why?" *Teaching Children Mathematics* 14 (January 2008): 315–316.

Glidden, P. "Build Your Own Fraction Computer." *Mathematics Teaching in the Middle School* 8 (December 2002): 204–208.

Gregg, J., and D. Gregg. "Measurement and Fair-Sharing Models for Dividing Fractions." *Mathematics Teaching in the Middle School* 12 (May 2007): 390–396.

Hope, J., and D. Owens. "An Analysis of the Difficulty of Learning Fractions." *Focus on Learning Problems in Mathematics* 1987, 9(4): 25–40.

Kalder, R. "Teaching Preservice Secondary Teachers How to Teach Elementary Mathematics Concepts." *Mathematics Teacher* 2007, 101(2): 146–149.

Lamon, S. *Teaching Fractions and Ratios for Understanding*. Mahwah, NJ: Lawrence Erlbaum Associates, 1999.

Ma, L. *Knowing and Teaching Elementary Mathematics*. Mahwah, N.J.: Lawrence Erlbaum Associates, 1999.

Mack, N. "Connecting to Develop Computational Fluency with Fractions." *Teaching Children Mathematics* 11 (November 2004): 226–231.

Middleton, J., M. van den Heuvel-Panhuizen, and J. Shew. "Using Bar Representations as a Model for Connecting Concepts of Rational Numbers." *Mathematics Teaching in the Middle School* 3 (January 1998): 302–312.

Moss, J., and R. Case. "Developing Children's Understanding of the Rational Numbers: A New Model and an Experimental Curriculum." *Journal For Research in Mathematics Education* 1999, 30(2): 122–147.

Moyer, P., and E. Mailley. "Inchworm and a Half: Developing Fraction and Measurement Concepts Using Mathematical Representations." *Teaching Children Mathematics* 10 (January 2004): 244–252.

Perlwitz, M. "Dividing Fractions: Reconciling Self-Generated Solutions with Algorithmic Answers." *Mathematics Teaching in the Middle School* 10 (October 2005): 278–283.

Perlwitz, M. "Two Students' Constructed Strategies to Divide Fractions." *Mathematics Teaching in the Middle School* 10 (February 2004): 121–126.

Roddick, C., and C. Silvas-Centeno. "Developing Understanding of Fractions through Pattern Blocks and Fair Trade." *Teaching Children Mathematics* 14 (October 2007): 140–145.

Tirosh, D. "Enhancing Prospective Teacher's Knowledge of Children's Conceptions: The Case of Division of Fractions." *Journal For Research in Mathematics Education* 2000 31(5): 5–25.

Tzur, R. "An Integrated Study of Children's Construction of Improper Fractions and the Teacher's Role in Promoting Learning." *Journal For Research in Mathematics Education* 1999, 30(4): 390–416.

Wearne, D., and J. Hiebert, "Constructing and Using Meaning for Mathematical Symbols: The Case of Decimal Fractions." In *Number Concepts and Operations in the Middle Grades*, edited by J. Hiebert and M. Behr. Hillsdale, NJ: LEA, 1988.

Wu, Z. "Multiplying Fractions." *Teaching Children Mathematics* 8 (November 2001): 174–177.

Yang, D., and R. Reys, "One Fraction Problem, Many Solution Paths." *Mathematics Teaching in the Middle School* 7 (September 2001): 10–14.

CHAPTER 7

Decimals and Real Numbers

Preliminary Problem

Jennie challenged Ryan to a contest in which all the digits 0, 1, 2, 3, . . . , 9 are used to create two decimals whose sum is 1. No other restrictions were given. How do you think that Ryan might have answered this challenge?

I n *Principles and Standards* (NCTM 2000), we find that students in grades 6–8 should

- work flexibly with fractions, decimals, and percents to solve problems;
- compare and order fractions, decimals, and percents efficiently and find their approximate locations on a number line; . . .
- develop an understanding of large numbers and recognize and appropriately use exponential, scientific, and calculator notation; . . .
- understand the meaning and effects of arithmetic operations with fractions, decimals, and integers. (p. 214)

In *Focal Points*, as early as grade 4 students are expected to begin relating "their understanding of fractions to reading and writing decimals, comparing and ordering decimals, and estimating decimal or fractional amounts in problem solving. They connect equivalent fractions and decimals by comparing models to symbols and locating equivalent symbols on the number line" (p. 31).

In this chapter, we develop the necessary understanding that teachers must have to help students reach these expectations. To begin, we look at a brief history of decimals.

Although the Hindu-Arabic numeration system discussed in Chapter 2 was used in many places around the sixth century, the extension of the system to decimals by the Dutch scientist Simon Stevin did not take place in European mathematics until about a thousand years later. The only significant improvement in the system since Stevin's time has been in notation. Even today there is no universally accepted form of writing a decimal. For example, in the United States, we write 6.75; in England, this number is written $6 \cdot 75$; and in Germany and France, it is written 6,75.

Decimals play an important part in the mathematics education of students in grades 4 through 8. In Section 7-1, we explore relationships between fractions and decimals and see how decimals are an extension of the base-ten system. Later in the chapter, we consider operations on decimals, properties of decimals, real numbers, and notions of algebra with real numbers.

Historical Note

In 1584, Simon Stevin (1548–1620), a quartermaster general in the Dutch army, wrote *La Thiende (The Tenth)*, a work that gave rules for computing with decimals. He not only stated rules for decimal computations but also suggested practical applications for decimals and recommended that his government adopt them. To show place value, Stevin used circled numerals between digits. For example, he wrote 0.4789 as 4 ① 7 ② 8 ③ 9 ④. ◆

THIENDE.   13
HET ANDER DEEL
DER THIENDE VANDE
WERCKINCHE.

I. VOORSTEL VANDE
VERGADERINGHE.

Wesende ghegeven Thiendetalen te vergaderen: hare Somme te vinden.

T'GHEGHEVEN. Het sijn drie oirdens van Thiendetalen, welcker eerste 27 ③ 8 ① 4 ③ 7 ③ , de tweede, 37 ③ 6 ① 7 ③, 5 ③ , de derde, 875 ③ 7 ① 8 ② 2 ③ , TBEGHEERDE. Wy moeten haer Somme vinden . WERCKING. Men sal de ghegheven ghetalen in oirden stellen als hier neven, die vergaderende naer de ghemeene manie re der vergaderinghe van heelegetalen aldus:

⓪	①	②	③
2 7	8	4	7
3 7	6	7	5
8 7	5	7	8	2
9 4 1	3	0	4

7-1   Introduction to Decimals

The word *decimal* comes from the Latin *decem*, meaning "ten." Most people first encounter decimals when dealing with money. For example, on a sign that says a bike costs $128.95, the dot in $128.95 is the **decimal point**. Because $0.95 is $\frac{95}{100}$ of a dollar, we have $128.95 = 128 + \frac{95}{100}$ dollars. Because $0.95 is 9 dimes and 5 cents, 1 dime is $\frac{1}{10}$ of a dollar, and 1 cent is $\frac{1}{100}$ of a dollar, $0.95 is $9 \cdot \frac{1}{10} + 5 \cdot \frac{1}{100}$, or $\frac{95}{100}$ of a dollar.

Consequently,

$$128.95 = 1 \cdot 10^2 + 2 \cdot 10 + 8 \cdot 1 + 9 \cdot \frac{1}{10} + 5 \cdot \frac{1}{10^2}$$

The digits in 128.95 correspond, respectively, to the following place values: 10^2, 10, 1, $\frac{1}{10}$, and $\frac{1}{10^2}$. (Each term in the sequence of place values after the first is $\frac{1}{10}$ of the previous term.) Thus, 12.61843 represents

$$1 \cdot 10 + 2 \cdot 1 + \frac{6}{10^1} + \frac{1}{10^2} + \frac{8}{10^3} + \frac{4}{10^4} + \frac{3}{10^5}, \quad \text{or} \quad 12\frac{61,843}{100,000}$$

The decimal 12.61843 is read "twelve and sixty-one thousand eight hundred forty-three hundred-thousandths." Each place of a decimal may be named by its power of 10. For example, the places of 12.61843 can be named as shown in Table 7-1.

Table 7-1

1	2	.	6	1	8	4	3
Tens	Units	*and*	Tenths	Hundredths	Thousandths	Ten-thousandths	Hundred-thousandths

> **REMARK** Note that $\frac{61,843}{100,000}$ represents a division of two whole numbers when the divisor is a power of 10; in this case, 10^5. The decimal for $\frac{61,843}{100,000}$ is 0.61843.

Decimals can be introduced with concrete materials. For example, suppose that a long in the base-ten block set represents 1 unit (instead of letting the cube represent 1 unit).

If that is the case, then the cube represents $\frac{1}{10}$. Then 5.4 could be represented as seen in Figure 7-1(a).

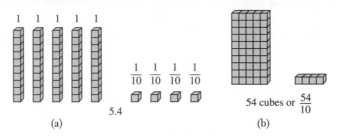

Figure 7-1

If we wanted a different interpretation of 5.4, we could use Figure 7-1(b), which shows 5.4 is also equivalent to 54 tenths, or $\frac{54}{10}$. This equivalence can be stated symbolically as

$$5.4 = 5 + 0.4 = 5 + \frac{4}{10} = \frac{50}{10} + \frac{4}{10} = \frac{54}{10}$$

This approach gives students a concrete connection between fractions and decimals.

NOW TRY THIS 7-1 In a set of base-ten blocks, let 1 flat represent 1 unit.

a. What will 1 long represent?
b. What will 1 cube represent?
c. Represent 1.23 using the blocks with these equivalences.

To represent a decimal such as 2.235, we can think of a block shown in Figure 7-2(a) as a unit. Then a flat represents $\frac{1}{10}$, a long represents $\frac{1}{100}$, and a cube represents $\frac{1}{1000}$. Using these objects, we show a representation of 2.235 in Figure 7-2(b).

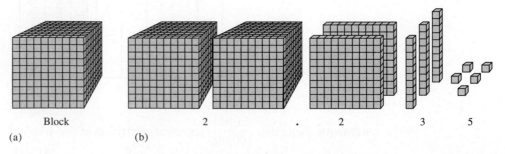

Figure 7-2

Table 7-2 shows other examples of decimals, their fractional meanings, and their common fraction notations.

Table 7-2

Decimal	Expanded Fractional Meaning	Common Fraction Notation
5.3	$5 + \dfrac{3}{10}$	$5\dfrac{3}{10}$, or $\dfrac{53}{10}$
0.02	$0 + \dfrac{0}{10} + \dfrac{2}{100}$	$\dfrac{2}{100}$
2.0103	$2 + \dfrac{0}{10} + \dfrac{1}{100} + \dfrac{0}{1000} + \dfrac{3}{10,000}$	$2\dfrac{103}{10,000}$, or $\dfrac{20,103}{10,000}$
$^{-}3.6$	$-\left(3 + \dfrac{6}{10}\right)$	$^{-}3\dfrac{6}{10}$, or $-\dfrac{36}{10}$

Decimals can also be written in expanded (or standard) form using place value and negative exponents. Thus,

$$12.61843 = 1 \cdot 10^1 + 2 \cdot 10^0 + 6 \cdot 10^{-1} + 1 \cdot 10^{-2} + 8 \cdot 10^{-3} + 4 \cdot 10^{-4} + 3 \cdot 10^{-5}$$

NOW TRY THIS 7-2 Decimals can sometimes cause confusion. In the *Blondie* cartoon, the cartoonists intended the price of tomatoes to be increased. Is that what the cartoon says? Give other examples of this type of mistake involving dollars and cents.

Example 7-1 shows how to convert rational numbers whose denominators are powers of 10 to decimals.

Example 7-1

Convert each of the following to decimals:

a. $\dfrac{25}{10}$ b. $\dfrac{56}{100}$ c. $\dfrac{205}{10,000}$

Solution a. $\dfrac{25}{10} = \dfrac{2 \cdot 10 + 5}{10} = \dfrac{2 \cdot 10}{10} + \dfrac{5}{10} = 2 + \dfrac{5}{10} = 2.5$

b. $\dfrac{56}{100} = \dfrac{5 \cdot 10 + 6}{10^2} = \dfrac{5 \cdot 10}{10^2} + \dfrac{6}{10^2} = 0 + \dfrac{5}{10} + \dfrac{6}{10^2} = 0.56$

c. $\dfrac{205}{10,000} = \dfrac{2 \cdot 10^2 + 0 \cdot 10 + 5}{10^4} = \dfrac{2 \cdot 10^2}{10^4} + \dfrac{0 \cdot 10}{10^4} + \dfrac{5}{10^4}$

$= \dfrac{2}{10^2} + \dfrac{0}{10^3} + \dfrac{5}{10^4} = 0 + \dfrac{0}{10^1} + \dfrac{2}{10^2} + \dfrac{0}{10^3} + \dfrac{5}{10^4} = 0.0205$

We reinforce the ideas in Example 7-1 through the use of a calculator. In Example 7-1(b), press ⑤ ⑥ ÷ ① ⓪ ⓪ = and watch the display. Divide by 10 and look at the new placement of the decimal point. Once more, divide by 10 (which amounts to dividing the original number, 56, by 10,000) and note the placement of the decimal point. This leads to the following general rule for dividing a decimal by a positive integer power of 10:

> To divide a decimal by 10^n, start at the decimal point, count n place values left, annexing zeros if necessary, and insert the decimal point to the left of the nth place value counted.

The fractions in Example 7-1 are easy to convert to decimals because the denominators are powers of 10. If the denominator of a fraction is not a power of 10, as in $\frac{3}{5}$, we use the problem-solving strategy of *converting the problem to a simpler one that we already know how to do*. First, we rewrite $\frac{3}{5}$ as a fraction in which the denominator is a power of 10, and then we convert the fraction to a decimal.

$$\frac{3}{5} = \frac{2 \cdot 3}{2 \cdot 5} = \frac{6}{10} = 0.6$$

The reason for multiplying the numerator and the denominator by 2 is apparent when we observe that $10 = 5 \cdot 2$ or $2 \cdot 5$.

NOW TRY THIS 7-3 Consider the following fractions. How could you convert them to decimals?

a. $\frac{5}{2}$

b. $\frac{3}{8}$ (*Hint:* $\frac{3}{8} = \frac{3}{2^3}$. By what power of 5 must one multiply to make the denominator a power of 10?)

c. $\frac{3}{20}$ (*Hint:* $\frac{3}{20} = \frac{3}{2^2 \cdot 5}$. By what power of 5 must one multiply to make the denominator a power of 10?)

In Now Try This 7-3, it appears that as long as we have the denominator of a fraction as a power of 10, then converting the fraction to a decimal is a simple use of division of a whole number by a power of 10.

Note that any power of 10 is of the form 10^n, and in general, because $10^n = (2 \cdot 5)^n = 2^n \cdot 5^n$, the prime factorization of the denominator that is a power of 10^n must be $2^n \cdot 5^n$. We use these ideas to write each fraction in Example 7-2 as a decimal.

Example 7-2

Express each of the following as decimals:

a. $\frac{7}{2^6}$ b. $\frac{1}{2^3 \cdot 5^4}$ c. $\frac{1}{125}$ d. $\frac{7}{250}$

Solution   a. $\frac{7}{2^6} = \frac{7 \cdot 5^6}{2^6 \cdot 5^6} = \frac{7 \cdot 15{,}625}{(2 \cdot 5)^6} = \frac{109{,}375}{10^6} = 0.109375$

b. $\frac{1}{2^3 \cdot 5^4} = \frac{1 \cdot 2^1}{2^3 \cdot 5^4 \cdot 2^1} = \frac{2}{2^4 \cdot 5^4} = \frac{2}{(2 \cdot 5)^4} = \frac{2}{10^4} = 0.0002$

$$\textbf{c. } \frac{1}{125} = \frac{1}{5^3} = \frac{1 \cdot 2^3}{5^3 \cdot 2^3} = \frac{8}{(5 \cdot 2)^3} = \frac{8}{10^3} = 0.008$$

$$\textbf{d. } \frac{7}{250} = \frac{7}{2 \cdot 5^3} = \frac{7 \cdot 2^2}{2 \cdot 5^3 \cdot 2^2} = \frac{28}{(2 \cdot 5)^3} = \frac{28}{10^3} = 0.028$$

  A calculator can quickly convert some fractions whose denominators can be written as powers of 10 to decimals. For example, to find $\frac{7}{2^6}$, press $\boxed{7}\boxed{\div}\boxed{2}\boxed{y^x}\boxed{6}\boxed{=}$; to convert $\frac{1}{125}$ to a decimal, press $\boxed{1}\boxed{\div}\boxed{1}\boxed{2}\boxed{5}\boxed{=}$ or press $\boxed{1}\boxed{2}\boxed{5}\boxed{1/x}\boxed{=}$. The display for $\frac{1}{125}$ on some calculators may show $\boxed{8-03}$, which is the calculator's notation for $\frac{8}{10^3}$, or $8 \cdot 10^{-3}$. (This *scientific notation* is discussed in more detail later in the chapter.)

The answers in Example 7-2 are illustrations of **terminating decimals**—*decimals that can be written with only a finite number of places to the right of the decimal point.* Not every rational number can be written as a terminating decimal. For example, if we attempt to rewrite $\frac{2}{11}$ as a terminating decimal using the method just developed, we first try to find a natural number b such that the following holds:

$$\frac{2}{11} = \frac{2b}{11b}, \quad \text{where } 11b \text{ is a power of 10}$$

By the Fundamental Theorem of Arithmetic, the only prime factors of a power of 10 are 2 and 5. Because $11b$ has 11 as a factor, we cannot write $11b$ as a power of 10. Therefore it seems that $\frac{2}{11}$ cannot be written as a terminating decimal. A similar argument using the Fundamental Theorem of Arithmetic holds in general, so we have the following result.

Theorem 7–1

A rational number $\frac{a}{b}$ in simplest form can be written as a terminating decimal if, and only if, the prime factorization of the denominator contains no primes other than 2 or 5.

Example 7-3

Which of the following fractions can be written as terminating decimals?

a. $\frac{7}{8}$ **b.** $\frac{11}{250}$ **c.** $\frac{21}{28}$ **d.** $\frac{37}{768}$ **e.** $\frac{3.85}{1000}$

Solution   **a.** Because the denominator, 8, is 2^3 and the fraction is in simplest form, $\frac{7}{8}$ can be written as a terminating decimal.

b. Because the denominator, 250, is $2 \cdot 5^3$ and the fraction is in simplest form, then $\frac{11}{250}$ can be written as a terminating decimal.

c. $\dfrac{21}{28}$ can be written in simplest form as $\dfrac{21}{28} = \dfrac{3 \cdot 7}{2^2 \cdot 7} = \dfrac{3}{2^2}$. The denominator is now 2^2, so $\dfrac{21}{28}$ can be written as a terminating decimal.

d. In simplest form, the fraction $\dfrac{37}{768} = \dfrac{37}{2^8 \cdot 3}$ has a denominator with a factor of 3, so $\dfrac{37}{768}$ cannot be written as a terminating decimal.

e. $\dfrac{3.85}{1000} \cdot \dfrac{100}{100} = \dfrac{385}{10^5} = 0.00385$, a terminating decimal.

REMARK As Example 7-3(c) shows, to determine whether a rational number $\dfrac{a}{b}$ can be represented as a terminating decimal, we consider the prime factorization of the denominator *only* if the fraction is in simplest form.

Ordering Terminating Decimals

A terminating decimal is easily located on a number line because it can be represented as a rational number $\dfrac{a}{b}$, where $b \neq 0$, and b is a power of 10. For example, consider 0.56, or $\dfrac{56}{100}$. One way to think about $\dfrac{56}{100}$ is as the rightmost endpoint of the 56th part of a unit segment divided into 100 equal parts, as in Figure 7-3.

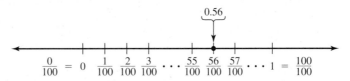

Figure 7-3

To aid in the ordering of decimals, we know that

$$\frac{5}{10} = \frac{5 \cdot 10}{10 \cdot 10} = \frac{50}{100} < \frac{56}{100} < \frac{60}{100} = \frac{6 \cdot 10}{10 \cdot 10} = \frac{6}{10}$$

so that $\dfrac{56}{100}$, or 0.56, is between $\dfrac{5}{10}$, or 0.5, and $\dfrac{6}{10}$, or 0.6. Thus, $0.5 < 0.56 < 0.6$. In a similar manner, two terminating decimals can be ordered by converting each to rational numbers in the form $\dfrac{a}{b}$, where a and b are integers with $b \neq 0$, and determining which is greater. For example, because $0.36 = \dfrac{36}{100}$, $0.9 = 0.90 = \dfrac{90}{100}$, and $\dfrac{90}{100} > \dfrac{36}{100}$, it follows that $0.9 > 0.36$. One could also tell that $0.9 > 0.36$ because \$0.90 is 90¢ and \$0.36 is 36¢ and 90¢ > 36¢.

School Book Page DECIMAL PLACE VALUE

Lesson 11-2

Key Idea
There are many ways to represent decimal numbers.

Think It Through
I can **use objects**, **draw pictures**, or **make a chart** to represent 1.48.

Decimal Place Value

LEARN

What are some ways to represent decimals?

Here are different ways to represent 1.48.

Number line:

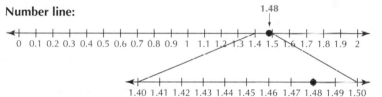

Grids:

1 one   4 tenths 8 hundredths

Place-value chart:

tens	ones		tenths	hundredths
	1	.	4	8

Expanded form: 1 + 0.4 + 0.08

Standard form: 1.48

Word form: One and forty-eight hundredths

> **Example**
>
> Write the word form and the expanded form for 5.0**2**. Then, tell the value of the red digit.
>
> *Word form:* five and two hundredths
>
> *Expanded form:* 5 + 0.02
>
> The red digit is in the hundredths place, so its value is 2 hundredths, 0.02.

*I write the word **and** for the decimal point.*

✓ **Talk About It**

1. Which digit is in the tenths place in 1.48?

2. Explain how to locate 1.48 on a number line.

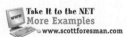
Take It to the NET
More Examples
www.scottforesman.com

628

Source: Scott Foresman-Addison Wesley Mathematics, Grade 4, 2008 (p. 628).

Comparing 0.36 and 0.9 by converting them to "money" suggests a method for comparing any two decimals. For example, consider the decimals 0.345 and 0.1474. Note that 0.345 is a decimal with thousandths as the smallest place value, and 0.1474 has ten-thousandths as its smallest place value. To compare the decimals, we could write both with ten-thousandths as the smallest place value. Note that 0.345 could be written as 0.3450. Now we have 0.3450 and 0.1474 both with ten-thousandths as the smallest place value. These decimals could be written as follows:

$$\frac{3450}{10,000} \quad \text{and} \quad \frac{1474}{10,000}$$

Research Note

Place value connections between whole numbers and decimal numbers are useful for learning, but children often fail to adjust for the decimal aspects. A common error is for a student to claim that 0.56 is greater than 0.6 because 56 is greater than 6 (Hiebert 1992).

As in Chapter 6, these two fractions can be compared simply by looking at the numerators. Because 3450 is greater than 1474, $\frac{3450}{10,000} > \frac{1474}{10,000}$, and hence $0.345 > 0.1474$.

This process is frequently shortened by looking only at the decimals lined up as follows:

$$0.3450 = 0.345$$
$$0.1474$$

And with the decimals lined up in this fashion, 3 tenths is greater than 1 tenth. Thus, $0.345 > 0.1474$. This suggests a way to order decimals without conversion to fractions. The steps we use to compare terminating decimals are like those for comparing whole numbers:

1. Line up the numbers by place value.
2. Start at the left and find the first place where the face values are different.
3. Compare these digits. The digit with the greater face value in this place represents the greater of the two numbers.

As noted by Hiebert, students do not always consider decimals when trying to order numbers.

From the student page (page 417) on Decimal Place Value, we see that even in grade 4, students learn different representations of decimals. Do number 2 in the Talk About It section.

Assessment 7-1 A

1. Write each of the following as a sum in expanded place value form:
   a. 0.023
   b. 206.06
   c. 312.0103
   d. 0.000132

2. Rewrite each of the following as decimals:
   a. $4 \cdot 10^3 + 3 \cdot 10^2 + 5 \cdot 10 + 6 + 7 \cdot 10^{-1} + 8 \cdot 10^{-2}$
   b. $4 \cdot 10^3 + 6 \cdot 10^{-1} + 8 \cdot 10^{-3}$
   c. $4 \cdot 10^4 + 3 \cdot 10^{-2}$
   d. $2 \cdot 10^{-1} + 4 \cdot 10^{-4} + 7 \cdot 10^{-7}$

3. Write each of the following as numerals:
   a. Five hundred thirty-six and seventy-six ten-thousandths
   b. Three and eight thousandths
   c. Four hundred thirty-six millionths
   d. Five million and two tenths

4. Write each of the following in words:
   a. 0.34
   b. 20.34
   c. 2.034
   d. 0.000034

5. Write each of the following terminating decimals in $\frac{a}{b}$ simplest form where $a, b \in Q$ and $b \neq 0$.
   a. 0.436   b. 25.16
   c. ⁻316.027   d. 28.1902
   e. ⁻4.3   f. ⁻62.01

6. Mentally determine which of the following represent terminating decimals:
   a. $\frac{4}{5}$ b. $\frac{61}{2^2 \cdot 5}$
   c. $\frac{3}{6}$ d. $\frac{1}{2^5}$

e. $\dfrac{36}{5^5}$ **f.** $\dfrac{133}{625}$ **g.** $\dfrac{1}{3}$

h. $\dfrac{2}{25}$ **i.** $\dfrac{1}{13}$ **j.** $\dfrac{26}{65}$

7. Where possible, write each of the numbers in exercise 6 as a terminating decimal.
8. Seven minutes is part of an hour. If 7 min were to be expressed as a decimal part of an hour, explain whether or not it would be a terminating decimal.
9. A unit segment has length of 1. If a unit segment had one endpoint at 3, the other at 4 on a number line, and we considered the $\dfrac{56}{100}$ part of that unit segment as in Figure 7-3, what decimal would the rightmost endpoint of the $\dfrac{56}{100}$ segment represent?
10. Given the U.S. monetary system, what decimal reason can you think of for having coins only for a penny, nickel, dime, quarter, and half-dollar as coins less than $1.00?
11. In each of the following, order the decimals from greatest to least:
    a. 13.4919, 13.492, 13.49183, 13.49199
    b. ⁻1.453, ⁻1.45, ⁻1.4053, ⁻1.493
12. Write the numbers in each of the following sentences as decimals in symbols:
    a. A mite has body length about fourteen thousandths of an inch.
    b. The Earth goes around the Sun once every three hundred sixty-five and twenty-four hundredths days.

13. Use a grid with 100 squares and represent 0.32. Explain your representation.
14. If the decimals 0.804, 0.84, and 0.8399 are arranged on a typical number line, which is furthest to the right?
15. Write a decimal number that has a ten-thousandths place and is between 8.34 and 8.341.
16. **a.** Show that between any two terminating decimals, there is another terminating decimal.
    b. Argue that part (a) can be used to show that there are infinitely many terminating decimals between any two terminating decimals.
17. **a.** Describe the decimal 0.613 using base-ten blocks.
    b. Explain whether you can describe a decimal such as 0.61345 using base-ten blocks.
18. A baseball player's batting average was reported as "three-twenty-two." A batting average is essentially determined when you divide the number of times that the player has hits by the number of times at bat. Explain why the reported batting average is not mathematically correct.
19. If "decimals" in other number bases work the same as in base ten, explain the meaning of the following: 3.145_{six}.
20. The five top swimmers in an event had the following times:
    Emily   64.54 sec   Kathy   64.02 sec
    Molly   64.46 sec   Rhonda   63.54 sec
    Martha   63.59 sec
    List them in the order they placed with the "best" time first.

Assessment 7-1 B

1. Write each of the following as a sum in expanded place value form:
    a. 0.045   **b.** 103.03
    c. 245.6701   **d.** 0.00034
2. Rewrite each of the following as decimals:
    a. $5 \cdot 10^3 + 2 \cdot 10^2 + 4 \cdot 10^{-1}$
    b. $4 \cdot 10^{-3} + 2 \cdot 10^4$
    c. $2 \cdot 10^2 + 3 \cdot 10^4$
3. Write each of the following as numerals:
    a. Two thousand twenty-seven thousandths
    b. Two thousand and twenty-seven thousandths
    c. Two thousand twenty and seven thousandths
    d. Four hundred-thousandths
4. Write each of the following in words:
    a. 0.45
    b. 2.035
    c. 45.0006
    d. 0.0000445

5. Write each of the following terminating decimals in $\dfrac{a}{b}$ simplest form where $a, b \in Q$ and $b \neq 0$.
    a. 28.32   **b.** 34.1736   **c.** ⁻27.32
6. Mentally determine which of the following represent terminating decimals:
    a. $\dfrac{4}{8}$ **b.** $\dfrac{1}{2^6}$ **c.** $\dfrac{137}{625}$
    d. $\dfrac{1}{17}$ **e.** $\dfrac{3}{25}$
7. Where possible, write each of the numbers in exercise 6 as a terminating decimal.
8. What whole number of minutes (less than 60) could be expressed as terminating decimal parts of an hour?
9. If in a set of base-ten blocks, one block represented $\dfrac{1}{10}$, what is the value of each of the following?
    a. 1 cube
    b. 1 flat

c. 1 long
d. 3 blocks, 1 long, and 4 cubes

10. In each of the following, order the decimals from least to greatest:
    a. 24.9419, 24.942, 24.94189, 24.94199
    b. ⁻34.25, ⁻34.251, ⁻34.205, ⁻34.2519

11. Use a grid with 100 squares and represent 0.23.

12. Write a decimal that is between 8.345 and 8.3456.

13. A normal carpenter's rule is marked off in sixteenths. What would be the values at each of the marks if the rule were done in decimal form?

14. Suppose the rule in exercise 13 was marked in thirty-seconds, what would be the decimal marking the rule between $\frac{1}{16}$ and $\frac{2}{16}$?

15. Use the results of exercises 13 and 14 to describe how you might argue that there are infinitely many terminating decimals between any two specific terminating decimals such as 0.0625 and 0.125.

16. Explain the mathematical meaning of a sign on a copy machine reading ".05¢ a copy."

17. If "decimals" in other number bases work the same as in base ten, explain the meaning of the following: 0.00334_{seven}.

18. If a swimmer cut 0.2 sec off of his time in a 50 m backstroke every day for the first 10 days of practice, how much time would he cut off?

Mathematical Connections 7-1

Communication

1. Using Simon Stevin's notation, how would you write the following:
   a. 0.3256   **b.** 0.0032

2. If 1 mL is 0.001 L, how should 18 mL be expressed as a terminating decimal number of liters?

3. Explain whether 1 day can be expressed as a terminating decimal part of a 365-da year.

4. Using the number-line model to depict terminating decimals, explain whether you think that there is a greatest terminating decimal less than 1.

5. Look up the meaning of *tithe*. How is that related to decimals?

6. Explain how you could use base-ten blocks to represent two and three hundred forty-five thousandths.

7. Explain why in Theorem 7–1 the rational number must be in simplest form before examining the denominator.

8. In your own words, explain what a decimal point is and what it does.

Open-Ended

9. Determine how decimal notation is symbolized in different countries.

10. Examine three elementary school textbooks and report how the introductions of the topics of exponents and decimals differ, if they do. Report the grade level that decimals are introduced.

Cooperative Learning

11. Based on a system similar to that of base ten, determine how you might introduce a "decimal" notation in base five.

12. Simon Stevin is credited with propagating the decimal system. In small groups, research the history of decimals to find contributions of the Arabs, the Chinese, and Renaissance mathematicians. Explain whether or not you believe Stevin "invented" the decimal system.

Questions from the Classroom

13. A student claims that 0.36 is greater than 0.9 because 36 is greater than 9. How do you respond?

14. A student argues that fractions should no longer be taught once students learn how to work with decimal numbers. How do you respond?

National Assessment of Educational Progress (NAEP) Question

What is 4 hundredths written in decimal notation?
   a. 0.004
   b. 0.04
   c. 0.400
   d. 4.00

NAEP, 2007, Grade 8

BRAIN TEASER Find two base-two "decimals" that can be represented by terminating decimals in base ten. Argue that any terminating base-two decimals can be represented by a base-ten terminating decimal.

7-2   Operations on Decimals

To develop an algorithm for addition of terminating decimals, consider the sum 2.16 + 1.73. In elementary school, base-ten blocks are recommended to demonstrate such an addition problem. Figure 7-4 shows how the addition can be performed.

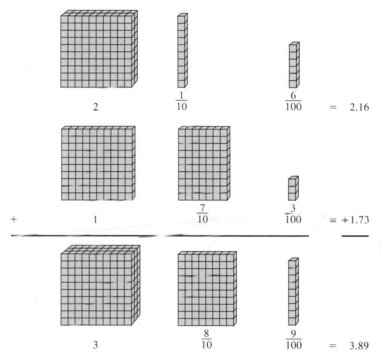

Figure 7-4

The computation in Figure 7-4 can be explained by *changing it to a problem we already know how to solve*, that is, to a sum involving fractions. We then use the commutative and associative properties of addition to aid in the computation, as follows:

$$2.16 + 1.73 = \left(2 + \frac{1}{10} + \frac{6}{100} \right) + \left(1 + \frac{7}{10} + \frac{3}{100} \right)$$

$$= (2 + 1) + \left(\frac{1}{10} + \frac{7}{10} \right) + \left(\frac{6}{100} + \frac{3}{100} \right)$$

$$= 3 + \frac{8}{10} + \frac{9}{100}$$

$$= 3.89$$

A different way of looking at the addition 2.16 + 1.73 links to a different representation of the decimals as fractions with denominators that are powers of 10 in the following:

$$2.16 = \frac{216}{100} \text{ and } 1.73 = \frac{173}{100} \text{ so that } 2.16 + 1.73 = \frac{216}{100} + \frac{173}{100} = \frac{216 + 173}{100} = \frac{389}{100} = 3.89.$$

Similarly, $2.1 + 1.73 = \frac{21}{10} + \frac{173}{100} = \frac{210}{100} + \frac{173}{100} = \frac{383}{100} = 3.83.$

In each of these cases, the decimals are converted to fractions with the same denominator that is a power of 10. Once this is done, the numerators can be added as whole numbers to form the numerator of the answer fraction, which can then be converted back to a decimal. If these decimals are written as shown below, one essentially "lines up the decimal points" and adds as one does with whole numbers, placing the decimal point in the appropriate place in the answer.

$$\begin{array}{r} 2.16 \\ +\ 1.73 \\ \hline 3.89 \end{array} \quad \text{and} \quad \begin{array}{r} 2.1 \\ +\ 1.73 \\ \hline 3.83 \end{array}$$

The second computation could be written as shown here:

$$\begin{array}{r} 2.10 \\ +\ 1.73 \\ \hline 3.83 \end{array}$$

In the computations above, we add units to units, tenths to tenths, and hundredths to hundredths. This can be accomplished most efficiently by keeping the numbers in their decimal forms, lining up the decimal points, and adding as if the numbers were whole numbers. This technique works for both addition and subtraction, as demonstrated on the student page (page 423). Note also the use of a calculator on the decimal addition, Example C.

Multiplying Decimals

Just as in the presented algorithms for adding terminating decimals by representing them as fractions, we develop and explain an algorithm for multiplication of decimals. Consider the product $4.62 \cdot 2.4$:

$$(4.62)(2.4) \ = \ \frac{462}{100} \cdot \frac{24}{10} \ = \ \frac{462}{10^2} \cdot \frac{24}{10^1} \ = \ \frac{462 \cdot 24}{10^2 \cdot 10^1} \ = \ \frac{11{,}088}{10^3} \ = \ 11.088$$

The answer to this computation was obtained by multiplying the whole numbers 462 and 24 and then dividing the result by 10^3.

On student page (page 424), observe how multiplication of a whole number times a decimal can be accomplished by first shading 0.6 three times, once each in salmon, green, and purple. And then 0.04 is shaded three times in similar colors, as seen on the rightmost part of the Activity. The product is the sum of all the shaded portions: $0.6 + 0.6 + 0.04 + 0.04 + 0.04$ or 1.92. In this manner, multiplication of a whole number times a decimal is interpreted almost exactly as the multiplication of a whole number by a whole number. Consider the shading method in the Activity at the bottom of the page to show why 0.5×0.7 is 0.35. Try part (d) on your own. Do you think that this method would be effective with the multiplication of decimals with more places?

An algorithm for multiplying decimals can be stated as follows:

If there are n digits to the right of the decimal point in one number and m digits to the right of the decimal point in a second number, multiply the two numbers, ignoring the decimals, and then place the decimal point so that there are $n + m$ digits to the right of the decimal point in the product.

REMARK There are $n + m$ digits to the right of the decimal point in the product because $10^n \cdot 10^m = 10^{n+m}$.

School Book Page   ADDING AND SUBTRACTING WHOLE NUMBERS AND DECIMALS

Lesson 2-5

Key Idea
Adding and subtracting decimals is similar to adding and subtracting whole numbers.

Adding and Subtracting Whole Numbers and Decimals

WARM UP

Estimate.

1. 80.6 + 501 + 102.4

2. 6,289.867 − 284.499

3. 8.592 − 4.635

4. 0.198 + 1.18

● LEARN

How can you add decimals?

When the starting gun is fired, relay runners can lose 0.2 second in reaction time, and another 1.8 seconds to go from a standing start to running speed. About how many seconds pass before running speed is reached?

Think It Through
I **remember** that annexing zeros does not change the decimal's value.

Example A

Find 0.2 + 1.8.
Estimate: 0 + 2 = 2

	What You **Think**	What You **Write**

STEP 1 Write the numbers, lining up decimal points.

$$\begin{array}{r} 0.2 \\ + 1.8 \\ \hline \end{array}$$

STEP 2 Add the tenths. Regroup if necessary. Write the decimal point in your answer.

$$\begin{array}{r} 0.2 \\ + 1.8 \\ \hline 2.0 \end{array}$$

The answer and estimate match, so the answer is reasonable.

Example B

Find 5.6 + 2.973.
Estimate: 6 + 3 = 9

$$\begin{array}{r} 5.600 \\ + 2.973 \\ \hline 8.573 \end{array}$$

Write the numbers, lining up the decimal points. Add thousandths, hundredths, tenths. Regroup if necessary. Write the decimal point in your answer.

Since 8.573 is close to 9, the answer is reasonable.

Example C

Use a calculator to find 338.09 + 517.3.

Estimate: 300 + 500 = 800

Press: 338.09 [+] 517.3 [ENTER =]

Display: 855.39

Since 855.39 is close to 800, the answer in the display is reasonable.

86

Source: Scott Foresman-Addison Wesley Mathematics, Grade 6, 2008 (p. 86).

School Book Page

MULTIPLYING WHOLE NUMBERS AND DECIMALS

Lesson 2-6

Key Idea
Multiplying decimals is similar to multiplying whole numbers. You just need to know where to place the decimal point in the product.

Materials
• 10-by-10 decimal grids or
⚙ **tools**

• colored pencils or markers

Think It Through
I can **use models** to show multiplication of decimals.

Multiplying Whole Numbers and Decimals

LEARN

How can you multiply a whole number by a decimal?

You can use the same methods for multiplying decimals as you use for multiplying whole numbers.

✓ **WARM UP**
Estimate.
1. 5.88×3.03
2. 725×4
3. 172×9.9

Activity

a. To find 3×0.64, shade the tenths for the decimal number on a grid. Do this 3 times, using a different color each time.

b. Using the same 3 colors, shade three groups of the hundredths for the decimal number.

c. Count all of the shaded hundredths.

d. Use grids to find 1.2×2 and 2×0.18.

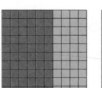

$3 \times 0.64 = 192$ hundredths $= 1.92$

How can you multiply a decimal by a decimal?

Activity

a. To find 0.5×0.7, shade 5 columns using the same color to show 5 tenths.

b. Shade 7 rows in another color to show 7 tenths.

c. Count the hundredths in the overlapping shaded area.

d. Use grids to find 0.4×0.7 and 1.6×0.3.

$0.5 \times 0.7 = 35$ hundredths $= 0.35$

90

Source: Scott Foresman-Addison Wesley Mathematics, Grade 6, 2008 (p. 90).

Example 7-4

Compute each of the following:

a. $(6.2)(1.43)$ **b.** $(0.02)(0.013)$ **c.** $(1000)(3.6)$

Solution a.   1.4 3   (2 digits after the decimal point)
             $\times$ 6.2   (1 digit after the decimal point)
             ———
              286
            8 5 8
            ———
            8.8 6 6   (2 + 1, or 3 digits after the decimal point)

b.   0.0 1 3
       $\times$ 0.0 2
       ————
       0.0 0 0 2 6

c.   3.6
       $\times$ 1 0 0 0
       ————
       3 6 0 0.0

NOW TRY THIS 7-4 Example 7-4(c) suggests that multiplication by 1000, or 10^3, results in moving the decimal point in the product three places to the right. **(a)** Explain why this is true using expanded notation and the distributive property of multiplication over addition. **(b)** In general, how does multiplication by 10^n, where n is a positive integer, affect the product? Why?

Scientific Notation

Many calculators display the decimals for the fractions $\dfrac{3}{45{,}689}$ and $\dfrac{5}{76{,}146}$ as $\boxed{6.5661319 \quad -05}$ and $\boxed{6.566333 \quad -05}$, respectively. The displays are in **scientific notation**. The first display is a notation for $6.5661319 \cdot 10^{-5}$ and the second for $6.566333 \cdot 10^{-5}$.

Scientists use scientific notation to handle either very small or very large numbers. For example, "the Sun is 93,000,000 mi from Earth" is expressed as "the Sun is $9.3 \cdot 10^7$ mi from Earth." A micron, a metric unit of measure that is 0.000001 m, is written $1 \cdot 10^{-6}$ m.

Definition of Scientific Notation

In **scientific notation**, a positive number is written as the product of a number greater than or equal to 1 and less than 10 and an integer power of 10. To write a negative number in scientific notation, treat the number as a positive number and adjoin the negative sign in front of the result.

The following numbers are in scientific notation:

$$8.3 \cdot 10^8, \quad 1.2 \cdot 10^{10}, \quad {}^{-}73.2 = {}^{-}(7.32 \cdot 10^8), \quad \text{and} \quad 7.84 \cdot 10^{-6}$$

The numbers $0.43 \cdot 10^9$ and $12.3 \cdot 10^{-6}$ are not in scientific notation because 0.43 and 12.3 are not greater than or equal to 1 and less than 10. To write a number like 934.5 in scientific notation, we divide by 10^2 to get 9.345 and then multiply by 10^2 to retain the value of the original number:

$$934.5 = \left(\frac{934.5}{10^2}\right)10^2 = 9.345 \cdot 10^2$$

This amounts to moving the decimal point two places to the left (dividing by 10^2) and then multiplying by 10^2. Similarly, to write 0.000078 in scientific notation, we first multiply by 10^5 to obtain 7.8 and then divide by 10^5 (or multiply by 10^{-5}) to keep the original value:

$$0.000078 = (0.000078 \cdot 10^5)10^{-5} = 7.8 \cdot 10^{-5}$$

This amounts to moving the decimal point five places to the right and multiplying by 10^{-5}.

Example 7-5

Write each of the following in scientific notation:

a. 413,682,000   b. 0.0000231   c. 83.7   d. $^-$10,000,000

Solution   a. $413,682,000 = \left(\dfrac{413,682,000}{10^8}\right)10^8 = 4.13682 \cdot 10^8$

b. $0.0000231 = (0.0000231 \cdot 10^5)10^{-5} = 2.31 \cdot 10^{-5}$

c. $83.7 = \left(\dfrac{83.7}{10^1}\right)10^1 = 8.37 \cdot 10^1$

d. $^-10,000,000 = {}^-\left(\dfrac{10,000,000}{10^7}\right)10^7 = {}^-(1 \cdot 10^7)$

Example 7-6

Convert the following to standard numerals:

a. $6.84 \cdot 10^{-5}$ b. $3.12 \cdot 10^7$ c. $^-(4.08 \cdot 10^4)$

Solution   a. $6.84 \cdot 10^{-5} = 6.84\left(\dfrac{1}{10^5}\right) = 0.0000684$

b. $3.12 \cdot 10^7 = 31,200,000$

c. $^-(4.08 \cdot 10^4) = {}^-40,800$

Numbers in scientific notation are easy to manipulate using the laws of exponents. For example, $(5.6 \cdot 10^5)(6 \cdot 10^4)$ can be rewritten $(5.6 \cdot 6)(10^5 \cdot 10^4) = 33.6 \cdot 10^9$, which is $3.36 \cdot 10^{10}$ in scientific notation. Also,

$$(2.35 \cdot 10^{-15})(2 \cdot 10^8) = (2.35 \cdot 2)(10^{-15} \cdot 10^8) = 4.7 \cdot 10^{-7}$$

 Calculators with an [EE] key can be used to represent numbers in scientific notation. For example, to find $(5.2 \cdot 10^{16})(9.37 \cdot 10^4)$, press

$$\boxed{5}\,\boxed{.}\,\boxed{2}\,\boxed{\text{EE}}\,\boxed{1}\,\boxed{6}\,\boxed{\times}\,\boxed{9}\,\boxed{.}\,\boxed{3}\,\boxed{7}\,\boxed{\text{EE}}\,\boxed{4}\,\boxed{=}$$

Dividing Decimals

With division, some sets are closed and others not. For example, the set of whole numbers was not closed under division. Consider that $1 \div 3$ does not yield a whole number because there is no whole number to multiply times 3 to get 1. Neither is the set of integers closed under division, by the same counterexample. However, the set of rational numbers without 0 is closed under division. Recall that $\dfrac{a}{b} \div \dfrac{c}{d}$, where b, c, and $d \ne 0$ gives $\dfrac{ad}{bc}$, a rational number.

Should we expect the set of decimals without 0 to be closed under division? The answer for the uninitiated is not automatic. We have shown that some fractions can be represented by terminating decimals. We saw earlier that $\frac{2}{11}$ could not be represented as a terminating decimal. And the set of rational numbers without 0 being closed under division does not imply that the set of decimals without 0 is closed under division.

However, like the set of whole numbers without 0 which is not closed under division, there are algorithms for dividing decimals that allow for remainders. Some divisions have 0 remainder, as seen in the following. Consider $75.45 \div 3$:

$$75.45 \div 3 = \frac{7545}{100} \div \frac{3}{1} = \frac{7545}{100} \cdot \frac{1}{3} = \frac{7545}{3} \cdot \frac{1}{100} = 2515 \cdot \frac{1}{100} = 25.15$$

By writing the dividend as a fraction whose denominator is a power of 10 and using the commutative property of multiplication and algorithms, the division is changed into a division of whole numbers $(7545 \div 3)$ times a power of 10; that is, $\frac{1}{100}$, or 10^{-2}. A different method of accomplishing the change of the division into whole-number division is to multiply the dividend and the divisor by the same power of 10, as follows:

$$75.45 \div 3 = \frac{75.45}{3} = \frac{75.45 \cdot 10^2}{3 \cdot 10^2} = \frac{7545}{3 \cdot 10^2} = \frac{7545}{3} \cdot \frac{1}{10^2}$$

Why does this work?

The process shown here is seen in the activity at the bottom of the student page (page 428). Complete the activity.

When the divisor is a whole number, the division can be handled as with whole numbers and the decimal point placed directly over the decimal point in the dividend. When the divisor is not a whole number, as in $1.2032 \div 0.32$, we can obtain a whole-number divisor by expressing the quotient as a fraction and then multiplying the numerator and denominator of the fraction by a power of 10. This corresponds to rewriting the division problem in form (a) as an equivalent problem in form (b), as follows:

$$\text{(a) } 0.32\overline{)1.2032} \qquad \text{(b) } 32\overline{)120.32}$$

In elementary school texts, this process is usually described as "moving" the decimal point two places to the right in both the dividend and the divisor. This process is usually indicated with arrows, as shown in the following.

$$
\begin{array}{r}
3.76 \\
0.32\overline{)1.2032} \\
96 \\
243 \\
224 \\
192 \\
192 \\
\hline
0
\end{array}
$$

Multiply divisor and dividend by 100.

School Book Page   DIVIDING BY A DECIMAL

Key Idea
To divide by a decimal, you need to change the divisor to a whole number.

Materials
• decimal models
  or tools

Think It Through
I can **use models** to show division by decimals.

Dividing by a Decimal

LEARN

How can you divide by a decimal?

Nihad and his friends divide the cost of a $2.70 bag of nuts. Each person contributes $0.90. How many people will share the nuts?

Activity

You can use the same methods for dividing decimals as you use for dividing whole numbers.

a. Find 2.7 ÷ 0.9.

b. Circle groups of 0.9.

c. Count the circled groups.

d. Use decimal models to find 3.0 ÷ 0.3 and 2.5 ÷ 0.5.

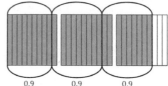

0.9   0.9   0.9

2.7 ÷ 0.9 = 3

dividend   divisor   quotient

How can you change a decimal divisor into a whole number divisor without changing the quotient?

Activity

a. Study the patterns below. Compare the first and second columns.

Pattern 1	
80 ÷ 0.4 = 200	800 ÷ 4 = 200
8 ÷ 0.4 = 20	80 ÷ 4 = 20
0.8 ÷ 0.4 = 2	8 ÷ 4 = 2
0.08 ÷ 0.4 = 0.2	0.8 ÷ 4 = 0.2

Pattern 2	
80 ÷ 0.04 = 2,000	8,000 ÷ 4 = 2,000
8 ÷ 0.04 = 200	800 ÷ 4 = 200
0.8 ÷ 0.04 = 20	80 ÷ 4 = 20
0.08 ÷ 0.04 = 2	8 ÷ 4 = 2

b. For Pattern 1, what happens to the quotient when you multiply the dividend and divisor by 10?

c. What number were the dividend and divisor multiplied by in Pattern 2?

d. Complete the table at the right.

Dividend	Divisor	Multiply Dividend and Divisor by:	Quotient
5.2	0.013	1,000	
5.2	0.13		40
5.2	1.3		

100

Source: Scott Foresman-Addison Wesley Mathematics, Grade 6, 2008 (p. 100).

Example 7-7

Compute each of the following:

a. $13.169 \div 0.13$ **b.** $9 \div 0.75$

Solution **a.**
$$
\begin{array}{r}
1\ 01.3 \\
0.13\overline{)13.169} \\
\underline{13} \\
16 \\
\underline{13} \\
39 \\
\underline{39} \\
0
\end{array}
$$

b.
$$
\begin{array}{r}
12 \\
0.75\overline{)9.00} \\
\underline{7\ 5} \\
1\ 50 \\
\underline{1\ 50} \\
0
\end{array}
$$

In Example 7-7(b), we appended two zeros in the dividend because $\dfrac{9}{0.75} = \dfrac{9 \cdot 100}{0.75 \cdot 100} = \dfrac{900}{75}$.

Example 7-8

An owner of a gasoline station must collect a gasoline tax of $0.11 on each gallon of gasoline sold. One week, the owner paid $1595 in gasoline taxes. The pump price of a gallon of gas that week was $3.35 without tax.

a. How many gallons of gas were sold during the week?
b. What was the revenue after taxes for the week?

Solution **a.** To find the number of gallons of gas sold during the week, we must divide the total gas tax bill by the amount of the tax per gallon:

$$\frac{1595}{0.11} = 14{,}500$$

Thus, 14,500 gallons were sold.
  b. To obtain the revenue after taxes, first determine the revenue before taxes. Then multiply the number of gallons sold by the cost per gallon:

$$(14{,}500)(\$3.35) = \$48{,}575$$

Next, subtract the gasoline taxes from the total revenue:

$$\$48{,}575 - \$1{,}595 = \$46{,}980$$

Thus, the revenue after gasoline taxes is $46,980.

NOW TRY THIS 7-5 Use decimal division to help Hi, in the *Hi and Lois* cartoon, determine the best buy.

Now consider a division as shown below:

a.
$$0.33\overline{)1.20\,3\,2} \quad 3.6\,4 \text{ Quotient}$$

```
            3.6 4  Quotient
  0.33)1.20 3 2
        99
        21 3
        19 8
         1 5 2
         1 3 2
           2 0  Remainder
```

b.
```
          364  Quotient
  33)12032
      99
      213
      198
      152
      132
       20  Remainder
```

Note that in part (a), all digits in the dividend are used and 20 is shown at the bottom of the division. This division could be treated as in the division of part (b), where the quotient is 364 and the remainder is 20. In part (a), the quotient is 3.64 and the remainder shown as 20 actually represents 0.0020, as can be seen by the original decimal alignment. This could be verified by checking: $3.64 \times 0.33 + 0.0020 = 1.2032$.

Mental Computation

Some of the tools used for mental computations with whole numbers can be used to perform mental computations with decimals, as seen in the following:

1. *Breaking and bridging*

$$1.5 + 3.7 + 4.48$$

$$= 4.5 + 0.7 + 4.48$$

$$= 5.2 + 4.48$$

$$= 9.2 + 0.48 = 9.68$$

$1.5 + 3$

$4.5 + 0.7$

$5.2 + 4$

$9.2 + 0.48$

2. *Using compatible numbers*
(Decimal numbers are compatible when they add up to a whole number.)

$$
\begin{array}{ll}
7.91 & \longrightarrow \quad 12 \\
3.85 & \\
4.09 & \longrightarrow \quad +\ 4 \\
+\ 0.15 & \\
& \qquad\quad 16
\end{array}
$$

$7.91 + 4.09$

$3.85 + 0.15$

$12 + 4$

3. *Making compatible numbers*

$$
\begin{array}{rcl}
9.27 & = & 9.25 + 0.02 \\
+\,3.79 & = & 3.75 + 0.04 \\
\hline
 & & 13.00 + 0.06 = 13.06
\end{array}
$$

4. *Balancing with decimals in subtraction*

$$
\begin{array}{rcl}
4.63 = & 4.63 + 0.03 & = \quad 4.66 \\
-\,1.97 = & -\,(1.97 + 0.03) & = \ -\,2.00 \\
\hline
 & & \qquad 2.66
\end{array}
$$

5. *Balancing with decimals in division*

$$0.25\overline{)8}$$

$$\times 4 \qquad \times 4$$

$$32$$
$$1\overline{)32}$$

REMARK Balancing with decimals in division uses the property $\dfrac{a}{b} = \dfrac{a \cdot c}{b \cdot c}$ if $c \neq 0$.

Rounding Decimals

Frequently, it is not necessary to know the exact numerical answer to a question. For example, if we want to know the distance to the Moon or the population of metropolitan New York City, the approximate answers of 239,000 mi and 21,200,000 people, respectively, may be adequate.

Often a situation determines how you should round. For example, suppose a purchase came to $38.65 and the cashier used a calculator to figure out the 6% sales tax by multiplying $0.06 \cdot 38.65$. The display showed 2.319. Because the display is between 2.31 and 2.32 and it is closer to 2.32, the cashier rounds up the sales tax to $2.32.

Suppose a display of 8.7345649 needs to be reported to the nearest hundredth. The display is between 8.73 and 8.74 but is closer to 8.73, so we round it down to 8.73. Next suppose the number 6.8675 needs to be rounded to the nearest thousandth. Notice that 6.8675 is exactly halfway between 6.867 and 6.868. *In such cases, it is common practice to round up* and therefore the answer to the nearest thousandth is 6.868. We write this as $6.8675 \approx 6.868$ and say 6.8675 is approximately equal to 6.868.

Example 7-9

Round each of the following numbers:

a. 7.456 to the nearest hundredth
b. 7.456 to the nearest tenth
c. 7.456 to the nearest unit
d. 7456 to the nearest thousand
e. 745 to the nearest ten
f. 74.56 to the nearest ten

Solution   **a.** $7.456 \approx 7.46$
      b. $7.456 \approx 7.5$
      c. $7.456 \approx 7$
      d. $7456 \approx 7000$
      e. $745 \approx 750$
      f. $74.56 \approx 70$

Rounding can also be done on some calculators using the $\boxed{\text{FIX}}$ key. If you want the number 2.3669 to be rounded to thousandths, you enter $\boxed{\text{FIX}}\boxed{3}$. The display will show 0.000. If you then enter 2.3669 and press the $\boxed{=}$ key, the display will show 2.367.

Estimating Decimal Computations Using Rounding

Rounded numbers can be useful for estimating answers to computations. For example, consider each of the following:

1. Karly goes to the grocery store to buy items that cost the following amounts. She estimates the total cost by rounding each amount to the nearest dollar and adding the rounded numbers.

$$
\begin{array}{rcr}
\$2.39 & \rightarrow & \$2 \\
0.89 & \rightarrow & 1 \\
6.13 & \rightarrow & 6 \\
4.75 & \rightarrow & 5 \\
+\ 5.05 & \rightarrow & 5 \\
\hline
& & \$19
\end{array}
$$

   Thus, Karly's estimate for her grocery bill is $19.

2. Karly's bill for car repairs was $72.80, and she has a coupon for $17.50 off. She can estimate her total cost by rounding each amount to the nearest 10 dollars and subtracting.

$$
\begin{array}{rcr}
\$72.80 & \rightarrow & \$70 \\
-\ 17.50 & \rightarrow & -\ 20 \\
\hline
& & \$50
\end{array}
$$

   Thus, an estimate for the repair bill is $50.

3. Karly sees a flash of lightning and hears the thunder 3.2 sec later. She knows that sound travels at 0.33 km/sec. She may estimate the distance she is from the lightning by rounding the time to the nearest unit and the speed to the nearest tenth and multiplying.

$$
\begin{array}{rcr}
0.33 & \rightarrow & 0.3 \\
\times\ 3.2 & \rightarrow & \times\ 3 \\
\hline
& & 0.9
\end{array}
$$

   Thus, Karly estimates that she is approximately 0.9 km from the lightning.

   An alternative approach is to recognize that $0.33 \approx \frac{1}{3}$ and 3.2 is close to 3.3, so an approximation using compatible numbers is $\left(\frac{1}{3}\right)3.3$, or 1.1 km.

4. Karly wants to estimate the number of miles she gets per gallon of gas. If she had driven 298 mi and it took 12.4 gal to fill the gas tank, she rounds and divides as follows:

$$
12.4\overline{)298} \quad \rightarrow \quad 12\overline{)300}^{\,25}
$$

   Thus, she got about 25 mi/gal.

5. Johanna wanted to place the decimal point in a product resulting when she bought 21.45 lb of sesame seeds to grind for tahini at $3.40 per pound. The multiplication (without a decimal) resulted in 7293. She knew that very rough estimates could be obtained by rounding 21.45 to 20 and $3.40 to $3. The product had to be in the neighborhood of 20 × 3, or 60. So her placement of the decimal point was $72.93.

When computations are performed with rounded numbers, the results may be significantly different from the actual answer.

NOW TRY THIS 7-6 Other estimation strategies, such as front-end, clustering, and grouping to nice numbers, that you investigated with whole numbers also work with decimals. Take the grocery store bill in part 1 on the preceding page and use a front-end-with-adjustment strategy to estimate the bill.

Round-off Errors

Round-off errors are typically compounded when computations are involved. For example, if two distances are 42.6 mi and 22.4 mi rounded to the nearest tenth, then the sum of the distances appears to be $42.6 + 22.4$, or 65 mi. To the nearest hundredth, the distances might have been more accurately reported as 42.55 and 22.35 mi, respectively. The sum of these distances is 64.9 mi. Alternatively, the original distances may have been rounded from 42.64 and 22.44 mi. The sum now is 65.08, or 65.1 rounded to the nearest tenth. The original sum of 65 mi is between 64.9 and 65.1 mi, but the simple rounding process used is not precise.

Similar errors arise in other arithmetic operations. *When computations are done with approximate numbers, the final result should not be reported using more significant digits than the number used with the fewest significant digits.* Non-zero digits are always significant. Zeroes before other digits are non-significant. Zeroes between other non-zero digits are significant. Zeroes to the right of a decimal point are significant. To avoid uncertainty, zeroes at the end of a number are significant only if to the right of a decimal point.

NOW TRY THIS 7-7 In the *Hi and Lois* cartoon that follows,

a. Estimate whether Trixie's calculation is accurate.
b. To the nearest hundred-millionth, how many times is Trixie older than her twin brother?

Assessment 7-2A

1. If Maura went to the store and bought a chair for $17.95, a lawn rake for $13.59, a spade for $14.86, a lawn mower for $179.98, and two six-packs of mineral water for $2.43 each, what was the bill?

2. **a.** Complete the following magic square; that is, make the sum of every row, column, and diagonal the same:

8.2		
3.7	5.5	
	9.1	2.8

   b. If each cell of the magic square has 0.85 added to it,
      (i) Is the square still magic?
      (ii) If the answer to part (i) is "yes," what is the sum of each row?
   c. If each cell of the original magic square is multiplied by 0.5, is the square still magic? If "yes," what is the sum of each row?

3. A stock rose 0.24 in the market on Thursday. If the resulting price was 73.245, what was the price of the stock before the rise?

4. A U.S. \$1 bill was valued at 1.0156 Canadian dollars on February 22, 2008. What was the value of \$27.32 American dollars in Canadian dollars that day?

5. A kilowatt hour means 1000 watts of electricity are being used continuously for 1 hr. The electric utility company in Laura's town charges \$0.06715 for each kilowatt hour used. Laura heats her house with three electric wall heaters that use 1200 watts per hour each.
   a. How much does it cost to heat her house for 1 day?
   b. How many hours would a 75-watt lightbulb have to stay on to result in \$1 for electricity charges?

6. One liter is 4.224 c. How many liters are in 36.5 c?

7. Florence Griffith-Joyner set a world record for the women's 100 m dash at the 1988 Summer Olympics in Seoul, South Korea. She covered the distance in 10.49 sec. If 1 m is equivalent to 39.37 in., express Griffith-Joyner's speed in terms of miles per hour.

8. If each of the following sequences is either arithmetic or geometric, continue the decimal patterns:
   a. 0.9, 1.8, 2.7, 3.6, 4.5, ____, ____, ____
   b. 0.3, 0.5, 0.7, 0.9, 1.1, ____, ____, ____

9. If the first term of a finite geometric sequence is 0.9 and its ratio is 0.2, what is the sum of the first five terms?

10. Interpret 0.2222 as a sum of a finite geometric sequence whose first term is 0.2. (*Hint:* Write 0.2222 as the sum of fractions whose denominators are powers of 10.)

11. In a finite geometric sequence, the first term is 0.2 and the sixth term is 0.000486. What are the second through fifth terms?

12. Estimate the placement of each of the following on the given number line by placing the letter for each computation in the appropriate box:
   a. $0.3 \div 0.31$ b. $0.3 \cdot 0.31$

13. A bank statement from a local bank shows that a checking account has a balance of \$83.62. The balance recorded in the checkbook shows only \$21.69. After checking the canceled checks against the record of these checks, the customer finds that the bank has not yet recorded six checks in the amounts of \$3.21, \$14.56, \$12.44, \$6.98, \$9.51, and \$7.49. Is the bank record correct? (Assume the person's checkbook records *are* correct.)

14. Convert each of the following to standard numerals:
   a. $3.2 \cdot 10^{-9}$
   b. $3.2 \cdot 10^{9}$
   c. $4.2 \cdot 10^{-1}$
   d. $6.2 \cdot 10^{5}$

15. Write the numerals in each of the following sentences in scientific notation:
   a. The diameter of Earth is about 12,700,000 m.
   b. The distance from Pluto to the Sun is about 4,486,000,000 km.
   c. Each year, about 50,000,000 cans are discarded in the United States.

16. Write the numerals in each of the following sentences in standard form:
   a. The mass of a dust particle is 7.53×10^{-10} g.
   b. The speed of light is approximately 2.98×10^{5} km per second.
   c. Jupiter is approximately 7.78570000×10^{8} km from the Sun.

17. Write the results of each of the following in scientific notation:
   a. $(8 \cdot 10^{12})(6 \cdot 10^{15})$
   b. $(16 \cdot 10^{12}) \div (4 \cdot 10^{5})$
   c. $(5 \cdot 10^{8})(6 \cdot 10^{9}) \div (15 \cdot 10^{15})$

18. Round each of the following numbers as specified:
   a. 203.651 to the nearest hundred
   b. 203.651 to the nearest ten
   c. 203.651 to the nearest unit
   d. 203.651 to the nearest tenth
   e. 203.651 to the nearest hundredth

19. Jane's car travels 224 mi on 12 gal of gas. Rounded to the nearest mile per gallon, how many miles to the gallon does her car get?

20. Audrey wants to buy some camera equipment to take pictures on her daughter's birthday. To estimate the total cost, she rounds each price to the nearest dollar and adds the rounded prices. What is her estimate for the items listed?

Camera	\$54.56
Film	\$4.50
Case	\$17.85

21. Estimate the sum or difference in each of the following by using (i) rounding and (ii) front-end estimation. Then perform the computations to see how close your estimates are to the actual answers.

   a.   65.84   b.   89.47
         24.29   − 32.16
         12.18
       + 19.75

   c.   5.85   d.   223.75
         6.13   − 87.60
         9.10
       + 4.32

22. Write two numbers in scientific notation such that one number is greater than 1, one number is less than 1, and the product of the two numbers is 1.

23. Find the least and the greatest possible products for the expression using the digits 1 through 9. Each digit may be used only once in each part.

24. Some digits in the following number are covered by squares:

   4 ☐☐ 3 ☐ . ☐☐ 8 ☐

   If each of the digits 1 through 9 is used exactly once in the number, determine the greatest possible number.

25. Iris worked a 40 hr week at $8.25/hr. Mentally compute her salary for the week and explain how you did it.

26. Mentally compute the number to fill in the blank in each of the following:
  a. $8.4 \cdot 6 = 4.2 \cdot$ ____
  b. $10.2 \div 0.3 = 20.4 \div$ ____
  c. $ab = (a/2) \cdot$ ____
  d. $a \div b = 2a \div$ ____

27. Which of the following result in equal quotients?
  a. $7 \div 0.25$ **b.** $70 \div 2.5$
  c. $0.7 \div 0.25$ **d.** $700 \div 25$

28. Use estimation to place the decimal point in the correct position in each of the following products:
  a. $534 \cdot 0.34 = 18156$
  b. $5.07 \cdot 29.3 = 148551$

Assessment 7-2 B

1. A stock's price dropped from $63.28 per share to $27.45. What was the loss on a single share of the stock?

2. Keith bought 30 lb of nuts at $3.00/lb and 20 lb of nuts at $5.00/lb. If he wanted to buy 10 more pounds of a different kind of nut to make the average price per pound equal to $4.50, what price should he pay for the additional 10 lb?

3. Automobile engines were once measured in cubic inches (in.^{3}) but are now usually measured in cubic centimeters (cm^3). If 2.54 cm is equivalent to 1 in., answer the following:
  a. Susan's 1963 Thunderbird has a 390 in.^{3} engine. Approximately how many cubic centimeters is this?
  b. Dan's 1991 Taurus has a 3000 cm^3 engine. Approximately how many cubic inches is this?

4. The javelin throw in the 2004 Olympics, Heat 1, gave the following results (with distances in meters): Carolina Klüft, SWE, 48.89; Svetlana Sokolova, RUS, 47.86; Shelia Burrell, USA, 47.69. By what lengths did Klüft defeat Sokolova and Burrell?

5. Complete the following magic square; that is, make the sum of every row, column, and diagonal the same:

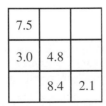

6. At a store, Samuel bought a bouquet for $4.99, a candy bar for $0.79, a memory stick for $49.99, and a bottle of water for $1.49. What was his bill?

7. Mississippi Electric charges $0.0485 for each kilowatt-hour used. Terry used her computer for 3.4 hr while the computer itself used 45 watts per hour and the computer monitor used 35 watts per hour. What was the charge for the computer usage?

8. If each of the following sequences is either arithmetic or geometric, continue the decimal patterns:
  a. 1, 0.5, 0.25, 0.125, _____, _____, _____
  b. 0.2, 1.5, 2.8, 4.1, 5.4, ____, ____, ____

9. If the first term of a finite geometric sequence is 0.4 and its ratio is 0.3, what is the sum of the first five terms?

10. Interpret the decimal 0.3333333 as a sum of a finite geometric sequence whose first term is 0.3. (*Hint:* Write 0.3333333 as the sum of fractions whose denominators are powers of 10.)

11. Estimate the placement of each of the following on the given number line by placing the letter for each computation in the appropriate box:
  a. $0.3 + 0.31$ **b.** $0.3 - 0.31$

12. Convert each of the following to standard numerals:
  a. $3.5 \cdot 10^7$ **b.** $3.5 \cdot 10^{-7}$
  c. $^-(2.4 \cdot 10^{-3})$

13. Write each of the numerals in each of the following sentences in scientific notation:
  a. The population of one state is 999,347 people.
  b. The area of North America is 24,490,000 mi^2.
  c. The nucleus of the diameter of an atom is estimated to be 10^{-15} m.

14. Write the results of each of the following in scientific notation:
  a. $(5 \cdot 10^7)(7 \cdot 10^{12})$
  b. $(^-13 \cdot 10^4) \div 65$
  c. $(3 \cdot 10^7)(4 \cdot 10^5) \div (6 \cdot 10^{-7})$

15. Round each of the following numbers as specified:
  a. 715.04 to the nearest hundred
  b. 715.04 to the nearest tenth
  c. 715.04 to the nearest unit
  d. 715.04 to the nearest ten
  e. 715.04 to the nearest thousand

16. Jane drives at a constant speed of 55.5 mph. How far should she expect to drive in $\frac{3}{4}$ hr?

17. Solve the following for x where x is a decimal:
    a. $2x + 1.3 = 4.1$ b. $4.2 - 3x = 10.2$
    c. $8.56 = 3 - 2x$

18. Write the numerals in each of the following sentences in standard form:
    a. A computer requires $4.4 \cdot 10^{-6}$ sec to do an addition problem.
    b. There are about $1.99 \cdot 10^4$ km of coastline in the United States.
    c. Earth has existed for approximately $3 \cdot 10^9$ yr.

19. Mary Kim invested $964 in 18 shares of stock. A month later, she sold the 18 shares at $61.48 per share. She also invested in 350 shares of stock for a total of $27,422.50. She sold this stock for $85.35 a share and paid $495 in total commissions. What was Mary Kim's profit or loss on the transactions to the nearest dollar?

20. Find the least and the greatest possible products for the expression using the digits 1 through 9. Each digit may be used only once in each part.

$$\square . \square \times \square . \square$$

21. Some digits in the following number are covered by squares:

$$4\,\square\,\square\,3\,\square\,.\,\square\,\square\,8\,\square$$

If each of the digits 1 through 9 is used exactly once in the number, determine the least possible number.

22. Iris worked a 40-hr week at $6.25/hr. Mentally compute her salary for the week and explain how you did it.

23. Estimate the sum or difference in each of the following by using (i) rounding and (ii) front-end estimation. Then perform the computation to see how close your estimates are to the actual answers.

a. 47.62
    27.99
    13.14
  + 7.61

b. 79.86
  − 27.37

c. 5.85
    6.17
    9.1
  + 4.23

d. 232.65
  − 78.92

24. Which of the following result in equal quotients?
    a. $9 \div 0.35$ b. $90 \div 3.5$
    c. $900 \div 35$

25. Is subtracting 0.3 the same as
    a. subtracting 0.30?
    b. subtracting 0.03? Explain.

26. Use estimation to choose a decimal to multiply by 9 in order to get within 1 of 93. Explain how you made your choice and check your estimate.

27. a. Fill in the parentheses in each of the following to write a true equation:

$$1 \cdot 2 + 0.25 = (\ )^2$$
$$2 \cdot 3 + 0.25 = (\ )^2$$

Conjecture what the next two equations in this pattern will be.

    b. Do the computations to determine if your next two equations are correct.
    c. Generalize your answer in part (a) by filling in an appropriate expression in the equation

$$n(n + 1) + 0.25 = (\ )^2$$

where n is the number of the terms in this sequence of equations.

Mathematical Connections 7-2

Communication

1. Give an example of a balanced checkbook where entries could be incorrect.
2. How is multiplication of decimals like multiplication of whole numbers? How is it different?
3. Why are estimation skills important in dividing decimals?
4. In the text, multiplication and division were done using both fractional and decimal forms. Discuss the advantages and disadvantages of each.
5. Explain why subtraction of terminating decimals can be accomplished by lining up the decimal points, subtracting as if the numbers were whole numbers, and then placing the decimal point in the difference.

Open-Ended

6. Find several examples of the use of decimals in the newspaper. Tell whether you think the numbers are exact or estimates. Also tell why you think decimals were used instead of fractions.
7. How could a calculator be used to develop or reinforce the understanding of multiplication of decimals?
8. How might certain calculators impede the learning of multiplication of decimals if they display the simplest form of a terminating decimal, for example, 0.3 instead of 0.30?
9. Wearne and Hiebert (1988) reported that students who connect physical representations of decimals with decimal

notation are more likely to create their own procedures for converting a fraction to its decimal notation. Do you think that this research supports or does not support the conversion of decimals to fractions to explain computational procedures?

Cooperative Learning

10. In your group, decide on all the prerequisite skills that students need before learning to perform arithmetic operations on decimals.

11. You will need a calculator and a partner to play the game described on page 438 on the student page from McDougal Littell *MATH Thematics, New Edition Book 1,* 2008.

Questions from the Classroom

12. A student multiples $(6.5)(8.5)$ to obtain the following:

$$
\begin{array}{r}
8.5 \\
\times\,6.5 \\
\hline
425 \\
510 \\
\hline
55.25
\end{array}
$$

However, when the student multiplies $8\frac{1}{2}\cdot 6\frac{1}{2}$, she obtains the following:

$$
\begin{array}{r}
8\frac{1}{2} \\
\times\,6\frac{1}{2} \\
\hline
4\frac{1}{4}\quad\left(\frac{1}{2}\cdot 8\frac{1}{2}\right) \\
48\phantom{\frac{1}{4}}\quad(6\cdot 8) \\
\hline
52\frac{1}{4}
\end{array}
$$

How is this possible?

13. A student tries to calculate $0.999^{10,000}$ on a calculator and finds the answer to be $4.5173346\cdot 10^{-5}$. The student wonders how it could be that a number like 0.999, so close to 1, when raised to some power could result in a number close to 0. How do you respond?

14. How would you respond to the following:
   a. A student claims that $\frac{9443}{9444}$ and $\frac{9444}{9445}$ are equal because both display 0.9998941 on his scientific calculator when the divisions are performed.
   b. Another student claims that the fractions are not equal and wants to know if there is any way the same calculator can determine which is greater.

Review Problems

15. Write 14.0479 in expanded place value form.
16. Without dividing, determine which of the following represent terminating decimals:
   a. $\dfrac{24}{36}$
   b. $\dfrac{49}{56}$
17. If the denominator of a fraction is 26, is it possible that the fraction could be written as a terminating decimal? Why or why not?
18. $\dfrac{35}{56}$ can be written as a decimal that terminates. Explain why.

Third International Mathematics and Science Study (TIMSS) Questions

A rubber ball rebounds to half the height it drops. If the ball is dropped from a rooftop 18 m above the ground, what is the total distance traveled by the time it hits the ground the third time?
   a. 31.5 m
   b. 40.5 m
   c. 45 m
   d. 63 m

In a discus-throwing competition, the winning throw was 61.60 m. The second-place throw was 59.72 m. How much longer was the winning throw than the second-place throw?
   a. 1.18 m
   b. 1.88 m
   c. 1.98 m
   d. 2.18 m

TIMSS, Grade 8, 1994

National Assessment of Educational Progress (NAEP) Question

It costs $0.25 to operate a clothes dryer for 10 minutes at a laundromat. What is the total cost to operate one clothes dryer for 30 minutes, a second for 40 minutes, and a third for 50 minutes?
   a. $3.25
   b. $3.00
   c. $2.75
   d. $2.00
   e. $1.20

Add the numbers $\dfrac{7}{10}, \dfrac{7}{100}$, and $\dfrac{7}{1,000}$. Write this sum as a *decimal*.

NAEP, Grade 8, 2007

School Book Page · ESTIMATING DECIMAL PRODUCTS

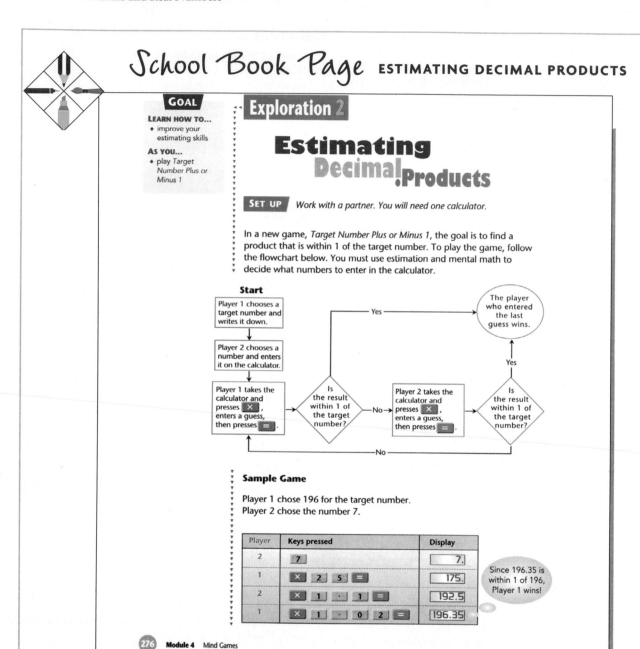

7-3 Nonterminating Decimals

In grade 7 *Focal Points*, we find:

> Students now use division to express any fraction as a decimal, including fractions that they must represent with infinite decimals. (p. 38)

Earlier in the chapter, we developed procedures for converting some rational numbers to decimals. For example, $\frac{7}{8}$ can be written as a terminating decimal as follows:

$$\frac{7}{8} = \frac{7}{2^3} = \frac{7 \cdot 5^3}{2^3 \cdot 5^3} = \frac{875}{1000} = 0.875$$

The decimal for $\frac{7}{8}$ can also be found by division:

$$
\begin{array}{r}
0.875 \\
8\overline{)7.000} \\
\underline{6\ 4} \\
60 \\
\underline{56} \\
40 \\
\underline{40} \\
0
\end{array}
$$

However, we showed that $\frac{2}{11}$ was not a terminating decimal. We investigate non-terminating decimals now.

Repeating Decimals

 If we use a calculator to find a decimal representation for $\frac{2}{11}$, the calculator may display 0.1818181. It seems that 18 repeats itself. To examine what digits, if any, the calculator did not display, consider the following division:

$$
\begin{array}{r}
0.18 \\
11\overline{)2.00} \\
\underline{1\ 1} \\
90 \\
\underline{88} \\
2
\end{array}
$$

At this point, if the division is continued, the division pattern would repeat, since the remainder 2 divided by 11 repeats the division. Thus the quotient is 0.181818.... A decimal of this type is a **repeating decimal**, and the repeating block of digits is the **repetend**. The repeating decimal is written $0.\overline{18}$, where the bar indicates that the block of digits underneath is repeated continuously.

Example 7-10

Convert the following to decimals:

a. $\dfrac{1}{7}$ **b.** $\dfrac{2}{13}$

Solution   We use a calculator to divide, and it seems that the division pattern repeats.

a. $\dfrac{1}{7} = 0.\overline{142857}$ **b.** $\dfrac{2}{13} = 0.\overline{153846}$

To see why in Example 7-10 the division pattern repeats as predicted, consider the following divisions:

a.
$$
\begin{array}{r}
0.142857 \\
7\overline{)1.000000} \\
7 \\
\overline{30} \\
28 \\
\overline{20} \\
14 \\
\overline{60} \\
56 \\
\overline{40} \\
35 \\
\overline{50} \\
49 \\
\overline{1}
\end{array}
$$

b.
$$
\begin{array}{r}
0.153846 \\
13\overline{)2.000000} \\
1\,3 \\
\overline{70} \\
65 \\
\overline{50} \\
39 \\
\overline{110} \\
104 \\
\overline{60} \\
52 \\
\overline{80} \\
78 \\
\overline{2}
\end{array}
$$

In $\dfrac{1}{7}$, the remainders obtained in the division are 3, 2, 6, 4, 5, and 1. These are all the possible nonzero remainders that can be obtained when dividing by 7. If we had obtained a remainder of 0, the decimal would terminate. Consequently, the seventh division cannot produce a new remainder. Whenever a remainder recurs, the process repeats itself. Using similar reasoning, we could predict that the repetend for $\dfrac{2}{13}$ could not have more than 12 digits, because there are only 12 possible nonzero remainders. However, one of the remainders could repeat sooner than that, which was actually the case in part (b). In general, if $\dfrac{a}{b}$ is any rational number in simplest form with $b \neq 0$ and $b > a$, and it does not represent a terminating decimal, the repetend has at most $b - 1$ digits. Therefore, a *rational number may always be represented either as a terminating decimal or as a repeating decimal.*

NOW TRY THIS 7-8

a. Write $\dfrac{1}{9}$ as a decimal.

b. Based on your answer in part (a), mentally compute the decimal representation for each of the following.

(i) $\dfrac{2}{9}$ (ii) $\dfrac{3}{9}$ (iii) $\dfrac{5}{9}$ (iv) $\dfrac{8}{9}$

Example 7-11

Use a calculator to convert $\frac{1}{17}$ to a repeating decimal.

Solution   In using a calculator, if we press $\boxed{1}$ $\boxed{\div}$ $\boxed{1}$ $\boxed{7}$ $\boxed{=}$, we obtain the following, shown as part of a division problem:

$$17\overline{)1.}^{0.0588235}$$

Without knowing whether the calculator has an internal round-off feature and with the calculator's having an eight-digit display, we find the greatest number of digits to be trusted in the quotient is six following the decimal point. (Why?) If we use those six places and multiply 0.058823 times 17, we may continue the operation as follows:

$$\boxed{.}\,\boxed{0}\,\boxed{5}\,\boxed{8}\,\boxed{8}\,\boxed{2}\,\boxed{3}\,\boxed{\times}\,\boxed{1}\,\boxed{7}\,\boxed{=}$$

We then obtain 0.999991, which we may place in the preceding division:

$$\begin{array}{r} 0.058823 \\ 17\overline{)1.000000} \\ \underline{999991} \\ 9 \end{array}$$

Next, we divide 9 by 17 to obtain 0.5294118. Again ignoring the rightmost digit, we continue as before, completing the division as follows, where the repeating pattern is apparent:

$$\begin{array}{r} 0.05882352941176470588235 \\ 17\overline{)1.0000000000000000000000000} \\ \underline{999991} \\ 9000000 \\ \underline{8999987} \\ 13000000 \\ \underline{12999985} \\ 15 \end{array}$$

Thus, $\frac{1}{17} = 0.\overline{0588235294117647}$, and the repetend is 16 digits long.

Example 7-11 suggests a method to show that every rational number can be written as a decimal. Additionally, Example 7-11 illustrates how a calculator with only a finite display of digits can be used to do division beyond what the calculator was designed to do. A common suggestion for elementary students is that they only use the division algorithm selectively to learn the process. The calculator example given and similar ones should be used to see whether you truly understand the division process and the place values involved along the way.

Writing a Repeating Decimal in the Form $\frac{a}{b}$, Where $a, b \in I, b \neq 0$

We have already considered how to write terminating decimals in the form $\frac{a}{b}$, where a, b are integers and $b \neq 0$. For example,

$$0.55 = \frac{55}{10^2} = \frac{55}{100}$$

To write $0.\overline{5}$ in a similar way, we see that because the repeating decimal has infinitely many digits, the denominator cannot be written as a single power of 10. To overcome this difficulty, we must somehow eliminate the infinitely repeating part of the decimal. If we let $n = 0.\overline{5}$, then our *subgoal* is to write an equation for n without a repeating decimal. It can be shown that $10(0.555\ldots) = 5.555\ldots = 5.\overline{5}$. Hence, $10n = 5.\overline{5}$. Using this information, we subtract the corresponding sides of the equations to obtain an equation whose solution can be written without a repeating decimal.

$$
\begin{array}{r}
10n = 5.\overline{5} \\
-\quad n = 0.\overline{5} \\
\hline
9n = 5 \\
n = \dfrac{5}{9}
\end{array}
$$

Thus, $0.\overline{5} = \dfrac{5}{9}$. This result can be checked by performing the division $5 \div 9$. (*Note:* Performing the subtraction gives an equation that contains only integers. The repeating blocks "cancel" each other.)

Infinite geometric sequences provide a general method to change repeating decimals to rational numbers in the form $\dfrac{a}{b}$, where a and b are integers and $b \neq 0$. Consider $0.\overline{5}$ as a sum of the geometric sequence $0.5, 0.05, 0.005, 0.0005\ldots$. With this geometric sequence the first term is 0.5 and the ratio is 0.1. Thus, we have the following equality:

$$0.\overline{5} = 0.5 + 0.05 + 0.005 + 0.0005 + \ldots$$

To determine a method of finding the sum of an infinite geometric sequence, use the following:

$$\text{Let } S = 0.5 + 0.05 + 0.005 + 0.0005 + \ldots$$

If we multiply both sides of the equation by 0.1, we have the following pair of equations:

$$
\begin{array}{l}
S = 0.5 + 0.05 + 0.005 + 0.0005 + \ldots \\
0.1S = \quad\quad\ \ 0.05 + 0.005 + 0.0005 + \ldots
\end{array}
$$

Subtracting the bottom equation from the top, we have

$$0.9S = 0.5$$

so that $S = \dfrac{0.5}{0.9}$, or $\dfrac{5}{9}$, the same result as before.

The process for *finding the sum of an infinite geometric sequence whose ratio is r, where* $0 < r < 1$ or $-1 < 0 < r$, follows:

$$
\begin{array}{r}
S = a + ar + ar^2 + ar^3 + \ldots \\
{}^{-}rS = {}^{-}(ar + ar^2 + ar^3 + \ldots) \\
\hline
S - rS = a \\
S(1 - r) = a \\
S = \dfrac{a}{1 - r}
\end{array}
$$

NOW TRY THIS 7-9

a. If $r = 1$, how do you know that S does not exist?
b. If $r > 1$, what do we know about r^{n+1} as n gets greater and greater?
c. Find a formula for the sum of a finite geometric sequence.

Repeating decimals are written as "fractions" in student texts, as seen on the following partial student page. Answer number 4.

School Book Page   ENRICHMENT

34. $10\frac{1}{5} - 5\frac{3}{4}$ **35.** $\frac{4}{7} + \frac{8}{9}$ **36.** $6 - 3\frac{2}{5}$

37. Algebra Find $4 + 2 \times 3 - 4 \div 2$.

  A. 3   **B.** 7   **C.** 8   **D.** 16

Enrichment

Writing Repeating Decimals as Fractions

A short way of writing the repeating decimal $0.323232\ldots$ is $0.\overline{32}$. You can write this repeating decimal as a fraction.

Write $1n = 0.323232\ldots$ Since two digits repeat, multiply both sides by 100 to get $100n = 32.323232\ldots$

$$100n = 32.323232\ldots$$
$$- \quad 1n = 0.323232\ldots$$
$$\overline{99n = 32}$$

Subtract and solve for n.

$$n = \frac{32}{99}$$

To write $0.1\overline{643}$ as a fraction, write $1n = 0.1643643\ldots$

$$1{,}000n = 164.3643643\ldots$$
$$- \quad 1n = 0.1643643\ldots$$
$$\overline{999n = 164.2}$$

Three digits repeat, so multiply both sides by 1,000 to get $1{,}000n = 164.3643643\ldots$

$$n = \frac{164.2}{999} = \frac{1{,}642}{9{,}990} = \frac{821}{4{,}995}$$

Subtract and solve for n. Simplify.

For 1–4, write each decimal as a fraction with whole numbers in the numerator and denominator.

  1. $0.\overline{4}$ **2.** $0.\overline{25}$ **3.** $0.\overline{153}$ **4.** $0.2\overline{73}$

 All text pages available online and on CD-ROM.   Section A Lesson 5-1   **251**

Source: Scott Foresman-Addison Wesley Mathematics, Grade 6, 2008 (p. 251).

Suppose a decimal has a repetend of more than one digit, such as $0.\overline{235}$. We can use the formula just developed where the first term of a geometric sequence is 0.235, $r = \dfrac{1}{1000}$, and find the sum as follows:

$$S = \frac{0.235}{1 - \dfrac{1}{1000}} = \frac{0.235}{\dfrac{999}{1000}} = \frac{235}{999}$$

Thus, $0.\overline{235} = \dfrac{235}{999}$.

An equivalent approach is to multiply the decimal $0.\overline{235}$ by 10^3, since there is a three-digit repetend. Let $n = 0.\overline{235}$. Our *subgoal* is to write an equation for n without the repeating decimal:

$$1000n = 235.\overline{235}$$
$$\underline{-\quad n = \quad\ 0.\overline{235}}$$
$$999n = 235$$
$$n = \frac{235}{999}$$

Again, $0.\overline{235} = \dfrac{235}{999}$.

We generalize the latter method by first noticing that $0.\overline{5}$ repeats in one-digit blocks. Therefore, to write it in the form $\dfrac{a}{b}$, we first multiply by 10^1. Because $0.\overline{235}$ repeats in three-digit blocks, we first multiply by 10^3. In general, *if the repetend is immediately to the right of the decimal point, first multiply by 10^m, where m is the number of digits in the repetend, and then continue as in the preceding cases.*

Now, suppose the repeating block does *not* occur immediately after the decimal point. For example, let $n = 2.3\overline{45}$. A strategy for solving this problem is to *change it to a related problem* we know how to solve; that is, change it to a problem where the repeating block immediately follows the decimal point. This becomes a *subgoal*. To accomplish this, we multiply both sides by 10:

$$n = 2.3\overline{45}$$
$$10n = 23.\overline{45}$$

We now proceed as with previous problems. Because $10n = 23.\overline{45}$ and the number of digits in the repetend is 2, we multiply by 10^2 as follows:

$$100(10n) = 2345.\overline{45}$$

Thus,

$$1000n = 2345.\overline{45}$$
$$\underline{-\ 10n = \quad 23.\overline{45}}$$
$$990n = 2322$$
$$n = \frac{2322}{990}, \text{ or } \frac{129}{55}$$

Hence, $2.3\overline{45} = \dfrac{2322}{990}$, or $\dfrac{129}{55}$.

NOW TRY THIS 7-10 Use the sum of an infinite geometric sequence to find an equivalent fraction for $0.\overline{235}$ and $2.3\overline{45}$.

A Surprising Result

To find the $\dfrac{a}{b}$ form of $0.\overline{9}$, we proceed as follows. Let $n = 0.\overline{9}$, then $10n = 9.\overline{9}$. Next we subtract and continue to solve for n:

$$10n = 9.\overline{9}$$
$$- \ n = 0.\overline{9}$$
$$\overline{9n = 9}$$
$$n = 1$$

Hence, $0.\overline{9} = 1$. This approach to the problem may not be convincing. Another approach to show that $0.\overline{9}$ is really another name for 1 is shown next:

(a) $\dfrac{1}{3} = 0.33333333\ldots$ **(b)** $\dfrac{2}{3} = 0.66666666\ldots$

Adding equations (a) and (b), we have $1 = 0.99999999\ldots$. or $0.\overline{9}$. This last decimal represents the infinite sum $\dfrac{9}{10} + \dfrac{9}{10^2} + \dfrac{9}{10^3} + \ldots$.

Some may prefer a visual approach to show that $0.\overline{9} = 1$. Consider the number line in Figure 7-5(a). Most would agree that $0.\overline{9}$ would be between 0.9 and 1.0, so we start there. Then $0.\overline{9}$ is between 0.99 and 1.0, as in Figure 7-5(b).

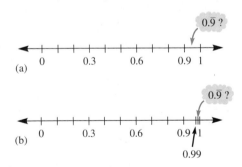

Figure 7-5

Then we can proceed similarly to argue that $0.\overline{9}$ would be between 0.999 and 1, and so forth. Next it is reasonable to ask if there is any tiny amount a such that $0.\overline{9} + a = 1$? The answer has to be no. (Why?) If there is no such number a, then $0.\overline{9}$ cannot be less than 1. With a comparable argument that $0.\overline{9}$ cannot be greater than 1, we know that $0.\overline{9} = 1$. In more advanced mathematics courses, sums like $0.\overline{9}$, or $0.9 + 0.09 + 0.009 + \ldots$, are defined as the *limits of finite sums*.

Ordering Repeating Decimals

We now know that any repeating decimal can be written as a rational number in the form $\dfrac{a}{b}$, where $b \neq 0$. Thus, any repeating decimal can be represented on a number line in a manner similar to the way that terminating decimals could be placed on the number line. Also, because we know that between any two rational numbers in $\frac{a}{b}$ form there are infinitely many more rational numbers of that form, it is reasonable that there should be infinitely many repeating decimals between any two other decimals.

To order repeating decimals, we could consider where a repeating decimal might lie on a number line or we could compare them using place value in a manner similar to how we ordered terminating decimals. For example, to order repeating decimals such as $1.\overline{3478}$ and $1.34\overline{7821}$, write the decimals one under the other, in their equivalent forms without the bars, and line up the decimal points (or place values) as follows:

$$1.34783478\ldots$$
$$1.34782178\ldots$$

The digits to the left of the decimal points and the first four digits after the decimal points are the same in each of the numbers. However, since the digit in the hundred-thousandths place of the top number, which is 3, is greater than the digit 2 in the hundred-thousandths place of the bottom number, $1.\overline{3478}$ is greater than $1.34\overline{7821}$.

It is easy to compare two fractions, such as $\frac{21}{43}$ and $\frac{37}{75}$, using a calculator. We convert each to a decimal and then compare the decimals.

$$\boxed{2}\,\boxed{1}\,\boxed{\div}\,\boxed{4}\,\boxed{3}\,\boxed{=}\;\rightarrow 0.4883721$$
$$\boxed{3}\,\boxed{7}\,\boxed{\div}\,\boxed{7}\,\boxed{5}\,\boxed{=}\;\rightarrow 0.4933333$$

Examining the digits in the hundredths place, we see that

$$\frac{37}{75} > \frac{21}{43}$$

Example 7-12

Find a rational number in decimal form between $0.\overline{35}$ and $0.\overline{351}$.

Solution   First, line up the decimals.
$$0.353535\ldots$$
$$0.351351\ldots$$

Then, to find a decimal between these two, observe that starting from the left, the first place at which the two numbers differ is the thousandths place. Clearly, one decimal between these two is 0.352. Others include 0.3514, $0.35\overline{15}$, and 0.35136. In fact, there are infinitely many others.

NOW TRY THIS 7-11 Given the terminating decimal 0.36, find two repeating decimals, one less than 0.36 but no more than 0.01 less, and one greater than 0.36, but no more than 0.01 greater.

Assessment 7-3A

1. Find the decimal representation for each of the following:
   a. $\frac{4}{9}$
   b. $\frac{2}{7}$
   c. $\frac{3}{11}$
   d. $\frac{1}{15}$
   e. $\frac{2}{75}$
   f. $\frac{1}{99}$
   g. $\frac{5}{6}$
   h. $\frac{1}{13}$
   i. $\frac{1}{21}$
   j. $\frac{3}{19}$

2. Convert each of the following repeating decimals to $\frac{a}{b}$ form, where a, b are integers and $b \neq 0$.
   a. $0.\overline{4}$
   b. $0.6\overline{1}$
   c. $1.3\overline{96}$
   d. $0.\overline{55}$
   e. $^{-}2.3\overline{4}$
   f. $^{-}0.0\overline{2}$

3. Express 1 min as a repeating decimal part of an hour.

4. Order the following decimals from greatest to least:
   $$^{-}1.4\overline{54}, \ ^{-}1.\overline{454}, \ ^{-}1.\overline{45}, \ ^{-}1.45\overline{4}, \ ^{-}1.454$$

5. Describe possible terms for the following pattern sequence:
   $$0, 0.5, 0.\overline{6}, 0.75, 0.8, 0.8\overline{3}, \underline{\quad}, \underline{\quad}, \underline{\quad}$$

6. In the repeating decimal $0.\overline{45}$ and with an interpretation of it as a geometric sequence, what is the common ratio?

7. Give an argument why $3\frac{1}{7}$ must be a repeating decimal.

8. Suppose $a = 0.\overline{32}$ and $b = 0.\overline{123}$.
   a. Find $a + b$ by adding from left to right. How many digits are in the repetend of the sum?
   b. Find $a + b$ if $a = 1.2\overline{34}$ and $b = 0.\overline{1234}$. Is the answer a rational number? How many digits are in the repetend?

9. Explain whether a terminating decimal could ever be written as a repeating decimal.

10. Find three decimals between each of the two following pairs of decimals:
    a. $3.\overline{2}$ and 3.22
    b. $462.\overline{24}$ and 462.243

11. Find the decimal halfway between the two following decimals: $0.\overline{4}$ and 0.5.

12. a. Find three rational numbers between $\frac{3}{4}$ and $0.7\overline{5}$.
    b. Find three rational numbers between $\frac{1}{3}$ and $0.3\overline{4}$.

13. a. What is the 21st digit in the decimal expansion of $\frac{3}{7}$?
    b. What is the 5280th digit in the decimal expansion of $\frac{1}{17}$?

14. a. Write each of the following as a fraction in the form $\frac{a}{b}$, where a and b are integers and $b \neq 0$.
      (i) $0.\overline{1}$ (ii) $0.\overline{01}$ (iii) $0.\overline{001}$
    b. What fraction would you expect for $0.\overline{0001}$?
    c. Mentally compute the decimal equivalent for $\frac{1}{90}$.

15. Use the fact that $0.\overline{1} = \frac{1}{9}$ to mentally convert each of the following into fractions:
    a. $0.\overline{2}$
    b. $0.\overline{3}$
    c. $0.\overline{5}$
    d. $2.\overline{7}$
    e. $9.\overline{9}$

16. Use the fact that $0.\overline{01} = \frac{1}{99}$ and $0.\overline{001} = \frac{1}{999}$ to mentally convert each of the following into fractions:
    a. $0.\overline{05}$
    b. $0.\overline{003}$
    c. $3.\overline{25}$
    d. $3.\overline{125}$

17. Find the sum of the finite geometric sequence whose first term is 0.4, whose ratio is 0.5, and which has five terms.

18. Use the formula for the sum of an infinite geometric series to find a rational number in the form $\frac{a}{b}$, where a and b are integers and $b \neq 0$, for the following repeating decimals:
    a. $0.2\overline{9}$
    b. $2.02\overline{9}$

19. In this section, you saw that $0.\overline{9}$ was equal to 1. In other words, $0.\overline{9}$ is another representation for 1, as is $\frac{1}{1}$. How would you write 1 as a repeating decimal in a different form from $0.\overline{9}$? Explain your representation.

20. a. Determine whether the sum of any two repeating decimals must be a repeating decimal. Assume each repetend is neither all 9s nor all 0s.
    b. Explain why you can always write the sum of a terminating decimal and a repeating decimal as a repeating decimal.
    c. Explain why the sum of any two terminating decimals can be written as a repeating decimal.
    d. Explain why it is not always possible to write the sum of any two repeating decimals as a terminating decimal.

Assessment 7-3B

1. Find the decimal representation for each of the following:

   a. $\frac{2}{3}$ **b.** $\frac{7}{9}$ **c.** $\frac{1}{24}$

   d. $\frac{3}{60}$ **e.** $\frac{2}{99}$ **f.** $\frac{7}{6}$

   g. $\frac{2}{21}$ **h.** $\frac{4}{19}$

2. Convert each of the following repeating decimals to $\frac{a}{b}$ form, where a and b are integers and $b \neq 0$:

   a. $0.\overline{7}$ **b.** $0.\overline{46}$ **c.** $2.\overline{37}$
   d. $2.3\overline{4}$ **e.** $^-4.3\overline{4}$ **f.** $^-0.0\overline{3}$

3. Express 1 sec as a repeating decimal part of an hour.

4. Order the following decimals from least to greatest:

   $$^-4.34, \, ^-4.\overline{34}, \, ^-4.3\overline{4}, \, ^-4.3\overline{43}, \, ^-4.4\overline{34}$$

5. Describe possible terms for the following arithmetic sequence:

   $$0, 0.\overline{3}, 0.\overline{6}, 1, 1.\overline{3}, \underline{\quad}, \underline{\quad}, \underline{\quad}$$

★ 6. Find an infinite geometric sequence whose sum is $2.3\overline{4}$.

7. Explain whether a repeating decimal could ever be written as a terminating decimal.

8. Give an argument why $\frac{1}{17}$ must be a repeating decimal.

9. Find three decimals between each of the two following pairs of decimals.

   a. $4.\overline{3}$ and 4.3
   b. $203.\overline{76}$ and $203.\overline{7}$

10. Find the decimal halfway between the two following decimals: $0.\overline{9}$ and 1.1.

11. Find three rational numbers between the following pairs of numbers:

   a. $\frac{2}{3}$ and 0.67 **b.** $\frac{2}{3}$ and $0.6\overline{7}$

12. What is the 23rd decimal in the expansion of $\frac{1}{17}$?

13. **a.** Write each of the following as a fraction in the form $\frac{a}{b}$, where a and b are integers and $b \neq 0$:

   (i) $0.\overline{2}$ **(ii)** $0.0\overline{2}$ **(iii)** $0.00\overline{2}$
   b. What fraction would you expect for $0.000\overline{2}$?
   c. Mentally compute the decimal equivalent for $\frac{4}{90}$.

14. Based on the fact that $0.\overline{9} = 1$, what are the equivalents of each of the following:

   a. $1.\overline{9}$ **b.** $2.\overline{9}$ **c.** $3.\overline{9}$

15. Use the fact that $0.\overline{1} = \frac{1}{9}$, $0.\overline{01} = \frac{1}{99}$, and $0.\overline{001} = \frac{1}{999}$ to convert each of the following mentally to the form $\frac{a}{b}$, where a and b are integers and $b \neq 0$:

   a. $0.\overline{4}$ **b.** $0.\overline{12}$ **c.** $0.\overline{111}$

16. Find the sum of the finite geometric sequence whose first term is 0.1, whose ratio is 0.3, and which has four terms.

17. Find the sum in $\frac{a}{b}$ form, where a and b are integers and $b \neq 0$ of the geometric sequence represented by the following:

   a. $0.\overline{29}$ **b.** $0.000\overline{29}$

18. Consider the repeating decimals, $0.\overline{23}$ and $0.\overline{235}$, how many places do you expect in the repetend of the sum of the two decimals? Why?

19. Find values of x so that each of the following is true. Write answers both in $\frac{a}{b}$ form, where a and b are integers and $b \neq 0$, and as decimals.

   a. $3x = 8$ **b.** $3x + 1 = 8$
   c. $3x - 1 = 8$ **d.** $1 - 3x = 8$
   e. $1 = 3x + 8$ **f.** $1 = 8 - 3x$

Mathematical Connections 7-3

Communication

1. **a.** If a grocery store advertised three lemons for $2.00, what is the cost of one lemon?
   b. If you choose to buy exactly one lemon at the cost given in part (a), what are you charged?
   c. How is the store treating the repeating decimal cost of one lemon?

   d. Explain whether or not a grocery store would ever use a repeating decimal as a cost for an item.
   e. Explain whether you think cash registers ever work with repeating decimals.

2. A friend claims that every finite decimal is equal to some infinite decimal. Is the claim true? Explain why or why not.

3. Some addition problems are easier to compute with fractions and some are easier to do with decimals. For example, $\frac{1}{7} + \frac{5}{7}$ is easier to compute than $0.\overline{142857} + 0.\overline{714285}$ and $0.4 + 0.25$ is easier to compute than $\frac{2}{5} + \frac{1}{4}$. Describe situations in which you think it would be easier to compute the additions with fractions than with decimals, and vice versa.

Open-Ended

4. Notice that $\frac{1}{7} = 0.\overline{142857}$, $\frac{2}{7} = 0.\overline{285714}$, $\frac{3}{7} = 0.\overline{428571}$, $\frac{4}{7} = 0.\overline{571428}$, $\frac{5}{7} = 0.\overline{714285}$, and $\frac{6}{7} = 0.\overline{857142}$.

   a. Describe a common property that all of these repeating decimals share.
   b. Suppose you memorized the decimal form for $\frac{1}{7}$. How could you quickly find the answers for the decimal expansion of the rest of the preceding fractions? Describe as many ways as you can.
   c. Find some other fractions that behave like $\frac{1}{7}$. In what way is the behavior similar?
   d. Based on your answer in (a), describe shortcuts for writing $\frac{k}{14}$ as a repeating decimal for $k = 1, 2, 3, \ldots, 13$.

5. a. Demonstrate that every integer can be written as a decimal.
   b. Multiply each of the following, giving your answer as a decimal in simplest form. (Remember your answer to part (a)).
      (i) $2 \cdot 0.\overline{3}$ **(ii)** $3 \cdot 0.\overline{3}$ **(iii)** $3 \cdot 0.\overline{35}$
   c. Explain whether the traditional algorithm for multiplying decimals can be used for multiplying repeating decimals.
   d. Explain whether repeating decimals can be multiplied.

6. a. Does your calculator allow you to enter repeating decimals? If so, in what form?
   b. Explain whether repeating decimal arithmetic can be performed on your calculator.

7. Explain whether you had rather have the solution to $3x = 7$ expressed as a fraction or as a repeating decimal.

Cooperative Learning

8. Choose a partner and play the following game. Write a repeating decimal of the form $0.\overline{abcdef}$. Tell your partner that the decimal is of that form but do not reveal the specific values for the digits. Your partner's objective is to find your repeating decimal. Your partner is allowed to ask you for the values of six digits that are at the 100th or greater places after the decimal point but not the digits in consecutive places. For example, your opponent may ask for the 100th, 200th, 300th, . . . digits but may not ask for the 100th and 101st digit. Switch roles at least once. After playing the game, discuss in your group a strategy for

asking your partner the least number of questions that will allow you to find your partner's repetend.

Questions from the Classroom

9. A student argues that repeating decimals are of little value because no calculator will handle computations with repeating decimals and in real life, decimals do not have an infinite number of digits. How do you respond?

10. Samantha says that base two could have no decimals, terminating or repeating. How would you convince her that is untrue?

Review Problems

11. John is a payroll clerk for a small company. Last month, the employee's gross earnings (earnings before deductions) totaled $27,849.50. John deducted $1520.63 for social security, $723.30 for state income tax, and $2843.62 for federal income tax. What was the employee's net pay (earnings after deductions)?

12. The speed of light is approximately 186,000 mi/sec. It takes light from the nearest star, Alpha Centauri, approximately 4 yr to reach Earth. How many miles away is Alpha Centauri from Earth? Express the answer in scientific notation.

13. Find the product of 0.22 and 0.35 on a calculator. How does the placement of the decimal point in the answer on the calculator compare with the placement of the decimal point using the rule in this chapter? Explain.

14. Answer the following:
   a. Find a number to add to $^-0.023$ to obtain a sum greater than 3 but less than 4.
   b. Find a number to subtract from $^-0.023$ to obtain a difference greater than 3 but less than 4.
   c. Find a number to multiply times 0.023 to obtain a product greater than 3 but less than 4.
   d. Find a number to divide into 0.023 to obtain a quotient greater than 3 but less than 4.

Third International Mathematics and Science Study (TIMSS) Questions

In which list are the numbers ordered from greatest to least?
   a. 0.233, 0.3, 0.32, 0.332
   b. 0.3, 0.32, 0.332, 0.233
   c. 0.32, 0.233, 0.332, 0.3
   d. 0.332, 0.32, 0.3, 0.233

TIMSS, Grade 8, 2003

In which of these pairs of numbers is 2.25 larger than the first number but smaller than the second number?

   a. 1 and 2   **b.** 2 and $\frac{5}{2}$

   c. $\frac{5}{2}$ and $\frac{11}{4}$ **d.** $\frac{11}{4}$ and 3

TIMSS, Grade 8, 2003

National Assessment of Educational Progress (NAEP) Question

Mrs. Jones bought 6 pints of berries. Each pint cost 87¢. Mrs. Jones used her calculator to find the cost of the berries and the display showed 522. What was the cost of the berries?

    a. $522
    b. $52.20
    c. $5.22
    d. $0.52

NAEP, Grade 4, 2003

7-4   Real Numbers

Every rational number can be expressed either as a repeating decimal or as a terminating decimal. The ancient Greeks discovered numbers that are not rational. Such numbers must have a decimal representation that neither terminates nor repeats. To find such decimals, we focus on the characteristics they must have:

1. There must be an infinite number of nonzero digits to the right of the decimal point.
2. There cannot be a repeating block of digits (a repetend).

One way to construct a nonterminating, nonrepeating decimal is to devise a pattern of infinite digits in such a way that there will definitely be no repeated block. Consider the number 0.1010010001 . . . . If the pattern continues, the next groups of digits are four zeros followed by 1, five zeros followed by 1, and so on. It is possible to describe a pattern for this decimal, but there is no repeating block of digits. Because this decimal is nonterminating and nonrepeating, it cannot represent a rational number. Such numbers that are not rational numbers are **irrational numbers**.

In the mid-eighteenth century, it was proved that the ratio of the circumference of a circle to its diameter, symbolized by π **(pi)**, is an irrational number. The numbers $\dfrac{22}{7}$, 3.14, or 3.14159 are rational number approximations of π. The value of π has been computed to over a trillion decimal places with no apparent pattern.

Square Roots

Irrational numbers occur in the study of area. For example, to find the area of a square, we use the formula $A = s^2$, where A is the area and s is the length of a side of the square. If a

---

Ten researchers, including Professor Yasumasa Kanada, at the Information Technology Center at Tokyo University calculated the value of pi to 1.2411 trillion places on a Hitachi supercomputer in September 2002. It took the computer 600 hours to do the computation.

Pi is the subject of several books for young students, including *Sir Cumference and the Dragon of Pi: A Math Adventure* (Neuschwander 1999). ◆

side of a square is 3 cm long, then the area of the square is $9\,cm^2$ (square centimeters). Conversely, we can use the formula to find the length of a side of a square, given its area. If the area of a square is $25\,cm^2$, then $s^2 = 25$, so $s = 5$ or $^-5$. Each of these solutions is a **square root** of 25. However, because lengths are always nonnegative, 5 is the only possible solution. The positive solution of $s^2 = 25$ (namely, 5) is the **principal square root** of 25 and is denoted $\sqrt{25}$. Similarly, the principal square root of 2 is denoted $\sqrt{2}$. Note that $\sqrt{16} \neq {}^-4$ because $^-4$ is not the principal square root of 16. Can you find $\sqrt{0}$?

Definition of the Principal Square Root

If a is any nonnegative number, the **principal square root** of a (denoted $\sqrt{a}$) is the nonnegative number b such that $b^2 = a$.

Example 7-13

Find the following:

a. The square roots of 144
b. The principal square root of 144
c. $\sqrt{\dfrac{4}{9}}$

Solution **a.** The square roots of 144 are 12 and $^-12$.
b. The principal square root of 144 is 12.
c. $\sqrt{\dfrac{4}{9}} = \dfrac{2}{3}$

Other Roots

We have seen that the positive solution to $s^2 = 25$ is denoted $\sqrt{25}$. Similarly, the positive solution to $s^4 = 25$ is denoted $\sqrt[4]{25}$. In general, if n is even, the positive solution to $x^n = 25$ is $\sqrt[n]{25}$ and is the principal **nth root** of 25. The number n is the **index**. Note that in the expression $\sqrt{25}$, the index 2 is understood and not expressed. In general, *the positive solution to $x^n = b$, where b is nonnegative, is $\sqrt[n]{b}$.*

Substituting $\sqrt[n]{b}$ for x in the equation $x^n = b$ gives the following:

$$(\sqrt[n]{b})^n = b$$

If b is negative, $\sqrt[n]{b}$ may not be a real number. For example, consider $\sqrt[4]{-16}$. If $\sqrt[4]{-16} = x$, then $x^4 = {}^-16$. Because any nonzero real number raised to the fourth power

---

Historical Note

Evaluating square roots may have been known by Vedic Hindu scholars before 600 BCE. The *Sulbasutra* (rule of chords), Sanskrit texts, contain approximations for some square roots that are incredibly accurate. The discovery of irrational numbers by members of the Pythagorean Society was disturbing, so they decided to keep the matter secret. One legend has it that a society member was drowned because he relayed the secret to persons outside the society. In 1525, Christoff Rudolff, a German mathematician, became the first to use the symbol $\sqrt{}$ for a radical or a root. ♦

is positive, there is no real-number solution to $x^4 = {}^-16$ and therefore $\sqrt[4]{{}^-16}$ is not a real number. Similarly, if we are restricted to real numbers, it is not possible to find *any* even root of a negative number. However, the value $^-2$ satisfies the equation $x^3 = {}^-8$. Hence, $\sqrt[3]{{}^-8} = {}^-2$. *In general, the odd root of a negative number is a negative number.*

Because $\sqrt{a}$, if it exists, is positive by definition, $\sqrt{(-3)^2} = \sqrt{9} = 3$ and not $^-3$. Many students think that $\sqrt{a^2}$ always equals a. This is true if $a \geq 0$, but false if $a < 0$. *In general,* $\sqrt{a^2} = |a|$.

Similarly, $\sqrt[4]{a^4} = |a|$ and $\sqrt[6]{a^6} = |a|$, but $\sqrt[3]{a^3} = a$ for all a. (Why?) Notice that when n is even and $b > 0$, the equation $x^n = b$ has two real-number solutions, $\sqrt[n]{b}$ and $^-\sqrt[n]{b}$. If n is odd, the equation has only one real-number solution, $\sqrt[n]{b}$, for any real number b.

Irrationality of Square Roots and Other Roots

Some square roots are rational numbers. Others, like $\sqrt{2}$, are irrational numbers. To see this, note that $1^2 = 1$ and $2^2 = 4$ and that there is no whole number s such that $s^2 = 2$. Is there a rational number $\dfrac{a}{b}$ such that $\left(\dfrac{a}{b}\right)^2 = 2$? To decide, we use the strategy of *indirect reasoning*. If we assume there is such a rational number $\dfrac{a}{b}$, then the following must be true:

$$\left(\frac{a}{b}\right)^2 = 2$$

$$\frac{a^2}{b^2} = 2$$

$$a^2 = 2b^2$$

If $a^2 = 2b^2$, then by the Fundamental Theorem of Arithmetic, the prime factorizations of a^2 and $2b^2$ are the same. In particular, the prime 2 appears the same number of times in the prime factorization of a^2 as it does in the factorization of $2b^2$. Because $b^2 = bb$, no matter how many times 2 appears in the prime factorization of b, it appears twice as many times in bb.

Also, a^2 has an even number of 2s for the same reason b^2 does. In $2b^2$, another factor of 2 is introduced, resulting in an odd number of 2s in the prime factorization of $2b^2$ and, hence, of a^2. But 2 cannot appear both an odd number of times and an even number of times in the same prime factorization of a^2. We have a contradiction.

This contradiction could have been caused only by the assumption that $\sqrt{2}$ is a rational number. Consequently, $\sqrt{2}$ must be an irrational number. We can use a similar argument to show that $\sqrt{3}$ is irrational or $\sqrt{n}$ is irrational, where n is a whole number but not the square of another whole number.

Many irrational numbers can be interpreted geometrically. For example, we can find a point on a number line to represent $\sqrt{2}$ by using the **Pythagorean Theorem**. That is, if a and b are the lengths of the shorter sides (legs) of a right triangle and c is the length of the longer side (hypotenuse), then $a^2 + b^2 = c^2$, as shown in Figure 7-6.

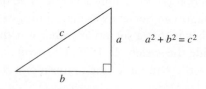

Figure 7-6

Figure 7-7 shows a segment 1 unit long constructed perpendicular to a number line at point P. Thus two sides of the triangle shown are each 1 unit long. If $a = b = 1$, then $c^2 = 2$ and $c = \sqrt{2}$. To find a point on the number line that corresponds to $\sqrt{2}$, we need to find a point Q on the number line such that the distance from 0 to Q is $\sqrt{2}$. Because $\sqrt{2}$ is the length of the hypotenuse, the point Q can be found by marking an arc with center 0 and radius c. The intersection of the positive number line with the arc is point Q.

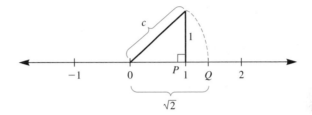

Figure 7-7

Similarly, other square roots can be constructed, as shown in Figure 7-8.

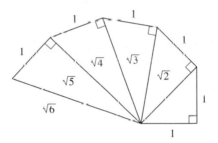

Figure 7-8

Estimating a Square Root

From Figure 7-7, we see that $\sqrt{2}$ must have a value between 1 and 2; that is, $1 < \sqrt{2} < 2$. To obtain a closer approximation of $\sqrt{2}$, we attempt to "squeeze" $\sqrt{2}$ between two numbers that are between 1 and 2. Because $(1.5)^2 = 2.25$ and $(1.4)^2 = 1.96$, it follows that $1.4 < \sqrt{2} < 1.5$. Because a^2 can be interpreted as the area of a square with side of length a, this discussion can be pictured geometrically, as in Figure 7-9.

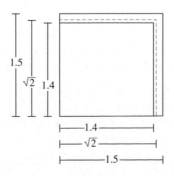

Figure 7-9

If we desire a more accurate approximation for $\sqrt{2}$, we can continue this squeezing process. We see that $(1.4)^2$, or 1.96, is closer to 2 than is $(1.5)^2$, or 2.25, so we choose numbers closer to 1.4 in order to find the next approximation. We find the following:

$$(1.42)^2 = 2.0164$$
$$(1.41)^2 = 1.9981$$

Thus, $1.41 < \sqrt{2} < 1.42$. We can continue this process until we obtain the desired approximation. Note that if a calculator has a square-root key, we can obtain an approximation directly.

NOW TRY THIS 7-12 One algorithm for calculating square roots is sometimes attributed to Archimedes. The algorithm makes use of making closer and closer estimates to the square root. To find the square root of a positive number, n, first make a guess. Call the first guess *Guess1*. Now compute as follows:

Step 1: Divide n by *Guess1*.
Step 2: Now add *Guess1* to the quotient obtained in step 1.
Step 3: Divide the sum in step 2 by 2. The quotient becomes *Guess2*.

Repeat the steps using *Guess2* to obtain successive guesses or until the desired accuracy is achieved.

a. Use the method described to find the square root of 13 to four decimal places.
b. Write the steps for the algorithm in a recursive formula.

The System of Real Numbers

The set of **real numbers**, R, is the union of the set of rational numbers and the set of irrational numbers. Real numbers represented as decimals can be terminating, repeating, or nonterminating and nonrepeating.

Every integer is a rational number as well as a real number. Every rational number is a real number, but not every real number is rational, as has been shown with $\sqrt{2}$. The relationships among sets of numbers are summarized in the tree diagram in Figure 7-10.

$$\text{Reals} = \{x | x \text{ is a decimal}\}$$

Irrationals $= \{x | x \text{ is a decimal that neither repeats nor terminates}\}$

Rationals $= \left\{ x | x = \dfrac{a}{b} \text{ where } a, b \in I \text{ and } b \neq 0 \right\}$
$= \{x | x \text{ is a repeating or terminating decimal}\}$

Integers $= \{ \ldots, {}^-3, {}^-2, {}^-1, 0, 1, 2, 3, \ldots \}$

Whole Numbers $= \{0, 1, 2, 3, 4, \ldots\}$

Natural Numbers $= \{1, 2, 3, 4, 5, \ldots\}$

Figure 7-10

The concept of fractions can now be extended to include all numbers of the form $\dfrac{a}{b}$, where a and b are real numbers with $b \neq 0$, such as $\dfrac{\sqrt{3}}{5}$. Addition, subtraction, multiplication, and division are defined on the set of real numbers in such a way that all the properties of these operations on rationals still hold. The properties are summarized next.

Theorem 7–2:   Properties of Real Numbers

Closure properties   For real numbers a and b, $a + b$ and ab are unique real numbers.

Commutative properties   For real numbers a and b, $a + b = b + a$ and $ab = ba$.

Associative properties   For real numbers a, b, and c, $a + (b + c) = (a + b) + c$ and $a(bc) = (ab)c$.

Identity properties   The number 0 is the unique additive identity and 1 is the unique multiplicative identity such that, for any real number a, $0 + a = a = a + 0$ and $1 \cdot a = a = a \cdot 1$.

Inverse properties   (1) For every real number a, ^-a is its unique additive inverse; that is, $a + {^-a} = 0 = {^-a} + a$. (2) For every nonzero real number a, $\dfrac{1}{a}$ is its unique multiplicative inverse; that is, $a\left(\dfrac{1}{a}\right) = 1 = \left(\dfrac{1}{a}\right)a$.

Distributive property of multiplication over addition   For real numbers a, b, and c, $a(b + c) = ab + ac$.

Denseness property   For real numbers a and b, there exists a real number c such that $a < c < b$.

Radicals and Rational Exponents

Scientific calculators have a $\boxed{y^x}$ key with which we can find the values of expressions like $3.41^{2/3}$ and $4^{1/2}$. What does $4^{1/2}$ mean? By extending the properties of exponents previously developed for integer exponents,

$$4^{1/2} \cdot 4^{1/2} = 4^{(1/2 + 1/2)}$$

Thus,

$$(4^{1/2})^2 = 4^1 = 4$$

and

$$4^{1/2} = \sqrt{4}$$

The number $4^{1/2}$ is assumed to be the principal square root of 4, that is, $4^{1/2} = \sqrt{4}$. *In general, if x is a nonnegative real number, then $x^{1/2} = \sqrt{x}$.* Similarly, $(x^{1/3})^3 = x^{(1/3)3} = x^1$ and $x^{1/3} = \sqrt[3]{x}$. This discussion leads to the following:

1. $x^{1/n} = \sqrt[n]{x}$, where $\sqrt[n]{x}$ is meaningful.
2. $(x^m)^{1/n} = \sqrt[n]{x^m}$ if $\gcd(m, n) = 1$.
3. $x^{m/n} = \sqrt[n]{x^m}$, if $\gcd(m, n) = 1$

◆ Historical Note

In 1874 Georg Cantor showed that the set of all real numbers cannot be put in a one-to-one correspondence with the set of natural numbers and in the process continued to compare infinities.   ◆

More Properties of Exponents

It can be shown that the properties of integer exponents also hold for rational exponents. These properties are equivalent to the corresponding properties of radicals *if the expressions involving radicals are meaningful.*

Let r and s be any rational numbers, x and y be any real numbers, and n be any nonzero integer. What are restrictions on the variables in each of the following?

a. $x^{-r} = \dfrac{1}{x^r}$

b. $(xy)^r = x^r y^r$ implies $(xy)^{1/n} = x^{1/n} y^{1/n}$ and $\sqrt[n]{xy} = \sqrt[n]{x}\,\sqrt[n]{y}$

c. $\left(\dfrac{x}{y}\right)^r = \dfrac{x^r}{y^r}$ implies $\left(\dfrac{x}{y}\right)^{1/n} = \dfrac{x^{1/n}}{y^{1/n}}$ and $\sqrt[n]{\dfrac{x}{y}} = \dfrac{\sqrt[n]{x}}{\sqrt[n]{y}}$

d. $(x^r)^s = x^{rs}$ implies $(x^{1/n})^s = x^{s/n}$ and hence, $(\sqrt[n]{x})^s = \sqrt[n]{x^s}$

The preceding properties can be used to write equivalent expressions for the roots of many numbers. For example, $\sqrt{96} = \sqrt{16 \cdot 6} = \sqrt{16} \cdot \sqrt{6} = 4 \cdot \sqrt{6}$. Similarly, $\sqrt[3]{54} = \sqrt[3]{27 \cdot 2} = \sqrt[3]{27} \cdot \sqrt[3]{2} = 3 \cdot \sqrt[3]{2}$.

Example 7-14

Simplify each of the following if possible:

a. $16^{1/4}$ **b.** $16^{5/4}$ **c.** $(-8)^{1/3}$ **d.** $125^{-4/3}$ **e.** $(-16)^{1/4}$

Solution

a. $16^{1/4} = (2^4)^{1/4} = 2^1 = 2$, or $16^{1/4} = \sqrt[4]{16} = 2$

b. $16^{5/4} = 16^{(1/4)5} = (16^{1/4})^5 = 2^5 = 32$

c. $(-8)^{1/3} = ((-2)^3)^{1/3} = (-2)^1 = -2$ or $(-8)^{1/3} = \sqrt[3]{-8} = -2$

d. $125^{-4/3} = (5^3)^{-4/3} = 5^{-4} = \dfrac{1}{5^4} = \dfrac{1}{625}$

e. Because every real number raised to the fourth power is positive, $\sqrt[4]{-16}$ is not a real number. Consequently, $(-16)^{1/4}$ is not a real number.

NOW TRY THIS 7-13 Compute $\sqrt[8]{10}$ on a calculator using the following sequence of keys:

a. Explain why this approach works.
b. For what values of n can $\sqrt[n]{10}$ be computed using only the $\sqrt{}$ key? Why?

Assessment 7-4 A

1. Write an irrational number whose digits are 2s and 3s.

2. Use the Pythagorean Theorem to find x.

a.

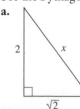

b.

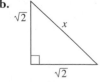

c.

3. Arrange the following real numbers in order from greatest to least:

$$0.9, \quad 0.\overline{9}, \quad 0.\overline{98}, \quad 0.9\overline{88}, \quad 0.9\overline{98}, \quad 0.8\overline{98}, \quad \sqrt{0.98}$$

4. Determine which of the following represent irrational numbers:

   a. $\sqrt{51}$ **b.** $\sqrt{64}$ **c.** $\sqrt{324}$
   d. $\sqrt{325}$ **e.** $2 + 3\sqrt{2}$ **f.** $\sqrt{2} \div 5$

5. If possible, find the square roots for each of the following without using a calculator:
   a. 225
   b. 169
   c. $^-81$
   d. 625

6. Find the approximate square roots for each of the following, rounded to hundredths, by using the squeezing method:
   a. 7
   b. 0.0120

7. Classify each of the following as true or false. If false, give a counterexample.
   a. The sum of any rational number and any irrational number is a rational number.
   b. The sum of any two irrational numbers is an irrational number.
   c. The product of any two irrational numbers is an irrational number.
   d. The difference of any two irrational numbers could be a rational number.

8. Find three irrational numbers between 1 and 3.

9. Find an irrational number between $0.\overline{53}$ and $0.\overline{54}$.

10. Between the terminating decimals 0.5 and 0.6, find an irrational number.

11. If one adds a rational number and an irrational number, what type of number results? Prove your answer.

12. Based on your answer in exercise 11, argue that there are infinitely many irrational numbers.

13. If R is the set of real numbers, Q is the set of rational numbers, I is the set of integers, W is the set of whole numbers, and S is the set of irrational numbers, find each of the following:
   a. $Q \cup S$
   b. $Q \cap S$
   c. $Q \cap R$
   d. $S \cap W$
   e. $W \cup R$
   f. $Q \cup R$

14. If the following letters correspond to the sets listed in exercise 13, complete the table by placing checkmarks in the appropriate columns. (N is the set of natural numbers.)

	N	I	Q	R	S
a. 6.7			✔	✔	
b. 5					
c. $\sqrt{2}$					
d. $^-5$					
e. $3\frac{1}{7}$					

15. If the following letters correspond to the sets listed in problem 13, put a checkmark under each set of numbers for which a solution to the problem exists. (N is the set of natural numbers.)

	N	I	Q	R	S
a. $x^2 + 1 = 5$					
b. $2x - 1 = 32$					
c. $x^2 = 3$					
d. $x^2 = 4$					
e. $\sqrt{x} = ^-1$					
f. $\frac{3}{4}x = 4$					

16. Determine for what real values of x, if any, each of the following statements is true:
   a. $\sqrt{x} = 8$
   b. $\sqrt{x} = ^-8$
   c. $\sqrt{^-x} = 8$
   d. $\sqrt{^-x} = ^-8$
   e. $\sqrt{x} > 0$
   f. $\sqrt{x} < 0$

17. A diagonal brace is placed in a $4\,\text{ft} \times 5\,\text{ft}$ rectangular gate. What is the length of the brace to the nearest tenth of a foot? (*Hint:* Use the Pythagorean Theorem.)

18. Write each of the following square roots in the form $a\sqrt{b}$, where a and b are integers and b has the least value possible:
   a. $\sqrt{180}$
   b. $\sqrt{363}$
   c. $\sqrt{252}$

19. Write each of the following in the simplest form or as $a\sqrt[n]{b}$, where a and b are integers, $b > 0$, and b has the least value possible:
   a. $\sqrt[3]{^-54}$
   b. $\sqrt[5]{96}$
   c. $\sqrt[3]{250}$
   d. $\sqrt[5]{^-243}$

20. In each of the following geometric sequences, find the missing terms:
   a. 5, _____, _____, 10
   b. 2, _____, _____, _____, 1

21. The following exponential function approximates the number of bacteria after t hr: $E(t) = 2^{10} \cdot 16^t$.
   a. What is the initial number of bacteria, that is, the number when $t = 0$?
   b. After $\frac{1}{4}$ hr, how many bacteria are there?
   c. After $\frac{1}{2}$ hr, how many bacteria are there?

22. Solve for x in the following, where x is a rational number:
   a. $3^x = 81$
   b. $4^x = 8$
   c. $128^{^-x} = 16$
   d. $\left(\frac{4}{9}\right)^{3x} = \frac{32}{243}$

23. Classify each of the following numbers as rational or irrational:
   a. $\sqrt{2} - \frac{2}{\sqrt{2}}$
   b. $(\sqrt{2})^{^-4}$

24. a. Is it ever true that $\sqrt{a^2} = ^-a$? Explain your answer.
   b. For which values of x, if any, is
      (i) $x^2 \geq 2$?
      (ii) $x^2 \geq ^-2$

25. Figure 7-8 showed a way to construct the square root of any positive integer. Figure 7-7 showed how the square root of a positive integer could be placed on a number line. Describe where $\sqrt{342}$ would be on a number line.

★26. Use the steps for the recursive formula found in Now Try This 7-12 to show that the algorithm could produce the square root of a positive real number.

Assessment 7-4B

1. Write an irrational number whose digits are 4s and 5s.
2. Use the Pythagorean Theorem to find x.

   a.

   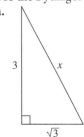

   b.

   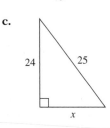

   c.

3. Arrange the following real numbers in order from greatest to least:

$$0.8, 0.\overline{8}, 0.\overline{89}, 0.8\overline{89}, \sqrt{0.7744}$$

4. Determine which of the following represent irrational numbers:

   a. $\sqrt{78}$ b. $\sqrt{81}$ c. $\sqrt[3]{343}$

   d. $3 + \sqrt{81}$ e. $2 \div \sqrt{2}$

5. If possible, find whole numbers that are the square roots for each of the following without using a calculator:

   a. 256   b. 324

   c. $^-25$ d. 1024

6. Find the approximate square roots for each of the following, rounded to hundredths, by using the squeezing method:

   a. 20.3   b. 1.64

7. Classify each of the following as true or false. If false, give a counterexample.

   a. The sum of any two rational numbers is a rational number.

   b. The difference of any two irrational numbers is an irrational number.

   c. The product of any rational number and any irrational number is an irrational number.

8. Find three irrational numbers between each of the following pairs:

   a. 3 and 4   b. $7.0\overline{5}$ and $7.0\overline{6}$

   c. 0.7 and 0.8

9. Knowing that $\sqrt{2}$ is an irrational number, argue that $\dfrac{\sqrt{2}}{2}$ is also an irrational number.

10. If R is the set of real numbers, Q is the set of rational numbers, I is the set of integers, W is the set of whole numbers, N is the set of natural numbers, and S is the set of irrational numbers, find each of the following:

    a. $Q \cap I$ b. $S - Q$ c. $R \cup S$

    d. Which of the sets could be a universal set for the rest of the sets?

11. If R is the set of real numbers, how would you describe $\overline{S}$, where S is the set of irrational numbers?

12. If the following letters correspond to the sets listed in exercise 10, complete the following table by placing checkmarks in the appropriate columns:

	N	I	Q	R	S
a. $\sqrt{3}$					
b. $4\frac{1}{2}$					
c. $^-3\frac{1}{7}$					

13. If the following letters correspond to the sets listed in exercise 10, put a checkmark under each set of numbers for which a solution to the problem exists:

	N	I	Q	R	S
a. $x^2 + 2 = 4$					
b. $1 - 2x = 32$					
c. $x^3 = 4$					
d. $\sqrt{x} = ^-2$					
e. $0.\overline{7}x = 5$					

14. Determine for what real values of x, if any, each of the following statements is true:

    a. $\sqrt{x} = 7$ b. $\sqrt{x} = ^-7$

    c. $\sqrt{^-x} = 7$ d. $^-\sqrt{x} = 7$

    e. $^-\sqrt{x} = ^-7$

15. Write each of the following square roots in the form $a\sqrt{b}$, where a and b are integers and b has the least value possible:

    a. $\sqrt{360}$ b. $\sqrt{40}$

    c. $\sqrt{240}$

16. Write each of the following in the simplest form or as $a\sqrt[n]{b}$, where a and b are integers and $b > 0$ and b is as small as possible:

    a. $\sqrt[3]{^-102}$ b. $\sqrt[6]{64}$

    c. $\sqrt[3]{64}$

17. In each of the following geometric sequences, find the missing terms:

    a. 4, _____, _____, 8

    b. 1, _____, _____, 2

18. If the two shorter sides of a right triangle have lengths 5 and 12, how long is the hypotenuse (the longest side)?

19. In an exponential function, $E(t) = 8^t$, where t represents time in hours.
   a. Is the exponential function's graph linear? Make a sketch to show your answer.
   b. What is the value of $E(0)$?
   c. After $\frac{1}{3}$ of an hour, what is the value of $E\left(\frac{1}{3}\right)$?

20. Solve for x in the following, where x is a real number:
   a. $2^x = 64$
   b. $4^x = 64$
   c. $2^{-x} = 64$
   d. $\sqrt{\frac{3x}{2}} = 36$

21. Classify each of the following numbers as rational or irrational:
   a. $\dfrac{1}{1 + \sqrt{2}}$
   b. $\dfrac{4}{\sqrt{2}} - \sqrt{2}$

22. Explain whether $\sqrt{a^2} = {}^-|a|$.

23. If $a > 2$, describe the approximate location of $\dfrac{1}{\sqrt{a}}$ on a number line.

24. Describe how you would decide whether $\dfrac{2}{\sqrt{3}} < \dfrac{3}{\sqrt{5}}$.

Mathematical Connections 7-4

Communication

1. A mathematician once described the set of rational numbers as the stars and the set of irrational numbers as the black in the night sky. What do you think the mathematician meant?

2. Find the value of $\sqrt{3}$ on a calculator. Explain why this can't be the exact value of $\sqrt{3}$.

3. Is it true that $\sqrt{a + b} = \sqrt{a} + \sqrt{b}$ for all a and b? Explain.

4. Pi (π) is an irrational number. Could $\pi = \dfrac{22}{7}$? Why or why not?

5. Without using a calculator or doing any computation, determine if $\sqrt{13} = 3.60\overline{5}$. Explain why or why not.

6. Is $\sqrt{x^2 + y^2} = x + y$ for all real number values of x and y? Explain your reasoning.

7. Without using a calculator, arrange the following in increasing order. Explain your reasoning.

$$(4/25)^{-1/3}, \ (25/4)^{1/3}, \ (4/25)^{-1/4}$$

Open-Ended

8. The sequence $1, 1.01, 1.001, 1.0001, \ldots$ is an infinite sequence of rational numbers.
   a. Write several other infinite sequences of rational numbers.
   b. Write an infinite sequence of irrational numbers.

9. a. Place five irrational numbers between $\frac{1}{2}$ and $\frac{3}{4}$.
   b. Write an infinite sequence of irrational numbers all of whose terms are between $\frac{1}{2}$ and $\frac{3}{4}$.

Cooperative Learning

10. Let each member of a group choose a number between 0 and 1 on a calculator and check what happens when the $\boxed{x^2}$ key is pressed In succession until it is clear that there is no reason to go on.
   a. Compare your answers and write a conjecture based on what you observe.
   b. Use other keys on the calculator in a similar way. Describe the process and state a corresponding conjecture.
   c. Why do you get the result you do in parts (a) and (b)?

11. A calculator displays the following: $(3.7)^{2.4} = 23.103838$. In your group, discuss the meaning of the expression $(3.7)^{2.4}$ in view of what you know about exponents. Compare your findings with those of other groups.

Questions from the Classroom

12. A student argues that the only solution to $x^2 = 4$ is 2 because $\sqrt{4}$ is equal to 2. How do you respond?

13. Jim asked, if $\sqrt{2}$ can be written as $\dfrac{\sqrt{2}}{1}$, why isn't it rational? How would you answer him?

14. Maria says that $1 + \sqrt{2}$ is not a number because the sum cannot be completed. How do you respond?

15. A student claims that $5 = {}^-5$ because both are square roots of 25. How do you help her?

16. Jose says that the equation $\sqrt{{}^-x} = 3$ has no solution, since the square root of a negative number does not exist. Why is this argument wrong?

17. A student argues that in the real world, irrational numbers are never used. How do you respond?

18. Alfonso says that there are infinitely many irrational numbers between 0 and 1 because $\frac{\pi}{n}$, where $n \geq 4$ and n is a natural number, is in the interval. Is Alfonso incorrect?

19. M. Brown's research in 1981 argued that students have significantly more difficulty with computations in word problems if there are decimals in the problems. What is your reaction to that research?

Review Problems

20. **a.** Human bones make up 0.18 of a person's body weight. How much do the bones of a 120 lb person weigh?
    b. Muscles make up about 0.4 of a person's body weight. How much do the muscles of a 120 lb person weigh?

21. Write each of the following decimals in the form $\frac{a}{b}$, where $a, b \in I, b \neq 0$:
    a. 16.72   **b.** 0.003   **c.** ⁻5.07   **d.** 0.123

22. Write a repeating decimal equal to each of the following without using more than one zero:
    a. 5   **b.** 5.1   **c.** $\frac{1}{2}$

23. Write 0.00024 as a fraction in simplest form.
24. Write $0.\overline{24}$ as a fraction in simplest form.
25. Write each of the following as a standard numeral:
    a. $2.08 \cdot 10^5$ **b.** $3.8 \cdot 10^{-4}$

National Assessment of Educational Progress (NAEP) Question

Term	1	2	3	4
Fraction	1/2	2/3	3/4	4/5

If the list of fractions above continues in the same pattern, which term will be equal to 0.95?
 a. The 100th
 b. The 95th
 c. The 20th
 d. The 19th
 e. The 15th

NAEP, Grade 8, 2003

BRAIN TEASER Many schools celebrate Pi Day on March 14. Use the cartoon below to suggest why the day is used and at what time the celebration might start.

*7-5   Using Real Numbers in Equations

In *PSSM*, we find the following:

> Electronic technologies—calculators and computers—are essential tools for teaching, learning and doing mathematics. They furnish visual images of mathematical ideas, they facilitate organizing and analyzing data, and they compute efficiently and accurately. They can support investigation by students in every area of mathematics, including geometry, statistics, algebra, measurement and number. (p. 24)

Additionally, in *PSSM* we find:

> Computer and calculators change what students can do with conventional representations and expand the set of representations with which they can work. . . . They can use computer algebra systems to manipulate expressions, and they can investigate complex data sets using spreadsheets. As students learn to use these new, versatile tools, they also can consider ways in which some representations used in electronic technology differ from conventional representations. (pp. 68–69)

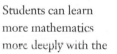

Students can learn more mathematics more deeply with the appropriate use of technology (Dunham and Dick, 1994; Sheets, 1993; Boers-van Oosterum 1990; Rojano 1996; Groves, 1994). ◆

In this section, we extend what students have studied in previous sections and begin to explore some topics algebraically using the technology of spreadsheets. Students studying this section may do the work without technology, but as the research note indicates, some learning can be more in-depth with the use of the technology.

In previous chapters, we have seen linear relations whose domain was the set of whole numbers or the set of integers but not when the domain was the entire set of real numbers. For example in Figure 7-11 if using only integers, a portion of the graph of the line that depicts the relation $y = 2x + 3$ would appear as follows.

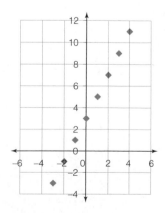

Figure 7-11

With all real numbers, the entire line of Figure 7-11 can be depicted as in Figure 7-12.

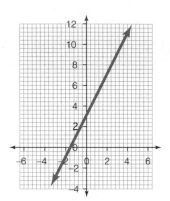

Figure 7-12

Open a new spreadsheet and in column A, enter 1 in cell A1. In cell A2, enter =A1+1, and use the fill down feature to display the first 20 elements in that column. In cell B1, enter =2*A1+3 and press return. The use of the function feature allows the value of the variable A1 to be picked up from cell A1, inserted into the given formula, and output 5 as 2(1) + 3. Click on cell B1. Use the fill down feature to give as many values in this column as needed to match those of column A.

   To produce a graph of the data in the completed columns, highlight the elements in both columns and use the tool bar to locate the Charts section of the bar. Choose the type of graph that you would like to use to display the data. In Figure 7-12 above, the line option was chosen. Once the option is selected, the graph will be drawn. Use this process to graph $y = {}^-2x + 5$.

Plotting Other Functions on a Spreadsheet

In Figure 7-13, we see the graph of $y = x^2$ and $y = x$ produced on a spreadsheet. Observe that the graph of $y = x^2$ is not linear and "grows" much faster than the linear graph of $y = x$ for $x > 1$.

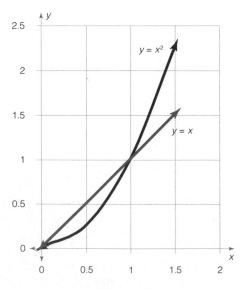

Figure 7-13

Graphs with the set of real numbers as the domain allow us to compare different functions. For example in Figure 7-14, consider the graph of $y = x^2$, $y = x$, and $y = \sqrt{x}$ when $x \geq 0$.

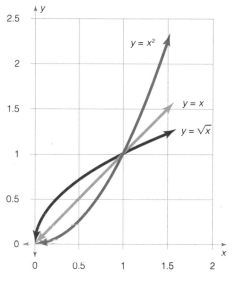

Figure 7-14

In Figure 7-14, it appears that the two graphs are symmetric across the "diagonal line" through the origin in the graph. In fact, the two functions are mirror images of each other across the line $y = x$. Does this work for all values of x for which the functions are defined?

The functions $y = x^3$ and $y = \sqrt[3]{x}$ have a similar relationship to that seen in Figure 7-14. Now Try This 7-14 explores another pair of functions with this type of relationship.

NOW TRY THIS 7-14 Graph the functions $y = 2x + 1$ and $y = (x - 1)/2$ on a graph with the function $y = x$. What do you observe?

NOW TRY THIS 7-15 Graphing technology can be used to investigate many different types of relationships. For example, plot the following on the same graph.

a. $y = 2x$ **b.** $y = 2x + 1$ **c.** $y = 2x + 2$ **d.** $y = 2x - 1$ **e.** $y = 2x - 2$

What do you observe about this set of lines? What do you observe about this set of equations?

Estimating Solutions to Linear Equations

Before students learn to solve systems of equations algebraically, a spreadsheet may be used to estimate solutions to the system. For example, consider the graph in Figure 7-15 where the lines $y = \left(\dfrac{3}{4}\right)x + 5$ and $y = {}^{-}x + 10$ are drawn.

It appears that the two lines cross (or intersect) when x has a value between 2 and 3. Recreate the graph using values of x at intervals of 0.1 and see if you can predict the coordinates of the point of intersection, as seen in Figure 7-16.

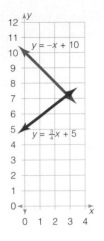

Figure 7-15

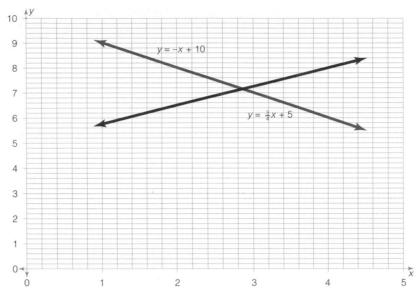

Figure 7-16

Now the solution appears to be a point with coordinates near (2.8, 7.2). Substituting 2.8 for x in each equation yields $y = \left(\dfrac{3}{4}\right)2.8 + 5 = 7.1$ in the first and $y = {}^-2.8 + 10$, or 7.2, in the second. So a better estimate for the solution would need to be an x-value slightly more than 2.8 which would yield a value between 7.1 and 7.2 for y. If you continued this process, you could estimate closer and closer to a solution. A further examination of the solution using systems of equations will be seen in a later chapter in this book.

Table 7-3

	A	B
1	1	−1
2	2	1
3	3	3
4	4	5
5	5	7
6	6	9
7	7	11
8	8	13
9	9	15
10	10	17
11	11	19
12	12	21
13	13	23

Solving One Equation in One Unknown

Just as the solution of a system of equations could be estimated as shown above, we could estimate (or find) the solution to one equation in one unknown using a spreadsheet. This technique could be useful when an equation is difficult. We demonstrate using a simple equation, $2x - 3 = 17$. An exploratory solution to the equation might consider values of x for which $2x - 3$ give results close to 17. See Table 7-3 produced by a spreadsheet, where values of x are depicted in column A and values of $2x - 3$ are depicted in column B. In Table 7-3, we see that when x has a value of 10, then the value of $2x - 3$ is equal to 17. Thus, the solution to the equation is $x = 10$. We could have deduced this algebraically as follows:

$$2x - 3 = 17$$
$$2x = 17 + 3$$
$$2x = 20$$
$$x = 10$$

Consider a nonlinear equation such as we saw earlier in this section, $y = x^2$. There are infinitely many solutions including (0, 0). However, it is not so easy for some to deduce for what values of x, if any, is $y = 0$ in the equation $y = x^2 - 5x + 6$.

Consider the spreadsheet in Table 7-4, where column A represents values of x and column B was created using the formula =A1^2−5*A1+6 to represent $x^2 - 5x + 6$.

Table 7-4

	A	B
1	1	2
2	2	0
3	3	0
4	4	2
5	5	6
6	6	12
7	7	20
8	8	30
9	9	42
10	10	56
11	11	72
12	12	90
13	13	110

In Table 7-4, we see that when the value of x in column A is 2 and when it is 3, then the corresponding values of y are 0 in each case. Thus we predict that $y = 0$ in the equation $y = x^2 - 5x + 6$ when $x = 2$ or 3. This can also be exhibited in the graph of Figure 7-17, drawn with the spreadsheet where the graph is restricted to the area of interest and the values of x are stepped at 0.1.

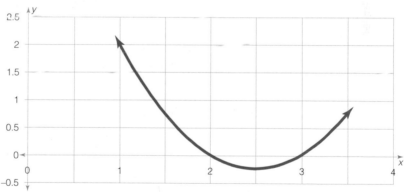

Figure 7-17

Table 7-5

x	$x^2 - 2$
$^-2$	2
$^-1.9$	1.61
$^-1.8$	1.24
$^-1.7$	0.89
$^-1.6$	0.56
$^-1.5$	0.25
$^-1.4$	$^-0.04$
$^-1.3$	$^-0.31$
$^-1.2$	$^-0.56$
$^-1.1$	$^-0.79$

Although the use of spreadsheets may not always determine an exact answer when solving equations, using a graph to depict data and a spreadsheet to focus on continually finer values of x can be useful in helping you predict solutions.

A spreadsheet uses only finite decimals in its calculations. Thus, in solving an equation such as $x^2 - 2 = 0$, which we have seen has solutions $x = \sqrt{2}$ and $x = {}^-\sqrt{2}$, the solutions can only be estimated. In Table 7-5, we see that as x increases in value by 0.1, the value of y changes from positive 0.25 when $x = {}^-1.5$ to $^-0.04$ when $x = {}^-1.4$. Between the values of $x = {}^-1.5$ and $x = {}^-1.4$, there should be a value of x that produces the value of 0 for y. Continual refinement of the values of x could produce an approximation of $\sqrt{2}$.

 The following partial student page depicts the solution of an equation that has an irrational number solution. Do number 3 in the Quick Check section. On the student page, note that the solution suggests that to solve $51 = t^2$, one should take the positive square root of each side of the equation. For the student example, that is appropriate because the solution is a unit of time

School Book Page APPLICATION: SKYDIVING

EXAMPLE   Application: Skydiving

3 The formula $d = 16t^2$ represents the approximate distance d in feet a skydiver falls in t seconds before opening the parachute. The formula assumes there is no air resistance. Find the time a skydiver takes to fall 816 feet before opening the parachute.

GO for Help

For help in using formulas, go to Lesson 2-6, Example 1.

$d = 16t^2$ ← Use the formula for distance and time.

$816 = 16t^2$ ← Substitute 816 for d.

$\frac{816}{16} = t^2$ ← Divide each side by 16 to isolate t.

$51 = t^2$ ← Simplify.

$\sqrt{51} = \sqrt{t^2}$ ← Find the positive square root of each side.

$\sqrt{}\ 51\ \boxed{=}\ 7.141428429$ ← Use a calculator.

$7.1 \approx t$ ← Round to the nearest tenth.

The skydiver takes about 7.1 seconds to fall 816 feet.

✓ Quick Check

3. Find the time a skydiver takes to fall each distance. Round to the nearest tenth of a second.
  a. 480 ft   **b.** 625 ft

Irrational numbers are numbers that cannot be written in the form $\frac{a}{b}$, where a is any integer and b is any nonzero integer. Rational and irrational numbers form the set of **real numbers.**

3-1 Exploring Square Roots and Irrational Numbers   **107**

Source: Prentice Hall Mathematics, Course 3, 2008 (p. 107).

and must be a positive number. There are no negative-number solutions. We develop an alternative approach to solving a similar equation next and then return to this problem again.

Consider the equation $x^2 = 9$. One can see that $x = 3$ or $^-3$ are solutions. Some might take the square root of both sides of the equation. A different method recalls the difference of squares and solves as follows:

$$x^2 = 9$$
$$x^2 - 9 = 0$$
$$x^2 - 3^2 = 0$$
$$(x - 3)(x + 3) = 0 \quad \text{(by factoring)}$$

Now one of $(x - 3)$ or $(x + 3)$ must be 0. So $x - 3 = 0$ or $x + 3 = 0$. Thus, $x = 3$ or $x = \ ^-3$.

This process could be applied to the equation $51 = t^2$. From the student page on page 466:

$$51 = t^2$$
$$0 = t^2 - 51$$
$$t^2 - 51 = 0$$
$$t^2 - (\sqrt{51})^2 = 0$$
$$(x - \sqrt{51})(x + \sqrt{51}) = 0 \qquad \text{By factoring}$$

Now one of $(x - \sqrt{51})$ or $(x + \sqrt{51})$ must be 0. So $x - \sqrt{51} = 0$ or $x + \sqrt{51} = 0$. Thus, $x = \sqrt{51}$ or $x = {}^-\sqrt{51}$. But because time cannot be negative in the example, we must use the positive solution.

A natural question is whether this process could be adapted to solving slightly more complicated equations such as $x^2 + 4x + 3 = 0$. Could this equation be written as a difference of two squares equal to 0 as in the example above? One potential answer comes from a historical approach by actually looking at two squares. We know that x^2 is a perfect square whose side length is x, but we have an additional $4x$. Figure 7-18(a) shows one depiction of $x^2 + 4x$ and Figure 7-18(b) shows a different configuration of the $4x$ portion of the figure in an effort to create a square using the information that we have. The $4x$ portion was split into two portions, each with an area of $2x$, and placed as pictured. If an additional portion could be found to "fill in the gap," or to "complete the square," as in Figure 7-18(c), then we would have completed one part of the task needed.

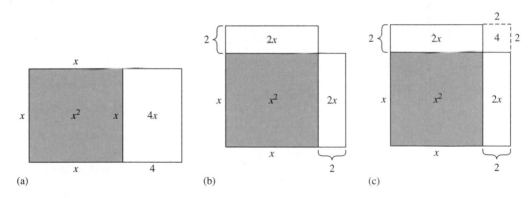

Figure 7-18

By appending a small 2×2 square in the upper right corner of Figure 7-18(c), we would have a square with a side length of $x + 2$. To translate this in terms of our original equation, we would have changed the left side to be $x^2 + 4x + 4 + 3$. To keep the equation "balanced," we would also need to subtract 4, because 4 was added. We do this as follows: $(x^2 + 4x + 4) + 3 - 4 = 0$, or $(x^2 + 4x + 4) - 1 = 0$. In this case, the expression in parentheses is $(x + 2)^2$ and 1 can be written as 1^2, giving us:

$$(x + 2)^2 - 1^2 = 0$$
$$[(x + 2) - 1][(x + 2) + 1] = 0$$

Now $(x + 2) - 1 = 0$ or $(x + 2) + 1 = 0$. So $x = {}^-1$, or $x = {}^-3$ are the solutions to the original equation. Note that we used x as a length in the drawing for a visualization of the process. But for the solution x is a negative number.

In Now Try This 7-16, we investigate how to solve a general quadratic equation of the form $ax^2 + bx + c = 0$, where $a \neq 0$.

NOW TRY THIS 7-16 To find solutions, if they exist, for equations of the form $ax^2 + bx + c = 0$, where $a \neq 0$, we could use the method of *completing the square* to make the equation fit the difference-of-squares form above. To solve this equation, we relate it to a problem that we already know how to solve, that is, a *difference-of-squares* equation.

The question then becomes can we rewrite $ax^2 + bx + c = 0$ in the form $m^2 - n^2 = 0$, which can then be rewritten as $(m - n)(m + n) = 0$ that we can solve? To think about how to rewrite $ax^2 + bx + c$ in the desired form, we first simplify the equation by dividing both sides of the equation by a to give

$x^2 + \dfrac{b}{a}x + \dfrac{c}{a} = 0$ and then use notions of geometry. Consider x^2 as the area of a square whose side has

length x, as in Figure 7-19(a).

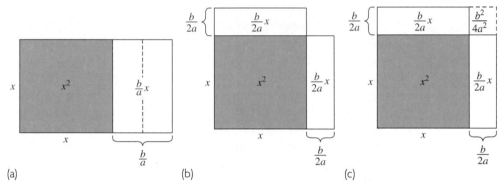

(a)   (b)   (c)

Figure 7-19

Similarly, $\dfrac{b}{a}x$ can be considered the area of a rectangle whose length is x and whose width is $\dfrac{b}{a}$, as in

Figure 7-19(a). To try and build a square with area $x^2 + \dfrac{b}{a}x$, we can split $\dfrac{b}{a}x$ into two equal parts and place

them as seen in Figure 7-19(b). Then to complete a square with area close to $x^2 + \dfrac{b}{a}x$, we need another

small square, as pictured in Figure 7-19(c). The needed small square has sides of length $\dfrac{b}{2a}$ and thus has area

$\left(\dfrac{b}{2a}\right)^2$, or $\dfrac{b^2}{4a^2}$. With this small square appended, the length of the side of the *completed square* is $x + \dfrac{b}{2a}$.

Now consider our simplified expression, $x^2 + \dfrac{b}{a}x + \dfrac{c}{a} = 0$. If we added $\dfrac{b^2}{4a^2}$ to the first two terms, we

would have $x^2 + \dfrac{b}{a}x + \dfrac{b^2}{4a^2}$ and would have to adjust the equation to keep it equivalent. We do this by

subtracting $\dfrac{b^2}{4a^2}$ as follows:

$$x^2 + \frac{b}{a}x + \frac{b^2}{4a^2} - \frac{b^2}{4a^2} + \frac{c}{a} = 0$$
$$\left(x + \frac{b}{2a}\right)^2 + \left(-\frac{b^2}{4a^2} + \frac{c}{a}\right) = 0$$

or equivalently,

$$\left(x + \frac{b}{2a}\right)^2 - \left(\frac{b^2}{4a^2} - \frac{c}{a}\right) = 0$$

The first expression is a perfect square $\left(x + \dfrac{b}{2a}\right)^2$ and the rest of the left side of the equation is $\dfrac{b^2}{4a^2} - \dfrac{c}{a}$,

or $\dfrac{b^2 - 4ac}{4a^2}$. If we wanted to write this latter expression as a perfect square, we could do that by taking the square root of the expression and squaring it as $\left(\sqrt{\dfrac{b^2 - 4ac}{4a^2}}\right)^2$. Combining the parts, we have

$\left(x + \dfrac{b}{2a}\right)^2 - \left(\sqrt{\dfrac{b^2 - 4ac}{4a^2}}\right)^2 = 0$. The left side of this last equation is in the form of $m^2 - n^2$ and can

be rewritten using the difference of squares as

$$\left(x + \frac{b}{2a} - \sqrt{\frac{b^2 - 4ac}{4a^2}}\right)\left(x + \frac{b}{2a} + \sqrt{\frac{b^2 - 4ac}{4a^2}}\right) = 0$$

Because the product is 0, then one of the factors must be 0, so we have $x + \dfrac{b}{2a} - \sqrt{\dfrac{b^2 - 4ac}{4a^2}} = 0$ or $x + \dfrac{b}{2a} + \sqrt{\dfrac{b^2 - 4ac}{4a^2}} = 0$, yielding the following:

$$x = -\frac{b}{2a} + \sqrt{\frac{b^2 - 4ac}{4a^2}} \quad \text{or} \quad x = -\frac{b}{2a} - \sqrt{\frac{b^2 - 4ac}{4a^2}}.$$

Thus,

$$x = \frac{-b + \sqrt{b^2 - 4ac}}{2a} \quad \text{or} \quad x = \frac{-b - \sqrt{b^2 - 4ac}}{2a}$$

This final result is often written in combined form, as seen below, and is called the *quadratic equation*:

$$x = \frac{-b \pm \sqrt{b^2 - 4ac}}{2a}$$

a. Use the "complete-the-square" process to solve $x^2 + 5x + 6 = 0$.
b. Use the quadratic equation to find solutions to $2x^2 - 14x + 24 = 0$.
c. Consider the expression $b^2 - 4ac$ to decide when a quadratic equation has no real number solution.

In the solution to $x^2 + 4x + 3 = 0$, we saw that the left side of the equation eventually factored into $(x + 1)(x + 3)$. Could those factors have been found without using the difference of squares or the quadratic equation of Now Try This 7-16?

To factor an expression like $x^2 + 4x + 3$, we want to write it with two factors of the form $(x + a)(x + b)$. If we had factors of that form and multiplied them together using the distributive property, we would have the following:

$$(x + a)(x + b) = (x + a)x + (x + a)b$$
$$= x^2 + ax + xb + ab$$
$$= x^2 + (a + b)x + ab$$

So, to factor $x^2 + 4x + 3$ we needed to look for two numbers a and b whose sum is 4 and whose product is 3. That method could have led us to $(x + 1)(x + 3)$.

Example 7-15

Solve $x^2 + 12x + 20 = 0$

a. by factoring.
b. by using the quadratic equation in Now Try This 7-16.

Solution   **a.** Two numbers whose sum is 12 and whose product is 20 are 10 and 2. Thus the factors are $(x + 10)(x + 2)$. Thus,

$$x^2 + 12x + 20 = 0$$
$$(x + 10)(x + 2) = 0$$
$$x = {}^-10 \quad \text{or} \quad x = {}^-2$$

b. $x = \dfrac{-12 \pm \sqrt{12^2 - 4 \cdot 1 \cdot 20}}{2 \cdot 1}$; $x = {}^-10$ or $x = {}^-2$.

Assessment 7-5A

1. Graph each of the following lines:

   a. $y = 3x - 7$

   b. $y = {}^-\left(\dfrac{3}{4}\right)x + 5$

   c. $y = x\sqrt{2}$

2. Graph each of the following:
   a. $y = x^2$
   b. $y = 2^x$

3. Explain whether you can graph $y = \sqrt{{}^-x}$. If so, graph it.

4. Find a function whose graph mirrors the graph of $y = {}^-x$ in the line $y = x$.

5. Use graphs from a spreadsheet to estimate the solution of the following system of linear equations:

$$y = {}^-2x + 7$$
$$y = 3x - 5$$

6. Solve the following equations:

   a. $2x - 13 = 23$ **b.** $\dfrac{2}{3}x + 13 = 41$

   c. $2x^3 = {}^-16$ **d.** $14x^2 - 25 = 3$
   e. $(x - 7)(x + 3) = 0$ **f.** $x^2 + 4x = 5$
   g. $2x^2 + 3x = 5$

7. One version of the formula for calculating the distance that a bullet travels when shot at an initial velocity of 1000 ft/sec over a period of time, t in seconds, is

$$d = \left(\dfrac{1}{2}\right)({}^-32)t^2 + 1000t.$$

  Use this formula to find the following:
   a. the distance the bullet travels after 3 sec
   b. the distance the bullet travels after 5 sec

8. A circle has circumference, C, directly proportional to its diameter, d; in particular, $C = \pi d$. What does the graph of this equation look like?

9. In the book *The Number Devil: A Mathematical Adventure* by Hans Magnus Enzensberger, the irrational number $\dfrac{1 + \sqrt{5}}{2}$, with decimal approximation 1.618 was discussed.
   a. Find the reciprocal of this number, to three decimal places.
   b. How is your answer in part (a) related to the solution to the equation $x^2 - x - 1 = 0$?

10. Two variables, x and y, have a product of 1. Sketch the graph of this relation.

Assessment 7-5B

1. Graph each of the following lines:

   a. $y = 2x - 5$ **b.** $y = {}^-\left(\dfrac{3}{5}\right)x + 7$

   c. $y = 2x\sqrt{2}$

2. Graph each of the following:
   a. $y = 2x^2$ **b.** $y = 3^x$

3. Does the graph of $y = {}^-\sqrt{{}^-x}$ exist? If so, graph it.

4. What is the equation of a function whose graph mirrors $y = {}^-x + b$, where b is a real number, in the line $y = x$?

5. Use graphs to estimate the solution of the following system of linear equations:

$$y = {}^-3x + 5$$
$$y = 2x - 7$$

6. Solve the following equations:

   a. $3x - 12 = 23$ **b.** $\frac{2}{3}x + 15 = 45$

   c. $3x^3 = {}^-24$ **d.** $14x^2 - 39 = 3$

   e. $(x - 7)(x + 2) = 0$ **f.** $2x^2 - 5 = 0$

   g. $2x^2 - 5x = {}^-3$

7. One version of the formula for calculating the distance that a bullet travels when shot at an initial velocity of 1000 ft/sec over a period of time, t in seconds, is $d = {}^-16t^2 + 1000t$. Use this formula to find the following:

   a. the distance the bullet travels after 2 sec

   b. the distance the bullet travels after 6 sec

8. Write the following expression in simplest form: $\left(\frac{2}{3}\right)(2x) \div \left(\frac{7\pi}{100}\right)$.

9. The diameter, d, of a circle is directly proportional to its circumference, C; in particular, $d = \frac{c}{\pi}$. Graph d as a function of C.

10. The function with an equation $y = x^2 - 2x + 1$ has a graph that is a *parabola*. Find the values of x where the graph crosses the x-axis.

Mathematical Connections 7-5

Communication

1. Read a chapter of *The Number Devil: A Mathematical Adventure* by Hans Enzensberger (1997) and write a paragraph to share the algebra in this fantasy mathematics book with classmates.

2. Debate the following statement with classmates, where one group takes the "pro" side and the other takes the "con" side: *A middle school student can fully understand the concept of an irrational number to solve an equation like $x^2 - 2 = 0$.*

Open-Ended

3. Research an approximation of π and explain the approximation.

4. Search the Internet to consider Georg Cantor's argument that there are as many rational numbers as there are natural numbers. Write an explanation of this argument for a classmate.

Cooperative Learning

5. In a group, investigate grade 6–8 textbooks and determine how irrational numbers are treated, if at all. Write a rationale about whether or not your group believes these numbers should be taught at this level.

6. In a group, research at least three articles involving the teaching of mathematics with spreadsheets and compare the findings.

Questions from the Classroom

7. A student argues that $y = (1 + \sqrt{2})x$ has no graph because $1 + \sqrt{2}$ cannot be written as an exact number. How do you help?

8. A student questions whether a spreadsheet or a graphing calculator is more useful to teach algebra concepts. How do you respond?

Review Questions

9. John Wallis, English mathematician, in 1650 discovered the following curious ratio:

$$\frac{2 \cdot 2 \cdot 4 \cdot 4 \cdot 6 \cdot 6 \cdot 8 \ldots}{1 \cdot 3 \cdot 3 \cdot 5 \cdot 5 \cdot 7 \cdot 7 \ldots}$$

To approximate the ratio, consider the following sequence:

$$\frac{2}{1}, \frac{2 \cdot 2}{1 \cdot 3}, \frac{2 \cdot 2 \cdot 4}{1 \cdot 3 \cdot 3}, \frac{2 \cdot 2 \cdot 4 \cdot 4}{1 \cdot 3 \cdot 3 \cdot 5}, \ldots$$

Write the decimal equivalents of these given terms.

10. In 1914, the following mnemonic appeared in the *Scientific American:* "May I have a large container of coffee?" Consider the number of letters in each word. What well-known number does this mnemonic represent?

11. **a.** Find $0.8 \div 0.32$ by converting these numbers to rational numbers in the form $\frac{a}{b}$, where a and b are integers and $b \neq 0$, dividing the rational numbers, and then changing the quotient back to decimal form.

    b. Find $0.8 \div 0.32$ using decimal division and compare your answer with the answer from part (a).

12. Change $7.27\overline{1}$ to a rational number in the form $\frac{a}{b}$, where a and b are integers and $b \neq 0$.

13. Find a repeating decimal between 0.21 and $0.\overline{2}$.

14. Order the following from least to greatest:

$$2.5, \frac{5}{3}, 2.0\overline{5}, 2.\overline{15}, \frac{7}{3}$$

National Assessment of Educational Progress (NAEP) Question

Mr. Hardt bought a square piece of carpet with an area of 39 square yards. The length of each side of this carpet is between which of the following?

a. 4 yards and 5 yards
b. 5 yards and 6 yards
c. 6 yards and 7 yards
d. 7 yards and 8 yards
e. 9 yards and 10 yards

NAEP, 2007, Grade 8

Hint for Solving the Preliminary Problem

Think about repeating decimals as one possible route to finding a solution.

Chapter Outline

I. Decimals

A. Every rational number can be represented as a terminating or repeating decimal.

B. A rational number $\frac{a}{b}$, in simplest form, whose denominator is of the form $2^m \cdot 5^n$, where m and n are whole numbers, can be expressed as a **terminating decimal**.

C. A **repeating decimal** is a decimal with a block of digits, called the **repetend**, that repeat infinitely many times.

D. A number is in **scientific notation** if it is written as the product of a number n that is greater than or equal to 1 and less than 10 and an integer power of 10. If a number is negative, follow the rules for a positive number and adjoin the negative sign.

E. An **infinite geometric sequence** with first term a and ratio r, where $0 < r < 1$, has sum

$$S = \frac{a}{1-r}.$$

II. Real numbers

A. An **irrational number** is represented by a nonterminating, nonrepeating decimal.

B. The set of **real numbers** is the set of all decimals, namely, the union of the set of rational numbers and the set of irrational numbers.

C. If a is any nonnegative real number, then the **principal square root** of a, denoted by $\sqrt{a}$,

is the nonnegative number b such that $bb = b^2 = a$.

D. Square roots can be found by using the **squeezing method** or the Archimedean method.

E. Computation methods with terminating or repeating decimals can be verified by converting the decimals to rational numbers in the form $\frac{a}{b}$, where a and b are integers and $b \neq 0$.

F. Estimations for decimal arithmetic can be accomplished following the same general rules as for other sets of numbers.

III. Radicals and rational exponents

A. $\sqrt[n]{x}$ or $x^{1/n}$ is the **nth root** of x and n is the **index**.

B. The following properties hold for all radicals if the expressions involving radicals are meaningful:

a. $\sqrt[n]{xy} = \sqrt[n]{x} \cdot \sqrt[n]{y}$

b. $\sqrt[n]{\dfrac{x}{y}} = \dfrac{\sqrt[n]{x}}{\sqrt[n]{y}}$

c. $x^{m/n} = \left(\sqrt[n]{x}\right)^m = \sqrt[n]{x^m}$, if m/n is in simplest form.

★ **IV.** The solution to a quadratic equation in the form $ax^2 + bx + c = 0$, where $a \neq 0$, is

$$x = \frac{-b \pm \sqrt{b^2 - 4ac}}{2a}.$$ For real number solutions, $b^2 - 4ac \geq 0$.

Chapter Review

1. a. On the number line, find the decimals that correspond to points A, B, and C.
   b. Indicate by D the point that corresponds to 0.09 and by E the point that corresponds to 0.15.

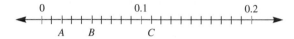

2. Write each of the following as a rational number in the form $\frac{a}{b}$, where a and b are integers and $b \neq 0$:
   a. 32.012   **b.** 0.00103

3. Describe a test to determine if a fraction can be written as a terminating decimal without one's actually performing the division.

4. A board is 142.4 cm long. How many shelves can be cut from it if each shelf is 55.3 cm long? (Disregard the width of the cuts.)

5. Write each of the following as a decimal:
   a. $\frac{4}{7}$ **b.** $\frac{1}{8}$ **c.** $\frac{2}{3}$ **d.** $\frac{5}{8}$

6. Write each of the following as a fraction in simplest form:
   a. 0.28   **b.** $^{-}6.07$ **c.** $0.\overline{3}$ **d.** $2.0\overline{8}$

7. Round each of the following numbers as specified:
   a. 307.625 to the nearest hundredth
   b. 307.625 to the nearest tenth
   c. 307.625 to the nearest unit
   d. 307.625 to the nearest hundred

8. Rewrite each of the following in scientific notation:
   a. 426,000
   b. $324 \cdot 10^{-6}$
   c. 0.00000237
   d. $^{-}0.325$

9. Order the following decimals from greatest to least:
   $1.45\overline{19}$, $1.451\overline{9}$, 1.4519, $1.45\overline{19}$, $^{-}0.134$, $^{-}0.13\overline{401}$, $0.13\overline{401}$

10. Each of the following is a geometric sequence. Find the missing terms.
   a. 5, _____, 10
   b. 1, _____, _____, _____, 1/4

11. Write each of the following in scientific notation without using a calculator:
   a. 1783411.56
   b. $\frac{347}{10^8}$
   c. $49.3 \cdot 10^8$
   d. $29.4 \cdot \frac{10^{12}}{10^{-4}}$

   e. $0.47 \cdot 1000^{12}$
   f. $\frac{3}{5^9}$

12. a. Find five decimals between 0.1 and 0.11 and order them from greatest to least.
   b. Find four decimals between 0 and 0.1 listed from least to greatest so that each decimal starting from the second is twice as large as the preceding one.
   c. Find four decimals between 0.1 and 0.2 and list them in increasing order so that the first one is halfway between 0.1 and 0.2, the second halfway between the first and 0.2, the third halfway between the second and 0.2, and similarly for the fourth one.

13. Classify each of the following as rational or irrational (assume the patterns shown continue):
   a. 2.191199911999119999119 . . .
   b. $\frac{1}{\sqrt{2}}$
   c. $\frac{4}{9}$
   d. 0.0011001100110011 . . .
   e. 0.001100011000011 . . .

14. Write each of the following in the form $a\sqrt{b}$ or $a\sqrt[n]{b}$, where a and b are positive integers and b has the least value possible:
   a. $\sqrt{242}$ **b.** $\sqrt{288}$
   c. $\sqrt{180}$ **d.** $\sqrt[3]{162}$

15. Answer each of the following and explain your answers:
   a. Is the set of irrational numbers closed under addition?
   b. Is the set of irrational numbers closed under subtraction?
   c. Is the set of irrational numbers closed under multiplication?
   d. Is the set of irrational numbers closed under division?

16. Find an approximation for $\sqrt{23}$ correct to three decimal places without using the $\boxed{y^x}$ or the $\boxed{\sqrt{}}$ keys.

17. Approximate $\sqrt[3]{2}$ to two decimal places by a method similar to the squeezing method.

18. Graph each of the following lines:
   a. $y = {}^{-}2x + 5$ **b.** $y = {}^{-}\left(\frac{1}{5}\right)x + 7$
   c. $y = {}^{-}x\sqrt{2}$

19. Graph each of the following:
   a. $y = {}^{-}x^2$ **b.** $y = 3^{-x}$

20. Use graphs to estimate the solution of the following system of linear equations:

$$y = {}^-1x + 6$$
$$y = 2x - 7$$

21. Solve the following equations for real number solution(s):

a. $^-3x + 12 = 23$

b. $\frac{2}{3}x + 18 = 42$

c. $x^3 = {}^-1$

d. $4x^2 - 33 = 3$

e. $(2x - 7)(3x + 2) = 0$

f. $x^2 - x = 6$

g. $2x^2 - 3x + 1 = 0$

22. What would be the third term of a geometric sequence whose first term is π and whose ratio is $\frac{1}{\pi}$?

23. Explain whether or not you would ever expect a price in a store to be an irrational number.

Selected Bibliography

Behr, M., I. Wachsmuth, T. Post, and R. Lesh. "Order and Equivalence of Rational Numbers: A Clinical Teaching Experiment." *Journal for Research in Mathematics Education* 15 (July 1984): 323–341.

Beigie, D. "Integrating Content to Create Problem-Solving Opportunities." *Mathematics Teaching in the Middle School* 13 (February 2008): 352–360.

Beigie, D. "Investigating Limits in Number Patterns." *Mathematics Teaching in the Middle School* 7 (April 2002): 438–443.

Boers-van Oosterum, M. "Understanding of Variables and Their Uses Acquired by Students in Traditional and Computer-Intensive Algebra." Ph.D. diss., University of Maryland College Park, 1990.

Brown, M. "Place Value and Decimals." In *Children's Understanding of Mathematics*. London: John Murray, 1981, pp. 11–16.

Drum, R., and W. Petty, Jr. "2 Is Not the Same as 2.0!" *Mathematics Teaching in the Middle School* 6 (September 2000): 34–38.

Dunham, P., and T. Dick. "Research on Graphing Calculators." *Mathematics Teacher* 87 (September 1994): 440–445.

Enzensberger, H. *The Number Devil: A Mathematical Adventure*. New York: Henry Holt and Company, 1997.

Flores, A. "On My Mind: The Finger and the Moon." *Mathematics Teaching in the Middle School* 13 (October 2007): 132.

Glasgow, R., G. Ragan, W. Fields, R. Reys, and D. Wasman. "The Decimal Dilemma." *Teaching Children Mathematics* 7 (October 2000): 89–93.

Groves, S. "Calculators: A Learning Environment to Promote Number Sense." Paper presented at the annual meeting of the American Educational Research Association, New Orleans, April 1994.

Hiebert, J. "Mathematical, Cognitive, and Instructional Analyses of Decimal Fractions." In *Analysis of Arithmetic for Mathematics Teaching*, edited by G. Leinhardt, R. Putman, and R. Hattrup. Hillsdale, NJ: LEA, 1992.

Irwin, K. "Using Everyday knowledge of Decimals to Enhance Learning." *Journal of Research in Mathematics Education* 32 (July 2001): 399–420.

Judd, W. "Instructional Games with Calculators." *Mathematics Teaching in the Middle School* 13 (February 2007): 312–314.

Leonard, J., and L. Campbell. "Using the Stock Market for Relevance in Teaching Number Sense." *Mathematics Teaching in the Middle School* 9 (February 2004): 294.

Lewis, L. "Irrational Numbers Can 'In-Spiral' You." *Mathematics Teaching in the Middle School* 13 (April 2007): 442–445.

Martine, S., and J. Bay-Williams. "Investigating Students' Conceptual Understanding of Decimal Fractions Using Multiple Representations." *Mathematics Teaching in the Middle School* 8 (January 2003): 244–247.

Neuschwander, C. *Sir Cumference and the Dragon of Pi: A Math Adventure*. Watertown, MA: Cambridge Publishing, Inc., 1999.

Oppenheimer, L., and R. Hunting. "Relating Fractions and Decimals: Listening to Students Talk." *Mathematics Teaching in the Middle School* 4 (February 1999): 318–321.

Reeder, S. "Are We Golden? Investigations with the Golden Ratio." *Mathematics Teaching in the Middle School* 13 (October 2007): 150–154.

Reeves, A., and M. Beasley. "Sudorku." *Mathematics Teaching in the Middle School* 13 (August 2007): 30–31, 36.

Reys, B., and F. Arbaugh. "Clearing Up the Confusion over Calculator Use in Grades K–5." *Teaching Children Mathematics* 8 (October 2001): 90–94.

Reys, R., B. Reys, N. Nohda, and H. Emori. "Mental Computation Performance and Strategy Use of Japanese Students in Grades 2, 4, 6, and 8." *Journal for Research in Mathematics Education* 26 (July 1995): 304–326.

Rojano, T. "Developing Algebraic Aspects of Problem Solving within a Spreadsheet Environment." In *Approaches to Algebra: Perspectives for Research and Teaching,* edited by Nadine Bednarz, Carolyn Kieran, and Lesley Lee. Boston: Kluwer Academic Publishers, 1996.

Shaw, K., and L. Aspinwall. "Mathematics Detective: Mission Possible: Exploring the Mysterious 1/7." *Mathematics Teaching in the Middle School* 11(November 2005): 178.

Sheets, C. *Effects of Computer Learning and Problem-solving Tools on the Development of Secondary School Students' Understanding of Mathematical Functions.* Ph.D. Diss., University of Maryland College Park, 1993.

Sowder, J., D. Wearne, W. Martin, and M. Strutchens. "What Do 8th-Grade Students Know about Mathematics?" In *Results and Interpretations of the 1990–2000 Mathematics Assessments of the National Assessment of Educational Progress,* edited by P. Kloosterman and F. Lester. Reston, VA: NCTM, 2004, pp. 105–144.

Thompson, A., and S. Sproule. "Deciding When to Use Calculators." *Mathematics Teaching in the Middle School* 6 (October 2000): 12–129.

Thompson, C., and V. Walker. "Connecting Decimals and Other Mathematical Content." *Teaching Children Mathematics* 2 (April 1996): 496–502.

Vinogradova, N. "Solving Quadratic Equations by Completing Squares." *Mathematics Teaching in the Middle School* 12 (March 2007): 403–405.

Wearne, D., and J. Hiebert. "A Cognitive Approach to Meaningful Mathematics Instruction: Testing a Local Theory Using Decimal Numbers." *Journal for Research in Mathematics Education* 19 (November 1988): 371–384.

Williams, S., and J. Copley. "Using Calculators to Discover Patterns in Dividing Decimals." *Mathematics Teaching in the Middle School* 1 (April 1994): 72–75.

Yang, D., and R. Reys. "One Fraction Problem, Many Solution Paths." *Mathematics Teaching in the Middle School* 7 (November 2001): 164–166.

8

Proportional Reasoning, Percents, and Applications

Preliminary Problem

A car dealer sold two cars. On the first car he made a 10% profit and on the second car he lost 10%. He sold the cars for $9999 each. What was the overall dollar amount of profit or loss on the two transactions, or did the dealer break even?

Proportional reasoning is an extremely important concept taught in grades K–8 that in many cases does not receive enough attention. As pointed out by Hoffer and Hoffer (1988), "Proportional reasoning is generally regarded as one of the important components of formal thought acquired in adolescence. . . . Failure to develop in this area by early to middle adolescence precludes study in a variety of disciplines requiring quantitative thinking and understanding including algebra, geometry, some aspects of biology, chemistry and physics." (p. 303) Proportionality has connections to most, if not all, of the other foundational middle-school topics and can provide a context to study these topics.

People face proportional reasoning problems in everyday life, for example, in changing recipes or determining a saving plan. These types of problems require students to identify the variables involved and then identify the relationship between these variables. An example of creating ratios to make comparisons in situations that involve pairs of numbers is given in this passage from *Principles and Standards:*

> Working with proportions is a major focus proposed in these Standards for the middle grades. Students should become proficient in creating ratios to make comparisons in situations that involve pairs of numbers, as in the following problem:
> If three packages of cocoa make fifteen cups of hot chocolate, how many packages are needed to make sixty cups? (p. 34)

Further, in the grade 7 *Focal Points* we find the following:

> Students extend their work with ratios to develop an understanding of proportionality that they apply to solve single and multistep problems in numerous contexts. (p. 19)

In this chapter, we cover ratios and proportions, proportional reasoning, and percents and then include an optional section on computing interest.

8-1   Ratios, Proportions, and Proportional Reasoning

As seen in the quotes above from *Principles and Standards* and the *Focal Points*, ratios and proportions are a very important part of the middle-school curriculum. Ratios are encountered in everyday life. For example, there may be a 2-to-3 ratio of Democrats to Republicans on a certain legislative committee, a friend may be given a speeding ticket for driving 69 miles per hour, or eggs may cost $1.40 a dozen. Each of these illustrates a **ratio**. Ratios are written $\frac{a}{b}$ or $a:b$ and are usually used to compare quantities.

A ratio of $1:3$ for boys to girls in a class means that the number of boys is $\frac{1}{3}$ that of girls; that is, there is 1 boy for every 3 girls. Notice we could also say that the ratio of girls to boys is $3:1$, or that there are 3 times as many girls as boys. Ratios can represent **part-to-whole** or **whole-to-part** comparisons. For example, if the ratio of boys to girls in a class is $1:3$, then the ratio of boys (part) to children (whole) is $1:4$. If there are b boys and g girls, then $g = 3b$; that is, $\frac{1}{3} = \frac{b}{3b}$. Also, the ratio of boys to the entire class is $\frac{b}{b+g} = \frac{b}{b+3b} = \frac{b}{4b} = \frac{1}{4}$.

We could also say that the ratio of all children (whole) to boys (part) is $4 : 1$. Some ratios give **part-to-part** comparisons, such as the ratio of the number of boys to girls or the number of students to one teacher. For example, a school might say that the average ratio of students to teachers cannot exceed $24 : 1$.

Notice that the ratio of $1 : 3$ for boys to girls in a class does not tell us how many boys and how many girls there are in the class. It only tells us the relative size of the groups. There could be 2 boys and 6 girls or 3 boys and 9 girls or 4 boys and 12 girls or some other numbers that give a fraction equivalent to $\frac{1}{3}$.

Example 8-1

There were 7 males and 12 females in the Dew Drop Inn on Monday evening. In the game room next door were 14 males and 24 females.

a. Express the number of males to females at the inn as a ratio (part-to-part).
b. Express the number of males to females at the game room as a ratio (part-to-part).
c. Express the number of males in the game room to the number of people in the game room as a ratio (part-to-whole).

Solution   **a.** The ratio is $\frac{7}{12}$.   **b.** The ratio is $\frac{14}{24}$, or $\frac{7}{12}$.   **c.** The ratio is $\frac{14}{38}$, or $\frac{7}{19}$.

Proportions

In a study of sixth graders conducted by Harel and colleagues (1994), children were shown a picture of a carton of orange juice and were told that the orange juice was made from orange concentrate and water. Then they were shown two glasses—a large glass and a small glass—and they were told that both glasses were filled with orange juice from the carton. They were then asked if the orange juice from each of the two glasses would taste equally "orangey" or if one would taste more "orangey." About half the students said the larger glass would be more "orangey" and about half said the smaller glass would be more "orangey." These students were thinking of only one quantity—the water alone or the orange concentrate alone. For example, one student said that the larger glass would be more "orangey" because the glass is bigger and it would hold more orange concentrate, while another said the small glass would be more "orangey" because it has less water but more orange per ounce.

Suppose Recipe A for an orange drink calls for 2 cans of orange concentrate for every 3 cans of water. We could say that the ratio of cans of orange concentrate to cans of water is $2 : 3$. We represent this pictorially as in Figure 8-1(a), where O represents a can of orange concentrate and W represents a can of water. In Figure 8-1(b) and (c), we continue the process of adding 2 cans of orange concentrate for every 3 cans of water.

(a) O O   (b) O O   O O   (c) O O   O O   O O
   W W W   W W W   W W W   W W W   W W W   W W W

Figure 8-1 (Recipe A)

From Figure 8-1 we could develop and continue the **ratio table**, as shown in Table 8-1.

Table 8-1

Cans of Orange Concentrate	2	4	6	8	10	12
Cans of Water	3	6	9	?	?	?

In Table 8-1, the ratios $2/3$ and $4/6$ are equal. The equation $2/3 = 4/6$ is a proportion. In general, we have the following.

Definition

A **proportion** is a statement that two given ratios are equal.

If Recipe B calls for 4 cans of orange concentrate for every 8 cans of water, then the ratio of cans of orange concentrate to cans of water for this recipe is $4:8$. We picture this as in Figure 8-2(a).

(a) O O O O   (b) O O | O O
    W W W W   W W | W W
    W W W W   W W | W W

Figure 8-2 (Recipe B)

Which of the two recipes produces a drink that tastes more "orangey"? In Figure 8-1(a), we see that in Recipe A there are 2 cans of orange concentrate for every 3 cans of water. In Figure 8-2(a), we see that in Recipe B there are 4 cans of orange concentrate for every 8 cans of water. To compare the two recipes, we need either the same number of cans of orange concentrate or the same number of cans of water. Either is possible. Figure 8-1(b) shows that for Recipe A there are 4 cans of orange concentrate for every 6 cans of water. In Recipe B, for 4 cans of orange concentrate there are 8 cans of water. Recipe B calls for more water per 4 cans of orange concentrate, so it is less "orangey." An alternative is to observe that in Figure 8-2(b), Recipe B could be divided to show that there are 2 cans of orange concentrate for every 4 cans of water. We could then compare this with Figure 8-1(a), showing 2 cans of orange concentrate for every 3 cans of water, and reach the same conclusion.

From our work in Section 6.1, we know that $2/3 = 4/6$ because $2 \cdot 6 = 3 \cdot 4$. Hence $2/3 = 4/6$ is a proportion. Also $2/3 \neq 4/8$ because $2 \cdot 8 \neq 3 \cdot 4$ and this is not a proportion. In general, we have the following theorem that follows from the property of *equal fractions* developed in Section 6.1.

Theorem 8–1

If a, b, c, and d are all real numbers and $b \neq 0$ and $d \neq 0$, then

$$\frac{a}{b} = \frac{c}{d} \text{ is a proportion if, and only if,} \quad ad = bc$$

NOW TRY THIS 8-1 Justify Theorem 8-1 by multiplying each side of the proportion by *bd*.

Students in the lower grades typically experience problems that are *additive*. Consider the problem below.

Allie and Bente type at the same speed. Allie started typing first. When Allie had typed 8 pages, Bente had typed 4 pages. When Bente has typed 10 pages, how many has Allie typed?

This is an example of an *additive* relationship. Students should reason that since the two people type at the same speed, when Bente has typed an additional 6 pages, Allie should have also typed an additional 6 pages, so she should have typed $8 + 6$, or 14, pages.

Next consider the following problem:

Carl can type 8 pages for every 4 pages that Dan can type. If Dan has typed 12 pages, how many pages has Carl typed?

If students try an *additive* approach, they will conclude that since Dan has typed 8 more pages than in the original relationship, then Carl should have typed an additional 8 pages. However, the correct reasoning is that since Carl types twice as fast as Dan he will type twice as many pages as Dan. Therefore, when Dan has typed 12 pages, Carl has typed 24 pages. The relationship between the ratios is *multiplicative*. Another way to solve this problem is to set up the proportion $\dfrac{8}{4} = \dfrac{x}{12}$, where x is the number of pages that Carl will type, and solve for x. Because $\dfrac{8}{4} = \dfrac{8 \cdot 3}{4 \cdot 3} = \dfrac{24}{12}$, then $x = 24$ pages.

In the problem introduced in the quote from the *Principles and Standards*, one term in the proportion is missing:

$$\text{packages} \rightarrow \dfrac{3}{15} = \dfrac{x}{60}$$
$$\text{cups} \quad \rightarrow$$

One way to solve the equation is to multiply both sides by 60, as follows:

$$\frac{3}{15} \cdot 60 = \frac{x}{60} \cdot 60$$
$$3 \cdot 4 = x$$
$$12 = x$$

Therefore, 12 packages of cocoa are needed to make 60 cups of hot chocolate.

We could also solve this by noticing $\dfrac{3}{15} = \dfrac{3 \cdot 4}{15 \cdot 4} = \dfrac{12}{60}$, so $x = 12$. Another method of solution uses Theorem 8–1. This is often called the *cross-multiplication method* or *cross-product method*. This equation is a proportion if, and only if,

$$3 \cdot 60 = 15x$$
$$180 = 15x$$
$$12 = x$$

The cross-product method (Theorem 8–1) is very efficient, but notice the research concerning this method pointed out in the Research Note.

Research Note

The cross-product algorithm for evaluating a proportion (using equality of fractions) is (1) an extremely efficient algorithm but rote and without meaning, (2) usually misunderstood, (3) rarely generated by students independently, and (4) often used as a "means of avoiding proportional reasoning rather than facilitating it" (Cramer and Post, 1993; Post et al. 1988; Hart 1984; Lesh et al. 1988). ◆

Example 8-2

If there are 3 cars for every 8 students at a high school, how many cars are there for 1200 students?

Solution   We use the strategy of *setting up a table*, as shown in Table 8-2

Table 8-2

Number of cars	3	x
Number of students	8	1200

The ratio of cars to students is always the same:

$$\text{Cars} \rightarrow \frac{3}{8} = \frac{x}{1200}$$
$$\text{Students} \rightarrow$$

$$3 \cdot 1200 = 8x$$
$$3600 = 8x$$
$$450 = x$$

Thus, there are 450 cars.

Next consider the student page. It shows the cost to rent a car from two different companies. The ratios for the What-A-Deal company do not form a proportion because $\frac{1}{20} \neq \frac{2}{35}$. The Value Vehicle ratios do form a proportion because $\frac{1}{20} = \frac{2}{40} = \frac{3}{60} = \frac{4}{80}$. To check for proportions, the student page uses *cross products*. Read through the example to see how attention is paid to the units in the problem. Work the *Talk About It* on the bottom of the student page.

On the student page, we find "We say that these quantities **vary proportionally**." What does this mean? Consider Table 8-3.

Table 8-3

Days (d)	1	2	3	4
Cost (c)	20	40	60	80

The ratios $\frac{d}{c}$ are all equal, that is, $\frac{1}{20} = \frac{2}{40} = \frac{3}{60} = \frac{4}{80}$. Thus, each pair of ratios forms a proportion. In this case, $\frac{d}{c} = \frac{1}{20}$ for all values of c and d. This is also expressed by saying that *d is proportional to c* or *d varies proportionally to c* or *d varies directly with c*. In this case, $d = \frac{1}{20} c$ for every c and d. The number $\frac{1}{20}$ is the **constant of proportionality.** For example, we can say that *gas used by a car is proportional to the miles traveled* or *lottery profits vary directly with the number of tickets sold.*

Definition

If the variables x and y are related by the equality $y = kx$, $\left(k = \frac{y}{x} \right)$, then **y is said to be proportional to x** and k is the **constant of proportionality** between y and x.

School Book Page UNDERSTANDING PROPORTIONS

Lesson 6-5

Key Idea
A proportion is a statement that two ratios are equal.

Vocabulary
• proportion
• cross products

316

Understanding Proportions

▶ LEARN

What is a proportion?

A **proportion** states that two ratios are equal. You can use the first two Value Vehicle ratios to write a proportion: $\frac{1 \text{ day}}{\$20} = \frac{2 \text{ days}}{\$40}$.

In a proportion, the units must be the same across the top and bottom, or down the left and right sides. In this case, days are across the top and dollars across the bottom.

For Value Vehicle, the ratios of days to the cost of renting a car are all equal. We say that these quantities **vary proportionally.** The What-A-Deal Wheels ratios are NOT all equal, so these quantities do NOT vary proportionally.

In the proportion $\frac{1}{20} = \frac{2}{40}$, 1×40 and 20×2 are cross products. The **cross products** of the terms in a proportion are equal.

✓ WARM UP

Write = or ≠.

1. $\frac{3}{4} \boxed{} \frac{9}{12}$ 2. $\frac{8}{16} \boxed{} \frac{4}{8}$

3. $\frac{20}{12} \boxed{} \frac{5}{3}$ 4. $\frac{3}{1} \boxed{} \frac{1}{3}$

5. $\frac{24}{6} \boxed{} \frac{4}{1}$ 6. $\frac{12}{15} \boxed{} \frac{3}{5}$

Value Vehicle

Days	Cost
1	$20
2	$40
3	$60
4	$80

What-A-Deal Wheels

Days	Cost
1	$20
2	$35
3	$48
4	$59

Example

Decide if the ratios $\frac{3 \text{ ft}}{8 \text{ sec}}$ and $\frac{9 \text{ ft}}{24 \text{ sec}}$ form a proportion.

What You Do

Look at the units.

$\frac{3 \text{ ft}}{8 \text{ sec}} \stackrel{?}{=} \frac{9 \text{ ft}}{24 \text{ sec}}$ The units are the same across the top and bottom.

Look at the cross products.

$3 \times 24 \stackrel{?}{=} 8 \times 9$ ← The cross products

$72 = 72$ are equal.

Why It Works

Multiply both sides of the proportion by 24×8 and simplify.

$\frac{24 \times 8 \times 3}{8} = \frac{9 \times 24 \times 8}{24}$

$24 \times 3 = 9 \times 8$ ← cross products

$72 = 72$

Since the units are the same and the cross products are equal, the ratios form a proportion.

✓ Talk About It

1. Do the ratios $\frac{4 \text{ ft}}{6 \text{ ft}}$ and $\frac{12 \text{ sec}}{18 \text{ sec}}$ form a proportion? Why or why not?

Source: Scott Foresman-Addison Wesley Mathematics, Grade 6, 2008 (p. 316).

> **REMARK** A central idea in proportional reasoning is that a relationship between two quantities is such that the ratio of one quantity to the other remains unchanged as the numerical values of both quantities change.

It is important to remember that in the ratio $a:b$ or $\dfrac{a}{b}$, a and b do not have to be integers.

For example, if in Eugene, Oregon, $\dfrac{7}{10}$ of the population exercise regularly, then $\dfrac{3}{10}$ of the population do not exercise regularly, and the ratio of those who do to those who do not is $\dfrac{7}{10} : \dfrac{3}{10}$. This ratio can be written $7:3$.

It is also important to notice units of measure when we work with proportions. For example, if a turtle travels 5 in. every 10 sec, how many feet does it travel in 50 sec? If units of measure are ignored, we might set up the following proportion:

$$\frac{5}{10} = \frac{x}{50}$$

In this proportion the units of measure are not listed. A more informative proportion that often prevents errors is the following:

$$\frac{5 \text{ in.}}{10 \text{ sec}} = \frac{x \text{ in.}}{50 \text{ sec}}$$

This implies that $x = 25$. Consequently, since 12 in. = 1 ft, the turtle travels $\dfrac{25}{12}$ ft, or $2\dfrac{1}{12}$ ft, or 2 ft 1 in.

Principles and Standards points out the following concerning the cross-multiplication method of solving proportions.

> Instruction in solving proportions should include methods that have a strong intuitive basis. The so-called cross-multiplication method can be developed meaningfully if it arises naturally in students' work, but it can also have unfortunate side effects when students do not adequately understand when the method is appropriate to use. Other approaches to solving proportions are often more intuitive and also quite powerful. For example, when trying to decide which is the better buy—12 tickets for $15.00 or 20 tickets for $23.00—students might choose to use a scaling strategy (finding the cost for a common number of tickets) or a unit-rate strategy (finding the cost for one ticket). (p. 221)

The **scaling strategy** for solving the problem would involve finding the cost for a common number of tickets. Because LCM(12, 20) = 60, we could choose to find the cost of 60 tickets under each plan.

In the first plan, since 12 tickets cost $15, then 60 tickets cost $75.

In the second plan, since 20 tickets cost $23, then 60 tickets cost $69.

Therefore, the second plan is a better buy.

The **unit-rate strategy** for solving this problem involves finding the cost of one ticket under each plan and then comparing the unit cost.

In the first plan, since 12 tickets cost $15, then 1 ticket costs $1.25.

In the second plan, since 20 tickets cost $23, then 1 ticket costs $1.15.

Therefore, the second plan is the better buy. Note the suggestion in the Research Note below for using the unit-rate strategy.

Research Note   The unit-rate method is perhaps the best method for working with problems involving ratios and proportions. The unit-rate method is strongly suggested as "scaffolding" for building proportional reasoning (Post et al. 1988). ♦

Example 8-3

Kai, Paulus, and Judy made $2520 for painting a house. Kai worked 30 hr, Paulus worked 50 hr, and Judy worked 60 hr. They divided the money in proportion to the number of hours worked. If they all earn the same rate of pay, how much did each earn?

Solution   Let x be the unit rate or the rate of pay per hour. Then, $30x$ denotes the amount of money that Kai received. Paulus received $50x$ because then, and only then, will the ratios of the amounts be the same as $30:50$, as required. Similarly, Judy received $60x$. Because the total amount of money received is $30x + 50x + 60x$, we have

$$30x + 50x + 60x = 2520$$
$$140x = 2520$$
$$x = 18$$

Hence,

$$\text{Kai received } 30x = 30 \cdot 18, \text{ or } \$540$$
$$\text{Paulus received } 50x = 50 \cdot 18, \text{ or } \$900$$
$$\text{Judy received } 60x = 60 \cdot 18, \text{ or } \$1080$$

Dividing each of the amounts by 18 shows that the proportion is as required.

Consider the proportion $\dfrac{15}{30} = \dfrac{3}{6}$. Because the ratios in the proportion are equal and because equal nonzero fractions have equal reciprocals, it follows that $\dfrac{30}{15} = \dfrac{6}{3}$. Also notice that the proportions are true because each results in $15 \cdot 6 = 30 \cdot 3$. In general, we have the following theorem.

Theorem 8–2

For any rational numbers $\dfrac{a}{b}$ and $\dfrac{c}{d}$, with $a \neq 0$ and $c \neq 0$, $\dfrac{a}{b} = \dfrac{c}{d}$ if, and only if, $\dfrac{b}{a} = \dfrac{d}{c}$.

Consider $\dfrac{15}{30} = \dfrac{3}{6}$ again. Notice that $\dfrac{15}{3} = \dfrac{30}{6}$; that is, the ratio of the numerators is equal to the ratio of the corresponding denominators. In general, we have the following theorem.

Theorem 8–3

For any rational numbers $\dfrac{a}{b}$ and $\dfrac{c}{d}$, with $b, c, d \neq 0$, $\dfrac{a}{b} = \dfrac{c}{d}$ if, and only if, $\dfrac{a}{c} = \dfrac{b}{d}$.

Scale Drawings

Ratio and proportions are used in scale drawings. For example, if the scale is 1 : 300, then the length of 1 cm in such a drawing represents 300 cm, or 3 m in true size. The **scale** is the ratio of the size of the drawing to the size of the object. The following example shows the use of scale drawings.

Example 8-4

The floor plan of the main floor of a house in Figure 8-3 is drawn in the scale of 1 : 300. Find the dimensions in meters of the living room.

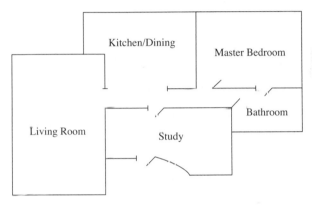

Figure 8-3

Solution   In Figure 8-3, the dimensions of the living room measured with a centimeter ruler are approximately 3.7 cm by 2.5 cm. Because the scale is 1 : 300, 1 cm in the drawing represents 300 cm, or 3 m in true size. Hence, 3.7 cm represents 3.7 · 3, or 11.1 m, and 2.5 cm represents 2.5 · 3, or 7.5 m. Hence, the dimensions of the living room are approximately 11.1 m by 7.5 m.

Assessment 8-1A

1. Answer the following regarding the English alphabet:
   a. Determine the ratio of vowels to consonants.
   b. What is the ratio of consonants to vowels?
   c. What is the ratio of consonants to letters in the English alphabet?
   d. Write a word that has a ratio of 2 : 3 of vowels to consonants.

2. **a.** If the ratio of boys to girls in a class is 2 : 3, what is the ratio of boys to all the students in the class? Why?
   b. If the ratio of boys to girls in a class is $m : n$, what is the ratio of boys to all the students in the class?
   c. If $\dfrac{3}{5}$ of the class are girls, what is the ratio of girls to boys?

3. Solve for x in each of the following proportions:
   a. $\dfrac{12}{x} = \dfrac{18}{45}$ **b.** $\dfrac{x}{7} = \dfrac{-10}{21}$
   c. $\dfrac{5}{7} = \dfrac{3x}{98}$ **d.** $3\dfrac{1}{2}$ is to 5 as x is to 15.

4. There are approximately 2 lb of muscle for every 5 lb of body weight. For a 90-lb person, approximately how much of the weight is muscle?

5. Which is a better buy—4 grapefruits for 80¢ or 12 grapefruits for $1.80?

6. On a map, $\dfrac{1}{3}$ in. represents 5 mi. If New York and Aluossim are 18 in. apart on the map, what is the actual distance between them?

7. David read 40 pages of a book in 50 min. How many pages should he be able to read in 80 min if he reads at a constant rate?

8. Two numbers are in the ratio 3 : 4. Find the numbers if
   a. their sum is 98.
   b. their product is 768.

9. Gary, Bill, and Carmella invested in a corporation in the ratio of 2 : 4 : 5, respectively. If they divide the profit of $82,000 proportionally to their investment, approximately how much will each receive?

10. Sheila and Dora worked $3\frac{1}{2}$ hr and $4\frac{1}{2}$ hr, respectively, on a programming project. They were paid $176 for the project. How much did each earn if they are both paid the same rate?

11. Vonna scored 75 goals in her soccer kicking practice. If her success-to-failure rate is 5 : 4, how many times did she attempt a goal?

12. Express each of the following as a ratio $\frac{a}{b}$, where a and b are whole numbers:
   a. $\frac{1}{6} : 1$
   b. $\frac{1}{3} : \frac{1}{3}$
   c. $\frac{1}{6} : \frac{2}{7}$

13. Use Theorems 8–2 and 8–3 to write three other proportions that follow from the following proportion:
$$\frac{12¢}{36 \text{ oz}} = \frac{16¢}{48 \text{ oz}}$$

14. The rise and span for a house roof are identified as shown on the drawing. The pitch of a roof is the ratio of the rise to the half-span.
   a. If the rise is 10 ft and the span is 28 ft, what is the pitch?

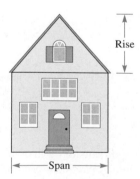

Rise

Span

   b. If the span is 16 ft and the pitch is $\frac{3}{4}$, what is the rise?

15. Gear ratios are used in industry. A gear ratio is the comparison of the number of teeth on two gears. When two gears are meshed, the revolutions per minute (rpm) are inversely proportional to the number of teeth; that is,
$$\frac{\text{rpm of large gear}}{\text{rpm of small gear}} = \frac{\text{Number of teeth on small gear}}{\text{Number of teeth on large gear}}$$

   a. The rpm ratio of the large gear to the small gear is 4 : 6. If the small gear has 18 teeth, how many teeth does the large gear have?
   b. The large gear revolves at 200 rpm and has 60 teeth. How many teeth are there on the small gear that has an rpm of 600?

16. A Boeing 747 jet is approximately 230 ft long and has a wingspan of 195 ft. If a scale model of the plane is about 40 cm long, what is the model's wingspan?

17. Jennifer weighs 160 lb on Earth and 416 lb on Jupiter. Find Amy's weight on Jupiter if she weighs 120 lb on Earth.

18. A recipe calls for 1 tsp of mustard seeds, 3 c of tomato sauce, $1\frac{1}{2}$ c of chopped scallions, and $3\frac{1}{4}$ c of beans. If one ingredient is altered as specified, how must the other ingredients be changed to keep the proportions the same? Explain your reasoning.
   a. 2 c of tomato sauce
   b. 1 c of chopped scallions
   c. $1\frac{3}{4}$ c of beans

19. The electrical resistance of a wire, measured in ohms (Ω), is proportional to the length of the wire. If the electrical resistance of a 5-ft wire is 4.2 Ω, what is the resistance of 18 ft of the same wire?

20. In a photograph of a father and his daughter, the daughter's height is 2.3 cm and the father's height is 5.8 cm. If the father is actually 188 cm tall, how tall is the daughter?

21. The amount of gold in jewelry and other products is measured in karats (K), where 24K represents pure gold. The mark 14K on a chain indicates that the ratio between the mass of the gold in the chain and the mass of the chain is 14 : 24. If a gold ring is marked 18K and it weighs 0.4 oz, what is the value of the gold in the ring if pure gold is valued at $300 per oz?

22. If Amber is paid $8.00 per hour for typing, the table shows how much she earns.

Hours (h)	1	2	3	4	5
Wages (w)	8	16	24	32	40

   a. How much did Amber make for a 40-hr work week?
   b. What is the constant of proportionality?

23. **a.** In Room A there are 1 man and 2 women; in Room B there are 2 men and 4 women; and in Room C there are 5 men and 10 women. If all the people in Rooms B and C go to Room A, what will be the ratio of men to women in Room A?
   b. Prove the following generalization of the proportions used in (a):
$$\text{If } \frac{a}{b} = \frac{c}{d} = \frac{e}{f}, \text{ then } \frac{a}{b} = \frac{c}{d} = \frac{e}{f} = \frac{a + c + e}{b + d + f}$$

Assessment 8-1B

1. Answer the following regarding the letters in the word *Mississippi*.
   a. Determine the ratio of vowels to consonants.
   b. What is the ratio of consonants to vowels?
   c. What is the ratio of consonants to letters in the word?

2. Solve for x in each of the following proportions:
   a. $\dfrac{5}{x} = \dfrac{30}{42}$ b. $\dfrac{x}{8} = \dfrac{^{-}12}{32}$
   c. $\dfrac{7}{8} = \dfrac{3x}{48}$ d. $3\dfrac{1}{2}$ is to 8 as x is to 24

3. There are 5 adult drivers to each teenage driver in Aluossim. If there are 12,345 adult drivers in Aluossim, how many teenage drivers are there?

4. A candle is 30 in. long. After burning for 12 min, the candle is 25 in. long. How long will it take for the whole candle to burn at the same rate?

5. A rectangular yard has a width-to-length ratio of $5:9$. If the distance around the yard is 2800 ft, what are the dimensions of the yard?

6. A grasshopper can jump 20 times its length. If jumping ability in humans (height) were proportional to a grasshopper's (length), how far could a 6-ft-tall person jump?

7. Jim found out that after working for 9 months he had earned 6 days of vacation time. How many days per year does he earn at this rate?

8. At Rattlesnake School the teacher–student ratio is $1:30$. If the school has 1200 students, how many additional teachers must be hired to reduce the ratio to $1:20$?

9. At a particular time, the ratio of the height of an object that is perpendicular to the ground to the length of its shadow is the same for all objects. If a 30-ft tree casts a shadow of 12 ft, how tall is a tree that casts a shadow of 14 ft?

10. The following table shows several possible widths W and corresponding lengths L of a rectangle whose area is $10\,\text{ft}^2$.

Width (W) (Feet)	Length (L) (Feet)	Area (Square Feet)
0.5	20	$0.5 \cdot 20 = 10$
1	10	$1 \cdot 10 = 10$
2	5	$2 \cdot 5 = 10$
2.5	4	$2.5 \cdot 4 = 10$
4	2.5	$4 \cdot 2.5 = 10$
5	2	$5 \cdot 2 = 10$
10	1	$10 \cdot 1 = 10$
20	0.5	$20 \cdot 0.5 = 10$

L
$\boxed{\text{Area} = 10\,\text{ft}^2}\;W$

   a. Use the values in the table and some additional values to graph the length L on the vertical axis versus the width W on the horizontal axis.

   b. What is the algebraic relationship between L and W?
   c. Write W as a function of L; that is, express W in terms of L.
   d. Write L as a function of W; that is, express L in terms of W.

11. Find three sets of x- and y-values for the following:

$$\frac{4 \text{ tickets}}{\$20} = \frac{x \text{ tickets}}{\$y}$$

12. If rent is $850 for each 2 weeks, how much is the rent for 7 weeks?

13. Leonardo da Vinci in his drawing *Vitruvian Man* showed that the man's armspan was equal to the man's height. Some other ratios are listed below.

$$\frac{\text{Length of hand}}{\text{Length of foot}} = \frac{7}{9}$$

$$\frac{\text{Distance from elbow to end of hand}}{\text{Distance from shoulder to elbow}} = \frac{8}{5}$$

$$\frac{\text{Length of hand}}{\text{Length of big toe}} = \frac{14}{3}$$

   Using the ratios above, answer the following:
   a. If the length of a big toe is 6 cm, how long should the hand be?
   b. If a hand is 21 cm, how long is the foot?
   c. If the distance from the elbow to the end of the hand is 20 in., what is the distance from the shoulder to the elbow?

14. On a city map, a rectangular park has a length of 4 in. If the actual length and width of the park are 500 ft and 300 ft, respectively, how wide is the park on the map?

15. Jim's car will travel 240 mi on 15 gal of gas. How far can he expect to go on 3 gal of gas?

16. Some model railroads use an O scale in their replicas of actual trains. The O scale uses the ratio 1 in./48 in. How many feet long is the actual locomotive if an O scale replica is 18 in. long?

17. a. On an American flag, what is the ratio of stars to stripes?
    b. What is the ratio of stripes to stars?

18. On an American flag, the ratio of the length of the flag to its width must be $19:10$.
    a. If a flag is to be $9\frac{1}{2}$ ft long, how wide should it be?
    b. The flag that was placed on the Moon measured 5 ft by 3 ft. Does this ratio form a proportion with the official length-to-width ratio? Why?

19. If $\dfrac{x}{y} = \dfrac{a}{b}$ is true, what other proportions do you know are true?

20. If a certain recipe takes $1\frac{1}{2}$ c of flour and 2 c of milk, how much milk should be added if the cook only has 1 c of flour?

★21. Prove that if $\frac{a}{b} = \frac{c}{d}$, $a \neq {}^{-}b$, and $a \neq b$, then the following are true:

a. $\dfrac{a + b}{b} = \dfrac{c + d}{d}$ $\left(Hint: \dfrac{a}{b} + 1 = \dfrac{c}{d} + 1\right)$

b. $\dfrac{a}{a + b} = \dfrac{c}{c + d}$ c. $\dfrac{a - b}{a + b} = \dfrac{c - d}{c + d}$

Mathematical Connections 8-1

Comunication

1. Iris has found some dinosaur bones and a fossil footprint. The length of the footprint is 40 cm, the length of the thigh bone is 100 cm, and the length of the body is 700 cm.
   a. What is the ratio of the footprint's length to the dinosaur's length?
   b. Iris found a new track that she believes was made by the same species of dinosaur. If the footprint was 30 cm long and if the same ratio of foot length to body length holds, how long is the dinosaur?
   c. In the same area, Iris also found a 50-cm thigh bone. Do you think this thigh bone belonged to the same dinosaur that made the 30-cm footprint that Iris found? Why or why not?

2. Suppose a 10-in. pizza costs $4. For you to find the price x of a 14-in. pizza, is it correct to set up the proportion $\dfrac{x}{4} = \dfrac{14}{10}$? (Assume the ratios of areas remain the same.) Why or why not?

3. When you use a proportion to solve a scale-drawing problem, can the proportion be set up in more than one way? Explain.

4. Nell said she can tell just by looking at the ratios $15 : 7$ and $15 : 8$ that these do not form a proportion. Is she correct? Why?

5. Sol had photographs that were 4 in. by 6 in., 5 in. by 7 in., and 8 in. by 10 in. Do the dimensions vary proportionately? Explain why.

6. Can $\dfrac{a}{b}$ and $\dfrac{a + b}{b}$ ever form a proportion? Why?

★7. In a condo complex, $\dfrac{2}{3}$ of the men were married to $\dfrac{3}{4}$ of the women. What is the ratio of married people to the total adult population of the condo complex? Explain how you can obtain this ratio without knowing the actual number of men or women.

Open-Ended

8. Write a paragraph in which you use the terms *ratio* and *proportion* correctly.

9. List three real-word situations that involve ratio and proportion.

10. Find examples of ratios in a newspaper.

11. Boyle's Law states that at a given temperature, the product of the volume V of a gas and the pressure P is a constant c, as follows:

$$PV = c$$

   a. If at a given temperature, a pressure of 48 lb/in.^{2} compresses a certain gas to a volume of 960 in.^{3}, what pressure would be necessary to compress the gas to a volume of 800 in.^{3} at the same temperature?
   b. Find three other real-world situations in which the variables are related mathematically like the variables in Boyle's Law. In each case, describe how the variables are related using ratio and proportion.

12. Research the Golden Ratio that the Greeks used in the design of the Parthenon. Write a report on this ratio and include a drawing of a golden rectangle.

Cooperative Learning

13. In *Gulliver's Travels* by Jonathan Swift we find the following:

   The seamstresses took my measure as I lay on the ground, one standing at my neck and another at mid-leg, with a strong cord extended, that each held by the end, while the third measured the length of the cord with a rule of an inch long. Then they measured my right thumb and desired no more; for by a mathematical computation, that twice around the thumb is once around the wrist, and so on to the neck and the waist; and with the help of my old shirt, which I displayed on the ground before them for a pattern, they fitted me exactly.

   a. Explore the measurements of those in your group to see if you believe the ratios mentioned for Gulliver.
   b. Suppose the distance around a person's thumb is 9 cm. What is the distance around the person's neck?
   c. What ratio could be used to compare a person's height to his or her double armspan? Does this ratio have anything to do with da Vinci's famous painting *Vitruvian Man*?
   d. Do you think there is a ratio between foot length and height? If so, what might it be?
   e. Estimate other body ratios and then see how close you are to actual measurements.

Questions from the Classroom

14. Mary is working with measurements and writes the following proportion:

$$12 \text{ in.}/1 \text{ ft} = 5 \text{ ft}/60 \text{ in.}$$

How would you help her?

15. Nora said she can use division to decide whether two ratios form a proportion; for example, $32:8$ and $40:10$ form a proportion because $32 \div 8 = 4$ and $40 \div 10 = 4$. Is she correct? Why?

16. Joe reported that if $\frac{a}{b}$ is a proper fraction and $\frac{a}{b} = \frac{c}{d}$, then $\frac{d}{c-d} = \frac{b}{a-b}$. Joan said she didn't believe it. How do you respond?

17. Al is 5 ft tall and has a shadow that is 18 in. long. At the same time, a tree has a shadow that is 15 ft long. Al sets up and solves the proportion as follows:

$$\frac{5 \text{ ft}}{15 \text{ ft}} = \frac{18 \text{ in.}}{x \text{ in.}}, \quad \text{so } x = 54 \text{ in.}$$

How would you help him?

18. Amy's friend told her the ratio of girls to boys in her new class is $5:6$. Amy was very surprised to think her class would have only 11 students. What do you tell her?

Third International Mathematics and Science Study (TIMSS) Questions

For every soft drink bottle that Fred collected, Maria collected 3. Fred collected a total of 9 soft drink bottles. How many did Maria collect?

a. 3   **b.** 12   **c.** 13   **d.** 27

TIMSS 2003, Grade 4

Three brothers, Bob, Dan, and Mark, receive a gift of 45,000 zeds from their father. The money is shared between the brothers in proportion to the number of children each one has. Bob has 2 children, Dan has 3 children, and Mark has 4 children. How many zeds does Mark get?

a. 5,000
b. 10,000
c. 15,000
d. 20,000

TIMSS 2003, Grade 8

National Assessment of Educational Progress (NAEP) Question

Sarah has a part-time job at Better Burgers restaurant and is paid $5.50 for each hour she works. She has made the chart below to reflect her earnings but needs your help to complete it.

a. Fill in the missing entries in the chart.

Hours Worked	Money Earned (in dollars)
1	$5.50
4	
	$38.50
$7\frac{3}{4}$	$42.63

b. If Sarah works h hours, then in terms of h, how much will she earn?

NAEP, 2007, Grade 8

BRAIN TEASER Manday read that the arm of the Statue of Liberty is 42 ft long. She would like to know how long the Statue of Liberty's nose is. How would you advise her to proceed?

8-2   Percents

Percents are very useful in conveying information. People hear that there is a 60% chance of rain or that their savings account is drawing 6% annual interest. Percents are special kinds of fractions, namely, fractions with a denominator of 100. The word **percent** comes from the Latin phrase *per centum*, which means *per hundred*. The root *cent* appears in words such as century (100 years), centimeter $\left(\frac{1}{100} \text{ m}\right)$, and centipede (100 legs). A bank that pays 6% annual simple interest on a savings account pays $6 for each $100 in the account for one year; that is, it pays 6/100 of whatever amount is in the account for 1 year. The symbol % indicates percent; 6% means 6 for each 100. Hence, to find 6% of $400, we determine how many hundreds are in 400. There are 4 hundreds in 400, so 6% of 400 is $6 \cdot 4 = 24$. Therefore, 6% of $400 = $24.

E-Manipulatives

For more help with percents using a hundreds grid, see the *Percents* module.

> **Definition of Percent**
>
> $$n\% = \frac{n}{100}$$

Figure 8-4

Thus, $n\%$ of a quantity is $\frac{n}{100}$ of the quantity. Therefore, 1% is one hundredth of a whole and 100% represents the entire quantity; whereas 200% represents $\frac{200}{100}$, or 2 times, the given quantity. Percents can be illustrated by using a hundreds grid. For example, what percent of the grid is shaded in Figure 8-4? Because 30 out of the 100, or $\frac{30}{100}$, of the squares are shaded, we say that 30% of the grid is shaded (or similarly, 70% of the grid is not shaded).

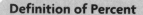

NOW TRY THIS 8-2 Write the fraction in lowest terms and the percent that represents the shaded portion in each part in Figure 8-5.

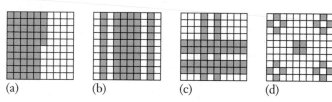

    (a)   (b)   (c)   (d)

Figure 8-5

Because $n\% = \frac{n}{100}$, one way to convert a number to a percent is to write it as a fraction with denominator 100; the numerator gives the amount of the percent. For example, $\frac{3}{4} = \frac{3 \cdot 25}{4 \cdot 25} = \frac{75}{100}$. Hence, $\frac{3}{4} = 75\%$.

Example 8-5

Write each of the following as a percent:

a. 0.03   **b.** $0.\overline{3}$ **c.** 1.2   **d.** 0.00042

e. 1   **f.** $\frac{3}{5}$ **g.** $\frac{2}{3}$ **h.** $2\frac{1}{7}$

Solution **a.** $0.03 = 100\left(\frac{0.03}{100}\right) = \frac{3}{100} = 3\%$

b. $0.\overline{3} = 100\left(\frac{0.\overline{3}}{100}\right) = \frac{33.\overline{3}}{100} = 33.\overline{3}\%$

c. $1.2 = 100\left(\frac{1.2}{100}\right) = \frac{120}{100} = 120\%$

d. $0.00042 = 100\left(\dfrac{0.00042}{100}\right) = \dfrac{0.042}{100} = 0.042\%$

e. $1 = 100\left(\dfrac{1}{100}\right) = \dfrac{100}{100} = 100\%$

f. $\dfrac{3}{5} = 100\left[\dfrac{\left(\dfrac{3}{5}\right)}{100}\right] = \dfrac{60}{100} = 60\%$

g. $\dfrac{2}{3} = 100\left[\dfrac{\left(\dfrac{2}{3}\right)}{100}\right] = \dfrac{\left(\dfrac{200}{3}\right)}{100} = \dfrac{66.\overline{6}}{100} = 66.\overline{6}\%$

h. $2\dfrac{1}{7} = 100\left[\dfrac{\left(2\dfrac{1}{7}\right)}{100}\right] = \dfrac{\left(\dfrac{1500}{7}\right)}{100} = \dfrac{1500}{7}\%,\ \text{or}\ 214\dfrac{2}{7}\%$

A number can also be converted to a percent by using a *proportion*. For example, to write $\dfrac{3}{5}$ as a percent, find the value of n in the following proportion:

$$\frac{3}{5} = \frac{n}{100}$$

$$\left(\frac{3}{5}\right)100 = n$$

$$n = 60$$

Therefore,

$$\frac{3}{5} = 60\%$$

Still another way to convert a number to a percent is to recall that $1 = 100\%$. Thus, for example, $\dfrac{3}{4} = \dfrac{3}{4}$ of $1 = \dfrac{3}{4} \cdot 1 = \dfrac{3}{4} \cdot 100\% = 75\%$.

REMARK The % symbol is crucial in identifying the meaning of a number. For example, $\dfrac{1}{2}$ and $\dfrac{1}{2}\%$ are different numbers: $\dfrac{1}{2} = 50\%$, which is not equal to $\dfrac{1}{2}\%$. Similary, 0.01 is different from 0.01%, which is 0.0001.

In the following student page on finding percents, an algorthmic method for converting a proportion to a percentage is seen in the example. Complete the example.

In the *Principles and Standards*, we find the following:

> As with fractions and decimals, conceptual difficulties need to be carefully addressed in instruction. In particular, percents less than 1% and greater than 100% are often challenging. (p. 217)

It is sometimes useful to convert percents to decimals. This can be done by writing the percent as a fraction in the form $\dfrac{n}{100}$ and then converting the fraction to a decimal.

School Book Page FINDING PERCENTS

GOAL

LEARN HOW TO...
- find percents

AS YOU...
- find Audience Approval ratings

Exploration 2

Finding Percents

SET UP *You will need the frequency table from Question 2 in the Setting the Stage.*

12 **Try This as a Class** Suppose a class gave two movies $3\frac{1}{2}$ or more stars as shown in the ratios below.

Movie A	**Movie B**
$\frac{19}{30}$	$\frac{15}{24}$

a. Estimate the Audience Approval rating for each movie. Explain the reasoning you used.

b. Using estimation, can you determine which movie had a higher Audience Approval rating? Explain.

▶ **A percent bar model can help you see how to set up a proportion to find the exact percent equivalent for a ratio.**

EXAMPLE

Set up a proportion to find the Audience Approval rating for Movie A, represented by the ratio $\frac{19}{30}$.

SAMPLE RESPONSE

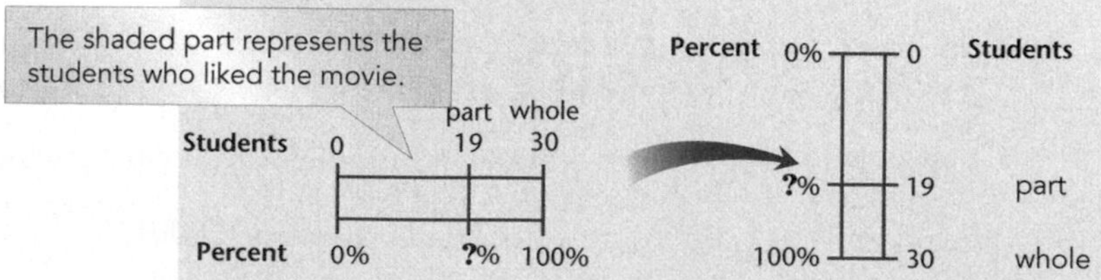

The shaded part represents the students who liked the movie.
Students
0
part 19
whole 30
Percent
0%
?%
100%
Percent
0%
0
Students
?%
19
part
100%
30
whole

Rotate the bar 90° so the parts are over the wholes.

Proportion

$$\text{part} \rightarrow \frac{x}{100} = \frac{19}{30} \leftarrow \text{part}$$
$$\text{whole} \rightarrow \qquad \leftarrow \text{whole}$$

Source: Math Thematics, Book 2 (Grade 7), McDougal Littell, 2008, (p. 362).

Example 8-6

Write each of the following percents as a decimal:

a. 5%   **b.** 6.3%   **c.** 100%

d. 250%   **e.** $\frac{2}{3}$%   **f.** $33\frac{1}{3}$%

Solution   **a.** $5\% = \frac{5}{100} = 0.05$ **b.** $6.3\% = \frac{6.3}{100} = 0.063$

c. $100\% = \frac{100}{100} = 1$ **d.** $250\% = \frac{250}{100} = 2.5$

e. $\frac{2}{3}\% = \frac{\frac{2}{3}}{100} = \frac{0.\overline{6}}{100} = 0.00\overline{6}$ **f.** $33\frac{1}{3}\% = \frac{33\frac{1}{3}}{100} = \frac{33.\overline{3}}{100} = 0.3\overline{3}$

Another approach to writing a percent as a decimal is first to convert 1% to a decimal. Because $1\% = \frac{1}{100} = 0.01$, we conclude that $5\% = 5 \cdot 0.01 = 0.05$ and that $6.3\% = 6.3 \cdot 0.01 = 0.063$.

NOW TRY THIS 8-3

a. Investigate how your calculator handles percents and tell what the calculator does when the % key is pushed.

b. Use your calculator to change $\frac{1}{3}$ to a percent.

Applications Involving Percent

In the grade 7 *Focal Points* we find the following:

> They (students) use ratio and proportionality to solve a wide variety of percent problems, including problems involving discounts, interest, taxes, tips, and percent increase or decrease. (p. 19)

Application problems that involve percents usually take one of the following forms:

1. Finding a percent of a number
2. Finding what percent one number is of another
3. Finding a number when a percent of that number is known

Before we consider examples illustrating these forms, recall what it means to find a fraction "of" a number. For example, $\frac{2}{3}$ of 70 means $\frac{2}{3} \cdot 70$. Similarly, to find 40% of 70, we have $\frac{40}{100}$ of 70, which means $\frac{40}{100} \cdot 70$, or $0.40 \cdot 70 = 28$.

A different way to think about 40% of 70 is to consider that 70 represents 100 parts (or the whole) and 40% requires only 40 of those 100 parts. For example, if

$$100 \text{ parts } = 70$$

$$1 \text{ part } = \left(\frac{70}{100}\right), \text{ or } 0.7$$

$$40 \text{ parts } = 40(0.7), \text{ or } 28$$

Thus, 40% of 70 = 28.

Research Note

In a study of grades 4, 6, and 8 Japanese students, a low percentage of students were able to correctly answer "What is 100% of 48?" Further research was called for to determine if this performance was representative of Japanese students' knowledge of percent. None of the students made any comments to suggest any conceptual links, connections, or similarities between the fraction and decimal computations (Reys et al. 1995).   ◆

As noted in the Research Note, Japanese students had trouble computing 100% of 48. The percent bar introduced on the student page on page 492 can be used as a model for understanding what 100% of a number means as well as understanding other percents. In Figure 8-6, consider the percent bar that represents 100% of the whole with 40% of the whole shaded. Note that 100% of the bar represents 70.

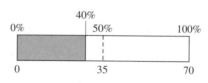

Figure 8-6

Table 8-4

0%	0
10%	
20%	
30%	
40%	?
50%	35
60%	
70%	
80%	
90%	
100%	70

Also, half of the percent bar (50% denoted by the dotted segment) represents half of 70, or 35. Thus, we know that 40% of the bar (or 40% of 70) is less than 35. In fact, if the top of the bar is thought of as being marked off in 1% intervals, there are 100 intervals marking whole numbers of percentages. If at the same time the bottom of the bar is considered to be marked in intervals of 1, there would be only 70 intervals marked at the bottom. Where would you expect the two sets of intervals to align?

Suppose that we know that as in Table 8-4, 0% corresponds to 0; 50% corresponds to 35; and 100% corresponds to 70. What percentages correspond to 10%, 20%, 30%, and so on? If there are 100 intervals marking percentages compared to only 70 intervals marking the corresponding length, there must be a ratio of $\frac{100}{70}$, or $\frac{10}{7}$. Thus, 10% should compare to 7; 20% to $2 \cdot 7$, or 14; and so on. Hence, 40% corresponds to $4 \cdot 7$, or 28.

Percents can be greater than 100%. For example, if your resting heart rate is considered the base unit, then this would be 100% of your resting heart rate. Increasing your heart rate would lead to a rate that is greater than 100%. Work through the following partial student page to investigate percents greater than 100%.

School Book Page

▶ **Percents Greater than 100%** Suppose Jane's resting heart rate is 60 beats per min and her active heart rate is 96 beats per min. One way to compare these two rates is to ask: "Jane's active heart rate is what percent of her resting heart rate?"

One way to visualize the comparison between Jane's active heart rate and her resting rate is with a percent bar model like the one in the Example below.

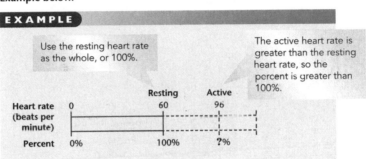

EXAMPLE

Use the resting heart rate as the whole, or 100%.

The active heart rate is greater than the resting heart rate, so the percent is greater than 100%.

| Heart rate (beats per minute) | 0 | | Resting 60 | Active 96 |
| Percent | 0% | | 100% | ?% |

18 a. How many heart beats is 50% of 60?

    b. How many heart beats is 100% of 60?

19 a. 150% = 100% + 50%. How many heartbeats is 150% of 60?

    b. How many heartbeats is 200% of 60?

20 Use the percent bar model above and your answers to Question 19 to estimate what percent Jane's active heart rate is of her resting heart rate.

✔ **QUESTION 21**

...checks that you can estimate percents greater than 100%.

21 ✔ **CHECKPOINT** Maura's resting heart rate is 80 beats per min and her active heart rate is 140 beats per min. Estimate what percent Maura's active heart rate is of her resting heart rate.

 Module 5 Recreation

Source: Math Thematics, New Edition, Book 2, McDougal Littell, 2008 (p. 378).

Example 8-7

A house that sells for $92,000 requires a 20% down payment. What is the amount of the down payment?

Solution The down payment is 20% of $92,000, or $0.20 \cdot \$92,000 = \$18,400$. Hence, the amount of the down payment is $18,400.

Example 8-8

If Alberto has 45 correct answers on an 80-question test, what percent of his answers are correct?

Solution   Alberto has $\dfrac{45}{80}$ of the answers correct. To find the percent of correct answers, we need to convert $\dfrac{45}{80}$ to a percent. We can do this by multiplying the fraction by 100 and attaching the % symbol, as follows:

$$\frac{45}{80} = 100 \cdot \frac{45}{80}\%$$
$$= 56.25\%$$

Thus, 56.25% of the answers are correct.

An alternative solution uses proportion. Let n be the percent of correct answers and proceed as follows:

$$\frac{45}{80} = \frac{n}{100}$$
$$\frac{45}{80} \cdot 100 = n$$
$$n = \frac{4500}{80} = 56.25$$

Thus, 56.25% of the answers are correct.

Example 8-9

Forty-two percent of the parents of the schoolchildren in the Paxson School District are employed at Di Paloma University. If the number of parents employed by the university is 168, how many parents are in the school district?

Solution   Let n be the number of parents in the school district. Then 42% of n is 168. We *translate this information into an equation and solve for n.*

$$42\% \text{ of } n = 168$$
$$\frac{42}{100}n = 168$$
$$0.42n = 168$$
$$n = \frac{168}{0.42} = 400$$

There are 400 parents in the school district.

Example 8-9 can be solved using a proportion. Forty-two percent, or $\dfrac{42}{100}$, of the parents are employed at the university. If n is the total number of parents, then $\dfrac{168}{n}$ also represents the fraction of parents employed there. Thus,

$$\frac{42}{100} = \frac{168}{n}$$
$$42n = 100 \cdot 168$$
$$n = \frac{16{,}800}{42} = 400$$

We can also solve the problem as follows:

$$42\% \text{ of } n \text{ is } 168$$

$$1\% \text{ of } n \text{ is } \frac{168}{42}$$

$$100\% \text{ of } n \text{ is } 100\left(\frac{168}{42}\right)$$

Therefore,

$$n = 100\left(\frac{168}{42}\right), \text{ or } 400$$

Example 8-10

Kelly bought a bicycle and a year later sold it for 20% less than what she paid for it. If she sold the bike for $144, what did she pay for it?

Solution We are looking for the original price, P, that Kelly paid for the bike. We know that she sold the bike for $144 and that this included a 20% loss. Thus, we can *write the following equation:*

$$\$144 = P - \text{ Kelly's loss}$$

Because Kelly's loss is 20% of P, we proceed as follows:

$$\$144 = P - 20\% \cdot P$$
$$\$144 = P - 0.20P$$
$$\$144 = (1 - 0.20)P$$
$$\$144 = 0.80P$$
$$\frac{\$144}{0.80} = P$$
$$\$180 = P$$

Thus, she paid $180 for the bike.

Example 8-11

Westerner's Clothing Store advertised a suit for 10% off, for a savings of $15. Later, the manager marked the suit at 30% off the original price. What is the amount of the current discount?

Solution A 10% discount amounts to a $15 savings. We could find the amount of the current discount if we knew the original price P. Thus, finding the original price becomes our *subgoal.* Because 10% of P is $15, we have the following:

$$10\% \cdot P = \$15$$
$$0.10P = \$15$$
$$P = \$150$$

To find the current discount, we calculate 30% of $150. Because $0.03 \cdot \$150 = \$45$, the amount of the 30% discount is $45.

In the *Looking Back* stage of problem solving, we check the answer and look for other ways to solve the problem. A different approach leads to a more efficient solution and confirms the answer. If 10% of the price is $15, then 30% of the price is 3 times $15, or $45.

NOW TRY THIS 8-4 In the cartoon below, compute the percentage and number of slices for the portions with olives, plain, and with onions and green peppers.

Mental Math with Percents

Mental math may be helpful when working with percents. Two techniques follow:

1. *Using fraction equivalents*

   Knowing fraction equivalents for some percents can make some computations easier. Table 8-5 gives several fraction equivalents.

Table 8-5

Percent	25%	50%	75%	$33\frac{1}{3}$%	$66\frac{2}{3}$%	10%	1%
Fraction Equivalent	$\frac{1}{4}$	$\frac{1}{2}$	$\frac{3}{4}$	$\frac{1}{3}$	$\frac{2}{3}$	$\frac{1}{10}$	$\frac{1}{100}$

These equivalents can be used in such computations as the following:

$$50\% \text{ of } \$80 = \left(\frac{1}{2}\right)80 = \$40$$

$$66\frac{2}{3}\% \text{ of } 90 = \left(\frac{2}{3}\right)90 = 60$$

2. *Using a known percent*

   Frequently, we may not know a percent of something, but we know a close percent of it. For example, to find 55% of 62, we might do the following:

$$50\% \text{ of } 62 = \left(\frac{1}{2}\right)(62) = 31$$

$$5\% \text{ of } 62 = \left(\frac{1}{2}\right)(10\%)(62) = \left(\frac{1}{2}\right)(6.2) = 3.1$$

   Adding, we see that 55% of 62 is $31 + 3.1 = 34.1$.

Estimations with Percents

Estimations with percents can be used to determine whether answers are reasonable. Following are two examples:

1. To estimate 27% of 598, note that 27% of 598 is a little more than 25% of 598, but 25% of 598 is approximately the same as 25% of 600, or $\frac{1}{4}$ of 600, or 150. Here, we have adjusted 27% downward and 598 upward, so 150 should be a reasonable estimate. A better estimate might be obtained by estimating 30% of 600 and then subtracting 3% of 600 to obtain 27% of 600, giving $180 - 18$, or 162.

2. To estimate 148% of 500, note that 148% of 500 should be slightly less than 150% of 500. 150% of 500 is $1.5(500) = 750$. Thus, 148% of 500 should be a little less than 750.

Example 8-12

Laura wants to buy a blouse originally priced at $26.50 but now on sale at 40% off. She has $17 in her wallet and wonders if she has enough cash. How can she mentally find out? (Ignore the sales tax.)

Solution   It is easier to find 40% of $25 (versus $26.50) mentally. One way is to find 10% of $25, which is $2.50. Now, 40% is 4 times that much, that is, $4 \cdot \$2.50$, or $10. Thus, Laura estimates that the blouse will cost $26.50 − $10, or $16.50. Since the actual discount is greater than $10 (40% of 26.50 is greater than 40% of 25), Laura will have to pay less than $16.50 for the blouse and, hence, she has enough cash.

Sometimes it may not be clear which operations to perform with percent. The following example investigates this.

Example 8-13

Which of the following statements could be true and which are false? Explain your answers.

a. Leonardo got a 10% raise at the end of his first year on the job and a 10% raise after another year. His total raise was 20% of his original salary.

b. Jung and Dina paid 45% of their first department store bill of $620 and 48% of the second department store bill of $380. They paid 45% + 48% = 93% of the total bill of $1000.

c. Bill spent 25% of his salary on food and 40% on housing. Bill spent 25% + 40% = 65% of his salary on food and housing.

d. In Bordertown, 65% of the adult population works in town, 25% works across the border, and 15% is unemployed.

e. In Clean City, the fine for various polluting activities is a certain percentage of one's monthly income. The fine for smoking in public places is 40%, for driving a polluting car is 50%, and for littering is 30%. Mr. Schmutz committed all three polluting crimes in one day and paid a fine of 120% of his monthly income.

Solution   a. In applications, percent has meaning only when it represents part of a quantity. For example, 10% of a quantity plus another 10% of the same quantity is 20% of that quantity. In Leonardo's case, the first 10% raise was calculated based on his original salary and the second 10% raise was calculated on his new salary. Consequently, the percentages cannot be added, and the statement is false. He received a 21% raise.

b. The last statement does not make sense; 45% of one bill plus 48% of the other bill is not 93% of the total bill because the bills are different.

c. Because the percentages are of the same quantity, the statement could be true.

d. Because the percentages are of the same quantity, that is, the number of adults, we can add them: 65% + 25% + 15% = 105%. But 105% of the population accounts for more (5% more) than the town's population, which is impossible. Hence, the statement is false.

> **e.** Again, the percentages are of the same quantity; that is, the individual's monthly income. Hence, we can add them: 120% of one's monthly income is a stiff fine, but possible.

Assessment 8-2A

1. Express each of the following as a percent:
   a. 7.89 **b.** 193.1 **c.** $\frac{5}{6}$
   d. $\frac{1}{8}$ **e.** $\frac{5}{8}$ **f.** $\frac{4}{5}$

2. Convert each of the following percents to a decimal:
   a. 16% **b.** $\frac{1}{5}$%
   c. $13\frac{2}{3}$% **d.** $\frac{1}{3}$%

3. Fill in the following blanks to find other expressions for 4%:
   a. ____ for every 100 **b.** ____ for every 50
   c. 1 for every ____ **d.** 8 for every ____
   e. 0.5 for every ____

4. Different calculators compute percents in various ways. To investigate this, consider $5 \cdot 6\%$.
   a. If the following sequence of keys is pressed, is the correct answer of 0.3 displayed on your calculator?

   $$\boxed{5}\;\boxed{\times}\;\boxed{6}\;\boxed{\%}\;\boxed{=}$$

   b. Press $\boxed{6}\;\boxed{\%}\;\boxed{\times}\;\boxed{5}\;\boxed{=}$. Is the answer 0.3?

5. Answer each of the following:
   a. What is 6% of 34?
   b. 17 is what percent of 34?
   c. 18 is 30% of what number?
   d. What is 7% of 49?

6. **a.** Write a fraction representing 5% of x.
   b. If 10% of an amount is a, what is the amount in terms of a?

7. Marc had 84 boxes of candy to sell. He sold 75% of the boxes. How many did he sell?

8. Gail made $16,000 last year and received a 6% raise. How much does she make now?

9. Gail received a 7% raise last year. If her salary is now $27,285, what was her salary last year?

10. Joe sold 180 newspapers out of 200. Bill sold 85% of his 260 newspapers. Ron sold 212 newspapers, 80% of those he had.
    a. Who sold the most newspapers? How many?
    b. Who sold the greatest percentage of his newspapers? What percent?
    c. Who started with the greatest number of newspapers? How many?

11. If a dress that normally sells for $35 is on sale for $28, what is the "percent off"? (This could be called a *percent of decrease*, or a *discount*.)

12. Mort bought his house in 2008 for $159,000. It was recently appraised at $195,000. What is the approximate *percent of increase* in value to the nearest percent?

13. Xuan weighed 9 1b when he was born. At 6 mo, he weighed 18 1b. What was the percent of increase in Xuan's weight?

14. Sally bought a dress marked 20% off. If the regular price was $28.00, what was the sale price?

15. An airline ticket costs $320 without the tax. If the tax rate is 5%, what is the total bill for the airline ticket?

16. Bill got 52 correct answers on an 80-question test. What percent of the questions did he answer incorrectly?

17. A real-estate broker receives 4% of an $80,000 sale. How much does the broker receive?

18. A survey reported that $66\frac{2}{3}$% of 1800 employees favored a new insurance program. How many employees favored the new program?

19. Which represents the greater percent: $\frac{325}{500}$ or $\frac{600}{1000}$? How can you tell?

20. An advertisement reads that if you buy 10 items, you get 20% off your total purchase price. You need 8 items that cost $9.50 each.
    a. How much would 8 items cost? 10 items?
    b. Is it more economical to buy 8 items or 10 items?

21. Soda is advertised at 45¢ a can or $2.40 a six-pack. If 6 cans are to be purchased, what percent is saved by purchasing the six-pack?

22. John paid $330 for a new mountain bicycle to sell in his shop. He wants to price it so that he can offer a 10% discount and still make 20% of the price he paid for it. At what price should the bike be marked?

23. Solve each of the following using mental mathematics:
    a. 15% of $22 **b.** 20% of $120
    c. 5% of $38 **d.** 25% of $98

24. For people to be safe but still achieve a cardiovascular training effect, they should monitor their heart rates while exercising. The maximum heart rate can be approximated by subtracting your age from 220. You can achieve a safe training effect if you maintain your heart rate (beats per min.) between 60% and 80% of that number for at least 20 min 3 times a week.

a. Determine the range for your age.

b. At the top of a long hill, Jeannie slows her bike and takes her pulse. She counts 41 beats in 15 sec.
   (i) Express in decimal form the amount of time in seconds between successive beats.
   (ii) Express the amount in terms of minutes.

25. A crew consists of one apprentice, one journeyman, and one master carpenter. The crew receives a check for $4200 for a job they just finished. A journeyman makes 200% of what an apprentice makes, and a master makes 150% of what a journeyman makes. How much does each person in the crew earn?

26. a. In an incoming freshman class of 500 students, only 20 claimed to be math majors. What percent of the freshman class is this?
    b. When the survey was repeated the next year, 5% of nonmath majors had decided to switch and become math majors.
       (i) How many math majors are there now?
       (ii) What percent of the former freshman class do they represent?

27. Ms. Price has received a 10% raise in salary in each of the last 2 yr. If her annual salary this year is $100,000, what was her salary 2 yr ago, rounded to the nearest penny?

28. *USA Today* (2005) reported that the U.S. Congress was sent a $2.57 trillion budget for fiscal year 2006. It further reported that one would have to purchase a $100 item every second for 815 years to spend that much money.
    a. Decide whether or not you believe these reports agree with each other.

b. Assuming that exactly 815 years were required to spend the entire $2.57 trillion, what percentage of the money was spent each year?

29. If you wanted to spend 25% of your monthly salary on entertainment and 56% of the salary on rent, could those amounts be $500 and $950? Why or why not?

30. An organization has 100,000 members. A bylaw change can be made at the annual business meeting held once each year, and a bylaw change must be approved by a majority of those attending the meeting. The chair of the meeting cannot vote unless there is a tie vote but does count as an attendee at the meeting.
    a. With these rules, what is the minimum number required at the meeting to make a bylaw change?
    b. Based on your answer to part (a), what percentage of the membership can change the bylaws of the organization?

31. A tip in a restaurant has been typically figured at 15% of the total bill.
    a. If the bill is $30, what would be the typical tip?
    b. If the patron receiving the bill gave a tip that was half the bill, what is the percentage of the tip?
    c. If the patron receiving the bill gave a tip that was equal to the bill, what is the percentage of the tip?

32. Suppose the percent bar below shows the number of students in a school who do not favor dress codes. How many students are in the school?

| Number of students | 0 | 374 | |
| Percent | 0% 50% 68% | 100% |

Assessment 8-2B

1. Express each of the following as a percent:
   a. 0.032   b. 0.2   c. $\frac{3}{20}$
   d. $\frac{13}{8}$ e. $\frac{1}{6}$ f. $\frac{1}{40}$

2. Convert each of the following percents to a decimal:
   a. $4\frac{1}{2}\%$ b. $\frac{2}{7}\%$
   c. 125%   d. $\frac{1}{4}\%$

3. Fill in the following blanks to find other expressions for 5%:
   a. ____ for every 100
   b. ____ for every 50
   c. 1 for every ____
   d. 8 for every ____
   e. 0.5 for every ____

4. Answer each of the following:
   a. 63 is 30% of what number?
   b. What is 7% of 150?
   c. 61.5 is what percent of 20.5?
   d. 16 is 40% of what number?

5. A used car originally cost $1700. One year later, it was worth $1400. What is the percentage of depreciation?

6. On a certain day in Glacier Park, 728 eagles were counted. Five years later, 594 were counted. What was the percentage of decrease in the number of eagle-scounted?

7. What is the sale price of a softball if the regular price is $6.80 and there is a 25% discount?

8. If a $\frac{1}{4}$-c serving of Crunchies breakfast food has 0.5% of the minimum daily requirement of vitamin C, how many cups would you have to eat to obtain the minimum daily requirement of vitamin C?

9. **a.** How can an estimate of 10% of a number help you estimate 35% of the number?
   b. Mentally compute 35% of $8.00.

10. If 30 is 150% of a number, is the number greater than or less than 30? Why?

11. What is 40% of 50% of a number?

12. If you add 20% of a number to the number itself, what percent of the result would you have to subtract to get the original number back?

13. The price of a suit that sold for $200 was reduced by 25%. By what percent must the price of the suit be increased to bring the price back to $200?

14. The car Elsie bought 1 yr ago has depreciated by $1116.88, which is 12.13% of the price she paid for it. How much did she pay for the car, to the nearest cent?

15. Solve each of the following using mental mathematics:
   a. 15% of $42   **b.** 20% of $280
   c. 5% of $28   **d.** 25% of $84

16. If we build a 10 × 10 model with blocks, as shown in the following figure, and paint the entire model, what percent of the cubes will have each of the following?
   a. Four faces painted
   b. Three faces painted
   c. Two faces painted

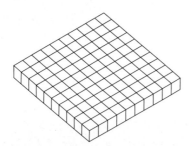

17. If 70% of the girls in a class wanted to have a prom and 40% of the boys wanted a prom, is it possible that only 50% of the students in the class wanted a prom? Explain your answer.

18. If 70% of the girls in a class wanted to have a prom and 60% of the boys wanted a prom, is it possible that only 50% of the students in the class wanted a prom? Explain your answer.

19. Draw a line segment that is 3 in. long and call it X. This segment represents 50% of another segment Y. Draw each of the following.
   a. A segment that represents 100% of segment Y.
   b. A segment that represents 25% of segment Y.
   c. A segment that represents 150% of segment Y.

20. Order these numbers from least to greatest.

   65%, 3/5, 0.70, 50%, 2/3, 0.55

21. **a.** It is recommended that no more than 30% of your calorie intake should be from fat. If you consumed about 2400 calories daily, what is the maximum amount of fat calories you should consume?
   b. If one cookie contains 140 calories and 70 calories in the cookie are fat calories, could you eat 3 cookies and not exceed the recommended amount of fat calories for the day?

22. If you buy a new bicycle for $380 and the sales tax is 9%, what is your total bill?

23. The number of known living species is about 1.7 million. About 4500 species are mammals. What percent of known living species are mammals?

24. There are 80 coins in a piggy bank of which 20% are quarters. What is the least possible amount of money that could be in the piggy bank?

25. A ski area reports 80% of its runs are open. If there are 60 runs open, how many runs are at the ski area?

26. A salesperson earns a weekly salary of $900 plus a commission rate of 4% on all sales. What did the person make for a week with total sales of $1800?

27. Jim bought two shirts that were originally marked at $40 each. One shirt was discounted 20% and the other was discounted 25%. The sales tax was 4.5%. How much did he spend in all?

28. If the sides of a square are increased by 80%, does this increase the area of the square by 80% ($A = s^2$)? If not, what happens to the area? Use percentages in your solution.

29. According to a *TV Guide* survey, 46% of people in the United States said they would not stop watching television for anything less than a million dollars. Use the percent bar for the U.S. population to estimate the number of people who would not stop watching television for anything less than a million dollars.

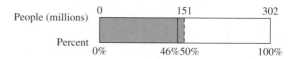

Mathematical Connections 8-2

Communication

1. Use mental math to find 11% of 850. Explain your method.
2. Does 0.4 = 0.4%? Explain.
3. What does it mean to reach 125% of your savings goal?
4. Is 4% of 98 the same as 98% of 4? Explain.
5. **a.** If 25% of a number is 55, is the number greater than or less than 55? Explain.
   b. If 150% of a number is 55, is the number greater than or less than 55? Explain.
6. Can 35% of one number be greater than 55% of another number? Explain.
7. Why does one picture have so much more shaded area when they both show 50%?

8. Why is it possible to have an increase of 150% in price but not a 150% decrease in price?
9. Two equal amounts of money were invested in two different stocks. The value of the first stock increased by 15% the first year and then decreased by 15% the second year. The second stock decreased by 15% the first year and increased by 15% the second year. Was one investment better than the other? Explain your reasoning.

Open-Ended

10. Write and solve a word problem whose solution involves the following. If one of these tasks is impossible, explain why.
    a. Addition of percent
    b. Subtraction of percent
    c. Multiplication of percent
    d. Division of percent
    e. A percent whose decimal representation is raised to the second power
    f. A percent greater than 100
11. Look at newspapers and magazines for information given in percents.
    a. Based on your findings, write a problem that involves social science as well as mathematics.
    b. Write a clear solution to your problem in (a).
12. Write a percentage problem whose answer is the solution of each of the following equations:
    a. $\dfrac{37}{100} = \dfrac{115}{x}$
    b. $\dfrac{p}{100} = \dfrac{a}{x}$

Cooperative Learning

13. Find the percentage of students in your class who engage in each of the following activities:

a. Studying and Doing Homework

Number of Hours per Week (h)	Percent
$h < 1$	
$1 < h < 3$	
$3 \leq h < 5$	
$5 \leq h < 10$	
$h \geq 10$	
Total	

b. Watching TV

Number of Hours per Week (h)	Percent
$h < 1$	
$1 \leq h < 5$	
$5 \leq h < 10$	
$h \geq 10$	
Total	

c. Did your totals add up to 100% in each table? Why or why not?

Questions from the Classroom

14. A student asks if 90% means 90 out of 100, how can she possibly score 90% on a test that has only 20 questions? How do you answer her?
15. A student says that $3\dfrac{1}{4}\% = 0.03 + 0.25 = 0.28$. Is this correct? Why?
16. A student reports that it is impossible to mark a product up 150% because 100% of something is all there is. What is your response?
17. A student argues that a $p\%$ increase in salary followed by a $q\%$ decrease is equivalent to a $q\%$ decrease followed by a $p\%$ increase because of the commutative property of multiplication. How do you respond?
18. A student argues that 0.01% = 0.01 because in 0.01%, the percent is already written as a decimal. How do you respond?

Review Problems

19. Jim's lawn mower engine needs 5 fl oz of oil mixed with every 2 gal of gasoline. A container has 12 gal of gasoline and Jim put in 34 fl oz of oil. Is this correct? Why?
20. A certain shade of green requires 4 parts blue to 5 parts yellow. If you have 25 qt of yellow, how much blue should be added?

21. Tell whether $4x/3$ and $12x/9$ *always*, *sometimes*, or *never* form a proportion. Explain.

22. Solve each proportion.

a. $\dfrac{x}{18} = \dfrac{21}{63}$ **b.** $\dfrac{27}{21} = \dfrac{36}{x}$

c. $\dfrac{15}{33} = \dfrac{x}{22}$ **d.** $\dfrac{19}{x} = \dfrac{152}{4}$

Third International Mathematics and Science Study (TIMSS) Questions

Experts say that 25% of all serious bicycle accidents involve head injuries and that, of all head injuries, 80% are fatal. What percentage of all serious bicycle accidents involve fatal head injuries?

a. 16%   **b.** 20%
c. 55%   **d.** 105%

TIMSS, Grade 8, 1995

Last year there were 1172 students at Beaton High School. This year there are 15 percent more students than last year. Approximately how many students are at Beaton High School this year?

a. 1800   **b.** 1600
c. 1500   **d.** 1400
e. 1200

TIMSS, Grade 8, 1995

National Assessment of Educational Progress (NAEP) Question

$4.95

$2.45

Sales Tax Table

Amount of Sales	Amount of Tax
$6.00	$0.36
6.20	0.37
6.40	0.38
6.60	0.40
6.80	0.41
7.00	0.42
7.20	0.43
7.40	0.44
7.60	0.46
7.80	0.47
8.00	0.48

Carlos bought the cereal and milk shown. Use the table to find out the total amount Carlos spent, including tax.
Total amount spent: _____
Show how you found your answer.

NAEP, Grade 4, 2007

BRAIN TEASER   The crust of a certain pumpkin pie is 25% of the pie. By what percent should the amount of crust be reduced in order to make it constitute 20% of the pie?

TECHNOLOGY CORNER   We can use spreadsheets to solve mixture problems. For example, consider the problem of finding out how many liters of water must be added to 5 L of pure lemon juice to change its concentration from 100% to less than 30% lemon juice.

Six lemonade mixtures were prepared starting from 5 L of pure lemon juice and adding water in 2 L increments. At each step, the percent of lemon juice in the mixture was calculated. The results of the process are summarized in the spreadsheet in Figure 8-7. The formulas used to obtain the results in a particular column are given in row 12.

a. Explain how this spreadsheet can be used to help students solve the problem.

b. Explain the formulas in row 12.

	A	B	C	D
	Liters of	Liters of	Total Liters	% Lemon Juice
1	Lemon Juice	Water Added	in Mixture	in Mixture
2	(L)	(L)	(L)	
3	5	0	5	100.00
4	5	2	7	71.43
5	5	4	9	55.56
6	5	6	11	45.45
7	5	8	13	38.46
8	5	10	15	33.33
9	5	12	17	29.41
10				
11	5	x	5+x	5/(5+x)*100
12		(where x is a		
13		multiple of 2)		

Sheet1   Sh

Ready

Figure 8-7

*8-3   Computing Interest

When a bank advertises a $5\frac{1}{2}$% interest rate on a savings account, the **interest** is the amount of money the bank will pay for using your money. The original amount deposited or borrowed is the **principal**. The percent used to determine the interest is the **interest rate**. Interest rates are given for specific periods of time, such as years, months, or days. Interest computed on the original principal is **simple interest**. For example, suppose we borrow $5000 from a company at an annual simple interest rate of 9% for 1 yr. The interest we owe on the loan for 1 yr is 9% of $5000, or $0.09 \cdot \$5000$. In general, if a principal, P, is invested at an annual interest rate of r, then the simple interest after 1 yr is $Pr \cdot 1$; after t years, it is Prt. Thus, if I represents simple interest, we have

$$I = Prt$$

The amount needed to pay off a $5000 loan at 9% annual simple interest for 1 yr is the $5000 borrowed plus the interest on the $5000; that is, $5000 + 5000 \cdot 0.09$, or $5450. In general, *an* **amount** (*or* **balance**) *A is equal to the principal P plus the interest I*; that is,

$$A = P + I = P + Prt = P(1 + rt)$$

Example 8-14

Vera opened a saving account that pays simple interest at the rate of $5\frac{1}{4}$% per year. If she deposits $2000 and makes no other deposits, find the interest and the final amount for the following time periods:

a. 1 yr   **b.** 90 days

Solution **a.** To find the interest for 1 yr, we proceed as follows:

$$I = \$2000 \cdot 5\frac{1}{4}\% \cdot 1 = \$2000 \cdot 0.0525 \cdot 1 = \$105$$

Her final amount at the end of 1 yr is

$$\$2000 + \$105 = \$2105$$

b. When the interest rate is annual and the interest period is given in days, we represent the time as a fractional part of a year by dividing the number of days by 365. Thus,

$$I = \$2000 \cdot 5\frac{1}{4}\% \cdot \frac{90}{365}$$

$$= \$2000 \cdot 0.0525 \cdot \frac{90}{365} \doteq \$25.89$$

Hence,

$$A \doteq \$2000 + \$25.89$$
$$A \doteq \$2025.89$$

Thus, Vera's amount after 90 days is approximately \$2025.89.

Example 8-15

Find the annual interest rate if a principal of \$10,000 increased to \$10,900 at the end of 1 yr.

Solution   Let the annual interest rate be $x\%$. We know that $x\%$ of \$10,000 is the increase. Because the increase is \$10,900 − \$10,000 = \$900, we use the strategy of *writing an equation* for x as follows:

$$x\% \text{ of } 10{,}000 = 900$$

$$\frac{x}{100} \cdot 10{,}000 = 900$$

$$x = 9$$

Thus, the annual interest rate is 9%. We can also solve this problem mentally by asking, "What percent of 10,000 is 900?" Because 1% of 10,000 is 100, to obtain 900, we take 9% of 10,000.

Compound Interest

In business transactions, interest is sometimes calculated daily (365 times a year). In the case of savings, the earned interest is added daily to the principal, and each day the interest is earned on a different amount; that is, it is earned on the previous interest as well as the principal. Interest earned in this way is **compound interest**. Compounding usually is done annually (once a year), semiannually (twice a year), quarterly (4 times a year), or monthly (12 times a year). However, even when the interest is compounded, it is given as an annual rate. For example, if the annual rate is 6% compounded monthly, the interest per month is $\frac{6}{12}\%$, or 0.5%. If it is compounded daily, the interest per day is $\frac{6}{365}\%$.

In general, *the interest rate per period is the annual interest rate divided by the number of periods in a year.*

We can use a spreadsheet to compare various compound interest rates. Work through the student page on page 508 and answer the questions in the *TRY IT* and the *ON YOUR OWN* sections.

If you invest $100 at 8% annual interest compounded quarterly, how much will you have in the account after 1 yr? The quarterly interest rate is $\frac{1}{4} \cdot 8\%$, or 2%. It seems that we would have to calculate the interest 4 times. But we can also reason as follows. If at the beginning of any of the four periods there are x dollars in the account, at the end of that period there will be

$$x + 2\% \text{ of } x = x + 0.02x$$
$$= x(1 + 0.02)$$
$$= x(1.02) \text{ dollars}$$

Hence, to find the amount at the end of any period, we need only multiply the amount at the beginning of the period by 1.02. From Table 8-6, we see that the amount at the end of the fourth period is $100 \cdot 1.02^4$. On a scientific calculator, we can find the amount using $\boxed{1}\ \boxed{0}\ \boxed{0}\ \boxed{\times}\ \boxed{1}\ \boxed{\cdot}\ \boxed{0}\ \boxed{2}\ \boxed{y^x}\ \boxed{4}\ \boxed{=}$. The calculator displays 108.24322. Thus, the amount at the end of 1 yr is approximately $108.24.

Table 8-6

Period	Initial Amount	Final Amount
1	100	$100 \cdot 1.02$
2	$100 \cdot 1.02$	$(100 \cdot 1.02)1.02$, or $100 \cdot 1.02^2$
3	$100 \cdot 1.02^2$	$(100 \cdot 1.02^2)1.02$, or $100 \cdot 1.02^3$
4	$100 \cdot 1.02^3$	$(100 \cdot 1.02^3)1.02$, or $100 \cdot 1.02^4$

Finding the final amount at the end of the nth period amounts to finding the nth term of a geometric sequence whose first term is $100 \cdot 1.02$ (amount at the end of the first period) and whose ratio is 1.02. Thus, the amount at the end of the nth period is given by $(100 \cdot 1.02)(1.02)^{n-1} = 100 \cdot 1.02^n$. We can generalize this discussion. If the principal is P and the interest rate per period is r, then the amount A after n periods is $P(1 + r)(1 + r)^{n-1}$, or $P(1 + r)^n$. Therefore, the *formula for computing the amount at the end of the nth period is*

$$A = P(1 + r)^n$$

Example 8-16

Suppose you deposit $1000 in a savings account that pays 6% annual interest compounded quarterly.

a. What is the balance at the end of 1 yr?
b. What is the *effective annual yield* on this investment; that is, what is the simple interest rate that would after 1 yr pay the same amount as the given compound interest rate?

Solution   a. An annual interest rate of 6% earns $\frac{1}{4}$ of 6%, or an interest rate of $\frac{0.06}{4}$, in 1 quarter. Because there are 4 periods, we have the following:

$$A = 1000\left(1 + \frac{0.06}{4}\right)^4 \doteq \$1061.36$$

The balance at the end of 1 yr is approximately $1061.36.

School Book Page COMPOUND INTEREST RATES

TECHNOLOGY

Using a Spreadsheet • Compound Interest

Problem: Which investment strategy will cause your $100 investment to increase the most in 4 years: if you earn 4% interest compounded annually or if you earn 3.75% interest compounded monthly?

A spreadsheet can help you find the answer to this problem.

❶ Enter the following information in your spreadsheet as shown.

	A	B	C	D	E	F
1	Year	Amount	Annual Rate	Month	Amount	Monthly Rate
2	0	100		0	100	

❷ Enter the following formulas.
In cell C2, enter =.04.
In cell F2, enter =.0375/12.
In cell A3, enter =A2+1.
In cell B3, enter =B2+B2*C$2.
In cell D3, enter =D2+1.
In cell E3, enter =E2+E2*F$2.

❸ Select cells A3 to F50 and use the **Fill Down** command.

	A	B	C	D	E	F
1	Year	Amount	Annual Rate	Month	Amount	Monthly Rate
2	0	100	0.04	0	100	0.003125
3	1	104		1	100.3125	

	A	B	C	D	E	F
1	Year	Amount	Annual Rate	Month	Amount	Monthly Rate
2	0	100	0.04	0	100	0.003125
3	1	104		1	100.3125	
4	2	108.16		2	100.6259	
5	3	112.4864		3	100.9404	
6	4	116.9858		4	101.2558	
50	48	657.0528		48	116.1562	

Solution: $100 at 4% interest compounded annually is $116.99; $100 at 3.75% interest compounded monthly is only $116.16.

TRY IT

Which investment strategy will cause your $100 investment to increase the most in 4 years: if you earn 5% interest compounded annually or if you earn 5% interest compounded quarterly (4 times a year)?

ON YOUR OWN

► Which is easier, computing compound interest by calculator or by using a spreadsheet? Explain.

► Why must you divide the interest rate by the number of compounding periods?

► Why do you have to enter formulas by using the "=" symbol?

312

Source: Scott Foresman-Addison Wesley Middle School Math, Course 3, 2002 (p. 312).

b. Because the interest earned is $1061.36 - $1000.00 = $61.36, the effective annual yield can be computed by using the simple interest formula, $I = Prt$.

$$61.36 = 1000 \cdot r \cdot 1$$

$$\frac{61.36}{1000} = r$$

$$0.06136 = r$$

$$6.136\% = r$$

The effective annual yield is 6.136%.

Example 8-17

To save for their child's college education, a couple deposits $3000 into an account that pays 7% annual interest compounded daily. Find the amount in this account after 8 yr.

Solution   The principal in the problem is $3000, the daily rate i is 0.07/365, and the number of compounding periods is $8 \cdot 365$, or 2920. Thus we have

$$A = \$3000\left(1 + \frac{0.07}{365}\right)^{2920} \doteq \$5251.74$$

Thus, the amount in the account is approximately $5251.74.

Assessment 8-3A

You will need a calculator to do most of the following problems.
1. Complete the following compound interest chart.

	Compounding Period	Principal	Annual Rate	Length of Time (Years)	Interest Rate per Period	Number of Periods	Amount of Interest Paid	Total Amount in Account
a.	Semiannual	$1000	6%	2				
b.	Quarterly	$1000	8%	3				
c.	Monthly	$1000	10%	5				
d.	Daily	$1000	12%	4				

2. Ms. Jackson borrowed $42,000 at 8.75% annual simple interest. If exactly 1 yr later she was able to repay the loan without penalty, how much interest would she owe?

3. Carolyn borrowed $125. If the interest rate is 1.5% per month on the unpaid balance and she does not pay this debt for 1 yr, how much interest will she owe at the end of the year?

4. Burger Queen will need $50,000 in 5 yr for a new addition. To meet this goal, the company deposits money in an account today that pays 3% annual interest compounded quarterly. Find the amount that should be invested to total $50,000 in 5 yr.

5. A company is expanding its line to include more products. To do so, it borrows $320,000 at 13.5% annual simple interest for a period of 18 mo. How much interest must the company pay?

6. To save for their retirement, a couple deposits $4000 in an account that pays 5.9% annual interest compounded quarterly. What will be the value of their investment after 20 yr?

7. A car company is offering car loans at a simple interest rate of 4.7%. Find the interest charged to a customer who finances a car loan of $7200 for 3 yr.

8. Johnny and Carolyn have three savings plans, which accumulated the following amounts of interest for 1 yr:

   (i) A passbook savings account that accumulated $53.90 on a principal of $980

   (ii) A certificate of deposit that accumulated $55.20 on a principal of $600

   (iii) A money market account that accumulated $158.40 on a principal of $1200

   Which of these accounts paid the best interest rate for the year?

9. A hamburger costs $1.35 and the price continues to rise at a rate of 11% a year for the next 6 yr. What will the price of a hamburger be at the end of 6 yr?

10. Adrien and Jarrell deposit $300 on January 1 in a holiday savings account that pays 1.1% per month interest. What is the effective annual yield?

11. An amount of $3000 was deposited in a bank at a rate of 2% annual interest compounded quarterly for 3 yr. The rate then increased to 3% annual interest and was compounded quarterly for the next 3 yr. If no money was withdrawn, what was the balance at the end of this time?

12. The New Age Savings Bank advertises 4% annual interest rates compounded daily, while the Pay More Bank pays 5.2% annual interest compounded annually. Which bank offers a better rate for a customer who plans to leave her money in for exactly 1 yr?

13. If a fixed simple interest rate is paid on a savings account for a period of several years and enough money is withdrawn from the account that allows the principal to remain fixed but no other money is withdrawn, what type of sequence does the money earned every year represent?

Assessment 8-3B

You will need a calculator to do most of the following problems.

1. Complete the following compound interest chart.

Compounding Period	Principal	Annual Rate	Length of Time (Years)	Interest Rate per Period	Number of Periods	Amount of Interest Paid	Total Amount in Account
a. Semiannual	$1000	4%	2				
b. Quarterly	$1000	6%	3				
c. Monthly	$1000	18%	5				
d. Daily	$1000	18%	4				

2. A man collected $28,500 on a loan of $25,000 he made 4 yr ago. If he charged simple interest, what was the rate he charged?

3. If college tuition is $10,000 this year, what will it be 10 yr from now, assuming a constant inflation rate of 9% a year?

4. Sara invested money at a bank that paid 3.5% annual interest compounded quarterly. If she had $4650 at the and of 4 yr, what was her initial investment?

5. The number of trees in a rain forest decreases each month by 0.5%. If the forest has approximately $2.34 \cdot 10^9$ trees, how many trees will be left after 20 yr?

6. A money market fund pays 14% annual interest compounded daily. What is the value of $10,000 invested in this fund after 15 yr?

7. A car is purchased for $15,000. If each year the car depreciates by 10% of its value the preceding year, what will its value be at the end of 3 yr?

8. Interest is compounded annually at 4% on a savings account for a period of n years and the interest remains in the account, the amount every year would represent what type of sequence?

9. If a publishing company signed an agreement to allow a textbook (1st edition with 500 pages) to expand over several editions to 1000 pages and the book was growing at approximately 10% in the number of pages over each edition, how many editions could be published before it reached the contractual limit?

10. Amy is charged 12% annual interest compounded monthly on the unpaid balance of a $2000 loan. She did

not make any payments for 2 yr. Her friend said the amount she owed had more than doubled. Is this correct? How much does she now owe?

11. If the price of a new car is expected to rise 2.5% per year, what would be the price of a new car in 5 yr if it now sells for $32,400?

12. Al invests $1000 at 6% annual interest compounded daily and Betty invests $1000 at 7% simple interest. After how many whole years will Al's investments be worth more than Betty's investment?

Mathematical Connections 8-3

Communication

1. Because of a recession, the value of a new house depreciated 10% each year for 3 yr in a row. Then, for the next 3 yr, the value of the house increased 10% each year. Did the value of the house increase or decrease after 6 yr? Explain.

2. Determine the number of years (to the nearest tenth) it would take for any amount of money to double if it were deposited at a 10% annual interest rate compounded annually. Explain your reasoning.

3. Each year a car's value depreciated 20% from the previous year. Mike claims that after 5 yr the car would depreciate 100% and would not be worth anything. Is Mike correct? Explain why or why not. If not, find the actual percent the car would depreciate after 5 yr.

Open-Ended

4. The effect of depreciation can be computed using a formula similar to the formula for compound interest.
   a. Assume depreciation is the same each month. Write a problem involving depreciation and solve it.
   b. Develop a general formula for depreciation defining what each variable in the formula stands for.

5. Find four large cities around the world and an approximate percentage rate of population growth for the cities. Estimate the population in each of the four cities in 25 yr.

6. State different situations that do not involve money in which a formula like the one for compound interest is used. In each case, state a related problem and write its solution.

Cooperative Learning

7. The federal Truth in Lending Act, passed in 1969, requires lending institutions to quote an annual percentage rate (APR) that helps consumers compare the true cost of loans regardless of how each lending institution computes the interest and adds on costs.
   a. Call different banks and ask for their APR on some loans and the meaning of APR.
   b. Based on your findings in (a), write a definition of APR.
   c. Use the information given by your credit card (you may need to call the bank) and compute the APR on cash advances. Is your answer the same as that given by the bank? Compare the APR for different credit cards.

Questions from the Classroom

8. Jen said she did not understand the difference between simple interest and compound interest. She said the higher the interest rate the bank pays, the more money she would get no matter what kind of interest it is. How would you respond?

9. Noel read that women make 75¢ for every dollar that men make. She says that this means that men are paid 25% more than women. Is she correct? Why?

10. A student claims that if the value of an item increases by 100% each year from its value the previous year and if the original price is d dollars, then the value after n years will be $d \cdot 2^n$ dollars. Is the student correct? Why or why not?

Hint for Solving the Preliminary Problem

Set up equations to find the worth of each car. Then compare the total worth of the two cars with the amount that he sold them for and answer the question.

Chapter Outline

I. Ratio and proportion

  A. A fraction $\frac{a}{b}$ is a **ratio**.

  B. A **proportion** is a statement that two given ratios are equal.

  C. If a, b, c, and d are all real numbers and $b \neq 0$ and $d \neq 0$, then $\frac{a}{b} = \frac{c}{d}$ if, and only if, $ad = bc$.

  D. If the variables x and y are related by the equality $y = kx, \left(k = \frac{y}{x} \right)$, then **$y$ is said to be proportional to x** and **k is the constant of proportionality**.

  E. Properties of proportions

    1. If $\frac{a}{b} = \frac{c}{d}$, then $\frac{b}{a} = \frac{d}{c}$, where $a \neq 0$ and $c \neq 0$.

    2. If $\frac{a}{b} = \frac{c}{d}$, then $\frac{a}{c} = \frac{b}{d}$, where $c \neq 0$.

II. Percent and interest

  A. **Percent** means *per hundred*. Percent is written using the % symbol: $x\% = \frac{x}{100}$.

  *__B.__ **Simple interest** is computed using the formula $I = Prt$, where I is the interest, P is the principal, r is the annual interest rate, and t is the time in years.

  *__C.__ **Compound interest** is computed using the formula $A = P(1 + r)^n$, where A is the balance, P is the principal, r is the interest rate per period, and n is the number of periods.

Chapter Review

1. Tom tossed a coin 30 times and got 17 heads.
  a. What is the ratio of heads to coin tosses?
  b. What is the ratio of heads to tails?
  c. What is the ratio of tails to heads?

2. Which bottle of juice is a better buy (cost per ounce): 48 fl oz for $3.05 or 64 fl oz for $3.60?

3. Eighteen-karat gold contains 18 parts (grams) gold and 6 parts (grams) other metals. Amy's new ring contains 12 parts gold and 3 parts other metals. Is the ring 18-karat gold? Why?

4. Solve for x in each of the following:

  a. $\dfrac{15}{12} = \dfrac{21}{x}$ **b.** $\dfrac{20}{35} = \dfrac{110}{x}$

  c. $\dfrac{\frac{1}{2}}{\frac{1}{3}} = \dfrac{\frac{3}{2}}{x}$

5. A recipe for fruit salad serves 4 people. It calls for 3 oranges and 16 grapes. How many oranges and grapes do you need to serve 11 people?

6. If the scale on a drawing of a house is 1 cm = 2.5 m, what is the length of the house if it measures 3 cm on the scale drawing?

7. In water (H_2O), the ratio of the weight of oxygen to the weight of hydrogen is approximately $8 : 1$. How many ounces of hydrogen are in 1 lb of water?

8. To estimate the number of fish in a lake, scientists use a tagging and recapturing technique. A number of fish are captured, tagged, and then released back into the lake. After a while, some fish are captured and the number of tagged fish is counted.

  Let T be the total number of fish captured, tagged, and released into the lake, n the number of fish in a recaptured sample, and t the number of fish found tagged in that sample. Finally, let x be the number of fish in the lake. The assumption is that the ratio between tagged fish and the total number of fish in any sample is approximately the same and hence scientists assume $\dfrac{t}{n} = \dfrac{T}{x}$. Suppose 173 fish were captured, tagged, and released. Then 68 fish were recaptured and among them 21 were found to be tagged. Estimate the number of fish in the lake.

9. A manufacturer produces the same kind of computer chip in two plants. In the first plant, the ratio of defective chips to good chips is $15 : 100$ and in the second plant, that ratio is $12 : 100$. A buyer of a large number of chips is aware that some come from the first plant and some from the second. However, she is not aware of how many come from each. The buyer would like to know the ratio of defective chips to good chips in any given order. Can she determine that ratio? If so, explain how. If not, explain why not.

10. Suppose the ratio of the lengths of the sides in two squares is $1 : r$. What is the ratio of their areas? ($A = s^2$.)

11. The Grizzlies won 18 games and lost 7.
    a. What is the ratio of games won to games lost?
    b. What is the ratio of games won to games played?

12. Express each of the following as a ratio $\dfrac{a}{b}$ where a and b are whole numbers:
    a. $\dfrac{1}{5} : 1$ **b.** $\dfrac{2}{5} : \dfrac{3}{4}$

13. The ratio of boys to girls in Mr. Good's class is 3 to 5, the ratio of boys to girls in Ms. Garcia's is the same, and you know that there are 15 girls in Ms. Garcia's class. How many boys are in Ms. Garcia's class?

14. Answer each of the following:
    a. 6 is what percent of 24?
    b. What is 320% of 60?
    c. 17 is 30% of what number?
    d. 0.2 is what percent of 1?

15. Change each of the following to a percent:
    a. $\dfrac{1}{8}$ **b.** $\dfrac{3}{40}$
    c. 6.27   **d.** 0.0123
    e. $\dfrac{3}{2}$

16. Change each of the following percents to a decimal:
    a. 60%   **b.** $\dfrac{2}{3}$%   **c.** 100%

17. Sandy received a dividend that equals 11% of the value of her investment. If her dividend was $1020.80, how much was her investment?

18. Five computers in a shipment of 150 were found to be defective. What percent of the computers were defective?

19. On a mathematics examination, a student missed 8 of 70 questions. What percent of the questions, rounded to the nearest tenth of a percent, did the student do correctly?

20. A laptop computer costs $3450 at present. This is 60% of the cost 4 yr ago. What was the cost of the system 4 yr ago? Explain your reasoning.

21. If, on a purchase of one new suit, you are offered successive discounts of 5%, 10%, or 20% in any order you wish, what order should you choose?

22. Jane bought a bicycle and sold it for 30% more than she paid for it. She sold it for $104. How much did she pay for it?

23. The student bookstore had a textbook for sale at $89.95. A student found the book on eBay for $62.00. If the student bought the book on eBay, what percentage of the cost of the bookstore book did she save?

24. When a store had a 60% off sale, Dori had a coupon for an additional 40% off any item and thought she should be able to obtain the dress that she wanted for free. If you were the store manager, how would you explain the mathematics of the situation to her?

25. Explain whether or not you could have each of the following as mathematically meaningful percentages:
    a. π%
    b. $\sqrt{2}$%
    c. $0.3\overline{4}$%
    d. $(1 + 0.\overline{3})$%

*26. A company was offered a $30,000 loan at a 12.5% annual simple interest rate for 4 yr. Find the simple interest due on the loan at the end of 4 yr.

*27. A fund pays 14% annual interest compounded quarterly. What is the value of a $10,000 investment after 3 yr?

Selected Bibliography

Abrahamson, D., and C. Cigan. "A Design for Ratio and Proportion Instruction." *Mathematics Teaching in the Middle School* 8 (May 2003): 493–501.

Beckman, C., D. Thompson, and R. Austin. "Exploring Proportional Reasoning Through Movies and Literature." *Mathematics Teaching in the Middle School* 9 (January 2004): 256–261.

Billings, E. "Problems That Encourage Proportion Sense." *Mathematics Teaching in the Middle School* 5 (January 2000): 310–313.

Chapin, S., and N. Anderson. "Crossing the Bridge to Formal Proportional Reasoning." *Mathematics Teaching in the Middle School* 8 (April 2003): 420–425.

Cramer, K., and T. Post. "Connecting Research to Teaching Proportional Reasoning." *Mathematics Teacher* 86 (May 1993): 404–407

Curcio, F., and N. Bezuk. "Understanding Rational Numbers and Proportions." *Curriculum and Evaluation Standards for School Mathematics Addenda Series Grades 5–8*. Reston, Va.: National Council of Teachers of Mathematics, 1994.

Harel, G., M. Behr, R. Lesh, and T. Post. "Invariance of Ratio: The Case of Children's Anticipatory Scheme for Constancy of Taste." *Journal of Research in Mathematics Education* 25 (July 1994): 324–345.

Hart, K. *Ratio: Children's Strategies and Errors.* Windsor, England: NFER-Nelson Pub. Co., 1984.

Hoffer, A., and S. Hoffer. "Ratios and Proportional Thinking." *Teaching Mathematics in Grades K–8*, edited by T. Post. Boston, MA: Allyn & Bacon, 1988, pp. 285–312.

Horak, V. "A Science Application of Area and Ratio Concepts." *Mathematics Teaching in the Middle School* 11 (April 2006): 360–365.

Lanius, C., and S. Williams. "Proportionality: A Unifying Theme for the Middle Grades." *Mathematics Teaching in the Middle School* 8 (April 2003): 392–396.

Lembke, L., and B. Reys. "The Development of, and Interaction between, Intuitive and School-taught Ideas about Percent. " *Journal of Research in Mathematics Education* 25 (May 1994): 237–259.

Lesh, R., T. Post, and M. Behr. "Proportional Reasoning." In *Number Concepts and Operations in the Middle Grades*, edited by J. Hiebert and M. Behr. Reston, VA: National Council of Teachers of Mathematics, 1988.

Lo, J., T. Watanabe, and J. Cai. "Developing Ratio Concepts: An Asian Perspective." *Mathematics Teaching in the Middle School* 9 (March 2004): 362–367.

Martine, S., and J. Bay-Williams. "Using Literature to Engage Students in Proportional Reasoning." *Mathematics Teaching in the Middle School* 9 (November 2003): 142–148.

Miller, J., and J. Fey. "Proportional Reasoning." *Mathematics Teaching in the Middle School* 5 (January 2000): 310–313.

Moss, J., and B. Caswell. "Building Percent Dolls: Connecting Linear Measurement to Learning Ratio and Proportion." *Mathematics Teaching in the Middle School* 10 (September 2004): 68–74.

Post, T., M. Behr, and R. Lesh. "Proportionality and the Development of Pre-Algebra Understandings." In *The Ideas of Algebra, K–12*, edited by A. Coxford and A. Shulte. Reston, VA: National Council of Teachers of Mathematics, 1988.

Reys, R., B. Reys, N. Nohda, and H. Emori. "Mental Computation Performance and Strategy Use of Japanese Students in Grades 2, 4, 6, and 8." *Journal for Research in Mathematics Education* 26 (July 1995): 304–326.

Scaptura, C., J. Suh, and G. Mahaffey. "Masterpieces to Mathematics: Using Art to Teach Fraction, Decimal, and Percent Equivalents." *Mathematics Teaching in the Middle School* 13 (August 2007): 24–28.

Seeley, C., and J. Schielack. "A Look at the Development of Ratios, Rates, and Proportionality." *Mathematics Teaching in the Middle School* 13 (October 2007): 140–142.

Sharp, J., and B. Adams. "Using a Pattern Table to Solve Contextualized Proportion Problems." *Mathematics Teaching in the Middle School* 8 (April 2003): 432–439.

Sowder, J., D. Wearne, W. Martin, and M. Strutchens. "What Do 8th-Grade Students Know about Mathematics?" In *Results and Interpretations of the 1990–2000 Mathematics Assessments of the National Assessment of Educational Progress*, edited by P. Kloosterman and F. Lester. Reston, VA.: NCTM, 2004, pp. 105–144.

Sweeney, E., and R. Quinn. "Concentration: Connecting Fractions, Decimals, & Percents." *Mathematics Teaching in the Middle School* 5 (January 2000): 324–328.

Thompson, C., and W. Bush. "Improving Middle School Teachers' Reasoning about Proportional Reasoning." *Mathematics Teaching in the Middle School* 8 (April 2003): 398–403.

VanDooren, W., D. DeBock. L. Verschaffel, and D. Janssens. "Improper Applications of Proportional Reasoning." *Mathematics Teaching in the Middle School* 9 (December 2003): 204–209.

Watson, J., M. Shaughnessy, and M. Perlwitz. "Proportional Reasoning: Lessons from Research in Data and Chance." *Mathematics Teaching in the Middle School* 10 (September 2004): 104–109.

CHAPTER 9

Probability

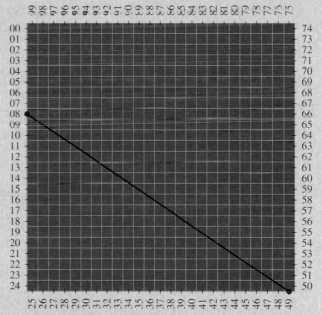

Preliminary Problem

The art on the left is reminiscent of art work of Frenchman François Morellet (1926–) who used random numbers in his work. Using a grid similar to the one on the above right, Morellet used two-digit numbers in pairs and connected them to form line segments. To simulate his process, choose any directory page of phone numbers, randomly choose a phone number to start, and then form 2 two-digit numbers from the last 4 digits of the chosen number. For example, 4908 is separated to form 49 and 08. Mark dots 49 and 08 on the grid and draw a segment connecting the dots. If Morellet used 10 phone numbers in this manner to determine the segments, what is the probability that at least 1 segment cannot be drawn? For this to happen, 1 four-digit number would have to determine only 1 dot, not 2.

Probability, with its roots in gambling, is used in such areas as predicting sales, planning political campaigns, determining insurance premiums, making investment decisions, and testing experimental drugs. Some examples of uses of probability in everyday conversations include the following:

What is the probability that the Chicago Cubs will win the World Series?

There is no chance you will get a raise.

There is a 50% chance of rain today.

The grades 6–8 *Principles and Standards* call for an increased emphasis on probability. In the grades 6–8 *Standards*, we find the following:

> Although the computation of probabilities can appear to be simple work with fractions, students must grapple with many conceptual challenges in order to understand probability. Misconceptions about probability have been held not only by many students but also by many adults (Konold 1989). To correct misconceptions, it is useful for students to make predictions and then compare the predictions with actual outcomes. (p. 254)

Probability Expectations

The *Principles and Standards* Data Analysis and Probability Standard describes its expectations for students with regard to the study of probability. The expectations include the following:

> In grades K–2, children should discuss events related to their experience as likely or unlikely. (p. 400)
>
> In grades 3–5, children should be able to "describe events as likely or unlikely and discuss the degree of likelihood using words such as *certain*, *equally likely*, and *impossible*." They should be able to "predict the probability of outcomes of simple experiments and test the predictions." They should "understand that the measure of the likelihood of an event can be represented by a number from 0 to 1." (p. 400)
>
> In grades 6–8, children should "understand and use appropriate terminology to describe complementary and mutually exclusive events." They should be able "to make and test conjectures about the results of experiments and simulations." They should be able to "compute probabilities of compound events using methods such as organized lists, tree diagrams, and area models." (p. 401)

In *Focal Points*, grade eight students are expected to "Use proportionality and a basic understanding of probability to make and test conjectures about the results of experiments and simulations" (p. 39). Additionally in *Focal Points*, we find that students in pre-K–8 are expected to "Compute probabilities for simple compound events, using such methods as organized lists, tree diagrams, and models" (p. 39).

In this chapter, we introduce all of the topics in the preceding expectations along with several others. We use tree diagrams and geometric probabilities (area models) to solve problems and to analyze games that involve spinners, cards, and dice. We introduce counting techniques and discuss the role of simulations in probability.

9-1   How Probabilities Are Determined

Probabilities are ratios, expressed as fractions, decimals, or percents, determined by considering results or outcomes of experiments. An **experiment** is an activity whose results can be observed and recorded. Each of the possible results of an experiment is an **outcome**. If we toss a typical coin that cannot land on its edge, there are two distinct possible outcomes: heads (H) and tails (T).

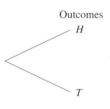

Figure 9-1

A set of all possible outcomes for an experiment is a **sample space**. The outcomes in the sample space cannot overlap. In a single coin toss of a typical coin, the sample space S is given by $S = \{H, T\}$. The sample space can be modeled by a **tree diagram**, as shown in Figure 9-1. Each outcome of the experiment is designated by a separate branch in the tree diagram. The sample space S for rolling a standard die as in Figure 9-2(a), is $S = \{1, 2, 3, 4, 5, 6\}$. Figure 9-2(b) gives a tree diagram for that sample space.

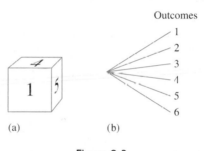

(a)   (b)

Figure 9-2

Any subset of a sample space is an **event**. For example, the set of all even-numbered rolls $\{2, 4, 6\}$ is a subset of all possible rolls of a die $\{1, 2, 3, 4, 5, 6\}$ and is an event.

Historical Note

The first time probability appears to have been mentioned in the Western history of mathematics is in a 1477 commentary on Dante's *Divine Comedy*, but most historians think that it originated in an unfinished dice game. The French mathematician Blaise Pascal (1623–1669) received a letter from his friend Chevalier de Méré, a professional gambler, who asked how to divide the stakes if two players start, but fail to complete, a game consisting of five matches in which the winner is the one who wins three out of five matches. The players decided to divide the stakes according to their chances of winning the game. Pascal shared the problem with Pierre de Fermat (1601–1665) and together they solved the problem, which prompted the development of probability. Since the work of the French mathematician Pierre Simon de Laplace (1749–1827), probability theory has become a major mathematical tool in science. ◆

Example 9-1

Suppose an experiment consists of drawing 1 slip of paper from a jar containing 12 slips of paper, each with a different month of the year written on it. Find each of the following:

a. the sample space S for the experiment
b. the event A consisting of outcomes having a month beginning with J
c. the event B consisting of outcomes having the name of a month that has exactly four letters
d. the event C consisting of outcomes having a month that begins with M or N

Solution **a.** $S = \{$January, February, March, April, May, June, July, August, September, October, November, December$\}$
b. $A = \{$January, June, July$\}$
c. $B = \{$June, July$\}$
d. $C = \{$March, May, November$\}$

Determining Probabilities

Around 1900, the English statistician Karl Pearson tossed a coin 24,000 times and recorded 12,012 heads. During World War II, John Kerrich, a Dane and a prisoner of war, tossed a coin 10,000 times. A subset of his results is in Table 9-1. The *relative frequency* column on the right is obtained by dividing the number of heads by the number of tosses of the coin.

Table 9-1

Number of Tosses	Number of Heads	Relative Frequency (rounded)
10	4	0.400
50	25	0.500
100	44	0.440
500	255	0.510
1,000	502	0.502
5,000	2,533	0.507
8,000	4,034	0.504
10,000	5,067	0.507

As the number of Kerrich's tosses increased, he obtained heads close to half the time. The relative frequency for Pearson's 24,000 tosses gives a similar result of 12,012/24,000, or approximately $\frac{1}{2}$. Kerrich was using the relative frequency interpretation of probability. In this interpretation, *the probability of an event is the long-run fraction of times that an event will occur given many repetitions under identical circumstances.*

When a probability is determined by observing outcomes of experiments, it is determined **experimentally**, or **empirically**. The exact number of heads that occurs when a fair coin is tossed a few times cannot be predicted accurately. A *fair coin* is a coin that is just as likely to land "heads" as it is to land "tails on each toss." When a fair coin is tossed many times and the fraction (or ratio) of heads is near $\frac{1}{2}$, we say that the probability of heads occurring is $\frac{1}{2}$ and write $P(\{H\}) = \frac{1}{2}$, or we shorten the symbolism to $P(H) = \frac{1}{2}$.

Probabilities only suggest what will happen in the "long run." This concept is called *The Law of Large Numbers* or sometimes *Bernoulli's Theorem*, and is given here as Theorem 9–1.

Theorem 9–1 : Law of Large Numbers (Bernoulli's Theorem)

If an experiment is repeated a large number of times, the *experimental* or *empirical* probability of a particular outcome approaches a fixed number as the number of repetitions increases.

In this text, by "probability" we mean *theoretical probability*. We assign **theoretical probabilities** to the outcomes under ideal conditions. For example, we could argue that since an ideal coin marked head and tail is symmetric and has two sides, then each side should appear about the same number of times if the coin is tossed many times. Again we would conclude that

$$P(H) = P(T) = \frac{1}{2}$$

When one outcome is just as likely as another, as in coin tossing, the outcomes are **equally likely**. If an experiment is repeated many times, the experimental probability of the event's occurring should be approximately equal to the theoretical probability of the event's occurring.

REMARK Unless otherwise noted a fair coin is considered to have a "head" side and a "tail" side. A die is considered to be a typical die with faces marked 1, 2, 3, 4, 5, or 6.

Research Note

Students tend to categorize events as equally likely because of their listing in a sample space. (Lecoutre 1992). ◆

A *fair* die is a die that is just as likely to land showing any of the numerals 1 through 6 on any toss. Its sample space S is given by $S = \{1, 2, 3, 4, 5, 6\}$, and $P(1) = P(2) = P(3) = P(4) = P(5) = P(6) = \frac{1}{6}$. The probability of rolling an even number, that is, the probability of the event $E = \{2, 4, 6\}$, is $\frac{3}{6}$, or $\frac{1}{2}$. For a sample space with equally likely outcomes, the probability of an event A can be defined as follows.

Definition of Probability of an Event with Equally Likely Outcomes

For an experiment with sample space S with equally likely outcomes, the **probability of an event A** is given by

$$P(A) = \frac{\text{Number of elements of } A}{\text{Number of elements of } S} = \frac{n(A)}{n(S)}$$

Historical Note

The Bernoulli family, of Belgian origin, fled Amsterdam for Basel, Switzerland, to escape religious persecution. The eldest son, Jakob (1654–1705), was the first of several family members who, against the wishes of their parents, pursued the study of mathematics. After taking his degree in theology, Jakob traveled for 7 years studying mathematics. He returned to Basel in 1683 and in 1687 became the chair of mathematics at the University of Basel. His greatest and most original work, *Ars Conjectandi* (*The Art of Conjecturing*), laid the foundation for the modern theory of probability. ◆

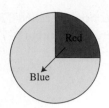

Figure 9-3

Note that the given definition of the probability of an event applies only to a sample space that has equally likely outcomes. If each possible outcome of the sample space is equally likely, the sample space is a **uniform sample space**. Thus, in a uniform sample space with n outcomes, the probability of each outcome is $\dfrac{1}{n}$. Trying to apply the definition of probability given to a space with outcomes that are not equally likely (nonuniform) leads to incorrect conclusions. For example, the sample space for spinning the spinner in Figure 9-3 is given by $S = \{\text{Red, Blue}\}$, but the outcome Blue is more likely to occur than is the outcome Red, so $P(\text{Red})$ is not equal to $\dfrac{1}{2}$ but $\dfrac{1}{4}$ (why?). If the spinner were spun 100 times, we could reasonably expect that about $\dfrac{1}{4}$ of 100, or 25, of the outcomes would be Red, whereas about $\dfrac{3}{4}$ of 100, or 75, of the outcomes would be Blue. See the Research Note by Lecoutre (page 519).

NOW TRY THIS 9-1

a. In an experiment of tossing a fair coin once, what is the sum of the probabilities of all the distinct outcomes in the sample space?

b. In an experiment of tossing a fair die once, what is the sum of the probabilities of all the distinct outcomes in the sample space?

c. Does the sum of the probabilities of all the distinct outcomes of any sample space always result in the same number?

Example 9-2

Let $S = \{1, 2, 3, 4, 5, \ldots, 25\}$. If a number is chosen **at random**, that is, with the same chance of being drawn as all other numbers in the set, calculate each of the following probabilities:

a. the event A that an even number is drawn
b. the event B that a number less than 10 and greater than 20 is drawn
c. the event C that a number less than 26 is drawn
d. the event D that a prime number is drawn
e. the event E that a number both even and prime is drawn

Solution   Each of the 25 numbers in set S has an equal chance of being drawn.

a. $A = \{2, 4, 6, 8, 10, 12, 14, 16, 18, 20, 22, 24\}$, so $n(A) = 12$. Thus,
$$P(A) = \frac{n(A)}{n(S)} = \frac{12}{25}.$$

b. $B = \varnothing$, so $n(B) = 0$. Thus, $P(B) = \dfrac{0}{25} = 0$.

c. $C = S$ and $n(C) = 25$. Thus, $P(C) = \dfrac{25}{25} = 1$.

d. $D = \{2, 3, 5, 7, 11, 13, 17, 19, 23\}$, so $n(D) = 9$. Thus,
$$P(D) = \frac{n(D)}{n(S)} = \frac{9}{25}.$$

e. $E = \{2\}$, so $n(E) = 1$. Thus, $P(E) = \dfrac{1}{25}$.

In Example 9-2(b), event B is the empty set. An event such as B that has no outcomes is an **impossible event** *and has probability* 0. If the word *and* were replaced by *or* in Example 9-2(b), then event B would no longer be the empty set. In Example 9-2(c), event C consists of drawing a number less than 26 on a single draw. Because every number in S is less than 26, $P(C) = \dfrac{25}{25} = 1$. An event that has probability 1 is a **certain event**.

If event A is the empty set, $\varnothing$, then it has no elements and its probability is the ratio $P(A) = \dfrac{n(\varnothing)}{n(S)} = \dfrac{0}{n(S)} = 0$. (Note that an assumption is made that the sample space does not have 0 elements.) Because A is a subset of the sample space S, the greatest number of elements that A can have is the number of elements in the sample space, S. So the probability of A in this case would be $P(A) = \dfrac{n(A)}{n(S)} = \dfrac{n(S)}{n(S)} = 1$. For any subset A of S with some but not all elements of S, $n(A)$ is greater than 0 but less than $n(S)$. Hence $P(A) = \dfrac{n(A)}{n(S)} < 1$. We summarize this discussion in the following theorem.

Theorem 9–2

If A is any event and S is the sample space, then $0 \le P(A) \le 1$.

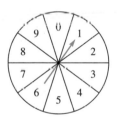

Figure 9-4

Consider one spin of the number wheel shown in Figure 9-4 where each sector of the circle has equal area. For this experiment, $S = \{0, 1, 2, 3, 4, 5, 6, 7, 8, 9\}$. If A is the event of spinning a number in the set $\{0, 1, 2, 3, 4\}$ and B is the event of spinning a number in the set $\{5, 7\}$, then using the definition of probability for equally likely events, $P(A) = \dfrac{n(A)}{n(S)} = \dfrac{5}{10}$ and $P(B) = \dfrac{n(B)}{n(S)} = \dfrac{2}{10}$. The probability of an event can be found by adding the probabilities of disjoint events representing the various outcomes in the set. For example, event $B = \{5, 7\}$ can be represented as the union of two disjoint events, that is, spinning a 5 or spinning a 7. Then $P(B)$ can be found by adding the probabilities of each event.

$$P(B) = P(5) + P(7) = \frac{1}{10} + \frac{1}{10} = \frac{2}{10}$$

Likewise,

$$P(A) = \frac{1}{10} + \frac{1}{10} + \frac{1}{10} + \frac{1}{10} + \frac{1}{10} = \frac{5}{10}$$

These are special cases of the following theorem, which holds for all finite probability spaces.

Theorem 9–3

The probability of an event is equal to the sum of the probabilities of the disjoint outcomes making up the event.

Example 9-3

If we draw a card at random from an ordinary deck of playing cards, what is the probability that

a. the card is an ace?
b. the card is an ace or a queen?

Solution   **a.** There are 52 cards in a deck, 4 of which are aces. If event A is drawing an ace, then A = $\left\{ \boxed{\spadesuit}, \boxed{\clubsuit}, \boxed{\diamondsuit}, \boxed{\heartsuit} \right\}$. We use the definition of probability for equally likely outcomes to compute the following:

$$P(A) = \frac{n(A)}{n(S)} = \frac{4}{52}$$

An alternative approach is to find the sum of the probabilities of obtaining each of the outcomes in the event, where the probability of drawing any single ace from the deck is $\frac{1}{52}$:

$$P(A) = \frac{1}{52} + \frac{1}{52} + \frac{1}{52} + \frac{1}{52} = \frac{4}{52}$$

b. The event E of getting an ace or a queen consists of 8 cards: 4 aces and 4 queens. Hence,

$$P(E) = \frac{n(E)}{n(S)} = \frac{8}{52}$$

Mutually Exclusive Events

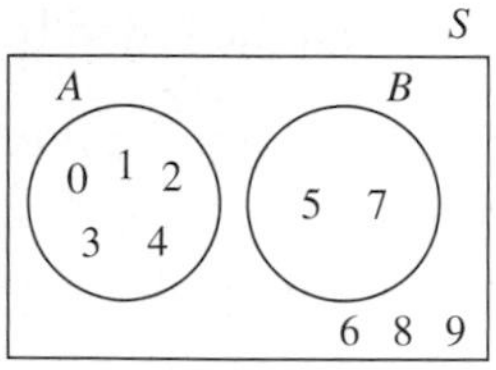

$A \cap B = \emptyset$

Figure 9-5

Consider one spin of the wheel in Figure 9-4. For this experiment, we have $S = \{0, 1, 2, 3, 4, 5, 6, 7, 8, 9\}$. If $A = \{0, 1, 2, 3, 4\}$ and $B = \{5, 7\}$, then $A \cap B = \emptyset$. See Figure 9-5. Two such events are **mutually exclusive** events. If event A occurs, then event B cannot occur, and we have the following definition.

Definition of Mutually Exclusive Events

Events A and B are **mutually exclusive** if they have no elements in common; that is, $A \cap B = \emptyset$.

Each outcome in the space $S = \{0, 1, 2, 3, 4, 5, 6, 7, 8, 9\}$ is equally likely, with probability $\frac{1}{10}$. Now with mutually exclusive events A and B given, if we write the probability of A or B as $P(A \cup B)$, we have the following:

$$P(A \cup B) = \frac{n(A \cup B)}{n(S)} = \frac{7}{10} = \frac{5 + 2}{10} = \frac{5}{10} + \frac{2}{10}$$

$$= \frac{n(A)}{n(S)} + \frac{n(B)}{n(S)} = P(A) + P(B)$$

The result developed in this example is true for all mutually exclusive events. In general, we have the following theorem.

> **Theorem 9–4**
>
> If events A and B are mutually exclusive, then $P(A \text{ or } B) = P(A \cup B) = P(A) + P(B)$.

For a sample space with equally likely outcomes, this property follows immediately from the fact that if $A \cap B = \varnothing$, then $n(A \cup B) = n(A) + n(B)$. In general, *the probability of the union of events such that any two are mutually exclusive is the sum of the probabilities of those events.*

Complementary Events

If the weather forecaster tells us that the probability of rain is 25%, what is the probability that it will not rain? These two events—rain and not rain—are **complements** of each other. Therefore, if the probability of rain is 25%, or $\frac{1}{4}$, the probability it will not rain is $100\% - 25\% = 75\%$, or $1 - \frac{1}{4} = \frac{3}{4}$. Notice that $P(\text{no rain}) = 1 - P(\text{rain})$. The two events rain and no rain are mutually exclusive because if one happens, the other cannot. Two mutually exclusive events whose union is the sample space are **complementary events**. If A is an event, the complement of A, written $\overline{A}$, is also an event. For example, consider the event $A = \{2, 4\}$ of tossing a 2 or a 4 using a standard die. The complement of A is the set $\overline{A} = \{1, 3, 5, 6\}$. Because the sample space is $S = \{1, 2, 3, 4, 5, 6\}$, we have $P(A) = \frac{2}{6}$ and $P(\overline{A}) = \frac{4}{6}$. Notice that $P(A) + P(\overline{A}) = \frac{2}{6} + \frac{4}{6} = 1$. This is true in general for any set A and its complement, $\overline{A}$.

> **Theorem 9–5**
>
> If A is an event and $\overline{A}$ is its complement, then
> $$P(A) + P(\overline{A}) = 1$$
> $$P(\overline{A}) = 1 - P(A) \ \text{ or } \ P(A) = 1 - P(\overline{A})$$

Non-Mutually Exclusive Events

Consider the spinner in Figure 9-6.

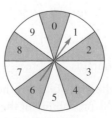

Figure 9-6

Let E be the event of spinning an even number and T the event of spinning a number divisible by 3; that is,

$$E = \{0, 2, 4, 6, 8\}$$
$$T = \{0, 3, 6, 9\}$$

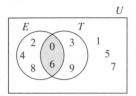

Figure 9-7

The event of spinning an even number and a number divisible by 3 is denoted $E \cap T$. Because $E \cap T = \{0, 6\}$, as seen in Figure 9-7, E and T are not mutually exclusive and $P(E \cup T) \neq P(E) + P(T)$. However, because each outcome is equally likely, we can still use the definition of the probability of an event to compute the probability of E or T as follows:

$$P(E \cup T) = \frac{n(E \cup T)}{n(S)}$$

Because $E \cup T = \{0, 2, 4, 6, 8, 3, 9\}$, $n(E \cup T) = 7$. Because, $n(S) = 10$, we have $P(E \cup T) = \dfrac{7}{10}$.

In general, we can also compute the probability of E or T by using the following result from Chapter 2:

$$n(E \cup T) = n(E) + n(T) - n(E \cap T)$$

Therefore, for a sample space with equally likely outcomes,

$$\begin{aligned}
P(E \cup T) &= \frac{n(E \cup T)}{n(S)} \\
&= \frac{n(E) + n(T) - n(E \cap T)}{n(S)} \\
&= \frac{n(E)}{n(S)} + \frac{n(T)}{n(S)} - \frac{n(E \cap T)}{n(S)} \\
&= P(E) + P(T) - P(E \cap T)
\end{aligned}$$

This result, although given for events in a sample space with equally likely outcomes, is true in general. This and other results of probability are summarized next.

Summary of Probability Properties

1. $P(\varnothing) = 0$ (impossible event).
2. $P(S) = 1$, where S is the sample space (certain event).
3. For any event A, $0 \leq P(A) \leq 1$.
4. If A and B are events and $A \cap B = \varnothing$, then $P(A \cup B) = P(A) + P(B)$.
5. If A and B are any events, then $P(A \cup B) = P(A) + P(B) - P(A \cap B)$.
6. If A is an event, then $P(\overline{A}) = 1 - P(A)$.

Example 9-4

A golf bag contains 2 red tees, 4 blue tees, and 5 white tees.

a. What is the probability of the event R that a tee drawn at random is red?
b. What is the probability of the event "not R;" that is, that a tee drawn at random is not red?
c. What is the probability of the event that a tee drawn at random is either red (R) or blue (B); that is, $P(R \cup B)$?

Solution **a.** Because the bag contains a total of $2 + 4 + 5$, or 11, tees and 2 tees are red,
$$P(R) = \frac{2}{11}.$$

b. The bag contains 11 tees and 9 are not red, so the probability of "not R" is $\frac{9}{11}$. Also, notice that $P(\overline{R}) = 1 - P(R) = 1 - \frac{2}{11} = \frac{9}{11}$.

c. The bag contains 2 red tees and 4 blue tees and $R \cap B = \emptyset$, so $P(R \cup B)$
$$= \frac{2}{11} + \frac{4}{11}, \text{ or } \frac{6}{11}.$$

 On the following student page, events are categorized as "less likely" if the probability of the event is near 0 and "more likely" if the event has probability close to 1. After reading the page, answer Number 4 using your name.

Example 9-5

Find the probability of rolling a sum of 7 or 11 when rolling a pair of fair dice.

Solution To solve this problem, we use the strategy of *making a table*. Figure 9-8(a) shows all possible outcomes of tossing the dice. We know that there are 6 possible results from tossing the first die and 6 from tossing the second die, so by the Fundamental Counting Principle, there are $6 \cdot 6$, or 36, entries in the table. It may be easier to read the results when they are recorded as ordered pairs, as in Figure 9-8(b), where the first component represents the number on the first die and the second component represents the number on the second die. We show the possible sums from rolling the pair of dice in Figure 9-8(c). In Figure 9-8(c) we see that a sum of 7 appears 6 times. Hence, the event "a sum of 7" arises from the following subset of the set of ordered pairs in Figure 9-8(b):

$$\{(6, 1), (5, 2), (4, 3), (3, 4), (2, 5), (1, 6)\}$$

Each outcome in this set is equally likely and hence $P(\text{a sum of } 7) = \frac{6}{36}$. Similarly, $P(\text{a sum of } 11) = \frac{2}{36}$. The probabilities of each of the possible sums can be calculated in

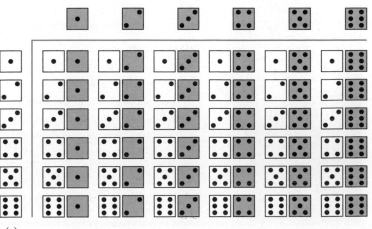

(a)

School Book Page — EXPRESSING PROBABILITY AS A FRACTION

Lesson 5-12

Key Idea
Probability can be expressed as a fraction.

Vocabulary
• probability

Materials
• spinner or

tools

Expressing Probability as a Fraction

LEARN

What is the probability of an event?

The **probability** of an event is a number that describes the chance that an event will occur.

Probability of an event = $\dfrac{\text{number of favorable outcomes}}{\text{number of possible outcomes}}$

The probability of an event ranges from 0 for an event that is impossible to 1 for an event that is certain.

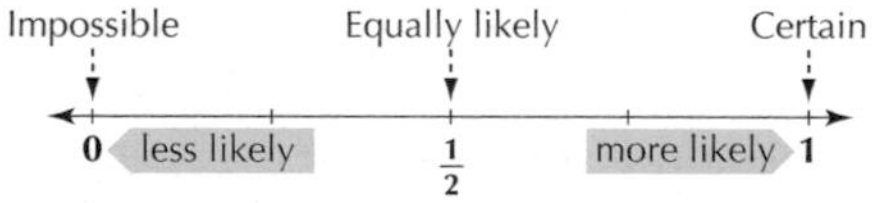

Impossible	Equally likely	Certain

0 ← less likely $\frac{1}{2}$ more likely → **1**

WARM UP

1. 29×53

2. 2.15×8

3. $456 \div 3$

4. $\$41.68 \div 2$

Example A

Beatrice wrote each letter of her name on a slip of paper and put the slips in a bag. If she draws one slip of paper from the bag, what is the probability of drawing a vowel?

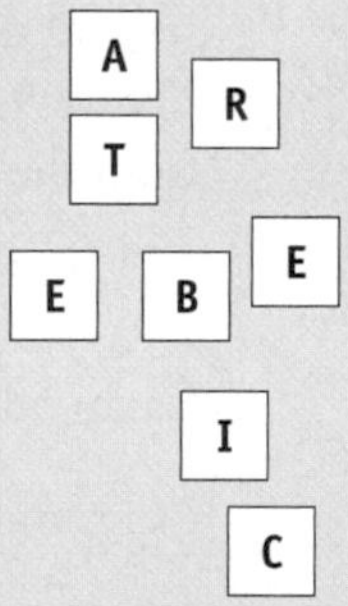

There are 8 possible outcomes (letters) and 4 favorable outcome (vowels).

$$\text{Probability of drawing a vowel} = \frac{\text{number of vowels}}{\text{number of letters}} = \frac{4}{8}$$

A   R   T   E   B   E   I   C

The probability of drawing a vowel is $\frac{4}{8}$, or $\frac{1}{2}$.

The probability of drawing a vowel can be written as $P(\text{vowel}) = \frac{1}{2}$.

✔ Talk About It

Use the letters in BEATRICE. Find the probability of choosing each letter or letters out of a bag.

1. a consonant   **2.** the letter E   **3.** not E

4. Reasoning If you put the letters of your name in a bag and draw one letter, what is $P(\text{vowel})$?

Source: Scott Foresman-Addison Wesley, Mathematics, Grade 5, 2008 (p. 302).

	Number on Second Die					
	1	2	3	4	5	6
1	(1, 1)	(1, 2)	(1, 3)	(1, 4)	(1, 5)	(1, 6)
2	(2, 1)	(2, 2)	(2, 3)	(2, 4)	(2, 5)	(2, 6)
3	(3, 1)	(3, 2)	(3, 3)	(3, 4)	(3, 5)	(3, 6)
4	(4, 1)	(4, 2)	(4, 3)	(4, 4)	(4, 5)	(4, 6)
5	(5, 1)	(5, 2)	(5, 3)	(5, 4)	(5, 5)	(5, 6)
6	(6, 1)	(6, 2)	(6, 3)	(6, 4)	(6, 5)	(6, 6)

(b)

Number on First Die

	Number on Second Die					
	1	2	3	4	5	6
1	2	3	4	5	6	7
2	3	4	5	6	7	8
3	4	5	6	7	8	9
4	5	6	7	8	9	10
5	6	7	8	9	10	11
6	7	8	9	10	11	12

Possible Sums

(c)

Number on First Die

Figure 9-8

the same way. From Figure 9-8(c), we see that the sample space for the experiment here is $\{2, 3, 4, 5, 6, 7, 8, 9, 10, 11, 12\}$ but that the sample space is not uniform; that is, the probabilities of the given sums are not equal. To summarize these probabilities, we construct a **probability distribution** as in Table 9-2

Table 9-2

Outcome	2	3	4	5	6	7	8	9	10	11	12
Probability	$\frac{1}{36}$	$\frac{2}{36}$	$\frac{3}{36}$	$\frac{4}{36}$	$\frac{5}{36}$	$\frac{6}{36}$	$\frac{5}{36}$	$\frac{4}{36}$	$\frac{3}{36}$	$\frac{2}{36}$	$\frac{1}{36}$

The probability of rolling a sum of 7 or 11 is given by $P(\text{sum of 7 or sum of 11}) = P(\text{sum of 7}) + P(\text{sum of 11}) = \frac{6}{36} + \frac{2}{36} = \frac{8}{36}$.

Example 9-6

A fair pair of dice is rolled. Let E be the event of rolling a sum that is an even number and F the event of rolling a sum that is a prime number. Find the probability of rolling a sum that is even *or* prime, that is, $P(E \cup F)$.

Solution   To solve this problem, we use Table 9-2. Note that $E \cup F = \{2, 4, 6, 8, 10, 12, 3, 5, 7, 11\}$. Therefore,

$$P(E \cup F) = P(2) + P(4) + P(6) + P(8) + P(10) + P(12) + P(3) + P(5) + P(7) + P(11)$$

$$= \frac{1}{36} + \frac{3}{36} + \frac{5}{36} + \frac{5}{36} + \frac{3}{36} + \frac{1}{36} + \frac{2}{36} + \frac{4}{36} + \frac{6}{36} + \frac{2}{36}$$

$$= \frac{32}{36}$$

Another approach is to use a property of probabilities (see page 525): We know that events E and F are not mutually exclusive because $E = \{2, 4, 6, 8, 10, 12\}$, $F = \{2, 3, 5, 7, 11\}$, and $E \cap F = \{2\}$. Therefore,

$$P(E \cup F) = P(E) + P(F) - P(E \cap F)$$
$$= \frac{18}{36} + \frac{15}{36} - \frac{1}{36}$$
$$= \frac{32}{36}$$

A third approach to finding $P(E \cup F)$ is to find $P(\overline{E \cup F})$ and subtract this probability from 1. Because $E \cup F = \{2, 3, 4, 5, 6, 7, 8, 10, 11, 12\}$, then $\overline{E \cup F} = \{9\}$ and $P(\overline{E \cup F}) = \frac{4}{36}$. Hence, $P(E \cup F) = 1 - P(\overline{E \cup F}) = 1 - \frac{4}{36} = \frac{32}{36}$.

NOW TRY THIS 9-2 Consider a normal phone. Suppose a number button is pushed at random.

a. What is the probability of pressing a 9?
b. What is the probability of pressing a 5?
c. What is the probability of pressing a button with three letters?
d. What is the probability of pressing a button with four letters?
e. What is the probability of pressing a number button with no letters?
f. Is the probability distribution for getting 0, 3, or 4 letters on a random key uniform? Explain your answer.

Assessment 9-1A

1. Consider the experiment of drawing a single card from a standard deck of cards and determine which of the following are uniform sample spaces (equally likely outcomes):
   a. {face card, not face card} NO
   b. {club, diamond, heart, spade} yes
   c. {black, red} yes
   d. {king, queen, jack, ace, even card, odd card} NO

2. Each letter of the alphabet is written on a separate piece of paper and placed in a box and then one piece is drawn at random.
   a. What is the probability that the selected piece of paper has a vowel written on it?
   b. What is the probability that it has a consonant written on it?
   c. What is the probability that the paper has a vowel on it or a letter from the word *probability*? $\frac{11}{26}$

3. The following spinner is spun:

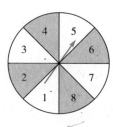

Find the probabilities of obtaining each of the following:
   a. P(factor of 35) $\frac{3}{8}$ b. P(multiple of 3) $\frac{1}{4}$
   c. P(even number) $\frac{1}{2}$ d. P(6 or 2) $\frac{1}{4}$
   e. P(11) 0 f. P(composite number) $\frac{3}{8}$
   g. P(neither a prime nor a composite) $\frac{1}{8}$

4. A card is selected from an ordinary deck of 52 cards. Find the probabilities for each of the following:
   a. A red card
   b. A face card
   c. A red card or a 10
   d. A queen
   e. Not a queen
   f. A face card or a club
   g. A face card and a club
   h. Not a face card and not a club

5. Suppose the drawer in the cartoon contains six black socks, four brown socks, and two green socks. Suppose the cat draws one sock from the drawer and that it is equally likely that any one of the socks is drawn. Find the probabilities for each of the following:
   a. The sock is brown.
   b. The sock is either black or green.
   c. The sock is red.
   d. The sock is not black.
   e. A cat reaches in the drawer without looking to pull out four socks. What is the probability that the cat gets two socks of the same color?

6. Riena has six unmarked CDs in a box, where each is dedicated to exactly one of English, mathematics, French, American history, chemistry, and computer science. Answer the following questions:
   a. If she chooses a CD at random, what is the probability she chooses the English CD?
   b. What is the probability that she chooses a CD that is neither mathematics nor chemistry?

7. According to a weather report, there is a 30% chance it will rain tomorrow. What is the probability it will not rain tomorrow? Explain your answer.

8. A roulette wheel has 38 slots around the rim. The first 36 slots are numbered from 1 to 36. Half of these 36 slots are red, and the other half are black. The remaining 2 slots are numbered 0 and 00 and are green. As the roulette wheel is spun in one direction, a small ivory ball is rolled along the rim in the opposite direction. The ball has an equally likely chance of falling into any one of the 38 slots. Find each of the following:
   a. The probability that the ball lands in a black slot
   b. The probability that the ball lands on 0 or 00
   c. The probability that the ball does not land on a number from 1 through 12
   d. The probability that the ball lands on an odd number or on a green slot

9. If the roulette wheel in exercise 8 is spun 190 times, predict about how many times the ball can be expected to land on 0 or 00.

10. Suppose a fair coin is tossed twice. Find the probability for each of the following:
    a. Exactly one head
    b. At least one head
    c. At most one head

11. In Sentinel High School, there are 350 freshmen, 320 sophomores, 310 juniors, and 400 seniors. If a student is chosen at random from the student body to represent the school, what is the probability that the chosen student is a freshman?

12. If A and B are mutually exclusive and if $P(A) = 0.3$ and $P(B) = 0.4$, what is $P(A \cup B)$?

13. A calculus class is composed of 35 men and 45 women. There are 20 business majors, 30 biology majors, 10 computer science majors, and 20 mathematics majors. No person has a double major. If a single student is chosen from the class, what is the probability that the student is the following:
    a. Female
    b. A computer science major
    c. Not a mathematics major
    d. A computer science major or a mathematics major

14. A box contains five white balls, three black balls, and two red balls.
    a. How many red balls must be added to the box so that the probability of drawing a red ball is $\frac{3}{4}$?
    b. How many black balls must be added to the original box so that the probability of drawing a white ball is $\frac{1}{4}$?

15. Zoe is playing a game in which she draws one ball from one of the boxes shown. She wins if she draws a white ball from either box #1 or box #2. She says that in order to maximize her chances of winning she will always pick box #2 because it has more white balls. Is she correct? Why?

#1

#2

16. If you flipped a fair coin 15 times and got 15 heads, what is the probability of getting a head on the 16th toss? Explain your answer.

17. Consider the letters in the word *dictionary*. If one letter is drawn at random, what is the probability of the following:

    a. The letter is a vowel?

    b. The letter is a consonant?

18. Answer each of the following as likely or unlikely:

    a. The probability that a current U.S. Supreme Court Justice chosen at random is a man.

    b. The probability that a current member of the U.S. Senate chosen at random is African-American.

    c. The probability that a person in the world chosen at random has Asian roots.

19. Explain whether you think that when dialing a phone number with its area code, each digit 0–9 has an equal chance of being chosen as the lead number.

Assessment 9-1B

1. When a thumbtack is dropped, it will land point up ($\perp$) or point down ($\wedge$). This experiment was repeated 80 times with the following results: point up: 56 times; point down: 24 times.

    a. What is the experimental probability that the thumbtack will land point up?

    b. What is the experimental probability that the thumbtack will land point down?

    c. If you were to try this experiment another 80 times, would you get the same results? Why?

    d. Would you expect to get nearly the same results on a second trial? Why?

2. An experiment consists of selecting the last digit of a telephone number from a telephone book. Assume that each of the 10 digits is equally likely to appear as a last digit. List each of the following:

    a. The sample space

    b. The event consisting of outcomes that the digit is less than 5

    c. The event consisting of outcomes that the digit is odd

    d. The event consisting of outcomes that the digit is not 2

    e. Find the probability of each of the events in (b) through (d).

3. In an open refrigerator, there are seven different types of diet soda, four different types of regular soda, and two types of bottles of water.

    a. Suppose Evan chose a bottle from the refrigerator at random. Could we realistically say that the probability of choosing a diet soda is $\frac{7}{13}$? Why or why not?

    b. If there are 16 total bottles of diet soda, 8 total bottles of regular soda, and 4 total bottles of water, what is the probability of each of the following:

       (i) Choosing a bottle of diet soda when a bottle is chosen at random

       (ii) Choosing a bottle of regular soda when a bottle is chosen at random

       (iii) Choosing a bottle of water when a bottle is chosen at random

4. A bag contains n cards each having one of the numbers $(1, 2, 3, 4, \ldots, n)$ written on it. If all numbers are used the probability of drawing a card with a number less than or equal to 10 is $\frac{4}{10}$. How many cards are in the bag?

5. Determine if each player has an equal probability of winning each of the following games:

    a. Toss a fair coin. If heads appears, I win; if tails appears, you lose.

    b. Toss a fair coin. If heads appears, I win; otherwise, you win.

    c. Toss a fair die numbered 1 through 6. If 1 appears, I win; if 6 appears, you win.

    d. Toss a fair die numbered 1 through 6. If an even number appears, I win; if an odd number appears, you win.

    e. Toss a fair die numbered 1 through 6. If a number greater than or equal to 3 appears, I win; otherwise, you win.

    f. Toss two fair dice numbered 1 through 6. If a 1 appears on each die, I win; if a 6 appears on each die, you win.

    g. Toss two fair dice numbered 1 through 6. If the sum is 3, I win; if the sum is 2, you win.

h. Toss two dice numbered 1 through 6; one die is red and one is white. If the number on the red die is greater than the number on the white die, I win; otherwise, you win.

6. The following questions refer to a very popular dice game, craps, in which a player rolls two dice:

a. Rolling a sum of 7 or 11 on the first roll of the dice is a win. What is the probability of winning on the first roll?

b. Rolling a sum of 2, 3, or 12 on the first roll of the dice is a loss. What is the probability of losing on the first roll?

c. Rolling a sum of 4, 5, 6, 8, 9, or 10 on the first roll is neither a win nor a loss. What is the probability of neither winning nor losing on the first roll?

d. After rolling a sum of 4, 5, 6, 8, 9, or 10, a player must roll the same sum again before rolling a sum of 7. Which sum—4, 5, 6, 8, 9, or 10—has the highest probability of occurring again?

e. What is the probability of rolling a sum of 1 on any roll of the dice?

f. What is the probability of rolling a sum less than 13 on any roll of the dice?

g. If the two dice are rolled 60 times, predict about how many times a sum of 7 will be rolled.

7. In each of the following, sketch a single spinner with the following characteristics.

a. The outcomes are M, A, T, and H, each with equally likely probability.

b. The outcomes are R, A, and T with $P(R) = \frac{3}{4}$, $P(A) = \frac{1}{8}$, and $P(T) = \frac{1}{8}$.

8. In the game of "Between," two cards are dealt and not replaced in the deck. You then pick a third card from the deck. To win, you must pick a card that has a value between the other two cards. The order of values is 2, 3, 4, 5, 6, 7, 8, 9, 10, J, Q, K, A, where the letters represent a jack, queen, king, and ace, respectively. Determine the probability of your winning if the first two cards dealt are the following:

a. a 5 and a jack   **b.** a 2 and a king

c. a 5 and a 6

9. Calculators, watches, scoreboards, and many other devices display numbers using arrays like the following. The device lights up different parts of the array (any segment lettered a through g), to display any single digit 0 through 9. Suppose a digit is chosen at random. Determine the probability for each of the following. (*Hint:* Use a digital watch to see which segments are lit up for the different numbers.)

a. Segment a will be lit.

b. Segment b will be lit.

c. Segments e and b will be lit.

d. Segment e or b will be lit.

10. If A is the set of students taking algebra and C is the set of students taking chemistry at Central High School, describe in words what is meant by each of the following probabilities:

a. $P(A \cup C)$ **b.** $P(A \cap C)$ **c.** $1 - P(C)$

11. A box contains 25% black balls and 75% white balls. The same number of black balls as were in the box are added (so the new number of black balls is twice the original number). A ball is now drawn from the box at random. What is the probability that it is black?

12. Suppose we have a box containing five white balls, three black balls, and two red balls. Is it possible to add the same number of balls of each color to the box so that when a ball is drawn at random the probability that it is a black ball is the following (explain):

a. $\frac{1}{3}$ **b.** 0.32

13. Nanci has 15 files in a filing cabinet labeled from A to O, one file per label. If she chooses one file at random, what is the probability of each of the following:

a. Nanci choose a file with the label B.

b. Nanci does not choose a file with the label B.

14. A teacher gave students three questions and wrote that she would randomly choose one of the three questions for an exam. If Harry prepared for only one of the questions, what is the probability that he did not study for the question that the teacher put on the exam?

15. Sylvia decided to simulate the probability of the birth of a boy or a girl by spinning a spinner with the numbers 1–7 on the seven equally divided sectors of the spinner. Suppose the birth of a boy is represented by spinning an odd number and a girl is represented by spinning an even number. If the chance of the birth of a boy is approximately equal to the chance for the birth of a girl, explain whether the spinner described is a good tool to simulate the birth.

16. There are approximately 47 words listed on each page of a certain dictionary. If you chose a page on which you knew that a certain word was listed, explain whether you think the probability of choosing the specific word is equal to the probability of not choosing the word if one word from the page is chosen at random.

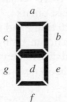

Mathematical Connections 9-1

Communication

1. If events A and B are from the same sample space, and if $P(A) = 0.8$ and $P(B) = 0.9$, can events A and B be mutually exclusive? Explain.

2. Bobbie says that when she shoots a free throw in basketball, she will either make it or miss it. Because there are only two outcomes and one of them is making a basket, Bobbie claims the probability of her making a free throw is $\frac{1}{2}$. Explain whether Bobbie's reasoning is correct.

3. If the following spinner is spun 100 times, Joe claims that the number 4 will occur most often because the greatest area of the spinner is covered by the number 4. What would you tell Joe about his conjecture? What is the probability that a 4 will occur on any spin?

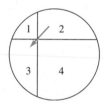

Open-Ended

4. Prepare a card with the numbers printed on it as shown.

$$\boxed{1 \quad 2 \quad 3 \quad 4}$$

Show the card to 10 people and ask each of them to pick a single digit listed on the card and tell you what number they picked. Record the results. Pool your results with 4 other people and answer the following questions:

   a. What is the probability of selecting a 3 from the card if the choice was made at random?
   b. Based on the data you gathered, what is the experimental probability of selecting a 3 based on your sample of 10? Based on the sample of 50?
   c. Why do you think the theoretical probability might not agree with the experimental probability?
   d. Conjecture what will happen if the numbers 1, 2, 3, 4, 5 are printed on the card and a similar experiment is repeated. Test your conjecture.

5. Select any book, go to the first complete paragraph in it, and count the number of words the paragraph has. Now count the number of words that start with a vowel. If the paragraph has fewer than 100 words, continue to count words until there are more than 100 words from where you started. What is the experimental probability that a word chosen at random from the book starts with a vowel? Open the book to any page and choose a paragraph with more

than 100 words. Count the words. Predict how many words on the page start with a vowel, and then count to see how close you were.

6. List three real-world situations that do not involve weather or gambling where probability might be used.

7. For each of the following letters, describe an event, if possible, that has the approximate probability marked by the letter on the probability line:

Cooperative Learning

8. Form groups of three or four students. Each group has a pair of dice. Each player needs 18 markers and a sheet of paper with a gameboard drawn on it similar to the following:

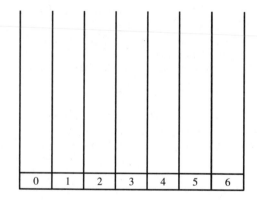

At the beginning of each game, each player places his or her markers on the boards in any arrangement above the numbers. Two players (or teams) then take turns rolling the dice. The result of each roll is the difference between the greater and the smaller of the two numbers. For example, if a 6 and a 4 were rolled, the difference would be $6 - 4$, or 2. All players who have a marker on the number that represents the difference can remove one marker. Only one marker can be removed each roll. The first player to remove all of the markers is the winner. The game may be stopped if it is clear that there can be no winner.

   a. Play the game twice. What differences seem to occur most often? Least often?
   b. Roll the dice 20 times and record how often the various differences occur. Using this information, explain how you would distribute your 18 markers to win.
   c. Compute the theoretical probabilities for each possible difference.

d. Use your answers to (c) to explain how you would arrange the markers in order to give yourself the best chance of winning.

9. The following game is played with two players and a single fair die numbered 1 through 6. The die is rolled and one person receives a score that is the square of the number appearing on the die. The other person will receive a score of 4 times the value showing on the die. The person with the greater score wins.
   a. Play the game several times to see if it appears to be a *fair game*, that is, a game in which each player has an equal chance of winning.
   b. Determine if this is a fair game. If it is not fair, who has the advantage? Explain how you arrived at your answer.

Questions from the Classroom

10. A student claims that if a fair coin is tossed and comes up heads 5 times in a row, then, according to the law of averages, the probability of tails on the next toss is greater than the probability of heads. What is your reply?

11. A student observes the following spinner and claims that the color red has the highest probability of appearing, since there are two red areas on the spinner. What is your reply?

12. A student tosses a coin 3 times, and a head appears each time. The student concludes that the coin is not fair. What is your response?

13. A student wonders why probabilities cannot be negative. What is your response?

14. A student claims that "if the probability of an event is $\frac{3}{5}$, then there are three ways the event can occur and only five elements in the sample space." How do you respond?

Third International Mathematics and Science Study (TIMSS) Questions

In an eighth-grade class of 30 students, the probability that a student chosen at random will be less than 13 years old is $\frac{1}{5}$. How many students in the class are less than 13 years old?
a. Two   **b.** Three   **c.** Four   **d.** Five   **e.** Six

TIMSS, Grade 8, 2003

In a school there were 1200 students (boys and girls). A sample of 100 students was selected at random, and 45 boys were found in the sample. Which of these is most likely to be the number of boys in the school?
a. 450   **b.** 500   **c.** 540   **d.** 600

TIMSS, Grade 8, 2003

National Assessment of Educational Progress (NAEP) Question

Each of the 6 faces of a certain cube is labeled either R or S. When the cube is tossed, the probability of the cube landing with an R face up is $\frac{1}{3}$.

How many faces are labeled R?
a. Five   **b.** Four   **c.** Three   **d.** Two   **e.** One

NAEP, Grade 8, 2005

LABORATORY ACTIVITY

1. Suppose a paper cup is tossed in the air. The different ways it can land are shown in Figure 9-9. Toss a cup 100 times and record each result. From this information, calculate the experimental probability of each outcome. Do the outcomes appear to be equally likely? Using experimental probabilities, predict how many times the cup will land on its side if tossed 100 times.

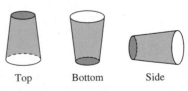

Figure 9-9

2. Toss a fair coin 200 times and record the results. From this information, calculate the experimental probability of getting heads on a particular toss. Does the experimental result agree with the expected theoretical probability of $\frac{1}{2}$?

3. Hold a coin upright on its edge under your forefinger on a hard surface and then spin it with your other finger so that it spins before landing. Repeat this experiment 100 times and calculate the experimental probability of the coin's landing on heads on a particular spin. Compare your experimental probabilities with those in activity 2.

9-2   Multistage Experiments with Tree Diagrams and Geometric Probabilities

In Section 9-1, we considered **one-stage experiments**, that is, experiments that were over after one step. For example, drawing one ball at random from the box containing a red, white, and green ball in Figure 9-10(a) is a one-stage experiment. A tree diagram for this experiment is shown in Figure 9-10(b).

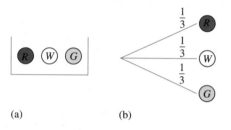

(a)   (b)

Figure 9-10

Next we consider a **two-stage experiment**. For example, a ball is drawn from the box in Figure 9-10(a) and its color is recorded. Then the ball is *replaced*, and a second ball is drawn and its color is recorded. A tree diagram for this experiment is given in Figure 9-11.

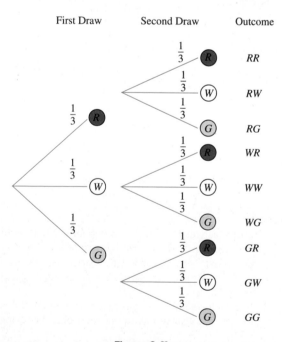

Figure 9-11

Notice that the tree diagram in Figure 9-11 can be used to generate the sample space. A sample space for this experiment may be written $\{RR, RW, RG, WR, WW, WG, GR, GW, GG\}$. Each of the outcomes in the sample space is equally likely and there are nine total outcomes, so the probability of each outcome is $\frac{1}{9}$.

An alternative way to generate the sample space is to use a table, as shown in Table 9-3.

Table 9-3

	Second Draw		
	R	W	G
R	RR	RW	RG
W	WR	WW	WG
G	GR	GW	GG

First Draw (labels R, W, G on rows)

NOW TRY THIS 9-3

1. For each of the following, use the tree diagram in Figure 9-11 to determine the probability that
    a. both balls are red.
    b. no ball is red.
    c. at least one ball is red.
    d. at most one ball is red.
    e. both balls are the same color.
2. Answer each part of question 1 using Table 9-3.

Example 9-7

Suppose we toss a fair coin 3 times and record the results. Find each of the following:

a. The sample space for this experiment
b. The event A of tossing 2 heads and 1 tail
c. The event B of tossing no tails
d. The event C of tossing a head on the last toss

Solution a. $S = \{HHH, HHT, HTH, THH, TTT, TTH, THT, HTT\}$
  b. $A = \{HHT, HTH, THH\}$
  c. $B = \{HHH\}$
  d. $C = \{HHH, HTH, THH, TTH\}$

More Multistage Experiments

The box in Figure 9-12(a) contains one colored ball and two white balls. If a ball is drawn at random and the color recorded, a tree diagram for the experiment might look like the one in Figure 9-12(b). Because each ball has the same chance of being drawn, we may combine the branches and obtain the tree diagram shown in Figure 9-12(c). Combining branches in this way is a common practice because it simplifies tree diagrams.

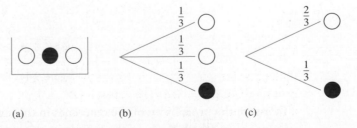

(a)   (b)   (c)

Figure 9-12

A probability model that uses geometric shapes is an *area model*. When area models are used to determine probabilities geometrically (a process sometimes called finding **geometric probabilities**), outcomes are associated with points chosen at random in a geometric region that represents a sample space. A way to show the probabilities of choosing a ball at random from the bin in Figure 9-12(a) is shown in Figure 9-13.

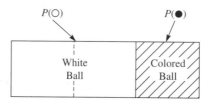

Figure 9-13

If the area of the entire rectangle is 1, the area of the white portion depicting the probability of drawing a white ball is represented as $\frac{2}{3}$ of the entire rectangle. The cross-hatched portion of the figure represents the probability of choosing a colored ball, or $\frac{1}{3}$. We interpret this as $P(\bigcirc) = \frac{2}{3}$ and $P(\bullet) = \frac{1}{3}$.

 NOW TRY THIS 9-4 Use a circle of area 1 to depict the probabilities for drawing a single ball in Figure 9-12(a).

Now suppose a ball is drawn at random from the box in Figure 9-12(a) and its color recorded. The ball is then *replaced*, and a second ball is drawn and its color recorded. The sample space for this two-stage experiment may be recorded using ordered pairs as $\{(\bullet, \bullet), (\bullet, \bigcirc), (\bigcirc, \bullet), (\bigcirc, \bigcirc)\}$ or, more commonly, as $\{\bullet\ \bullet, \bullet\ \bigcirc, \bigcirc\ \bullet, \bigcirc\ \bigcirc\}$, as shown in the tree diagram in Figure 9-14.

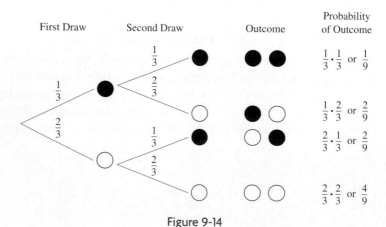

Figure 9-14

To assign the probability of the outcomes in this experiment, consider, for example, the path for the outcome $\bullet\ \bigcirc$. In the first draw, the probability of obtaining a colored ball is

$\frac{1}{3}$. Then, the probability of obtaining a white ball in the second draw is $\frac{2}{3}$. Thus we expect to obtain a colored ball on the first draw $\frac{1}{3}$ of the time and then on the second draw, to obtain a white ball $\frac{2}{3}$ of those times that we obtained a colored ball on the first draw, that is, $\frac{2}{3}$ of $\frac{1}{3}$, or $\frac{2}{3} \cdot \frac{1}{3}$. Observe that this product can be obtained by multiplying the probabilities along the branches used for the path leading to ● ○; that is, $\frac{1}{3} \cdot \frac{2}{3}$, or $\frac{2}{9}$. The probabilities shown in Figure 9-14 are obtained by following the paths leading to each of the four outcomes and multiplying the probabilities along the paths. This procedure is an instance of the following general theorem.

Theorem 9–6: Multiplication Rule for Probabilities for Tree Diagrams

For all multistage experiments, the probability of the outcome along any path of a tree diagram is equal to the product of all the probabilities along the path.

REMARK The sum of the probabilities on all the branches from any point always equals 1, and the sum of the probabilities for all the possible outcomes must also be 1.

We can use geometric probabilities to depict Figure 9-14 in a manner similar to that used for multiplying fractions in an earlier chapter. We depict the first draw as in Figure 9-15(a) where the $P(●) = \frac{1}{3}$ and the $P(○) = \frac{2}{3}$. Next, for the second draw we use horizontal lines to separate the large rectangles again into thirds and use cross-hatching to aid in showing the probabilities of the given sections.

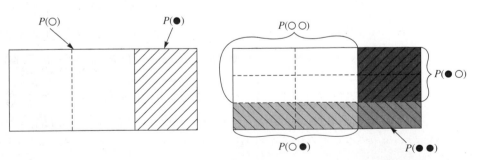

Figure 9-15

Thus we have the following: $P(● \, ●) = \frac{1}{9}$; $P(○ \, ●) = \frac{2}{9}$; $P(● \, ○) = \frac{2}{9}$; and $P(○ \, ○) = \frac{4}{9}$.

(Note: Geometric probabilities lose some usefulness as problems get more complicated.)

Look again at the box pictured in Figure 9-12(a). This time, suppose two balls are drawn one-by-one *without replacement*. A tree diagram for this experiment, along with the set of possible outcomes, is shown in Figure 9-16. Compare this figure to the one in Figure 9-14. How are they different?

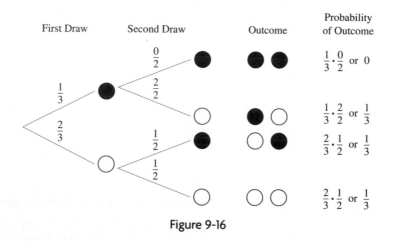

Figure 9-16

REMARK The branch showing the second draw, with probability $\dfrac{0}{2}$, is added here for completeness. As seen later, the branch could be deleted, as could any unneeded branches of the tree diagram.

The denominators of the fractions in the second draw are all 2. Because the draws are made without replacement, only two balls remain for the second draw.

Consider event A, consisting of the outcomes for drawing exactly one colored ball in the two draws without replacement. This event is given by $A = \{● ○, ○ ●\}$. Since the outcome ● ○ appears $\dfrac{1}{3}$ of the time, and the outcome ○ ● appears $\dfrac{1}{3}$ of the time, then either ● ○ or ○ ● will appear $\dfrac{2}{3}$ of the time. Thus, $P(A) = \dfrac{1}{3} + \dfrac{1}{3} = \dfrac{2}{3}$.

Event B, consisting of outcomes for drawing *at least* one colored ball, could be recorded as $B = \{● ○, ○ ●, ● ●\}$. Because $P(● ○) = \dfrac{1}{3}, P(○ ●) = \dfrac{1}{3}$, and $P(● ●) = 0$, we have $P(B) = \dfrac{1}{3} + \dfrac{1}{3} + 0 = \dfrac{2}{3}$. Because $\overline{B} = \{○, ○\}$ and $P(\overline{B}) = \dfrac{1}{3}$, the probability of B could have been computed as follows: $P(B) = 1 - P(\overline{B}) = 1 - \dfrac{1}{3} = \dfrac{2}{3}$.

Independent Events

In Figure 9-16, the fact that the first ball was not replaced affects the probability of the color of the second ball drawn. When the occurrence of one event has no influence on the outcome of a second event, the events are **independent**. For example, if two coins are flipped and event E_1 is obtaining a head on the first coin and E_2 is obtaining a tail on the second coin, then E_1 and E_2 are independent events because one event has no influence on the second. Figure 9-17(a) depicts flipping the coins. In Figure 9-17(b), the tree diagram is abbreviated to show the branches of interest. Notice that $P(E_1) = \dfrac{1}{2}, P(E_2) = \dfrac{1}{2}$, and $P(E_1 \cap E_2) = \dfrac{1}{4}$. So in this case, $P(E_1 \cap E_2) = P(E_1) \cdot P(E_2)$.

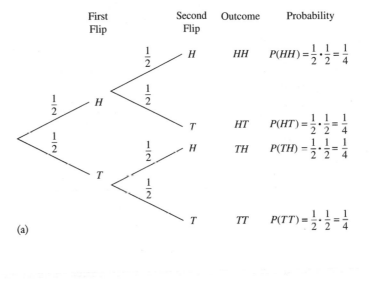

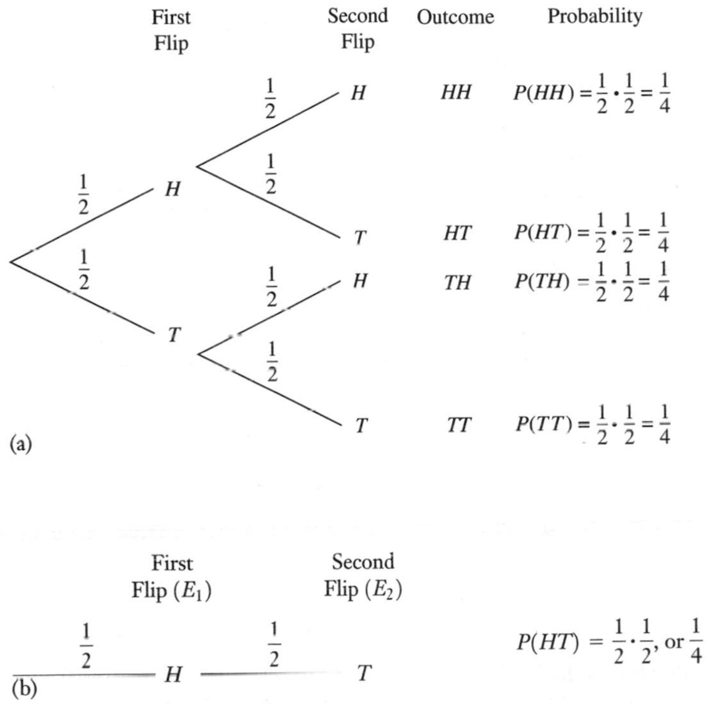

Figure 9-17

Next, consider two boxes as in Figure 9-18. Box 1 contains two white and two black balls, and box 2 contains two white balls and three black balls. Let B_1 be the event of drawing a black ball from box 1 and B_2 the event of drawing a black ball from box 2. These events are independent. Suppose a ball is drawn from each box and we are interested in the probability that each ball is black; that is, $P(B_1 \cap B_2)$. We know that $P(B_1) = \dfrac{2}{4}$, or $\dfrac{1}{2}$, and $P(B_2) = \dfrac{3}{5}$.

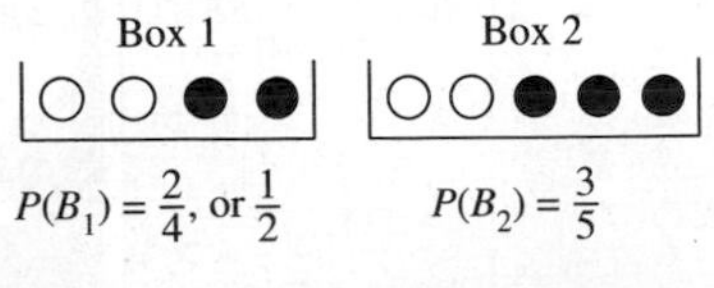

Box 1 **Box 2**

$P(B_1) = \dfrac{2}{4}$, or $\dfrac{1}{2}$ $P(B_2) = \dfrac{3}{5}$

Figure 9-18

Is $P(B_1 \cap B_2) = P(B_1) \cdot P(B_2)$ in this case as well? We can answer this question by computing $P(B_1 \cap B_2)$ using a familiar approach. If we consider all the black balls to be different and all the white balls to be different, there are $4 \cdot 5$ different pairs of balls. (Why?) Among these pairs, we are interested in pairs consisting only of black balls. There are $2 \cdot 3$ such pairs. Why? Hence,

$$P(B_1 \cap B_2) = \frac{2 \cdot 3}{4 \cdot 5} = \frac{2}{4} \cdot \frac{3}{5} = \frac{1}{2} \cdot \frac{3}{5}$$

Thus we see that for the independent events B_1 and B_2, we have $P(B_1 \cap B_2) = P(B_1) \cdot P(B_2)$. This discussion can be generalized as follows.

Theorem 9–7

For any independent events E_1 and E_2,

$$P(E_1 \cap E_2) = P(E_1) \cdot P(E_2)$$

The accompanying research note discusses student errors with a conjunction of two events.

Research Note

Students often will assign a higher probability to the conjunction of two events than to either of the two events individually. This conjunction fallacy occurs even if students have had a probability course. For example, students rate the probability of "being 55 and having a heart attack" as more likely than the probability of either "being 55" or "having a heart attack" (Kahneman and Tversky 1983). ◆

The student page on page 541 gives examples of problems that involve independent and dependent events where **dependent events** are defined to be not independent events. Answer question 1 at the bottom of the student page.

NOW TRY THIS 9-5 In the following cartoon, assume that the events are independent and that there is a 30% chance of rain tonight and a 60% chance of rain tomorrow.

a. What is the probability that it will rain both times?
b. What is the probability that it will not rain either of the times?
c. What is the probability that it will rain exactly one of the times?
d. What is the probability that it will rain at least one of the times?
e. In real life, do you think that the events of rain tonight and rain tomorrow are independent events? Why or why not?

WIZARD OF ID

School Book Page   INDEPENDENT EVENTS

Lesson 11-15

Key Idea
To find the probability that two events will occur, you need to know whether the occurrence of one event has an effect on the occurrence of the other.

Vocabulary
• independent events
• dependent events

Think It Through
I need to **remember vocabulary terms.** Mutually exclusive events cannot happen at the same time. Independent and dependent events can happen at the same time.

Multiplying Probabilities

LEARN

How do you find the probability that two events will occur?

Roxanne draws one marble from this bag without looking. Then she draws a second marble.

Example A

What is the probability that Roxanne will get a red marble each time if she replaces the first marble before she draws the second one?

Since she replaces the first marble before the second draw, the outcome of the first draw has no effect on the outcome of the second draw. The two draws are **independent events.**

To find P(red, red), find the probability of each event and multiply.

P(red, red) = P(red on 1st draw) × P(red on 2nd draw)

$$= \frac{3}{5} \times \frac{3}{5} = \frac{9}{25} = 36\%$$

1st draw

P(red) = $\frac{3}{5}$

2nd draw

P(red) = $\frac{3}{5}$

Example B

What is the probability that Roxanne will get a red marble each time if she does not replace the first marble before she draws second one?

Since the outcome of the first draw affects the outcome of the second draw, the two draws are **dependent events.**

To find P(red, red), find the probability of each event and multiply.

P(red, red) = P(red on 1st draw) × P(red on 2nd draw)

$$= \frac{3}{5} \times \frac{1}{2} = \frac{3}{10} = 30\%$$

1st draw

P(red) = $\frac{3}{5}$

2nd draw

P(red) = $\frac{1}{2}$

✔ Talk About It

1. In Example B, why is the probability of the first draw different from the second draw?

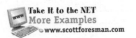
Take It to the NET
More Examples
www.scottforesman.com

672

Source: Scott Foresman-Addison Wesley Mathematics, Grade 6, 2008 (p. 672).

Example 9-8

Figure 9-19 shows a box with 11 letters. Some letters are repeated. Suppose 4 letters are drawn at random from the box one-by-one without replacement. What is the probability of the outcome *BABY*, with the letters chosen in exactly the order given?

PROBABILITY

Figure 9-19

Solution   We are interested in only the tree branch leading to the outcome *BABY*. The portion needed is shown in Figure 9-20.

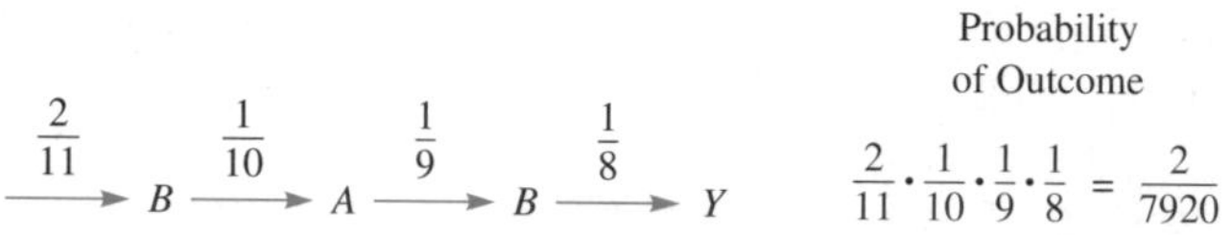

Probability
of Outcome

$$\xrightarrow{\frac{2}{11}} B \xrightarrow{\frac{1}{10}} A \xrightarrow{\frac{1}{9}} B \xrightarrow{\frac{1}{8}} Y \qquad \frac{2}{11} \cdot \frac{1}{10} \cdot \frac{1}{9} \cdot \frac{1}{8} = \frac{2}{7920}$$

Figure 9-20

The probability of the first B is $\frac{2}{11}$ because there are 2 B's out of 11 letters. The probability of the second B is $\frac{1}{9}$ because there are 9 letters left after 1 B and 1 A have been chosen. Then, $P(BABY)$ is $\frac{2}{7920}$, as shown.

In Example 9-8, suppose 4 letters are drawn one-by-one from the box and the letters are replaced after each drawing. In this case, the branch needed to find $P(BABY)$ in the order drawn is pictured in Figure 9-21. Then, $P(BABY) = \frac{2}{11} \cdot \frac{1}{11} \cdot \frac{2}{11} \cdot \frac{1}{11}$, or $\frac{4}{14,641}$.

Probability
of Outcome

$$\xrightarrow{\frac{2}{11}} B \xrightarrow{\frac{1}{11}} A \xrightarrow{\frac{2}{11}} B \xrightarrow{\frac{1}{11}} Y \qquad \frac{2}{11} \cdot \frac{1}{11} \cdot \frac{2}{11} \cdot \frac{1}{11} \text{ or } \frac{4}{14,641}$$

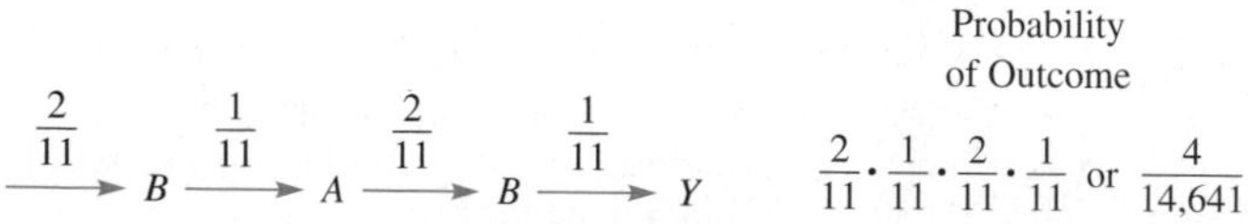

Figure 9-21

Example 9-9

Consider the three boxes in Figure 9-22. A letter is drawn from box 1 and placed in box 2. Then, a letter is drawn from box 2 and placed in box 3. Finally, a letter is drawn from box 3. What is the probability that the letter drawn from box 3 is *B*?

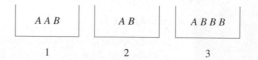

| A A B | A B | A B B B |
| 1 | 2 | 3 |

Figure 9-22

Solution   A tree diagram for this experiment is given in Figure 9-23. Notice that the denominators in the second stage are 3 rather than 2 because in this stage, there are now three letters in box 2. The denominators in the third stage are 5 because in this stage, there are five letters in box 3. To find the probability that a *B* is drawn from box 3, add the probabilities for the outcomes *AAB*, *ABB*, *BAB*, and *BBB* that make up this event.

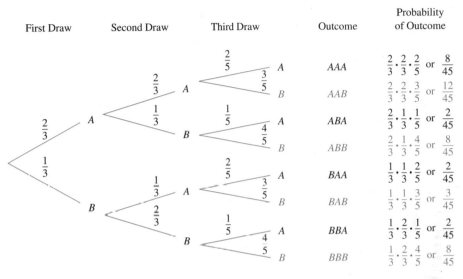

Figure 9-23

Thus, the probability of obtaining a *B* on the draw from box 3 in this experiment is

$$\frac{12}{45} + \frac{8}{45} + \frac{3}{45} + \frac{8}{45} = \frac{31}{45}$$

NOW TRY THIS 9-6 Suppose that in Example 9–9 it is known that the letter *A* was drawn on the first draw. What is the probability that

a. the last letter drawn is a *B*?
b. the last letter drawn is an *A*?
c. the last two letters drawn will match, that is, 2 *A*'s or 2 *B*'s?

Modeling Games

We can use models to analyze games that involve probability. Consider the following game, which Arthur and Gwen play: There are two colored marbles and one white marble in a box, as in Figure 9-24(a). Gwen mixes the marbles, and Arthur draws two marbles at random without replacement. If the two marbles match, Arthur wins; otherwise, Gwen wins. Does each player have an equal chance of winning? We *develop a model* to analyze the game. One possible model is a tree diagram, as shown in Figure 9-24(b). Because the outcome ○ ○ cannot happen, the tree diagram could also be shortened as in Figure 9-24(c).

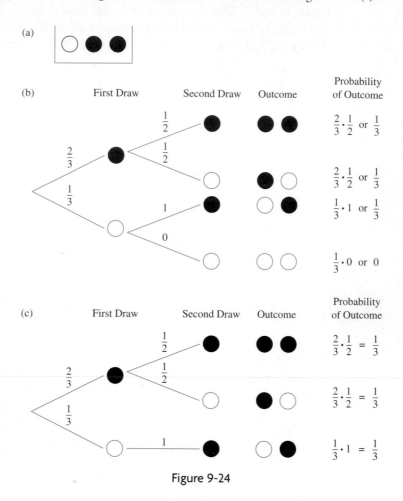

Figure 9-24

The probability that the marbles are the same color is $\frac{1}{3} + 0$, or $\frac{1}{3}$, and the probability that they are not the same color is $\frac{1}{3} + \frac{1}{3}$, or $\frac{2}{3}$. Because $\frac{1}{3} \neq \frac{2}{3}$, the players do not have the same chance of winning.

An alternative model for analyzing this game is given in Figure 9-25(a), where the colored and white marbles are shown along with the possible ways of drawing two marbles. Each line segment in the diagram represents one pair of marbles that could be drawn. S indicates that the marbles in the pair are the same color, and D indicates that the marbles are different colors. Because there are two D's in Figure 9-25(a), we see that the probability of drawing two different-colored marbles is $\frac{2}{3}$. Likewise, the probability of drawing two marbles of the same color is $\frac{1}{3}$. Because $\frac{2}{3} \neq \frac{1}{3}$, the players do not have an equal chance of winning.

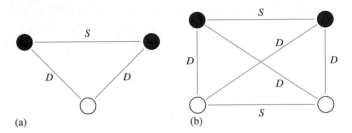

Figure 9-25

Will adding another white marble give each player an equal chance of winning? With two white and two colored marbles, we have the model in Figure 9-25(b). Therefore, $P(D) = \frac{4}{6}$, or $\frac{2}{3}$, and $P(S) = \frac{2}{6}$, or $\frac{1}{3}$. We see that adding a white marble does not change the probabilities.

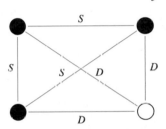

Figure 9-26

Next, consider a game with the same rules but using three colored marbles and one white marble. Figure 9-26 shows a model for this situation.

Thus, the probability of drawing two marbles of the same color is $\frac{3}{6}$, and the probability of drawing two marbles of different colors is $\frac{3}{6}$. Finally, we have a game in which each player has an equal chance of winning.

NOW TRY THIS 9-7 Referring to the games described above, answer the following:

a. Does each player have an equal chance of winning if only one white marble and one colored marble are used and the marble is replaced after the first draw?
b. Find games that involve different numbers of marbles in which each player has an equal chance of winning.
c. Find a pattern for the numbers of colored and white marbles that allow each player to have an equal chance of winning.

Problem Solving　A String-Tying Game

In a party game, a child is handed six strings, as shown in Figure 9-27(a). Another child ties the top ends two at a time, forming three separate knots, and the bottom ends, forming three separate knots, as in Figure 9-27(b). If the strings form one closed ring, as in Figure 9-27(c), the child tying the knots wins a prize. What is the probability that the child wins a prize on the first try? Guess the probability before working the problem.

(a)

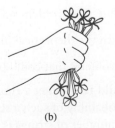

(b)

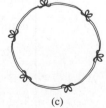

(c)

Figure 9-27

Understanding the Problem   The problem is to determine the probability that one closed ring will be formed. One closed ring means that all six pieces are joined end-to-end to form one, and only one, ring, as shown in Figure 9-27(c).

Devising a Plan   Figure 9-28(a) shows what happens when the ends of the strings of one set are tied in pairs at the top. Notice that no matter in what order those ends are tied, the result appears as in Figure 9-28(a).

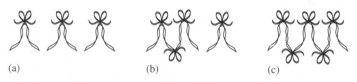

(a)   (b)   (c)

Figure 9-28

Then, the other ends are tied in a three-stage experiment. If we pick any string in the first stage, then there are five choices for its mate. Four of these choices are favorable choices for forming a ring. (Why?) Thus, the probability of forming a favorable first tie is $\frac{4}{5}$. Figure 9-28(b) shows a favorable tie at the first stage.

For any one of the remaining four strings, there are three choices for its mate. Two of these choices are favorable ones. (Why?) Thus, the probability of forming a favorable second tie is $\frac{2}{3}$. Figure 9-28(c) shows a favorable tie at the second stage.

Now, two ends remain. Since nothing can go wrong at the third stage, the probability of making a favorable tie is 1. If we use the probabilities completed at each stage and a single branch of a tree diagram, we can calculate the probability of performing three successful ties in a row and hence the probability of forming one closed ring.

Carrying Out the Plan   If we let S represent a successful tie at each stage, then the branch of the tree with which we are concerned is the one shown in Figure 9-29.

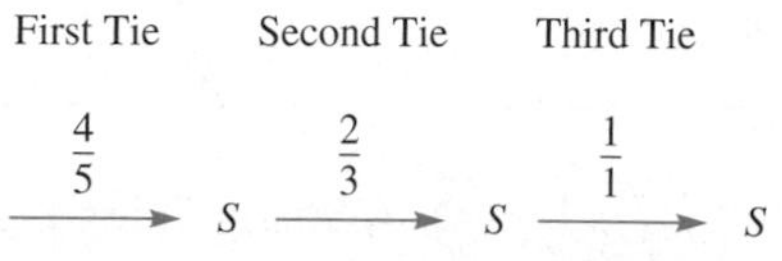

First Tie   Second Tie   Third Tie

$$\xrightarrow{\frac{4}{5}} S \xrightarrow{\frac{2}{3}} S \xrightarrow{\frac{1}{1}} S$$

Figure 9-29

Thus, the probability of forming one ring is $P(\text{ring}) = \frac{4}{5} \cdot \frac{2}{3} \cdot \frac{1}{1} = \frac{8}{15} = 0.5\overline{3}$.

Looking Back   The probability that a child will form a ring on the first try is $\frac{8}{15}$. A class might simulate this problem several times with strings to see how the fraction of successes compares with the theoretical probability of $\frac{8}{15}$.

Related problems that could be posed for solution include the following:

1. If a child fails to get a ring 10 times in a row, the child may not play again. What is the probability of such a streak of bad luck?
2. If the number of strings is reduced to three and the rule is that an upper end must be tied to a lower end, what is the probability of a single ring?

3. If the number of strings is three, but an upper end can be tied to either an upper or a lower end, what is the probability of a single ring?
4. What is the probability of forming three rings in the original problem?
5. What is the probability of forming two rings in the original problem?

Geometric Probability (Area Models) Revisited

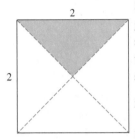

Figure 9-30

Suppose we throw darts at a square target 2 units long on a side and divided into four congruent triangles, as shown in Figure 9-30. If the dart must hit the target somewhere and if all spots can be hit with equal probability, what is the probability that the dart will land in the shaded region? The entire target, which has an area of 4 square units, represents the sample space. The shaded area is the event of a successful toss. The area of the shaded part is $\frac{1}{4}$ of the sample space. The probability of the dart's landing in the shaded region is the ratio of the area of the event to the area of the sample space, or $\frac{1}{4}$. In general when computing the probability of an event geometrically, we find the ratio: area of region representing event : area of region representing the sample space.

| Problem Solving | A Quiz-Show Game |

On a quiz show, a contestant stands at the entrance to a maze that opens into two rooms, labeled A and B in Figure 9-31. The master of ceremonies' assistant is to place a new car in one room and a donkey in the other. The contestant must walk through the maze into one of the rooms and will win whatever is in that room. If the contestant makes each decision in the maze at random, in which room should the assistant place the car to give the contestant the best chance to win?

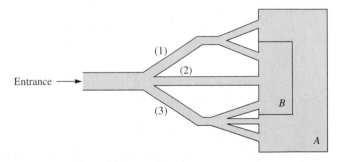

Figure 9-31

Understanding the Problem   The contestant must first choose one of the paths marked 1, 2, or 3 and then choose another path as she proceeds through the maze. To determine the room the contestant is most likely to choose, the assistant must be able to determine the probability of the contestant's reaching each room.

Devising a Plan   One way to determine where the car should be placed is to *model the choices with a tree diagram* and to compute the probabilities along the branches of the tree.

Carrying Out the Plan   A tree diagram for the maze is shown in Figure 9-32, along with the possible outcomes and the probabilities of each branch.

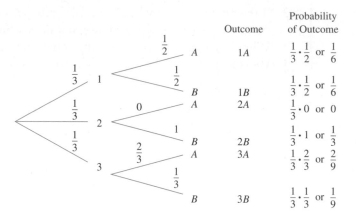

Figure 9-32

The probability that room A is chosen is $\frac{1}{6} + 0 + \frac{2}{9} = \frac{7}{18}$. Hence the probability that room B is chosen is $1 - \frac{7}{18} = \frac{11}{18}$. Thus room B has the greater probability of being chosen. This is where the car should be placed for the contestant to have the best chance of winning it.

Looking Back   An alternative model for this problem and for many probability problems is an area model. The rectangle in Figure 9-33(a) represents the first three choices that the contestant can make. Because each choice is equally likely, each is represented by an equal area. If the contestant chooses the upper path, then rooms A and B have an equal chance of being chosen. If she chooses the middle path, then only room B can be entered. If she chooses the lower path, then room A is entered $\frac{2}{3}$ of the time. This can be expressed in terms of the area model shown in Figure 9-33(b). Dividing the rectangle into pieces of equal area, we obtain the model in Figure 9-33(c), in which the area representing room B is shaded. Because the area representing room B is greater than the area representing room A, room B has the greater probability of being chosen. Figure 9-33(c) can enable us to find the probability of choosing room B. Because the shaded area consists of 11 rectangles out of a total of 18 rectangles, the probability of choosing room B is $\frac{11}{18}$. We can vary the problem by changing the maze or by changing the locations of the rooms.

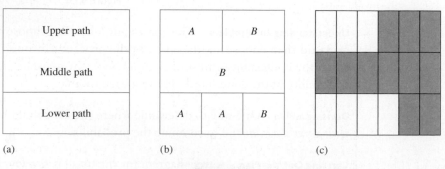

Figure 9-33

Assessment 9-2A

1. a. A box contains three white balls and two red balls. A ball is drawn at random from the box and not replaced. Then a second ball is drawn from the box. Draw a tree diagram for this experiment and find the probability that the two balls are of different colors.

b. Suppose that a ball is drawn at random from the box in part (a), its color is recorded, and then the ball is put back in the box. Draw a tree diagram for this experiment and find the probability that the two balls are of different colors.

2. A box contains six letters, shown as follows. What is the probability of the outcome *DAN* in that order if three letters are drawn one-by-one

a. with replacement?

b. without replacement?

$$\boxed{R\,A\,N\,D\,O\,M}$$

3. An executive committee consisted of 10 members: 4 women and 6 men. Three members were selected at random to be sent to a meeting in Hawaii. A blindfolded woman drew 3 of the 10 names from a hat to determine who would go. All 3 names drawn were women's. What was the probability of such luck?

4. Following are three boxes containing balls. Draw a ball from box 1 and place it in box 2. Then draw a ball from box 2 and place it in box 3. Finally, draw a ball from box 3.

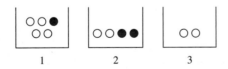

a. What is the probability that the last ball, drawn from box 3, is white?

b. What is the probability that the last ball drawn is colored?

5. Assume the probability is $\frac{1}{2}$ that a child born is a boy and births are independent. What is the probability that if a family will have four children (no twins), they will all be boys?

6. A box contains five slips of paper. Each slip has one of the numbers 4, 6, 7, 8, or 9 written on it and all numbers are used. The first player reaches into the box and draws two slips and adds the two numbers. If the sum is even, the player wins. If the sum is odd, the player loses.

a. What is the probability that the player wins?

b. Does the probability change if the two numbers are multiplied? Explain.

7. The following shows the numbers of symbols on each of the three dials of a standard slot machine:

Symbol	Dial 1	Dial 2	Dial 3
Bar	1	3	1
Bell	1	3	3
Plum	5	1	5
Orange	3	6	7
Cherry	7	7	0
Lemon	3	0	4
Total	20	20	20

Find the probability for each of the following:

a. Three plums   **b.** Three oranges

c. Three lemons   **d.** No plums

8. You play a game in which you first choose one of the two spinners shown. You then spin your spinner and a second person spins the other spinner. The one with the greater number wins.

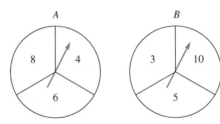

a. Which spinner should you choose? Why?

b. Notice that the sum of the numbers on each spinner is 18. Design two spinners with unequal sums so that choosing the spinner with the lesser sum will give the player a greater probability of winning.

9. If a person takes a five-question true-false test, what is the probability that the score is 100% correct if the person guesses on every question?

10. Rattlesnake and Paxson Colleges play four games against each other in a chess tournament. Rob Fisher, the chess whiz from Paxson, withdrew from the tournament, so the probabilities that Rattlesnake and Paxson will win each game are $\frac{2}{3}$ and $\frac{1}{3}$, respectively. Determine the following probabilities:

a. Paxson loses all four games.

b. The match is a draw with each school winning two games.

11. Consider the following dartboard: (Assume that all quadrilaterals are squares and that the x's represent equal measures.)

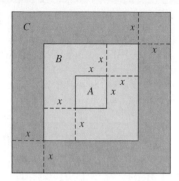

If a dart may land in any region on the board with probability proportional to the area of that region, what is the probability that it will land in
   a. Section A?   b. Section B?   c. Section C?

12. In the following square dartboard, suppose a dart is equally likely to land in any small square of the board.

Points are given as follows:

Region	Points
A	10
B	8
C	6
D	4
E	2

   a. What is the total area of the board?
   b. What is the probability of a dart's landing in each region of the board?
   c. If two darts are tossed, what is the probability of scoring 20 points?
   d. What is the probability that the dart will land in neither D nor E?

13. An electric clock is stopped by a power failure. What is the probability that the second hand is stopped between the 3 and the 4?

14. Let $A = \{x \mid {}^{-}1 < x < 1\}$ and $B = \{x \mid {}^{-}3 < x < 2\}$. If a real number is picked at random from set B, what is the probability it will be in set A?

15. There are 40 employees in a certain firm. We know that 28 of these employees are males, 2 of these males are secretaries, and 10 secretaries are employed by the firm. What is the probability that an employee chosen at random is a secretary, given that the person is a male?

16. Four blue socks, four white socks, and four gray socks are mixed in a drawer. You pull out two socks, one at a time, without looking.
   a. Draw a tree diagram along with the possible outcomes and the probabilities of each branch.
   b. What is the probability of getting a pair of socks of the same color?
   c. What is the probability of getting two gray socks?
   d. Suppose that, instead of pulling out two socks, you pull out four socks. What is the probability now of getting two socks of the same color?

17. When you toss a quarter 4 times, what is the probability that you get at least
   a. as many heads as tails?
   b. as many tails as heads?

18. A manufacturer found that among 500 randomly selected smoke detectors only 450 worked properly. Based on this information, how many smoke detectors would you have to install to be sure that the probability that at least one of them will work will be at least 99.99%?

★19. Carolyn will win a large prize if she wins two tennis games in a row out of three games. She is to play alternately against Billie and Bobby. She may choose to play Billie-Bobby-Billie or Bobby-Billie-Bobby. She wins against Billie 50% of the time and against Bobby 80% of the time. Which alternative should she choose, and why?

Assessment 9-2B

1. Suppose an experiment consists of spinning X and then spinning Y, as follows:

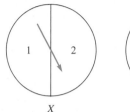

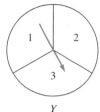

Find the following:
a. The sample space S for the experiment
b. The event A consisting of outcomes from spinning an even number followed by an even number
c. The event B consisting of outcomes from spinning at least one 2
d. The event C consisting of outcomes from spinning exactly one 2

2. Following are three boxes containing letters.

|MATH| |AND| |HISTORY|
   1   2   3

a. From box 1, three letters are drawn one-by-one without replacement and recorded in order. What is the probability that the outcome is *HAT*?
b. From box 1, three letters are drawn one-by-one with replacement and recorded in order. What is the probability that the outcome is *HAT*?
c. One letter is drawn at random from box 1, then another from box 2, and then another from box 3, with the results recorded in order. What is the probability that the outcome is *HAT*?
d. If a box is chosen at random and then a letter is drawn at random from the box, what is the probability that the outcome is *A*?

3. Two boxes with letters follow. You are to choose a box and draw three letters at random, one-by-one, without replacement. If the outcome is SOS, you win a prize.

|SOS| |SOSSOS|
  1   2

a. Which box should you choose?
b. Which box would you choose if the letters are to be drawn with replacement?

4. An assembly line has two inspectors. The probability that the first inspector will miss a defective item is 0.05.

If the defective item passes the first inspector, the probability that the second inspector will miss it is 0.01. What is the probability that a defective item will pass by both inspectors?

5. Following are two boxes containing colored and white balls. A ball is drawn at random from box 1. Then a ball is drawn at random from box 2, and the colors of balls from both boxes are recorded in order.

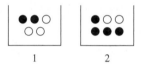

Find each of the following:
a. The probability of two white balls
b. The probability of at least one colored ball
c. The probability of at most one colored ball
d. The probability of ● ○ or ○ ●

6. A penny, a nickel, a dime, and a quarter are tossed. What is the probability of obtaining at least three heads on the tosses?

7. Brittany is going to ascend a four step staircase. At any time, she is just as likely to stride up one step or two steps. Find the probability that she will ascend the four steps in
a. two strides   b. three strides   c. four strides

8. Suppose we spin the following spinner with the first spin giving the numerator and the second spin giving the denominator of a fraction. What is the probability that the fraction will be greater than $1\frac{1}{2}$?

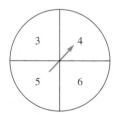

9. An experiment consists of spinning the spinner shown and then flipping a coin with sides numbered 1 and 2.

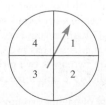

What is the probability that
  a. the number on the spinner will be greater than the number on the coin?
  b. the outcome will consist of two consecutive integers in any order?

10. The combinations on the lockers at the high school consist of three numbers, each ranging from 0 to 39. If a combination is chosen at random, what is the probability that the first two numbers are each multiples of 9 and the third number is a multiple of 4?

11. The following box contains the 11 letters shown. The letters are drawn one-by-one without replacement, and the results are recorded in order. Find the probability of the outcome MISSISSIPPI.

$$\boxed{\text{M I I I I P P S S S S}}$$

12. The land area of Earth is approximately $57,500,000 \text{ mi}^2$. The water area of Earth is approximately $139,600,000 \text{ mi}^2$. If a meteor lands at random on the planet, what is the probability, to the nearest tenth, that it will hit water?

13. A husband and wife discover that there is a 10% probability of their passing on a hereditary disease to any of their children. If they plan to have three children, what is the probability that at least one child will inherit the disease?

14. At a certain hospital, 40 patients have lung cancer, 30 patients smoke, and 25 have lung cancer and smoke. Suppose the hospital contains 200 patients. If a patient chosen at random is known to smoke, what is the probability that the patient has lung cancer?

15. In a certain population of caribou, the probability of an animal's being sickly is $\frac{1}{20}$. If a caribou is sickly, the probability of its being eaten by wolves is $\frac{1}{3}$. If a caribou is not sickly, the probability of its being eaten by wolves is $\frac{1}{150}$. If a caribou is chosen at random from the herd, what is the probability that it will be eaten by wolves?

16. Solve the Quiz-Show Game in this section by replacing Figure 9-31 with the following maze:

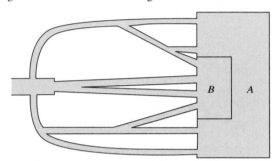

★17. Jane has two tennis serves, a hard serve and a soft serve. Her hard serve is in (a good serve) 50% of the time, and her soft serve is in (good) 75% of the time. If her hard serve is in, she wins 75% of her points. If her soft serve is in, she wins 50% of her points. Since she is allowed to reserve one time if her first serve is out, what should her serving strategy be? That is, should she serve hard followed by soft; both hard; soft followed by hard; or both soft? Note that two bad serves is a lost point.

Mathematical Connections 9-2

Communication

1. Jim rolled a fair die 5 times and obtained a 3 every time. He concluded that on the next roll, a 3 is more likely to occur than the other numbers. Explain whether this is true.

2. A witness to a crime observed that the criminal had blond hair and blue eyes and drove a red car. When the police look for a suspect, is the probability greater that they will arrest someone with blond hair and blue eyes or that they will arrest someone with blond hair and blue eyes who drives a red car? Explain your answer.

3. José says that if a family has two children, then the only possibilities are two boys, two girls, or one boy and one girl. Because there are only three possibilities, José claims the probability of having two boys is $\frac{1}{3}$. How would you respond to José?

4. You are given three white balls, one red ball, and two identical boxes. You are asked to distribute the balls in the boxes in any way you like. You then are asked to select a box (after the boxes have been shuffled) and to pick a ball at random from that box. If the ball is red, you win a prize. How should you distribute the balls in the boxes to maximize your chances of winning? Justify your reasoning.

Open-Ended

5. Make up a game in which the players have an equal chance of winning and that involves rolling two regular dice.

★6. How can the faces of two cubes be numbered so that when they are rolled, the resulting sum is a number 1 to 12 inclusive and each sum has the same probability?

7. Use graph paper to design a dartboard such that the probability of hitting a certain part of the board is $\frac{3}{5}$. Explain your reasoning.

Cooperative Learning

8. Use two spinners divided into four equal areas with the areas numbered 1–4. A player spins both spinners and computes the product of the two numbers. If the product is 1, 2, 3, or 4, player A wins and places a marker on the appropriate square on the gameboard shown. In the same way, player B wins if the product is 6, 8, 9, 12, or 16 and places a marker in the appropriate square. The game ends when one player crosses the finish line or 20 plays have been made. Each player receives 1 point for each marker, and the game winner is the person with the most points when the game ends.

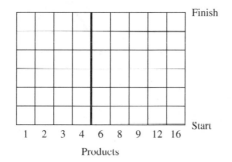

a. Do you think the game looks fair? Why or why not? If not, who do you think has the best chance of winning?
b. Play the game to see if it seems fair. Do you think it is fair based on playing it? Why?
c. Complete the following table to determine possible products.

	2nd spin			
×	1	2	3	4
1				
2				
3			12	
4				

(1st spin)

d. Based on the table in part (c), is this a fair game? Explain why.
e. Replace the spinners in the preceding game with ones divided into six equal areas with the numbers 1, 2, 3, 4, 5, and 6 in the regions. Design a similar game and decide what products each player should use so that this is a fair game. Explain how your game is fair.

9. Play the following game in pairs. One player chooses one of four equally likely outcomes from the sample space {*HH, HT, TH, TT*}, obtained by tossing a fair coin twice. The other player then chooses one of the other outcomes. A coin is flipped until either player's choice appears. For ex-

ample, the first player chooses *TT* and the second player chooses *HT*. If the first two flips yield *TH*, then no one wins and the game continues. If, after five flips, the string *THHHT* appears, the second player is the winner because the sequence *HT* finally appeared. Play the game 10 times. Does each player appear to have the same chance of winning? Analyze the game for the case in which the first player chooses *TT* and the second *HT*, and explain whether the game is fair for all choices made by the two players. (*Hint*: Find the probability that the first player wins the game by showing that the first player will win if, and only if, "tails" appears on the first and on the second flip.)

10. Consider the three spinners *A*, *B*, and *C* shown in the following figure:

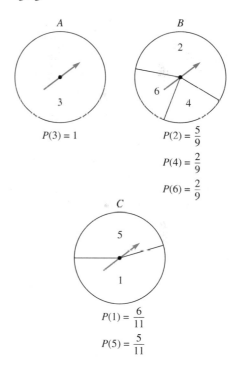

a. Suppose there are only two players and that the first player chooses a spinner, then the second player chooses a different spinner, and each person spins his or her spinner, with the greater number winning. Play the game several times to get a feeling for it. Determine if each player appears to have the same chances of winning the game. If not, which spinner should you choose in order to win?
b. This time play the same game with three players. If each player must choose a different spinner, is the winning strategy the same as it was in (a)? Why or why not?

Questions from the Classroom

11. An experiment consists of tossing a fair coin twice. The student reasons that there are three possible outcomes: two heads, one head and one tail, or two tails. Thus,

$$P(HH) = \frac{1}{3}.$$ What is your reply?

12. A student would like to know the difference between two events being independent and two events being mutually exclusive. How would you answer her?

13. In response to the question "If a fair die is rolled twice, what is the probability of rolling a pair of 5s?" a student replies, "One-third, because $\frac{1}{6} + \frac{1}{6} = \frac{1}{3}$." How do you respond?

14. A student is not sure when to add and when to multiply probabilities. How do you respond?

15. Alberto is to spin the spinners shown and compute the probability of two blacks. He looks at the spinners and says the answer is $\frac{1}{2}$ because $\frac{1}{2}$ of the areas of the circles are black. How do you respond?

16. There are two bags each containing red balls and yellow balls. Bag A contains 1 red and 4 yellow balls. Bag B contains 3 red and 13 yellow balls. Alva says that you should always choose bag B if you are trying to draw a red ball because it has more red balls than bag A. How do you respond?

Review Problems

17. Match the following phrase to the probability that describes it:

    a. A certain event   **(i)** $\dfrac{1}{1000}$

    b. An impossible event   **(ii)** $\dfrac{999}{1000}$

    c. A very likely event   **(iii)** 0

    d. An unlikely event   **(iv)** $\dfrac{1}{2}$

    e. A 50% chance   **(v)** 1

18. A date in the month of April is chosen at random. Find the probability of the date's being each of the following:
    a. April 7
    b. April 31
    c. Before April 20

19. Three men were walking down a street talking when they met a fourth man. If the fourth man knew that only two of the men always lied and the third tells the truth and he asked the three a question, what is the probability that he got a truthful answer when one man answered?

20. If there were 16 people at a party and one of the group decided to conduct a poll of who the president of the United States was, explain whether you would expect the answer to be accurate and why.

Third International Mathematics and Science Study (TIMSS) Question

The figure below shows a spinner with 24 sectors. When someone spins the arrow, it is equally likely to stop on any sector.

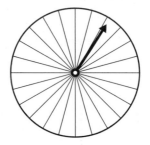

$\frac{1}{8}$ of the sectors are blue, $\frac{1}{24}$ are purple, $\frac{1}{2}$ are orange, and $\frac{1}{3}$ are red. If a person spins the arrow, on which color sector is the spinner LEAST likely to stop?
    a. blue
    b. purple
    c. orange
    d. red

TIMSS, Grade 8, 2003

National Assessment for Educational Progress (NAEP) Question

A package of candies contained only 10 red candies, 10 blue candies, and 10 green candies. Bill shook up the package, opened it, and started taking out one candy at a time and eating it. The first 2 candies he took out and ate were blue. Bill thinks the probability of getting a blue candy on his third try is $\frac{10}{30}$ or $\frac{1}{3}$. Is Bill correct or incorrect?

NAEP, Grade8, 2005

BRAIN TEASER Bradley Efron, a Stanford University statistician, designed a set of nonstandard dice whose faces are numbered as shown in Figure 9-34. The dice are to be used in a game in which each player chooses a die and then rolls it. Whoever rolls the greater number is the winner. What strategy should you use so that you have the best chance of winning this game?

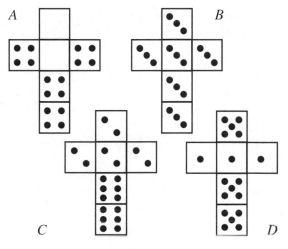

Figure 9-34

9-3　Using Simulations in Probability

Students may use simulations to study phenomena too complex to analyze by other means. A **simulation** is a technique used to act out a problem by conducting experiments whose outcomes are analogous to the original problem. Using simulations, students typically estimate a probability using many trials rather than determine probabilities theoretically. Simulations can take various forms. For example, consider the simulation on the student page on page 556. In this case, the students use a simulation with a globe to find the probability that a meteorite will hit land. Follow the instructions on the student page and then answer the questions in Now Try This 9-8

NOW TRY THIS 9-8

a. After doing the simulation on the student page, what percentage of Earth do you think is covered by land?

b. What is the probability that a meteorite will hit water?

c. It is estimated that approximately 500 meteorites hit Earth each year. How many would you expect to hit land each year?

School Book Page

Exploration 1

GEOMETRIC Probability

SET UP   *Work as a class. You will need an inflatable globe.*

▶ **A meteorite has an equally likely chance of hitting anywhere on Earth. To find the probability that a meteorite will hit land you will conduct an experiment with a globe.**

2 **Try This as a Class** Follow the steps below to simulate a meteorite falling to Earth.

First   Make a table like the one shown to record the results of your experiment.

Next   Carefully toss an inflated globe from one person to another. Each time the globe is caught, make a tally mark to record whether the left index finger is touching land or touching water before tossing it again. Record the results of 30 tosses.

Then   Complete your table by calculating the percent of the tosses that hit land and the percent that hit water.

FOR ◀HELP
with *finding probabilities*, see
MODULE 4, p. 242

	Tally	Number of tosses	Percent of tosses
Land			
Water			
Total		30	

  566   **Module 8** Our Environment

Source: Math Thematics New Edition, Book 1, McDougal Littell, 2008 (p. 566).

In the *Principles and Standards*, we find the following:

> If simulations are used, teachers need to help students understand what the simulation data represent and how they relate to the problem situation, such as flipping coins. (p. 254)

In Section 9-1 and in Table 9-1, we saw partial results from John Kerrich's coin tosses. It would take a considerable amount of time to toss a coin 10,000 times as he did. But suppose we could simulate those tosses easily without having to flip a coin that many times. If that could be done, then we could make predictions based on the simulation, which, if the simulation was done correctly, would be very much like the experimental results that he found. So what makes sense to simulate a coin toss?

Because the probabilities of tossing a fair coin and obtaining a head or a tail are each $\frac{1}{2}$, it is sensible to devise a system that has two outcomes, each of which is expected to have a probability of $\frac{1}{2}$. Consider the following possibilities (and there are many more):

Scenario 1: Take an ordinary deck of playing cards. There are 52 cards (without jokers) in the deck, half of which are red and half of which are black. One simulation for the coin toss is simply to draw a card at random from the deck. Assign red to mean "a head appears," and assign black to mean "a tail appears." The probability of drawing a red (or obtaining a head) is $\frac{1}{2}$, and the probability of obtaining a black (or a tail) is $\frac{1}{2}$. No coin has to be used, but the flipping of a coin could be simulated by drawing a card.

Scenario 2: Take an ordinary die with six faces, half of which are even and half of which are odd. We could assign an even number as a "head" and an odd number as a "tail." Rolling the die then becomes a simulation for flipping a coin.

Scenario 3: We could use a **random-number table** as in Table 9-4. The digits in this table form a list of digits selected at random, often by a computer or a calculator. To simulate the coin toss, we pick a number at random to start, and then read across the table, letting an even digit represent "heads" and an odd digit represent "tails." By keeping a record of what we obtain as we read, we simulate the probability of heads as the ratio of the number of even digits found divided by the total number read.

Scenario 4: Use a spreadsheet to generate random numbers much like that of scenario 3. We can do this in Excel 2007 by assigning in cell A1 the following: =randbetween (1, 100) and pressing return. An alternate is to use = Rounddown (Rand (), 2*100 + 1) in some earlier versions. A random number between 1 and 100 will be placed in cell A1. By using the fill down command, we generate as many random numbers as desired. Then we can determine how many are even and how many are odd to simulate the probabilities as in Scenario 3. (Note that we could make the spreadsheet "count" the even and odd numbers obtained for us.)

Having a simulation to toss a coin is not especially meaningful in the real world, but suppose you wanted to estimate how many boys and how many girls are in an expected population boom of 1,000,000 children. If we assume that the probability of a boy or a girl is approximately the same, then we could use any of the simulations in scenarios 1–4 to estimate the probability of each. Based on the probabilities, a proportionate number of the 1,000,000 births could be predicted to be girls. The remainder would be boys.

Table 9-4   Random Digits

36422	93239	76046	81114	77412	86557	19549	98473	15221	87856
78496	47197	37961	67568	14861	61077	85210	51264	49975	71785
95384	59596	05081	39968	80495	00192	94679	18307	16265	48888
37957	89199	10816	24260	52302	69592	55019	94127	71721	70673
31422	27529	95051	83157	96377	33723	52902	51302	86370	50452
07443	15346	40653	84238	24430	88834	77318	07486	33950	61598
41348	86255	92715	96656	49693	99286	83447	20215	16040	41085
12398	95111	45663	55020	57159	58010	43162	98878	73337	35571
77229	92095	44305	09285	73256	02968	31129	66588	48126	52700
61175	53014	60304	13976	96312	42442	96713	43940	92516	81421
16825	27482	97858	05642	88047	68960	52991	67703	29805	42701
84656	03089	05166	67571	25545	26603	40243	55482	38341	97782
03872	31767	23729	89523	73654	24626	78393	77172	41328	95633
40488	70426	04034	46618	55102	93408	10965	69744	80766	14889
98322	25528	43808	05935	78338	77881	90139	72375	50624	91385
13366	52764	02407	14202	74172	58770	65348	24115	44277	96735
86711	27764	86789	43800	87582	09298	17880	75507	35217	08352
53886	50358	62738	91783	71944	90221	79403	75139	09102	77826
99348	21186	42266	01531	44325	61042	13453	61917	90426	12437
49985	08787	59448	82680	52929	19077	98518	06251	58451	91140
49807	32863	69984	20102	09523	47827	08374	79849	19352	62726
46569	00365	23591	44317	55054	99835	20633	66215	46668	53587
09988	44203	43532	54538	16619	45444	11957	69184	98398	96508
32916	00567	82881	59753	54761	39404	90756	91760	18698	42852
93285	32297	27254	27198	99093	97821	46277	10439	30389	45372
03222	39951	12738	50303	25017	84207	52123	88637	19369	58289
87002	61789	96250	99337	14144	00027	43542	87030	14773	73087
68840	94259	01961	42552	91843	33855	00824	48733	81297	80411
88323	28828	64765	08244	53077	50897	91937	08871	91517	19668
55170	71062	64159	79364	53088	21536	39451	95649	65256	23950

The examples above are for expected single outcomes of an experiment. Suppose we simulate the probability of a couple having two girls (*GG*) in an expected family. We could use the random-digit table with an even digit representing a girl and an odd digit representing a boy. Because there are two children, we need to consider pairs of digits. If we examine 100 pairs, then the simulated probability of *GG* will be the number of pairs of even digits divided by 100, the total number of pairs considered.

Example 9-10

A baseball player, Reggie, has a batting average of .400. Assume that this is the theoretical probability of him getting a hit on any particular time at bat. Estimate the probability that he will get at least one hit in his next three times at bat.

Solution   We use a random-digit table to simulate this example. We choose a starting point and place the random digits in groups of three. Because Reggie's probability of getting a hit on any particular time at bat is 0.400, we could use the occurrence of four numbers from 0 through 9 to represent a hit. Suppose a hit is represented by the digits 0, 1, 2, and 3. At least one hit is obtained in three times at bat if, in any sequence of three digits, a 0, 1, 2, or 3 appears. Data for 50 trials are given next:

780	862	760	580	783	720	590	506	021	366
848	118	073	077	042	254	063	667	374	153
377	883	573	683	780	115	662	591	685	274
279	652	754	909	754	892	310	673	964	351
803	034	799	915	059	006	774	640	298	961

We see that a 0, 1, 2, or 3 appears in 42 out of the 50 trials. Thus, an estimate for the probability of at least one hit on Reggie's next three times at bat is $\frac{42}{50}$.

NOW TRY THIS 9-9

a. Use the random-digit table to estimate the probability that in a family of three, there are two girls and one boy.

b. Determine the theoretical probability of the family's having two girls and one boy and compare the answer to the simulated probability in (a).

c. Should the answers in (a) and (b) always be exactly the same? Why? How can you make sure that the answers are approximately the same?

From a random sample, we may be able to deduce information about the population from which the sample was taken. To see how this can be done, consider Example 9-11.

Example 9-11

To determine the number of fish in a certain pond, suppose we capture 300 fish, mark them, and throw them back into the pond. Suppose that the next day, 200 fish are caught and 20 of these are already marked. These 200 fish are then thrown back into the pond. Estimate how many fish are in the pond.

Solution Because 20 of the 200 fish are marked, we assume that $\frac{20}{200}$, or $\frac{1}{10}$, of the fish are marked. Thus, $\frac{1}{10}$ of the population is marked. If n represents the population, then $\frac{1}{10} n = 300$ and $n = 300 \cdot 10 = 3000$. Hence, an estimate for the fish population of the pond is 3000.

The following simulation problem concerns how many chips there are in chocolate chip cookies.

Example 9-12

Suppose Lucy makes enough batter for exactly 100 chocolate chip cookies and mixes 100 chocolate chips into the batter. If the chips are distributed at random and Charlie chooses a cookie at random from the 100 cookies, estimate the probability that it will contain exactly one chocolate chip.

Solution   We can use a simulation to estimate the probability of choosing a cookie with exactly one chocolate chip. We construct a 10×10 grid, as shown in Figure 9-35(a), to represent the 100 cookies Lucy made. Each square (cookie) can be associated with some ordered pair, where the first component is for the horizontal scale and the second is for the vertical scale. For example, the squares (0, 2) and (5, 3) are pictured in Figure 9-35(a). Using the random-digit Table 9-4, close your eyes and then take a pencil and point to one number to start. Look at the number and the number immediately following it. Consider these numbers an ordered pair and continue until you obtain 100 ordered pairs to represent the 100 chips. For example, suppose we start at a 3 and the numbers following 3 are as follows:

$$39968 \qquad 80495 \qquad 00192\ldots$$

Then the ordered pairs would be given as (3, 9), (9, 6), (8, 8), (0, 4), and so on. Use each pair of numbers as the coordinates for the square (cookie) and place a tally on the grid to represent each chip, as shown in Figure 9-35(b). We estimate the probability that a cookie has exactly one chip by counting the number of squares with exactly one tally and dividing by 100.

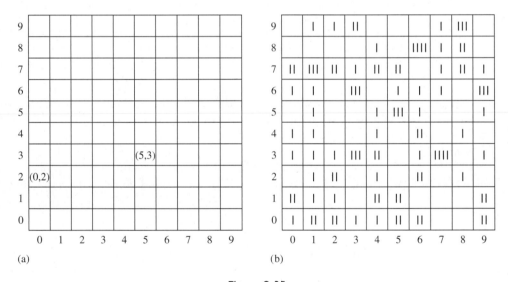

Figure 9-35

Table 9-5 shows the results of one simulation. Thus the estimate for the probability of Charlie receiving a cookie with exactly one chip is $\dfrac{34}{100}$.

Table 9-5

Number of Chips	Number of Cookies
0	38
1	34
2	20
3	6
≥ 4	2

Try a simulation on your own and compare your results with the preceding ones and with the results given in Table 9-6, obtained by theoretical methods.

Table 9-6

Number of Chips	Number of Cookies
0	36.8
1	36.8
2	18.4
3	6.1
≥ 4	1.9

Assessment 9-3A

1. Could you use a thumb tack to simulate the birth of boys and girls?
2. How could you use a random-number table to estimate the probability that two cards drawn from a standard deck of cards with replacement will be of the same suit?
3. How might you use a random-digit table to simulate each of the following?
   a. Tossing a single die
   b. Choosing three people at random from a group of 20 people
   c. Spinning the spinner, where the probability of each color is as shown in the following figure:

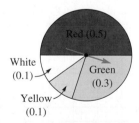

4. A school has 500 students. The principal is to pick 30 students at random from the school to go to the Rose Bowl. How can this be done by using a random-digit table?
5. In a certain city, the probability that it will rain on a certain day is 0.8 if it rained the day before. The probability that it will be dry on a certain day is 0.3 if it was dry the day before. It is now Sunday, and it is raining. Use the random-digit table to simulate the weather for the rest of the week.

6. It is reported that 15% of people who come into contact with a person infected with strep throat contract the disease. How might you use the random-digit table to simulate the probability that at least one child in a three-child family will catch the disease, given that each child has come into contact with the infected person?
7. Pick a block of two digits from the random-digit table. What is the probability that the block picked is less than 30?
8. An estimate of the fish population of a certain pond was found by catching 200 fish and marking and returning them to the pond. The next day, 300 fish were caught, of which 50 had been marked the previous day. Estimate the fish population of the pond.
9. Suppose that in the World Series, the two teams are evenly matched. The two teams play until one team wins four games, and no ties are possible.
   a. What is the maximum number of games that could be played?
   b. Use simulation to approximate the probabilities that the series will end in (i) four games and in (ii) seven games.
10. Assume Carmen Smith, a basketball player, makes free throws with 80% probability of success and is placed in a one-and-one situation where she is given a second foul shot only if the first shot goes through the basket. Simulate the 25 attempts from the foul line in one-and-one situations to determine how many times we would expect Carmen to score 0 points, 1 point, and 2 points.

Assessment 9-3B

1. How could you use a spinner as shown below to simulate the birth of boys and girls?

2. How could you use a random-digit generator or random-number table to simulate rain if you knew that 60% of the time with conditions as you have today, it will rain?

3. How could you use a random-digit generator or a random-number table to simulate choosing 4 people at random from a group of 25 people?

4. In a school with 200 students, the band director chooses four people at random to carry banner in front of the marching band. If she uses a random-digit table or a random-number generator, how could this be done?

5. Simulate the probability of tossing two fair coins 25 times and obtaining the outcome TT, where T represents tails. What is the simulated probability and what method did you use?

6. An estimate of the frog population in a certain pond was found by catching 30 frogs, marking them, and returning them to the pond. The next day, 50 frogs were caught, of which 14 had been marked the previous day. Estimate the frog population of the pond.

7. Suppose a dot is placed at random in a 10 × 10 graph grid in which squares have been numbered from 1 to 100 with no number repeated. Now simulate 100 students choosing a number at random between 1 and 100 inclusive. How many students might you expect to choose the square with the dot?

8. In a class of seven students, a teacher spins a seven-sectored spinner (with the same size sectors) to determine how to ask students questions. About how many questions would an individual student be asked out of 100 questions total?

Mathematical Connections 9-3

Communication

1. In an attempt to reduce the growth of its population, China instituted a policy limiting a family to one child. Rural Chinese suggested revising the policy to limit families to one son. Assuming the suggested policy is adopted and that any birth is as likely to produce a boy as a girl, explain how to use simulation to answer the following:
   a. What would be the average family size?
   b. What would be the ratio of newborn boys to newborn girls?

2. Consider a "walk" on the following grid starting out at the origin 0 and "walking" 1 unit (block) north, and at each intersection turning left with probability $\frac{1}{2}$, turning right with probability $\frac{1}{6}$, and moving straight with probability $\frac{1}{3}$. Explain how to simulate the "walk" using a regular six-sided die.

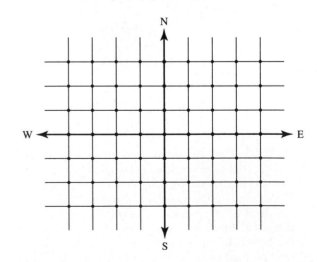

Open-Ended

3. What is the probability that in a group of five people chosen at random, at least two will have birthdays in the same

month? Design a simulation for this problem and try your simulation 10 times.

4. The probability of the home team's winning a basketball game is 80%. Describe a simulation of the probability that the home team will win three home games in a row.

5. Montana duck hunters are all perfect shots. Ten Montana hunters are in a duck blind when 10 ducks fly over. All 10 hunters pick a duck at random to shoot at, and all 10 hunters fire at the same time. How many ducks could be expected to escape, on the average, if this experiment were repeated a large number of times? How could this problem be simulated?

Cooperative Learning

6. The sixth-grade class decided that the ideal number of children in a family is four: two boys and two girls.
   a. As a group, design a simulation to determine the probability of two boys and two girls in a family with four kids.
   b. Have each person in your group try the simulation 25 times and compare the probabilities.
   c. Combine the results of all the members of your group and use this information to find a simulated probability.
   d. Compute the theoretical probability of having two boys and two girls in a family of four and compare your answer to the simulated probability.

7. Have you ever wondered how you would score on a 10-item true-false test if you guessed at every answer?
   a. Simulate your score by tossing a coin 10 times, with heads representing true and tails representing false. Check your answers by using the following key and score yourself:

1	2	3	4	5	6	7	8	9	10
F	T	F	F	T	F	T	T	T	F

   b. Combine your results with those of others in your group to find the average (mean) number of correct answers for your group.
   c. How many items would you expect to get correct if the number of items were 30 instead of 10?
   d. What is the theoretical probability of your getting all the answers correct on a 10-item true-false test if all answers were chosen by flipping a coin?

8. a. Estimate how many cards you would expect to have to turn over on the average in an ordinary playing deck before an ace appears.
   b. Have each person in your group repeat the experiment 10 times or simulate it for 10 trials. Then find the average number of cards turned before an ace apppears.
   c. Combine the results of all the members of your group and find the average number of cards that are turned before an ace appears. How does this average compare to your individual estimate?

9. A cereal company places a coupon bearing a number from 1 to 9 in each box of cereal. If the numbers are distributed at random in the boxes, estimate the number of boxes you would have to purchase in order to obtain all nine numbers. Explain how the random-digit table could be used to estimate the number of coupons. Each person in the group should simulate 10 trials, and the results in the group should be combined to find the estimate.

Questions from the Classroom

10. A student said that he had heard of Monte Carlo methods and wanted to know how those methods related to simulations. What do you say?

11. Maximilian said that he could make up an experimental probability answer to a homework question and no one would ever know. How do you react?

Review Problems

12. In a two-person game, four coins are tossed. If exactly two heads come up, you win. If anything else comes up, you lose. Does each player have an equal chance of winning the game? Explain why or why not.

13. A single card is drawn from an ordinary deck. What is the probability of obtaining each of the following?
   a. A club
   b. A queen and a spade
   c. Not a queen
   d. Not a heart
   e. A spade or a heart
   f. The 6 of diamonds
   g. A queen or a spade
   h. Either red or black

14. From a sack containing seven red marbles, eight blue marbles, and four white marbles, marbles are drawn at random for several experiments. Determine the probability of each of the following events:
   a. One marble drawn at random is either red or blue.
   b. The first draw is red and the second is blue, where one marble is drawn at random, its color is recorded, the marble is replaced, and another marble is drawn.
   c. The event in (b) where the first marble is not replaced.

15. A basketball player shoots free throws and makes them with probability $\frac{1}{3}$. What is the probability the player will miss three in a row?

TECHNOLOGY CORNER

1. On some graphing calculators, if we choose the MATH menu and then select PRB, which stands for PROBABILITY, we find **RANDINT(** or **RANDI(**, the random integer feature. RANDINT(generates a random integer within a specified range. It requires two inputs which are the upper and lower boundaries for the integers. For example, RANDINT(1, 10) generates a random integer from 1 through 10.
   a. How could you use RANDINT(to simulate tossing a single die?
   b. How could you use RANDINT(to simulate the sum of the numbers when tossing two dice?
2. Some graphing calculators have a **RAND** function, a random-digit generator. RAND generates a random number greater than 0 and less than 1. For example, RAND might produce the numbers .5956605, .049599836, or .876572691. To have RAND produce random numbers from 1 to 10 as in (1), we enter int (10 *RAND) + 1. The **int** (greatest integer) feature is found in the MATH menu under NUM. The feature int returns the greatest integer less than or equal to a number.
   a. How could you use RAND to simulate tossing a single die?
   b. How could you use RAND to simulate tossing two dice and taking the sum of the numbers?
3. Use one of these random features to simulate tossing two dice 30 times. Based on your simulation, what is the probability that a sum of 7 will occur?

9-4   Odds, Conditional Probability, and Expected Value

Computing Odds

People talk about the *odds in favor of* and the *odds against* a particular event's happening. In the *Grass Roots* cartoon, a teacher has reported to have said that the odds against winning the lottery are greater than 55 million to 1. What does this mean?

Before answering that question, we think about a simpler set of odds to see what is meant. Suppose that an event has odds that are stated as 4 : 1; this means that for every 4 times the event occurs 1 time, it is expected not to occur. So one way to translate odds could be thought of as follows:

$$\text{Odds in favor of an event} = \frac{\text{Number of ways that an event could occur}}{\text{Number of ways that the event could not occur}}$$

Similarly, we could write the following:

$$\text{Odds against an event} = \frac{\text{Number of ways that an event could not occur}}{\text{Number of ways that the event could occur}}$$

 NOW TRY THIS 9-10 Use the information in the *Grass Roots* cartoon to interpret the odds in favor of winning the lottery.

Frequently, there are other interpretations of odds. When the **odds in favor** of the president's being reelected are 4 to 1, this refers to how likely the president is to win the election relative to how likely the president is to lose. The probability of the president's winning is 4 times the probability of losing. If W represents the event the president wins the election and L represents the event the president loses, then

$$P(W) = 4P(L)$$
$$\frac{P(W)}{P(L)} = \frac{4}{1}, \text{ or } 4:1$$

Because W and L are complements of each other, $L = \overline{W}$, and $P(\overline{W}) = 1 - P(W)$ so

$$\frac{P(W)}{P(\overline{W})} = \frac{P(W)}{1 - P(W)} = \frac{4}{1}, \text{ or } 4:1$$

The **odds against** the president's winning are how likely the president is to lose relative to how likely the president is to win. Using the preceding information, we have

$$\frac{P(L)}{P(W)} = \frac{P(L)}{P(\overline{L})} = \frac{P(\overline{W})}{P(W)} = \frac{1 - P(W)}{P(W)} = \frac{1}{4}, \text{ or } 1:4$$

Formally, odds are defined as follows.

Definition of Odds

Let $P(A)$ be the probability that event A occurs and $P(\overline{A})$ be the probability that event A does not occur. Then the **odds in favor** of event A are

$$\frac{P(A)}{P(\overline{A})} \quad \text{or} \quad \frac{P(A)}{1 - P(A)}$$

and the **odds against** event A are

$$\frac{P(\overline{A})}{P(A)} \quad \text{or} \quad \frac{1 - P(A)}{P(A)}$$

When we say that the odds in favor of an event are 4 : 1, remember that this is a ratio that can be used to determine the probability that the event occurs. We are saying that if circumstances surrounding an event are to occur many times, we would expect that for every 5 times the event could occur, we would expect 4 of those times for it to happen and 1 time that it would not. This interpretation gives us the following intuitive results:

$$P(\text{event happens}) = \frac{4}{5}$$

$$P(\text{event does not happen}) = \frac{1}{5}$$

We could have started with probabilities and worked to get odds as follows:

$$\frac{P(\text{event happens})}{P(\text{event does not happen})} = \frac{\frac{4}{5}}{\frac{1}{5}} = \frac{4}{5} \div \frac{1}{5} = \frac{4}{5} \cdot \frac{5}{1} = \frac{4}{1}$$

This process can be generalized. Note that the probabilities are ratios. We can determine the odds in favor of or against an event without knowing the number of outcomes in the event. When odds are calculated for equally likely outcomes, we have the following:

Odds in favor of an event A: $\dfrac{P(A)}{P(\overline{A})} = \dfrac{n(A)}{n(S)} \div \dfrac{n(\overline{A})}{n(S)} = \dfrac{n(A)}{n(S)} \cdot \dfrac{n(S)}{n(\overline{A})} = \dfrac{n(A)}{n(\overline{A})}$

Odds against an event A: $\dfrac{n(\overline{A})}{n(A)}$

Thus, in the case of equally likely outcomes, we have

$$\text{Odds in favor} = \frac{\text{Number of favorable outcomes}}{\text{Number of unfavorable outcomes}}$$

$$\text{Odds against} = \frac{\text{Number of unfavorable outcomes}}{\text{Number of favorable outcomes}}$$

When you roll a die, the number of favorable ways of rolling a 4 in one throw of a die is 1, and the number of unfavorable ways is 5. Thus the odds in favor of rolling a 4 are 1 to 5.

Example 9-13

For each of the following, find the odds in favor of the event's occurring:

a. rolling a number less than 5 on a die
b. tossing heads on a fair coin
c. drawing an ace from an ordinary 52-card deck
d. drawing a heart from an ordinary 52-card deck

Solution   a. The probability of rolling a number less than 5 is $\frac{4}{6}$; the probability of rolling a number not less than 5 is $\frac{2}{6}$. The odds in favor of rolling a number less than 5 are $\left(\frac{4}{6}\right) \div \left(\frac{2}{6}\right)$, or 4 : 2, or 2 : 1.

b. $P(H) = \dfrac{1}{2}$ and $P(\overline{H}) = \dfrac{1}{2}$. The odds in favor of getting heads are

$$\left(\dfrac{1}{2}\right) \div \left(\dfrac{1}{2}\right), \text{ or } 1:1.$$

c. The probability of drawing an ace is $\dfrac{4}{52}$, and the probability of not drawing

an ace is $\dfrac{48}{52}$. The odds in favor of drawing an ace are $\left(\dfrac{4}{52}\right) \div \left(\dfrac{48}{52}\right)$, or $4:48$,

or $1:12$.

d. The probability of drawing a heart is $\dfrac{13}{52}$, or $\dfrac{1}{4}$, and the probability of not

drawing a heart is $\dfrac{39}{52}$, or $\dfrac{3}{4}$. The odds in favor of drawing a heart are

$$\left(\dfrac{13}{52}\right) \div \left(\dfrac{39}{52}\right) = \dfrac{13}{39}, \text{ or } 13:39, \text{ or } 1:3.$$

NOW TRY THIS 9-11 In Example 9-12(a), there are four ways to roll a number less than 5 on a die (favorable outcomes) and two ways of not rolling a number less than 5 (unfavorable outcomes), so the odds in favor of rolling a number less than 5 are 4:2, or 2:1. Work the other three parts of Example 9-12 using this approach.

Example 9-14

In the following cartoon, find the probability of making totally black copies if the odds are 3 to 1 against making totally black copies:

Solution   The odds against making totally black copies are 3 to 1. Let B represent the event of making a totally black copy and $\overline{B}$ represent not making a totally black copy. We have

$$\frac{P(\overline{B})}{1 - P(\overline{B})} = \frac{3}{1}$$

$$P(\overline{B}) = 3(1 - P(\overline{B}))$$

$$P(\overline{B}) = 3 - 3P(\overline{B})$$

$$4P(\overline{B}) = 3$$

$$P(\overline{B}) = \frac{3}{4}$$

$$P(B) = 1 - P(\overline{B}) = 1 - \frac{3}{4}$$

$$P(B) = \frac{1}{4}$$

In Example 9-13, the odds in favor of making totally black copies are $1:3$. Therefore, the ratio of favorable outcomes to unfavorable outcomes is $\frac{1}{3}$. Thus, in a sample space modeling this situation with four elements, there would be 1 favorable outcome and 3 unfavorable outcomes and $P(\text{black copies}) = \frac{1}{1 + 3} = \frac{1}{4}$. In general, we have the following.

Theorem 9–8

If the odds in favor of event E are $m:n$, then $P(E) = \dfrac{m}{m + n}$.

If the odds against event E are $m:n$, then $P(E) = \dfrac{n}{m + n}$.

Conditional Probabilities

When the sample space of an experiment is affected by additional information, the new sample space often is reduced in size. For example, suppose we toss a fair coin 3 times and consider the following events:

A: getting a tail on the first toss

B: getting a tail on all three tosses

In Figure 9-36 we see that $P(A) = \dfrac{4}{8} = \dfrac{1}{2}$ and $P(B) = \dfrac{1}{8}$.

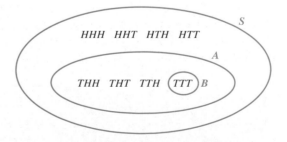

(*S* represents the sample space)

Figure 9-36

What if we were told that event A has occurred (that is, a tail occurred on the first toss), and we are now asked to find $P(B)$? The sample space is now reduced to {*THH, THT, TTH, TTT*}. Now the probability that all three tosses are tails given that the first toss is a tail is $\frac{1}{4}$. (Note that $A \cap B = \{TTT\}$.) The notation that we use for this situation is $P(B|A)$, read "the probability of B given A," and we write $P(B|A) = \frac{1}{4}$. Notice that

$$P(B|A) = \frac{1}{4} = \frac{\frac{1}{8}}{\frac{4}{8}} = \frac{P(A \cap B)}{P(A)}$$

This is true in general and we have the following.

Definition of Conditional Probability

If A and B are events in sample space S and $P(A) \neq 0$, then the **conditional probability** that event B occurs given that event A has occurred is given by

$$P(B|A) = \frac{P(A \cap B)}{P(A)}$$

Example 9-15

What is the probability of rolling a 6 on a fair die if you know that the roll is an even number?

Solution   If event B is rolling a 6 and event A is rolling an even number, then

$$P(B|A) = \frac{P(A \cap B)}{P(A)} = \frac{\frac{1}{6}}{\frac{1}{2}} = \frac{1}{3}$$

Note that it is not really necessary to use the formula for conditional probability to answer the question, because the new sample space is {2, 4, 6}, and so $P(6) = \frac{1}{3}$.

The accompanying research note outlines six fundamental concepts for a student to reason with probability.

Research Note   Six concepts are fundamental to a young child trying to reason in a probability context. These six are sample space, experimental probability, theoretical probability, probability comparisons, conditional probability, and independence (Jones et al. 1999).

Expected Value

Racetracks use odds for betting purposes. If the odds against Fast Jack are $3:1$, this means the track will pay $3 for every $1 you bet. If Fast Jack wins, then for a $5 bet, the track will return your $5 plus $3 \cdot \$5$, or $15, for a total of $20. The $3:1$ odds means the track expects Fast Jack to lose 3 out of 4 times in this situation. If the odds at racetracks were accurate, bettors would receive an even return for their money; that is, bettors would not expect to win or lose money in the long run. For example, if the stated odds of 3 : 1 against Fast Jack were accurate, then the probability of Fast Jack's losing the race would be $\frac{3}{4}$ and of Fast Jack's winning, $\frac{1}{4}$. If we compute the expected average winnings (expected value) over the long run, the gain is $3 for every $1 bet for a win, and a loss of $1 otherwise. The expected value, E, is computed as follows:

$$E = 3\left(\frac{1}{4}\right) + {}^{-}1\left(\frac{3}{4}\right) = 0$$

Therefore, the expected value is $0. If the racetrack gave accurate odds, it could not stay in business because it could not cover its expenses and make a profit. This is why the track overestimates the horses' chances of winning by about 20% and uses "house odds."

As another example of using odds, consider the spinner in Figure 9-37; the payoff is in each sector of the circle. Using area models, we can assign the following probabilities to each region:

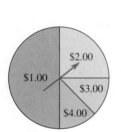

$$P(\$1.00) = \frac{1}{2} \quad P(\$2.00) = \frac{1}{4} \quad P(\$3.00) = \frac{1}{8} \quad P(\$4.00) = \frac{1}{8}$$

Figure 9-37

Should the owner of this spinner expect to make money over an extended period of time if the charge is $2.00 per spin?

To determine the average payoff over the long run, we find the product of the probability of landing on the payoff and the payoff itself and then find the sum of the products. This computation is given by

$$E = \left(\frac{1}{2}\right)1 + \left(\frac{1}{4}\right)2 + \left(\frac{1}{8}\right)3 + \left(\frac{1}{8}\right)4 = 1.875$$

The owner can expect to pay out about $1.88 per spin. This is less than the $2.00 charge, so the owner should make a profit if the spinner is used many times. The sum of the products in this example, $1.875, is the **expected value**, or **mathematical expectation**, of the experiment of spinning the wheel in Figure 9-37 once. The owner's expected average earnings are $2.00 − $1.875 = $0.125 per spin, and the player's expected average earnings (loss) are ${}^{-}$0.125. Note that the expected value may be none of the individual possible outcomes so might never be the actual amount won on any trial.

The expected value is an average of winnings over the long run. Expected value can be used to predict the average result of an experiment when it is repeated many times. But *an expected value cannot be used to determine the outcome of any single experiment.*

Definition of Expected Value

If, in an experiment, the possible outcomes are numbers $a_1, a_2, \ldots, a_n$, occurring with probabilities $p_1, p_2, \ldots, p_n$, respectively, then the **expected value (mathematical expectation)** E is given by the equation

$$E = a_1 \cdot p_1 + a_2 \cdot p_2 + a_3 \cdot p_3 + \ldots + a_n \cdot p_n$$

NOW TRY THIS 9-12 In the popular television game show *Deal or No Deal*, a contestant had six cases left to choose from with amounts for the cases having values $0.01, $1.00, $100, $1000, $10,000, and $1,000,000. Each case has equal probability of being chosen. The player has to choose two cases, which are removed. What is the mathematical expectation that the player could expect to be offered to deal and not play any more?

When payoffs are involved and the expected value minus cost to play a game of chance is $0, the game is a **fair game**. Do you think gambling casinos and lotteries make sure that the games are not fair?

> **REMARK** Previously, a *fair game* was defined as a game in which each player has an equal chance of winning. If there are two players and each has the same probability of winning a given number of dollars from the other, it follows that the expected value minus the cost to play (net winning) for each player is $0.

Example 9-16

Suppose you pay $5.00 to play the following game. Two coins are tossed. You receive $10 if two heads occur, $5 if exactly one head occurs, and nothing if no heads appear. Is this a fair game? That is, are the net winnings $0?

Solution   Before we determine the net winning, recall that $P(HH) = \dfrac{1}{4}$, $P(HT \text{ or } TH) = \dfrac{1}{2}$, and $P(TT) = \dfrac{1}{4}$. To find the expected value, we perform the following computation:

$$E = \left(\frac{1}{4}\right)(\$10) + \left(\frac{1}{2}\right)(\$5) + \left(\frac{1}{4}\right)(\$0) = \$5$$

The net winnings are $0 so this is a fair game.

Problem Solving   A Coin-Tossing Game

Al and Betsy played a coin-tossing game in which a fair coin was tossed until a total of either three heads or three tails occurred. Al was to win when a total of three heads were tossed, and Betsy was to win when a total of three tails were tossed. Each bet $50 on the game. If the coin was lost when Al had two heads and Betsy had one tail, how should the stakes be fairly split if the game is not continued?

Understanding the Problem   Al and Betsy each bet $50 on a coin-tossing game in which a fair coin was to be tossed. Al was to win when a total of three heads was obtained; Betsy was to win when a total of three tails was obtained. When two heads and one tail had occured, the coin was lost. The problem is how to split the stakes fairly.

Different people could have different interpretations of what "splitting fairly" means. Possibly, though, the best is to split the pot in proportion to the probabilities of each player's winning the game when play was halted. We must calculate the expected value for each player and split the pot accordingly.

Devising a Plan   A third head would make Al the winner, whereas Betsy needs two more tails to win. A tree diagram that simulates the completion of the game allows us to find the probability of each player's   winning the game. Once we find the probabilities, all we need to do is multiply the probabilities by the amount of the pot, $100, to determine each player's fair share.

Carrying Out the Plan   The tree diagram in Figure 9-38 shows the possibilities for game winners if the game is completed. We can find the probabilities of each player's winning as follows:

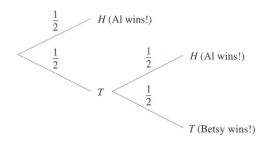

Figure 9-38

$$P(\text{Betsy wins}) = \frac{1}{2} \cdot \frac{1}{2} = \frac{1}{4}$$

$$P(\text{Al wins}) = 1 - \frac{1}{4} = \frac{3}{4}$$

Hence, the fair way to split the stakes is for Al to receive $\frac{3}{4}$ of $100, or $75, and Betsy should receive $\frac{1}{4}$ of $100, or $25.

Looking Back   The problem could be made even more interesting by assuming that the coin is not fair so that the probability is not $\frac{1}{2}$ for each branch in the tree diagram. Other possibilities arise if the players have unequal amounts of money in the pot or if more tosses are required in order to win.

 BRAIN TEASER Al tosses one quarter and at the same time Betty tosses two quarters. What is the probability that Betty gets the same number of heads as Al?

 ## Assessment 9-4A

1. a. What are the odds in favor of drawing a face card from an ordinary deck of playing cards?
   b. What are the odds against drawing a face card?
2. On a single roll of a pair of dice, what are the odds against rolling a sum of 7?

3. If the probability of a boy's being born is $\frac{1}{2}$, and a family plans to have four children, what are the odds against having all boys?

4. If the odds against Deborah's winning first prize in a chess tournament are 3 to 5, what is the probability that she will win first prize?

5. On a tote board at a racetrack, the odds for Gameylegs are listed as $26:1$. Tote boards list the odds that the horse will lose the race. If this is the case, what is the probability of Gameylegs's winning the race?

6. If the probability that a randomly chosen household has a cat is 0.27, what are the odds against a chosen household having a cat?

7. A container has three white balls and two red balls. A first ball is drawn at random and not replaced. Then a second ball is drawn. Given the following conditions, what is the probability that the second ball was red?
   a. The first ball was white.
   b. The first ball was red.

8. Suppose five quarters, five dimes, five nickels, and ten pennies are in a box. One coin is selected at random. What is the expected value of this experiment?

9. If the odds in favor of Fast Leg's winning a horse race are 5 to 2 and the first prize is $14,000, what is the expected value of Fast Leg's winning the race?

10. Sweepstakes are required by law to display the odds of winning as well as the payoffs. That information is sufficient to calculate the expected value of a sweepstakes. Suppose that mailing an entry for the sweepstakes costs $0.42 and the odds in favor of winning the various prizes are as follows:

Odds for each prize	Prize	Quantity
1 to 20,000,000	$1,000,000	1
1 to 20,000,000	$100,000	10
1 to 1,000,000	$1,000	100

   a. What is the expected value of the sweepstakes for any individual?
   b. At what postage rate would the drawing be fair?

11. Suppose it costs $8 to roll a pair of dice. You get paid the sum of the numbers in dollars that appear on the dice. Is it a fair game?

12. If the probability of an event happening is $\frac{88}{93}$, what are the odds against it happening?

13. From a set of eight marbles, five red and three white, you choose one at random. What are the odds in favor of choosing a red marble?

14. In exercise 13, what are the odds of not choosing a red marble?

15. If the odds of achieving a Grand Slam golf sweep are $2:9$, what is the probability in favor of achieving the sweep?

16. A regular die has the numbers 1–6, respectively, on its faces. When a die is tossed, what are the odds of a prime number showing?

Assessment 9-4B

1. **a.** Susan said that the odds in favor of drawing a black card from a normal deck of 52 cards was $2:1$. Do you agree?
   b. Explain the odds in favor of drawing a black card from the normal deck Susan used.

2. If a family has one girl and plans to have another child, answer the following if the probability of a girl is 1/2:
   a. What is the probability of the second child's being a boy?
   b. What is the probability of the family having one girl and one boy given their current family?

3. Diane tossed a coin 9 times and got 9 tails. Assume that Diane's coin is fair and answer each of the following questions:
   a. What is the probability of tossing a tail on the 10th toss?
   b. What is the probability of tossing 10 more tails in a row?
   c. What are the odds against tossing 10 more tails in a row?

4. What are the odds in favor of tossing at least two heads if a fair coin is tossed 3 times?

5. If the probability of rain for the day is 60%, what are the odds against its raining?

6. On an American roulette wheel, half of the slots numbered 1 through 36 are red and half are black. Two slots, numbered 0 and 00, are green. What are the odds against a red slot's coming up on any spin of the wheel?

7. If a whole number less than 10 is chosen at random, what are the odds that the number is greater than 5?

8. If the probability of an event is $0.\overline{3}$, what are the odds against the event?

9. What are the odds in favor of randomly drawing the letter S from the letters in the word *MISSISSIPPI* ?

10. The following spinner is spun. Given the conditions listed below, what is the probability that

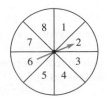

a. it lands on an odd number?
b. it lands on a number divisible by 3?
c. it does not land on 5, 6, or 7?
d. it lands on a number less than 4?

11. If $P(A) = \frac{2}{3}$, $P(B) = \frac{1}{2}$, and $P(B|A) = \frac{1}{3}$, find $P(A|B)$.

12. You play a game in which two dice are rolled. If a sum of 7 appears, you win $10; otherwise, you lose $2.00. If you intend to play this game for a long time, should you expect to make money, lose money, or come out about even? Explain.

13. On a roulette wheel are 36 slots numbered 1 through 36 and 2 slots numbered 0 and 00. You can bet on a single number. If the ball lands on your number, you receive 35 chips plus the chip you played.
a. What is the probability that the ball will land on 17?
b. What are the odds against the ball landing on 17?
c. If each chip is worth $1, what is the expected value for a player who plays the number 17 for a long time?

14. Suppose a standard six-sided die is rolled and you receive $1 for every dot showing on the top of the die. What should the cost of playing the game be in order to make it a fair game?

15. In the cartoon at the beginning of the chapter, suppose the odds against winning the lottery were 55 million to 1. What is the probability of winning the lottery given those odds?

16. If the lottery with odds as in exercise 15 has a prize of $13,567,290, is buying a ticket for $1 a fair deal? Explain your answer.

17. In the television show *Deal or No Deal*, a contestant is down to two cases with equal probabilities of being chosen; the money amounts associated with the cases are $250,000 and $0.01. The banker has offered the contestant $120,000. What would be your advice about taking the deal for the offered money? Back up your answer with expected value.

Mathematical Connections 9-4

Communication

1. Explain the difference between odds and probability.

2. A game involves tossing two coins. A player wins $1.00 if both tosses result in heads. What should you pay to play this game in order to make it a fair game? Explain your answer.

Open-Ended

3. Suppose you flip two fair coins. Design a fair game with a payoff based on the number of heads that are shown.

4. An insurance company sells a policy that pays $50,000 in case of accidental death. According to company figures, the rate of accidental death is 47 per 100,000 population. What annual premium should the company charge for this coverage? Explain how much profit the company will make under your plan, how you determined the amount of profit needed for the company, and how the annual premium was computed.

5. Write a game-type problem about odds and payoffs so that the odds in favor of an event are 2 : 3 and the game is a fair game.

Cooperative Learning

6. As a group, design a game that involves cards, dice, or spinners.
a. Write the rules so that any person who wants to play can understand the game.

b. Write a description explaining whether the game is fair and how you arrived at your conclusion.
c. Calculate the odds of each player's winning.
d. If betting is involved, discuss expected values.
e. Exchange a game with another group and compare your analysis of their game with their analysis of your group's game.

Questions from the Classroom

7. A student claims that if the odds in favor of winning a game are $a : b$, then out of every $a + b$ games she would win a games. Hence, the probability of winning the game is $\frac{a}{a + b}$. Is the student's reasoning correct? Why or why not?

8. A student wants to know why if the odds in favor of an event are $3 : 4$, the probability of the event occurring is not $\frac{3}{4}$? How do you respond?

9. Maria wants 51% of her class to vote for her. To achieve this, she decides that there is a probability of $\frac{1}{3}$ that an individual student will not vote. Also, she promises 24 of the 48 students that she will do what they individually want if she is elected so that she is reasonably sure they will vote for her. Would you advise her that she is on safe ground for winning with this strategy?

Review Problems

10. Refer to the following spinners and write the sample space for each of the following experiments:

Spinner 1

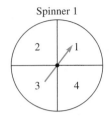

Spinner 2

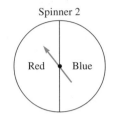

a. Spin spinner 1 once.
b. Spin spinner 2 once.

c. Spin spinner 1 once and then spin spinner 2 once.
d. Spin spinner 2 once and then roll a die.
e. Spin spinner 1 twice.
f. Spin spinner 2 twice.

11. Draw a spinner with two sections, red and blue, such that the probability of getting (Blue, Blue) on two spins is $\frac{25}{36}$.

12. Find the probability of getting two vowels when someone draws two letters from the English alphabet with replacement.

 BRAIN TEASER It is your first day of class; your class has 40 students. A friend who does not know any students in your class bets you that at least 2 of them share a birthday (month and day). What are your friend's chances of winning the bet?

9-5   Using Permutations and Combinations in Probability

Permutations of Unlike Objects

An arrangement of things in a definite order with no repetitions is a **permutation**. For example, different arrangements of the three letters R, A, and T are seen in Figure 9-39.

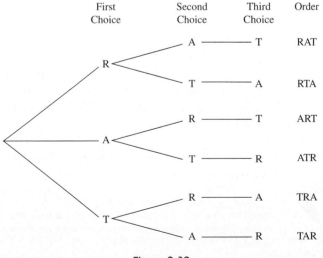

Figure 9-39

Notice that order is important and there are no repetitions. Determining the number of possible arrangements of the three letters without making a list can be done using the Fundamental Counting Principle. (Recall that the Fundamental Counting Principle states that if an event M can occur in m ways and, after M has occurred, event N can occur in n ways, then event M followed by event N can occur in mn ways.) Because there are three ways to choose the first letter, two ways to choose the second letter, and one way to choose the third letter, there are $3 \cdot 2 \cdot 1$, or 6, ways to arrange the letters. It is common to record the number of permutations of three objects taken three at a time as $_3P_3$. Therefore, $_3P_3 = 6$.

Consider how many ways the owner of an ice cream parlor can display 10 flavors in a row along the front of the display case. The first position can be filled in 10 ways, the second position in 9 ways, the third position in 8 ways, and so on. By the Fundamental Counting Principle, there are $10 \cdot 9 \cdot 8 \cdot 7 \cdot 6 \cdot 5 \cdot 4 \cdot 3 \cdot 2 \cdot 1$, or 3,628,800, ways to display the flavors. If there were 16 flavors, there would be $16 \cdot 15 \cdot 14 \cdot 13 \cdot \ldots \cdot 3 \cdot 2 \cdot 1$ ways to arrange them.

In general, *if there are n objects, then the number of possible ways to arrange the objects in a row is the product of all the natural numbers from n to 1, inclusive.* This expression, **n factorial**, is denoted **n!** as shown next.

$$n! = n(n - 1)(n - 2) \cdot \ldots \cdot 3 \cdot 2 \cdot 1$$

For example, $5! = 5 \cdot 4 \cdot 3 \cdot 2 \cdot 1$, $3! = 3 \cdot 2 \cdot 1$, and $1! = 1$. Using factorial notation is helpful in counting and probability problems.

Many calculators have a factorial key such as $\boxed{x!}$. To use this key, enter a whole number and then press the factorial key. For example, to compute 5!, press $\boxed{5}\;\boxed{x!}$ and 120 will appear on the display.

Consider the set of people in a small club, {Al, Betty, Carl, Dan}. For them to elect a president and a secretary, order is important and no repetitions are possible. How many ways are there to elect a committee of two in which one person is president and the other secretary.

One way to answer the question is to agree that the choice "Al, Betty" denotes Al as president and Betty as secretary, while the choice "Betty, Al" indicates that Betty is president and Al is secretary. Thus, order is important and no repetitions are possible. Consequently, counting the number of possibilities is a permutation problem. Since there are four ways of choosing a president and then three ways of choosing a secretary, by the Fundamental Counting Principle, there are $4 \cdot 3$, or 12, ways of choosing a president and a secretary. Choosing two officers from a club of four is a permutation of four people chosen two at a time. The number of possible permutations of four objects taken two at a time, denoted $_4P_2$, may be counted using the Fundamental Counting Principle. Therefore, we have $_4P_2 = 4 \cdot 3$, or 12.

NOW TRY THIS 9-13

a. Write $_nP_2$, $_nP_3$, and $_nP_4$ in terms of n.
b. Based on your answers in part (a), write $_nP_r$ in terms of n and r.
c. In the club mentioned above, how many ways are there to choose a president, vice president, and secretary?

The number of permutations can be written in terms of factorials. Consider the number of permutations of 20 objects chosen 3 at a time:

$$\begin{aligned} _{20}P_3 &= 20 \cdot 19 \cdot 18 \\ &= \frac{20 \cdot 19 \cdot 18 \cdot (17 \cdot \ldots \cdot 3 \cdot 2 \cdot 1)}{(17 \cdot \ldots \cdot 3 \cdot 2 \cdot 1)} \\ &= \frac{20!}{17!} \\ &= \frac{20!}{(20-3)!} \end{aligned}$$

This can be generalized as follows:

Permutation of Objects in a Set

In a set of n elements, the number of ways to choose r elements from the set in order, the permutations of n objects taken r at a time, is given by

$$_nP_r = \frac{n!}{(n-r)!}$$

Recall that $_nP_n$ is the number of permutations of n objects chosen n at a time; that is, the number of ways of rearranging n objects in a row. We have seen that this number is $n!$. If we use the formula for $_nP_r$ to compute $_nP_n$, we obtain

$$_nP_n = \frac{n!}{(n-n)!} = \frac{n!}{0!}$$

Consequently, $n! = n!/0!$. To make this equation true, *we define* $0!$ *to be* 1.

 Many calculators, especially graphing calculators, can calculate the number of permutations of n objects taken r at a time. This feature is usually denoted $\boxed{_nP_r}$. To use this key, enter the value of n, then press $\boxed{_nP_r}$, followed by the value of r. If you then press $\boxed{=}$ or $\boxed{\text{ENTER}}$, the number of permutations is displayed.

NOW TRY THIS 9-14

a. Try to use a factorial key $\boxed{x!}$ on your calculator to compute $\dfrac{100!}{98!}$. What happens? Why?

b. Without using a calculator, use the definition of factorials to compute the expression in part (a).

Example 9-17

a. A baseball team has nine players. Find the number of ways the manager can arrange the batting order.

b. Find the number of ways of choosing three initials from the alphabet if none of the letters can be repeated.

Solution   a. Because there are nine ways to choose the first batter, eight ways to choose the second batter, and so on, there are $9 \cdot 8 \cdot 7 \cdot \ldots \cdot 2 \cdot 1 = 9!$, or 362,880, ways of arranging the batting order. Using the formula for permutations, we have $_9P_9 = \dfrac{9!}{0!} = 362{,}880$.

b. There are 26 ways of choosing the first letter, 25 ways of choosing the second letter, and 24 ways of choosing the third letter. Hence, there are $26 \cdot 25 \cdot 24$, or 15,600, ways of choosing the three letters. Alternatively, if we use the formula for permutations, we have

$$_{26}P_3 = \frac{26!}{23!} = \frac{26 \cdot 25 \cdot 24 \cdot (23 \cdot 22 \cdot 21 \cdot \ldots \cdot 1)}{23 \cdot 22 \cdot 21 \cdot \ldots \cdot 1}$$
$$= 26 \cdot 25 \cdot 24$$
$$= 15{,}600$$

NOW TRY THIS 9-15 If the digits 1 through 9 inclusive are used with no repeats to form a four-digit code number, what is the probability that a code number selected at random starts with 9?

Permutations Involving Like Objects

In the previous counting examples, each object to be counted was distinct. Suppose we wanted to arrange the letters in the word *ZOO*. How many choices would we have? A tree diagram, as in Figure 9-40, suggests that there might be $3 \cdot 2 \cdot 1 = 3!$, or 6, possibilities. However, looking at the list of possibilities shows that *ZOO*, *OZO*, and *OOZ* each appear twice because the *O*'s are not different. We need to determine how to remove the duplication in arrangements such as this where some objects are the same. To eliminate the duplication, we divide the number of arrangements shown by the number of ways the two *O*'s can be rearranged, which is 2!. Consequently, there are $\dfrac{3!}{2!}$, or 3, ways of arranging the letters in *ZOO*. The arrangements are *ZOO*, *OZO*, and *OOZ*.

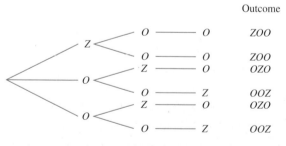

Figure 9-40

This discussion is generalized in the following:

Permutations of Like Objects

In a set of n elements, of which r_1 are alike, r_2 are alike, and so on through r_k, then the number of different arrangements of all n elements, where alike elements are indistinguishable, is equal to

$$\frac{n!}{r_1! \cdot r_2! \cdot r_3! \cdot \ldots \cdot r_k!}$$

Example 9-18   Find the number of rearrangements of the letters in each of the following words:

a. *bubble*   **b.** *statistics*

Solution   **a.** There are 6 letters with *b* repeated 3 times. Hence, the number of arrangements is

$$\frac{6!}{3!} = 6 \cdot 5 \cdot 4 = 120$$

b. There are 10 letters in the word *statistics,* with three *s*'s, three *t*'s, and two *i*'s in the word. Hence, the number of arrangements is

$$\frac{10!}{3! \cdot 3! \cdot 2!} = \frac{10 \cdot 9 \cdot 8 \cdot 7 \cdot 6 \cdot 5 \cdot 4 \cdot 3 \cdot 2 \cdot 1}{3 \cdot 2 \cdot 1 \cdot 3 \cdot 2 \cdot 1 \cdot 2 \cdot 1} = 50,400$$

Combinations

Reconsider the club {Al, Betty, Carl, Dan}. Suppose a two-person committee is selected with no chair. In this case, order is not important, and an Al-Betty choice is the same as a Betty-Al choice. An arrangement of objects in which the order makes no difference is a **combination**. A comparison of the results of electing a president and a secretary for the club and the results of simply selecting a two-person committee are shown in Figure 9-41. Because each two permutations "shrink" into one combination, we see that the number of combinations is the number of permutations divided by 2, or

$$\frac{4 \cdot 3}{2} = 6$$

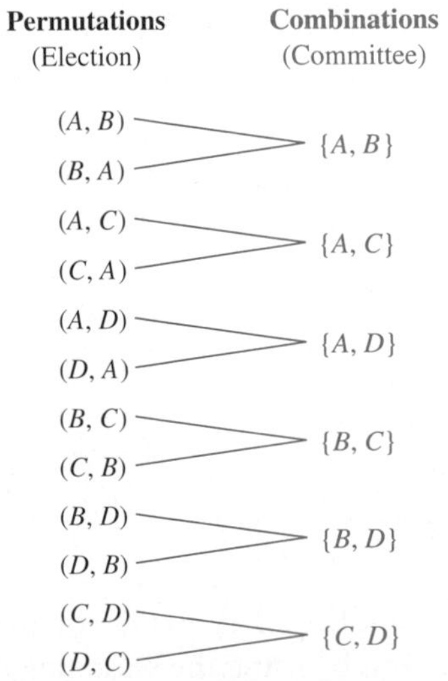

Permutations (Election)   **Combinations** (Committee)

(A, B)
(B, A)   {A, B}

(A, C)
(C, A)   {A, C}

(A, D)
(D, A)   {A, D}

(B, C)
(C, B)   {B, C}

(B, D)
(D, B)   {B, D}

(C, D)
(D, C)   {C, D}

Figure 9-41

In how many ways can a committee of three people be selected from the club {Al, Betty, Carl, Dan}? To solve this problem, we proceed as we did earlier and find the number of ways to select three people from a group of four for three offices, say president, vice president, and secretary (a permutation problem) and then use this result to see how many combinations of people are possible for the committee. Figure 9-42 shows a partial list for both problems.

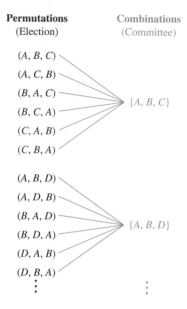

Figure 9-42

By the Fundamental Counting Principle, if order is important, the number of ways to choose three people from the list of four is $4 \cdot 3 \cdot 2$, or 24. However, with each triple chosen, there are 3!, or 6, ways to rearrange the triple, as seen in Figure 9-42. Therefore, there are 3! times as many permutations as combinations, or equivalently, each 3!, or 6, permutations "shrink" into one combination. Therefore, to find the number of combinations, we divide the number of permutations, 24, by 3!, or 6, to obtain 4. The four committees are $\{A, B, C\}$, $\{A, B, D\}$, $\{A, C, D\}$, and $\{B, C, D\}$.

In general, we use the following rule to count combinations.

Counting Combinations

To find the number of combinations possible in a counting problem, first use the Fundamental Counting Principle to find the number of permutations and then divide by the number of ways in which each choice can be arranged.

Symbolically, the number of combinations of n objects taken r at a time is denoted $_nC_r$ and can be computed as follows:

$$_nC_r = \frac{_nP_r}{_rP_r} = \frac{\dfrac{n!}{(n-r)!}}{r!} = \frac{n!}{r!(n-r)!}$$

Note: It is not necessary to memorize the formula above; we can always find the number of combinations by using the reasoning developed in the committee example.

Example 9-19

The Library of Science Book Club offers 3 free books from a list of 42. If you circle 3 choices from a list of 42 numbers representing the book on a postcard, how many possible choices are there?

Solution   Order is not important, so this is a combination problem. By the Fundamental Counting Principle, there are $42 \cdot 41 \cdot 40$ ways to choose the 3 free books. Because each set of 3 circled numbers could be rearranged $3 \cdot 2 \cdot 1$ ways, there is an extra factor of 3! in the original $42 \cdot 41 \cdot 40$ ways. Therefore, the number of combinations possible for 3 books is

$$\frac{42 \cdot 41 \cdot 40}{3!} = 11{,}480$$

If we use the formula for $_nC_r$, we have $_{42}C_3 = \dfrac{42!}{3! \cdot 39!}$, or 11,480.

Example 9-20

At the beginning of the second quarter of a mathematics class for elementary school teachers, each of the class's 25 students shook hands with each of the other students exactly once. How many handshakes took place?

Solution   Since the handshake between persons A and B is the same as that between persons B and A, this is a problem of choosing combinations of 25 people 2 at a time. There are

$$_{25}C_2 = \frac{25 \cdot 24}{2!} = 300$$

different handshakes.

Example 9-21

Given a class of 12 girls and 10 boys, answer each of the following:

a. In how many ways can a committee of 5 consisting of 3 girls and 2 boys be chosen?
b. What is the probability that a committee of 5, chosen at random from the class, consists of 3 girls and 2 boys?
c. How many of the possible committees of 5 have no boys?
d. What is the probability that a committee of 5, chosen at random from the class, consists only of girls?

Solution   a. Based on the information given, we do not assign special functions to members on a committee and, hence, the order of the children on a committee does not matter. From 12 girls we can choose 3 girls in $_{12}C_3$ ways. Each of these choices can be paired with $_{10}C_2$ combinations of boys. By the Fundamental Counting Principle, the total number of committees is

$$_{12}C_3 \cdot {}_{10}C_2 = \frac{12 \cdot 11 \cdot 10}{3!} \cdot \frac{10 \cdot 9}{2} = 9900$$

b. The total number of committees of 5 is $_{22}C_5$, or 26,334. Using part (a), we find the probability that a committee of 5 will consist of 3 girls and 2 boys to be

$$\frac{_{12}C_3 \cdot {}_{10}C_2}{_{22}C_5} = \frac{9900}{26{,}334} \approx 0.3759$$

c. The number of ways to choose 0 boys and 5 girls from the 12 girls in the class is

$$_{10}C_0 \cdot {}_{12}C_5 = 1 \cdot {}_{12}C_5 = \frac{12 \cdot 11 \cdot 10 \cdot 9 \cdot 8}{5 \cdot 4 \cdot 3 \cdot 2 \cdot 1} = 792$$

d. $\dfrac{_{12}C_5}{_{22}C_5} = \dfrac{792}{26{,}334} \approx 0.030$

Problem Solving   A True-False Test Problem

In the following *Peanuts* cartoon, suppose Peppermint Patty took a six-question true-false test. If she answered each question true or false at random, what is the probability that she answered exactly 50% of the questions correctly?

Understanding the Problem   A score of 50% indicates that Peppermint Patty answered $\frac{1}{2}$ of the six questions, or three questions, correctly. She answered the questions true or false at random, so the probability that she answered a given question correctly is $\frac{1}{2}$. We are asked to determine the probability that Patty answered exactly three of the questions correctly.

Devising a Plan   We do not know which three questions Patty missed. She could have missed any three out of six on the test. Suppose she answered questions 2, 4, and 5 incorrectly. In this case, she would have answered questions 1, 3, and 6 correctly. We can compute the probability of this set of answers by *using a branch of a tree diagram*, as in Figure 9-43, where C represents a correct answer and I represents an incorrect answer.

$$\text{Question:} \quad 1 \qquad 2 \qquad 3 \qquad 4 \qquad 5 \qquad 6 \qquad \begin{array}{c}\text{Probability}\\\text{of Outcome}\end{array}$$

$$\xrightarrow{\frac{1}{2}} C \xrightarrow{\frac{1}{2}} I \xrightarrow{\frac{1}{2}} C \xrightarrow{\frac{1}{2}} I \xrightarrow{\frac{1}{2}} I \xrightarrow{\frac{1}{2}} C \qquad \left(\frac{1}{2}\right)^6$$

Figure 9-43

Multiplying the probabilities along the branches, we obtain $\left(\frac{1}{2}\right)^6$ as the probability of answering questions 1 through 6 in the following way: $C\,I\,C\,I\,I\,C$. There are other ways to answer exactly three questions correctly: for example, $C\,C\,C\,I\,I\,I$. The probability of answering questions 1 through 6 in this way is also $\left(\frac{1}{2}\right)^6$. The number of ways to answer exactly three of the questions correctly is simply the number of ways of arranging three C's and three I's in a row, which is also the number of ways of choosing three correct questions out of six, that is, $_6C_3$. Because all these arrangements give Patty a score of 50%, the desired probability is the sum of the probabilities for each arrangement.

Carrying Out the Plan   There are $_6C_3$, or 20, sets of answers similar to the one in Figure 9-43, with three correct and three incorrect answers. The product of the probabilities for each of these sets of answers is $\left(\frac{1}{2}\right)^6$, so the sum of the probabilities for all 20 sets is $20\left(\frac{1}{2}\right)^6$, or 0.3125. Thus, Peppermint Patty has a probability of 0.3125 of obtaining a score of exactly 50% on the test.

Looking Back It seems paradoxical to learn that the probability of obtaining a score of 50% on a six-question true-false test is not close to $\frac{1}{2}$. As an extension of the problem, suppose a passing score is a score of at least 70%. Now what is the probability that Peppermint Patty will pass? What is the probability of her obtaining a score of at least 50% on the test? If the test is a six-question multiple-choice test with five alternative answers for each question, what is the probability of obtaining a score of at least 50% by random guessing?

Problem Solving   Matching Letters to Envelopes

Stephen placed three letters in envelopes while he was having a telephone conversation. He addressed the envelopes and sealed them without checking if each letter was in the correct envelope. What is the probability that each of the letters was inserted correctly?

Understanding the Problem Stephen sealed three letters in addressed envelopes without checking to see if each was in the correct envelope. We are to determine the probability that each of the three letters was placed correctly. This probability could be found if we knew the sample space, or at least how many elements are in the sample space.

Devising a Plan To aid in solving the problem, we represent the respective letters as a, b, and c and the respective addressed envelopes as A, B, and C. For example, a correctly placed letter a would be in envelope A. To construct the sample space, we use the strategy of *making a table*. The table should show all the possible permutations of letters in envelopes. Once the table is completed, we can determine the probability that each letter is placed correctly.

Carrying Out the Plan Table 9-10 is constructed by using the envelope labels A, B, and C as headings and listing all ways that letters a, b, and c could be placed in the envelopes. The first arrangement is the only case out of six in which each of the envelopes is labeled correctly, so the probability that each envelope is labeled correctly is $\frac{1}{6}$.

Table 9-10

A	B	C	
a	b	c	} Envelope labels
a	b	c	
a	c	b	
b	a	c	
b	c	a	} Letter content
c	a	b	
c	b	a	

Looking Back Is the probability of having each letter placed incorrectly the same as the probability of having each letter placed correctly? A first guess might be that the probabilities are the same, but that is not true. Why?

We also could have used a counting argument to solve the problem. Given an envelope, there is only one correct letter to place in the envelope. Thus, there is one correct way to place the letters in the envelopes. By the Fundamental Counting Principle, there are $3 \cdot 2 \cdot 1$ ways of choosing the letters to place in the envelopes, so the probability of having the letters correctly placed is $\frac{1}{6}$.

Assessments 9-5A

1. The eighth-grade class at a grade school has 16 girls and 14 boys. How many different boy-girl dates can be arranged?

2. The telephone prefix for a university is 243. The prefix is followed by four digits. How many different telephone numbers are possible before a new prefix is needed?

3. Carlin's Pizza House offers 3 kinds of salad, 15 kinds of pizza, and 4 kinds of dessert. How many different three-course meals can be ordered?

4. Decide whether each of the following is true:
   a. $6! = 6 \cdot 5!$
   b. $3! + 3! = 6!$
   c. $\dfrac{6!}{3!} = 2!$

5. In how many ways can the letters in the word *SCRAMBLE* be rearranged?

6. How many two-person committees can be formed from a group of six people?

7. Assume a class has 30 members.
   a. In how many ways can a president, a vice president, and a secretary be selected?
   b. How many committees of three people can be chosen?

8. A five-volume numbered set of books is placed randomly on a shelf. What is the probability that the books will be numbered in the correct order from left to right?

9. Take 10 points in a plane, no 3 of them on a line. How many straight lines can be drawn if each line is drawn through a pair of points?

10. Sally has four red flags, three green flags, and two white flags. How many nine-flag signals can she run up a flagpole?

11. At a party, 28 handshakes took place. Each person shook hands exactly once with each of the others present. How many people were at the party?

12. How many different 5-card hands can be dealt from a standard deck of 52 playing cards?

13. In a certain lottery game, 54 numbers are randomly mixed and 6 are selected. A person must pick all 6 numbers to win. Order is not important. What is the probability of winning?

14. From a group of 10 boys and 12 girls, a committee of 4 students is chosen at random. What is the probability that
   a. all 4 members on the committee will be girls?
   b. all 4 members of the committee will be boys?
   c. there will be at least 1 girl on the committee?

15. From a group of 20 Britons, 21 Italians, and 4 Danes, a committee of 8 people is chosen at random. Find the probability that (express your answers using notation for combinations)
   a. the committee will consist of 2 Britons, 4 Italians, and 2 Danes.
   b. the committee will have no Britons.
   c. there will be at least one Briton on the committee.
   d. all members of the committee will be Britons.

16. Stephen placed five letters in envelopes while he was watching television and addressed the envelopes without checking if each letter was in the correct envelope.
   a. What is the probability that all five letters were in the correct envelopes?
   b. What is the probability that exactly four letters were addressed correctly?
   c. If n letters were put in n envelopes at random, what is the probability that exactly $n - 1$ of them were in the correct envelopes?

17. A company is setting up four-digit ID numbers for employees.
   a. How many four-digit numbers are there if numbers can start with 0 and digits can be repeated?
   b. How many four-digit numbers are there if numbers can start with 0 and all the digits must be different?
   c. If you randomly assign a four-digit ID number that can start with 0 and digits can be repeated, what is the probability that all the digits are even?

18. A company president presents identical awards to four out of six finalists for the awards. In how many ways can she present the awards?

19. April wonders how many ways she could record nine songs on a CD. What is the answer?

20. Your English teacher asks that you read any three of the eight books on his reading list. How many choices do you have for the set of three books you read?

Assessment 9-5B

1. If a coin is tossed 5 times, in how many different ways can the sequence of heads and tails appear?

2. Radio stations in the United States have call letters that begin with either K or W. Some have three letters; others have four letters. How many three-letter call letters are possible? How many four-letter call letters are possible?

3. Decide whether each of the following is true or false:
   a. $\dfrac{6!}{3} = 2!$
   b. $\dfrac{6!}{5!} = 6!$
   c. $\dfrac{6!}{4!2!} = 15$
   d. $n!(n+1) = (n+1)!$

4. Find the number of ways to arrange the letters in the following words:
   a. *OHIO*
   b. *ALABAMA*
   c. *ILLINOIS*
   d. *MISSISSIPPI*
   e. *TENNESSEE*

5. In a car race, there are 6 Chevrolets, 4 Fords, and 2 Pontiacs. In how many ways can the 12 cars finish if we consider only the makes of the cars?

6. A basketball coach was criticized in the newspaper for not trying out every combination of players. If the team roster has 12 players, how many 5-player combinations are possible?

7. Find the number of shortest paths from point A to point B along the edges of the cubes in each of the following. (For example, in (a) one shortest path is A-C-D-B.)

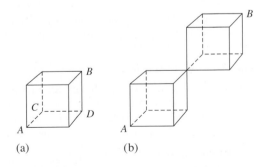

(a)   (b)

8. A committee of three people is selected at random from a set consisting of seven Americans, five French people, and three English people.
   a. What is the probability that the committee consists of all Americans?
   b. What is the probability that the committee has no Americans?

9. License plates in a certain state have three letters followed by three digits. How many different plates are possible if no repetitions of letters or digits are allowed?

10. Social Security numbers are in the form ###-##-####, where each symbol represents a number 0 through 9. How many Social Security numbers are possible using this format?

11. The probability of a basketball player's making a free throw successfully at any time in a game is $\dfrac{2}{3}$. If the player attempts ten free throws in a game, what is the probability that exactly six are made?

12. A fair die is rolled 8 times. What is the probability of getting
    a. 1 on each of the eight rolls?
    b. 6 exactly twice in the eight rolls?
    c. 6 at least once in the eight rolls?

13. Two fair dice are rolled 5 times and the sum of the numbers that come up is recorded. Find the probability of getting
    a. a sum of 7 on each of the five rolls.
    b. a sum of 7 exactly twice in the five rolls.

14. What are the odds against a royal flush in poker; that is, a 10, jack, queen, king, and ace all of the same suit?

15. How many different 12-person juries can be selected from a pool of 24 people?

16. How many ways are there to arrange the letters in the word *permutation*?

17. At the American Kennel Club, there are 36 dogs entered in a show where there will be a first-, second-, and third-place award. How many possibilities are there for these awards?

18. A club selects an executive committee of 5 and from the 5 one of the group becomes the president. From a membership of 32, how many different choices are possible?

Mathematical Connections 9-5

Communication

1. The terms *Fundamental Counting Principle*, *permutations*, and *combinations* are all used to work with counting problems. In your own words, explain how all these terms are related and how they are used.

2. **a.** A bicycle lock has three reels, each of which contains the numbers 0 through 9. To open the lock, you must enter the numbers in the correct order, such as 369 or 455, where one number is chosen from each reel. How many different possibilities are there for the numbers to open the lock? Explain how you arrived at your answer.
   b. The lock in the cartoon is a *combination* lock. Explain why this is probably not a good name for this lock for someone who has studied counting problems.

REAL LIFE ADVENTURES by Gary Wise and Lance Aldrich

The hacksaw industry would be in serious trouble
without the combination-lock industry.

3. **a.** Ten people are to be seated on 10 chairs in a line. Among them is a family of 3 that does not want to be separated. How many different seating arrangements are possible? Explain how you arrived at your answer.
   b. How many possible seating arrangements are there in part (a) in which the family members do not all sit together? Explain how you arrived at your answer.

4. In how many ways can five couples be seated in a row of 10 chairs if no couple is separated? Explain how you arrived at your answer.

Open-Ended

5. Suppose the Department of Motor Vehicles uses only six spaces and the numbers 0 through 9 to create its license plates. Numbers can be repeated.
   a. How many license plates are possible?
   b. Based on the 2000 census, determine whether there are any states in which the answer in (a) might provide enough license plates.
   c. If you were in charge of making license plates for the state of California, describe the method you would use to ensure you would have enough license plates.

Cooperative Learning

6. The following triangular array of numbers is a part of **Pascal's triangle:**

							Row
			1				(0)
		1		1			(1)
	1		2		1		(2)
1		3		3		1	(3)
1	4		6		4	1	(4)
1	5	10		10	5	1	(5)
1	6	15	20	15	6	1	(6)

   a. In your group, decide how the triangle was constructed and complete the next two rows.
   b. Describe at least three number patterns in Pascal's triangle.
   c. Find the sum of the numbers in each row. Predict the sum of the numbers in row 10.
   d. The entries in row 2 are just $_2C_0, _2C_1$, and $_2C_2$. Have different members of your group investigate whether a similar pattern holds for other rows in Pascal's triangle.
   e. Describe how you could use combinations to find any entry in Pascal's triangle.

Questions from the Classroom

7. A student does not understand the meaning of $_4P_0$. He wants to know how we can consider permutations of four objects chosen zero at a time. How do you respond?

8. A student wants to know why, if we can define 0! as 1, we cannot define $\frac{1}{0}$ as 1. How do you respond?

9. A student claims that one does not need to define permutations when he has the Fundamental Counting Principle. What do you say?

Review Problems

10. Two cards are drawn at random without replacement from a deck of 52 cards. What is the probability that
  a. at least 1 card is an ace?
  b. exactly 1 card is red?

11. If two regular dice are tossed, what is the probability of tossing a sum greater than 10?

12. Two coins are tossed. You win $5.00 if both coins are heads and $3.00 if both coins are tails and lose $4.00 if the coins do not match. What is the expected value of this game? Is this a fair game?

13. On a roulette wheel, the probability of winning when you pick a particular number is $\frac{1}{38}$. Suppose you bet $1.00 to play the game, and if your number is picked, you get back $36.
  a. Is this a fair game?
  b. What would happen if you played this game a large number of times?

National Assessment of Educational Progress (NAEP) Question

There are 15 girls and 11 boys in a mathematics class. If a student is selected at random to run an errand, what is the probability that a boy will be selected?

  a. $\frac{4}{26}$

  b. $\frac{11}{26}$

  c. $\frac{15}{26}$

  d. $\frac{11}{15}$

  e. $\frac{15}{11}$

NAEP, Grade 8, 1990

BRAIN TEASER An airplane can complete its flight if at least $\frac{1}{2}$ of its engines are working. If the probability that an engine fails is 0.01 and all engine failures do not depend on each other, what is the probability of a successful flight if the plane has
  a. two engines?   **b.** four engines?

Hint for Solving the Preliminary Problem

The two dots chosen to draw a segment are determined by the randomly chosen phone number's last four digits. The four digits are separated to form 2 two-digit numbers. Consider any such number *abcd*. For a single segment not to be drawn, then for some *abcd*, *ab* = *cd*. The complement of at least one not being drawn is that all can be drawn. So find the probability that all can be drawn. There are no restrictions on *ab* but *cd* cannot be equal to *ab* if a segment is to be drawn. What is the probability of *cd* being equal to *ab*? If they match, then only one dot is determined and no segment can be drawn.

Chapter Outline

  I. Probability
    A. Probabilities can be determined **experimentally (empirically)** or **theoretically**.
    B. A **sample space** is the set of all possible **outcomes** of an **experiment**.
    C. An **event** is a subset of a sample space.
    D. Outcomes are **equally likely** if each outcome is as likely to occur as another.

  E. If all outcomes of an experiment are *equally likely*, the **probability of an event** *A* from sample space *S* is given by

$$P(A) = \frac{n(A)}{n(S)}$$

F. If each possible outcome of the sample space is equally likely, the sample space is a **uniform sample space**.

G. Law of Large Numbers If an experiment is repeated a large number of times, the experimental (empirical) probability of a particular outcome approaches a fixed number as the number of repetitions increases.

H. An **impossible event** is an event with a probability of 0. An impossible event can never occur.

I. A **certain event** is an event with a probability of 1. A certain event is sure to happen.

J. If A is an event, then $0 \leq P(A) \leq 1$.

K. Two events are **mutually exclusive** if, and only if, only one of the events can occur at any given time—that is, if, and only if, the events are disjoint.

L. The probability of the **complement of an event** is given by $P(\overline{A}) = 1 - P(A)$, where A is the event and $\overline{A}$ is its complement.

M. A **probability distribution** assigns a probability to each outcome in a sample space.

N. Multiplication rule for probabilities For all **multistage experiments**, the probability of the outcome along any path of a tree diagram is equal to the product of all the probabilities along the path.

O. If A and B are two events, then $P(A \cup B) = P(A) + P(B) - P(A \cap B)$.

P. If events E_1 and E_2 are **independent**—that is, the occurrence of one has no influence on the occurrence of the other—then $P(E_1 \cap E_2) = P(E_1) \cdot P(E_2)$.

Q. Simulations can play an important part in probability. Fair coins, dice, spinners, and random-digit tables are useful in performing simulations.

II. Odds, expected value, and conditional probability

A. The **odds in favor** of an event A are given by

$$\frac{P(A)}{P(\overline{A})} = \frac{P(A)}{1 - P(A)} = \frac{\text{Number of ways } A \text{ can occur}}{\text{Number of ways } A \text{ cannot occur}}$$

B. The **odds against** an event A are given by

$$\frac{P(\overline{A})}{P(A)} = \frac{P(\overline{A})}{1 - P(\overline{A})} = \frac{\text{Number of ways } \overline{A} \text{ can occur}}{\text{Number of ways } \overline{A} \text{ cannot occur}}$$

C. If A and B are events and $P(A) \neq 0$, then the **conditional probability** that event B occurs given that event A occurs is

$$P(B|A) = \frac{P(A \cap B)}{P(A)}$$

D. If, in an experiment, the possible outcomes are numbers $a_1, a_2, \ldots, a_n$, occurring with probabilities $p_1, p_2, \ldots, p_n$, respectively, then the **expected value (mathematical expectation)** E is defined as

$$E = a_1 \cdot p_1 + a_2 \cdot p_2 + a_3 \cdot p_3 + \ldots + a_n \cdot p_n$$

E. A **fair game** is a game in which the expected net winnings or expected value is \$0.

III. Counting principles

A. Fundamental Counting Principle If an event M can occur in m ways and, after it has occurred, event N can occur in n ways, then event M followed by event N can occur in mn ways.

B. Permutations are arrangements in which order is important. The number of permutations of r elements chosen from n elements is given by

$$_nP_r = \frac{n!}{(n - r)!}$$

C. The expression $n!$, called n **factorial**, represents the product of all the natural numbers less than or equal to n. $0!$ is defined as 1.

D. Permutations of like objects If a set contains n elements, of which r_1 are of one kind, r_2 are of another kind, and so on through r_k, then the number of different arrangements of all n elements is equal to

$$\frac{n!}{r_1! \cdot r_2! \cdot r_3! \cdot \ \cdots \ \cdot r_k!}$$

E. Combinations are arrangements in which order is *not* important. To find the number of combinations possible, first use the Fundamental Counting Principle to find the number of permutations and then divide by the number of ways in which each choice can be arranged:

$$_nC_r = \frac{_nP_r}{_rP_r}$$

Chapter Review

1. A coin is flipped 3 times and heads (*H*) or tails (*T*) are recorded.
   a. List all the elements in the sample space.
   b. List the elements in the event "at least two heads appear."
   c. Find the probability that the event in part (b) occurs.

2. Suppose the names of the days of the week are placed in a box and one name is drawn at random.
   a. List the sample space for this experiment.
   b. List the event consisting of outcomes that the day drawn starts with the letter *T*.
   c. What is the probability of drawing a day that starts with *T*?

3. If you have a jar of 1000 jelly beans and you know that $P(\text{Blue}) = \frac{4}{5}$ and $P(\text{Red}) = \frac{1}{8}$, list several things you can say about the number of beans in the jar.

4. In the 2000 presidential election, George W. Bush received 50,460,110 votes and Albert A. Gore received 51,003,926. If a 2000-voter for either Bush or Gore is chosen at random, answer the following:
   a. What is the probability that the voter opted for Bush?
   b. What is the probability that the voter opted for Gore?
   c. What are the odds that a voter chosen at random did not vote for Bush?

5. A box contains three red balls, five black balls, and four white balls. Suppose one ball is drawn at random. Find the probability of each of the following events:
   a. A black ball is drawn.
   b. A black or a white ball is drawn.
   c. Neither a red nor a white ball is drawn.
   d. A red ball is not drawn.
   e. A black ball and a white ball are drawn.
   f. A black or white or red ball is drawn.

6. One card is selected at random from an ordinary set of 52 cards. Find the probability of each of the following events:
   a. A club is drawn.
   b. A spade and a 5 are drawn.
   c. A heart or a face card is drawn.
   d. A jack is not drawn.

7. A box contains five colored balls and four white balls. If three balls are drawn one-by-one, find the probability that they are all white if the draws are made as follows:
   a. With replacement
   b. Without replacement

8. Consider the following two boxes. If a letter is drawn from box 1 and placed into box 2 and then a letter is drawn from box 2, what is the probability that the letter is an *L*?

9. Use the following boxes for a two-stage experiment. First select a box at random and then select a letter at random from the box. What is the probability of drawing an *A*?

10. Consider the following boxes. Draw a ball from box 1 and put it into box 2. Then draw a ball from box 2 and put it into box 3. Finally, draw a ball from box 3. Construct a tree diagram for this experiment and calculate the probability that the last ball chosen is colored.

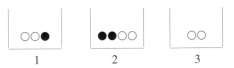

11. What are the odds in favor of drawing a jack when one card is drawn from an ordinary deck of playing cards?

12. A die is rolled once. What are the odds against rolling a prime number?

13. If the odds in favor of a particular event are 3 to 5, what is the probability that the event will occur?

14. A game consists of rolling two dice. Rolling double 1s pays $7.20. Rolling double 6s pays $3.60. Any other roll pays nothing. What is the expected value for this game?

15. A total of 3000 tickets have been sold for a drawing. If one ticket is drawn for a single prize of $1000, what is a fair price for a ticket?

16. In a special raffle, a ticket costs $2. You mark any four digits on a card (repetition and 0 are allowed). If you select the winning number, you win $15,000. What is the expected value?

17. How many four-digit numbers can be formed if the first digit cannot be 0 and the last digit must be 2?

18. A club consists of 10 members. In how many different ways can a group of 3 people be selected to go on a European trip?

19. Find the number of ways that 4 flags can be displayed on a flagpole, one above the other, if 10 different flags are available.

20. Five women live together in an apartment. Two have blue eyes. If two of the women are chosen at random, what is the probability that they both have blue eyes?

21. Five evenly matched horses (Applefarm, Bandy, Cash, Deadbeat, and Egglegs) run in a race.
    a. In how many ways can the first-, second-, and third-place horses be determined?
    b. Find the probability that Deadbeat finishes first and Bandy finishes second in the race.
    c. Find the probability that the first-, second-, and third-place horses are Deadbeat, Egglegs, and Cash, in that order.

22. Al and Ruby each roll an ordinary die once. What is the probability that the number of Ruby's roll is greater than the number of Al's roll?

23. Amy has a quiz on which she is to answer any three of the five questions. If she is equally well versed on all questions and chooses three questions at random, what is the probability that question 1 is not chosen?

24. How many batting lineups are there for the nine players of a baseball team if the center fielder must bat fourth and the pitcher last?

25. On a certain street are three traffic lights. At any given time, the probability that a light is green is 0.3. What is the probability that a person will hit all three lights when they are green if they represent independent events?

26. A three-stage rocket has the following probabilities for failure: The probability for failure at stage one is $\frac{1}{6}$; at stage two, $\frac{1}{8}$; and at stage three, $\frac{1}{10}$. What is the probability of a successful flight, given that the first stage was successful and all stages are independent?

27. How could each of the following be simulated by using a random-digit table?
    a. Tossing a fair die
    b. Picking 3 months at random from the 12 months of the year
    c. Spinning the spinner shown

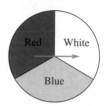

28. Otto says that if you toss three coins you get $3H$, $2H$, $1H$, or $0H$, so the probability of getting three heads is $\frac{1}{4}$. How do you respond?

29. If a dart is thrown at the following tangram dartboard and we assume the dart lands at random on the board, what is the probability of its landing in each of the following areas?
    a. Area A b. Area B c. Area C

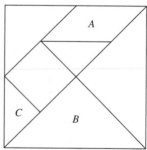

30. The points M, N, O, P, and Q in the following figure represent exits on a highway (the numbers represent miles). An accident occurs at random between points M and Q. What is the probability that it has occurred between N and O?

Selected Bibliography

Beck, S., and V. Huse. "A Virtual Spin on the Teaching of Probability." *Teaching Children Mathematics* 13 (May 2007): 482–486.

Bintz, W., and S. Moore. "Using Literature to Teach Factorials." *Mathematics Teaching in the Middle School* 8 (May 2003): 461–465.

Caulfield, R., S. Harkness, and R. Riley. "Surprise! Turn Routine Problems into Worthwhile Tasks." *Mathematics Teaching in the Middle School* 9 (December 2003): 198–202.

Coffey, D., and M. Richardson. "Rethinking Fair Games." *Mathematics Teaching in the Middle School* 10 (February 2005): 298–303.

Colgan, M. "March Math Madness: The Mathematics of the NCAA Basketball Tournament." *Mathematics Teaching in the Middle School* 11 (March 2006): 334–342.

Edwards, T., and S. Hensien. "Using Probability Experiments to Foster Discourse." *Teaching Children Mathematics* 6 (April 2000): 524–529.

Ewbank, W., and J. Ginther. "Probability on a Budget." *Mathematics Teaching in the Middle School* 7 (January 2002): 280–283.

Hardy, M. "Burgers, Graphs, and Combinations." *Mathematics Teaching in the Middle School* 7 (October 2001): 72–76.

Hoiberg, K., J. Sharp, T. Hodgson, and J. Colbert. "Geometric Probability and the Areas of Leaves." *Mathematics Teaching in the Middle School* 10 (March 2005): 326–332.

http://radicalart.info/AlgorithmicArt/scatter/index.html (accessed April 18, 2008).

Hylton-Lindsay, A. "Problem Solving, Patterns, Probability, Pascal, and Palindromes." *Mathematics Teaching in the Middle School* 8 (February 2003): 288–293.

Jardine, D. "Looking at Probability through a Historical Lens." *Mathematics Teaching in the Middle School* 6 (September 2000): 50–54.

Jones, G., C. Thornton, C. Langrall, and J. Tarr. "Understanding Students' Probabilistic Reasoning." In *Developing Mathematical Reasoning in Grades K–12*, edited by L. Stiff. Reston, VA: NCTM, 1999.

Kahneman, D., and A. Tversky. "Extensional versus Intuitive Reasoning: The Conjunction Fallacy in Probability Judgement." *Psychological Review*, 1983, 90 (4): 293–315.

Lappan, G., et al. "Area Models for Probability." *Mathematics Teacher* 80 (November 1987): 650–654.

Lawrence, A. "From *The Giver* to *The Twenty-One Balloons:* Explorations with Probability." *Mathematics Teaching in the Middle School* 4 (May 1999): 504–509.

Lecoutre, M. "Cognitive Models and Problem Spaces in 'Purely Random' Situations." *Educational Studies in Mathematics*, 1992, 23: 557–568.

McCoy, L., S. Buckner, and J. Munley. "Probability Games from Diverse Cultures." *Mathematics Teaching in the Middle School* 12 (March 2007): 394–402.

McMillen, S. "What Are the Odds? Taxed to the Max!" *Mathematics Teaching in the Middle School* 10 (March 2005): 342, 347.

Metz, M. "Playing Fair at the Fair." *Mathematics Teaching in the Middle School* 8 (February 2003): 303, 306–307.

Mittag, K., and S. Taylor. "Hitting the Bull's Eye: A Dart Game Simulation Using Graphing Calculator." *Mathematics Teaching in the Middle School* 12 (September 2006): 116–122.

Nicholson, C. "Is Chance Fair? One Student's Thoughts on Probability." *Teaching Children Mathematics* 12 (September 2005): 83–89.

Norton, R. "Determining Probabilities by Examining Underlying Structure." *Mathematics Teaching in the Middle School* 7 (October 2001): 78–82.

Pagni, D. "The Coat Check Problem: A S(t)imulating Lesson." *Mathematics Teaching in the Middle School* 13 (October 2007): 182–187.

Quinn, R. "Using Attribute Blocks to Develop a Conceptual Understanding of Probability." *Mathematics Teaching in the Middle School* 6 (January 2001): 290–294.

Rubel, L. "Middle School and High School Student's Probabilistic Reasoning on Coin Tasks." *Journal of Research in Mathematics Education* 38 (November 2007): 531–557.

Tarr, J. "Providing Opportunities to Learn Probability Concepts." *Teaching Children Mathematics* 8 (April 2002): 482–487.

Tayek, C. "Solutions to the Dicey Situation Problem." *Teaching Children Mathematics* 13 (May 2007): 474–477.

Wiest, L., and R. Quinn. "Exploring Probability through an Evens-Odds Dice Game." *Mathematics Teaching in the Middle School* 4 (March 1999): 358–362.

CHAPTER 10

Data Analysis/Statistics: An Introduction

Preliminary Problem

A headline in a Mississippi news-paper read, "Women Earn 68% of What Men Do." In the article's supporting evidence, the following was displayed. What is your reaction to the headline and the evidence?

Percentage of Men's Salaries that Women Make

Job Category	Percentages
Executive/Managerial	81
Faculty	80
Professional Non-faculty	79
Clerical	107
Technical	80
Skilled Crafts	84

Source: Daily Mississippian, Tuesday, September 18, 2007, Vol. 98, No. 196, p. 1.

I n *Principles and Standards*, we find the following:

> Instructional programs from prekindergarten through grade 12 should enable all students to
>
> • formulate questions that can be addressed with data and collect, organize, and display relevant data to answer them;
>
> • select and use appropriate statistical methods to analyze data;
>
> • develop and evaluate inferences and predictions that are based on data . . . . (p. 48)

In addition, we find the following:

> Students need to know about data analysis and related aspects of probability in order to reason statistically—skills necessary to becoming informed citizens and intelligent consumers. (p. 48)

Curriculum Focal Points recommends that from the earliest grades students will "pose questions and gather data about themselves and their surroundings" (kindergarten, p. 26), "represent data using concrete objects, pictures, and graphs" (grade 1, p. 26), and "describe parts of the data and the set of data as a whole to determine what the data show" (grades kindergarten and 1, p. 26). Students in grades 3–5 are expected to "design investigations to address a question and consider how data-collection methods affect the nature of the data set," "collect data using observations, surveys, and experiments," and "represent data using tables and graphs such as line plots, bar graphs, and line graphs" (p. 33). Additionally students in grade 5 are expected to "recognize the differences in representing categorical and numerical data" (p. 33). In grade 7, students are expected to "formulate questions, design studies, and collect data about a characteristic shared by two populations or different characteristics within one population" (p. 39).

Data analysis usually refers to a more informal approach to statistics. It is a relatively new term in mathematics. *Statistics* once referred to numerical information about state or political territories; it comes from the Latin *statisticus*, meaning "of the state." Today, much of statistics involves making sense of data.

In *Guidelines for Assessment and Instruction in Statistics Education (GAISE) Report: A PreK–12 Curriculum Framework* (March 2005) (hereafter referred to as the *Statistics Framework*), presented to the American Statistical Association, recommendations are made for statistics education that complement the *Principles and Standards for School Mathematics* (NCTM 2000). In the *Statistics Framework*, specific data analysis recommendations are divided into three parts—A, B, and C—with the more intuitive parts for early grades being in

---

◆ Historical Note

The seventeenth-century work of John Graunt (1620–1674) and the nineteenth-century work of Adolph Quetelet (1796–1874) involved making predictions on the collection of data. Graunt dealt with birth and death records; Quetelet dealt with crime and mortality rates. Florence Nightingale (1820–1910) worked with mortality tables during the Crimean War to improve hospital care. Other notables in data collection and analysis include Sir Francis Galton (1822–1911) and Gregor Mendel (1822–1884). In the twentieth century, work continued by Ronald Fisher (1890–1962) in genetics and Andrei Nikolaevich Kolmogorov (1903–1987), chairman of the Commission for Mathematical Education, the Presidium of the Academy of Sciences of the former Soviet Union. ◆

part A and the more advanced ideas being in part C. According to the *Statistics Framework*, "Sound statistical reasoning skills take a long time to develop.... The surest way to reach the necessary skill level is to begin the educational process in the elementary grades and keep strengthening and expanding these skills throughout the middle and high school years" (p. 3). The combination of documents recommends a solid background for both elementary students and prospective teachers of elementary students.

In both *Principles and Standards* and *Curriculum Focal Points*, we see that students in the early grades should explore the basic ideas of statistics by collecting data, organizing the data pictorially, and then interpreting information from their displays. The ideas of gathering, representing, and analyzing data are expanded in the later grades. In this chapter, we deal with categorical and numerical data, representations of data, and key statistical concepts including measures of central tendency and of variation. Additionally, some uses and misuses of statistics are shown. Formulating questions and designing studies at an elementary level can be found online at www.pearsonhighered.com/Billstein10einfo.

10-1   Displaying Data: Part I

Visual illustrations are an important part of statistics. Such visual illustrations or graphs take many forms: pictographs, circle graphs, pie charts, dot plots, line plots, scatterplots, stem and leaf plots, box plots, frequency tables, histograms, bar graphs, and frequency polygons or line graphs. A *graph* is a picture that displays data. Graphs are used to try to tell a story. In the *Herman* cartoon, we see a graph being used to display some particular data to make a point to an audience. What message do you think the presenter is trying to get across? What labels might appear on the vertical and horizontal axes?

"That's the last time I go on vacation."

Research Note

Describing data frequently involves reading information from graphical displays, tables, lists, and so on. The majority of students in elementary and middle-school grades can read data from these representations accurately (Beaton, Mullis, Martin et al. 1996; Bright and Friel 1998; Jones, Thornton, Langrall, et al. 1999; Jones, Thornton, Langrall, et al. 2000; Periera-Mendoza and Mellor 1991). ◆

As seen in the research note, students typically can read information from graphs. In *Principles and Standards*, we find the following:

> A fundamental idea in prekindergarten through grade 2 is that data can be organized or ordered and that this "picture" of the data provides information about the phenomenon or question. In grades 3–5, students should develop skill in representing their data, often using bar graphs, tables, or line plots. . . . Students in grades 6–8 should begin to compare the effectiveness of various types of displays in organizing the data for further analysis or in presenting the data clearly to an audience. (p. 49)

In particular, we find the following by grade level.

In prekindergarten through grade 2 all students should

- pose questions and gather data about themselves and their surroundings;
- sort and classify objects according to their attributes and organize data about the objects;
- represent data using concrete objects, pictures, and graphs;
- describe parts of the data and the set of data as a whole to determine what the data show.

In grades 3–5 all students should

- design investigations to address a question and consider how data-collection methods affect the nature of the data set;
- collect data using observations, surveys, and experiments;
- represent data using tables and graphs such as line plots, bar graphs, and line graphs;
- recognize the differences in representing categorical and numerical data;
- describe the shape and important features of a set of data and compare related data sets, with an emphasis on how the data are distributed. . . .

In grades 6–8 all students should

- formulate questions, design studies, and collect data about a characteristic shared by two populations or different characteristics within one population;
- select, create, and use appropriate graphical representations of data, including histograms, box plots, and scatterplots; . . .
- discuss and understand the correspondence between data sets and their graphical representations, especially histograms, stem-and-leaf plots, box plots, and scatterplots; . . .
- make conjectures about possible relationships between two characteristics of a sample on the basis of scatterplots of the data and approximate lines of fit; . . . . (p. 401)

Data: Categorical and Numerical

Statistical thinking begins in early grades with a need to know such things as "the most popular" pet, the favorite type of shoe, the most used color to paint, and so on. **Data** may be collected to find answers to such questions. Data collected may be either **categorical** or **numerical** depending on the questions being answered. For example, according to the *Statistics Framework*, in the elementary grades, students may

> be interested in the favorite type of music among students at a certain grade level. . . . The class might investigate the question: *What type of music is most popular among students?* . . . The characteristic, favorite music type is a categorical variable—each child in that grade would be placed in a particular nonnumerical category based on his or her favorite music type. The resulting data are often called Categorical Data. (p. 22)

Categorical data are data that represent characteristics of objects or individuals in groups (or categories), such as black or white, inside or outside, male or female. **Numerical data** are data collected on numerical variables. For example, in grade school, students may ask whether there is a difference in the distance that girls and boys can jump. The distance jumped is a numerical variable and the collected data is numerical.

Frequently, a variety of representations are needed for the different types of data, but in common usage, there are some representations that are used for both types of data. We will see different representations for both categorical and numerical data in the following subsections, with the graphic representations of two variables delayed until Section 10-2. Consider the research note that emphasizes the value of technology in representations.

Research Note

Explorations with categorical and numerical data in instruction that uses technology with primary-aged children produce more focused and less idiosyncratic descriptions of the data (Jones, Thornton, Langrall, et al. 1999). ◆

Pictographs

An elementary student might use a picture graph, or **pictograph**, to represent tallies of categories. For example, categorical data might be seen in the determination of the month in which the most newspapers were recycled. The month is the *category* and the data might be depicted in a pictograph. Pictographs are often seen in newspapers and magazines. In the *Frank and Ernest* cartoon, the projected results of the third quarter's hunts are seen in a pictograph. With only what is pictured, what is the most common animal that the hunters bag?

Frank and Ernest

©2007 Thaves. Reprinted with permission. Newspaper dist. by NEA, Inc.

The information in the pictograph could be represented in a **frequency table**, as seen in Table 10-1, that gives data counts.

Table 10-1

Type	Frequency
🦗	2
🐃	5
🐕	3
🦆	1

In a pictograph, a symbol or an icon is used to represent a quantity of items. A *legend* tells what the symbol represents. Pictographs are used frequently to show comparisons of outputs, as in Figure 10-1. A major disadvantage of pictographs is also evident in Figure 10-1. The month of September contains a partial bundle of newspapers. It is impossible to tell from the graph the weight of that bundle with any accuracy.

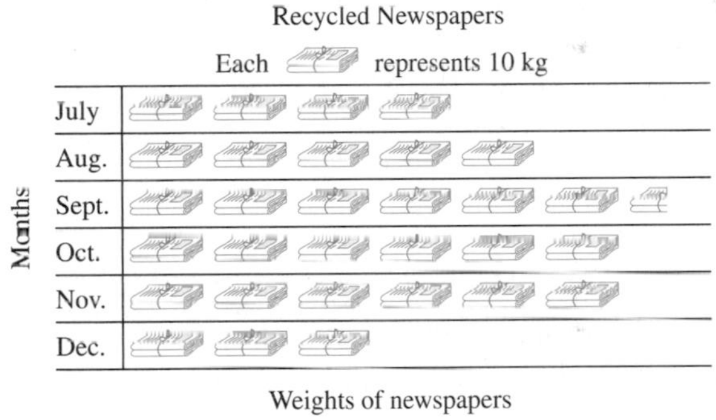

Recycled Newspapers

Each ▭ represents 10 kg

Weights of newspapers

Figure 10-1

REMARK All graphs need titles; if needed, legends should be shown.

The number of students in each teacher's fifth-grade class at Hillview is depicted in the tabular representation in Table 10-2. Each teacher is a category and the frequency/count table provides a method to summarize the categorical data. Figure 10-2 depicts the information in a pictograph.

Table 10-2

Teacher	Frequency or Count
Ames	20
Ball	20
Cox	15
Day	25
Eves	15
Fagin	10

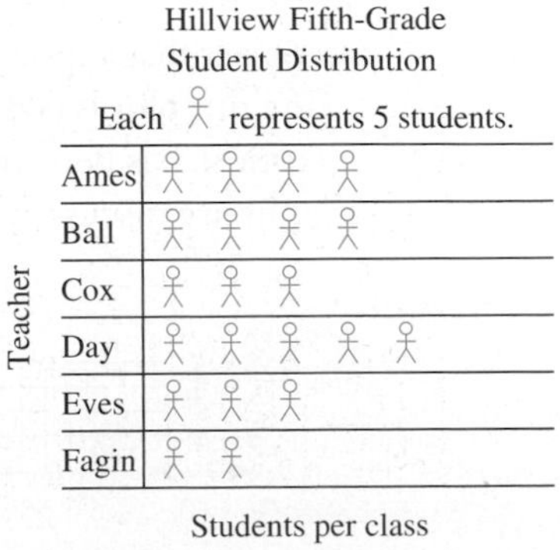

Hillview Fifth-Grade
Student Distribution

Each 👤 represents 5 students.

Students per class

Figure 10-2

Dot Plots

Next we examine a **dot plot**, sometimes called a **line plot**. Dot plots provide a quick, simple way of organizing data. Typically, we use them when there is only one group of data with fewer than 50 values.

> **REMARK** The names *dot plots* and *line plots* have been used synonymously in the past, but current usage by statistics educators favors dot plots.

Suppose the 30 students in Abel's class received the following test scores:

82 97 70 72 83 75 76 84 76 88 80 81 81 52 82

82 73 98 83 72 84 84 76 85 86 78 97 97 82 77

A dot plot for the class scores consists of a horizontal number line on which each score is denoted by a dot, or an ×, above the corresponding number-line value, as shown in Figure 10-3. The number of ×'s above each score indicates how many times each score occurred.

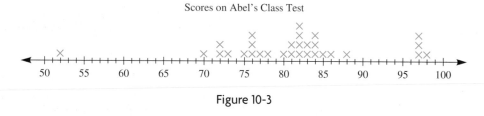

Figure 10-3

Figure 10-3 yields information about Abel's class exam scores. For example, three students scored 76 and four scored greater than 90. We also see that the low score was 52 and the high score was 98. Several features of the data become more obvious when dot plots are used. For example, outliers, clusters, and gaps are apparent. An **outlier** is a data point whose value is significantly greater or less than other values, such as the score of 52 in Figure 10-3. (Outliers are discussed in greater detail in a later section.) A **cluster** is an isolated group of points, such as the one located at the scores 97 and 98. A **gap** is a large space between points, such as the one between 88 and 97.

Another feature of the data is the score that appears most often in the data set. In Figure 10-3, 82 is the score that appears the most number of times. The count or measurement of an object that appears the most often is the **mode**. The mode is discussed in more detail later.

If a dot plot is constructed on grid paper, then shading in the squares with ×'s and adding a vertical axis depicting the scale allows the formation of a *bar graph*, as in Figure 10-4. (Bar graphs are discussed in more detail later in this section.)

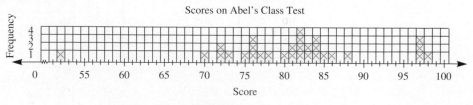

Figure 10-4

> **REMARK** The break in the horizontal axis is denoted by a squiggle and indicates that a part of the number line is missing.

    The following student page shows a grade 6 approach to frequency tables and dot plots. Answer questions *d*, *e*, and *f*.

Stem and Leaf Plots

The **stem and leaf plot** is akin to the dot plot, but the number line is usually vertical, and digits are used rather than ×'s. A stem and leaf plot of test scores for Abel's class is shown in Figure 10-5. The numbers on the left side of the vertical segment are the **stems**, and the numbers on the right side are the **leaves**. In Figure 10-5, the stems are the tens digits of the scores on the test and the leaves are the units digits. In this case, the legend, "9|7 represents 97," shows how to read the plot.

Scores for Abel's Class

```
5 | 2
6 |
7 | 0223566678
8 | 011222233444568        9|7 represents 97
9 | 7778
```

Figure 10-5

The data in a stem and leaf plot, as in Figure 10-5, are in order from least to greatest on a given row. This is an **ordered stem and leaf plot**. Stem and leaf plots are sometimes first constructed with data in the order given. To make them useful, most users then order the data.

There is no unique way to construct stem and leaf plots. Smaller numbers are usually placed at the top so that when the plot is turned counterclockwise 90°, it resembles a bar graph or a histogram (discussed later in this section). Important advantages of stem and leaf plots are that they can be created by hand rather easily and they do not become unmanageable when the number of values becomes large. Moreover, no original values are lost in a stem and leaf plot. For example, in the stem and leaf plot of Figure 10-5, we know that one score was 75 and that exactly three students scored 97. A disadvantage of stem and leaf plots is that we do lose some information; for example, we know from such a plot that a student scored 88, but we do not know which one.

Following is a summary of how to construct a stem and leaf plot:

1. Find the high and low values of the data.
2. Decide on the stems.
3. List the stems in a column from least to greatest.
4. Use each piece of data to create leaves to the right of the stems on the appropriate rows.
5. If the plot is to be ordered, list the leaves in order from least to greatest.
6. Add a legend identifying the values represented by the stems and leaves.
7. Add a title explaining what the graph is about.

School Book Page   FREQUENCY TABLES AND LINE PLOTS

Lesson 11-3

Key Idea
Making a frequency table or a line plot helps to organize and display data.

Vocabulary
- frequency table
- interval
- line plot
- outlier

Think It Through
I can **make a table** to show the frequency of various data values.

Frequency Tables and Line Plots

LEARN

Activity

How can you make a frequency table and a line plot?

The manager of a shoe store recorded the ages of customers who purchased athletic shoes recently.

a. Copy the **frequency table** at the right. Then follow the steps below to complete the table.

Step 1 List age groups, or **intervals.**
Step 2 Make a **tally mark** for each data value.
Use ⊞ for every five.
Step 3 In the far right column, show the frequency by writing the number of tally marks for that age group.

b. How many people 50 years of age or older bought shoes?

A **line plot** is another way to display data. Instead of tally marks, a line plot uses Xs.

c. Copy and complete the line plot of the shoe data.

d. How are the frequency table and line plot alike? How are they different?

e. What does the tallest column of Xs signify?

f. Could you have used different intervals for the frequency table or line plot? Do you think you would always need to use intervals? Explain.

Ages of Customers

14	37	30
25	19	31
8	20	
10	47	
12	61	
35	26	
26	38	
67	21	
23	20	
6	52	
55	11	
13	23	
34	41	
29	42	
9	62	
54	56	
43	34	
27	50	

Age Group	Tally	Frequency
0-9	|||	3
10-19	⊞		6
20-29	⊞ ⊞	10
30-39		
40-49		
50-59		
60-69		

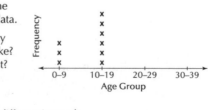

Take It to the NET
More Examples
www.scottforesman.com

Source: Scott Foresman-Addison Wesley Mathematics, Grade 6, 2008 (p. 628).

NOW TRY THIS 10-1 Construct a stem and leaf plot using the data in Table 10-3, which lists the presidents of the United States and their ages at death.

Table 10-3

President	Age at Death	President	Age at Death	President	Age at Death
George Washington	67	Franklin Pierce	64	Woodrow Wilson	67
John Adams	90	James Buchanan	77	Warren Harding	57
Thomas Jefferson	83	Abraham Lincoln	56	Calvin Coolidge	60
James Madison	85	Andrew Johnson	66	Herbert Hoover	90
James Monroe	73	Ulysses Grant	63	Franklin Roosevelt	63
John Q. Adams	80	Rutherford Hayes	70	Harry Truman	88
Andrew Jackson	78	James Garfield	49	Dwight Eisenhower	78
Martin Van Buren	79	Chester Arthur	57	John Kennedy	46
William H. Harrison	68	Grover Cleveland	71	Lyndon Johnson	64
John Tyler	71	Benjamin Harrison	67	Richard Nixon	81
James K. Polk	53	William McKinley	58	Gerald Ford	93
Zachary Taylor	65	Theodore Roosevelt	60	Ronald Reagan	93
Millard Fillmore	74	William Taft	72		

Back-to-Back Stem and Leaf Plots

If two sets of related data with a similar number of data values are to be compared, a *back-to-back stem and leaf plot* can be used. In this case, two plots are made: one with leaves to the right, and one with leaves to the left. For example, if Abel gave the same test to two classes, he might prepare a back-to-back stem and leaf plot, as shown in Figure 10-6, where the data for the first class is on the left and for the second class is on the right.

Abel's Class Test Scores

Second-period Class		Fifth-period Class
20	5	2
531	6	24
99987542	7	1257
875420	8	4456999
1	9	2457
	10	0

0|5| represents a score of 50 |5|2 represents a score of 52

Figure 10-6

NOW TRY THIS 10-2 In the stem and leaf plots in Figure 10-6, which class do you think did better on the test? Why?

Example 10-1   Group the presidents in Table 10-3 into two groups, the first consisting of George Washington to Rutherford B. Hayes and the second consisting of James Garfield to Ronald Reagan.

a. Create a back-to-back stem and leaf plot of the two groups and see if there appears to be a difference in ages at death between the two groups.

b. Which group of presidents seems to have lived longer?

Solution   **a.** Because the ages at death vary from 46 to 93, the stems vary from 4 to 9. In Figure 10-7, the first 19 presidents are listed on the left and the remaining 19 on the right.

Ages of Presidents at Death

Early Presidents		Later Presidents
	4	96
63	5	787
364587	6	707034
0741983	7	128
3|8| represents 053	8	81 |6|7 represents
83 years old 0	9	033 67 years old

Figure 10-7

b. The early presidents seem, on average, to have lived longer because the ages at the high end, especially in the 70s through 90s, come more often from the early presidents. The ages at the lower end come more often from the later presidents. For the stems in the 50s and 60s, the numbers of leaves are about equal.

A stem and leaf plot shows how wide a range of values the data cover, where the values are concentrated, whether the data have any symmetry, where gaps in the data are, and whether any data points are decidedly different from the rest of the data.

Grouped Frequency Tables

The stem and leaf plot in Figure 10-5 naturally groups scores into intervals or **classes**. For the data in Figure 10-5, the following classes are used: 40–49, 50–59, 60–69, 70–79, 80–89, and 90–99. Each class has interval size 10; for example, 10 different scores can fall within the interval 50 and 59. Students often incorrectly report the interval size as 9 because $59 - 50 = 9$.

A **grouped frequency table** shows how many times data occurs in a range. For example, consider the data in Table 10-3 for the ages of the presidents at death. These results are summarized in Table 10-4.

Table 10-4

Ages at Death	Tally	Frequency
40–49	||	2
50–59	⊬	5
60–69	⊬ ⊬ ||	12
70–79	⊬ ⊬	10
80–89	⊬	5
90–99	||||	4
	Total	38

Table 10-4 shows that 12 ages appear in the interval 60 through 69, but it does not show the particular ages in the interval. As the interval size increases, information is lost. Choices of interval size may vary. Classes should be chosen to accommodate all the data and each item should fit into only one class; that is, the classes should not overlap. Data from frequency tables can be graphed, as will be shown next.

Histograms and Bar Graphs

The data in Table 10-4 could be pictured graphically using a **histogram**. Figure 10-8(a) shows a histogram of the frequencies in Table 10-4. A histogram is made up of adjoining rectangles, or bars. In this case, the death ages are shown on the horizontal axis and the numbers along the vertical axis give the scale for the frequency. The frequencies of the death ages are shown by the bars, which are all the same width. The scale on the vertical axis must be uniform.

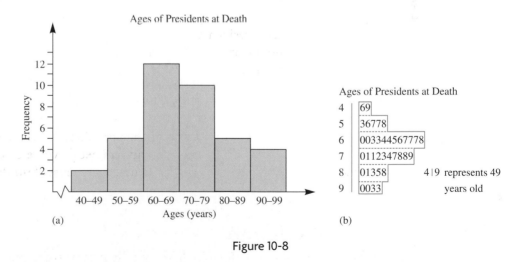

Figure 10-8

REMARK In some reports the midpoints of the bars in a histogram are marked instead of the intervals.

Histograms can be made easily from single-sided stem and leaf plots. For example, if we take the stem and leaf plot in Figure 10-8(b) and enclose each row (set of leaves) in a bar, we will have what looks like a histogram. We can make Figure 10-8(b) resemble Figure 10-8(a) by rotating the graph 90° counterclockwise. Histograms show gaps and clusters just as stem and leaf plots do. However, with a histogram we cannot retrieve data as we can in a stem and leaf plot. Another disadvantage of a histogram is that it is often necessary to estimate the heights of the bars.

The partial student page shows a histogram in grade 6. Read the page and try problems 11 and 12. Note the definition of histogram is slightly different from this book's.

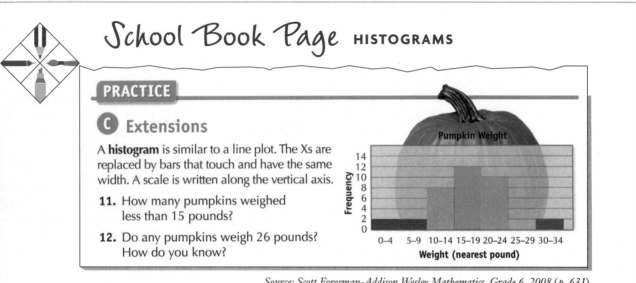

School Book Page HISTOGRAMS

PRACTICE

C Extensions

A **histogram** is similar to a line plot. The Xs are replaced by bars that touch and have the same width. A scale is written along the vertical axis.

11. How many pumpkins weighed less than 15 pounds?

12. Do any pumpkins weigh 26 pounds? How do you know?

Source: Scott Foresman-Addison Wesley Mathematics, Grade 6, 2008 (p. 631).

E-Manipulative Activity

Use the activity *Bar Graphs* to create and analyze data in a bar graph.

A **bar graph** typically has spaces between the bars and is used to depict categorical data. A bar graph showing the heights in centimeters of five students is given in Figure 10-9. The height of each bar represents the height in centimeters of each student named on the horizontal axis. Each space between bars is usually one-half the width of a bar.

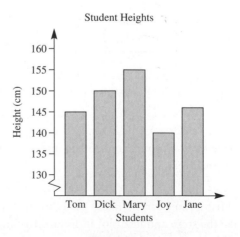

Figure 10-9

Note that in Figure 10-9 the bars representing Tom, Dick, Mary, Joy, and Jane could be placed in any order, whereas in Figure 10-8(a), the order cannot be changed without losing the continuity of the numbering along the horizontal axis. A distinguishing feature between histograms and bar graphs is that there is no ordering that has to be done among the bars of the bar graph, whereas there is an order for a histogram.

Double-bar graphs can be used to make comparisons in data. For example, the data in the back-to-back stem and leaf plot of Figure 10-7 can be pictured as shown in Figure 10-10. The green bars represent the later presidents, and the blue bars represent the earlier presidents.

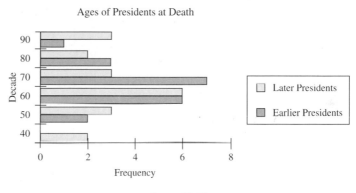

Figure 10-10

Other Bar Graphs

Table 10-5 shows various types of shoes worn by students in one class and the approximate percentages of students who wore them. Figure 10-11 depicts a **percentage bar graph** with that same data.

Table 10-5

Shoe Type	Frequency	*Percentage
Dress	1	4%
Flip-Flops	4	14%
Crocs	7	24%
Loafers	3	10%
Tennis	14	48%

*Note: percentages are approximate to sum to 100%.

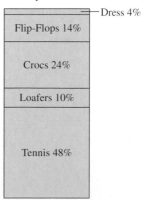

Figure 10-11

Table 10-6 shows information about the expenditures of a business over a period of years. This data is depicted in a **stacked bar graph** of the data in Figure 10-12.

Table 10-6

Years	Materials	Labor
1970–1979	$ 795,000.00	$1,500,000.00
1980–1989	$ 950,000.00	$1,900,000.00
1990–1999	$1,230,000.00	$2,400,000.00
2000–2009	$1,500,000.00	$2,400,000.00

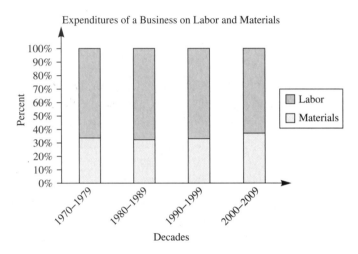

Expenditures of a Business on Labor and Materials

Figure 10-12

Circle Graphs (Pie Charts)

Another type of graph used here to represent categorical data is the circle graph. A **circle graph**, or **pie chart**, consists of a circular region partitioned into disjoint sections, with each section representing a part or percentage of the whole. A circle graph shows how parts are related to the whole. An example of a circle graph is given in Figure 10-13. Example 10-2 shows how a circle graph can be constructed from given data.

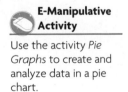

E-Manipulative Activity

Use the activity *Pie Graphs* to create and analyze data in a pie chart.

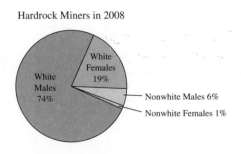

Hardrock Miners in 2008

Figure 10-13

Example 10-2

Construct a circle graph for the information in Table 10-7, which is based on information taken from a U.S. Bureau of the Census Report (2006).

Table 10-7

Age	Number of U.S. People (nearest million)
Under 5	20
5–19	62
20–34	61
35–54	64
55–65	42
Over 65	50
Total	299

Solution   The entire circle represents the total 299 million people. The measure of the central angle (an angle whose vertex is at the center of the circle) of each sector of the graph is proportional to the fraction or percentage of the population the section represents. For example, the measure of the angle for the sector for the under-5 group is $\frac{20}{299}$, or approximately 7% of the circle. Because the entire circle is 360°, then $\frac{20}{299}$ of 360°, which is approximately 24°, should be devoted to the under-5 group. Similarly, we can compute the number of degrees for each age group, as shown in Table 10-8. (What formulas could be used in a spreadsheet to create columns 3 and 4?)

Table 10-8

Age	Ratio	Approximate Percent	Approximate Degrees
Under 5	20/299	7	24
5–19	62/299	21	75
20–34	61/299	20	73
35–54	64/299	21	77
55–65	42/299	14	51
Over 65	50/299	17	60
Total	299/299	100	360

The percents and degrees in Table 10-8 are only approximate. Appropriate software or a compass can be used to draw a circle and a protractor can be used to draw the sectors in the circle graph, as shown in Figure 10-14.

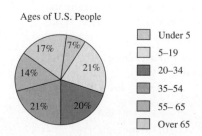

Figure 10-14

Assessment 10-1A

1. The following pictograph shows the approximate number of people who speak the six most common languages on Earth.
   a. About how many people speak Spanish?
   b. About how many people speak English?
   c. About how many more people speak Mandarin than Arabic?

Number of People Speaking the Six
Most Common Languages

Arabic	
English	
Hindi	
Mandarin	
Russian	
Spanish	

Each ⬤ represents 100 million people.

2. The Automobile Club of America considers that its holiday weekends start on Friday afternoon at 6:00 P.M. They last until Monday night at midnight.
   a. If 150 people died on one such weekend by Saturday at 6:00 P.M. and the death rate is constant, draw a graphical representation showing how many people might die on highways in the entire weekend.
   b. Discuss the reasonableness of your graph and the estimated total.

3. If a 3-in. long rectangular bar represents 100% of a population, use a percentage bar graph to represent the budget of a family whose total monthly income is $4500 and who spends the following:

Rent	$1800
Food	$1500
Transportation	$200
Entertainment	$400
Utilities	$300
Savings	All that is left

4. The following stem and leaf plot gives the weight in pounds of all 15 students in the Algebra 1 class at East Junior High:

Weights of Students in East Junior
High Algebra 1 Class

7	24	
8	112578	
9	2478	
10	3	10|3 represents
11		103 lb
12	35	

   a. Write the weights of the 15 students.
   b. What is the weight of the lightest student in the class?
   c. What is the weight of the heaviest student in the class?

5. Draw a histogram based on the stem and leaf plot in exercise 4.

6. Toss a coin 30 times.
   a. Construct a dot plot for the data.
   b. Draw a bar graph for the data.

7. The following figure shows a bar graph of the rainfall in centimeters during the last school year:

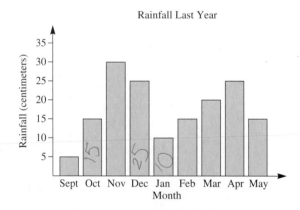

   a. Which month had the most rainfall, and how much did it have?
   b. How much total rain fell in October, December, and January?

8. HKM Company employs 40 people of the following ages:

37	58	21	63	48	52	24	52	37	23
23	34	45	46	23	26	21	18	41	27
23	45	32	63	20	19	21	23	54	62
41	32	26	41	25	18	23	34	29	26

   a. Draw a stem and leaf plot for the data.
   b. Are more employees in their 40s or in their 50s?
   c. How many employees are less than 30 years old?
   d. What percentage of the people are 50 years or older?

9. Given the following bar graph, estimate the length of the following rivers:
   a. Mississippi   **b.** Columbia

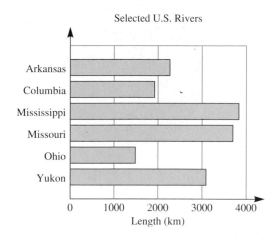

Selected U.S. Rivers

10. Five coins are tossed 64 times. A distribution for the number of heads obtained is shown in the following table. Draw a bar graph for the data.

Number of Heads	0	1	2	3	4	5
Frequency	2	10	20	20	10	2

11. **a.** What type of graph might be used to depict the changing life expectancies for women in the United States over the past 200 years?
    b. Justify your use of the graph listed in part (a).

12. In the following graph, a book club of 21 people chose their favorite book type, such as detective, romance, and so on.
    a. How many people are represented in each sector of the circle graph?
    b. What do you think are the types of books in the 5% sectors and why?

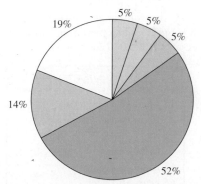

Book Choices of 21 People

13. If the number of people reading mysteries in different age groups in a survey are depicted as in the bar graph given, explain whether or not you think that the graph is accurate and why.

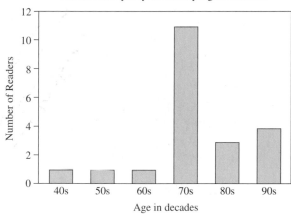

Number of Mystery Readers by Age in Decades

14. Based on the data in the graph of exercise 13, in what age decade do people read the most?

15. The following are the amounts (to the nearest dollar) paid by 25 students for textbooks during the fall term:

35	42	37	60	50
42	50	16	58	39
33	39	23	53	51
48	41	49	62	40
45	37	62	30	23

   a. Draw an ordered stem and leaf plot to illustrate the data.
   b. Construct a grouped frequency table for the data, starting the first class at $15.00 with intervals of $5.00 each.
   c. Draw a bar graph of the data.

16. The following bar graph shows the life expectancies for men and women:

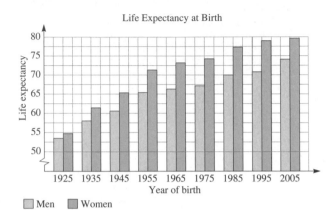

Life Expectancy at Birth

Men   Women

a. Whose life expectancy has changed the most since 1925?
b. In 1925, about how much longer was a woman expected to live than a man?
c. In 2005, about how much longer was a woman expected to live than a man?

17. Use the circle graph to answer the following questions:

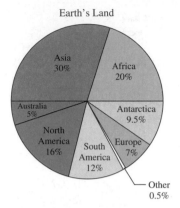

Earth's Land

a. Which is the largest continent?
b. Which continent is about twice the size of Antarctica?
c. How does Africa compare in size to Asia?
d. Which two continents make up about half of Earth's surface?
e. What is the ratio of the size of Australia to North America?
f. If Europe has approximately 4.1 million mi² of land, what is the total area of the land on Earth?

Assessment 10-1B

1. Make a pictograph to represent the categorical data in the following table. Use to represent 10 glasses of lemonade sold.

Glasses of Lemonade Sold

Day	Tally	Frequency
Monday	HHT HHT HHT	15
Tuesday	HHT HHT HHT HHT	20
Wednesday	HHT HHT HHT HHT HHT HHT	30
Thursday	HHT	5
Friday	HHT HHT	10

2. Draw a double-bar graph showing that in a certain city, the *New York Times* is more popular than the *Wall Street Journal*. Show the title, legend, and any other needed information to make the graph understandable to the reader.

3. Following are the ages of the 30 students from Washington School who participated in the city track meet. Draw a dot plot to represent these data.

10 10 11 10 13 8 10 13 14 9
14 13 10 14 11 9 13 10 11 12
11 12 14 13 12 8 13 14 9 14

4. Display the following information about households with computers in a pictograph where a computer icon represents 2,000,000 computers.

Age of Householder	
15 to 24 years	4,034
25 to 34 years	13,543
35 to 44 years	17,482
45 to 54 years	16,464
55 to 64 years	10,405
65 years and over	8,005
Numbers reported in thousands	

Source: U.S. Census Bureau. Current Population Survey, October 2003.

5. **a.** Use the characteristics below in the Reasons for No Internet Access to draw a bar graph of the number data.

b. Describe how you handled the smaller number in your graph in relation to the much larger ones to be depicted.

Characteristic	Total Number in 1000s	Percentage
INTERNET ACCESS		
Total households	113,126	100.0
Internet access	61,852	54.7
No Internet access	51,274	45.3
REASONS FOR NO INTERNET ACCESS		
Total households	51,274	100.0
Don't need it, not interested	20,185	39.4
Costs are too high	11,950	23.3
No computer or computer inadequate	11,777	23.0
Lack of confidence or skills	2,282	4.5
Lack of time to use the Internet	1,177	2.3
Have access to Internet elsewhere	1,064	2.1
Concern that children will access inappropriate sites	451	0.9
Privacy and security concerns	402	0.8
Language barriers	266	0.5
Other reason	1,720	3.4

Source: U.S. Census Bureau, Current Population Survey, October 2003.

6. Use the percentages in the table in exercise 5 to construct a percentage bar showing the data representing no Internet access.

7. The following spreadsheet shows the educational attainment of the U.S. population in 2006 from age 15 on.
   a. Use the data given for the number of high school graduates in a stem and leaf plot to show the number

of graduates in the different age groups. (Round data as needed.)
   b. Does a stem and leaf plot help illustrate the data?
   c. What type of graph might be more informative?
8. Use the data from exercise 7 to observe any differences between those people completing an eighth-grade education and those who are high school graduates.

	A	B	C	D	E	F	G	H	I	J
5										Educational Attainment
6	All Races and Both sexes	Total	None	1st - 4th grade	5th - 6th grade	7th - 8th grade	9th grade	10th grade	11th grade/2	High school graduate
7	15 years and over	233,194	996	2,119	3,950	7,857	8,576	9,891	13,574	69,548
8	15 to 17 years	13,344	19	15	41	2,352	4,245	4,128	2,337	147
9	18 to 19 years	7,572	7	20	27	106	168	440	2,463	2,287
10	20 to 24 years	20,393	31	72	216	274	408	473	1,224	6,216
11	25 to 29 years	20,138	50	129	384	320	447	445	959	5,768
12	30 to 34 years	19,343	60	120	389	317	390	359	775	5,534
13	35 to 39 years	20,771	88	165	409	289	376	366	805	6,055
14	40 to 44 years	22,350	88	142	394	341	338	480	836	7,081
15	45 to 49 years	22,518	87	153	316	355	369	485	759	7,340
16	50 to 54 years	20,279	77	165	350	315	292	434	734	6,160
17	55 to 59 years	17,827	76	145	246	394	239	402	494	5,414
18	60 to 64 years	13,153	73	167	182	354	266	393	503	4,511
19	65 to 69 years	10,231	57	198	210	462	273	336	457	3,752
20	70 to 74 years	8,323	66	215	218	480	204	323	405	3,160
21	75 years and over	16,951	217	414	568	1,496	562	828	823	6,124
22	18 years and over	219,849	977	2,104	3,910	5,505	4,331	5,763	11,237	69,401
23	15 to 24 years	41,309	57	107	283	2,732	4,821	5,041	6,025	8,651
24	25 years and over	191,884	939	2,012	3,667	5,124	3,755	4,850	7,549	60,898
25	15 to 64 years	197,689	656	1,293	2,954	5,418	7,537	8,404	11,888	58,513
26	65 years and over	35,505	340	827	996	2,438	1,039	1,487	1,686	13,035

Source: U.S. Census Bureau, Current Population Survey, 2006 Annual Social and Economic Supplement, Internet Release date: March 15, 2007 (numbers in thousands).

9. In the double-bar graph below, the percentage of the population 25 years and over from Michigan with a high school diploma or more is depicted by race.

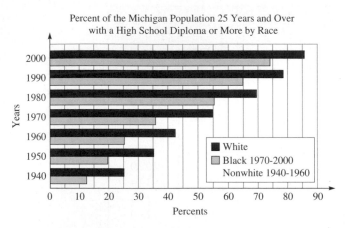

Percent of the Michigan Population 25 Years and Over with a High School Diploma or More by Race

Source: U.S. Census Bureau, *A Half-Century of Learning: Historical Statistics on Educational Attainment in the United States, 1940 to 2000,* January 18, 2007.

a. What conclusions, if any, can be drawn from the graph?

b. What questions might the legend raise?

10. The following table shows the grade distribution for the final examination in the mathematics course for elementary teachers. Draw a circle graph for the data.

Grade	Frequency
A	4
B	10
C	37
D	8
F	1

11. In a circle graph, a sector containing 32° represents what percentage of the data?

12. A percentage bar graph is sometimes used like a circle graph. If the bar is 8 cm, how long is each piece?

Savings	Rent	Food	Auto Payment	Tuition
10%	30%	12%	27%	x%

13. The following horizontal bar graph gives the top speeds of several animals:

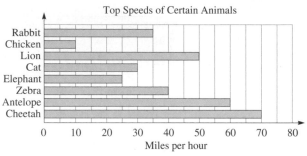

Top Speeds of Certain Animals

a. Which is the slowest animal shown?

b. How fast can a chicken run?

c. Which animal can run twice as fast as a rabbit?

d. Can a lion outrun a zebra?

14. A list of presidents, with the number of children for each, follows:

1. Washington, 0	**2.** J. Adams, 5
3. Jefferson, 6	**4.** Madison, 0
5. Monroe, 2	**6.** J. Q. Adams, 4
7. Jackson, 0	**8.** Van Buren, 4
9. W. H. Harrison, 10	**10.** Tyler, 15
11. Polk, 0	**12.** Taylor, 6
13. Fillmore, 2	**14.** Pierce, 3
15. Buchanan, 0	**16.** Lincoln, 4
17. A. Johnson, 5	**18.** Grant, 4
19. Hayes, 8	**20.** Garfield, 7
21. Arthur, 3	**22.** Cleveland, 5
23. B. Harrison, 3	**24.** McKinley, 2
25. T. Roosevelt, 6	**26.** Taft, 3
27. Wilson, 3	**28.** Harding, 0
29. Coolidge, 2	**30.** Hoover, 2
31. F. D. Roosevelt, 6	**32.** Truman, 1
33. Eisenhower, 2	**34.** Kennedy, 4
35. L. B. Johnson, 2	**36.** Nixon, 2
37. Ford, 4	**38.** Carter, 4
39. Reagan, 4	**40.** G. Bush, 5
41. Clinton, 1	**42.** G. W. Bush, 2

a. Construct a dot plot for these data.

b. Make a frequency table for these data.

c. What is the most frequent number of children?

15. The table at right depicts the number of deaths in the United States from Acquired Immunodeficiency Syndrome (AIDS) by age in 2003 and in 2005:
 a. Choose and construct a graph to display the data.
 b. Are any patterns of difference evident in the comparison of the two groups of data?

Age (yr)	2003	2005
<13	23	7
13–19	45	56
20–29	694	614
30–39	4217	3231
40–49	7037	6632
50–59	3851	4164
≥60	1537	1613

Source: Centers for Disease Control and Prevention, 2005.

Mathematical Connections 10-1

Communication

1. a. Discuss when a pictograph might be more appropriate than a circle graph.
 b. Discuss when a circle graph might be more appropriate than a bar graph.
 c. Give an example of a set of data for which a stem and leaf plot would be more informative than a histogram.
2. Explain whether a circle graph would change if the amount of data in each category was doubled.
3. Explain why the sum of the percents in a circle graph should always be 100%. How could it happen that the sum is only close to 100%?
4. The federal budget for 1 yr is typically depicted with one type of visual representation. Which one is used and why?
5. Tell whether it is appropriate to use a bar graph for each of the following. If so, draw the appropriate graph.
 a. U.S. population   **b.** Continents of the world

Year	U.S. Population
1920	105,710,620
1930	122,775,046
1940	131,669,275
1950	150,697,361
1960	179,323,175
1970	203,302,031
1980	226,542,203
1990	248,765,170
2000	281,421,906

Continent	Area in Square Miles (mi^2)
Africa	11,694,000
Antarctica	5,100,000
Asia	16,968,000
Australia	2,966,000
Europe	4,066,000
North America	9,363,000
South America	6,886,000

6. Discuss the trend in U.S. car sales from 1985 to 2003 based on the information in the given circle graphs.

U.S. Car Sales by Vehicle Size and Type

Data taken from *The World Almanac and Book of Facts 2005,* World Almanac Books, 2005.

7. Find six recent examples of different types of visual representations of data in your newspaper. Explain whether you think the representations are appropriate.
8. Choose a topic, describe how you would go about collecting data on the topic, and then explain how you would display your data in a graph. Tell why you chose the particular graph.
9. A graph similar to the following one was depicted on a package of cigarettes in Canada.

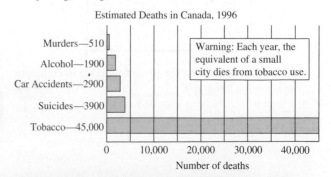

a. Make a circle graph of the data.

b. Explain which graph you think is more effective.

c. Explain whether you would like to see this labeling on cigarette packages in the United States.

Cooperative Learning

10. Find data regarding the cost of running for the U.S. Senate. Use the data to argue that only wealthy people may run for the Senate.

11. Decide on and give a rationale for the type of graph that you would use to show the percentage of time that professors spend on teaching, service, and research.

12. Choose one page of this text. Find the word length of every word on the page. Draw a graph depicting this data. If you chose another page of the book at random, what would you expect the most common word length to be? Why?

Questions from the Classroom

13. Jackson asks if a stem and leaf plot has to be constructed with the largest numbers at the bottom and the least ones at the top and why this should or should not be done. How do you respond?

14. Aliene says that she constructed the following graph on a spreadsheet and wants to know if it is an acceptable type of graph to be used to depict data regarding the types of cars owned by a certain subset of the population. How do you react?

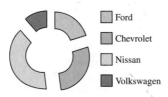

	Ford
	Chevrolet
	Nissan
	Volkswagen

National Assessment of Educational Progress (NAEP) Questions

FINAL TEST SCORES	
Score	Number of Students
95	50
90	120
85	170
80	60
75	10

Use the information in the preceding table to complete the following bar graph.

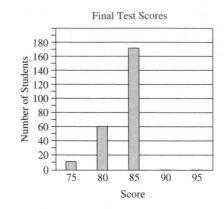

Final Test Scores

NAEP, Grade 4, 2003

The graph below and written summary on the next page [sic] present information about the sleep habits of newborn babies, one year olds, four year olds, and ten year olds. Each solid bar represents a period of sleep.

Some of the information presented in the summary does not agree with the information in the graph.

For example, there is an error in sentence 1 that has already been identified and corrected for you.

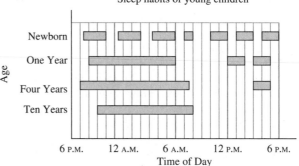

Sleep habits of young children

- In sentences 2 and 3 below, underline the information that is not correct based on the graph. There is an error in each statement.
- Then, write the correct information above the errors in sentences 2 and 3.

1. According to research that has been done on sleep habits and patterns of sleep in children, the number of hours that a newborn baby sleeps in a 24-hour period of time is more than that of a ten year old.

2. From the time a child is born until it reaches age ten, the number of different time periods of sleep increases as the child grows older.

3. Newborns need 2 more hours of sleep than ten year olds between 6 A.M. and 6 P.M.

NAEP, Grade 8, 2007

10-2   Displaying Data: Part II

In the previous section, we examined various graphical depictions of data. In this section, we explore other graphical displays: line graphs, scatterplots, and trend lines.

Line Graphs

A graphical form used to present numerical data is a line graph (or a broken line graph). A **line graph** typically shows the trend of a variable over time. Time is usually marked on the horizontal axis, with the variable being considered marked on the vertical axis. An example is seen in Figure 10-15, where Sanna's weight over 10 years is depicted. Observe that consecutive data points are connected by line segments.

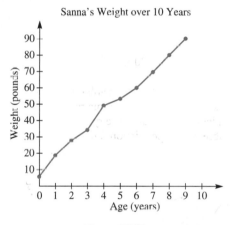

Figure 10-15

Figure 10-15 is the first example of a depiction of **continuous data**. Though the dots mark only Sanna's weight at given years, we know that she had a weight at every age along the horizontal axis. It is for this reason that it makes sense to connect the given data points with a set of line segments, as depicted. All other data examined thus far have been **discrete data**. For example, in Figure 10-9 of the previous section, it would make no sense to connect the points at the tops of the bars representing the heights of five students. These are discrete data that essentially stand as individual pieces of information. They are graphed together to examine the set of information, but connecting them would have no mathematical or statistical meaning. It is sometimes done to "prettify" a graph.

On the grade 5 student page, a line graph is drawn that depicts the number of calories burned when a person is standing over a period of time. This type of information would be of particular interest to guards, for example at Buckingham Palace in England, where they stand at attention for long periods of time. Observe that in this graph, the line is continuous because some calories are being burned the entire time one stands guard. Notice also that this line is not "broken." Answer questions 1–3.

 NOW TRY THIS 10-3 Because the line depicted on the student page contains no "breaks," what do the data imply?

School Book Page   LINE GRAPHS

Lesson 5-3

Key Idea
Line graphs are used to compare data over time.

Vocabulary
• line graph
• trend
• axes (p. 263)

Materials
• grid paper or
 tools

Think It Through
I can look at **the way the graph rises or falls** to determine the trend.

Line Graphs

◀ **LEARN**

How do you read a line graph?

A **line graph** uses data points connected with line segments to represent data collected over time.

A line graph is often used to show a **trend** or general direction in data. The appearance of the graph indicates if the data numbers are increasing or decreasing.

The graph below shows how many calories are burned over a period of 4 hours. To read the graph, locate a point and read the numbers on both **axes.**

To Determine a Trend
• If the part of a line between two points is rising from left to right, the data numbers are increasing.
• Similarly, if the part of the line between the two points is falling from left to the right, the data numbers are decreasing.

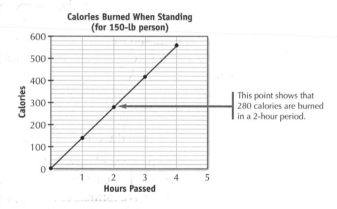

Calories Burned When Standing
(for 150-lb person)

This point shows that 280 calories are burned in a 2-hour period.

✓ Talk About It

1. About how many calories are burned by a 150-lb person who stands for 3 hours?

2. Is the graph rising or falling from left to right? Describe the trend that is shown.

3. **Representations** A 150-lb person burns 100 calories each hour when sitting. If after 2 hours of standing up, a person of this weight were to then sit for 2 hours, how would the graph change?

Take It to the NET
More Examples
www.scottforesman.com

266

Source: Scott Foresman-Addison Wesley, Grade 5, 2008 (p. 266).

Scatterplots

Sometimes a relationship between variables cannot be easily depicted by even a broken line. Frequently, a **scatterplot** is used. Figure 10-16 shows a scatterplot depicting the relation between the number of hours studied and quiz scores. The highest score is a 10 and the lowest is 1.

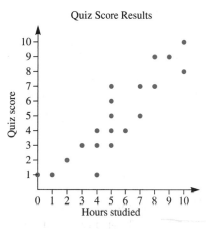

Figure 10-16

Table 10-9

Number of the Term	Term
1	2
2	4
3	6
4	8
5	10
6	12
7	14
8	16
9	18
10	20

In a scatterplot like that in Figure 10-16, we frequently try to discern patterns, but before that problem is approached, let's consider a simpler discrete set of data that second or third graders recognize. Imagine an arithmetic sequence obtained by skip counting by 2 to obtain 2, 4, 6, 8, 10, ... We know that 2 is the first term in the sequence, 4 is the second term, 6 is the third term, and we could continue. We summarize the data as in Table 10-9.

In Chapter 1, we described the value of the term using at least two different methods: a recursive method, and a closed-form method. In the recursive method, we recognized the following:

First term is 2.

Next term is 2 more than the previous term.

In the closed-form method, we observed the constant difference of 2 in the terms and learned that this was an arithmetic sequence such that the value of the term was twice the number of the term. We wrote this as a formula as $y = 2n$, where n was the number of the term.

Now consider what might happen if we plotted the same data with a scatterplot, as in Figure 10-17.

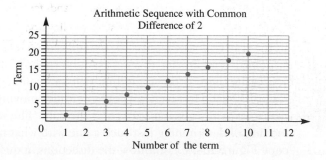

Figure 10-17

Note that the graph of Figure 10-17 is depicted as a set of discrete data in a scatterplot. But this scatterplot is unlike the one of Figure 10-16. All of the points in this scatterplot appear to lie along a line. If we draw a representation of the line, as in Figure 10-18, then we have depicted the data as a set of continuous data.

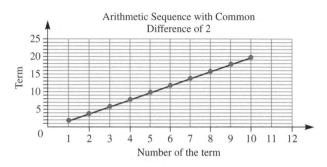

Figure 10-18

Putting together what we know from Chapter 1 with what we have learned about graphs, we could say that a point would lie along this line (or be on this line) if the term is twice the number of its term. The second, or "*y*-coordinate," is twice the first, or "*x*-coordinate." (Note that this description makes sense only if the vertical axis is labeled as a "*y*-axis" and the horizontal axis is labeled as the "*x*-axis.") Generalizing to continuous data, we have $y = 2x$ for all *x* that are real numbers. With this final representation, we moved from the discrete case, and in the process lost the sequence notation, but for any value of *x* given, we could predict the corresponding value of *y*. The line itself is a **trend line** for the data given. It is a line that closely fits the data and can be used to describe it. In this case, all the given data fall on this line and we have a complete fit of the data by the line.

NOW TRY THIS 10-4 Consider a sequence of data points formed using the following recursive formula:

First term ⁻10.
Next term is 4 more than the previous term.

   a. Construct a table of values for this recursive relation.
   b. Draw a scatterplot for the relation with the trend line.
   c. Identify by formula the "best" trend line for the data.

In the scatterplot of Figure 10-16 (and repeated as Figure 10-19), is it possible to identify a "good" trend line for this data? And if so, how might it be done?

Clearly, there is no single line that can contain all of the data points in the graph of Figure 10-19. However, we could consider moving a line across the data to define a line that follows the general pattern of the data. The student page on page 619 describes one method of determining just such a line with what might be called a "spaghetti fit." Photocopy Figure 10-19 and follow the directions in question 18. If we used a spaghetti fit with Figure 10-19, we might produce a trend line as seen in Figure 10-20. There could be a better trend line as seen in Figure 10-21(a). Could we find an equation of this line?

School Book Page **PREDICTING WITH A GRAPH**

Exploration 3

Predicting **GRAPH** with a

GOAL

LEARN HOW TO...
◆ fit a line to data in a scatter plot
◆ use a scatter plot to make predictions

AS YOU...
◆ analyze the body ratio data you collected

KEY TERMS
◆ fitted line
◆ scatter plot

SET UP *Work in a group. You will need: • completed Labsheet 3A • Labsheet 3B • uncooked spaghetti • graph paper*

▶ In Explorations 1 and 2 you used ratios to make predictions. You can also use a graph to make predictions from data.

Use Labsheet 3B for Questions 18–20 and 22.

18 **Try This as a Class** Follow the steps below to make predictions about height and reach using a graph.

 Step 1

Place a piece of uncooked spaghetti on the *Reach Compared to Height Graph* so it lies close to most of the points. Try to have about the same number of points on one side of the spaghetti as on the other side.

Step 2

Draw a line segment on the *Reach Compared to Height Graph* along the edge of your spaghetti. The segment you drew on the graph is a **fitted line**. It can be used to predict unknown measurements using known ones.

19 **a.** How can the fitted line help you predict a person's height if you know the person's reach is 135 cm?

   b. What do you expect the reach of a person 152 cm tall to be?

Source: McDougal Littell Math Thematics, New Edition, Book 1, 2008 (p. 387).

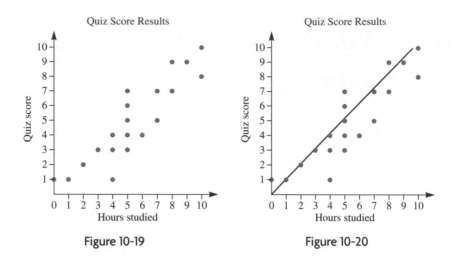

Figure 10-19   Figure 10-20

One method of attempting to find an equation of the line in Figure 10-20 is to realize that the line approximately goes through points with coordinates (2, 2) and (7, 8). When the *x*-coordinate in this case goes from 2 to 7, an increase of 5, the *y*-coordinate goes from 2 to 8, an increase of 6. We might interpret that as the *y*-coordinates increase at $\frac{6}{5}$ the rate of the corresponding *x*-coordinates. That is, $y = \frac{6}{5}x$.

However, if we plotted $y = \frac{6}{5}x$ on the graph, we would see that it is a bit higher than a "better" trend line, as seen in Figure 10-21. Thinking about the spaghetti-fit technique, we could "lower" the line in Figure 10-20 by "moving" it down approximately $\frac{1}{2}$ unit, yielding an equation $y = \frac{6}{5}x - \frac{1}{2}$, as seen in Figure 10-21.

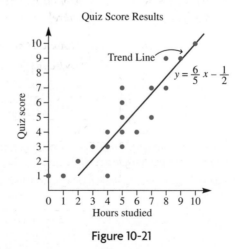

Figure 10-21

On some scatterplots the data points fall near a trend line. The trend line can be used to make predictions. If a trend line slopes up from left to right, as in Figure 10-21, then we say there is a *positive association* between the number of hours studied and the quiz score. From the trend line in Figure 10-21, we would predict that students who studied 7 hours might score about 8.

In Figure 10-21, note that the student who studied 4 hours and received a score of 1 did no better than the one who did not study. Although there is a possible association between studying and scoring well on a quiz, we cannot deduce cause and effect based on scatterplots and trend lines.

Later in the book, we will again study equations of trend lines that are considered *lines-of-best-fit*, that is, lines that appear to most closely fit the data. If the trend line slopes downward to the right, we can also make predictions; we say there is a *negative association*. If the points do not approximately fall about any line, we say there is *no association*. Scatterplots also show clusters of points and outliers. Examples of various associations are given in Figure 10-22.

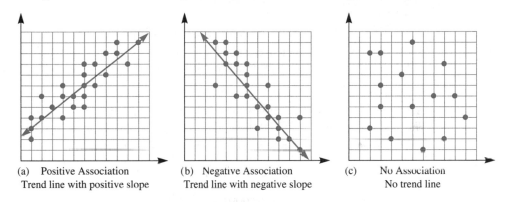

(a) Positive Association
Trend line with positive slope

(b) Negative Association
Trend line with negative slope

(c) No Association
No trend line

Figure 10-22

> **REMARK** Some books choose to discuss "correlation" instead of "association" as we have done here. In this regard, we are following the recommendations of the *Statistics Framework* for pre-high school and have chosen not to discuss correlation coefficients.

Frequently, broken line graphs are used together to demonstrate different sets of data where comparisons may be made. For example, Figure 10-23 shows the "crowdedness" of housing from 1940 to 2000, with one broken line showing "crowded" with more than 1 person per room and the other showing "severely crowded" with more than 1.5 persons per room.

Historical Census of Housing Crowding

Crowded and Severely Crowded: 1940 to 2000

Figure 10-23
Source: U.S. Census Bureau, Housing and Household Economic Statistics Division; Last Revised December 2, 2004.

The student page reveals that students in the sixth grade also consider several representations on the same graph to compare data. Answer questions 1 and 2 on the page.

School Book Page   UNDERSTAND GRAPHIC
SOURCES: GRAPHS

Reading For Math Success

Understand Graphic Sources: Graphs

Understanding graphic sources such as graphs when you read in math can help you use the **problem-solving strategy,** *Make a Graph,* in the next lesson.

In reading, understanding graphs can help you understand what you read. In math, understanding graphs can help you solve problems.

Average High and Low Temperatures in Washington, D.C.

● average high
● average low

The labels along the axes of the <u>graph</u> *describe the data that are in the graph.*

This point indicates that the average high temperature in Washington, D.C. in November is 50°F.

The key tells what data set each color represents.

Each point on the <u>graph</u> *corresponds to a month and either a high or low temperature.*

1. What is the average high temperature in Washington, D.C., in May? the average low temperature in March?

2. Describe the trend you see in the temperatures in Washington, D.C., from January to July.

646

Source: Scott Foresman-Addison Wesley, Grade 6, 2008 (p. 646).

Choosing a Data Display

◆ *Research Note*

Many elementary students have difficulty creating visual displays of data (Mullis, Martin, et al. 1997). Middle-school students have substantial gaps in abilities to construct graphs from given data (Berg and Phillips 1994). ◆

Choosing an appropriate data display is not always easy, as seen in the accompanying research note. Each type of graph is suitable for presenting certain kinds of data. In this chapter, you have seen pictographs, dot plots, stem and leaf plots, histograms, bar graphs, line graphs, scatterplots, and circle graphs (pie charts). Some types of display and uses follow:

Bar graph—Used to compare numbers of data items in grouped categories; order of data does not matter except for convenience.

Histogram—Used to compare numbers of data items grouped in numerical intervals; order matters in the data depicted.

Stem and leaf plot—Used to show each value in a data set and to group values into intervals.

Scatterplot—Used to show the relationship between two variables.

Line graph—Used to show how data values change over time; normally used for continuous data.

Circle graph—Used to show the division of a whole into parts.

NOW TRY THIS 10-5

a. Which graph in Figure 10-24(a) or (b) displays the data more effectively? Why?

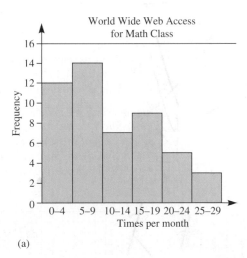

(a)

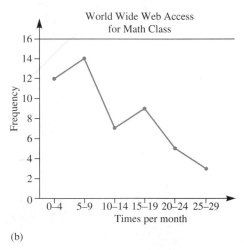

(b)

Figure 10-24

Note: Figure 10-24(b) could have been formed by connecting the midpoints of the tops of the bars in Figure 10-24(a).

b. Explain whether connecting the dots with line segments as in Figure 10-24(b) is meaningful.

c. To show each of the following, which graph is the best choice: line graph, bar graph, or circle graph? Why?

   (i) The percentage of a college student's budget devoted to housing, clothing, food, tuition and books, taxes

   (ii) Showing the change in the cost of living over the past 12 months

Assessment 10-2A

1. The following graph shows how the value of a car depreciates each year. This graph allows us to find the trade-in value of a car for each of 5 years. The percents given in the graph are based on the selling price of the new car.

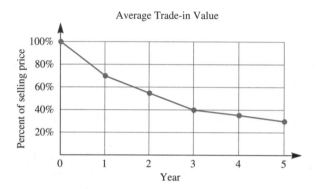

Average Trade-in Value

   a. What is the approximate trade-in value of a $12,000 car after 1 yr?
   b. How much has a $20,000 car depreciated after 5 yr?
   c. What is the approximate trade-in value of a $20,000 car after 4 yr?
   d. Dani wants to trade in her car before it loses half its value. When should she do this?

2. Coach Lewis kept track of the basketball team's jumping records for a 10-year period, as follows:

Year	2000	2001	2002	2003	2004	2005
Record (nearest in.)	65	67	67	68	70	74

Year	2006	2007	2008	2009
Record (nearest in.)	77	78	80	81

   a. Draw a scatterplot for the data.
   b. What kind of association is there for these data, if any?

3. Refer to the following scatterplot regarding movie attendance in a certain city:

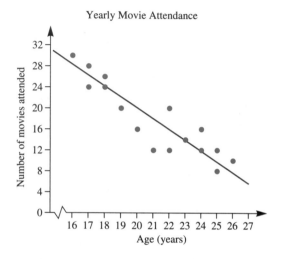

Yearly Movie Attendance

   a. What type of association exists for these data?
   b. About how many movies does an average 25-year-old attend?
   c. From the data in the scatterplot, conjecture how old you think a person is who attends 16 movies a year.

4. Given an "add 4" sequence with a first term of 2, do the following:
   a. Plot this sequence as a scatterplot.
   b. Sketch a trend line.
   c. Find an equation for the trend line you sketched.

Note: In exercises 5 and 6, numbers marking axes are moved to show dots.

5. For each of the scatterplots below, answer the following:
   (i) What type of association, if any, can you identify?
   (ii) Is there an identifiable trend line? If so, sketch it.
   a.

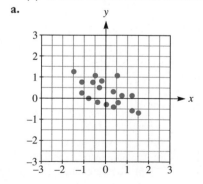

b.

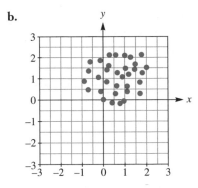

6. In the following scatterplots, estimate an equation of the trend line pictured:

a.

b.

c.

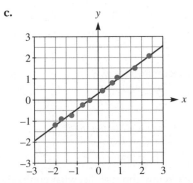

7. The following data show the Olympic year and the length of the discus throw to the nearest inch. Use the data to draw a scatterplot and predict the length of the throw in 1996.

1964	2402
1968	2551
1972	2535
1976	2657
1980	2624
1984	2622
1988	2709
1992	2564

8. The following data show the cost of various diamonds per carat:

Carats	Cost($)	Carats	Cost($)
0.17	355	0.17	353
0.16	328	0.18	438
0.17	350	0.17	318
0.18	325	0.18	419
0.25	642	0.17	346
0.16	342	0.15	315
0.15	322	0.17	350
0.19	485	0.32	918
0.21	483	0.32	919
0.15	323	0.15	298
0.18	462	0.16	339
0.28	823	0.16	338
0.16	336	0.23	595
0.2	498	0.23	553
0.23	595	0.17	345
0.29	860	0.33	945
0.12	223	0.25	655
0.26	663	0.35	1086
0.25	750	0.18	443
0.27	720	0.25	678
0.18	468	0.25	675

Source: Singfat Chu. "Diamond Ring Pricing Using Linear Regression." *Journal of Statistics Education* 4 (1996).

Draw a scatterplot of the data and determine if there is a trend line that could be used to predict the cost of 0.5 carat of a diamond.

9. The following data show the year, the name of the actress, and the age at which she won an Oscar since 1950. Plot the data as a scatterplot, see if there is a trend line, and if so, predict the age of the winner in 1997.

1950	Judy Holliday	28
1951	Vivien Leigh	38
1952	Shirley Booth	45
1953	Audrey Hepburn	24
1954	Grace Kelly	26
1955	Anna Magnani	47
1956	Ingrid Bergman	41
1957	Joanne Woodward	27
1958	Susan Hayward	39
1959	Simone Signoret	38
1960	Elizabeth Taylor	28
1961	Sophia Loren	27
1962	Anne Bancroft	31
1963	Patricia Neal	37
1964	Julie Andrews	30
1965	Julie Christie	24

1966	Elizabeth Taylor	34
1967	Katharine Hepburn	60
1968	Katharine Hepburn	61
1968	Barbra Streisand	26
1969	Maggie Smith	35
1970	Glenda Jackson	34
1971	Jane Fonda	34
1972	Liza Minnelli	26
1973	Glenda Jackson	37
1974	Ellen Burstyn	42
1975	Louise Fletcher	41
1976	Faye Dunaway	35
1977	Diane Keaton	31
1978	Jane Fonda	41
1979	Sally Field	33
1980	Sissy Spacek	30

1981	Katharine Hepburn	74
1982	Meryl Streep	33
1983	Shirley MacLaine	49
1984	Sally Field	38
1985	Geraldine Page	61
1986	Marlee Matlin	21
1987	Cher	41
1988	Jodie Foster	26
1989	Jessica Tandy	81
1990	Kathy Bates	42
1991	Jodie Foster	29
1992	Emma Thompson	33
1993	Holly Hunter	35
1994	Jessica Lange	45
1995	Susan Sarandon	49
1996	Frances McDormand	39

10. On the partial student page shown, answer questions 8–9.

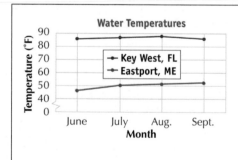

8. During which month is the water temperature in Eastport the highest?

A. June   **C.** August

B. July   **D.** September

9. Which statement is TRUE?

A. The water temperature is higher in Maine than in Florida.

B. The water temperature in both cities increases from June to September.

C. In the summer, the water temperature in Eastport is greater than 50°F.

D. The water temperature in Key West decreases from August to September.

Source: Scott Foresman-Addison Wesley Mathematics, Grade 6, 2008 (p. 684).

11. If a trend line had equation $y = 3x + 5$, what y-value would you expect to obtain when x has the following values?

a. $^-2$ **b.** 14   **c.** 0   **d.** 5

12. If a trend line had equation $y = {}^-2x - 5$, what type of association would you expect the data to have?

13. If a trend line had equation $y = 12$, what could you say about the data?

Assessment 10-2B

1. A learning specialist tracked the average reported study times of her students and had advice for faculty members as a result. Graph the data she tracked and write advisory comments to the faculty.

Monday	1.5 hr
Tuesday	$1\frac{3}{4}$ hr
Wednesday	$\frac{3}{4}$ hr
Thursday	2 hr
Friday	15 min
Saturday	15 min
Sunday	1.5 hr

2. Use the following data to draw a line graph. Use the graph to see what might be determined about the temperature of a patient in the clinic.

Time	Temperature in Degrees Fahrenheit
6:00 A.M.	96.8
7:00 A.M.	98.8
8:00 A.M.	99.9
9:00 A.M.	99.8
10:00 A.M.	100.2
11:00 A.M.	101.2
12:00 P.M.	102.0

Note: In exercises 3 and 4, numbers marking the axes are more to show dots.

3. Use the scatterplots shown to answer the following:
   (i) What type of association, if any, can you identify?
   (ii) Identify a trend line for the data if one exists. If one exists, sketch it.

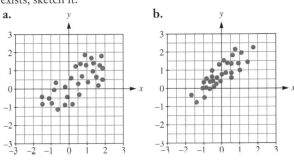

4. Given an "add 5" sequence with a first term of 3, do the following:
   a. Plot this sequence as a scatterplot.
   b. Sketch a trend line.
   c. Find an equation for the trend line you sketched.

5. In the following scatterplots, estimate an equation of the trend line pictured:

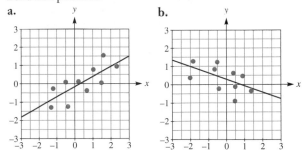

6. The data shown are for the Olympic long jump.

Year	Long Jump to the Nearest Inch
1900	283
1904	289
1908	295
1912	299
1920	282
1924	293
1928	305
1932	301
1936	317
1948	308
1952	298
1956	308
1960	320
1964	318
1968	351
1972	325
1976	329
1980	336
1984	336
1988	343
1992	343
1996	335
2000	331
2004	327

a. Draw a line graph of the data.
b. If you were predicting the length achieved by the 2008 summer Olympic long jump winner, what would be your prediction? Why?

7. On the partial student page shown, answer questions 7–9.

8. The data shown estimate the number of flea eggs produced over a period of days.

Day	No. of Eggs	Day	No. of Eggs
1	436	15	550
2	495	16	487
3	575	17	585
4	444	18	549
5	754	19	475
6	915	20	435
7	945	21	523
8	655	22	390
9	782	23	425
10	704	24	415
11	590	25	450
12	411	26	395
13	547	27	405
14	584		

Source: Moore D., and McCabe G. (2005). *Introduction to the Practice of Statistics*, p. 27.

a. Draw a line graph of the given data.
b. Predict how many eggs there might be on day 28.

9. If a trend line has equation $y = 2x - 7$, what y-value would you expect to obtain when x has the following values?

a. $^-2$ **b.** 14   **c.** 0   **d.** 5

10. If a trend line had equation $y = {}^-2x + 5$, what type of association would you expect the data to have?

11. If a trend line had an equation $y = 15$, what could you say about the data?

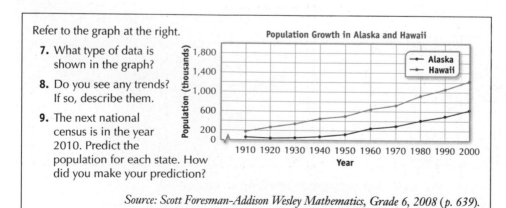

Refer to the graph at the right.

7. What type of data is shown in the graph?

8. Do you see any trends? If so, describe them.

9. The next national census is in the year 2010. Predict the population for each state. How did you make your prediction?

Population Growth in Alaska and Hawaii

Source: Scott Foresman-Addison Wesley Mathematics, Grade 6, 2008 (p. 639).

Mathematical Connections 10-2

Communications

1. Write an explanation of *association* in relation to scatterplots that could be used with eighth graders.
2. Explain a relation between a trend line and a function.

Cooperative Learning

3. With a group, examine a set of sixth- through eighth-grade textbooks to determine the ways that scatterplots, associations, and line graphs are presented.
4. Consider the women's 100-yd dash (and the subsequent 100-m dash) in the Olympics since records have been kept. Determine the best way to combine all data and to depict the data.

Open-Ended

5. Compare the equations of trend lines and linear equations from an eighth-grade algebra book.
6. Use the Internet to identify some of the modern developers of methods of depicting data with graphs.

Questions from the Classroom

7. Jacquie argued that scatterplots had little value because rarely does a single trend line fit the data. How do you respond?
8. Merle says that there is no reason to practice drawing graphs because spreadsheets can be used to draw any images desired. How do you respond?

9. Arthur says that one can determine a country's economic stability by looking on a scatterplot at the value of its currency as related to the American dollar. He continues by saying that the trend line for the data will be flat, or horizontal. How do you react?

Review Problems

10. The Smith family drew a circle graph of their budget that contained the following:

    Taxes, 20%
    Rent, 32%
    Food, 20%
    Utilities, 5%
    Gas, 13%
    Miscellaneous, 12%

    What advice could you give the family concerning the data?
11. Adjust the miscellaneous percentage in exercise 10 and draw a percentage graph of the entire data set.
12. a. Use the data in exercise 10-2B, 6, to draw a back-to-back stem and leaf plot. Separate the data into two categories, "old" and "new," by using 1952 as the division.
    b. What differences do you notice between the two sets?
13. Use the "new" data from exercise 12 to draw a histogram of the data.

BRAIN TEASER In the *FoxTrot* cartoon, Peter describes one situation where a line-of-best fit might be used. Decide whether you think this is appropriate or not.

10-3   Measures of Central Tendency and Variation

In *Principles and Standards*, we find the following by grade level:

In grades 3–5 all students should . . .

- use measures of center, focusing on the median, and understand what each does and does not indicate about the data sets;

- compare different representations of the same data and evaluate how well each representation shows important aspects of the data; . . . (p. 400)

In grades 6–8 all students should . . .

- find, use, and interpret measures of center and spread, including mean and interquartile range; . . . (p. 401)

◆ *Research Note*

Fourth graders tend to interpret the average as a mode (Mokros and Russell 1995). Students tend to use the mode to summarize data because it is so easily seen in a graph (Bright and Friel 1998). Some students tend to conceptualize the average as a median, the center of the data (Strauss and Bichler 1988). However, primary students tend not to have the balance-point idea for the mean and cannot use an algorithmic procedure to find it (Mokros and Russell 1995).   ◆

The media present us with a variety of data and statistics. For example, we find in the *World Almanac* that the average person's lifetime includes 6 years of eating, 4 years of cleaning, 2 years of trying to return telephone calls to people who never seem to be in, 6 months waiting at stop lights, 1 year looking for misplaced objects, and 8 months opening junk mail. In the previous section, we examined data by looking at graphs to display the overall distribution of values. In this section, we describe specific aspects of data by using a few carefully chosen numbers. These numbers will help in the analysis of data. Two important aspects of data are its *center* and its *spread*. The mean and median are **measures of central tendency** or of **center** that describe where data are centered. Each of these measures is a single number that describes the data. However, each does it slightly differently. The *range, interquartile range, variance, mean absolute deviation*, and *standard deviation* introduced later in this section describe the spread of data and should be used with measures of central tendency. Though the mean and median are the primary measures of central tendency discussed here, the research note shows that they may not be the numbers students identify first.

A word that is often used in statistics is *average*. For example, suppose that, as the *Far Side* cartoon on page 631 suggests, the average number of children in a family is 1.5. What does this mean? How can the average number of children be 1.5?

To explore more about averages, examine the following set of data for three teachers, each of whom claims that his or her class scored better *on the average* than the other two classes did:

Mr. Smith: 62, 94, 95, 98, 98
Mr. Jones: 62, 62, 98, 99, 100
Ms. Rivera: 40, 62, 85, 99, 99

All of these teachers are correct in their assertions because each has used a different number to characterize the scores in the class. In the following, we examine how each teacher can justify the claim.

THE FAR SIDE® BY GARY LARSON

"Bob and Ruth! Come on in. ... Have you met
Russell and Bill, our 1.5 children?"

Computing Means

The number commonly used to characterize a set of data is the **arithmetic mean**, frequently
called the **average**, or the **mean**. To find the mean of scores for each of the teachers given
previously, we find the sum of the scores in each case and divide by 5, the number of scores.

$$\text{Mean (Smith):} \quad \frac{62 + 94 + 95 + 98 + 98}{5} = \frac{447}{5} = 89.4$$

$$\text{Mean (Jones):} \quad \frac{62 + 62 + 98 + 99 + 100}{5} = \frac{421}{5} = 84.2$$

$$\text{Mean (Rivera):} \quad \frac{40 + 62 + 85 + 99 + 99}{5} = \frac{385}{5} = 77$$

In terms of the mean, Mr. Smith's class scored better than the others. In general, we define
the *arithmetic mean* as follows.

Definition of Mean

The **arithmetic mean** of the numbers $x_1, x_2, \ldots, x_n$, denoted $\bar{x}$ and read "x bar," is given by

$$\bar{x} = \frac{x_1 + x_2 + x_3 + \ldots + x_n}{n}$$

Understanding the Mean as a Balance Point

Because the mean is the most widely used measure of central tendency, we provide a model for thinking about it. Suppose a student at a rural school reports that the mean number of pets for the six students in a group is 5. Do we know anything about the distribution of these pets? All six students could have exactly five pets, as shown in the dot plot in Figure 10-25(a).

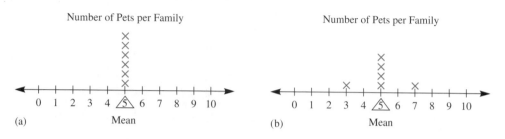

Figure 10-25

If we change the dot plot as shown in Figure 10-25(b), the mean is still 5. Notice that the new dot plot could be obtained by moving one value from Figure 10-25(a) 2 units to the right and then balancing this by moving one value 2 units to the left. We can think of the mean as a *balance point*.

Consider Figure 10-26, which shows the number of children for each family in a group. The mean of 5 is the balance point where the sum of the distances from the mean to the data points above the mean equals the sum of the distances from the mean to the data points below the mean. The sum of the distances above the mean is 3 + 5, or 8. The sum of the distances below the mean is 1 + 2 + 2 + 3, or 8. In this case, we see that the data are centered about the mean, but the mean does not belong to the set of data.

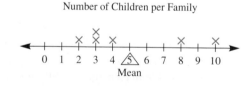

Figure 10-26

Knowing that the mean is a balance point as depicted in Figure 10-26 also shows the importance of not reporting the mean as a single number to summarize data. Using the data in Figure 10-26, it would be possible to rearrange the data to have the same mean but be spread very differently. For example, the data, independent of context, is given in the left column of Table 10-10. The right column of Table 10-10 has 5 subtracted from each of the data points to the right of the mean while 10 is added to each of the data points to the left of the mean. Note that the mean of each of the sets of data are exactly the same but the data are differently spread. In the next subsection, we consider the mean absolute deviation and the standard deviation, either of which could be used with the mean to describe the spread of the data.

Table 10-10

	2 3 3 4 8 10	$^-3$ $^-2$ $^-2$ $^-1$ 18 20
Mean	$(2 + 3 + 3 + 4 + 8 + 10)/6 = 5$	$(^-3 + {}^-2 + {}^-2 + {}^-1 + 18 + 20)/6 = 5$

NOW TRY THIS 10-6

a. A litter of six puppies was born with a mean weight of 7 lb. List two possibilities for the weights of the pups.

b. Could the mean of a set of scores ever be equal to the greatest score? The least score? Explain your answers.

c. Using the data in Figure 10-26, move the data points until a "balance point" (the mean) can be found or estimated. For example, move the value 10 two units left to 8 and counterbalance that by moving the value 2 two units right to 4. Continue until a single balance point is determined. What is the mean?

Computing Medians

The value exactly in the middle of an ordered set of numbers is the **median**. To find the median for the teachers' scores, we arrange each of their scores in increasing or decreasing order and pick the middle score. Intuitively, we know that half the scores are greater than the median and half are less.

Median (Smith):   62, 94, ⟨95,⟩ 98, 98   median = 95

Median (Jones):   62, 62, ⟨98,⟩ 99, 100   median = 98

Median (Rivera):   40, 62, ⟨85,⟩ 99, 99   median = 85

In terms of the median, Mr. Jones's class scored better than the others.

With an odd number of scores, as in the present example, the median is the middle score. With an even number of scores, however, the median is defined as the mean of the middle two scores. Thus, to find the median, we add the middle two scores and divide by 2. For example, the median of the scores

$$64, 68, \boxed{70, 74,} \ 82, 90$$

is given by

$$\frac{70 + 74}{2}, \text{ or } 72$$

In general, to find the median for a set of n numbers, proceed as follows:

1. Arrange the numbers in order from least to greatest.
2. **a.** If n is odd, the median is the middle number.
   b. If n is even, the median is the mean of the two middle numbers.

Just as the mean needs a measure of spread to help it describe a set of data, the median needs such a measure as well. For example, if you consider the data of Jones's students' scores as 62, 62, 98, 99, and 100 with median of 98, suppose the scores were 32, 32, 98, 99, and 100. The median is still 98 but the two sets of scores are very different. A median is often reported with the **interquartile range**, a measure of spread that shows where the middle 50% of the scores lie with the median in that range. The interquartile range is discussed later in this section. The two together form a much better pair to describe the data than the median alone.

Finding Modes

The **mode** of a set of data is sometimes reported as a measure of central tendency, but when it is reported in that form, it is frequently being misused. The mode of a set of data is the number that appears most frequently, if there is one, but the mode does not have to be in any way a measure of central tendency. Examples of this follow in the discussion. A mode is frequently reported with categorical data. Note that in some distributions, no number appears more than once. In other distributions, there may be more than one mode. Saying that there is no mode does not say that there are no measures of central tendency. *We include the mode here because of past practice not because it is a current statistical use of the term.*

The set of scores 64, 79, 80, 82, 90 has no mode. (Some would say that this set of data has five modes, but because no one score appears more than another, we prefer to say there is no mode.) The set of scores 64, 75, 75, 82, 90, 90, 98 is **bimodal** (two modes) because both 75 and 90 are modes. It is possible for a set of data to have too many modes for this type of number to be useful in describing the data. For the three classes listed previously, if the mode were used as the criterion for an average, Ms. Rivera's class scored better than the others.

Mode (Smith):	62, 94, 95, 98, 98	mode = 98
Mode (Jones):	62, 62, 98, 99, 100	mode = 62
Mode (Rivera):	40, 62, 85, 99, 99	mode = 99

> **REMARK** The preceding example shows a misuse of both *mode* and *average*.

Example 10-3

Find (a) the mean, (b) the median, and (c) the mode for the following collection of data:

$$60 \quad 60 \quad 70 \quad 95 \quad 95 \quad 100$$

Solution a. $\bar{x} = \dfrac{60 + 60 + 70 + 95 + 95 + 100}{6} = \dfrac{480}{6} = 80$

b. The median is $\dfrac{70 + 95}{2}$, or 82.5.

c. The set of data is bimodal and has both 60 and 95 as modes.

NOW TRY THIS 10-7 When the data values are all the same, the mean, median, and mode are all the same. Describe a situation, in which not all the data points are the same and the mean, median, and mode are still the same.

NOW TRY THIS 10-8

a. Suppose the average number of children per family for the employees of the university in a certain
city is 2.58. Could this be a mean? Median? Mode? Explain why.

b. Answer the questions in (a) if the average number of children was reported to be 2.5.

Choosing the Most Appropriate Average

Although the *mean* is the most commonly used "average" to describe a set of data, it may
not always be the most appropriate choice.

Example 10-4

Suppose a company employs 20 people. The president of the company earns $200,000, the
vice president earns $75,000, and 18 employees earn $10,000 each. Is the mean the best
number to choose to represent the "average" salary for the company?

Solution   The mean salary for this company is

$$\frac{\$200,000 + \$75,000 + 18(\$10,000)}{20} = \frac{\$455,000}{20} = \$22,750$$

In this case, the mean salary of $22,750 is not representative. Either the median or mode,
both of which are $10,000, would describe the typical salary better.

In Example 10-4, notice that *the mean is affected by extreme values*. In most cases, the
median is not affected by extreme values. The median, however, can also be misleading, as
shown in the following example.

Example 10-5

Suppose nine students make the following scores on a test:

$$30, 35, 40, 40, 92, 92, 93, 98, 99$$

Is the median the best "average" to represent the set of scores?

Solution   The median score is 92. From that score, one might infer that the individuals all
scored very well, yet 92 is certainly not a typical score. In this case, the mean of approximately 69 might be more appropriate than the median. However, with the spread of the
scores, neither is very appropriate for this distribution.

The *mode*, too, can be misleading in describing a set of data with very few items that
occur frequently, as shown in the following example.

Example 10-6

Is the mode an appropriate "average" for the following test scores?

$$40, 42, 50, 62, 63, 65, 98, 98$$

Solution   The mode of the set of scores is 98 because this score occurs most frequently.
The score of 98 is not representative of the set of data because of the large spread of scores
and the much lower mean (and median).

The choice of which number to use to represent a particular set of data is not always easy. In the example involving the three teachers, each teacher chose the average that best suited his or her claim. The type of average should always be specified along with a measure of spread (*Statistics Framework*, p. 29).

As stated in the research note, the body of research on organizing data at the early grades is limited.

BRAIN TEASER   Mr. Ramirez and Ms. Jonsey gave tests to their classes with the results seen in the table.

	Overall Mean	Mean for Females	Mean for Males	Percent Females
Ramirez	218	230	205	
Jonsey	221	224		88

What are the missing entries? For an overall mean, Ms. Jonsey's class is higher, but for females, the mean is higher in Mr. Ramirez's class. Is it possible that the mean for males is higher in Mr. Ramirez's class as well?

♦ *Research Note*

First and second graders have some conception of mode and median and a limited conception of spread (Jones, Thornton, Langrall, et al. 1999; Jones, Thornton, Langrall, et al. 2000). Most elementary students understand that the mean is between extreme values (Strauss and Bichler 1988). ♦

Problem Solving   The Missing Grades

Students of Dr. Van Horn were asked to keep track of their own grades. One day, Dr. Van Horn asked the students to report their grades. One student had lost the papers but claims to remember the grades on four of six assignments: 100, 82, 74, and 60. In addition, the student remembered that the mean of all six papers was 69, and the other two papers had identical grades. What were the grades on the other two homework papers?

Understanding the Problem   The student had scores of 100, 82, 74, and 60 on four of six papers. The mean of all six papers was 69, and two identical scores were missing. The missing scores must be less than 60; otherwise, from observation, the mean could not be less than three of the four known scores and greater than the fourth one given.

Devising a Plan   To find the missing grades, we use the strategy of *writing an equation*. The mean is obtained by finding the sum of the scores and then dividing by the number of scores, which is 6. So if we let x stand for each of the two missing grades, we have

$$69 = \frac{100 + 82 + 74 + 60 + x + x}{6}$$

Carrying Out the Plan   We now solve the equation as follows:

$$69 = \frac{100 + 82 + 74 + 60 + x + x}{6}$$

$$69 = \frac{316 + 2x}{6}$$

$$49 = x$$

Since the solution to the equation is $x = 49$, each of the two missing scores was 49.

Looking Back   The answer of 49 seems reasonable. We can check this by computing the mean of the scores 100, 82, 74, 60, 49, 49 and showing that it is 69.

Measures of Spread

The mean and median provide limited information about a whole distribution of data. For example, if you sit in a sauna for 30 minutes and then in a refrigerated room for 30 minutes, an average temperature of your surroundings for that hour might sound comfortable. To tell how much the data are scattered, we develop measures of *spread* or *dispersion*. Perhaps the easiest way to measure spread is the **range**, the difference between the greatest and the least values in a data set. For example, the range in the set of data 1, 3, 7, 8, 10 is $10 - 1 = 9$. However, just because the ranges of two sets of data are the same, the data do not have to have the same dispersion. For example, the data set 1, 10, 10, 10, 10 also has a range of 9 and is spread quite differently from the first collection of data. For this reason we need other measures of spread besides the range.

Another measure of spread is the *interquartile range (IQR)*. The IQR is the range of the middle half of the data. Consider the following set of test scores:

$$20 \quad 25 \quad 40 \quad 50 \quad 50 \quad 60 \quad 70 \quad 75 \quad 80 \quad 80 \quad 90 \quad 100 \quad 100$$

The range for this set of scores is $100 - 20 = 80$. The median score for this set of data is 70. We mark this location with a vertical bar between the 7 and the 0 and circle the data point for emphasis, as shown.

$$20 \quad 25 \quad 40 \quad 50 \quad 50 \quad 60 \quad \boxed{7 \mid 0} \quad 75 \quad 80 \quad 80 \quad 90 \quad 100 \quad 100$$

Next, we consider only the data values to the left of the **vertical bar** and draw another vertical bar where the median of those values is located:

$$20 \quad 25 \quad 40 \mid 50 \quad 50 \quad 60$$

The score of $45 = (40 + 50)/2$ is the median of the lower half of the scores and is the **lower quartile**. The lower quartile, or the **first quartile**, is denoted Q_1. One quarter, or 25%, of the scores lie at or below Q_1. Similarly, we can find the upper, or third, quartile (Q_3), which is $(80 + 90)/2$, or 85. The **upper quartile** (Q_3) is the median of the upper half of all scores. Three-quarters, or 75%, of the scores lie at or below Q_3. Thus we have divided the scores into four groups of three scores each:

$$\underbrace{20 \quad 25 \quad 40} \mid \underbrace{50 \quad 50 \quad 60} \quad 7 \mid 0 \quad \underbrace{75 \quad 80 \quad 80} \mid \underbrace{90 \quad 100 \quad 100}$$

Lower extreme	Lower quartile, Q_1		Median, Q_2	Upper quartile, Q_3	Upper extreme

The *interquartile range* **(IQR)** is the difference between the upper quartile and the lower quartile. In this case, IQR $= 85 - 45 = 40$. The IQR is itself another useful measure of spread because it is less influenced by extreme values. The IQR represents the interval between Q_1 and Q_3 containing the middle 50% of the values.

The interquartile range is the measure of spread most often reported with the median. This is done because the median marks the exact center of the data. However, as we saw earlier, the same median could be reported with two sets of data that have very different spreads. With the interquartile range reported along with the median, not only do we know the middle, we know how spread out the middle 50% of the data are. If they have a wide spread, then we are aware of it and can assess and use the data accordingly. For example, in the data set just presented, the median of 70 should be reported with the interquartile range of 40 or the interval $[Q_1, Q_3]$. In Now Try This 10-9, we are asked to create a distorted set of data that has these two measures associated with it.

NOW TRY THIS 10-9 Create a set of data that ranges from 0 to 100 and is widely spread yet has median 70 and an interquartile range of 50.

Box Plots

Displaying data in a **box plot** (or a **box-and-whisker plot**) is a way to avoid issues such as those found in Now Try This 10-9. A box plot is a way to display data visually and draw informal conclusions. Box plots show only certain data; they are visual representations of the *five-number summary* of the data. The five numbers are the median, the upper and lower quartiles (and, hence, the interquartile range information), and the least and greatest values in the distribution. The center, the spread, and the overall range are immediately evident by looking at the plot.

To construct a box plot, we need data's median, upper and lower quartiles, and extremes. To construct the box, we draw bars at the quartiles. We draw segments from each end of the box to the extreme values to form the whiskers. The box plot can be either vertical or horizontal. A vertical version of the box plot for the given data is shown in Figure 10-27.

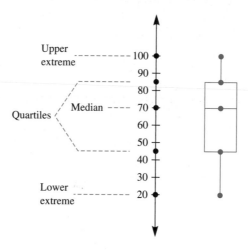

Figure 10-27

The box plot gives a fairly clear picture of the spread of the data. If we look at the graph in Figure 10-27, we can see that the median is 70, the maximum value is 100, the minimum value is 20, and the upper and lower quartiles are 45 and 85.

Example 10-7

What are the minimum and maximum values, the median, and the lower and upper quartiles of the box plot shown in Figure 10-28?

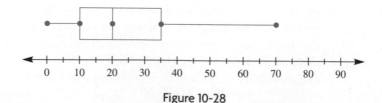

Figure 10-28

Solution   The minimum value is 0, the maximum value is 70, the median is 20, the lower quartile is 10, and the upper quartile is 35.

Outliers

An *outlier* is a value that is widely separated from the rest of a group of data. For example, in a set of scores such as

$$91 \quad 92 \quad 92 \quad 93 \quad 93 \quad 93 \quad 94$$

all data are grouped close together and no values are widely separated. However, in a set of scores such as

$$21 \quad 92 \quad 92 \quad 93 \quad 93 \quad 93 \quad 95 \quad 150$$

both 21 and 150 are widely separated from the rest of the data. These values are potential outliers. The upper and lower extreme values are not necessarily outliers. In data such as

$$75 \quad 90 \quad 91 \quad 92 \quad 92 \quad 93 \quad 93$$

it is not easy to decide, so we develop a convention for determining outliers. *An **outlier** is any value that is more than 1.5 times the interquartile range above the upper quartile or below the lower quartile.* Statisticians sometimes use values different from 1.5 to determine outliers.

It is common practice to indicate outliers with asterisks. Whiskers are then drawn to the extreme points that are not *outliers.* To investigate how this works, consider Example 10-8.

Example 10-8

Draw a box plot of the data in Table 10-11 and identify possible outliers.

Table 10-11 Final Medal Standings for Top 20 Countries—2004 Summer Olympics

United States	103
Russia	92
China	63
Australia	49
Germany	48	$Q_3 = 42.5$
Japan	37
France	33
Italy	32
Korea	30
Britain	30	Median $(Q_2) = 28.5$
Cuba	27
Ukraine	23
Netherlands	22
Romania	19
Spain	19	$Q_1 = 18$
Hungary	17
Greece	16
Belarus	15
Canada	12
Bulgaria	12

Solution   The extreme scores are 103 and 12, the median is 28.5, $Q_1 = 18$, and $Q_3 = 42.5$. The IQR is $42.5 - 18$, or 24.5. Outliers are scores that are greater than $42.5 + 1.5(24.5)$, or 79.25, or less than $18 - 1.5(24.5)$, or $^-18.75$. Therefore, in this data 92 and 103 are the only outliers. A box plot is given in Figure 10-29. Notice that the whisker stops at the extreme point 12 on the lower end and at 63 on the upper end. Outliers are indicated with asterisks.

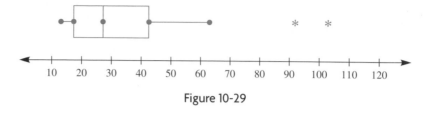

Figure 10-29

TECHNOLOGY CORNER   A graphing calculator can be used to create a box plot from collected data investigating student heights. Consider the following:

a. Collect student heights in a class with measurements made in centimeters. Collect data to the near-est centimeter.
b. Enter all the data into the calculator using the **LIST** key and entering the data into L1.
c. With all data entered, set up the plot for the data using 2nd **PLOT**. Press **ENTER** to select Plot1.
d. Select the box plot as the type; use Xlist as L1.
e. Use the **WINDOW** key to set up an approximate window for the box plot. Make the Xmin 5 cm less than the least data point and Xmax 5 cm greater than the greatest measure. Press **Y =** to remove any equations and press **GRAPH** to explore the plot.
f. Use the **TRACE** key with the left and right arrows to record the minimum point, the quartiles and the maximum point.

Source: This activity was adapted from the following web site: http://timiddlegrades.com/activitiesmath/290.

Comparing Sets of Data

Box plots are used primarily for large sets of data or for comparing several distributions. The stem and leaf plot is usually a much clearer display for a single distribution. Parallel box plots drawn using the same number line give us the easiest comparison of medians, extreme scores, and the quartiles for the sets of data. As an example, we construct parallel box plots comparing the data in Table 10-12.

Table 10-12 Mean Earnings of Males and Females, 1967–2003 (full-time workers age 15 and older; income in current dollars)

Year	Male	Female	Year	Male	Female	Year	Male	Female	Year	Male	Female
2003	$ 41,483	$ 24,630	1993	$ 28,939	$ 15,761	1983	$18,109	$8,780	1974	$9,861	$4,161
2002	41,057	23,619	1992	26,810	14,922	1982	17,381	8,195	1973	9,289	3,799
2001	40,859	23,602	1991	26,369	14,449	1981	16,515	7,440	1972	8,635	3,577
2000	40,254	22,428	1990	26,041	13,913	1980	15,340	6,772	1971	7,892	3,333
1999	38,384	21,312	1989	25,746	13,226	1979	14,311	6,026	1970	7,537	3,138
1998	36,315	20,462	1988	24,054	12,311	1978	13,113	5,599	1969	7,202	2,945
1997	34,794	19,511	1987	22,798	11,538	1977	12,063	5,291	1968	6,626	2,732
1996	32,800	18,369	1986	21,822	10,741	1976	11,165	4,875	1967	6,054	2,483
1995	31,454	17,265	1985	20,652	10,173	1975	10,429	4,513			
1994	30,367	16,478	1984	19,438	9,584						

Source: Data taken from U.S. Census Data, 2004.

Before constructing parallel box plots, we find the five important values for each group of data. These values are given in Table 10-13.

Table 10-13

Value	Female	Male
Maximum	$24,630.00	$41,483.00
Upper quartile	16,871.50	30,910.50
Median	10,173.00	20,652.00
Lower quartile	4,694.00	10,797.00
Minimum	2,483.00	6,054.00

In this example, the IQR for males is $20,113.50; the IQR for females is $12,177.50. Checking reveals there are no outliers.

Next we draw the horizontal scale and construct the box plots for the female and male populations using the data in Table 10-13, as shown in Figure 10-30.

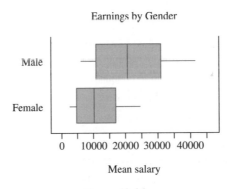

Figure 10-30

From the box plot in Figure 10-30, we can see that the mean salaries for males have been higher than those for females, since the extreme values, median, and quartiles for the males are greater than those for females. Also, more than 50% of the mean salaries for males are greater than those for the mean salaries of females over the time period.

REMARK Some may choose to display the data in Table 10-12 with two lines (or two line graphs), one depicting male salaries and the other female salaries.

Although we cannot spot clusters or gaps in box plots as we can with stem and leaf or line plots, we can more easily compare data from different sets. With box plots, we do not need to have sets of data that are approximately the same size, as we did for stem and leaf plots. To compare data from two or more sets using their box plots, we first study the boxes to see if they are located in approximately the same places. Next, we consider the lengths of the boxes to see if the variability of the data is about the same. We also check whether the median, the quartiles, and the extreme values in one set are greater than those in another set. If they are, the data in the first set are greater than those of the other set, no matter how we compare them. If they are not, we can continue to study the data for other similarities and differences.

NOW TRY THIS 10-10 Using only the data from 1983 through 1992 in Table 10-12, decide whether there are outliers. How do you think the two box plots would compare for this data with the comparison made in Figure 10-30?

 Read carefully the box plots from the following student text and complete problem 1.

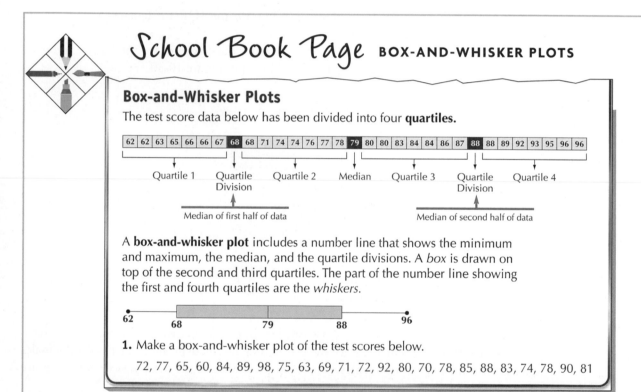

School Book Page **BOX-AND-WHISKER PLOTS**

Box-and-Whisker Plots

The test score data below has been divided into four **quartiles**.

| 62 | 62 | 63 | 65 | 66 | 66 | 67 | **68** | 68 | 71 | 74 | 74 | 76 | 77 | 78 | **79** | 80 | 80 | 83 | 84 | 84 | 86 | 87 | **88** | 88 | 89 | 92 | 93 | 95 | 96 | 96 |

Quartile 1   Quartile Division   Quartile 2   Median   Quartile 3   Quartile Division   Quartile 4

Median of first half of data   Median of second half of data

A **box-and-whisker plot** includes a number line that shows the minimum and maximum, the median, and the quartile divisions. A *box* is drawn on top of the second and third quartiles. The part of the number line showing the first and fourth quartiles are the *whiskers*.

62   68   79   88   96

1. Make a box-and-whisker plot of the test scores below.

72, 77, 65, 60, 84, 89, 98, 75, 63, 69, 71, 72, 92, 80, 70, 78, 85, 88, 83, 74, 78, 90, 81

Source: Scott Foresman-Addison Wesley, Mathematics, Grade 6, 2008 (p. 631).

Variation: Mean Absolute Deviation, Variance, and Standard Deviation

Earlier in the section, we noted that a measure of spread is needed when data are summarized with a single number, such as the mean or median. There are three measures of spread that are generally discussed and each has different uses. The most sophisticated is the standard deviation, as we will see in the following paragraphs. To examine these measures of spread, we start again with a simple test-score example.

Suppose Professors Abel and Babel each taught a section of a graduate statistics course and each had six students. Both professors gave the same final exam. The results, along with the means for each group of scores, are given in Table 10-14, with stem and leaf plots in Figure 10-31(a) and (b), respectively. As the stem and leaf plots show, the sets of data are very different. The first is more spread out, or varies more, than the second. However, each set has 60 as the mean. Each median also equals 60. Although the mean and the median for these two groups are the same, the two distributions of scores are very different.

Table 10-14

Abel's Class Scores	Babel's Class Scores
100	70
80	70
70	60
50	60
50	60
10	40
$\bar{x} = \dfrac{360}{6} = 60$	$\bar{x} = \dfrac{360}{6} = 60$

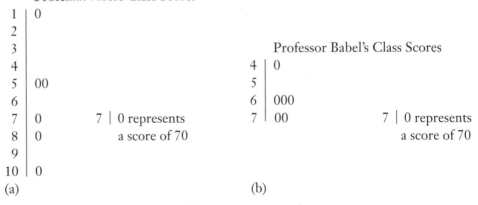

Professor Abel's Class Scores

```
 1 | 0
 2 |
 3 |
 4 |
 5 | 00
 6 |
 7 | 0          7 | 0 represents
 8 | 0              a score of 70
 9 |
10 | 0
```
(a)

Professor Babel's Class Scores

```
 4 | 0
 5 |
 6 | 000
 7 | 00         7 | 0 represents
                    a score of 70
```
(b)

Figure 10-31

As we have seen, there are many ways to measure the spread of data. The simplest way is to find the range. The range for Professor Abel's class is $100 - 10$, or 90. The range for Professor Babel's class is $70 - 40$, or 30. If we use the range as our measure of dispersion, we see that Abel's class is much more spread out than Babel's class. If we use the interquartile range, the IQR for Abel's class is 30, and for Babel's class, 10. Again, these measures of spread show more of a spread for Abel's class than for Babel's. Remember the disadvantage of the range is that it uses only extreme values.

Mean Absolute Deviation

One of the most basic ways to measure the spread of data is to measure the distance that each data point is away from the mean. The absolute value is one method used for finding distance. The **mean absolute deviation (MAD)** makes use of the absolute value to find the distance each data point is away from the mean; then the mean of those distances is found to give an "average distance from the mean" for each of the points. For example, we find the mean absolute deviation of the scores from the mean by using the following steps:

- Measure the distance from the mean by calculating the score minus the mean.
- Find the absolute value of each difference.
- Sum those absolute values (the absolute deviation).
- Find the mean absolute deviation (MAD) by dividing the sum by the number of scores.

In Table 10-15, we give a sample set of data and compute the mean absolute deviation.

Table 10-15

Test Scores	$\lvert$Test Score $-$ Mean$\rvert$
99	$\lvert 99 - 83.2 \rvert$, or 15.8
67	$\lvert 67 - 83.2 \rvert$, or 16.2
84	$\lvert 84 - 83.2 \rvert$, or 0.8
99	$\lvert 99 - 83.2 \rvert$, or 15.8
67	$\lvert 67 - 83.2 \rvert$, or 16.2
$\bar{x} = \dfrac{(99 + 67 + 84 + 99 + 67)}{5} = 83.2$	Sum of absolute deviations $= 64.8$
	$\text{MAD} = \dfrac{64.8}{5}$, or 12.96

The MAD may be summarized as follows for the numbers $x_1, x_2, x_3, \ldots, x_n$, where $\bar{x}$ is the mean of the numbers:

$$\text{MAD} = \frac{\lvert x_1 - \bar{x} \rvert + \lvert x_2 - \bar{x} \rvert + \ldots + \lvert x_n - \bar{x} \rvert}{n}$$

Visual pictures of the mean absolute deviation for the given set of test scores are given in Figure 10-32(a) and (b).

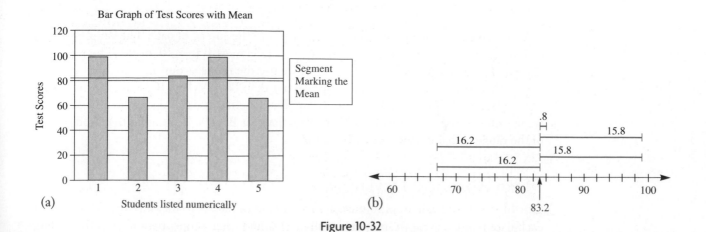

Figure 10-32

The mean absolute deviation is a rough measure of the "average" distance that a score is from the mean and hence gives an idea of how far away from the mean the test scores are. The MAD is a relatively easy measure for students to find, is recommended for use in the *Statistics Framework* (p. 43), and serves as a precursor to the standard deviation. The MAD does give a measure of spread that works well for some data sets. It handles fairly large deviations in some sets well. Consider the data for Abel from Table 10-14 and displayed in Table 10-16.

Table 10-16

Abel's Scores Class	|Abel's Scores − 60|	Babel's Class Scores	|Babel's Scores − 60|
100	40	70	10
80	20	70	10
70	10	60	0
50	10	60	0
50	10	60	0
10	50	40	20
Mean = 60	MAD = $\dfrac{140}{6}$, or approximately 23.3	Mean = 60	MAD = $\dfrac{40}{6}$, or approximately 6.7

Using the mean absolute deviation on the two sets of scores, it is easy to see that even though the means are 60, the spread is very different. For Abel's class, the scores are an average of 23.3 points from the mean, while in Babel's class, the scores are an average of 6.7 points from the mean. Babel's test scores are "less spread out."

Statisticians commonly use two other measures of spread: the *variance* and the *standard deviation*. These measures are also based on how far the scores are from the mean. To find out how far each value differs from the mean, we subtract each value in the data from the mean to obtain the deviation. Some of these deviations may be positive, and others may be negative. Because the mean is the balance point, the total of the deviations above the mean equals the total of the deviations below the mean. (The mean of the deviations is 0 because the sum of the deviations is 0.) Squaring the deviations makes them all positive. The mean of the squared deviations is the **variance**. Because the variance involves squaring the deviations, it does not have the same units of measurement as the original observations. For example, lengths measured in feet have a variance measured in square feet. To obtain the same units as the original observations, we take the square root of the variance and obtain the **standard deviation**.

The steps involved in calculating the variance, v, and standard deviation, s, of n numbers are as follows:

1. Find the mean of the numbers.
2. Subtract the mean from each number.
3. Square each difference found in step 2.
4. Find the sum of the squares in step 3.
5. Divide by n to obtain the variance, v.
6. Find the square root of v to obtain the standard deviation, s.

These six steps can be summarized for the numbers $x_1, x_2, x_3, \ldots, x_n$ as follows, where $\bar{x}$ is the mean of these numbers:

$$s = \sqrt{v} = \sqrt{\frac{(x_1 - \bar{x})^2 + (x_2 - \bar{x})^2 + (x_3 - \bar{x})^2 + \ldots + (x_n - \bar{x})^2}{n}}$$

> **REMARK** In some textbooks, this formula involves division by $n - 1$ instead of by n. Division by $n - 1$ is more useful for advanced work in statistics.

The variances and standard deviations for the final exam data from the classes of Professors Abel and Babel are calculated by using Tables 10-17 and 10-18, respectively.

Table 10-17 Abel's Scores

x	$x - \bar{x}$	$(x - \bar{x})^2$
100	40	1600
80	20	400
70	10	100
50	$^-10$	100
50	$^-10$	100
10	$^-50$	2500
Totals: 360	0	4800

$\bar{x} = \dfrac{360}{6} = 60$

$v = \dfrac{4800}{6} = 800$

$s = \sqrt{800} \approx 28.3$

Table 10-18 Babel's Scores

x	$x - \bar{x}$	$(x - \bar{x})^2$
70	10	100
70	10	100
60	0	0
60	0	0
60	0	0
40	$^-20$	400
Totals: 360	0	600

$\bar{x} = \dfrac{360}{6} = 60$

$v = \dfrac{600}{6} = 100$

$s = \sqrt{100} = 10$

Values far from the mean on either side will have greater positive squared deviations, whereas values close to the mean will have smaller positive squared deviations. Therefore, the standard deviation is a greater number when the values from a set of data are widely spread and a lesser number (close to 0) when the data values are close together.

Example 10-9

Professor Abel gave two group exams. Exam A had grades of 0, 0, 0, 100, 100, 100, and exam B had grades of 50, 50, 50, 50, 50, 50. Find the following for each exam:

a. Mean   **b.** Range   **c.** Mean absolute deviation
d. Standard deviation   **e.** Median   **f.** Interquartile range

Solution   **a.** The means for exams A and B are each 50.
b. The range for exams A and B are 100 and 0, respectively.
c. The mean absolute deviations for the two exams are as follows:

$$\text{MAD}_A = \frac{|0 - 50| + |0 - 50| + |0 - 50| + |100 - 50| + |100 - 50| + |100 - 50|}{6} = 50$$

$$\text{MAD}_B = \frac{|50 - 50| + |50 - 50| + |50 - 50| + |50 - 50| + |50 - 50| + |50 - 50|}{6} = 0$$

d. The standard deviations for exams A and B are as follows:

$$s_A = \sqrt{\frac{3(0 - 50)^2 + 3(100 - 50)^2}{6}} = 50$$

$$s_B = \sqrt{\frac{6(50 - 50)^2}{6}} = 0$$

e. The medians for the exams are each 50.
f. The interquartile range for exams A and B, respectively, are 100 and 0.

In Example 10-9, exam A has mean 50, mean absolute deviation of 50, and standard deviation of 50. Exam A also had median 50 and interquartile range 100. Reporting the three measures together in each case demonstrates that the scores are widely spread away from the mean and median. Exam B has mean 50, mean absolute deviation 0, and standard deviation of 0. Its median is 50 with interquartile range 0. All of the descriptors for Exam B show that there is very little spread of data from the mean and median. (In this case there is none.) Example 10-9 was chosen with extreme scores to illustrate what can happen with measures of central tendency and measures of spread. Normally, one can tell more about the distribution of data points when the mean, mean absolute deviation, or standard deviation are reported together, as well as when the median and interquartile range are reported together.

Normal Distributions

To better understand how standard deviations are used as measures of spread, we next consider normal distributions. The graphs of normal distributions are the bell-shaped curves called *normal curves*. These curves often describe distributions such as IQ scores for the population of the United States.

A **normal curve** is a smooth, bell-shaped curve that depicts frequency values distributed symmetrically about the mean. (Also, the mean, median, and mode all have the same value.) The normal curve is a theoretical distribution that extends infinitely in both directions. It gets closer and closer to the *x* axis but never reaches it. On a normal curve, about 68% of the values lie within 1 standard deviation of the mean, about 95% lie within 2 standard deviations, and about 99.8% are within 3 standard deviations. The percentages represent approximations of the total percent of area under the curve. The curve and the percentages are illustrated in Figure 10-33.

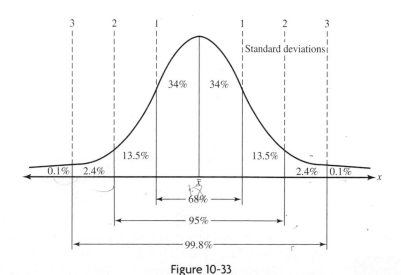

Figure 10-33

REMARK The normal curve, or normal-probability curve, has equation

$y = \dfrac{1}{s\sqrt{2\pi}} e^{(-1/2)(x-\bar{x})^2/s^2}$, where $-\infty < x < \infty$, s is the standard deviation, and

$\bar{x}$ is the mean; e is approximately 2.718.

Suppose the area under the curve represents the population of the United States. Psychologists claim that the mean IQ is 100 and the standard deviation is 15. They also claim that an IQ score of over 130 represents a superior score. Because 130 is equal to the mean plus 2 standard deviations, we see from Figure 10-33 that only 2.5% of the population fall into this category.

Example 10-10

When a standardized test was scored, there was a mean of 500 and a standard deviation of 100. Suppose that 10,000 students took the test and their scores had a bell-shaped distribution, making it possible to use a normal curve to approximate the distribution.

a. How many scored between 400 and 600?
b. How many scored between 300 and 700?
c. How many scored between 200 and 800?
d. How many scored above 800?

Solution **a.** Since 1 standard deviation on either side of the mean is from 400 to 600, about 68% of the scores fall in this interval. Thus, 0.68(10,000), or 6800, students scored between 400 and 600.
b. About 95% of 10,000, or 9500, students scored between 300 and 700.
c. About 99.8% of 10,000, or 9980, students scored between 200 and 800.
d. About 0.1% of 10,000, or 10, students scored above 800.

REMARK About 0.2%, or 20, students' scores in Example 10-10 fall outside 3 standard deviations. About 10 of these students did very well on the test, and about 10 students did very poorly.

Application of the Normal Curve

Suppose that a group of students asked their teacher to grade "on a curve." If the teacher gave a test to 200 students and the mean on the test was 71, with a standard deviation of 7, the graph in Figure 10-34 shows how the grades could be assigned. In Figure 10-34, the teacher has used the normal curve in grading. (The use of the normal curve presupposes that the teacher had a bell-shaped distribution of scores and also that the teacher arbitrarily decided to use the lines marking standard deviations to determine the boundaries of the A's, B's, C's, D's, and F's.) Thus, based on the normal curve in Figure 10-34, Table 10-19 shows the range of grades that the teacher might assign if the grades are rounded. Students who ask their teachers to grade on the curve may wish to reconsider if the normal curve is to be used.

Historical Note

Abraham De Moivre (1667–1754), a French Huguenot, was the first to develop and study the normal curve. He was one of the first to study actuarial information, in his book *Annuities upon Lives*. De Moivre's work with the normal curve went essentially unnoticed and was developed independently by Pierre de Laplace (1749–1827) and Carl Friedrich Gauss (1777–1855), who found so many applications for the normal curve that it is referred to as the *Gaussian curve*. ♦

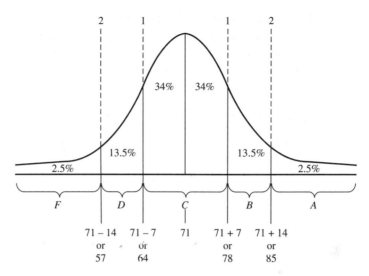

Figure 10-34

Table 10-19

Test Score	Grade	Number of People per Grade	Percentage Receiving Grade
85 and above	A	5	2.5%
78–84	B	27	13.5%
64–77	C	136	68%
57–63	D	27	13.5%
Below 57	F	5	2.5%

Percentiles

When students take a standardized test such as the ACT or SAT, their scores are often reported in **percentiles**. A percentile shows a person's score relative to other scores. For example, if a student's score is at the 82nd percentile, this means that approximately 82% of those taking the test scored lower than the student and approximately 18% had higher scores.

One application of percentiles is to make comparisons. For example, consider Kristy and Kim as they applied for a job. On the application, Kristy wrote that she finished 15th in her class. Kim reported that she finished 40th in her class. Does Kristy's class rank imply that she is a better student than Kim? Suppose that there were 50 students in Kristy's class and 400 students in Kim's class. In such a case, we need a common method of comparing their rankings. In order to make a balanced comparison, we can find Kristy's and Kim's percentile ranks, or percentiles, in their respective classes. In Kristy's class, there were $50 - 15$, or 35, students ranked below her, so we say that Kristy ranked at the 70th percentile. Similarly, $(400 - 40)/400$, or 90%, of the students ranked below Kim. We say that Kim ranked at the 90th percentile for her school. Thus, comparatively, Kim ranked higher in her class than Kristy did in hers.

As the name implies, percentiles divide the set of data into 100 equal parts. For example, if Tomas reports that he scored at the 50th percentile (the median, or Q_2) on the SAT, he is saying that he scored better than 50% of the people taking the test. In general, the rth percentile is denoted by P_r. In a case where the distribution is normal, we could use the percentages in

Figure 10-33 to determine percentiles. For example, someone scoring 2 standard deviations above the median is at the $(0.1 + 2.4 + 13.5 + 34 + 34 + 13.5)$ percentile, or better than 97.5% of the population. Because percentiles are typically reported using whole percentages, that places the person at the 98th percentile.

The discussion of percentiles has centered on distributions that can be represented by a normal curve. Percentiles can also be found from distributions that are not normal. This can be done by constructing a frequency table, determining the number of scores less than the score for which we are trying to determine a percentile, and then dividing that number by the total number of scores in the set of data. This is investigated further in the assessment set.

Deciles are points that divide a distribution into 10 equally spaced sections. There are nine deciles, denoted $D_1, D_2, \ldots, D_9$ and $D_1 = P_{10}, D_2 = P_{20}, \ldots, D_9 = P_{90}$.

REMARK In some finite sets of test scores, a particular percentile may not be represented. This does not stop their being used with small data sets, but it does inhibit their usefulness.

Example 10-11

A standardized test that was distributed along a normal curve had a median (and mean) of 500 and a standard deviation of 100. The 16th percentile, P_{16}, is 400 because 400 is 1 standard deviation below the median. Find each of the following:

a. P_{50} **b.** P_{84}

Solution a. Since 500 is the median, 50% of the distribution is less than 500. Thus, $P_{50} = 500$.

b. Since 600 is 1 standard deviation above the median, 84% of the distribution is less than 600. Thus, $P_{84} = 600$.

Example 10-12

a. Ossie was ranked 25th in a class of 250. What was his percentile rank?

b. In a class of 50, Cathy has a percentile rank of 60. What is her class standing?

Solution a. There were $250 - 25$, or 225, students ranked below Ossie. Hence, $225/250$, or 90%, of the class ranked below him. Therefore, Ossie ranked at the 90th percentile.

b. Cathy's percentile rank is 60. Thus 60% of the class ranks below her. Because 60% of 50 is 30, 30 students ranked below Cathy. Therefore, Cathy is 20th in her class.

BRAIN TEASER At a birthday party, the honoree would not tell his age but agreed to give hints. He computed and announced that the mean age of his seven party guests was 21. When 29-year-old Jill arrived at the party, the honoree announced that the mean age of the eight people was now 22. Jack, another 29-year-old, arrived next. The honoree then added his age to the set of ages of the other nine people and announced that the mean was now 27. How old was the honoree?

Assessment 10-3A

1. Calculate the mean, the median, and the mode for each of the following data sets:
   a. 2, 8, 7, 8, 5, 8, 10, 5
   b. 10, 12, 12, 14, 20, 16, 12, 14, 11
   c. 18, 22, 22, 17, 30, 18, 12
2. Write an example of a data set with seven data points for which the mode is not a good descriptor of the center of the data.
3. The mean score on a set of 20 tests is 75. What is the sum of the 20 test scores?
4. The mean for a set of 28 scores is 80. Suppose two more students take the test and score 60 and 50. What is the new mean?
5. Selina claimed that in her class all of the scores on a test were either 100 or 50, so the mean must be 75.
   a. Explain with an example how Selina may have reasoned.
   b. Explain whether this reasoning is valid.
6. Suppose in Selina's class there were three students who scored 100 and nine who scored 50. What is the mean of the scores in the class?
7. Suppose there were m students in a class who scored 100 and n students who scored 50. Write an algebraic expression for the mean in terms of m and n.
8. A table showing Jon's fall-quarter grades follows. Find his grade point average for the term (A = 4, B = 3, C = 2, D = 1, F = 0).

Course	Credits	Grade
Math	5	B
English	3	A
Physics	5	C
German	3	D
Handball	1	A

9. If the mean weight of seven linesmen on a team is 230 lb and the mean weight of the four backfield members is 190 lb, what is the mean weight of the 11-person team?
10. The following table gives the annual salaries of the 40 dancers in a certain troupe.
    a. Find the mean annual salary for the troupe.
    b. Find the median annual salary.
    c. Find the mode.

Salary	Number of Dancers
$ 18,000	2
22,000	4
26,000	4
35,000	3
38,000	12
44,000	8
50,000	4
80,000	2
150,000	1

11. Use the data in exercise 10 to find the following:
    a. Range
    b. Mean absolute deviation
    c. Standard deviation
    d. Interquartile range
12. Use the data in exercise 10 and the values in exercise 11 to answer the following:
    a. Which values from exercise 11 should be reported to best describe the data in exercise 10?
    b. Explain why you made the choice you did in part (a).
13. Maria filled her car's gas tank. The mileage odometer read 42,800 mi. When the odometer read 43,030, Maria filled the tank with 12 gal. At the end of the trip, she filled the tank with 18 gal and the odometer read 43,390 mi. How many miles per gallon did she get for the entire trip?
14. The youngest person in a company is 24 years old. The range of ages is 34 yr. How old is the oldest person in the company?
15. To receive an A in a class, Willie needs at least a mean of 90 on five exams. Willie's grades on the first four exams were 84, 95, 86, and 94. What minimum score does he need on the fifth exam to receive an A in the class?
16. Ginny's median score on three tests was 90. Her mean score was 92 and her range was 6. What were her three test scores?
17. Following are box plots comparing the ticket prices of two performing arts theaters:

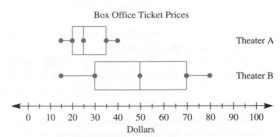

Box Office Ticket Prices

a. What is the median ticket price for each theater?
b. Which theater has the greatest range of prices?
c. What is the highest ticket price at either theater?
d. Make some statements comparing the ticket prices at the two theaters.

18. What is the standard deviation of the heights of seven trapeze artists if their heights are 175 cm, 182 cm, 190 cm, 180 cm, 192 cm, 172 cm, and 190 cm?

19. In a Math 131 class at DiPaloma University, the grades on the first exam were as follows:

$$96 \quad 71 \quad 43 \quad 77 \quad 75 \quad 76 \quad 61$$
$$83 \quad 71 \quad 58 \quad 97 \quad 76 \quad 74 \quad 91$$
$$74 \quad 71 \quad 77 \quad 83 \quad 87 \quad 93 \quad 79$$

a. Find the mean.
b. Find the median.
c. Find the mode.
d. Find the IQR.
e. Find the variance of the scores.
f. Find the standard deviation of the scores.
g. Find the mean absolute deviation of the scores.

20. For certain workers, the mean wage is $5.00/hr, with a standard deviation of $0.50. If a worker is chosen at random, what is the probability that the worker's wage is between $4.50 and $5.50? Assume a normal distribution of wages.

21. Assume a normal distribution and that the average phone call in a certain town lasted 4 min, with a standard deviation of 2 min. What percentage of the calls lasted less than 2 min?

22. In a normal distribution, how are the mean and the median related? Why?

23. On a certain exam, the mean is 72 and the standard deviation is 9. If a grade of A is given to any student who scores at least 2 standard deviations above the mean, what is the lowest score that a person could receive and still get an A?

24. Assume that the heights of American women are approximately normally distributed, with a mean of 65.5 in. and a standard deviation of 2.5 in. Within what range are the heights of 95% of American women?

25. If a standardized test with scores distributed normally has a mean of 65 and a standard deviation of 12, find the following:
   a. Q_2 **b.** P_{16} **c.** P_{84} **d.** D_5

26. Listed in the table is a set of children's weights and the cumulative total of children.
a. Find the percentiles (rounded to the nearest tenth) for each weight by dividing the cumulative total in each case by the total number of children.
b. Sketch the "cumulative" curve, in which weights are marked on a horizontal axis and percentiles are marked on a vertical axis.
c. Should "cumulative" curves or percentile curves always have the shape you found in part (b)? Why or why not?

Cumulative Totals and Percentiles of 39 Children's Weights

Weights	Frequency	Cumulative Frequency	Percentile
90	0	0	
91	1	1	
92	0	1	
93	3	4	
94	2	6	
95	6	12	
96	5	17	
97	7	24	
98	5	29	
99	8	37	
100	2	39	

Assessment 10-3B

1. Calculate the mean, the median, and the mode for each of the following data sets:
  a. 82, 80, 63, 75, 92, 80, 92, 90, 80, 80
  b. 5, 5, 5, 5, 5, 10

2. a. If each of six students scored 80 on a test, find each of the following for the scores:
    (i) Mean   **(ii)** Median   **(iii)** Mode
  b. Make up another set of six scores that are not all the same but in which the mean, median, and mode are all 80.

3. The tram at a ski area has a capacity of 50 people with a load limit of 7500 lb. What is the mean weight of the passengers if the tram is loaded to its weight limit and capacity?

4. The names and ages for each person in a family of five follow:

Name	Dick	Jane	Kirk	Jean	Scott
Age	40	36	8	6	2

a. What is the mean age?
b. Find the mean of the ages 5 yr from now.
c. Find the mean 10 yr from now.
d. Describe the relationships among the means found in (a), (b), and (c).

5. Suppose you own a hat shop and decide to order hats in only *one* size for the coming season. To decide which size to order, you look at last year's sales figures, which are itemized according to size. Should you find the mean, median, or mode for the data? Why?

6. Suppose there were n students in a class; h of them scored 100 and the rest scored 50. Write an algebraic expression for the mean for the class in terms of n and h.

7. **a.** Mr. Alberto wanted to count the score on a term paper as 60% of a final grade, homework as 25% of the grade, and the final exam as the remainder of the grade. If you were in the class, made 85 on the term paper, had a 78 average on homework, and scored 90 on the final exam, what number could Mr. Alberto use to determine your grade *if* he used his grading scheme?

   b. Write an algebraic expression to generalize the scoring procedure for Mr. Alberto's class to allow him to use any percentages he likes for the scoring scheme. Clearly describe what the variables represent.

8. If 99 people had a mean income of $12,000, how much is the mean income increased by the addition of a single income of $200,000?

9. Refer to the following chart. In a gymnastics competition, each competitor receives six scores. The highest and lowest scores are eliminated, and the official score is the mean of the four remaining scores.

Gymnast	Scores					
Balance Beam						
Meta	9.2	9.2	9.1	9.3	9.8	9.6
Lisa	9.3	9.1	9.4	9.6	9.9	9.4
Olga	9.4	9.5	9.6	9.6	9.9	9.6
Uneven Bars						
Meta	9.2	9.1	9.3	9.2	9.4	9.5
Lisa	10.0	9.8	9.9	9.7	9.9	9.8
Olga	9.4	9.6	9.5	9.4	9.4	9.4
Floor Exercises						
Meta	9.7	9.8	9.4	9.8	9.8	9.7
Lisa	10.0	9.9	9.8	10.0	9.7	10.0
Olga	9.4	9.3	9.6	9.4	9.5	9.4

   a. If the only events in the competition are the balance beam, the uneven bars, and the floor exercise, find the winner of each event.

   b. Find the overall winner of the competition if the overall winner is the person with the highest combined official scores.

10. Choose the data set(s) that fit the descriptions given in each of the following:

   a. The mean is 6.
   The range is 6.
   Set A: 3, 5, 7, 9
   Set B: 2, 4, 6, 8
   Set C: 2, 3, 4, 15

   b. The mean is 11.
   The median is 11.
   The mode is 11.

Set A: 9, 10, 10, 11, 12, 12, 13
Set B: 11, 11, 11, 11, 11, 11, 11
Set C: 9, 11, 11, 11, 11, 12, 12

   c. The mean is 3.
   The median is 3.
   It has no mode.

   Set A: 0, $2\frac{1}{2}$, $6\frac{1}{2}$
   Set B: 3, 3, 3, 3
   Set C: 1, 2, 4, 5

   d. The box plot shown next.

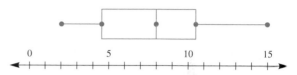

Set A: 2, 3, 4, 4, 6, 6, 7, 15
Set B: 2, 3, 6, 6, 8, 9, 12, 14, 15
Set C: 2, 4.5, 8, 13, 15
Set D: 2, 3, 6, 6, 8, 9, 10, 11, 15

11. The mean of five numbers is 6. If one of the five numbers is removed, the mean becomes 7. What is the value of the number that was removed?

12. **a.** Find the mean and the median of the following arithmetic sequences:
   (i) 1, 3, 5, 7, 9
   (ii) 1, 3, 5, 7, 9, ..., 199
   (iii) 7, 10, 13, 16, ..., 607

   b. Based on your answers in (a), make a conjecture about the mean and the median of any arithmetic sequence.

13. Construct a box plot for the following gas mileages per gallon of various company cars:

   22 18 14 28 30 12 38 22
   30 39 20 18 14 16 10

14. Construct a box plot for the following set of test scores. Indicate outliers, if any, with asterisks.

   20 95 40 70 90 70 80 80 90 95

15. The following table shows the heights in feet of the tallest 10 buildings in Los Angeles and in Minneapolis:

Los Angeles	Minneapolis
858	950
750	775
735	668
699	579
625	561
620	447
578	440
571	416
534	403
516	366

a. Draw horizontal box plots to compare the data.

b. Are there any outliers in this data? If so, which values are they?

c. Based on your box plots from (a), make some comparisons of the heights of the buildings in the two cities.

16. The following shows box plots for the piano practice times in hours per week for Tom and Dick. Make some comparisons for their practice times.

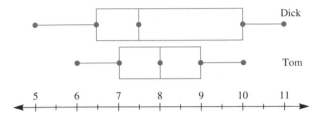

17. a. If all the numbers in a set are equal, what is the standard deviation?

   b. If the standard deviation of a set of numbers is zero, must all the numbers in the set be equal?

18. In a school system, teachers start at a salary of $25,200 and have a top salary of $51,800. The teachers' union is bargaining with the school district for next year's salary increment.

   a. If every teacher is given a $1000 raise, what happens to each of the following?

   (i) Mean

   (ii) Median

   (iii) Extremes

   (iv) Quartiles

   (v) Standard deviation

   (vi) IQR

   b. If every teacher received a 5% raise, what does this do to the following?

   (i) Mean

   (ii) Standard deviation

19. The mean IQ score for 1500 students is 100, with a standard deviation of 15. Assuming the scores have a normal curve,

   a. how many have an IQ between 85 and 115?

   b. how many have an IQ between 70 and 130?

   c. how many have an IQ over 145?

20. Sugar Plops boxes say they hold 16 oz. To make sure they do, the manufacturer fills the box to a mean weight of 16.1 oz, with a standard deviation of 0.05 oz. If the weights have a normal curve, what percentage of the boxes actually contain 16 oz or more?

21. According to psychologists, IQs are normally distributed, with a mean of 100 and a standard deviation of 15.

   a. What percentage of the population has IQs between 100 and 130?

   b. What percentage of the population has IQs lower than 85?

22. The weights of newborn babies are distributed normally, with a mean of approximately 105 oz and a standard deviation of 20 oz. If a newborn is selected at random, what is the probability that the baby weighs less than 125 oz?

23. A tire company tests a particular model of tire and found the tires to be normally distributed with respect to wear. The mean was 28,000 mi and the standard deviation was 2500 mi. If 2000 tires are tested, about how many are likely to wear out before 23,000 mi?

24. A standardized mathematics test given to 10,000 students had the scores normally distributed. The mean was 500 and the standard deviation was 60. A student scoring below 440 points was deficient in mathematics. About how many students were rated deficient?

25. In a class of 25 students, Bill has a rank of 25th. At what percentile is he?

26. Jill was 80th in a class of 200, whereas Nathan, who is in the same class, has a percentile rank of 80. Which student has the higher standing in the class?

Mathematical Connections 10-3

Communication

1. If you were considering ages and wanted one number to represent the age at which a person can get a driver's license, which "average" would you use and why?

2. A movie chain conducts a popcorn poll in which each person entering a theater and buying a box of popcorn is asked a yes-no question. Which "average" do you think is used to report the result and why?

3. When a government agency reports the average rainfall for a state for a year, which "average" do you think it uses and why?

4. Carl had scores of 90, 95, 85, and 90 on his first four tests.
   a. Find the median, mean, and mode.

   b. Carl scored a 20 on his fifth exam. Which of the three averages would Carl want the instructor to use to compute his grade? Why?

   c. Which measure is affected most by an extreme score?

5. The mean of the five numbers given is 50:

$$20 \quad 35 \quad 50 \quad 60 \quad 85$$

   a. Add four numbers to the list so that the mean of the nine numbers is still 50.

   b. Explain how you could choose the four numbers to add to the list so that the mean did not change.

   c. How does the mean of the four numbers you added to the list compare to the original mean of 50?

6. Sue drives 5 mi at 30 mph and then 5 mi at 50 mph. Is the mean speed for the trip 40 mph? Why or why not?
7. Explain why the mode could be a less-than-adequate measure of center for a data set.
8. If a custodian were to report the amount of tissue in a set of bathroom stalls, what type of measure would you want reported to convince you that there was an adequate supply until the stalls were checked again? Explain your answer.
9. If you were to make an argument that there were not enough women's bathrooms in a theater, what type of data would you use and why? Explain the types of measures of center and spread that you would report for your data.
10. The mean of 5, 7, 9 is 7. The mean of 67, 72, 77 is 72.
    a. Find two more examples where the mean of an ordered set of data is the "middle" data point.
    b. Suppose that a data set consists of $a_1, a_2, a_3, \ldots, a_n$, an arithmetic sequence. Explain why the mean of this data set is $\frac{a_1 + a_n}{2}$.

Open-Ended

11. In 1991, students from countries around the world took the International Assessment of Educational Progress test. The following table gives the average mathematics scores for 15 countries, along with the number of days that students in each country spend in school each year.
    a. Make a conjecture about the relationship between the country's mathematics score and the number of days spent in school.
    b. Test your conjecture by using a box plot to compare students who spend 190 days or fewer in school with those who spend more than 190 days in school.
    c. Explain how your graph supports or disproves your conjecture.

Country	Days of School	Math Score
Canada	188	62
France	174	64
Hungary	177	68
Ireland	173	61
Israel	215	63
Italy	204	64
Jordan	191	40
Korea	222	73
Scotland	191	61
Slovenia	190	57
Soviet Union	198	70
Spain	188	55
Switzerland	207	71
Taiwan	222	73
United States	178	55

    d. Draw a scatterplot for the data and see if there seems to be an association between mathematics scores and days in school. Is this the same message your box plot

gave you? Which plot would you use to support your conjecture?
12. Use the data in the following table to compare the number of people living in the United States from 1810 through 1900 and the number living in the United States from 1910 through 2000. Use any form of graphical representation to make the comparison and explain why you chose the representation that you did.

Year	Number (thousands)	Year	Number (thousands)
1810	7,239	1910	92,228
1820	9,638	1920	106,021
1830	12,866	1930	123,202
1840	17,068	1940	132,164
1850	23,191	1950	151,325
1860	31,443	1960	179,323
1870	38,558	1970	203,302
1880	50,189	1980	226,542
1890	62,979	1990	248,765
1900	76,212	2000	281,422

Data taken from *The World Almanac and Book of Facts 2002*, World Almanac Books, 2002.

Cooperative Learning

13. In small groups, determine a method of finding the number and types of graphs and statistical representations used in at least two newspapers in your campus library. Based on your findings, write a report defending which type(s) of representation should be emphasized in a journalistic statistics class.

Questions from the Classroom

14. A student asks, "If the average income of 10 people is $10,000 and one person gets a raise of $10,000, is the median, the mean, or the mode changed and, if so, by how much?"
15. Jose asks, "Why can a median number of children not be 3.8?" How do you respond?
16. Suppose the class takes a test and the following averages are obtained: mean, 80; median, 90; mode, 70. Tom, who scored 80, would like to know if he did better than half the class. What is your response?
17. A student asks if it is possible to find the mode for data in a grouped frequency table. What is your response?
18. A student asks if she can draw any conclusions about a set of data if she knows that the mean for the data is less than the median. How do you answer?
19. A student asks if it is possible to have a standard deviation of $^-5$. How do you respond?
20. Mel's mean on 10 tests for the quarter was 89. She complained to the teacher that she should be given an A because she missed the cutoff of 90 by only a single point. Explain whether it is clear that she really missed an A by only a single point if tests scored based on 100 pts.

Review Problems

21. Given the following double-bar graph, make some comparisons of the number of men and women in the labor force over the years.

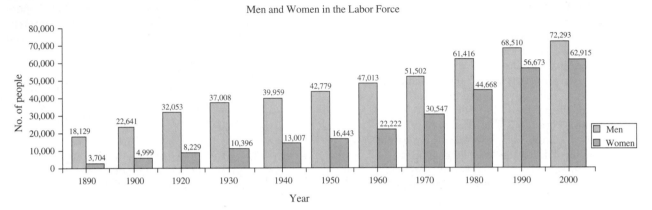

Men and Women in the Labor Force

Data taken from *The World Almanac and Book of Facts 2002,* World Almanac Books, 2002.

22. Consider the following circle graph. What is the number of degrees in each sector of the graph?

U.S. Car Sales by Vehicle Size and Type
for Ali Auto Sales

2008

Large 7% Luxury 17%

Midsize 48% Small 28%

23. Given the bar graph shown, answer the following:
   a. Which mountain is the highest? Approximately how high is it?
   b. Which mountains are higher than 6000 m?

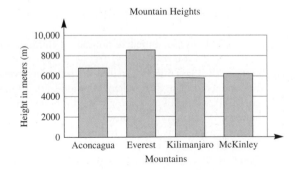

Mountain Heights

24. Following are raw test scores from a history test:

86 85 87 96 55
90 94 82 68 77
88 89 85 74 90
72 80 76 88 73
64 79 73 85 93

a. Construct an ordered stem and leaf plot for the given data.
b. Construct a grouped frequency table for these scores with intervals of 5, starting the first class at 55.
c. Draw a histogram of the data.
d. If a circle graph of the grouped data in (b) were drawn, how many degrees would be in the section representing the 85 through 89 interval?

National Assessment of Educational Progress (NAEP) Questions

Score	Number of Students
90	1
80	3
70	4
60	0
50	3

The table above shows the scores of a group of 11 students on a history test. What is the average (mean) score of the group to the nearest whole number?

NAEP, 2003, Grade 8

Tetsu rides his bicycle x miles the first day, y miles the second day, and z miles the third day. Which of the following expressions represents the average number of miles per day that Tetsu travels?

a. $x + y + z$ **b.** xyz
c. $3(x + y + z)$ **d.** $3(xyz)$
e. $(x + y + z)/3$

NAEP, 2003, Grade 8

The prices of gasoline in a certain region are $1.41, $1.36, $1.57, and $1.45 per gallon. What is the median price per gallon for gasoline in this region?

a. $1.41
b. $1.43

c. $1.44
d. $1.45
e. $1.47

NAEP, Grade 8, 2005

LABORATORY ACTIVITY We can model the mean as a measure of central tendency using a strip of cardboard that is 1 in. wide and 1 ft long with holes 1 in. apart and $\frac{1}{8}$ in. from the edge, as in Figure 10-35(a). Use string and tape to suspend the strip from a desk with the string tied through a hole punched between 56 and 57. Then use paper clips of equal size to investigate means, as in Figure 10-35(b).

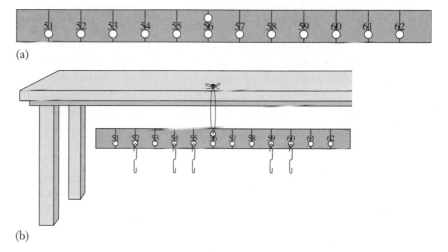

(a)

(b)

Figure 10-35

1. a. If paper clips are hung in the holes at 51, 53, and 60, where should an additional clip be hung in order to achieve a balance?
    b. If paper clips are hung at 51, 54, and 60, where should two additional paper clips be hung to achieve a balance?
    c. If paper clips are hung at 51, 53, 54, and 55, where should four additional clips be hung to achieve a balance?
2. Find the mean of the data of all the numbers in each part in (1) and compare this answer with the number in the center of the strip.
3. Would any of the means in (2) change if we hung an additional paper clip in the center of the strip?
4. Find the median and mode for 51, 53, 54, 56, 58, 58, and 59.
5. Hang paper clips in each hole in (4). If a number appears more than once, hang that number of paper clips in the hole. Is there a balance around the median? Is there a balance around the mode?
6. Under what conditions do you think there would be a balance around (a) the mean, (b) the median, and (c) the mode? Test your conjecture using the cardboard strip.

TECHNOLOGY CORNER Use a graphing calculator or create a spreadsheet to find the mean absolute deviation, the variance, and standard deviation of the set of scores 32, 41, 47, 53, and 57. If you use a spreadsheet, make column A be the set of scores, column B the scores minus the mean, column C the absolute value of the differences, and column D the squares of the difference of the scores and the mean. Compute the mean absolute deviation, the variance, and the standard deviation.

10-4   Abuses of Statistics

In *Principles and Standards*, we find the following:

> The amount of data available to help make decisions in business, politics, research, and everyday life is staggering. Consumer surveys guide the development and marketing of products. Polls help determine political-campaign strategies, and experiments are used to evaluate the safety and efficacy of new medical treatments. Statistics are often misused to sway public opinion on issues or to misrepresent the quality and effectiveness of commercial products. (p. 48)

As noted, statistics are frequently abused. Benjamin Disraeli (1804–1881), an English prime minister, once remarked, "There are three kinds of lies: lies, damned lies, and statistics." People sometimes deliberately use statistics to mislead others. This can be seen in advertising. More often, however, the misuse of statistics is the result of misinterpreting what the data and statistics mean. For example, if we were told that the "average" depth of water in a lily pond was 2 feet, most of us would presume that a heron could stand up in any part of the pond. This is not necessarily the case, as seen in the following cartoon.

THE FAR SIDE®   BY GARY LARSON

Consider an advertisement reporting that of the people responding to a recent survey, 98% said that Buffepain is the most effective pain reliever of headaches and arthritis. To certify that the statistics are not being misused, the following information should have been reported:

1. The questions being asked
2. The number of people surveyed
3. The number of people who responded

4. How the people who participated in the survey were chosen
5. The number and type of pain relievers tested
6. How the answers were interpreted

Without the information listed, the following situations are possible, all of which could cause the advertisement to be misleading:

1. Suppose 1,000,000 people nationwide were sent the survey, and only 50 responded. This would mean that there was only a 0.005% response, which would certainly cause us to mistrust the ad.
2. Suppose a survey sentence read, "Buffepain is the best pain reliever I've tried for headaches and arthritis," and there were no questions about the kind and type of other pain relievers tried.
3. Of the 50 responding in (1), suppose 49 responses were affirmative. The 98% claim is true, but 999,950 people did not respond.
4. Suppose all the people who received the survey were chosen from a town in which the major industry was the manufacture of Buffepain. It is very doubtful that the survey would represent an unbiased sample.
5. Suppose only two "pain relievers" were tested: Buffepain, whose active ingredient is 100% aspirin, and a placebo containing only powdered sugar.

This is not to say that advertisements of this type are all misleading or dishonest, but simply that data interpretations are only as honest as their reporters.

In the Buffepain report, a primary issue deals with the survey conducted, the number of people involved in the survey, how they were chosen, and how results of the survey were interpreted. Surveys are common for gaining information from a population. However, there are classic examples in history of how surveys and survey information have been either misused or misinterpreted. A well-known example was seen in the predictions of the winner of the Harry S. Truman/Thomas E. Dewey 1948 U.S. presidential election. A leading pollster, Elmo Roper, was so confident of a loss by Truman that on September 9, 1948, approximately 2 months before the election, he announced that there would be no more polls on the election. Additionally, while Truman was still on the campaign trail, *Newsweek*, after polling 50 key political experts, stated on October 11, 1948, approximately 1 month before the election, that Dewey would win.

In the *Newsweek* survey, a very select group was surveyed. Questions that should have been asked include (1) How was the group chosen? (2) Was the group representative of the voters in the United States? (3) Could the result of this survey appropriately have been generalized to all voters? All of these types of questions are important to surveyors. What is the population to which the results are being generalized? Is the entire population to be surveyed, or is the survey given only to a sample of the population? If only a sample will be used, how is the sample chosen? Is the sample of the population randomly chosen so that each person being surveyed has an equally likely chance of being chosen? How large is the population and how large a sample must be used so that the sample is representative of the population?

A version of a survey is sometimes seen in the early grades when students try to determine the favorite pet of the students in their school. A very simple version of this type of survey uses only the students in one class. If that is done, then the answer is known for that entire population. Note that the answer is not generalizable to the whole school. How might the sample be taken if students wanted to find the favorite pet of all the students in the school?

NOW TRY THIS 10-11 Based on the graph in Figure 10-36, write arguments to determine whether or not the following inferences are correct:

a. 82% of the parents surveyed say that there is not too much focus on preparing for tests.
b. Only 9% of the parents say that learning is thwarted.
c. 12% of the parents say test questions are too difficult and that expectations are unreasonable.
d. At least 89% of the parents believe that schools require too many tests.
e. The vast majority of parents say that teachers are not putting too much academic pressure on their children.

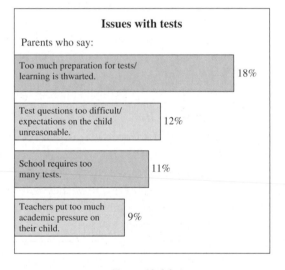

Figure 10-36
Source: Data from *USA Today*. November 21, 2000.

A different type of misuse of data and statistics involves graphs. Among the things to look for in a graph are the following. If they are not there, then the graph may be misleading.

1. Title
2. Labels on both axes of a line or bar chart and on all sections of a pie chart
3. Source of the data
4. Key to a pictograph
5. Uniform size of symbols in a pictograph
6. Scale: Does it start with zero? If not, is there a break shown?
7. Scale: Are the numbers equally spaced?

To see an example of a misleading use of graphs, consider how graphs can be used to distort data or exaggerate certain pieces of information. Graphs using a break in the vertical axis can be used to create different visual impressions, which are sometimes misleading. For example, consider the two graphs in Figure 10-37, which represent the number of girls trying out for basketball at each of three middle schools. As we can see, the graph in Figure 10-37(a) portrays a different picture from the one in Figure 10-37(b).

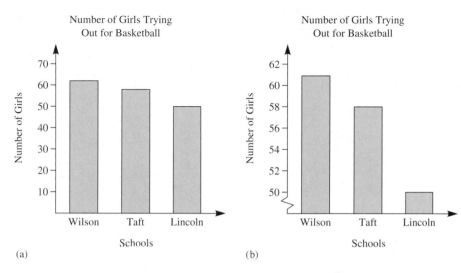

Figure 10-37

The research note indicates that young students have difficulty interpreting data. How do you think they would respond to Figure 10-37?

A line graph, histogram, or bar graph can be altered by changing the scale of the graph. For example, consider the data in Table 10-20 for the number of graduates from a community college for the years 2004 through 2008.

Table 10-20

Year	2004	2005	2006	2007	2008
Number of Graduates	140	180	200	210	160

The graphs in Figure 10-38(a) and (b) represent the same data, but different scales are used in each. The statistics presented are the same, but these graphs do not convey the same psychological message. In Figure 10-38(b), the spacing of the years on the horizontal axis of the graph is more spread out and that for the numbers on the vertical axis is

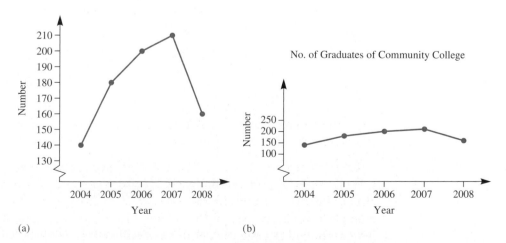

Figure 10-38

more condensed than in Figure 10-38(a). Both of these changes minimize the variability of the data. A college administrator might use a graph like the one in Figure 10-38(b) to convince people that the college was not in serious enrollment trouble.

Another error that frequently occurs is the use of continuous graphs, as in Figure 10-39, to depict data that are discrete (a finite number of data values). In Figure 10-39, we see the enrollment in Math 206 for five semesters. The bar graph accurately depicts the enrollment data, but if this graph is replaced with the continuous line in Figure 10-39 but without the bars, then many points along the line are meaningless.

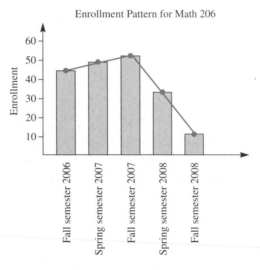

Figure 10-39

Other ways to distort graphs include omitting a scale, as in Figure 10-40(a). The scale is given in Figure 10-40(b).

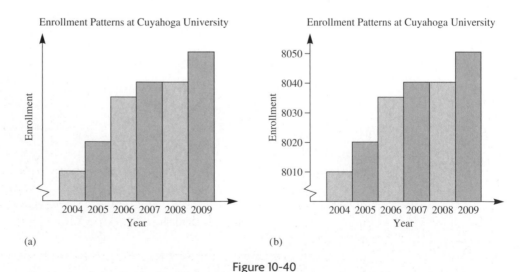

Figure 10-40

Other graphs can also be misleading. Suppose, for example, that the number of boxes of cereal sold by Sugar Plops last year was 2 million and the number of boxes of cereal sold by

Korn Krisps was 8 million. The Korn Krisps executives prepared the graph in Figure 10-41 to demonstrate the data. The Sugar Plops people objected. Do you see why?

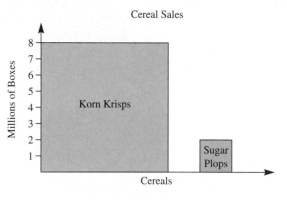

Figure 10-41

The graph in Figure 10-41 clearly distorts the data, since the figure for Korn Krisps is both 4 times as high and 4 times as wide as the bar for Sugar Plops. Thus, the area of the bar representing Korn Krisps is 16 times the comparable area representing Sugar Plops, rather than 4 times the area, as would be justified by the original data. To depict the data accurately, the length of a side of the Sugar Plops bar should be 4 units. Then four of these bars would "fit" in the Korn Krisps bar. Figure 10-42 shows how the comparison of Sugar Plops and Korn Krisps cereals might look if the figures were made three-dimensional. The figure for Korn Krisps has a volume 64 times the volume of the Sugar Plops figure.

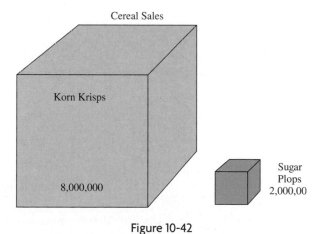

Figure 10-42

Circle graphs easily become distorted when attempts are made to depict them as three-dimensional. Many graphs of this type do not acknowledge either the variable thickness of the depiction or the distortion due to perspective. Observe that the 27% sector pictured in Figure 10-43 looks far greater than the 23% sector, although they should be very nearly the same size.

The final examples of the misuses of statistics involve misleading uses of mean, median, and mode. All these are "averages" and can be used to suit a person's purposes. As discussed in Section 10-3 in the example involving the teachers Smith, Jones, and Rivera, each teacher had reported that his or her class had done better than the other two. Each of the teachers used a different number to represent the test scores.

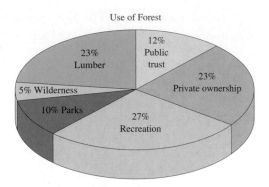

Figure 10-43

As another example, company administrators wishing to portray to prospective employees a rosy salary picture may find a mean salary of $58,000 for line workers as well as upper management in the schedule of salaries. At the same time, a union that is bargaining for salaries may include part-time employees as well as line workers and will exclude management personnel in order to present a mean salary of $29,000 at the bargaining table. The important thing to watch for when a mean is reported is disparate cases in the reference group. If the sample is small, then a few extremely high or low scores can have a great influence on the mean. Suppose Figure 10-44 shows the salaries of both management and line workers of the company. If the median is being used as the average, then the median might be $43,500, which is representative of neither major group of employees. The bimodal distribution means the median is nonrepresentative of the distribution.

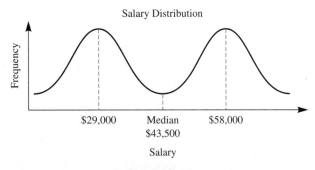

Figure 10-44

To conclude our comments on the misuse of statistics, consider a quote from Darrell Huff's book *How to Lie with Statistics* (p. 8):

So it is with much that you read and hear. Averages and relationships and trends and graphs are not always what they seem. There may be more in them than meets the eye, and there may be a good deal less.

The secret language of statistics, so appealing in a fact-minded culture, is employed to sensationalize, inflate, confuse, and oversimplify. Statistical methods and statistical terms are necessary in reporting the mass data of social and economic trends, business conditions, "opinion" polls, the census. But without writers who use the words with honesty and understanding and readers who know what they mean, the result can be semantic "nonsense."

Assessment 10-4A

This entire set of assessment items is appropriate for communication and cooperative learning. Many items are open-ended and several lend themselves to further investigation.

1. Discuss whether the following claims could be misleading. Explain why and how.
   a. A car manufacturer claims its car is quieter than a glider.
   b. A motorcycle manufacturer claims that more than 95% of its cycles sold in the United States in the last 15 yr are still on the road.
   c. A company claims its fruit juice has 10% more fruit solids than is required by U.S. government standards. (The government requires 10% fruit solids.)
   d. A brand of bread claims to be 40% fresher.

2. The city of Podunk advertised that its temperature was the ideal temperature in the country because its mean temperature was 25°C. What possible misconceptions could people draw from this advertisement?

3. What is wrong with the following line graph?

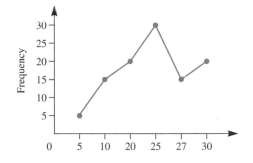

4. Can you draw any valid conclusions about a set of data in which the mean is less than the median?

5. A student read that 9 out of 10 pickup trucks sold in the last 10 yr are still on the road. She concluded that the average life of a pickup is around 10 yr. Is she correct?

6. In the *Shoe* cartoon shown, explain the intended meaning of the second panel.

7. Doug's Dog Food Company wanted to impress the public with the magnitude of the company's growth. Sales of Doug's Dog Food had doubled from 2008 to 2009, so the company displayed the following graph, in which the radius of the base and the height of the 2009 can are double those of the 2008 can. What does the graph really show with respect to the growth of the company? (*Hint:* The volume of a cylinder is given by $V = \pi r^2 h$, where r is the radius of the base and h is the height.)

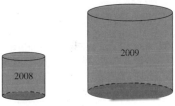

Doug's Dog Food Sales

8. Refer to the following circle graph.

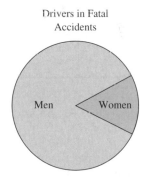

Ms. McNulty claims that on the basis of this information, we can conclude that men are worse drivers than women. Discuss whether you can reach that conclusion from the pictograph or you need more information. If more information is needed, what would you like to know?

9. The following graph was prepared to compare prices of washing machines at three stores:

Which of the following statements is true? Explain why or why not.
a. Prices vary widely at the three stores.
b. The price at Discount House is 4 times as great as that at J-Mart.
c. The prices at J-Mart and Super Discount differ by less than $10.

10. The following table gives the number of accidents per year on a certain highway for a 5-yr period:

Year	2004	2005	2006	2007	2008
Number of Accidents	24	26	30	32	38

a. Draw a bar graph to convince people that the number of accidents is on the rise and that something should be done about it.
b. Draw a bar graph to show that the rate of accidents is almost constant, and that nothing needs to be done.

11. Write a list of scores for which the mean and median are not representative of the list.

12. If you knew that 43% and 57% of mathematics teachers of grades K–4 and 5–8, respectively, considered themselves well prepared to use calculators/computers to demonstrate mathematical principles, and you assume each of the following roles, what should you do based on these data? In each case, how might you be misusing the data?
a. Teacher   b. Principal
c. Parent

13. The No Child Left Behind Act (2000) signed into law in the United States demands that teachers be highly qualified. One aspect of that law said that teachers of middle-school mathematics should have a major in the subject they teach. In January 2003, a survey revealed that 61% of middle-school mathematics classes are led by teachers who lack even a minor in mathematics.
a. Using these data, what actions should a state superintendent take?
b. How might you be misusing the data to justify your actions?

14. In 2008, it was reported that an average teacher's salary in the United States was $32,488. This was about $20,000 less than the average salary for a field engineer.
a. As a state legislator, how would you propose using these data in your state?
b. How might you be misusing the data if you were the state legislator?

15. According to the U.S. Department of Education in 2007, at least one-third of all entry-level teachers will quit within 5 yr of starting teaching. In urban schools, the percentage is higher.
a. As a collegiate advisor for prospective teachers, what advice would you give a student who wants to become a teacher of mathematics?
b. How might you be misusing the data in framing your advice?

16. Consider a state such as Montana that has both mountains and prairies. What numbers might you report to depict the "average" height above sea level of such a state? Why?

17. Is it possible for a state or country to have a mean sea level height that is negative? If so, what might such a region look like?

18. Describe how you might pick a random sample of adults that is representative of the members of your town.

19. A very large and successful manufacturer of computer chips released data in the early 1990s stating that approximately 65% of its chips were defective when they came off the assembly line. Give some reasons why this statistic could be accurate and yet the company could still be successful.

20. A student read in the newspaper that the pill form of a drug taken once per day is up to 92% effective in warding off the flu. She concludes that if she takes the pill 8% more times per day, she will be 100% safe from contracting the flu. Explain whether you agree with her and why.

21. In 2003, slightly more than 5% of the adult population in Haiti was infected with the HIV virus. Because almost 95% of the adult population was not infected at the time, Sam concluded that the virus was not a serious threat to the country. Do you agree with him? Why or why not?

Assessment 10-4B

1. Discuss whether the following claims could be misleading. Explain why and how.
   a. A used-car dealer claims that a car she is trying to sell will get up to 30 mpg.
   b. Sudso claims that its detergent will leave your clothes brighter.
   c. A sugarless gum company claims that 8 of every 10 dentists responding to the survey recommend sugarless gum.
   d. Most accidents occur in the home. Therefore, to be safer, you should stay out of your house as much as possible.
   e. More than 95% of the people who fly to a certain city do so on Airline A. Therefore, most people prefer Airline A to other airlines.

2. Jenny averaged 70 on her quizzes during the first part of the quarter and 80 on her quizzes during the second part of the quarter. When she found out that her final average for the quarter was not 75, she went to argue with her teacher. Give a possible explanation for Jenny's misunderstanding.

3. Suppose the following circle graphs are used to illustrate the fact that the number of elementary teaching majors at teachers' colleges has doubled between 2000 and 2009, while the percentage of male elementary teaching majors has stayed the same. What is misleading about the way the graphs are constructed?

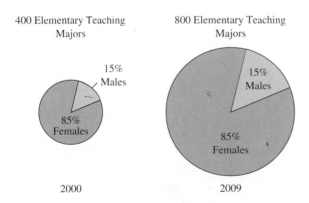

4. General Cooster once asked a person by the side of a river if the river was too deep to ride his horse across. The person responded that the average depth was 2 ft. If General Cooster rode out across the river, what assumptions did he make on the basis of the person's information?

5. Which of the following pieces of information would not be helpful in deciding the type of automobile that is the most economical to drive?
   a. Range of insurance costs
   b. Modes of drivers' ages for specific vehicle types
   c. Mean miles per gallon
   d. Typical cost of repairs per year
   e. Cost of routine maintenance

6. Explain what is wrong with the following graph:

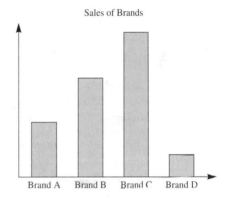

7. The following graph depicts the mean center of population of the United States and shows how the center has shifted from 1790 to 2000. Based solely on this graph, could you conclude that the population of the West Coast has increased since 1790?

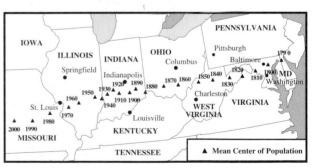

Source: U.S. Bureau of the Census, *2000 Census of Population and Housing, Population and Housing Unit Counts, United States* (2000 CPH-2-1).

8. The data in the graph of exercise 7 roughly follows a line. If the mean center of the population continues to move westward in the next 200 years, where would you expect it to be at the end of that period? At the end of 400 years?

9. In a recent survey, teachers rated their mathematics textbooks as follows:

National Survey of Teachers of Mathematics

	Grades K–4 (percentage)	Grades 5–8 (percentage)
Very poor	1	2
Poor	3	5
Fair	18	16
Good	35	33
Very good	36	33
Excellent	8	10

At about the same time that the survey was being done, newspapers printed articles saying that national experts in mathematics have been very critical of mathematics textbooks at all levels.

  a. What is your reaction to the survey and the reports?

  b. Does the national survey agree with the newspaper articles? Explain your thinking as you react.

10. **a.** Use the following data to justify the amount of time that you expect to assign for weekly homework to classes in grades K–4 and grades 5–8.

National Survey of Teachers Concerning the Amount of Homework Assigned per Week

	K–4 (percentages)	5–8 (percentages)
0–30 min	48	8
31–60 min	27	21
61–90 min	13	26
91–120 min	8	24
2–3 hr	3	17
More than 3 hr	1	5

  b. How might the survey data be misused to justify assigning at least 2 hr of homework per week?

11. The two headlines shown here were written based on exactly the same data set:

"Obesity has increased over 30%."

"On the average, Americans' weight has increased by less than 10 pounds."

  a. As a reporter, which headline would you use?

  b. As a reader, which headline would have the greatest impact?

  c. As an educated reader, which headline might you consider the most accurate knowing that both were written based on the same data set? Explain your answer.

12. In a British study around 1950, a group of 649 men with lung cancer were surveyed. A control group of the same size was established from a set of men who did not have lung cancer. The groups were matched according to ethnicity, age, and socioeconomic status. The statistics from the survey follow.*

	Lung Cancer Cases	Controls	Totals
Smokers	647	622	1269
Nonsmokers	2	27	29

  a. What is the fraction of smoking in the group that has lung cancer in the study?

  b. What is the fraction of smoking in the control group?

  c. If one person is chosen at random from each of the two groups (smokers and nonsmokers), what is the probability that a randomly chosen smoker; a randomly chosen non-smoker has cancer?

  d. Do you think that this data present a strong association between lung cancer and smoking? Explain your answer.

  e. Do you think that this evidence is conclusive that smoking causes lung cancer?

  (* *Problem taken from* Statistics Framework)

13. What are the characteristics that you think a sample might have to have to be representative of an entire population?

14. **a.** Suppose you are a student doing a survey of eating habits in a high school. If you sit in the hall and ask each student who passes you in 1 hr questions about their eating habits, explain whether or not you think you have a representative sample of the school population.

  b. If you are conducting the survey mentioned in part (a) and interview students in the cafeteria at noon, explain whether or not you think that you have a representative sample of the school population.

  c. For this survey, explain how you could choose a sample to be reasonably sure that you got a representative sample of the population.

15. A prospective homeowner considered the dropping interest rates for house loans in the early 1990s and decided to wait until the year 2000 to buy. Explain what type of statistics might be used in making this decision and whether you consider the prospective homeowner's decision to be wise.

16. A student got 99 on a quiz and was ecstatic over the grade. What other information might you need in order to decide if the student was justified in being happy?

17. A school administrator reports to you, a school board member, that the "average" number of students in a class in a school is 32. What other information might you need in order to predict whether any single classroom was overcrowded?

18. Alf used the graph shown to assert that 16-year-olds are better drivers than people in their twenties, and that people driving in their eighties are the safest. Explain whether or not you believe this assertion.

Number of Drivers in Fatal Crashes 1988

Source: Williams, A., and O. Carston, "Driver Age and Crash Involvement." *American Journal of Public Health* 79 (March 1989): 326–327.

19. The following chart lists the number of complaints received about airlines as reported in *U.S. News and World Report* on February 5, 2001:

Most complaints, Nov. '00	
United Airlines	252
American Airlines	162
Delta Air Lines	119
Fewest	
Alaska Airlines	13
Southwest Airlines	22
Continental Airlines	60

Discuss whether you think that United, American, and Delta are the worst airlines and Alaska, Southwest, and Continental are the best based on the above information.

20. In the graphic shown, two states are compared using their students' SAT scores and the spending ranking per pupil by those states in 1998. What are some erroneous conclusions that could be reached?

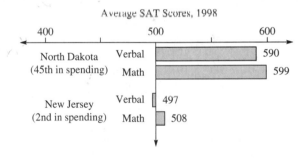

Average SAT Scores, 1998

Mathematical Connections 10-4

Review Problems

1. On the English 100 exam, the scores were as follows:

$$43 \quad 91 \quad 73 \quad 65$$
$$56 \quad 77 \quad 84 \quad 91$$
$$82 \quad 65 \quad 98 \quad 65$$

  a. Find the mean.
  b. Find the median.
  c. Find the mode.
  d. Find the variance.

  e. Find the standard deviation.
  f. Find the mean absolute deviation.

2. If the mean of a set of 36 scores is 27 and two more scores of 40 and 42 are added, what is the new mean?

3. On a certain exam, Tony corrected 10 papers and found the mean for his group to be 70. Alice corrected the remaining 20 papers and found that the mean for her group was 80. What is the mean of the combined group of 30 students?

4. Following are the men's gold-medal times for the 100 m run in the Olympic games from 1896 to 2003, rounded to the nearest tenth. Construct an ordered stem and leaf plot for the data.

Year	Time (sec), (rounded)
1896	12.0
1900	11.0
1904	11.0
1908	10.8
1912	10.8
1920	10.8
1924	10.6
1928	10.8
1932	10.3
1936	10.3
1948	10.3
1952	10.4
1956	10.5
1960	10.2
1964	10.0
1968	10.0
1972	10.1
1976	10.1
1980	10.3
1984	10.0
1988	9.9
1992	9.7
1996	9.8
2000	9.9

5. Following are the record swimming times of the women's 100-m freestyle and 100-m butterfly in the Olympics from 1960 to 2004. Draw parallel box plots of the two sets of data to compare them.

Year	Time—100 m Freestyle (sec)	Time—100 m Butterfly (sec)
1960	61.20	69.50
1964	59.50	64.70
1968	60.00	65.50
1972	58.59	63.34
1976	55.65	60.13
1980	54.79	60.42
1984	55.92	59.26
1988	54.93	59.00
1992	54.64	58.62
1996	54.50	59.13
2000	53.83	56.61
2004	53.84	57.72

National Assessment of Educational Progress (NAEP) Questions

Hamburger Prices 1985–1990

According to the graph, how many times did the yearly increase of the price of a hamburger exceed 10 cents?

a. None   **b.** One   **c.** Two

d. Three   **e.** Four

NAEP, Grade 8, 2003

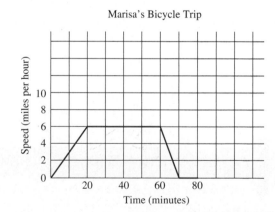

Marisa's Bicycle Trip

The graph above represents Marisa's riding speed throughout her 80-minute bicycle trip. Use the information in the graph to describe what could have happened on the trip, including her speed throughout the trip.

During the first 20 minutes, Marisa _____

_____

From 20 minutes to 60 minutes, she _____

_____

From 60 minutes to 80 minutes, she _____

_____

NAEP, Grade 8, 2003

Elements That Make Up the Earth's Crust

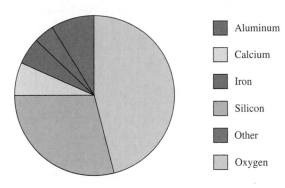

■ Aluminum

□ Calcium

■ Iron

▨ Silicon

■ Other

▨ Oxygen

According to the graph at left, which element forms the second greatest portion of the earth's crust?
a. Oxygen
b. Silicon
c. Aluminum
d. Iron
e. Calcium

NAEP, Grade 8, 2007

BRAIN TEASER Ouida had test grades of 91, 89, 83, 78, 76, and 75. Her teacher uses the mean of the test scores as the final grade of the class. Ouida has the option of keeping her current test average or replacing the highest and lowest scores that she currently has with a single score on a final exam. If Ouida chooses to take the final exam, find the lowest score she can receive and still maintain at least the mean that she currently has.

Hint for Solving the Preliminary Problem

A major issue with the set of data is that we do not know the bases of the percentages. Think about this as you frame your answer. Another possibility is to use a spreadsheet and try to create a numerical example that fits the data in the table of the Preliminary Problem.

Chapter Outline

I. Descriptive statistics

  A. Types of data

    1. Categorical data

    2. Numerical data

    3. Discrete data

    4. Continuous data

  B. Information can be summarized in each of the following forms:

    1. Pictographs

    2. Dot plots/line plots

    3. Stem and leaf plots

    4. Frequency tables

    5. Histograms

    6. Bar graphs/stacked bar graphs

    7. Line graphs/broken line graphs

    8. Circle graphs/pie charts

    9. Box plots/box-and-whisker plots

    10. Scatterplots

    11. Percentage bar graphs

II. Measures of central tendency

  A. The **mean** of n given numbers is the sum of the numbers divided by n.

  B. The **median** of a set of numbers is the middle number if the numbers are arranged in numerical order; if there is no middle number, the median is the mean of the two middle numbers.

  C. The **mode** of a set of numbers is the number or numbers that occur most frequently in the set.

III. Measures of variation

  A. The **range** is the difference between the greatest and least numbers in the data.

  B. The **mean absolute deviation (MAD)** of a data set is the mean of the absolute values of the differences of the data points and the mean.

  C. The **variance (v)** of a data set is found by subtracting the mean from each value, squaring each of these differences, finding the sum of these squares, and dividing by n, where n is the number of observations.

D. The **standard deviation (s)** is equal to the square root of the variance.

E. **Box plots**, or **box-and-whisker plots**, focus attention on the median, the quartiles, and the extremes and invite comparisons among them.

1. The **lower quartile** is the median of the items in the dataset less than the median.

2. The **upper quartile** is the median of the items in the dataset greater than the median.

3. The **interquartile range (IQR)** is calculated as the difference between the upper quartile and the lower quartile.

4. An **outlier** is any value more than 1.5 IQR above the upper quartile or more than 1.5 IQR below the lower quartile.

IV. Other data topics

A. Scatterplots are graphs of ordered pairs that allow us to examine the relationship (**association**) between two sets of variables.

B. A **trend line** or a line-of-best-fit is a line that approximates data in a scatterplot.

C. A **normal curve** is a smooth, bell-shaped curve that depicts frequency values distributed symmetrically about the mean, as shown:

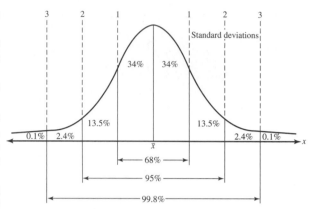

D. A **percentile** shows a person's score relative to other scores.

V. Types of variability

A. Measurement

B. Natural

C. Induced

D. Sampling

Chapter Review

1. Suppose you read that "the average family in Rattlesnake Gulch has 2.41 children." What average is being used to describe the data? Explain your answer. Suppose the sentence had said 2.5? Then what are the possibilities?

2. At Bug's Bar-B-Q restaurant, the average (mean) weekly wage for full-time workers is $250. There are 10 part-time employees whose average weekly salary is $100, and the total weekly payroll is $4000. How many full-time employees are there?

3. Find the mean, the median, and the mode for each of the following groups of data:
   a. 10, 50, 30, 40, 10, 60, 10
   b. 5, 8, 6, 3, 5, 4, 3, 6, 1, 9

4. Find the range, mean absolute deviation, interquartile range, variance, and standard deviation for each set of scores in exercise 3.

5. The mass, in kilograms, of each child in Ms. Rider's class follows:

   40 49 43 48 46 42 49 39 47 49

   42 41 42 39 41 40 45 43 44 42

   a. Make a dot plot for the data.
   b. Make an ordered stem and leaf plot for the data.
   c. Make a frequency table for the data.
   d. Make a bar graph of the data.

6. The grades on a test for 30 students follow:

   96 73 61   76 77 84

   78 98 98   80 67 82

   61 75 79   90 73 80

   85 63 86 100 94 77

   86 84 91   62 77 64

   a. Make a grouped frequency table for these scores, using four classes and starting the first class at 61.
   b. Draw a histogram of the grouped data.

7. The budget for the Wegetem Crime Co. is $2,000,000. Draw a circle graph to indicate how the company spends its money, where $600,000 is spent on bribes, $400,000 for legal fees, $300,000 for bail

money, $300,000 for contracts, and $400,000 for public relations. Indicate percentages on your graph.

8. What, if anything, is wrong with the following bar graph?

Monthly Health Club Costs

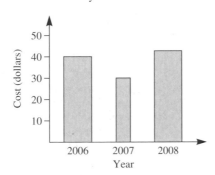

9. The mean salary of 24 people is $9000. How much will one additional salary of $80,000 increase the mean salary?

10. A cheetah can run 70 mph, a lion can run 30 mph, and a human can run 28 mph. Draw a bar graph to represent these data.

11. The life expectancies at birth for males and females are approximated in the following table:

Year	Male	Female	Year	Male	Female
1970	67.1	74.7	1989	71.7	78.5
1971	67.4	75.0	1990	71.8	78.8
1972	67.4	75.1	1991	72.0	78.9
1973	67.6	75.3	1992	72.1	78.9
1974	68.2	75.9	1993	72.1	78.9
1975	68.8	76.6	1994	72.4	79.0
1976	69.1	76.8	1995	72.5	78.9
1977	69.5	77.2	1996	73.1	79.1
1978	69.6	77.3	1997	73.6	79.2
1979	70.0	77.8	1998	73.8	79.5
1980	70.0	77.5	1999	73.9	79.4
1981	70.4	77.8	2000	74.1	79.5
1982	70.9	78.1	2001	74.4	79.8
1983	71.0	78.1	2002	74.7	79.9
1984	71.2	78.2	2003	74.8	80.1
1985	71.2	78.2	2004	75.2	80.4
1986	71.3	78.3	2005	75.2	80.4
1987	71.5	78.4	2006	75.2	80.4
1988	71.5	78.3	2007	75.4	83.2

a. Draw back-to-back ordered stem and leaf plots to compare the data.
b. Draw box plots to compare the data.
c. Describe the spread of the data using the inter-quartile range.

12. Larry and Marc took the same courses last quarter. Each bet that he would receive the better grades. Their courses and grades are as follows:

Course	Larry's Grades	Marc's Grades
Math (4 credits)	A	C
Chemistry (4 credits)	A	C
English (3 credits)	B	B
Psychology (3 credits)	C	A
Tennis (1 credit)	C	A

Marc claimed that the results constituted a tie, since both received 2 A's, 1 B, and 2 C's. Larry said that he won the bet because he had the higher grade-point average for the quarter. Who is correct? (Allow 4 points for an A, 3 points for a B, 2 points for a C, 1 point for a D, and 0 points for an F.)

13. Following are the lengths in yards of the nine holes of the University Golf Course:

$$160 \quad 360 \quad 330$$
$$350 \quad 180 \quad 460$$
$$480 \quad 450 \quad 380$$

Find each of the following measures with respect to the lengths of the holes:
a. Median   b. Mode
c. Mean   d. Standard deviation
e. Range   f. Interquartile range
g. Variance   h. Mean absolute deviation

14. The speeds in miles per hour of 30 cars were checked by radar. The data are as follows:

$$62 \quad 67 \quad 69 \quad 72 \quad 75 \quad 60 \quad 58 \quad 86 \quad 74 \quad 68$$
$$56 \quad 67 \quad 82 \quad 88 \quad 90 \quad 54 \quad 67 \quad 65 \quad 64 \quad 68$$
$$74 \quad 65 \quad 58 \quad 75 \quad 67 \quad 65 \quad 66 \quad 64 \quad 45 \quad 64$$

a. Find the median.
b. Find the upper and lower quartiles.
c. Draw a box plot for the data and indicate outliers (if any) with asterisks.
d. What percentage of the speeds is in the interval $[Q_1, Q_3]$?
e. If every person driving faster than 70 mph received a ticket, what percentage of the drivers received speeding tickets?
f. Describe the spread of the data.
g. Find P_{25}, P_{75}, and D_5.

15. The following scattergram was developed with information obtained from the girls trying out for the high school basketball team:

Height and Weights
of Girls' Basketball Team Players

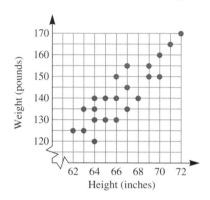

a. What kind of association exists between the heights and weights that are listed?
b. What is the weight of the girl who is 72 in. tall?
c. How tall is the girl who weighs 145 lb?
d. What is the mode of the heights?
e. What is the range of the weights?

16. The following table shows statistics for two grade-school basketball players. Collette is the leading scorer for the girls, while Rudy is the leading scorer for the boys.

	Collette	Rudy
Mean points per game	24	24
Mean absolute deviation (in points per game)	6	14
Points scored in the final game	36	36

a. Explain whether Collette or Rudy is the more consistent scorer.
b. For each player, identify the range of values within 1 MAD of the mean.
c. Explain which player's performance was most impressive in the last game.

17. If the test scores were distributed normally, what is the probability that a student scores more than 2 standard deviations above the mean?

18. If every person on an academic team had exactly the same score, describe the following:
a. The mean of the scores
b. The median of the scores
c. The mode of the scores
d. The standard deviation of the scores
e. The mean absolute deviation of the scores

19. If one tossed a fair die 500 times and drew a bar graph showing the frequencies of the resulting tosses, describe what you think the graph would look like.

20. A cereal company has an advertisement on one of its boxes that says, "Lose up to 6 lb in 2 weeks."
a. How much might a person who weighs 175 lb when he starts the plan expect to weigh in 1 yr if this rate could be continued?
b. Explain whether or not you believe your mathematical answer to part (a) is possible for weight loss.
c. A disclaimer on the box says that the "average weight loss is 5.0 lb." On what basis do you think that the cereal company could make the two claims?

21. A mathematics department has three women out of a tenure-track faculty of 20. One of the faculty members claimed that the statistics showed that the department was biased against tenure-track women faculty. How would you respond to that claim?

22. Explain whether or not it is reasonable to use a single number to describe the "average" depth of a swimming pool that includes a diving pool as a part of the swimming pool.

23. A box of cereal bars claims to have 138 g of contents. How might the company have determined the weight to make that claim on all comparable boxes to ensure reasonable accuracy of weight?

24. a. In box plots showing the ages of students at two different universities, the interquartile range for one was 3 (from 17 to 20) and the other was 5 (from 24 to 29). If you were a typical high school graduating senior and were trying to decide between the two universities using these data, which would you choose and why?
b. If you were a student at one of the two universities and were 23 years old, at which university might you expect to appear as a whisker in the box plot?

25. The *Weekly Reader* has conducted a poll of kids aged 9 and up that has proven right in the last 11 elections. In 2004, kids picked Bush by a 66% to 33% margin. What is your reaction to the validity of this type of poll?

26. The Nielsen Television Index rating of 30 means that an estimated 30% of American televisions are tuned to the show with that rating. The ratings are based on the preferences of a scientifically selected sample of 1200 homes.
a. Discuss possible ways in which viewers could bias this sample.
b. How could networks attempt to bias the results?

27. List and give examples of several ways to misuse statistics graphically.

28. Explain whether you think it is reasonable for a ski resort to advertise excellent skiing because the runs have 64 in. of snow at the top of the hill and 23 in. at the bottom.

29. The following bar graph shows the number of students per computer in the United States from 1983 to 2000.

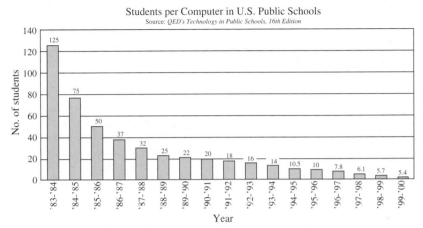

Students per Computer in U.S. Public Schools
Source: *QED's Technology in Public Schools, 16th Edition*

Data taken from *The World Almanac and Book of Facts 1999,* World Almanac Books, 2002.

a. Predict the results for the years 00–01, 01–02, and 02–03.

b. Between what two years did the greatest decrease in the number of students per computer occur?

c. Is a bar graph the most appropriate graph to use for these data? Why?

d. How could you distort these data to make it appear that the change has been very dramatic over the years?

30. The heights of 1000 girls at East High School were measured, and the mean was found to be 64 in., with a standard deviation of 2 in. If the heights are approximately normally distributed, about how many of the girls are

a. over 68 in. tall?

b. between 60 and 64 in. tall?

c. If a girl is selected at random at East High School, what is the probability that she will be over 66 in. tall?

31. A standardized test has a mean of 600 and a standard deviation of 75. If 1000 students took the test and their scores approximated a normal curve, how many scored between 600 and 750?

32. Use the information in problem 31 to find

a. P_{16} **b.** D_5 **c.** P_{84}

Selected Bibliography

Basile, C., "Collecting Data Outdoors: Making Connections to the Real World." *Teaching Children Mathematics* 6 (September 1999): 8–12.

Beaton, A., I. Mullis, M. Martin, E. Gonzalez, D. Kelly, and T. Smith. *Mathematics Achievement in the Middle School Years: IEA's Third International Mathematics and Science Study (TIMSS).* Chestnut Hill, MA: Boston College, Center for the Study of Testing, Evaluation, and Educational Policy, 1996.

Berg, C., and D. Phillips. "An Investigation of the Relationship between Logical Thinking Structures and the Ability to Construct and Interpret Line Graphs." *Journal of Research in Science Teaching* 31 (1994): 323–344.

Bremigan, B. "Developing a Meaningful Understanding of the Mean." *Mathematics Teaching in the Middle School* 9 (September 2003): 23–26.

Bright, G., and S. Friel. "Graphical Representations: Helping Students Interpret Data." In S. Lajoie (ed.), *Reflections on Statistics: Learning, Teaching, and Assessment in Grades K–12*. Mahwah, N.J.: Erlbaum, 1998, pp. 63–88.

Colgan, M. "March Math Madness: The Mathematics of the NCAA Basketball Tournament." *Mathematics Teaching in the Middle School* 11 (March 2006): 334.

Doerr, H., and L. English. "A Modeling Perspective on Students' Mathematical Reasoning about Data." *Journal of Research in Mathematics Education* 34 (March 2003): 110–136.

Franklin, C., and J. Garfield. "The GAISE Project: Developing Statistics Education Guidelines for Grades Pre-K–12 and College Courses." In *Thinking and Reasoning with Data and Chance, 68th Yearbook*. Reston, VA: National Council of Teachers of Mathematics, 2006, pp. 345–375.

Franklin, C., and D. Mewborn. "The Statistical Education of Grades Pre-K–12 Teachers." In *Thinking and Reasoning with Data and Chance, 68th Yearbook*. Reston, VA: National Council of Teachers of Mathematics, 2006, pp. 335–344.

Friel, S., W. O'Connor, and J. Manner. "More than 'Meanmedianmode' and a Bar Graph: What's Needed to Have a Statistical Conversation?" In *Thinking and Reasoning with Data and Chance, 68th Yearbook*. Reston, VA: National Council of Teachers of Mathematics, 2006, pp. 117–137.

Goldsby, D. "Lollipop Statistics." *Mathematics Teaching in the Middle School* 9 (September 2003): 12–15.

Jacobs, V. "How Do Students Think about Statistical Sampling?" *Mathematics Teaching in the Middle School* 7 (December 1999): 240–246, 263.

Johnson, A. "Now & Then Community Planning through Data Analysis." *Mathematics Teaching in the Middle School* 7 (March 2000): 458–464.

Jones, G., C. Thornton, C. Langrall, E. Mooney, A. Wares, B. Perry, and I. Putt. "A Framework for Characterizing Children's Statistical Thinking." *Mathematical Thinking and Learning* 2 (2000): 269–307.

Jones, G., C. Thornton, C. Langrall, E. Mooney, A. Wares, B. Perry, and I. Putt. "Using Students' Statistical Thinking to Inform Instruction." Paper presented at the research presession of the meeting of the National Council of Teachers of Mathematics, San Francisco, 1999.

Kader, G. "Means and MADs." *Mathematics Teaching in the Middle School* 4 (March 1999): 398–403.

Marshall, P., and L. Varro. "Cracking the Code." *Mathematics Teaching in the Middle School* 10 (August 2004): 54.

Martinie, S. "Families Ask: Data Analysis and Statistics in the Middle School?" *Mathematics Teaching in the Middle School* 12 (August 2006): 48.

Maus, J. "Every Story Tells a Picture." *Mathematics Teaching in the Middle School* 10 (April 2005): 375–379.

McClain, K., J. Leckman, P. Schmitt, and T. Regis. "Changing the Face of Statistical Data Analysis in the Middle Grades." In *Thinking and Reasoning with Data and Chance, 68th Yearbook*. Reston, VA: National Council of Teachers of Mathematics, 2006, pp. 229–240.

McClain, K., and P. Schmitt. "Teachers Grow Mathematically Together: A Case Study from Data Analysis." *Mathematics Teaching in the Middle School* 9 (January 2004): 274.

McDuffie, A., and J. Morrison. "Supporting Teacher Learning: Teachers Learning about Data Display: Connecting Mathematics and Science Inquiry. *Teaching Children Mathematics* 14 (February 2008): 375.

McMillen, S., and M. Sullivan. "Cartoon Corner: What's Your Chance? The Cookie Contest." *Mathematics Teaching in the Middle School* 10 (May 2005): 469.

Mokros, J., and S. Russell. "Children's Concepts of Average and Representation." *Journal for Research in Mathematics Education* 26 (January 1995): 20–39.

Mullis, I., Martin, M., Beaton, A., Gonzalez, E., Kelly, D., and Smith, T. *Mathematics Achievement in the Primary School Years. IEA's Third International Mathematics and Science Study*. Chestnut Hill, MA: Boston College, 1997.

Pereira-Mendoza, L., and J. Mellor. "Students' Concepts of Bar Graphs: Some Preliminary Findings." In D. Vere-Jones (ed.), *Proceedings of the Third International Conference on Teaching Statistics*, vol. 1, pp. 150–157. Amsterdam, The Netherlands: International Statistical Institute, 1991.

Putt, I., G. Jones, C. Thornton, B. Perry, C. Langrall, and E. Mooney. "Young Students' Informal Statistical Knowledge." *Teaching Statistics* 21 (1999): 74–77.

Russell, S. "What Does It Mean that '5 Has a Lot'? From the World to Data and Back." In *Thinking and Reasoning with Data and Chance, 68th Yearbook*. Reston, VA: National Council of Teachers of Mathematics, 2006, pp. 17–29.

Schwartz, S., and D. Whitin. "Graphing with Four-year Olds: Exploring the Possibilities through Staff Development." In *Thinking and Reasoning with Data and Chance, 68th Yearbook*. Reston, VA: National Council of Teachers of Mathematics, 2006, pp. 5–16.

Shaughnessy, J. "Research on Students' Understanding of Some Big Concepts in Statistics." In *Thinking and Reasoning with Data and Chance, 68th Yearbook*. Reston, VA: National Council of Teachers of Mathematics, 2006, pp. 77–98.

Strauss, S., and E. Bichler. "The Development of Children's Concepts of the Arithmetic Average." *Journal for Research in Mathematics Education* 19 (March 1988): 64–80.

Wallace, E. "Short Takes: Another Good Idea: More than Meets the Eye: An Interdisciplinary Exploration." *Mathematics Teaching in the Middle School* 12 (March 2007): 380.

Watson, J., and J. Shaughnessy. "Proportional Reasoning: Lessons from Research in Data and Chance." *Mathematics Teaching in the Middle School* 10 (September 2004): 104–109.

Whitin, D., and P. Whitin. "Talk Counts: Discussing Graphs with Young Children." *Teaching Children Mathematics* 10 (November 2003): 142–149.

Wilson, M., and C. Krapfl. "Exploring Mean, Median, and Mode with a Spreadsheet." *Mathematics Teaching in the Middle School* 1 (September–October 1995): 490–495.

Zawojewski, J., and J. Shaughnessy. "Mean and Median: Are They Really So Easy?" *Mathematics Teaching in the Middle School* 7 (March 2000): 436–440.

Introductory 11
Geometry

Preliminary Problem

One day Ms. Ha brought to her class sheets of paper printed with 3-by-3 arrays of dots as shown. She showed the students two concave quadrilaterals with vertices on four of the nine dots and pointed out that quadrilaterals in different positions should be considered different. She then challenged the class with the following questions. How would you answer these questions?

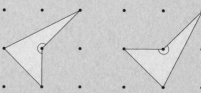

1. How many concave kites can be created on 4 of the 9 dots?
2. How many different concave quadrilaterals can be created with their vertices on 4 of the 9 dots?

The origins of geometry date back some 4000 years to ancient Egypt, when yearly floods of the fertile Nile River valley required regular resurveys of the surrounding land for purposes of taxation. Later, the Babylonians added to geometrical knowledge with an approximation of π. Centuries later, two Greek words—*geo* ("earth") and *metron* ("measure")—were combined to give this practical science of land measurement a name. The consummate Greek geometer Euclid (325–~265 BCE) pushed geometry from the realm of the practical into a more theoretical one with his *Elements* (see Historical Note).

Since Euclid, geometry has progressed from giving us the knowledge that the world is not flat, to measurements between planets, to Mandelbrot's recent work with fractals; it has evolved into a subject that we no longer consider in isolation from the rest of mathematics.

According to the *Principles and Standards*,

> The study of geometry in grades 3–5 requires thinking and doing. As students sort, build, draw, model, trace, measure, and construct, the capacity to visualize geometric relationships will develop. At the same time they are learning to reason and to make, test, and justify conjectures about these relationships. This exploration requires access to a variety of tools, such as graph paper, rulers, pattern blocks, geoboards, and geometric solids, and is greatly enhanced by electronic tools that support exploration, such as dynamic geometry software. (p. 165)

The following are some of the expectations from the *Principles and Standards* for different grade levels that will be covered in this chapter.

> In grades pre-K–2, all students should
>
> • recognize, name, build, draw, compare, and sort two- and three-dimensional shapes;
>
> • describe attributes and parts of two- and three-dimensional shapes;
>
> • investigate and predict the results of putting together and taking apart two- and three-dimensional shapes. (p. 96)

> In grades 3–5, all students should
>
> • identify, compare, and analyze attributes of two- and three-dimensional shapes and develop vocabulary to describe the attributes;
>
> • classify two- and three-dimensional shapes according to their properties and develop definitions of classes of shapes such as triangles and pyramids;

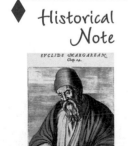

Historical Note

Little is known about the personal life of Euclid of Alexandria (ca. 300 BCE), but legend tells us that he studied geometry for its beauty and logic. Euclid is best known for the *Elements*, a work so systematic and encompassing that upon its completion, many earlier mathematical works were discarded and lost to future generations. The *Elements*, comprising 13 books, includes not only geometry but also arithmetic and topics in algebra. In the *Elements*, Euclid set up a deductive system by starting with a set of statements that he assumed to be true and then showed that geometric discoveries followed logically from these assumptions. Euclid's work was a major influence on the development of modern mathematics and science. ◆

- investigate, describe, and reason about the results of subdividing, combining, and transforming shapes;

- make and test conjectures about geometric properties and relationships and develop logical arguments to justify conclusions. (p. 164)

In grades 6–8, all students should

- precisely describe, classify, and understand relationships among types of two- and three-dimensional objects using their defining properties;

- use visual tools such as networks to represent and solve problems. (p. 240)

Euclidean geometry has been taught to schoolchildren since Euclid's time and has been one of the foundations of a liberal arts education. Many people have found great pleasure in the study of geometry, among them the eminent British philosopher and Nobel Prize winner (in literature), Sir Bertrand Russell. In his autobiography he wrote:

At the age of eleven, I began Euclid, with my brother as my tutor. This was one of the great events of my life, as dazzling as first love. I had not imagined there was anything so delicious in the world . . . from that moment until . . . I was thirty-eight, mathematics was my chief interest and my chief source of happiness.

11-1   Basic Notions

The fundamental building blocks of geometry are *points*, *lines*, and *planes*. Ironically, the building blocks are *undefined terms* in order to avoid circular definitions. An example of a circular definition and the frustration involved in starting without some basic, undefined notions is given in the following *B.C.* cartoon.

Because other geometric concepts are developed from undefined terms, we present an intuitive notion of these terms in the illustrations of Table 11-1.

Table 11-1 Notation

Term/Symbolism	Illustration
point A point B point C	
line ℓ	
line m, line AB, $\overleftrightarrow{AB}$, or $\overleftrightarrow{BA}$	
plane α	
plane ABC or plane γ	

A line has no thickness and it extends forever in two directions. *It is uniquely determined by two points*, that is, given two points, there is one and only one line that connects these points (notice that we have not defined what lines and points are, but rather have described some of their properties). A basic concept in geometry is that of a *distance*. If P and Q are any two points, we can create a number line on the line PQ such that there is one-to-one correspondence between the points on the line and real numbers. Moreover, we can correspond to P the real number 0 and to Q a positive real number. The distance between P and Q, written PQ, is the real number that corresponds to Q. Table 11-2 illustrates many commonly used terms and their meanings in geometry.

Table 11-2

Term	Illustration and Symbolism
Collinear points are points on the same line. (Any two points are collinear but not every three points have to be collinear.)	ℓ contains points A, B, and C. Line ℓ contains points A, B, and C. Points A, B, and C belong to line ℓ. Points A, B, and C are collinear. Points A, B, and D are not collinear.
When A, B, and C are three collinear points, then B is **between** A and C if $AB + BC = AC$. Point D is not between A and C.	
A **line segment**, or **segment**, is a subset of a line that contains two points of the line and all points between those two points.	$\overline{AB}$ or $\overline{BA}$
A ray $\overrightarrow{AB}$ is a subset of the line $\overleftrightarrow{AB}$ that contains the endpoint A, the point B, all the points between A and B, and all points C on the line such that B is between A and C.	$\overrightarrow{AB}$

NOW TRY THIS 11-1

a. In Table 11-2, it is stated that B is between points A and C if A, B, and C are collinear and $AB + BC = AC$. If A, B, and C are three different points, is it possible to have $AB + BC = AC$ and to have the points not collinear?

b. If we think of lines, segments, and rays as sets of points, find
  i. $\overrightarrow{AB} \cup \overrightarrow{BA}$
  ii. $\overrightarrow{AB} \cap \overrightarrow{BA}$

Planar Notions

A plane has no thickness and it extends indefinitely. *A plane is uniquely determined by three noncollinear points.* In other words, given three noncollinear points, there is one and only one plane that contains these points. Table 11-3 illustrates intuitive planar notions.

Table 11-3

Term	Illustration and Symbolism
Points in the same plane are **coplanar**.	Points D, E, and G are coplanar.
Noncoplanar points cannot be placed in a single plane.	Points D, E, F, and G are noncoplanar.
Lines in the same plane are **coplanar lines**.	Lines DE, DF, and FE are coplanar. Lines DE and EG are coplanar.

(continues)

Table 11-3 *continued*

Term	Illustration and Symbolism
Intersecting lines are lines with exactly one point in common.	Lines *DE* and *GE* are intersecting lines; they intersect at point *E*.
Skew lines are lines for which there is no plane that contains them (they are not coplanar).	Lines *GF* and *DE* are skew lines.
Concurrent lines are three or more lines that intersect in the same point.	Lines *DE*, *EG*, and *EF* are concurrent.
Parallel lines are coplanar lines that have no points in common. That is, a line is parallel to itself.*	*m* is parallel to *n*, written $m \parallel n$; also, $m \parallel m$ and $n \parallel n$.

♦ *Research Note*

Students build false interpretations of geometric terms from their exposure to a limited number of static pictures. For example, many students claim that two lines cannot be parallel unless they are the same length or are oriented vertically or horizontally (Kerslake 1981). ♦

Notice that in Table 11-3 the parallel lines *m* and *n* are neither horizontal nor vertical. As pointed out in the Research Note, it is important that students see parallel lines that are drawn in different positions.

NOW TRY THIS 11-2

a. Can skew lines have a point in common? Why?
b. Can skew lines be parallel? Why?
c. On a globe, a "line" is a great circle, that is, a circle the same size as the equator. How many different lines can be drawn through two different points on a globe?

Euclid considered the following statements concerning points, lines, and planes. He stated the first four as **axioms**—statements assumed to be true without proof—and showed further that the last three follow logically from the first four. Consequently, he called the latter statements *theorems*.

1. Axiom: There is exactly one line that contains any two distinct points.
2. Axiom: If two points lie in a plane, then the line containing the points lies in the plane.

*This definition is not standard. Sometimes it is convenient to define a line to be parallel to itself.

3. **Axiom:**   If two distinct planes intersect, then their intersection is a line.
4. **Axiom:**   There is exactly one plane that contains any three distinct noncollinear points.
5. **Theorem:**   A line and a point not on the line determine a unique plane.
6. **Theorem:**   Two distinct parallel lines determine a unique plane.
7. **Theorem:**   Two distinct intersecting lines determine a unique plane.

Problem Solving   Lines Through Points

Given 15 points, no 3 of which are collinear, how many lines can be drawn through the 15 points?

Understanding the Problem   Because 2 points determine exactly one line, we must consider ways to find out how many lines are determined by the 15 points.

Devising a Plan   We use the strategy of *examining related simpler cases* of the problem in order to think through the original problem. Figure 11-1 shows that 3 noncollinear points determine 3 lines, 4 determine 6 lines, 5 determine 10 lines, and 6 determine 15 lines.

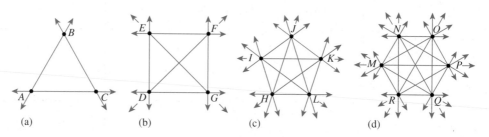

(a)   (b)   (c)   (d)

Figure 11-1

Examining Figure 11-1(d), we see that through any one of the 6 points, we can draw only 5 lines to connect to the other 5 points. If 5 lines are drawn through each point, there seem to be 6 times 5, or 30, lines. But in this approach, each line is counted twice; for example, line NO is counted as 1 of the 5 lines through N and also as 1 of the 5 lines through O. Thus, there are $\dfrac{6 \cdot 5}{2}$, or 15, lines in the figure. We use this information to determine the solution for 15 points.

Carrying Out the Plan   Each of the 15 points can be paired with all points other than itself to determine a line; that is, each point can be paired with the other 14 points to determine 14 lines. If we do this and account for counting each line twice, we see that there should be $\dfrac{15 \cdot 14}{2}$, or 105, lines.

Looking Back   Using this reasoning, we conclude that the number of lines determined by n points, no three of which are collinear, is $\dfrac{n(n-1)}{2}$.

An alternative solution to this problem uses the notion of combinations found in Chapter 9. The number of ways that n points, no 3 of which are collinear, can be chosen 2 at a time to form lines is $_nC_2$, or $\dfrac{n(n-1)}{2}$.

A somewhat different approach also uses a strategy of *solving a simpler problem*. Two points determine exactly 1 line. When we add a third point so that the 3 points are not collinear, as in Figure 11-1(a), we can connect that point to the 2 existing points and create 2 new lines, for a total of 1 + 2 lines. When a fourth point is added (so that no 3 of the 4 points are collinear), it can be connected to the 3 existing points to obtain 3 additional lines, for a total of 1 + 2 + 3 lines. Continuing in this way so that no 3 points are collinear, when the 15th point is added to the existing 14 points, 14 new lines are created for a total of 1 + 2 + 3 + ... + 14 lines. Using what we know about the sum of an arithmetic sequence (see Chapter 1), we get

$$1 + 2 + 3 + \ldots + 14 = \frac{14(14 + 1)}{2} = \frac{14 \cdot 15}{2}$$

or 105 lines.

In general, when the nth point is added, we can connect it to the $n - 1$ existing points to obtain $n - 1$ new lines. Thus the number of lines determined by n points, no 3 of which are collinear, is $1 + 2 + 3 + \ldots + (n - 1) = \dfrac{(n - 1)(1 + n - 1)}{2}$, or $\dfrac{(n - 1)n}{2}$.

NOW TRY THIS 11-3 Use a geometric model to find the number of handshakes that take place at a party of 20 people if each person shakes hands with everybody else at the party. (*Hint.* Think about people as points and handshakes as segments.)

Other Planar Notions

Two distinct planes either intersect in a line or are parallel. In Figure 11-2(a), planes α and β are parallel; that is, they have no points in common. Figure 11-2(b) shows planes that intersect in $\overleftrightarrow{AB}$.

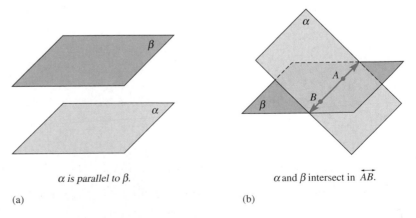

α *is parallel to* β. α *and* β *intersect in* $\overleftrightarrow{AB}$.

(a)   (b)

Figure 11-2

A line and a plane can be related in one of three ways. If a line and a plane have no points in common, the line is parallel to the plane, as in Figure 11-3(a). If two points of a line are in the plane, then the entire line containing the points is contained in the plane, as in Figure 11-3(b). In Figure 11-3(b), line AB separates plane α into two **half-planes**, denoted AB-C and AB-D. Plane α is the union of the three mutually disjoint sets: the two half-planes AB-C and AB-D, and the line AB. If a line intersects a plane but is not contained in the plane, it intersects the plane at only one point, as in Figure 11-3(c).

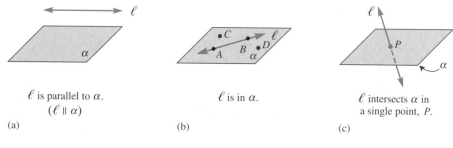

ℓ is parallel to α.
($\ell \parallel \alpha$)

(a)

ℓ is in α.

(b)

ℓ intersects α in
a single point, P.

(c)

Figure 11-3

Angles

When two rays share an endpoint, an **angle** is formed, as shown in Figure 11-4(a). The rays of an angle are the **sides** of the angle, and the common endpoint is the **vertex** of the angle. An angle can be named by three different points: the vertex and a point on each ray, with the vertex always listed between the other two points. Thus the angle in Figure 11-4(a) may be named $\angle CBA$ or $\angle ABC$. When there is no risk of confusion, it is customary simply to name an angle by its vertex, by a number, or by a lowercase Greek letter. The angle in Figure 11-4(a) therefore also can be named $\angle B$ or $\angle 1$. In Figure 11-4(b), however, more than one angle has vertex P, namely, $\angle QPR$, $\angle RPS$, and $\angle QPS$. Thus the notation $\angle P$ is inadequate for naming any one of the angles α, β, or $\angle QPS$. The **interior of an angle** is the region in the plane between the two sides of the angle. More formally, the interior of $\angle ABC$ is the set of all points in the plane, which are in the half plane AB-C and also in the half plane BC-A Angles such as $\angle QPR$ (or α) and $\angle RPS$ (or β) in Figure 11-4(b) are **adjacent angles**, which share a common vertex and a common side and their interiors do not overlap.

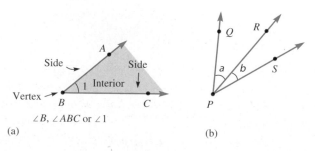

$\angle B$, $\angle ABC$ or $\angle 1$

(a)

(b)

Figure 11-4

Angle Measurement

An angle is measured according to the amount of "opening" between its sides. The **degree** is commonly used to measure angles. A complete rotation about a point has a measure of 360 degrees, written 360°. One degree is $\frac{1}{360}$ of a complete rotation. Figure 11-5 shows that $\angle BAC$ has a measure of 30 degrees, written $m(\angle BAC) = 30°$. The measuring device pictured in the figure is a **protractor**. A degree is subdivided into 60 equal parts—**minutes**—and each minute is further divided into 60 equal parts—**seconds**. The measurement 29 degrees, 47 minutes, 13 seconds is written 29°47′13″.

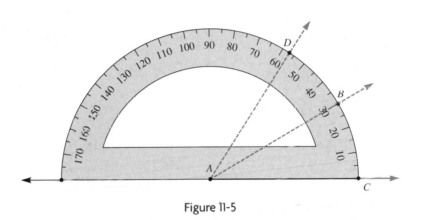

Figure 11-5

In calculus, the use of radians simplifies many formulas and calculations. An angle of 1 **radian** is the angle whose vertex is at the center of the circle and that intercepts an arc equal in length to the radius of the circle.* Thus, if in Figure 11-5 the arc CD is as long as the radius AC, then $\angle CAD$ measures 1 radian. One radian is approximately 57.296 degrees.

 NOW TRY THIS 11-4 Convert 8.42° to degrees, minutes, and seconds.

 **Historical Note** The French mathematician and astronomer Pierre Herigone (1580–1643) used a symbol for angle in 1634, but the Mathematical Association of America recommended $\angle$ as the standard symbol for angle in the United States in 1923. The use of 360° to measure angles seems to date to the Babylonians (4000–3000 BCE). ◆

---

*A *circle* is the set of all points in a plane that are the same distance from a given point, the *center*. An *arc* of a circle is any part of the circle that can be drawn without lifting a pencil (circles and arcs will be discussed in more detail in Chapter 12).

Example 11-1

a. In Figure 11-6, find the measure of $\angle BAC$ if $m(\angle 1) = 47°45'$ and $m(\angle 2) = 29°58'$.
b. Express $47°45'36''$ as a number of degrees.

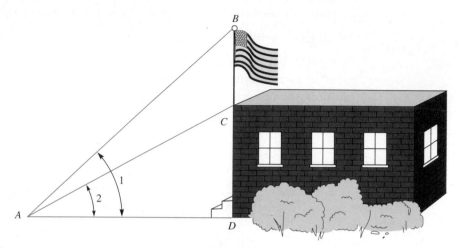

Figure 11-6

Solution **a.** $m(\angle BAC) = 47°45' - 29°58'$
$= 46°(60 + 45)' - 29°58'$
$= 46°105' - 29°58'$
$= (46 - 29)° + (105 - 58)'$
$= 17°47'$

b. $45' = \left(\dfrac{45}{60}\right)° = 0.75°$

$36'' = \left(\dfrac{36}{60}\right)' = \left(\dfrac{1}{60} \cdot \dfrac{36}{60}\right)°$

$= 0.01°$

Thus,

$$47°45'36'' = 47° + 0.75° + 0.01°$$
$$= 47.76°$$

REMARK Notice that the arithmetic with degrees, minutes, and seconds as seen in Example 11-1 is analogous to doing arithmetic in different number bases.

Types of Angles

We can create different types of angles by paper folding. Consider the folds shown in Figure 11-7(a) and (b). A piece of paper is folded in half and then reopened. If any point on the fold line, labeled ℓ, is chosen as the vertex O, then the measure of the angle pictured is 180°. If the paper is refolded and folded once more, as shown in Figure 11-7(c), and then is reopened, as shown in Figure 11-7(d), four angles of the same size are created. Each angle has measure 90° and is a right angle. The symbol ⌐ denotes a right angle. (Why is the measure of each of these angles 90°?)

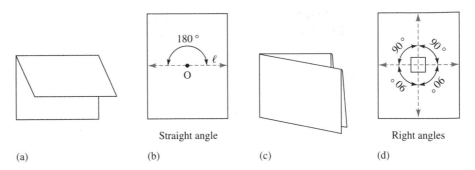

Figure 11-7

If the paper is folded as shown in Figure 11-8 and reopened, then angles α and β are formed, with measures that are less than 90° and greater than 90°, respectively. (Note that β has measure less than 180°.) Angle α is an *acute* angle, and β is an *obtuse* angle. In Figure 11-8(b), the measures of α and β add up to 180°. Any two angles the sum of whose measures is 180° are **supplementary** angles. We say that each angle is a **supplement** of the other.

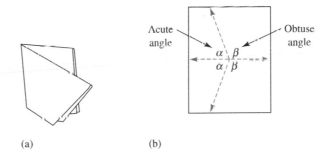

Figure 11-8

The types of planar angles are shown in Figure 11-9, along with their definitions.

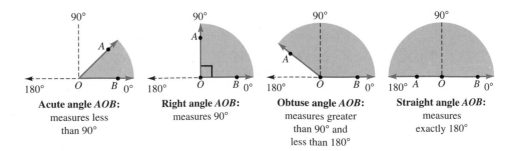

Figure 11-9

REMARK Notice that a straight angle can be described without referring to degree measure: A straight angle is an angle whose sides are opposite rays forming a straight line. A model of a straight angle can be obtained by folding a page in half.

On the student page students are asked to sketch angles with a given measure and work applications of angle measure. Answer the questions on the student page.

> **REMARK** In higher mathematics and in scientific applications, it is important to view an angle as being created by a ray rotating about its endpoint. If the ray makes one full rotation, we say that it sweeps an angle of 360°. Angles with positive measure are created by a counterclockwise rotation, and angles with negative measure by a clockwise rotation. Angles whose measures are greater than 360° are created when the ray makes more than one full rotation about its endpoint.

School Book Page

B Reasoning and Problem Solving

Estimation Without using a protractor, try to sketch an angle with the given measure. Then use a protractor to check your estimate.

19. 90°   **20.** 45°   **21.** 30°   **22.** 60°   **23.** 15°

Math and Social Studies

The grade of a highway refers to how steep the pavement is. It is usually given as a percent. Highway grades of 5% or greater require extreme caution by truck drivers. A grade steeper than 10% is rarely found on highways.

A grade of 6% is modeled in the scale drawing below.

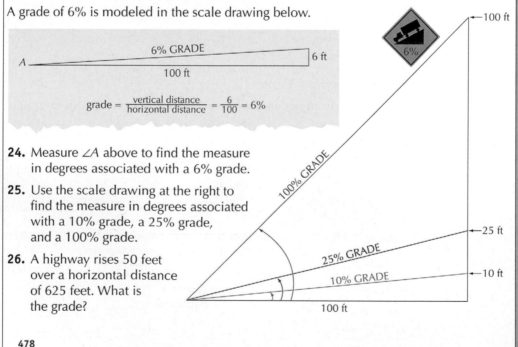

grade = $\dfrac{\text{vertical distance}}{\text{horizontal distance}}$ = $\dfrac{6}{100}$ = 6%

24. Measure ∠A above to find the measure in degrees associated with a 6% grade.

25. Use the scale drawing at the right to find the measure in degrees associated with a 10% grade, a 25% grade, and a 100% grade.

26. A highway rises 50 feet over a horizontal distance of 625 feet. What is the grade?

478

Source: Scott Foresman-Addison Wesley Math, Grade 6, 2008 (p. 478).

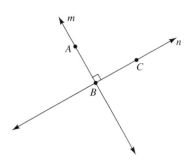

Figure 11-10

Perpendicular Lines

When two lines intersect so that the angles formed are right angles, as in Figure 11-7(d) and Figure 11-10, the lines are **perpendicular lines**. In Figure 11-10, lines m and n are perpendicular, and we write $m \perp n$. Two intersecting segments, two intersecting rays, one segment and one ray, a segment and a line, or a ray and a line that intersect are perpendicular if they lie on perpendicular lines. For example, in Figure 11-10, $\overline{AB} \perp \overline{BC}$, $\overrightarrow{BA} \perp \overrightarrow{BC}$, and $\overleftrightarrow{AB} \perp \overrightarrow{BC}$.

A Line Perpendicular to a Plane

If a line and a plane intersect, they can be perpendicular. For example, consider Figure 11-11, where planes β and γ represent two walls whose intersection is $\overleftrightarrow{AB}$. The edge $\overleftrightarrow{AB}$ is perpendicular to the floor. Also, every line in the plane of the floor (plane α) passing through point A is perpendicular to $\overleftrightarrow{AB}$. This discussion should help you understand what we mean by "a line perpendicular to a plane." A **line perpendicular to a plane** is a line that is perpendicular to every line in the plane through its intersection with the plane.

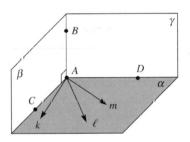

Figure 11-11

NOW TRY THIS 11-5

a. Is it possible for a line intersecting a plane to be perpendicular to exactly one line in the plane through its intersection with the plane? Explain by making an appropriate drawing.

b. Is it possible for a line intersecting a plane to be perpendicular to two distinct lines in a plane going through its point of intersection with the plane, and yet not be perpendicular to the plane?

c. Can a line be perpendicular to infinitely many lines?

d. If a line ℓ intersecting a plane α at point A is perpendicular to two distinct lines in the plane through A, what seems to be true about ℓ and α?

Because spatial ability is crucial to students' mathematical development (see Research Note), we now extend the concept of an angle between two lines to an angle between two planes.

Perpendicular Planes and Dihedral Angles

Figure 11-12(a) shows two perpendicular planes α and β, which can be modeled by two pages of a book opened at 90°. If $\overline{AB}$ and $\overline{AC}$ represent the edges of the book, then each is perpendicular to the binding $\overline{AD}$. Notice that $\angle BAC$ is a right angle. If P is any point on

$\overline{AD}$, Q in plane α, and S in plane β so that $\overrightarrow{PQ} \perp \overleftrightarrow{AD}$ and $\overrightarrow{PS} \perp \overleftrightarrow{AD}$, then $\angle QPS$ is also a right angle. Since $\angle QPS$ measures 90°, we say that the **planes are perpendicular**. If we view Figure 11-12(a) as the union of two half-planes and the common line AD, we have a **dihedral angle** that measures 90°.

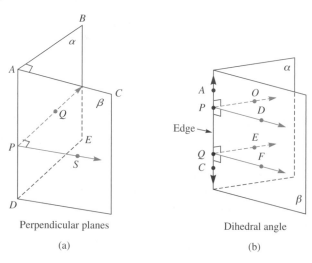

Perpendicular planes

(a)

Dihedral angle

(b)

Figure 11-12

In general, a dihedral angle is formed by the union of two half-planes and the common line defining the half-planes. In Figure 11-12(b), the dihedral angle O-AC-D is formed by the intersecting planes α and β. Note that the point O is in the plane α, $\overleftrightarrow{AC}$ is the edge of the dihedral angle, and point D is in plane β. A dihedral angle is measured by any of the associated planar angles such as $\angle OPD$, where $\overrightarrow{PO} \perp \overleftrightarrow{AC}$ and $\overrightarrow{PD} \perp \overleftrightarrow{AC}$. If any of the four dihedral angles created by the intersecting planes measures 90°, then all four dihedral angles measure 90° and the planes are said to be perpendicular. Notice the analogy between the definition of a planar angle and a dihedral angle. A planar angle is a union of two rays and a common endpoint, while a dihedral angle is the union of two half-planes and a common line.

Assessment 11-1A

1. a. Points A, B, C, and D are collinear. In how many ways can the line be named using only these points? (Assume that different order means different name.)
   b. Points A, B, C, D, and E are coplanar and no three are collinear. In how many ways can the plane be named using only these points?
2. The following figure is a rectangular box in which $EFGH$ and $ABCD$ are rectangles and $\overline{BF}$ is perpendicular to planes $EFGH$ and $ABCD$. Answer the following:
   a. Name two pairs of skew lines.
   b. Are $\overleftrightarrow{BD}$ and $\overleftrightarrow{FH}$ parallel, skew, or intersecting lines?
   c. Are $\overleftrightarrow{BD}$ and $\overleftrightarrow{GH}$ parallel?

d. Find the intersection of $\overleftrightarrow{BD}$ and plane EFG.
e. Explain why planes BDH and FHG are perpendicular.

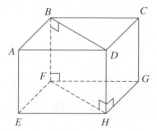

3. Use the following drawing of one of the Great Pyramids of Egypt (with square base) to find the following:
   a. The intersection of $\overline{AD}$ and $\overline{CE}$
   b. The intersection of planes ABC, ACE, and BCE
   c. The intersection of $\overleftrightarrow{AD}$ and $\overrightarrow{CA}$
   d. A pair of skew lines
   e. A pair of parallel lines
   f. A plane not determined by one of the triangular faces or by the base

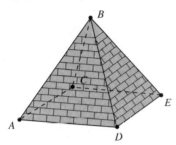

4. Determine how many acute angles are in the following figure:

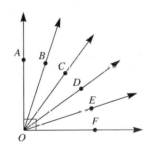

5. Identify a physical model for each of the following:
   a. Perpendicular lines
   b. An acute angle
   c. An obtuse angle

6. Find the measure of each of the following angles:
   a. $\angle EAB$ b. $\angle EAD$
   c. $\angle GAF$ d. $\angle CAF$

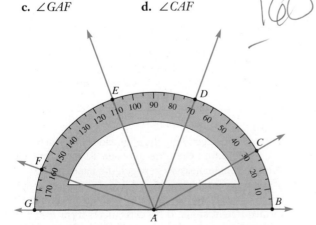

7. a. Perform each of the following operations. Leave your answers in simplest form.
   (i) $18°35'29'' + 22°55'41''$
   (ii) $15°29' - 3°45'$
   b. Express the following without decimals:
   (i) $0.9°$
   (ii) $15.13°$

8. Consider a correctly set clock that starts ticking at noon and answer the following:
   a. Find the measure of the angle swept by the hour hand by the time it reaches
   (i) 3 P.M.
   (ii) 12:25 P.M.
   (iii) 6:50 P.M.
   b. Find the exact angle measure between the minute and the hour hands at 1:15 P.M.

9. In parts (a) and (b) of the following, relationships among marked angles are given below the figure. Find the measure of the marked angles. In part (c), find only the measure of $\angle BOC$ and tell why the exact values of x and y cannot be determined.

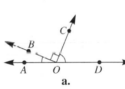

$m(\angle AOB) = \frac{1}{3}m(\angle COD)$ $m(\angle AOB)$ is $35°$ less than $3 \cdot m(\angle BOC)$

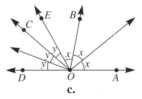

10. Given three collinear points A, B, C, (B is between A and C), four different rays can be named using these points: $\overrightarrow{AB}$, $\overrightarrow{BA}$, $\overrightarrow{BC}$, and $\overrightarrow{CB}$. Determine how many different rays can be named given each of the following:
   a. Four collinear points
   b. Five collinear points
   c. n collinear points

11. Refer to the following table.

Number of Intersection Points of Coplanar Lines

	0	1	2	3	4	5
2			Not possible	Not possible	Not possible	Not possible
3					Not possible	Not possible
4						
5						
6						

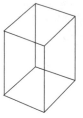
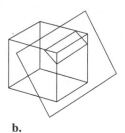

a.   **b.**

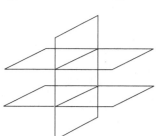

c.   **d.**

a. Sketch the possible intersections of the given number of lines. Three sketches are given for you.
b. Given *n* lines, find a formula for determining the greatest possible number of intersection points.
12. Trace each of the following drawings. In your tracings, use dashed lines for segments that would not be seen and solid lines for segments that would be seen. (Different people may see different perspectives.)

13. Draw pictures illustrating a real-world example of the following:
a. Three planes intersecting in a common line
b. Three planes intersecting in a common point

Assessment 11-1B

1. The following figure is a box in which the top and bottom are rectangles and $\overline{BF}$ perpendicular to plane *FGH*. Answer the following:
a. Find the intersection of $\overleftrightarrow{BH}$ and plane *DCG*.
b. Name two pairs of perpendicular planes.
c. Name two lines that are perpendicular to plane *EFH*.
d. What is the measure of dihedral angle *D-HG-F*?

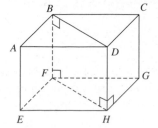

2. a. Perform the following operations. Leave your answers in simplest form.
   (i) $21°35'31'' + 49°51'32''$
   (ii) $93°38'14'' - 13°49'27''$
b. Express the following in degrees, minutes, and seconds, without decimals:
   (i) $10.3°$
   (ii) $15.14°$
3. Consider a correctly functioning clock that starts ticking at noon and find the time between 12 noon and 1 P.M. when the angle measure between the hands is $180°$.
4. Mario was studying right angles and wondered if during his math class the minute and hour hands of the clock formed a right angle. If his class meets from 2:00 P.M. to 2:50 P.M., is a right angle formed? If it is, figure out to the nearest second when the hands form the right angle.

5. In each of the following, relationships among marked angles are given below the figure. Find the measures of the marked angles.

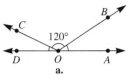

a.

$$m(\angle DOC) = \frac{3}{4}m(\angle BOA)$$

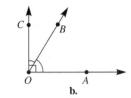

b.

$m(\angle AOB)$ is 30° less than $2 \cdot m(\angle BOC)$

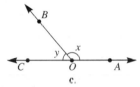

c.

$$m(\angle AOB) - m(\angle BOC) = 50°$$

6. a. Referring to the figure below, construct a line and point A on the line. Choose an arbitrary point B not on the line and draw $\overrightarrow{AB}$. Use a protractor (or geometry utility such as GSP) to measure $\angle DAB$ and $\angle BAC$. Find half the measure of each angle to draw $\overrightarrow{AE}$ and $\overrightarrow{AF}$, which divide $\angle DAB$ and $\angle BAC$ in half as shown. Now measure $\angle EAF$.
  b. Repeat part (a) for different locations of point B and conjecture what the measure of $\angle EAF$ will always be.
  c. Prove or disprove your conjecture in part (b).

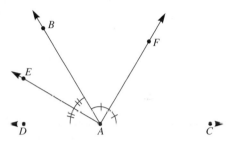

7. In each of the following pairs, determine whether the symbols name the same geometric figure:
  a. $\overrightarrow{AB}$ and $\overrightarrow{BA}$
  b. $\overline{AB}$ and $\overline{BA}$
  c. $\overleftrightarrow{AB}$ and $\overleftrightarrow{BA}$

8. a. How many planes are determined by three non-collinear points?
  b. How many planes are determined by four noncoplanar points?
  c. How many planes are determined by n points, no four of which are coplanar?

Mathematical Connections 11-1

Communication

1. Forest rangers use degree measures to identify directions and locate critical spots such as fires. In the following drawing, a forest ranger at tower A observes smoke at a bearing of 149° (clockwise from the north), while another forest ranger at tower B observes the same source of smoke at a bearing of 250° (clockwise from the north).

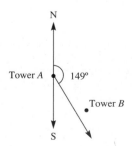

  a. Choose two locations for A and B and use a protractor and a straightedge to locate the source of the smoke.
  b. Explain how the forest rangers could find the location of the fire.
  c. Describe other situations in which location can be determined by similar methods.

2. Is it possible to locate four points in a plane such that the number of lines determined by the points is not exactly 1, 4, or 6? Explain.

3. Given n lines in the plane, no two of which are parallel and no three of which intersect in a single point, find the number of nonoverlapping regions that are created by the n lines. Explain your reasoning.

4. Angles COB and COA are supplementary, and points A, O, B are collinear. $\angle COA$ is divided into n angles each of measure x, and $\angle COB$ is divided into n angles each of measure y. Find $x + y$ in terms of n.

5. A line n is perpendicular to plane α, and plane β contains n (n is in plane β). Must planes α and β be perpendicular? Explain why or why not.

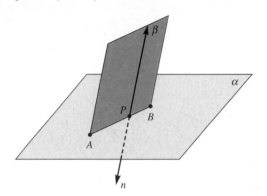

Cooperative Learning

6. Let each member of your group use a protractor to make a triangle out of cardboard that has one angle measuring 30° and another 50°. Answer the following and compare your solutions with other members of your group:

a. Show how to use the triangle (without a protractor) to draw an angle with measure 40°.

b. Is there more than one way to draw an angle of 40° using the triangle? Explain.

c. What other angles can be drawn with the triangle? Why?

Open-Ended

7. Within the classroom, identify a physical object with the following:

   a. Parallel lines   **b.** Parallel planes

   c. Skew lines   **d.** Right angles

8. On a sheet of dot paper or on a geoboard like the one shown, create the following:

   a. Right angle   **b.** Acute angle

   c. Obtuse angle   **d.** Adjacent angles

   e. Parallel segments   **f.** Intersecting segments

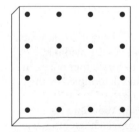

Questions from the Classroom

9. Henry claims that a line segment has a finite number of points because it has two endpoints. How do you respond?

10. A student claims that if any two planes that do not intersect are parallel, then any two lines that do not intersect should also be parallel. How do you respond?

11. A student says that it is actually impossible to measure an angle, since each angle is the union of two rays that extend infinitely and therefore continue forever. What is your response?

12. Maggie claims that to make the measure of an angle greater, you just extend the rays. How do you respond?

13. A student says that a line is parallel to itself. How do you reply?

14. A student says there can be only 360 different rays emanating from a point, since there are only 360° in a circle. How do you respond?

15. A student would like to know how a carpenter can tell if a wall is perpendicular to the floor. How do you respond?

Third International Mathematics and Science Study (TIMSS) Questions

On the grid, draw a line parallel to line L.

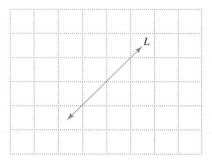

TIMSS, Grade 4, 2003

In the figure, the measure of $\angle POR$ is 110°, the measure of $\angle QOS$ is 90°, and the measure of $\angle POS$ is 140°.

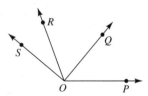

What is the measure of $\angle QOR$?

TIMSS, Grade 8, 2003

National Assessment of Educational Progress (NAEP) Question

What is the intersection of rays PQ and QP in the figure above?

   a. Segment PQ

   b. Line PQ

   c. Point P

   d. Point Q

   f. The empty set

NAEP, Grade 8, 2007

LABORATORY ACTIVITY   Many geometric ideas can be experienced through paper folding, as in Figure 11-13. Consider the following construction of perpendicular lines and answer the questions that follow:

Fold one corner
of a sheet of
paper over and
crease it along
the fold.

Fold any part
of the crease
onto itself.

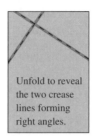

Unfold to reveal
the two crease
lines forming
right angles.

Figure 11-13

a. Explain why the two crease lines are perpendicular.
b. Use a sheet of paper and follow the preceding instructions to create two pairs of perpendicular lines.
c. Use paper folding to create angles with the following measures:
   i. 45°   ii. 135°   iii. 22°30′
   (*Mathematics Through Paper Folding* by Olson and *Geometric Exercises in Paper Foldings* by Row are listed in the bibliography as resource books.)
d. Cut out a large circle; fold it in half by creasing along a line through the center of the circle. Fold the resulting half circle in half again. Unfold and tell what is the measure of the smallest angle created by the creases.
e. Suppose you continue folding in half as in part (d); what is the measure of the smallest angle created after three folds?
f. What is the measure of the smallest angle created after *n* folds?

11-2   Polygons

With a pencil, draw a path on a piece of paper without lifting the pencil and without retracing any part of the path except single points. The resulting drawing is restricted to the plane of the paper, and not lifting the pencil implies that there are no breaks in the drawing. The drawing is **connected** and is a **curve**.

In Table 11-4 on page 704 each curve is in the plane and is connected, but there are some obvious differences among the curves. Table 11-4 shows sample curves and their classifications. A check is placed in a box if the curve has the attribute listed at the top.

Research Note

Dina and Pierre van Hiele, after years of extensive research, contend that students develop an understanding of geometry by progressing through five distinctive levels in a hierarchical manner similar to those associated with Piaget (Carpenter 1980; Clements and Battista 1992):
Level I—Recognition and Visualization: Students can name and perceive geometric figures.
Level II—Analysis: Students can identify and isolate specific attributes of a figure.
Level III—Order: Students understand the role of a definition and recognize that specific properties follow from others.
Level IV—Deduction: Students are able to work within a deduction system—postulates, theorems, and proofs.
Level V—Rigor: Students understand both rigor in proofs and abstract geometric systems such as non-Euclidean geometries. ◆

Table 11-4

Curve	Simple	Closed	Polygon	Convex	Concave
	✔				
	✔				
	✔	✔			✔
	✔	✔		✔	
	✔	✔			✔
	✔	✔	✔	✔	
	✔	✔	✔	✔	
	✔	✔	✔		✔

A **simple** curve does not intersect itself, except that if you draw it with a pencil, the starting and stopping points may be the same. A **closed** curve can be drawn starting and stopping at the same point. A curve can be classified as simple, nonsimple, closed, nonclosed, and so on. **Polygons** are simple closed curves with *sides* that are only segments. A point where two sides of a polygon meet is a *vertex*. **Convex** curves are simple and closed, such that *the segment connecting any two points in the interior of the curve is wholly contained in the interior of the curve*. **Concave** curves are simple, closed, and not convex; that is, it is possible for a line segment connecting two interior points to cross outside the interior of the curve.

 NOW TRY THIS 11-6 Draw a curve that is neither simple nor closed.

As in Figure 11-14(a), every simple closed curve separates the plane into three disjoint subsets: the interior of the curve, the exterior of the curve, and the curve itself. Of specific interest to us are polygons and their interiors, together called **polygonal regions**. Figure 11-14(b) shows a polygonal region.

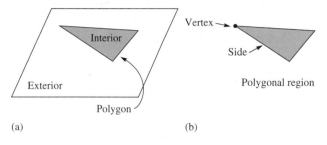

(a)   (b)

Figure 11-14

Whether a point is inside or outside a curve is not always obvious. This is explored in Now Try This 11-7.

NOW TRY THIS 11-7 Determine whether point X is inside or outside the simple closed curve of Figure 11-15. Explain your reasoning so that it can be generalized to other simple closed curves.

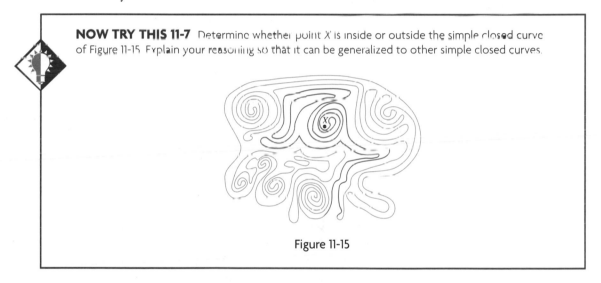

Figure 11-15

More About Polygons

Polygons are classified according to the number of sides or vertices they have. For example, consider the polygons listed in Table 11-5.

Table 11-5

Polygon	Number of Sides or Vertices
Triangle	3
Quadrilateral	4
Pentagon	5
Hexagon	6
Heptagon	7
Octagon	8
Nonagon	9
Decagon	10
.	.
.	.
.	.
n-gon	n

A polygon is referred to by the capital letters that represent its consecutive vertices, such as *ABCD* or *CDAB* (but not *BCAD*) shown in Figure 11-16(a). Any two sides of a convex polygon having a common vertex determine an **interior angle**, or **angle of the polygon**, such as ∠1 of polygon *ABCD* in Figure 11-16(a). An **exterior angle of a convex polygon** is determined by a side of the polygon and the extension of a contiguous side of the polygon. An example is ∠2 in Figure 11-16(b). Any line segment connecting nonconsecutive vertices of a polygon, such as $\overline{AC}$ in Figure 11-16(a), is a **diagonal** of the polygon.

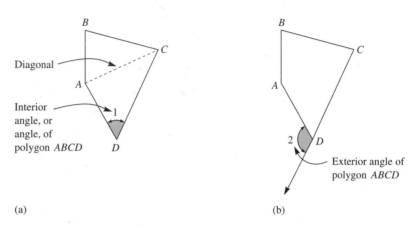

(a)   (b)

Figure 11-16

Congruent Segments and Angles

Most modern industries operate on the notion of creating **congruent parts**, parts that are of the same size and shape. For example, the specifications for all cars of a particular model are the same, and all parts produced for that model are basically the same. Usually congruent figures refer to figures in a plane. For example, two line **segments** are **congruent** (≅) if a tracing of one line segment can be fitted exactly on top of the other line segment. If $\overline{AB}$ is congruent to $\overline{CD}$, we write $\overline{AB} \cong \overline{CD}$. Two **angles** are **congruent** if they have the same measure. Congruent segments and congruent angles are shown in Figure 11-17(a) and (b), respectively.

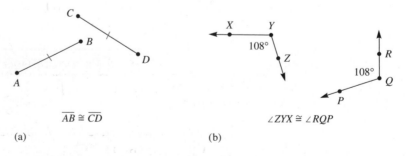

$\overline{AB} \cong \overline{CD}$ ∠*ZYX* ≅ ∠*RQP*

(a)   (b)

Figure 11-17

Notice in Figure 11-17(a), $\overline{AB} = \overline{BA}$, but $\overline{AB} \neq \overline{CD}$ because the segments contain different sets of points. However, $\overline{AB} \cong \overline{CD}$. In geometry, we use the equal sign "=" for *exactly* the same shape, size, and location. Notice that $\overline{AB} \cong \overline{CD}$ if, and only if, the segments have the same length, that is, if, and only if, *AB* = *CD*.

Regular Polygons

Convex polygons in which all the interior angles are congruent and all the sides are congruent are **regular polygons**. A regular polygon is both *equiangular* and *equilateral*. A regular triangle is an equilateral triangle. A regular pentagon and a regular hexagon are illustrated in Figure 11-18. The congruent sides and congruent angles are marked.

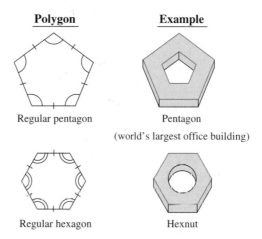

Polygon   **Example**

Regular pentagon   Pentagon
                    (world's largest office building)

Regular hexagon   Hexnut

Figure 11-18

NOW TRY THIS 11-8 Sketch the following:

1. A polygon that is equiangular but not equilateral.
2. A polygon that is equilateral but not equiangular.

Triangles and Quadrilaterals

Triangles and quadrilaterals may be classified according to their angle measures or according to their side length, as shown in Table 11-6.

Table 11-6

Definition	Illustration	Example
A triangle containing one right angle is a **right triangle**.		
A triangle in which all the angles are acute is an **acute triangle**.		YIELD
A triangle containing one obtuse angle is an **obtuse triangle**.		
A triangle with no congruent sides is a **scalene triangle**.		

(continues)

Table 11-6 *continued*

Definition	Illustration	Example
A triangle with at least two congruent sides is an **isosceles triangle**.		
A triangle with three congruent sides is an **equilateral triangle**.		
A **trapezoid** is a quadrilateral with at least one pair of parallel sides.		
A **kite** is a quadrilateral with two adjacent sides congruent and the other two sides also congruent.		
An **isosceles trapezoid** is a trapezoid with congruent base angles.		
A **parallelogram** is a quadrilateral in which each pair of opposite sides is parallel.		
A **rectangle** is a parallelogram with a right angle.		
A **rhombus** is a parallelogram with two adjacent sides congruent.		
A **square** is a rectangle with two adjacent sides congruent.		

REMARK In chapter 12 we will be able to prove that a quadrilateral is:

    a. a rectangle if, and only if, it has four right angles,

    b. a rhombus if, and only if, all its sides are congruent,

    c. a square if, and only if, it has four right angles and four congruent sides.

Some texts give different definitions for trapezoids and other figures. Many elementary texts define a trapezoid as a quadrilateral with *exactly* one pair of parallel sides. Note the definition of a trapezoid on the following partial student page. What additional arrows would have been drawn if our definition of trapezoid were used?

REMARK **a.** Notice that our definitions are mathematically convenient. For example, because we define a rectangle as a parallelogram with a right angle, any theorem proved for parallelograms will also be valid for rectangles.
b. In mathematics, when we say that a triangle has *at least* two sides congruent it means that it may have exactly two sides or all three sides congruent. Thus, an equilateral triangle is isosceles, but an isosceles triangle is not necessarily equilateral.

The many names for geometric figures can be overwhelming, as depicted in the following cartoon.

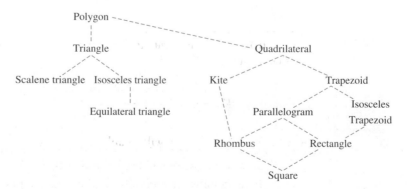

www.cartoonroom.com

WITH SO MANY GEOMETRIC FIGURES, I'M GONNA RUN OUT OF ALPHABET NAMING THE PARTS!

Hierarchy Among Polygons

Using set concepts, we can say that the set of all triangles is a proper subset of the set of all polygons. Also, the set of all equilateral triangles is a proper subset of the set of all isosceles triangles. This hierarchy is shown in Figure 11-19, where more general terms appear above more specific ones.

Figure 11-19

School Book Page — CLASSIFYING QUADRILATERALS

Lesson 6-6

Key Idea
Quadrilaterals can be classified by the properties of their angles and sides.

Vocabulary
• parallelogram
• trapezoid
• rectangle
• rhombus
• square

Materials
• ruler
• protractor

Think It Through
I can measure the angles in various quadrilaterals to **try, check, and revise** my prediction.

346

Classifying Quadrilaterals

LEARN

How do you classify quadrilaterals?

Quadrilaterals can be classified by their angles or pairs of sides. Remember that a quadrilateral is any polygon with 4 sides.

Parallelogram
A quadrilateral with both pairs of opposite sides parallel and equal in length

Trapezoid
A quadrilateral with only one pair of parallel sides

Rectangle
A parallelogram with four right angles

Rhombus
A parallelogram with all sides the same length

Rectangles and rhombuses are special parallelograms.

Square
A rectangle with all sides the same length

A square is a special rectangle. It is also a special rhombus.

WARM UP
Find the missing angle measure.
1. 82°, 60°, ?
2. 64°, ?

Activity

What is the sum of the measures of the angles of a quadrilateral?

a. Which quadrilaterals have all right angles? What is the sum of the measures of the four angles in each of those quadrilaterals?

b. Draw a large quadrilateral that has no right angles. What do you predict is the sum of the measures of the four angles?

c. Check your prediction by using a protractor to measure each of the angles. What is the sum of the measures?

d. Draw two more large quadrilaterals. In each quadrilateral, measure the angles. What is the sum of the measures?

Source: Scott Foresman-Addison Wesley, Mathematics, Grade 5, 2008 (p. 346).

NOW TRY THIS 11-9 Use the definitions in Table 11-6 to experiment with several drawings to decide which of the following are true:

1. An equilateral triangle is isosceles.
2. A square is a regular quadrilateral.
3. If one angle of a rhombus is a right angle, then all the angles of the rhombus are right angles.
4. A square is a rhombus with a right angle.
5. All the angles of a rectangle are right angles.
6. A rectangle is an isosceles trapezoid.
7. Some isosceles trapezoids are kites.
8. If a kite has a right angle, then it must be a square.

Assessment 11-2A

1. Determine for each of the following which of the figures labeled (1) through (10) can be classified under the given term:
   a. Simple closed curve
   b. Polygon
   c. Convex polygon
   d. Concave polygon

(1)   (2)   (3)   (4)   (5)

(6)   (7)   (8)   (9)   (10)

2. What is the maximum number of intersection points between a quadrilateral and a triangle (where no sides of the polygons are on the same line)?
3. What type of polygon must have a diagonal such that part of the diagonal falls in the exterior of the polygon?
4. Which of the following figures are convex and which are concave? Why?

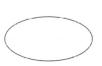

a.   **b.**

c.   **d.**

5. If possible, draw the following triangles. If it is not possible, state why.
   a. An obtuse scalene triangle
   b. An acute scalene triangle
   c. A right scalene triangle
   d. An obtuse equilateral triangle
   e. A right equilateral triangle
   f. An obtuse isosceles triangle
   g. An acute isosceles triangle
   h. A right isosceles triangle
6. Determine how many diagonals each of the following has:
   a. Pentagon   **b.** Decagon
   c. 20-gon   **d.** n-gon

7. Identify each of the following triangles as scalene, isosceles, or equilateral (there may be more than one term that applies to these triangles):

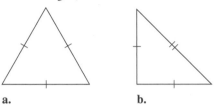

a.   **b.**

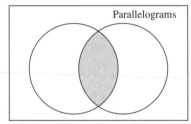

$AC > BC > AB$

c.

8. Describe the shaded region in the following Venn diagram, where the universal set is the set of all parallelograms, the left circle represents all rhombuses and the right circle all the rectangles.

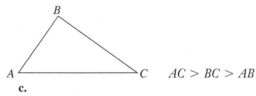

Parallelograms

9. Draw any angle BAC where $AB = AC$. Then construct an isosceles triangle ABC. Crease the triangle along the vertex A so that vertex B falls on C. List properties of an isosceles triangle that seem to be true.

10. Draw any triangle ABC and trace it on a second sheet of paper labeling the traced vertices A', B', and C'. Next rotate the second sheet of paper by $180°$ about M, the midpoint of $\overline{AB}$. Then trace the original triangle ABC on the second sheet so that A falls on B' and B on A', as shown.

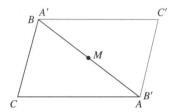

a. What kind of figure does $ACBC'$ seem to be? List some of the properties of this figure that seem to be true.

b. What kind of triangle must ABC in part (a) be so that the quadrilateral $ACBC'$ is
  (i) a rhombus?
  (ii) a rectangle?
  (iii) a square?

Assessment 11-2B

1. Determine for each of the following which of the figures (if any) labeled (1) through (10) can be classified under the given terms:
  a. Isosceles triangle
  b. Isosceles but not equilateral triangle
  c. Equilateral but not isosceles triangle
  d. Parallelogram but not a trapezoid
  e. A trapezoid but not a parallelogram
  f. A rectangle but not a square
  g. A square but not a rectangle
  h. A square but not a trapezoid
  i. A rhombus but not a kite
  j. A rhombus
  k. A kite

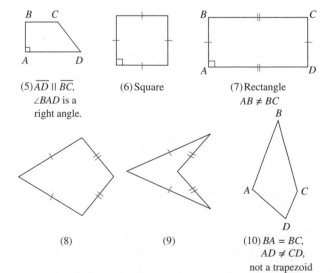

(5) $\overline{AD} \parallel \overline{BC}$,
$\angle BAD$ is a
right angle.

(6) Square

(7) Rectangle
$AB \neq BC$

(8)   (9)   (10) $BA = BC$,
$AD \neq CD$,
not a trapezoid

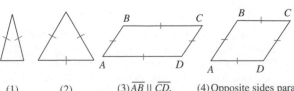

(1)   (2)   (3) $\overline{AB} \parallel \overline{CD}$,
$\overline{AD} \parallel \overline{BC}$

(4) Opposite sides parallel,
all sides congruent.

2. Notice that a pentagon has only two diagonals that intersect at a given vertex. Determine how many diagonals intersect at a given vertex in each of the following polygons:
   a. Hexagon   **b.** Decagon
   c. 20-gon   **d.** *n*-gon
3. Describe the shaded region as simply as possible.

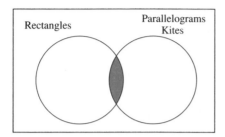

4. Informally, *a line of symmetry* of a figure is a line with the property that if the figure is folded along that line, the part of the figure on one side of the line matches perfectly the part on the other side of the line. Notice that a rectangular sheet of paper can be folded onto itself horizontally and vertically and hence a rectangle has two lines of symmetry. (More on lines of symmetry in chapter 14.) Draw all the lines of symmetry (if any exist) for each of the following:

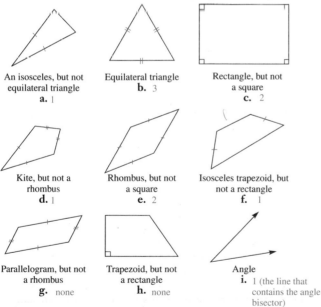

An isosceles, but not equilateral triangle
a. 1

Equilateral triangle
b. 3

Rectangle, but not a square
c. 2

Kite, but not a rhombus
d. 1

Rhombus, but not a square
e. 2

Isosceles trapezoid, but not a rectangle
f. 1

Parallelogram, but not a rhombus
g. none

Trapezoid, but not a rectangle
h. none

Angle
i. 1 (the line that contains the angle bisector)

5. **a.** How can you find the center of a circle by paper folding?
   b. How many lines of symmetry does a circle have? What are they?
6. Describe all the lines of symmetry (if any) of each of the following:
   a. A segment
   b. A ray
   c. A line
   d. Two distinct parallel lines
   e. Two intersecting lines
   f. Two intersecting segments
7. What is the minimum number of small squares that must be colored blue so that the large square has diagonal $\overline{BD}$ as a line of symmetry? Color the appropriate squares.

Mathematical Connections 11-2

Communication

1. **a.** Fold a rectangular piece of paper to create a square. Describe your procedure in writing and orally with a classmate. Explain why your approach creates a square.

   b. Crease the square in (a) so that the two diagonals are shown. Use paper folding to show that the diagonals of a square are congruent and perpendicular and divide each other into congruent parts. Describe your procedure and explain why it works.

Open-Ended

2. On a geoboard or dot paper, construct each of the following:
   a. A scalene triangle
   b. An obtuse triangle
   c. An isosceles trapezoid
   d. A trapezoid that is not isosceles
   e. A convex hexagon
   f. A concave quadrilateral
   g. A parallelogram
   h. A rhombus that is not a square.

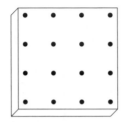

Cooperative Learning

3. Work with a partner. One of you constructs a figure on a geoboard or draw it on a piece of paper and identify it. Do not show the figure to your partner but tell your partner sufficient properties of the figure to identify it. Have your partner identify the figure you constructed. Your partner earns 1 point if the figure is correctly identified and 2 points if a figure is found that has all the required attributes but is different from the one you drew. Each of you should take the same number of turns. Try this with each of the following types of figure:
   a. Scalene triangle
   b. Isosceles triangle
   c. Square
   d. Parallelogram
   e. Trapezoid
   f. Rectangle
   g. Regular polygon
   h. Rhombus
   i. Isosceles trapezoid
   j. A kite that is not a rhombus
4. Work with partners to create a Venn diagram with the universal set being all triangles and the subsets being isosceles, equilateral, and right triangles.
5. a. Investigate the meaning and uses of Reuleaux triangles.
   b. Explain the similarities and differences between a Reuleaux triangle and an equilateral triangle.
6. a. With partners, decide how you might define a triangle on a globe.
   b. With partners, a globe, string, and a ruler, decide if you think a square can exist on a globe.

Questions from the Classroom

7. A student asks whether a polygon whose sides are congruent is necessarily a regular polygon and whether a polygon with all angles congruent is necessarily a regular polygon. How do you answer?

8. A student asks how to find the shortest path between two points A and B on a right circular cylinder. How do you respond?

9. One student says, "My sister's high school geometry book talked about equal angles. Why don't we use the term 'equal angles' instead of 'congruent angles'?" How do you reply?
10. Jodi identifies figure (a) as a rectangle and figure (b) as a square. She claims that figure (b) is not a rectangle because it is a square. How do you respond?

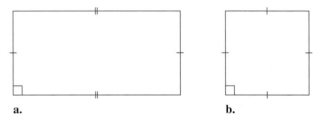

a.   b.

11. Millie claims that a rhombus is regular because all of its sides are congruent. How do you respond?
12. A student asks if geometry is worth studying if lines have different meaning on a sphere. What do you say?

Review Problems

13. If three distinct rays with the same vertex are drawn as shown in the following figure, then three angles are formed: $\angle AOB$, $\angle AOC$, and $\angle BOC$.

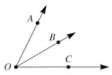

What is the maximum number of angles (measuring less than 180°) formed by using
   a. 10 distinct noncollinear rays with the same vertex?
   b. n distinct noncollinear rays with the same vertex?
14. Determine the possible intersection sets of a line and an angle.
15. Classify the following as true or false. If false, tell why.
   a. A ray has two endpoints.
   b. For any points M and N, $\overleftrightarrow{MN} = \overleftrightarrow{NM}$.
   c. Skew lines are coplanar.
   d. $\overrightarrow{MN} = \overrightarrow{NM}$
   e. A line segment contains an infinite number of points.
   f. If two distinct planes intersect, their intersection is a line segment.

Third International Mathematics and Science Study (TIMSS) Questions

a. Draw 1 straight line on this rectangle to divide it into 2 triangles.

b. Draw 1 straight line on this rectangle to divide it into 2 rectangles.

c. Draw 2 straight lines on this rectangle to divide it into 1 rectangle and 2 triangles.

TIMSS, Grade 4, 2003

In the picture there are a number of geometric shapes, like circles, squares, rectangles, and triangles. For example, the sun looks like a circle.

Draw lines to three other different objects in the picture and write what shapes they look like.

TIMSS, Grade 4, 2003

National Assessment of Educational Progress (NAEP) Question

Nick has a square piece of paper. He draws the two diagonals of the square, finds the point where they intersect, and labels that point A. Then he folds each of the four corners of the paper onto point A. What geometric shape is produced?

a. A square
b. A right triangle
c. An isosceles triangle
d. A pentagon
f. A hexagon

NAEP, Grade 8, 2007

LABORATORY ACTIVITY Use paper folding, tape or glue, and any other tools to create a shape larger than but like that in Figure 11-20.

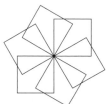

Figure 11-20

11-3   More About Angles

In Figure 11-21 two lines intersect and form the angles marked 1, 2, 3, and 4.

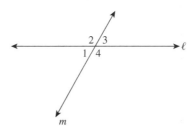

Figure 11-21

Vertical angles are pairs of angles such as $\angle 1$ and $\angle 3$ and appear any time two lines intersect. Another pair of vertical angles in Figure 11-21 is $\angle 2$ and $\angle 4$. Assuming we know the meaning of opposite rays (two different rays that share a common endpoint and whose union is a line), we can define vertical angles as follows:

> **Definition**
>
> **Vertical angles** created by intersecting lines are a pair of angles whose sides are two pairs of opposite rays.

REMARK Notice that the word *definition* in mathematics is quite different from its meaning in everyday language. For example, the definition of vertical angles uses as little information as possible to describe what we mean by vertical angles. We don't put all we know about vertical angles in the definition because these properties can be proved. Recall, for example, that we defined a parallelogram as a quadrilateral in which opposite sides are parallel. Although it is true that opposite sides of a parallelogram are also congruent, we do not put this fact in the definition since it can be proved (this will be done in Chapter 12) and therefore is a theorem.

You have probably noticed that vertical angles seem to be congruent. We state this in the following theorem and provide a proof.

> **Theorem 11–1**   Vertical angles are congruent.

Proof: Referring to Figure 11-21, we need to show that $\angle 1 \cong \angle 3$ and $\angle 2 \cong \angle 4$. We prove the first congruence. For that purpose, notice that $m(\angle 1) + m(\angle 2) = 180°$ and $m(\angle 3) + m(\angle 2) = 180°$. Thus,

$$m(\angle 1) = 180° - m(\angle 2)$$
$$m(\angle 3) = 180° - m(\angle 2)$$

Consequently, $m\angle 1 = m\angle 3$. In a similar way we can show that $m(\angle 2) = m(\angle 4)$.

Other pairs of angles appear frequently enough that it is convenient to refer to them by specific names. Table 11-7 shows several types of angle.

Table 11-7

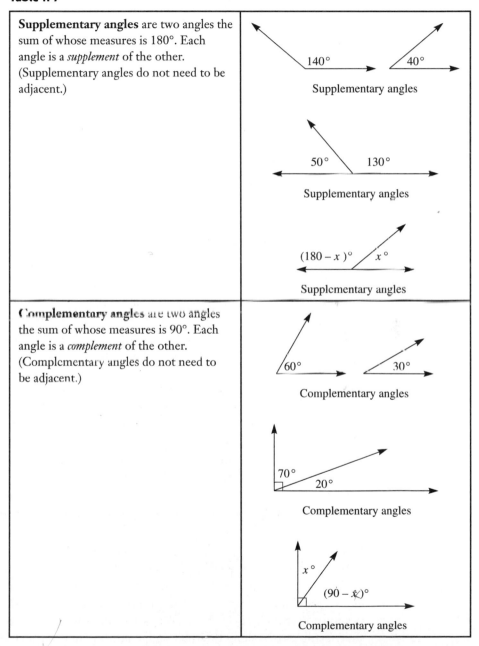

Supplementary angles are two angles the sum of whose measures is 180°. Each angle is a *supplement* of the other. (Supplementary angles do not need to be adjacent.)	140° 40° Supplementary angles 50° 130° Supplementary angles $(180 - x)°$ $x°$ Supplementary angles
Complementary angles are two angles the sum of whose measures is 90°. Each angle is a *complement* of the other. (Complementary angles do not need to be adjacent.)	60° 30° Complementary angles 70° 20° Complementary angles $x°$ $(90 - x)°$ Complementary angles

Angles are also formed when a line intersects two distinct lines. Any line that intersects a pair of lines in a plane is a **transversal** of those lines. In Figure 11-22(a), line p is a transversal of lines m and n, all in the same plane. Angles formed by these lines are named according to their placement in relation to the transversal and the two given lines. They are listed in Table 11-8.

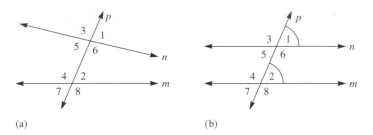

(a)   (b)

Figure 11-22

Table 11-8

Interior angles	$\angle 2, \angle 4, \angle 5, \angle 6$
Exterior angles	$\angle 3, \angle 1, \angle 7, \angle 8$
Alternate interior angles	$\angle 5$ and $\angle 2$, $\angle 4$ and $\angle 6$
Alternate exterior angles	$\angle 1$ and $\angle 7$, $\angle 3$ and $\angle 8$
Corresponding angles	$\angle 3$ and $\angle 4$, $\angle 5$ and $\angle 7$, $\angle 1$ and $\angle 2$, $\angle 6$ and $\angle 8$

Suppose corresponding angles such as $\angle 1$ and $\angle 2$ in Figure 11-22(b) are respectively congruent. With this assumption, and because $\angle 1$ and $\angle 5$ are congruent vertical angles, we know that the pair of alternate interior angles, $\angle 2$ and $\angle 5$, are also congruent. Similarly, each pair of corresponding angles, alternate interior angles, and alternate exterior angles are congruent.

If we examine Figure 11-22(b) further, we see that lines m and n appear to be parallel when $\angle 1$ is congruent to $\angle 2$. Conversely, if the lines are parallel, the pairs of angles mentioned previously are congruent. This is true and is summarized in the following statement.

Angles and Parallel Lines Property

If any two distinct coplanar lines are cut by a transversal, then a pair of corresponding angles, alternate interior angles, or alternate exterior angles are congruent if, and only if, the lines are parallel.

REMARK   The if part of the above property is only true in Euclidean geometry.

Constructing Parallel Lines

A method commonly used by architects to construct a line ℓ through a given point P parallel to a given line m is shown in Figure 11-23. Place the side $\overline{AB}$ of triangle ABC on line m, as shown in Figure 11-23(a). Next, place a ruler on side $\overline{AC}$. Keeping the ruler stationary, slide triangle ABC along the ruler's edge until its side $\overline{AB}$ (marked $\overline{A'B'}$) contains point P, as in Figure 11-23(b). Use the side $\overline{A'B'}$ to draw the line ℓ through P parallel to m.

To show that the construction produces parallel lines, notice that when triangle ABC slides, the measures of its angles are unchanged. The angles of triangle ABC and triangle $A'B'C'$ in Figure 11-23(b) are correspondingly congruent angles. $\angle A$ and $\angle A'$ are corresponding angles formed by m and ℓ and the transversal $\overline{EF}$. Because corresponding angles are congruent, Theorem 11-3 implies that $\ell \parallel m$. In Chapter 12, we show how to construct parallel lines using only a compass and straightedge.

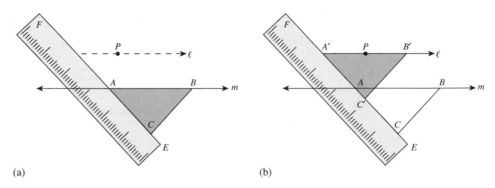

(a)   (b)

Figure 11-23

The Sum of the Measures of the Angles of a Triangle

The sum of the measures of the angles in a triangle can be observed to be 180°. We see this by using a torn triangle, as shown in Figure 11-24. Angles 1, 2, and 3 of triangle *ABC* in Figure 11-24(a) are torn as pictured and then replaced as shown in Figure 11-24(b). The three angles seem to lie along a single line ℓ.

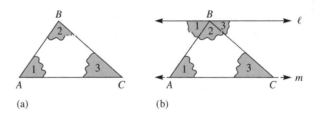

(a)   (b)

Figure 11-24

Research Note

Studies that have attempted to involve students in mathematical discovery and discourse—conjecture, careful reasoning, and building of validating arguments that can be scrutinized by others—have shown improvements in learning (Clements & Battista, 1992, p. 442).   ◆

Repeating the procedure for several different triangles, we find that the angle measures seem to sum to 180° or a straight angle. This conclusion, which is based on observation, is an example of *inductive reasoning*—discussed in Chapter 1—and is only a conjecture. In contrast, *deductive reasoning*, or proof, shows that a statement is true using the given information, previously defined and undefined terms, theorems or statements assumed to be true, and logic. A conclusion based on *deductive reasoning* must be true if the hypothesis is true. Notice, however, the importance of conjecturing as pointed out in the Research Note. We now state the following theorem and its proof.

Theorem 11–2   The sum of the measures of the interior angles of a triangle is 180°.

Proof: In Figure 11-25(a), $\triangle ABC$ has interior angles 1, 2, and 3; we need to prove that $m(\angle 1) + m(\angle 2) + m(\angle 3) = 180°$. Motivated by the experiment in Figure 11-24, in Figure 11-25(b) we place $\angle 4$ so that $\angle 4 \cong \angle 1$. We next extend $\overrightarrow{BD}$ to form line ℓ. In this way, $\angle 5$ is formed. If we could show that $\angle 5 \cong \angle 3$, the proof would be completed. Because we constructed $\angle 4$ so that $\angle 4 \cong \angle 1$, we have a pair of congruent alternate angles created by lines ℓ and m and transversal $\overline{AB}$. Thus, $\ell \| m$ and therefore $\angle 5 \cong \angle 3$, as these are alternate interior angles created by the parallel lines ℓ and m and the transversal $\overleftrightarrow{BC}$. Consequently,

$$m(\angle 1) + m(\angle 2) + m(\angle 3) = m(\angle 4) + m(\angle 2) + m(\angle 5) = 180°$$

so

$$m(\angle 1) + m(\angle 2) + m(\angle 3) = 180°$$

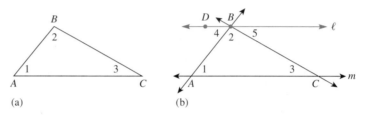

(a)   (b)

Figure 11-25

Example 11-2

In the framework for a tire jack, shown in Figure 11-26(a), $ABCD$ is a parallelogram. If $\angle ADC$ of the parallelogram measures 50°, what are the measures of the other angles of the parallelogram?

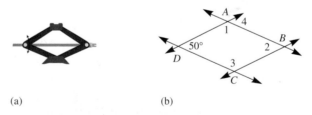

(a)   (b)

Figure 11-26

Solution   Refer to Figure 11-26(b). We draw the lines containing the sides of parallelogram $ABCD$. $\angle ADC$ has a measure of 50°, and $\angle 4$ and $\angle ADC$ are corresponding angles formed by parallel lines $\overleftrightarrow{AB}$ and $\overleftrightarrow{CD}$ cut by transversal $\overleftrightarrow{AD}$. So it follows from the *Angles and Parallel Lines Property* that $m(\angle 4) = 50°$. Because $\angle 1$ and $\angle 4$ are supplementary, $m(\angle 1) = 180° - 50° = 130°$. Using similar reasoning, we find that $m(\angle 2) = 50°$ and $m(\angle 3) = 130°$.

Example 11-3

In Figure 11-27, $m \| n$ and k is a transversal. Explain why $m(\angle 1) + m(\angle 2) = 180°$.

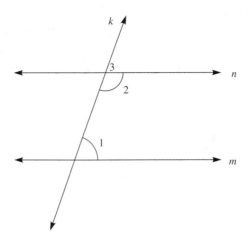

Figure 11-27

Solution   Because $\angle 1$ and $\angle 3$ are corresponding angles and $m \| n$ by the property on page 712, $m(\angle 1) = m(\angle 3)$. Also because $\angle 2$ and $\angle 3$ are supplementary angles, $m(\angle 2) + m(\angle 3) = 180°$. Substituting $m(\angle 1)$ for $m(\angle 3)$, we have $m(\angle 2) + m(\angle 1) = 180°$.

NOW TRY THIS 11-10   When a plane flies directly from New York to London and then to Nairobi and back to New York, it flies along the sides of a spherical triangle, as shown in Figure 11-28(a), approximately following the surface of Earth. On a sphere, the shortest path between two points is along an arc of a great circle. A great circle is obtained when a plane through the center of the sphere intersects the sphere. (An example of a great circle is the equator.) To obtain a great circle through two given points on the sphere, we need consider only a plane through the two points and the center of the sphere. A great circle through points A and B is shown in Figure 11-28(b). A *spherical triangle* consists of three points on the sphere and arcs of great circles (the shortest path) connecting the points.

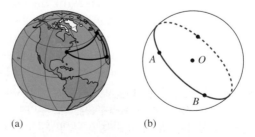

(a)   (b)

Figure 11-28

a. How many great circles go through the North Pole and the South Pole on the globe?
b. Describe a spherical triangle that has one vertex at the North Pole and the other two vertices on the equator.
c. Is there a spherical triangle with three right angles? If so, find one.

The Sum of the Measures of the Interior Angles of a Convex Polygon with *n* Sides

To answer the question in the cartoon, we find the sum of the measures of all the interior angles in any convex *n*-gon by first considering several special cases. From any vertex of a polygon, diagonals can be drawn from the vertex to form adjacent, nonoverlapping triangular regions. In the quadrilateral in Figure 11-29(a), the diagonal from *B* partitions the quadrilateral into two triangles.

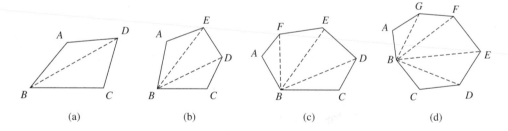

Figure 11-29

In the pentagon in Figure 11-29(b), the diagonals from *B* partition the pentagon into three triangles. In the hexagon in Figure 11-29(c), the diagonals from *B* partition the hexagon into four triangles. Notice that in each of the polygons in Figure 11-29, the number of triangles is two less than the number of sides. This will always be the case, because when we increase the number of sides by 1, we are adding one triangle and hence the number of triangles is still 2 less than the number of sides. This is shown in Figure 11-29(d), which has one more side than the hexagon in Figure 11-29(c).

The number of triangles in a convex *n*-gon created by all the diagonals from a single vertex is $n - 2$. Now to find the sum of the measures of all the interior angles in any convex polygon, we add the measures of all the interior angles in the triangles. Since the sum of the measures of the angles in any triangle is 180°, the sum of the measures of the interior angles in any convex *n*-gon is $(n - 2)180°$. We have proved the following theorem.

Theorem 11–3

The sum of the measures of the interior angles of any convex *n*-gon is $(n - 2)180°$.

> **REMARK** Other ways to derive the formula in Theorem 11–5 will be explored in problem 3 of the Mathematical Connections 11-3.

The Sum of the Measures of the Exterior Angles of a Convex n-gon

Figure 11-30 shows the interior and exterior angles of a pentagon (the exterior angles are marked in blue). The measures of the interior angles are α_1, α_2, α_3, α_4, α_5 and of the exterior angles β_1, β_2, β_3, β_4, β_5. Since the exterior and interior angle at a vertex are supplementary and the sum of the measure of the interior angles in a pentagon is $(5-2) \cdot 180°$, we have:

$$(\alpha_1 + \beta_1) + (\alpha_2 + \beta_2) + (\alpha_3 + \beta_3) + (\alpha_4 + \beta_4) + (\alpha_5 + \beta_5) = 5 \cdot 180°$$
$$(\alpha_1 + \alpha_2 + \alpha_3 + \alpha_4 + \alpha_5) + (\beta_1 + \beta_2 + \beta_3 + \beta_4 + \beta_5) = 5 \cdot 180°$$
$$(5-2) \cdot 180° + (\beta_1 + \beta_2 + \beta_3 + \beta_4 + \beta_5) = 5 \cdot 180°$$
$$\beta_1 + \beta_2 + \beta_3 + \beta_4 + \beta_5 = 5 \cdot 180° - 3 \cdot 180°$$
$$= 2 \cdot 180° = 360°$$

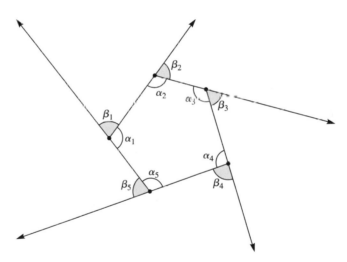

Figure 11-30

Thus the sum of the measures of the exterior angles in a pentagon is 360°. An analogous approach works for any convex n-gon. Let S be the sum of the measures of the interior angles and E the sum of the measures of the exterior angles. We know that $S + E = n \cdot 180°$. Because $S = (n-2) \cdot 180°$, we get

$$(n-2) \cdot 180° + E = n180°$$
$$E = n \cdot 180° - (n-2) \cdot 180°$$
$$= [n - (n-2)] \cdot 180°$$
$$= 2 \cdot 180° = 360°$$

Thus we have proved the following theorem.

Theorem 11–4

The sum of the measures of the exterior angles of a convex n-gon is 360°.

A more intuitive way to justify Theorem 11–4 is shown in Figure 11-31(a). Imagine walking clockwise around the convex pentagon starting at vertex A. At each vertex we need to turn by an exterior angle. At the end of the walk we are at A heading in the same direction we started; thus it seems that our total turn is through 360°, whether we walk around a convex pentagon or any other convex polygon. This may be even more clear if we extend the sides of the exterior angles and look at the figure from afar. Because the sides are infinite and the pentagon is small, the figure will look like the one in Figure 11-31(b).

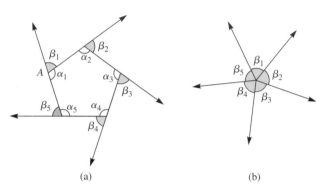

(a)   (b)

Figure 11-31

NOW TRY THIS 11-11

a. Using the fact that the sum of the measures of the exterior angles in a convex n-gon is 360°, derive the formula for the sum of the interior angles.

b. Express the measure of a single interior angle of a regular n-gon in terms of n.

Example 11-4

a. Find the measure of each interior angle of a regular decagon.

b. Find the number of sides of a regular polygon each of whose interior angles has a measure of 175°.

Solution
a. Because a decagon has 10 sides, the sum of the measures of the angles of a decagon is $10 \cdot 180 - 360$, or 1440°. A regular decagon has 10 angles, all of which are congruent, so each one has a measure of $\dfrac{1440°}{10}$, or 144°. As an alternative solution using Now Try This 11-10, each exterior angle is $\dfrac{360°}{10}$, or 36°. Hence, each interior angle is $180 - 36$, or 144°.

b. Each interior angle of the regular polygon is 175°. Thus, the measure of each exterior angle of the polygon is $180° - 175°$, or 5°. Because the sum of the measures of all exterior angles of a convex polygon is 360°, the number of exterior angles is $\dfrac{360}{5}$, or 72. Hence, the number of sides is 72.

Example 11-5

In Figure 11-32, lines k and l are parallel. The angles at A and B are as shown. Find x, the measure of $\angle BCA$.

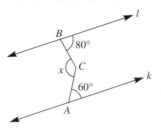

Figure 11-32

Solution We use the strategy of *examining a related problem*. If we had a transversal intersecting the parallel lines, we could use the fact that corresponding or alternate interior angles are congruent. For that purpose we need to extend either $\overline{BC}$ or $\overline{AC}$. In Figure 11-32(a), we extend $\overline{BC}$ and obtain the transversal $\overleftrightarrow{BC}$ that intersects line k at D. Because the marked angles with vertices at B and D are alternate interior angles created by the parallel lines and the transversal, they are congruent, and hence the marked angle at D measures $80°$. We can now find the measure of the third angle in $\triangle ACD$; it is $180° - (60° + 80°)$, or $40°$. Thus, $x = 180° - 40°$, or $140°$.

Alternative approaches are suggested in Figure 11-33(b), where line m is constructed parallel to ℓ, and Figure 11-33(c), where a line perpendicular to k is drawn through point C (because $\ell \parallel k$, that line will be perpendicular to ℓ as well). Solve the problem using each of these figures.

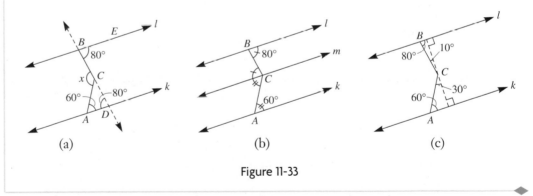

(a)   (b)   (c)

Figure 11-33

Walks Around Stars

We have seen how walking around a polygon can justify the fact that the sum of the measures of the exterior angles of any convex polygon is $360°$. We can use the same idea to find the sum of the measures of the exterior angles in the regular five-pointed star in Figure 11-34 with congruent angles at A, B, C, D, and E. The star in Figure 11-34 can be obtained from a regular convex pentagon by finding its vertices as intersections of the lines containing the non-adjacent sides of the pentagon. We want to find the measure of each interior angle of the star.

We begin a trip on the star at any vertex A, walk along the line to B, and then turn by the exterior angle whose measure is x, walk to C, turn by x again, walk to D, turn by x, walk to E, turn by x, walk back to A, and finally turn again by x. In this way we not only started at A and ended there but we also face in the same direction as when

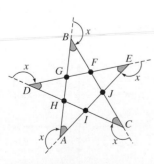

Figure 11-34

we started. Our total turning is $5x$, but because we face in the same direction as we started, we must have turned a multiple of $360°$. Thus, $5x = k \cdot 360°$ or $x = k \cdot 72°$, for some positive integer k. If $k = 1$, then $x = 72°$, which is not reasonable because it is clear from the figure that $90° < x < 180°$. If $k = 2$, then $x = 144°$, and each shaded interior angle of the five-pointed star is $180° - 144°$, or $36°$.

We can also find the measure of each interior angle in a different way. In $\triangle HGD$, $\angle DGH$ and $\angle DHG$ are exterior angles of the pentagon $FGHIJ$ and hence by Theorem 11–4 each measures $\dfrac{360°}{5}$ or $72°$. Thus the measure of $\angle GDH$ is $180° - 2 \cdot 72°$ or $360°$.

BRAIN TEASER   Sylvia, a graphic designer, needs to know the measurement of the shaded angles at the vertices of the 12-pointed star in Figure 11-35, in which all the sides and all the marked angles are congruent. How can she find the exact measurement of the angles without using a protractor?

Figure 11-35

Assessment 11-3A

1. If three lines all meet in a single point, how many pairs of vertical angles are formed?

2. Find the measure of the third angle in each of the following triangles:

a.

b.

c.

d.

3. For each of the following figures, determine whether m and n are parallel lines. Justify your answers.

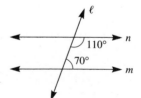
a.

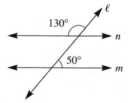
b.

c.

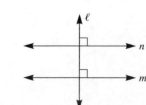
d.

4. Two angles are complementary and the ratio of their measures is 7 : 2. What are the angle measures?

5. **a.** In a regular polygon, the measure of each interior angle is 162°. How many sides does the polygon have?
   b. Find the measure of each of the interior angles of a regular dodecagon (ten-sided polygon).

6. In the following figure, $\overleftrightarrow{DE} \parallel \overleftrightarrow{BC}$, $\overleftrightarrow{EF} \parallel \overleftrightarrow{AB}$, and $\overleftrightarrow{DF} \parallel \overleftrightarrow{AC}$. Also, $m(\angle 1) = 45°$ and $m(\angle 2) = 65°$. Find each of the following values:
   a. $m(\angle 3)$ **b.** $m(\angle D)$
   c. $m(\angle E)$ **d.** $m(\angle F)$

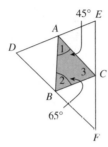

7. Lines a, b, and c are parallel. Find the measures of the numbered angles.

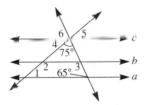

8. For each of the following, determine x if the lines a and b are parallel.

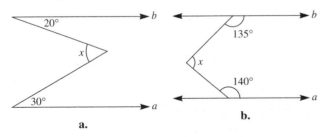

   a.   **b.**

9. In the following figures, find the measures of the angles marked x.

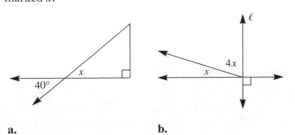

   a.   **b.**

10. In each of the following, a relationship between the marked angles is given. In each case, prove that $k \parallel l$.

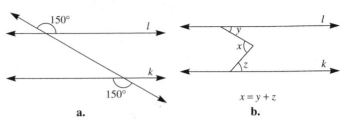

   a.   **b.**
                                     $x = y + z$

11. **a.** Determine the measure of an angle whose measure is twice that of its complement.
    b. If two angles of a triangle are complementary, what is the measure of the third angle?

12. Find the sum of the measures of the marked angles in the following figure:

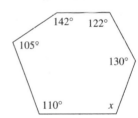

13. Find the measure of angle x in the following figure.

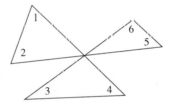

14. Find the measures of angles 1, 2, and 3 given that $TRAP$ is a trapezoid with $\overline{TR} \parallel \overline{PA}$.

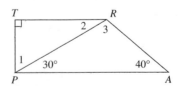

15. Home plate on a baseball field has three right angles and two other congruent angles. Refer to the following figure and find the measures of each of these two other congruent angles:

16. Refer to the following figure and answer the following:

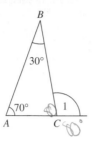

a. Find $m(\angle 1)$.

b. $\angle 1$ is an exterior angle of $\triangle ABC$. Use your answer in (a) to make a conjecture concerning the measure of an exterior angle of a triangle. Justify your conjecture.

17. If one interior angle of a parallelogram measures 120°, what are the measures of the other three interior angles?

18. In the figure, x, y, z, and w are measures of the angles as shown. If $y = 2x = \frac{1}{2}z = \frac{1}{3}w$, find x, y, z, and w.

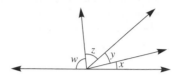

19. In each of the following figures, find the measures of the unknown marked angles.

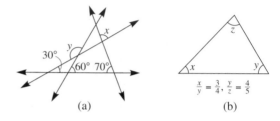

$$\frac{x}{y} = \frac{3}{4}, \frac{y}{z} = \frac{4}{5}$$

(a)   (b)

Assessment 11-3B

1. For each of the following, determine x if the lines a and b are parallel.

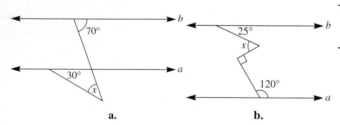

a.   b.

2. In the following figures, find the measures of the angles marked x and y:

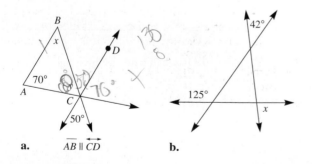

a. $\overleftrightarrow{AB} \parallel \overleftrightarrow{CD}$ b.

3. In each of the following, a relationship between the marked angles is given. In each case, prove that $k \parallel l$.

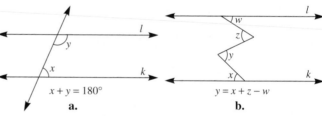

$x + y = 180°$ $y = x + z - w$
a.   b.

4. a. Determine the measure of an angle whose measure is $\frac{2}{3}$ of its complement.

b. Can two angles of a triangle be supplementary? Why or why not?

5. Find the sum of the measures of the marked angles in each of the following figures:

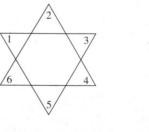

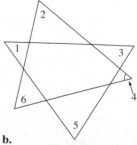

a.   b.

6. Calculate the measure of each angle of a pentagon, where the measures of the angles form an arithmetic sequence and the least measure is 60°.

7. Two sides of a regular octagon are extended as shown in the following figure. Find the measure of $\angle 1$.

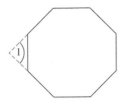

8. Refer to the following figure and answer (a) and (b):

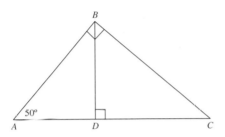

a. If $m(\angle ABC) = 90°$ and $\overline{BD} \perp \overline{AC}$ and $m(\angle A) = 50°$, find the measure of all the angles of $\triangle ABC$, $\triangle ADB$, and $\triangle CDB$.

b. If $m(\angle A) = \alpha$ in (a), find the measures of all the angles of $\triangle ABC$, $\triangle ADB$, and $\triangle CDB$ in terms of α.

9. In a quadrilateral the smallest angle has measure $49°30'$. If the measures of the four angles form an arithmetic sequence, find the measures of the other angles.

10. In each of the following figures, find the measures of the marked angles.

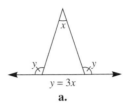

$y = 3x$

a.

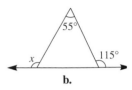

b.

The sequence x, y, z, u, v, w is an arithmetic sequence and $v = 136°$

c.

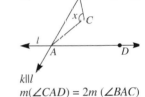

$k\|l$
$m(\angle CAD) = 2m(\angle BAC)$
$m(\angle CBF) = 2m(\angle ABC)$

d.

Mathematical Connections 11-3

Communication

1. a. If one angle of a triangle is obtuse, can another also be obtuse? Why or why not?
   b. If one angle in a triangle is acute, can the other two angles also be acute? Why or why not?
   c. Can a triangle have two right angles? Why or why not?
   d. If a triangle has one acute angle, is the triangle necessarily acute? Why or why not?

2. In the following figure, A is a point not on line ℓ. Is it possible to have two distinct perpendicular segments from A to ℓ in a plane? Why or why not?

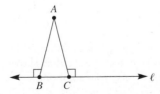

3. a. Explain how to find the sum of the measures of the interior angles of any convex pentagon by choosing any point P in the interior and constructing triangles, as shown in the following figure:

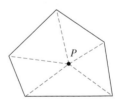

b. Using the method suggested by the diagram in (a), explain how to find the sum of the measures of the angles of any convex n-gon. Is your answer in (b) the same as the one already obtained in this section, $(n - 2) \cdot 180°$?

4. In the following figure, the legs of the ladder are congruent. If the ladder makes an angle of $120°$ with the ground, what is x? Explain your reasoning.

5. Explain how, through paper folding, you would show each of the following:

   a. In an isosceles triangle, the angles opposite congruent sides are congruent.

   b. If two angles of a triangle are congruent, the triangle is isosceles.

   c. In an equilateral triangle, all the interior angles are congruent.

   d. An isosceles trapezoid has two pairs of congruent angles.

6. a. Explain how to find the sum of the measures of the marked interior angles of a concave quadrilateral like the following:

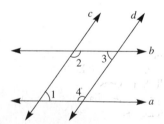

   b. Conjecture whether the formula for the sum of the measures of the angles of a convex polygon is true for concave polygons.

   c. Justify your conjecture in (b) for pentagons and hexagons and explain why your conjecture is true in general.

7. Regular hexagons have been used to tile floors. Can a floor be tiled using only regular pentagons? Why or why not?

8. Lines *a* and *b* are cut by a transversal *c*, creating angles 1, 2 and 4. If $m(\angle 1) = m(\angle 3)$ and $m(\angle 2) = m(\angle 4)$, can you conclude that *a* and *b* are parallel? Justify your answer.

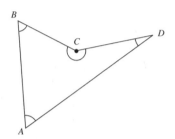

9. Express β in terms of *x*, *y*, *z*, and *w*. Justify each step in your reasoning.

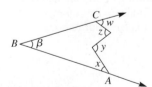

10. A beam of light from *A* hits the surface of a mirror at point *B*, is reflected, and then hits a perpendicular mirror and is reflected again. If $\angle 1 \cong \angle 2$ and $\angle 3 \cong \angle 4$, prove that the reflected beam is parallel to the incoming beam, that is, that the rays $\overrightarrow{AB}$ and $\overrightarrow{CD}$ are parallel.

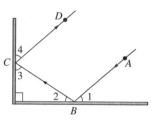

11. *ABCD* is a quadrilateral in which opposite angles have the same measure as indicated in the following figure. What kind of quadrilateral is *ABCD*? Justify your answer.

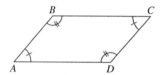

12. *ABCD* is a parallelogram. The angle bisectors of each of the interior angles of the parallelogram intersect to form the shaded figure shown. Answer the questions that follow.

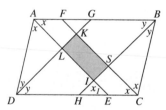

   a. Draw several different parallelograms and conjecture the kind of figure the shaded quadrilateral is.

 ★ b. Justify your conjecture.

13. The sides of △*DEF* are parallel to the sides of △*BCA*. If the measures of two angles of △*ABC* are 60° and 70° as shown, find the measures of angles of △*DEF*. Justify your answer.

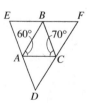

14. The measure of $\angle A$ in △*ABC* is α. The exterior angles at *B* and *C* have been bisected as shown. Find $m(\angle D)$ in terms of α. Justify your solution.

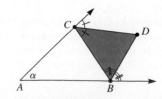

15. In the accompanying figure, $a \| b$. The angles at P and Q have been trisected. The trisection rays form a quadrilateral $ABCD$.

a. Find the measures of $\angle CBA$ and $\angle CDA$.

★ **b.** Show that the measures of A and C of the quadrilateral $ABCD$ cannot be uniquely determined with the given information.

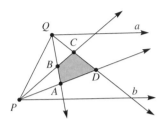

Open-Ended

16. Draw three different nonconvex polygons. When you walk around a polygon, at each vertex you need to turn either right (clockwise) or left (counterclockwise). A turn to the left is measured by a positive number of degrees and a turn to the right by a negative number of degrees. Find the sum of the measures of the turn angles of the polygons you drew. Assume you start at a vertex facing in the direction of a side, walk around the polygon, and end up at the same vertex facing in the same direction as when you started.

17. In *Now Try This 11-9*, you may have found that on a sphere a triangle with three right angles is possible. List other geometric properties of figures on a sphere that are different from corresponding properties in the plane.

Cooperative Learning

18. In $\triangle ABC$, $\overrightarrow{AD}$ and $\overrightarrow{BD}$ are *angle bisectors*; that is, they divide the angles at A and B respectively into congruent angles.

a. If the measures of $\angle A$ and $\angle B$ are known, then $m(\angle D)$ can be found. Explain how.

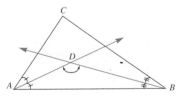

b. Suppose the measures of $\angle A$ and $\angle B$ are not known but that of $\angle C$ is. Can $m(\angle D)$ be found? To answer this question, assign each member of your group a triangle with different angles but with the same measure for $\angle C$. Each person should compute $m(\angle D)$ for his or her triangle. Use the results to make a conjecture related to the previous question.

c. Discuss a strategy for answering the question in (b) and write a solution to be distributed to the entire class.

19. Each person in your group is to draw a large triangle like $\triangle ABC$ in the following figure and cut it out. Obtain the crease $\overline{BB'}$ by folding the triangle at B so that A falls on some point A' on $\overline{AC}$. Next, unfold and fold the top B along $\overline{BB'}$ so that B falls on B'. Then fold vertices A and C along $\overline{AC}$ to match point B', as shown in the following figures:

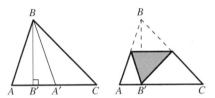

a. Why is $\overline{BB'}$ perpendicular to $\overline{AC}$?

b. What theorem does the folded figure illustrate? Why?

c. The folded figure seems to be a rectangle. Explain why.

d. What is the length of $\overline{GF}$ in terms of the length of the base $\overline{AC}$ of $\triangle ABC$? Why?

Questions from the Classroom

20. A student asks why great circles (circles obtained as a result of intersecting the sphere with a plane through the center of the sphere) are considered lines on a sphere. The same student asks if it is possible to have parallel lines on a sphere. How do you respond?

21. Jan, a tile designer, wants to make tiles in the shape of a convex polygon with all interior angles acute. She is wondering if there are any such polygons besides triangles. How do you respond?

22. A student wonders if there exists a convex decagon with exactly four right angles. How do you respond?

Review Problems

23. In each of the following, conjecture the required properties. If this is not possible, explain why.

a. Two properties that hold true for all rectangles but not for all rhombuses

b. Two properties that hold true for all squares but not for all isosceles trapezoids

c. Two properties that hold true for all parallelograms but not for all squares

24. Sort the shapes shown according to the following attributes:
  a. Number of parallel sides
  b. Number of right angles
  c. Number of congruent sides

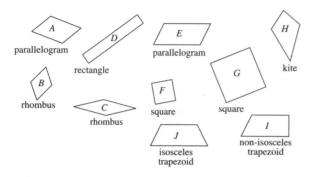

parallelogram

rectangle

parallelogram

kite

rhombus

rhombus

square

square

isosceles trapezoid

non-isosceles trapezoid

25. Use the figures in problem 24 to conjecture properties characteristic of different classes of figures. For example, "Congruent opposite sides describe a parallelogram."

Third International Mathematics and Science Study (TIMSS) Question

In the figure, PQ and RS are intersecting straight lines.

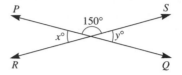

What is the value of $x + y$?
  a. 15   **b.** 30   **c.** 60   **d.** 180   **e.** 300

TIMSS, Grade 4, 2003

National Assessment of Educational Progress (NAEP) Question

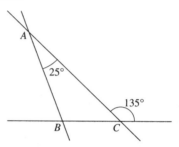

In the triangle, what is the degree measure of $\angle ABC$?
  a. 45   **b.** 100   **c.** 110   **d.** 135   **e.** 160

NAEP, Grade 8, 2003

11-4 Geometry in Three Dimensions

Simple Closed Surfaces

A visit to the grocery store exposes us to many three-dimensional objects that have simple closed surfaces. Figure 11-36 shows some examples.

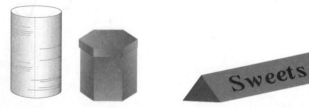

Figure 11-36

A **simple closed surface** has exactly one interior, has no holes, and is hollow. An example is a sphere, as shown in Figure 11-37(c). A **sphere** is defined as the set of all points at a given distance from a given point, the **center**. The set of all points on a simple closed surface with all interior points is a **solid**. Figures 11-37(a), (b), (c), and (d) are examples of simple closed surfaces; (e) and (f) are not. A **polyhedron** (*polyhedra* is the plural) is a simple closed surface made up of polygonal regions, or **faces**. The vertices of the polygonal regions are the **vertices** of the polyhedron, and the sides of each polygonal region are the **edges** of the polyhedron. Figures 11-37(a) and (b) are examples of polyhedra but (c), (d), (e), and (f) are not.

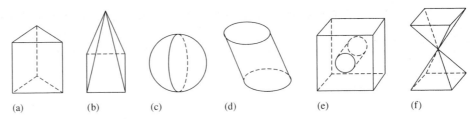

Figure 11-37

A **prism** is a polyhedron in which two congruent faces lie in parallel planes and the other faces are bounded by parallelograms. Figure 11-38 shows four different prisms. The shaded parallel faces of a prism are the **bases** of the prism. A prism usually is named after its bases, as the figure suggests.

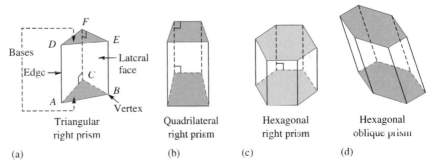

Figure 11-38

The faces other than the bases are the **lateral faces** of a prism. If the lateral faces of a prism are all bounded by rectangles, the prism is a **right prism**, as in Figure 11-38(a)–(c). Figure 11-38(d) is an **oblique prism** because some of its lateral faces are *not* bounded by rectangles.

Students often have trouble drawing three-dimensional figures. Figure 11-39 gives an example of how to draw a right pentagonal prism.

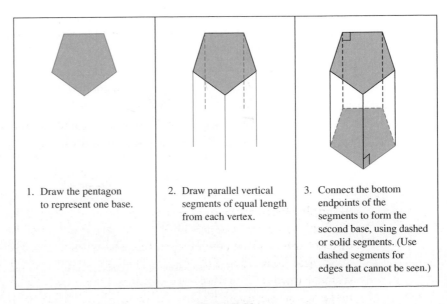

1. Draw the pentagon to represent one base.

2. Draw parallel vertical segments of equal length from each vertex.

3. Connect the bottom endpoints of the segments to form the second base, using dashed or solid segments. (Use dashed segments for edges that cannot be seen.)

Figure 11-39

A **pyramid** is a polyhedron determined by a polygon and a point not in the plane of the polygon. The pyramid consists of the triangular regions determined by the point and each pair of consecutive vertices of the polygon and the polygonal region determined by the polygon. The polygonal region is the **base** of the pyramid, and the point is the **apex**. As with a prism, the faces other than the base are **lateral faces**. Pyramids are classified according to their bases, which are shaded in Figure 11-40. A pyramid is a **right pyramid** if all its lateral faces are *congruent isosceles* triangles. (Congruent triangles will be defined in chapter 12. Intuitively congruent triangles have the same shape and size.)

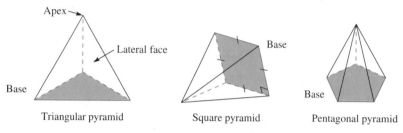

Triangular pyramid   Square pyramid   Pentagonal pyramid

Figure 11-40

To draw a pyramid, follow the steps in Figure 11-41.

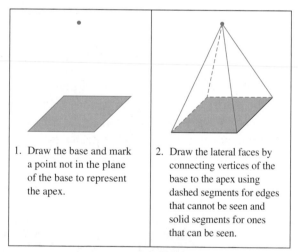

1. Draw the base and mark a point not in the plane of the base to represent the apex.

2. Draw the lateral faces by connecting vertices of the base to the apex using dashed segments for edges that cannot be seen and solid segments for ones that can be seen.

Figure 11-41

If one or more corners of a polyhedron is removed by an intersecting plane or planes, we say that the polyhedron is a **truncated polyhedron**. Figure 11-42 shows several truncated polyhedra.

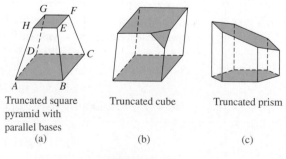

Truncated square pyramid with parallel bases   Truncated cube   Truncated prism
(a)   (b)   (c)

Figure 11-42

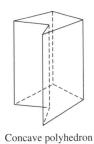

Concave polyhedron

Figure 11-43

Regular Polyhedra

A polyhedron is a **convex polyhedron** if, and only if, the segment connecting any two points in the interior of the polyhedron is itself in the interior. Figure 11-43 shows a concave polyhedron. A **regular polyhedron** is a convex polyhedron whose faces are congruent regular polygonal regions such that the number of edges that meet at each vertex is the same for all the vertices of the polyhedron.

Regular polyhedra, shown in Figure 11-44, have fascinated mathematicians for centuries. At least three of the polyhedra were identified by the Pythagoreans (ca. 500 BCE). Two other polyhedra were known to the followers of Plato (ca. 350 BCE). Three of the five polyhedra occur in nature in the form of crystals of sodium sulphantimoniate, sodium chloride (common salt), and chrome alum. The other two do not occur in crystalline form but have been observed as skeletons of microscopic sea animals called *radiolaria*.

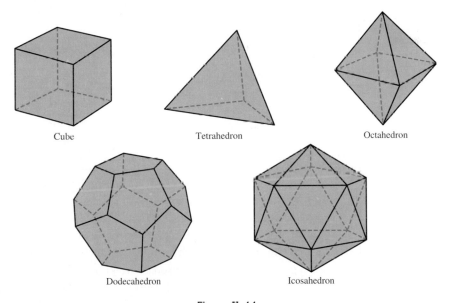

Cube   Tetrahedron   Octahedron

Dodecahedron   Icosahedron

Figure 11-44

Problem Solving   Regular Polyhedra?

How many regular polyhedra are there?

Understanding the Problem   Each face of a regular polyhedron is congruent to each of the other faces of that polyhedron, and each face is bounded by a regular polygon. We are to find the number of different regular polyhedra.

◆ *Historical Note*

The regular solid polyhedra are known as the **Platonic solids**, after the Greek philosopher Plato (ca. 350 BCE). Plato attached a mystical significance to the five regular polyhedra, associating them with what he believed were the four elements (earth, air, fire, and water) and the universe. Plato suggested that the smallest particles of earth have the form of a cube, those of air an octahedron, those of fire a tetrahedron, those of water an icosahedron, and those of the universe a dodecahedron (see Figure 11-44).   ◆

Devising a Plan   The sum of the measures of all the angles of the faces at a vertex of a regular polyhedron must be less than 360°. We next examine the measures of the interior angles of regular polygons to determine which of the polygons could be faces of a regular polyhedron. Then we try to determine how many types of polyhedra there are.

Carrying Out the Plan   We determine the size of an angle of some regular polygons, as shown in Table 11-9. Could a regular heptagon be a face of a regular polyhedron? At least three figures must fit together at a vertex to make a polyhedron. (Why?) If three angles of a regular heptagon were together at one vertex, then the sum of the measures of these angles would be $\dfrac{3 \cdot 900°}{7}$, or $\dfrac{2700°}{7}$, which is greater than 360°. Similarly, more than three angles cannot be used at a vertex. Thus, a heptagon cannot be used to make a regular polyhedron.

Table 11-9

Polygon	Measure of an Interior Angle
Triangle	60°
Square	90°
Pentagon	108°
Hexagon	120°
Heptagon	$\left(\dfrac{900}{7}\right)°$

The measure of an interior angle of a regular polygon increases as the number of sides of the polygon increases. (Why?) Thus any polygon with more than six sides has an interior angle greater than 120°. So if three angles were to fit together at a vertex, the sum of the measures of the angles would be greater than 360°. This means that the only polygons that might be used to make regular polyhedra are equilateral triangles, squares, regular pentagons, and regular hexagons. Consider the possibilities given in Table 11-10.

Table 11-10

Polygon	Measure of an Interior Angle	Number of Polygons at a Vertex	Sum of the Angles at the Vertex	Polyhedron Formed	Model
Triangle	60°	3	180°	**Tetrahedron**	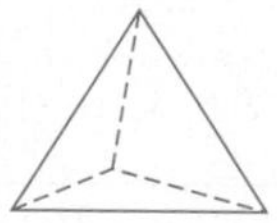
Triangle	60°	4	240°	**Octahedron**	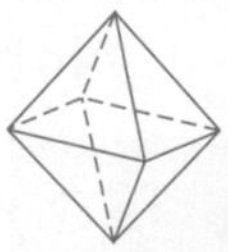

(continues)

Table 11-10 *continued*

Polygon	Measure of an Interior Angle	Number of Polygons at a Vertex	Sum of the Angles at the Vertex	Polyhedron Formed	Model
Triangle	60°	5	300°	**Icosahedron**	
Square	90°	3	270°	**Cube**	
Pentagon	108°	3	324°	**Dodecahedron**	

Notice that we were not able to use six equilateral triangles to make a polyhedron because $6(60°) = 360°$ and the triangles would lie in a plane. Similarly, we could not use four squares or any hexagons. We also could not use more than three pentagons because if we did, the sum of the measures of the angles would be more than 360°. However, equilateral triangles, squares, and regular hexagons are used to tile floors. More on tiling is covered in Chapter 14.

Looking Back Interested readers may want to investigate **semiregular polyhedra**. These are also formed by using regular polygons as faces, but the regular polygons used need not have the same number of sides. In addition at each vertex semiregular polyhedra have the same number of faces of each kind. Two semiregular polyhedra are shown in Figure 11-45(a) and (b).

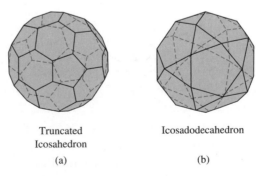

Truncated
Icosahedron

(a)

Icosadodecahedron

(b)

Figure 11-45

The patterns in Figure 11-46, called *nets*, can be used to construct the five regular polyhedra. It is left as an exercise to determine other patterns for constructing the regular polyhedra.

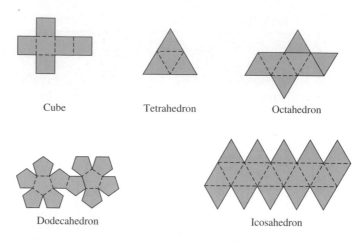

Cube   Tetrahedron   Octahedron

Dodecahedron   Icosahedron

Figure 11-46

Compare these nets, with the nets introduced on the student page.

NOW TRY THIS 11-12 A simple relationship among the number of faces, the number of edges, and the number of vertices of any polyhedron was discovered by the French mathematician and philosopher René Descartes (1596–1650) and rediscovered by the Swiss mathematician Leonhard Euler (1707–1783). Table 11-11 suggests a relationship among the numbers of vertices (V), edges (E), and faces (F). This relationship is known as **Euler's formula:** $V + F - E = 2$. Check that the relationship holds for the truncated polyhedron in Figure 11-45.

Table 11-11

Name	V	F	E
Tetrahedron	4	4	6
Cube	8	6	12
Octahedron	6	8	12
Dodecahedron	20	12	30
Icosahedron	12	20	30

◆ *Historical Note*

Leonhard Euler went blind in 1766 and for the remaining 17 years of his life continued to do mathematics by dictating to a secretary and by writing formulas in chalk on a slate for his secretary to copy down. He published 530 papers in his lifetime and left enough work to supply the *Proceedings of the St. Petersburg Academy* for the next 47 years.   ◆

School Book Page VIEWS OF SOLID FIGURES

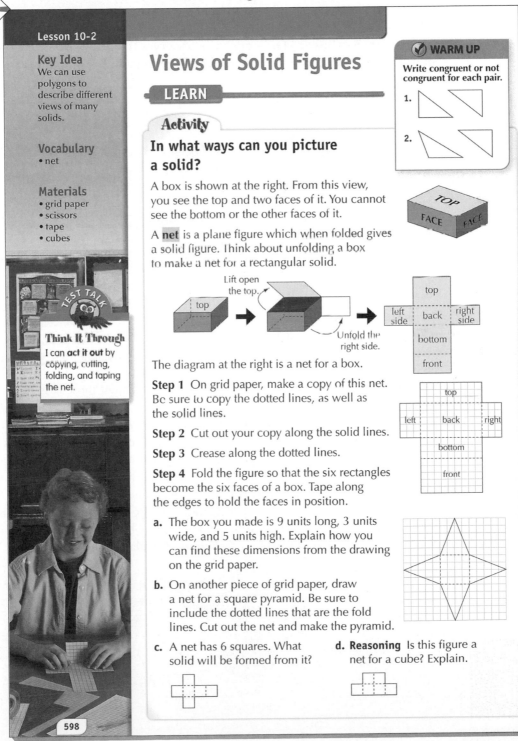

Lesson 10-2

Key Idea
We can use polygons to describe different views of many solids.

Vocabulary
• net

Materials
• grid paper
• scissors
• tape
• cubes

TEST TALK

Think It Through
I can **act it out** by copying, cutting, folding, and taping the net.

598

Views of Solid Figures

LEARN

Activity

In what ways can you picture a solid?

A box is shown at the right. From this view, you see the top and two faces of it. You cannot see the bottom or the other faces of it.

A **net** is a plane figure which when folded gives a solid figure. Think about unfolding a box to make a net for a rectangular solid.

The diagram at the right is a net for a box.

Step 1 On grid paper, make a copy of this net. Be sure to copy the dotted lines, as well as the solid lines.

Step 2 Cut out your copy along the solid lines.

Step 3 Crease along the dotted lines.

Step 4 Fold the figure so that the six rectangles become the six faces of a box. Tape along the edges to hold the faces in position.

a. The box you made is 9 units long, 3 units wide, and 5 units high. Explain how you can find these dimensions from the drawing on the grid paper.

b. On another piece of grid paper, draw a net for a square pyramid. Be sure to include the dotted lines that are the fold lines. Cut out the net and make the pyramid.

c. A net has 6 squares. What solid will be formed from it?

d. Reasoning Is this figure a net for a cube? Explain.

WARM UP
Write congruent or not congruent for each pair.

1.

2.

Source: Scott Foresman-Addison Wesley, Mathematics, Grade 5, 2008 (p. 598).

Cylinders and Cones

A cylinder is an example of a simple closed surface that is not a polyhedron. Consider line segment $\overline{AB}$ and a line ℓ as shown in Figure 11-47. When $\overline{AB}$ moves so that it always remains parallel to line ℓ and points A and B trace simple closed planar curves other than polygons, the surface generated by $\overline{AB}$, along with the simple closed curves and their interiors, form a **cylinder**. The simple closed curves traced by A and B, along with their interiors, are the **bases** of the cylinder, and the remaining points constitute the *lateral surface of the cylinder*.

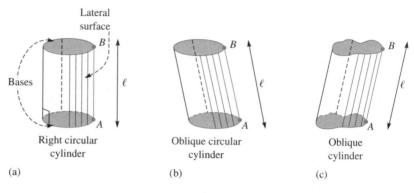

Figure 11-47

If a base of a cylinder is a circular region, as in Figure 11-47(a) and (b), the cylinder is a **circular cylinder**. If the line segment forming a circular cylinder is perpendicular to a base, the cylinder is a **right cylinder**. Circular cylinders that are not right cylinders are **oblique cylinders**. The cylinder in Figure 11-47(a) is a right cylinder; those in Figures 11-47(b) and (c) are oblique cylinders.

Suppose we have a simple closed curve, other than a polygon, in a plane and a point P not in the plane of the curve. The union of line segments connecting point P to each point of a simple closed curve, the simple closed curve, and the interior of the curve is a **cone**. Cones are pictured in Figure 11-48. Point P is the **vertex** of the cone. The points of the cone not in the base constitute the *lateral surface of the cone*. A line segment from vertex P perpendicular to the plane of the base is the **altitude**. A **right circular cone**, such as the one in Figure 11-48(a), is a cone whose altitude intersects the base (a circular region) at the center of the circle. Figure 11-48 illustrates an **oblique circular cone**.

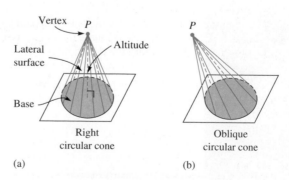

Figure 11-48

Assessment 11-4A

1. Identify each of the following polyhedra. If a polyhedron can be described in more than one way, give as many names as possible.

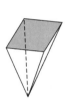 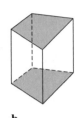 

  a.   **b.**   **c.**

2. Given the tetrahedron shown, name the following:
   a. Vertices
   b. Edges
   c. The interior faces
   d. Intersection of face DRW and edge $\overline{RA}$
   e. Intersection of face DRW and face DAW

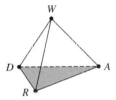

3. Identify four shapes of containers that can be found in a grocery store.
4. Determine the minimum number of faces possible for each of the following:
   a. Prism   **b.** Pyramid   **c.** Polyhedron
5. Classify each of the following as true or false:
   a. If the lateral faces of a prism are rectangles, it is a right prism.
   b. Every pyramid is a prism.
   c. Some pyramids are polyhedra.
   d. The bases of a prism lie in perpendicular planes.
   e. The bases of all cones are circles.
   f. A cylinder has only one base.
   g. All lateral faces of an oblique prism are rectangular regions.
   h. All regular polyhedra are convex.
6. If possible, sketch each of the following:
   a. An oblique square prism
   b. An oblique square pyramid

7. For each of the following, draw a prism and a pyramid that have the given region as a base:
   a. Triangle
   b. Pentagon
   c. Regular hexagon
8. The following are pictures of stacks of solid cubes. In each case, determine the number of cubes in the stack and the number of faces that are glued together.

    a.   **b.**

9. Name each polyhedron that can be constructed using the following nets made of figures as described.

   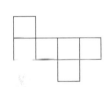

a. regular hexagon **b.** square and   **c.** squares
and isosceles triangles   equilateral
                 triangle

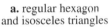 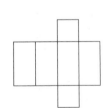

d. rectangles   **e.** rectangles and regular hexagons

10. The following nets along with shaded tabs can be used to construct polyhedra. Copy the nets (or construct similar ones), fold them, and glue them together using the tabs. What kind of polyhedra do you obtain?

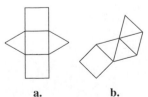

      a.   **b.**

11. Draw nets like the ones in problem 10 and as few tabs as possible for each net so that when folded the nets can be glued with the tabs to create polyhedra.

12. The figure on the left in each of the following represents a card attached to a wire, as shown. Match each figure on the left with what it would look like if you were to revolve it by spinning the wire between your fingers.

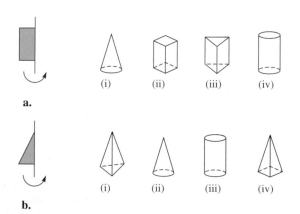

a.

b.

13. Which of the following three-dimensional figures could be used to make a shadow like in (a)? In (b)?

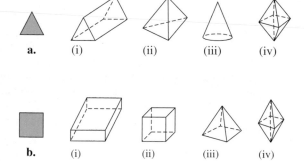

a.   (i)   (ii)   (iii)   (iv)

b.   (i)   (ii)   (iii)   (iv)

14. Consider a jar with a lid, as illustrated in the following figure. The jar is half filled with water. In drawings (a) and (b), sketch the water.

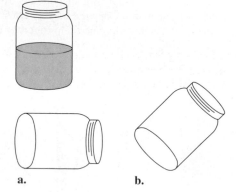

a.   b.

15. On the left of each of the following figures is a net for a three-dimensional object. On the right are several objects. Which object will the net fold to make?

a.

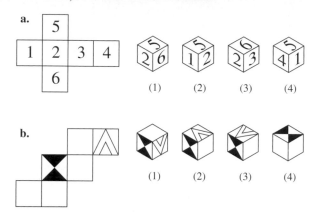

(1)   (2)   (3)   (4)

b.

(1)   (2)   (3)   (4)

16. Name the intersection of each of the following with the plane shown:

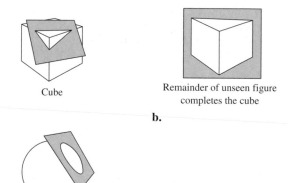

Cube

Remainder of unseen figure completes the cube

a.   b.

Sphere

c.

17. When a plane intersects a cube, different-sized rectangles can be obtained. Answer the following questions and explain your reasoning:
   a. Show the largest possible rectangle that can be obtained.
   b. Is a smallest rectangle? Explain

18. The following is a net that can be used to create a tetrahedron. M, N, P, Q, S are midpoints of sides of the triangles (a midpoint divides a segment into two congruent segments).
   a. Predict what kind of figure points M, N, P, Q, S and like segments MN, NP, PQ, and QS will form when the net is folded into a tetrahedron.

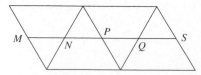

b. Draw a larger net similar to the one shown (make angles of 60°). Mark the corresponding midpoints (use a marked ruler) and fold it into a tetrahedron. Do the midpoints create a figure you predicted in part (a)?

19. The right hexagonal prism shown has regular hexagons as bases. Answer the following questions:

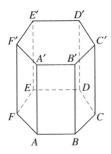

a. Name all the pairs of parallel lateral faces. (Faces are parallel if the planes containing the faces are parallel.)

b. What is the measure of the dihedral angle between two adjacent lateral faces? Why?

20. For each of the following figures, find $V + F - E$, where V, E, and F stand, respectively, for the number of vertices, edges, and faces:

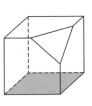

a.   **b.**

Assessment 11-4B

1. Two prisms are sketched on dot paper, as in the following figure. Complete the drawings by using dashed segments for the hidden edges.

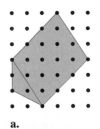
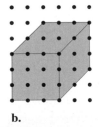

a.   **b.**

2. The following are pictures of solid cubes lying on a flat surface. In each case, determine the number of cubes in the stack and the number of faces that are glued together.

a.   **b.**

3. Copy the given net (enlarge if desired) and construct the polyhedron from the net. (The shaded flaps are not part of the net.) Name the polyhedron.

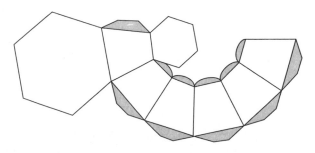

4. The following figures represent cards attached to a wire, as shown. For each card, sketch and name the three-dimensional figure resulting from revolving it about the wire.

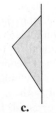
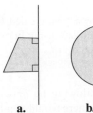
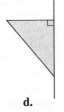

a.   **b.**   **c.**   **d.**

5. A diagonal of a prism is any segment determined by two vertices that do not lie in the same face, as shown in the following figure.

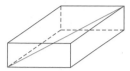

Complete the following table showing the total number of diagonals for various prisms:

Prism	Vertices per Base	Diagonals per Vertex	Total Number of Diagonals
Quadrilateral	4	1	4
Pentagonal	5		
Hexagonal			
Heptagonal			
Octagonal			
.			
.			
.			
n-gonal			

6. Sketch the intersection of each of the following with the plane shown:

Right pentagonal prism

a.

Right circular cone
(plane parallel to base)

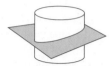

Right circular cylinder
(plane not parallel to base)

b.   **c.**

7. For each of the following three-dimensional figures, draw or name possible cross sections when the three-dimensional figure is sliced by a plane.
  a. Right pentagonal prism
  b. Right circular cylinder

8. Given a tetrahedron, explain how to obtain each of the following shapes by slicing the tetrahedron by a plane. Describe a procedure for getting each shape.
  a. An equilateral triangle
  b. A scalene triangle
  c. A rectangle

9. Draw a right pentagonal prism with bases, $ABCDE$ and $A'B'C'D'E'$ and answer the following questions:
  a. Name all the pairs of parallel lateral faces. (Faces are parallel if the planes containing the faces are parallel.)
  b. What is the measure of the dihedral angle between two adjacent lateral faces? Why?

10. Answer each of the following questions about a pyramid and a prism, each having an n-gon as a base:
  a. How many faces does each have?
  b. How many vertices does each have?
  c. How many edges does each have?
  d. Use your answers to (a), (b), and (c) to verify Euler's formula $V + F - E = 2$ for all pyramids and all prisms.

Mathematical Connections 11-4

Communication

1. How many possible pairs of bases does a rectangular prism have? Explain.

2. A circle can be approximated by a "many-sided" polygon. Use this notion to describe the relationship between each of the following:
  a. A pyramid and a cone
  b. A prism and a cylinder

3. Can either or both of the following be drawings of a quadrilateral pyramid? If yes, where would you be standing in each case? Explain why.

a.   **b.**

4. When a plane intersects a three-dimensional object, a cross section is created. Sketch a cube and show how each of the following cross sections can be obtained. Explain your reasoning:

a. An equilateral triangle
b. A scalene triangle
c. A rectangle
d. A square
e. A pentagon
f. A hexagon
g. A parallelogram that is not a rectangle
h. A rhombus that is not square

5. The following is a picture of a right rectangular prism. M and N are points on two edges such that $\overline{MN} \parallel \overline{AD}$. Answer the following questions and explain your reasoning:

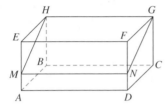

a. Is $HGNM$ a rectangle or a parallelogram that is not a rectangle?
b. If the vertex H is connected to each of the vertices A, B, C, and D, a rectangular pyramid is formed. Is it a right pyramid?
c. A pyramid is formed by connecting the intersection of the diagonals $\overline{HF}$ and $\overline{EG}$ with the vertices of the base $ABCD$. Is it a right pyramid?

Open-Ended

6. When a box in the shape of a right rectangular prism, like the one in the following figure, is cut by a plane halfway between the opposite sides and parallel to these sides, that plane is a *plane of symmetry*. If a mirror is placed at the plane of symmetry, the reflection of the front part of the box will look just like the back part. Draw several space figures and find the number of planes of symmetry for each. Summarize your results in a table. Can you identify any figures with infinitely many planes of symmetry?

Cooperative Learning

7. In a two-person game, draw a three-dimensional figure without showing it to your partner. Tell your partner the shape of all possible cross sections of your figure sufficient to identify the figure. If your partner can identify your figure, 1 point is earned by your partner. If a figure is identified that has all the cross sections listed but that figure is not your figure, 2 points are earned by your partner. Each of you should take an equal number of turns.

8. Some of the following nets can be folded into cubes. Have each person in your group draw all the nets that can be folded into cubes. Share your findings with the group and decide how many different such nets there

are. Discuss what "different" means in this case. Finally, compare your group's answers with those of other groups.

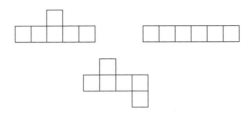

9. Choose different triples of points on the edges of each of the following figures. Then find the cross section of where a plane through each triple of points intersects the figure. (The following figures show only a few examples.) List possible figures that can be obtained in this manner.

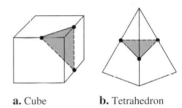

a. Cube   **b.** Tetrahedron

Questions from the Classroom

10. A student asks how to find the shortest path between two points A and B on two different faces which are neither the top or bottom of a right rectangular prism, without leaving the prism. How do you respond?

11. Jodi has a model of a tetrahedron (shown below) and would like to know how many different nets exist for the tetrahedron. How do you respond?

Review Problems

12. If two angles of one triangle are congruent to two angles of another triangle, must the third angles of both be congruent? Why or why not?

13. Triangles ABC and CDE are equilateral triangles with points A, C, and E collinear. Find the measure of $\angle BCD$.

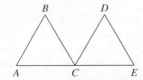

14. What is the measure of each angle in a regular nonagon?

15. Classify the following as true or false. If false, tell why.

    a. Every rhombus is a parallelogram.

    b. Every polygon has at least three sides.

    c. Triangles can have at most two acute angles.

16. Assume that lines ℓ, m, and n, as shown, are in the same plane. What can you conclude? Why?

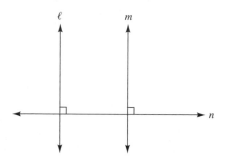

Which of the following could NOT be folded into a cube?

a.   **b.**   **c.**   **d.**

NAEP, Grade 8, 2003

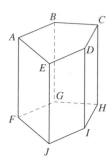

In the figure above, points A, E, and H are on a plane that intersects a right prism. What is the intersection of the plane with the right prism?

    a. A line

    b. A triangle

    c. A quadrilateral

    d. A pentagon

    e. A hexagon

NAEP, Grade 12, 2003

BRAIN TEASER   A rectangular region can be rolled to form the lateral surface of a right circular cylinder. What shape of paper is needed to make an oblique circular cylinder?

LABORATORY ACTIVITY   Consider a structure made of cubes with the front view, the side view looking from the right, and the top view as in Figure 11-49. Build the structure.

    Top   Front   Right

Figure 11-49

 **Hint for Solving the Preliminary Problem**

    1. Notice that there are four positions for the kite shown, since any of the four corners of the array can be chosen as the vertex, which in the figure is in the bottom left corner.

    2. Draw first all possible concave quadrilaterals with a vertex at the bottom left corner.

Chapter Outline

I. Basic geometric notions

  A. Points, lines, and planes

    1. Points, lines, and **planes** are basic, but undefined, terms.

    2. Collinear points are points that belong to the same line.

    3. Segments and **rays** are important subsets of lines.

    4. Coplanar points are points that lie in the same plane. **Coplanar lines** are defined similarly.

    5. Two lines with exactly one point in common are **intersecting lines**.

    6. Concurrent lines are lines that contain a common point.

    7. Two distinct coplanar lines with no points in common are **parallel**.

    8. Skew lines are lines that cannot be contained in the same plane.

    9. Parallel planes are planes with no points in common.

    10. Space is the set of all points.

    11. An **angle** is the union of two rays with a common endpoint.

    12. Angles are classified according to size as **acute, obtuse, right,** or **straight**.

    13. Two lines that meet to form a right angle are **perpendicular**.

    14. A **dihedral angle** is the union of two half-planes and the common line defining the half-planes.

    15. A **line** is **perpendicular** to a **plane** if it is perpendicular to every line in the plane through its intersection with the plane.

    16. Two intersecting **planes** are **perpendicular** if any of the four dihedral angles created by the planes measures 90°.

  B. Plane figures

    1. A **closed curve** is a curve that, when traced, has the same starting and stopping points and may cross itself at individual points.

    2. A **simple curve** is a curve that does not cross itself when traced, although the starting and stopping points may be the same.

    3. A **polygon** is a simple closed curve with segments as sides.

      a. A **diagonal** is any line segment connecting two nonconsecutive vertices of a polygon.

      b. A **convex polygon** is one such that if any two points of the polygonal region are connected by a segment, the segment is a subset of the polygonal region.

      c. A **concave polygon** is a nonconvex polygon.

      d. A **regular polygon** is a polygon in which all the interior angles are congruent and all the sides are congruent.

    4. A **polygonal region** is the union of a polygon and its interior.

    5. Triangles are classified according to the lengths of their sides as **scalene, isosceles,** or **equilateral** and according to the measures of their angles as **acute, obtuse,** or **right**.

    6. Quadrilaterals with special properties are **trapezoids, parallelograms, rectangles, kites, isosceles trapezoids, rhombuses,** and **squares**.

II. Theorems involving angles

  A. Supplements of the same angle, or of congruent angles, are congruent.

  B. Complements of the same angle, or of congruent angles, are congruent.

  C. Vertical angles formed by intersecting lines are congruent.

  D. If any two distinct coplanar lines are cut by a transversal, then a pair of **corresponding angles, alternate interior angles,** or **alternate exterior angles** are congruent if, and only if, the lines are parallel.

  E. The sum of the measures of the interior angles of a triangle is 180°.

  F. The sum of the measures of the interior angles of any convex polygon with n sides is $180°n - 360°$, or $180°(n - 2)$. One interior angle of a regular n-gon measures $\dfrac{180°n - 360°}{n}$, or $\dfrac{180°(n - 2)}{n}$.

  G. The sum of the measures of the exterior angles of any convex polygon is 360°.

III. Three-dimensional figures

  A. A **polyhedron** is a simple closed surface formed by polygonal regions.

  B. Three-dimensional figures with special properties are **prisms, pyramids, regular polyhedra, cylinders, cones,** and **spheres**.

  C. Euler's formula: $V + F - E = 2$, where V stands for the number of vertices, F for the number of faces, and E for the number of edges

Chapter Review

1. Refer to the following line *m*:

   a. Other than *m* list three names for the line.
   b. Name two rays on *m* that have endpoint *B*.
   c. Find a simpler name for the intersection of rays $\overrightarrow{AB}$ and $\overrightarrow{BA}$.
   d. Find a simpler name for the intersection of rays $\overrightarrow{BA}$ and $\overrightarrow{AC}$.

2. In the figure below, $\overrightarrow{PQ}$ is perpendicular to the plane α. Answer the following:
   a. Name a pair of skew lines.
   b. Using only the letters in the figure, name as many planes as possible that are perpendicular to α.
   c. What is the intersection of planes *APQ* and β?
   d. Is there a single plane containing *A*, *B*, *P*, and *Q*? Explain your answer.

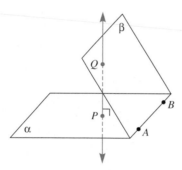

3. Draw each of the following curves:
   a. A simple closed curve
   b. A closed curve that is not simple
   c. A concave hexagon
   d. A convex decagon

4. a. Can a triangle have two obtuse angles? Justify your answer.
   b. Can a parallelogram have four acute angles? Justify your answer.

5. Find each of the following:
   a. $113°57' + 18°14'$
   b. $84°13' - 27°45'$
   c. $113°57' + 18.4°$
   d. $0.75°$ in minutes and seconds

6. In a certain triangle, the measure of one angle is twice the measure of the smallest angle. The measure of the third angle is 7 times as great as the measure of the smallest angle. Find the measures of each of the angles in the triangle.

7. a. Explain how to derive an expression for the sum of the measures of the interior angles in a convex *n*-gon.
   b. In a certain regular polygon, the measure of each interior angle is 176°. How many sides does the polygon have?

8. Sketch each of the following:
   a. Three planes that intersect in a point
   b. A plane and a cone that intersect in a circle
   c. A plane and a cylinder that intersect in a segment
   d. Two pyramids that intersect in a triangle

9. Sketch drawings to illustrate different possible intersections of a square pyramid and a plane.

10. In a periscope, a pair of mirrors are parallel. If the dotted line in the following figure represents a path of light and $m(\angle 1) = m(\angle 2) = 45°$, find $m(\angle 3)$ and $m(\angle 4)$:

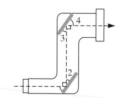

11. Find $6°48'59'' + 28°19'36''$. Write your answer in simplest terms.

12. In the figure, ℓ is parallel to *m*, and $m(\angle 1) = 60°$. Find each of the following:
    a. $m(\angle 3)$
    b. $m(\angle 6)$
    c. $m(\angle 8)$

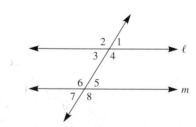

13. Draw a large triangle *ABC* and tear off $\angle B$. Fit the torn-off angle as $\angle BAD$, as shown in the following figure:

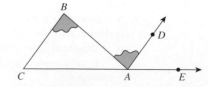

a. Why is $\overleftrightarrow{AD} \parallel \overleftrightarrow{CB}$?

b. Why is $\angle DAE \cong \angle C$?

c. Use parts (a) and (b) to show that the sum of the measures of the interior angles of a triangle is 180°.

14. Prove the transitive property for parallel lines, that is, if $a \parallel b$ and $b \parallel c$, then $a \parallel c$ (assume $c \neq a$).

15. Wafiq claims that he can easily prove that the sum of measures of the interior angles in any triangle is 180°. He says that any right triangle is half of a rectangle. Because a rectangle has four right angles, its angles add up to 360°, so the angles of a right triangle add up to half of 360°, or 180°. Next he claims that he can divide any triangle into two right triangles and use what he proved for right triangles to prove the theorem for any triangle. Is Wafiq's approach correct? Explain why or why not.

16. Prove that if $x = \alpha + \beta$, then $a \parallel b$.

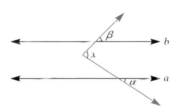

17. The polygon $ABCDE$ contains a point P in its interior that can be joined to all the vertices by segments entirely in the interior of the polygon. In this way, the polygon has been "divided" into five triangles. Use such a division into triangles to answer the following:

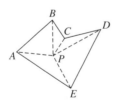

a. Find the sum of the measures of the interior angles of the polygon $ABCDE$.

b. If an n-gon can be subdivided into triangles in a similar way, explain how to derive the formula for the sum of the measures of the interior angles of the n-gon in terms of n.

c. Explain how to find the sum of the measures of the interior angles of the following polygon:

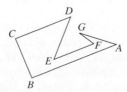

18. In $\triangle ABC$, line DE is parallel to line AB, and line DF is parallel to line AC. If $m(\angle C) = 70°$ and $m(\angle B) = 45°$, find the measures of the angles labeled 1, 2, 3, 4, and 5.

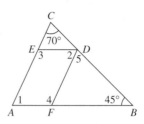

19. If $\angle ACB$ is a right angle as shown, and the other angles are as indicated in the figure, find x, the measure of $\angle DCB$.

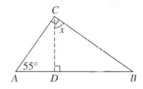

20. In the following figure, $\angle 1 \cong \angle 2$.

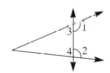

a. Prove that $\angle 3 \cong \angle 4$.

b. Suppose $m(\angle 1) < m(\angle 2)$. What can you conclude about $m(\angle 3)$ and $m(\angle 4)$? Justify your answer.

21. In the accompanying figure, planes α and β intersect in line AB. The dihedral angle $S\text{-}AB\text{-}Q$ measures 90°. The lines PS and PQ determine a new plane, γ.

a. Why is plane γ perpendicular to α as well as to β?

b. Why is line AB perpendicular to γ?

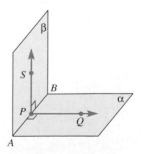

22. If a pyramid has an octagon for a base, how many lateral faces does it have?

23. a. Given 10 points in the plane, no 3 of which are collinear, how many triangles can be drawn whose vertices are the given points? Explain your reasoning.

   b. Generalize and answer part (a) for n points.

24. In each of the following, determine the number of sides of a regular polygon with the stated property. If such a regular polygon does not exist, explain why.

   a. Each exterior angle measures 20°.

   b. Each exterior angle measures 25°.

   c. The sum of all the exterior angles is 3600°.

   d. The total number of diagonals is 4860.

25. For each of the convex polygons of n sides, find the measure of the largest interior angle.

   a. $n = 10$, the smallest interior angle measures 100° and the angles increase in an arithmetic sequence.

   b. $n = 5$, the measures of the angles are labeled x, y, z, u, v, where $y = 2x$, z is $\frac{3}{5}$ of y, u is $\frac{4}{5}$ of z, and $v = z$.

26. Carefully draw nets that can be folded into each of the following. Cut the nets out and using glue or tape, fold them into the given polyhedron.

   a. Tetrahedron

   b. Square pyramid

   c. Right rectangular prism

   d. Right circular cylinder with a top and a bottom

27. a. Draw as many different nets as possible that can be folded into an open cube (a cube without a top).

   b. How many different nets are possible in part (a)? Explain what "different" means to you.

28. If ABC is a right triangle and $m(\angle A) = 42°$, what is the measure of the other acute angle?

29. In the following, draw in dotted lines for the unseen segments:

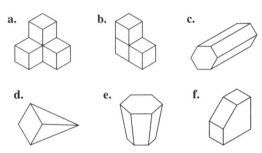

a.　　**b.**　　**c.**

d.　　**e.**　　**f.**

30. If you are drawing a two-dimensional representation of a large $n \times n \times n$ cube, will the edges of the square faces appear parallel as they do in a normal square on a piece of paper? Why or why not?

31. Explain whether or not a cylinder has to have a circular base.

32. Draw a curve that is neither simple nor closed.

Selected Bibliography

Andrews, A. G. "Solving Geometric Problems by Using Unity Blocks." *Teaching Children Mathematics* 5 (February 1999): 318–323.

Bartels, B. H. "Truss(t)ing Triangles." *Mathematics Teaching in the Middle School* 3 (March–April 1998): 394–396.

Britton, B., and S. Stump. "Unexpected Riches from a Geoboard Quadrilateral Activity." *Mathematics Teaching in the Middle School* 6 (April 2001): 490–493.

Burger, W. F., and J. M. Shaughnessy. "Characterizing the van Heile Levels of Development in Geometry." *Journal for Research in Mathematics Education* (1986), 17(1): 31–48.

Carpenter, T. "Research in Cognitive Development." In *Research in Mathematics Education*, edited by R. Shumways. Reston, Va.: NCTM, 1980.

Carroll, W. "Middle School Students' Reasoning about Geometric Situations." *Mathematics Teach-*

ing in the Middle School 3 (March–April 1998): 398–403.

Clements, D., and M. Battista. "Geometry and Special Reasoning." In *Handbook of Research on Mathematics Teaching and Learning*, edited by D. Grouws. New York: MacMillan, 1992.

Clements, D., and J. Sarama. "The Earliest Geometry." *Teaching Children Mathematics* 7 (October 2000): 82–86.

Clements, D., and J. Sarama. "The Role of Technology in Early Childhood Learning." *Teaching Children Mathematics* 8 (February 2002): 340–343.

Ellis, M., and C. Yeh. "Coloring Maps: How Many Colors Are Necessary?" *Teaching Children Mathematics* 8 (April 2008): 485–488.

Ellis, M., and C. Yeh. "How Many Triangles?" *Teaching Children Mathematics* 4 (November 2007): 214–216.

Franco, B. *Unfolding Mathematics with Unit Origami.* Berkeley, CA: Key Curriculum Press 1999.

Grouws, D., ed. *Handbook of Research on Mathematics Teaching and Learning.* Charlotte, NC: Information Age Publishing, 2006.

Hannibal, M. A. "Young Children's Developing Understanding of Geometric Shapes." *Teaching Children Mathematics* 5 (February 1999): 353–357.

Hoffer, A. "Geometry and Visual Thinking." In *Teaching Mathematics in Grades K–8: Research Based Methods*, edited by R. R. Post. Columbus, OH: Allyn and Bacon, 1988.

Hoffer, A. "Making a Better Beer Glass." *Mathematics Teacher* 75 (May 1982): 378–379.

Kennedy, J., and E. McDowell. "Geoboard Quadrilaterals." *Mathematics Teacher* 91 (April 1998): 288–290.

Kerslake, D. "Graphs." In K. M. Hart (Ed.) *Children's Understanding of Mathematics.* London: John Murray, 1981, 120–136.

Libeskind, S. *Euclidean and Transformational Geometry.* Sudbury, MA: Jones and Bartlett Publishers, Inc., 2008.

Lipka, J., S. Wildfeuer, N. Wahlberg, M. George, and D. Ezran. "Elastic Geometry and Storyknifing: A Yup'ik Eskimo Example." *Teaching Children Mathematics* 7 (February 2001): 337–343.

Long, V. "From Polygons to Poetry." *Mathematics Teaching in the Middle School* 6 (April 2001): 436–438.

Moyer, P. S., and J. J. Bolyard. "Classify and Capture: Using Venn Diagrams and Tangrams to Develop Abilities in Mathematical Reasoning and Proof." *Mathematics Teaching in Middle School* 3 (February 2003): 325–330.

Olson, A. T. *Mathematics Through Paper Folding.* Washington, D.C.: NCTM (1975).

Olson, J. "What Shapes Can You Make?" *Teaching Children Mathematics* 5 (February 1999): 330–331.

Olson, M. "Coloring Tetras with Two Colors." *Teaching Children Mathematics* 7 (January 2001): 294–295.

Peterson, B. "From Tessellations to Polyhedra: Big Polyhedra." *Mathematics Teaching in the Middle School* 5 (February 2000): 348–356.

Renne, C. G. "Is a Rectangle a Square? Developing Mathematical Vocabulary and Conceptual Understanding." *Teaching Children Mathematics* 10 (January 2004): 258–263.

Rich, W., and J. Joyner. "Using Interactive Web Sites to Enhance Mathematics Learning." *Teaching Children Mathematics* 8 (February 2002): 380–383.

Row, T. S. *Geometric Exercises in Paper Folding.* Mineola, NY: Dover Publications, Inc. 1966.

Sundberg, S. "A Plethora of Polyhedra." *Mathematics Teaching in the Middle School* 3 (March–April 1998): 388–391.

van Hiele, P. M. "Developing Geometric Thinking through Activities That Begin with Play." *Teaching Children Mathematics* 5 (February 1999): 310–316.

CHAPTER 12

Constructions, Congruence, and Similarity

Preliminary Problem

Rosa needed a wooden triangle that had one 18-in. side and angle measures of 36°, 84°, and 60°. Rosa had a carpenter make the triangle. Later, she needed another triangle with the same specifications. When the second triangle arrived, it was not congruent to the first one. Explain how this is possible.

T he *Principles and Standards* state that students should be able to do the following:

Pre-K–2

• relate ideas in geometry to ideas in number and measurement. (p. 396)

Grades 3–5

• explore congruence and similarity;

• make and use coordinate systems to specify locations and to describe paths;

• build and draw geometric objects;

• use geometric models to solve problems in other areas of mathematics, such as number and measurement;

• recognize geometric ideas and relationships and apply them to other disciplines and to problems that arise in the classroom or in everyday life. (p. 396)

Grades 6–8

• use coordinate geometry to represent and examine the properties of geometric shapes;

• use coordinate geometry to examine special geometric shapes, such as regular polygons or those with pairs of parallel or perpendicular sides;

• draw geometric objects with specified properties, such as side lengths or angle measures. (p. 397)

The *Focal Points* for grade 8 emphasize the use of similar triangles:

Students use fundamental facts about distance and angles to describe and analyze figures and situations in two- and three-dimensional space and to solve problems, including those with multiple steps. They prove that particular configurations of lines give rise to similar triangles because of the congruent angles created when a transversal cuts parallel lines. Students apply this reasoning about similar triangles to solve a variety of problems, including those that ask them to find heights and distances. (p. 20)

In this chapter, we introduce, through constructions and visualization, the concepts of congruence and similarity. We study equations of lines, investigate systems of equations both geometrically and algebraically, and consider basic trigonometry.

12-1   Congruence Through Constructions

In mathematics, **similar** (∼) objects have the same shape but not necessarily the same size; **congruent** (≅) objects have the same size as well as the same shape. Whenever two figures are congruent, they are also similar. However, the converse is not true. Examples of similar and congruent objects are seen in Figure 12-1.

A 100% photocopy of a two-dimensional object is **congruent** to the original. When we photocopy and either reduce or enlarge the image, we get a **similar** object.

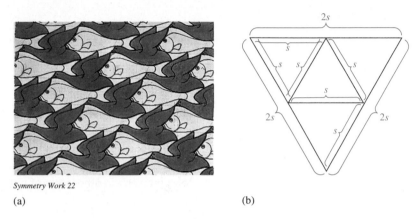

Symmetry Work 22

(a)   (b)

Figure 12-1

The Escher print in Figure 12-1(a) shows congruent fish and congruent birds. Figure 12-1(b) contains both congruent and similar equilateral triangles. The smaller triangles are both similar to and congruent to each other. They also are similar to the large triangle that contains them. Figure 12-1(b) is an example of a **rep-tile**, a figure that is used to construct a larger, similar figure. In Figure 12-1(b), one of the smaller equilateral triangles is a rep-tile. Figure 12-2 shows two circles with the same shape but different sizes. The size of a circle is determined by its radius. In fact, a **circle** is the set of all points in a plane at the same distance (**radius**) from a given point, its **center**. In Figure 12-2, the larger circle (circle O) has center O and radius OA, and the smaller circle (circle O') has center O' and radius $O'A'$. Note that *radius* is used to describe both the length OA and the segment $\overline{OA}$. Circles with equal radii are congruent; however, all circles are similar to each other.

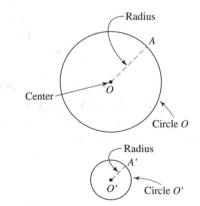

Figure 12-2

Before studying congruent and similar figures, we first consider some notation and review definitions from Chapter 11. For example, *any two line segments are congruent if they have the same length, and two angles are congruent if they have the same measure.* The length of line segment $\overline{AB}$ is denoted AB. Symbolically, we may write the following about congruent segments and angles.

> **Definition of Congruent Segments and Angles**
>
> $\overline{AB} \cong \overline{CD}$ if, and only if, $AB = CD$
> $\angle ABC \cong \angle DEF$ if, and only if, $m(\angle ABC) = m(\angle DEF)$

Geometric Constructions

Ancient Greek mathematicians constructed geometric figures with a straightedge (no markings on it) and a collapsible compass. Figure 12-3 shows a more modern compass that can be used to mark off and duplicate lengths and to construct circles or arcs with a radius of a given measure. To draw a circle when given the radius PQ of a circle, we follow the steps illustrated in Figure 12-3.

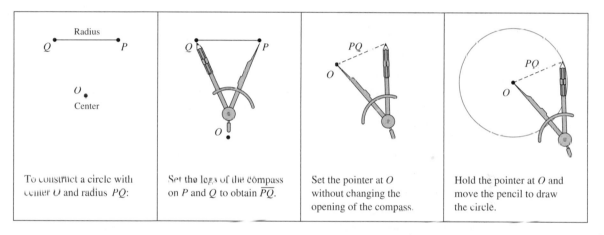

To construct a circle with center O and radius PQ:	Set the legs of the compass on P and Q to obtain $\overline{PQ}$.	Set the pointer at O without changing the opening of the compass.	Hold the pointer at O and move the pencil to draw the circle.

Figure 12-3 Constructing a circle given its radius

An **arc** of a circle is any part of the circle that can be drawn without lifting a pencil. (The entire circle could be considered an arc.) The **center of an arc** is the center of the circle containing the arc. If there is no danger of ambiguity in a discussion, the endpoints of the arc are used to name the arc, otherwise three points are used. In Figure 12-4, $\overparen{ACB}$ is a **minor arc** and $\overparen{ADB}$ is a **major arc**. If the two arcs determined by a pair of points on the circle are the same size, each is a **semicircle**. If only two points are used in the arc name, then the minor arc is intended. Thus, in Figure 12-4, $\overparen{AB}$ is another name for $\overparen{ACB}$. A segment connecting two points on a circle is a **chord** of the circle. If a chord contains the center, it is a **diameter**. In Figure 12-4, $\overline{AC}$ and $\overline{AD}$ are chords, and $\overline{AD}$ is also a diameter.

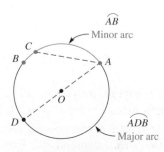

Figure 12-4

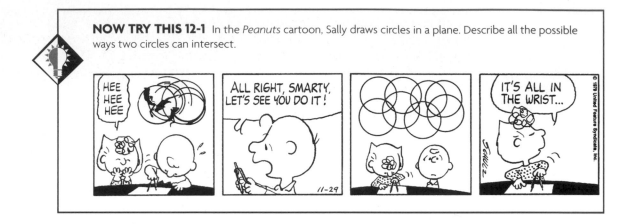

NOW TRY THIS 12-1 In the *Peanuts* cartoon, Sally draws circles in a plane. Describe all the possible ways two circles can intersect.

Constructing Segments

There are many ways to draw a segment congruent to a given segment $\overline{AB}$. A natural approach is to use a ruler, measure $\overline{AB}$, and then draw a congruent segment. A different way is to trace $\overline{AB}$ onto a piece of paper. A third method is to use a straightedge and a compass, as in Figure 12-5.

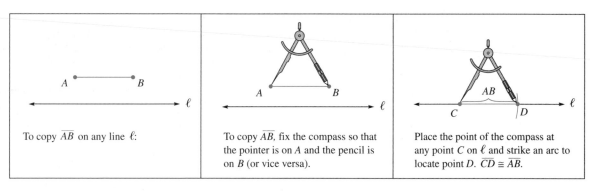

To copy $\overline{AB}$ on any line ℓ:

To copy $\overline{AB}$, fix the compass so that the pointer is on A and the pencil is on B (or vice versa).

Place the point of the compass at any point C on ℓ and strike an arc to locate point D. $\overline{CD} \cong \overline{AB}$.

Figure 12-5 Constructing a line segment congruent to a given segment

The straight line and circle were considered the basic geometric figures by the Greeks, and the straightedge and compass are their physical analogs. It is believed that the Greek philosopher Plato (427–347 BCE) rejected the use of mechanical devices other than the straightedge and compass for geometric constructions because use of other tools emphasized practicality rather than "ideas." ◆

Triangle Congruence

Informally, two figures are congruent if it is possible to fit one figure onto the other so that all matching parts coincide. In Figure 12-6, $\triangle ABC$ and $\triangle A'B'C'$ have corresponding congruent parts. Tick marks are used to show congruent segments and angles in the tri-

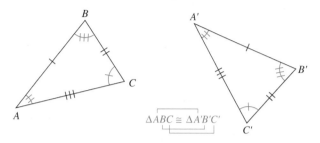

Figure 12-6

angles. If we were to trace $\triangle ABC$ in Figure 12-6 and put the tracing over $\triangle A'B'C'$ so that the tracing of A is over A', the tracing of B is over B', and the tracing of C is over C', $\triangle ABC$ would coincide with $\triangle A'B'C'$. This suggests the following definition of congruent triangles.

> **Definition of Congruent Triangles**
> $\triangle ABC$ is congruent to $\triangle A'B'C'$, written $\triangle ABC \cong \triangle A'B'C'$, if, and only if, $\angle A \cong \angle A'$, $\angle B \cong \angle B'$, $\angle C \cong \angle C'$, $\overline{AB} \cong \overline{A'B'}$, $\overline{BC} \cong \overline{B'C'}$, and $\overline{AC} \cong \overline{A'C'}$.

In the definition of congruent triangles $\triangle ABC$ and $\triangle A'B'C'$, the order of the vertices is such that the listed congruent angles and congruent sides correspond, leading to the statement: "Corresponding parts of congruent triangles are congruent," abbreviated as CPCTC.

NOW TRY THIS 12-2 In Figure 12-6, $\triangle ABC \cong \triangle A'B'C'$. List all possible ways that the congruence can be symbolized.

Example 12-1

Write an appropriate symbolic congruence for each of the pairs.

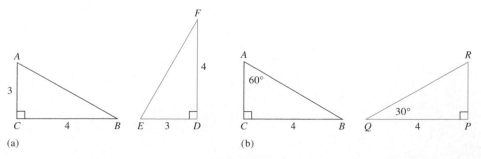

Figure 12-7

Solution   **a.** Vertex C corresponds to D because the angles at C and D are right angles. Also, because $\overline{CB} \cong \overline{DF}$ and C corresponds to D, B corresponds to F. Consequently, A corresponds to E. Thus, $\triangle ABC \cong \triangle EFD$.

        b. $\overline{BC} \cong \overline{QP}$. Vertex C corresponds to P because the angles at C and P are both right angles. To establish the other correspondences, we first find the missing angle measures in the triangles. We see that $m(\angle B) = 90° - 60° = 30°$ and $m(\angle R) = 90° - 30° = 60°$. Consequently, A corresponds to R because $m(\angle A) = m(\angle R) = 60°$, and B corresponds to Q because $m(\angle B) = m(\angle Q) = 30°$. Thus, $\triangle ABC \cong \triangle RQP$.

◆

Side, Side, Side Congruence Condition (SSS)

Calibration experts work to ensure that car parts on an automotive assembly line are interchangeable so that the same part fits on all basic models of the same car. For the cars to be congruent, the parts must be congruent. In design for automotive production, decisions have to be made about the minimal set of items to consider for eventual congruency. In considering congruence of figures in geometry, we also look for a minimal number of conditions that assure congruence.

If three sides and three angles of one triangle are congruent to the corresponding three sides and three angles of another triangle, then the triangles are congruent. However, do we need to know that all six parts of one triangle are congruent to the corresponding parts of the second triangle in order to conclude that the triangles are congruent? What if we know only five parts, or four parts, or just three parts to be congruent? In this section and in the next section, we will show when three parts are sufficient to determine congruency and when they are not.

Consider the triangle formed by attaching three segments, as in Figure 12-8. Such a triangle is *rigid*; that is, its size and shape cannot be changed. Because of this property, a manufacturer can make duplicates if the lengths of the sides are known. Note that once the sides are known, all the angles are automatically determined.

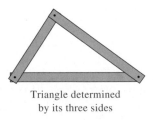

Triangle determined
by its three sides

Figure 12-8

Many bridges and other structures that have exposed frameworks demonstrate the practical use of the rigidity of triangles, as seen in Figure 12-9.

Figure 12-9

Because a triangle is completely determined by its three sides, we have the following theorem.

Theorem 12–1:   Side, Side, Side (SSS)

If the three sides of one triangle are congruent, respectively, to the three sides of a second triangle, then the triangles are congruent.

REMARK Under certain assumptions this statement can be proved, and therefore we list it as a theorem.

Example 12-2

For each part in Figure 12-10, use SSS to explain why the given triangles are congruent:

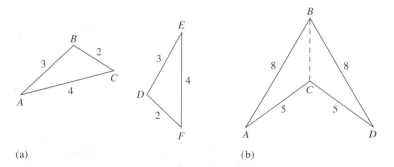

(a)   (b)

Figure 12-10

Solution a. $\triangle ABC \cong \triangle EDF$ by SSS because $\overline{AB} \cong \overline{ED}$, $\overline{BC} \cong \overline{DF}$, and $\overline{AC} \cong \overline{EF}$.
b. $\triangle ABC \cong \triangle DBC$ by SSS because $\overline{AB} \cong \overline{DB}$, $\overline{AC} \cong \overline{DC}$, and $\overline{BC} \cong \overline{BC}$.

Geometric constructions are useful in drafting and design. Designers most frequently use computers and computer software in their work. In today's world—other than in school—manual tools such as the compass, ruler, triangle, and protractor are used infrequently. In Euclid's time, mathematicians required that constructions be done using only a compass and a straightedge—an unmarked ruler. Euclidean constructions follow these rules:

1. Given two points, a unique straight line can be drawn containing the points as well as a unique segment connecting the points. (This is accomplished by aligning the straightedge across the points.)
2. It is possible to extend any part of a line.
3. A circle can be drawn given its center and radius.
4. Any number of points can be chosen on a given line, segment, or circle.
5. Points of intersection of two lines, two circles, or a line and a circle can be used to construct segments, lines, or circles.
6. No other instruments (such as a marked ruler, triangle, or protractor) or procedures can be used to perform Euclidean constructions.

In reality, compass and ruler constructions are subject to error. For example, a geometric line is an ideal line with zero width. However, a drawing of a line, no matter how sharp the pencil and how good the straightedge is, has a nonzero width. Then why insist on performing constructions using the preceding rules? There are many good reasons. We can look at this activity as a game, and games have rules. Playing games can be a source of enjoyment, and so can performing Euclidean constructions—but with the additional benefit of learning geometry in the process. Also, to use computer graphics, or even to use computer software such as *Geometer's Sketchpad* (*GSP*), one needs to be able to use strategies similar to or the same as those used in Euclidean constructions.

Constructing a Triangle Given Three Sides

Using the SSS theorem, we can construct a triangle $A'B'C'$ congruent to a given triangle ABC if we know the lengths of the three sides. We illustrate this in Figure 12-11.

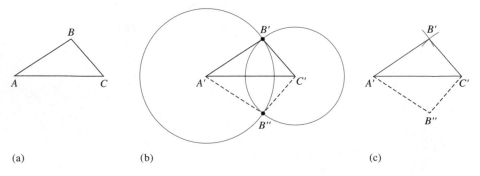

(a)   (b)   (c)

Figure 12-11

First, we construct in Figure 12-11 a segment congruent to one of the three segments: for example, $\overline{A'C'}$ congruent to $\overline{AC}$. To complete the triangle construction, we must locate the other vertex, B'. The distance from A' to B' is AB. All points at a distance AB from A'

are on a circle with center at A' and radius of length AB. Similarly, B' must be on a circle with center C' and radius of length BC. The only possible locations for B' are at the points where the two circles intersect. Either point B' or B'' is acceptable. Usually, a picture of the construction shows only one possibility and the construction uses only arcs, as seen in Figure 12-11(c). By SSS, $\triangle A'B'C' \cong \triangle A'B''C'$.

> **REMARK** Starting the construction with a segment $\overline{A'B'}$ congruent to $\overline{AB}$ or with $\overline{B'C'}$ congruent to $\overline{BC}$ would also result in triangles congruent to $\triangle ABC$.

NOW TRY THIS 12-3 From using the SSS construction as in Figure 12-11, we could think that given any three segments, we could construct a triangle whose sides are congruent to the given segments. To determine if this is true, cut a soda straw into several pieces of different lengths. Make possible triangles and answer the following questions:

a. Could a triangle be constructed from every three pieces?
b. If three pieces were exactly the same length, what type of triangle could be constructed?
c. If the length of one piece is the exact sum of the lengths of the other two pieces, can a triangle be constructed from the three pieces? Why or why not?
d. If one piece is longer than the two other pieces put together, can a triangle be constructed from the three pieces? Why or why not?

TECHNOLOGY CORNER Work through *GSP* Lab 4 in the Technology Manual to investigate congruent triangles.

Now Try This 12-3 lends credence to the following property.

> **Theorem 12–2: Triangle Inequality**
> The sum of the measures of any two sides of a triangle must be greater than the measure of the third side.

This theorem says that given any three segments it is not always possible to construct a triangle whose sides will be congruent to the given segments. For example, if the sides are 8, 14, and 5 units long, then $8 + 14 > 5$ and $14 + 5 > 8$, but since $8 + 5 < 14$, a triangle with such side lengths does not exist.

Constructing Congruent Angles

We use the SSS notion of congruent triangles to construct an angle congruent to a given angle, $\angle B$, by making $\angle B$ a part of an isosceles triangle and then reproducing this triangle, as in Figure 12-12.

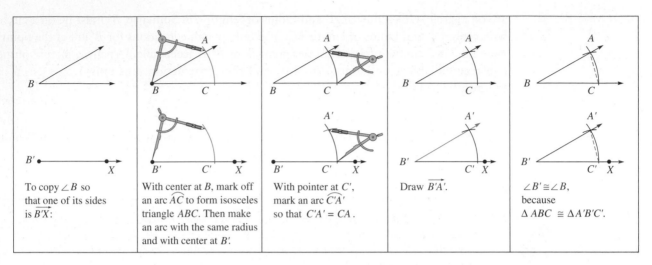

Figure 12-12 Copying an angle

> **REMARK** With the compass and straightedge alone, it is impossible in general to construct an angle if given only its measure. For example, an angle of measure 20° cannot be constructed with a compass and a straightedge only. Instead, a protractor or a geometry drawing utility or some other measuring tool must be used.

Side, Angle, Side Property (SAS)

We have seen that, given three segments, no more than one triangle can be constructed. Could more than one triangle be constructed when given only two segments? Consider Figure 12-13(b), which shows three different triangles with sides congruent to the segments given in Figure 12-13(a). The length of the third side depends on the measure of the **included angle** between the two given sides.

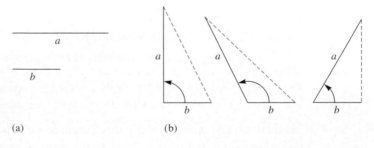

(a)   (b)

Figure 12-13

Historical Note
Three geometry compass and straightedge construction problems that concerned mathematicians for centuries were (1) trisecting a general angle, (2) duplicating a cube (constructing a cube whose volume is twice a given cube), and (3) squaring a circle (constructing a square with exactly the same area as a given circle). These constructions were proved to be impossible only in the 1800s. ♦

Constructions Involving Two Sides and an Included Angle of a Triangle

Figure 12-14 shows how to construct a triangle congruent to $\triangle ABC$ by using two sides $\overline{AB}$ and $\overline{AC}$ and the included angle, $\angle A$, formed by these sides. First, a ray with an arbitrary endpoint A' is drawn, and $\overline{A'C'}$ is constructed congruent to $\overline{AC}$. Then, $\angle A'$ is constructed so that $\angle A' \cong \angle A$, and B' is marked on the side of $\angle A'$ not containing C' so that $\overline{A'B'} \cong \overline{AB}$. Connecting B' and C' completes $\triangle A'B'C'$ so that $\triangle A'B'C'$ appears to be congruent to $\triangle ABC$.

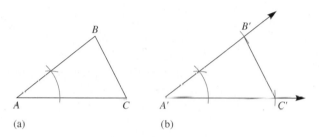

(a)   (b)

Figure 12-14

It is true that if we knew the lengths of two sides and the measure of the angle included between them, we could construct a unique triangle. We express this fact as **Side, Angle, Side (SAS)**.

Axiom 12–1:   Side, Angle, Side (SAS)

If two sides and the included angle of one triangle are congruent to two sides and the included angle of another triangle, respectively, then the two triangles are congruent.

REMARK In many approaches to Euclidean geometry, this SAS congruence condition is an axiom; in other approaches, it is proved as a theorem. In fact, if SSS is taken as an axiom in place of SAS axiom, other congruency conditions can be proved and therefore become theorems.

 NOW TRY THIS 12-4 If, in two triangles, two sides and an angle not included between these sides are congruent, respectively, show the triangles don't have to be congruent. State as few additional conditions as possible that can be placed on the sides or angles to make the triangles congruent.

Example 12-3

For each part of Figure 12-15, use SAS to show that the given pair of triangles are congruent.

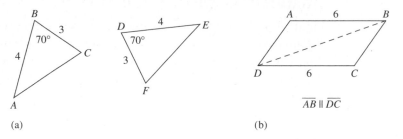

(a)   (b)

$\overline{AB} \parallel \overline{DC}$

Figure 12-15

Solution **a.** $\triangle ABC \cong \triangle EDF$ by SAS because $\overline{AB} \cong \overline{ED}$, $\angle B \cong \angle D$ (the included angles), and $\overline{BC} \cong \overline{DF}$.

        b. Because $\overline{AB} \cong \overline{CD}$ and $\overline{DB} \cong \overline{BD}$, we need either another side or another angle to show that the triangles are congruent. We know that $\overline{AB} \parallel \overline{DC}$. Since parallel segments $\overline{AB}$ and $\overline{DC}$ are cut by transversal $\overline{BD}$, we have alternate interior angles $\angle ABD$ and $\angle BDC$ congruent. Now $\triangle ABD \cong \triangle CDB$ by SAS.

We have just seen that two pairs of congruent sides and a pair of included angles (SAS) determine congruent triangles. But what if the angle is not an included angle? In Figure 12-16, $\triangle ABC$ and $\triangle ABD$ have two pairs of congruent sides and a common angle. They share $\overline{AB}$ and $\overline{BC} \cong \overline{BD}$, and also they share $\angle A$. However, $\triangle ABC$ is not congruent to $\triangle ABD$. Thus, SSA (Side-Side-Angle) which specifies two sides and a non-included angle does not assure congruence.

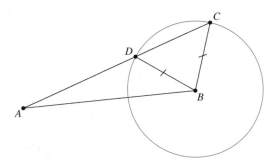

Figure 12-16

Nevertheless (as you may have discovered in Now Try This 12-4), if the angle in SSA is a right angle, then we have two right triangles with two pairs of congruent sides (one of which is opposite the right angle), as shown in Figure 12-17, and the triangles are congruent. The side opposite the right angle is the hypotenuse. The other sides of a right triangle are called legs.

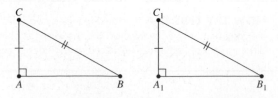

Figure 12-17

It can be proved that the triangles are congruent and therefore the congruence condition, referred to as the **Hypotenuse-Leg** (HL) congruence theorem, can be stated as a theorem.

Theorem 12–3: Hypotenuse-Leg (HL)

If the hypotenuse and a leg of one right triangle are congruent to the hypotenuse and a leg of another right triangle, then the triangles are congruent.

REMARK The SSA condition is also valid when the non-included angle is obtuse.

Other congruence and noncongruence conditions for triangles are explored in Section 12-2.

Selected Triangle Properties

The SAS and other conditions for congruence allow us to investigate a number of properties of triangles. For example, in Figure 12-18, consider isosceles triangle ABC with $\overline{AB} \cong \overline{AC}$. We use *paper folding* to construct the angle bisector of $\angle A$. For that purpose, crease the triangle through vertex A so that vertex B "falls" on vertex C (this can be done because $\overline{AB} \cong \overline{AC}$). If point D is the intersection of the crease and $\overline{BC}$, then our folding assures that $\angle BAD \cong \angle CAD$. Thus, by SAS, $\triangle ABD \cong \triangle ACD$. Because corresponding parts must be congruent, we have $\angle B \cong \angle C$, $\overline{BD} = \overline{CD}$, $\angle BDA \cong \angle CDA$. Because the last pair of angles are supplementary, it follows that each is a right angle. Consequently, $\overline{AD}$ is perpendicular to $\overline{BC}$ and bisects $\overline{BC}$. A line that is perpendicular to a segment and bisects it is the **perpendicular bisector** of the segment.

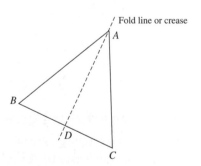

Figure 12-18

In Figure 12-18, notice that if point A is equidistant from the endpoints B and C, then A is on the perpendicular bisector of $\overline{BC}$. The converse of this statement is also true. The statement and its converse are given in Theorem 12–4.

Theorem 12–4

a. Any point equidistant from the endpoints of a segment is on the perpendicular bisector of the segment.
b. Any point on the perpendicular bisector of a segment is equidistant from the endpoints of the segment.

The findings related to Figure 12-18 are summarized in Theorem 12–5.

> **Theorem 12–5**
>
> For every isosceles triangle:
>
> **a.** the angles opposite the congruent sides are congruent. (Base angles of an isosceles triangle are congruent.)
> **b.** the angle bisector of an angle formed by two congruent sides is the perpendicular bisector of the third side of the triangle.

An **altitude** of a triangle is the perpendicular segment from a vertex of the triangle to the line containing the opposite side of the triangle. In Figure 12-18 we see that the perpendicular bisector is also the altitude to $\overline{BC}$. Figure 12-19 (a) and (b) shows the three altitudes $\overline{AE}$, $\overline{BF}$, and $\overline{CD}$ of an acute triangle and an obtuse triangle. Notice that in Figure 12-19(b), by definition the altitude $\overline{AE}$ is the perpendicular segment from the vertex A to the line containing the opposite side of the triangle. Similarly, the altitude $\overline{CD}$ is the perpendicular segment from vertex C to the line containing the opposite side, $\overline{AB}$. In Figure 12-19(a), we see that the three altitudes intersect in a single point in the interior of the triangle. In Figure 12-19(b), the three lines containing the altitudes intersect in a single point P in the exterior of the triangle.

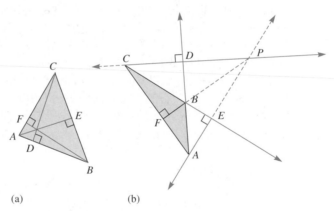

(a)   (b)

Figure 12-19

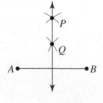

Figure 12-20

Construction of the Perpendicular Bisector of a Segment

Theorem 12–5 can be used to construct the perpendicular bisector of a segment by constructing any two points equidistant from the endpoints of the segment. Each point is a vertex of an isosceles triangle, and the two points determine the perpendicular bisector of the segment. In Figure 12-20 we have constructed point P equidistant from A and B by drawing intersecting arcs with the same radius—one with center at A, and the other with center at B. Point Q is constructed similarly with two intersecting arcs. By Theorem 12–4(a), each point is on the perpendicular bisector of $\overline{AB}$. Because two points determine a unique line, $\overleftrightarrow{PQ}$ must be the perpendicular bisector of $\overline{AB}$.

NOW TRY THIS 12-5 In Figure 12-20 we constructed the perpendicular bisector of $\overline{AB}$ by drawing four arcs. An alternative construction of the perpendicular bisector of $\overline{AB}$ can be accomplished by drawing one large-enough arc (or circle) with center at A and another one with the same radius and centered at B. The two points where the arcs intersect determine the perpendicular bisector of the segment. Draw any segment and construct its perpendicular bisector using only two arcs.

Construction of a Circle Circumscribed About a Triangle

In Figure 12-21(a), a circle is **circumscribed** about a given $\triangle ABC$; that is, every vertex of the triangle is on the circle. How can such a **circumcircle** be constructed?

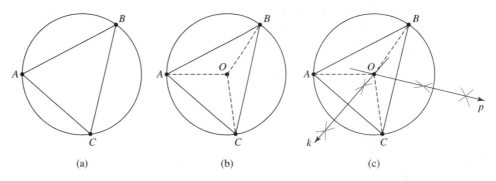

(a)   (b)   (c)

Figure 12-21

Suppose we know the location of center O of the circle, as in Figure 12-21(b). The properties of the center of a circle should enable us to find its location. Because O is the center of the circle circumscribed about $\triangle ABC$, $OA = OC = OB$. The fact that $OA = OC$ implies that O is equidistant from the endpoints of segment $\overline{AC}$. Hence, by Theorem 12-5, O is on the perpendicular bisector, k, of $\overline{AC}$, as shown in Figure 12-21(c). Similarly, because $OC = OB$, O is on the perpendicular bisector, p, of $\overline{BC}$. Because O is on k and on p, it is the point of intersection of the two perpendicular bisectors. Thus, given $\triangle ABC$, we can construct the center of the circumscribed circle by constructing perpendicular bisectors of any two sides of the triangle. The point where the perpendicular bisectors intersect is the **circumcenter**. Thus, the required circle is the circle with the center at O and radius OA (the **circumradius**). The construction is shown in Figure 12-21(c).

Problem Solving   Archaeological Find

At the site of an ancient settlement, archaeologists found a fragment of a saucer, as shown in Figure 12-22. To restore the saucer, the archaeologists needed to determine the radius of the original saucer. How can they do this?

Figure 12-22

Understanding the Problem   The border of the shard shown in Figure 12-22 was part of a circle. To reconstruct the saucer, we are to determine the center and the radius of the circle of which the shard is a part.

Devising a Plan   We can use a *model* to determine the radius. We trace an outline of the circular edge of the three-dimensional shard on a piece of paper. The result is an arc of a circle, as shown in Figure 12-23. To determine the radius, we must find the center O. We know that by connecting O to points of the circle, we obtain congruent radii. Also, each pair of radii determines an isosceles triangle. Consider points A, B, and O. Triangle ABO is isosceles and O is on the perpendicular bisector of $\overline{AB}$.

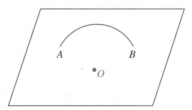

Figure 12-23

Carrying Out the Plan   To find a line containing point O, construct a perpendicular bisector of $\overline{AB}$, as in Figure 12-24(a). Similarly, any other segment (for example, $\overline{AC}$) with endpoints on $\overline{AB}$ has a perpendicular bisector containing O, as in Figure 12-24(b). The point of intersection of the two perpendicular bisectors is point O. To complete the problem, measure the length of either $\overline{OB}$ or $\overline{OA}$.

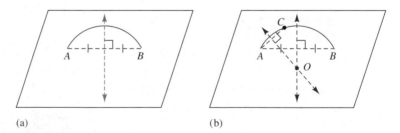

(a)   (b)

Figure 12-24

Looking Back   A related problem is, What would happen if the piece of pottery had been part of a sphere? Would the same ideas still work?

NOW TRY THIS 12-6 In the photograph we see congruent hexagons. Do you think that there is a Side-Side-Side-Side-Side-Side property for determining hexagons congruent? Prepare a pro or con argument.

 ## Assessment 12-1A

1. In the definition of a circle, why are the points restricted to a plane?
2. Where do all the points in space equidistant from the end points of $\overline{AB}$ lie?
3. **a.** Use any tool to draw triangle ABC in which BC is greater than AC. Measure the angles opposite $\overline{BC}$ and $\overline{AC}$. Compare the angle measures. What did you find?
   b. Based on your finding in (a), make a conjecture concerning the lengths of sides and the measures of angles of a triangle.
4. Use any tools to construct each of the following, if possible:
   a. A segment congruent to $\overline{AB}$ and an angle congruent to $\angle CAB$

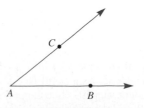

   b. A triangle with sides of lengths 2 cm, 3 cm, and 4 cm
   c. A triangle with sides of lengths 4 cm, 3 cm, and 5 cm (What kind of triangle is it?)
   d. A triangle with sides 4 cm, 5 cm, and 10 cm
   e. An equilateral triangle with sides 5 cm
   f. A triangle with sides 6 cm and 7 cm and an included angle of measure 75°

g. A triangle with sides 6 cm and 7 cm and a nonincluded angle of measure 40°

h. A triangle with sides 6 cm and 6 cm and a nonincluded angle of measure 40°

i. A right triangle with legs 4 cm and 8 cm (the legs include the right angle)

5. For each of the conditions in problem 4(b) through (i), does the given information determine a unique triangle? Explain why or why not.

6. How many different triangles can be constructed with toothpicks by connecting the toothpicks only at their ends if each triangle can contain at most five toothpicks per side?

7. For each of the following, determine whether the given conditions are sufficient to prove that $\triangle PQR \cong \triangle MNO$. Justify your answers.
   a. $\overline{PQ} \cong \overline{MN}, \overline{PR} \cong \overline{MO}, \angle P \cong \angle M$
   b. $\overline{PQ} \cong \overline{MN}, \overline{PR} \cong \overline{MO}, \overline{QR} \cong \overline{NO}$
   c. $\overline{PQ} \cong \overline{MN}, \overline{PR} \cong \overline{MO}, \angle Q \cong \angle N$

8. Is it possible to construct a right triangle with all sides congruent? Justify your answer.

9. If two students are constructing equilateral triangles with a side of length 4 in., explain whether or not the triangles constructed must necessarily be congruent.

10. If you have a paper cutout of a circle and no other construction tools, explain how you can find the center of the circle.

11. In the following drawing, the triangles are equilateral. If the circumcircles are constructed around the largest triangle and around the "middle" triangle, what can be said of the two circumcircles? Why?

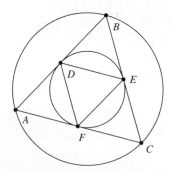

12. Given a scalene triangle, use it and any tools to construct a rep-tile by finding the midpoints of the sides of the triangle.

13. A group of students on a hiking trip wants to find the distance AB across a pond (see the following figure). One student suggests choosing any point C, connecting it with B, and then finding point D such that $\angle DCB \cong \angle ACB$ and $\overline{DC} \cong \overline{AC}$. How and why does this help in finding the distance AB?

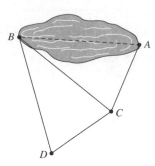

14. Using only a compass and a straightedge, perform each of the following:
   a. Construct an equilateral triangle with the following side $\overline{AB}$:

   b. Construct a 60° angle.
   c. Construct an isosceles triangle with $\angle A$ (see the following figure) as the angle included between the two congruent sides:

   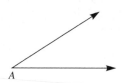

15. Refer to the figure shown and, using only a compass and a straightedge, perform each of the following:

   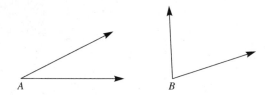

   a. Construct $\angle C$ so that $m(\angle C) = m(\angle A) + m(\angle B)$.
   b. Using the angles in (a), construct $\angle C$ so that $m(\angle C) = m(\angle B) - m(\angle A)$.

16. **a.** Draw a circle and choose an arbitrary point A on the circle, use any method to mark off points B, C, D, E and F so that $\overline{AB}, \overline{BC}, \overline{CD}, \overline{DE}$ and $\overline{EF}$ are congruent to the radius of the circle, as shown in the following figure:

   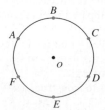

b. Connect each of the points to point O and connect the points in order around the circle.

c. Explain why all of the six triangles are congruent.

d. If the radius of the circle is r, what is the length of the side of the hexagon?

e. Construct any circle, then use your answer to part (d) to inscribe a regular hexagon in the circle; that is, the sides of the hexagon are congruent and all the vertices of the hexagons are on the circle. Describe your construction in words.

f. Construct any circle. Use the construction in part (e) to construct an equilateral triangle inscribed in the circle.

17. Write a definition for congruent arcs.

18. How many different one-to-one correspondences could be listed between the following:
  a. Vertices of two triangles
  b. Vertices of two quadrilaterals
  c. Vertices of two n-gons

19. In a pair of right triangles, suppose two legs of one are congruent, respectively, to two legs of the other. Explain whether the triangles are congruent and why.

20. Construct an obtuse triangle similar to the one shown. Then, using only a compass and a straightedge, construct the circle that circumscribes the triangle. Describe your construction.

21. For which of the following figures is it possible to find a circle that circumscribes the figure? If it is possible to find such a circle, draw the figure and construct the circumscribing circle.
  a. A right triangle
  b. A regular hexagon

22. Construct any obtuse triangle ABC. Then construct the three altitudes $\overline{AD}$, $\overline{BE}$, and $\overline{CF}$. Mark the points D, E, and F and point P where the lines containing the altitudes intersect

23. Consider a normal stop sign on a highway. Explain whether or not a circle could be circumscribed about this sign.

24. In a circle, consider the lines containing two perpendicular radii. The lines intersect the circle in four points. If the four points are connected,
  a. What type of figure is formed?
  b. Prove your answer in part (a).

Assessment 12-1B

1. In space, where are all the points that are the same distance from a given point?

2. Use only a compass and a straightedge to construct each of the following, if possible:
  a. A segment congruent to $\overline{CB}$ and an angle congruent to $\angle ACB$

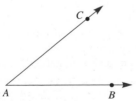

  b. A triangle with sides of lengths 2, 3, and 4 units
  c. An isosceles right triangle
  d. An equilateral triangle with any side length

3. A rancher designed a wooden gate as illustrated in the following figure. Explain the purpose of the diagonal boards on the gate.

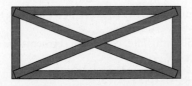

4. A rural homeowner had his television antenna held in place by three guy wires, as shown in the following figure.

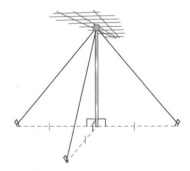

  a. If the stakes are on level ground and the distances from the stakes to the base of the antenna are the same, what is true about the lengths of the wires? Why?
  b. If the stakes are not on level ground yet are the same distance from the base of the antenna, explain whether you can make the same conclusion regarding the lengths of the wires.

5. What is true of the perpendicular bisectors of each of the sides of a regular pentagon? Justify your answer.

6. Using only a compass and a straightedge, perform each of the following:

a. Use the fact that the perpendicular bisector of the base of an isosceles triangle is also the angle bisector of the opposite angle, to construct a 30° angle.

b. Construct an isosceles triangle and the three angle bisectors of the angles.

7. Refer to the figure shown and, using only a compass and a straightedge, perform each of the following:

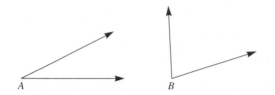

a. Construct ∠C so that $m(\angle C) = 2m(\angle A) + m(\angle B)$.
b. Construct ∠C so that $m(\angle C) = 2m(\angle B) - m(\angle A)$.

8. An equilateral triangle ABC is congruent to itself.

a. Write all possible true correspondences between the triangle and itself.

b. Use one of your answers in (a) to show that an equilateral triangle is also equiangular.

9. If $ABCDEF$ is a regular hexagon inscribed in a circle of radius r, prove that the length of each side of the hexagon equals r.

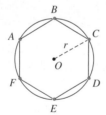

10. Let $ABCD$ be a square with diagonals $\overline{AC}$ and $\overline{BD}$ intersecting in point F, as shown in the figure.

a. What is the relationship between point F and the diagonals $\overline{BD}$ and $\overline{AC}$? Why?

b. What are the measures of angles BFA and AFD? Why?

11. What is the least number of congruent parts that one must know in order to be sure that square pyramids are congruent?

12. How many different one-to-one correspondences could be listed between the following:

a. Vertices of two regular pentagons

b. Vertices of two regular n-gons

13. Use any instruments to construct a square. Then, using only a compass and a straightedge, construct the circle that circumscribes the square.

14. For which of the following figures is it possible to find a circle that circumscribes the figure? If it is possible to find such a circle, draw the figure using any tools, and construct the circumscribing circle.

a. A rhombus

b. A regular decagon

15. In the following drawing, a compass is used to draw circle O. A point P is marked on the circle. Using the compass with the same setting that was used to draw the circle and using point P as the center, draw an arc that intersects the original circle in two points A and B. Repeat the process using points A and B on the circle. Continue this process.

a. What do you observe?

b. If you connected the points on the circle in order with line segments, starting at P and going in a clockwise fashion around the circle, what figure would you draw? Why?

16. a. If $\overline{MN}$ is any chord of circle O, what can you say about the perpendicular bisector of $\overline{MN}$?

b. Prove your answer in part (a)

Mathematical Connections 12-1

Communication

1. In a circle with center A and radius AB, let P be a point of $\overline{AB}$ that is not an endpoint. Explain whether or not P is on the circle.

2. Write an argument to convince the class that Theorem 12–5 is true.

3. In the following kite, congruent segments are shown with tick marks.

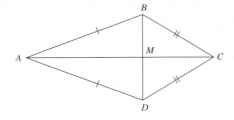

a. Argue that the diagonal $\overline{AC}$ bisects $\angle A$ and $\angle C$.
b. Let M be the point on which the diagonals of kite $ABCD$ intersect. Measure $\angle AMD$ and make a conjecture concerning the angle between the diagonals of a kite. Justify your conjecture.
c. Show that $\overline{BM} \cong \overline{MD}$.

4. If the kite in problem 3 were concave, do the same answers to parts (a) through (c) hold? Explain your answer.

5. Televisions are measured by the lengths of the diagonals across the screen. Are all 19 in. television screens congruent? Justify your answer.

6. If given any regular polygon, describe how you can find the center of the circle that circumscribes the polygon.

7. Explain why the quadrilateral $ABCD$ is a kite.

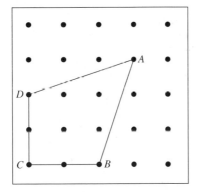

8. To construct the perpendicular to a line ℓ through a point P on the line, use a drafting triangle to align one of the legs of the triangle with the line, as shown in the figure. Next move the triangle along the leg until the other leg passes through the given point, P. Then answer the following questions.
a. Explain whether or not the construction described above will work if the point P is not on line ℓ.
b. Explain whether or not the above constructions are valid Euclidean constructions.

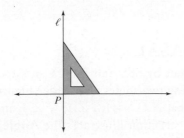

Open-Ended

9. a. Find at least five examples of congruent objects.
b. Find at least five examples of similar objects that are not congruent.

10. Design a quilt pattern that involves rep-tiles or find a pattern and describe the rep-tiles involved.

11. Pose and solve a problem in which one needs to show that two angles are congruent and in whose solution the Hypotenuse-Leg theorem can be used.

12. Use $\triangle AFB$ and $\triangle CED$ shown on the following dot paper to verify $\overline{AB}$ and $\overline{CD}$ are
a. congruent
b. perpendicular to each other (consider the angles in $\triangle BGC$.)

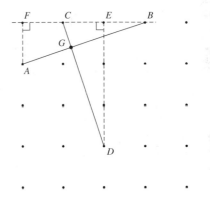

Cooperative Learning

13. Make a set of drawings of congruent figures and write an argument to convince a group of fellow students that the correspondence among vertices is necessary for determining that two triangles are congruent.

14. Use the theorem (not in the text) that in a triangle the side opposite the greater angle is greater than the side opposite the smaller angle to answer the following:
a. Explain why the hypotenuse of a right triangle is longer than a leg of the triangle.
b. Define the distance between a line and a point not on the line. Use part (a) to show that it is the length of the shortest segment among all the segments connecting the given point to points on the line.

Questions from the Classroom

15. On a test, a student wrote $AB \cong CD$ instead of $\overline{AB} \cong \overline{CD}$. Is this answer correct? Why?

16. A student asks if there are any constructions that cannot be done with a compass and a straightedge. How do you answer?

17. A student asks for a mathematical definition of congruence that holds for all figures. How do you respond?

18. A student constructed a line and three congruent segments *AD*, *DE*, and *EB* on the line. Then she constructed point *C* so that *AC* = *BC*. She claimed that by drawing $\overrightarrow{CD}$ and $\overrightarrow{CE}$, as shown in the following figure, she has trisected (divided into three congruent angles) ∠*ACB*. What is your response?

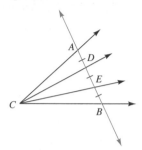

19. Claire claims that she discovered the following procedure for constructing a line through a given point *P* parallel to a given line ℓ. She would like to know how to prove that her construction is correct. How do you respond?

Claire: I choose any point *Q* on line ℓ so that $\overline{PQ}$ is not perpendicular to ℓ and construct the perpendicular bisector of $\overline{PQ}$, intersecting ℓ at *S* and $\overline{PQ}$, at *M*. I find point *T* (*T* ≠ *S*) such that *MT* = *MS*. The line *PT* is parallel to ℓ.

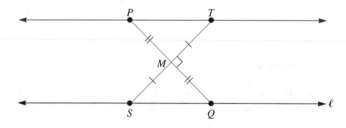

20. **a.** Zara claims that in spite of the fact that every triangle is congruent to itself, the statement △*ABC* ≅ △*BAC* is generally false. However, if *AC* = *BC*, the statement is true. Is she correct? If so, why?

    b. Zara wants to know if it is possible using part (a) to prove that in an isosceles triangle the base angles are congruent. How do you respond?

Third International Mathematics and Science Study (TIMSS) Questions

In square *EFGH*, which of these is FALSE?
   a. △*EIF* and △*EIH* are congruent.
   b. △*GHI* and △*GHF* are congruent.
   c. △*EFH* and △*EGH* are congruent.
   d. △*EIF* and △*GIH* are congruent.

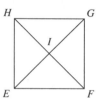

TIMSS, Grade 8, 2003

ABCD is a trapezoid.

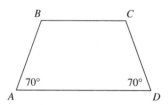

Another trapezoid, *GHIJ* (not shown), is congruent (the same size and shape as) to *ABCD*. Angles *G* and *J* each measure 70°. Which of these could be true?
   a. *GH* = *AB*
   b. Angle *H* is a right angle.
   c. All sides of *GHIJ* are the same length.
   d. The perimeter of *GHIJ* is 3 times the perimeter of *ABCD*.
   e. The area of *GHIJ* is less than the area of *ABCD*.

TIMSS, Grade 8, 2003

12-2   Other Congruence Properties

Angle, Side, Angle Property (ASA)

Triangles can be determined to be congruent by SSS and SAS. Can a triangle be constructed congruent to a given triangle by using two angles and a side? Figure 12-25 shows the construction of a triangle *A′B′C′* such that $\overline{A'C'} \cong \overline{AC}$, ∠*A′* ≅ ∠*A*, and ∠*C′* ≅ ∠*C*. It seems that △*A′B′C′* ≅ △*ABC*. This construction illustrates the **Angle, Side, Angle (ASA)** property of congruence.

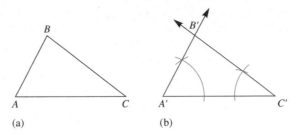

Figure 12-25

> **Theorem 12–6:   Angle, Side, Angle (ASA)**
>
> If two angles and the included side of one triangle are congruent to two angles and the included side of another triangle, respectively, then the triangles are congruent.

Example 12-4

In Figure 12-26, $\triangle ABC$ and $\triangle DEF$ have two pairs of angles congruent and a pair of sides congruent. Show that the triangles are congruent.

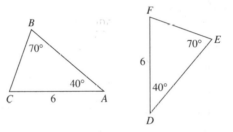

Figure 12-26

Solution   Notice that $\angle A \cong \angle D$ and $\angle B \cong \angle E$, which implies that $\angle C \cong \angle F$ because the measure of each is $180° - (70 + 40)°$. Then, since $\overline{AC} \cong \overline{DF}$, we have $\triangle ABC \cong \triangle DEF$ by ASA.

Example 12-4 is a special case of the following theorem (which can be justified in a similar way).

> **Theorem 12–7 Angle, Angle, Side (AAS)**
>
> If two angles and a *side opposite one of these two angles* of a triangle are congruent to the two corresponding angles and the corresponding side in another triangle, then the two triangles are congruent.

Notice in Theorem 12–7 the emphasis on the fact that the side in one triangle must correspond to the side in the other triangle. If that is not the case, two angles and a side do not assure congruence, as shown in Figure 12-27 where a side and two angles in $\triangle ABC$ are congruent to a side and two angles in $\triangle DEF$. In that figure, $\overline{AB} \cong \overline{EF}$, $\angle A \cong \angle D$, and $\angle E \cong \angle B$, but the triangles are not congruent. Side $\overline{AB}$ is not opposite one of the angles mentioned. Consequently, the condition AAS does not assure congruence unless the side is opposite one of the two angles.

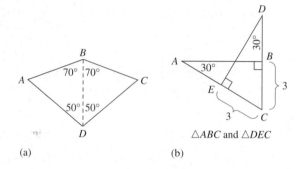

Figure 12-27

Example 12-5

Show that the triangles in each part of Figure 12-28 are congruent.

(a)

(b)

△ABC and △DEC

Figure 12-28

Solution   **a.** ∠ABD ≅ ∠CBD, $\overline{BD} \cong \overline{BD}$, and ∠ADB ≅ ∠CDB. Consequently, by ASA, △ABD ≅ △CBD.

b. Since ∠C ≅ ∠C, ∠ABC ≅ ∠DEC, and $\overline{EC} \cong \overline{BC}$, by AAS, △ABC ≅ △DEC.

NOW TRY THIS 12-7   Draw a parallelogram and one of its diagonals. Use the definition of a parallelogram given in Chapter 11 and the ASA congruence property to prove that opposite sides of a parallelogram are congruent.

Example 12-6

a. Using the definition of a parallelogram and the property that opposite sides in a parallelogram are congruent (see Now Try This 12-7), prove that the diagonals of a parallelogram bisect each other; that is, in Figure 12-29(a), show that $AO = OC$ and $BO = OD$.

b. Draw a line through the point O where the diagonals of a parallelogram intersect, as in Figure 12-29(b). The line intersects the opposite sides of the parallelogram at points P and Q. Prove that $\overline{OP} \cong \overline{OQ}$.

c. Prove that $\overline{SQ} \cong \overline{PT}$ in Figure 12-29(c).

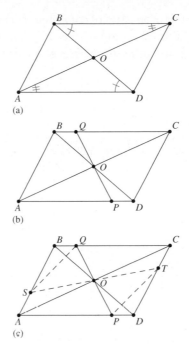

Figure 12-29

Solution **a.** In Figure 12-29(a), we need to show that $\overline{BO} \cong \overline{DO}$ and $\overline{AO} \cong \overline{CO}$. These segments appear in two pairs of triangles that seem to be congruent. One pair is $\triangle AOD$ and $\triangle COB$. To prove that these triangles are congruent, we notice that $\overline{AD} \cong \overline{CB}$ (see Now Try This 12-7). We have no information about the other corresponding sides; however, we know that $\overline{BC} \parallel \overline{AD}$. Consequently, $\angle CBO \cong \angle ADO$, as these are alternate interior angles formed by the parallel lines and the transversal $\overleftrightarrow{BD}$. Similarly, using the transversal $\overleftrightarrow{AC}$, we have $\angle OAD \cong \angle OCB$ (congruent angles have been similarly marked in Figure 12-29(a)). Now $\triangle AOD \cong \triangle COB$ by ASA. Because corresponding parts in these triangles are congruent, we have $\overline{AO} \cong \overline{OC}$ and $\overline{BO} \cong \overline{OD}$.

b. As in part (a), there are two pairs of triangles that have segments and $\overline{OQ}$ and $\overline{OP}$ as sides; one such pair is $\triangle AOP$ and $\triangle COQ$. In these triangles, $\overline{AO} \cong \overline{CO}$ (as was proved in part (a)). As in part (a), $\angle OAP \cong \angle OCQ$. Also, $\angle AOP \cong \angle COQ$ because these are vertical angles. Thus, $\triangle AOP \cong \triangle COQ$ (why?) and $\overline{OP} \cong \overline{OQ}$ because these are corresponding parts in these congruent triangles.

c. $\overline{OS} \cong \overline{OT}$ and $\overline{OQ} \cong \overline{OP}$ (using the results of part (b)). $\angle SOQ \cong \angle TOP$ because these are vertical angles. These imply $\triangle SQO \cong \triangle TPO$ by SAS so $\overline{SQ} \cong \overline{PT}$ by CPCTC.

NOW TRY THIS 12-8 If the diagonals of a quadrilateral bisect each other, must the quadrilateral be a parallelogram? Explain why or why not.

Using properties of congruent triangles, we can deduce various properties of quadrilaterals. Table 12-1 summarizes the definitions and lists some properties of six quadrilaterals. These and other properties of quadrilaterals are further investigated in Assessment 12-2.

Table 12-1

Quadrilateral and Its Definition	Theorems about the Quadrilateral
 *Trapezoid:* A quadrilateral with at least one pair of parallel sides	Consecutive angles between parallel sides are supplementary.
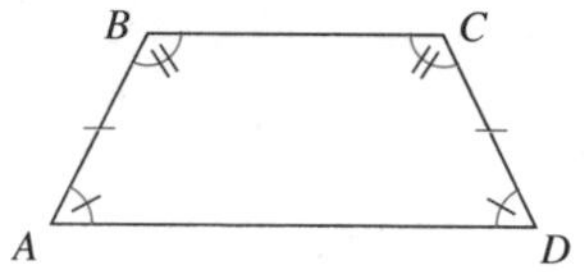 *Isosceles trapezoid:* A trapezoid with a pair of congruent base angles	**a.** Each pair of base angles (marked) are congruent. **b.** A pair of opposite sides are congruent. **c.** If a trapezoid has congruent diagonals it is isosceles.
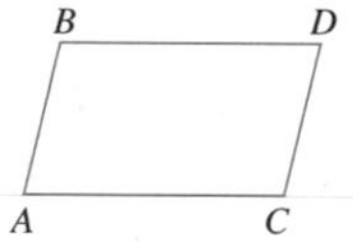 *Parallelogram:* A quadrilateral in which each pair of opposite sides is parallel	**a.** A parallelogram has all the properties of a trapezoid. **b.** Opposite sides are congruent. **c.** Opposite angles are congruent. **d.** Diagonals bisect each other. **e.** A quadrilateral in which the diagonals bisect each other is a parallelogram.
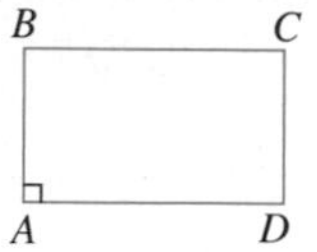 *Rectangle:* A parallelogram with a right angle	**a.** A rectangle has all the properties of a parallelogram. **b.** All the angles of a rectangle are right angles. **c.** A quadrilateral in which all the angles are right angles is a rectangle. **d.** The diagonals of a rectangle are congruent and bisect each other. **e.** A quadrilateral in which the diagonals are congruent and bisect each other is a rectangle.
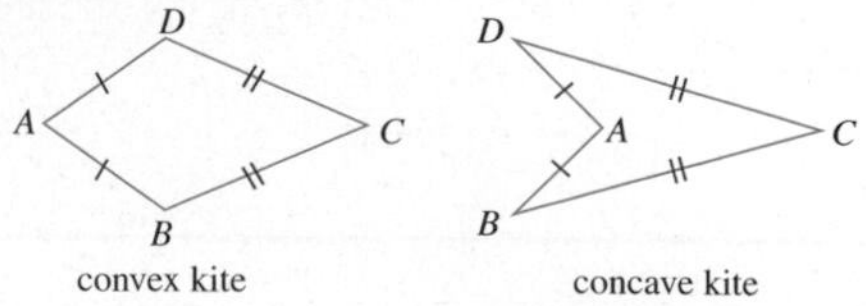 *Kite:* A quadrilateral with two disjoint pairs of congruent adjacent sides	**a.** Lines containing the diagonals are perpendicular to each other. **b.** A line containing one diagonal ($\overline{AC}$ in the figure) is a bisector of the other diagonal. **c.** One diagonal ($\overline{AC}$ in the figure) bisects nonconsecutive angles.

(continues)

Table 12-1 (continued)

Quadrilateral and Its Definition	Theorems about the Quadrilateral
Rhombus: A parallelogram with two adjacent sides congruent	**a.** A rhombus has all the properties of a parallelogram and a kite. **b.** A quadrilateral in which all the sides are congruent is a rhombus. **c.** The diagonals of a rhombus are perpendicular to and bisect each other. **d.** Each diagonal bisects opposite angles.
Square: A rectangle with all sides congruent	A square has all the properties of a parallelogram, a rectangle, and a rhombus.

The proofs of some of the theorems listed in Table 12-1 will be explored in the Assessments; meanwhile, in the following example we assume the theorems about parallelograms and prove a theorem for rectangles.

Example 12-7

Prove that a quadrilateral in which the diagonals are congruent and bisect each other is a rectangle.

Solution If *ABCD* is a quadrilateral whose diagonals bisect each other, we know from Table 12-1 that it is a parallelogram. It remains to be proved that *ABCD* in Figure 12-30 is a rectangle, that is, that one of its angles is a right angle (see definition of a rectangle). Let's try to show that ∠*BAD* in Figure 12-30 is a right angle. What do we know about the angles of *ABCD*? Because it is a parallelogram, ∠*BAD* and ∠*CDA* are supplementary (see property (a) of a parallelogram in Table 12-1). Thus, to show that each is a right angle, it will suffice to prove that the angles are congruent. These angles are in △*ABD* and △*DCA*, which are congruent by SSS since $\overline{AB} \cong \overline{DC}$ (in a parallelogram opposite sides are congruent), $\overline{AD} \cong \overline{DA}$, and $\overline{BD} \cong \overline{CA}$ (given). Therefore, ∠*BAD* ≅ ∠*CDA*. Since these angles are supplementary and congruent, each must be a right angle.

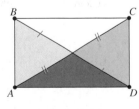

Figure 12-30

Example 12-8

In Figure 12-31(a), △*ABC* is a right triangle, and $\overline{CD}$ is a **median** (segment connecting a vertex to the midpoint of the opposite side) to the hypotenuse $\overline{AB}$. Prove that the median to the hypotenuse in a right triangle is half as long as the hypotenuse.

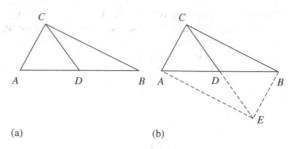

(a)   (b)

Figure 12-31

Solution   We are given that $\angle C$ is a right triangle and D is the midpoint of $\overline{AB}$. We need to prove that $2CD = AB$. In Figure 12-31(b), we extend $\overline{CD}$ to obtain $CE = 2CD$. Because D is the midpoint of $\overline{AB}$ and also the midpoint of $\overline{CE}$ (by our construction), the diagonals $\overline{AB}$ and $\overline{CE}$ of quadrilateral $ACBE$ bisect each other. Thus $ACBE$ is a parallelogram (see Table 12-1). We also know that $\angle C$ is a right angle (given). Thus, $ACBE$ is a rectangle. We know (Table 12-1) that the diagonals of a rectangle are congruent. Thus, $CE = AB$ and therefore $2CD = AB$, or $CD = \dfrac{1}{2}AB$.

Congruence of Quadrilaterals and Other Figures

What minimum conditions determine congruence of quadrilaterals? For example, does the SSS condition for triangles extend to the SSSS condition for quadrilaterals? Figure 12-32 shows a parallelogram with hinges at the vertices. Pulling any two vertices will give a new parallelogram with the same sides but different angles, and therefore infinitely many non-congruent parallelograms.

One way to determine a quadrilateral is to give directions for drawing it. Start with a side and tell by what angle to turn at the end of each side to draw the next side (the turn is by an exterior angle). Thus, SASAS seems to be a valid congruence condition for quadrilaterals (this condition can be proved by dividing a quadrilateral into triangles and using what we know about congruence of triangles).

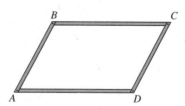

Figure 12-32

Assessment 12-2A

1. Use any tools to construct each of the following, if possible:
   a. A triangle with angles measuring 60° and 70° and an included side of 8 in.

   b. A triangle with angles measuring 60° and 70° and a nonincluded side of 8 cm on a side of the 60° angle
   c. A triangle with angles measuring 30°, 70°, and 80°

2. For each of the conditions in problem 1(a) through (c), is it possible to construct two noncongruent triangles? Explain why or why not.

3. For each of the following, determine whether the given conditions are sufficient to prove that $\triangle PQR \cong \triangle MNO$. Justify your answers.
   a. $\angle Q \cong \angle N$, $\angle P \cong \angle M$, $\overline{PQ} \cong \overline{MN}$
   b. $\angle R \cong \angle O$, $\angle P \cong \angle M$, $\overline{QR} \cong \overline{NO}$

4. A parallel ruler, shown as follows, can be used to draw parallel lines. The distance between the parallel segments $\overline{AB}$ and $\overline{DC}$ can vary. The ruler is constructed so that the distance between A and B equals the distance between D and C. The distance between A and C is the same as the distance between B and D. Explain why $\overline{AB}$ and $\overline{DC}$ are always parallel.

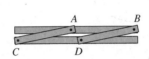
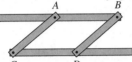

5. In each of the following, identify the word in the set {parallelogram, rectangle, rhombus, trapezoid, kite, or square} needed to make the resulting sentence true. If none of the words makes the sentence true, answer "none" and justify your answer.
   a. A quadrilateral is a _____ if, and only if, its diagonals bisect each other.
   b. A quadrilateral is a _____ if, and only if, its diagonals are congruent.
   c. A quadrilateral is a _____ if, and only if, its diagonals are perpendicular.

6. Create several trapezoids that have a pair of opposite nonparallel sides congruent. Measure all angles and make a conjecture about the relationships among pairs of angles.

7. Classify each of the following statements as true or false. If the statement is false, provide a counterexample.
   a. The diagonals of a square are perpendicular bisectors of each other.
   b. If all sides of a quadrilateral are congruent, the quadrilateral is a rhombus.
   c. If a rhombus is a square, it must also be a rectangle.
   d. A trapezoid is a parallelogram.

8. a. Construct quadrilaterals having exactly one, two, or four right angles.
   b. Can a quadrilateral have exactly three right angles? Why?
   c. Can a parallelogram have exactly two right angles? Why?

9. Quadrilateral $ABCD$ pictured is a kite. Prove that $\overleftrightarrow{BD} \perp \overleftrightarrow{AC}$.

10. Classify each of the following figures formed from regular hexagon $ABCDEF$.

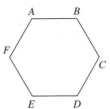

   a. Quadrilateral $ABDE$
   b. Quadrilateral $ABEF$

11. What is the shape that is formed by taking the inner piece of cardboard in a paper towel roll, carefully cutting along the seam, and flattening? How do you know?

12. The game of Triominoes has equilateral-triangular playing pieces with numbers at each vertex, shown as follows:

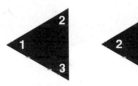

   If two pieces are placed together as shown in the following figure, explain what type of quadrilateral is formed:

13. A **sector** of a circle is a pie-shaped section bounded by two radii and an arc. What is a minimal set of conditions for determining that two sectors of the same circle are congruent?

14. In the rectangle $ABCD$ shown, X and Y are midpoints of the given sides, $DP = AQ$.

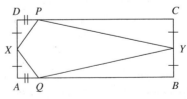

   a. What type of quadrilateral is $PYQX$? Prove your answer.
   b. If points P and Q are moved at a constant rate and in the same direction along $\overline{DC}$ and $\overline{AB}$, respectively, does this change your answer in part (a)? Why or why not?

15. Draw any triangle and make a copy of it. Can the two triangles be "put together" to form a parallelogram? Explain your answer.

16. a. Use a square to create a rep-tile.
    b. Could your pattern from part (a) be used with any rectangle to create a rep-tile?

17. A compact disc (CD) is circular. Why do you think that most CD cases are square?

18. Draw two quadrilaterals such that two angles and an included side in one quadrilateral are congruent, respectively, to two angles and an included side in the other quadrilateral. However, the quadrilaterals are not congruent.

19. Using a straightedge and a compass, construct any convex kite. Then construct a second convex kite that is not congruent to the first but whose sides are congruent to the corresponding sides of the first kite.

20. What type of figure is formed by joining the midpoints of a rectangle (see the following figure)? Justify your answer.

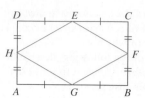

21. What minimum information is sufficient to determine congruency for each of the following?
    a. Two squares
    b. Two rectangles
    c. Two parallelograms

22. Describe a set of minimal conditions to determine if two regular polygons are congruent.

Assessment 12-2B

1. Use any tools to construct each of the following, if possible:
   a. A right triangle with one acute angle measuring 75° and a leg of 5 cm on a side of the 75° angle
   b. A triangle with angles measuring 30°, 60°, and 90°

2. For each of the conditions in problem 1(a) and (b), is it possible to construct two noncongruent triangles? Explain why or why not.

3. In both the ASA and AAS properties of congruence, if two angles of one triangle are congruent, respectively, to two angles of another triangle, what must be true about the third angles of the triangles? Justify your answer.

4. a. For two right triangles, give a minimal set of conditions based on ASA and AAS to argue that the two triangles are congruent.
   b. Using your answer in (a), write two theorems that can be used to show that right triangles are congruent.

5. For each of the following, determine whether the given conditions are sufficient to prove that $\triangle PQR \cong \triangle MNO$. Justify your answers.
   a. $\overline{PQ} \cong \overline{MN}$, $\overline{PR} \cong \overline{MO}$, $\angle N \cong \angle Q$
   b. $\angle P \cong \angle M$, $\angle Q \cong \angle N$, $\angle R \cong \angle O$

6. In each of the following, choose one word, *parallelogram*, *rectangle*, *rhombus*, *trapezoid*, *kite*, or *square*, so that the resulting sentence is true. If none of the words makes the sentence true, answer "none."
   a. A quadrilateral is a _____ if, and only if, its diagonals are congruent and bisect each other.
   b. A quadrilateral is a _____ if, and only if, its diagonals are perpendicular and bisect each other.
   c. A quadrilateral is a _____ if, and only if, its diagonals are congruent and perpendicular and they bisect each other.

   d. A quadrilateral is a _____ if, and only if, a pair of opposite sides is parallel and congruent.

7. *ABCD* in the following figure is a trapezoid in which $\overline{AD}\|\overline{BC}$ and $AD > BC$. Prove that if $AB = CD$ then $\angle A \cong \angle D$. (*Hint:* through *C* draw a line parallel to $\overline{AB}$.)

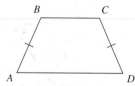

8. Classify each of the following statements as true or false. If the statement is false, provide a counterexample.
   a. No rectangle is a rhombus.
   b. No trapezoid is a square.
   c. Some squares are trapezoids.
   d. A rhombus is a parallelogram.
   e. A square is a rhombus.

9. a. Construct a pentagon with all sides congruent and exactly three right angles.
   b. Can a quadrilateral have four acute angles? Why?
   c. Can a kite have exactly two right angles? Why?

10. Are $\angle BAD$ and $\angle DCB$ congruent? Explain why or why not.

11. Classify each of the figures formed from regular hexagon *ABCDEF*.

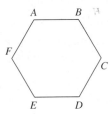

   a. Quadrilateral *ABCD*
   b. Quadrilateral *ABCE*

12. a. What is the maximum number of 3" × 5" rectangular pieces of paper that can be cut from an 8 1/2" × 11" sheet of paper?
   b. Are all 3" × 5" rectangular cards congruent? Why or why not?

13. Each fourth grader is given a protractor, two 30-cm sticks, and two 20-cm sticks and is asked to form a quadrilateral with an interior 75° angle. Sketch all possibilities.

14. Draw a quadrilateral such that no circle can circumscribe the quadrilateral. Explain why no such circle exists.

15. Can a quadrilateral have three obtuse angles? Why or why not?

16. In the following quadrilateral, *AB = AC = AD*. Prove that $m(\angle 1) + m(\angle 4) = m(\angle 2) + m(\angle 3)$.

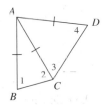

17. Draw two quadrilaterals such that two 45° angles and an included side in one quadrilateral are congruent, respectively, to two angles and an included side in the other quadrilateral. However, the quadrilaterals are not congruent.

18. a. What type of figure is formed by joining the midpoints of the sides of a parallelogram?
   b. Justify your answer to part (a).

19. What minimum information is sufficient to determine congruency for each of the following?
   a. Two rectangles
   b. Two kites
   c. Two isosceles triangles

20. Suppose polygon *ABCD* shown in the following figure is any parallelogram:

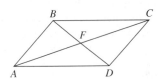

Use congruent triangles to justify each of the following:
   a. $\angle A \cong \angle C$ and $\angle B \cong \angle D$ (opposite angles are congruent).
   b. $\overline{BC} \cong \overline{AD}$ and $\overline{AB} \cong \overline{CD}$ (opposite sides are congruent).
   c. $\overline{BF} \cong \overline{DF}$ and $\overline{AF} \cong \overline{CF}$ (the diagonals bisect each other).
   d. $\angle DAB$ and $\angle ABC$ are supplementary.

Mathematical Connections 12-2

Communication

1. Stan is standing on the bank of a river wearing a baseball cap. Standing erect and looking directly at the other bank, he pulls the bill of his cap down until it just obscures his vision of the opposite bank. He then turns around, being careful not to disturb the cap, and picks out a spot that is just obscured by the bill of his cap. He then paces off the distance to this spot and claims that the distance across the river is approximately equal to the distance he paced. Is Stan's claim true? Why?

2. Most ironing boards are collapsible for storage and can be adjusted to fit the height of the person using them. The surface of the board, though, remains parallel to the floor regardless of the height. Explain how to construct the legs

of an ironing board to ensure the surface is always parallel to the floor.

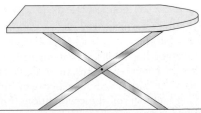

3. Using the fact that the perpendicular bisector of the base of an isosceles triangle is also the angle bisector, explain how to construct the angle bisector of a given angle using only a straightedge and a compass.

4. The marked angles and a side in the two triangles are congruent. Can you conclude that the triangles are congruent? Why or why not?

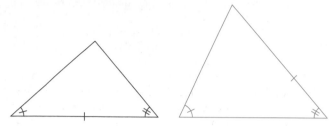

5. In the following figure, point P is on the angle bisector of $\angle A$. The distance from P to one of the sides of the angle is PB and to the other side it is PC. It seems that $PB = PC$. Answer the following:

a. Pick another point on the angle bisector and construct the distances from that point to the sides of the angle. What seems to be true about the distances?

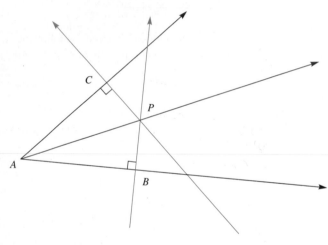

b. State a conjecture about the points on the angle bisector of an angle by completing the following sentence:

The distances from every point on the angle bisector of an angle . . .

c. Prove your conjecture in part (b).
d. State and prove the converse of your statement in part (b).

Open-Ended

6. By definition, two quadrilaterals are congruent if there is a one-to-one correspondence between the vertices such that corresponding angles are congruent and corresponding sides are congruent. Thus, there are eight relations—four between sides and four between angles. If the following number of these conditions hold, will the quadrilaterals be congruent?
a. 7   **b.** 4

7. On a 4 by 4 geoboard or dot paper (with 16 dots), pose and solve questions concerning the congruence of the following:

a. Triangles.
b. Isosceles trapezoids.
c. Quadrilaterals that are neither squares nor rectangles.

Cooperative Learning

8. a. Record the definitions of *trapezoid* and *kite* given in different Grade 6–8 and secondary-school geometry textbooks.
b. Compare the definitions found with those in this text and with those other groups found.
c. Defend the use of one definition over another.

Questions from the Classroom

9. A student claims that polygon $ABCD$ in the following drawing is a parallelogram if $\angle 1 \cong \angle 2$. Is he correct? Why?

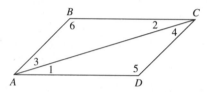

10. A student asks why "congruent" rather than "equal" is used to discuss triangles that have the same size and shape. What do you say?

11. A student asks why triangles often appear in structures such as bridges, building frames, and trusses. How do you respond?

12. A student says that she knows that a parallelogram has two altitudes but when it is long and narrow like the one shown, she does not know how to construct with any tools the altitude to the shorter sides. How do you respond?

13. A student claims that if a line ℓ is drawn through vertex A of $\triangle ABC$ parallel to $\overline{BC}$ and a new point A' is chosen anywhere on ℓ, then the height of $\triangle A'BC$ to side $\overline{BC}$ stays the same. He would like to know if this is true and if so why. How do you respond?

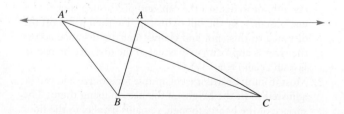

Review Problems

14. In the following regular pentagon, use the existing vertices to find all the triangles congruent to $\triangle ABC$. Show that the triangles actually are congruent.

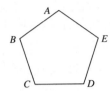

15. If possible, construct a triangle that has the three segments a, b, and c shown here as its sides; if not possible, explain why not.

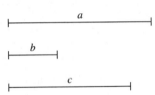

16. Construct an equilateral triangle whose sides are congruent to the following segment:

17. For each of the following pairs of triangles, determine whether the given conditions are sufficient to show that the triangles are congruent. If the triangles are congruent, tell which property can be used to verify this fact.

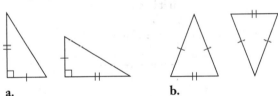

a.   **b.**

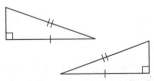

c.

Third International Mathematics and Science Study (TIMSS) Questions

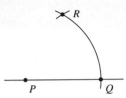

In the figure above, an arc of a circle with center P has been drawn to cut the line at Q. Then an arc with the same radius and center Q was drawn to cut the first arc at R. What would be the size of angle PRQ?

a. 30°
b. 45°
c. 60°
d. 75°

TIMSS, Grade 8, 2003

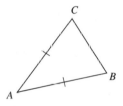

The triangle ABC has $AB = AC$.
Draw a line to divide triangle ABC into two congruent triangles

TIMSS, Grade 8, 2003

TECHNOLOGY CORNER

a. Use *The Geometer's Sketchpad* and inductive reasoning to test the truth of the following statement: "If one pair of sides of a quadrilateral is congruent and parallel, then the quadrilateral is a parallelogram."
b. Prove the conjecture in part (a).

12-3   Other Constructions

We use the definition of a rhombus and the following properties (also listed in Table 12-1) to accomplish basic compass-and-straightedge constructions:

1. A rhombus is a parallelogram in which all the sides are congruent.
2. A quadrilateral in which all the sides are congruent is a rhombus.
3. Each diagonal of a rhombus bisects the opposite angles.
4. The diagonals of a rhombus are perpendicular.
5. The diagonals of a rhombus bisect each other.

Constructing a Parallel Line

To construct a line parallel to a given line ℓ through a point P not on ℓ, as in the leftmost panel of Figure 12-33, our strategy is to construct a rhombus (using property 2 listed above) with one of its vertices at P and one of its sides on line ℓ. Because the opposite sides of a rhombus are parallel, one of the sides through P will be parallel to ℓ. This construction is shown in Figure 12-33.

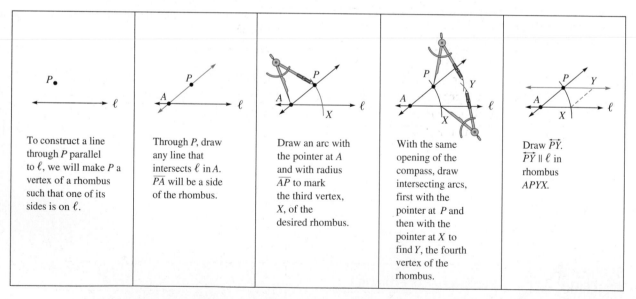

Figure 12-33 Constructing parallel lines (rhombus method)

Figure 12-34 shows another way to construct parallel lines. If congruent corresponding angles are formed by a transversal cutting two lines, then the lines are parallel. Thus, the first step is to draw a transversal through P that intersects ℓ. The angle marked α is formed by the transversal and line ℓ. By constructing an angle with a vertex at P congruent to α, we create congruent corresponding angles; therefore, $\overleftrightarrow{PQ} \parallel \ell$.

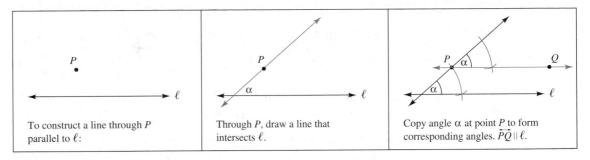

Figure 12-34 Constructing parallel lines (corresponding angle method)

NOW TRY THIS 12-9 Parallel lines are frequently constructed using either a ruler and one triangle or two triangles. If a ruler and a triangle are used, the ruler is left fixed and the triangle is slid so that one side of the triangle touches the ruler at all times. In Figure 12-35, the hypotenuses of the right triangles are all parallel (also the legs not on the ruler are all parallel). How can this method be used to construct a line through a given point parallel to a given line?

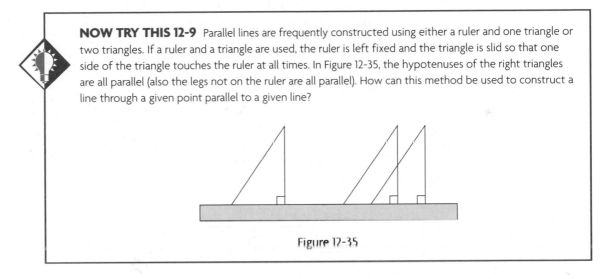

Figure 12-35

Paper folding can be used to construct parallel lines. For example, in Figure 12-36(a) to construct a line m parallel to line p through point Q, we can fold a perpendicular to line p so that the fold line does not contain point Q, as shown in Figure 12-36(b). Then by marking the image of point Q and connecting point Q and its image, Q', we have $\overleftrightarrow{QQ'}$ parallel to line p (why?).

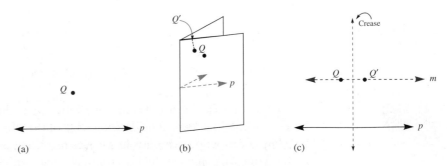

Figure 12-36

Constructing Angle Bisectors

Another construction based on a property of a rhombus is a construction of an **angle bisector**, a ray that separates an angle into two congruent angles. The diagonal of a rhombus with vertex A bisects $\angle A$, as shown in Figure 12-37.

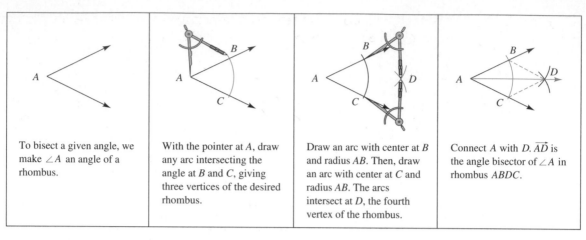

| To bisect a given angle, we make ∠A an angle of a rhombus. | With the pointer at A, draw any arc intersecting the angle at B and C, giving three vertices of the desired rhombus. | Draw an arc with center at B and radius AB. Then, draw an arc with center at C and radius AB. The arcs intersect at D, the fourth vertex of the rhombus. | Connect A with D. $\overrightarrow{AD}$ is the angle bisector of ∠A in rhombus ABDC. |

Figure 12-37 Bisecting an angle (rhombus method)

◆ *Research Note*

Students who study only two-dimensional figures in early grades may have difficulty discriminating among terms for two-dimensional and three-dimensional figures in later years. Constructions involving folding may help (Nieuwoudt and van Niekerk, 1997). ◆

REMARK Another way to bisect an angle can be devised by using the fact that a perpendicular bisector of the base of an isosceles triangle contains the vertex and is the angle bisector of the opposite angle. Actually, the steps of the construction can be the same as in Figure 12-37, where △BAC is an isosceles triangle and $\overleftrightarrow{AD}$ is the perpendicular bisector of $\overline{BC}$, because both A and D are equidistant from the endpoints of $\overline{BC}$.

We can bisect an angle by folding a line through the vertex so that one side of the angle folds onto the other side. For example, in Figure 12-38 we bisect ∠ABC by folding and creasing the paper through the vertex B so that $\overrightarrow{BC}$ coincides with $\overrightarrow{BA}$.

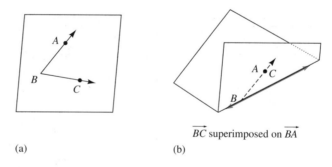

$\overrightarrow{BC}$ superimposed on $\overrightarrow{BA}$

(a)   (b)

Figure 12-38

A Mira is a plastic device that acts as a reflector so that the image of an object can be seen behind the Mira. The drawing edge of the Mira acts as a folding line on paper. Any construction demonstrated in this text using paper folding can also be done with a Mira. To construct the bisector of an angle with a Mira, we place the drawing edge of the Mira on the vertex of the angle and reflect one side of the angle onto the other, as shown in Figure 12-39.

Angle bisector

Figure 12-39

Constructing Perpendicular Lines

To construct a line through P perpendicular to line ℓ, where P is not a point on ℓ, as in Figure 12-40, recall that the diagonals of a rhombus are perpendicular to each other. If we construct a rhombus with a vertex at P and a diagonal on ℓ, as in Figure 12-40, the segment connecting the fourth vertex Q to P is perpendicular to ℓ.

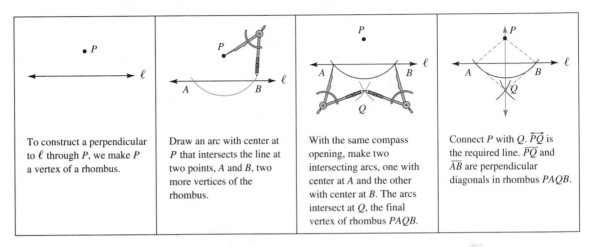

| To construct a perpendicular to ℓ through P, we make P a vertex of a rhombus. | Draw an arc with center at P that intersects the line at two points, A and B, two more vertices of the rhombus. | With the same compass opening, make two intersecting arcs, one with center at A and the other with center at B. The arcs intersect at Q, the final vertex of rhombus $PAQB$. | Connect P with Q. $\overleftrightarrow{PQ}$ is the required line. $\overleftrightarrow{PQ}$ and $\overline{AB}$ are perpendicular diagonals in rhombus $PAQB$. |

Figure 12-40 Constructing a perpendicular to a line from a point not on the line

In Section 12-1 we saw how to construct the perpendicular bisector of a segment using a property of a perpendicular bisector stated in Theorem 12–5. Here we show how a property of a rhombus can also be used for constructing the perpendicular bisector of a segment.

To construct the perpendicular bisector of a line segment, as in Figure 12-41, we use the fact that the diagonals of a rhombus are perpendicular bisectors of each other.

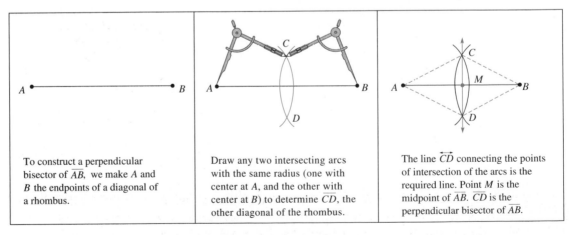

| To construct a perpendicular bisector of $\overline{AB}$, we make A and B the endpoints of a diagonal of a rhombus. | Draw any two intersecting arcs with the same radius (one with center at A, and the other with center at B) to determine $\overline{CD}$, the other diagonal of the rhombus. | The line $\overleftrightarrow{CD}$ connecting the points of intersection of the arcs is the required line. Point M is the midpoint of $\overline{AB}$. $\overline{CD}$ is the perpendicular bisector of $\overline{AB}$. |

Figure 12-41 Bisecting a line segment

The construction yields a rhombus such that the original segment is one of the diagonals of the rhombus and the other diagonal is the perpendicular bisector, as in Figure 12-41.

Constructing a perpendicular to a line ℓ at a point M on ℓ is based on the fact that the diagonals of a rhombus are perpendicular bisectors of each other. Observe in Figure 12-41 that $\overline{CD}$ is a perpendicular to $\overline{AB}$ through M. Thus we construct a rhombus whose diagonals intersect at point M, as in Figure 12-42.

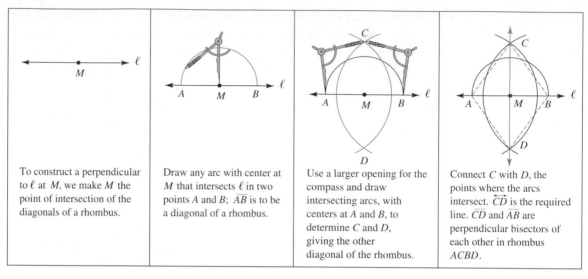

To construct a perpendicular to ℓ at *M*, we make *M* the point of intersection of the diagonals of a rhombus.	Draw any arc with center at *M* that intersects ℓ in two points *A* and *B*; $\overline{AB}$ is to be a diagonal of a rhombus.	Use a larger opening for the compass and draw intersecting arcs, with centers at *A* and *B*, to determine *C* and *D*, giving the other diagonal of the rhombus.	Connect *C* with *D*, the points where the arcs intersect. $\overleftrightarrow{CD}$ is the required line. $\overleftrightarrow{CD}$ and $\overline{AB}$ are perpendicular bisectors of each other in rhombus *ACBD*.

Figure 12-42 Constructing a perpendicular to a line from a point on the line

Perpendicularity constructions can also be completed by means of paper folding or by using a Mira. To use paper folding to construct a perpendicular to a given line ℓ at a point *P* on the line, we fold the line onto itself, as shown in Figure 12-43(a). The fold line is perpendicular to ℓ. To perform the construction with a Mira, we place the Mira with the drawing edge on *P*, as shown in Figure 12-43(b), so that ℓ is reflected onto itself. The line along the drawing edge is the required perpendicular.

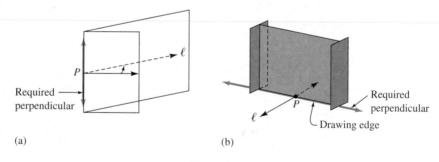

(a)   (b)

Figure 12-43

Constructing perpendiculars is useful in locating altitudes of a triangle. The construction of altitudes is described in Example 12-9.

Example 12-9   Given triangle *ABC*, construct an altitude from vertex *A* in each part of Figure 12-44.

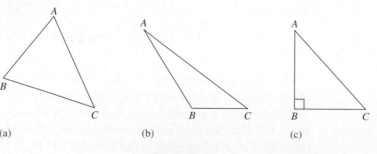

(a)   (b)   (c)

Figure 12-44

Solution   **a.** An altitude is the segment perpendicular from a vertex to the line containing the opposite side of a triangle, so we need to construct a perpendicular from point A to the line containing $\overline{BC}$. Such a construction is shown in Figure 12-45. $\overline{AD}$ is the required altitude.

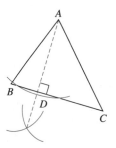

Figure 12-45

b. The construction of the altitude from vertex A is shown in Figure 12-46. Notice that the required altitude $\overline{AD}$ does not intersect the interior of $\triangle ABC$.

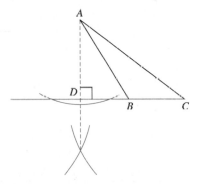

Figure 12-46

c. Triangle ABC is a right triangle. The altitude from vertex A is the side $\overline{AB}$. No construction is required.

TECHNOLOGY CORNER   Use *The Geometer's Sketchpad* in each of the following:

a. Draw all the altitudes of a triangle. Make conjectures about the lines containing the altitudes of each of the following types of triangles: acute, right, and obtuse.
b. Draw an angle and its angle bisector. Pick a point on the angle bisector and calculate the distance from the point to each side of the angle.
c. Move the point to different positions on the angle bisector. What do you notice about the distance?

Properties of Angle Bisectors

Consider the angle bisector in Figure 12-47. It seems that any point P on the angle bisector is equidistant from the sides of the angle; that is, $PD = PE$. (*The distance from a point to a line is the length of the perpendicular segment from the point to the line.*)

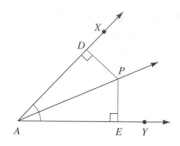

Figure 12-47

To justify this, we find two congruent triangles that have these segments as corresponding sides. The only triangles pictured are $\triangle ADP$ and $\triangle AEP$. Because $\overrightarrow{AP}$ is the angle bisector, $\angle DAP \cong \angle EAP$. Also, $\angle PDA$ and $\angle PEA$ are right angles and are thus congruent. $\overline{AP}$ is congruent to itself, so $\triangle PDA \cong \triangle PEA$ by AAS. Thus, $\overline{PD} \cong \overline{PE}$ because they are corresponding parts of congruent triangles PDA and PEA. Consequently, we have proved the first part of the following theorem.

Theorem 12–8

a. Any point P on an angle bisector is equidistant from the sides of the angle.
b. Any point that is equidistant from the sides of an angle is on the angle bisector of the angle.

> **REMARK** Both parts of the theorem can be stated as: A point is on an angle bisector of an angle if, and only if, it is equidistant from the sides of the angle.

Notice that we have proved only Theorem 12–8(a). You are asked to demonstrate part (b) in Now Try This 12-10.

 NOW TRY THIS 12-10 Prove that if a point is in the interior of an angle and is equidistant from the sides of the angle, the point must be on the angle bisector of that angle.

The Incenter of a Triangle

The intersection of any two angle bisectors of a triangle is the **incenter** of the triangle. In Figure 12-48, P is the intersection of the angle bisectors of $\angle A$ and of $\angle B$. We will show that the third angle bisector, the bisector of $\angle C$, also goes through P, which will imply that the three angle bisectors intersect in a single point; that is, are *concurrent*. From P we

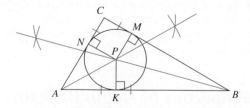

Figure 12-48

construct the perpendiculars to the three sides of the triangle, $\overline{PN}$, $\overline{PK}$, and $\overline{PM}$. Because P is on the angle bisector of $\angle A$, from Theorem 12–8(a) we know that

1. $PN = PK$

Similarly, because P is on the angle bisector of $\angle B$, we know that

2. $PK = PM$

Now from (1) and (2) we conclude that $PN = PM$. But this means that P is equidistant from the sides of $\angle C$. Thus, by Theorem 12–8(b), P is on the angle bisector of $\angle C$. We have proved the following theorem.

> **Theorem 12–9**
>
> The angle bisectors of a triangle are concurrent and the three distances from the point of inter-section to the sides are equal.

Notice that if in Figure 12-48 we draw a circle with the center P and radius PM (or PK or PN), the circle seems to fit exactly inside the triangle. In fact each line segment "touches" the circle at one point only. The sides of the triangle are **tangent** to the circle. *A line is tangent to a circle if it intersects the circle in one and only one point.* It is possible to show that a line perpendicular to a radius at the endpoint of the radius that is not the center is tangent to the circle. Thus, the sides of the $\triangle ABC$ in Figure 12-48 are tangent to the circle. Such a circle is **inscribed** in the triangle. The center of the inscribed circle (the **incircle**) is the **incenter** of the triangle.

REMARK To find the radius of the inscribed circle, it is sufficient to construct just one perpendicular (to one of the sides). The point of intersection of the perpendicular with the corresponding side when connected with P determines the radius.

 NOW TRY THIS 12-11 Use any tools to investigate whether a circle can be inscribed in a square.

Assessment 12-3A

1. Refer to the following figure and use a compass and a straightedge to construct a line m through P parallel to ℓ, using each of the following:
  a. Alternate interior angles
  b. Alternate exterior angles

$P \bullet$

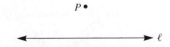

ℓ

2. For each of the following items, compare these methods of construction: paper folding, Mira, compass and straight-edge, and *The Geometer's Sketchpad* (if available).

  a. Bisector of $\angle A$
  b. Perpendicular bisector of $\overline{AB}$
  c. Perpendicular from point P to line m

3. Construction companies avoid vandalism at night by hanging expensive pieces of equipment from the boom of a crane, as shown in the following figure:

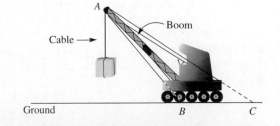

a. If you consider a triangle with two vertices A and B as marked in the figure and the intersection of a line through the cable holding the equipment and the ground as the third vertex, what type of triangle is formed?

b. If you consider the triangle formed by points A, B, and C, describe where the altitude containing vertex A of the triangle is.

4. Construct the perpendicular bisectors of each of the sides of the following triangles. Use any desired method.

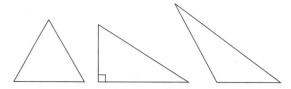

a. Make a conjecture about the perpendicular bisectors of the sides of an acute triangle.

b. Make a conjecture about the perpendicular bisectors of the sides of a right triangle.

c. Make a conjecture about the perpendicular bisectors of the sides of an obtuse triangle.

d. For each of the three triangles, construct the circle that circumscribes the triangle.

5. a. Given triangle ABC, as in the following, construct a point P that is equidistant from the three vertices of the triangle. Explain why your construction is correct.

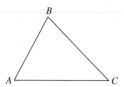

b. Repeat (a) for the following obtuse triangle:

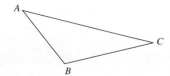

6. Draw an acute triangle and an obtuse triangle and in each case construct the circle inscribed in the triangle. Explain how to find the radius of the circle

7. For which of the following figures is it possible to construct a circle that is inscribed in the figure (each side of the figure must be tangent to the circle)? If it is possible to find such a circle, draw the figure and construct the inscribed circle using only a straightedge and a compass; if not, explain how you decided that it is impossible to find an inscribed circle.

a. A rectangle

b. A rhombus

c. A regular hexagon

8. Given $\overline{AB}$ in the following figure, construct a square with $\overline{AB}$ as a side:

9. Given A, B, and C as vertices, use a compass and a straightedge to construct a parallelogram $ABCD$:

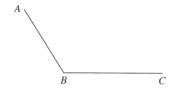

10. In right triangle ABC, point O is the incenter; $CD = 2$"; $AC = 8$"; $AB = 10$" Find y if $DB = y$.

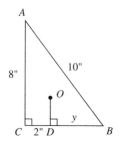

11. a. Show that any parallelogram can be dissected (cut apart) to form two non-isosceles trapezoids.

b. Prove that any two congruent trapezoids can always be put together to form a parallelogram.

12. Describe how to construct the incircle of a regular pentagon.

13. In the parallelogram $ABCD$ shown, find x if $x = EF$.

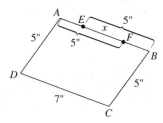

14. Some car jacks are constructed like a collapsed rhombus. When used to raise a car, the rhombus is easily seen.

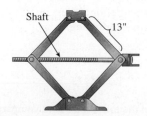

If the sides of the rhombus are 13", what is the maximum length of the shaft?

15. If *INTO* is a rhombus and *X* is any point on $\overline{NT}$ (except *N* and *T*), what type of quadrilateral is *IXTO*? Why?

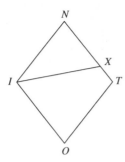

16. Using any tools, construct each of the following, if possible. If the construction is not possible, explain why.
    a. A square, given one side
    b. A rectangle, given one diagonal
    c. A triangle with two obtuse angles
    d. A parallelogram with exactly three right angles

17. Using only a compass and a straightedge, construct angles with each of the following measures:
    a. 30°
    b. 45°
    c. 75°

18. Copy $\overline{AB}$ in the following figure close to the bottom of a blank page. Use a compass and a straightedge to construct the perpendicular bisector of $\overline{AB}$. You are not allowed to put any marks below $\overline{AB}$.

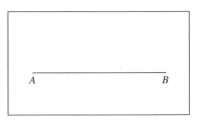

19. **a.** Explain why the hypotenuses in Figure 12-35 are all parallel.
    b. Use the "sliding triangle" method described in Figure 12-35 to construct a line through a point *P* parallel to a line *ℓ*.

20. Draw a convex quadrilateral similar to the one shown below and construct each of the following points, if possible. If it is not possible, explain why. Describe each construction in words and explain why it produces the required point.
    a. The point that is equidistant from $\overline{AB}$ and $\overline{AD}$ and also from points *B* and *C*
    b. The point that is equidistant from $\overline{AB}$, $\overline{AD}$, and $\overline{BC}$

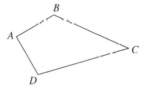

21. Describe how to inscribe a square in any circle with a given center. Explain why your construction produces a square.

Assessment 12-3B

1. Draw a figure like the one below, then use a compass and a straightedge to construct line *m* through *P* parallel to *ℓ*, using each of the following:
   a. Perpendicular lines
   b. A quadrilateral whose diagonals bisect each other.

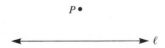

2. Construct an obtuse triangle and the perpendicular bisectors of two of its sides. Then construct the circle that circumscribes the triangle.

3. Draw a large obtuse triangle and construct the circle inscribed in the triangle. Why is the center of the circle equidistant from the sides of the triangle?

4. For which of the following figures is it possible to construct a circle that circumscribes the figure (each vertex of the figure must be on the circle)? If it is possible to find

such a circle, draw the figure and construct the circumscribed circle using only a straightedge and a compass; if not, explain how you decided that it is impossible to find the circumscribing circle.
   a. An isosceles trapezoid
   b. A parallelogram that is not a rectangle
   c. A rectangle

5. Given $\overline{AB}$ in the following figure, construct a square with $\overline{AB}$ as a diagonal:

6. Suppose you are "charged" 10¢ each time you use your straightedge to draw a line segment and 10¢ each time you use your compass to draw an arc. Using only a compass and a straightedge, determine the cheapest way to construct an equilateral triangle.

7. In the drawing, $\overline{AX}$ bisects $\angle BAC$; $AB = AC$; $m(\angle BCA) = 72°$, find *BX*.

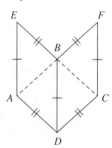

8. In the picture A, B, F are collinear, as are E, B, C. What is the most that can be concluded about the type of figure $ABCD$ is? Prove your answer.

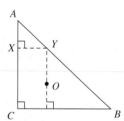

9. In the isosceles right triangle shown, O is the incenter; the radius of the inscribed circle is r. Find AX.

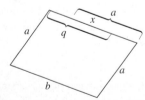

10. Describe how to construct the incircle of a regular decagon.

11. In the parallelogram shown, find x in terms of a and b.

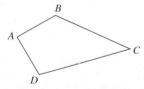

12. In the following concave quadrilateral $APBQ$, $\overline{PQ}$ and $\overline{AB}$ are the diagonals; $\overline{AP} \cong \overline{BP}$, $\overline{AQ} \cong \overline{BQ}$, and $\overline{PQ}$ has been extended until it intersects $\overline{AB}$ at C:

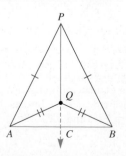

a. Make a conjecture concerning $\overline{PQ}$ and $\overline{AB}$.
b. Justify your conjecture in (a).
c. Make conjectures concerning the relationships between $\overrightarrow{PQ}$ and $\angle APB$ and between $\overrightarrow{QC}$ and $\angle AQB$.
d. Justify your conjectures in (c).

13. If two pieces of tape of the same width cross each other, what type of parallelogram is $ABCD$? Why?

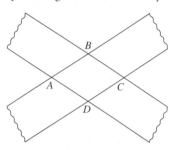

14. Using any tools, construct each of the following, if possible. If the construction is not possible, explain why.
a. A square, given one diagonal
b. A parallelogram, given two of its adjacent sides
c. A rhombus, given its diagonals
d. A parallelogram given a side and all the angles

15. Given an equilateral triangle and using only a compass and a straightedge, construct angles with each of the following measures:
a. 15°
b. 105°
c. 120°

16. Draw a line ℓ and a point P not on the line and use the sliding triangle method described in Figure 12-35 to construct a perpendicular to ℓ through P using a straightedge and a right triangle.

17. Draw a convex quadrilateral similar to the one shown below and construct each of the following points, if possible. If it is not possible, explain why. Describe each construction in words and explain why it produces the required point.
a. The point that is equidistant from A, B, and C
b. The point that is equidistant from A, B, C, and D
c. The point that is equidistant from the four sides

18. Describe how to inscribe an equilateral triangle in any circle. Explain why your construction produces an equilateral triangle.

19. Describe how to inscribe a regular octagon in a given circle with a given center.

Mathematical Connections 12-3

Communication

1. Given an angle and a roll of tape, describe how you might construct the bisector of the angle.

2. Write a letter from you, a curriculum developer, to parents explaining whether or not the geometry curriculum in Grades 5–8 should include construction problems that use only a compass and straightedge.

3. Patty paper constructions are accomplished using the waxed paper that is sometimes applied between hamburger patties by commercial meat companies. Research patty paper constructions and organize a presentation on them.

4. If a line is tangent to a circle at point P, and A is any point on the line, then $\overline{AP}$ is a *tangent segment*. Use this definition to prove that the two tangent segments from a point A outside the circle to the circle are congruent. That is, prove that $\overline{AP} \cong \overline{AQ}$.

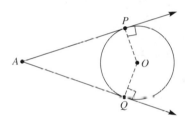

Open-Ended

5. The outer edge of a CD and the inner edge of a CD are concentric circles.
   a. Define *concentric circles*.
   b. Find at least six illustrations of the use of concentric circles.

6. Explain whether or not there is a single perpendicular from a point to a line on a sphere if a line is defined as a great circle of the sphere (a circle obtained by intersecting the sphere with a plane through its center).

7. Explain whether or not there are parallel lines on a sphere.

Cooperative Learning

8. In your group, perform the paper-folding construction shown and answer the questions that follow. Let P and Q be two opposite vertices of a rectangular piece of paper, as shown in (i). Fold P onto Q so that a crease is formed. This results in (ii), where A and B are the endpoints of the crease. Next crease again so that point A folds onto point B. This results in (iii). Next, unfold to obtain two creases, as in (iv).

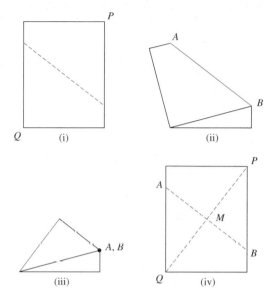

a. Individually write an explanation of why every point on $\overline{AB}$ is equidistant from the endpoints of $\overline{PQ}$. Compare the explanations in your group and prepare one explanation to be presented to the class.

b. Discuss in your group whether $APBQ$ is a rhombus.

Questions from the Classroom

9. A student wonders whether the following construction of a tangent to a circle from an exterior point P is a valid construction. How do you respond?

Rotate the ruler about point P counterclockwise until it just touches the circle as shown in the figure below:

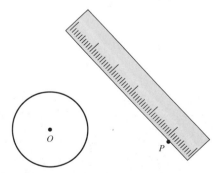

10. A student wants to know how to bisect a given angle drawn on a large sheet of paper using only an unmarked ruler. How do you respond?

11. Gail is asked to draw a triangle and to construct its incircle. She finds the point O where two angle bisectors intersect and point D where the angle bisector $\overrightarrow{BO}$ intersects the opposite side and draws the circle with center O and radius OD. The teacher marks the construction "wrong." Gail wants to know why. How do you respond?

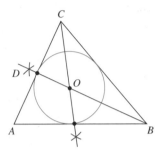

Review Problems

12. In the following figure, $\overleftrightarrow{AB} \parallel \overleftrightarrow{ED}$ and $\overline{BC} \cong \overline{CE}$. Explain why $DE = AB$.

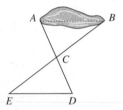

13. Draw $\triangle ABC$. Then construct $\triangle PQR$ congruent to $\triangle ABC$ using each of the following combinations:
    a. Two sides of $\triangle ABC$ and an angle included between these sides
    b. The three sides of $\triangle ABC$
    c. Two angles and a side included between these angles
14. In two right triangles, $\triangle ABC$ and $\triangle DEF$, if $\angle A$ and $\angle D$ are congruent and $\overline{AC}$ and $\overline{DF}$ are congruent, what can you conclude about the two triangles. Why?

Third International Mathematics and Science Study (TIMSS) Question

In the $\triangle ABC$ the altitudes BN and CM intersect at point S. The measure of $\angle MSB$ is $40°$ and the measure of $\angle SBC$ is $20°$. Write a PROOF of the following statement: "$\triangle ABC$ is isosceles." Give geometric reasons for statements in your proof.

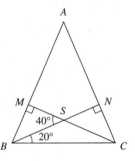

TIMSS, Grade 12, 1995

TECHNOLOGY CORNER Use *The Geometer's Sketchpad* to do each of the following:

1. a. Draw a circle. Label its center and draw any angle whose vertex is the center and whose sides intersect the circle, as in Figure 12-49. Such an angle is a **central angle** of the circle. A central angle and its intercepted arc are considered to have the same measure.

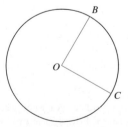

Figure 12-49

   b. Choose any point P on the circumference of the circle that is not B and not C and not in the interior of $\angle BOC$. Draw $\angle BPC$ and find its measure. $\angle BPC$ is an **inscribed angle** of the circle.
   c. Make a conjecture about the measures of $\angle BOC$ and $\angle BPC$.
2. Draw an inscribed triangle in a circle in such a way that one side of the triangle is a diameter of the circle. Find the measure of the angle that does not have the diameter as a side. Make a conjecture about the measure of any such angle.

12-4   Similar Triangles and Similar Figures

When a germ is examined under a microscope or when a slide is projected on a screen, the shapes in each case usually remain the same, but the sizes are altered. Informally, such figures are **similar**. For example, see the student page and do Problem 3. The ratio of the corresponding side lengths is the **scale factor**. On the student page the scale factor in the top figures is $\frac{4}{3}$.

It seems that in any enlargement or reduction the resulting figure is similar to the original; that is, the corresponding angle measures remain the same and the corresponding sides are proportional. In general, we have the following definition of similar triangles and similar polygons.

Definition of Similar Triangles

$\triangle ABC$ is similar to $\triangle DEF$, written $\triangle ABC \sim \triangle DEF$, if, and only if, $\angle A \cong \angle D$, $\angle B \cong \angle E$, $\angle C \cong \angle F$, and $\dfrac{AB}{DE} = \dfrac{AC}{DF} = \dfrac{BC}{EF}$.

Definition of Similar Polygons

Two polygons with the same number of vertices are similar if there is a one-to-one correspondence between the vertices of one and the vertices of the other such that the corresponding interior angles are congruent and corresponding sides are proportional.

As with congruence of triangles, so with similarity—we do not need all the conditions in the definition to conclude that the triangles are similar. In fact, the following conditions suffice to conclude that the triangles are similar: AAA (or AA), SSS, and SAS. We state these conditions in the following theorems, which are proved in more advanced geometry texts.

Theorem 12–10:   SSS Similarity for Triangles

If corresponding sides of two triangles are proportional, then the triangles are similar.

Theorem 12–11:   SAS Similarity for Triangles

Given two triangles, if two sides are proportional and the included angles are congruent, then the triangles are similar.

Theorem 12–12:   Angle, Angle (AA) Similarity for Triangles

If two angles of one triangle are congruent, respectively, to two angles of a second triangle, then the triangles are similar.

School Book Page SIMILAR FIGURES

Exploration 1 ▸▸▸▸▸▸▸▸▸▸▸▸▸▸▸▸▸▸▸▸▸▸▸

Similar Figures
Similar Figures

GOAL

LEARN HOW TO...
  ◆ tell whether triangles are similar
  ◆ make indirect measurements

AS YOU...
  ◆ estimate the height of a cliff

KEY TERMS
  ◆ similar
  ◆ corresponding angles
  ◆ corresponding sides

SET UP   *You will need:* • *Labsheet 4A* • *centimeter ruler* • *protractor*

▸ As you will see in this exploration, Charlotte Lopez uses *similar* figures to estimate the cliff height. Two figures are **similar** if they have the same shape, but not necessarily the same size. The figures below are similar.

∠*P* and ∠*T* are *corresponding angles*.
Corresponding angles have the same measure.

$m\angle P = m\angle T$

Read $m\angle T$ as "the measure of angle *T*."

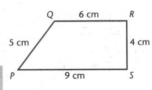

$\overline{PQ}$ and $\overline{TU}$ are *corresponding sides*. The lengths of **corresponding sides** are in proportion.

$$\frac{PQ}{TU} = \frac{QR}{UV}$$

Read $\overline{TU}$ as "segment *TU*." Read *TU* as "the length of segment *TU*."

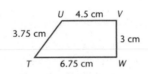

3 a. Try This as a Class Name all pairs of corresponding angles and corresponding sides in the figures above.

  b. Find the ratios $\frac{PQ}{TU}$, $\frac{QR}{UV}$, $\frac{RS}{VW}$, and $\frac{PS}{TW}$. What do you notice?

4 ✔ **CHECKPOINT** The figures below are similar. Copy and complete each statement.

  a. $m\angle C = m\ \underline{\ ?\ }$

  b. $\frac{BC}{FG} = \frac{?}{GH}$

  c. $\frac{?}{AD} = \frac{EF}{?}$

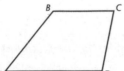

✔ **QUESTION 4**

...checks your understanding of the definition of similar figures.

Source: Math Thematics, Book 3, New Edition, McDougal Littell, 2008 (p. 193).

In what follows we give examples in which the similarity conditions are used.

Example 12-10

Given the pairs of similar triangles in Figure 12-50, find a one-to-one correspondence among the vertices of the triangles such that the corresponding angles are congruent. Then write the proportions for the corresponding sides that follow from the definition.

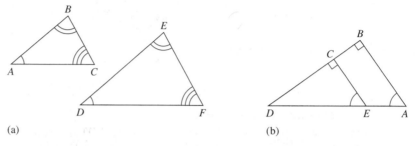

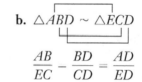

Figure 12-50

Solution   **a.** $\triangle ABC \sim \triangle DEF$

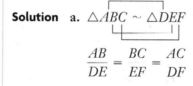

$$\frac{AB}{DE} = \frac{BC}{EF} = \frac{AC}{DF}$$

b. $\triangle ABD \sim \triangle ECD$

$$\frac{AB}{EC} = \frac{BD}{CD} = \frac{AD}{ED}$$

Example 12-11

For each part of Figure 12-51, find a pair of similar triangles.

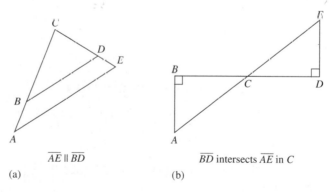

$\overline{AE} \parallel \overline{BD}$ $\overline{BD}$ intersects $\overline{AE}$ in C

(a)   (b)

Figure 12-51

Solution   **a.** Because $\overline{AE} \parallel \overline{BD}$, congruent corresponding angles are formed by a transversal cutting the parallel segments. Thus, $\angle CBD \cong \angle CAE$ and $\angle CDB \cong \angle CEA$. Also, $\angle C \cong \angle C$, so $\triangle CBD \sim \triangle CAE$ by AAA.

b. $\angle B \cong \angle D$ because both are right triangles. Also, $\angle ACB \cong \angle ECD$ because they are vertical angles. Thus, $\triangle ACB \sim \triangle ECD$ by AA.

NOW TRY THIS 12-12 Congruency of corresponding angles is sufficient to prove that two triangles are similar. Is the same condition sufficient to prove other polygons are similar to each other? Explain your answer.

Example 12-12   In Figure 12-52, solve for x.

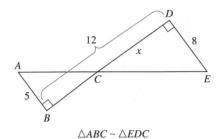

$$\triangle ABC \sim \triangle EDC$$

Figure 12-52

Solution $\triangle ABC \sim \triangle EDC$, so

$$\frac{AB}{ED} = \frac{AC}{EC} = \frac{BC}{DC}$$

Now, $AB = 5$, $ED = 8$, and $CD = x$, so $BC = 12 - x$. Thus,

$$\frac{5}{8} = \frac{12 - x}{x}$$

$$5x = 8(12 - x)$$

$$5x = 96 - 8x$$

$$13x = 96$$

$$x = \frac{96}{13}$$

Properties of Proportion

Similar triangles give rise to various properties that involve proportions. For example, in

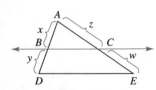

Figure 12-53

Figure 12-53 if $\overline{BC} \parallel \overline{DE}$, then $\frac{AB}{BD} = \frac{AC}{CE}$. This can be justified as follows: $\overline{BC} \parallel \overline{DE}$, so $\triangle ADE \sim \triangle ABC$ (why?). Consequently, $\frac{AD}{AB} = \frac{AE}{AC}$, which may be written as follows:

$$\frac{x + y}{x} = \frac{z + w}{z}$$

$$\frac{x}{x} + \frac{y}{x} = \frac{z}{z} + \frac{w}{z}$$

$$1 + \frac{y}{x} = 1 + \frac{w}{z}$$

$$\frac{y}{x} = \frac{w}{z}$$

$$\frac{x}{y} = \frac{z}{w}$$

The statement is not true for polygons. For example, a rectangle and a square have congruent angles but do not have to be similar.

This result is summarized in the following theorem.

Theorem 12–13

If a line parallel to one side of a triangle intersects the other sides, then it divides those sides into proportional segments.

The converse of Theorem 12–13 is also true; that is, if in Figure 12-53 we know that $\dfrac{AB}{BD} = \dfrac{AC}{CE}$, then we can conclude that $\overline{BC} \parallel \overline{DE}$.

We can prove this fact using the SAS similarity theorem. We can reverse the steps in the proof of Theorem 12–13 and from $\dfrac{x}{y} = \dfrac{z}{w}$ get $\dfrac{x + y}{x} = \dfrac{z + w}{z}$ and hence that $\dfrac{AD}{AB} = \dfrac{AE}{AC}$. Using this proportion and the fact that $\triangle ABC$ and $\triangle ADE$ share $\angle A$, we conclude by the SAS similarity theorem that $\triangle ABC \sim \triangle ADE$. Consequently, $\angle ABC \cong \angle ADE$ and therefore $\overline{BC} \parallel \overline{DE}$. We summarize this result in the following theorem.

Theorem 12–14

If a line divides two sides of a triangle into proportional segments, then the line is parallel to the third side.

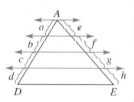

Figure 12-54

Similarly, if lines parallel to $\overline{DE}$ intersect $\triangle ADE$, as shown in Figure 12-54, so that $a = b = c = d$, it can be shown that $e = f = g = h$. This result is stated in the following theorem.

Theorem 12–15

If parallel lines cut off congruent segments on one transversal, then they cut off congruent segments on any transversal.

Theorem 12–15 can be used to divide a given segment into any number of congruent parts. For example, using only a compass and a straightedge, we can divide segment $\overline{AB}$ in Figure 12-55 into three congruent parts by making the construction resemble Figure 12-54.

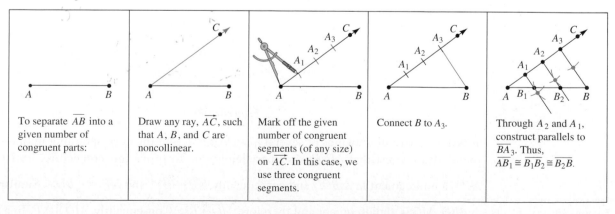

To separate $\overline{AB}$ into a given number of congruent parts:	Draw any ray, $\overrightarrow{AC}$, such that A, B, and C are noncollinear.	Mark off the given number of congruent segments (of any size) on $\overrightarrow{AC}$. In this case, we use three congruent segments.	Connect B to A_3.	Through A_2 and A_1, construct parallels to $\overline{BA_3}$. Thus, $\overline{AB_1} \cong \overline{B_1B_2} \cong \overline{B_2B}$.

Figure 12-55 Separating a segment into congruent parts

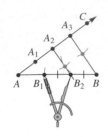

Figure 12-56

REMARK It is necessary to construct only $\overline{A_2B_2}$. We can then use a compass to mark off point B_1 such that $B_1B_2 = BB_2$, as in Figure 12-56.

Midsegments of Triangles and Quadrilaterals

The segment connecting the midpoints of two sides of a triangle or two adjacent sides of a quadrilateral is a **midsegment**. In Figure 12-57, M and N are midpoints of $\overline{AB}$ and $\overline{BC}$, respectively, and $\overline{MN}$ is a midsegment. Because $\dfrac{MB}{MA} = \dfrac{BN}{CN} = 1$, by Theorem 12–14 $\overline{MN} \parallel \overline{AC}$. How is MN related to AC? Notice that $\triangle MBN \sim \triangle ABC$ (why?). Therefore,

$$\frac{MN}{AC} = \frac{MB}{AB} = \frac{MB}{2MB} = \frac{1}{2}$$

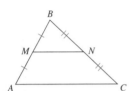

Figure 12-57

Consequently, we have the following theorem.

Theorem 12–16:   The Midsegment Theorem

The midsegment (segment connecting the midpoints of two sides of a triangle) is parallel to the third side of the triangle and half as long.

The following theorem is an immediate consequence of Theorem 12–15.

Theorem 12–17

If a line bisects one side of a triangle and is parallel to a second side, then it bisects the third side and therefore is a midsegment.

Example 12-13

In the quadrilateral $ABCD$ in Figure 12-58, M, N, P, and Q are the midpoints of the sides. What kind of quadrilateral is $MNPQ$?

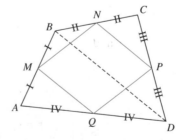

Figure 12-58

Solution   Careful drawings of several quadrilaterals and the corresponding midpoints of their sides suggest that $MNPQ$ is a parallelogram. To prove this conjecture, notice that NP is a midsegment in $\triangle BCD$ and consequently $\overline{NP} \parallel \overline{BD}$ $\left(\text{and } NP = \dfrac{1}{2}BD\right)$. Similarly, in $\triangle ABD$, $\overline{MQ}$ is a midsegment and therefore $\overline{MQ} \parallel \overline{BD}$. Consequently, $\overline{MQ} \parallel \overline{NP}$. In a similar way we could show that $\overline{MN} \parallel \overline{QP}$ (consider midsegments in $\triangle ABC$ and in $\triangle ADC$) and therefore by the definition of a parallelogram, $MNPQ$ is a parallelogram.

A **median** of a triangle is a segment connecting a vertex of the triangle to the midpoint of the opposite side. A triangle has three medians. A careful drawing of the three medians shown in Figure 12-59 suggests that the three medians are concurrent, which they are. The point of intersection, G, is the **center of gravity**, or the **centroid**, of the triangle.

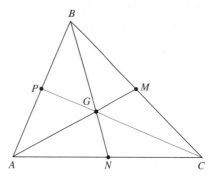

Figure 12-59

If a triangular piece of somewhat thick uniform material such as wood or metal is placed at its centroid on a sharp object like a pencil tip, it will balance. In the next example, we prove an interesting property of the centroid.

Example 12-14

Show that the centroid of a triangle divides each median in the ratio 1:2.

Solution In Figure 12-60, we have two medians of $\triangle ABC$ and their point of intersection G. We need to prove that $AG = 2MG$ and $BG = 2NG$. The fact that M and N are midpoints of two sides suggests constructing MN, the midsegment. The midsegment theorem tells us that $\overline{MN} \parallel \overline{AB}$ and $MN = \frac{1}{2}AB$. The fact that $\overline{MN} \parallel \overline{AB}$ implies that the equally marked angles are congruent because they are alternate interior angles formed by a transversal of the parallel lines. Thus, $\triangle ABG \sim \triangle MNG$ by the AA similarity condition. Using the fact that $AB = 2MN$ we have

$$\frac{BG}{NG} = \frac{AG}{MG} = \frac{AB}{MN} = 2$$

Thus, $BG = 2GN$ and $AG = 2MG$.

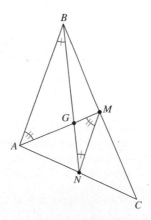

Figure 12-60

Indirect Measurements

Similar triangles have long been used to make indirect measurements. Thales of Miletus (ca. 600 BCE) is believed to have determined the height of the Great Pyramid of Egypt by using ratios involving shadows, similar to those pictured in Figure 12-61. The Sun is so far away that it should make approximately congruent angles at B and B'. Because the angles at C and C' are right angles, $\triangle ABC \sim \triangle A'B'C'$. Hence,

$$\frac{AC}{A'C'} = \frac{BC}{B'C'}$$

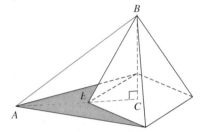

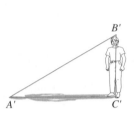

Figure 12-61

And because $AC = AE + EC$, the following proportion is obtained:

$$\frac{AE + EC}{A'C'} = \frac{BC}{B'C'}$$

The person's height and shadow can be measured. Also, the length AE of the shadow of the pyramid can be measured, and EC can be found because the base of the pyramid is a square. Each term of the proportion except the height of the pyramid is known. Thus, the height BC of the pyramid can be found by solving the proportion.

Example 12-15

On a sunny day, a tall tree casts a 40-m shadow (Figure 12-62). At the same time, a meter-stick held vertically casts a 2.5-m shadow. How tall is the tree?

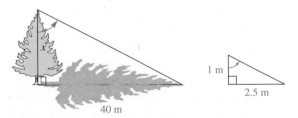

Figure 12-62

Solution   In Figure 12-62, the triangles are similar by AA because the tree and the stick both meet the ground at right angles and the angles formed by the Sun's rays are congruent because the shadows are measured at the same time.

$$\frac{x}{40} = \frac{1}{2.5}$$
$$2.5x = 40$$
$$x = 16$$

The tree is 16 m tall.

BRAIN TEASER Two neighbors, Smith and Wheeler, plan to erect flagpoles in their yards. Smith wants a 10 ft pole, and Wheeler wants a 15 ft pole. To keep the poles straight while the concrete bases harden, guy wires are to be tied from the tops of the flagpoles to a fence post on the property lines and to the bases of the flagpoles, as shown in Figure 12-63. How high should the fence post be and how far apart should they erect the flagpoles for this scheme to work?

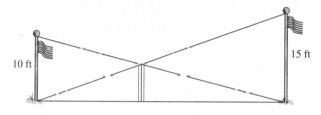

10 ft

15 ft

Figure 12-63

LABORATORY ACTIVITY The device pictured in Figure 12-64 is a pantograph. It is used to draw enlarged versions of figures. In Figure 12-64, the red dots represent either brads or nuts and bolts. The strips are made of lath or cardboard and are rigid. A pointer at *D* is used to trace along an original figure, which causes the pencil at *F* to draw an enlarged version of the figure. Make or obtain a pantograph and experiment with enlarging figures. Explain how and why this works.

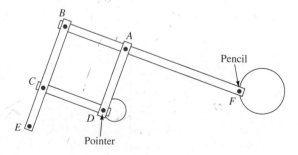

Figure 12-64

Assessment 12-4A

1. In the cartoon, the human character's size is increased. If we assume that the human figures are similar and the person's height has increased from 5'10" to 6'11", what is the scale factor?

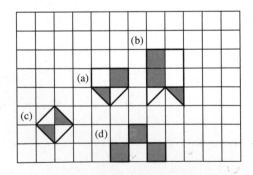

2. Which of the following are always similar? Why?
   a. Any two equilateral triangles
   b. Any two squares
   c. Any two rectangles
   d. Any two rectangles in which one side is twice as long as the other.

3. Use grid paper to draw figures that have sides three times as large as the ones in the following figure and that are colored similarly.

4. a. Sketch two nonsimilar polygons for which corresponding angles are congruent.
   b. Sketch two nonsimilar polygons for which corresponding sides are proportional.

5. a. Examine several examples of similar polygons and make a conjecture concerning the ratio of their perimeters.
   b. Prove your conjecture in part (a).

6. a. Which of the following pairs of triangles are similar? If they are similar, explain why.

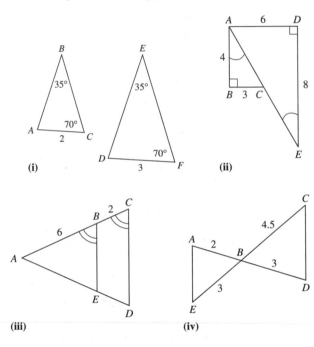

   b. For each pair of similar triangles, find the scale factor of the sides of the triangles.

7. In the following figures, find the measure of the sides marked *x*.

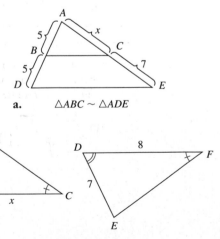

a.   △*ABC* ~ △*ADE*

b.   △*ABC* ~ △*EDF*

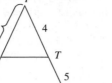
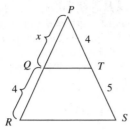

c. △PQT ~ △PRS

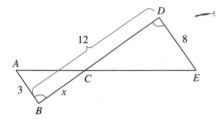

d. △ABC ~ △EDC

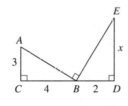

e. △ABC ~ △BED

8. Draw a segment $\overline{AB}$ and then use a compass and a straightedge to separate $\overline{AB}$ into five congruent pieces.

9. Given segments of length a, b, and c as shown, construct a segment of length x so that $\frac{a}{b} = \frac{c}{x}$. (*Hint:* Use the idea shown in Figure 12-54.)

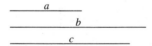

10. In the cartoon shown, if a smaller map were obtained and its scale were half the size of the original scale, would the distance to be traveled be different? Why?

11. Sketch with approximate measures two hexagons with corresponding sides proportional but so that they are not similar.

12. In the figure shown, prove that $ac = bd$.

13. Sketch a pair of hexagons with corresponding angles congruent but so that the hexagons are not similar.

14. Rep-tiles from Section 12-1 involved similar figures. Draw a rep-tile involving at least six parallelograms.

15. If one ruler is marked off in inches and another is marked off in centimeters, explain whether or not the two rulers must be similar.

16. Times is a computer font. Explain whether or not the Times font with sizes 16 and 36 are similar.

17. If you copy a page on a machine at 75%, you should get a similar copy of the page. What is the corresponding setting to obtain the original from the copy? Why?

18. In the following figure, find the distance AB across the pond using the similar triangles shown:

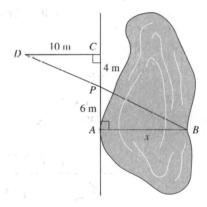

19. Samantha wants to know how far above the ground the top of a leaning flagpole is. At high noon, when the Sun is almost directly overhead, the shadow cast by the pole is 7 ft long. Samantha holds a plumb bob with a string 3 ft long up to the flagpole and determines that the point of the plumb bob touches the ground 13 in. from the base of the flagpole. How far above the ground is the top of the pole?

3 ft

13 in.

7 ft

20. The midpoints M, N, P, Q of the sides of a quadrilateral $ABCD$ have been connected and an interior quadrilateral is obtained. We have shown that $MNPQ$ is a parallelogram. What is the most you can say about the kind of parallelogram $MNPQ$ is if $ABCD$ is
  a. A rhombus?
  b. A kite?
  c. An isosceles trapezoid?
  d. A quadrilateral that is neither a rhombus nor a kite but whose diagonals are perpendicular to each other?

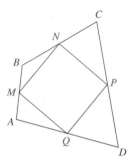

21. Explain why all pairs of regular octagons are similar.
22. If you took cross sections of a typical ice cream cone parallel to the circular opening where the ice cream is usually placed, explain whether the cross sections would be similar.

Assessment 12-4B

1. In parallelograms $ABCD$ and $A_1B_1C_1D_1$, $\angle BAD \cong \angle B_1A_1D_1$. Explain whether or not the parallelograms are similar.
2. Which of the following are always similar? Why?
  a. Any two rectangles in which the diagonal in one is twice as long as in the other.
  b. Any two rhombuses
  c. Any two circles
  d. Any two regular polygons
  e. Any two regular polygons with the same number of sides

3. Which of the following pairs of triangles are similar? If they are similar, explain why and find the scale factor if possible.

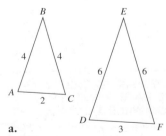

a.

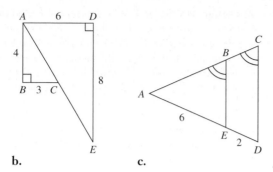

b.   **c.**

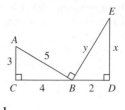

d.

5. Given the following figure, use a compass and a straight-edge to separate $\overline{AB}$ into five congruent pieces:

6. In the following right triangle ABC, $\overline{CD} \perp \overline{AB}$:

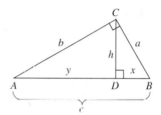

a. Find three pairs of similar triangles. Justify your answers.
b. Write the corresponding proportions for each set of similar triangles.
c. Use part (b) and the figure to show that $a^2 = xc$. Also argue that $b^2 = yc$.
d. Use part (c) to show that $a^2 + b^2 = c^2$. State this result in words using the legs and hypotenuse of a right triangle.
e. Use part (b) to show that $h^2 = xy$.
f. Show that x, h, and y form a geometric sequence.

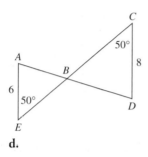

d.

4. Prove that in the following figures triangles in each part are similar and find the measure of the sides marked with x, and in part (d) x and y.

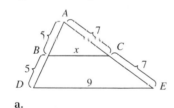

a.

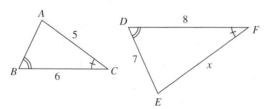

b.

7. Given segments of length a, b, and c as shown, construct a segment of length x so that $\dfrac{a}{b} = \dfrac{x}{c}$. (*Hint:* Use the idea shown in Figure 12-54.)

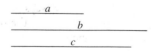

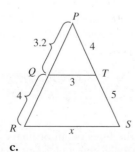

c.

8. Triangle ABC is similar to triangle DEF with a side of triangle ABC that is 75% of its corresponding side in triangle DEF. Also, triangle GHI is similar to triangle DEF with a side of triangle GHI that is 32% of its corresponding side in triangle DEF.
   a. Are triangles ABC and GHI similar to each other? Why, or why not?
   b. What are the possible ratios of corresponding sides of triangles ABC and GHI?

9. Sketch with approximate measures two pentagons with corresponding sides proportional but so that they are not similar.

10. If you copy a document on a machine, you should get a similar copy. Suppose you want to reduce a document to $\frac{1}{6}$ of its original size and you first reduce it by 25%. By what percent do you need to reduce the copy to obtain the desirable size?

11. To find the height of a tree, a group of Girl Scouts devised the following method. A girl walks toward the tree along its shadow until the shadow of the top of her head coincides with the shadow of the top of the tree. If the girl is 150 cm tall, her distance to the foot of the tree is 15 m, and the length of her shadow is 3 m, how tall is the tree?

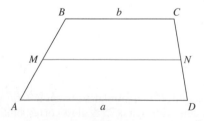

12. The angle bisector of one of the angles in an isosceles triangle is constructed. This angle bisector partitions the original triangle into two isosceles triangles.
    a. What are the angle measures of the original triangle? (There are two possibilities.)
    b. Which, if any, triangles are congruent? Similar? Explain your answers.

13. a. In the accompanying figure, $ABCD$ is a trapezoid. M is the midpoint of $\overline{AB}$. Through M, a line parallel to the bases has been drawn, intersecting $\overline{CD}$ at N. (i) Explain why N must be the midpoint of $\overline{CD}$ and (ii) express MN in terms of a and b, the lengths of the parallel sides of the trapezoid.

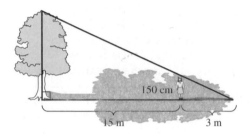

b. Denote MN by c. Use your answer to part (a) to show that b, c, and a form an arithmetic sequence.

c. In the trapezoid $ABCD$, the lengths of the bases are a and b as shown. Side $\overline{AB}$ has been divided into 9 congruent segments. Through the endpoints of the segments, lines parallel to $\overline{AD}$ have been drawn. In this way, 8 new segments connecting the sides $\overline{AB}$ and $\overline{CD}$ have been created. Show that the sequence of 10 terms, starting with b, proceeding with the lengths of the parallel segments, and ending with a, is an arithmetic sequence and find the sum of the sequence in terms of a and b.

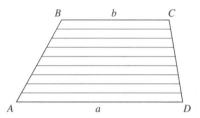

14. $ABCD$ is a convex quadrilateral and N, P, Q, M are the midpoints of its sides as shown. The diagonals $\overline{BD}$ and $\overline{AC}$ intersect at T, $\overline{NP}$ intersects $\overline{AC}$ at S, and $\overline{BD}$ intersects $\overline{MN}$ at V.
    a. Prove that $NSTV$ is a parallelogram
    b. Complete the following statements and prove them.
       (i) $MNPQ$ is a rectangle if, and only if, $\overline{AC}$ and $\overline{BD}$ are _____.
       (ii) $MNPQ$ is a square if, and only if, $\overline{AC}$ and $\overline{BD}$ are _____.

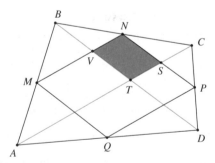

15. Explain why all pairs of regular n-gons are similar.
16. Must all cross sections of a circular cylinder be similar? Why? Draw a sketch to illustrate your answer.

Mathematical Connections 12-4

Communication

1. Do you think any two cubes are similar? Why or why not?
2. Architects frequently build three-dimensional scale drawings to construct scale models of projects. Are all the models similar to the finished products? Why or why not?

3. Assuming the lines on an ordinary piece of notebook paper are parallel and equidistant, describe a method for using the paper to divide a piece of licorice evenly among 2, 3, 4, or 5 children. Explain why it works.

Open-Ended

4. Build two similar towers out of blocks.
  a. What is the ratio of the heights of the towers?
  b. What is the ratio of the perimeters of the bases of the towers?

5. On a sunny day, go outside and measure the heights of objects and their accompanying shadows. Use the data gathered by an entire class and plot graphs. Plot all data points on the same graph, with shadow lengths on the horizontal axis and object heights on the vertical axis. What do you observe? Why?

Cooperative Learning

6. A building was to be built on a triangular piece of property. The architect was given the approximate measurements of the angles of the triangular lot as 54°, 39°, and 87° and the lengths of two of the sides as 100 m and 80 m. When the architect began the design on drafting paper, she drew a triangle to scale with the corresponding measures and found that the lot was considerably smaller than she had been led to believe. It appeared that the proposed building would not fit. The surveyor was called. He confirmed each of the measurements and could not see a problem with the size. Neither the architect nor surveyor could understand the reason for the other's opinion.
  a. Have one person in your group play the part of the architect and explain why she felt she was correct.
  b. Have one person in the group explain the reason for the miscommunication.
  c. Have the group suggest a way to provide an accurate description of the lot.

Questions from the Classroom

7. A student asks if, for the same n, all convex n-gons with all angles congruent are similar. How do you respond?
8. A student says she thinks all circles are similar but would like to know why. How do you answer her?
9. A student argues that in the following figure triangles *ABC* and *DBE* are similar because $\overline{DE}$ divides the sides proportionally. One side is divided in the ratio $\frac{2}{4}$, while the other is divided in the ratio $\frac{3}{6}$. How do you respond?

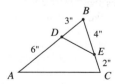

10. A student asks if there is an ASA similarity condition for triangles. How do you respond?

Review Problems

11. Given the following base of an isosceles triangle and the altitude to that base, construct the triangle:

_____

Base

_____

Altitude

12. Given the following length of a side of an equilateral triangle, construct an altitude of the equilateral triangle.

_____

13. Write a paragraph describing how you could construct an isosceles right triangle when given the length of the hypotenuse of the 45°-45°-90° triangle.
14. Use a compass and a straightedge to draw a pair of obtuse vertical angles and the angle bisector of one of these angles. Extend the angle bisector. Does the extended angle bisector bisect the other vertical angle? Justify your answer.
15. Describe a minimal set of conditions that can be used to argue that two quadrilaterals are congruent.

Third International Mathematics and Science Study (TIMSS) Question

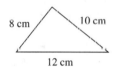

Which of the following triangles is similar to the triangle shown above?

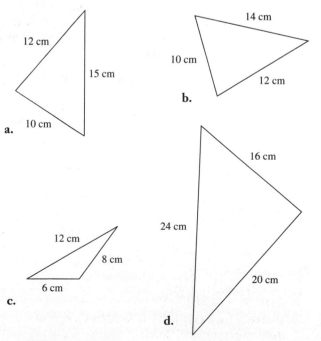

National Assessment of Educational Progress (NAEP) Question

The figure at right shows two right angles. The length of AE is x and the length of DE is 40.

Show all of the steps that lead to finding the value of x. Your last step should give the value of x.

NAEP, Grade 8, 2007

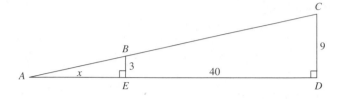

TECHNOLOGY CORNER   Work through *GSP* Lab 6 in the Technology Manual.

TECHNOLOGY CORNER   In 1975, Benoit Mandelbrot invented the word *fractal* to describe certain irregular and fragmented shapes. These shapes are such that if one looks at a small part of the shape, the small part resembles the larger shape. These shapes are somewhat like rep-tiles, although the smaller structures of fractals are not necessarily identical to the larger structure. (Recall that with rep-tiles, the smaller structures are similar and are identical in all aspects except size.) Fractal geometry was introduced for the purpose of modeling natural phenomena such as irregular coastlines, arteries and veins, the branching structure of plants, the thermal agitation of molecules in a fluid, and sponges. Figure 12-65 shows an example of a fractal—a computer-generated picture of the Mandelbrot set known as the "Tail of the Seahorse."

Figure 12-65

Earlier, in 1906, Helge von Koch came up with a curve that has infinite perimeter. To visualize this curve, we construct in Figure 12-66 a sequence of polygons $S_1, S_2, S_3, \ldots$ as follows:

a. S_1 is an equilateral triangle.
b. S_2 is obtained from S_1 by dividing each side of the triangle into three congruent parts and constructing on the middle part an equilateral triangle with the base removed.
c. S_3 is obtained from S_2 as S_2 was obtained from S_1.

We continue in a similar way to obtain the other polygons of the sequence in Figure 12-66(d) and (e). These polygons come closer and closer to a curve, called the *snowflake curve*, which is another example of a fractal. Use *The Geometer's Sketchpad* to construct Figure 12-66(a)–(c).

(continued)

TECHNOLOGY CORNER *(continued)*

If S_1 has perimeter 3 units, what is the perimeter of S_2 and S_3? What do you think happens to the perimeter of the snowflakes as the number of sides of the polygons increases?

d. Suppose Figure 12-66(a) is a regular tetrahedron and Figure 12-66(b) is a regular tetrahedron with proportionately smaller regular tetrahedrons placed on the faces of the original similarly to the placement of the equilateral triangles shown. If this process continues, what do you think the resulting figure approaches?

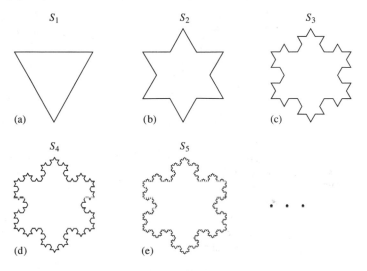

S_1 S_2 S_3

(a)   (b)   (c)

S_4 S_5

(d)   (e)   . . .

Figure 12-66

Note: GSP's Iterate command, in the Transform menu, is useful for constructing fractals.

BRAIN TEASER A particular kaleidoscope is a right prism with an equilateral triangle as a base. A beam of light is reflected at a 60° angle from a point P on a side of the triangular base, as shown in Figure 12-67. The beam is reflected in the plane of the base to the different mirrored surfaces and continues bouncing off at 60° angles. Find the length of the path of the reflected light when it reaches the point at which it originated.

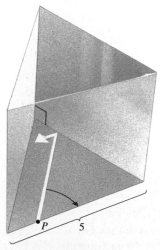

P 5

Figure 12-67

12-5   Lines and Linear Equations in a Cartesian Coordinate System

The Cartesian coordinate system (named for René Descartes) enables us to study geometry using algebra and to interpret algebraic phenomena geometrically. A Cartesian coordinate system is constructed by placing two number lines perpendicular to each other, as shown in Figure 12-68. The intersection point of the two lines is the **origin**, the horizontal line is the **x-axis**, and the vertical line is the **y-axis**. The location of any point P can be described by an ordered pair of numbers (a, b), where a perpendicular from P to the x-axis intersects at a point with coordinate a and a perpendicular from P to the y-axis intersects at a point with coordinate b; the point is identified as $P(a, b)$. The first component in the ordered pair (a, b) is the **abscissa**, or **x-coordinate**, of P. The second component is the **ordinate**, or **y-coordinate**, of P. There is a one-to-one correspondence between all the points in the plane and all the ordered pairs of real numbers. For example, in Figure 12-68, R has coordinates $(^-4, ^-3)$, written $R(^-4, ^-3)$.

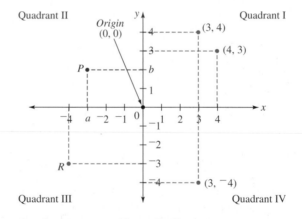

Figure 12-68

Equations of Vertical and Horizontal Lines

Every point on the x-axis has a y-coordinate of zero. Thus, the x-axis can be described as the set of all points (x, y) such that $y = 0$. We say that the x-axis has equation $y = 0$. Similarly, the y-axis can be described as the set of all points (x, y) such that $x = 0$ and y is an arbitrary real number. Thus, $x = 0$ is the equation of the y-axis. If we plot the set of all points that satisfy a given condition, the resulting picture on the Cartesian coordinate system is the **graph** of the set.

Example 12-16

Sketch the graph for each of the following:

a. $x = 2$
b. $y = 3$
c. $x < 2$ and $y = 3$

Solution   **a.** The equation $x = 2$ represents the set of all points (x, y) for which $x = 2$ and y is any real number. This set is the line perpendicular to the x-axis at $(2, 0)$, as in Figure 12-69(a).

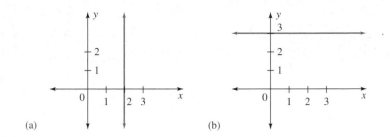

Figure 12-69

b. The equation $y = 3$ represents the set of all points (x, y) for which $y = 3$ and x is any real number. This set is the line perpendicular to the y-axis at $(0, 3)$, as in Figure 12-69(b).

c. The statements in part (c) represent the set of all points (x, y) for which $x < 2$, but y is always 3. The set is part of a line, as shown in Figure 12-70. Note that the hollow dot at $(2, 3)$ indicates that this point is not included in the solution set.

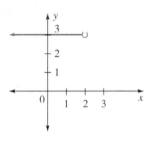

Figure 12-70

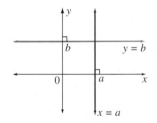

Figure 12-71

In Example 12-20 we found the graphs of the equations $x = 2$ and $y = 3$. In general, *the graph of the equation $x = a$, where a is some real number, is the line perpendicular to the x-axis through the point with coordinates $(a, 0)$, as shown* in Figure 12-70. Similarly, *the graph of the equation $y = b$ is the line perpendicular to the y-axis through the point with coordinates $(0, b)$.*

Equations of Lines

Consider the arithmetic sequence 4, 7, 10, 13, . . . in Table 12-2 whose xth term is $3x + 1$. If the number of the term is the x-coordinate and the corresponding term the y-coordinate, the set of points appear to lie on a line that is parallel to neither the x-axis nor the y-axis, as in Figure 12-72.

Table 12-2

Number of Term	Term
1	4
2	7
3	10
4	13
.	.
.	.
.	.
x	$3x + 1$

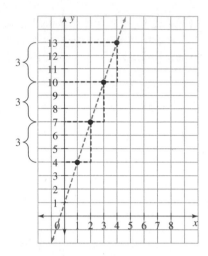

Figure 12-72

In Table 12-2, we see that the common difference in the arithmetic sequence is 3. Likewise, there is a difference of 3 in the y-coordinates of the marked points. Correspondingly, there is a difference of 1 in the number of the term. Hence we might say that the rate of change of the term with respect to the number of the term is 3, or the ratio is 3 to 1.

When we think about the rate of change and the ratio, several rhetorical questions arise: Does one find comparable similarities with any arithmetic sequence? What is the relation between the general term $3x + 1$ and the given line in Figure 12-72? The graph in Figure 12-72 crosses the y-axis at the point with coordinates $(0, 1)$; if there were a "zeroth" term for the arithmetic sequence, then the term value would be $3 \cdot 0 + 1$, or 1; does something comparable happen with other arithmetic sequences?

Do all points corresponding to arithmetic sequences lie along lines? To help answer this question, we consider the following sequences in Table 12-3.

Table 12-3

Number of Term x	$1x$	$2x$	$\frac{1}{2}x$	$(^-1)x$	$(^-2)x$
1	1	2	$\frac{1}{2}$	$^-1$	$^-2$
2	2	4	1	$^-2$	$^-4$
3	3	6	$\frac{3}{2}$	$^-3$	$^-6$
4	4	8	2	$^-4$	$^-8$
5	5	10	$\frac{5}{2}$	$^-5$	$^-10$
6	6	12	3	$^-6$	$^-12$
.	.	.	.	.	.
.	.	.	.	.	.
.	.	.	.	.	.
x	x	$2x$	$\frac{1}{2}x$	^-x	^-2x

In Figure 12-73, the sets of ordered pairs (x, y) are plotted on a graph so that the number of the term is the x-coordinate and the corresponding term appearing in a particular column in Table 12-3, the y-coordinate. The sets of points appear to determine straight lines.

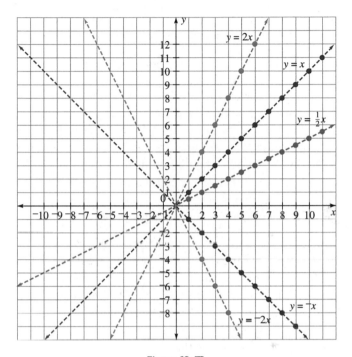

Figure 12-73

All five lines in Figure 12-73 have equations of the form $y = mx$, where m takes the values $2, 1, \frac{1}{2}, {}^-1,$ and $^-2$. If all points along a given dashed line are connected, then all the points on that line satisfy the corresponding equation. The number m is a measure of steepness and is the **slope** of the line whose equation is $y = mx$. The graph goes up from left to right (increases) if m is positive, and it goes down from left to right (decreases) if m is negative.

NOW TRY THIS 12-13

a. In the equation $y = mx$, if m is 0, what happens to the line?
b. What happens to the line as m increases?
c. What happens to the line when m decreases?

   All lines in Figure 12-73 pass through the origin. This is true for any line whose equation is $y = mx$. If $x = 0$, then $y = m \cdot 0 = 0$ and $(0, 0)$ is a point on the graph of $y = mx$. Conversely, it is possible to show that any nonvertical line passing through the origin has an equation of the form $y = mx$ for some value of m.

Example 12-17

Find the equation of the line that contains $(0, 0)$ and $(2, 3)$.

Solution   The line goes through the origin; therefore its equation has the form $y = mx$. To find the equation of the line, we must find the value of m. The line contains $(2, 3)$, so we substitute 2 for x and 3 for y in the equation $y = mx$ to obtain $3 = m \cdot 2$, and thus $m = \dfrac{3}{2}$. Hence, the required equation is $y = \dfrac{3}{2}x$.

Next, we consider equations of the form $y = mx + b$, where b is a real number. To do this, we examine the graphs of $y = x + 2$ and $y = x$. Given the graph of $y = x$, we can obtain the graph of $y = x + 2$ by "raising" each point on the first graph by 2 units. This is because for a certain value of x, the corresponding y value is 2 units greater. This is shown in Figure 12-74(a). Similarly, to sketch the graph of $y = x - 2$, we first draw the graph of $y = x$ and then lower each point vertically by 2 units, as shown in Figure 12-74(a).

The graphs of $y = x + 2$ and $y = x - 2$ are parallel lines. In general, *for a given value of m, the graph of $y = mx + b$ is a straight line through $(0, b)$ and parallel to the line whose equation is $y = mx$.*

Further, the graph of the line $y = mx + b$, can be obtained from the graph of $y = mx$ by sliding $y = mx$ up (or down) b units depending on the value of b, as shown in Figure 12-74(b). In general, *any two parallel lines have the same slope or are vertical lines with no slope.*

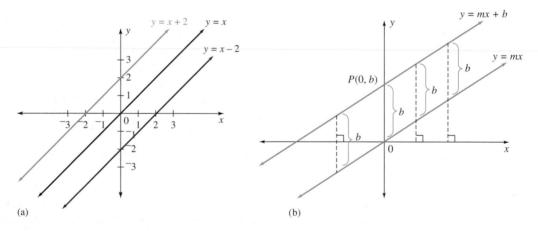

(a)   (b)

Figure 12-74

The graph of $y = mx + b$ in Figure 12-74(b) crosses the y-axis at point $P(0, b)$. The value of y at the point of intersection of any line with the y-axis is the **y-intercept**. Thus, b is the y-intercept of $y = mx + b$; this form of linear equation is the **slope-intercept form**. Similarly, the value of x at the point of intersection of a line with the x-axis is the **x-intercept**.

Example 12-18

Given the equation $y - 3x = {}^-6$, do the following:

a. Find the slope of the line.
b. Find the y-intercept.
c. Find the x-intercept.
d. Sketch the graph of the equation.

Solution a. To write the equation in the form $y = mx + b$, we add $3x$ to both sides of the given equation to obtain $y = 3x + (^-6)$. Hence, the slope is 3.

b. The form $y = 3x + (^-6)$ shows that $b = -6$, which is the y-intercept. (The y-intercept can also be found directly by substituting $x = 0$ in the equation and finding the corresponding value of y.)

c. The x-intercept is the x-coordinate of the point where the graph intersects the x-axis. At that point, $y = 0$. Substituting 0 for y in $y = 3x - 6$ gives 2 as the x-intercept.

d. Knowing the y-intercept and the x-intercept gives us two points, $(0, {}^-6)$ and $(2, 0)$, on the line. We may plot these points and draw the line through them to obtain the desired graph in Figure 12-75.

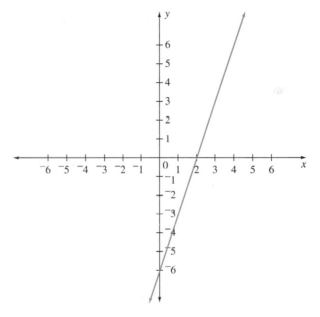

Figure 12-75

Recall that any vertical line has equation $x = a$ for some real number a. This equation cannot be written in slope-intercept form. In general, *every line has an equation of the form either $y = mx + b$ or $x = a$*. Any equation that can be put in one of these forms is a **linear equation**.

> **Theorem 12–18: Equation of a Line**
>
> Every line has an equation of the form either $y = mx + b$ or $x = a$, where m is the slope and b is the y-intercept.

Using Similar Triangles to Determine Slope

We have defined the slope of a line with equation $y = mx + b$ to be m. The slope is a measure of steepness of a line. A different way to discuss the steepness of a line is to consider the rate of change in y-values in relation to their corresponding x-values. In Figure 12-76, line k has a greater rate of change than line ℓ. In other words, line k rises higher than line ℓ for the same horizontal run.

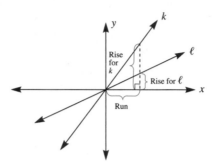

Figure 12-76

Thus, we could express the rate of change, the steepness, or the slope, as the ratio $\dfrac{\text{Change in } y\text{-values}}{\text{Corresponding change in } x\text{-values}}$. The rate is frequently generalized as $\dfrac{\text{Rise}}{\text{Run}}$. In Figure 12-77, right triangles have been constructed and shaded on several lines. In each triangle, the horizontal side is the run and the vertical side is the rise.

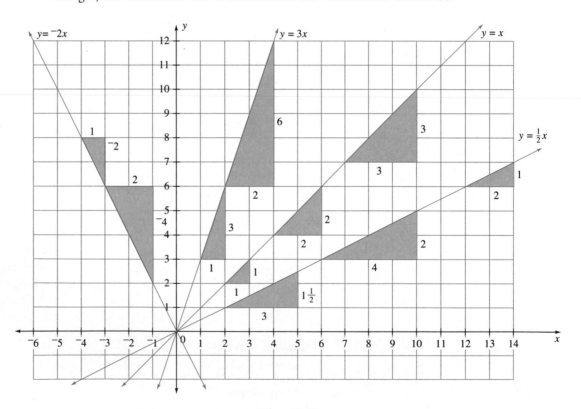

Figure 12-77

The slope of each line in Figure 12-77 can be calculated as the rise over the run in any of the shaded triangles, with hypotenuse (the side opposite the right angle) along the given line. To test this fact, notice that

$$\text{for } y = \frac{1}{2}x, \quad m = \frac{\text{Rise}}{\text{Run}} = \frac{1\frac{1}{2}}{3} = \frac{2}{4} = \frac{1}{2}$$

$$\text{for } y = x, \quad m = \frac{\text{Rise}}{\text{Run}} = \frac{1}{1} = \frac{2}{2} = \frac{3}{3} = 1$$

$$\text{for } y = 3x, \quad m = \frac{3}{1} = \frac{6}{2} = 3$$

$$\text{for } y = {}^-2x, \quad m = \frac{{}^-4}{2} = \frac{{}^-2}{1} = {}^-2$$

Using the previous notion, we find that the slope of a line $\overleftrightarrow{AB}$ is the change in y-coordinates divided by the corresponding change in x-coordinates of any two points on $\overleftrightarrow{AB}$. The difference $x_2 - x_1$ is the *run*, and the difference $y_2 - y_1$ is the *rise*. The slope formula can be interpreted with coordinates, as shown in Figure 12-78. Notice that the ratio $\frac{y_2 - y_1}{x_2 - x_1}$ is always the same, regardless of which two points on a given nonvertical line are chosen. This is so because the right triangles created on the same line are similar.

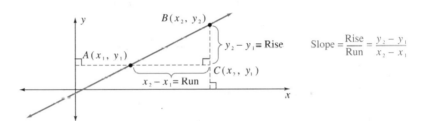

Figure 12-78

The discussion of slope is summarized in the following formula.

Slope Formula

Given two points $A(x_1, y_1)$ and $B(x_2, y_2)$ with $x_1 \neq x_2$, the slope m of the line $\overleftrightarrow{AB}$ is

$$m = \frac{y_2 - y_1}{x_2 - x_1}$$

REMARK By multiplying both the numerator and the denominator on the right side of the slope formula by $^-1$, we obtain

$$m = \frac{y_2 - y_1}{x_2 - x_1} = \frac{(y_2 - y_1)(^-1)}{(x_2 - x_1)(^-1)} = \frac{y_1 - y_2}{x_1 - x_2}$$

This shows that although it does not matter which point is named (x_1, y_1) and which is named (x_2, y_2), *the order of the coordinates in the subtraction must be the same.*

NOW TRY THIS 12-14

a. Use the slope formula to find the slope of any horizontal line.
b. What happens when we attempt to use the slope formula for a vertical line? What is your conclusion about the slope of a vertical line?

When a line is inclined downward from the left to the right, the slope is negative. This is illustrated in Figure 12-79, where the graph of the line $y = {}^-2x$ is shown. The slope of line $y = {}^-2x$ can be calculated as $\dfrac{\text{Rise}}{\text{Run}} = \dfrac{{}^-4}{2} = \dfrac{{}^-2}{1}$.

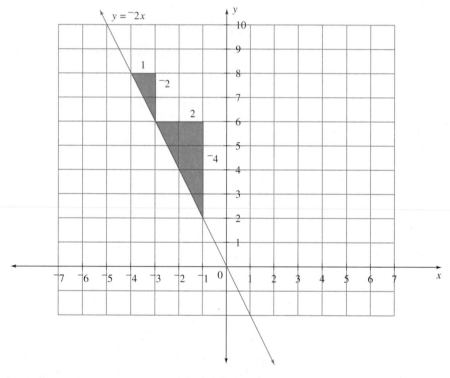

Figure 12-79

Example 12-19

a. Given $A(3, 1)$ and $B(5, 4)$, find the slope of $\overleftrightarrow{AB}$.
b. Find the slope of the line passing through the points $A(-3, 4)$ and $B(-1, 0)$.

Solution   a. $m = \dfrac{4 - 1}{5 - 3} = \dfrac{3}{2}$,   or $\dfrac{1 - 4}{3 - 5} = \dfrac{{}^-3}{{}^-2} = \dfrac{3}{2}$

b. $m = \dfrac{4 - 0}{{}^-3 - ({}^-1)} = \dfrac{4}{{}^-2} = {}^-2$,   or $\dfrac{0 - 4}{{}^-1 - ({}^-3)} = \dfrac{{}^-4}{2} = {}^-2$

Given two points on a nonvertical line, we can use the slope formula to find the slope of the line and its equation. This is demonstrated in the following example.

Example 12-20

In Figure 12-80, the points $(^-4, 0)$ and $(1, 4)$ are on the line ℓ. Find:

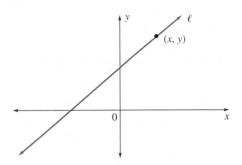

Figure 12-80

a. The slope of the line
b. The equation of the line

Solution   **a.** $m = \dfrac{4 - 0}{1 - (^-4)} = \dfrac{4}{5}$

b. Point (x, y) represents any point on the line different from $(^-4, 0)$ if, and only if, the slope determined by points $(^-4, 0)$ and (x, y) is $\dfrac{4}{5}$, the slope of the line we found in part (a). Thus,

$$\frac{y - 0}{x - (^-4)} = \frac{4}{5}$$

$$y = \frac{4}{5}(x + 4)$$

The slope-intercept form of the equation of the line is $y = \dfrac{4}{5}x + \dfrac{16}{5}$.

Systems of Linear Equations

The mathematical descriptions of many problems involve more than one equation, each having more than one unknown. To solve such problems, we must find a common solution to the equations, if it exists. An example is given on the following student page. Answer question 21.

Example 12-21

May Chin ordered lunch for herself and several friends by phone without checking prices. She paid $18.00 for three soyburgers and twelve orders of fries. Another time she paid $12 for four soyburgers and four orders of fries. Set up a system of equations with two unknowns representing the prices of a soyburger and an order of fries, respectively.

Solution   Let x be the price in dollars of a soyburger and y be the price of an order of fries. Three soyburgers cost $3x$ dollars, and twelve orders of fries cost $12y$ dollars. Because May paid $18.00 for her entire order, we have $3x + 12y = 18$, or, after dividing each side by 3, $x + 4y = 6$. Similarly, $4x + 4y = 12$, or $x + y = 3$.

Section 2

Key Term

linear
equation

Key Concepts

Modeling Linear Change (pp. 483–485)

When a quantity changes by the same amount at regular intervals, the quantity shows linear change. You can use a linear equation to model linear change. You can also use a table or a graph.

Example Lynda borrowed $175 from her parents. She pays them back $10 a week. Her sister Maria borrowed $200 and pays back $15 a week. Who will finish paying off her loan first?

Write an equation for each person. Let y = the amount the person owes after x weeks.

Lynda	Maria
$y = 175 - 10x$	$y = 200 - 15x$

Number of weeks	Amount Lynda owes	Amount Maria owes
0	175	200
1	165	185
2	155	170
3	145	155
4	135	140
5	125	125
6	115	110
...	...	...
13	45	5
14	35	0

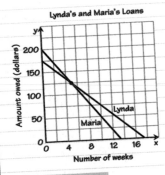

Lynda's and Maria's Loans

Maria only has to pay $5 in the 14th week.

After five weeks, they owe the same amount of money. After that, Maria is paying more per week than Lynda, so she will pay off her loan first.

21 **Key Concepts Question** In the Example above, how does the graph show when the sisters owe the same amount of money?

An ordered pair satisfying the two linear equations in Example 12-25 is a point that belongs to each of the lines. Figure 12-81 shows the graphs of $x + 4y = 6$ and $x + y = 3$. We are looking for x and y that satisfy each equation; that is, a point on each of the lines. The two lines appear to intersect at $(2, 1)$. Thus, $(2, 1)$ seems to be the solution of the given system of equations. This solution can be checked by substituting 2 for x and 1 for y in each equation. Because two distinct lines intersect in only one point, $(2, 1)$ is the only solution to the system. Therefore, in Example 12-21 a soyburger cost $2 and fries cost $1.

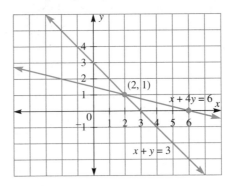

Figure 12-81

Substitution Method

Drawbacks to estimating a solution to a system of equations graphically include an inability to read some coordinates of points. However, there are algebraic methods for solving systems of linear equations. Example 12-22 demonstrates one such algebraic method: the **substitution method**.

Example 12-22   Solve the following system:

$$3x - 4y = 5$$
$$2x + 5y = 1$$

Solution   First, rewrite each equation, expressing y in terms of x.

$$y = \frac{3x - 5}{4} \quad \text{and} \quad y = \frac{1 - 2x}{5}$$

Because we are looking for x and y that satisfy each equation, we can equate the expressions for y and solve the resulting equation for x.

$$\frac{3x - 5}{4} = \frac{1 - 2x}{5}$$
$$5(3x - 5) = 4(1 - 2x)$$
$$15x - 25 = 4 - 8x$$
$$23x = 29$$
$$x = \frac{29}{23}$$

Substituting $\frac{29}{23}$ for x in $y = \frac{3x - 5}{4}$ gives $y = \frac{-7}{23}$. Hence, $x = \frac{29}{23}$ and $y = \frac{-7}{23}$. This can be checked by substituting the values for x and y in the original equations.

Sometimes when using the substitution method, it is more convenient to solve a system of equations by expressing x in terms of y in one of the equations and substituting the obtained expression for x in the other equation.

Elimination Method

The **elimination method** for solving two equations with two unknowns is based on eliminating one of the variables by adding or subtracting the original or equivalent equations. For example, consider the following system:

$$x - y = {}^-3$$
$$x + y = 7$$

By adding the two equations, we can eliminate the variable y. The resulting equation can then be solved for x.

$$
\begin{array}{rcl}
x - y &=& {}^-3 \\
x + y &=& 7 \\
\hline
2x &=& 4 \\
x &=& 2
\end{array}
$$

Substituting 2 for x in the first equation (either equation may be used) gives $y = 5$. Checking this result shows that $x = 2$ and $y = 5$, or $(2, 5)$, is the solution to the system. Notice that the solution gives the point where the graphs intersect.

Often, another operation is required before equations are added so that an unknown can be eliminated. For example, consider the following system:

$$3x + 2y = 5$$
$$5x - 4y = 3$$

Adding the equations does not eliminate either unknown. However, if the first equation contained $4y$ rather than $2y$, the variable y could be eliminated by adding. To obtain $4y$ in the first equation, we multiply both sides of the equation by 2 to obtain the equivalent equation $6x + 4y = 10$. Adding the equations in the equivalent system gives the following:

$$
\begin{array}{rcl}
6x + 4y &=& 10 \\
5x - 4y &=& 3 \\
\hline
11x &=& 13 \\
x &=& \dfrac{13}{11}
\end{array}
$$

To find the corresponding value of y, we substitute $\dfrac{13}{11}$ for x in either of the original equations and solve for y. We may also use the elimination method again and solve for y. To eliminate the x-values from the original system, we multiply the first equation by 5 and the second by ${}^-3$ (or the first by ${}^-5$ and the second by 3). Then we add the two equations and solve for y.

$$
\begin{array}{rcl}
15x + 10y &=& 25 \\
{}^-15x + 12y &=& {}^-9 \\
\hline
22y &=& 16 \\
y &=& \dfrac{16}{22}, \text{ or } \dfrac{8}{11}
\end{array}
$$

Consequently, $\left(\dfrac{13}{11}, \dfrac{8}{11}\right)$ is the solution of the original system. This solution, as always, should be checked by substitution in the *original* equations.

Solutions to Various Systems of Linear Equations

All examples thus far have had unique solutions. However, other situations may arise. Geometrically, a system of two linear equations can be characterized as follows:

1. *The system has a unique solution if, and only if, the graphs of the equations intersect in a single point.*
2. *The system has no solution if, and only if, the equations represent distinct parallel lines.*
3. *The system has infinitely many solutions if, and only if, the equations represent the same line.*

Consider the following system.

$$2x - 3y = 1$$
$$^-4x + 6y = 5$$

In an attempt to solve for x, we multiply the first equation by 2 and then add as follows:

$$4x - 6y = 2$$
$$\underline{^-4x + 6y = 5}$$
$$0 = 7$$

A false statement results. Logically, a false result can occur only on the basis of a false assumption or an incorrect procedure. Because our procedure is correct in this case, there must be a false assumption. We assumed that the system has a solution. That assumption caused a false statement; therefore, the assumption itself must be false. Hence, the system has no solution. In other words, the solution set is $\varnothing$. This situation arises if, and only if, the corresponding lines are parallel.

Next, consider the following system:

$$2x - 3y = 1$$
$$^-4x + 6y = ^-2$$

To solve this system, we multiply the first equation by 2 and add as follows:

$$4x - 6y = 2$$
$$\underline{^-4x + 6y = ^-2}$$
$$0 = 0$$

The resulting statement, $0 = 0$, is always true. Rewriting the equation as $0 \cdot x + 0 \cdot y = 0$ shows that all values of x and y satisfy this equation. The values of x and y that satisfy both $0 \cdot x + 0 \cdot y = 0$ and $2x - 3y = 1$ are those that satisfy $2x - 3y = 1$. There are infinitely many such pairs of x and y that correspond to points on the line $2x - 3y = 1$ and hence to $^-4x + 6y = ^-2$.

One way to check that a system has infinitely many solutions is by observing whether each of the original equations represents the same line. In the preceding system, both equations may be written as $y = \dfrac{2}{3}x - \dfrac{1}{3}$. Another way to check that a system has infinitely many solutions is to observe whether one equation can be multiplied by some number to obtain the second equation. For example, multiplying the equation $2x - 3y = 1$ by $^-2$ yields the second equation, $^-4x + 6y = ^-2$.

Example 12-23

Identify each of the following systems as having a unique solution, no solution, or infinitely many solutions:

a. $2x - 3y = 5$
$\frac{1}{2}x - y = 1$

b. $\frac{x}{3} - \frac{y}{4} = 1$
$3y - 4x + 12 = 0$

c. $6x - 9y = 5$
$^{-}8x + 12y = 7$

Solution   One method is to attempt to solve each system. Another method is to write each equation in the slope-intercept form and interpret the system geometrically.

a. *First method.*   To eliminate x, multiply the second equation by $^{-}4$ and add the equations.

$$
\begin{array}{rcr}
2x - 3y &=& 5 \\
{}^{-}2x + 4y &=& {}^{-}4 \\
\hline
y &=& 1
\end{array}
$$

Substituting 1 for y in either equation gives $x = 4$. Thus, $(4, 1)$ is the unique solution of the system.

Second method.   In slope-intercept form, the first equation is $y = \frac{2}{3}x - \frac{5}{3}$. The second equation is $y = \frac{1}{2}x - 1$. The slopes of the corresponding lines are $\frac{2}{3}$ and $\frac{1}{2}$, respectively. Consequently, the lines are distinct and are not parallel and, therefore, intersect in a single point whose coordinates are the unique solution to the original system.

b. *First method.*   Multiply the first equation by 12 and rewrite the second equation as $^{-}4x + 3y = ^{-}12$. Then, adding the resulting equations gives the following:

$$
\begin{array}{rcr}
4x - 3y &=& 12 \\
{}^{-}4x + 3y &=& {}^{-}12 \\
\hline
0 &=& 0
\end{array}
$$

This implies that the two equations represent the same line, and the original system has infinitely many solutions (all the points on the line).

Second method.   In slope-intercept form, both equations have the form

$$y = \frac{4}{3}x - 4$$

Thus, the two lines are identical, so the system has infinitely many solutions.

c. *First method.*   To eliminate y, multiply the first equation by 4 and the second by 3; then, add the resulting equations.

$$
\begin{array}{rcr}
24x - 36y &=& 20 \\
{}^{-}24x + 36y &=& 21 \\
\hline
0 &=& 41
\end{array}
$$

No pair of numbers satisfies $0 \cdot x + 0 \cdot y = 41$, so this equation has no solutions, and, consequently, the original system has no solutions.

Second method.   In slope-intercept form, the first equation is $y = \frac{2}{3}x - \frac{5}{9}$. The second equation is $y = \frac{2}{3}x + \frac{7}{12}$. The corresponding lines have the same slope, $\frac{2}{3}$, but different y-intercepts. Consequently, the lines are parallel and the original system has no solution.

NOW TRY THIS 12-15 Find all the solutions (if any) of each of the following systems by graphing the equations in each system. Then justify your answers by an algebraic approach.

a. $x - y = 1$
   $2x - y = 5$

b. $2x - y = 1$
   $2y - 4x = 3$

c. $2x - 3y = 1$
   $6y - 4x = 2$

Fitting a Line to Data

As noted in Chapter 10, in many practical situations, a relationship between two variables comes from collected data such as from population or business surveys. When the data are graphed, there may not be a single line that goes through all of the points, but the points may appear to approximate, or "follow," a straight line. In such cases, it is useful to find the equation of what seems to be the **trend line**. Knowing the equation of such a line enables us to predict an outcome without actually performing the experiment.

There are several approaches to define, and hence find, the trend line. We take a graphical approach as follows:

1. Choose a line that seems to follow the given points so that there are about an equal number of points below the line as above the line.
2. Determine two convenient points on the line and approximate the x- and y-coordinates of these points.
3. Use the points in (2) to determine the equation of the line.

Example 12-24

A shirt manufacturer noticed that the number of units sold depends on the price charged. The data in Table 12-4 show the number of units sold for a given price per unit.

a. Find the equation of a line that seems to fit the data best.
b. Use the equation in (a) to predict the number of units that will be sold if the price per unit is $60.

Table 12-4

Price per Unit (Dollars)	Number of Units Sold (Thousands)
50	200
44	250
41	300
33	380
31	400
24.5	450
20	500
14.5	550

Solution **a.** Figure 12-82(a) shows the graph of the data displayed in Table 12-4. Figure 12-82(b) shows a line that seems to fit the data so that approximately the same number of points are below the line as above the line. We choose the points (50, 200) and (20, 500), which are on the line in Figure 12-82(b).

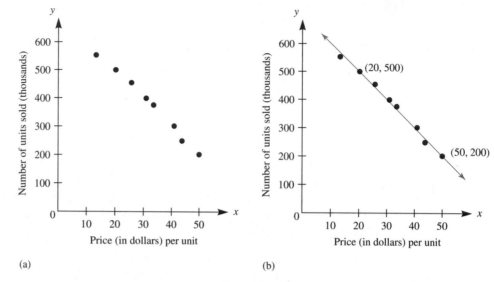

(a)   (b)

Figure 12-82

To find the equation of the line, we may find m and b in the equation $y = mx + b$. Substituting the points (50, 200) and (20, 500) into this equation, we obtain the following:

$$200 = 50m + b$$
$$500 = 20m + b$$

One way to solve the equations is to express b in terms of m for each equation:

$$b = 200 - 50m$$
$$b = 500 - 20m$$

We equate the expressions for b and solve for m:

$$200 - 50m = 500 - 20m$$
$$200 - 500 = 50m - 20m$$
$$^{-}300 = 30m$$
$$m = {}^{-}10$$

Substituting this value for m, we obtain

$$b = 200 - 50(^{-}10) = 700$$

Consequently, the equation of the trend line is $y = {}^{-}10x + 700$.

b. Using the equation in (a), substitute $x = 60$ to obtain $y = {}^{-}10(60) + 700$, or $y = 100$. Thus, we predict that 100,000 units will be sold if the price per unit is $60.

Assessment 12-5A

1. The graph of $y = mx$ is given in the following figure. Sketch the graphs for each of the following on the same figure. Explain your answers.
   a. $y = mx + 3$
   b. $y = mx - 3$

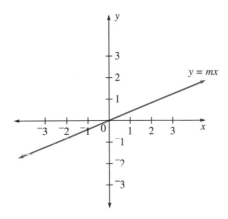

2. Sketch the graphs for each of the following equations:
   a. $y = \dfrac{-3}{4}x + 3$
   b. $y = 3$
   c. $y = 15x - 30$

3. Find the x-intercept and y-intercept for the equations in problem 2, if they exist.

4. In the following figure, part (i) shows a dual-scale thermometer and part (ii) shows the corresponding points plotted on a graph:

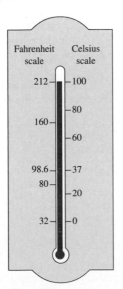

(i)

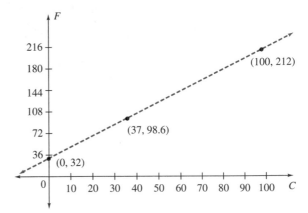

(ii)

   a. Use two of the points on the graph to develop a formula for conversion from degrees Celsius (C) to degrees Fahrenheit (F).
   b. Use your answer in (a) to find a formula for converting from degrees F to degrees Celsius.

5. Write each of the following equations in slope-intercept form and identify the slope and y-intercept:
   a. $3y - x - 0$ b. $x + y - 3$ c. $x - 3y$

6. For each of the following, write the equation of the line determined by the given pair of points in slope-intercept form or in the form $x = a$:
   a. $(-4, 3)$ and $(1, -2)$
   b. $(0, 0)$ and $(2, 1)$
   c. $\left(0, \dfrac{-1}{2}\right)$ and $\left(\dfrac{1}{2}, 0\right)$

7. Find the coordinates of two other points collinear (on the same line) with each of the following pairs of given points:
   a. $P(2, 2)$, $Q(4, 2)$ b. $P(0, 0)$, $Q(0, 1)$

8. For each of the following, give as much information as possible about x and y:
   a. The ordered pairs $(-2, 0)$, $(-2, 1)$, and (x, y) represent collinear points.
   b. The ordered pair (x, y) is in the fourth quadrant.

9. Consider the lines through $P(2, 4)$ and perpendicular to the x- and y-axes, respectively. Find the area and the perimeter of the rectangle formed by these lines and the axes.

10. Find the equations for each of the following:
   a. The line containing $P(3, 0)$ and perpendicular to the x-axis
   b. The line containing $P(-4, 5)$ and parallel to the x-axis

11. For each of the following, find the slope, if it exists, of the line determined by the given pair of points:
   a. $(4, 3)$ and $(^-5, 0)$ **b.** $(\sqrt{5}, 2)$ and $(1, 2)$
   c. (a, a) and (b, b)

12. Write the equation of each line in problem 11.

13. Wildlife experts found that the number of chirps a cricket makes in a 15-sec interval is related to the temperature T in degrees Fahrenheit, as shown in the following graph:

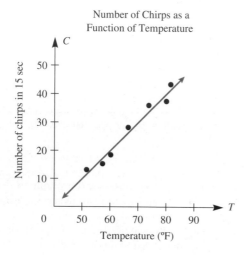

Number of Chirps as a Function of Temperature

   a. If C is the number of chirps in 15 sec, write a formula for C in terms of T (temperature in degrees Fahrenheit) that seems to fit the data best.
   b. Use the equation in (a) to predict the number of chirps in 15 sec when the temperature is 90°.
   c. If N is the number of chirps per minute, write a formula for N in terms of T.

14. An equilateral triangle whose side length is 6 is placed so that one side lies along the x-axis of a coordinate system and one vertex is at the origin.
   a. Write the coordinates of the midpoint of the side along the x-axis.
   b. Write the equation of the perpendicular bisector of the side along the x-axis.

15. Write the equation of lines meeting each of the following conditions:
   a. A horizontal line containing the point with coordinates $(3, 4)$
   b. A vertical line containing the point with coordinates $(^-7, ^-8)$

16. Consider a trapezoid with vertices whose coordinates are $(0, 0)$, $(4, 6)$, $(10, 6)$, and $(15, 0)$.
   a. What are the coordinates of the endpoints of the midsegment of the trapezoid?
   b. What is the equation of the midsegment?

17. Suppose a car is traveling at a constant speed of 60 mph.
   a. Draw a graph showing the relation between the distance traveled and the amount of time that it takes to travel the distance.
   b. What is the slope of the line that you drew in part (a)?

18. a. Graph the following data and find the equation of the trend line.
   b. Use your answer in (a) to predict the value of y when $x = 10$.

x	y
1	8.1
2	9.9
3	12
4	14.1
5	15.9
6	18
7	19.9

19. Solve each of the following systems, if possible. Indicate whether the system has a unique solution, infinitely many solutions, or no solution.
   a. $y = 3x - 1$
       $y = x + 3$
   b. $3x + 4y = ^-17$
       $2x + 3y = ^-13$

20. The owner of a 5000-gal oil truck loads the truck with gasoline and kerosene. The profit on each gallon of gasoline is 13¢ and on each gallon of kerosene it is 12¢. How many gallons of each fuel did the owner load if the profit was $640?

21. Josephine's bank contains 27 coins. If all the coins are either dimes or quarters and the value of the coins is $5.25, how many of each kind of coin are there?

22. If a triangle has vertices with coordinates $(0, 32)$, $(20, 0)$, and $(100, 212)$ and a similar triangle has vertices $(0, 16)$ and $(10, 0)$, find a possible set of coordinates for the third vertex.

23. If two sets of data have a negative association, what can be said about the slope of the trend line for the data?

Assessment 12-5B

1. Sketch the graphs for each of the following equations:

   a. $x = {}^-2$ **b.** $y = 3x - 1$ **c.** $y = \dfrac{1}{20}x$

2. Find the x-intercept and y-intercept for the equations in problem 1, if they exist.

3. Write each of the following equations in slope-intercept form and identify the slope and y-intercept:

   a. $\dfrac{x}{3} + \dfrac{y}{4} = 1$

   b. $3x - 4y + 7 = 0$

   c. $x - y = 4(x - y)$

4. For each of the following, write the equation of the line determined by the given pair of points in slope-intercept form or in the form $x = a$:

   a. $(0, 1)$ and $(2, 1)$

   b. $(2, 1)$ and $(2, {}^-1)$

   c. $({}^-a, 0)$ and $(a, 0), a \neq 0$

5. Find the coordinates of two other points collinear (on the same line) with each of the following pairs of given points:

   a. $P({}^-1, 0), Q({}^-1, 2)$ **b.** $P(0, 0), Q(1, 1)$

6. Give as much information as possible about x and y in the following:

   a. The ordered pairs $({}^-2, 1), (0, 1),$ and (x, y) represent collinear points.

   b. $(x, y), (0, 0)$ and $({}^-1, {}^-1)$ are collinear.

7. Find the equations for each of the following:

   a. The line containing $P(0, {}^-2)$ and parallel to the x-axis

   b. The line containing $P({}^-4, 5)$ and parallel to the y-axis

8. For each of the following, find the slope, if it exists, of the line determined by the given pair of points:

   a. $({}^-4, 1)$ and $(5, 2)$ **b.** $({}^-3, 81)$ and $({}^-3, 198)$

   c. $(1.0001, 12)$ and $(1, 10)$

9. Write the equation of each line in problem 8.

10. A rectangle has two vertices on the x-axis, two vertices on the y-axis, and one vertex at the point with coordinates $(4, 6)$.

    a. Make a sketch showing that there is a rectangle with these characteristics.

    b. Write the coordinates of each of the other three vertices.

    c. Write the equations of the diagonals of the rectangle.

    d. Find the coordinates of the point of intersection of the diagonals.

11. A triangle has vertices with coordinates $({}^-4, 6), (2, {}^-6),$ and $(8, 10)$.

    a. Write the equations of the lines containing each of the sides of the triangle.

    b. Prove that your equations are correct by finding the coordinates of the points of intersection of each pair of sides of the triangle.

12. If three vertices of a kite have coordinates $(0, 0), (6, 14),$ and $(6, {}^-14)$, write an equation of a line that contains the other vertex. Is there more than one solution to the problem?

13. Solve each of the following systems, if possible. Indicate whether the system has a unique solution, infinitely many solutions, or no solution.

    a. $2x - 6y = 7$ **b.** $4x - 6y = 1$
       $3x - 9y = 10$ $6x - 9y = 1.5$

14. The vertices of a triangle are given by $(0, 0), (10, 0),$ and $(6, 8)$. Show that the segments connecting $(5, 0)$ and $(6, 8), (10, 0)$ and $(3, 4)$, and $(0, 0)$ and $(8, 4)$ intersect at a common point.

15. At the end of 10 mo, the balance of an account earning annual simple interest is $2100.

    a. If, at the end of 18 mo, the balance is $2180, how much money was originally in the account?

    b. What is the annual rate of interest?

16. **a.** Solve each of the following systems of equations. What do you notice about the answers?

    i. $x + 2y = 3$
       $4x + 5y = 6$

    ii. $2x + 3y = 4$
        $5x + 6y = 7$

    iii. $31x + 32y = 33$
         $34x + 35y = 36$

    b. Write another system similar to those in (a). What solution did you expect? Check your guess.

    c. Write a general system similar to those in (a). What solution does this system have? Why?

17. If a line ℓ forms a 45° angle with the x-axis, what is the slope of

    a. Line ℓ?

    b. Line m, which is perpendicular to line ℓ?

Mathematical Connections 12-5

Communication

1. An arithmetic sequence having a linear graph was depicted in this chapter. Explain whether a geometric sequence has a linear graph.

2. Dahlia wanted to find what temperature has the same measure when measured in degrees Fahrenheit or degrees Celsius. To do that, she graphed the equation of the conversion formula $C = \dfrac{5}{9}(F - 32)$. On the same coordinate system, she also graphed the equation $C = F$ and found the intersection point $(^-40, ^-40)$. Explain why this procedure answers the question Dahlia asked.

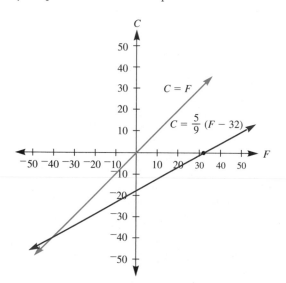

3. Explain in more than one way why two lines with the same slope and different y intercepts are parallel.

Open-Ended

4. Look for data in newspapers, magazines, or books whose graphs appear to be close to linear and find the equations of the lines that you think best fit the data.

5. **a.** Write equations of two lines that intersect but when graphed look parallel.
   b. At what point do those two lines intersect?

Cooperative Learning

6. Play the following game between your group and another group. Each group makes up four linear equations that have a common property and presents the equations to the other group. For example, one group could present the equations $2x - y = 0$, $4x - 2y = 3$, $y - 2x = 3$, and $3y - 6x = 5$. If the second group discovers a common property that the equations share, such as the graphs of the

equations are four parallel lines, they get 1 point. Each group takes a specified number of turns.

Questions from the Classroom

7. A student would like to know why it is impossible to find the slope of a vertical line. How do you respond?

8. A student argues that there is no trend line for a set of data. Could the student be correct? Give an example to make your point to the student.

9. One student says that a tangent is another expression for a slope. Is the student correct?

10. Jonah tried to solve the equation $^-5x + y = 20$ by adding 5 to both sides. He wrote $5 - 5x + y = 5 + 20$ or $0 \cdot x + y = 25$ and finally $y = 25$. How would you help Jonah?

11. Jill would like to know why two lines with an undefined slope are parallel. How would you respond?

Review Problems

12. $ABCD$ is a trapezoid with $\overline{BC} \parallel \overline{AD}$. Solve for x.

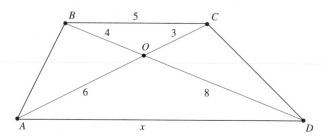

13. Draw an arbitrary triangle ABC. Let M, N, and P be the corresponding midpoints of the sides $\overline{AB}$, $\overline{BC}$, and $\overline{AC}$. Connect the points M, N, and P. Four triangles are created. Is it true that all of the smaller triangles are congruent and each is similar to $\triangle ABC$? Why or why not?

14. In triangle ABC, a square has been inscribed as shown. The lengths of the sides of $\triangle ABC$ are 3, 4, and 5 as shown. Find the length of a side of the square. (*Hint:* First show that $\triangle AED \sim \triangle ABC$.)

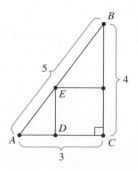

15. In the figure shown, find *BE*.

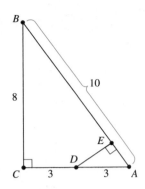

Third International Mathematics and Science Study (TIMSS) Question

A straight line passes through the points $(2, 3)$ and $(4, 7)$. Which of these points is also on the line?
  a. $(0, 2)$
  b. $(1, 2)$
  c. $(2, 4)$
  d. $(3, 5)$
  e. $(4, 5)$

TIMSS, Grade 8, 2003

National Assessment of Educational Progress (NAEP) Question

Which of the following is the graph of the line with equation $y = {}^{-}2x + 1$?

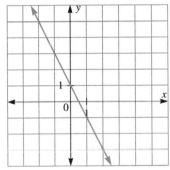

a.

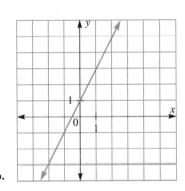

b.

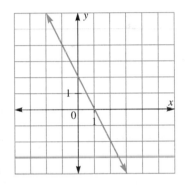

c.

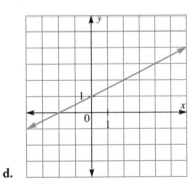

d.

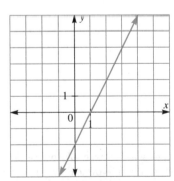

e.

NAEP, Grade 8, 2007

Hint for Solving the Preliminary Problem

Construct a triangle with a side 18 in. long and the given angles. Then construct a new triangle in the interior of the first triangle by drawing a line parallel to one of the sides. There are infinitely many such triangles. Choose one with a side of 18 in.

Chapter Outline

I. Congruence

    A. Two geometric figures are **congruent** if, and only if, they have the same size and shape.

    B. Two triangles are congruent if they satisfy any of the following properties:

        1. Side, Side, Side (SSS)

        2. Side, Angle, Side (SAS)

        3. Angle, Side, Angle (ASA)

        4. Angle, Angle, Side (AAS)

    C. **Triangle inequality:** The sum of the measures of any two sides of a triangle must be greater than the measure of the third side.

    D. Corresponding parts of congruent figures are congruent.

    E. A **rep-tile** is a figure that is used in the construction of a similar figure.

II. Circles

    A. An **arc** of a circle is part of the circle that can be drawn without lifting a pencil. The **center of an arc** is the center of the circle containing the arc.

    B. A **chord** is a segment whose endpoints lie on a circle.

III. Proportion

    A. If a line parallel to one side of a triangle intersects the other sides, it divides those sides into proportional segments.

    B. If parallel lines cut off congruent segments on one transversal, they cut off congruent segments on any transversal.

IV. Constructions that can be accomplished using a compass and a straightedge

    A. Copy a line segment.

    B. Copy a circle.

    C. Copy an angle.

    D. Bisect a segment.

    E. Bisect an angle.

    F. Construct a perpendicular from a point to a line.

    G. Construct a perpendicular bisector of a segment.

    H. Construct a perpendicular to a line through a point on the line.

    I. Construct a parallel to a line through a point not on the line.

    J. Divide a segment into congruent parts.

    K. Inscribe some regular polygons in a circle.

    L. Circumscribe a circle about a triangle.

    M. Inscribe a circle in a triangle.

V. Similarity and proportion

    A. Two polygons are **similar** if, and only if, their corresponding angles are congruent and their corresponding sides are proportional.

    B. **AA:** If two angles of one triangle are congruent to two angles of a second triangle, respectively, the triangles are similar.

    C. **SSS:** If three sides of one triangle are proportional to corresponding sides of another triangle, then the triangles are similar.

    D. **SAS:** If two sides of one triangle are proportional to two corresponding sides of another triangle, and the included angles are congruent, then the triangles are similar.

    E. If a line divides two sides of a triangle into proportional segments, then the line is parallel to the third side.

VI. Lines in a Cartesian coordinate system

    A. Every nonvertical line has an equation of the form $y = mx + b$, where m is the slope and b is the y-intercept.

    B. The equation of any vertical line can be written in the form $x = a$, and any horizontal line in the form $y = b$.

    C. **Slope formula:** Given two points $A(x_1, y_1)$ and $B(x_2, y_2)$, the slope m of line AB is given by the following:

$$m = \frac{y_2 - y_1}{x_2 - x_1} = \frac{y_1 - y_2}{x_1 - x_2} = \frac{\text{Rise}}{\text{Run}}$$

    D. The equation of a line with slope m through a given point with coordinates (x_1, y_1) is $y - y_1 = m(x - x_1)$.

    E. Parallel nonvertical lines have the same slope.

    F. If two lines have the same slope, they are parallel.

    G. The slope of a horizontal line is 0, and the slope of a vertical line is undefined.

    H. A system of **linear equations** can be solved graphically by drawing the graphs of the equations.

    I. A system of linear equations can be solved algebraically by either the **substitution method** or the **elimination method**.

    J. The **trend line** is a straight fitted line that lies close to the points in the graph of a set of data.

Chapter Review

1. Each of the following figures contains at least one pair of congruent triangles. Identify them and tell why they are congruent.

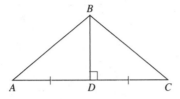

a.

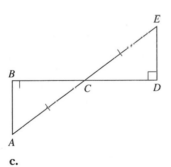

$AC = BD$

b.

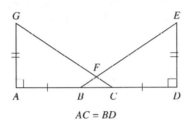

c.

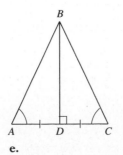

d.

e.

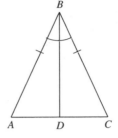

f.

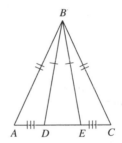

g.

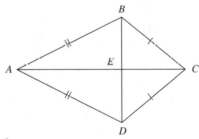

h.

2. In the following figure, $ABCD$ is a square and $\overline{DE} \cong \overline{BF}$. What kind of figure is $AECF$? Justify your answer.

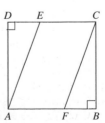

3. Construct each of the following by (1) using a compass and straightedge and (2) paper folding:

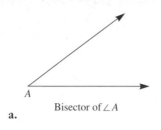

a. Bisector of ∠A

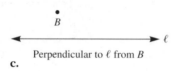

b. Perpendicular to ℓ at B

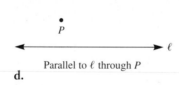

c. Perpendicular to ℓ from B

d. Parallel to ℓ through P

4. For each of the following pairs of similar triangles, find the missing measures:

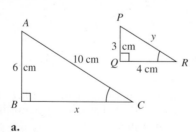

a.

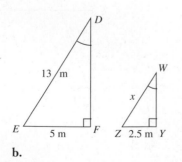

b.

5. Divide the following segment into five congruent parts:

_____

6. If ABCD is a trapezoid, $\overline{EF} \parallel \overline{AD}$ and $\overline{AC}$ is a diagonal, then what is the relationship between $\dfrac{a}{b}$ and $\dfrac{c}{d}$?

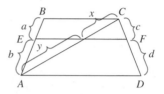

7. Given the following figure, construct a circle that contains A and B and has its center on ℓ:

8. For each of the following figures, show that appropriate triangles are similar and find x and y:

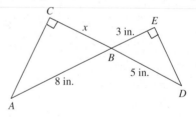

a.

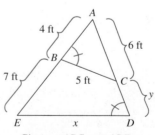

Given: ∠ABC ≅ ∠ADE

b.

9. Determine if each of the following is true or false. If false, explain why.
   a. A radius of a circle is a chord of the circle.
   b. If a radius bisects a chord of a circle, then it is perpendicular to the chord.

10. A person 2 m tall casts a shadow 1 m long when a building has a 6 m shadow. How high is the building?

11. a. Which of the following polygons can be inscribed in a circle? Assume that all sides of each polygon are congruent and that all the angles of polygons (ii) and (iii) are congruent.

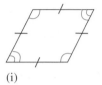

(i)   (ii)   (iii)

   b. Based on your answer in (a), make a conjecture about what kinds of polygons can be inscribed in a circle.

12. Determine the vertical height of the playground slide shown in the following figure:

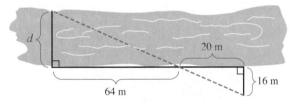

13. Find the distance d across the river sketched.

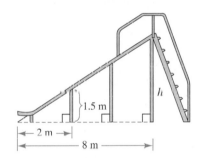

14. Describe how any parallelogram can be dissected and reassembled to form a rectangle.

15. A regular hexagon $ABCDEF$ can be divided into two congruent trapezoids by drawing $\overline{AD}$.
    a. What is the relation of $\overline{AD}$ to the circumcircle of the regular hexagon?
    b. Prove that $\angle ABD$ is a right angle.

16. If two lines on a coordinate plane are parallel, justify whether or not their slopes must always be equal.

17. In an equilateral triangle, what must be true about all midsegments?

18. In any triangle, a midsegment is drawn. The triangle is divided into two parts, a triangle and a quadrilateral.
    a. Prove that the triangle is similar to the original.
    b. Prove that the quadrilateral is a trapezoid.

19. Is the following statement always true, always false, or true in some cases and false in others? Explain your answer.

   A quadrilateral whose diagonals are congruent and perpendicular is a square.

20. For each of the following, write the equation of the line determined by the given pair of points.
    a. $(2, {}^-3)$ and $({}^-1, 1)$ b. $({}^-3, 0)$ and $(3, 2)$

21. Use slope to determine if there is a single line through the points with coordinates $(4, 2)$, $(0, {}^-1)$, and $(7, {}^-5)$. Explain your reasoning.

22. Solve each of the following systems, if possible. If the system does not have a unique solution, explain why not.
    a. $x + 2y = 3$
       $2x - y = 9$
    b. $\dfrac{x}{2} + \dfrac{y}{3} = 1$
       $4y - 3x = 2$
    c. $x - 2y = 1$
       $4y - 2x = 0$

23. Why are two congruent triangles ABC and $A'B'C'$ similar?

Selected Bibliography

Beigie, D. "Coordinate Plane Set Detective," *Mathematics Teaching in the Middle School* 9 (January 2004): 251–255.

Beigie, D. "Computer-Generated Fractal Art," *Mathematics Teaching in the Middle School* 10 (February 2005): 262–269.

Carpenter, T., M. Corbitt, H. Kepner, M. Lindquist, and R. Reys. *Results from the Second Mathematics Assessment of the National Assessment of Educational Progress.* Reston, VA: National Council of Teachers of Mathematics, 1981.

Drum, R. "Miniature Toys Introduce Ratio and Proportion with a Real-world Connection." *Mathematics Teaching in the Middle School* 7 (September 2001): 50–54.

Dwyer, N., B. Causey-Lee, and N. Irby. "Conceptualizing Ratios with Look-alike Polygons," *Mathematics Teaching in the Middle School* 8 (April 2005): 426–431.

Johnston-Wilder, S., and J. Mason, Eds. *Developing Thinking in Geometry.* London: The Open University, 2005.

Kastberg, S. "Euclidean Tools and the Creation of Euclidean Geometry." *Mathematics Teaching in the Middle School* 7 (January 2002): 294–296.

Kennedy, J., and E. McDowell. "Geoboard Quadrilaterals." *Mathematics Teacher* 91 (April 1998): 288–290.

Libeskind, S. *Euclidean and Transformational Geometry.* Sudbury, MA: Jones and Bartlett Publishers, Inc., 2008.

Nieuwoudt, H., and R. van Niekerk. "The Spatial Competence of Young Children through the Development of Solids." Paper presented at the meeting of the American Educational Research Association, Chicago, March, 1997.

Piatek-Jimenez, K. "Building Intuitive Arguments for the Triangle Congruence Conditions." *Mathematics Teacher* 101 (February 2008): 463–467.

Sharp, J., and K. Hoiberg. "And Then There Was Luke: The Geometric Thinking of a Young Mathematician." *Teaching Children Mathematics* 7 (March 2001): 432–439.

Slavit, D. "Above and Beyond AAA: The Similarity and Congruence of Polygons." *Mathematics Teaching in the Middle School* 3 (January 1998): 276–280.

Slovin, H. "Moving to Proportional Reasoning." *Mathematics Teaching in the Middle School* 6 (September 2000): 58–60.

Tripathi, P. N. "Developing Mathematical Understanding Through Multiple Representations." *Mathematics Teaching in the Middle School* 13 (April 2008): 438–445.

Welchman, R., and J. Urso. "Midpoint Shapes." *Teaching Children Mathematics* 6 (April 2000): 506–509.

Zykova, V. "The Psychology of Sixth-Grade Pupils' Mastery of Geometric Concepts," In *The Learning of Mathematical Concepts*, edited by J. Kilpatrick and I. Wirzup. Soviet Studies in the Psychology of Learning and Teaching Mathematics (vol. 1). Palo Alto, CA: SMSG, 1969.

13

Concepts of Measurement

Preliminary Problem

Bobbie makes a circular pizza that is 14 in. in diameter. This includes a 1-in. crust all around the outside. She wants to make a 10-in. circular pizza with the same percentage of crust around it as the 14-in. pizza. Approximately how wide will the crust be on the 10-in. pizza to the nearest tenth of an inch?

P*rinciples and Standards* discusses the measurable attributes of objects as follows:

> A measurable attribute is a characteristic of an object that can be quantified. Line segments have length, plane regions have area, and physical objects have mass. As students progress through the curriculum from preschool through high school, the set of attributes they can measure should expand. Recognizing that objects have attributes that are measurable is the first step in the study of measurement. Children in prekindergarten through grade 2 begin by comparing and ordering objects using language such as longer and shorter. Length should be the focus in this grade band, but weight, time, area, and volume should also be explored. In grades 3–5, students should learn about area more thoroughly, as well as perimeter, volume, temperature, and angle measure. In these grades, they learn that measurements can be computed using formulas and need not always be taken directly with a measuring tool. Middle-grade students build on these earlier measurement experiences by continuing their study of perimeter, area, and volume and by beginning to explore derived measurements, such as speed. (p. 44)

In the United States, two measurement systems are used regularly: the English system and the metric system. In *Principles and Standards*, we find the following concerning the use of the two measurement systems:

> Since the customary English system of measurement is still prevalent in the United States, students should learn both customary and metric systems and should know some rough equivalences between the metric and customary systems—for example, that a two-liter bottle of soda is a little more than half a gallon. The study of these systems begins in elementary school, and students at this level should be able to carry out simple conversions within both systems. Students should develop proficiency in these conversions in the middle grades and should learn some useful benchmarks for converting between the two systems. (pp. 45–46)

Many concepts of measurement are more confusing for children than for adults because students lack everyday measurement experiences. In this chapter, we use both systems of measurement for length, area, volume, mass, and temperature with the philosophy that students should learn to think within a system. We also develop formulas for the area of plane figures and for surface areas and volumes of solids. We use the concept of area in discussing the Pythagorean theorem.

13-1   Linear Measure

To measure a segment, we must decide on a unit of measure. In the cartoon, the parents decided to start measuring Jeremy in sofa lengths as their unit of measure. Early attempts at measurement lacked a standard unit and used fingers, hands, arms, and feet as units of measure. "Hands" are still used to measure the heights of horses. These early crude measurements were refined eventually and standardized by the English.

Zits *by Jerry Scott and Jim Borgman*

NOW TRY THIS 13-1 The *Principles and Standards* state that in grades PreK–2, students should begin their study of measurement using nonstandard units such as paper clips (p. 45).

a. Estimate the length of this textbook page in terms of paper clips.
b. Measure the textbook page in paper clips.
c. How close was your estimate to the measure?
d. Repeat parts (a–c) using *hands*.

The English System

Originally, in the English system, a yard was the distance from the tip of the nose to the end of an outstretched arm of an adult person, and a foot was the length of a human foot. In 1893, the United States defined the yard and other units in terms of metric units. Some units of length in the English system and relationships among them are summarized in Table 13-1.

Table 13-1

Unit	Equivalent in Other Units
yard (yd)	3 ft
foot (ft)	12 in.
mile (mi)	1760 yd, or 5280 ft

◆ *Historical Note*

The yard has been defined in many different ways in the United States over the history of the country. In 1832, it was defined as the distance between the 27th and 63rd inches on a certain brass bar made by Troughton of London. In 1856, the yard was redefined in terms of the British Bronze Yard No. 11 (which was 0.00087 in. longer than the Troughton yard). In 1893, the yard was redefined in terms of the international meter as $\frac{3600}{3937}$ of a meter through use of an 1866 U.S. law making the use of the metric system permissible in the United States. In 1960, the meter was redefined in terms of the wavelength of light from krypton-86 and still later in terms of the distance light travels in $\frac{1}{299,792,458}$ sec. Effective July 1, 1959, the yard is defined in terms of the international yard, which in turn was based on the international definition of a meter. ◆

Converting Units of Measure

To convert from one unit of length to another, different processes can be used. For example, to convert 5.25 mi to yards we use the facts that 1 mi = 5280 ft and 1 ft = $\frac{1}{3}$ yd. Therefore,

$$5.25 \text{ mi} = 5.25(5280) \text{ ft} = 5.25(5280)\left(\frac{1}{3}\right) \text{ yd, or } 9240 \text{ yd}$$

Likewise, to convert 432 in. to yards we use the facts that 1 ft = $\frac{1}{3}$ yd and 1 in. = $\frac{1}{12}$ ft. Therefore,

$$432 \text{ in.} = 432\left(\frac{1}{12}\right) \text{ ft} = 432\left(\frac{1}{12}\right)\left(\frac{1}{3}\right) \text{ yd, or } 12 \text{ yd}$$

Dimensional Analysis (Unit Analysis)

A different process for converting units of measure is known as **dimensional analysis**. This process works with *unit ratios* (ratios equivalent to 1). Since 1 yd = 3 ft, then $\frac{1 \text{ yd}}{3 \text{ ft}} = 1$ and $\frac{3 \text{ ft}}{1 \text{ yd}} = 1$ and these are unit ratios. Also, $\frac{5280 \text{ ft}}{1 \text{ mi}} = 1$ and is a unit ratio. Therefore, to convert 5.25 mi to yards, we have the following:

$$5.25 \text{ mi} = 5.25 \text{ mi} \cdot \frac{5280 \text{ ft}}{1 \text{ mi}} \cdot \frac{1 \text{ yd}}{3 \text{ ft}} = 9240 \text{ yd}$$

Example 13-1   If a cheetah is clocked at 60 miles per hour (mph), what is its speed in feet per second?

Solution

$$60 \frac{\text{mi}}{\text{hr}} = 60 \frac{\text{mi}}{\text{hr}} \cdot \frac{5280 \text{ ft}}{1 \text{ mi}} \cdot \frac{1 \text{ hr}}{60 \text{ min}} \cdot \frac{1 \text{ min}}{60 \text{ sec}} = 88 \frac{\text{ft}}{\text{sec}}$$

REMARK The use of *dimensional analysis* will be seen in the solution to many parts of the examples in this text that involve conversions. It will not be the only technique used.

Example 13-2   Complete each of the following:

a. 219 ft = _____ yd
b. 8432 yd = _____ mi
c. 0.2 mi = _____ ft
d. 64 in. = _____ yd

Solution   **a.** Because $1 \text{ ft} = \frac{1}{3} \text{ yd}$, $219 \text{ ft} = 219 \cdot \frac{1}{3} \text{ yd} = 73 \text{ yd}$. Alternatively, $219 \text{ ft} = 219 \text{ ft} \cdot \frac{1 \text{ yd}}{3 \text{ ft}} = 73 \text{ yd}$.

b. Because $1 \text{ yd} = \frac{1}{1760} \text{ mi}$, $8432 \text{ yd} = 8432 \cdot \frac{1}{1760} \text{ mi} \approx 4.79 \text{ mi}$. Alternatively, $8432 \text{ yd} = 8432 \text{ yd} \cdot \frac{3 \text{ ft}}{1 \text{ yd}} \cdot \frac{1 \text{ mi}}{5280 \text{ ft}} \approx 4.79 \text{ mi}$.

c. $1 \text{ mi} = 5280 \text{ ft}$. Hence, $0.2 \text{ mi} = 0.2 \cdot 5280 \text{ ft} = 1056 \text{ ft}$. Alternatively, $0.2 \text{ mi} = 0.2 \text{ mi} \cdot \frac{5280 \text{ ft}}{1 \text{ mi}} = 1056 \text{ ft}$.

d. We first find a connection between yards and inches. We have $1 \text{ yd} = 3 \text{ ft}$ and $1 \text{ ft} = 12 \text{ in.}$ Hence, $1 \text{ yd} = 3 \text{ ft} = 3 \cdot 12 \text{ in.} = 36 \text{ in.}$ Hence, $1 \text{ in.} = \frac{1}{36} \text{ yd}$; therefore, $64 \text{ in.} = 64 \cdot \frac{1}{36} \text{ yd} \approx 1.78 \text{ yd}$. Alternatively, $64 \text{ in.} = 64 \text{ in.} \cdot \frac{1 \text{ ft}}{12 \text{ in.}} \cdot \frac{1 \text{ yd}}{3 \text{ ft}} \approx 1.78 \text{ yd}$.

The Metric System

At this time, the United States is the only major industrial nation in the world that continues to use the English system. However, the use of the **metric system** in the United States has been increasing, particularly in the scientific community and in industry.

Different units of length in the metric system are obtained by multiplying a power of 10 times the base unit. Table 13-2 gives some of the prefixes for these units, their symbols, and the multiplication factors.

◆ Historical Note

The metric system, a decimal system, was proposed in France in 1670 by Gabriel Mouton. However, not until the French Revolution in 1790 did the French Academy of Sciences bring various groups together to develop the system. In France at that time, over 800 different measures were in use. There were 13 distinct measures for the *foot* ranging from 10.6 in. to 13.4 in. The Academy recognized the need for a standard base unit of linear measurement. The members chose $\frac{1}{10,000,000}$ of the distance from the equator to the North Pole on a meridian through Paris as the base unit of length and called it the **meter (m)**, from the Greek word *metron*, meaning "to measure."

The U.S. Congress included encouragement for U.S. industrial metrication in the Omnibus Trade and Competitiveness Act of 1988 by designating the metric system as the preferred system of weights and measures for U.S. trade and commerce and by requiring each federal agency to be metric by the end of fiscal year 1992. By December 31, 2009, all products sold in Europe (with limited exceptions) are required to have only metric units on their labels. Dual labeling will not be permitted.   ◆

Table 13-2

Prefix	Symbol	Factor
kilo	k	1000 (one thousand)
*hecto	h	100 (one hundred)
*deka	da	10 (ten)
*deci	d	0.1 (one tenth)
centi	c	0.01 (one hundredth)
milli	m	0.001 (one thousandth)

*Not commonly used

Metric prefixes, combined with the base unit meter, name different units of length. Table 13-3 gives units, the symbol for each, and their relationship to the meter.

Table 13-3

Unit	Symbol	Relationship to Base Unit
kilometer	km	1000 m
*hectometer	hm	100 m
*dekameter	dam	10 m
meter	**m**	**base unit**
*decimeter	dm	0.1 m
centimeter	cm	0.01 m
millimeter	mm	0.001 m

*Not commonly used

REMARK Two other prefixes, mega (1,000,000) and micro (0.000001), are used for very large and very small units, respectively. The symbols for mega and micro are M and μ, respectively.

NOW TRY THIS 13-2 If our money system used metric prefixes and the base unit was a dollar, give metric names to each of the following:

a. dime   b. penny   c. $10 bill   d. $100 bill   e. $1000 bill

Benchmarks that can be used for estimations for a meter, a decimeter, a centimeter, and a millimeter are shown in Figure 13-1. The kilometer is commonly used for measuring longer distances: 1 km = 1000 m. Nine football fields, including end zones, laid end to end are approximately 1 km long.

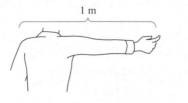

1 m   1 dm   1 cm   1 mm

Figure 13-1

In the grade 2 *Focal Points*, we find the following:

> They (students) understand linear measure as an iteration of units and use rulers and other measurement tools with that understanding. They understand the need for equal-length units, the use of standard units of measure (centimeter and inch), and the inverse relationship between the size of a unit and the number of units used in a particular measurement (i.e., children recognize that the smaller the unit, the more iterations they need to cover a given length). (p. 14)

As mentioned in the *Focal Points* and in the Research Note, the smaller the unit the more iterations we need to cover a given length. This motivates the need for larger units of length and the ability to convert one unit to another.

Conversions among metric lengths are accomplished by multiplying or dividing by powers of 10. As with converting dollars to cents or cents to dollars, we simply move the decimal point to the left or right, depending on the units. For example,

$$0.123 \text{ km} = 1.23 \text{ hm} = 12.3 \text{ dam} = 123 \text{ m} = 1230 \text{ dm} = 12{,}300 \text{ cm} = 123{,}000 \text{ mm}$$

It is possible to convert units by using the chart in Figure 13-2. We count the number of steps from one unit to the other and move the decimal point that many steps in the same direction.

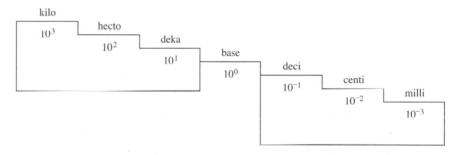

Figure 13-2

Research Note

Research is inconclusive about prerequisite relationships between conservation of measure and a child's ability to measure attributes. Conservation is, however, a prerequisite for understanding the relationship between the size of a unit and the number of units involved in a measurement. For example, the number of inches in a measurement will be greater than the number of yards in the same measurement (Hiebert 1981). ◆

Example 13-3

Complete each of the following:

a. $1.4 \text{ km} = $ _____ m
b. $285 \text{ mm} = $ _____ m
c. $0.03 \text{ km} = $ _____ cm

Solution **a.** Because 1 km = 1000 m, to change kilometers to meters, we multiply by 1000. Hence, 1.4 km = 1.4 (1000 m) = 1400 m. Alternatively,

$$1.4 \text{ km} = 1.4 \text{ km} \cdot \frac{1000 \text{ m}}{1 \text{ km}} = 1400 \text{ m}.$$

b. Because 1 mm = 0.001 m, to change from millimeters to meters, we multiply by 0.001. Thus, 285 mm = 285(0.001 m) = 0.285 m. Alternatively,

$$285 \text{ mm} = 285 \text{ mm} \cdot \frac{1 \text{ m}}{1000 \text{ mm}} = 0.285 \text{ m}.$$

c. To change kilometers to centimeters, we first multiply by 1000 to convert kilometers to meters and then multiply by 100 to convert meters to centimeters. Therefore, we move the decimal point five places to the right to obtain 0.03 km = 3000 cm. An alternative approach is to use Figure 13-2. To go from kilo to centi on the steps, we move five places to the right, so we need to move the decimal point in 0.03 five places to the right resulting in 3000.

Linear units of length are commonly measured with rulers. Figure 13-3 shows part of a centimeter ruler. Rulers can be used to measure distance.

Figure 13-3

In the grade 5 *Focal Points*, we find the following, "Students' experiences connect their work with solids and volumes to their earlier work with capacity and mass. They solve problems that require attention to both approximation and precision of measurement" (p. 17).

Measuring distances in the real world frequently results in errors. Thus, many industrial plants using parts from a variety of sources rely on portable calibration units taken from plant to plant to test measuring instruments used in constructing the parts. This calibration helps assure that the final assembly plant can fit all parts together. To calibrate the measuring instruments, technicians establish the greatest possible error (GPE) allowable in order to obtain the final fit. The **greatest possible error (GPE)** of a measurement is one-half the unit used. For example, if the width of a piece of board was measured as 5 cm, the actual width must be between 4.5 cm and 5.5 cm. Therefore, the GPE for this measurement is 0.5 cm. If the width of a button is measured as 1.2 cm, then the actual width is between 1.15 cm and 1.25 cm, and so the GPE is 0.05 cm or 0.5 mm.

When drawings are given, we assume that the listed measurements are accurate. When actually measuring objects in the real world, we find that such accuracy is usually impossible.

Distance Properties

A person using the expression "the shortest distance between two points is a straight line" may have good intentions, but the statement is actually false. (Why?) A correct statement is "the shortest among all the paths in a plane connecting two points A and B is the segment $\overline{AB}$." (The length of $\overline{AB}$ is denoted by AB.) This fact is one of the basic properties of distance, listed next.

Properties of Distance

1. The distance between any two points A and B is greater than or equal to 0, written $AB \geq 0$.
2. The distance between any two points A and B is the same as the distance between B and A, written $AB = BA$.
3. For any three points A, B, and C, the distance between A and B plus the distance between B and C is greater than or equal to the distance between A and C, written $AB + BC \geq AC$.

In the special case where A, B, and C are collinear and B is between A and C, as in Figure 13-4(a), we have $AB + BC = AC$. Otherwise, if A, B, and C are not collinear, as in Figure 13-4(b), then they form the vertices of a triangle and $AB + BC > AC$. This inequality, $AB + BC > AC$, is the **Triangle Inequality**.

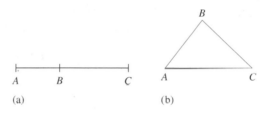

(a)   (b)

Figure 13-4

Theorem 13–1:   Triangle Inequality

The sum of the lengths of any two sides of a triangle is greater than the length of the third side.

REMARK In more rigorous treatments of geometry, the Triangle Inequality is proved and is therefore a theorem.

NOW TRY THIS 13-3 If two sides of a triangle are 31 cm and 85 cm long and the measure of the third side must be a whole number of centimeters,

a. what is the longest the third side can be?
b. what is the shortest the third side can be?

REMARK Notice the difference between AB, $\overline{AB}$, and $\overleftrightarrow{AB}$. AB is the distance between two points A and B and therefore a nonnegative real number. $\overline{AB}$ is the segment connecting points A and B and therefore a set of points. $\overleftrightarrow{AB}$ is the line through points A and B.

Distance Around a Plane Figure

The **perimeter** of a simple closed curve is the length of the curve. If a figure is a polygon, its perimeter is the sum of the lengths of its sides. A perimeter has linear measure.

Example 13-4   Find the perimeter of each of the shapes in Figure 13-5.

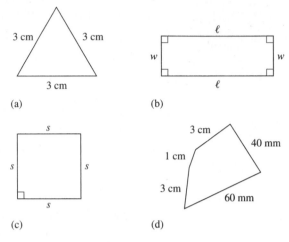

(a)   (b)

(c)   (d)

(e) A regular n-gon with side s

Figure 13-5

Solution **a.** The perimeter is $3(3) = 9$ cm.
   b. The perimeter is $2w + 2\ell$.
   c. The perimeter is $4s$.
   d. Because 40 mm = 4 cm and 60 mm = 6 cm, the perimeter is $1 + 3 + 4 + 6 + 3 = 17$ cm, or 170 mm.
   e. Because all sides of a regular n-gon with side length s have the same length, the perimeter is ns.

REMARK When letters are used for distances, as in Example 13-4(b), (c), and (e), we do not usually attach dimensions to an answer. If the perimeter is $4s$, we actually mean that it is $4s$ units of length.

Problem Solving   Roping a Square

Given a square of unknown size, stretch a thin rope tightly around it. Now take the rope off, add 100 ft to it, and put the extended rope back around the square so that it is evenly spaced and the new rope makes a square around the original square, as in Figure 13-6. Find d, the distance between the squares.

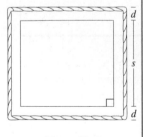

Figure 13-6

Understanding the Problem We are to determine the distance between a square and a new square formed by adding 100 ft of rope to a rope that was stretched around the original square. Figure 13-6 shows the situation if the length of the side of the original square is s and the unknown distance is d.

Devising a Plan If we use variables to represent the unknowns, we can *write an equation* to model the problem. The perimeter of the new square is $4s + 100$. Another way to represent this perimeter is $4(s + 2d)$. Therefore, we have $4s + 100 = 4(s + 2d)$. We must solve this equation for d.

Carrying Out the Plan   We solve the equation as shown next.

$$4s + 100 = 4(s + 2d)$$
$$4s + 100 = 4s + 8d$$
$$100 = 8d$$
$$12.5 = d$$

Therefore, the distance between the squares is 12.5 ft.

Looking Back   A different way to think about the problem is to consider the individual parts that must sum to 100. Note that the perimeter of the original square (the sum of the marked congruent segments) does not change in Figure 13-7, but the eight red segments sum to 100 ft so one red segment has length $\frac{100}{8}$, or 12.5 ft. Notice that the answer is the same regardless of the size of the original square. The distance is always 12.5 ft. This problem can be extended to figures other than squares. Try the Brain Teaser on page 871.

Figure 13-7

Circumference of a Circle

A **circle**, as shown in Figure 13-8, is defined as the set of all points in a plane that are the same distance from a given point, the **center**. The perimeter of a circle is its **circumference**. The ancient Greeks discovered that if they divided the circumference of any circle by the length of its diameter, they always obtained approximately the same number. It can be proved that the ratio of circumference C to diameter d is always the same for all circles. It is symbolized as π (**pi**). In the late eighteenth century, mathematicians proved that the ratio $\frac{C}{d}$, or π, is neither a terminating nor a repeating decimal. Rather, it is an irrational number. For most practical purposes, π is approximated by $\frac{22}{7}$, $3\frac{1}{7}$, or 3.14. These values are not exact values of π.

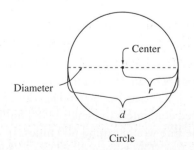

$C = \pi d$, where d is the length of the diameter of the circle.

$C = 2\pi r$, where r is the length of the radius of the circle.

Circle

Figure 13-8

The relationship $\dfrac{C}{d} = \pi$ is used for finding the circumference of a circle and normally is written $C = \pi d$ or $C = 2\pi r$ because the length of the diameter d is twice the radius (r) of the circle. The exact circumference of a circle with diameter 6 cm is 6π cm, and an approximation might be $6(3.14) = 18.84$ cm.

LABORATORY ACTIVITY To approximate the value of π, you need string, a marked ruler, and several different-sized round tin cans or jars. Pick a can and wrap the string tightly around the can. Use a pen to mark a point on the string where the beginning of the string meets the string again. Unwrap the string and measure its length. Next, determine the diameter of the can by tracing the bottom of the can on a piece of paper. Fold the circle onto itself to find a line of symmetry. The chord determined by the line is a diameter of the circle. Measure the diameter and determine the ratio of the circumference to the diameter. (Use the same units in all of your measurements.) Repeat the experiment with at least three cans and find the average of the corresponding ratios.

Do this activity with a class and record diameters and corresponding circumferences in columns A and B. Plot the graph of this data, with circumference on the vertical axis and the diameter on the horizontal axis. What do you observe? Find a fitted line for the data. What is the slope of this line?

BRAIN TEASER Suppose a wire is stretched tightly around Earth. (The radius of Earth is approximately 6400 km.) Then suppose the wire is cut and its length is increased by 20 m. It is then placed back around the planet so that it is the same distance from Earth at every point. Could you walk under the wire?

NOW TRY THIS 13-4 Explain the following cartoon.

Historical Note

$\pi = 3.14159\ 26535\ 89793\ 23846\ 26433\ 83279\ 50288\ 41971\ 69399\ 37510\ 58209\ldots$
Archimedes (b. 287 BCE) found an approximation for π given by the inequality $3\dfrac{10}{71} < \pi < 3\dfrac{10}{70}$. A Chinese astronomer thought that $\pi = \dfrac{355}{113}$. Ludolph van Ceulen (1540–1610), a German mathematician, calculated π to 35 decimal places. The approximation was engraved on his tombstone. Leonhard Euler adopted the symbol π in 1737. In 1761, Johann Lambert, an Alsatian mathematician, proved that π is an irrational number. In 1989, Columbia University mathematicians and Soviet émigré brothers David and Gregory Chudnovsky used computers to establish 480 million digits of π. If these digits were printed along a line, the line would extend 600 miles. In 1995, the brothers were able to reach the billionth digit of pi. The known digits of pi have now exceeded 1 trillion and no pattern among the digits was found. ◆

Arc Length

The length of an arc depends on the radius of the circle and the central angle determining the arc. If the central angle has a measure of 180°, as in Figure 13-9(a), the arc is a **semicircle**. The length of a semicircle is $\frac{1}{2} \cdot 2\pi r$, or πr. The length of an arc whose central angle is $\theta°$ can be developed as in Figure 13-9(b) by using proportional reasoning. Since a circle has 360°, an angle of $\theta°$ determines $\frac{\theta}{360}$ of a circle. Because the circumference of a circle is $2\pi r$, an arc of $\theta°$ has length $\frac{\theta}{360} \cdot 2\pi r$, or $\frac{\pi r\theta}{180}$.

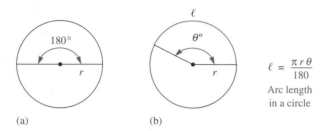

$$\ell = \frac{\pi r \theta}{180}$$

Arc length in a circle

(a)   (b)

Figure 13-9

Example 13-5

Find each of the following:

a. the circumference of a circle if the radius is 2 m
b. the radius of a circle if the circumference is 15π m
c. the length of a 25° arc of a circle of radius 10 cm
d. the radius of an arc whose central angle is 87° and whose arc length is 154 cm

Solution   a. $C = 2\pi(2) = 4\pi$; thus the circumference is 4π m.

b. $C = 2\pi r$ implies $15\pi = 2\pi r$. Hence, $r = \frac{15}{2}$ and the radius is 7.5 m.

c. The arc length is $\frac{\pi r\theta}{180} = \frac{\pi \cdot 10 \cdot 25}{180}$ cm, or $\frac{25\pi}{18}$ cm, or approximately 4.36 cm.

d. The arc length ℓ is $\frac{\pi r\theta}{180}$, so that $154 = \frac{\pi r \cdot 87}{180}$. Thus, $r \approx 101.4$ cm.

Assessment 13-1A

1. Use the following picture of a ruler to find each of the lengths in centimeters:

a. AB
b. DE
c. CJ
d. EF

e. IJ
f. AF
g. IC
h. GB

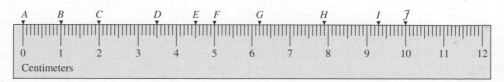

2. Estimate and then measure each of the following in terms of the units listed:
   a. The length of your desktop in cubits (elbow to outstretched fingers)
   b. The length of this page in pencil lengths
   c. The width of this book in pencil widths

3. The United Kingdom used the following money system in the past: 1 pound = 20 shillings, 1 shilling = 12 pence, 1 penny = 2 half-pence (pence is the plural of penny), 1 penny = 4 farthings. Use dimensional analysis to convert each of the following:
   a. 1 pound = _____ farthings
   b. 30 shillings = _____ pounds
   c. 12 pounds = _____ pence

4. On a 4 × 4 geoboard as shown, find examples of the following:

   a. The square with greatest perimeter
   b. A square with least perimeter
   c. A triangle with least perimeter

5. Complete each of the following:
   a. 100 in. = _____ yd   b. 400 yd = _____ in.
   c. 300 ft = _____ yd   d. 372 in. = _____ ft

6. Draw segments that you estimate to be of the given lengths. Use a metric ruler to check the estimates.
   a. 10 mm   b. 100 mm   c. 1 cm   d. 10 cm

7. Estimate the length of the following segment and then measure it:

   ├──────────────────────────────────┤

   Express the measurement in each of the following units:
   a. Millimeters   b. Centimeters

8. Choose an appropriate metric unit and estimate each of the following measures (check your estimates):
   a. The length of a pencil
   b. The diameter of a nickel
   c. The width of the top of a desk

9. Redo exercise 8 using English measures.

10. Complete the following table:

Item	m	cm	mm
a. Length of a piece of paper		35	
b. Height of a woman	1.63		
c. Width of a filmstrip			35
d. Length of a cigarette			100
e. Length of two meter-sticks laid end to end	2		

11. For each of the following, place a decimal point in the number to make the sentence reasonable:
    a. A stack of 10 dimes is 1000 mm high.
    b. The desk is 770 m high.
    c. The distance from one side of a street to the other is 100 m.
    d. A dollar bill is 155 cm long.

12. List the following in decreasing order: 8 cm, 5218 mm, 245 cm, 91 mm, 6 m, 700 mm.

13. Draw each of the following as accurately as possible:
    a. A regular polygon whose perimeter is 12 cm
    b. A circle whose circumference is 4 in.

14. a. What is the perimeter of a semicircle of a circle whose radius is 1 unit?
    b. What is the perimeter of a semicircle of a circle whose radius is 1/2 unit?
    c. Are the two semicircles in parts (a) and (b) similar? If so, what is the ratio of their perimeters? If not, why not?

15. Guess the perimeter in centimeters of each of the following figures and then check the estimates using a ruler:

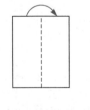

    a.   b.

16. Complete each of the following:
    a. 10 mm = _____ cm   b. 262 m = _____ km
    c. 3 km = _____ m   d. 30 mm = _____ m

17. Draw a triangle *ABC*. Measure the length of each of its sides in millimeters. For each of the following, tell which is greater and by how much:
    a. $AB + BC$ or AC
    b. $BC + CA$ or AB
    c. $AB + CA$ or BC

18. Can the following be the lengths of the sides of a triangle? Why or why not?
    a. 23 cm, 50 cm, 60 cm
    b. 10 cm, 40 cm, 50 cm

19. Take an $8\frac{1}{2}$ × 11-in. piece of paper, fold it as shown in the following figure, and then cut the folded paper along the diagonal as shown.

a. Rearrange the two smaller pieces to find a triangle with the minimum perimeter.

b. Arrange the two smaller pieces to form a triangle with the maximum perimeter.

20. For each of the following circumferences, find the radius of the circle:

    a. 12π cm   **b.** 6 m

21. For each of the following, if a circle has the dimensions given, determine its circumference:

    a. 6 cm diameter

    b. $\dfrac{2}{\pi}$ cm radius

22. What happens to the circumference of a circle if the length of the radius is doubled?

23. Jet planes can exceed the speed of sound, so a new measurement called *Mach number* was invented to describe the speed of such planes. Mach 2 is twice the speed of sound. (Mach number is a number that indicates the ratio of the speed of an object through a medium to the speed of sound in the medium.) The speed of sound in air is approximately 344 m/sec.

    a. If the speed of a plane is described as Mach 2.5, what is its speed in kilometers per hour (km/hr)?

    b. If the speed of a plane is described as Mach 3, what is its speed in meters per second?

    c. Describe the speed of 5000 km/hr as a Mach number.

24. Refer to the following figure and determine the perimeter of the paint lane (shaded area) and the semicircle above it determined by the free-throw line on a basketball court:

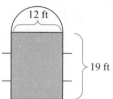

25. Find the length of the side of a square that has the same perimeter as a rectangle that is 66 cm by 32 cm.

26. Give the greatest possible error for each of the following measurements:

    a. 23 m

    b. 3.6 cm

    c. 3.12 m

Assessment 13-1B

1. Use the following picture of a ruler to find each of the lengths in inches:

   a. AB **e.** $I\!\mathcal{J}$
   b. DE **f.** AF
   c. $C\!\mathcal{J}$ **g.** IC
   d. EF **h.** GB

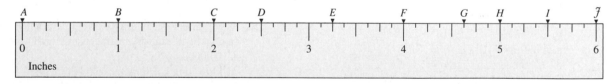

2. Estimate and then measure each of the following in terms of the units listed:

   a. The length of your desktop in mathematics book lengths

   b. The length of this page in paper dollar widths

   c. The width of this page in paper dollar lengths

3. Convert each of the following:

   a. 100 in. = _____ ft   **b.** 400 yd = _____ ft
   c. 300 ft = _____ in.   **d.** 372 in. = _____ yd

4. Draw segments that you estimate to be of the following lengths. Use a metric ruler to check the estimates.

   a. 0.01 m   **b.** 15 cm   **c.** 35 mm   **d.** 150 mm

5. Estimate the length of the following segment and then measure it:

   Express the measurement in each of the following units:

   a. Millimeters   **b.** Centimeters

6. Choose an appropriate metric unit and estimate each of the following measures (check your estimates):

   a. The thickness of the top of a desk

   b. The length of this sheet of paper

   c. The height of a door

7. Redo exercise 6 using English measures.

8. Complete the following table:

Item	m	cm	mm
a. Width of a piece of paper		20	
b. Height of a woman	1.52		
c. Length of a pencil			90
d. Length of a baseball bat	1.1		

9. For each of the following, place a decimal point in the number to make the sentence reasonable:
  a. The basketball player is 1950 cm tall.
  b. A new piece of chalk is about 8100 cm long.
  c. The speed limit in town is 400 km/hr.
10. List the following in decreasing order: 8 m, 5218 cm, 245 cm, 91 m, 6 m, 925 mm.
11. Draw each of the following as accurately as possible:
  a. A triangle whose perimeter is 4 in.
  b. A nonconvex quadrilateral whose perimeter is 8 cm
12. Guess the perimeter in centimeters of each of the following figures and then check the estimates using a ruler:

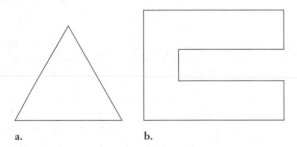

a.   b.

13. Complete each of the following:
  a. 35 m = _____ cm
  b. 359 mm = _____ m
  c. 647 mm = _____ cm
  d. 0.1 cm = _____ mm
14. Can the following be the lengths of the sides of a triangle? Why or why not?
  a. 20 cm, 40 cm, 50 cm
  b. 20 cm, 40 cm, 60 cm
  c. 41 cm, 250 mm, 12 cm
15. a. How do the ratios of the perimeters of similar triangles compare to the ratios of the corresponding sides? Why?
  b. Does the relationship in part (a) hold for all similar polygons?
16. The following figure made of 6 unit squares has a perimeter of 12 units. The figure is made in such a way that each square must share at least one complete side with another square.

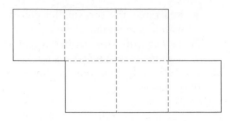

  a. Add more squares to the figure above so that the perimeter of the new figure is 18.
  b. Consider any figure made of squares where each square must share at least one complete side with another square, what is the minimum number of squares required to build a figure of perimeter 18?
  c. Under the same condition as part (b), what is the maximum number of squares possible to build a figure of perimeter 18?
17. For each of the following circumferences, find the radius of the circle:
  a. 0.67 m   **b.** 92π cm
18. For each of the following, if a circle has the dimensions given, determine its circumference:
  a. 3-cm radius
  b. 6π-cm diameter
19. The following figure is a circle whose radius is r units. The diameters of the two semicircular regions inside the large circle are also r units long. Compute the length of the curve that separates the shaded and white regions.

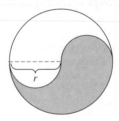

20. Astronomers use a light-year to measure distance. A light-year is the distance light travels in 1 yr. The speed of light is 300,000 km/sec.
  a. How long is 1 light-year in kilometers?
  b. The nearest star (other than the sun) is Alpha Centauri. It is 4.34 light-years from Earth. How far is that in kilometers?
  c. How long will it take a rocket traveling 60,000 km/hr to reach Alpha Centauri?
  d. How long will it take the rocket in (c) to travel to the sun if it takes approximately 8 min 19 sec for light from the sun to reach the earth?
21. On a circular merry-go-round, one horse is 3 m from the center and another is 6 m from the center. The merry-go-round makes 3 revolutions per minute. Are the bases of the merry-go-round that the two horses are standing on traveling at the same speed? If not, how fast is each traveling?
22. Give the greatest possible error for each of the following:
  a. 136 m   **b.** 3.5 ft   **c.** 3.62 cm

Mathematical Connections 13-1

Communication

1. Howard Eves (1911–2004) described the creation of the metric system as one of the greatest accomplishments of the eighteenth century. Do you agree or disagree? Explain.

2. How could you define the perimeter of a nonsimple closed curve? Give an example in your explanation.

3. There has been considerable debate about whether the United States should change to the metric system.
   a. Based on your experiences with linear measure, what do you see as the advantages of changing?
   b. Which system do you think would be easier for children to learn? Why?
   c. What things will probably not change if the United States adopts the metric system?

4. In track, the second lane from the inside of the track is longer than the inside lane. Use this information to explain why, in running events that require a complete lap of the track, runners are lined up at the starting blocks as shown in the following figure:

Starting blocks

Open-Ended

5. Observe that it is possible to build a triangle with toothpicks that has sides of 3, 4, and 5 toothpicks, as shown, and answer the questions that follow.

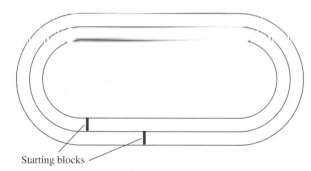

   a. Find two other triples of toothpicks that can be used as sides of a triangle and two other triples that cannot be used to create a triangle.

   b. Describe how to tell whether a given triple of numbers a, b, c can be used to construct a triangle with sides of a, b, and c toothpicks. Explain why your rule is valid.

Cooperative Learning

6. a. Help each person in the group find his or her height in centimeters.
   b. Help each person in the group find the length of his or her outstretched arms (horizontal) from fingertip to fingertip.
   c. Compare the difference between the two measurements in parts (a) and (b). Compare the results of the group members and make a conjecture about the relationship between the two measurements.
   d. Compare your group's results with other groups to determine if they have similar findings.

7. Jerry wants to design a gold chain 60 cm long made of thin gold wire circles, each of which is the same size. He wants to use the least amount of wire and wonders what the radius of each circle should be.

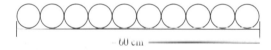

   60 cm

   a. Each member of the group should choose a specific number of circles and find the length of wire needed to make a 60 cm chain with the chosen number of circles.
   b. Compare your results and make a conjecture based on the results.
   c. Justify your conjecture.

Questions from the Classroom

8. A student claims that the circumference of a semicircle is the same as that of a circle from which the semicircle was taken because one has to measure the "outer part" and the "inner part." How do you respond?

9. A student has a tennis ball can with a flat top and bottom containing three tennis balls. To the student's surprise, the circumference of the top of the can is longer than the height of the can. The student wants to know if this fact can be explained without performing any measurements. How do you respond?

Third International Mathematics and Science Study (TIMSS) Questions

Which of these is closest to the length of the pencil in the figure?

a. 9 cm   **b.** 10.5 cm   **c.** 12 cm   **d.** 13.5 cm

TIMSS, Grade 8, 1994

What is the ratio of the length of a side of a square to its perimeter?

a. $\dfrac{1}{1}$ **b.** $\dfrac{1}{2}$ **c.** $\dfrac{1}{3}$ **d.** $\dfrac{1}{4}$

TIMSS, Grade 8, 1994

National Assessment of Educational Progress (NAEP) Question

Which of these units would be the best to use to measure the length of a school building?

a. Millimeters
b. Centimeters
c. Meters
d. Kilometers

NAEP, Grade 4, 2007

LABORATORY ACTIVITY Use a piece of ordinary $8\frac{1}{2}$ by 11-in. paper to answer the following:

a. What is its perimeter?
b. *Without* cutting any of the edges of the piece of paper, cut a shape from the page large enough so that you can walk through it. What is the approximate perimeter of the new shape?

13-2   Areas of Polygons and Circles

In the grade 4 *Focal Points*, we find the following:

> Students recognize area as an attribute of two-dimensional regions. They learn that they can quantify area by finding the total number of same-sized units of area that cover the shape without gaps or overlaps. They understand that a square that is 1 unit on a side is the standard unit for measuring area. They select appropriate units, strategies (e.g., decomposing shapes), and tools for solving problems that involve estimating or measuring area. Students connect area measure to the area model that they have used to represent multiplication, and they use this connection to justify the formula for the area of a rectangle. (p. 16)

In this section, we quantify area using same-sized square units and we develop formulas for the area of various polygons.

Area is measured using square units and the area of a region is the number of nonoverlapping square units that covers the region. A square measuring 1 ft on a side has an area of 1 square foot, denoted $1\,\text{ft}^2$. A square measuring 1 cm on a side has an area of 1 square centimeter, denoted $1\,\text{cm}^2$.

> **REMARK** Students sometimes confuse the area of $5\,\text{cm}^2$ with the area of a square 5 cm on each side. The area of a square 5 cm on each side is $(5\,\text{cm})^2$, or $25\,\text{cm}^2$. Five squares each 1 cm by 1 cm have the area of $5\,\text{cm}^2$. Thus, $5\,\text{cm}^2 \ne (5\,\text{cm})^2$.

Areas on a Geoboard

Addition Method

In teaching the concept of area, intuitive activities should precede the development of formulas. Many such activities can be accomplished using a *geoboard* or *dot paper*. Notice that the square unit is defined in the upper left corner of the dot paper in Figure 13-10(a). The area of the pentagon can be found by finding the sum of the areas of smaller pieces. Finding the area in this way uses the *addition method*. The region in Figure 13-10(b) has been divided into smaller pieces in Figure 13-10(c). What is the area of this shape?

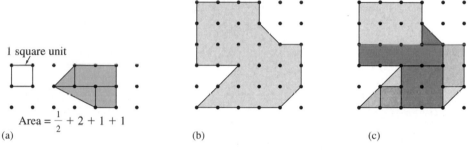

1 square unit

$\text{Area} = \dfrac{1}{2} + 2 + 1 + 1$

(a)   (b)   (c)

Figure 13-10

Rectangle Method

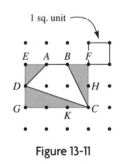

1 sq. unit

Figure 13-11

Another method of finding the area of shapes on the dot paper is the *rectangle method*. To find the area of quadrilateral *ABCD* in Figure 13-11, we construct the rectangle *EFCG* around the quadrilateral and then subtract the areas of the shaded triangles *EAD*, *BFC*, and *DGC*. The area of the rectangle *EFCG* can be counted to be 6 square units. The area of $\triangle EAD$ is $\dfrac{1}{2}$ square unit, and the area of $\triangle BFC$ is half the area of rectangle *BFCK*, or $\dfrac{1}{2}$ of 2, or 1 square unit. Similarly, the area of $\triangle DGC$ is half the area of rectangle *DHCG*; that is, $\dfrac{1}{2} \cdot 3$, or $\dfrac{3}{2}$ square units. Consequently, the area of *ABCD* is $6 - \left(\dfrac{1}{2} + 1 + \dfrac{3}{2}\right)$, or 3 square units.

Example 13-6

Using a geoboard, find the area of each of the shaded parts of Figure 13-12.

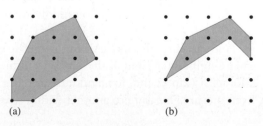

(a)   (b)

Figure 13-12

Solution **a.** We construct a rectangle around the hexagon and then subtract the areas of regions *a*, *b*, *c*, *d*, and *e* from the area of this rectangle, as shown in Figure 13-13. Therefore, the area of the hexagon is $16 - (3 + 1 + 1 + 1 + 1)$, or 9 square units. The addition method could also be used in this problem.

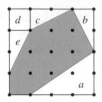

Figure 13-13

b. The area of the hexagon equals the area of the surrounding rectangle shown in Figure 13-14 minus the sum of the areas of figures *a*, *b*, *c*, *d*, *e*, *f*, and *g*. Thus the area of the hexagon is $12 - \left(3 + 1 + \dfrac{1}{2} + \dfrac{1}{2} + 1 + 1 + 1 \right)$, or 4 square units.

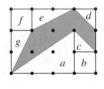

Figure 13-14

NOW TRY THIS 13-5 How many of each of the following nonstandard shapes are contained in Figure 13-15?

a. Triangles

b. Rhombuses

c. Trapezoids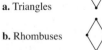

Figure 13-15

Converting Units of Area

The price of new carpet for a house is quoted in terms of dollars per square yard, for example, $12.50/yd^2$. The most commonly used units of area in the English system are the square inch (in.^{2}), the square foot (ft^2), the square yard (yd^2), the square mile (mi^2), and, for land measure, the acre (A). In the metric system, the most commonly used units are the square millimeter (mm^2), the square centimeter (cm^2), the square meter (m^2), the square kilometer (km^2), and, for land measure, the hectare (ha). It is often necessary to convert from one area measure to another within a system.

To determine how many 1-cm squares are in 1 m^2, look at Figure 13-16(a). There are 100 cm in 1 m, so each side of the square meter has a measure of 100 cm. Thus, it takes 100 rows of 100 1-cm squares each to fill a square meter; that is, $100 \cdot 100$, or 10,000 1-cm squares. Because the area of each centimeter square is $1 \text{ cm} \cdot 1 \text{ cm}$, or 1 cm^2, there are 10,000 cm^2 in 1 m^2. In general, the area A of a square that is s units on a side is s^2, as shown in Figure 13-16(b).

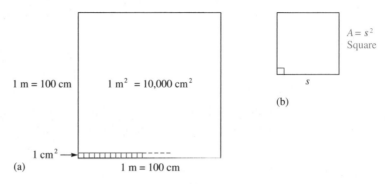

Figure 13-16

Other metric conversions of area measure can be developed similarly. For example, Figure 13-17(a) shows that 1 m^2 = 10,000 cm^2 = 1,000,000 mm^2. Likewise, Figure 13-17(b) shows that 1 m^2 = 0.000001 km^2. Similarly, 1 cm^2 = 100 mm^2 and 1 km^2 = 1,000,000 m^2.

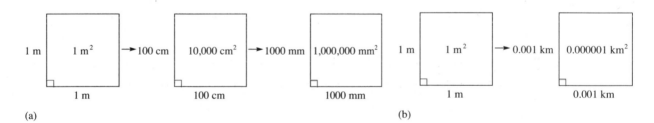

Figure 13-17

Table 13-4 shows symbols for metric units of area and their relationship to the square meter.

Table 13-4

Unit	Symbol	Relationship to Square Meter
square kilometer	km^2	1,000,000 m^2
*square hectometer	hm^2	10,000 m^2
*square dekameter	dam^2	100 m^2
square meter	**m^2**	**1 m^2**
*square decimeter	dm^2	0.01 m^2
square centimeter	cm^2	0.0001 m^2
square millimeter	mm^2	0.000001 m^2

*Not commonly used

Example 13-7

▶▶▶▶▶▶▶▶▶▶

Complete each of the following:

a. $5 \, cm^2 = $ _____ mm^2
b. $124{,}000{,}000 \, m^2 = $ _____ km^2

Solution   **a.** $1 \, cm^2 = 100 \, mm^2$ implies $5 \, cm^2 = 5 \cdot 1 \, cm^2 = 5 \cdot 100 \, mm^2 = 500 \, mm^2$.
        b. $1 \, m^2 = 0.000001 \, km^2$ implies $124{,}000{,}000 \, m^2 = 124{,}000{,}000 \cdot 1 \, m^2 = $
           $124{,}000{,}000 \cdot 0.000001 \, km^2 = 124 \, km^2$.

◆

Based on the relationship among units of length in the English system, it is possible to convert among English units of area. For example, because $1 \, yd = 3 \, ft$, it follows that $1 \, yd^2 = (1 \, yd)^2 = 1 \, yd \cdot 1 \, yd = 3 \, ft \cdot 3 \, ft = 9 \, ft^2$. Similarly, because $1 \, ft = 12 \, in.$, $1 \, ft^2 = (1 \, ft)^2 = 1 \, ft \cdot 1 \, ft = 12 \, in. \cdot 12 \, in. = 144 \, in.^2$. Table 13-5 summarizes various relationships among units of area in the English system.

Table 13-5

Unit of Area	Equivalent of Other Units
$1 \, ft^2$	$\frac{1}{9} \, yd^2$, or $144 \, in.^2$
$1 \, yd^2$	$9 \, ft^2$, or $1296 \, in.^2$
$1 \, mi^2$	$3{,}097{,}600 \, yd^2$, or $27{,}878{,}400 \, ft^2$

Land Measure

The concept of area is widely used in land measure. The common unit of land measure in the English system is the **acre**. Historically, an acre was the amount of land a man with one horse could plow in one day. There are $4840 \, yd^2$ in 1 acre. For very large land measures in the English system, the **square mile** (mi^2), or 640 acres, is used.

In the metric system, small land areas are measured in terms of a square unit 10 m on a side, called an **are** (pronounced "air") and denoted by **a**. Thus, $1 \, a = 10 \, m \cdot 10 \, m$, or $100 \, m^2$. Larger land areas are measured in **hectares**. A hectare is 100 a. A hectare, denoted by **ha**, is the amount of land whose area is $10{,}000 \, m^2$. It follows that 1 ha is the area of a square that is 100 m on a side. Therefore, $1 \, ha = 1 \, hm^2$. For very large land measures, the **square kilometer**, denoted by km^2, is used. One square kilometer is the area of a square with a side 1 km, or 1000 m, long. Land area measures are summarized in Table 13-6.

Table 13-6

Unit of Area	Equivalent of Other Units
1 a	$100 \, m^2$
1 ha	100 a, or $10{,}000 \, m^2$, or $1 \, hm^2$
$1 \, km^2$	$1{,}000{,}000 \, m^2$ or 100 ha
1 acre	$4840 \, yd^2$
$1 \, mi^2$	640 acres

Example 13-8

a. A square field has a side of 400 m. Find the area of the field in hectares.

b. A square field has a side of 400 yd. Find the area of the field in acres.

Solution   **a.** $A = (400 \text{ m})^2 = 160{,}000 \text{ m}^2 = \dfrac{160{,}000}{10{,}000} \text{ ha} = 16 \text{ ha}$

　　　　　　 b. $A = (400 \text{ yd})^2 = 160{,}000 \text{ yd}^2 = \dfrac{160{,}000}{4840} \text{ acre} \approx 33.1 \text{ acres}$

---

*Research
Note*

Middle-school students find areas by counting visual units rather than using formulas learned in the past, even if counting is tedious and complex (Figueras and Waldegg 1984).   ◆

Area of a Rectangle

As mentioned in the Research Note, students may count the number of units of area contained in any given region. For example, suppose the square in Figure 13-18(a) represents 1 square unit. Then, the rectangle *ABCD* in Figure 13-18(b) contains $3 \cdot 4$, or 12 square units.

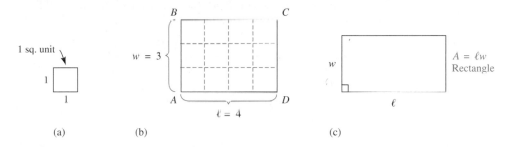

Figure 13-18

If the unit in Figure 13-18(a) is 1 cm^2, then the area of rectangle *ABCD* is 12 cm^2. In general, the area A of any rectangle can be found by multiplying the lengths of two adjacent sides ℓ and w, or $A = \ell w$, as given in Figure 13-18(c).

TECHNOLOGY CORNER *GSP* Lab 8 in the Technology Manual can be used at this point to motivate the formulas for finding the areas of a rectangle, parallelogram, and trapezoid.

Example 13-9

Find the area of each rectangle in Figure 13-19.

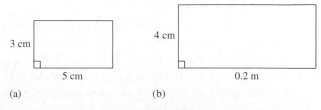

Figure 13-19

Solution   **a.** $A = (3 \text{ cm})(5 \text{ cm}) = 15 \text{ cm}^2$

b. First, write the lengths of the sides in the same unit of length. Because $0.2 \text{ m} = 20 \text{ cm}$, $A = (4 \text{ cm})(20 \text{ cm}) = 80 \text{ cm}^2$. Alternatively, $4 \text{ cm} = 0.04 \text{ m}$, so $A = (0.04 \text{ m})(0.2 \text{ m}) = 0.008 \text{ m}^2$.

NOW TRY THIS 13-6 Estimate the area in square centimeters of a dollar bill. Measure and calculate how close your estimate is to the actual area.

Area of a Parallelogram

The area of a parallelogram can be found by *reducing the problem to one that we already know how to solve*, in this case, finding the area of a rectangle. One important strategy for deriving the formula for the area of a parallelogram is **dissection**. In dissection, we cut a figure with unknown area into a number of pieces. By reassembling these pieces we obtain a figure whose area we know how to find.

Informally, to find the area of the parallelogram in Figure 13-20(a), we can cut the shaded triangular piece of the parallelogram and move it to obtain the rectangle in Figure 13-20(b). Because the shaded areas are congruent, the area of the parallelogram in Figure 13-20(a) is the same as the area of the rectangle in Figure 13-20(b). The area of the rectangle is bh so the area of the parallelogram is bh.

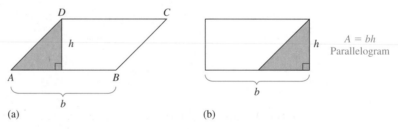

Figure 13-20

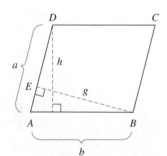

Figure 13-21

In general, any side of a parallelogram can be designated as a **base**. The **height** (*h*) is the distance between the bases and is always the length of a segment perpendicular to the lines containing the bases. The area of parallelogram $ABCD$ is given by $A = bh$; that is, the length of the base times the corresponding height. Similarly, in Figure 13-21, EB, or g, is the height that corresponds to the bases $\overline{AD}$ and $\overline{BC}$, each of which has measure a. Consequently, the area of the parallelogram $ABCD$ is ag. Therefore, $A = ag = bh$.

Area of a Triangle

The formula for the area of a triangle can be derived from the formula for the area of a parallelogram. To explore this, suppose $\triangle BAC$ in Figure 13-22(a) has base of length b and height h. Let $\triangle BAC'$ be the image of $\triangle BAC$ when $\triangle BAC$ is rotated 180° about M, the midpoint of $\overline{AB}$, as in Figure 13-22(b). Proving that quadrilateral $BCAC'$ is a parallelogram is left as an exercise. Parallelogram $BCAC'$ has area bh and is constructed of congruent triangles BAC and BAC'. So the area of $\triangle ABC$ is $\frac{1}{2}bh$. In general, the area of a triangle is equal to half the product of the length of a side and the height to that side.

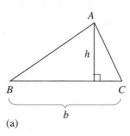
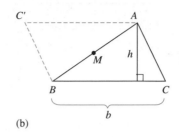

$A = \frac{1}{2} bh$
Triangle

(a)   (b)

Figure 13-22

In Figure 13-23, $\overline{BC}$ is a base of $\triangle ABC$, and the corresponding height h_1, or AE, is the distance from the opposite vertex A to the line containing $\overline{BC}$. Similarly, $\overline{AC}$ can be chosen as a base. Then h_2, or BG, the distance from the opposite vertex B to the line containing $\overline{AC}$, is the corresponding height. If $\overline{AB}$ is chosen as a base, then the corresponding height is h_3, or CF. Thus, the area of $\triangle ABC$ is

$$\text{Area}(\triangle ABC) = \frac{bh_1}{2} = \frac{ah_2}{2} = \frac{ch_3}{2}$$

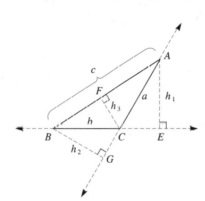

Figure 13-23

NOW TRY THIS 13-7 To see how the formula for the area of a triangle can be derived from the formula for the area of a rectangle, cut out any triangle as in Figure 13-24(a); fold to find the altitude h; and fold the altitude in half as shown in Figure 13-24(b). Next fold along the colored segments in the trapezoid in Figure 13-24(c) to obtain the rectangle in Figure 13-24(d).

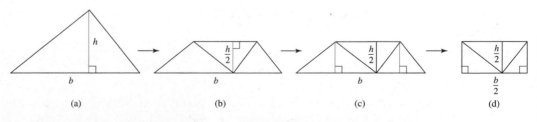

(a)   (b)   (c)   (d)

Figure 13-24

a. What is the area of the rectangle in Figure 13-24(d)?
b. How can the formula for the area of a triangle be developed from your answer in part (a)?

NOW TRY THIS 13-8   In Figure 13-25, $\ell \parallel \overleftrightarrow{AB}$. How are the areas of $\triangle ABP$, $\triangle ABQ$, $\triangle ABR$, $\triangle ABS$, $\triangle ABT$, and $\triangle ABU$ related? Explain your answer.

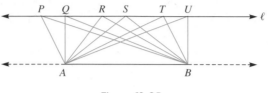

Figure 13-25

Problem Solving   Fence Problem

Two farmers own large fields divided by a common border consisting of two sections of fence as shown in Figure 13-26. They want to replace the two sections of fence with a new straight fence so that the areas of the new regions are the same as the old areas. In other words, each farmer should have the same amount of land as before the border was changed. Where should the new "straight" fence be placed?

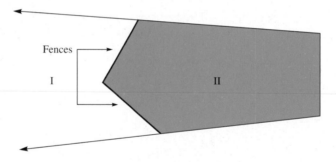

Figure 13-26

Solution   One of the farmers owns the region marked I in Figure 13-27, and the other owns the region marked II. The farmer who owns region II owns $\triangle ABC$. The area of that triangle equals the area of any triangle with $\overline{AC}$ as a base and vertex on the line ED parallel to the base $\overline{AC}$. Thus, if point B "moves" along $\overleftrightarrow{ED}$, we get a variety of triangles whose areas are the same as the area of $\triangle ABC$. If we choose $\triangle ADC$, the area of region II does not change and the new border $\overline{CD}$ is straight. Making the border become $\overline{AE}$ is another possibility.

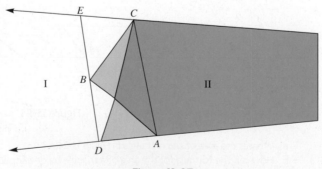

Figure 13-27

Example 13-10   Find the areas of the figures in Figure 13-28. Assume the quadrilaterals *ABCD* in (a) and (b) are parallelograms. (Figures are not drawn to scale.)

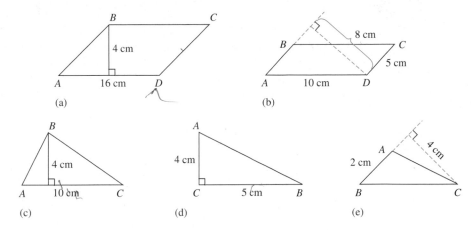

(a)   (b)

(c)   (d)   (e)

Figure 13-28

Solution **a.** $A = bh = (16 \text{ cm})(4 \text{ cm}) = 64 \text{ cm}^2$

b. $A = bh = (5 \text{ cm})(8 \text{ cm}) = 40 \text{ cm}^2$

c. $A = \frac{1}{2}bh = \frac{1}{2}(10 \text{ cm})(4 \text{ cm}) = 20 \text{ cm}^2$

d. $A = \frac{1}{2}bh = \frac{1}{2}(5 \text{ cm})(4 \text{ cm}) = 10 \text{ cm}^2$

e. $A = \frac{1}{2}bh = \frac{1}{2}(2 \text{ cm})(4 \text{ cm}) = 4 \text{ cm}^2$

(handwritten) ⓐ $A = bh$ $= (12)(13)$

(handwritten) //

Area of a Kite

The area of a kite can be found by relating it to the area of a rectangle. Consider the kite shown in Figure 13-29(a), where d_1 and d_2 are the lengths of the diagonals. If the kite is dissected and reassembled as shown in Figure 13-29(b), then a rectangle is formed with length d_1 and width $d_2/2$. The rectangle in Figure 13-29(b) has area $(d_1 d_2)/2$, or $\frac{1}{2}(d_1 d_2)$.

Therefore, the area of a kite is equal to half the product of the lengths of its diagonals. Partitioning the kite into triangles and finding the areas of the triangles can also be used to find the area of a kite.

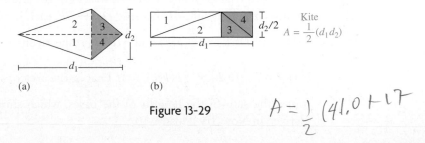

(a)   (b)

Figure 13-29

(handwritten) $A = \frac{1}{2}(41.0 + 17$

> **REMARK** Since rhombuses and squares are also kites, the area formula for kites developed above can be used to find the areas of rhombuses and squares.

NOW TRY THIS 13-9 We can find the area of a kite using the lengths of the two diagonals, as seen in Figure 13-29. Prove that the same formula works for any quadrilateral whose diagonals are perpendicular to each other, as seen in Figure 13-30.

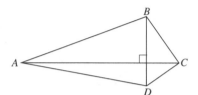

Figure 13-30

Area of a Trapezoid

The formula for the area of a trapezoid can be developed using the formula for the area of a parallelogram, as shown in Now Try This 13-10.

NOW TRY THIS 13-10 Cut out a trapezoid *ABCD* as shown in Figure 13-31. Copy and place as shown by the dashed trapezoid. Use the total figure obtained to derive the formula for the area of a trapezoid.

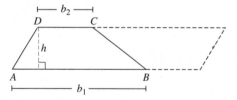

Figure 13-31

   Areas of general polygons can also be found by partitioning the polygons into triangles, finding the areas of the triangles, and summing those areas. In Figure 13-32(a), trapezoid *ABCD* has bases b_1 and b_2 and height h. By connecting points *B* and *D*, as in Figure 13-32(b), we create two triangles: one with base of length *AB* and height *DE* and the other with base of length *CD* and height *BF*. Because $\overline{DE} \cong \overline{BF}$, each has length h. Thus, the areas of triangles *ADB* and *DCB* are $\frac{1}{2}(b_1h)$ and $\frac{1}{2}(b_2h)$, respectively. Hence, the area of trapezoid *ABCD* is $\frac{1}{2}(b_1h) + \frac{1}{2}(b_2h) = \frac{1}{2}h(b_1 + b_2)$. That is, the area of a trapezoid is equal to half the height times the sum of the lengths of the bases, which should be the same as the formula you derived in Now Try This 13-10.

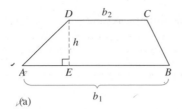
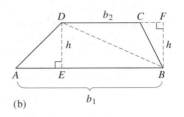

Figure 13-32

Example 13-11

Find the areas of the trapezoids in Figure 13-33.

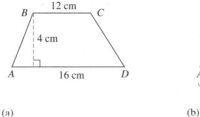

(a)

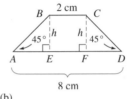

(b)

Figure 13-33

Solution   **a.** $A = \dfrac{1}{2}h(b_1 + b_2) = \dfrac{1}{2}(4 \text{ cm})(12 \text{ cm} + 16 \text{ cm}) = 56 \text{ cm}^2$

b. To find the area of trapezoid $ABCD$, we use the strategy of *determining a subgoal* of finding the height, h. In Figure 13-33(b), $BE = CF = h$. Also, $\overline{BE}$ is a side of $\triangle ABE$, which has angles with measures of 45° and 90°. Consequently, the third angle in triangle ABE is $180 - (45 + 90)$, or 45°. Therefore, $\triangle ABE$ is isosceles and $AE = BE = h$. Similarly, it follows that $FD = h$. Because $AD = 8 \text{ cm} = h + EF + h$, we could find h if we knew the value of EF. From Figure 13-33(b), $EF = BC = 2 \text{ cm}$ because $BCFE$ is a rectangle (why?) and opposite sides of a rectangle are congruent. Now $h + EF + h = h + 2 + h = 8 \text{ cm}$.

Thus, $h = 3 \text{ cm}$ and the area of the trapezoid is $A = \dfrac{1}{2}(3 \text{ cm})(2 \text{ cm} + 8 \text{ cm})$, or 15 cm^2.

Area of a Regular Polygon

The area of a triangle can be used to find the area of any regular polygon, as illustrated *using a simpler case* strategy involving a regular hexagon in Figure 13-34(a). The hexagon can be separated into six congruent triangles with side s and height a as shown in Figure 13-34(a). The height of such a triangle of a regular polygon is the *apothem* and is denoted a. The area of each triangle is $\dfrac{1}{2}as$. Because six triangles make up the hexagon, the area of the hexagon is $6\left(\dfrac{1}{2}as\right)$, or $\dfrac{1}{2}a(6s)$. However, $6s$ is the perimeter, p, of the hexagon, so the area of the hexagon is $\frac{1}{2}ap$. The same process can be used to develop the formula for the area of any regular polygon. That is, the area of any regular polygon is $\frac{1}{2}ap$, where a is the height of one of the triangles involved and p is the perimeter of the polygon, as shown in Figure 13-34(b).

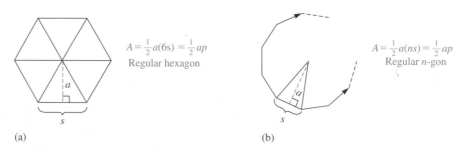

$$A = \tfrac{1}{2}a(6s) = \tfrac{1}{2}ap$$
Regular hexagon

$$A = \tfrac{1}{2}a(ns) = \tfrac{1}{2}ap$$
Regular n-gon

(a)

(b)

Figure 13-34

Area of a Circle

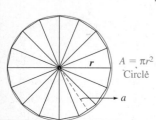

$A = \pi r^2$
Circle

Figure 13-35

We use the strategy of *examining a related problem* to find the area of a circle. The area of a regular polygon inscribed in a circle, as in Figure 13-35, approximates the area of the circle, and we know that the area of any regular n-gon is $\frac{1}{2}ap$, where a is the height of a triangle of the n-gon and p is the perimeter. If the number of sides n is made very large, then the perimeter and the area of the n-gon are close to those of the circle.

Also, the apothem a is getting closer and closer to the radius r of the circle, and the perimeter p is approaching the circumference $2\pi r$. Because the area of the circle is approximately equal to the area of the n-gon, $\frac{1}{2}ap \approx \frac{1}{2}r(2\pi r) = \pi r^2$. In fact, the area of the circle is precisely πr^2.

Another approach for leading students to discover the formula for finding the area of a circle is mentioned in the grade 7 *Focal Points*:

Students see that the formula for the area of a circle is plausible by decomposing a circle into a number of wedges and rearranging them into a shape that approximates a parallelogram. (p. 19)

An example of this approach is seen in Now Try This 13-11 and the accompanying student page.

NOW TRY THIS 13-11

a. Draw a circle of radius 4 cm and divide it into eight equal-sized sections.
b. Using the circle you constructed, answer questions 25–28 on the student page.

School Book Page   AREA OF A CIRCLE

24 **Use Labsheet 2A.** Follow the directions below.

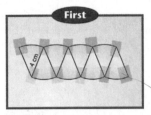

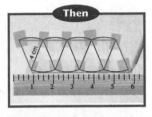

Cut out the *Circle*. Then cut apart the eight sectors and arrange them to form the figure shown. Tape the figure to a sheet of paper.

Use a ruler to draw segments across the top and bottom of your figure. Extend the sides of the figure to meet the bottom segment.

25 What kind of polygon is the new figure you drew in Question 24?

26 **Try This as a Class** Use the figure you made in Question 24.

　　a. **Estimation** Explain how you could use the new figure you drew to estimate the area of the circle. Then estimate the area. Do you think this is a good estimate? Why or why not?

　　b. How is the length of the base of the figure related to the circumference of the circle?

　　c. How is the height of the figure related to the radius of the circle?

27 **Discussion** Examine the drawings shown.

　　a. As a circle is cut into more and more sectors and put back together as shown, what begins to happen to its shape?

　　b. The area A of a parallelogram is found by multiplying the length of its base b by the height h, or $A = bh$. Use your figure to explain why this formula can be written as $A = \frac{1}{2}Cr$ to find the area of a circle, where C is the circumference and r is the length of the radius of the circle.

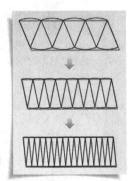

28 The circumference C of a circle is equal to $2\pi r$, where r is the length of the radius. Rewrite the formula $A = \frac{1}{2}Cr$ by substituting $2\pi r$ for C.

Source: Math Thematics, Book 2, McDougal Littell, 2005 (p. 402).

Area of a Sector

A **sector** of a circle is a pie-shaped region of the circle determined by an angle whose vertex is the center of the circle. This angle is a **central angle**. The area of a sector depends on the radius of the circle and the measure of the central angle determining the sector. If the angle has a measure of 90°, as in Figure 13-36(a), the area of the sector is one-fourth the area of the circle, or $\frac{90}{360}\pi r^2$. The area of a sector with central angle of 1° is $\frac{1}{360}$ of the area of the circle, and a sector with central angle $\theta°$ has area $\frac{\theta}{360}$ of the area of the circle, or $\frac{\theta}{360}(\pi r^2)$, as shown in Figure 13-36(b).

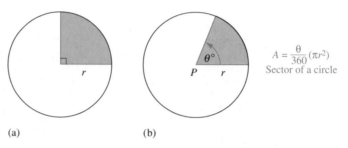

(a)   (b)

Figure 13-36

Assessment 13-2A

1. As an example of measuring area with a nonstandard measure, estimate the area of your desktop in terms of a piece of notebook paper as your unit of area. Then measure the area of your desktop with the paper and compare it to your estimate.

2. Choose the most appropriate metric units (cm², m², or km²) and English units (in.², yd², or mi²) for measuring each of the following:
   a. Area of a sheet of notebook paper
   b. Area of a quarter
   c. Area of a desktop
   d. Area of a classroom floor

3. Estimate and then measure each of the following using cm² or m².
   a. Area of a door
   b. Area of a desktop

4. Complete the following conversion table:

Item	m²	cm²	mm²
a. Area of a sheet of paper		588	
b. Area of a cross section of a crayon			192
c. Area of a desktop	1.5		
d. Area of a dollar bill		100	
e. Area of a postage stamp		5	

5. Complete the following:
   a. $4000 \text{ ft}^2 = $ _____ yd²
   b. $10^6 \text{ yd}^2 = $ _____ mi²
   c. $10 \text{ mi}^2 = $ _____ acre
   d. $3 \text{ acres} = $ _____ ft²

6. Complete each of the following:
  a. A football field (with endzones included) is about 49 m × 100 m, or _____ m².
  b. About _____ a are in two football fields with endzones.
  c. About _____ ha are in two football fields with endzones.

7. Find the areas of each of the following figures if the distance between two adjacent dots in a row or a column is 1 unit:

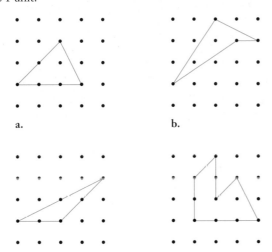

  a.　　　　**b.**

  c.　　　　**d.**

8. Find the area of △*ABC* in each of the following:

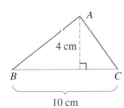

  a.

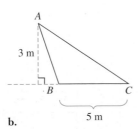

  b.

9. If a triangle is inscribed in a circle so that one of the triangle's sides is a diameter of the circle, what is the greatest area that the triangle can have in terms of the radius, *r*, of the circle?

10. If two different squares have sides in the ratio $\frac{a}{b}$,
  a. are the squares similar? Why?
  b. what is the ratio of the areas of the squares?

11. Find the area of each of the following quadrilaterals:

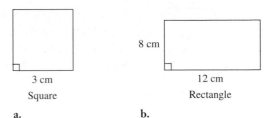

  a.　　　　**b.**

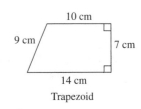

  c.　　　　**d.**

12. a. A rectangular piece of land is 1300 m × 1500 m.
    (i) What is the area in square kilometers?
    (ii) What is the area in hectares?
  b. A rectangular piece of land is 1300 yd × 1500 yd.
    (i) What is the area in square miles?
    (ii) What is the area in acres?
  c. Explain which measuring system you would rather use to solve problems like those in (a) and (b).

13. For a parallelogram whose sides are 6 cm and 10 cm, which of the following is true?
  a. The data are insufficient to enable us to determine the area.
  b. The area equals 60 cm².
  c. The area is greater than 60 cm².
  d. The area is less than 60 cm².

14. If the diagonals of a rhombus are 12 cm and 5 cm units long, find the area of the rhombus.

15. Find the cost of carpeting the following rectangular rooms:
  a. Dimensions: 6.5 m × 4.5 m; cost = $13.85/m²
  b. Dimensions: 15 ft × 11 ft; cost = $30/yd²

16. Find the area of each of the following. Leave your answers in terms of π.

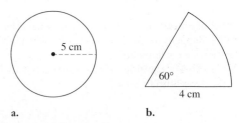

  a.　　　　**b.**

17. Joe uses stick-on square carpet tiles to cover his 3 m × 4 m bathroom floor. If each tile is 10 cm on a side, how many tiles does he need?

18. Find the area of each of the following regular polygons:

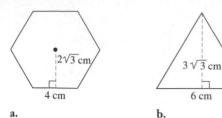

a.

b.

19. a. If a circle has a circumference of 8π cm, what is its area?

b. If a circle of radius r and a square with a side of length s have equal areas, express r in terms of s.

20. Find the area of each of the following shaded parts. Assume all arcs are circular with centers marked. Leave all answers in terms of π.

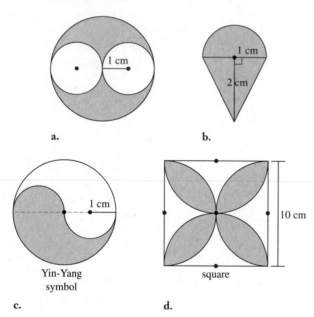

a.

b.

Yin-Yang symbol

square

c.

d.

21. A circular flower bed is 6 m in diameter and has a circular sidewalk around it 1 m wide. Find the area of the sidewalk in square meters.

22. a. If the area of a square is 144 cm², what is its perimeter?

b. If the perimeter of a square is 32 cm, what is its area?

23. a. What happens to the area of a square when the length of each side is doubled?

b. If the ratio of the sides of two squares is 1 to 5, what is the ratio of their areas?

24. Find the shaded area in the following figure:

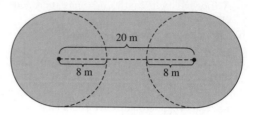

25. Quadrilateral *MATH* has been dissected into squares. The area of the red square is 64 square units and the area of the blue square is 81 square units. Determine the dimensions and the area of quadrilateral *MATH*.

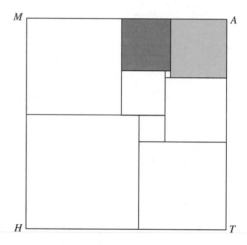

26. Complete and explain how to use geometric shapes to find an equivalent algebraic expression not involving parentheses for each of the following:

a. $a(b + c)$

b. $(a + b)(c + d)$

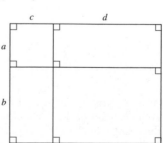

27. In the following figure, $\ell \parallel \overleftrightarrow{AB}$. If the area of $\triangle ABP$ is 10 cm^2, what are the areas of $\triangle ABQ$, $\triangle ABR$, $\triangle ABS$, $\triangle ABT$, and $\triangle ABU$? Explain your answers.

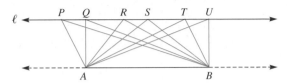

28. Heron's formula can be used to find the area of a triangle if the lengths of the three sides are known. If the lengths of the three sides are a, b, and c units and the semiperimeter $s = \dfrac{a + b + c}{2}$, then the area of the triangle is given by $\sqrt{s(s - a)(s - b)(s - c)}$. Use Heron's formula to find the areas of the right triangles with the sides given.
   a. 3 cm, 4 cm, 5 cm
   b. 5 cm, 12 cm, 13 cm

Assessment 13-2B

1. As an example of measuring area with a nonstandard measure, estimate the area of your desktop in terms of your hand as your unit of area. Then measure the area of your desktop with your hand and compare it to your estimate.

2. Choose the most appropriate metric units (cm^2, m^2, or km^2) and English units (in.^2, yd^2, or mi^2) for measuring each of the following:
   a. Area of a parallel parking space
   b. Area of an airport runway
   c. Area of your mathematics book cover

3. Estimate and then measure each of the following using cm^2, m^2, or km^2:
   a. Area of a chair seat
   b. Area of a whiteboard or chalkboard

4. Complete the following conversion table:

	m^2	cm^2	mm^2
a.	52		
b.			105
c.		86	
d.			10,000
e.	8.2		

5. Complete the following:
   a. $99 \text{ ft}^2 = $ _____ yd^2
   b. $10^6 \text{ yd}^2 = $ _____ ft^2
   c. $6.5 \text{ mi}^2 = $ _____ acres
   d. 3 acres $= $ _____ yd^2

6. Find the areas of each of the following figures if the distance between two adjacent dots in a row or a column is 1 unit:

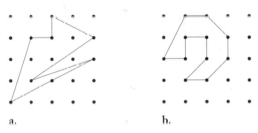

a.   b.

7. If all vertices of a polygon are points on square-dot paper, the polygon is a **lattice polygon**. In 1899, G. Pick discovered a surprising theorem involving I, the number of dots *inside* the polygon, and B, the number of dots that lie *on* the polygon. The theorem states that the area of any lattice polygon is $I + \dfrac{1}{2}B - 1$. Check that this is true for the polygons in exercise 6.

8. Find the area of $\triangle ABC$ in each of the following. (Drawings not to scale.)

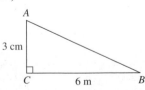

a.

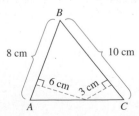

b.

9. If a triangle and a square have one side in common, where could the third vertex of the triangle lie if the area of the triangle is exactly equal to area of the square?

10. **a.** If triangle ABC is similar to triangle DEF and $\dfrac{AB}{DE} = \dfrac{2}{3}$, what is the ratio of the heights of the triangles?
   b. What is the ratio of the areas of the two triangles? Why?

11. Find the area of each of the following quadrilaterals:

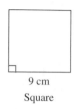

9 cm

Square

a.

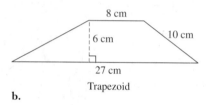

8 cm

6 cm

10 cm

27 cm

Trapezoid

b.

12. Find the area of each of the following.

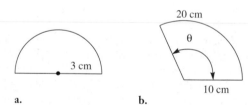

20 cm

θ

3 cm

10 cm

a.   **b.**

13. A rectangular plot of land is to be seeded with grass. If the plot is 22 m × 28 m and a 1 kg bag of seed is needed for 85 m² of land, how many bags of seed are needed?

14. Suppose the largest square peg possible is placed in a circular hole as shown in the following figure and that the largest circular peg possible is placed in a square hole. In which case is there a smaller percentage of space wasted?

15. Find the area of each of the following shaded parts. Assume all arcs are circular. Leave all answers in terms of π.

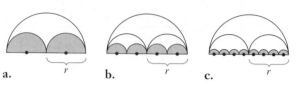

a. r **b.** r **c.** r

16. **a.** What happens to the area of a circle if its diameter is doubled?
   b. What happens to the area of a circle if its radius is increased by 10%?
   c. What happens to the area of a circle if its circumference is tripled?

17. A rectangular field is 64 m × 25 m. Shawn wants to fence a square field that has the same area as the rectangular field. How long are the sides of the square field?

18. A store has wrapping paper on sale. One package is 3 rolls of $2\dfrac{1}{2}$ ft × 8 ft for $6.00. Another package is 5 rolls of $2\dfrac{1}{2}$ ft × 6 ft for $8.00. Which is the better buy per square foot?

19. The following figure consists of five congruent squares. Find a segment through point P that divides the figure into two parts of equal area.

P

20. **a.** Sketch a graph showing the relationship between the length and width of all rectangles with perimeters of 12 cm.
   b. Sketch a graph showing the relationship between the length and width of all rectangles with areas of 12 cm².

21. For a dartboard (see the following figure), Joan is trying to determine how the area of the outside shaded region compares with the sum of the areas of the 3 inside shaded regions so that she can determine payoffs. How do they compare?

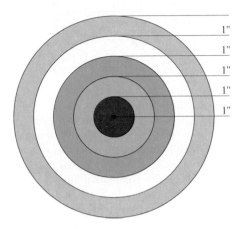

1"
1"
1"
1"
1"

22. Draw two rectangles that have the same perimeter but different areas.

23. **a.** If the area of a rectangle remains constant but its perimeter increases, how has the shape of the rectangle changed?
    b. If the perimeter of a rectangle remains constant but its area has increased, how has the shape of the rectangle changed?

24. Given △*ABC* with parallel lines dividing *AB* into three congruent segments as shown, how does the area of △*BDE* compare with the area of △*ABC*?

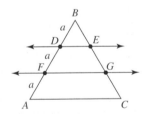

25. Use exercise 25 and compare the following areas:
    a. Triangle *DBE* and trapezoid *DEGF*
    b. Triangle *DBE* and trapezoid *FGCA*
    c. Trapezoids *DEGF* and *FGCA*
    d. Trapezoid *DEGF* and triangle *ABC*
    e. Trapezoid *FGCA* and triangle *ABC*
    f. Triangle *ABC* and trapezoid *DECA*

★ 26. In the following figure, quadrilateral *ABCD* is a parallelogram and *P* is any point on *AC*. Prove that the area of △*BCP* is equal to the area of △*DPC*.

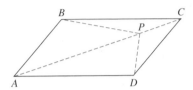

27. Squares *A* and *B* are congruent. One vertex of *B* is at the center of square *A*. What is the ratio of the shaded area to the area of square *A*?

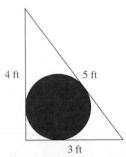

28. Rheba wanted a piece of red circular glass inset into her right triangular window. If the window had sides with measures of 3 ft, 4 ft, and 5 ft, and the glass was to be an inscribed circle as shown, what is the radius of the circular glass? (Areas may be used to solve this problem.)

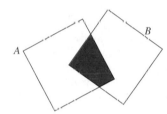

4 ft   5 ft

3 ft

Mathematical Connections 13-2

Communication

1. Suppose a triangle has sides of lengths 6 in., 11 in., and 13 in. Explain how you can find the area of this triangle when the height is not given.

2. John claimed he had a garden twice as large as Al's rectangular-shaped garden that measured 15 ft by 30 ft. When they visited John's rectangular-shaped garden, they found it measured 18 ft by 50 ft. Al claimed that it could not be twice as large since neither the length nor the width were twice as large. Who was correct and why?

3. a. If a 10 in. (diameter) pizza costs $10, how much should a 20 in. pizza cost? Explain the assumptions you made in your answer.

   b. If the ratio between the diameters of two pizzas is 1 : *k*, what should the ratio be between the prices? Explain the assumptions you made in your answer.

4. a. Explain how the following drawing can be used to determine a formula for the area of △*ABC*:

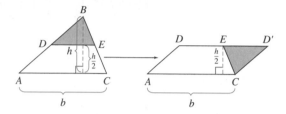

   b. Use paper cutting to reassemble △*ABC* in (a) into parallelogram *ADD'C*.

5. The area of a parallelogram can be found by using the concept of a half-turn (a turn by 180°). Consider the parallelogram *ABCD*, and let *M* and *N* be the midpoints of $\overline{AB}$ and $\overline{CD}$, respectively. Rotate the shaded triangle with vertex *M* about *M* by 180° clockwise and rotate the shaded triangle with vertex *N* about *N* by 180° clockwise. What kind of figure do you obtain? Now complete the argument to find the area of the parallelogram.

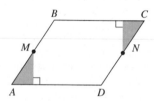

6. If the length of a rectangle is increased by 10% and the width of the rectangle is decreased by 10%, is the area changed? If so, does it increase or decrease and by what percent?

Open-Ended

7. a. Estimate the area in square centimeters that your handprint will cover.

   b. Trace the outline of your hand on square-centimeter grid paper and use the outline to obtain an estimate for the area. Explain how you arrived at your estimate.

8. a. Give dimensions of a square and a rectangle that have the same perimeter but the square has the greater area.

   b. Give dimensions of a square and a rectangle that have equal area but the rectangle has greater perimeter.

Cooperative Learning

9. Use five 1 × 1 squares to build the cross shape shown and discuss the questions that follow.

   a. What is the area of this shape?
   b. What is the perimeter of this shape?
   c. Add squares to the shape so that each square added touches a complete edge with at least one other square.
      (i) What is the minimum number of squares that can be added so that the shape has a perimeter of 18?
      (ii) What is the maximum number?
      (iii) What is the maximum area the new shape could have and still have a perimeter of 18?
   d. Using the five squares, have members of the group start with shapes different from the original cross shape and answer the questions in (c). Discuss your results.
   e. Explore shapes that are made up of more than five squares.

Questions from the Classroom

10. On a field trip, Glenda, a sixth-grade student, was looking at a huge dinosaur footprint and wondering about its area. Glenda said all you have to do is place a string around the border of the print and then take the string off and form it into a square and compute the area of the square. How would you help her?

11. A student asks, "Can I find the area of an angle?" How do you respond?

12. A student claims that because *are* and *hectare* are measures of area, we should say "square are" and "square hectare." How do you respond?

13. Larry and Gary are discussing whose garden has the most area to plant flowers. Larry claims that all they have to do is walk around the two gardens to get the perimeter and the one with the greatest perimeter has the greatest area. How would you help these students?

14. Jimmy claims that to find the area of a parallelogram he just has to multiply length times width. In the figure he multiplies $25 \times 20 = 500$ in.^{2}. What would you tell him?

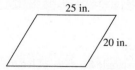

Review Problems

15. A glass coffee tabletop is essentially square but has rounded circular corners. Find its perimeter.

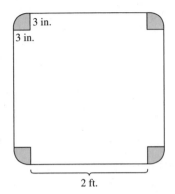

16. a. The earth has a circumference of approximately 39,750 km. With this circumference, what is its radius?
   b. Use the measurement in part (a) to determine the length of an arc from the North Pole to the equator.

20. Compare the perimeter of a regular hexagon to the circumference of its circumscribed circle.

Third International Mathematics and Science Study (TIMSS) Questions

A rectangular picture is pasted to a sheet of white paper as shown.

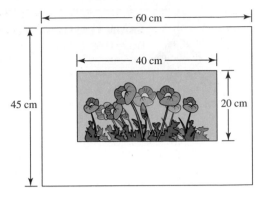

What is the area of the white paper not covered by the picture?

a. 165 cm^2 **b.** 500 cm^2
c. 1900 cm^2 **d.** 2700 cm^2

TIMSS, Grade 8, 1994

The length of a rectangle is 6 cm, and its perimeter is 16 cm. What is the area of the rectangle in square centimeters?

TIMSS, Grade 8, 1994

National Assessment of Educational Progress (NAEP) Questions

Mark's room is 12 feet wide and 15 feet long. Mark wants to cover the floor with carpet. How many square feet of carpet does he need?

Answer: _____ square feet

The carpet costs $2.60 per square foot. How much will the carpet cost?

Answer: $ _____

NAEP, Grade 4, 2007

LABORATORY ACTIVITY

1. On a 5 × 5 geoboard, make △DEF as shown in Figure 13-37. Keep the rubber band around D and E fixed and move the vertex F to all the possible locations so that the triangles formed will have the same area as the area of △DEF. How do the locations for the third vertex relate to $\overline{DE}$?

Figure 13-37

2. On a 5 × 5 geoboard, construct, if possible, squares of areas 1, 2, 3, 4, 5, 6, and 7 square units.

3. On a 5 × 5 geoboard, construct triangles that have areas $\frac{1}{2}$, 1, $1\frac{1}{2}$, 2, . . . , until the maximum-sized triangle is reached.

BRAIN TEASER The rectangle in Figure 13-38(b) was apparently formed by cutting the square in Figure 13-38(a) along the dotted lines and reassembling the pieces as pictured.

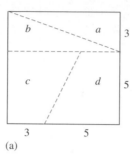

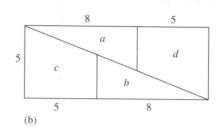

(a)

(b)

Figure 13-38

1. What is the area of the square in 13-38(a)?
2. What is the area of the rectangle in 13-38(b)?
3. How do you explain the discrepancy between the areas?

13-3   The Pythagorean Theorem, Distance Formula, and Equation of a Circle

Surveyors often have to calculate distances that cannot be measured directly, such as distances across water, as illustrated in Figure 13-39.

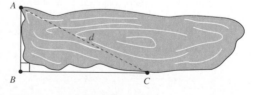

Figure 13-39

To measure the distance from point A to point C, they could use one of the most remarkable and useful theorems in geometry: the Pythagorean theorem. This theorem was illustrated on a Greek stamp in 1955, as shown in Figure 13-40(a), to honor the 2500th anniversary of the founding of the Pythagorean School.

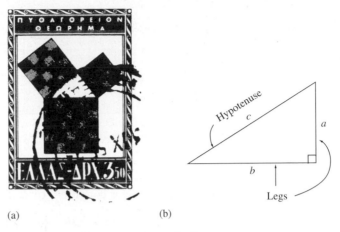

(a)   (b)

Figure 13-40

In the triangle on the stamp and in triangle ABC in Figure 13-40(b), the side opposite the right angle is the **hypotenuse**. The other two sides are **legs**. Interpreted in terms of area, the Pythagorean theorem states that the area of a square with the hypotenuse of a right triangle as a side is equal to the sum of the areas of the squares with the legs as sides.

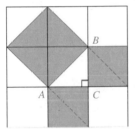

Figure 13-41

Because the Pythagoreans affirmed geometric results on the basis of special cases, mathematical historians believe it is possible they may have discovered the theorem by looking at a floor tiling like the one illustrated in Figure 13-41. Each square can be divided by its diagonal into two congruent isosceles right triangles, so we see that the shaded square constructed with $\overline{AB}$ as a side consists of four triangles, each congruent to $\triangle ABC$. Similarly, each of the shaded squares with legs $\overline{BC}$ and $\overline{AC}$ as sides consists of two triangles congruent to $\triangle ABC$. Thus, the area of the larger square is equal to the sum of the areas of the two smaller squares. The theorem is true in general and is stated as follows using Figure 13-40(b).

Theorem 13–2:   Pythagorean Theorem

If a right triangle has legs of lengths a and b and hypotenuse of length c, then $c^2 = a^2 + b^2$.

TECHNOLOGY CORNER *GSP* Lab X in the Technology Manual can be used at this point to investigate the Pythagorean theorem.

There are hundreds of known proofs for the Pythagorean theorem. The classic book *The Pythagorean Proposition*, by E. Loomis, contains many of these proofs. Some proofs involve the strategy of *drawing diagrams* with a square area c^2 equal to the sum of the areas a^2 and b^2 of two other squares. One such proof is given in Figure 13-42. In Figure 13-42(a), the measures of the legs of a right triangle ABC are a and b and the measure of the hypotenuse is c. We draw a square with sides of length $a + b$ and subdivide it, as shown in Figure 13-42(b). In Figure 13-42(c), another square with side of length $a + b$ is drawn and each of its sides is divided into two segments of length a and b, as shown.

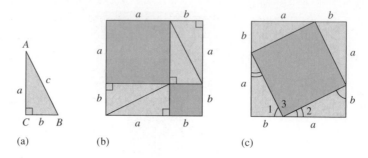

Figure 13-42

Each cream-colored triangle is congruent to $\triangle ABC$ (why?). Consequently, each triangle has hypotenuse c and the same area, $\frac{1}{2}ab$. Thus, the length of each side of the inside blue quadrilateral in Figure 13-42(c) is c and so the figure is a rhombus. Because the triangles are right triangles, their acute angles are complementary. Hence $m(\angle 1) + m(\angle 2) = 90°$ so $m(\angle 3) = 90°$. Therefore, the blue quadrilateral is a square whose area is c^2. To complete the proof, we consider the four triangles in Figure 13-42(b) and (c). Because the areas of the sets of four triangles in both Figure 13-42(b) and (c) are equal, the sum of the areas of the two shaded squares in Figure 13-42(b) equals the area of the shaded square in Figure 13-42(c), that is, $a^2 + b^2 = c^2$.

The grade 8 *Focal Points* call for using the type of proof shown above, as seen in the following:

> Students explain why the Pythagorean theorem is valid by using a variety of methods—for example, by decomposing a square in two different ways. (p. 20)

Another example of this method is given in Now Try This 13-12.

NOW TRY THIS 13-12 In 1873, Henry Perigal, a London stockbroker, printed what has been called the "paper and scissors" proof of the Pythagorean theorem. It is illustrated in Figure 13-43. Explain how this figure could be used to justify the theorem.

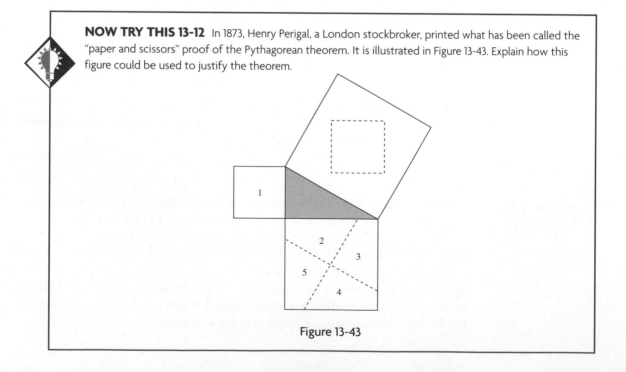

Figure 13-43

Example 13-12

a. For the drawing in Figure 13-44, find the value of x.

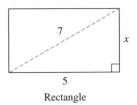

Rectangle

Figure 13-44

b. The size of a rectangular television screen is given as the length of the diagonal of the screen. If the length of the screen is 24 in. and the width is 18 in. as shown in Figure 13-45, what is the diagonal length?

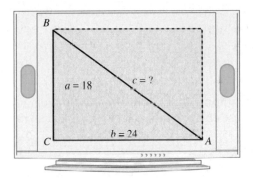

Figure 13-45

Solution **a.** In the rectangle, the diagonal partitions the rectangle into two right triangles, each with length 5 units and width x units. Thus we have the following:

$$5^2 + x^2 = 7^2$$
$$25 + x^2 = 49$$
$$x^2 = 24$$
$$x = \sqrt{24}, \text{ or approximately } 4.9 \text{ units}$$

b. A right triangle is formed with the diagonal as the hypotenuse and the legs of measure 24 in. and 18 in. The Pythagorean theorem can be used to find the length of the diagonal.

$$c^2 = 18^2 + 24^2$$
$$c^2 = 324 + 576$$
$$c^2 = 900$$
$$c = 30$$

Because all the measurements are inches, the diagonal has length 30 in.

Example 13-13

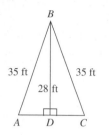

35 ft   35 ft

28 ft

A   D   C

Figure 13-46

A pole $\overline{BD}$, 28 ft high, is perpendicular to the ground. Two wires $\overline{BC}$ and $\overline{BA}$, each 35 ft long, are attached to the top of the pole and to stakes A and C on the ground, as shown in Figure 13-46. If points A, D, and C are collinear, how far are the stakes A and C from each other?

Solution $\overline{AC}$ is not a side in any known right triangle, but we want to find AC. Because a point equidistant from the endpoints of a segment must be on a perpendicular bisector of the segment, it follows that $AD = DC$. Therefore, AC is twice as long as DC. Our *subgoal* is to find DC. We may find DC by applying the Pythagorean theorem in triangle BDC. This results in the following:

$$28^2 + (DC)^2 = 35^2$$
$$(DC)^2 = 35^2 - 28^2$$
$$DC = \sqrt{441}, \text{ or } 21 \text{ ft}$$
$$AC = 2 \cdot DC = 42 \text{ ft}$$

Example 13-14

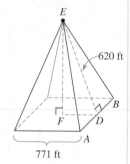

E

620 ft

B

F   D

A

771 ft

Figure 13-47

How tall is the Great Pyramid of Cheops, a right regular square pyramid, if the base has a side 771 ft and the slant height (altitude of $\triangle EAB$) is 620 ft?

Solution In Figure 13-47, $\overline{EF}$ is a leg of a right triangle formed by $\overline{FD}$, $\overline{EF}$, and $\overline{ED}$. Because the pyramid is a right regular pyramid, $\overline{EF}$ intersects the base at its center. Thus, $DF = \left(\frac{1}{2}\right) AB$, or $\left(\frac{1}{2}\right) 771$, or 385.5 ft. Now ED, the slant height, has length 620 ft, and we can apply the Pythagorean theorem as follows:

$$(EF)^2 + (DF)^2 = (ED)^2$$
$$(EF)^2 + (385.5)^2 = (620)^2$$
$$(EF)^2 = 235,789.75$$
$$EF \approx 485.6 \text{ ft}$$

Thus, the Great Pyramid is approximately 485.6 ft tall.

Special Right Triangles

An isosceles right triangle has two legs of equal length and two 45° angles. Any such triangle is a **45°-45°-90° right triangle**. Drawing a diagonal of a square forms two of these triangles, as shown in Figure 13-48.

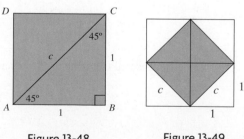

Figure 13-48   **Figure 13-49**

In Figure 13-49, we see several 45°-45°-90° triangles. Each side of the shaded square is the hypotenuse of a 45°-45°-90° triangle. The area of the shaded square is 2 square units (why?). Therefore, $c^2 = 2$ and $c = \sqrt{2}$. Another way to see that $c = \sqrt{2}$ is to apply the Pythagorean theorem to one of the nonshaded triangles. Because $c^2 = 1^2 + 1^2 = 2$, then $c = \sqrt{2}$.

In one isosceles right triangle in Figure 13-49, each leg is 1 unit long and the hypotenuse is $\sqrt{2}$ units long. This is generalized when the isosceles right triangle has a leg of length a, as follows.

Theorem 13–3:   45°-45°-90° Triangle Relationships

In an isosceles right triangle with the length of each leg a, the hypotenuse has length $a\sqrt{2}$.

NOW TRY THIS 13-13 In the following cartoon, Jason runs a different pattern than he was told. Explain how he arrived at this pattern.

Figure 13-50(a) shows that in terms of area a 30°-60°-90° triangle is half of an equilateral triangle. When the equilateral triangle has side 2 units long, then in the 30°-60°-90° triangle, the leg opposite the 30° angle is 1 unit long and the leg opposite the 60° angle has a length of $\sqrt{3}$ units. (Why?)

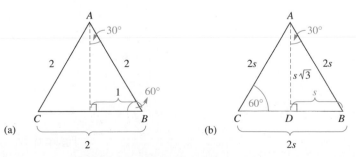

Figure 13-50

This example may also be generalized using the triangle in Figure 13-50(b). When the side of the equilateral triangle ABC is $2s$, then in triangle ABD, the side opposite the 30° angle, $\overline{BD}$, is s units long, and AD may be found using the Pythagorean theorem to have a length of $s\sqrt{3}$ units. This discussion is summarized in the following.

> **Theorem 13–4:   30°-60°-90° Triangle Relationships**
>
> In a 30°-60°-90° triangle, the length of the hypotenuse is twice the length of the leg opposite the 30° angle, and the leg length opposite the 60° angle is $\sqrt{3}$ times the length of the shorter leg.

Converse of the Pythagorean Theorem

The converse of the Pythagorean theorem is also true. It provided a useful way for early surveyors—in particular, the Egyptian rope stretchers—to determine right angles. Figure 13-51(a) shows a knotted rope with 12 equally spaced knots. Figure 13-51(b) shows how the rope might be held to form a triangle with sides of lengths 3, 4, and 5. The triangle formed is a right triangle and contains a 90° angle. Note that $5^2 = 3^2 + 4^2$.

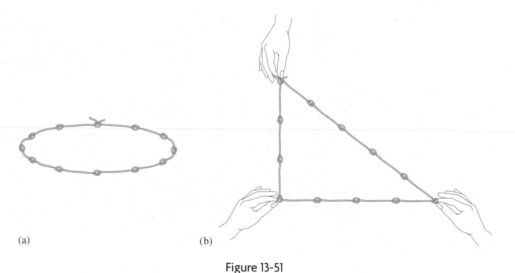

(a)   (b)

Figure 13-51

Given a triangle with sides of lengths a, b, and c such that $a^2 + b^2 = c^2$ as shown in Figure 13-52(a), must the triangle be a right triangle?

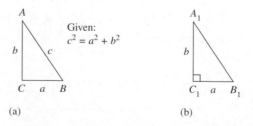

(a)   (b)

Figure 13-52

To investigate this question, we construct a right triangle with two sides a and b, as shown in Figure 13-52(b). By the Pythagorean theorem, $(A_1B_1)^2 = a^2 + b^2$. Therefore, $(A_1B_1)^2 = c^2$ and $A_1B_1 = c_1$. By the side-side-side (SSS) property, $\triangle ABC \cong \triangle A_1B_1C_1$ and hence $m(\angle C) = 90°$. Therefore, we have the following theorem, which is the converse of the Pythagorean theorem.

> ### Theorem 13–5:   Converse of the Pythagorean Theorem
>
> If $\triangle ABC$ is a triangle with sides of lengths a, b, and c such that $a^2 + b^2 = c^2$, then $\triangle ABC$ is a right triangle with the right angle opposite the side of length c.

Example 13-15

Determine if the following can be the lengths of the sides of a right triangle:

a. $51, 68, 85$ **b.** $2, 3, \sqrt{13}$ **c.** $3, 4, 7$

Solution **a.** $51^2 + 68^2 = 7225 = 85^2$, so $51, 68$, and 85 can be the lengths of the sides of a right triangle.
     b. $2^2 + 3^2 = 4 + 9 = 13 = (\sqrt{13})^2$, so $2, 3$, and $\sqrt{13}$ can be the lengths of the sides of a right triangle.
     c. $3^2 + 4^2 \neq 7^2$, so the measures cannot be the lengths of the sides of a right triangle. In fact, since $3 + 4 - 7$, segments with these lengths do not form a triangle.

NOW TRY THIS 13-14

a. Draw three segments that could be used to form the sides of a right triangle and discuss how you would show that these three lengths determine a right triangle.
b. Multiply the lengths of the three segments in (a) by a fixed number and determine if the resulting three lengths can be sides of a right triangle.
c. Using three new numbers, repeat the experiment in (a) and (b). Form a conjecture based on your experiments.

The Distance Formula: An Application of the Pythagorean Theorem

Given the coordinates of two points A and B, we can find the distance AB. We first consider the special case in which the two points are on one of the axes. For example, in Figure 13-53(a), $A(2, 0)$ and $B(5, 0)$ are on the x-axis. The distance between these two points is 3 units:

$$AB = OB - OA = 5 - 2 = 3$$

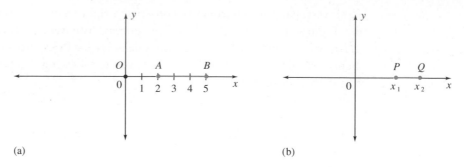

(a)   (b)

Figure 13-53

In general, if two points P and Q are on the x-axis, as in Figure 13-53(b), with x-coordinates x_1 and x_2, respectively, and $x_2 > x_1$, then $PQ = x_2 - x_1$. In fact, *the distance between two points on the x-axis is always the absolute value of the difference between the x-coordinates of the points* (why?). A similar result holds for any two points on the y-axis.

Figure 13-54 shows two points in the plane: $C(2, 5)$ and $D(6, 8)$. The distance between C and D can be found by using the strategy of *looking at a related problem.* We obtain a right triangle by drawing perpendiculars from the points to the x-axis and to the y-axis, as shown in Figure 13-54, thus defining triangle CDE. The lengths of the legs of triangle CDE are found by using horizontal and vertical distances and properties of rectangles.

$$CE = |6 - 2| = 4$$
$$DE = |8 - 5| = 3$$

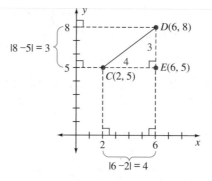

Figure 13-54

The distance between C and D can be found by applying the Pythagorean theorem to the triangle.

$$CD^2 = DE^2 + CE^2$$
$$= 3^2 + 4^2$$
$$= 25$$
$$CD = \sqrt{25}, \text{ or } 5$$

The method can be generalized to find a formula for the distance between any two points $A(x_1, y_1)$ and $B(x_2, y_2)$. Construct a right triangle with $\overline{AB}$ as its hypotenuse by drawing segments through A parallel to the x-axis and through B parallel to the y-axis, as shown in Figure 13-55. The lines containing the segments intersect at point C, forming right triangle ABC. Now, apply the Pythagorean theorem.

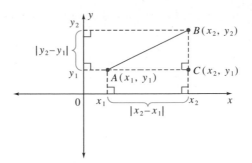

Figure 13-55

In Figure 13-55, we see that $AC = |x_2 - x_1|$ and $BC = |y_2 - y_1|$. By the Pythagorean theorem, $(AB)^2 = |x_2 - x_1|^2 + |y_2 - y_1|^2$, and consequently $AB = \sqrt{|x_2 - x_1|^2 + |y_2 - y_1|^2}$. Because $|x_2 - x_1|^2 = (x_2 - x_1)^2$ and $|y_2 - y_1|^2 = (y_2 - y_1)^2$, $AB = \sqrt{(x_2 - x_1)^2 + (y_2 - y_1)^2}$. This result is known as the **distance formula**.

> **Theorem 13–6:   Distance Formula**
>
> The distance between the points $A(x_1, y_1)$ and $B(x_2, y_2)$ is given by
> $$AB = \sqrt{(x_2 - x_1)^2 + (y_2 - y_1)^2}.$$

NOW TRY THIS 13-15 Investigate whether it makes any difference in the distance formula if $(x_1 - x_2)$ and $(y_1 - y_2)$ are used instead of $(x_2 - x_1)$ and $(y_2 - y_1)$, respectively.

Example 13-16

  a. Show that $A(7, 4)$, $B(^-2, 1)$, and $C(10, ^-5)$ are the vertices of an isosceles triangle.
  b. Show that $\triangle ABC$ in (a) is a right triangle.

Solution   **a.** Using the distance formula, we find the lengths of the sides.
$$AB = \sqrt{(^-2 - 7)^2 + (1 - 4)^2} = \sqrt{(^-9)^2 + (^-3)^2} = \sqrt{90}$$
$$BC = \sqrt{[10 - (^-2)]^2 + (^-5 - 1)^2} = \sqrt{12^2 + (^-6)^2} = \sqrt{180}$$
$$AC = \sqrt{(10 - 7)^2 + (^-5 - 4)^2} = \sqrt{3^2 + (^-9)^2} = \sqrt{90}$$

Thus, $AB = AC$, and so the triangle is isosceles.
  b. Because $(\sqrt{90})^2 + (\sqrt{90})^2 = (\sqrt{180})^2$, $\triangle ABC$, is a right triangle with $\overline{BC}$ as hypotenuse and $\overline{AB}$ and $\overline{AC}$ as legs.

Example 13-17

Determine whether the points $A(0, 5)$, $B(1, 2)$, and $C(2, ^-1)$ are collinear.

Solution   It is hard to tell by graphing the points whether the points are collinear (on the same line). If they are not collinear, they would be the vertices of a triangle, and hence $AB + BC$ would be greater than AC (triangle inequality). If $AB + BC = AC$, a triangle could not be formed and the points would be collinear. Using the distance formula, we find the lengths of the sides:

$$AB = \sqrt{(0-1)^2 + (5-2)^2} = \sqrt{1+9} = \sqrt{10}$$

$$BC = \sqrt{(2-1)^2 + (^-1-2)^2} = \sqrt{1+9} = \sqrt{10}$$

$$AC = \sqrt{(0-2)^2 + [5-(^-1)]^2} = \sqrt{4+36} = \sqrt{40}$$

Thus, $AB + BC = 2\sqrt{10}$. Is this sum equal to $\sqrt{40}$? Because $\sqrt{40} = \sqrt{4 \cdot 10} = \sqrt{4} \cdot \sqrt{10} = 2\sqrt{10}$, we have $AB + BC = AC$, and consequently, A, B, and C are collinear.

> **REMARK** An alternative solution to Example 13-17 is to show that the slopes of each of the segments that can be formed are the same.

Using the Distance Formula to Develop the Equation of a Circle

Recall that a *circle* is the set of all points in a given plane equidistant from a given point, the center. Suppose that (x, y) is any point on a circle with the center at the origin, $(0, 0)$, as shown in Figure 13-56(a). Any point on the circle is the same distance from the center and this distance is the *radius, r*. The triangle formed in Figure 13-56(a) is a right triangle with side lengths equal to x and y and hypotenuse equal to r.

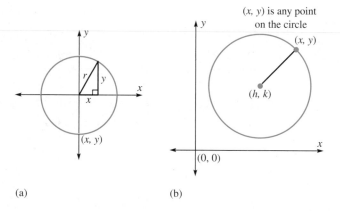

(a)   (b)

Figure 13-56

From the distance formula, we have $r = \sqrt{x^2 + y^2}$. Squaring both sides, we have $r^2 = x^2 + y^2$. We can rewrite this formula to obtain the general equation of a circle with center at the origin, as given below.

Theorem 13–7:   Equation of a Circle with Center at the Origin

An equation of a circle with the center at the origin and radius r is $x^2 + y^2 = r^2$.

If the center is not at the origin, how would the equation change? Suppose the center is at (h, k) and r is the radius as shown in Figure 13-56(b). Then we can use the distance formula to obtain the following.

$$\sqrt{(x - h)^2 + (y - k)^2} = r \quad \text{or} \quad (x - h)^2 + (y - k)^2 = r^2$$

Thus, we have the following.

Theorem 13–8:   Equation of a Circle with Center at (h, k)

An equation of a circle with center (h, k) and radius r is $(x - h)^2 + (y - k)^2 = r^2$.

Example 13-18

a. Find an equation of a circle with its center at $(2, {}^-5)$ and radius of 3.

b. Given the equation of a circle $(x - 3)^2 + (y + 4)^2 = 3$, find the radius and the center.

Solution   **a.** Let $(h, k) = (2, {}^-5)$ and $r = 3$; then using $(x - h)^2 + (y - k)^2 = r^2$ we have $(x - 2)^2 + (y + 5)^2 = 9$.

   b. Rewriting $(x - 3)^2 + (y + 4)^2 = 3$ as $(x - 3)^2 + [y - ({}^-4)]^2 = (\sqrt{3})^2$ and using $(x - h)^2 + (y - k)^2 = r^2$, we have $r^2 = (\sqrt{3})^2$ and $(h, k) = (3, {}^-4)$. Therefore, the radius is $\sqrt{3}$ and the center is at $(3, {}^-4)$.

BRAIN TEASER A farmer has a square plot of land. An irrigation system can be installed with the option of one large circular sprinkler, as in Figure 13-57(a), or nine small sprinklers, as in Figure 13-57(b). The farmer wants to know which plan will provide water to the greater percentage of land in the field, regardless of the cost and the watering pattern. What advice would you give?

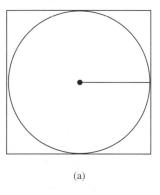

(a)

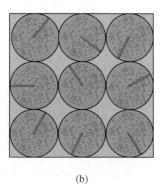

(b)

Figure 13-57

Assessment 13-3A

1. Find the length of the following segments. Assume that the horizontal and vertical distances between neighboring dots is 1 unit.

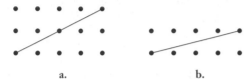

      a.   b.

2. Use the Pythagorean theorem to find x in each of the following:

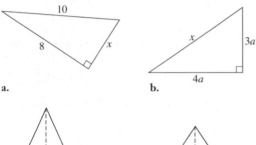

  a.   b.

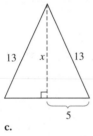
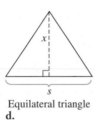

                    Equilateral triangle
  c.   d.

3. On a 5 × 5 geoboard, construct segments of the following lengths.
   a. $\sqrt{13}$
   b. $\sqrt{5}$

4. Find a square on a 4 × 4 geoboard that has an area of 2 square units and a perimeter of $4\sqrt{2}$ units.

5. What is the perimeter of the largest isosceles triangle on a 5 × 5 geoboard?

6. If the hypotenuse of a right triangle is 30 cm long and one leg is twice as long as the other, how long are the legs of the triangle?

7. For each of the following, determine whether the given numbers represent lengths of sides of a right triangle:
   a. 10, 24, 16
   b. 16, 34, 30
   c. $\sqrt{2}, \sqrt{2}, 2$

8. What is the longest line segment that can be drawn in the interior of a right rectangular prism that is 12 cm wide, 15 cm long, and 9 cm high?

9. Starting from point A, a boat sails due south for 6 mi, then due east for 5 mi, and then due south for 4 mi. How far is the boat from A?

10. A 15 ft ladder is leaning against a wall. The base of the ladder is 3 ft from the wall. How high above the ground is the top of the ladder?

11. In the following figure, two poles are 25 m and 15 m high. A cable 14 m long joins the tops of the poles. Find the distance between the poles.

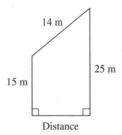

                    Distance

12. For each of the following, solve for the unknowns:

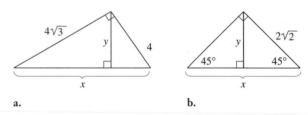

  a.   b.

13. A builder needs to calculate the dimensions of a regular hexagonal window. Assuming the height CD of the window is 1.3 m, find the width AB (O is the midpoint of $\overline{AB}$) in the following figure:

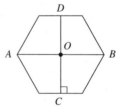

14. If a player standing on third base throws on a straight line to first base, how far is the ball thrown? (*Hint:* The distance from home plate to first base is 90 ft.)

15. To make a home plate for a neighborhood baseball park, we can cut the plate from a square, as shown in the following figure.

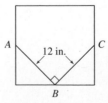

If A, B, and C are midpoints of the sides of the square, what are the dimensions of the square to the nearest tenth of an inch?

16. A company wants to lay a string of buoys across a lake. To find the length of the lake, they made the following measurements. What is the length of the lake?

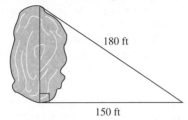

180 ft

150 ft

17. Before James Garfield was elected president of the United States, he discovered a proof of the Pythagorean theorem. He formed a trapezoid like the one that follows and found the area of the trapezoid in two ways. Can you discover his proof?

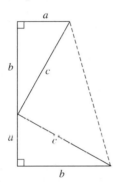

18. Use the following figure to prove the Pythagorean theorem by first proving that the quadrilateral with side c is a square. Then, compute the area of the square with side $a + b$ in two ways: (a) as $(a + b)^2$ and (b) as the sum of the areas of the four triangles and the square with side c.

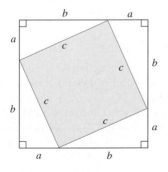

19. For each of the following, find the length of $\overline{AB}$:
    a. $A(0, 3)$, $B(0, 7)$
    b. $A(0, 3)$, $B(4, 0)$
    c. $A(^-1, 2)$, $B(3, ^-4)$
    d. $A(4, ^-5)$, $B\left(\dfrac{1}{2}, \dfrac{^-7}{4}\right)$

20. If the length of the hypotenuse in a 30°-60°-90° triangle is $c/2$ units, what is the length of the side opposite the 60° angle? Explain your answer.

21. In triangle ABC with vertices at $A(0, 0)$, $B(6, 0)$, and $C(0, 8)$,
    a. write the equations of the lines that contain the altitudes.
    b. what are the coordinates of the intersection of the altitudes?

22. a. Find the possible coordinates of the third vertex of an isosceles right triangle that has endpoints of one leg with coordinates $(0, 0)$ and $(8, 8)$ and the hypotenuse lies on the x-axis.
    b. What are the lengths of the sides of the triangle in part (a)?
    c. Show that the sides of the triangle satisfy the Pythagorean theorem.

23. Find an equation of the circle with the given center and radius:
    a. Center at $(^-3, 4)$, radius $= 4$
    b. Center at $(^-3, ^-2)$, radius $= \sqrt{2}$

24. Give the center of the circle and the radius.
    a. $x^2 + y^2 = 16$
    b. $(x - 3)^2 + (y - 2)^2 = 100$
    c. $(x + 2)^2 + (y - 3)^2 = 5$
    d. $x^2 + (y + 3)^2 = 9$

Assessment 13-3B

1. Find the length of the following segments. Assume that the horizontal and vertical distances between neighboring dots is 1 unit.

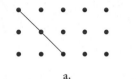

   a.   b.

2. Use the Pythagorean theorem to find x in each of the following:

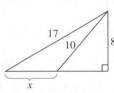

   a.

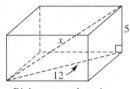

   Right rectangular prism
   b.

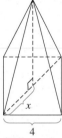
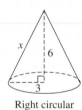

   Right square
   pyramid
   c.

   Right circular
   cone
   d.

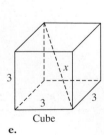

   Cube
   e.

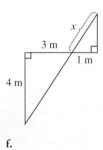

   f.

3. On a 5 × 5 geoboard, construct segments of the following lengths.
   a. $\sqrt{10}$ b. $\sqrt{17}$

4. On a 4 × 4 geoboard, the greatest square has a perimeter of 12 units. Find a polygon on the geoboard with a greater perimeter. Prove your answer.

5. For each of the following, determine whether the given numbers represent lengths of sides of a right triangle:
   a. $\frac{3}{2}, \frac{4}{2}, \frac{5}{2}$ b. $\sqrt{2}, \sqrt{3}, \sqrt{5}$
   c. 18, 24, 30

6. A door is 6 ft 6 in. tall and 36 in. wide. Can a piece of plywood that is 7 ft by 8 ft be carried through the door?

7. Find the area of each of the following:

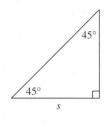
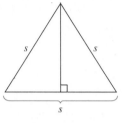

   a.   b.

8. The following two rhombuses have perimeters that are equal. Use the properties of a rhombus to find the area of each rhombus.

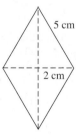
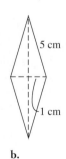

   a.   b.

9. What is the longest piece of straight dry spaghetti that will fit in a cylindrical can that has a radius of 2 in. and height of 10 in.?

10. If possible, draw a square with the given number of square units on a 5 × 5 geoboard grid.
    a. 5   b. 7   c. 8
    d. 14   e. 15

11. Use the following drawing to prove the Pythagorean theorem by using corresponding parts of similar triangles $\triangle ACD$, $\triangle CBD$, and $\triangle ABC$. Lengths of sides are

indicated by a, b, c, x, and y. (*Hint:* Show that $b^2 = cx$ and $a^2 = cy$.)

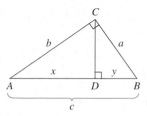

12. An access ramp enters a building 1 m above ground level and starts 3 m from the building. How long is the ramp?

13. A CB radio station C is located 3 mi from the interstate highway h. The station has a range of 6.1 mi in all directions from the station. If the interstate is along a straight line, how many miles of highway are in the range of this station?

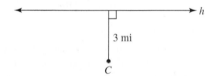

14. Show how the following figure could be used to prove the Pythagorean theorem:

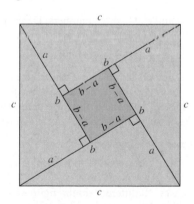

15. Construct semicircles on right triangle ABC with $\overline{AB}$, $\overline{BC}$, and $\overline{AC}$ as diameters, as shown below.

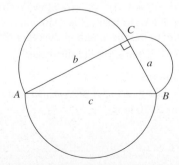

Is the area of the semicircle on the hypotenuse equal to the sum of the areas of the semicircles on the legs? Why?

16. In exercise 15, semicircles were constructed on each side of the right triangle ABC. Flip the semicircle on the hypotenuse over the hypotenuse, as shown in the figure below. Show that the area of the right triangle, t, is equal to the sum of the areas of the two crescents c_1 and c_2.

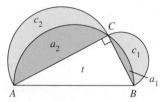

17. On each side of a right triangle, construct an equilateral triangle. Is the area of the triangle constructed on the hypotenuse always equal to the sum of the areas of the triangles constructed on the legs? Why?

18. Find the perimeter of the triangle with vertices at $A(0, 0)$, $B(^-4, ^-3)$, and $C(^-5, 0)$.

19. Show that the triangle whose vertices are $A(^-2, ^-5)$, $B(1, ^-1)$, and $C(5, 2)$ is isosceles.

20. In triangle ABC with vertices at $A(0, 0)$, $B(6, 0)$, and $C(0, 8)$,
   a. write the equations of the perpendicular bisectors of each of the sides of the triangle.
   b. what are the coordinates of the intersection of the perpendicular bisectors?
   c. what is the radius of the circumcircle for the triangle?
   d. what are the coordinates of the center of the circumcircle?

21. In an equilateral triangle with one vertex at the origin and one with coordinates $(8, 0)$, find the possible coordinates of the third vertex.

22. A boat starts at point A, moves 3 km due north, then 2 km due east, then 1 km due south, and then 4 km due east to point B. Find the distance AB.

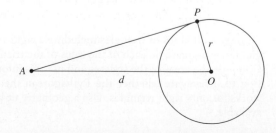

23. The distance from point A to the center of a circle with radius r is d. Express the length of the tangent segment $\overline{AP}$ in terms of r and d.

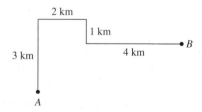

24. Find the equation of the circle with the given center and radius.
  a. Center at $(0, 0)$ and $r = \sqrt{5}$
  b. Center at $(^-6, ^-7)$ and $r = 6$

25. Give the center and radius for each circle.
  a. $(x - 3)^2 + (y + 2)^2 = 9$
  b. $3x^2 + 3y^2 = 9$

26. Find the area between two concentric (same center) circles whose equations are given below.

$$(x + 4)^2 + (y - 8)^2 = 9$$

and

$$(x + 4)^2 + (y - 8)^2 = 64$$

Mathematical Connections 13-3

Communication

1. Given the following square, describe how to use a compass and a straightedge to construct a square whose area is as follows:

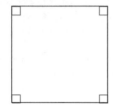

  a. Twice the area of the given square
  b. Half the area of the given square

2. Gail tried the Egyptian method of using a knotted rope to determine a right angle so that she could build a shed. She placed her knots so that each was 1 ft from the next. She stretched out her rope in the form of a triangle whose sides were of lengths 5, 12, and 13 ft. Did she have a right angle? Explain why or why not.

3. Explain how you would find the distance XY across the lake shown below, and then find XY.

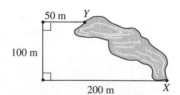

4. The sum of the squares of the lengths of all the sides of a right triangle is 200. Explain how to find the length of the hypotenuse and then find the length of the hypotenuse.

Open-Ended

5. Draw several kinds of triangles including a right triangle. Draw a square on each of the sides of the triangles. Compute the areas of the squares and use this information to investigate whether the Pythagorean theorem works for only right triangles. Use a geometry utility if available.

6. A **Pythagorean triple** is a sequence of three natural numbers a, b, and c that satisfy the relationship $a^2 + b^2 = c^2$. The least three numbers that form a Pythagorean triple are 3-4-5. Another triple is 5-12-13 because $5^2 + 12^2 = 13^2$.
  a. Find two other Pythagorean triples.
  b. Does doubling each number in a Pythagorean triple result in a new Pythagorean triple? Why or why not?
  c. Does adding a fixed number to each number in a Pythagorean triple result in a new Pythagorean triple? Why or why not?
  d. Suppose $a = 2uv$, $b = u^2 - v^2$, and $c = u^2 + v^2$, where u and v are whole numbers. Determine whether a-b-c is a Pythagorean triple.

Cooperative Learning

7. Have each person in the group use a 1-m string to make a different right triangle. Measure each side to the nearest centimeter. Use these measurements to see if the Pythagorean theorem holds for your measurements.

Questions from the Classroom

8. As part of the discussion of the Pythagorean theorem, squares were constructed on each side of a right triangle. A student asks, "If different similar figures are constructed on each side of the triangle, does the same type of relationship still hold?" How do you reply?

9. Amy says that the equation of the circle she just drew is $(x - 3)^2 + (y + 2)^2 = ^-16$. How do you respond?

10. One leg of a right triangle is 3 cm and the hypotenuse is 5 cm. Joni found the length of the other leg by evaluating $\sqrt{3^2 + 5^2}$. What error did she make?

Review Problems

11. Arrange the following in decreasing order: 3.2 m, 322 cm, 0.032 km, 3020 mm.

12. Find the area of each of the following figures:

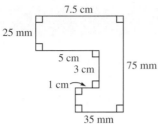

a.

6 cm

4 cm

10 cm

b.

600 cm

4 m

10 m

Trapezoid

c.

13. Complete the following table, which concerns circles:

	Radius	Diameter	Circumference	Area
a.	5 cm			
b.		24 cm		
c.				17π m^2
d.			20π cm	

14. A 10 m wire is wrapped around a circular region. If the wire fits exactly, what is the area of the region?

National Assessment of Educational Progress (NAEP) Question

The endpoints of a line segment are the points with coordinates $(2, 1)$ and $(8, 9)$. What are the coordinates of the midpoint of this line segment?

a. $\left(2, 3\frac{1}{2}\right)$

b. $(3, 4)$

c. $(5, 5)$

d. $\left(4\frac{1}{2}, 5\frac{1}{2}\right)$

e. $(10, 10)$

NAEP Grade 8, 2005

TECHNOLOGY CORNER Use *GSP* to determine the relationship between the length of the hypotenuse of a 45°-45°-90° triangle and the length of a leg.

a. Construct a 45°-45°-90° triangle, label the vertices as in Figure 13-59, and measure the lengths of the sides. Record the data for triangle 1 in the following table and compute the ratio:

Figure 13-59

	AC	CB	AB	AB/CB
Triangle 1				
Triangle 2				
Triangle 3				
Triangle 4				

b. Repeat (a) for three other triangles.
c. Make a conjecture about the relationship between the length of the hypotenuse and the length of a leg for these triangles.
d. Given a 30°-60°-90° triangle, determine the relationship between the lengths of the hypotenuse and the shorter leg and the relationships between the lengths of the longer and shorter legs.

13-4   Surface Areas

In the grade 5 *Focal Points* we find the following:

> They (*students*) decompose three-dimensional shapes and find surface areas and volumes of prisms. As they work with surface area, they find and justify relationships among the formulas for the areas of different polygons. They measure necessary attributes of shapes to use area formulas to solve problems. (p. 17)

In this section, we decompose three-dimensional shapes to find surface areas and we justify relationships among the formulas for the areas of different polygons. However, as mentioned in the Research Note, students have a difficult time translating information from three-dimensional shapes to two dimensions, and vice-versa.

Painting houses, buying roofing, seal-coating driveways, and buying carpet are among the common applications that involve computing areas. In many real-world problems, we must find the surface areas of such three-dimensional figures as prisms, cylinders, pyramids, cones, and spheres. Formulas for finding these areas are usually based on finding the area of two-dimensional pieces of the three-dimensional figures.

Surface Area of Right Prisms

Consider the cereal box shown in Figure 13-60(a). Ignore the flaps for gluing the box together. To find the amount of cardboard necessary to make the box, we cut the box along the edges and make it lie flat, as shown in Figure 13-60(b). When we do this we obtain a *net* for the box. The box is composed of a series of rectangles. We find the area of each rectangle and sum those areas to find the surface area of the box.

◆ *Research Note*

Adolescents have a difficult time translating visual information from a three-dimensional environment to two dimensions, and vice-versa (Ben-Chaim et al. 1984). This may cause problems when trying to find surface areas of three-dimensional figures using two-dimensional images. ◆

(a)

(b)

Figure 13-60

NOW TRY THIS 13-16

a. Find the surface area of the box in Figure 13-60.

b. Could the box be made from a rectangular piece of cardboard 21 in. by 15 in.? If not, what size rectangle could you use and how would you do it?

A similar process can be used for many three-dimensional figures. For example, the surface area of the cube in Figure 13-61(a) is the sum of the areas of the faces of the cube. A net for a cube is shown in Figure 13-61(b). Because each of the six faces is a square of area $16\,cm^2$, the surface area is $6(16\,cm^2)$, or $96\,cm^2$. In general, for a cube whose edge is e units as in Figure 13-61(c), the surface area is $6e^2$.

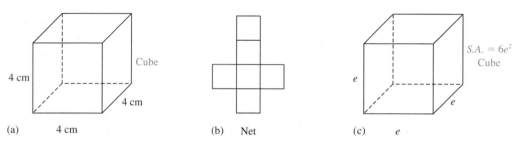

(a)   4 cm   (b)   Net   (c) e

Cube $S.A. = 6e^2$ Cube

Figure 13-61

To find the surface area of a right prism, like the cereal box in Figure 13-60(a), we find the sum of the areas of the rectangles that make up the lateral faces and the areas of the top and bottom. The sum of the areas of the lateral faces is the **lateral surface area**. The **surface area** $(S.A.)$ is the sum of the lateral surface area and the area of the bases.

NOW TRY THIS 13-17 Figure 13-62 shows a right pentagonal prism with a net for the prism. If B stands for the area of each of the prism's bases, show that the surface area of the prism could be computed as $S.A. = ph + 2B$, where p is the perimeter of the base of the prism and h is the height. Does this formula hold for all right prisms? Why or why not?

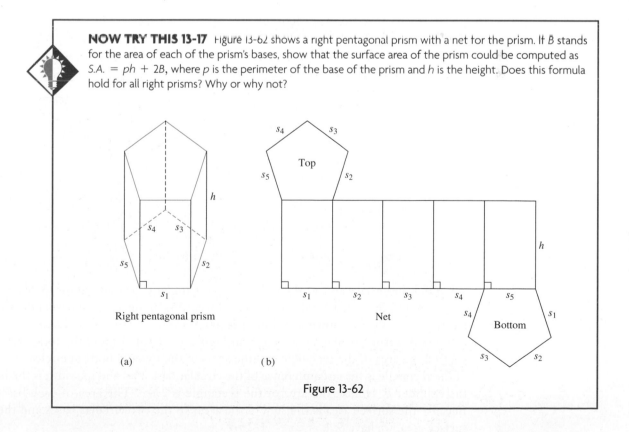

Right pentagonal prism

Net

(a)   (b)

Figure 13-62

Example 13-19

Find the surface area of each of the right prisms in Figure 13-63.

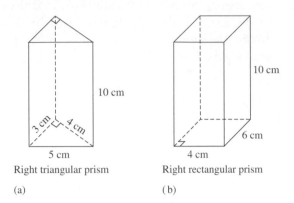

10 cm

3 cm 4 cm

5 cm

Right triangular prism

(a)

10 cm

6 cm

4 cm

Right rectangular prism

(b)

Figure 13-63

Solution   **a.** Each base is a right triangle. The area of the bases is $2\left[\frac{1}{2}(3 \cdot 4)\right]$, or 12 cm².

The area of the three lateral faces is $4 \cdot 10 + 3 \cdot 10 + 5 \cdot 10$, or 120 cm². Thus, the surface area is 12 cm² + 120 cm², or 132 cm².

b. The area of the bases is $2(4 \cdot 6)$, or 48 cm². The lateral surface area is $2 \cdot (10 \cdot 6) + 2 \cdot (10 \cdot 4)$, or 200 cm². Thus, the surface area is 248 cm².

Surface Area of a Cylinder

As the number of sides of a right regular prism increases, as shown in Figure 13-64, the figure approaches the shape of a right circular cylinder.

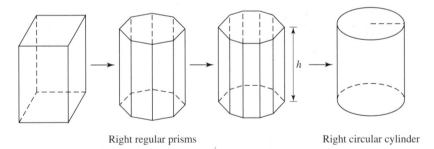

Right regular prisms   Right circular cylinder

Figure 13-64

To find the surface area of the right circular cylinder shown in Figure 13-65, we cut off the bases and slice the lateral surface open by cutting along any line perpendicular to the bases. Such a slice is shown as a dotted segment in Figure 13-65(a). Then we unroll the cylinder to form a rectangle, as shown in Figure 13-65(b). To find the total surface area, we find the area of the rectangle and the areas of the top and bottom circles. The length of the rectangle is the circumference of the circular base $2\pi r$, and its width is the height of the cylinder h. Hence, the area of the rectangle is $2\pi rh$. The area of each base is πr^2. Because the surface area is the sum of the areas of the two circular bases and the lateral surface area, we have $S.A. = 2\pi r^2 + 2\pi rh$.

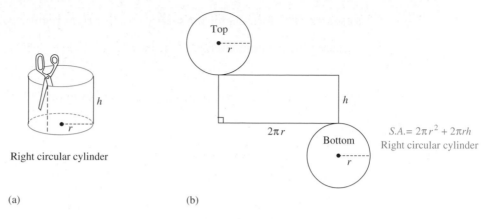

Figure 13-65

Surface Area of a Pyramid

The surface area of a pyramid is the sum of the lateral surface area of the pyramid and the area of the base. A **right regular pyramid** is a pyramid such that the segments connecting the apex to each vertex of the base are congruent and the base is a regular polygon. The lateral faces of the right regular pyramid pictured in Figure 13-66(a) are congruent isosceles triangles. Each triangular face has an altitude of length ℓ, the *slant height*. Because the pyramid is right regular, each side of the base has the same length b. To find the lateral surface area of a right regular pyramid, we need to find the area of one face, $\frac{1}{2}b\ell$, and multiply it by n, the number of faces. Adding the lateral surface area $n\left(\frac{1}{2}b\ell\right)$ to the area of the base B gives the surface area.

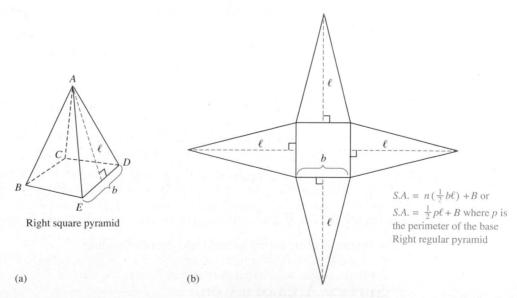

Figure 13-66

> **REMARK** The formula for the surface area of a right regular pyramid can be simplified because nb is the perimeter, p, of the base. Thus, $S.A. = \frac{1}{2}p\ell + B$.

Example 13-20

Find the surface area of the right square pyramid in Figure 13-67.

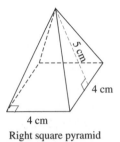

Right square pyramid

Figure 13-67

Solution   The surface area consists of the area of the square base plus the area of the four triangular faces. Hence, the surface area is

$$4 \text{ cm} \cdot 4 \text{ cm} + 4\left(\frac{1}{2} \cdot 4 \text{ cm} \cdot 5 \text{ cm}\right) = 16 \text{ cm}^2 + 40 \text{ cm}^2$$

$$= 56 \text{ cm}^2$$

Example 13-21

The Great Pyramid of Cheops is a right square pyramid with a height of 148 m and a square base with perimeter of 940 m. The altitude of each triangular face is 189 m. The basic shape of the Transamerica Building in San Francisco is a right square pyramid that has a height of 260 m and a square base with a perimeter of 140 m. The altitude of each triangular face is 261 m. How do the lateral surface areas of the two structures compare?

Solution   The length of one side of the square base of the Great Pyramid is $\frac{940}{4}$, or 235 m.

Likewise, the length of one side of the square base of the Transamerica Building is $\frac{140}{4}$, or 35 m. The lateral surface area (*L.S.A.*) of the two are computed as follows:

$$(\text{Great Pyramid}) \ L.S.A. = 4\left(\frac{1}{2} \cdot 235 \cdot 189\right) = 88,830 \text{ m}^2$$

$$(\text{Transamerica}) \ L.S.A. = 4\left(\frac{1}{2} \cdot 35 \cdot 261\right) = 18,270 \text{ m}^2$$

Therefore, the lateral surface area of the Great Pyramid is approximately $\frac{88,830}{18,270}$ or 4.9 times as great as that of the Transamerica Building.

Surface Area of a Cone

As the number of sides of a right regular pyramid increases, as shown in Figure 13-68, the figure approaches the shape of a right circular cone.

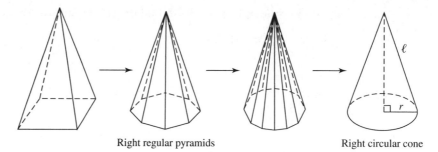

Right regular pyramids   Right circular cone

Figure 13-68

It is possible to find a formula for the surface area of a right circular cone by approximating the cone with a pyramid with many sides. We could inscribe in the circular base of the cone a regular polygon with many sides. The polygon can be used as the base of a right regular pyramid. The lateral surface area of the pyramid is close to the lateral surface area of the cone. The greater the number of faces of the pyramid, the closer the surface area of the pyramid is to that of the cone. In Figure 13-69(a), the lateral surface of the pyramid is $\frac{1}{2}ph$,

where p is the perimeter of the base and h is the height of each triangle. With many sides in the pyramid, the perimeter of its base is close to the perimeter of the circle, $2\pi r$. The height of each triangle of the pyramid is close to the slant height ℓ, a segment that connects the vertex of the cone with a point on the circular base. Consequently, it is reasonable that

the lateral surface area of the cone becomes $\frac{1}{2} \cdot 2\pi r \cdot \ell$, or $\pi r\ell$. To find the total surface

area of the cone, we add πr^2, the area of the base. Thus, $S.A. = \pi r^2 + \pi r\ell$, as given in Figure 13-69(b).

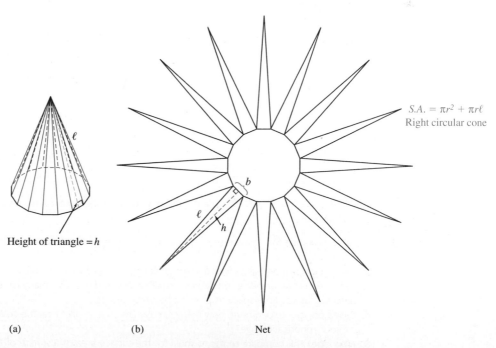

$S.A. = \pi r^2 + \pi r\ell$
Right circular cone

Height of triangle $= h$

(a)   (b)   Net

Figure 13-69

Example 13-22   Given the cone in Figure 13-70, find the surface area of that cone.

5 cm

4 cm

3 cm

Right circular cone

Figure 13-70

Solution   The base of the cone is a circle with radius 3 cm and area $\pi(3 \text{ cm})^2$, or $9\pi \text{ cm}^2$. The lateral surface has area $\pi(3 \text{ cm})(5 \text{ cm})$, or $15\pi \text{ cm}^2$. Thus we have the following surface area:

$$\begin{aligned} S.A. &= \pi(3 \text{ cm})^2 + \pi(3 \text{ cm})(5 \text{ cm}) \\ &= 9\pi \text{ cm}^2 + 15\pi \text{ cm}^2 \\ &= 24\pi \text{ cm}^2 \end{aligned}$$

Surface Area of a Sphere

A **great circle** of a sphere is a circle on the sphere whose radius is equal to the radius of the sphere. A great circle is obtained from the intersection of a plane through the center of the sphere and the sphere. Because there are infinitely many different planes through the center, there are infinitely many great circles on a sphere. However, they are all congruent. Finding a formula for the surface area of a sphere is a simple task using calculus, but it is not easy using only elementary mathematics. It can be shown that the surface area of a sphere is 4 times the area of a great circle of the sphere. Therefore, the formula for the surface area of a sphere is $S.A. = 4\pi r^2$, as given in Figure 13-71.

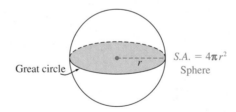

Great circle

r

$S.A. = 4\pi r^2$

Sphere

Figure 13-71

LABORATORY ACTIVITY   The following activity can be used to motivate the formula for the surface area of a sphere.

1. Take an orange that is as close to spherical as you can find and determine its radius. (This will be approximate.)
2. Take a compass and draw four disjoint circles that have the same radius as your orange.
3. Peel the orange, breaking the skin into small pieces, and see how many circles can be filled with the pieces. Leave as little space as possible between pieces.
4. Based on your experiment, how many circular areas ($A = \pi r^2$) can be covered by the surface (skin) of your spherical orange?
5. How does this compare to the formula for the surface area of a sphere of the same radius?

Assessment 13-4A

1. Which of these nets could be folded along the dotted segments to form a cube?

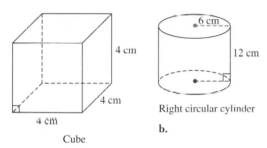

   a.   b.

   c.   d.

2. Find the surface area of each of the following:

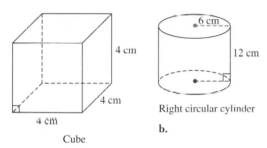

4 cm

4 cm

4 cm

Cube

a.

6 cm

12 cm

Right circular cylinder

b.

6 cm

5 cm

8 cm

Right rectangular prism

c.

4 cm

Sphere

d.

8 cm

6 cm

Right circular cone

e.

3. How many liters of paint are needed to paint the walls of a rectangular prism-shaped room that is 6 m × 4 m with a ceiling height of 2.5 m if 1 L of paint covers 20 m²? (Assume there are no doors or windows and paint comes in 1-L cans.)

4. The napkin ring pictured in the following figure is to be re-silvered. How many square millimeters must be covered?

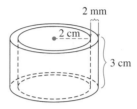

2 mm

2 cm

3 cm

5. Assume the radius of Earth is 6370 km and Earth is a sphere. What is its surface area?

6. Two cubes have sides of length 4 cm and 6 cm, respectively. What is the ratio of their surface areas?

7. The base of a right pyramid is a regular hexagon with sides of length 12 m. The altitude of the pyramid is 9 m. Find the total surface area of the pyramid.

8. A soup can has a $2\frac{5}{8}$ in. diameter and is 4 in. tall. What is the area of the paper that will be used to make the label for the can if the paper covers the entire lateral surface area?

9. A square piece of paper 10 cm on a side is rolled to form the lateral surface area of a right circular cylinder and then a top and bottom are added. What is the surface area of the cylinder?

10. The top of a right rectangular box has an area of 88 cm². The sides have areas 32 cm² and 44 cm². What are the dimensions of the box?

11. How does the lateral surface area of a right circular cone change if
    a. the slant height is tripled but the radius of the base remains the same?
    b. the radius of the base is tripled but the slant height remains the same?
    c. the slant height and the radius of the base are multiplied by 3?

12. Find the surface area of a right square pyramid if the area of the base is 100 cm² and the height of the pyramid is 20 cm.

13. A sector of a circle can be used to construct a right circular cone. The length of the arc of the sector becomes the circumference of the circular base of the cone.
    a. If the length of the arc is 6π, what is the radius of the base of the cone that can be constructed?
    b. In part (a), the radius of the sector is 5 units; what is the slant height of the cone that can be constructed?
    c. Using the information in parts (a) and (b), what is the height of the cone that can be constructed?
    d. Using the information in parts (a)–(c), what is the angle measure for the original sector?

14. The sector shown in the following figure is rolled into a cone so that the dashed edges just touch. Find the following:

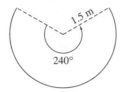

a. The lateral surface area of the cone
b. The total surface area of the cone

15. If the cardboard tube of a toilet paper roll has diameter of 2.5 in. and is 4 in. tall, what is the lateral surface area of the cardboard roll?

16. If two right circular cones are similar with radii of the bases in the ratio 1 : 2, what is the ratio of their surface areas?

17. If two cubes have total surface areas of 64 in.² and 36 in.², what is the ratio of their edges?

18. The total surface area of a cube is 10,648 cm². What is the length of each of the following?
a. One of the sides
b. A diagonal that is not a diagonal of a face

★ **19.** Find the total surface area of the following stand, which was cut from a right circular cone:

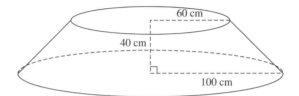

Assessment 13-4B

1. Which of the following nets could be folded to form a rectangular prism?

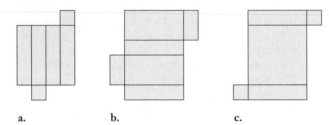

a.   **b.**   **c.**

2. Find the surface area of each of the following:

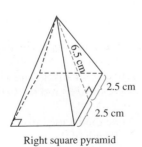

Right square pyramid

a.

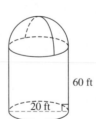

b. Right circular cylinder topped with hemisphere

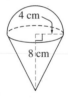

c. Right circular cone topped with hemisphere

3. How many liters of paint are needed to paint the walls of a rectangular prism-shaped room that is 8 m × 5 m with a ceiling height of 2.5 m if 1 L of paint covers 20 m²? (Assume there are no doors or windows and paint is sold in 1-L cans.)

4. Suppose one right circular cylinder has radius 2 m and height 6 m and another has radius 6 m and height 2 m.
a. Which cylinder has the greater lateral surface area?
b. Which cylinder has the greater total surface area?

5. Approximately how much material is needed to make the tent illustrated in the following figure (both ends and the bottom should be included)?

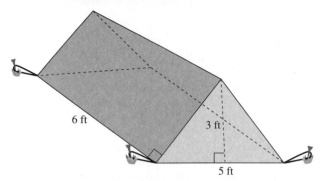

6. How does the surface area of a right rectangular box (including top and bottom) change if
   a. each dimension is doubled?
   b. each dimension is tripled?
   c. each dimension is multiplied by a factor of k?

7. What happens to the surface area of a sphere if the radius is
   a. doubled?   **b.** tripled?

8. Suppose a structure is composed of unit cubes with at least one face of each cube connected to the face of another cube, as shown in the following figure:
   a. If one cube is added, what is the maximum surface area the structure can have?

   b. If one cube is added, what is the minimum surface area the structure can have?
   c. Is it possible to design a structure so that one can add a cube and yet add nothing to the surface area of the structure? Explain your answer.

9. Suppose a paper cup is a frustum of a cone (that is, the cone is truncated).
   a. If the paper cup is rolled in a plane, with the paths of the top and bottom of the cup traced, what figure is formed?
   b. Describe the measures of the figure formed in relation to the original cup.

10. The region in each of the following figures revolves about the indicated axis. For each case, sketch the three-dimensional figure obtained and find its surface area.

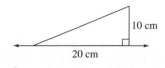

a.

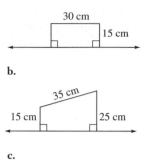

b.

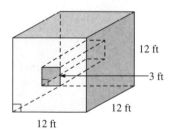

c.

11. A right square pyramid and a right circular cone are inscribed in a cube as shown. Find each of the following:
   a. Surface area of the pyramid
   b. Surface area of the cone

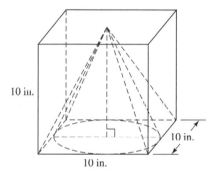

12. The diameter of Jupiter is about 11 times as great as the diameter of Earth. How do the surface areas compare?

13. Find the surface area of the figure formed by removing a section of the cube as shown below.

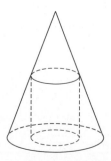

★ **14.** In the following figure, a right circular cylinder is inscribed in a right circular cone. Find the lateral surface area of the cylinder if the height of the cone is 40 cm, the height of the cylinder 30 cm, and the radius of the base of the cone is 25 cm.

Mathematical Connections 13-4

Communication

1. A student wonders if she doubles each measurement of a cereal box, will she need twice as much cardboard to make the new box. How would you help her decide?

2. Tennis balls are packed tightly three to a can that is shaped like a cylinder.
   a. Estimate how the surface area of the balls compares to the lateral surface area of the can. Explain how you arrived at your estimate.
   b. See how close your estimate in (a) was by actually computing the surface area of the balls and the lateral surface area of the can.

3. Which has the greater effect on the surface area of a right circular cylinder—doubling the radius or doubling the height? Explain.

4. In the drawing below, cube B was cut from a larger cube of surface area 216 cm^2 resulting in Figure A. The surface area of cube B is 24 cm^2. What effect did removing cube B from cube A have on the surface area? Explain.

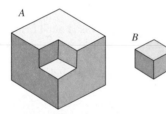

Open-Ended

5. One method of estimating body surface area in burn victims uses the fact that 100 handprints will approximately cover the whole body.
   a. What percentage of the body surface area is the surface area of two handprints?
   b. Estimate the percentage of the body surface area of one arm. Explain how you arrived at your estimate.
   c. Estimate your body surface area in square centimeters. Explain how you arrived at your estimate.
   d. Find the area of the flat part of your desk. How does the area of the desk compare with the surface area of your body?

6. Draw a net for a right prism that has surface area 80 cm^2.

Cooperative Learning

7. a. Shawn used small cubes to build a bigger cube that was solid and was three cubes long on each side. He then painted all the faces of the new, large cube red. He dropped the newly painted cube and all the little cubes came apart. He noticed that some cubes had only one face painted, some had two faces painted, and so on. Describe the number of cubes with 0, 1, 2, 3, 4, 5, or 6 faces painted. Have each member of the group choose a different number of sides and then combine your data to see if it makes sense. Look for any patterns that occur.
   b. What would the answers be if the large cube was four small cubes long on a side?
   c. Make a conjecture about how to count the cubes if the large cube were n small cubes long on a side.

Questions from the Classroom

8. Jodi says that if you double the radius of a right circular cone and divide the slant height by 2, then the surface area of the cone stays the same since the 2s cancel each other out. How do you respond?

9. Abi used the formula $S.A. = \pi r(r + \ell)$, where r is the radius and ℓ is the slant height, to find the surface area of a right circular cone. Is she correct? What will you tell her?

10. Jan says that if you double each of the dimensions of a rectangular box, it will take twice as much wrapping paper to wrap it. How do you respond?

Review Problems

11. Complete each of the following:
    a. 10 m^2 = ____ cm^2
    b. 13,680 cm^2 = ____ m^2
    c. 5 cm^2 = ____ mm^2
    d. 2 km^2 = ____ m^2
    e. 10^6 m^2 = ____ km^2
    f. 10^{12} mm^2 = ____ m^2

12. The sides of a rectangle are 10 cm and 20 cm long. Find the length of a diagonal of the rectangle.

13. The length of the side of a rhombus is 30 cm. If the length of one diagonal is 40 cm, find the length of the other diagonal.

14. Find the perimeters and the areas of the following figures:

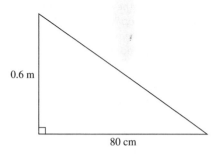

0.6 m

80 cm

a.

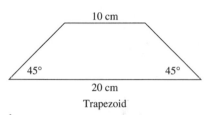

10 cm

45°   45°

20 cm

Trapezoid

b.

15. In the following, the length of the diagonal $\overline{AC}$ of rhombus $ABCD$ is 40 cm; $AE = 24$ cm. Find the length of a side of the rhombus and the length of the diagonal $\overline{BD}$.

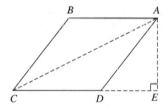

National Assessment of Educational Progress (NAEP) Question

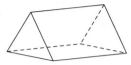

Which of the following can be folded to form the prism above?

a.

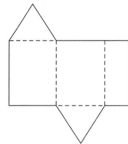

b.

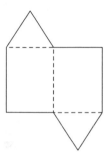

c.

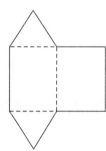

d.

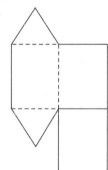

e.

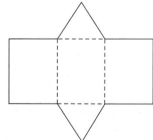

NAEP, Grade 8, 2005

TECHNOLOGY CORNER Find various-shaped cardboard containers and cut them apart to form nets. Make sure the nets are smaller than your computer screen.

a. Estimate the surface area of the nets.
b. Trace the perimeter of each net on a transparency sheet and hang the sheet on the monitor of your computer. Use *GSP* to trace the outline from the transparency and compute the area.
c. Compare your estimate in (a) with the answer in (b).

BRAIN TEASER A manufacturer of paper cups wants to produce paper cups in the form of truncated cones 16 cm high, with one circular base of radius 11 cm and the other of radius 7 cm, as shown in Figure 13-72. When the base of such a cup is removed and the cup is slit and flattened, the flattened region looks like a part of a circular ring. To design a pattern to make the cup, the manufacturer needs the data required to construct the flattened region. Find these data.

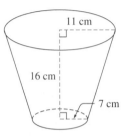

Figure 13-72

13-5   Volume, Mass, and Temperature

In Section 13-4, we investigated surface areas of various-shaped containers. In this section, we explore how much the containers will hold. This distinction is sometimes confused by elementary school students. Whereas surface area is the number of square units covering a three-dimensional figure, volume describes how much space a three-dimensional figure contains. The unit of measure for volume must be a shape that tessellates space. Cubes tessellate space; that is, they can be stacked so that they leave no gaps and fill space. Standard units of volume are based on cubes and are *cubic units*. A cubic unit is the amount of space enclosed within a cube that measures 1 unit on a side. The distinction between surface area and volume is demonstrated in Figure 13-73.

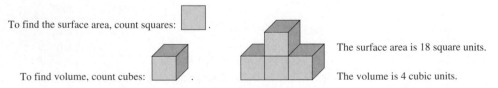

Figure 13-73

NOW TRY THIS 13-18 In Figure 13-74, the purple block is moved from one position to another.

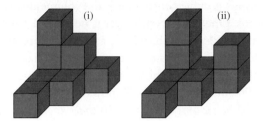

Figure 13-74

a. How does the volume in (i) compare with the volume in (ii)?
b. Do (i) and (ii) have the same surface area? If not, which has the greater surface area?
c. Find the surface area of each figure.

Volume of Right Rectangular Prisms

In the grade 7 *Focal Points* we find the following:

> By decomposing two- and three-dimensional shapes into smaller, component shapes, students find surface areas and develop and justify formulas for the surface areas and volumes of prisms and cylinders. As students decompose prisms and cylinders by slicing them, they develop and understand formulas for their volumes (*Volume = Area of base × Height*). They apply these formulas in problem solving to determine volumes of prisms and cylinders. (p. 19)

The quote from the *Focal Points* assumes the prisms and cylinders are right prisms and right circular cylinders. With this in mind, we investigate how to find the volume of a prism by using component shapes.

The volume of a right rectangular prism can be found by determining how many cubes are needed to build it as a solid. To find the volume, count how many cubes cover the base and then how many layers of these cubes are used to fill the prism. As shown in Figure 13-75(a), there are $8 \cdot 4$, or 32, cubes required to cover the base and there are five such layers. The volume of the rectangular prism is $(8 \cdot 4)5$, or 160 cubic units. For any right rectangular prism with dimension ℓ, w, and h measured in the same linear units, the volume of the prism is given by the area of the base, ℓw, times the height, h, or $V = \ell w h$, as shown in Figure 13-75(b).

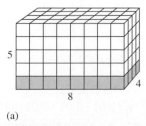

5

8

4

(a)

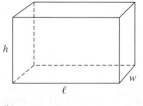

h

ℓ

w

$V = \ell w h$
Right rectangular prism

(b)

Figure 13-75

Converting Metric Measures of Volume

The most commonly used metric units of volume are the **cubic centimeter** and the **cubic meter**. A cubic centimeter is the volume of a cube whose length, width, and height are each 1 cm. One cubic centimeter is denoted 1 cm^3. Similarly, a cubic meter is the volume of a cube whose length, width, and height are each 1 m. One cubic meter is denoted 1 m^3. Other metric units of volume are symbolized similarly.

Figure 13-76 shows that since 1 dm = 10 cm, 1 dm^3 = (10 cm)(10 cm)(10 cm) = 1000 cm^3. Figure 13-77 shows that 1 m^3 = 1,000,000 cm^3 and that 1 dm^3 = 0.001 m^3. *Each metric unit of length is 10 times as great as the next smaller unit. Each metric unit of area is 100 times as great as the next smaller unit. Each metric unit of volume is 1000 times as great as the next smaller unit.* For example:

$$1 \text{ cm} = 10 \text{ mm}$$
$$1 \text{ cm}^2 = 100 \text{ mm}^2$$
$$1 \text{ cm}^3 = 1000 \text{ mm}^3$$

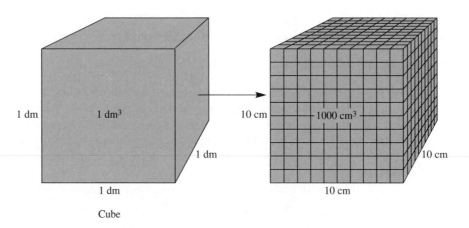

Figure 13-76

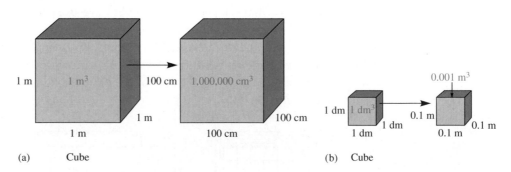

Figure 13-77

Because 1 cm = 0.01 m, then 1 cm^3 = (0.01 · 0.01 · 0.01) m^3, or 0.000001 m^3. To convert from cubic centimeters to cubic meters, we move the decimal point six places to the left.

NOW TRY THIS 13-19

a. Describe how to determine how many places to move the decimal point and in what direction in a metric area conversion if you know how many places and the direction to move in the corresponding length conversion.

b. Describe how to determine how many places to move the decimal point and in what direction in a metric volume conversion if you know how many places and the direction to move in the corresponding length conversion.

Example 13-23   Complete each of the following:

a. $5\,m^3 = $ _____ cm^3

b. $12{,}300\,mm^3 = $ _____ cm^3

Solution   **a.** $1\,m = 100\,cm$, so $1\,m^3 = (100\,cm)(100\,cm)(100\,cm)$, or $1{,}000{,}000\,cm^3$. Thus, $5\,m^3 = (5)(1{,}000{,}000\,cm^3) = 5{,}000{,}000\,cm^3$.

  b. $1\,mm = 0.1\,cm$, so $1\,mm^3 = (0.1\,cm)(0.1\,cm)(0.1\,cm)$, or $0.001\,cm^3$. Thus, $12{,}300\,mm^3 = 12{,}300(0.001\,cm^3) = 12.3\,cm^3$.

In the metric system, cubic units may be used for either dry or liquid measure, although units such as liters and milliliters are usually used for liquid measures. By definition, a **liter**, symbolized L, equals, or is the capacity of, a cubic decimeter; that is, $1\,L = 1\,dm^3$. (In the United States, L is the symbol for liter, but this is not universally accepted.)

Because $1\,L = 1\,dm^3$ and $1\,dm^3 = 1000\,cm^3$, it follows that $1\,L = 1000\,cm^3$ and $1\,cm^3 = 0.001\,L$. Also, $0.001\,L = 1\,milliliter = 1\,mL$. Hence, $1\,cm^3 = 1\,mL$. These relationships are summarized in Figure 13-78 and Table 13-7.

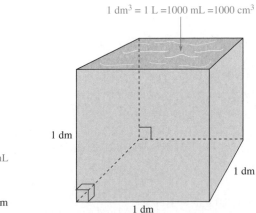

Cube

Cube

Figure 13-78

Table 13-7

Unit	Symbol	Relation to Liter
kiloliter	kL	1000 L
*hectoliter	hL	100 L
*dekaliter	daL	10 L
liter	**L**	**1 L**
*deciliter	dL	0.1 L
centiliter	cL	0.01 L
milliliter	mL	0.001 L

*Not commonly used

Example 13-24

Complete the following:

a. 27 L = _____ mL
b. 362 mL = _____ L
c. 3 mL = _____ cm^3
d. 3 m^3 = _____ L

Solution   **a.** 1 L = 1000 mL, so 27 L = 27 · 1000 mL = 27,000 mL. Alternatively, 27 L =

$$27 \text{ L} \cdot \frac{1000 \text{ mL}}{1 \text{ L}} = 27,000 \text{ mL}$$

b. 1 mL = 0.001 L, so 362 mL = 362(0.001 L) = 0.362 L

c. 1 mL = 1 cm^3, so 3 mL = 3 cm^3

d. 1 m^3 = 1000 dm^3 and 1 dm^3 = 1 L, so 1 m^3 = 1000 L and 3 m^3 = 3000 L

Alternatively, $3 \text{ m}^3 = 3 \text{ m}^3 \cdot \frac{1000 \text{ dm}^3}{1 \text{ m}^3} \cdot \frac{1 \text{ L}}{1 \text{ dm}^3} = 3000 \text{ L}.$

Converting English Measures of Volume

Basic units of volume in the English system are the cubic foot (1 ft^3), the cubic yard (1 yd^3), and the cubic inch (1 in.^{3}). In the United States, 1 gal = 231 in.^{3}, which is about 3.8 L, and 1 qt = $\frac{1}{4}$ gal, or about 58 in.^{3}.

Relationships among the one-dimensional units enable us to convert from one unit of volume to another, as shown in the following example.

Example 13-25

Convert each of the following, as indicated:

a. 45 yd^3 = _____ ft^3
b. 4320 in.^{3} = _____ yd^3
c. 10 gal = _____ ft^3
d. 3 ft^3 = _____ yd^3

Solution   **a.** Because 1 yd^3 = (3 ft)^3 = 27 ft^3, 45 yd^3 = 45 · 27 ft^3, or 1215 ft^3.

b. Because 1 in. = $\frac{1}{36}$ yd, 1 in.^{3} = $\left(\frac{1}{36}\right)^3$ yd^3. Consequently, 4320 in.^{3} =

$4320\left(\frac{1}{36}\right)^3$ yd^3 ≈ 0.0926 yd^3, or approximately 0.1 yd^3.

c. Because 1 gal = 231 in.^{3} and 1 in.^{3} = $\left(\frac{1}{12}\right)^3$ ft^3, 10 gal = 2310 in.^{3} =

$2310\left(\frac{1}{12}\right)^3$ ft^3 ≈ 1.337 ft^3, or approximately 1.3 ft^3.

Alternatively, $10 \text{ gal} = 10 \text{ gal} \cdot \frac{231 \text{ in.}^3}{1 \text{ gal}} \cdot \frac{\left(\frac{1}{12}\right)^3 \text{ ft}^3}{1 \text{ in.}^3} ≈ 1.3 \text{ ft}^3$

d. From (a), 1 ft^3 = $\frac{1}{27}$ yd^3. Hence, 3 ft^3 = 3 · $\frac{1}{27}$ yd^3 = $\frac{1}{9}$ yd^3 ≈ 0.1 yd^3.

Students typically depend on physical techniques for determining volumes of objects that may lead to errors in other situations. For example, counting the number of cubes to fill a box to find the volume may lead to problems when students are presented with a two-dimensional image of a box and asked to find its volume. The counting technique also is lost when the dimensions of the box are given in fractions or decimals (Hart 1981). ◆

Volumes of Prisms and Cylinders

As pointed out in the Research Note, students need practice finding volumes of prisms shown as two-dimensional images. We have shown that the volume of a right rectangular prism, as shown in Figure 13-79, involves multiplying the area of the base times the height. If we denote the area of the base by B and the height by h, then $V = Bh$.

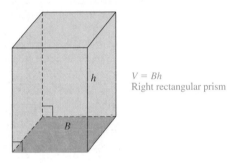

$V = Bh$
Right rectangular prism

Figure 13-79

Formulas for the volumes of many three-dimensional figures can be derived using the volume of a right prism. In Figure 13-80(a), a rectangular solid box has been sliced into thin layers. If the layers are shifted to form the solids in Figure 13-80(b) and (c), the volume of each of the three solids is the same as the volume of the original solid. This idea is the basis for **Cavalieri's Principle**.

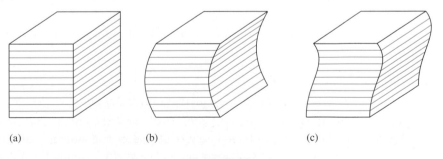

(a)   (b)   (c)

Figure 13-80

Cavalieri's Principle

Two solids each with a base in the same plane have equal volumes if every plane parallel to the bases intersects the solids in cross sections of equal area.

NOW TRY THIS 13-20

a. The two right prisms in Figure 13-81 have the same height. How do their volumes compare? Explain why.

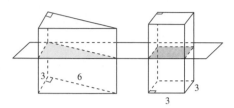

Figure 13-81

b. Consider the right prism and right circular cylinder in Figure 13-82(a) and (c) as stacks of papers. If the papers are shifted as shown in Figure 13-82(b) and (d), an oblique prism and an oblique cylinder, respectively, are formed.
  (i) Explain how the volume of the oblique prism is related to the volume of the right prism.
  (ii) Explain how the volume of the oblique cylinder is related to the volume of the right cylinder.

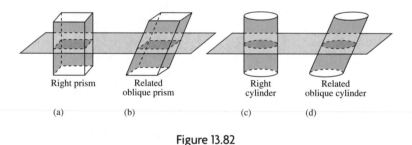

| Right prism | Related oblique prism | Right cylinder | Related oblique cylinder |
| (a) | (b) | (c) | (d) |

Figure 13.82

The volume of a cylinder can be approximated using prisms with increasing numbers of sides in their bases. The volume of each prism is the product of the area of the base and the height. Similarly, the **volume, V, of a cylinder** is the product of the area of the base B and the height h. If the base is a circle of radius r, and the height of the cylinder is h, then $V = Bh = \pi r^2 h$.

Historical Note

Bonaventura Cavalieri (1598–1647), an Italian mathematician and disciple of Galileo, contributed to the development of geometry, trigonometry, and algebra in the Renaissance. He became a Jesuit at an early age and later, after reading Euclid's *Elements*, was inspired to study mathematics. In 1629, Cavalieri became a professor at Bologna and held that post until his death. Cavalieri is best known for his principle concerning the volumes of solids.   ♦

Example 13-26   Find the volume of each figure in Figure 13-83.

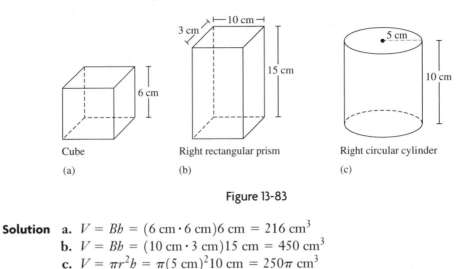

Figure 13-83

Solution a. $V = Bh = (6 \text{ cm} \cdot 6 \text{ cm})6 \text{ cm} = 216 \text{ cm}^3$
b. $V = Bh = (10 \text{ cm} \cdot 3 \text{ cm})15 \text{ cm} = 450 \text{ cm}^3$
c. $V = \pi r^2 h = \pi (5 \text{ cm})^2 10 \text{ cm} = 250\pi \text{ cm}^3$

Volumes of Pyramids and Cones

Figure 13-84(a) shows a right prism and a right pyramid with congruent bases and equal heights.

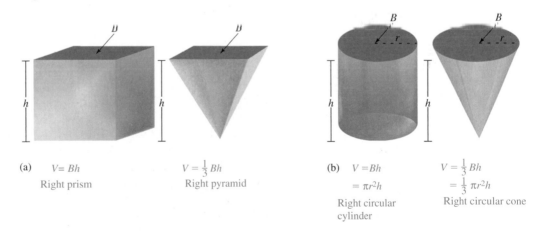

(a) $V = Bh$
Right prism

$V = \frac{1}{3} Bh$
Right pyramid

(b) $V = Bh$
$= \pi r^2 h$
Right circular cylinder

$V = \frac{1}{3} Bh$
$= \frac{1}{3} \pi r^2 h$
Right circular cone

Figure 13-84

How do we know how the volumes of the containers in Figure 13-84 are related? Students may explore the relationship by filling the pyramid with water, sand, or rice and pouring the contents into the prism. They should find that it takes three full pyramids to fill the prism. Therefore the volume of the pyramid is equal to one-third the volume of the prism. This relationship between prisms and pyramids with congruent bases and heights, respectively, is true in general; that is, for a pyramid $V = (1/3)Bh$, where B is the area of the base and h is the height. The same relationship holds between the volume of a cone and the volume of a cylinder, where they have congruent bases and equal heights, as shown in Figure 13-84(b). Therefore, the volume of a right circular cone is given by $V = (1/3)Bh$, or $V = (1/3)\pi r^2 h$.

This relationship between the volume of a cone and the volume of a cylinder with congruent bases and congruent heights is explored in Problem 4.2 on the student page on page 942. Answer parts A–D.

Another way to determine the area of a pyramid in terms of a prism is to start with a cube and three diagonals from one vertex drawn to other vertices of the "opposite face," as shown in Figure 13-85(a). We can see that there are three pyramids formed inside the cube, as shown in Figure 13-85(b), (c), and (d).

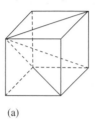
(a)

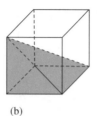
(b)

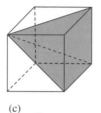
(c)

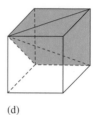
(d)

Figure 13-85

The three pyramids are identical in size and shape, do not overlap, and their union is the whole cube. Therefore, each of these pyramids has a volume one-third that of the cube. This result is true in general, and once again we see that for a pyramid $V = (1/3)Bh$, where B is the area of the base and h is the height. This can be demonstrated by building three paper models of the pyramids and fitting them together into a prism. A net that can be enlarged and used for the construction is given in Figure 13-86.

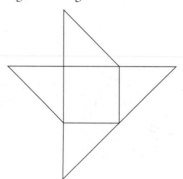

Figure 13-86

Example 13-27 Find the volume of each figure in Figure 13-87.

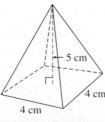

Right square pyramid

(a)

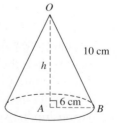

Right circular cone

(b)

Figure 13-87

School Book Page   CONES AND CYLINDERS

Problem 4.2 Cones and Cylinders, Pyramids and Cubes

- Roll a piece of stiff paper into a cone shape so that the tip touches the bottom of the cylinder you made in Problem 4.1.

- Tape the cone shape along the seam. Trim the cone so that it is the same height as the cylinder.

- Fill the cone to the top with sand or rice, and empty the contents into the cylinder. Repeat this as many times as needed to fill the cylinder completely.

A. What is the relationship between the volume of the cone and the volume of the cylinder?

B. Suppose a cylinder, a cone, and a sphere have the same radius and the same height. What is the relationship between the volumes of the three shapes?

C. Suppose a cone, a cylinder, and a sphere all have the same height, and that the cylinder has a volume of 64 cubic inches. How do you use the relationship in Question B to find

    1. the volume of a sphere whose radius is the same as the cylinder?

    2. the volume of a cone whose radius is the same as the cylinder?

D. Suppose the radius of a cylinder, a cone, and a sphere is 5 centimeters and the height of the cylinder and cone is 8 centimeters. Find the volume of the cylinder, cone, and sphere.

Investigation 4 Cones, Spheres, and Pyramids   **51**

Source: Filling and Wrapping, Connected Mathematics 2; Prentice Hall, 2006, (p. 51).

Solution   **a.** The figure is a pyramid with a square base, whose area is 4 cm · 4 cm and whose height is 5 cm. Hence, $V = \frac{1}{3}Bh = \frac{1}{3}(4 \text{ cm} \cdot 4 \text{ cm})(5 \text{ cm}) = \frac{80}{3} \text{ cm}^3$.

         b. The base of the cone is a circle of radius 6 cm. Because the volume of the cone is given by $V = \frac{1}{3}\pi r^2 h$, we need to know the height. In the right triangle OAB, $OA = h$ and by the Pythagorean theorem, $h^2 + 6^2 = 10^2$. Hence, $h^2 = 100 - 36$, or 64, and $h = 8$ cm. Thus, $V = \frac{1}{3}\pi r^2 h = \frac{1}{3}\pi(6 \text{ cm})^2(8 \text{ cm}) = 96\pi \text{ cm}^3$.

Example 13-28   Figure 13-88 is a net for a pyramid. If each triangle is equilateral, find the volume of the pyramid.

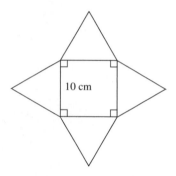

10 cm

Figure 13-88

Solution   The pyramid obtained from the folded model is shown in Figure 13-89. The volume of the pyramid is $V = \frac{1}{3}Bh = \frac{1}{3} \cdot 10^2 h$. We must find h. Notice that h is a leg in the right triangle EOF, where F is the midpoint of $\overline{CB}$. We know that $OF = 5$ cm. If we knew EF, we could find h by applying the Pythagorean theorem to $\triangle EOF$. To find the length of $\overline{EF}$, notice that $\overline{EF}$ is a leg in the right triangle EBF. ($\overline{EF}$ is the perpendicular bisector of $\overline{BC}$ in the equilateral triangle BEC.) In the right triangle EBF, we have $(EB)^2 = (BF)^2 + (EF)^2$. Because $EB = 10$ cm and $BF = 5$ cm, it follows that $10^2 = 5^2 + (EF)^2$, or $EF = \sqrt{75}$ cm $\approx$ 8.66 cm. In $\triangle EOF$, we have $h^2 + 5^2 = (EF)^2$, or $h^2 + 25 = 75$. Thus, $h = \sqrt{50}$ cm $\approx$ 7.07 cm, and $V \approx \frac{1}{3} \cdot 10^2 \cdot 7.07 \approx 235.7$ cm^3.

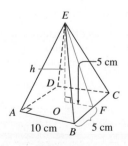

Figure 13-89

Volume of a Sphere

To find the volume of a sphere, imagine that a sphere is composed of a great number of right regular congruent pyramids with apexes at the center of the sphere and that the vertices of the base touch the sphere, as shown in Figure 13-90. If the pyramids have very small bases, then the height of each pyramid is nearly the radius r. Hence, the volume of each pyramid is $\frac{1}{3}Bh$ or $\frac{1}{3}Br$, where B is the area of the base. If there are n pyramids each with base area B, then the total volume of the pyramids is $V = \frac{1}{3}nBr$. Because nB is the total surface area of all the bases of the pyramids and because the sum of the areas of all the bases of the pyramids is very close to the surface area of the sphere, $4\pi r^2$, the volume of the sphere is given by $V = \frac{1}{3}(4\pi r^2)r = \frac{4}{3}\pi r^3$.

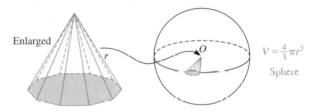

Figure 13-90

Example 13-29   Find the volume of a sphere whose radius is 6 cm.

Solution $V = \dfrac{4}{3}\pi(6 \text{ cm})^3 = \dfrac{4}{3}\pi(216 \text{ cm}^3) = 288\pi \text{ cm}^3$

NOW TRY THIS 13-21 A cylinder, a cone, and a sphere have the same radius and same height, as shown in Figure 13-91.

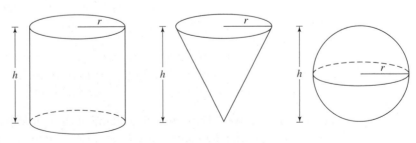

Figure 13-91

a. Find the volume of each figure in terms of r and h.
b. Use your answer to part (a) to show that the volumes are in the ratio $3 : 1 : 2$.

Problem Solving   Volume Comparisons: Cylinders and Boxes

A metal can manufacturer has a large quantity of rectangular metal sheets 20 cm × 30 cm. Without cutting the sheets, the manufacturer wants to make cylindrical pipes with circular cross sections from some of the sheets and box-shaped pipes with square cross sections from the other sheets. The volume of the box-shaped pipes is to be greater than the volume of the cylindrical pipes. Is this possible? If so, how would the pipes be made and what are their volumes?

Understanding the Problem   We are to use 20 cm × 30 cm rectangular sheets of metal to make some cylindrical pipes as well as some box-shaped pipes with square cross sections that have a greater volume than do the cylindrical pipes. Is this possible, and if so, how should the pipes be designed and what are their volumes?

Figure 13-92 shows a sheet of metal and two sections of pipe made from it, one cylindrical and the other box-shaped. A model for such pipes can be designed from a piece of paper by bending it into a right circular cylinder or by folding it into a right rectangular prism, as shown in the figure.

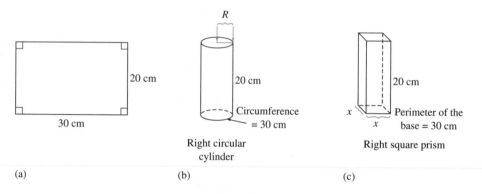

(a)   (b)   (c)

Figure 13-92

Devising a Plan   If we compute the volume of the cylinder in Figure 13-92(b) and the volume of the prism in Figure 13-92(c), we can determine which is greater. If the prism has a greater volume, the solution of the problem will be complete. Otherwise, we look for other ways to design the pipes before concluding that a solution is impossible.

To compute the volume of the cylinder, we find the area of the base. The area of the circular base is πr^2. To find r, we note that the circumference of the circle $2\pi r$ is 30 cm. Thus, $r = \frac{30}{2\pi} \approx 4.77$ cm, and the area of the circle is $\pi r^2 = \pi\left(\frac{30}{2\pi}\right)^2 = \frac{900\pi}{4\pi^2} = \frac{225}{\pi} \approx 71.62$ cm^2.

With the given information, we can also find the area of the base of the rectangular box. Because the perimeter of the base of the prism is $4x$, we have $4x = 30$, or $x = 7.5$ cm. Thus, the area of the square base is $x^2 = (7.5)^2$, or 56.25 cm^2.

Carrying Out the Plan   Denoting the volume of the cylindrical pipe by V_1 and the volume of the box-shaped pipe by V_2, we have $V_1 \approx 71.62 \cdot 20$, or approximately 1432.4 cm^3. For the volume of the box-shaped pipe, we have $V_2 = 56.25 \cdot 20$, or 1125 cm^3. We see that in the first design for the pipes, the volume of the cylindrical pipe is greater than the volume of the box-shaped pipe. This is not the required outcome.

Rather than bend the rectangular sheet of metal along the 30 cm side, we could bend it along the 20 cm side to obtain either pipe, as shown in Figure 13-93. Denoting the radius of the cylindrical pipe by r, the side of the box-shaped pipe by y, and their volumes by V_3 and V_4, respectively, we have $V_3 = \pi r^2 \cdot 30 = \pi[20/(2\pi)]^2 30 = (10^2 \cdot 30)/\pi$, or approximately

954.9 cm^3. Also, $V_4 = y^2 30 = \left(\dfrac{20}{4}\right)^2 30 = 25 \cdot 30$, or 750 cm^3. Because $V_2 = 1125$ cm^3 and $V_3 = 945.9$ cm^3, we see that the volume of the box-shaped pipe with an altitude of 20 cm is greater than the volume of the cylindrical pipe with an altitude of 30 cm. We could determine the order of volumes from greatest to least for the four shapes.

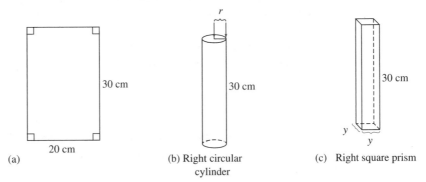

(a)   20 cm   (b) Right circular   (c) Right square prism
                                                cylinder

Figure 13-93

Looking Back   We could ask for the volumes of other three-dimensional objects that can be obtained by bending the rectangular sheets of metal. Also, because the lateral surface areas of the four types of pipes were the same but their volumes were different, we might want to investigate whether there are other cylinders and prisms that have the same lateral surface area and the same volume. Is it possible to find a circular cylinder with lateral surface area of 600 cm^2 and smallest possible volume? Similarly, is there a circular cylinder with the given surface area and greatest possible volume?

On the partial student page that follows, consider how to find the volume of the Great Plains tepee in Number 18.

✔ **QUESTION 18**

...checks that you can find the volume of a cone.

18 ✔ **CHECKPOINT** The shape of tepees used by Native American peoples on the Great Plains resembles a cone. A typical tepee might have stood 18 ft high and had a diameter of 15 ft at its base. Estimate the volume of such a tepee.

▶ **Most buildings are made up of a variety of space figures. The shape below is an alternate representation of a mat house. It is a triangular prism with half of a cone at either end.**

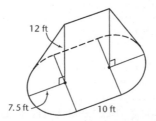

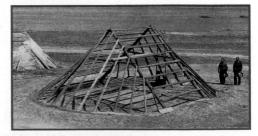

19 **Discussion** Describe how you would find the volume of a mat house shaped like the figure above. Then find the volume. Round your answer to the nearest tenth.

Source: Math Thematics, New Edition, Book 3, McDougal Littell, 2008 (p. 354).

Mass

Three centuries ago, Isaac Newton pointed out that in everyday life the word *weight* is used for what is really mass. *Mass* is a quantity of matter, as opposed to *weight*, which is a force exerted by gravitational pull. When astronauts are in orbit above Earth, their weights have changed even though their masses remain the same. In common parlance on Earth, *weight* and *mass* are used interchangeably. In the English system, weight is measured in avoirdupois units such as tons, pounds, and ounces. One pound (lb) equals 16 ounces (oz) and 2000 lb equals 1 English ton.

In the metric system, a fundamental unit for mass is the **gram**, denoted g. An ordinary paper clip or a thumbtack each has a mass of about 1 g. As with other base metric units, prefixes are added to *gram* to obtain other units. For example, a kilogram (kg) is 1000 g. Two standard loaves of bread have a mass of about 1 kg. A person's mass also is measured in kilograms. A newborn baby has a mass of about 3 kg. Another unit of mass is the metric ton (t), which is equal to 1000 kg. The metric ton is used to record the masses of objects such as cars and trucks. A small car has a mass of about 1 t. Mega (1,000,000) and micro (0.000001) are other prefixes used with gram.

Table 13-8 lists metric units of mass. Conversions that involve metric units of mass are handled in the same way as conversions that involve metric units of length.

Table 13-8

Unit	Symbol	Relationship to Gram
ton (metric)	t	1,000,000 g
kilogram	kg	1000 g
*hectogram	hg	100 g
*dekagram	dag	10 g
gram	**g**	**1 g**
*decigram	dg	0.1 g
*centigram	cg	0.01 g
milligram	mg	0.001 g

*Not commonly used

Example 13-30

Complete each of the following:

a. 34 g = _____ kg   **b.** 6836 kg = _____ t

Solution   **a.** 34 g = 34(0.001 kg) = 0.034 kg
            b. 6836 kg = 6836(0.001 t) = 6.836 t

Relationships Among Metric Units of Volume, Capacity, and Mass

The relationships among the units of volume, capacity, and mass in the metric system is illustrated in Figure 13-94. Does the English system have any relationships among units that are easy to remember?

1 cm³ (1 mL) of water   1 g

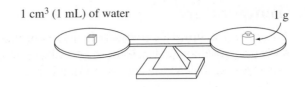

1 dm³ (1 L) of water

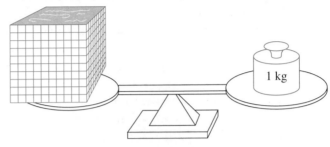

1 kg

Figure 13-94

NOW TRY THIS 13-22 Find the following relationships:

a. 1 cm³ of water has a mass of 1 _____.

b. 1 dm³ of water has a mass of 1 _____.

c. 1 L of water has a volume of 1 _____.

d. 1 cm³ of water has a capacity of 1 _____.

e. 1 mL of water has a mass of 1 _____.

f. 1 m³ of water has a capacity of 1 _____.

g. 1 m³ of water has a mass of 1 _____.

Example 13-31

A waterbed measures 180 cm × 210 cm × 20 cm.

a. Approximately how many liters of water can it hold?
b. What is its mass in kilograms when it is full of water?

Solution a. The volume of the waterbed is approximated by multiplying the length ℓ times the width w times the height h.

$$V = \ell wh$$
$$= 180 \text{ cm} \cdot 210 \text{ cm} \cdot 20 \text{ cm}$$
$$= 756{,}000 \text{ cm}^3, \text{ or } 756{,}000 \text{ mL}$$

Because 1 mL = 0.001 L, the volume is 756 L.

b. Because 1 L of water has a mass of 1 kg, 756 L of water has a mass of 756 kg, which is 0.756 t.

REMARK To see one advantage of the metric system, suppose the bed in Example 13-31 is 6 ft × 7 ft × 9 in. Try to approximate the volume in gallons and the weight of the water in pounds.

Temperature

For normal temperature measurements in the metric system, the base unit is the **degree Celsius**, named for Anders Celsius, a Swedish scientist. The Celsius scale has 100 equal divisions between 0 degrees Celsius (0°C), the freezing point of water, and 100 degrees Celsius (100°C), the boiling point of water, as seen in Figure 13-95. The **kelvin** (K) temperature scale is an extension of the degree Celsius scale down to *absolute zero*, a hypothetical temperature characterized by a complete absence of heat energy. The freezing point of water on this scale is 273.15 kelvins. In the English system, the Fahrenheit scale has 180 equal divisions between 32°F, the freezing point of water, and 212°F, the boiling point of water.

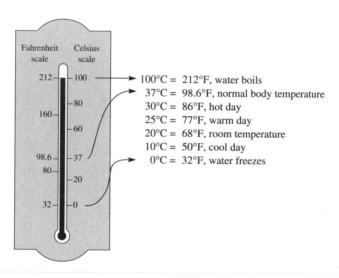

Figure 13-95

Figure 13-95 gives other temperature comparisons of the two scales and further illustrates the relationship between them. Because the Celsius scale has 100 divisions between the freezing point and the boiling point of water, whereas the Fahrenheit scale has 180 divisions, the relationship between the two scales is 100 to 180, or 5 to 9. For every 5 degrees on the Celsius scale, there are 9 degrees on the Fahrenheit scale, and for each degree on the Fahrenheit scale, there is $\frac{5}{9}$ degree on the Celsius scale. Because the ratio between the number of degrees above freezing on the Celsius scale and the number of degrees above freezing on the Fahrenheit scale remains the same and equals $\frac{5}{9}$, we may convert temperature from one system to the other.

For example, suppose we want to convert 50° on the Fahrenheit scale to the corresponding number on the Celsius scale. On the Fahrenheit scale, 50° is $50 - 32$, or 18°, above freezing, but on the Celsius scale, it is $\frac{5}{9} \cdot 18$, or 10°, above freezing. Because the freezing temperature on the Celsius scale is 0°, 10° above freezing is 10° Celsius. Thus, 50°F = 10°C. In general, F degrees is $F - 32$ above freezing on the Fahrenheit scale, but only $\frac{5}{9}(F - 32)$ above freezing on the Celsius scale. Thus, we have the relation $C = \frac{5}{9}(F - 32)$. If we solve the equation for F, we obtain $F = \frac{9}{5}C + 32$.

NOW TRY THIS 13-23 Does it ever happen that the temperature measured in Celsius degrees is the same if it is measured in Fahrenheit degrees? If so, when?

Assessment 13-5A

1. Complete each of the following:
   a. $8 \text{ m}^3 = $ _____ dm^3
   b. $675{,}000 \text{ m}^3 = $ _____ km^3
   c. $7000 \text{ mm}^3 = $ _____ cm^3
   d. $400 \text{ in.}^3 = $ _____ yd^3
   e. $0.2 \text{ ft}^3 = $ _____ in.^3

2. If a faucet is dripping at the rate of 15 drops/min and there are 20 drops/mL, how many liters of water are wasted in a 30-day month?

3. Find the volume of each of the following:

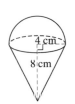

a. Right circular cone with hemisphere

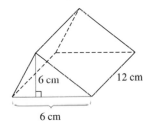

b. Right triangular prism

c. Right circular cone

d. Sphere

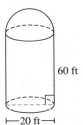

e. Right circular cylinder with hemisphere

4. Complete the following table:

	a.	b.	c.	d.	e.	f.
cm^3		500			750	4800
dm^3	2					
L			1.5			
mL				5000		

5. Place a decimal point in each of the following to make it an accurate sentence:
   a. A paper cup holds about 2000 mL.
   b. A regular soft-drink bottle holds about 320 L.
   c. A quart milk container holds about 10 L.
   d. A teaspoonful of cough syrup is about 500 mL

6. Two cubes have sides of lengths 4 cm and 6 cm, respectively. What is the ratio of their volumes?

7. What happens to the volume of a sphere if the radius is doubled?

8. Complete the following table for right rectangular prisms with the given dimensions:

	a.	b.	c.	d.
Length	20 cm	10 cm	2 dm	15 cm
Width	10 cm	2 dm	1 dm	2 dm
Height	10 cm	3 dm		
Volume (cm^3)				
Volume (dm^3)				7.5
Volume (L)			4	

9. Earth's diameter is approximately 4 times the Moon's and both bodies are spheres. What is the ratio of their volumes?

10. An Olympic-sized pool in the shape of a right rectangular prism is 50 m × 25 m. If it is 2 m deep throughout, how many liters of water does it hold?

11. A standard drinking straw is 25 cm long and 4 mm in diameter. How much liquid can be held in the straw at one time?

12. **a.** What happens to the volume of an aquarium that is in the shape of a right rectangular prism if the length, width, and height are all doubled?
   b. From your answer to (a), conjecture what happens to the volume of the aquarium if all the measurements are tripled.
   c. When you multiply each linear dimension of the aquarium by a positive value n, what happens to the volume?

13. The Great Pyramid of Cheops is a right square pyramid with height of 148 m and a square base with a perimeter of 940 m. The Transamerica Building in San Francisco has the basic shape of a right square pyramid that has a square base with a perimeter of 140 m and a height of 260 m. Which one has the greater volume and by how many times as great?

14. A right circular cone-shaped paper water cup has a height of 8 cm and a radius of 4 cm. If the cup is filled with water to half its height, what portion of the volume of the cup is filled with water?

15. If each edge of a cube is increased by 30%, by what percent does the volume increase?

16. One freezer measures 1.5 ft × 1.5 ft × 5 ft and sells for $350. Another freezer measures 2 ft × 2 ft × 4 ft and sells for $400. Which freezer is the better buy in terms of dollars per cubic foot?

17. A tennis ball can in the shape of a right circular cylinder holds three tennis balls snugly. If the radius of a tennis ball is 3.5 cm, what percentage of the tennis ball can is occupied by air?

18. A box is packed with six soda cans, as in the following figure. What percentage of the volume of the interior of the box is not occupied by the cans?

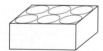

19. A right rectangular prism with base $ABCD$ as the bottom is shown in the following figure:

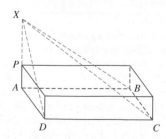

Suppose X is drawn so that $AX = 3AP$, where AP is the height of the prism, and X is connected to A, B, C, and D to form a pyramid. How do the volumes of the pyramid and the prism compare?

20. A right circular cylindrical can is to hold approximately 1 L of water. What should be the height of the can if the radius is 12 cm?

21. An engineer is to design a square-based pyramid whose volume is to be 100 m³.
   a. Find the dimensions (the length of a side of the square and the altitude) of one such pyramid.
   b. How many (noncongruent) such pyramids are possible? Why?

22. A square sheet of cardboard measuring y cm on a side is to be used to produce an open-top box when the maker cuts off a small square x cm × x cm from each corner and bends up the sides. Find the volume of the box if $y = 200$ and $x = 20$.

23. For each of the following, select the appropriate metric unit of measure (gram, kilogram, or metric ton):
   a. Car
   b. Adult
   c. Can of frozen orange juice
   d. Elephant

24. For each of the following, choose the correct unit (milligram, gram, or kilogram) to make each sentence reasonable:
   a. A staple has a mass of about 340 _____.
   b. A professional football player has a mass of about 110 _____.
   c. A vitamin tablet has a mass of about 1100 _____.

25. Complete each of the following:
   a. 15,000 g = _____ kg
   b. 0.036 kg = _____ g
   c. 4320 mg = _____ g
   d. 0.03 t = _____ kg
   e. 25 oz = _____ lb

26. A paper dollar has a mass of approximately 1 g. Is it possible to lift $1,000,000 in the following denominations:
   a. $1 bills   **b.** $10 bills
   c. $100 bills   **d.** $1000 bills
   e. $10,000 bills

27. A fish tank, which is a right rectangular prism, is 40 cm × 20 cm × 20 cm. If it is filled with water, what is the mass of the water?

28. Convert each of the following from degrees Fahrenheit to the nearest integer degree Celsius:
   a. 10°F
   b. 30°F
   c. 212°F

29. Answer each of the following:
   a. The thermometer reads 20°C. Can you go snow skiing?
   b. Your body temperature is 39°C. Are you ill?
   c. The temperature reads 35°C. Should you go water skiing?
   d. It's 30°C in the room. Are you comfortable, hot, or cold?

Assessment 13-5B

1. Complete each of the following:
   a. $500 \text{ cm}^3 = \underline{\hspace{1cm}} \text{ m}^3$
   b. $3 \text{ m}^3 = \underline{\hspace{1cm}} \text{ cm}^3$
   c. $0.002 \text{ m}^3 = \underline{\hspace{1cm}} \text{ cm}^3$
   d. $25 \text{ yd}^3 = \underline{\hspace{1cm}} \text{ ft}^3$
   e. $1200 \text{ in.}^3 = \underline{\hspace{1cm}} \text{ ft}^3$

2. Jeremy has a fish tank that has a 40 cm by 70 cm rectangular base. The water is 25 cm deep. When he drops rocks into the tank, the water goes up by 2 cm. What is the volume in liters of the rocks?

3. Maggie is planning to build a new one-story house with floor area of 2000 ft^2. She is thinking about putting in a 9 ft ceiling instead of an 8 ft ceiling. If she does this, how many more cubic feet of space will she have to heat and cool?

4. Find the volume of each of the following:

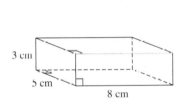

a. Right rectangular prism

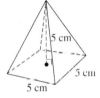

b. Right square pyramid

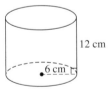

c. Right circular cylinder

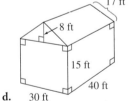

d.

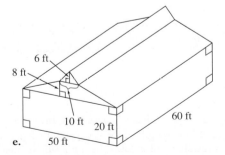

e.

5. Complete the following chart:

	a.	b.	c.	d.	e.	f.
cm³		200			202	6500
dm³	6					
L				3		
mL			1200			

6. Determine the volume of silver needed to make the napkin ring in the following figure out of solid silver. Give your answer in cubic millimeters.

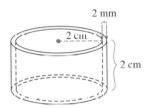

7. What happens to the volume of a sphere if the radius is tripled?

8. Complete the following chart for right rectangular prisms with the given dimensions:

	a.	b.	c.	d.
Length	5 cm	8 cm	2 dm	15 cm
Width	10 cm	6 dm	1 dm	2 dm
Height	20 cm	4 dm		
Volume (cm³)				
Volume (dm³)				12
Volume (L)			10	

9. Determine how many liters a right circular cylindrical tank holds if it is 6 m long and 13 m in diameter.

10. The Great Pyramid of Cheops has a square base of 771 ft on a side and a height of 486 ft. How many rooms 35 ft × 20 ft × 8 ft would be needed to have a volume equivalent to that of the Great Pyramid's?

11. A rectangular swimming pool with dimensions 10 m × 25 m is being built. The pool has a shallow end that is uniform in depth and a deep end that drops off as shown

in the following figure. What is the volume of this pool in cubic meters?

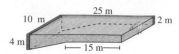

12. If 50 steel marbles that are 1 cm in diameter are melted down, will enough steel result to build a marble that is 4 cm in diameter? Explain.

13. A heavy metal sphere with radius 10 cm is dropped into a right circular cylinder with base radius of 10 cm. If the original cylinder has water in it that is 20 cm high, how high is the water after the sphere is placed in it?

14. What unit would you use to describe the volume of Earth? Justify your reasoning.

15. **a.** If two cubes have sides in the ratio 2 : 5, what is the ratio of their volumes?
    b. If two similar cones have heights in the ratio $a : b$, what is the ratio of their volumes?

16. A theater decides to change the shape of its popcorn container from a regular box to a right regular pyramid, as shown in the following figure, and charge only half as much.

If the containers are the same height and the tops are the same size, is this a bargain for the customer? Explain.

17. Two spherical cantaloupes of the same kind are sold at a fruit and vegetable stand. The circumference of one is 60 cm and that of the other is 50 cm. The larger melon is $1\frac{1}{2}$ times as expensive as the smaller. Which melon is the better buy and why?

★ 18. Half of the air is let out of a spherical balloon. If the balloon remains in the shape of a sphere, how does the radius of the smaller balloon compare to the original radius?

19. For each of the following, select the appropriate metric unit of measure (gram, kilogram, or metric ton):
    a. Jar of mustard
    b. Bag of peanuts
    c. Army tank
    d. Cat

20. For each of the following, choose the correct unit (milligram, gram, or kilogram) to make each sentence reasonable:
    a. A dime has a mass of 2 _____.
    b. The recipe said to add 4 _____ of salt.
    c. One strand of hair has a mass of 2 _____.

21. Complete each of the following:
    a. 8000 kg = _____ t
    b. 72 g = _____ kg
    c. 5 kg 750 g = _____ g
    d. 2.6 lb = _____ oz
    e. 3.8 lb = _____ oz

22. Convert each of the following from degrees Fahrenheit to the nearest integer degree Celsius:
    a. 0°F   **b.** 100°F
    c. ⁻40°F

23. Answer each of the following:
    a. The thermometer reads 26°C. Will the outdoor ice rink be open?
    b. It is 40°C. Will you need a sweater at the outdoor concert?
    c. Your bath water is 16°C. Will you have a hot, warm, or chilly bath?

24. **a.** Rainfall is usually measured in linear measure. Suppose St. Louis received 2 cm of rain on a given day. If a certain lot in St. Louis has measure 1 ha, how many liters of rainfall fell on the lot?
    b. What is the mass of the water that fell on the lot?

Mathematical Connections 13-5

Communication

1. **a.** Which will increase the volume of a circular cylinder more: doubling its height or doubling its radius? Explain.
   b. Is your answer the same for a circular cone? Why?

2. Explain how you would find the volume of an irregular shape.

3. Read the following problems (i) and (ii):
   (i) A tank in the shape of a cube 5 ft 3 in. on a side is filled with water. Find the volume in cubic feet, the capacity in gallons, and the weight of the water in pounds.
   (ii) A tank in the shape of a cube 2 m on a side is filled with water. Find the volume in cubic meters, the capacity in liters, and the mass of the water in kilograms.
   Discuss which problem is easier to work and why.

4. A furniture company gives an estimate for moving based upon the size of the rooms in an apartment. Write a rationale for why this is feasible. What assumptions are being made?

Open-Ended

5. A right circular cylinder has a 4-in. diameter, is 6 in. high, and is completely full of water. Design a right rectangular prism that will hold almost exactly the same amount of water.

6. Circular-shaped cookies are to be packaged 48 to a box. Each cookie is approximately 1 cm thick and has a diameter of 6 cm. Design a box that will hold this volume of cookies and has the least amount of surface area.

7. Design a cylinder that will hold 1 L of juice. Give the dimensions of your cylinder and tell why you designed the shape as you did.

Cooperative Learning

8. **a.** Find many different types of cans that are in the shape of a cylinder. Measure the height and diameter of each can.
   b. Compute the surface area and volume for each can.
   c. Based on the information collected, make a recommendation for designing an "ideal" can.

Questions from the Classroom

9. A student asks whether the volume of a prism can ever be the same number as its surface area. How do you answer?

10. A student interpreted 5 cm³ as shown below. What is wrong with this interpretation?

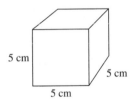

11. A student claims that it does not make any difference if his temperature is 2 degrees above normal Fahrenheit or 2 degrees above normal Celsius because in either case he is only 2 degrees above normal. How do you respond?

12. Andrea claims that if she doubles the length and width of the base of a rectangular prism and triples the height, she has increased the volume by a factor $2 \cdot 2 \cdot 3 = 12$. What would you tell her?

13. Jamie had $6.00 to spend on popcorn at a movie theater. She had to choose between buying two small cylindrical containers at $3.00 each or one large cylindrical container for $6.00. She noticed that the containers were about the same height and that the diameter of the large container looked about twice as long as the diameter of the small container. She bought two small containers and then asked in math class the next day if she made the right choice to get the most popcorn for her money. How would you help her?

Review Problems

14. Find the perimeter and the area of the following figures. Leave answers as exact values.

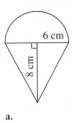

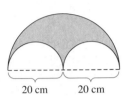

a.   **b.** The shaded portion only

15. Complete the following:
   a. 350 mm = _____ cm
   b. 1600 cm² = _____ m²
   c. 0.4 m² = _____ mm²
   d. 5.2 cm² = _____ m²

16. Determine whether each of the following is a right triangle:

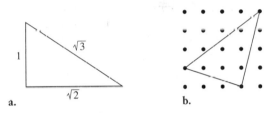

a.   **b.**

17. Find the surface area of each of the following. Leave answers as exact values.

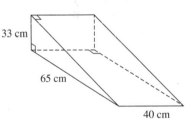

a. Right circular cone   **b.** Right triangular prism

Third International Mathematics and Science Study (TIMSS) Questions

All the small blocks are the same size. Which stack of blocks has a different volume from the others?

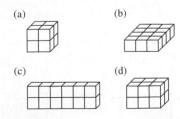

(a)   (b)   (c)   (d)

TIMSS, Grade 8, 2003

Oranges are packed in boxes. The average diameter of the oranges is 6 cm, and the boxes are 60 cm long, 36 cm wide, and 24 cm deep.

Which of these is the BEST approximation of the number of oranges that can be packed in a box?

a. 30   **b.** 240   **c.** 360   **d.** 1920

TIMSS, Grade 8, 2003

National Assessment of Educational Progress (NAEP) Question

How many 200-milliliter servings can be poured from a pitcher that contains 2 liters of juice?

a. 20   **b.** 15   **c.** 10   **d.** 5   **e.** 1

NAEP, Grade 8, 2007

Hint for Solving the Preliminary Problem

Find the area of the crust by finding the area of the 14-in. pizza minus the area with no crust and then convert this area to a percentage of the 14-in. pizza. Next find the area of the 10-in. pizza and use the percentage found for the 14-in. pizza to determine the area of the crust for the 10-in. pizza. You can use this area along with the area of the 10-in. pizza to find the radius of the pizza without the crust. From here you can determine the width of the crust.

Chapter Outline

I. The English system of measure
   A. Linear measure
      1 ft = 12 in.
      1 yd = 3 ft
      1 mi = 5280 ft = 1760 yd
   B. Area measure
      1. Units commonly used are the **square inch** (in.^{2}), **square yard** (yd^2), and **square foot** (ft^2).
      2. Land can be measured in **acres**.
   C. Volume and capacity measure
      Units commonly used are the **cubic inch** (in.^{3}), **cubic foot** (ft^3), **cubic yard** (yd^3), and **gallon.**
   D. Mass and weight
      Units of mass commonly used are **pound** (lb), **ounce** (oz) $\left(1\,oz = \dfrac{1}{16}\,lb\right)$, and **ton** (t) (1 ton = 2000 lb).
II. The metric system of measure
   A. A summary of relationships among prefixes and the base unit of linear measure follows:

Prefix	Unit	Relationship to Base Unit	Symbol
kilo	kilometer	1000 m	km
*hecto	hectometer	100 m	hm
*deka	dekameter	10 m	dam
	meter	**1 m**	**m**
*deci	decimeter	0.1 m	dm
centi	centimeter	0.01 m	cm
milli	millimeter	0.001 m	mm

*Not commonly used

   B. Area measure
      1. Units commonly used are the **square kilometer** (km^2), **square meter** (m^2), **square centimeter** (cm^2), and **square millimeter** (mm^2).
      2. Land can be measured using the **are** (a) (100 m^2) and the **hectare** (ha) (10,000 m^2).
   C. Volume and capacity measure
      1. Units commonly used are the **cubic meter** (m^3), **cubic decimeter** (dm^3), and **cubic centimeter** (cm^3).
      2. 1 dm^3 = 1 L (liter) and 1 cm^3 = 1 mL
   D. Mass and weight
      1. Units of mass commonly used are the **milligram** (mg), **gram** (g), **kilogram** (kg), and **metric ton** (t).
      2. 1 L and 1 mL of water have masses of 1 kg and 1 g, respectively.
   E. Temperature
      1. In the metric system the unit commonly used is the **degree Celsius** (°C). In the English system, the unit of temperature is the **degree Fahrenheit** (°F). The scientific unit used is the **kelvin** (K).
      2. Basic temperature reference points are the following:
         100°C—boiling point of water
         37°C—normal body temperature
         20°C—comfortable room temperature
         0°C—freezing point of water
      3. $C = \dfrac{5}{9}(F - 32)$ and $F = \dfrac{9}{5}C + 32$

III. Distance

    A. Distance properties. Given points A, B, and C,

        1. $AB \geq 0$

        2. $AB = BA$

        3. $AB + BC \geq AC$

    B. The distance around a two-dimensional figure is the **perimeter**. The distance C around a circle is the **circumference**. $C = 2\pi r = \pi d$, where r is the radius of the circle and d is the **diameter**.

    C. Distance formula. The distance between the points $A(x_1, y_1)$ and $B(x_2, y_2)$ is given by $AB = \sqrt{(x_2 - x_1)^2 + (y_2 - y_1)^2}$.

    D. Equation of a circle An equation of a circle with center (h, k) and radius r is $(x - h)^2 + (y - k)^2 = r^2$.

IV. Areas

    A. Formulas for areas

        1. Square: $A = s^2$, where s is the length of a side.

        2. Rectangle: $A = \ell w$, where ℓ is the length and w is the width.

        3. Parallelogram: $A = bh$, where b is the length of the base and h is the height.

        4. Triangle: $A = \frac{1}{2}bh$, where b is the length of the base and h is the altitude to that base.

        5. Trapezoid: $A = \frac{1}{2}h(b_1 + b_2)$, where b_1 and b_2 are the lengths of the bases and h is the height.

        6. Kite: $A = \frac{1}{2}(d_1 d_2)$, where d_1 and d_2 are the lengths of the diagonals of the kite.

        7. Regular polygon: $A = \frac{1}{2}ap$, where a is the apothem and p is the perimeter.

        8. Circle: $A = \pi r^2$, where r is the radius.

        9. Sector: $A = \theta \pi r^2 / 360$, where θ is the measure of the central angle forming the sector and r is the radius of the circle containing the sector.

    B. The Pythagorean theorem: In any right triangle, the square of the length of the hypotenuse is equal to the sum of the squares of the lengths of the legs.

    C. Converse of the Pythagorean theorem: If triangle ABC has sides of lengths a, b, and c such that $a^2 + b^2 = c^2$, then $\triangle ABC$ is a right triangle with the right angle opposite the side of length c.

    D. Triangle relations

        1. Property of 30°-60°-90° triangle: The length of the hypotenuse in a 30°-60°-90° triangle is 2 times the length of the leg opposite the 30° angle, and the length of the leg opposite the 60° angle is $\sqrt{3}$ times the length of the short leg.

        2. Property of 45°-45°-90° triangle: The length of the hypotenuse of a 45°-45°-90° triangle is $\sqrt{2}$ times the length of a leg.

V. Surface areas and volumes

    A. Formulas for areas

        1. Prism: $S.A. = 2B + ph$, where B is the area of a base, p is the perimeter of the base, and h is the height of the prism.

        2. Right circular cylinder: $S.A. = 2\pi r^2 + 2\pi rh$, where r is the radius of the circular base and h is the height of the cylinder.

        3. Right circular cone: $S.A. = \pi r^2 + \pi r\ell$, where r is the radius of the circular base and ℓ is the slant height.

        4. Right regular pyramid: $S.A. = B + \frac{1}{2}p\ell$, where B is the area of the base, p is the perimeter of the base, and ℓ is the slant height.

        5. Sphere: $S.A. = 4\pi r^2$, where r is the radius of the sphere.

    B. Formulas for volumes

        1. Prism: $V = Bh$, where B is the area of the base and h is the height.

        a. Right rectangular prism: $V = \ell wh$, where ℓ is the length, w is the width, and h is the height.

        b. Cube: $V = e^3$, where e is an edge.

        2. Right circular cylinder: $V = \pi r^2 h$, where r is the radius of the base and h is the height of the cylinder.

        3. Pyramid: $V = \frac{1}{3}Bh$, where B is the area of the base and h is the height of the pyramid.

        4. Circular cone: $V = \frac{1}{3}\pi r^2 h$, where r is the radius of the circular base and h is the height.

        5. Sphere: $V = \frac{4}{3}\pi r^3$, where r is the radius of the sphere.

Chapter Review

1. Complete the following.
   a. 50 ft = _____ yd
   b. 947 yd = _____ mi
   c. 0.75 mi = _____ ft
   d. 349 in. = _____ yd
   e. 5 km = _____ m
   f. 165 cm = _____ m
   g. 52 cm _____ mm
   h. 125 m = _____ km

2. Given three segments of length p, q, and r, where $p > q$, determine if it is possible to construct a triangle with sides of length p, q, and r in each of the following cases. Justify your answers.
   a. $p - q > r$
   b. $p - q = r$

3. Determine the area of the shaded region on each of the following geoboards if the unit of measure is 1 cm^2:

   a.

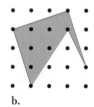

   b.

   c.

4. Explain how the formula for the area of a trapezoid can be found by using the following figures:

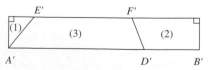

5. Lines a, b, and c are parallel to the line containing side $\overline{AB}$ of the triangles shown.

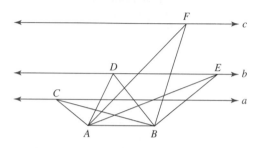

   List the triangles in order of size of their areas from least to greatest. Explain why your order is correct.

6. Use the figure shown to find each of the following areas:
   a. The area of the regular hexagon
   b. The area of the circle

7. Find the area of each shaded region in the following figures:

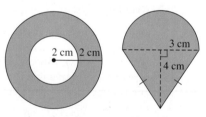

   a.   b.

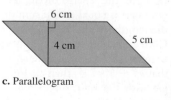

   c. Parallelogram

   d. Sector

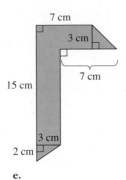

e.

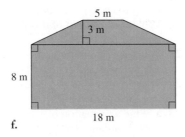

f.

8. Find the surface area of the following box (include the top and bottom):

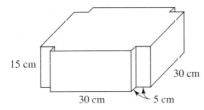

9. A baseball diamond is actually a square 90 ft on a side. What is the distance a catcher must throw from home plate to second base?

10. Find the length of segment AG in the spiral shown.

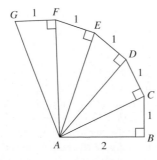

11. For each of the following, determine whether the measures represent sides of a right triangle.
  a. 5 cm, 12 cm, 13 cm
  b. 40 cm, 60 cm, 104 cm

12. Find the surface area and volume of each of the following figures:

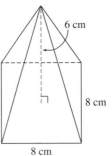   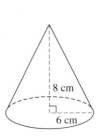

Right square pyramid   Right circular cone
a.   **b.**

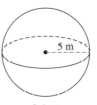   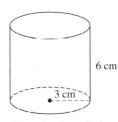

Sphere   Right circular cylinder
c.   **d.**

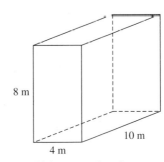

Right rectangular prism
e.

13. Find the lateral surface area of the following right circular cone:

14. Doug's Dog Food Company wants to impress the public with the magnitude of the company's growth. Sales of Doug's Dog Food doubled from 2000 to 2008, so the company is displaying the following graph, which shows the radius of the base and the height of the 2008 can to be double those of the 2000 can. What does the graph really show with respect to the company's growth? Explain your answer.

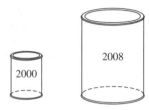

15. Find the area of the kite shown in the following figure:

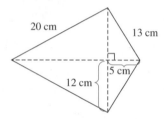

20 cm   13 cm
5 cm
12 cm

16. The diagonal of a rectangle has measure 1.3 m, and a side of the rectangle has measure 120 cm. Find the following:
   a. Perimeter of the rectangle
   b. Area of the rectangle

17. Find the area of a triangle that has sides of 3 m, 3 m, and 2 m.

18. A poster is to contain $0.25 \, m^2$ of printed matter, with margins of 12 cm at top and bottom and 6 cm at each side. Find the width of the poster if its height is 74 cm.

19. A right circular cylinder of height 10 cm and a right circular cone share a circular base and have the same volume. What is the height of the cone?

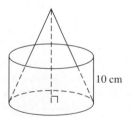

10 cm

20. Use a coordinate system to locate triangle *ABC* with vertices having coordinates $A(0, 0)$, $B(8, 0)$, and $C(6, 10)$.
   a. Write equations for each of the lines that contain the sides of the triangle.
   b. Find the midpoints of the sides of the triangle. (*Hint:* Use similar triangles and midsegments if you need to do so.)

c. Write the equations of each of the lines that contain the medians of the triangle. A median connects a vertex with the midpoint of the opposite side.
   d. Find the point of intersection of the medians (the **centroid**).
   e. The centroid of the triangle divides the medians in some given ratio. What is that ratio? Prove your answer.

21. On a 5 × 5 geoboard paper,
   a. draw a polygon whose perimeter is greater than 16.
   b. draw a polygon with least perimeter. What is the perimeter?
   c. draw a polygon with greatest area.

22. What is the length of the diagonal of a $8\frac{1}{2} \times 11$ in. piece of paper?

23. What is the circumference of the circumcircle of a triangle whose sides have length 6, 8, and 10 in.?

24. What is the radius of a circumcircle for an equilateral triangle of side 6 in.?

25. a. Find an equation of a circle with center at $(3, ^-4)$ and a radius of 5.
   b. Given that the equation of a circle is $36 = (x - 5)^2 + (y + 3)^2$, find the radius and the center.

26. Complete each of the following with the appropriate metric unit.
   a. A very heavy object has mass that is measured in _____.
   b. A cube whose length, width, and height are each 1 cm has a volume of _____.
   c. If the cube in (b) is filled with water, the mass of the water is _____.
   d. $1 \, L = $ _____ dm^3
   e. If a car uses 1 L of gas to go 12 km, the amount of gas needed to go 300 km is _____ L.
   f. $20 \, ha = $ _____ a
   g. $51.8 \, L = $ _____ cm^3
   h. $10 \, km^2 = $ _____ m^2
   i. $50 \, L$ _____ mL
   j. $5830 \, mL = $ _____ L
   k. $25 \, m^3 = $ _____ dm^3
   l. $75 \, dm^3 = $ _____ mL
   m. $52,813 \, g = $ _____ kg
   n. $4800 \, kg$ _____ t

27. a. A tank that is a right rectangular prism is $1 \, m \times 2 \, m \times 3 \, m$. If the tank is filled with water, what is the mass of the water in kilograms?
   b. Suppose the tank in (a) is exactly half full of water and then a heavy metal sphere of radius 30 cm is put into the tank. How high is the water now if the height of the tank is 3 m?

28. For each of the following, fill in the correct unit to make the sentence reasonable:
   a. Anna filled the gas tank with 80 _____.
   b. A man has a mass of about 82 _____.

c. The textbook has a mass of 978 _____.

d. A nickel has a mass of 5 _____.

e. A typical adult cat has a mass of about 4 _____.

f. A compact car has a mass of about 1.5 _____.

g. The amount of coffee in the cup is 180 _____.

29. For each of the following, decide if the situation is likely or unlikely:

a. Carrie's bath water has a temperature of 15°C.

b. Anne found 26°C too warm and so lowered the thermostat to 21°C.

c. Jim is drinking water that has a temperature of ⁻5°C.

d. The water in the teakettle has a temperature of 120°C.

e. The outside temperature dropped to 5°C, and ice appeared on the lake.

30. Complete each of the following:

a. 2 dm^3 of water has a mass of _____ g.

b. 1 L of water has a mass of _____ g.

c. 3 cm^3 of water has a mass of _____ g.

d. 4.2 mL of water has a mass of _____ kg.

e. 0.2 L of water has a volume of _____ m^3.

Selected Bibliography

Ameis, J. "Developing an Area Formula for a Circle with Goldilocks and the Three Bears." *Mathematics Teaching in the Middle School* 7 (November 2001): 140–142.

Austin, R., D. Thompson, and C. Beckmann. "Exploring Measurement Concepts Through Literature: Natural Links Across Disciplines." *Mathematics Teaching in the Middle School* 10 (January 2005): 218–224.

Bellasanta, B., B. Hunter, K. Irwin, M. Sheldon, C. Thompson, and C. Vistro-Yu. "By the Unit or Square Unit." *Mathematics Teaching in the Middle School* 7 (November 2001): 132–137.

Ben-Chaim, D., G. Lappan, and R. Houang. "Adolescents' Ability to Communicate Spatial Information: Analyzing and Affecting Students' Performance." *Educational Studies in Mathematics* 15 (5, 1984): 323–341.

Berry, R., and J. Wiggins. "Measurement in the Middle Grades." *Mathematics Teaching in the Middle School* 7 (November 2001): 154–156.

Chappell, M. "Geometry in the Middle Grades: From Past to the Present." *Mathematics Teaching in the Middle School* 9 (May 2001): 516–519.

Figueras, O., and G. Waldegg. "A First Approach to Measuring (Children between 11 and 13 Years Old)." *Proceedings of the Sixth Annual Meeting of the North American Branch of the International Group for the Psychology of Mathematics Education*, edited by J. Moser. Madison, WI: University of Wisconsin, 1984.

Hart, K. "Measurement," In *Children's Understanding of Mathematics: 11–16*. London: John Murray, 1981.

Hartzler, S. "Ratios of Linear, Area, and Volume Measures in Similar Solids." *Mathematics Teaching in the Middle School* 8 (January 2003): 228–232.

Hiebert, J. "Children's Thinking," In *Mathematics Education Research: Implications for the '80s*, edited by E. Fennema. Alexandria, VA: ASCD, 1981.

Johnson, A., and K. Norris. "The Beginnings of the Metric System." *Mathematics Teaching in the Middle School* 13 (October 2007): 172–181.

Johnston, D. "Measurement, Scale, and Theater Arts." *Mathematics Teaching in the Middle School* 13 (October 2007): 172–181

Joram, E., and V. Oleson. "Learning About Area by Working with Building Plans." *Mathematics Teaching in the Middle School* 9 (April 2004): 450–455

Kroon, C. "Metric Madness." *Mathematics Teaching in the Middle School* 9 (April 2004): 412–417.

Loomis, E. *The Pythagorean Proposition*. National Council of Teachers of Mathematics. Reston, VA: NCTM, 1976.

Martinie, S. "Measurement: What's the Big Idea?" *Mathematics Teaching in the Middle School* 9 (April 2004): 430–431.

Merz, A. "Hurry Up and Weight." *Teaching Children Mathematics* 10 (September 2003): 8–14.

Moyer, P., and E. Mailley. "*Inchworm and a Half*: Developing Fraction and Measurement Concepts Using Mathematical Representations." *Teaching Children Mathematics* 10 (January 2004): 244–252.

Preston, R., and T. Thompson. "Integrating Measurement Across the Curriculum." *Mathematics Teaching in the Middle School* 9 (April 2004): 436–441.

Pumala, V., and D. Klabunde. "Learning Measurement through Practice." *Mathematics Teaching in the Middle School* 10 (May 2005): 452–460.

Rozanski, K., C. Beckmann, and D. Thompson. "Exploring Size with the *Grouchy Ladybug*." *Teaching Children Mathematics* 10 (October 2003): 84–89.

Scanlon, G. "Sweet-Tooth Geometry." *Mathematics Teaching in the Middle School* 8 (May 2003): 466–469.

Stegemoller, W., and R. Stegemoller. "A Path to Discovery." *Mathematics Teaching in the Middle School* 9: (April 2004): 459–464.

Taylor, M. "Do Your Students Measure Up Metrically?" *Teaching Children Mathematics* 7 (January 2001): 282–287.

Tent, M. "Circles and the Number Π." *Mathematics Teaching in the Middle School* 6 (April 2001): 452–457.

Thompson, T., and R. Preston. "Measurement in the Middle Grades: Insights from NAEP and TIMSS." *Mathematics Teaching in the Middle School.* 9 (May 2004): 514–519.

Weinberg, S. "How Big Is Your Foot?" *Mathematics Teaching in the Middle School* 6 (April 2001): 476–481.

Weinberg, S., P. Hammrich, and M. Bruce. "The Giants Project." *Mathematics Teaching in the Middle School* 8 (April 2003): 406–413.

Young, S., and R. O'Leary. "Creating Numerical Scales for Measuring Tools." *Teaching Children Mathematics* 8 (March 2002): 400–405.

14

Motion Geometry and Tessellations

Preliminary Problem

Marilen challenged her classmates to find a regular polygon with rotational symmetry of 80°. Find all such polygons.

Euclid envisioned moving one geometric figure in a plane and placing it on top of another to determine if the two figures were congruent. Intuitively, we know this can be done by making a tracing of one figure, then sliding, turning, flipping the tracing, or using some combination of these motions, and finally placing to match atop the other figure. Additionally similarities (and/or scaling) can be accomplished through a combination of the motions and dilations.

Symmetries are fundamental in the study of geometry, nature, and shapes. Many symmetries are the results of sliding, flipping, and turning shapes. In *Focal Points*, we find the following expectations for students:

For grades 3–5:

- Predict and describe the results of sliding, flipping, and turning two-dimensional shapes. (p. 31)

- Describe a motion or a series of motions that will show that two shapes are congruent. (p. 32)

- Identify and describe line and rotational symmetry in two- and three-dimensional shapes and designs. (p. 32)

For grades 6–8:

- Describe sizes, positions, and orientations of shapes under informal transformations such as flips, turns, slides and scaling. (p. 37)

- Examine the congruence, similarity, and line or rotational symmetry of objects using transformations. (p. 37)

A study of motions and symmetries leads to tessellations of the plane or space. A **tessellation** is the filling of a plane or space with repetitions of congruent figures in such a way that none overlap and there are no gaps. This chapter contains sections on motions, symmetries, and tessellations.

In the *Principles and Standards*, we find the following:

Young children come to school with intuitions about how shapes can be moved. Students can explore motions such as slides, flips, and turns by using mirrors, paper folding, and tracing. Later, their knowledge about transformations should become more formal and systematic. In grades 3–5 students can investigate the effects of transformations and begin to describe them in mathematical terms. Using dynamic geometry software, they can begin to learn the attributes needed to

---

◆ Historical Note

In 1872, at age 23, Felix Klein (1849–1925) was appointed to a chair at the University of Erlangen, Germany. His inaugural address, the *Erlanger Programm*, described geometry as the study of properties of figures that do not change under a particular set of transformations. Specifically, Euclidean geometry was described as the study of such properties of figures as area and length, which remain unchanged under a set of transformations called *isometries*.   ◆

define a transformation. In the middle grades, students should learn to understand what it means for a transformation to preserve distance, as translations, rotations, and reflections do. . . . At all grade levels, appropriate consideration of symmetry provides insights into mathematics and into art and aesthetics. (p. 43)

The *Principles and Standards* also lists the following expectations for different grade levels:

In prekindergarten through grade 2, all students should

- recognize and apply slides, flips, and turns;
- recognize and create shapes that have symmetry. (p. 96)

14-1   Translations and Rotations

Any **motion** that preserves length or distance is an **isometry** (derived from Greek and meaning "equal measure"). A translation is an isometry and is a one-to-one correspondence between a plane and itself. Any function from a plane to itself that is a one-to-one correspondence is a **transformation** of the plane. So a translation is also a transformation. Because transformations are functions, they have properties of functions.

Translations

Figure 14-1(a) shows a two-dimensional representation of a child moving down a slide without twisting or turning. This type of motion is a **translation**, or **slide**. In Figure 14-1(a), the child (**preimage**) at the top of the slide moves a certain distance in a certain direction along a **slide line** to obtain the **image** at the bottom of the slide. Figure 14-1(a) shows a translation that takes the preimage to its image. In Figure 14-1(b) the translation is determined by the **slide arrow**, or **vector**, from M to N. The vector determines the image of any point in a plane in the following way: The image of a point A in the plane is the point A' obtained by sliding A along a line parallel to $\overrightarrow{MN}$ in the direction from M to N by the distance MN. (MN is also denoted by d in Figure 14-1(b).) Notice that $AA' = MN = d$ and $\overleftrightarrow{MN}$ is parallel to $\overleftrightarrow{AA'}$. It appears that under the translation, figures change neither their shapes nor their sizes. In fact, a translation preserves both length and angle size, and thus congruence of figures.

Figure 14-1(b) shows how segment AB is slid, or translated, by the slide arrow from M to N along slide line ℓ. Because $\overline{AB} \cong \overline{A'B'}$ and $AA' = BB' = MN = d$, quadrilateral $AA'B'B$ is a parallelogram, and thus $\overline{AB} \parallel \overline{A'B'}$.

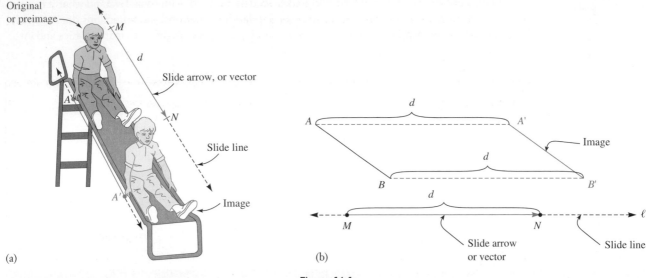

Figure 14-1

In succeeding paragraphs and activities, we define a translation, consider some of its properties in the Technology Corner, and consider constructions and representations of a translation.

Definition of a Translation

A **translation** is a motion (or transformation) of a plane that moves every point of the plane a specified distance in a specified direction along a straight line.

TECHNOLOGY CORNER *GSP* Lab 11 activities 6–8 can be used to investigate the following properties of translations:

• A figure and its image are congruent.
• The image of a line is a line parallel to it.

Constructions of Translations

The image of a figure under a translation can be constructed easily with tracing paper or by using only a compass and straightedge. Using tracing paper is a natural way to construct a "motion." Traditionally, most constructions in geometry have been accomplished with a compass and a straightedge. We consider first how the latter construction might be completed and leave the actual construction as an activity in Now Try This 14-1.

To construct the image of an object under a translation, we first need to know how to construct the image A' of a single point A. In Figure 14-1(b), by definition of translation $MN = AA'$ and $\overleftrightarrow{MN}$ is parallel to $\overleftrightarrow{AA'}$ Thus, $MAA'N$ is a parallelogram. (Why?)

Consequently, to construct the translation image A' of a point A with a compass and straightedge, we need only to construct a parallelogram $MAA'N$ so that $\overrightarrow{AA'}$ is in the same direction as $\overrightarrow{MN}$.

NOW TRY THIS 14-1 In Figure 14-2 use a compass and straightedge to construct the following:

a. The image of A under a translation that takes M to N.
b. The image of A under a translation that takes N to M.

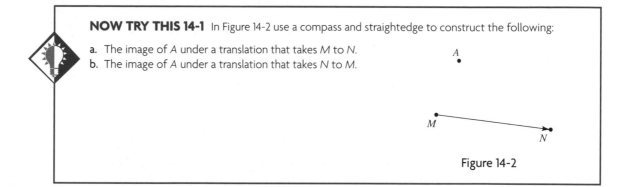

Figure 14-2

To find the image of a triangle under a translation, it suffices to find the images of the three vertices of the triangle and to connect these images with segments to form the triangle's image.

It is possible to use a geoboard, dot paper, or a grid to find an image of a segment. The following example uses dot paper.

Example 14-1   Find the image of $\overline{AB}$ under the translation from X to X' pictured on the dot paper in Figure 14-3.

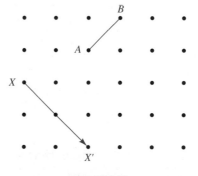

Figure 14-3

Solution   We want the image of $\overline{AB}$ to be parallel to $\overline{AB}$ so that $ABB'A'$ is a parallelogram. X', the image of X under the translation, could be obtained from X by shifting X, 2 units vertically down and then 2 units horizontally to the right, as shown in Figure 14-4. This shifting determines the slide arrow from X to X'. The image of each point on the dot paper can be obtained by first shifting it 2 units down and then 2 units to the right. Thus, $\overline{A'B'}$, the image of $\overline{AB}$, is found in this way in Figure 14-4.

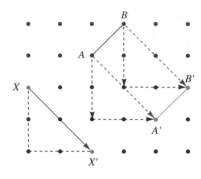

Figure 14-4

Coordinate Representation of Translations

In many applications of mathematics, such as computer graphics, it is necessary to use translations in a coordinate system. In Figure 14-5, $\triangle A'B'C'$ is the image of $\triangle ABC$ under the translation defined by the slide arrow from O to O', where O is the origin and O' has coordinates $(5, {}^-2)$. Point O' is the image of point O under the given translation. The point $O'(5, {}^-2)$ can be obtained by moving O horizontally to the right 5 units and then 2 units down. As each point in triangle ABC is translated in the direction from O to O' by the distance OO', we can obtain the image of any point by moving horizontally to the right 5 units and then vertically 2 units down. In Figure 14-5 the images A', B', and C' for points A, B, and C are shown. Table 14-1 shows how the coordinates of the image vertices A', B', and C' in Figure 14-5 are obtained from the coordinates of A, B, and C.

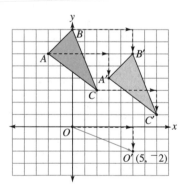

Figure 14-5

Table 14-1

Point (x, y)	Image Point $(x + 5, y - 2)$
$A\ ({}^-2, 6)$	$A'\ (3, 4)$
$B\ (0, 8)$	$B'\ (5, 6)$
$C\ (2, 3)$	$C'\ (7, 1)$

This discussion suggests that a translation is described by showing how the coordinates of any point (x, y) change. The translation in Table 14-1 can be written symbolically as $(x, y) \rightarrow (x + 5, y - 2)$, where "$\rightarrow$" denotes "moves to." In this notation, $(x + 5, y - 2)$ is the image of (x, y).

Theorem 14–1:   Translation in a Coordinate System

A translation is a function from the plane to the plane such that to every point (x, y) corresponds the point $(x + a, y + b)$, where a and b are real numbers.

Example 14-2

Find the coordinates of the image of the vertices of quadrilateral $ABCD$ in Figure 14-6 under the translations in parts (a) through (c). Draw the image in each case.

a. $(x, y) \rightarrow (x - 2, y + 4)$
b. A translation determined by the slide arrow from $A(0, 0)$ to $A'(^-2, 4)$
c. A translation determined by the slide arrow from $S(4, {}^-3)$ to $S'(2, 1)$

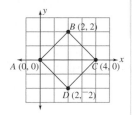

Figure 14-6

Solution **a.** Because $(x, y) \rightarrow (x - 2, y + 4)$, the images A', B', C', and D' of the corresponding points A, B, C, and D can be found as follows:

$$A(0, 0) \rightarrow A'(0 - 2, 0 + 4), \text{ or } A'(^-2, 4)$$
$$B(2, 2) \rightarrow B'(2 - 2, 2 + 4), \text{ or } B'(0, 6)$$
$$C(4, 0) \rightarrow C'(4 - 2, 0 + 4), \text{ or } C'(2, 4)$$
$$D(2, {}^-2) \rightarrow D'(2 - 2, {}^-2 + 4), \text{ or } D'(0, 2)$$

The square $ABCD$ and its image $A'B'C'D'$ are shown in Figure 14-7.

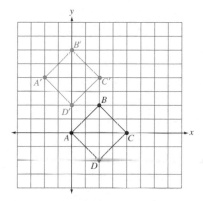

Figure 14-7

b. This is the same translation as in part (a) and hence the image of $ABCD$ is the same as in part (a).

c. The translation from $S(4, {}^-3)$ to $S'(2, 1)$ moves an x-coordinate $2 - 4$, or $^-2$ units and a y-coordinate $1 - {}^-3$, or 4 units. Thus, every point (x, y) moves to a point with coordinates $(x + (2 - 4), y + (1 - {}^-3))$, or $(x - 2, y + 4)$. Note that this is the same translation as in part (a).

Translations are useful in determining frieze patterns such as those appearing in wallpaper designs. An example is seen in Figure 14-8.

Figure 14-8

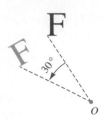

Figure 14-9

Rotations

A **rotation**, or **turn**, is another kind of isometry. Figure 14-9 illustrates congruent figures that resulted from a rotation about point O. The image of the letter **F** under a 30° counterclockwise rotation with center O is shown in green.

A rotation can be constructed by using tracing paper, as in Figure 14-10. In Figure 14-10(a), $\triangle ABC$ and point O are traced on tracing paper. Holding point O fixed, we turn the tracing paper to obtain the image $\triangle A'B'C'$, as shown in Figure 14-10(b). Point O is the **turn center**, and $\angle COC'$ is the **turn angle**.

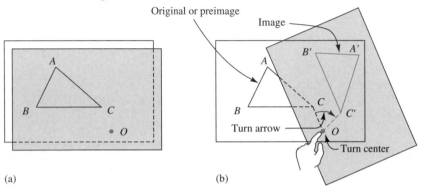

(a)   (b)

Figure 14-10 Construction of a rotation using tracing paper

To determine a rotation, we must know three pieces of data: the turn center; the direction of the turn, either clockwise (a negative measure) or counterclockwise (a positive measure); and the amount of the turn. The amount and the direction of the turn can be illustrated by a **turn arrow**, or they can be specified in numbers of degrees.

This discussion leads to the following definition.

Definition of Rotation

A **rotation** is a transformation of the plane determined by holding one point—the center—fixed and rotating the plane about this point by a certain amount in a certain direction (a certain number of degrees either clockwise or counterclockwise).

To construct an image of a figure under a rotation, observe that every point on the tracing-paper construction moves along a circle. Also the angle formed by any point (not the turn center), the center of the rotation O, and the image of the point is the angle of the turn. (Why?) With this in mind, we construct the image of a point under a rotation in Now Try This 14-2.

NOW TRY THIS 14-2 Use a compass and a straightedge to construct the image of point P under a rotation with center O through the angle and in the direction given in Figure 14-11. (*Hint*: Construct an isosceles triangle BAC with B on one side of the given angle and C on the other side so that $AB = AC = OP$. Then construct $\triangle POP'$ congruent to $\triangle BAC$.)

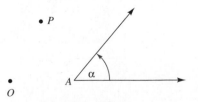

Figure 14-11

TECHNOLOGY CORNER *GSP* Lab 11 activities 9–11 can be used to investigate properties of rotations.

A rotation is an isometry, so the image of a figure under a rotation is congruent to the original figure. It can be shown that *under any isometry, the image of a line is a line, the image of a circle is a circle, and the images of parallel lines are parallel lines.*

For certain angles like 90°, rotations may be constructed on a geoboard or dot paper, as demonstrated in Example 14-3.

Example 14-3

In Figure 14-12, find the image of $\triangle ABC$ under the rotation shown with center O.

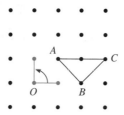

Figure 14-12

Solution $\triangle A'B'C'$, the image of $\triangle ABC$, is shown in Figure 14-13. The image of A is A' because $\angle A'OA$ is a right angle (why?) and $OA = OA'$. Similarly, B' is the image of B. To find the location of C', we use the fact that rotation is an isometry and hence $\triangle A'B'C' \cong \triangle ABC$. Thus, $\angle B \cong \angle B'$. The location of point C' shown makes $\angle B \cong \angle B'$ and $C'B' = CB$ (why?). The direction of the rotation is and counterclockwise as specified.

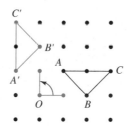

Figure 14-13

A rotation of 360° about a point moves any point and hence any figure onto itself. Such a transformation is an **identity transformation**. Any point may be the center of such a rotation.

A rotation of 180° about a point is a **half-turn**. Because a half-turn is a rotation, it has all the properties of rotations. Figure 14-14 shows some shapes and their images under a half-turn about point O.

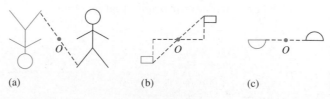

(a)   (b)   (c)

Figure 14-14

The half-screw in Figure 14-15 has a head that shows a practical example of a half-turn in hardware. To "lock" the screw, the head is turned 180°.

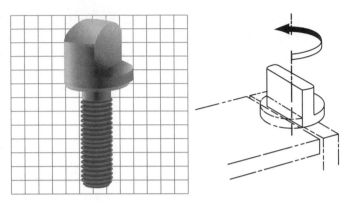

Figure 14-15

A different application of rotations appears when we consider the slopes of perpendicular lines, as seen in the following.

Slopes of Perpendicular Lines

Transformations can be used to investigate various mathematical relationships. For example, they can be used to determine the relationship between the slopes of two perpendicular lines, neither of which is vertical.

We first consider a special case in which the lines go through the origin. Suppose the two perpendicular lines are as pictured in Figure 14-16 with one point, B, of ℓ_1 having coordinates (a, b). The coordinates of point B determine the coordinates of point $A(a, 0)$ and triangle OAB is formed. If the plane containing the lines is rotated 90° counterclockwise about center O, then the image of ℓ_1 becomes ℓ_2. The image of triangle OAB becomes triangle $OA'B'$. Because the rotation takes a triangle to a congruent triangle, $\triangle OAB \cong \triangle OA'B'$. Thus, A' has coordinates $(0, a)$ and B' has coordinates $(^-b, a)$. (Why?)

Now consider the slopes of ℓ_1 and ℓ_2. The slope of ℓ_1 is b/a, and the slope of ℓ_2 is $a/(^-b)$. The slopes are negative reciprocals of each other and their product is $^-1$.

Any two perpendicular lines (neither vertical) that do not intersect at the origin can always be found using two lines parallel to the original lines but that pass through the origin. Because parallel lines have equal slopes, the relationship between the slopes (m_1 and m_2) of the perpendicular lines is the same as the relationship between the slopes of the perpendicular lines through the origin; that is, $m_1 m_2 = {}^-1$.

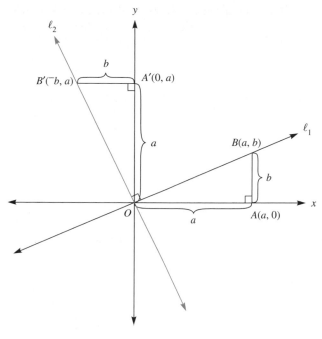

Figure 14-16

It is also possible to prove the converse statement; that is, if the slopes of two lines satisfy the condition $m_1 m_2 = {}^-1$, then the lines are perpendicular. We summarize these results in the following theorem.

Theorem 14–2:   Slopes of Perpendicular Lines

Two lines, neither of which is vertical, are perpendicular if, and only if, their slopes m_1 and m_2 satisfy the condition $m_1 m_2 = {}^-1$. Every vertical line has no slope but is perpendicular to a line with slope 0.

Example 14-4

Find the equation of line ℓ through point $({}^-1, 2)$ and perpendicular to the line $y = 3x + 5$.

Solution   If m is the slope of ℓ, as in Figure 14-17, then the equation of line ℓ is $y = mx + b$.

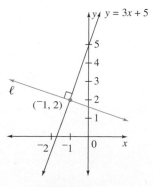

Figure 14-17

Because the line $y = 3x + 5$ has slope 3 and is perpendicular to ℓ, we have $3m = {}^-1$; therefore, $m = \dfrac{{}^-1}{3}$. Consequently, the equation of ℓ is

$$y = \dfrac{{}^-1}{3}x + b$$

Because the point $({}^-1, 2)$ is on ℓ, we can substitute $x = {}^-1, y = 2$ in $y = \dfrac{{}^-1}{3}x + b$ and solve for b as follows:

$$2 = \dfrac{{}^-1}{3}({}^-1) + b$$

$$\dfrac{5}{3} = b$$

Consequently, the equation of ℓ is

$$y = \dfrac{{}^-1}{3}x + \dfrac{5}{3}$$

Assessment 14-1A

1. For each of the following, find the image of the given quadrilateral under a translation from A to B:

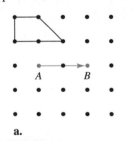

a.

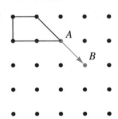

b.

2. Construct the image of $\overline{BC}$ under the translation pictured in the figure by using the following:
   a. Tracing paper
   b. Compass and straightedge

3. Find the coordinates of the image for each of the following points under the translation defined by $(x, y) \rightarrow (x + 3, y - 4)$:
   a. $(0, 0)$ **b.** $({}^-3, 4)$ **c.** $({}^-6, {}^-9)$

4. Find the coordinates of the points whose images under the translation $(x, y) \rightarrow (x - 3, y + 4)$ are the following:
   a. $(0, 0)$ **b.** $({}^-3, 4)$ **c.** $({}^-6, {}^-9)$

5. Consider the translation $(x, y) \rightarrow (x + 3, y - 4)$. In each of the following, draw the image of the figure under the translation and find the coordinates of the images of the labeled points:

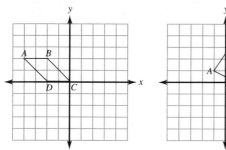

a.   **b.**

6. Consider the translation $(x, y) \rightarrow (x + 3, y - 4)$. In the following, draw the figure whose image is shown:

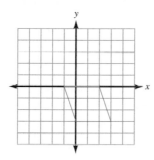

7. Find the image of the following quadrilateral in a 90° counterclockwise rotation about O:

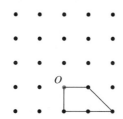

8. If ℓ is a line whose equation is $y = 2x - 1$, find the equation of the image of ℓ under each of the following translations:
   a. $(x, y) \rightarrow (x, y - 2)$
   b. $(x, y) \rightarrow (x + 3, y)$

9. If $y = {}^-2x + 3$ is the image of line k under the translation $(x, y) \rightarrow (x + 3, y - 2)$, find the equation of k.

10. Use a compass and a straightedge to find the image of line ℓ under a half-turn about point O as shown.

11. The image of $\overline{AB}$ under a rotation is given in the following figure. Find $\overline{AB}$.

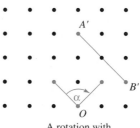

A rotation with
center O through α

12. The image of NOON is still NOON after a certain half-turn. List some other words that have the same property. What letters can such words contain?

13. Answer each of the following:
   a. Draw a line ℓ and any two points A and B so that $\overline{AB}$ is parallel to the line. Find the image of ℓ under a translation from A to B.
   b. Draw a line ℓ and any two points A and B so that $\overline{AB}$ is not parallel to ℓ. Construct ℓ', the image of ℓ under the translation from A to B.
   c. How are ℓ and ℓ' in part (b) related? Why?
   d. What is the image of $\angle ABC$ under 10 successive rotations about B if $m(\angle ABC) = 36°$.

14. The images of any point under a rotation by certain angles can be found with only a compass and straightedge (without the use of a protractor). Construct P', the image of P when it is rotated about O, as shown in the figure, for angles with the following measures and direction:
   a. 45° counterclockwise
   b. 60° clockwise

• P

• O

15. For each of the following points, find the coordinates of the image point under a half-turn about the origin:
   a. $(4, 0)$
   b. $(2, 4)$
   c. $({}^-2, {}^-4)$
   d. (a, b)

16. In the following figure, find the image of the figure under a half-turn about O:

• O

17. Draw any line and label it ℓ. Use tracing paper to find ℓ', the image of ℓ under each of the following rotations. In each case, describe in words how ℓ' is related to ℓ.

  a. Half-turn about point O on ℓ

  b. A 90° turn counterclockwise about point O, not on ℓ

18. a. Find the final image of $\triangle ABC$ by performing two rotations in succession each with center O, one by angle α and one by angle β, in the directions shown in the figure.

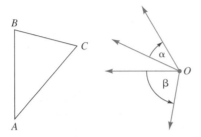

  b. Is the order of the rotations important?

  c. Could the result have been accomplished in one rotation?

19. Use a drawing similar to Figure 14-16 to find

  a. the images of the following points under a 90° rotation counterclockwise about the origin:

  (i) $(2, 3)$ **(ii)** $(^-1, 2)$

  (iii) (m, n) in terms of m and n.

  b. Show that under a half-turn with the origin as center, the image of a point with coordinates (a, b) has coordinates $(^-a, ^-b)$.

  c. Use what you found in part (a) to get the image of $P(a, b)$ under rotation clockwise by 90° about the origin. (*Hint*: Rotate first as in part (a), then apply a half-turn about the origin.)

20. Three vertices of a rhombus are at $O(0, 0), A(3, 4)$, $C(a, 0)$. Find the following:

  a. All possible coordinates of point C

  b. All possible coordinates for the fourth vertex B

  c. Verify that the diagonals of each of the rhombi you found in part (b) are perpendicular to each other.

21. The following figure is a rhombus with sides of length a and one of the vertices at (h, k):

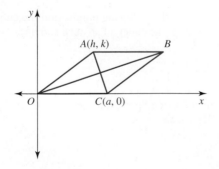

a. Find the coordinates of B in terms of a, h, and k.

b. Explain why $h^2 + k^2 = a^2$.

c. Prove that the diagonals of a rhombus are perpendicular to each other.

22. a. Draw two points O and M and a point P. Construct P', the image of P under a half-turn about O and then P'', the image of P' under a half-turn in M.

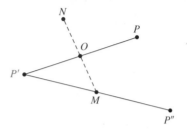

  b. Find the image of P under a single translation that takes N to M, where O is the midpoint of $\overline{MN}$. How does the image of P under this translation compare with P''?

  c. Repeat parts (a) and (b) for a new point Q (first find Q', the image of Q under a half-turn about O and then Q'', the image of Q' under half-turn in M).

  d. Conjecture what single transformation will have the same effect on any point in the plane as a half-turn in O followed by a half-turn in M.

  ★ **e.** Justify your conjecture in part (d).

23. a. What is the image of any point with coordinates (a, b) when two half-turns with the same center are completed successively? Explain.

  b. Explain how one half-turn can "undo" another.

24. Name four geometric figures that can be their own images under a half-turn and one that cannot.

25. In the drawing below, points A, B, and C are given. Find all locations of point D so that a parallelogram is formed with the four points.

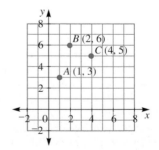

Assessment 14-1B

1. Find the figure whose image is given in each of the following under a translation from X to X':

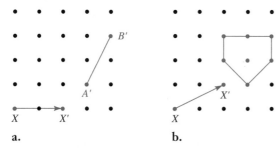

a.

b.

2. Construct $\overline{BC}$ whose image $\overline{B'C'}$ under the translation is pictured in the figure by using the following:
   a. Tracing paper
   b. Compass and straightedge

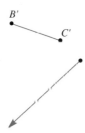

3. Find the coordinates of the image for each of the following points under the translation defined by $(x, y) \rightarrow (x + 3, y - 4)$:
   a. $(7, 14)$ b. $(^-3, ^-5)$
   c. (h, k)

4. Find the coordinates of the points whose images under the translation $(x, y) \rightarrow (x - 3, y + 4)$ are the following:
   a. $(7, 14)$ b. $(^-7, ^-10)$
   c. (h, k)

5. Consider the translation $(x, y) \rightarrow (x + 3, y - 4)$. In each of the following, draw the image of the figure under the translation, and find the coordinates of the images of the labeled points:

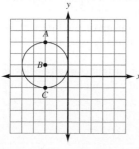

a.

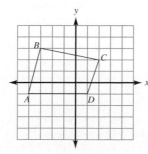

b.

6. Consider the translation $(x, y) \rightarrow (x + 3, y - 4)$. In the following, draw the figure whose image is shown:

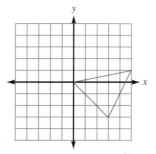

7. Find the figure whose image was found through a 90° clockwise rotation about point U and is shown below.

8. If ℓ is a line whose equation is $y = 2x - 1$, find the equation of the image of ℓ under each of the following translations:
   a. $(x, y) \rightarrow (x - 3, y + 2)$
   b. $(x, y) \rightarrow (x - 5, y - 4)$

9. If P' is the image of point P under a half-turn about O, what can be said about points P', P, and O? Why?

10. Use a compass and a straightedge to find the preimage of ℓ under a half-turn about point O as shown.

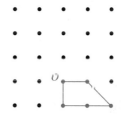

11. The image of $\overline{AB}$ under a rotation is given in the following figure. Find $\overline{AB}$.

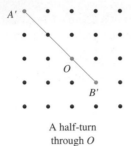

A half-turn
through O

12. The image of MOW is still MOW after a certain half-turn. Explain whether MOM has this property.

13. a. Refer to the following figure and use paper folding or any other method to show that if P' is the image of P under rotation about point O by a given angle, then O is on the perpendicular bisector of $\overline{PP'}$.

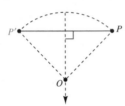

b. $\triangle A'B'C'$ shown in the following figure was obtained by rotating $\triangle ABC$ about a certain point O. Explain how to find the point O and the angle of rotation.

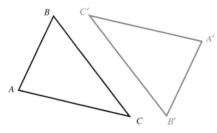

c. Triangles ABC and $A'B'C'$ shown are congruent. Trace them and explain why it is impossible to find a rotation under which $\triangle A'B'C'$ is the image of $\triangle ABC$.

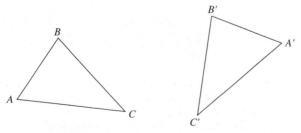

14. The images of any point under a rotation by certain angles can be found with only a compass and straightedge (without the use of a protractor). Construct P', the image of P when it is rotated about O, as shown in the figure, for angles with the following measures and direction:

a. 30° counterclockwise
b. 30° clockwise

15. For each of the following points, find the coordinates of the image point under a half-turn about the origin:
a. $(0, 3)$
b. $(^-2, 5)$
c. $(^-a, ^-b)$

16. In the following figure, find the image of the figure under a half-turn about O:

17. Draw any line and label it ℓ. Use tracing paper to find ℓ', the image of ℓ under each of the following rotations. In each case, describe in words how ℓ' is related to ℓ.
a. Half-turn about point O, not on ℓ
b. A 60° turn counterclockwise about point O, not on ℓ

18. When $\triangle ABC$ in the following figure is rotated about point O by 360°, each of the vertices traces a path:

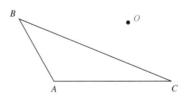

a. What geometric figure does each vertex trace?
b. Identify all points O for which two vertices trace an identical path. Justify your answer.
c. Given any $\triangle ABC$, is there a point O such that the three vertices trace an identical path? If so, describe how to find such a point.

19. Find the equation of the image of the line $y = 3x - 1$ under the following transformations:
a. Half-turn about the origin
b. A 90° rotation about the origin counterclockwise

20. Find the equation of the line through $(^-1, ^-3)$ and perpendicular to
a. $y = 2x + 1$
b. $y = ^-2x + 3$
c. $x = ^-4$
d. $y = ^-3$

21. Use any tools to construct a regular octagon (8-gon) inscribed in a circle and answer the following questions:

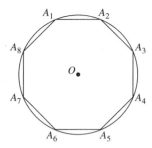

a. $\triangle A_5OA_6$ is the image of $\triangle A_1OA_2$ under a rotation about O clockwise by an angle whose measure is α. Find α, if $0 < \alpha \leq 180°$.

b. Find the smallest angle by which the octagon needs to be rotated so that its image is itself.

c. What is the image of each vertex under a half-turn in O? Why?

d. What is the image of the octagon under a half-turn in O? Why?

22. a. A translation from A to B is followed by a translation from B to C, as shown. Show that the same result can be accomplished by a single translation. What is that translation?

b. Find a single translation equivalent to a translation from A to B followed by a translation from D to E.

c. Suppose the order of the translations in part (b) is reversed; that is, a translation from D to E is followed by a translation from A to B. Is the result different from the one in part (b)?

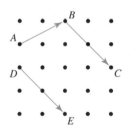

23. If $\overline{AB}$ is rotated 90° clockwise about point O, explain whether its image is perpendicular to $\overline{AB}$, regardless of the location of point O.

24. a. Points $A(1, 2)$, $B(^-2, 3)$, and $C(3, 8)$ are given. Find all possible coordinates of point D so that the four points form a rectangle.

b. For each location of point D found in part (a), find a translation that will take $\overline{AB}$ to the opposite side of the rectangle formed.

25. Suppose one of the vertices of a square whose diagonals intersect at the origin has coordinates $(^-2, 5)$. Find the coordinates of the other three vertices.

26. The image below is part of a net for a pyramid. The original is rotated counterclockwise through α with center O.

a. Sketch that image.

b. If the original and turn image together are the net for the lateral surface of a pyramid, name the pyramid.

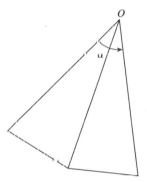

Mathematical Connections 14-1

Communication

1. If we are given two congruent nonparallel segments, is it always possible to find a rotation so that the image of one segment will be the other segment? Explain why or why not.

2. a. If you rotate an object 180° clockwise or counterclockwise, using the same center, is the image the same in both cases? Explain.

b. Answer part (a) if you rotate the object 360°.

3. For each of the following figures, trace the figure on tracing paper, rotate the tracing by 180° about the given point *O*, sketch the image, and then make a conjecture about the kind of figure that is formed by the union of the original figure and its image. In each case, explain why you think your conjecture is true.

O is the midpoint of $\overline{BC}$.

a.

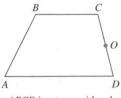

ABCD is a trapezoid and
O is the midpoint of $\overline{CD}$.

b.

ABCD is a square and
O is the intersection
of its diagonals.

c.

Open-Ended

4. A drawing of a cube, shown in the following figure, can be created by drawing a square *ABCD*, finding its image under translation defined by the slide arrow from *A* to *A′* so that $AA' = AB$, and connecting the points *A*, *B*, *C*, and *D* with their corresponding images.

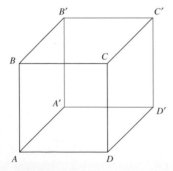

Draw several other perspective geometric figures using translations. In each case, name the figures and indicate the slide arrow that defines the translation.

5. Wall stenciling has been used to obtain an effect similar to that of wallpapering. The stencil pattern in the following figure can be used to create a border on a wall:

Measure the length of a wall of a room and design your own stencil pattern to create a border. Cut the pattern from a sheet of plastic or cardboard. Define the translation that will accomplish creating an appropriate border for the wall.

6. The following pattern can be created by rotating figure *A* about *O* clockwise by the indicated angle, then rotating the image *B* about *O* by the same angle, and then rotating the image *C* about *O* by the same angle, and so on until one of the images coincides with the original figure *A*:

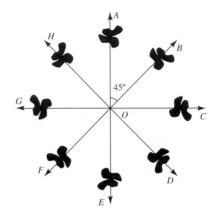

Make several designs with different numbers of congruent figures around a circle in which the image of each figure under the same rotation is the next figure and so that one of the images coincides with the original figure.

Cooperative Learning

7. Mark a point *A* on a sheet of paper and set a straightedge through *A* as shown in the following figure. Find a circular shape (a jar lid is a good choice) and mark point *P* on the edge of the shape. Place the shape on the straightedge so that *P* coincides with *A*. Consider the path traced by *P* as the circle rolls so that its edge stays in contact with the straightedge all the time and until *P* comes in contact with the straightedge again at point *B*.

a. Have one member of your group roll the circular shape and another draw the path traced by *P* as accurately as possible.

b. Discuss how to check if the path traced by point *P* is an arc of a circle.

c. Find the length of $\overline{AB}$.

Questions from the Classroom

8. A student asks if every translation on a grid can be accomplished by a translation along a vertical direction followed by a translation along a horizontal direction. How do you respond?

9. A student asks why, in a half-turn, the direction is never specified. How do you respond?

Third International Mathematics and Science Study (TIMSS) Questions

This figure will be turned to a different position.

Which of these could be the figure after it is turned?

(a)   (b)   (c)   (d)

TIMSS, Grade 4, 2003

Rectangle *PQRS* can be rotated (turned) onto rectangle *UVST*.

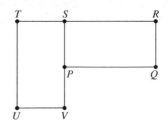

What point is the center of rotation?
a. *P*   **b.** *R*   **c.** *S*   **d.** *T*   **e.** *V*

TIMSS, Grade 8, 2003

TECHNOLOGY CORNER Use *GSP* to draw an equilateral triangle and two altitudes of the triangle. Let *O* be the point at which the altitudes intersect. Rotate the triangle by 120° about *O* in any direction. Make a conjecture based on this experiment. Do you think your conjecture may be true for some triangles that are not equilateral? Why?

BRAIN TEASER In Figure 14-18, a coin is shown above and touching another coin. Suppose the top coin is rotated around the circumference of the bottom coin without slipping until it rests directly below the bottom coin. Will the head be straight up or upside down?

Figure 14-18

14-2 Reflections and Glide Reflections

Reflections

Another isometry is a **reflection**, or **flip**. One example of a reflection often encountered in our daily lives is a mirror image. Figure 14-19 shows a figure with its mirror image.

Figure 14-19

Another reflection is shown in the following "B.C." cartoon.

We can obtain reflections in a line in various ways. Consider the half tree shown in Figure 14-20(a). Folding the paper along the **reflecting line** and drawing the image gives the **mirror image**, or *image*, of the half tree. In Figure 14-20(b), the paper is shown unfolded. Another way to simulate a reflection in a line involves using a Mira, as illustrated in Figure 14-20(c).

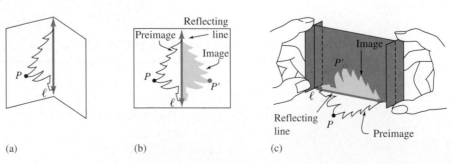

(a)   (b)   (c)

Figure 14-20

In Figure 14-21(a), the image of P under a reflection in line ℓ is P'. $\overline{PP'}$ is both perpendicular to and bisected by ℓ, or equivalently, ℓ is the perpendicular bisector of $\overline{PP'}$. In Figure 14-21(b), P is its own image under the reflection in line ℓ. If ℓ were a mirror, then P' would be the mirror image of P. This leads us to the following definition of a reflection.

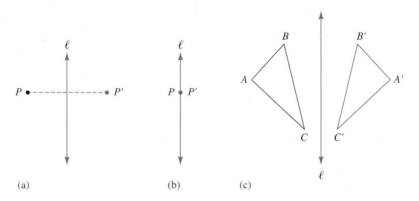

(a)   (b)   (c)

Figure 14-21

Definition of Reflection

A **reflection** in a line ℓ is a transformation of a plane that pairs each point P of the plane with a point P' in such a way that ℓ is the perpendicular bisector of $\overline{PP'}$, as long as P is not on ℓ. If P is on ℓ, then $P = P'$.

In Figure 14-21(c), we see another property of a reflection. In the original triangle ABC, if we walk clockwise around the vertices, starting at vertex A, we see the vertices in the order A-B-C. However, in the reflection image of triangle ABC, if we start at A' (the image of A) and walk clockwise, we see the vertices in the following order: A'-C'-B'. Thus a reflection does something that neither a translation nor a rotation does; it reverses the **orientation** of the original figure.

There are many methods of constructing a reflection image. We already illustrated such constructions with paper folding and a Mira. Next, we illustrate the construction of the image of a figure under a reflection in a line with tracing paper.

Constructing a Reflection by Using Tracing Paper

Figure 14-22(a) shows the use of tracing paper. We trace the original figure, the *reflecting line*, and a point on the reflecting line, which we use as a *reference point*. When we flip the tracing paper over to perform the reflection, we align the reflecting line and the reference point, as in Figure 14-22(b). Aligning the reference point ensures that no translating occurs along the reflecting line when the reflection is performed. If we wish the image to be on the paper with the original, we may indent the tracing paper or acetate sheet to mark the images of the original vertices.

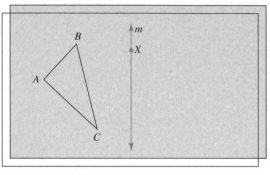

(a)

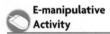
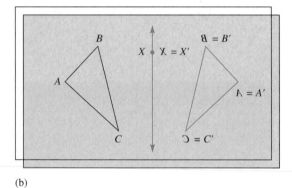

(b)

Figure 14-22

NOW TRY THIS 14-3 Use the definition of a reflection in a line and properties of a rhombus to construct the image P' of point P in Figure 14-23 under reflection in line m using only a compass and straightedge.

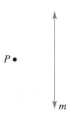

Figure 14-23

Reflections, rotations, and translations appear in the elementary curriculum, as shown in the following partial student page. Work through the activity and questions 1 through 5 on the page.

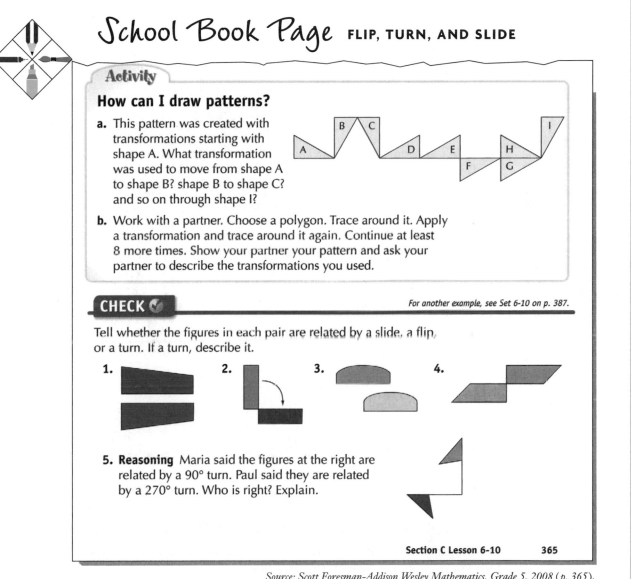

School Book Page FLIP, TURN, AND SLIDE

Activity

How can I draw patterns?

a. This pattern was created with transformations starting with shape A. What transformation was used to move from shape A to shape B? shape B to shape C? and so on through shape I?

b. Work with a partner. Choose a polygon. Trace around it. Apply a transformation and trace around it again. Continue at least 8 more times. Show your partner your pattern and ask your partner to describe the transformations you used.

CHECK ✓   *For another example, see Set 6-10 on p. 387.*

Tell whether the figures in each pair are related by a slide, a flip, or a turn. If a turn, describe it.

1.   **2.**   **3.**   **4.**

5. Reasoning Maria said the figures at the right are related by a 90° turn. Paul said they are related by a 270° turn. Who is right? Explain.

Section C Lesson 6-10   **365**

Source: Scott Foresman–Addison Wesley Mathematics, Grade 5, 2008 (p. 365).

As the Research Note (p. 958) relates, students may have difficulty operating on shapes. Example 14-5 deals with a reflection of a line.

Example 14-5

Describe how to construct the image of $\overleftrightarrow{AB}$ under a reflection in line m in Figure 14-24.

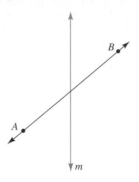

Figure 14-24

Solution   Under a reflection, the image of a line is a line. Thus, to find the image of $\overleftrightarrow{AB}$, it is sufficient to choose any two points on the line and find their images. The images determine the line that is the image of $\overleftrightarrow{AB}$. We choose two points whose images are easy to find. Point X, the intersection of $\overleftrightarrow{AB}$ and m, is its own image. If we choose point A and use the hint in Now Try This 14-3, we produce the construction shown in Figure 14-25.

♦ *Research Note*

Students have some informal understanding of geometric transformations such as reflections and rotations but have a difficult time operating on shapes using these transformations (Kuchemann 1981). ♦

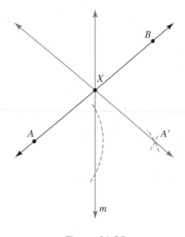

Figure 14-25

Constructing a Reflection on Dot Paper or a Geoboard

On dot paper or a geoboard, the images of figures under a reflection can sometimes be found by inspection, as seen in Example 14-6.

Example 14-6

Find the image of $\triangle ABC$ under a reflection in line m, as in Figure 14-26.

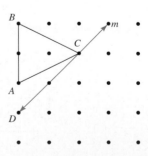

Figure 14-26

Solution   The image $A'B'C'$ is given in Figure 14-27. Note that C is the image of itself and the images of the vertices A and B are A' and B' such that m is the perpendicular bisector of $\overline{AA'}$ and $\overline{BB'}$.

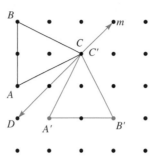

Figure 14-27

 NOW TRY THIS 14-4 Trace $\triangle ABC$ and $\triangle A'B'C'$ from Figure 14-27 on a sheet of paper (without tracing line m). Find the line of reflection using paper folding.

Reflections in a Coordinate System

For some reflecting lines, like the x-axis and y-axis and the line $y = x$, it is quite easy to find the coordinates of the image, given the coordinates of the point. In Figure 14-28, the line $y = x$ bisects the angle between the x-axis and y-axis. The image of $A(1, 4)$ is the point $A'(4, 1)$. Also the image of $B(^-3, 0)$ is $B'(0, ^-3)$.

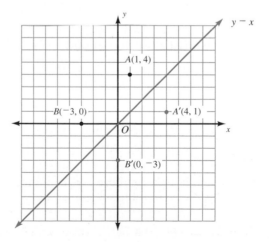

Figure 14-28

 NOW TRY THIS 14-5 Show that in general the image of $P(a, b)$ under reflection in line $y = x$ is the point $P'(b, a)$ and consequently that the reflection in the line $y = x$ exchanges the coordinates of the point.

TECHNOLOGY CORNER *GSP* Lab 11 activities 1–5 can be used to investigate properties of reflections.

Glide Reflections

Another basic isometry is a **glide reflection**. An example of a glide reflection is shown in the footprints of Figure 14-29. We consider the footprint labeled F_1 to have been translated to footprint F_2 and then reflected over line m (parallel to $\overline{F_1F_2}$) to yield F_3, the image of F_2. F_3 is the final image of F_1.

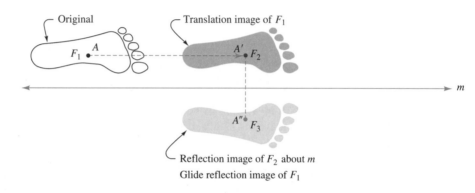

Figure 14-29

◆ *Research Note*

Children first discover that figures can be superimposed by motion when one glides the transparent sheet on the plane without lifting it, or by reversals when it is necessary to lift the transparent sheet, flip it in space to change sides, and then glide it (Demal 2004). ◆

Note that point A is translated to find point A' and then point A' is reflected over line m (parallel to $\overline{F_1F_2}$) to obtain point A''. Thus, A'' is the image of A in the glide reflection. The illustration in Figure 14-29 leads us to the following definition.

> **Definition of Glide Reflection**
>
> A **glide reflection** is a transformation consisting of a translation followed by a reflection in a line parallel to the slide arrow.

Because constructing a glide reflection involves constructing a translation and a reflection, the task of constructing a glide reflection is not a new problem. Exercises involving the construction of images of figures under glide reflections are left as assessments. As the research notes states, children are comfortable with glide reflections done on acetate sheets.

TECHNOLOGY CORNER *GSP* Lab 12, activity 1 can be used to investigate glide reflections.

Congruence via Isometries

We have seen that under an isometry, the image of a figure is a congruent figure. Also, given two congruent polygons, it is possible to show that one can be transformed to the other by using a sequence of isometries. In fact, *two geometric figures are congruent if and only if, one is an image of the other under a single isometry or under a composition of isometries.* The following example shows one illustration of this approach to congruence.

Example 14-7   *ABCD* in Figure 14-30 is a rectangle. Describe a sequence of isometries to show

a. $\triangle ADC \cong \triangle CBA$ **b.** $\triangle ADC \cong \triangle BCD$ **c.** $\triangle ADC \cong \triangle DAB$

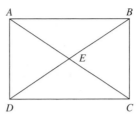

Figure 14-30

Solution   **a.** A half-turn of $\triangle ADC$ with center E is one such transformation.

b. A reflection in a line passing through E and parallel to $\overline{AD}$ is one such transformation.

c. A reflection of $\triangle ADC$ in a line passing through E and parallel to $\overline{DC}$ is one such transformation.

Light Reflecting from a Surface

When a ray of light bounces off a mirror or when a billiard ball bounces off the rail of a billiards table, the **angle of incidence**, the angle formed by the incoming ray in Figure 14-31 and a line perpendicular to the mirror, is congruent to the **angle of reflection**, the angle between the reflected ray and the line perpendicular to the mirror.

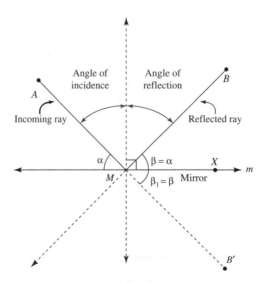

Figure 14-31

Because the angle of incidence is congruent to the angle of reflection, the respective complements of these angles must be congruent. If the measures of these complements are α and β, as indicated in Figure 14-31, then $\alpha = \beta$. Figure 14-31 shows B', the image of B under reflection in m, and hence $\angle XMB'$ is the image of $\angle XMB$. Notice that β, the measure of $\angle XMB$, must equal β_1, the measure of $\angle XMB'$, because reflection preserves angle measurement. Because $\alpha = \beta$ and $\beta = \beta_1$, we get $\alpha = \beta_1$. For that reason, points A, M, and B' are collinear (why?). We can show that these facts imply *Fermat's Principle:* Light follows the path of shortest distance; that is, the path *A-M-B* that light travels is the shortest among all the paths connecting A with a point in the mirror to B.

 NOW TRY THIS 14-6 Trace Figure 14-31 and mark on the mirror m a point P other than M. Prove that $AM + MB < AP + PB$ by showing that $MB = MB'$ and $BP = B'P$, and then by using the triangle inequality applied to $\triangle APB'$ and the fact that A, M, and B' are collinear.

Assessment 14-2A

1. For each of the following figures, find the image of the given quadrilateral in a reflection in ℓ:

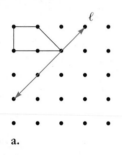

a.

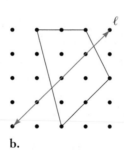

b.

2. Determine which of the following figures have a reflecting line such that the image of the figure under the reflecting line is the figure itself. In each case, find as many such reflecting lines as possible, sketching appropriate drawings.
   a. Circle
   b. Segment
   c. Ray
   d. Line
   e. Square
   f. Rectangle
   g. Scalene triangle
   h. Isosceles triangle
   i. Equilateral triangle
   j. Trapezoid whose base angles are not congruent
   k. Isosceles trapezoid
   l. Arc
   m. Kite
   n. Rhombus
   o. Regular hexagon
   p. Regular n-gon

3. Determine the final result when $\triangle ABC$ is reflected in line ℓ and then its image is reflected again in ℓ.

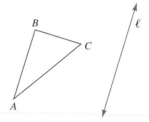

4. a. Refer to the following figure and suppose lines ℓ and m are parallel and $\triangle ABC$ is reflected in ℓ to obtain $\triangle A'B'C'$ and then $\triangle A'B'C'$ is reflected in m to obtain $\triangle A''B''C''$. Determine whether the same final image is obtained if $\triangle ABC$ is reflected first in m and then its image is reflected in ℓ.

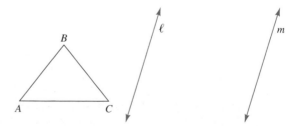

   b. Conjecture what single transformation will take $\triangle ABC$ directly to $\triangle A''B''C''$. Check your conjecture using tracing paper.

5. a. For the following figure, use any construction method to find the image of $\triangle ABC$ if $\triangle ABC$ is reflected in ℓ to obtain $\triangle A'B'C'$ and then $\triangle A'B'C'$ is reflected in m to obtain $\triangle A''B''C''$ (ℓ and m intersect at O):

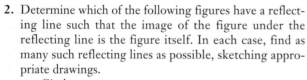

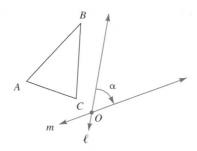

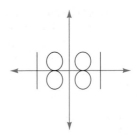

b. Conjecture what single transformation will take △ABC directly to △$A''B''C''$. Check your conjecture using tracing paper.

c. Answer the questions in (a) and in (b) for the case in which ℓ and m are perpendicular.

6. Use a Mira, if available, to investigate exercises 4 and 5.

7. Given △ABC and its reflection image △$A'B'C'$, find the line of reflection.

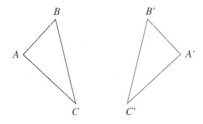

8. a. The word TOT is its own image when it is reflected through a vertical line through O, as shown in the following figure. List some other words that are their own images when reflected similarly.

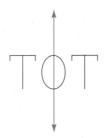

b. The image of BOOK is still BOOK when it is reflected through a horizontal line. List some other words that have the same property. Which uppercase letters can you use?

c. With an appropriate font, the image of 1881 is 1881 after reflection in either a horizontal or vertical line, as shown in the following figure. What other natural numbers less than 2000 have this property?

9. A glide reflection is determined by a translation parallel to a given line followed by a reflection in that line.

a. Determine whether the same final image is obtained if the reflection is followed by the translation.

b. Use your answer in (a) to determine whether the reflection and translation involved in the glide reflection commute.

10. For the following figure numbered 1, decide whether a reflection, a translation, a rotation, or a glide reflection will transform the figure into each of the other numbered figures (there may be more than one answer).

11. Given points $A(3, 4)$, $B(2, ^-6)$, and $C(^-2, 5)$, find the coordinates of the images of these points under each of the following transformations:

a. Reflection in the x-axis

b. Reflection in the line $y = x$

12. a. Conjecture what the image of a point with coordinates (x, y) will be under each of the transformations in exercise 11.

b. Suppose a point P with coordinates (x, y) is reflected in the x-axis and then its image P' is reflected in the y-axis to obtain P''. What are the coordinates of P'' in terms of x and y? Justify your answer.

13. Find the equations of the images of the following lines when reflected in the x-axis:

a. $y = ^-x + 3$

b. $y = 0$

14. Find the equation of the images of the following lines when the reflection line is the y-axis.

a. $y = ^-x + 3$ **b.** $y = 0$

15. A quadrilateral has vertices at $(a, 0)$, $(0, a)$, $(^-a, 0)$, and $(0, -a)$, where $a > 0$.

a. What is the most you can say about the quadrilateral?

b. Find the image of each vertex under the reflection in the line $y = x$.

16. The two circles "touch" each other at point P; that is, the tangent to one circle at P is also the tangent to the other circle. Find the line of reflection such that the image of the smaller circle will be in the interior of the larger circle, but still "touching" the larger circle.

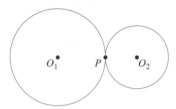

17. Two congruent circles with centers O_1 and O_2 intersect at points A and B.

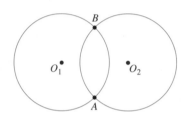

   a. In what line of reflection will the image of the circle with center O_1 be the circle with center O_2?

   b. Can the circle with center O_1 be transformed into the circle with center O_2 by a translation? If so, describe the translation.

18. Graph each pair of the following lines and for each construct the corresponding line of reflection so that the image of one line in the pair will be the second line. For each, identify the line of reflection.
   a. $y = {}^-x$ and $y = x$
   b. $y = 2x$ and $y = {}^-2x$

19. Two farm houses are located away from a road, as shown. A telephone company wants to construct a telephone pole at the edge of the road so that the telephone cables connecting the houses to the pole are as short as possible. Where should the pole be located?

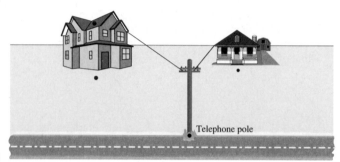

Telephone pole

Assessment 14-2B

1. In the following figure, find the images of $\triangle ABC$ under a reflection in line ℓ:

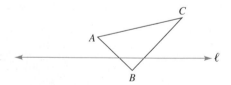

2. Draw a line and then draw a circle whose center is not on the line. Find the image of the circle under a reflection in the line.

3. Sketch the image of the figure below through a reflection in line ℓ.

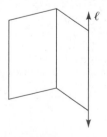

4. Find the equation of the image of the line with equation $y = 2x + 1$ when it is reflected in each of the following:
   a. x-axis
   b. y-axis
   c. $y = x$

5. Describe the set of all lines that are their own images when reflected in each of the following:
   a. x-axis
   b. y-axis
   c. $y = x$

6. a. Construct a "stylized" bow tie by using a reflection of the triangle below:

   b. Tell which reflecting line was used in part (a) and why.
   c. Is there more than one reflecting line that could have been used in part (a)? If so, where are others?

7. In the Beetle Bailey cartoon strip, Sarge's pet's name is Otto.
   a. If Otto is spelled in all capital letters, draw a reflecting line to show that the name can be its own image in a reflection.
   b. Where is the reflecting line?

8. Zuni art contains figures similar to the one partially drawn below. Use the pictured reflecting line to complete an abstract version of this image.

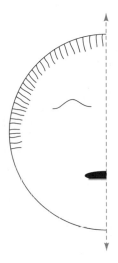

9. In the Pharlemina's Favorite quilt pattern below, describe a motion that will take part (a) to part (b).

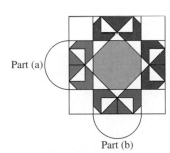

Part (a)

Part (b)

10. Follow the directions in exercise 9 for the Dutchman's Puzzle quilt pattern shown

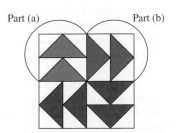

Part (a)   Part (b)

11. Given points $A(3, 4)$, $B(2, {}^{-}6)$, and $C({}^{-}2, 5)$, find the coordinates of the images of these points under each of the following transformations:
    a. Reflection in the y-axis
    b. Reflection in the line $y = {}^{-}x$

12. Conjecture what the image of a point with coordinates (x, y) will be under each of the transformations in exercise 11.

13. Find the equations of the images of the following lines when reflected in the x-axis:
    a. $y = 3x$ b. $y = {}^{-}x$
    c. $x = 0$

14. Find the equation of the images of the lines in exercise 13 when the reflection line is the y-axis.

15. In which line will the two intersecting circles reflect onto themselves; that is, the image of the circles will be the same two circles?

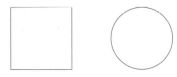

16. Construct a square and a circle of an appropriate size so that it will be possible to transform the circle by a reflection into the square such that the image of the circle will be tangent to all sides of the square. Identify the line of reflection.

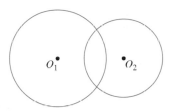

17. Graph each of the following pairs of lines and for each construct the corresponding line of reflection so that the image of one line in the pair will be the second line. For each, identify the line of reflection.
    a. $x = 0$ and $y = 0$
    b. $x = {}^{-}2$ and $x = 3$

18. A *fixed point* of a transformation is a point whose image is the point itself. List all the fixed points of the following isometries:
    a. Reflection in line ℓ
    b. Rotations by a given angle and direction about point O
    c. Translation with slide arrow from A to B
    d. Glide reflection determined by a translation with slide arrow $\overrightarrow{AB}$ followed by a reflection in line ℓ parallel to line AB

Mathematical Connections 14-2

Communication

1. When a billiard ball bounces off a side of a pool table, assume the angle of incidence is congruent to the angle of reflection. In the following figure showing a scale drawing of a pool table, a cue ball is at point *A*. Show how a player should aim to hit two sides of the table and then the ball at *B*. Justify your solution.

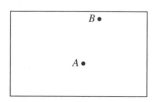

2. **a.** Draw an isosceles triangle *ABC* and then construct a line such that the image of △*ABC* when reflected in the line is △*ABC* though every point is not necessarily its own image. Explain why the line you constructed has the required property.
   b. For what kind of triangles is it possible to find more than one line with the property in (a)? Justify your answer.
   c. Given a scalene triangle *ABC*, is it possible to find a line ℓ such that when △*ABC* is reflected in ℓ, its image is itself? Explain your answer.
   d. Draw a circle with center *O* and a line with the property that the image of the circle, when reflected in the line, is the original circle. Identify all such lines. Justify your answer.

3. Use the following drawing to explain how a periscope works:

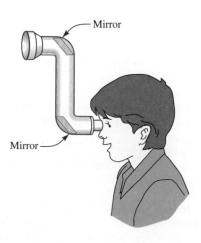

Open-Ended

4. In the following figure representing a miniature golf course hole, explain and justify the procedure showing how to aim the ball so that it gets in the hole if it is to bounce off
   a. one wall only.   **b.** two walls.

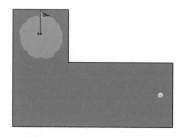

5. Design several wall stencil patterns using a reflection. In each case, explain how you would use the stencil in practice.
6. Design wall stencil patterns using a glide reflection.

Cooperative Learning

7. In the following figure representing a pool table, ball *B* is sent on a path that makes a 45° angle with the table wall, as shown. It bounces off the wall 5 times and returns to its original position.

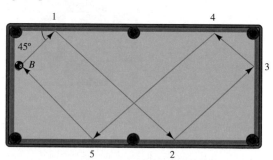

   a. Have each member of your group use graph paper to construct rectangular models of different-sized pool tables. Simulate the experiment using any tools (such as a straightedge, compass, and protractor) by choosing different positions for ball *B*.
   b. Share the results of your experiments with the rest of the group and together conjecture for which dimensions of the pool table and for what positions of *B* the experiment described in the problem will work.

Questions from the Classroom

8. A student asks, "If I have a point and its image, is that enough to determine whether the image was found using a translation, reflection, rotation, or glide reflection?" How do you respond?

9. Another student asks a question similar to question 9 but is concerned about a segment and its image. How do you respond to this student?

10. Sammi said that in the drawing below the image A' of point A under a reflection in line ℓ can also be found by a rotation of α with center O. Therefore, every reflection is actually a rotation. How do you respond?

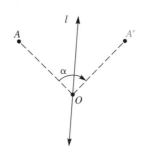

Review Problems

11. Which single digits are their own images under a rotation by an angle whose measure is less than $360°$?

12. What is the image of a point (a, b) under a half-turn in the origin?

13. Find all possible rotations that transform a circle onto itself.

14. Explain how an isometry can be used to construct a rectangle whose area is equal to that of the parallelogram $ABCD$ in the following figure:

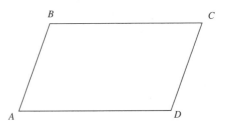

BRAIN TEASER   Two cities are on opposite sides of a river, as shown in Figure 14-32.

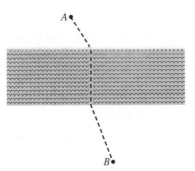

Figure 14-32

The cities' engineers want to build a bridge across the river that is perpendicular to the banks of the river and access roads to the bridge so that the distance between the cities is as short as possible. Where should the bridge and the roads be built?

14-3   Size Transformations

Isometries preserve distance. Consequently, the image of a figure under an isometry is a figure congruent to the original. A different type of transformation happens when a slide is projected on a screen. All objects on the slide are often enlarged on the screen by the same factor. Figure 14-33 is another example of such a transformation, a **size transformation**. (In *Focal Points*, the word *scaling* is used.) The point O is the *center* of the size transformation and 2 is the *scale factor*. Points O, A, and A' are collinear and $OA' = 2OA$; also, O, C, and C' are collinear and $OC' = 2OC$. Similarly, O, B, and B' are collinear and $OB' = 2OB$. It can be shown that $\triangle A'B'C'$ is similar to $\triangle ABC$ and hence that each side of $\triangle A'B'C'$ is twice as long as the corresponding sides of $\triangle ABC$.

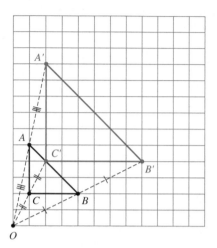

Figure 14-33

In general, we have the following definition.

Definition of Size Transformation

A size transformation from the plane to the plane with center O and scale factor r $(r > 0)$ is a transformation that assigns to each point A in the plane the point A' such that O, A, and A' are collinear and $OA' = rOA$ and such that O is not between A and A'.

REMARK It is possible to define a size transformation when the scale factor is negative in the same way as in the preceding definition for positive scale factor, except that O *must be between A and A'*. Also, note that a size transformation is sometimes called a **dilation**.

Example 14-8

a. In Figure 14-34(a), find the image of point P under a size transformation with center O and scale factor $\frac{2}{3}$.

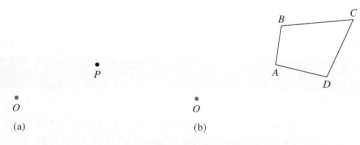

Figure 14-34

b. Find the image of the quadrilateral $ABCD$ in Figure 14-34(b) under the size transformation with center O and scale factor $\frac{2}{3}$.

Solution **a.** In Figure 14-35(a), we connect O with P and divide $\overline{OP}$ into three congruent

parts. The point P' is the image of P because $OP' = \frac{2}{3}OP$.

b. We find the image of each of the vertices and connect the images to obtain the quadrilateral $A'B'C'D'$, shown in Figure 14-35(b).

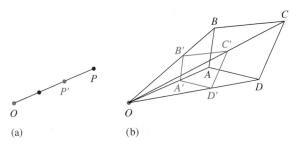

(a)   (b)

Figure 14-35

In Figure 14-35(b), the sides of the quadrilateral $A'B'C'D'$ are all parallel to the corresponding sides of the original quadrilateral, and the angles of the quadrilateral $A'B'C'D'$ are congruent to the corresponding angles of quadrilateral $ABCD$. Also, each side in the quadrilateral $A'B'C'D'$ is $\frac{2}{3}$ as long as the corresponding side of quadrilateral $ABCD$. These properties are true for any size transformation and are summarized in the following theorem.

Theorem 14–3

A size transformation with center O and scale factor r $(r > 0)$ has the following properties:

1. The image of a line segment is a line segment parallel to the original segment and r times as long.
2. The image of an angle is an angle congruent to the original angle.

Example 14-9

Show that $\triangle ABC$ in Figure 14-36 is the image of $\triangle ADE$ under a size transformation. Identify the center of the size transformation and the scale factor.

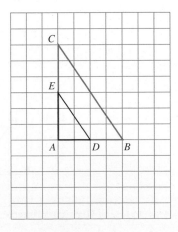

Figure 14-36

Solution   Because $\dfrac{AB}{AD} = \dfrac{AC}{AE} = 2$, we choose A as the center of the size transformation and 2 as the scale factor. Notice that under this transformation, the image of A is A itself. The image of D is B, and the image of E is C.

From Theorem 14–3, it follows that the image of a polygon under a size transformation is a similar polygon. (Why?) However, for any two similar polygons it is not always possible to find a size transformation so that the image of one polygon under the transformation is the other polygon. But, given two similar polygons, we can "move" one polygon to a place so that it will be the image of the other under a size transformation. The following example shows such an instance.

Example 14-10

Show that $\triangle ABC$ in Figure 14-37 is the image of $\triangle APQ$ under a succession of isometries with a size transformation.

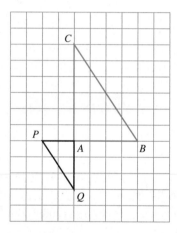

Figure 14-37

Solution   We use the strategy of *looking at a related problem*. In Example 14-9, the common vertex served as the center of the size transformation. This was possible because the corresponding sides of the triangles were parallel. To achieve a similar situation, we first transform $\triangle APQ$ by a half-turn in A and obtain $\triangle AP'Q'$, as shown in Figure 14-38.

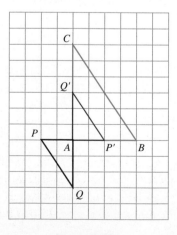

Figure 14-38

Now C is the image of Q' under a size transformation with center at A and scale factor 2. B is the image of P', and A is the image of itself under this transformation. Thus, $\triangle ABC$ can be obtained from $\triangle APQ$ by first finding the image of $\triangle APQ$ under a half-turn in A and then applying a size transformation with center A and scale factor 2 to that image.

◆

REMARK It was noted earlier that a size transformation could be defined using a negative scale factor. If that route had been taken, with A as the center, a size transformation with scale factor $^-2$ could be used to achieve the desired result.

Examples 14-9 and 14-10 are a basis for a definition of similar figures.

Definition of Similar Figures

Two figures are similar if it is possible to transform one onto the other by a sequence of isometries followed by a size transformation.

 Size transformations and similarity in the coordinate plane are discussed on the following student page. Answer the questions on the student page. Conjecture the coordinates of the image of the point (x, y) under a size transformation with center at the origin and scale factor r.

Applications of Size Transformations

One way to make an object appear three-dimensional is to use a **perspective drawing**. For example, to make a letter appear three-dimensional we can use a size transformation with an appropriate center O and a scale factor with selected images to create a three-dimensional effect, as shown in Figure 14-39 for the letter C.

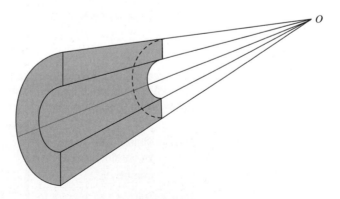

Figure 14-39

When a picture of an object is taken as with a box camera in Figure 14-40(a) with the pinhole lens at O, the object appears upside down on the negative. The picture of the object on the negative can be interpreted as an image under composition of a half-turn and a size transformation. Figure 14-40(b) illustrates the image of an arrow from A to B under a composition of a half-turn followed by a size transformation with scale factor $\frac{1}{2}$.

School Book Page TRANSFORMATIONS AND SIMILARITY

15. You can use an algorithm to model a change that involves stretching. The transformation shown is modeled by the algorithm below.

Step 1 Multiply the *x*- and *y*-coordinates of each point by 2.

Step 2 Translate each point up 1 unit.

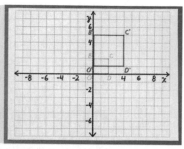

a. Draw *OBCD* with coordinates *O*(0, 0), *B*(0, 2), *C*(2, 2) and *D*(2, 0). Translate the figure up 1 unit. Then multiply each coordinate by 2. Draw the final image.

b. Compare your results from part (a) with the transformation shown. Does the order in which you perform the steps matter?

16. Copy the figure below. Then use the algorithm to transform the figure.

Step 1 Multiply the *x*- and *y*-coordinates of each point by 3.

Step 2 Reflect each point over the *x*-axis.

Step 3 Translate each point up 12 units.

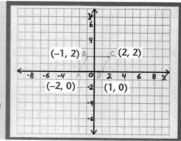

17. You already know how to stretch a figure using multiplication. You can also use multiplication to shrink a figure. Draw the figure below. Then use the algorithm to transform the figure.

Step 1 Multiply the *x*- and *y*-coordinates of each point by $\frac{1}{2}$.

Step 2 Translate each point to the left 1 unit.

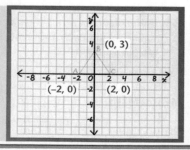

Source: Math Thematics New Edition, Book 3, McDougal Littell, 2008 (p. 439).

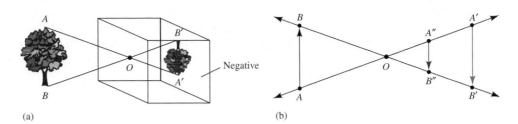

Figure 14-40

In Figure 14-40 the image A' of A under the half-turn with center O is found on the ray opposite $\overrightarrow{OA}$ so that $OA' = OA$. The point B', the image of B under the half-turn, is found similarly on the ray opposite $\overrightarrow{OB}$. The images of A' and B' under the size transformation are A'' and B'', respectively. Consequently, the image of the arrow from A to B under the composition of the half-turn followed by the size transformation is the arrow from A'' to B''.

A geometric construction application of size transformations is shown in Figure 14-41, where a square with two vertices on $\overline{AB}$ and the other two on the semicircular arc are desired.

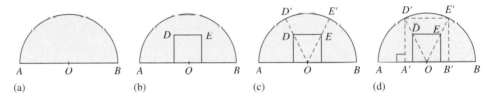

Figure 14-41

To solve the problem, construct any "small" square with one side on $\overline{AB}$ and centered at O, as in Figure 14-41(b). Draw rays $\overrightarrow{OD}$ and $\overrightarrow{OE}$ intersecting the semicircle at D' and E', respectively, as in Figure 14-41(c). Now drop perpendiculars to $\overline{AB}$ from D' and E' to obtain A' and B', respectively. Quadrilateral $A'D'E'B'$ shown in Figure 14-41(d) is the required square. (Why?)

Assessment 14-3A

1. In the following figures, describe a sequence of isometries followed by a size transformation so that the larger triangle is the final image of the smaller one:

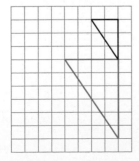

a.

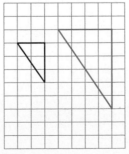

b.

2. In the following drawing, find the image of $\triangle ABC$ under the size transformation with center O and scale factor $\frac{1}{2}$:

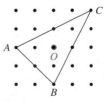

3. In each of the following drawings, find transformations that will take $\triangle ABC$ to its image, $\triangle A'B'C'$, which is similar:

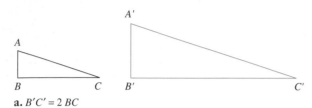

   a. $B'C' = 2\,BC$

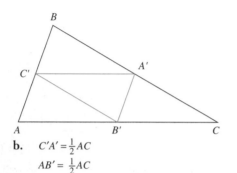

   b. $C'A' = \frac{1}{2}AC$

   $AB' = \frac{1}{2}AC$

4. In the following figure, the smaller triangle is the image of the larger under a size transformation centered at point O. Find the scale factor and the length of x and y as pictured.

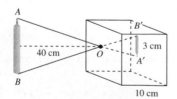

5. $\overline{A'B'}$ is the image of a candle $\overline{BA}$ produced by a box camera. Given the measurements in the figure, find the height of the candle.

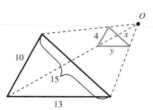

6. The following describes a size transformation with center O and image of the segment of length 4 in blue. Find the scale factor and the lengths designated by x and y.

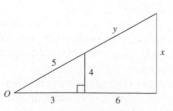

7. Find the coordinates of the images of $A(2, 3)$, and $B(^-2, 3)$ under the following transformations. Assume that all size transformations are centered at the origin.
   a. A size transformation with a scale factor 3 followed by a size transformation with a scale factor 2
   b. A size transformation with scale factor 2 followed by a size transformation with a scale factor 3

8. If a size transformation with center O and scale factor r takes a quadrilateral $ABCD$ to $A'B'C'D'$, what size transformation will take $A'B'C'D'$ back to $ABCD$?

9. **a.** Explain why in a coordinate system a size transformation with center at the origin and scale factor $r\,(r > 0)$ is given by $(x, y) \rightarrow (rx, ry)$.
   b. The transformation $(x, y) \rightarrow (^-2x, ^-2y)$ can be achieved by a size transformation followed by an isometry. Find that size transformation and the isometry.
   c. Find the equations of the images of each of the following lines under size transformation in part (a) with the given scale factor r:

   (i) $y = 2x, r = \dfrac{1}{2}$

   (ii) $y = 2x, r = 2$

   (iii) $y = 2x + 1, r = \dfrac{1}{3}$

   (iv) $y = {}^-x - 1, r = 3$

10. What sequence of transformations will transform a circle with center O_1 and radius 2 onto a circle with center O_2 and radius 3?

11. **a.** If a size transformation had a scale factor of 1, describe the net result.
    b. Why could a size transformation not have a scale factor of 0?

12. Consider a size transformation on a number line where the center of the transformation has O as its coordinate and the scale factor is 3. Describe the set of images of the points whose coordinates are integers.

13. If a figure had an area of 7 and was transformed by a size transformation with a scale factor of 3, what is the area of the image?

14. If a 2" × 3" photograph is enlarged to a 4" × 6" photograph, explain whether this can be represented by a size transformation.

Assessment 14-3B

1. In the following figures, describe a sequence of isometries followed by a size transformation so that the larger triangle is the final image of the smaller one:

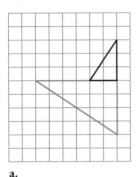

    a.   **b.**

2. Sketch the image of the Octagon Quilt pattern with the center of the pattern as the center of a size transformation with a scale factor of 2.

Octagon

3. In each of the following drawings, find transformations that will take $\triangle ABC$ to its image, $\triangle A'B'C'$, which is similar:

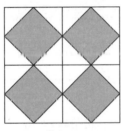

a. $AB = 2\,A'B'$

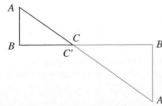

b. $\overline{AB} \parallel \overline{A'B'}$; points A, C, and A' are collinear; points B, C', and B' are collinear; $AB = \frac{3}{2}A'B'$

4. In the following figure, the smaller pentagon is the image of the larger under a size transformation centered at point O. Find the scale factor and the length of x and y as pictured.

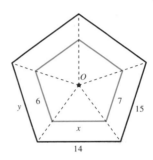

5. Find the image of $\triangle ABC$ under the size transformation with center O and scale factor $\frac{1}{2}$.

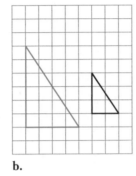

6. The following describes a size transformation with center O and image in blue. Find the scale factor and the lengths designated by x and y.

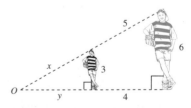

7. Find the coordinates of the images of $A(2, 3)$, and $B(^-2, 3)$ under the following transformations. Assume that all size transformations are centered at the origin.
   a. A size transformation with scale factor 2 followed by a translation with a slide arrow from $(2, 1)$ to $(3, 4)$
   b. A translation with a slide arrow from $(2, 1)$ to $(3, 4)$ followed by a size transformation with scale factor 2
   c. What can you conclude from your answers to parts (a) and (b) concerning the order in which the transformations are performed?

8. What size transformation will transform a circle with center at the origin and radius 4 onto a circle with the same center and radius 3?

9. Construct any triangle ABC and a square $DEFG$ so that D is on $\overline{AB}$ and the vertices F and G are on $\overline{AC}$, as shown, and answer the following:

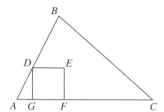

   a. Find the images of the square $DEFG$ under several size transformations with center A and different scale factors. What do all the images of vertex E have in common?

   b. Explain why the intersection of $\overleftrightarrow{AE}$ and $\overline{BC}$ is a vertex of a square. Construct that square.

10. Copy the following figure onto grid paper and determine the center and the scale factor of the size transformation. Why is there only one possibility for the center?

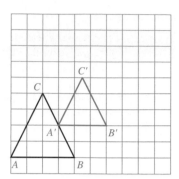

11. What size transformation will undo a size transformation with center O and scale factor $\dfrac{3}{4}$?

12. On the student page on page 972 coordinates are multiplied by 2 and these points are translated up 1 unit. Explain whether this operation describes a size transformation and, if so, determine its center.

Mathematical Connections 14-3

Communication

1. Which of the following properties do not change under a size transformation? Explain how you can be sure of your answers.

   a. Distance between points

   b. Angle measure

   c. Parallelism; that is, if two lines are parallel to each other, then their images are parallel to each other.

2. Given two similar figures, explain how to tell if there is a size transformation that transforms one of the figures onto the other.

3. **a.** Consider two consecutive size transformations, each with center O and corresponding scale factors $\dfrac{1}{2}$ and $\dfrac{1}{3}$, respectively. Suppose the image of figure F under the first transformation is F' and the image of F' under the second transformation is F''. What single transformation will map F directly onto F''? Explain why.

   b. What would be the answer to (a) if the scale factors were r_1 and r_2?

4. **a.** Is the image of a circle with center O under a size transformation with center O always a circle? Explain why or why not.

   b. Assume that under a size transformation with scale factor r, the image of a segment of length a is a segment of length ra and answer part (a) of this question in case the center of the circle is not at the origin.

Open-Ended

5. Describe several real-life situations other than the ones discussed in this section in which size transformations occur.

6. Use a sheet of graph paper with a coordinate system. Locate the origin as the center of a transformation with a scale factor of $^-1$ and find the image of some triangle located in the first quadrant. What other transformation that you've studied has the same image as this transformation?

Cooperative Learning

7. Have members of your group draw several figures and find their images under a size transformation with a scale factor of 3 and center of your choice.

   a. How does the perimeter of each image compare to the perimeter of the original figure? Compare your answers.

   b. How does the area of each image compare to the area of the original figure? Compare your answers.

   c. Make a conjecture concerning the relationship between the perimeter of each image and the perimeter of the original figure under a size transformation with a scale factor r.

   d. Repeat (c) for the area of each figure.

   e. Discuss your findings and come up with a group conjecture.

Questions from the Classroom

8. A student asks, "If I have two triangles that are not similar, is it possible to transform one onto the other by a sequence of isometries followed by a size transformation?" How do you respond?

9. A student says that a coordinate grid under a size transformation with the center at the origin and scale factor 2 does not change the grid. The image is still a coordinate grid. How do you respond?

Review Problems

10. Describe a transformation that would "undo" each of the following:
   a. A translation determined by slide arrow from M to N
   b. A rotation of $75°$ with center O in a clockwise direction
   c. A rotation of $45°$ with center A in a counterclockwise direction
   d. A glide reflection that is the composition of a reflection in line m and a translation that takes A to B where $\overleftrightarrow{AB} \parallel m$
   e. A reflection in line n

11. In the following coordinate plane, find the images of each of the given points in the transformation that is the composition of a reflection in line m followed by a reflection in line n.

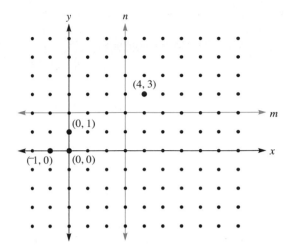

   a. $(4, 3)$ **b.** $(0, 1)$
   c. $(^-1, 0)$ **d.** $(0, 0)$

12. Find each of the following:
   a. Reflection image of an angle in its angle bisector
   b. Reflection image of a square in one of its diagonals

TECHNOLOGY CORNER   Use GSP to draw $\triangle ABC$ and three lines m, n, and p that intersect in a single point, as shown in Figure 14-42. Reflect $\triangle ABC$ in line m to obtain its image $\triangle A'B'C'$. Then reflect $\triangle A'B'C'$ in line n to obtain $\triangle A''B''C''$. Finally, reflect $\triangle A''B''C''$ in line p to obtain the final image $\triangle A'''B'''C'''$.

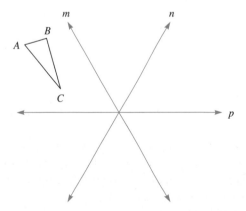

Figure 14-42

Find a single line q that could be used to reflect the original $\triangle ABC$ onto the final image.

14-4   Symmetries

Line Symmetries

The concept of a reflection can be used to identify line symmetries of a figure. All the drawings in Figure 14-43 have symmetries about the dashed lines.

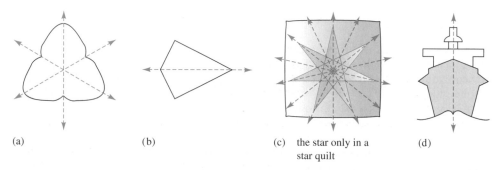

(a)   (b)   (c)   the star only in a   (d)
                                                     star quilt

Figure 14-43

Mathematically, a geometric figure has a **line of symmetry** ℓ if it is its own image under a reflection in ℓ. A method of creating a symmetrical figure is seen in Example 14-11.

Example 14-11   In Figure 14-44, we are given a figure and a line m. Do the minimum amount of drawing to create a figure from the given figure so that the result has line m as its line of symmetry.

Figure 14-44

Solution   For the resulting figure both to be symmetric about line m and to incorporate the existing figure, we reflect the existing figure about line m. The desired result of doing that is the combination of the original figure and the image. The resulting figure is shown in Figure 14-45.

Figure 14-45

Example 14-12

How many lines of symmetry does each drawing in Figure 14-46 have?

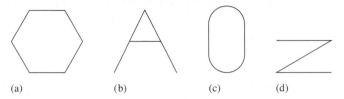

(a)   (b)   (c)   (d)

Figure 14-46

Solution **a.** 6   **b.** 1   **c.** 2   **d.** 0

 NOW TRY THIS 14-7   Create a figure with exactly 4 lines of symmetry.

Example 14-13

Given an arc AB in Figure 14-47, use symmetry and paper folding to find the center and the radius of the circle to which the arc belongs.

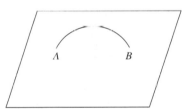

A B

Figure 14-47

Solution   The arc is part of a circle. The center of the circle is equidistant from the end-points of any chord of the circle and hence is on the perpendicular bisector of the chord. That perpendicular bisector is a line of symmetry of the circle. A circle has infinitely many lines of symmetry and each line passes through the center of the circle, where all the lines of symmetry intersect. To find a line of symmetry, fold the paper containing arc AB so that a portion of the arc is folded onto itself. Then unfold the paper and draw the line of symmetry on the fold mark, as shown in Figure 14-48(a). By refolding the paper in Figure 14-48(a) so that a different portion of the arc AB is folded onto itself, we can determine a second line of symmetry, as shown in Figure 14-48(b). The two dotted lines of symmetry intersect at O, the center of the circle of which $\overset{\frown}{AB}$ is an arc. Either $\overline{OB}$, $\overline{OA}$, or any segment connecting O to a point on the arc represents a radius of the circle.

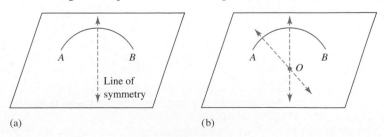

A B A B
           Line of O
           symmetry

(a)   (b)

Figure 14-48

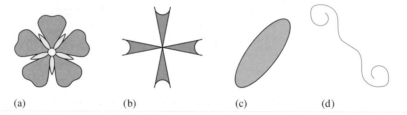

Figure 14-49

Rotational (Turn) Symmetries

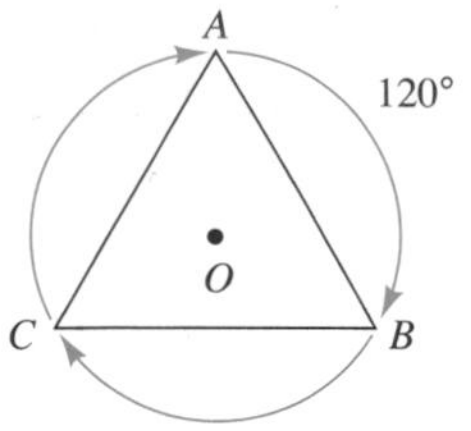

A figure has **rotational symmetry**, or **turn symmetry**, when the traced figure can be rotated less than 360° about some point so that it matches the original figure. Note that the condition "less than 360°" is necessary because any figure will coincide with itself if it is rotated 360° about any point. In Figure 14-49, the equilateral triangle coincides with itself after a rotation of 120° about point O. Hence, we say that the triangle has 120° rotational symmetry. Also in Figure 14-49, if we were to rotate the triangle another 120°, we would find again that it matches the original. So we can say that the triangle also has 240° rotational symmetry.

In general, if a figure has α degrees of rotational symmetry, it also will coincide with itself when rotated by $n\alpha$ degrees for any integer n. For this reason, in rotational symmetry it is sufficient to report the smallest possible positive angle measure that turns the figure onto itself. Notice that a circle has a rotational symmetry by any turn around its center. Thus, a circle has infinitely many rotational symmetries.

Other examples of figures that have rotational symmetry are shown in Figure 14-50. Figures 14-50(a), (b), (c), and (d) have 72°, 90°, 180°, and 180° rotational symmetries, respectively [(a) and (b) also have other rotational symmetries].

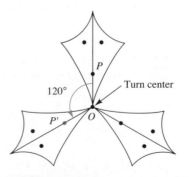

(a)	(b)	(c)	(d)

Figure 14-50

For "simple" figures, we can determine whether a figure has rotational symmetry by tracing it and turning the tracing about a point (the center of the figure) to see if it aligns on the figure before the tracing has turned in a complete circle, or 360°. The amount of the rotation can be determined by measuring the angle $\angle POP'$ through which a point P is rotated around a point O to match another point P' when the figures align. Such an angle, $\angle POP'$, is labeled with points P, O, and P' in Figure 14-51 and has measure 120°. Point O, the point held fixed when the tracing is turned, is the *turn center*.

Figure 14-51

NOW TRY THIS 14-8 Describe a figure that has 2° turn symmetry.

Example 14-14   Determine the amount of the turn for the rotational symmetries of each part of Figure 14-52.

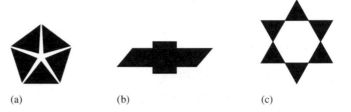

(a)   (b)   (c)

Figure 14-52

Solution   **a.** The amounts of the turns are $\dfrac{360°}{5}$ (or 72°), 144°, 216°, and 288°.

b. The amount of the turn is 180°.

c. The amounts of the turns are 60°, 120°, 180°, 240°, and 300°.

The rotations in Figure 14-52(b) and (c) exemplify yet another type of symmetry, namely, point symmetry.

Point Symmetry

Any figure that has 180° rotational symmetry is said to have **point symmetry** about the turn center. Some figures with point symmetry are shown in Figure 14-53(a). As illustrated in Figure 14-53(b), any figure with point symmetry is its own image under a half-turn. This makes the center of the half-turn the midpoint of a segment connecting a point and its image.

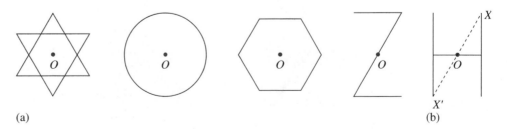

(a)   (b)

Figure 14-53

 Tracing and paper folding are commonly used in elementary school to investigate symmetries of figures, as shown in the following student page. Work through the questions on the student page.

School Book Page   INVESTIGATING SYMMETRY

Symmetry

LEARN

Activity

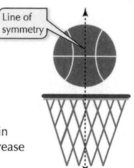

✓ **WARM UP**

Draw an example of
each figure. Then draw
a flip.

1. rectangle

2. trapezoid

3. right triangle

4. obtuse triangle

How can you describe and create symmetric figures?

An artist designed the trademark at
the right for a sporting goods
company. Many trademarks are
symmetric figures. This means they
can be folded into two congruent
parts that fit on top of each other.
The fold line is a **line of symmetry.**

Line of
symmetry

You can follow the steps below to create a
design with two lines of symmetry.

a. Fold a sheet of paper in half. Then fold it in
half again the other way (so the second crease
is perpendicular to the first).

b. Draw a path that starts on one folded edge
and ends at the other folded edge, as shown below.

c. Cut along the curve. Then open up the folded paper.

The figure you made should be symmetric with two lines
of symmetry.

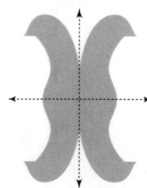

d. How many congruent parts are there in the figure you made?

e. Are the congruent parts related by slides, reflections, or turns?

Source: Scott Foresman-Addison Wesley Mathematics, Grade 5, 2008 (p. 368).

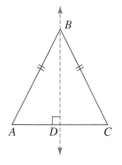

Figure 14-54

Classification of Figures by Their Symmetries

Geometric figures in a plane can be classified according to the number of symmetries they have. Consider a triangle described as having exactly one line of symmetry and no turn symmetries. What could the triangle look like? The only possibility is a triangle in which exactly two sides are congruent; that is, an isosceles triangle that is not equilateral. The line of symmetry passes through a vertex, as shown in Figure 14-54. We can describe equilateral and scalene triangles in terms of the number of lines of symmetry they have. This is left as an exercise.

A square, as in Figure 14-55, can be described as a four-sided polygon with four lines of symmetry—d_1, d_2, h, and v—and three turn symmetries about point O. In fact, we can use lines of symmetry and turn symmetries to define various types of quadrilaterals normally used in geometry. It is left as an exercise to see how these definitions differ from those in Chapter 11.

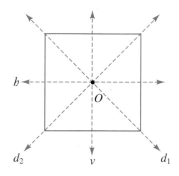

Figure 14-55

Symmetries of Three-Dimensional Figures

A three-dimensional figure has a **plane of symmetry** when every point of the figure on one side of the plane has a mirror image on the other side of the plane. Examples of figures with plane symmetry are shown in Figure 14-56. Solids can also have point symmetry, line symmetry, and turn symmetry. Notice that the right circular cylinder in Figure 14-56(a) has turn symmetries about the line n, which goes through the centers of the upper and lower bases and is perpendicular to them. The line n is the **axis of rotation**. If the cylinder is rotated by any number of degrees about its axis of rotation, its image is the cylinder itself. Hence, the cylinder has infinitely many rotational symmetries. The cube in Figure 14-56(b) also has rotational symmetries. For the cube, if n is a line connecting the centers of two opposite square faces, then a rotation of 90° in any direction will result in an image that coincides with the original cube. We can create an axis of rotation by connecting the centers of any two opposite faces; the cube has several axes of rotation, each with a 90° rotational symmetry. A figure such as in 14-56(c) may have no rotational symmetries other than 360° about any point.

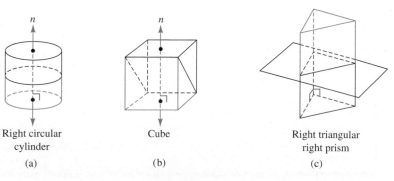

Right circular cylinder

(a)

Cube

(b)

Right triangular right prism

(c)

Figure 14-56

NOW TRY THIS 14-9

a. If we connect two opposite vertices of a cube that are farthest apart, like the pair in Figure 14-57, we obtain an axis of rotation. (i) What is the smallest angle by which the cube needs to be rotated about this axis so that its image will coincide with the original cube? (ii) How many such axes are there?

b. Figure 14-57(b) shows another axis of rotation for the cube. (i) By what angle does the cube need to be rotated about this axis for the image to be the same as the original cube? (ii) How many such axes of rotation are there for a cube?

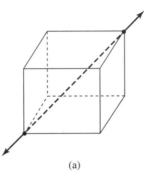
(a)

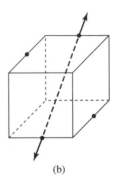
(b)

Figure 14-57

Assessment 14-4A

1. What geometric properties make "SOS" a good choice for the international distress symbol?

2. Various international signs have symmetries. Determine which of the following, if any, have (i) line symmetry, (ii) rotational symmetry, and/or (iii) point symmetry:

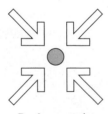

Rendezvous point

a.

Light switch

b.

3. Design symbols that have each of the following symmetries, if possible:
   a. Line symmetry but not rotational symmetry
   b. Rotational symmetry but not point symmetry
   c. Rotational symmetry but not line symmetry

4. In each of the following figures, complete the sketches so that they have line symmetry about ℓ:

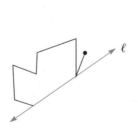

a.

b.

5. a. Determine the number of lines of symmetry, if any, in each of the following flags.
   b. Sketch the lines of symmetry for each, if they exist.

Switzerland
(i)

South Korea
(ii)

6. a. Determine how many lines of symmetry the following figure has:

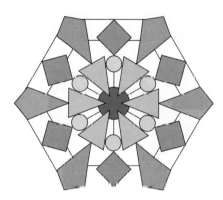

   b. Does the figure have rotational symmetry?
7. Find the lines of symmetry, if any, for each of the following trademarks:

Volkswagen
of America
a.

The Yellow Pages
b.

8. In each of the following figures, complete the sketches so that they have the indicated symmetry:

Point symmetry about O
a.

60° rotational symmetry about O
b.

9. Determine how many planes of symmetry, if any, each of the following three-dimensional vehicle controls has:

a.   Army vehicle
       fuel system

b.   Aircraft
       RPM

10. The following figures were drawn using the computer language Logo. Determine what kinds of symmetries, if any, each figure has.

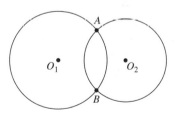

a.   **b.**

11. Describe all the symmetries of the following:
   a. A rectangle
   b. An isosceles trapezoid
   c. A non-square rhombus
   d. An angle
   e. A square
   f. A regular hexagon
   g. Two intersecting circles with centers O_1 and O_2 and intersection points A and B

   h. The colored ring between two circles with the same center O

12. If a figure has point symmetry, must it have line symmetry? Explain your answer.

13. Explain whether the following figures have plane symmetry:

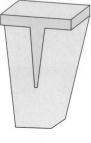

a.   b.

Assessment 14-4B

1. When a driver sees AMBULANCE on an emergency vehicle in a rearview mirror, the word is easy to read. However, looking at the "word" on the vehicle with no mirror involved, it appears very different. Explain whether AMBULANCE has any type of symmetry or describe the property being used to make the word readable in a mirror.

2. Explain why X is used to mark helicopter landing pads.

3. Various international signs have symmetries. Determine which of the following, if any, have (i) line symmetry, (ii) rotational symmetry, and/or (iii) point symmetry:

Bar

Observation deck

a.   b.

4. Determine the types of symmetry that each separate quilt pattern below has (line, rotational, point), if any.

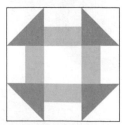

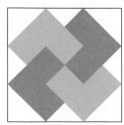

a.   Churn Dash   b.   Card Trick

c.   Friendship Star   d.   Linoleum

5. Redesign the quilt pattern below so that it has line symmetry.

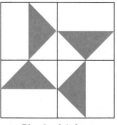

Pinwheel Askew

6. a. Determine the number of lines of symmetry, if any, in each of the following flags.
   b. Sketch the lines of symmetry for each, if any.

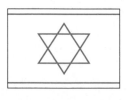

Israel

(i)

Barbados

(ii)

7. Explain whether the following quilt patterns have rotational symmetry, and, if so, identify the turn center.

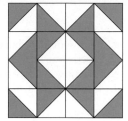

Devil's Puzzle

a.

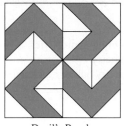

Empire Star

b.

8. Find the lines of symmetry, if any, for each of the following trademarks:

Chevrolet

a.

Chrysler Corporation

b.

9. Determine how many planes of symmetry, if any, each of the following three-dimensional vehicle controls has:

a. Army vehicle special-purpose equipment

b. Automotive finger-operated continuous multiturn

10. The following figures were drawn using the computer language Logo. Determine what kinds of symmetries, if any, each figure has.

a.

b.

11. Describe all the symmetries of the following:
  a. An isosceles triangle
  b. A parallelogram
  c. A kite
  d. An equilateral triangle
  e. A regular pentagon
  f. Two touching circles with centers O_1 and O_2

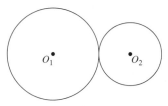

12. If a figure has line symmetries in each of two perpendicular lines, what other symmetry must it have?

13. Explain whether the following figures have plane symmetry:

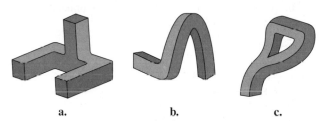

a.   **b.**   **c.**

14. Complete the following figures so that they have plane symmetry:

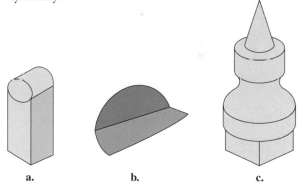

a.   **b.**   **c.**

Mathematical Connections 14-4

Communication

1. Answer each of the following. If your answer is no, provide a counterexample.
   a. If a figure has point symmetry, must it have rotational symmetry? Why?
   b. If a figure has rotational symmetry, must it have point symmetry? Why?
   c. Can a figure have point, line, and rotational symmetry? If so, sketch a figure that has these properties.
   d. If a figure has point symmetry, must it have line symmetry? Is the converse true? Why?
   e. If a figure has both point and line symmetry, must it have rotational symmetry? Why?

Open-Ended

2. If possible, sketch a triangle that satisfies each of the following:
   a. It has no lines of symmetry.
   b. It has exactly one line of symmetry.
   c. It has exactly two lines of symmetry.
   d. It has exactly three lines of symmetry.
3. Sketch a figure that has point symmetry but no line symmetry.
4. a. In the following figure, $ABCD$ is a rectangular sheet of paper. Fold the paper so that the opposite edges $\overline{AB}$ and $\overline{DC}$ coincide (the crease $\overline{EF}$ is created). Then fold the resulting rectangle $ABEF$ so that $\overline{BE}$ and $\overline{AF}$ coincide (the crease $\overline{HG}$ is created). The rectangle $AFGH$ is obtained. Now cut a curved piece of paper out of the corner G as shown and unfold the paper. Describe all the symmetries that the unfolded figure has.

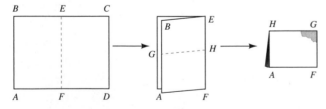

   b. Repeat the experiment in (a) by successively folding a new sheet of paper 3 times and cutting out a curved piece containing a corner that resulted from the three folds. Predict all the symmetries of the unfolded figure. Check your answer by unfolding the paper.
5. For each of the following, use paper folding or any other method to create, if possible, two nonsimilar figures that have the specified symmetries:

   a. Point symmetry but no other symmetries
   b. Exactly one line of symmetry
   c. Exactly two lines of symmetry but no rotational symmetry
   d. 45° rotational symmetry

Cooperative Learning

6. With a partner, design a three-mirror kaleidoscope by fastening three mirrors together, each perpendicular to a flat surface so that they form an equilateral triangular prism. Place colored paper with a pattern (or design your own pattern) in the base of the kaleidoscope. Peer over the edge of the kaleidoscope to view the generated figure. Repeat the experiment for different patterns.
7. With another person or group, play the following game several times. First, draw a polygon that has different kinds of symmetries. Without revealing your polygon to your partner, tell your partner all you know about the symmetries of the figure. Next, from this information, your opponent attempts to draw the type of figure you drew. If your opponent produces a figure of the type you drew, he or she earns 2 points. If your opponent produces a different figure having the symmetries that you reported or if you failed to reveal some of the symmetries of your polygon, your opponent gets 3 points. Alternate roles several times.

Questions from the Classroom

8. A student claims that a kite has no lines of symmetry. How do you respond?
9. A student says that every three-dimensional figure that has plane symmetry automatically has line symmetry. Do you agree? Why or why not?
10. A student wants to know if it is true that a line has infinitely many lines of symmetry. How do you respond?
11. A student wants to know all the symmetries of two intersecting lines. How do you respond?
12. A student asks if a semicircle (a circular arc consisting of half a circle) has half as many symmetries as the entire circle. The same student also wants to know how to find all the symmetries of any arc of a circle. How do you respond?
13. A student heard that the graph of $y = x^2$ as well as the graph of $y = |x|$ has a line symmetry but the graph of $y = x^3$ has point symmetry. The student wants to know why. How do you respond?

Review Problems

14. For each of the following cases, find the image of the given figure using paper folding:

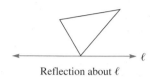

Reflection about ℓ

a.

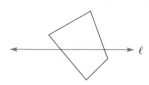

Reflection about ℓ

b.

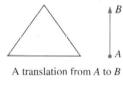

A translation from A to B

c.

15. Construct each image in exercise 14 using a compass and a straightedge.

Third International Mathematics and Science Study (TIMSS) Questions

There are several ways of arranging the tiles so that they form patterns. The grid below has been shaded to show how tiles can be placed on some of the squares. The pattern can be continued so that AB and CD are lines of symmetry.

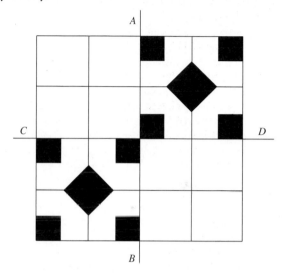

Shade in all the remaining squares on the grid so that the resulting pattern is symmetrical about line AB, and also is symmetrical about line CD.

TIMSS, Grade 8, 2003

Draw a line of symmetry on the triangle below.

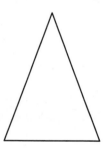

TIMSS, Grade 4, 2003

BRAIN TEASER Suppose you are looking in a mirror hung flat on a wall and your image goes from the top of the mirror to the bottom, as in Figure 14-58. How does the length of the part of your body that you see compare with the length of the mirror?

Figure 14-58

14-5   Tessellations of the Plane

In this section we use concepts from motion geometry to study *tessellations* of the plane. A **tessellation** of a plane is the filling of the plane with repetitions of figures in such a way that no figures overlap and there are no gaps. (Similarly, one can tessellate space.) The tiling of a floor and various mosaics are examples of tessellations. Maurits C. Escher was a master of tessellations. Many of his drawings have fascinated mathematicians for decades. An example from his work *Study of Regular Division of the Plane with Reptiles* (pen, ink, and watercolor), 1939, contains an exhibit of a tessellation of the plane by a lizardlike shape, as shown in Figure 14-59.

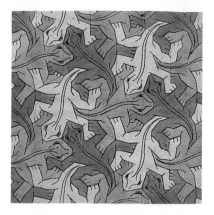

Figure 14-59

At the heart of the tessellation in Figure 14-59, we see a regular hexagon. But perhaps the simplest tessellation of the plane can be achieved with squares. Figure 14-60 shows two different tessellations of the plane with squares.

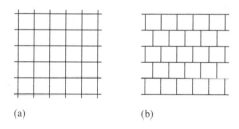

(a)   (b)

Figure 14-60

 The following section of a grade 4 student page shows that the ties between math and art in tessellations are studied early in a student's career. Answer questions 15–17.

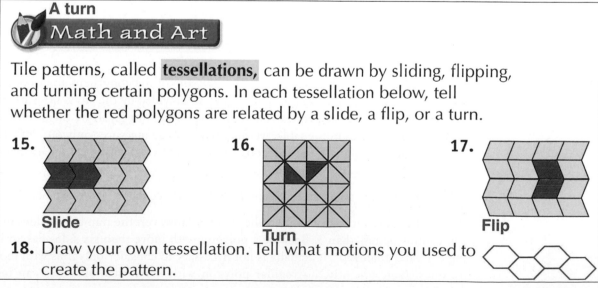

A turn

Math and Art

Tile patterns, called **tessellations,** can be drawn by sliding, flipping, and turning certain polygons. In each tessellation below, tell whether the red polygons are related by a slide, a flip, or a turn.

15. Slide

16. Turn

17. Flip

18. Draw your own tessellation. Tell what motions you used to create the pattern.

Source: Scott Foresman-Addison Wesley Mathematics, Grade 4, 2008 (p. 454).

Regular Tessellations

Tessellations with regular polygons are appealing and interesting because of their simplicity. Figure 14-61 shows portions of tessellations with equilateral triangles (a) and with regular hexagons (b).

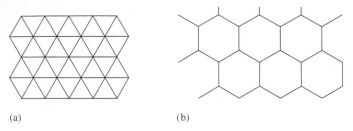

(a)   (b)

Figure 14-61

To determine other regular polygons that tessellate the plane, we investigate the possible size of the interior angle of a tessellating polygon. If n is the number of sides of a regular polygon, then because the sum of the measures of the exterior angles of the regular polygon is 360°, the measure of a single exterior angle of the polygon is $360°/n$. Hence, the measure of an interior angle is $180° - 360°/n$. Table 14-2 gives some values of n, the type of regular polygon related to each, and the angle measure of an interior angle found by using the expression $180° - 360°/n$.

Table 14-2

Number of Sides (*n*)	Regular Polygon	Measure of Interior Angle
3	Triangle	60°
4	Square	90°
5	Pentagon	108°
6	Hexagon	120°
7	Heptagon	900/7°
8	Octagon	135°
9	Nonagon	140°
10	Decagon	144°

If a regular polygon tessellates the plane, the sum of the congruent angles of the polygons around every vertex must be 360°. Thus, 360 divided by the angle measure gives the number of angles around a vertex and hence must be an integer. If we divide 360° by each of the angle measures in the table, we find that of these measures only 60°, 90°, and 120° divide 360°; hence of the listed polygons, only an equilateral triangle, a square, and a regular hexagon can tessellate the plane.

Can other regular polygons tessellate the plane? Notice that $\frac{360}{120} = 3$. Hence, 360 divided by a number greater than 120 is smaller than 3. However, the number of sides of a polygon cannot be less than 3. Because a polygon with more than six sides has an interior angle greater than 120°, it actually is not necessary to consider polygons with more than six sides. Consequently, no regular polygon with more than six sides can tessellate the plane.

Semiregular Tessellations

When more than one type of regular polygon is used and the arrangement of the polygons at each vertex is the same, the tessellation is *semiregular*. The partial student page below shows an example of a semiregular tessellation and one that is not semiregular.

School Book Page   SEMIREGULAR TESSELLATIONS

Extension ▶▶

Semi-Regular Tessellations

In Exercise 6 on page 573, you saw that one regular polygon can be used to create a tessellation. You can also create tessellations using more than one regular polygon. For example, an equilateral triangle and a regular hexagon were used to create the tessellation below.

17. a. Which pattern blocks are regular polygons?

b. Create two or three different tessellations that include more than one type of the pattern blocks you identified in part (a).

A *semi-regular tessellation* uses more than one kind of regular polygon, and has the same arrangement of polygons at every vertex. The tessellation shown above is an example of a semi-regular tessellation.

18. Two different arrangements of regular polygons appear in the tessellation below, so it is not a semi-regular tessellation.

a. Identify which of your tessellations from Exercise 17(b) are semi-regular tessellations.

b. Is the tessellation shown on page 565 a semi-regular tessellation? Explain.

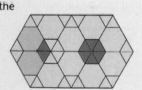

Source: Math Thematics, Book 2, McDougal Littell, 2005 (p. 575).

NOW TRY THIS 14-10 Create a semiregular tessellation consisting only of squares and equilateral triangles.

Tessellating with Other Shapes

Next, we consider tessellating the plane with arbitrary convex quadrilaterals. Before reading on, you may wish to investigate the problem yourself, with the help of cardboard quadrilaterals.

Figure 14-62 shows an arbitrary convex quadrilateral and a way to tessellate the plane with the quadrilateral. Successive half-turns of the quadrilateral about the midpoints P, Q, R, and S of its sides will produce four congruent quadrilaterals around a common vertex. Notice that the sum of the measures of the angles around vertex A is $a + b + c + d$. This is the sum of the measures of the interior angles of the quadrilateral, or $360°$.

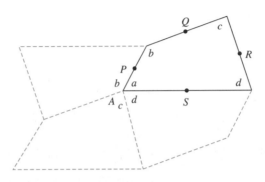

Figure 14-62

As we saw earlier in this section, a regular pentagon does not tessellate the plane. However, some nonregular pentagons do. One is shown in Figure 14-63, along with a tessellation of the plane by the pentagon.

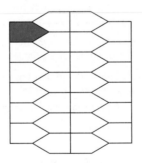

Figure 14-63

Historical Note

Determining which irregular pentagons tessellate is a surprisingly rich problem. Mathematicians thought they had solved it when they had classified eight types of pentagons that would tessellate. They believed they had all of them. But then in 1975, Marjorie Rice, with no formal training in mathematics, discovered a ninth type of tessellating pentagon. She went on to discover four more by 1977. Her interest was piqued by reading an article in *Scientific American* by Martin Gardner. Two of the pentagons she found are shown in Figure 14-64. The problem of how many types of pentagons tessellate remains unsolved.   ◆

Type 9 discovered in February 1976

Type 13 discovered in December 1977

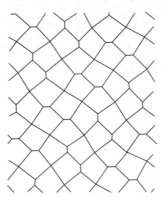
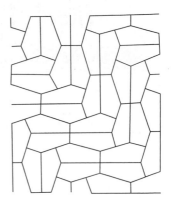

Figure 14-64

What other types of designs can be made that tessellate a plane? The plane geometry and motions studied earlier give some clues on how to design shapes that work. Consider one of the methods used in Chapter 13 to determine the area of a parallelogram. In Figure 14-65(a), triangle *ABE* is removed from the left of parallelogram *ABCD* and slid to the right, forming the rectangle *BB'E'E* of Figure 14-65(b). This same notion can be used to create a tessellating shape.

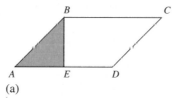
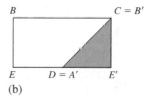

(a)

(b)

Figure 14-65

Consider any polygon known to tessellate a plane, such as rectangle *ABCD* in Figure 14-66(a). On the left side of the figure draw any shape in the interior of the rectangle, as in Figure 14-66(b). Cut this shape from the rectangle and slide it to the right by the slide that takes *A* to *B*, as shown in Figure 14-66(c). The resulting shape will tessellate the plane. (Why?)

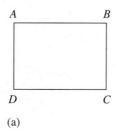
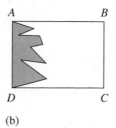
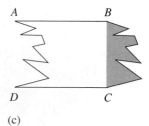

(a)

(b)

(c)

Figure 14-66

The following student pages show how a rectangle may be altered in two directions, shapes formed, and artistic details added to form "student art."

School Book Page TESSELLATIONS

Lesson 9-14

Key Idea
You can use what you know about transformations to create interesting tessellations.

Vocabulary
• tessellation

Materials
• construction paper
• scissors
• ruler
• tape
• markers or colored pencils

Tessellations

LEARN

Activity

How can you create a tessellation based on a rectangle?

A **tessellation** is a pattern of congruent figures that fills the plane without gaps or overlaps. A rectangle is one of the polygons that can tessellate the plane. You can alter the sides of a rectangle to produce a new shape that tessellates the plane.

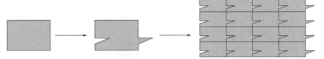

You can extend this idea to change a rectangle into an interesting shape that tessellates the plane.

✔ **WARM UP**

1. On grid paper, draw the triangle with vertices at (7, −5), (7, −1), and (1, −5). Slide the triangle to the left 8 units and up 6 units.

2. Repeat Exercise 1, but this time slide the triangle to the right 8 units and down 6 units.

a. Cut out a rectangular piece of paper about 6 cm by 12 cm. Label the sides as shown.

b. Draw and cut any shape out of side 1. Do <u>not</u> cut off a corner. If you cut off a numeral, simply remember which side is which.

c. Without turning the shape over, slide it straight across, and tape it onto side 3.

d. Similarly, cut into side 2, slide it straight across, and tape it onto side 4.

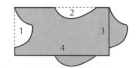

e. Repeat the process one or two more times, cutting into any <u>unaltered</u> part of the rectangle, except a corner. Remember to slide each piece straight across to the opposite side, and then tape it in place.

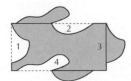

516

(continues)

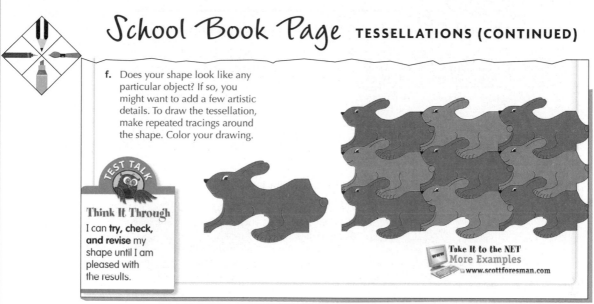

f. Does your shape look like any particular object? If so, you might want to add a few artistic details. To draw the tessellation, make repeated tracings around the shape. Color your drawing.

Think It Through

I can **try, check, and revise** my shape until I am pleased with the results.

Take It to the NET
More Examples
www.scottforesman.com

Source: Scott Foresman-Addison Wesley Mathematics, Grade 6, 2008 (pp. 516–517).

A second method of forming a tessellation involves a series of rotations of parts of a figure. In Figure 14-67(a), we start with an equilateral triangle *ABC*, choose the midpoint *O* of one side of the triangle, and cut out a shape, as in Figure 14-67(b), being careful not to cut away more than half of angle *B*, and then rotate the shape 180° clockwise around point *O*. If we continue this process on the other two sides, then we obtain a shape that can be rotated around point *A* to tessellate the plane. Complete the tessellating shape and tessellate the plane with it in Now Try This 14-11.

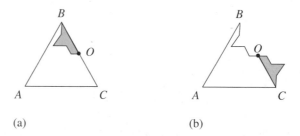

(a)   (b)

Figure 14-67

NOW TRY THIS 14-11 Continue the drawing of the tessellating shape in Figure 14-67. Cut out the shape and use it to draw a tessellation of the plane.

 The partial student page following shows a grade 6 tessellation construction involving rotations.

PRACTICE

C Extensions

20. Follow these steps to change a square into a shape that tessellates the plane.

a. Cut out a square piece of paper about 8 cm on a side. Label the sides and vertices as shown.

b. Draw and cut any shape out of side 1. Do <u>not</u> cut off a corner.

c. Rotate the piece about point A and tape it to side 2. Be sure to tape the piece the same distance from point A as it was before you cut it.

d. Similarly, cut into side 3, rotate the piece about point B, and tape the piece to side 4.

e. Repeat the process a few more times, cutting into any <u>unaltered</u> part of the square except a corner. Remember this plan:

- Rotate pieces from side 1 onto side 2, and from side 2 onto side 1. (Be sure to tape the piece the same distance from A.)

- Rotate pieces from side 3 onto side 4, and from side 4 onto side 3. (Be sure to tape the piece the same distance from B.)

f. Add artistic details and draw the tessellation.

Source: Scott Foresman–Addison Wesley Mathematics, Grade 6, 2008 (p. 519).

Assessment 14-5A

1. On dot paper, draw a tessellation of the plane using the following figure:

2. a. Tessellate the plane with the following quadrilateral:

  b. Is it possible to tessellate the plane with any quadrilateral? Why or why not?

3. On dot paper, use the following pentomino, to make a tessellation of the plane, if possible. (A *pentomino* is a polygon composed of five congruent, nonoverlapping squares.)

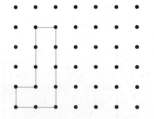

4. There are many ways to tessellate the plane by using combinations of regular polygons, as shown in the following figure:

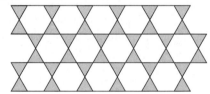

Try to produce other such tessellations by using only equilateral triangles, squares, and regular hexagons.

5. To determine if a shape created using a glide reflection will tessellate the plane, complete the following:

  a. Start with a rectangle. Determine some shape that you might use with a slide to form a tessellating shape. Slide it as shown. Determine the horizontal line of symmetry of the rectangle, and reflect as shown.

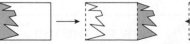
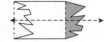

  b. Explain why the described series of motions is a glide reflection.

  c. Determine whether the final shape will tessellate the plane.

6. The **dual of a regular tessellation** is the tessellation obtained by connecting the centers of the polygons in the original tessellation that share a common side. The dual of the tessellation of equilateral triangles is the tessellation of regular hexagons, shown in color in the following figure:

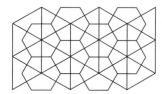

Describe and show the dual of each of the following:
   a. The regular tessellation of squares shown in Figure 14-60(a)
   b. The tessellation of squares in Figure 14-60(b)
   c. A tessellation of regular hexagons
7. A sidewalk is made of tiles of the type shown in the following figure:

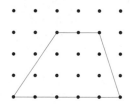

Each tile is made of three regular hexagons with some sides removed. Draw a partial tessellation composed of seven such figures.
8. Which of the following tessellations are semiregular and which are not? Justify your answers.

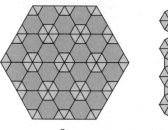 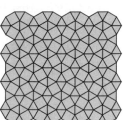

a.   **b.**

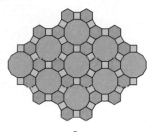

c.

Assessment 14-5B

1. On dot paper, draw a tessellation of the plane using the following figure:

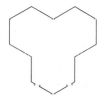

2. a. Tessellate the plane with the following triangle:

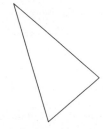

   b. Is it possible to tessellate a plane with any triangle? Why or why not?
3. On dot paper, use each of the following pentominoes, one at a time, to make a tessellation of the plane, if possible:

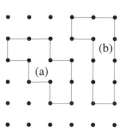

4. Produce a tessellation of a plane using a combination of regular octagons and squares.
5. A "bow-tie" figure is formed by right triangles (white) near the center of the Brown Goose quilt pattern. Explain whether this bow-tie figure will tessellate a plane.

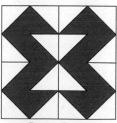

Brown Goose

6. The partial student page below shows "student work." Answer questions 17 and 18. *Source: Scott Foresman–Addison Wesley Mathematics, Grade 6, 2008 (p. 518).*

Math and Art

The Dutch artist M. C. Escher (1898 to 1972) is renowned for his interesting tessellations. Inspired by Escher's tessellating shapes of fish and lizards, Jon and Miguel created the art below.

17. What transformation would move fish *A* to fish *B*?

18. What transformation would move lizard *C* to lizard *D*? lizard *C* to lizard *E*?

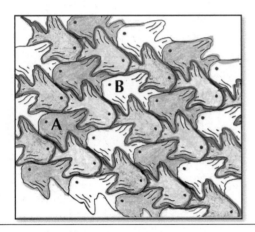

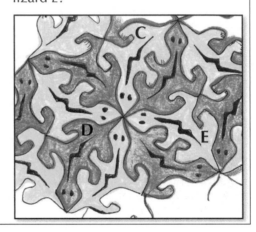

7. Which of the following tessellations are semiregular and which are not? Justify your answers.

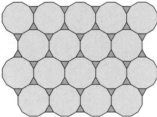

a.

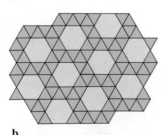

b.

8. Explain whether the "arrow" below will tessellate a plane.

9. A concrete tile similar to that shown below is used to construct a sound barrier wall for an interstate highway in Washington, DC. Sketch a tessellation using the figure to show that a wall could be built of the tiles.

10. A portion of a Pieced Star Quilt pattern is shown below. Will this hexagon tessellate a plane? Prove your answer.

Mathematical Connections 14-5

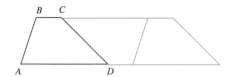

Communication

1. The following figure is a partial tessellation of the plane with the trapezoid *ABCD*:

   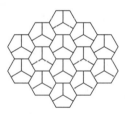

   a. Explain how the tessellation can be used to find a formula for the area of the trapezoid.
   b. Tessellate the plane with a triangle and show how the tessellation can be used to find the relationship between the length of the segment connecting the midpoints of the two sides of a triangle and the length of the third side.

2. Explain in your own words why only three types of regular polygons tessellate the plane.

3. The following figure shows how to tessellate the plane with irregular pentagons. Explain how the pentagons can be constructed and actually construct at least three such pentagons.

   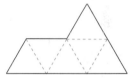

Open-Ended

4. There are endless numbers of figures that tessellate a plane. In the following drawing, the shaded figure is shown to tessellate the plane:

   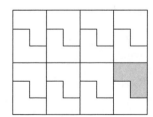

   Design several different polygons and show how each can tessellate the plane. What transformations are used in each of your designs? Explain how they are used to tessellate the plane.

5. Examine different quilt patterns or floor coverings and make a sketch of those you found that tessellate a plane.

6. A cube will tessellate space but a sphere will not. List several other solids that will tessellate space and several that will not.

Cooperative Learning

7. Each member of a small group is to find a drawing by M. C. Escher that does not appear in this text and in which the concept of tessellation is used. Each then shows the other members of the group, in detail, how he or she thinks Escher created the tessellation in the drawing.

8. a. Convince the members of your group that the following figure containing six equilateral triangles tessellates the plane:

   b. As a group, find different figures that contain six equilateral triangles. How many such figures can you find? Discuss the meaning of "different."
   c. Find some of the figures in (b) that are rep-tiles. (A rep-tile is a figure whose copies can be used to form a larger figure similar to itself.) Convince other members of your group that your figures are rep-tiles and that they tessellate the plane.

9. Trace each of the following shapes and cut out several copies of each. Each member of a small group should pick a shape, decide if it tessellates the plane, and convince other members of the group that it does or does not tessellate. The group should report the answers with figures or an argument why a shape does not tessellate.

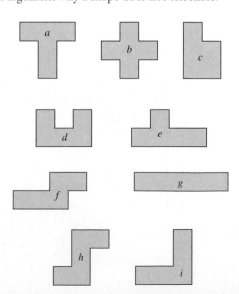

Questions from the Classroom

10. Isiah claims that any figure with plane symmetry could be thought of as an original with its image in a mirror. How do you respond?

11. Maria says that she can define a "rotational" motion in three dimensions with a line instead of a point as "center." How do you respond?

Review Questions

12. Find the image of the figure below in each of the following:
   a. A translation from M to N
   b. A 90° rotation counterclockwise through M

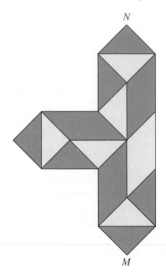

13. With the quilt patterns below, define all symmetries of each.

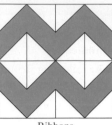

Dutchman's Puzzle
a.

Ribbons
b.

14. Determine whether the figure below has any symmetries.

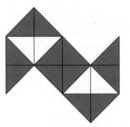

LABORATORY ACTIVITY Use pattern blocks to construct tessellations using each of the following types of pieces:

1. Squares
2. Equilateral triangles
3. Octagons and squares
4. Rhombuses

BRAIN TEASER An architect needs to determine the location of an airport T serving three cities A, B, and C so that the sum of the distances from the airport to the three cities is minimal and the airport is in the interior of triangle ABC. The architect's friend, a mathematician, suggests investigating the problem by picking any point for T and rotating $\triangle ATB$ clockwise about B by 60°. Assuming $\triangle ABC$ is acute, can you help the architect to find the true location of the airport?

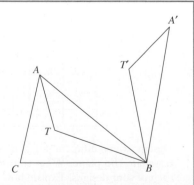

Hint for Solving the Preliminary Problem

To have turn symmetry of 80°, a figure must coincide with itself after a turn of 80°. For example, a regular 360-gon can be rotated onto itself after every 1° turn with its center as the center of the turn. After 80 such turns, the figure will have been rotated 80° and will fit itself. Thus, it has 80° rotational, or turn, symmetry. What other figures have this property? This problem is found in Martin (1996).

Chapter Outline

I. Motions of the plane

  A. A **transformation** of the plane is a function from the plane to itself that is a one-to-one correspondence. **Isometries** are transformations that preserve distance.

    1. A **translation** is a transformation of the plane that moves every point a specified distance in a specified direction along a straight line.

    In terms of coordinates, a translation of the plane is a transformation of the plane such that to every point (x, y) corresponds the point $(x + a, y + b)$ for real numbers a and b.

    2. A **rotation** is a transformation of the plane determined by holding one point (the center) fixed and rotating the plane about this point by a certain amount in a certain direction.

    3. A **half-turn** is a rotation of 180°.

    4. Properties of a rotation can be used to show that two lines, neither of which is vertical, are perpendicular if, and only if, their slopes m_1 and m_2 satisfy the condition $m_1 m_2 = {}^-1$.

    5. A **reflection** in a line m is a transformation among points of the plane that pairs each point P of the plane with a point P' in such a way that m is the perpendicular bisector of $\overline{PP'}$, as long as P is not on m. If P is on m, then $P = P'$.

    6. A **glide reflection** is the composition of a translation and a reflection in a line parallel to the slide arrow of the translation.

  B. Two figures are **congruent** if it is possible to transform one onto the other by a sequence of isometries.

  C. A **size transformation** S from the plane to the plane is defined as follows: Some point O, the center of the size transformation, is its own image. For any other point Q of the plane, its image Q' is such that $OQ'/OQ = r$, where r is a positive real number and O, Q, and Q' are collinear, with O not between Q and Q'.

  D. Two figures are **similar** if it is possible to transform one onto the other by a sequence of isometries followed by a size transformation.

II. Symmetries

  A. A figure has **line symmetry** if it is its own image under a reflection.

  B. A figure has **rotational symmetry** if it is its own image under a rotation of less than 360° about a turn center.

  C. A figure has **point symmetry** if it has 180° rotational symmetry.

  D. A three-dimensional figure has a **plane of symmetry** when every point of the figure on one side of the plane has a mirror image on the other side of the plane.

  E. A three-dimensional figure has **rotational symmetry** if when rotated by an amount less than 360° about an axis of rotation it coincides with itself.

III. Tessellations

  A **tessellation** of a plane is the filling of the plane with repetitions of figures in such a way that no figures overlap and there are no gaps.

Chapter Review

1. Complete each of the following motions:

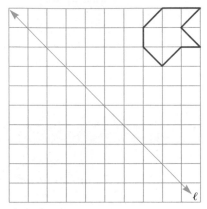

A reflection in ℓ

a.

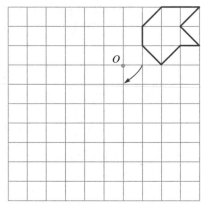

A rotation in *O* through the given arc

b.

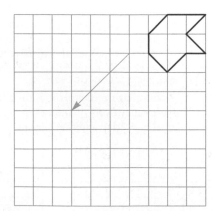

A translation, as pictured

c.

2. For each of the following figures, construct the image of △*ABC*.

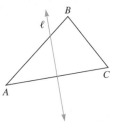

Through a reflection in ℓ

a.

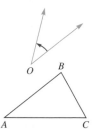

Through the given rotation in *O*

b.

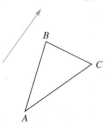

Through the translation arrow pictured

c.

3. Determine how many lines of symmetry, if any, each of the following figures has:

a.   **b.**

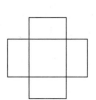

c.   **d.**

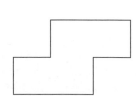

e.   **f.**

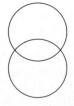

4. For each of the following figures, identify the types of symmetry (line, rotational, or point) it possesses:

a.

b.

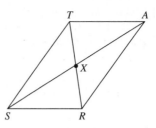

c.

5. Determine how many planes of symmetry each of the following has:
   a. A spherical rubber ball
   b. A right cylindrical water pipe
   c. A box that is a right rectangular prism, not a cube, and no square faces
   d. A cube

6. What type of symmetry (line, rotational, or point) does each of the lowercase letters of the printed English alphabet have?

7. In the following figure, $\triangle A'B'C'$ is the image of $\triangle ABC$ under a size transformation:

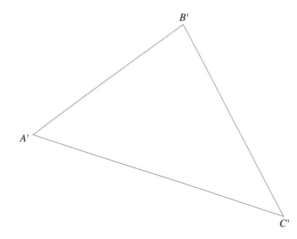

Locate points A, B, and C such that A' is the center of the size transformation and $BC = \dfrac{1}{2} B'C'$.

8. Given that *STAR* in the figure shown is a parallelogram, describe a sequence of isometries to show the following:
   a. $\triangle STA \cong \triangle ARS$
   b. $\triangle TSR \cong \triangle RAT$

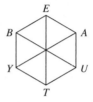

9. Given that *BEAUTY* in the figure shown is a regular hexagon, describe a sequence of isometries that will transform the following:
   a. *BEAU* onto *AUTY*
   b. *BEAU* onto *YTUA*

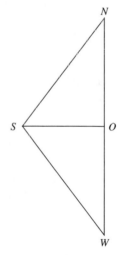

10. Given that $\triangle SNO \cong \triangle SWO$ in the following figure, describe one or more isometries that will transform $\triangle SNO$ onto $\triangle SWO$:

11. Show that $\triangle SER$ in the following figure is the image of $\triangle HOR$ under a succession of isometries with a size transformation:

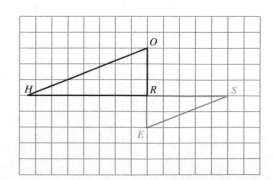

12. Show that $\triangle TAB$ in the following figure is the image of $\triangle PIG$ under a succession of isometries with a size transformation:

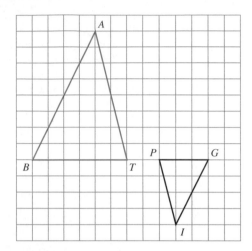

13. The triangle $A'B'C'$ with $A'(0, 7.91)$, $B'(^-5, ^-4.93)$, $C'(4.83, 0)$ is the image of triangle ABC under the translation $(x, y) \rightarrow (x + 3, y - 5)$.
   a. Find the coordinates of A, B, and C.
   b. Under what translation will the image of $\triangle A'B'C'$ be $\triangle ABC$?

14. If a translation determined by $(x, y) \rightarrow (x + 3, y - 2)$ is followed by another translation determined by $(x, y) \rightarrow (x - 3, y + 2)$, describe a transformation that would achieve the same thing.

15. Suppose $\triangle A''B''C''$ with $A''(0, 0)$, $B''(1, 5)$, $C''(^-2, 7)$ is the image of $\triangle ABC$ under the translation $(x, y) \rightarrow (x + 2, y - 1)$ followed by the translation $(x, y) \rightarrow (x + 1, y + 3)$.
   a. Find the coordinates of A, B, and C.
   b. Under what single transformation will the image of $\triangle ABC$ be $\triangle A''B''C''$?

16. Write each of the following as a single transformation:
   a. **(i)** A translation from A to B followed by a translation from B to C

   (ii) A translation from B to C followed by a translation from A to B
   b. A rotation about O by 90° counterclockwise, followed by a rotation about O by 30° clockwise
   c. **(i)** A size transformation with center O and scale factor 3 followed by a size transformation with center O and scale factor 2
   (ii) The size transformations in part (i) in reverse order

17. Which of the following figures has more symmetries? Justify your answer.
   a. A rhombus or a square
   b. An isosceles trapezoid or a parallelogram

18. Given the line $y = 2x + 3$ and the point $P(^-1, 3)$, find the equation of each of the following:
   a. The line through P parallel to the given line
   b. The line through P intersecting the given line at the point whose y-coordinate is 1
   c. The line through P perpendicular to the given line

19. Find the equation of the image of the line $y = ^-x + 3$ under each of the following transformations:
   a. The translation $(x, y) \rightarrow (x + 2, y - 3)$
   b. Reflection in the x-axis
   c. Reflection in the y-axis
   d. Reflection in the line $y = x$
   e. Half-turn about the origin
   f. Size transformation with center at the origin and scale factor 2

20. **a.** Consider the translation $(x, y) \rightarrow (x + 3, y - 5)$. The image of $(1, 2)$ is $(4, ^-3)$. What translation takes $(4, ^-3)$ to $(1, 2)$?
   b. What is the net result of following the translation $(x, y) \rightarrow (x + h, y + k)$ by the translation $(x, y) \rightarrow (x - h, y - k)$?

21. Explain why a regular octagon cannot tessellate the plane.

22. A square will tessellate a plane; can we always find a square that will tessellate a rectangle?

23. Can any tessellating shape be used as a unit for measuring area? If so, explain why.

Selected Bibliography

Britton, J. "Escher in the Classroom." *Mathematics Teaching in the Middle School* 12 (April 2007): 480.

Clason, R., D. Ericksen, and C. Ericksen. "Cross-and-Turn Tile Patterns." *Mathematics Teaching in the Middle Schools* 2 (May 1997): 430–437.

Clements, D., and J. Sarama. "The Earliest Geometry." *Teaching Children Mathematics* 7 (October 2000): 82–86.

Dayoub, I., and J. Lott. *Geometry: Constructions and Transformations*. White Plains, NY: Dale Seymour Publishing Company, 1977.

Demal, M., and D. Popeler. "*L'enseignement de la géométrie*." *Bulletin de l'apmep* Num 454 (2004), APMEP, Paris, pp. 717–728.

Fossnaugh, L., and M. Harrell. "Covering the Plane with Rep-Tiles." *Mathematics Teaching in the Middle Schools* 1 (January–February 1996): 666–670.

Harrell, M., and L. Fossnaugh. "Allium to Zircon: Mathematics & Nature." *Mathematics Teaching in the Middle Schools* 2 (May 1997): 380–389.

Harris, J. "Using Literature to Investigate Transformations." *Teaching Children Mathematics* 4 (May 1998): 510–513.

http://finitegeometry.org/sc/16/quiltgeometry.html. Accessed May 9, 2008.

Kuchemann, D. "Reflections and Rotations." In *Children's Understanding of Mathematics: 11–16*. London: John Murray, 1981.

Martin, G. *Geometry: A Moving Experience*. Honolulu, HI: Curriculum Research and Development Group, 1996.

Peterson, B. "From Tessellations to Polyhedra." *Mathematics Teaching in the Middle Schools* 5 (February 2000): 348–356.

Ranucci, E. "Master of Tessellations: M. C. Escher, 1898–1972." *Mathematics Teaching in the Middle School* 12 (April 2007): 476–479.

Sakshaug, L. "Patterns in Squares." *Teaching Children Mathematics* 7 (November 2000): 172–173.

Seidel, J. "Symmetry in Season." *Teaching Children Mathematics* 4 (January 1998): 244–249.

Sellke, D. "Geometric Flips via the Arts." *Teaching Children Mathematics* 4 (February 1999): 379–383.

Speer, W., and J. Dixon. "Reflections of Mathematics." *Teaching Children Mathematics* 2 (May 1996): 537–543.

Walter, M. *The Mirror Puzzle Book*. New York: Parkwest Publications (1985).

Wesslen, M., and S. Fernandez. "Transformational Geometry." *Mathematics Teaching* 191 (June 2005).

Westegaard, S. "Using Quilt Blocks to Construct Understanding." *Mathematics Teaching in the Middle School* 13 (February 2008): 361–365.

White, D. "Kenta, Kilts, and Kimonos: Exploring Cultures and Mathematics through Fabrics." *Teaching Children Mathematics* 7 (February 2001): 354–359.

Wilkes, G. "The Mathematics of Radial Designs." *Mathematics Teaching in the Middle School* 12 (April 2007): 474–475.

Zaslavsky, C. "Symmetry in American Folk Art." *Arithmetic Teacher* 37 (January 1990): 6–12.

Zilliox, J., and S. Lowrey. "Many Faces Have I." *Mathematics Teaching in the Middle Schools* 3 (November–December 1997): 180–183.

Copyright Acknowledgments

Answers to Assessment A exercises, odd-numbered Mathematical Connections problems, Chapter Review problems, Now Try This problems, Brain Teasers, Laboratory Activities, Technology Corners, and Preliminary Problems appear in the Student Edition.

Chapter 1

Assessment 1-1A

1. (a) 4950 (b) 251,001   2. 10,248   3. 12
4. 160 mi   5. Dandy, Cory, Alabababa, Bubba   6. 45
7. $1.19, (one 50¢, one 25¢, four 10¢, and four 1¢)
8. (a) (i) 541×72 (ii) divide 754 by 12
(b) (i) 257×14 (ii) divide 124 by 75   9. $5,256,000
10. 12   11. $2.45   12. 23 rungs   13. (a) 10,500
squares (b) $n^2 + 5n$ squares   14. width = 230 ft;
length = 310 ft   15. 26°F   16. Al—winter; Betty—
summer; Carl—spring; Dan—fall   17. $A = 9$.

Mathematical Connections 1-1

Communication

1. Answers vary; for example, problem-solving skills can help students to meet future challenges in work, life, and school. Problem-solving skills allow students to take on new tasks and problems and have the confidence to do so. If a first approach to a problem fails, good problem solvers can come up with alternative approaches. Much of the mathematics that students are taught is introduced through interesting problems. Students need to know how to problem solve to make progress on these problems and in turn learn the mathematics.   3. Answers vary from students having no idea of what to do and just trying a guess to see how close it is to using a guess as a way to start to close in on the exact answer. If intelligent guesses are made and something is learned from each guess, then the student can close in on the exact answer.

Open-Ended

5. Answers vary; for example, any of the investigation problems in this section are examples of problems that could be approached with the strategy discussed.

Cooperative Learning

7. Answers vary; for example, if the average reach of people in your group was 1.8 m, then it would take approximately 22,000,000 people.   9. (a) For a 100-page book, you need 25 sheets of paper.   (b) The sum of the page numbers on the same side of the sheet is 101.   (c) The sum of all the page numbers in a 100-page book is 5050.
(d) The general case is given in the following table.

Number of Sheets	Number of Book Pages	Sum of Two Page Numbers on Same Side of Sheet	Sum of All Page Numbers
1	4	5	$1 + 2 + 3 + 4 = \dfrac{4 \cdot 5}{2} = 10$
2	8	9	$1 + 2 + 3 + \ldots + 8 = \dfrac{8 \cdot 9}{2} = 36$
3	12	13	$1 + 2 + 3 + \ldots + 12 = \dfrac{12 \cdot 13}{2} = 78$
$\vdots$	$\vdots$	$\vdots$	$\vdots$
n	$4n$	$4n + 1$	$1 + 2 + 3 + \ldots + 4n = \dfrac{4n(4n + 1)}{2}$ $= 2n(4n + 1)$

Questions from the Classroom

11. Answers vary; for example, it is in the last step where students examine whether their answer is reasonable and whether it checks given the original conditions in the problem. Many times, students discover wrong answers at this point since they may have never bothered to check if the answer they arrived at makes sense. It is also at this step that students reflect on the mathematics that was used and determine whether there might be different ways of doing the problem. Also at this stage students reflect on any connections to other problems or generalizations that are

emphasized in the NCTM standards. **13.** Answers vary; for example, if these nine numbers are to be used in a magic square, then the sum in each of the three columns must be the same and must be a natural number. The natural number must be 1/3 of the sum of all nine numbers. However $1 + 3 + 4 + 5 + 6 + 7 + 8 + 9 + 10 = 53$ and $53/3 = 17\ 2/3$, which is not a natural number. Therefore, these numbers cannot be used for a magic square.

Assessment 1-2A

1. (a) ▢▢▢▢▢▢ **(b)** △▽△▽△▽ **(c)**

2. (a) 11, 13, 15; arithmetic **(b)** 250, 300, 350; arithmetic **(c)** 96, 192, 384; geometric **(d)** $10^6, 10^7, 10^8$ geometric **(e)** 33, 37, 41; arithmetic **(f)** $6^3, 7^3, 8^3$; neither **3. (a)** 199; $2n - 1$ **(b)** 4950; $50(n - 1)$ **(c)** $3 \cdot 2^{99}; 3 \cdot 2^{n-1}$ **(d)** $10^{100}; 10^n$ **(e)** 405; $5 + 4n$ or $9 + 4(n - 1)$ **(f)** $100^3 = 1,000,000; n^3$ **4.** 2, 7, 12 **5. (a)** 2, 4, 8 **(b)** 169, 256, 169 The rule is to square the sum of the digits in the previous term. **(c)** 4, 16, 37. If u_n is the units digit of a_n, t_n is the tens digit of a_n, and h_n is the hundreds digit of a_n, then $a_n = (u_{n-1})^2 + (t_{n-1})^2 + (h_{n-1})^2$ **(d)** The original sequence will be repeated indefinitely. **6. (a)** 30, 42, 56 **(b)** 10,100 **(c)** $n(n + 1)$ or $n^2 + n$ **7. (a)** 41 **(b)** $4n + 1$, or $5 + (n - 1)4$ **(c)** $12n + 4$ **8. (a)** 42 **(b)** $4n + 2$ or $6 + (n - 1)4$ **9.** 1200 students **10.** 23rd year **11. (a)** 3, 5, 9, 15, 23, 33 **(b)** 4, 6, 10, 16, 24, 34 **(c)** 15, 17, 21, 27, 35, 45 **12. (a)** 299, 447, 644 **(b)** 56, 72, 90 **13. (a)** 101 **(b)** 61 **(c)** 200 **(d)** 11 **14. (a)** 3, 6, 11, 18, 27 **(b)** 4, 9, 14, 19, 24 **(c)** 9, 99, 999, 9999, 99999 **(d)** 5, 8, 11, 14, 17 **15. (a)** Answers vary; for example, if $x = 5$, then $\frac{5 + 5}{5} \neq 5 + 1$. **(b)** Answers vary; for example, if $x = 2$, then $(2 + 4)^2 \neq 2 + 16$. **16. (a)** 41 **(b)** $n^2 + (n - 1) = n^2 + n - 1$ **(c)** yes, the 35th figure **17.** $a_3 = 8, a_4 = 11, a_5 = 14$ **18.** The 12th term of the geometric sequence is greater than the 12th term of the arithmetic sequence. **19. (a)** 1, 5, 9, 13, 17, 21, ... **(b)** $4n + 1$, or $S + (n - 1)4$ **20.** Two solutions are possible: 64, 128, 256, or $-64, 128, -256$.

21. (a) 51 **(b)** $n^2 + \dfrac{n(n - 1)}{2}$, or $\dfrac{3n^2 - n}{2}$, or $\dfrac{n(3n - 1)}{2}$

Mathematical Connections 1-2

Communication

1. (a) Answers vary; for example, both sequences start out the same but then the first is arithmetic with $d = 2$ and the second one is geometric with $r = 2$. **(b)** Answers vary; for example, both are arithmetic with $d = 2$, but they are different in that one generates even numbers and one generates odd numbers. **(c)** Answers vary; for example, both sequences are arithmetic with $d = 5$ in the first

sequence and $d = 50$ in the second sequence. The second sequence can be generated by multiplying each term in the first sequence by 10. **3. (a)** Yes. The difference between terms in the new sequence is the same as in the old sequence because a fixed number was added to each number in the sequence. **(b)** Yes. If the fixed number is k, the difference between terms of the second sequence is k times the difference between terms of the first sequence. **(c)** Yes. The difference of the new sequence is the sum of the original difference.

Open-Ended

5. Answers vary; for example, two more patterns follow:

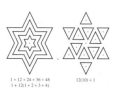

$1 + 12 + 24 + 36 + 48$
$1 + 12(1 + 2 + 3 + 4)$ $12(10) + 1$

7. Answers vary depending on sequence.

Cooperative Learning

9. (a) 81 **(b)** 40 **(c)** 3^{n-1} **(d)** $1 + 3 + 3^2 + 3^3 + \ldots + 3^{n-2}$; for $n \geq 2$.

Questions from the Classroom

11. Answers vary; for example, Joey could be told that he has noticed a pattern that works for many numbers but he must be careful in generalizing his observation to all numbers. He should be encouraged to look at more numbers to see if he can find a *counterexample* to his claim. If he starts investigating, he will find that 14 ends in 4 but is not divisible by 4. 34 is another number that ends in 4 but is not divisible by 4. These counterexamples prove that Joey's conjecture is false. **13.** The student should be asked to show examples of why this claim is true. Then the student needs to explore more possibilities to see if a counterexample can be found. Consider $\frac{5}{16}$ and $\frac{3}{4}$. The numerator of the first fraction is greater than the numerator of the second fraction and the denominator of the first fraction is greater than the denominator of the second fraction. However, $\frac{5}{16}$ is less than a half and $\frac{3}{4}$ is greater than a half, so clearly $\frac{5}{16}$ can't be greater than $\frac{3}{4}$. Thus we have a *counterexample* to show the conjecture is false. **15.** Two terms can lead to more than one sequence. For example, the terms 3, 6, ... could lead to 3, 6, 9, 12, ..., which is an arithmetic sequence with fixed difference 3. It might also lead to 3, 6, 9, 15, 24, 39, ..., in which each successive term is obtained by adding the two previous terms. Another sequence is 3, 6, 10, 15, 21, ..., in which the rule is add 3 to the first term, then add 4 to the second term, then add 5 to the third term, and so on. From these examples, we can see that two terms are not enough to determine any sequence.

Review Problems

17. 90   **19.** We need one 12-person tent and a combination of tents to hold 14 people. There are 10 ways: 662, 653, 6332, 62222, 5522, 5333, 53222, 33332, 332222, and 2222222.

Assessment 1-3A

1. (a) false statement   **(b)** false statement   **(c)** not a statement   **(d)** true statement   **(e)** not a statement
2. (a) There exists a natural number x such that $x + 8 = 11$.   **(b)** There exists a natural number x such that $x^2 = 4$.   **(c)** For all natural numbers x, $x + 3 = 3 + x$.   **(d)** For all natural numbers x, $5x + 4x = 9x$.   **3. (a)** For every natural number, x, $x + 8 = 11$.   **(b)** Every natural number x satisfies $x^2 = 4$.
(c) There is no natural number x such that $x + 3 = 3 + x$.
(d) There is no natural number x such that $5x + 4x = 9x$.
4. (a) This book does not have 500 pages.   **(b)** $3 \cdot 5 \neq 15$
(c) Some dogs do not have four legs.   **(d)** No rectangles are squares.   **(e)** All rectangles are squares.   **(f)** Some dogs have fleas.   **5. (a)** T   **(b)** T
6. (a)

p	$\sim p$	$\sim(\sim p)$
T	F	T
F	T	F

(b)

p	$\sim p$	$p \vee \sim p$	$p \wedge \sim p$
T	F	T	F
F	T	T	F

(c) yes   **(d)** no   **7. (a)** $q \wedge r$ **(b)** $r \vee \sim q$
(c) $\sim(q \wedge r)$ **(d)** $\sim q$ **8. (a)** false   **(b)** true   **(c)** false
(d) false   **(e)** false   **9. (a)** false   **(b)** true
(c) false   **(d)** false   **(e)** true   **10. (a)** no   **(b)** no
11.

p	q	$\sim p$	$\sim p \wedge q$
T	T	F	F
T	F	F	F
F	T	T	T
F	F	T	F

12. (a) $p \to q$ **(b)** $\sim p \to q$ **(c)** $p \to \sim q$ **(d)** $p \to q$
(e) $\sim q \to \sim p$ **(f)** $q \leftrightarrow p$ **13. (a)** Converse: If $2x = 10$, then $x = 5$. Inverse: If $x \neq 5$, then $2x \neq 10$.
Contrapositive: If $2x \neq 10$, then $x \neq 5$.   **(b)** Converse: If you do not like mathematics, then you do not like this book. Inverse: If you like this book, then you like mathematics. Contrapositive: If you like mathematics, then you like this book.   **(c)** Converse: If you have cavities, then you do not use Ultra Brush toothpaste. Inverse: If you use Ultra Brush toothpaste, then you do not have cavities. Contrapositive: If you do not have cavities, then you use Ultra Brush toothpaste.   **(d)** Converse: If your grades are high, then you are good at logic. Inverse: If you are not good at logic, then your grades are not high. Contrapositive: If your grades are not high, then you are not good at logic.

14. (a) no   **(b)** yes   **(c)** no   **15.** If a number is not a multiple of 4, then it is not a multiple of 8. (Contrapositive)
16. (a) valid   **(b)** valid   **(c)** invalid   **17. (a)** Some freshmen are intelligent.   **(b)** If I study for the final, then I will look for a teaching job.   **(c)** There exist triangles that are isosceles.   **18. (a)** If a figure is a square, then it is a rectangle.   **(b)** If a number is an integer, then it is a rational number.   **(c)** If a polygon has exactly three sides, then it is a triangle.
19. (a) $3 \cdot 2 \neq 6$ or $1 + 1 = 3$ **(b)** You cannot pay me now *and* you cannot pay me later.

Mathematical Connections 1-3

Communication

1. Commands, questions, and opinions are not statements because they can't be classified as true or false.   **3.** A compound statement may be formed by combining two or more statements. Connectives such as *and, or, if . . . then,* and *not* are used to form compound statements.   **5.** Given the disjunction *p or q*, the *inclusive* use of "or" means "*p or q or both.*" In logic, we use the inclusive "or." Lawyers sometimes use the phrase "and/or" to mean the inclusive use of "or." The *exclusive* use of "or" means "*either p or q but not both.*"   **7.** Dr. No is a male spy who is not poor and not tall.   **9.** Answers vary.

Cooperative Learning

11. Answers vary.

Questions from the Classroom

13. When $\sim(p \wedge q)$ is written, the negation sign operates on everything inside the parentheses; that is, it negates the conjunction $p \wedge q$. You would find the truth value for $p \wedge q$ and then negate it. When $\sim p \wedge q$ is written, the negation symbol operates only on the p statement and not the conjunction.
15. In the example,

Hypotheses:	All teachers are over 6 ft tall.
	Kay is a teacher.
Conclusion:	Kay is over 6 ft tall.

An Euler (Venn) diagram can be drawn to show that all teachers belong to the set of people over 6 ft tall. Kay belongs to the set of teachers. Thus, the argument is valid even though the hypothesis is false.

Chapter Review

1. (a) $15, 21, 28$ **(b)** $32, 27, 22$ **(c)** $400, 200, 100$
(d) $21, 34, 55$ **(e)** $17, 20, 23$ **(f)** $256, 1024, 4096$
(g) $16, 20, 24$ **(h)** $125, 216, 343$ **2. (a)** neither
(b) arithmetic **(c)** geometric **(d)** neither **(e)** arithmetic
(f) geometric **(g)** arithmetic **(h)** neither
3. (a) $3n + 2$ **(b)** $n^3 - 1$ **(c)** 3^n **4. (a)** $1, 4, 7, 10, 13$
(b) $2, 6, 12, 20, 30$ **(c)** $3, 7, 11, 15, 19$ **5. (a)** $10,100$
(b) $10,201$ **6. (a)** F; for example, $3 + 3 = 6$ and 6 is
not odd. **(b)** F; for example, 19 is odd and it ends in 9.
(c) T; the sum of any two even numbers is even because
$2m + 2n = 2(m + n)$, where m and n are natural numbers.
7.

16	3	2	13
5	10	11	8
9	6	7	12
4	15	14	1

8. 26 **9.** $2.00 **10.** 21 posts **11.** 128 matches
12. (a) $3 + 6 + 9 + 12 + 15 = \dfrac{15 \cdot 6}{2}; 3 + 6 + 9 +$
$12 + 15 + 18 = \dfrac{18 \cdot 7}{2}$ **(b)** $3 + 6 + 9 + 12 + \ldots +$
$3n = 3 \cdot 1 + 3 \cdot 2 + 3 \cdot 3 + 3 \cdot 4 + \ldots + 3n =$
$3 \cdot (1 + 2 + 3 + 4 + \ldots + n) = 3\left(\dfrac{n(n + 1)}{2}\right)$
$= \dfrac{(3n)(n + 1)}{2}$, where n is a natural number.
13. 44,000,000 turns **14.** 20 students **15.** 39 boxes
16. 48 triangles **17.** 9 hr **18.** 235 **19.** There will be
96,000 ants on the seventh day and 192,000 ants on the eighth
day, so it will certainly be full by then. **20.** 4 questions
21. Yes; cut off 10 cm leaving 80 cm. Next cut off 20 cm
leaving 60 cm. **22.** 4 or -4 **23.** In statement (i) each
and every student passed the final. In statement (ii) at least one
student passed the final and possibly all the students passed.
24. (a) yes **(b)** yes **(c)** no **(d)** yes **25. (a)** No
women smoke. **(b)** $3 + 5 \neq 8$ **(c)** Some heavy-metal
rock is not loud, or not all heavy-metal rock is loud.
(d) Beethoven wrote some music that is not classical.
26. Converse: If someone will faint, we will have a rock
concert. Inverse: If we do not have a rock concert, then no
one will faint. Contrapositive: If no one will faint, then we
will not have a rock concert.
27.

p	q	$\sim p$	$\sim q$	$p : \sim q$	$q : \sim p$
T	T	F	F	F	F
T	F	F	T	T	T
F	T	T	F	T	T
F	F	T	T	T	T

Therefore, $p \rightarrow \sim q \equiv q \rightarrow \sim p$.

28. (a)

p	q	$\sim q$	$(p \wedge \sim q)$	$(p \wedge q)$	$(p \wedge \sim q) \vee (p \wedge q)$
T	T	F	F	T	T
T	F	T	T	F	T
F	T	F	F	F	F
F	F	T	F	F	F

(b)

p	q	$\sim p$	$(p \vee q)$	$(p \vee q) \wedge \sim p$	$[(p \vee q) \wedge \sim p] \rightarrow q$
T	T	F	T	F	T
T	F	F	T	F	T
F	T	T	T	T	T
F	F	T	F	F	T

29. (a) Joe Czernyu loves Mom and apple pie.
(b) The structure of the Statue of Liberty will eventually
rust. **(c)** Albertina passed Math 100. **30.** Let the
following letters represent the given sentences:
p: You are fair-skinned.
q: You will sunburn.
r: You do not go to the dance.
s: Your parents want to know why you didn't go to the
dance.
Symbolically, $p \rightarrow q, q \rightarrow r, r \rightarrow s$. Using contrapositives we
have: $\sim s \rightarrow \sim r$, $\sim r \rightarrow \sim q$, $\sim q \rightarrow \sim p$. By the chain rule,
$\sim s \rightarrow \sim p$; that is, if your parents do not want to know why
you didn't go to the dance, then you are not fair-skinned.
31. (a) Valid, *modus tollens* **(b)** Valid, *modus ponens*

Answers to Now Try This

1-1. 11 pieces for 10 cuts; $(n + 1)$ pieces for n cuts
1-2. (a) 2500 **(b)** $\left(\dfrac{a_1 + a_n}{2}\right)n$ **1-3.** 120
1-4. 90 days **1-5.** Answers vary; for example, because
each person owes $13, Al could pay $4.25 to Betty and
$4.00 to Carl; Dani could pay $7.00 to Carl and everyone
would be even. **1-6.** 23 floors
1-7. Answers vary, for example,

$$\begin{array}{r} 132 \\ + 932 \\ \hline 1064 \end{array} \quad \text{or} \quad \begin{array}{r} 173 \\ + 873 \\ \hline 1046 \end{array}$$

1-8. 83 **1-9.** Al plays tennis; Bob plays baseball; Carl
plays basketball; Dan swims. **1-10. (a)** Answers vary; for
example, the next three terms could be $\triangle, \triangle, \bigcirc$. **(b)** The
pattern could be one circle, two triangles, one circle, two
triangles, and so on. **1-11. (a)** inductive reasoning
(b) The next several numbers also work. **(c)** Yes, if
$x = 11$, then $11^2 + 11 + 11$ is not prime because it is
divisible by 11. **1-12.** Because the 2nd term is 11, then
$11 = a_1 + d$. Because the 5th term is 23, then $23 = a_1 + 4d$.
Solving for a_1 and then equating the answers, we have
$11 - d = 23 - 4d$, which implies that $d = 4$. To find the
100th term, we substitute in $a_1 + (n - 1)d$ and the 100th
term is $7 + (100 - 1)4 = 7 + 99 \cdot 4 = 403$.

1-13. (a) 4 **(b)** 7 **(c)** 12 **(d)** 20 **(e)** 33 **(f)** The sum of the first n Fibonacci numbers is one less than the Fibonacci number two numbers later in the sequence. **(g)** $F_1 + F_2 + F_3 + F_4 + \ldots + F_n = F_{n+2} - 1$

1-14. (a) After 10 hours, there are $2 \cdot 3^{10} = 118{,}098$ bacteria, and after n hours, there are $2 \cdot 3^n$ bacteria. **(b)** After 10 hours, there are $2 + 10 \cdot 3 = 32$ bacteria, and after n hours, there are $2 + n \cdot 3$ bacteria. We can see that after only 10 hours geometric growth is much faster than arithmetic growth. In this case, 118,098 versus 32. This is true in general when $n > 1$.

1-15. (a)

(b)

1	2	3	4
4	12	24	40

(c) 4 12 24 40 60 84 112
   8 12 16 20 24 28
     4   4   4   4   4

(d) No, finding differences for a_{100} and a_n is very hard. It is easier to find a pattern involving the number of horizontal sticks and vertical sticks, that is, $a_{100} = 101 \cdot 100 + 101 \cdot 100 = 20{,}200$ and $a_n = (n+1)n + (n+1)n$ or $2[(n+1)n]$ or $2n^2 + 2n$.

1-16.

p	q	$\sim p$	$\sim q$	$p \vee q$	$\sim(p \vee q)$	$\sim p \wedge \sim q$
T	T	F	F	T	F	F
T	F	F	T	T	F	F
F	T	T	F	T	F	F
F	F	T	T	F	T	T

$\sim(p \vee q) \equiv \sim p \wedge \sim q$

1-17.

p	q	$p \rightarrow q$	$\sim(p \rightarrow q)$	$\sim q$	$p \wedge \sim q$
T	T	T	F	F	F
T	F	F	T	T	T
F	T	T	F	F	F
F	F	T	F	T	F

$\sim(p \rightarrow q) \equiv p \wedge \sim q$

1-18.

p	q	$p \rightarrow q$	$q \rightarrow p$	$(p \rightarrow q) \wedge (q \rightarrow p)$
T	T	T	T	T
T	F	F	T	F
F	T	T	F	F
F	F	T	T	T

Answers to Brain Teasers

Section 1-1

35 moves. This can be solved using the strategy of examining simpler cases and looking for a pattern. If one person is on each side, 3 moves are necessary. If two people are on each side, 8 moves are necessary. With 3 people on each

side, 15 moves are necessary. If n people are on each side, $(n + 1)^2 - 1$ moves are required.

Section 1-2

312211; the pattern counts the number of times a number occurs in the previous row. For example, to find the sixth row we examine the fifth row. There are three 1s, two 2s, and one 1, so the sixth row is 312211. The pattern continues using this rule.

Answer to Laboratory Activity

Section 1-1

The number of moves is $2^n - 1$ for n coins. This can be solved using the strategy of examining a simpler problem. If there is one coin, 1 move is necessary. If there are two coins, 3 moves are necessary. For three coins, the number of moves is 7. For four coins, the number of moves is 15.

At the rate of one move per second, it would take approximately 584,942,417,418 years to move 64 coins.

Answer to Preliminary Problem

It is possible to label each bowl correctly if you choose the bowl labeled APPLES AND ORANGES. If the piece of fruit drawn is an apple, then the APPLES sign must be placed under this bowl. Because each bowl is labeled incorrectly, the ORANGES sign must be moved to the bowl that was labeled APPLES and this leaves only one bowl for the APPLES AND ORANGES sign. If the selected piece of fruit was an orange, then similar reasoning could be used and the ORANGES sign would be shifted to the bowl and then the APPLES sign shifted, leaving only one spot for the APPLES AND ORANGES sign.

Note that if the bowl labeled APPLES was selected and you happened to pick an orange, then you would not know which sign to shift. Similar reasoning can be used to show that the bowl labeled ORANGES could not be selected first.

Chapter 2

Assessment 2-1A

1. (a) $\overline{\overline{\text{M}}}$CDXXIV; the double bar over M represents $1000 \cdot 1000 \cdot 1000$. **(b)** 46,032; the 4 in 46,032 represents 40,000 while the 4 in 4632 represents only 4000. **(c)** < ▼▼: the space in the latter number indicates < is multiplied by 60. **(d)** 𝕏ՈI; the 𝕏 represents 1000 while 𝟿 represents only 100. **(e)** ☺ represents three groups of 20 plus zero 1s and ⁙⁙⁙ represents three 5s and three 1s. **2. (a)** MCML; MCMXLVIII **(b)** << <▼▼ <<< **(c)** 𝕏991; 𝕏9ՈՈՈՈՈՈՈՈՈIIIIIIIIII **(d)** ⁙⁙⁙⁙; ⁙⁙ **3.** 1922 **4. (a)** CXXI **(b)** XLII **5. (a)** ▼ <▼▼; ՈՈՈՈՈՈII; LXXII; ⁙⁙⁙ **(b)** 602; 999ᵁ; DCII; ⁙⁙

(c) 1223; << <<ᵛᵛᵛ; MCCXXIII; ⋮⋮   **6.** **(a)** Hundreds
(b) Tens   **7.** **(a)** 3,004,005 **(b)** 20,001   **8.** 811 or 910
9. **(a)** 86 **(b)** 11   **10.** 2112_{four} **11.** **(a)** $(1, 10, 11,$
$100, 101, 110, 111, 1000, 1001, 1010, 1011, 1100, 1101,$
$1110, 1111)_{two}$ **(b)** $(1, 2, 3, 10, 11, 12, 13, 20, 21, 22, 23,$
$30, 31, 32, 33)_{four}$ **12.** 20   **13.** $2032_{four} = 2 \cdot 4^3 +$
$0 \cdot 4^2 + 3 \cdot 4^1 + 2 \cdot 1$ **14.** **(a)** 111_{two} **(b)** EEE_{twelve}
15. **(a)** ETE_{twelve}; $EE1_{twelve}$ **(b)** 11111_{two}; 100001_{two}
(c) 554_{six}; 1000_{six} **16.** **(a)** There is no numeral 4 in base
four.   **(b)** There are no numerals 6 or 7 in base five.
17. 3 blocks, 1 flat, 1 long, 2 units
18.

19. **(a)** 8 pennies can be traded for 1 nickel and 3 pennies.
After the trade, we have 2 quarters, 10 nickels and 3 pennies.
10 nickels can be traded for 2 quarters. After this trade, we
have 4 quarters, 0 nickels and 3 pennies.   **(b)** Assume that
you have 73 cents in any possible combination, for example,
10 nickels and 23 pennies. Because 23 pennies can be traded
for 4 nickels and 3 pennies, we have 14 nickels and 3 pennies.
14 nickels can be traded for 2 quarters and 4 nickels. After the
second trade, we should have 2 quarters, 4 nickels, and
3 pennies. We obtain: $73 = 243_{five}$.   **20.** **(a)** 10 flats = 1
block; 10 flats in base ten = 1000   **(b)** 20 flats = 1
block + 8 flats; 20 flats in base twelve = 1800_{twelve}
21. Methods vary. 100010_{two} **22.** **(a)** 117 **(b)** 45
(c) 1331   **23.** 1 prize of $625, 2 prizes of $125 and 1 of
$25   **24.** **(a)** 8 weeks, 2 days **(b)** 1 day, 5 hr
25. **(a)** 6 **(b)** 1   **26.** Above the bar are depicted 5s, 50s,
500s, and 5000s. Below the bar are 1s, 10s, 100s, and 1000s.
Thus there are $1 \cdot 5000, 1 \cdot 500, 3 \cdot 100, 1 \cdot 50, 1 \cdot 5,$ and $2 \cdot 1$
depicted for a total of 5857. The number 4869 could be de-
picted as follows:

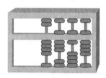

27. Assume an eight-digit display without scientific
notation. 98,765,432   **28.** **(a)** Answers vary; e.g., subtract
2020. **(b)** Answers vary; e.g., subtract 50.

Mathematical Connections 2-1

Communication

1. Answers will vary. Ben is incorrect. Zero is a place holder
in the Hindu-Arabic system. It is used to differentiate
between numbers like 54 and 504. If zero were nothing,
then we could eliminate it without changing our number
system.   **3.** **(a)** This is primarily for readability with
blocks grouped by 1000s and for naming. It has been

proposed with the metric system to drop the commas and
simply use spaces instead.   **(b)** Answers vary.

Open-Ended

5. 4; 1, 2, 4, 8; 1, 2, 4, 8, 16

Cooperative Learning

7. Answers vary; for example, in base two we have two digits, in base five we have five digits, and in base ten we have
ten digits. A negative base could be defined.   **9.** It is correct. However, Romans usually reserved the bar for numbers
greater than 4000. Because M is a special symbol for 1000, it
is preferable to write MI for 1001 rather than $\overline{\text{II}}$.

Assessment 2-2A

1. **(a)** $\{m, a, t, h, e, i, c, s\}$ **(b)** $\{x \mid x$ is a natural number
where $x > 20\}$ or $\{21, 22, 23, \ldots\}$ **2.** **(a)** $P = \{a, b, c, d\}$
(b) $\{1, 2\} \subset \{1, 2, 3, 4\}$ **(c)** $\{0, 1\} \not\subseteq \{1, 2, 3, 4\}$
(d) $0 \notin \{ \}$ or $0 \notin \varnothing$ **3.** **(a)** Yes **(b)** Yes **(c)** No
4. **(a)** 720 **(b)** $n(n - 1)(n - 2) \cdot \ldots \cdot 3 \cdot 2 \cdot 1$
5. **(a)** 24 **(b)** 6 **(c)** 12   **6.** $A = C, E = H, I = J$
7. **(a)** $1100 - 100,$ or 1000 if arithmetic **(b)** 501
(c) 11 **(d)** 100 **(e)** 5   **8.** $\overline{A}$ is the set of all college students
with at least one grade that is not an A; that is, those college
students who do not have a straight-A average.
9. **(a)** 7 **(b)** 0   **10.** **(a)** $n(D) = 5$ **(b)** $C = D$
11. **(a)** $\notin$ **(b)** $\notin$ **(c)** $\in$ **(d)** $\in$ **12.** **(a)** $\not\subseteq$ **(b)** $\subseteq$
(c) $\not\subseteq$ **(d)** $\not\subseteq$ **13.** **(a)** Yes **(b)** No. A may equal B.
(c) Yes **(d)** No. Consider $A = \{1\}$ and $B = \{1, 2\}$.
14. **(a)** Let $A = \{1, 2, 3\}$ and $B = \{1, 2, 3, 4, \ldots, 100\}$.
Since $A \subset B, n(A)$ is less than $n(B)$. So $3 < 100$. **(b)** Let
$A = \varnothing$ and $B = \{1, 2, 3\}$. Since $A \subset B, n(A) = 0$ is less than
$n(B) = 3,$ so $0 < 3$.   **15.** 35   **16.** 81

Mathematical Connections 2-2

Communication

1. A set is well defined if any object can be classified as
belonging to the set or not belonging to the set. For example, the set of U.S. presidents is well defined but the set of
rich U.S. presidents is not well defined because "rich" is a
matter of opinion.   **3.** Yes, $\varnothing \subset A$ for all sets A since A
contains at least one element and $\varnothing$ contains none; also
$\varnothing \subseteq A$.   **5.** To show $A \not\subseteq B$, we must be able to find at
least one element from set A that does not belong to set B.
7. If A and B are finite subsets, we say $n(A) \leq n(B)$ in case
A is a subset (not necessarily proper) of B.

Open-Ended

9. **(a)** Let A be the set of all natural numbers not equal to 1
with N as the universal set. Then $\overline{A} = \{1\}$ is finite.
(b) Answers vary. Let A be the set of even natural numbers
with N as the universal set. $\overline{A}$ is the set of odd natural numbers and so is infinite.

Cooperative Learning

11. (a) There are $2^{64} \approx 1.84 \times 10^{19}$ subsets of $\{1,2,3,\ldots,64\}$. If a computer can list one every millionth of a second, then it would take about

$$1.84 \times 10^{19} \times 0.000001 \sec \times \frac{1\,\text{yr}}{31{,}536{,}000\,\text{sec}} \approx 580{,}000\,\text{yr}$$

to list all the subsets. **(b)** There are $64 \cdot 63 \cdot 62 \cdot \ldots \cdot 2 \cdot 1 \approx 1.27 \times 10^{89}$ one-to-one correspondences between the two sets. So it would take about

$$1.27 \times 10^{89} \times 0.000001 \sec \times \frac{1\,\text{year}}{31{,}536{,}000\,\text{sec}} \approx 4 \times 10^{75}\,\text{yr}$$

to list all the one-to-one correspondences between the sets.

Questions from the Classroom

13. One way to designate the empty set is { }. Anything we enclose in the braces is an element of the set. Thus $\{\varnothing\}$ is a set having one element, the symbol of the empty set; so it is not empty. The difficulty usually arises from the reluctance to consider the empty set as an element. **15.** The set $A = \{1, \{1\}\}$ has two elements, 1 and $\{1\}$.

Review Problems

17. Answers will vary. The metric system is based on powers of 10. The common lengths and conversions are given below.

10 mm (millimeters) = 1 cm (centimeter)
10 cm = 1 dm (decimeter)
10 dm = 1 m (meter)
10 m = 1 dam (dekameter)
10 dam = 1 hm (hectometer)
10 hm = 1 km (kilometer)

The conversion plan will work much like base-ten conversions. **19.** 1410 **21. (a)** about 121 weeks **(b)** Approximately 3 years **(c)** Answers will vary. **(d)** Answers will vary.

Assessment 2-3A

1. (a) A or C **(b)** N **(c)** $\varnothing$ **2. (a)** Yes **(b)** Yes **(c)** Yes **(d)** Yes **3. (a)** True **(b)** False. Let $A = \{a,b,c\}$ and $B = \{a,b\}$. Then $A - B = \{c\}$, but $B - A = \varnothing$. **(c)** False. Let $U = \{a,b,c\}$, $A = \{a\}$ and $B = \{b\}$. Then $A \cap B = \varnothing$ and $\overline{A \cap B} = U$. $\overline{A} = \{b,c\}$; $\overline{B} = \{a,c\}$, and $\overline{A} \cap \overline{B} = \{c\}$. $\overline{A \cap B} \neq \overline{A} \cap \overline{B}$. **(d)** False. Let $A = \{a,b\}; B = \{b\}$. $A \cup B = \{a,b\}; (A \cup B) - A = \varnothing \neq B$. **(e)** False. Let $A = \{1,2,3\}, B = \{3,4,5\}$. Then $(A - B) \cup A = \{1,2,3\}$, but $(A - B) \cup (B - A) = \{1,2\} \cup \{4,5\} = \{1,2,4,5\}$. **4. (a)** $A \cap B = B$ **(b)** $A \cup B = A$ **5. (a)**

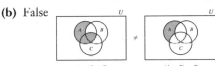

(b)

6. (a) $S \cup \overline{S} = U$ **(b)** $\overline{U} = \varnothing$ **(c)** $S \cap \overline{S} = \varnothing$ **(d)** $\varnothing \cap S = \varnothing$ **7. (a)** $A - B = A$ **(b)** $A - B = \varnothing$ **8.** Yes. By definition $A - B$ is the set of all elements in A that are not in B. If $A - B$ is the empty set, then this means that there are no elements in A that are not in B, which makes A a subset of B. **9.** Answers will vary. **(a)** $B - A$ **(b)** $\overline{A \cup B}$ **(c)** $(A \cap B) - C$ **10.**

$\overline{A \cap B}$

11. (a) False

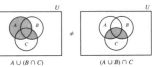

$A \cup (B \cap C)$ $(A \cup B) \cap C$

(b) False

$A - (B - C)$ $(A - B) - C$

12. (a) $(A \cap B \cap C) \subseteq (A \cap B)$ **(b)** $(A \cup B) \subseteq (A \cup B \cup C)$ **13. (a) (i)** 5 **(ii)** 2 **(iii)** 2 **(iv)** 3 **(b) (i)** $n + m$ **(ii)** The smaller of the two numbers m and n **(iii)** m **(iv)** n **14. (a)** Greatest is 15; least is 6. **(b)** Greatest is 4; least is 0. **15. (a)** The set of college basketball players more than 200 cm tall **(b)** The set of humans who are not college students or who are college students less than or equal to 200 cm tall **(c)** The set of humans who are college basketball players or who are college students taller than 200 cm **(d)** The set of all humans who are not college basketball players and who are not college students taller than 200 cm **(e)** The set of all college students taller than 200 cm who are not basketball players **(f)** The set of all college basketball players less than or equal to 200 cm tall **16.** 18 **17.** 4 **18. (a)** 20 **(b)** 10 **(c)** 10 **19.** 3. Using the following Venn diagram and the fact that the set of people who are O negative is $100 - n(A \cup B \cup C)$, we see that the answer is 3.

20. (a) False. Let $A = \{a,b,c\}$ and $B = \{1,2,3\}$. **(b)** False. Let $A = \{1,2,3\}$ and $B = \{1,2,3,4\}$. **(c)** True. **21.** Steelers versus Jets, Vikings versus Packers, Bills versus Redskins, Cowboys versus Giants **22. (a)** $A \times B = \{(x,a),(x,b),(x,c),(y,a),(y,b),(y,c)\}$ **(b)** $B \times A = \{(a,x),(a,y),(b,x),(b,y),(c,x),(c,y)\}$ **23. (a)** $C = \{a\}, D = \{b,c,d,e\}$ **(b)** $C = \{1,2\}$, $D = \{1,2,3\}$ **(c)** $C = D = \{0,1\}$

Mathematical Connections 2-3

Communication

1. (a) Yes. $(A \cap B) \subseteq (A \cup B)$ **(b)** No. For example, let $A = \{1,2,3\}$, and $B = \{4\}$. Then $2 \in A \cup B$, but $2 \notin A \cap B$.   **3.** No. Let $A = \{1\}$ and $B = \{a\}$. Then $A \times B = \{(1,a)\}$, but $B \times A = \{(a,1)\}$. These are not equal.

Open-Ended

5. Answers vary.

Cooperative Learning

7. Answers vary.

Questions from the Classroom

9. The student is right. If $A = \{1, 2\}$, $B = \{2, 3\}$ and $C = \{2, 4\}$, $A \cap B = A \cap C$ but $B \neq C$. Also, to show that the hypothesis implies $B = C$, we show that $B \subseteq C$ and $C \subseteq B$. To show that $B \subseteq C$, let $x \in B$, then $x \in A \cup B$ and because $A \cup B = A \cup C$, $x \in A \cup C$. Consequently, $x \in A$ or $x \in C$. If $x \in C$, then $B \subseteq C$. If $x \in A$, then since we started with $x \in B$, it follows that $x \in A \cap B$. Because $A \cap B = A \cap C$, we conclude that $x \in A \cap C$ and therefore $x \in C$. Thus, $B \subseteq C$. Similarly, starting with $x \in C$, it can be shown that $x \in B$ and hence that $C \subseteq B$.   **11.** Even though the Cartesian product of sets includes all pairings in which each element of the first set is the first component in a pair with each element of the second set, this is not necessarily a one-to-one correspondence. A one-to-one correspondence implies that there must be the same number of elements in each set. This is not the case in a Cartesian product. For example, consider sets $A = \{1\}$ and $B = \{a, b\}$.

Review Problems

13. The number "two" exists in base two but there is no single digit to represent "two."   **15. (a)** $\{x \mid 3 < x < 10$ where $x \in N\}$ **(b)** $\{15, 30, 45\}$ **17. (a)** These are all the subsets of $\{2,3,4\}$. There are $2^3 = 8$ such subsets.   **(b)** There are 8 subsets that contain the number 1.   **(c)** Twelve subsets contain 1 or 2 (or both). There are 4 subsets of $\{3,4\}$. We can form subsets that contain 1 or 2 or both by adjoining 1 to each, 2 to each, or 1 and 2 to each. By the Fundamental Counting Principle then, there are $3 \cdot 4 = 12$ possibilities. (It's also not hard to list them.)   **(d)** Four subsets contain neither 1 nor 2.   **(e)** 16; B has $2^5 = 32$ subsets. Half contain 5 and half do not.   **(f)** Every subset of A is a subset of B. The others can be listed by adjoining the element 5 to each subset of A. So there are twice as many subsets of B as subsets of A. A has 16 subsets and B has 32 subsets.   **19.** Answers vary.   **21.** 60

Chapter Review

1. (a) tens   **(b)** thousands   **(c)** hundreds
2. (a) 400,044   **(b)** 117   **(c)** 1704   **(d)** 11   **(e)** 1448

3. (a) CMXCIX   **(b)** ∩∩∩∩∩∩∩∩ⅢⅢⅢ   **(c)** ÷∴   **(d)** 2341_{five}
(e) 11011_{two} **4. (a)** 3^{17} **(b)** 2^{21} **5.** 2020_{three}
6. 1 block, 2 flats, 2 longs, 0 units
7. (a)   **(b)**

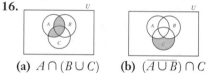

8. (a) 10^{10} **(b)** 5^8 **(c)** 2^{31} **9. (a)** 10,000,000,023
(b) $10,000,000,001_{\text{two}}$ **(c)** $10,000,000,001_{\text{five}}$
(d) 9,999,999,999   **(e)** $1,111,111,111_{\text{two}}$ **(f)** $EEEEE_{\text{twelve}}$
10. Answers vary. Selling pencils by the units, dozen, and gross is an example of the use of base 12.   **11. (a)** The Egyptian system had seven symbols. It was a *tally* system and a *grouping* system, and it used the *additive property*. It did not have a symbol for zero but this was not very important because it did not use place value.   **(b)** The Babylonian system used only two symbols. It was a place value system (base 60) and was additive within the positions. It lacked a symbol for zero until around 300 BCE.   **(c)** The Roman system used seven symbols. It was additive, subtractive, and multiplicative. It did not have a symbol for 0.   **(d)** The Hindu-Arabic system uses 10 symbols. It uses place value and it has a symbol for 0.   **12. (a)** 1003_{five} **(b)** 10000000_{two}
(c) $T8_{\text{twelve}}$ **13. (a)** 4210014_{five} **(b)** 10000001000_{two}
(c) $E0T018_{\text{twelve}}$ **(d)** 1100010_{eight} **14.** $\varnothing, \{m\}, \{a\},$ $\{t\}, \{b\}, \{m,a\}, \{m,t\}, \{m,b\}, \{a,t\}, \{a,b\}, \{t,b\}, \{m,a,t\},$ $\{m,a,b\}, \{m,t,b\}, \{a,t,b\}, \{m,a,t,b\}$ **15. (a)** $A \cup B = A$
(b) $C \cap D = \{l, e\}$ **(c)** $\overline{D} = \{u, n, i, v, r\}$
(d) $A \cap \overline{D} = \{r, v\}$ **(e)** $\overline{B \cup C} = \{s, v, u\}$
(f) $(B \cup C) \cap D = \{l, e, a\}$ **(g)** $\{i, n\}$ **(h)** $\{e\}$
(i) 5   **(j)** 16
16.

(a) $A \cap (B \cup C)$ **(b)** $(\overline{A \cup B}) \cap C$

17. 5040   **18. (a)** Answers will vary.   **(b)** 6
$t \leftrightarrow e$
$b \leftrightarrow n$
$e \leftrightarrow d$
19. It is not true that $A \cap (B \cup C) = (A \cap B) \cup C$ for all A, B, and C, as shown in the following diagrams.

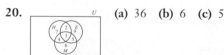

$A \cap (B \cup C)$ $\neq$ $(A \cap B) \cup C$

20.   **(a)** 36   **(b)** 6   **(c)** 5

21. Answers vary.   **(a)** $B \cup (C \cap A)$ **(b)** $B - C$
22. (a) False. Consider the sets $\{a\}$ and $\{2\}$.   **(b)** False. It is not a proper subset of itself.   **(c)** False. Consider the sets $\{t, b, e\}$ and $\{e, n, d\}$. They have the same number of elements, but they are not equal.   **(d)** False. This is in one-to-one correspondence with the set of natural numbers.

(e) False. The set $\{5,10,15,20, \dots \}$ is a proper subset of the natural numbers and is equivalent to the natural numbers, since there is a one-to-one correspondence between the two sets. **(f)** False. Let $B = \{1,2,3\}$ and A be the set of natural numbers. **(g)** True **(h)** False. Let $A = \{1,2,3\}$ and $B = \{a,b,c\}$. **23. (a)** Because $A \cup B$ is the union of disjoint sets, $A - B$, $B - A$, and $A \cap B$, the equation is true. **(b)** True, because $A - B$ and B are disjoint as well as $B - A$ and A; also, $A \cup B = (A - B) \cup B = (B - A) \cup A$. **24. (a)** 17 **(b)** 34 **(c)** 0 **(d)** 17 **25.** 7 **26.** Answers vary. The first question might be "Is the state or province one of the 48 contiguous states in the United States?" If the answer is *yes*, the second question could be "Does it begin with a vowel?" The third question can then be "Is it ___?" If the answer to the first question is *no*, the second question could be "Is it in Canada?" and the third question could be the same as above. **27. (a)** Let $A = \{1,2,3, \dots ,13\}$ and $B = \{1,2,3\}$; then B is a proper subset of A. Therefore, B has fewer elements than A, so $n(B)$ is less than $n(A)$; thus, $3 < 13$. **(b)** Let $A = \{1,2,3, \dots ,12\}$ and $B = \{1,2,3, \dots ,9\}$. B is a proper subset of A, so A has more elements than B; thus $n(A)$ is greater than $n(B)$, $12 > 9$. **28.** 12 outfits

Answers to Now Try This

2-1. 3 blocks 12 flats 11 longs 17 units
$=$ 3 blocks (1 block 2 flats) (10 long 1 long) (10 units 7 units)
$=$ 4 blocks 2 flats (1 flat 1 long) (1 long 7 units)
$=$ 4 blocks 3 flats 2 longs 7 units
$= 4327$
2-2. (a) 𝌆🄌🄌🄌𝍩𝍩𝍩 999∩∩‖ **(b)** 203,034
(c) Answers vary; for example, writing large numbers is very cumbersome as the system is additive and does not use place value. Performing operations involving addition, subtraction, multiplication, and division is hard because of the way that numbers are represented.
2-3. (a) ▼▼▼ <<▼▼▼▼▼ <<▼ **(b)** $2 \cdot 60^2 + 11 \cdot 60 + 1 = 7861.$ **(c)** Answers vary. The Hindu-Arabic system has a symbol for 0 and this is very important in a system that uses place value. Because it uses base sixty, the Babylonian system requires the use of many symbols to write numbers such as 59.
2-4. (a) The illustration indicates successive divisions by 5s. This shows that there are 164 fives in 824 with 4 as a remainder. Next there are 32 fives in 164 with 4 fives as a remainder. This process continues until we see that there is one 625 in 824 with one 125, two 25s, four 5s, and 4 units remaining.
(b) 5 ⌊728
$\quad$ 5 ⌊145 $\ \ $3
$\quad$ 5 ⌊29 $\ \ $0
$\quad$ 5 ⌊5 $\ \ $4
$\qquad$ 1→0

Thus, the answer is 10403_{five} **2-5. (a)** and **(b)**: Number the swimming lanes 1, 2, 3, 4 and name the people A, B, C, D. Then we represent the correspondence:

$$
\begin{array}{ccc}
1 & \leftrightarrow & A \\
2 & \leftrightarrow & B \\
3 & \leftrightarrow & C \\
4 & \leftrightarrow & D \\
\end{array}
\quad \text{as} \quad
\begin{array}{cccc}
1 & 2 & 3 & 4 \\
A & B & C & D \\
\end{array}
$$

The 24 one-to-one correspondences are

1 2 3 4	1 2 3 4	1 2 3 4	1 2 3 4
A B C D	B A C D	C A B D	D A B C
A B D C	B A D C	C A D B	D A C B
A C B D	B C A D	C B A D	D B A C
A C D B	B C D A	C B D A	D B C A
A D B C	B D A C	C D A B	D C A B
A D C B	B D C A	C D B A	D C B A

(c) We notice that $24 = 4 \cdot 3 \cdot 2 = 4 \cdot 3 \cdot 2 \cdot 1$. We also notice that we had four choices for people to swim in lane 1. After making a choice, we see that we had three choices to swim in lane 2, leaving us with two choices for lane 3 and, finally, one choice for lane 4. Extrapolating from this, we conjecture that there are

$$5 \cdot 4 \cdot 3 \cdot 2 \cdot 1 = 120$$

distinct one-to-one correspondences between a pair of five-element sets. **2-6.** If event M_1 can occur in m_1 ways and after it has occurred, event M_2 can occur in m_2 ways, and after it has occurred, event M_3 can occur in m_3 ways, and so on, where events $M_1, M_2, M_3, \dots , M_n$ can occur correspondingly in $m_1, m_2, m_3, \dots , m_n$ ways, then event M_1 followed by event M_2 followed by event $M_3, \dots$ followed by event M_n, can occur in $m_1 \cdot m_2 \cdot m_3 \cdot \ \dots \ \cdot m_n$ ways.
2-7. (a) No, two sets may be equivalent without being equal. To see this consider the following example:

$$A = \{a,b,c\}$$
$$B = \{1,2,3\}$$

Then,

$$a \leftrightarrow 1$$
$$b \leftrightarrow 2$$
$$c \leftrightarrow 3$$

is a one-to-one correspondence between A and B, and therefore $A \sim B$. However, $A \neq B$. **(b)** Yes, if two sets are equal, then each element can be paired with itself in a one-to-one correspondence to show equivalency. **2-8.** The set of natural numbers is $N = \{1,2,3,4,5, \dots \}$. If N were finite, then there would be some greatest element q. However, $q + 1$ is a natural number greater than q and would still be in the set N. Thus, there can be no greatest element in N and it cannot be finite. **2-9. (a)** Yes, by definition, $A \subseteq B$ means that every element of A is an element of B. Similarly, $A \subset B$ means that every element of A is an element of B but there exists an element in B which is not an element of A. Hence, if $A \subset B$, it is true that every element of A is in B. Consequently, $A \subseteq B$. Notice that if the more stringent condition $A \subset B$ is satisfied, then the weaker condition $A \subseteq B$ must also be satisfied. **(b)** No. To see this, consider the following counterexample:

$$A = \{a,b,c\}$$
$$B = \{a,b,c\}$$

Then, $A \subseteq B$. Notice that $A \not\subset B$ since $A = B$.

2-10. (a) Let two representations of the empty set be $\varnothing$ and { }. Suppose that $\varnothing \not\subseteq$ { }. Then there must be some element of $\varnothing$ that is not in { }. This cannot happen, so $\varnothing \subseteq$ { }. **(b)** Again use the two representations, $\varnothing$ and { }. From part (a) we know that $\varnothing \subseteq$ { }. Suppose $\varnothing \subset$ { }. If $\varnothing$ is a proper subset of { }, then there must be some element in { } that is not in $\varnothing$. There is no such element, so $\varnothing \not\subset$ { }.

2-11. (a) Assuming that a simple majority forms a winning coalition, we see that any subset consisting of three or more senators is a winning coalition. There are 16 such subsets. To see this, let $\{A, B, C, D, E\}$ be the set of five senators on the committee. Then the following are all possible winning coalitions:

$\{A,B,C\}$	$\{A,B,D\}$	$\{A,B,E\}$	$\{A,C,D\}$
$\{A,C,E\}$	$\{A,D,E\}$	$\{B,C,D\}$	$\{B,C,E\}$
$\{B,D,E\}$	$\{C,D,E\}$	$\{A,B,C,D\}$	$\{A,B,C,E\}$
$\{A,B,D,E\}$	$\{A,C,D,E\}$	$\{B,C,D,E\}$	$\{A,B,C,D,E\}$

From the list, we see that there are five subsets containing exactly four members. We also see that there are five senators on the committee. To understand why these numbers are the same, notice that creating a four-element subset is equivalent to deleting a single element from the total set. That is, we can give a one-to-one correspondence between the set of four-element subsets of $\{A, B, C, D, E\}$ and the set of senators by corresponding to each four-element subset the senator who is not in that subset, as shown:

$\{A,B,C,D\}$	$\leftrightarrow$	E
$\{A,B,C,E\}$	$\leftrightarrow$	D
$\{A,B,D,E\}$	$\leftrightarrow$	C
$\{A,C,D,E\}$	$\leftrightarrow$	B
$\{B,C,D,E\}$	$\leftrightarrow$	A

(b) We can give a one-to-one correspondence between the three-element subsets and the two-element subsets of $\{A, B, C, D, E\}$ by matching each three-element subset with the unique two-element subset that contains the senators on the committee but not in the subset; for example, $\{A, B, C\}$ 4 $\{D, E\}$. From part (a), we know that there are exactly 10 three-element subsets of the committee; hence, there must be 10 two-element subsets of the committee.

2-12. (a) 15 **(b)** $2^n - 1$ **2-13.** The formula is $n(A \cup B) = n(A) + n(B) - n(A \cap B)$. To justify this formula, notice that in $n(A \cup B)$, the elements of $A \cap B$ are counted only once. In $n(A) + n(B)$, the elements of $A \cap B$ are counted twice, once in A and once in B. Thus, subtracting $n(A \cap B)$ from $n(A) + n(B)$ makes the number equal to $n(A \cup B)$. For example, if $A = \{a, b, c\}$ and $B = \{c, d\}$, then $A \cup B = \{a, b, c, d\}$ and $n(A \cup B) = 4$. However, $n(A) + n(B) = 3 + 2 = 5$ since c is counted twice. Because $A \cap B = \{c\}$, $n(A \cap B) = 1$ and $n(A) + n(B) - n(A \cap B) = 4$. **2-14.** It is always true that $A \cap (B \cap C) = (A \cap B) \cap C$. The following figure gives Venn diagrams of each side of this equation. Because the Venn diagrams result in the same set, the equation is always true.

Similarly, it is always true that $A \cup (B \cup C) = (A \cup B) \cup C$. The following Venn diagrams justify this statement.

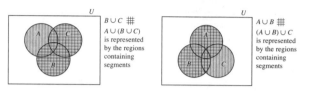

However, in general, $A - (B - C) \neq (A - B) - C$. To see this, consider the following counterexample:

$$A = \{1,2,3,4,5\}$$
$$B = \{1,2,3\}$$
$$C = \{3,4\}$$

Then, $A - (B - C) = A - \{1, 2\} = \{3, 4, 5\}$, but $(A - B) - C = \{4, 5\} - C = \{5\}$. Thus, for the preceding choice of A, B and C, we have that $A - (B - C) \neq (A - B) - C$. **2-15.** The following Venn diagrams show $A \cup (B \cap C) = (A \cup B) \cap (A \cup C)$.

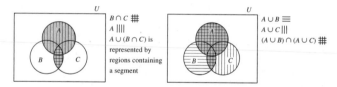

The property should be called the distributive property of set union over intersection.

2-16.

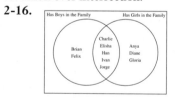

Answer to Brain Teaser

Section 2-2

Consider the set of students in Mr. Gonzales's class. This set has $2^{24} - 1$ nonempty subsets. The set of students in Ms. Chan's class has $2^{25} - 1$ nonempty subsets. Using the Fundamental Counting Principle, we find that $(2^{24} - 1) \cdot (2^{25} - 1)$, or approximately 563 trillion school committees could be formed to contain at least one student from each class. This number is greater than the population of the world, which is approximately 6 billion. Thus, Linda is right.

Answers to Laboratory Activity

Section 2-3

1. $A \cup B$ is the set of all blocks that are green, large, or both green and large. $B \cup A$ is the set of all blocks that are large, green, or both large and green. The sets are equal.
2. $\overline{A \cap B}$ is the set of all blocks that are not both large and green. $\overline{A} \cap \overline{B}$ is the set of all the small blocks that are not green. The sets are not equal.
3. $\overline{A \cap B}$ is the set of all blocks that are not both green and large. $\overline{A} \cup \overline{B}$ is the set of all the blocks that are not green or are not large. The sets are equal.
4. $A - B$ is the set of green blocks that are not large. $A \cap \overline{B}$ is the set of blocks that are green and not large. The sets are equal.

Answer to Preliminary Problem

Approaches to the problem may vary. Suppose A is the set of all adults, F is the set of all females and M is the set of all Mississippi residents. The diagram below shows variables for the unknown numbers of elements (people) in each region.

With this categorization, we have the following.
a – Mississippi men
b – Mississippi women
c – Tennessee men
d – Tennessee women
e – Tennessee girls
f – Mississippi girls
g – Mississippi boys
h – Tennessee boys
Also with the information in the problem we have

$$a + b = 24$$
$$b + d = 29$$
$$a + b + f + g = 44$$
$$e + f = 29$$
$$d = 17$$
$$b + f = 26$$
$$c + h = 17$$

We want $a + b + c + d + e + f + g + h$. Combining

$$a + b + f + g = 44$$
$$c + h = 17$$
$$d = 17$$

We know

$$a + b + c + d + f + g + h = 44 + 17 + 17 = 78.$$

If we had a value for e, we have the solution.
With $d = 17$ and $b + d = 29$, we know $b = 12$.
With $b = 12$ and $a + b = 24$, we know $a = 12$.
With $b = 12$ and $b + f = 26$, we know $f = 14$.
And with $f = 14$ and $e + f = 29$, we know $e = 15$.
Hence, the total number on the bus is $78 + 15 = 93$.

Chapter 3

Assessment 3-1A

1. For example, let $A = \{1, 2\}, B = \{2, 3\}$; then $A \cup B = \{1, 2, 3\}$. Thus, $n(A) = 2, n(B) = 2, n(A \cup B) = 3$, but $n(A) + n(B) = 2 + 2 = 4 \neq n(A \cup B)$. **2. (a)** true **(b)** false **(c)** true **3.** $n(A \cap B) = 2$ **4. (a)** 3, 4, 5, 6 **(b)** 3 **5. (a)** yes **(b)** yes **(c)** yes **(d)** No. $3 + 5 \notin V$. **(e)** yes **6. (a)** commutative property of addition **(b)** associative property of addition **(c)** commutative property of addition **(d)** identity property of addition **(e)** commutative property of addition **(f)** associative property of addition **7.** No. If $k = 0$, we would have $k = 0 + k$, implying $k < k$ which is false. **8. (a) (i)** For any whole numbers a and b, $a < b$ if, and only if, there exists a natural number k such that $b - k = a$, or equivalently if, and only if, $b - a$ is a natural number. **(ii)** For any whole numbers a and b, $a > b$ if, and only if, there exists a natural number k such that $a - k = b$, or equivalently $a - b$ is a natural number. **(b)** $a \geq b$ if, and only if, $a - b$ is a whole number. **9. (a)** 33, 38, 43 **(b)** 56, 49, 42 **10. (a)** 9 **(b)** 8 **(c)** 3 **(d)** 6 or 8 **(e)** 5 **(f)** 4 or 8 **(g)** 9 **11.** 0 **12. (a)**

8	1	6
3	5	7
4	9	2

(b)

17	10	15
12	14	16
13	18	11

13. (a) Kent is the shortest and Vera is the tallest. **(b)** Kent, 140 cm; Mischa, 142 cm; Sally, 148 cm; Vera, 152 cm **14. (a)** $9 = 7 + x$, or $9 = x + 7$ **(b)** $x = 6 + 3$ or $x = 3 + 6$ **(c)** $9 = x + 2$ or $9 = 2 + x$ **15. (a)** $8 + 3 = 11, 3 + 8 = 11, 11 - 3 = 8, 11 - 8 = 3$ **(b)** $13 - 8 = 5, 13 - 5 = 8, 8 + 5 = 13, 5 + 8 = 13$ **16. (a)** $a \geq b$ **(b)** $b \geq c$ and $a \geq b - c$ **17. (a)**

(b) $\Box + 5 = 8$, so $\Box = 3$ **(c)** ♀♀♀♀♀♀♀♀♀ ↕↕↕↕↕↕ difference is 3 **(d)** |8 – 5•|•— 5 —| 0 1 2 3 4 5 6 7 8 **18. (a)** 4 **(b)** 8 **(c)** 5 **(d)** 9

Mathematical Connections 3-1

Communication

1. No, sets A and B do not have to be disjoint. For example, suppose there are 11 students taking both algebra and biology. This implies the following Venn diagram:

From this Venn diagram we see that $n(A \cup B) = 41$ and there are not necessarily 52 students taking algebra or biology. **3.** An arrow starting at 0 and ending at 3 represents the same number as an arrow starting at 4 and ending at 7. One way to explain this to students is to make physical models of each and show that the lengths are the same by matching. **5.** It is very useful to have students learn more than one method to model addition and subtraction if the methods are to generalize to sets of numbers other than

whole numbers. The missing-addend approach is useful for all sets of numbers in solving subtraction problems. The method of counting to compute additions is not effective with sets of fractions or real numbers.   **7. (a)** When you put 9 and 4 together, it is equal to the same length as 13.   **(b)** If you put 9 and 4 together on top of 4 and 9 together, they both equal 13.   **(c)** Take the length 9 away from 13 and the length left is equal to 4.   **(d)** Take the length 4 away from 13 and the length left is equal to 9.
9. For 0 to be an identity for subtraction, the following would have to hold: For any integer a, $a - 0 = a = 0 - a$. In this case, $a - 0 = a$ works but $0 - a = a$ does not work.

Open-Ended

11. Answers vary; for example, let $A = \{a, b\}$ and $B = \{a, b, c, d\}$. Then $4 - 2 = n(B - A) = n(\{c, d\}) = 2$.

Cooperative Learning

13. (a) The table shows that if you add any two single-digit whole numbers, your answer is also a whole number.   **(b)** The table shows that if you add any two single-digit whole numbers, the order is not important; that is, if $a \in W$ and $b \in W$, $a + b = b + a$. Each row of answers has a corresponding column of identical answers.   **(c)** The first row and first column show that if you add any digit to the identity, 0, you get the identical digit back again.   **(d)** The properties reduce the number of facts to be remembered: for example, the 19 facts in the first row and column can be learned by just knowing that 0 is the additive identity. The commutative property also reduces the number of facts; for example, if you know $9 + 2$, then you know $2 + 9$.
15. Answers vary.

Questions from the Classroom

17. Any number can be represented by a directed arrow of a given length. In this case, the directed arrow represents 3 units. Any arrow 3 units in length can be used to represent 3, regardless of its starting point.   **19.** Answers vary; for example, the examples that the student is showing are all true but producing examples that work is not proof that the claim is true in general. All that is needed is to produce one *counterexample* to show that the claim is not true. If we consider $5 - 8$, we can see that there is no whole number that satisfies this subtraction. If the set were closed, then if any two elements were picked and one subtracted from the other, the answer would have to be a whole number. In the case of $5 - 8$ this does not happen.

Assessment 3-2A

1. (a)   981   **(b)**   2025   **2.**
$\underline{+421}$ 1196
      1402 $\underline{+3148}$
                    6369

3. (a) one possibility: 863   **(b)** one possibility: 368
            $\underline{+752}$ $\underline{+257}$
              1615   625
4. No, he can have either the fish or the salad.
5. yes, \$124   **6.**   3428
                    $\underline{+5631}$
                      9059

7. (a)   93 $93 + 3$ 96
       $\underline{-37}$ → $\underline{-(37 + 3)}$ → $\underline{-40}$
                                                56

(b)   321 $321 + 2$ 323 $323 + 60$ 383
     $\underline{-38}$ → $\underline{-(38+2)}$ → $\underline{-40}$ → $\underline{-(40+60)}$ → $\underline{-100}$
                                                                    283

8. (a) (i)   687   **(ii)**   359
         $\underline{+549}$ $\underline{+673}$
           16   12
           12   12
           11   9
         $\underline{}$ $\underline{}$
         1236   1032
(b) The algorithm works because the placement of partial sums still accounts for place value. It is fairly easy in the example because only two digits are added at a time. This process can be adapted if more than two numbers were added.   **9.** Answers vary; for example: **(a)** The student added $8 + 5 = 13$ and wrote down 13 with no regrouping. He then added $2 + 7 = 9$ and wrote down 9.   **(b)** The student added $8 + 5 = 13$ and instead of writing down 3 and regrouping with the 1, he wrote down 1 and regrouped with the 3.   **(c)** The student only recorded the difference in the units $(9 - 5)$, the tens $(5 - 0)$, and the hundreds $(3 - 2)$. The student always subtracted the smaller number from the larger number ignoring what number was "on top" and which was "on the bottom."   **(d)** The student regrouped 3 hundreds as 2 hundreds and 10 tens but then did not regroup the 10 tens as $9 \cdot 10 + 15$.
10. Step 1—place value
Step 2—commutative and associative properties of addition
Step 3—distributive property of multiplication over addition
Step 4—single-digit addition facts
Step 5—place value
11. (a) $68 + 23 = (6 \cdot 10 + 8) + (2 \cdot 10 + 3)$
$= (6 \cdot 10 + 2 \cdot 10) + (8 + 3)$
$= (6 + 2)10 + (8 + 3)$
$= 8 \cdot 10 + 11$
$= 8 \cdot 10 + (1 \cdot 10 + 1)$
$= (8 \cdot 10 + 1 \cdot 10) + 1$
$= (8 + 1) \cdot 10 + 1 = 9 \cdot 10 + 1 = 91$
(b) $174 + 285 = (1 \cdot 100 + 7 \cdot 10 + 4) + (2 \cdot 100$
$+ 8 \cdot 10 + 5)$
$= (1 \cdot 100 + 2 \cdot 100)$
$+ (7 \cdot 10 + 8 \cdot 10) + (4 + 5)$
$= (1 + 2) \cdot 100 + (7 + 8) \cdot 10 + (4 + 5)$
$= 3 \cdot 100 + 15 \cdot 10 + 9$
$= 3 \cdot 100 + (10 + 5) \cdot 10 + 9$
$= 3 \cdot 100 + 1 \cdot 100 + 5 \cdot 10 + 9$

$$= (3 + 1) \cdot 100 + 5 \cdot 10 + 9$$
$$= 4 \cdot 100 + 5 \cdot 10 + 9$$
$$= 459$$

(c) $2458 + 793 = (2 \cdot 1000 + 4 \cdot 100 + 5 \cdot 10 + 8)$
$\qquad\qquad\qquad + (7 \cdot 100 + 9 \cdot 10 + 3)$
$\qquad\quad = 2 \cdot 1000 + (4 \cdot 100 + 7 \cdot 100)$
$\qquad\qquad + (5 \cdot 10 + 9 \cdot 10) + (8 + 3)$
$\qquad\quad = 2 \cdot 1000 + (4 + 7) \cdot 100 +$
$\qquad\qquad (5 + 9) \cdot 10 + (8 + 3)$
$\qquad\quad = 2 \cdot 1000 + 11 \cdot 100 + 14 \cdot 10 + 11$
$\qquad\quad = 2 \cdot 1000 + (10 + 1) \cdot 100 +$
$\qquad\qquad (10 + 4) \cdot 10 + (1 \cdot 10 + 1)$
$\qquad\quad = 2 \cdot 1000 + 1 \cdot 1000 + 1 \cdot 100$
$\qquad\qquad + 1 \cdot 100 + 4 \cdot 10 + 1 \cdot 10 + 1$
$\qquad\quad = (2 + 1) \cdot 1000 + (1 + 1) \cdot 100$
$\qquad\qquad + (4 + 1) \cdot 10 + 1$
$\qquad\quad = 3 \cdot 1000 + 2 \cdot 100 + 5 \cdot 10 + 1$
$\qquad\quad = 3251$

12. (a)
```
     4  3  5  8
  +  3  8  6  4
  0/7/1/1/1/2
     8  2  2  2
```
(b)
```
     4  9  2  3
  +  9  8  9  7
  1/3/1/7/1/1/0
  1  4  8  2  0
```

13. (a) 121_{five} **(b)** 20_{five} **(c)** 1010_{five} **(d)** 14_{five} **(e)** 1001_{two} **(f)** 1010_{two}

14.

+	0	1	2	3	4	5	6	7
0	0	1	2	3	4	5	6	7
1	1	2	3	4	5	6	7	10
2	2	3	4	5	6	7	10	11
3	3	4	5	6	7	10	11	12
4	4	5	6	7	10	11	12	13
5	5	6	7	10	11	12	13	14
6	6	7	10	11	12	13	14	15
7	7	10	11	12	13	14	15	16

Base eight

15. (a) 9 hr 33 min 25 sec **(b)** 1 hr 39 min 40 sec

16. It appears that the calculator is performing the operation twice.

17. (a)
```
      1 1
     4 3 2
     9 7 6
    1 4 1
  + 1 4 1 8
   ‾‾‾‾‾‾‾‾
     2 8 2 6
```
(b)
```
     3 1 2
     1 3 0
     2 2
   4 3 3 0
   2 0 3
   1 2 0
  ‾‾‾‾‾‾
   3 1 0 five
```

18. (a) 3 gross 10 doz 9 ones **(b)** 6 gross 3 doz 4 ones

19. There is no numeral 5 in base five: $22_{\text{five}} + 33_{\text{five}} = 110_{\text{five}}$. **20. (a)**
$$\begin{array}{r} 230_{\text{five}} \\ -\ 22_{\text{five}} \\ \hline 203_{\text{five}} \end{array}$$
(b)
$$\begin{array}{r} 20010_{\text{three}} \\ -\ 2022_{\text{three}} \\ \hline 10211_{\text{three}} \end{array}$$

21. (a) 1241_{five} **(b)** 101_{two} **(c)** TET_{twelve} **(d)** 4000_{five}

22. (a) The method produces a palindrome in each case:
(i) 363 (ii) 9339 (iii) 5005. **(b)** for example, 89 or 97

Mathematical Connections 3-2

Communication

1. Answers vary. This approach emphasizes the meaning of place value of the digits and may be easier for young children than the standard algorithm. It can also serve as a transition to the standard algorithm. **3.** Answers vary. For example, you could talk about how adding 1 and subtracting 1 has the effect of adding 0 to the problem, which produces an equivalent problem. The reason for changing the original problem to the new problem is that the numbers are easier to work with. You could tell her that her technique works well and discuss how it works.

5. (a) Answers vary. **(b)** In the example, if 10 is added to the 4 ones to get 14 ones and nothing else is done, this changes the problem. So 10 must be taken away so that the net result is just adding 0, which does not change the original problem. When 1 ten is added to the 8 tens being subtracted on the bottom, this balances the 1 ten that was added in the minuend. Similarly, the algorithm is continued by adding values in the minuend and then adding a corresponding value in the proper place value position in the subtrahend to neutralize the addition on the top.

7. For example, the words *regroup* and *trade* more accurately reflect the actions that are taken when performing an addition or subtraction problem. The words *borrow* and *carry* seem to reflect mechanics and not the mathematical ideas used in the algorithm.

Open-Ended

9. Students are encouraged to use the bibliographies at the end of the chapters to seek mathematics education articles to answer this question.

Questions from the Classroom

11. You might first determine what Joe intends to do next and what answer he comes up with. You might then discuss ways to check if the answer using this technique is reasonable and correct. You could model the problem using base-ten blocks and start out with 6 longs and 8 units and ask how he might "take away" 19. Various models can be used to show that we need to consider taking 9 from 18 instead of $9 - 8$. **13.** Betsy is confused on what you should add to check a subtraction. She should try smaller numbers to get a feeling for what needs to be done—for example, $9 - 5 = 4$—to check we add $4 + 5$ to see that we get back to 9. Number lines or colored rods could be used to show that $9 - 5 = 4$ and that $4 + 5$ gets her back to the length of 9.

Review Problems

15. No, for example $2 + 3$ is not an element of set $\{1, 2, 3\}$.

Assessment 3-3A

1. (a) 5 **(b)** 4 **(c)** any whole number **2.** Each possible pairing of two of the sets is disjoint. **3. (a)** yes

(b) yes   **(c)** yes   **4. (a)** No, $2 + 3 = 5$.   **(b)** yes
(c) Not closed for either, $2 + 4 = 6$ and $2 \cdot 3 = 6$.
5. (a) $ac + ad + bc + bd$ **(b)** $\square \cdot \triangle + \square \cdot \bigcirc$
(c) $ab + ac - ac$ or ab **6. (a)** $(5 + 6) \cdot 3 = 33$ **(b)** no
parentheses needed   **(c)** no parentheses needed
(d) $(9 + 6) \div 3 = 5$ **7. (a)** $y(x + y)$ **(b)** $x(y + 1)$
(c) $ab(a + b)$ **8. (a)** 6   **(b)** 0   **(c)** 4   **9.** 72
10. (a) associative property of multiplication   **(b)** commutative property of multiplication   **(c)** commutative property
of multiplication   **(d)** identity property of multiplication
(e) zero multiplication property   **(f)** distributive property
of multiplication over addition   **11. (a)** closure property of
multiplication   **(b)** zero multiplication property
(c) identity property of multiplication   **12. (a)** distributive
property of multiplication over addition
(b) $32 \cdot 12 = 32(10 + 2)$
$\qquad\qquad = 32 \cdot 10 + 32 \cdot 2$
$\qquad\qquad = 320 + 64$
$\qquad\qquad = 384$
13. (a) $9(10 - 2) = 9 \cdot 10 - 9 \cdot 2 = 90 - 18 = 72$
(b) $20(8 - 3) = 20 \cdot 8 - 20 \cdot 3 = 160 - 60 = 100$
14. (a) $(a + b)^2 = (a + b)(a + b) = (a + b)a +$
$(a + b)b = a^2 + ba + ab + b^2 = a^2 + 2ab + b^2$
(b) The area of the square with side $a + b$ can be expressed
as $(a + b) \cdot (a + b)$ and also as the sum of areas of four
regions: two squares, a^2 and b^2, and two rectangles ab and
ba. Hence,

$$(a + b)^2 = a^2 + b^2 + ab + ba = a^2 + 2ab + b^2$$

	a	b
a	a^2	ab
b	ba	b^2

$\}\, a + b$

15. The area of the complete square is $(a + b)^2$. The area
of the small square is $(a - b)^2$. The area of each of the $a \times b$
rectangles is ab, so the area of the four rectangles is $4ab$.
Therefore, the area of the large square minus the small
square is $(a + b)^2 - (a - b)^2 = 4ab$.
16. (a) $(ab)c = c(ab)$ commutative property of
$\qquad\qquad\qquad\qquad$ multiplication
$\qquad\quad = (ca)b$ associative property of
$\qquad\qquad\qquad\qquad$ multiplication
(b) $(a + b)c = c(a + b)$ commutative property of
$\qquad\qquad\qquad\qquad\quad$ multiplication
$\qquad\qquad = c(b + a)$ commutative property of
$\qquad\qquad\qquad\qquad\quad$ addition
17. a. $y(x - y)$ **b.** $47(101 - 1)$ **(c)** $ab(b - a)$
18. (a) $40 = 8 \cdot 5$ **(b)** $326 = 2 \cdot x$ **19. (a)** $(8 \div 4) \div$
$2 \neq 8 \div (4 \div 2)$ **(b)** $8 \div (2 + 2) \neq (8 \div 2) + (8 \div 2)$
20. (a) Suppose we have two bags of marbles, in one bag a
marbles and in the other b marbles. We want to distribute the
marbles equally among c students. Then the number of marbles that each student gets can be found in two ways as
follows: Putting all the marbles in one bag, we have $a + b$
marbles and each student gets $(a + b) \div c$ marbles. We could
also divide the marbles in the first bag first, and then divide
the marbles in the second bag. In this way each student would

get $(a \div c) + (b \div c)$ marbles.   **(b)** Let $a \div c = x$ and
$b \div c = y$. Then $a = cx$ and $b = cy$. Consequently,

$$a + b = cx + cy$$
$$= c(x + y)$$

Now by definition of division, $x + y = (a + b) \div c$. When
you substitute for x and y, the property follows.
21. (a) 4   **(b)** 3   **(c)** 2   **22.** 5 months.   **23.** 2; 3 left
24. (a) $(1,36), (2,18), (3,12), (4,9), (6,6), (9,4), (12,3), (18,2),$
$(36,1)$ **(b)**

(c) The points in part (b) lie along a curve while the points
from the addition lie along a line.   **25.** A possible answer
is given, resulting in $4 \cdot 3$, or 12, color schemes.

Exterior	Interior
blue	red, blue, white
red	red, blue, white
green	red, blue, white
white	red, blue, white

26. (a) 3   **(b)** 2   **(c)** 2   **(d)** 6   **(e)** 4   **27.** yes, 64
28. (a) Subtract 18 from 45.   **(b)** Divide 54 by 9.
(c) Add 11 and 48.   **(d)** Add 6 and 8.   **29. (a)** A/l
(b) $f/3$ **(c)** $60b$ **(d)** $d/7$

Mathematical Connections 3-3

Communication

1. The remainder is 1. The number can be written as
$10q + 6$, where q is a whole number. $10q$ is divisible by 5.
When 6 is divided by 5, the remainder is 1.   **3.** Answers
vary; for example, you might think of $9 \cdot 7$ as $7 \cdot 9$ and see if
that helps. You might think of $9 \cdot 7$ as $9 \cdot 6 + 9$ or
$54 + 9 = 63$ or $9 \cdot 7 = 9 \cdot 5 + 9 \cdot 2 = 45 + 18 = 63$.
5. This is the case when x is either 0 or 1.

Open-Ended

7. Answers vary; for example, a taxi driver charges \$3 for en-
tering a cab and \$2 per minute for 6 minutes. (The prices here
are unrealistic, but this is the type of problem students may
suggest.)

Cooperative Learning

9. (a) Answers vary   **(b)** We know that if $a \div b = c$, then
$a = bc$, where a, b, and c are integers and $c \neq 0$. Therefore,
to find $35 \div 5$ we look in the table and go down to the 5
row, then over to 35, and then to find the other factor we go
up to 7. Therefore, $35 \div 5 = 7$.   **(c)** The only way for a
product of two numbers to be odd is if the two factors are
odd. The eight products surrounding each odd number all
involve an even factor and hence are even.

Questions from the Classroom

11. You could encourage Sue to substitute numbers for a and b to see if her claim is true. For example, if $a = 2$ and $b = 4$, then $3(2 \cdot 4) = 3 \cdot 8 = 24$ and $(3 \cdot 2)(3 \cdot 4) = b \cdot 12 = 72$. Therefore, we have a counter-example to show Sue's claim is false. At this point the correct associative and distributive properties could be demonstrated. **13.** In general, $a \div (b - c) \neq (a \div b) - (a \div c)$. For example, $100 \div (25 - 5) \neq (100 \div 25) - (100 \div 5)$. In fact, the right-hand side is $4 - 20$, which is not defined in the set of whole numbers. However, the right distributive property of division over subtraction does hold provided each expression is defined in the set of whole numbers; that is, $(b - c) \div a = (b \div a) - (c \div a)$.

Review Problems

15. for example, $\{0, 1\}$ **17. (a)** The student did not regroup $7 + 6 = 13$ as $1 \cdot 10 + 3$ and bring the 1 to the tens column. **(b)** The student added $5 + 7 = 12$ and just wrote it down and did not consider place value in the tens column. Likewise, the student added $3 + 4 = 7$ and wrote it in the hundreds column. **(c)** The student just took the absolute difference between 9 and 6 instead of regrouping to take 9 from 16. **(d)** The student did not decrease the number of tens after regrouping.

Assessment 3-4A

1. (a)
$$\begin{array}{r} 426 \\ \times\ 783 \\ \hline 1278 \\ 3408 \\ 2982 \\ \hline 333558 \end{array}$$
(b)
$$\begin{array}{r} 327 \\ \times\ 941 \\ \hline 327 \\ 1308 \\ 2943 \\ \hline 307707 \end{array}$$

2. (a)

7 2 8
6 | 6 3 1 8 7 2 | 9
8 | 2 8 0 8 3 2 | 4
4 3 2
$728 \cdot 94 = 68,432$

(b)

3 0 6
| 0 6 0 0 1 2 | 2
7 | 1 2 0 2 4 | 4
3 4 4
$306 \cdot 24 = 7,344$

3. Diagonals separate place value as value placement does in the traditional algorithm. **4. (a)** 5^{19} **(b)** 6^{15} **(c)** 10^{313} **(d)** 10^{12} or $2^{12} \cdot 5^{12}$ **5. (a)** 2^{100} because $2^{80} + 2^{80} = 2^{80}(1 + 1) = 2^{80} \cdot 2 = 2^{81}$ **(b)** 2^{102} because $2^{101} = 2^{100} \cdot 2$ and $2^{102} = 2^{100} \cdot 4$ **6.** The following partial products, which are obtained through the distributive property of multiplication over addition, are shown in the model.

(a)
$$\begin{array}{rl} 22 & \\ \times\ 13 & \\ \hline 6 & (3 \times 2) \\ 60 & (3 \times 20) \\ 20 & (10 \times 2) \\ 200 & (10 \times 20) \\ \hline 286 & \end{array}$$

(b)
$$\begin{array}{rl} 15 & \\ \times\ 21 & \\ \hline 5 & (1 \times 5) \\ 10 & (1 \times 10) \\ 100 & (20 \times 5) \\ 200 & (20 \times 10) \\ \hline 315 & \end{array}$$

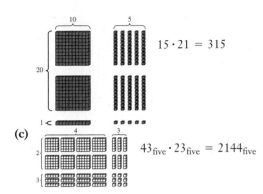

$15 \cdot 21 = 315$

(c)

$43_{\text{five}} \cdot 23_{\text{five}} = 2144_{\text{five}}$

To find $43_{\text{five}} \cdot 23_{\text{five}}$, count the number of flats, longs, and ones; we have $4 \cdot 2$ flats, $2 \cdot 3 + 3 \cdot 4$ longs, and $3 \cdot 3$ units. Remembering that 5 units = 1 long, 5 longs = 1 flat, and 5 flats = 1 block, we have $43_{\text{five}} \cdot 23_{\text{five}} = 2$ blocks, 1 flat, 4 longs, and 4 units = 2144_{five}. **7. (a)** $293 \cdot 476 = 139,468$ **(b)** Placement indicates place value.

(c)
$$\begin{array}{r} 363 \\ \times\ 84 \\ \hline 2904 \\ 1452 \\ \hline 30492 \end{array} \quad \begin{array}{l} (8 \times 363) \\ (4 \times 363) \end{array}$$

8.
$$\begin{array}{r|r} 17\times & 63 \\ 8 & 126 \\ 4 & 252 \\ 2 & 504 \\ \hline 1 & 1008 \end{array}$$
and $63 + 1008 = 1071$

9. (a) 1332 calories **(b)** Jane, 330 more calories **(c)** Maurice, 96 more calories **10.** No **11. (a)** 77 remainder 7 **(b)** 8 remainder 10 **(c)** 10 remainder 91 **12.** 3 **13.**

2	11
4	15
0	7
6	19
12	31

14. (a)
$$\begin{array}{r} 32 \\ \times\ 69 \\ \hline 2208 \end{array} \qquad \begin{array}{r} 23 \\ \times\ 96 \\ \hline 2208 \end{array}$$

(b) $(10a + b)(10c + d) = (10b + a)(10d + c)$ implies $100ac + 10bc + 10ad + bd = 100bd + 10ad + 10bc + ac$ or $99ac = 99bd$, which implies that $ac = bd$. Therefore this works whenever the product of the ones digits is the same as the product of the tens digits. **15.** 3 hr **16.** 1356, 2712, and 452 **17.** $142 **18. (a)** 5 was multiplied by 6 to obtain 30; the 3 was regrouped then 3 was multiplied by 2 to obtain 6; the "regroup" was added to obtain 9, which was recorded. **(b)** When 1 was brought down, the quotient of 0 was not recorded. **19. (a)** place value distributive property of multiplication over addition; associative property of multiplication; definition of a^n; identity property for addition and zero multiplication property, place value

(b) $34 \cdot 10^2 = (3 \cdot 10 + 4)10^2$ place value

$\qquad = (3 \cdot 10)10^2 + 4 \cdot 10^2$ distributive property of multiplication over addition

$\qquad = 3(10 \cdot 10^2) + 4 \cdot 10^2$ associative property of multiplication

$\qquad = 3 \cdot 10^3 + 4 \cdot 10^2$ definition of a^n

$\qquad = 3 \cdot 10^3 + 4 \cdot 10^2 + 0 \cdot 10$ property of addition identity and the zero multiplication property

$\qquad + 0 \cdot 1$

$\qquad = 3400$ place value

20. 58 buses needed, not all full

21. (a) 763   **(b)** 678   **22. (a)** nine   **(b)** six

$$\begin{array}{r} 763 \\ \times\ 8 \\ \hline 6104 \end{array} \qquad \begin{array}{r} 678 \\ \times\ 3 \\ \hline 2034 \end{array}$$

23. (a)

```
        3   2   3
   3  |2/1|1/3|2/2|  4
      |/2 |/ |/2 |
   0  |1/0|0/1|1/1|  2
      |/1 |/4 |/1 |
        2   2   1
```

(b) $a = 5, b = 7$

24. (a) 233_{five} **(b)** $4_{\text{five}} \text{R} 1_{\text{five}}$ **(c)** 1513_{six} **(d)** 31_{five}
(e) 110_{two} **(f)** 1101110_{two}

Mathematical Connections 3-4

Communication

1. Because $345 \cdot 678 = 345 \cdot (6 \cdot 10^2 + 7 \cdot 10 + 8)$, one explanation could be as follows: Using the distributive property of multiplication over addition, first multiply 345 by 6 and the result by 10^2; then multiply 345 by 7 and the result by 10; then multiply 345 by 8. Add all the numbers previously obtained. **3.** The result is always 4. Let the original number be x. The operation appears as follows:

$$\big[(2x)3 + 24\big]/6 - x = 4$$

5. Answers vary depending upon students' choices. Many students may favor the lattice multiplication algorithm because only single digits are multiplied and addition is accomplished later. **7.** $abba = a \cdot 10^3 + b \cdot 10^2 + b \cdot 10 + a = a \cdot 1001 + b \cdot 110 = 11(91 \cdot a + 10 \cdot b)$. Yes, $abccba = a \cdot 10^5 + b \cdot 10^4 + c \cdot 10^3 + c \cdot 10^2 + b \cdot 10 + a = a \cdot 100001 + b \cdot 10010 + c \cdot 1100 = 11(a \cdot 9091 + b \cdot 910 + c \cdot 100)$.

Cooperative Learning

9. Arguments vary depending upon groups. Some will argue that addition should be followed by subtraction because they are inverses of each other. Others will argue that addition should be followed by multiplication because multiplication is repeated addition. This would also postpone subtraction until students are more ready for it.

Questions from the Classroom

11. Evidently the student does not understand the process of long division. The repeated subtraction method should help in understanding the mistake.

$$\begin{array}{r} 6\overline{)36} \\ -\ 6 \quad \text{1 six} \\ \hline 30 \\ -\ 30 \quad \text{5 sixes} \\ \hline \text{6 sixes} \end{array}$$

Instead of adding 1 and 5, the student wrote 15. **13.** If the number has three digits with the unit digit 0, we have $ab0 \div 10 = ab$ since $ab \cdot 10 = ab0$. This will not work if the 0 digit is not the unit digit. **15. (a)** The first equation is true because $39 + 41 = 39 + (1 + 40) = (39 + 1) + 40 = 40 + 40$. Now, $39 \cdot 41 = (40 - 1)(40 + 1) = 40^2 - 1^2$, and $40^2 - 1 \neq 40^2$. **(b)** Yes, this pattern continues because the numbers being considered are in the form $(a - 1)(a + 1)$, which is equal to $a^2 - 1$.

Review Problems

17. (a) $(a + b + 2)x$ **(b)** $(3 + x)(a + b)$
19. (a) $36 = 4 \cdot 9$ **(b)** $112 = 2x$ **(c)** $48 = x \cdot 6$, or $48 = 6x$ **(d)** $x = 7 \cdot 17$

Assessment 3-5A

1. (a) 160   **(b)** 120   **2. (a)** $(9 \cdot 6) \cdot (2 \cdot 5) = 54 \cdot 10 = 540$ **(b)** $(8 \cdot 7) \cdot (25 \cdot 4) = 56 \cdot 100 = 5600$ **3. (a)** 605 **(b)** 963   **4. (a)** 36   **(b)** 120   **(c)** 46   **(d)** 97
5. 496 mi   **6. (a)** $28 + 2 = 30$; $30 + 20 = 50$; $50 + 3 = 53$; so the answer is $2 + 20 + 3 = 25$.
(b) $47 + 3 = 50$; $50 + 10 = 60$; $60 + 3 = 63$; so the answer is $3 + 10 + 3 = 16$.   **7. (a)** $86 + 37 = (80 + 30) + (7 + 6) = 123$ by adding the tens and the units separately.   **(b)** $97 + 54 = 97 + 3 + 54 - 3 = 100 + 51 = 151$ by adding 3 to 97 and subtracting it from 54.   **(c)** $230 + 60 + 70 + 44 + 40 + 6 = (230 + 70) + (60 + 40) + (44 + 6) = 450$ by using compatible numbers.   **8. (a)** 5300   **(b)** 100,000   **(c)** 120,000 **(d)** 2330   **9.** Answers vary; for example:
(a) $900 \div 30 = 30$ **(b)** $25,000 - 20,000 = 5000$
(c) $30 \cdot 30 = 900$ **(d)** $2000 + 3000 + 6000 + 1000 = 12,000$ **10.** Answers vary; for example: **(a)** $2 + 3 + 5 = 10$, so 10,000 is an initial estimate. $10,000 + 2000$ (adjustment) $= 12,000$ for the final estimate.   **(b)** The sum of the front digits is 22, so $2200 + 270$ adjustment gives 2470.   **11. (a)** The first set of numbers is not clustered. The second set is clustered about 500, so an estimate is 2500.   **(b)** Estimates vary.   **12. (a)** The range is 600 $(20 \cdot 30)$ to 1200 $(30 \cdot 40)$.   **(b)** The range is 700 $(100 + 600)$ to 900 $(200 + 700)$.   **(c)** The range is 230 $(200 + 30)$ to 340 $(300 + 40)$.   **13.** Answers vary; for example, $3300 - 100 - 300 - 400 - 500 = 2000$. The estimate is high because the amounts were rounded to the hundreds place and $8 was taken away from the check

amounts while \$13 was added to \$3287.   **14.** For example, $35 \cdot 20 = 700$ seats or $40 \cdot 25 = 1000$ seats; 700 will be low and 1000 high.   **15. (a)** Different answers since the estimates of 800 and 220 are way off.   **(b)** Same answers because 22 was divided by 2 to obtain 11 while 32 was multiplied by 2 to obtain 64. The result is multiplying the original computation by 2/2, or 1, which does not change it.   **(c)** Same answers because the first number was multiplied by 3 and the second number was divided by 3, which results in just multiplying the original computation by 1, which does not change it.   **16. (a)** false   **(b)** false   **(c)** false   **(d)** true   **17.** The clustering strategy gives $6 \cdot 70,000$, or 420,000.   **18. (a)** high; $299 \cdot 3 < 300 \cdot 3$ **(b)** low; $6,001 \div 299 > 6000 \div 300$ **(c)** low; $6,000 \div 299 > 6000 \div 300$ **(d)** low; $10 \cdot 99$ is only 990   **19.** One possibility is that to find $(10x + 5)^2$ we could write $(10x + 5)^2 = 100x^2 + 50x + 50x + 25 = 100x^2 + 100x + 25 = 100x(x + 1) + 25$. For example, in 65^2 we take $6 \cdot 7 = 42$ and append 25 to obtain 4225.

Mathematical Connections 3-5

Communication

1. Mental mathematics is the process of producing an exact answer to a computation without using external aids. Computational estimation is the process of forming an approximate answer to a numerical problem.   **3.** Answers vary; for example, students might suggest that mental mathematics and estimation are necessary every day in order to determine quickly whether their bills at various businesses are being computed correctly. Mental math and estimation help students to know whether the answers that appear on the calculator are reasonable. NCTM makes several important points about the importance of mental math and estimation in *Principles and Standards* and *Focal Points*.

Open-Ended

5. Answers vary; for example, when you are determining the amount of a tip for a waiter in a restaurant.

Cooperative Learning

7. Answers vary depending upon the grade level and textbook chosen.

Questions from the Classroom

9. Molly has a good idea and almost has it right. You just need to point out that she needs to add 2 at the end instead of subtract 2. The reasons for this should be explained. For example,

$$\begin{aligned}
261 - 48 &= 261 - (50 - 2) \\
&= 261 - 50 + 2 \\
&= (261 - 50) + 2 \\
&= 211 + 2 \\
&= 213
\end{aligned}$$

11. By using the calculator first, she is not learning the estimation skills that are so useful in making decisions. One important use of estimation is to determine if an answer is reasonable. For example, if the problem is $492 \cdot 63$, by estimating, the student would know that the answer should be approximately 30,000. If she gets an answer such as 17,712 ($492 \cdot 36$) or 59,346 ($942 \cdot 63$), she would know that there was an error in computation. (Maybe she hit the wrong key on the calculator.)

Review Problems

13. (a)

$$\begin{array}{r} 18\overline{)623} \\ -180 \leftarrow 10 \ (18\text{'s}) \\ \hline 443 \end{array} \qquad \begin{array}{r} 34 \\ 18\overline{)623} \\ -54 \\ \hline 83 \\ -72 \\ \hline 11 \end{array}$$

$$\begin{array}{r} 18\overline{)623} \\ -180 \leftarrow 10 \ (18\text{'s}) \\ \hline 443 \\ -180 \leftarrow 10 \ (18\text{'s}) \\ \hline 83 \\ -18 \leftarrow 1 \ (18\text{'s}) \\ \hline 65 \\ -18 \leftarrow 1 \ (18) \\ \hline 47 \\ -18 \leftarrow 1 \ (18) \\ \hline 29 \\ -18 \leftarrow 1 \ (18) \\ \hline 11 \quad 34 \ (18\text{'s}) \end{array}$$

(b)

$$\begin{array}{r} 21\overline{)493} \\ -210 \leftarrow 10 \ (21\text{'s}) \\ \hline 283 \\ -210 \leftarrow 10 \ (21\text{'s}) \\ \hline 73 \\ -21 \leftarrow 1 \ (21) \\ \hline 52 \\ -21 \leftarrow 1 \ (21) \\ \hline 31 \\ -21 \leftarrow 1 \ (21) \\ \hline 10 \quad 23 \ (21\text{'s}) \end{array} \qquad \begin{array}{r} 23 \\ 21\overline{)493} \\ -42 \\ \hline 73 \\ -63 \\ \hline 10 \end{array}$$

(c)

$$\begin{array}{r} 97\overline{)1000} \\ -970 \leftarrow 10 \ (97\text{'s}) \\ \hline 30 \end{array} \qquad \begin{array}{r} 10 \\ 97\overline{)1000} \\ -97 \\ \hline 30 \\ -0 \\ \hline 30 \end{array}$$

Chapter Review

1. (a) distributive property of multiplication over addition   **(b)** commutative property of addition   **(c)** identity property of multiplication   **(d)** distributive property of multiplication over addition   **(e)** commutative property of multiplication   **(f)** associative property of multiplication   **2. (a)** $3 < 13$ because there exists a natural number, namely 10, such that $3 + 10 = 13$ **(b)** $12 < 9$ or $9 < 12$ because there exists a natural number, namely 3, such that

$9 + 3 = 12$.   **3. (a)** 15, 14, 13, 12, 11, or 10   **(b)** 10
(c) any whole number   **(d)** whole numbers from 0
through 26   **4. (a)** $15a$ **(b)** $5x^2$ **(c)** $xa + xb + xy$
(d) $3x + 15 + xy + 5y$ or $(x + 5)(3 + y)$ **5.** 40 cans
6. 12 outfits   **7.** 26   **8.** $6000 for 80 people is cheaper.
9. $214   **10.** $400   **11. (a)** If n is the original number,
then each of the following lines shows the result of
performing the instruction:

$$n$$
$$n + 17$$
$$2(n + 17) = 2n + 34$$
$$2n + 30$$
$$4n + 60$$
$$4n + 80$$
$$n + 20$$
$$n$$

(b) Answers vary; for example, the next two lines could be
subtract 65 and then divide by 4.   **(c)** Answers vary.
12. 1119   **13.** 60,074   **14. (a)** 5 remainder 243
(b) 91 remainder 10   **(c)** 120_{five} remainder 2_{five} **(d)** 11_{two}
remainder 10_{two} **15. (a)** $912 \cdot 5 + 243 = 4803$
(b) $11 \cdot 91 + 10 = 1011$ **(c)** $23_{\text{five}} \cdot 120_{\text{five}} + 2_{\text{five}} =$
3312_{five} **(d)** $11_{\text{two}} \cdot 11_{\text{two}} + 10_{\text{two}} = 1011_{\text{two}}$
16. (a) $(19 \cdot 194)10 = 36,860$ **(b)** $(379 \cdot 193)100 =$
7,314,700   **(c)** $481 \cdot 73 \cdot (8 \cdot 125) = (481 \cdot 73)1000 =$
35,113,000   **(d)** $374 \cdot 893 \cdot (200 \cdot 50) = (374 \cdot 893) \cdot$
$10,000 = 3,339,820,000$ **17.** $395   **18.** $4380
19. 2600 cases   **20.** $3842   **21.** $2.16   **22.** 36 bikes
and 18 trikes   **23. (a)** 212_{five} **(b)** 101_{two} **(c)** 1442_{five}
(d) 101101_{two} **24. (a)** for example, $(26 + 24) +$
$(37 - 7) = 50 + 30 = 80$ **(b)** for example,
$(7 \cdot 9)(4 \cdot 25) = 63 \cdot 100 = 6300$ **25. (a)** 441   **(b)** 36
(c) 180   **(d)** 406   **26.** Answers vary; for example,
(a) $2300 + 300$ (adjustment) $= 2600$, (b) 2600.
27. $2400 \cdot 4 = 9600$ **28.** Answers vary; for example, we
first estimate how many 14s are contained in 322. There are
at least 20, so using place value we record a 2 above the 2 in
32 to indicate that we are removing 20 sets of 14. If we
remove 20 sets of 14, or 280, 42 is left. We next estimate how
many 14s are contained in 42. The answer is 3, so we write 3
next to the 2 in the quotient to indicate we are removing another 3 sets of 14. This leaves 0, so there are no more 14s to
remove and nothing left over. The 23 indicates that 322 contains 23 sets of 14 and the 0 indicates that there is nothing
left over.   **29. (a)** $999 \cdot 47 + 47 = 47(999 + 1) =$
$47 \cdot 1000 = 47,000$ **(b)** $43 \cdot 59 + 41 \cdot 43 = 43 \cdot (59 + 41)$
$= 43 \cdot 100 = 4300$ **(c)** $1003 \cdot 79 - 3 \cdot 79 = 79 \cdot$
$(1003 - 3) = 79 \cdot 1000 = 79,000$ **(d)** $1001 \cdot 113 - 113 =$
$113 \cdot (1001 - 1) = 113 \cdot 1000 = 113,000$ **(e)** $101 \cdot 35 =$
$(100 + 1)35 = 100 \cdot 35 + 1 \cdot 35 = 3500 + 35 = 3535$
(f) $98 \cdot 35 = (100 - 2)35 = 100 \cdot 35 - 2 \cdot 35 =$
$3500 - 70 = 3430$ **30. (a)** $3x^3 + 4x^2 + 7x + 8 +$
$+ (5x^2 + 2x + 1) = 3x^3 + 9x^2 + 9x + 9$
(b) Answers vary.   **(c)** Answers vary; for example,
$34 \cdot 10^2 = (3 \cdot 10 + 4) \cdot 10^2 = 3 \cdot 10^3 +$
$4 \cdot 10^2 + 0 \cdot 10 + 0 = 3400$ and $(3x + 4)x^2 = 3x^3 + 4x^2$.

Answers to Now Try This

3-1. For example, if the sets of elements are $\{a, b, c\}$ and
$\{a, d\}$, then the union of the sets of elements is $\{a, b, c, d\}$.
The union has only 4 elements while the original sets have
3 and 2, respectively. The sum of 3 and 2 is 5, not 4, the
number of elements in the set union.   **3-2.** Answers vary;
for example, students are used to starting with 1 when they
count, so they sometimes start with 1 on the number line. It
should be pointed out that we are working with whole numbers and the first whole number is 0. Next, to represent 3
on a number line an arrow (vector) of length 3 units must
be used. The length of the arrow in Figure 4-3 is 2 units.
3-3. (a) Closed; an even number plus an even number is
always even.   **(b)** Not closed; for example, $1 + 3 = 4$ and
4 is not an element of F.
3-4.

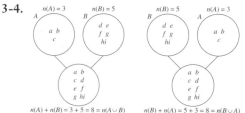

$$n(A) + n(B) = 3 + 5 = 8 = n(A \cup B) \qquad n(B) + n(A) = 5 + 3 = 8 = n(B \cup A)$$

3-5. Let $B \subseteq A$. If $n(A) = a$ and $n(B) = b$, then
$a - b = n(A - B)$. For example, $A = \{a, b, d, e\}$ and
$B = \{a\}, A - B = \{b, d, e\}, n(A) = 4, n(B) = 1$,
$n(A - B) = 3$.   **3-6. (a)** The set of whole numbers is not
closed under subtraction; for example, $2 - 5$ is not a whole
number.   **(b)** Subtraction is not associative for whole
numbers; for example, $9 - (7 - 2) \neq (9 - 7) - 2$.
(c) Subtraction is not commutative for whole numbers; for
example, $3 - 2 \neq 2 - 3$.   **(d)** There is no identity for
whole-number subtraction; for example, $5 - 0 = 5 \neq 0 - 5$.
3-7. (a) 5   **(b)** 7   **(c)** 10   **3-8.** $182 + 61$ can be
represented by:

1 block	8 flats	2 units
	6 flats	1 unit
1 block (10 flats + 4 flats)		3 units
or 1 block (1 block + 4 flats)		3 units
or 2 blocks	4 flats	3 units
or 243		

3-9. (i) The method is valid because subtracting and
adding the same number does not change the original sum.
(ii) $97 + 69 = (97 + 3) + (69 - 3)$
$$= 100 + 66$$
$$= 166$$
3-10. (a) 1000_{five} **(b)** 31_{five}
3-11. (a)

+	0_{two}	1_{two}
0_{two}	0_{two}	1_{two}
1_{two}	1_{two}	10_{two}

(b) (i) 1101_{two} **(ii)** 1111_{two}
-111_{two} $+ 111_{\text{two}}$
110_{two} 10110_{two}

3-12. One possible explanation is as follows: The first shirt
can be worn with each of the 5 pairs of pants for a total of
5 different outfits. Also, the second shirt can be worn with

each of the 5 pairs of pants for a total of 5 outfits different from the previous outfits. Similarly, the 3rd, 4th, 5th, and 6th shirts can be combined with the 5 pairs of pants for new outfits. In this way, each of the 6 shirts can be used for 5 new outfits for a total $5 + 5 + 5 + 5 + 5 + 5 = 6 \cdot 5$ outfits.
3-13. 91 members **3-14. (a)** The set of whole numbers is not closed under division; for example, $8 \div 5$ is not a whole number. Likewise, $8 \div 2 \neq 2 \div 8$ and $(8 \div 4) \div 2 \neq 8 \div (4 \div 2)$ show it is not commutative or associative.
(b) 1 is not the identity for whole-number division because $n \div 1 = 1$ for all whole numbers n, and $1 \div n \neq n$ except when $n = 1$.
3-15.
$$\begin{aligned}
10 \cdot 600 &= 10^1(6 \cdot 10^2) \\
&= (6 \cdot 10^2)10^1 \\
&= 6 \cdot (10^2 10^1) \\
&= 6 \cdot 10^3 \\
&= 6 \cdot 10^3 + 0 \cdot 10^2 + 0 \cdot 10 + 0 \cdot 1 \\
&= 6000 \\
20 \cdot 300 &= (2 \cdot 10^1)(3 \cdot 10^2) \\
&= (2 \cdot 3)(10^1 \cdot 10^2) \\
&= 6(10^1 \cdot 10^2) \\
&= 6 \cdot 10^3 \\
&= 6 \cdot 10^3 + 0 \cdot 10^2 + 0 \cdot 10 + 0 \cdot 1 \\
&= 6000
\end{aligned}$$
3-16.
$$\begin{aligned}
7 \cdot 4589 &= 7(4 \cdot 10^3 + 5 \cdot 10^2 + 8 \cdot 10 + 9) \\
&= (7 \cdot 4)10^3 + (7 \cdot 5)10^2 + (7 \cdot 8) \cdot 10 + 7 \cdot 9 \\
&= 28000 + 3500 + 560 + 63 \\
&= 32123
\end{aligned}$$
3-17. Answers vary; for example, **(a)** $40 + 160 = 200$ and $29 + 31 = 60$ so the sum is 260. **(b)** $3679 - 400 = 3279$ and $3279 - 74 = 3205$. **(c)** $75 + 25 = 100$ and $100 + 3 = 103$. **(d)** $2500 - 500 = 2000$ and $2000 - 200 = 1800$. **3-18.** Answers vary; for example, **(a)** $4 \cdot 25 = 100$ and $32 \cdot 100 = 3200$. **(b)** $123 \cdot 3 = 100 \cdot 3 + 23 \cdot 3 = 300 + 69 = 369$. **(c)** $25 \cdot 35 = (30 - 5)(30 + 5) = 30^2 - 5^2 = 900 - 25 = 875$. **(d)** $5075/25 = 5000/25 + 75/25 = 200 + 3 = 203$.
3-19. Answers vary; for example, **(a)** To estimate $4525 \cdot 9$, we know $4525 \cdot 10 = 45,250$ and since we have only 9 sets of 4525 we can take away approximately 5000 from our estimate and we have 40,250. **(b)** To estimate $3625/42$, we know the answer will be close to $3600/40$, or 90.

Answers to Brain Teasers

Section 3-1

Answers vary. For example,

1	2	3
8	9	4
7	6	5

Section 3-2

The license plate number is 10968.

Section 3-4

(a)
$$\begin{array}{r} 570,140 \\ \times\ 6 \\ \hline 3,420,840 \end{array}$$

(b)
$$\begin{array}{r} 38 \\ 38 \\ +\ 38 \\ \hline 114 \end{array} \quad \text{or} \quad \begin{array}{r} 39 \\ 39 \\ +\ 39 \\ \hline 117 \end{array}$$

Section 3-5

There are 85 members of a PTA to contact using a phone chain. We are to assume that each call takes 30 sec and that everyone is at home and answers the phone. The key to solving this problem is to realize that when a caller calls one of his/her two people, the first person being called does not wait for the second person to be called before this first person can begin calling his/her two people. A strategy is to build a model of this phone tree and a table to keep track of the number of people contacted. A model for the first 3 min. is given below. The table on the right of the model shows the number of *People called* in each 30-sec interval and it also shows the *Total number called*.

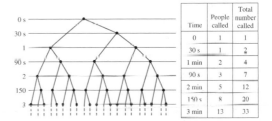

Time	People called	Total number called
0	1	1
30 s	1	2
1 min	2	4
90 s	3	7
2 min	5	12
150 s	8	20
3 min	13	33

If we examine the numbers in the *People called* column. we notice a familiar sequence, that is, the *Fibonacci* sequence where each number in the sequence is the sum of the previous two numbers. We can see from the model that this will continue to happen because the number of dots in each segment is the sum of the dots in the previous two segments. Therefore, we need to continue the table to see at what time the total number of people called reaches or exceeds 85. Continuing the table, we see that there would be $8 + 13 = 21$ more people called at 210 sec, giving a total of $33 + 21 = 54$. For 4 min., there would be $13 + 21 = 34$ more people called giving a total of $54 + 34 = 88$. Therefore, we can call 85 people in 4 min.

We can extend the problem by trying different numbers of members of the PTA or generalizing the problem to finding the length of time required to call n people. We could also investigate how the Fibonacci sequence shows up in other contexts, such as pine cones or the breeding habits of some animals.

Answers to Laboratory Activity

Section 3-2

Either abacus may be used. Both are comparable. Some may prefer the suan pan because there are two counters above the bar and five below the bar.

Section 3-3

1. yes   **2.** 18 and 19   **3.** In general, even numbers appear to reach 1 quicker.   **4.** Answers depend on choices. (Note: Question 1 is a famous unsolved problem in mathematics.)

Section 3-4

1. (a) computer   **(b)** Answers will vary.
2. (a) When a person tells his or her age by listing cards, the person is giving the base-two representation for his or her age. The number can then be determined by adding the numbers in the upper left-hand corners of the named cards.
(b) Card F would have 32 in the upper left corner. Each of the numbers 1 through 63 could be written in base two so you can tell where to place them on the cards.

Solution to the Preliminary Problem

Answers vary. One set of answers showing how to use five 5s and grouping symbols to obtain all the numbers 1–10 is given below.

$$(5 \div 5) + [(5 - 5) \cdot 5] = 1$$
$$[(5 + 5) \div 5] + (5 - 5) = 2$$
$$[(5 + 5) \div 5] + (5 \div 5) = 3$$
$$[(5 \cdot 5) \div 5] - (5 \div 5) = 4$$
$$5 + [(5 - 5) \cdot 55] = 5$$
$$[(5 \cdot 5) \div 5] + (5 \div 5) = 6$$
$$5 + (5 \div 5) + (5 \div 5) = 7$$
$$5 + [(5 + 5 + 5) \div 5] = 8$$
$$[55 - (5 + 5)] \div 5 = 9$$
$$5 + 5 + [(5 - 5) \cdot 5] = 10$$

Chapter 4

Assessment 4-1A

1. (a) $10 + 2d$ **(b)** $2n - 10$ **(c)** $10n^2$ **(d)** $n^2 - 2n$
2. (a) $\dfrac{7(n + 3) - 14}{7} - n$ **(b)** 1   **3. (a)** $2(n + 1)$
(b) $(n + 2)^2 - 2(n + 1)$ **4. (a)** $20 + 25b$ dollars
(b) $175d$ cents   **(c)** $3x + 3$ **(d)** $q \cdot 2^n$ **(e)** $40 - 3t°F$
(f) $4s + 15{,}000$ dollars   **(g)** $3x + 5$ **(h)** $3m$
5. $S = 20P$ **6.** $g = 5 + b$ **7.** $6n + 4$
8. (a) $P = 8t$ dollars   **(b)** $P = 15 + 10(t - 1)$ dollars
for $t \geq 1$ **9.** $5x + 1300$ dollars   **10.** Answers in
dollars: **(a)** youngest x, eldest $3x$, middle $\dfrac{3x}{2}$ dollars
(b) middle y, eldest $2y$, youngest $\dfrac{2y}{3}$ **(c)** eldest z, youngest $\dfrac{z}{3}$, middle $\dfrac{z}{2}$

Mathematical Connections 4-1

Communication

1. Both are correct. For the first student, x is the first of the three consecutive natural numbers. The second chose x to be the second of the three consecutive natural numbers.

Open-Ended

3. Answers will vary.

Questions from the Classroom

5. The student is correct. If the third integer is x, the five integers are $x - 2, x - 1, x, x + 1, x + 2$; their sum is $5x$. For any arithmetic sequence with difference d, the five integers are $x - 2d, x - d, x, x + d, x + 2d$. Their sum is $5x$.
7. Yes, for example in $\{A \mid A \subset W\}$, A is a variable set; any set that is a proper subset of W, the set of whole numbers.

Assessment 4-2A

1. If $\triangle\square = 12$, then $\bigcirc\bigcirc + 12 = 18$ and $\bigcirc = 3$. Then $\square\bigcirc\bigcirc = \square + 6 = 10$, so $\square = 4$. If $\square = 4$, then $4 + \triangle = 12$ and $\triangle = 8$. Therefore, $\bigcirc = 3, \square = 4$, and $\triangle = 8$.   **2. (a)** 24   **(b)** 20   **(c)** 9   **(d)** 3   **(e)** 3
3. 22   **4.** 524 student tickets   **5.** Let x be the amount the youngest receives. Then $x + 3x + x + 14{,}000 = 486{,}000$, or $5x = 472{,}000$. Youngest received \$94,400, oldest \$283,200, middle \$108,400.   **6.** 41, 41 and 38 in.
7. Let x be the number of nickels. Then $67 - x$ is the number of dimes. So $10(67 - x) + 5x = 420$, $x = 50$. Thus, 50 nickels and 17 dimes.   **8.** Ricardo 4, Miriam 14
9. 625   **10.** 350 yd by 175 yd

Mathematical Connections 4-2

Communication

1. Both are correct. For the first student, x is the first of the three consecutive whole numbers. The second chose x to be the second of the three consecutive whole numbers.

Open-Ended

3. Answers will vary. For example: **(a)** $(x + 1)(x + 2) = x^2 + 3x + 2$ **(b)** $2x + 3 = 2x - 1$
(c) $3x + 1 = 2x + 1$

Questions from the Classroom

5. $x = 0$ is a solution. The student is wrong because we cannot divide by 0.   **7.** Because there might be an error setting up the equation and hence the equation might not be a true modeling of the problem

Review Problems

9. $x = 3y$ **11.** Jack x, Julie $2x$, Tira $6x$

Assessment 4-3A

1. (a) Double the input number.   **(b)** Add 6 to the input number.   **2. (a)** This is not a function, since the input 1 is paired with two outputs (*a* and *d*).   **(b)** This is a function.
3. (a) Answers will vary. For example: **(b)** 30

4. (a)

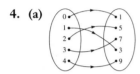

(b) $\{(0,1),(1,3),(2,5),(3,7),(4,9)\}$

(c)

x	*f(x)*
0	1
1	3
2	5
3	7
4	9

(d)

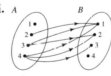

5. (a) This is a function.   **(b)** This is a function.
(c) This is a function.
6. (a) i.

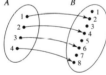   **ii.**

(b) Part (i) is a function from *A* to *B*. For each element in *A*, there is a unique element in *B*. The range of the function is $\{2,4,6,8\}$.
7.

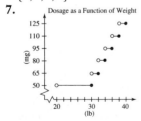

8. (a) 8 dollars   **(b)** $3n + 2$ dollars   **9. (a)** $L(n) = 2n + (n - 1)$, or $3n - 1$ **(b)** $L(n) = n^2 + 1$
10. (a)

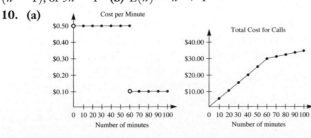

Notice that because we can't really depict 100 dots on the graphs a dot is drawn only for a multiple of 10 minutes. Because in part (a) we assumed a whole number of minutes the dots should not be connected; however, because of the scale, if we drew 100 dots, they would seem connected.
(b) That the company charges for each part of a minute at the rate of $0.50 per minute   **(c)** The two segments represent different charges per minute. The one representing the higher cost is steeper.   **(d)** $C(t) = 0.50t$ if $0 \le t \le 60, C(t) = 30 + 0.10(t - 60)$ if $t > 60$.
11. (a) $5n - 2$ **(b)** 3^n **12. (a)** 30   **(b)** 65
13. (a) The first three are.   **(b)** Only (ii) and (iii).
(c) The first three are.   **14. (a)** $2 \cdot 1 + 2 \cdot 7 = 16$; $2 \cdot 2 + 2 \cdot 6 = 16; 2 \cdot 6 + 2 \cdot 2 = 16; 2 \cdot 5 + 2 \cdot 5 = 20$
(b) $\{(1, 9), (2, 8),(3, 7), (4, 6), (5, 5), (6, 4), (7, 3), (8, 2), (9, 1)\}$ **(c)** The domain is $N \times N$, and the range is the set of all even numbers greater than or equal to 4.
15. (a) 50 cars   **(b)** between 6:00 A.M. and 6:30 A.M.
(c) 0   **(d)** between 8:30 A.M. and 9 A.M., by 100 cars
(e) Segments are used because the data are continuous rather than discrete. For example, there are a number of cars at 5:20 A.M.   **16. (a)** $H(2) = 192; H(6) = 192$; $H(3) = 240; H(5) - 240$. Some of the heights correspond to the ball going up, some to the ball coming down.
(b)

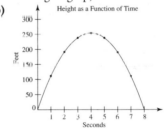

At $t = 4$ sec, the ball's height is $H(4) = 256$ ft above the ground.   **(c)** 8 sec   **(d)** $0 \le t \le 8$ **(e)** $0 \le H(t) \le 256$
17. (a) i. 40
    ii. 49
  (b) i. $S(n) = 2n(n + 1)$
     ii. $S(n) = (n + 1)^2 + (n + 2)n$ or $2n^2 + 4n + 1$
18. (a) $\dfrac{n(n + 1)}{2}$ **(b)** 4^{n-1} **19.** The converse is false.

For example, the set of ordered pairs $\{(1, 2), (1, 3)\}$ does not represent a function, since the element 1 is paired with two different second components.   **20.** Only (b) does not represent a function; if *x*, the input, is any given whole number, then *y* is not unique, as it could be any whole number *y* such that $y > x - 2$.   **21.** Only (b) is not. For $x = 1$ there are five values of *y*.   **22. (a)** boys: *B*, *H*; girls: *A*, *C*, *D*, *G*, *I*, *J*, *E*, *F*   **(b)** $\{(A, B), (A, C), (A, D), (C, A), (C, B), (C, D), (D, A),(D, B), (D, C), (F, G), (G, F), (I, J), (J, I), (E, H)\}$ **(c)** no

Mathematical Connections 4-3

Communication

1. yes, since each element of A is paired with exactly one element of B **3. (a)** This is not a function, since a faculty member may teach more than one class.   **(b)** This is a function (assuming only one teacher per class).   **(c)** This is not a function, since not every senator is paired with a committee. (Not every senator is the chairperson of a committee.)

Open-Ended

5. Answers will vary.   **7.** Answers will vary.
9. Answers will vary.

Cooperative Learning

11. Answers will vary.

Questions from the Classroom

13. Not every function is a sequence. Consider, for example, the function whose inputs are the students at a university and whose outputs are their student identification numbers.
15. One way is to show that for a given output y there is a unique input. If $y = 3x + 5$ and y is an output, then the unique input is $\dfrac{y - 5}{3}$. Another approach is to write the sequence of outputs 5, 8, 11, 14, 17, ... Then to 0 correspond 5, to 1 correspond 8, etc.

Review Problems

17. (a) 50   **(b)** 2   **(c)** 100   **(d)** 3

Chapter Review

1. $S = 13P$ **2.** There are 103 times as many girls as boys.   **3.** $f = 3y$ **4.** $10S - 10n$ **5.** 26   **6. (a)** If n is the original number, then each of the following lines shows the result of performing the instruction:

$$n$$
$$n + 17$$
$$2(n + 17) = 2n + 34$$
$$2n + 30$$
$$4n + 60$$
$$4n + 80$$
$$n + 20$$
$$n$$

(b) Answers will vary; for example, the next two lines could be subtract 65 and then divide by 4.   **(c)** Answers will vary.
7. (a) 12   **(b)** 29   **(c)** 3   **(d)** no solution   **(e)** Every number is a solution.   **8.** Paige 111, Jordan 222, and Mike 666   **9.** science books 17 days, other books 3 days   **10.** Rashid 50, Dahlia 150, Jacobo 300
11. (a) function   **(b)** not a function   **(c)** function
12. (a) range = $\{3, 4, 5, 6\}$ **(b)** range = $\{14, 29, 44, 59\}$
(c) range = $\{0, 1, 4, 9, 16\}$ **(d)** range = $\{5, 9, 15\}$

13. (a) This is not a function, since one student can have two majors.   **(b)** This is a function. The range is the subset of the natural numbers that includes the number of pages in each book in the library.   **(c)** This is a function. The range is $\{6, 8, 10, 12, \dots\}$.   **(d)** This is a function. The range is $\{0, 1\}$.   **(e)** This is a function. The range is N.
14. (a) $C(x) = 200 + 55(x - 1)$

(b)

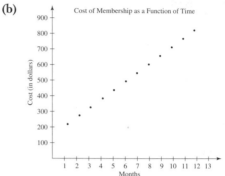

Cost of Membership as a Function of Time

(c) In the ninth month, the cost exceeds $600.   **(d)** in the 107th month   **15.** 5   **16. (a)** Yes, each input has exactly one output.   **(b)** No, for $x = 4$, there are two values for y.   **(c)** No, for $x = 5$, there are two values for y.   **17. (a)** 14, 18, 22, 26   **(b)** The graph consists of points on the line $y = 4x + 2$ for $x = 1, 2, 3, 4, \dots$.   **(c)** $y = 4x + 2$
(d) The graph consists of some points on a straight line but not all the points on the line, and hence is not a straight line.

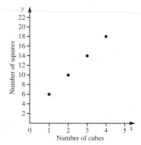

Answers to Now Try This

4-1. (a) If we remove the four white corner tiles, then the number of remaining white tiles on a side is the same as the number of shaded tiles, i.e., n. On four sides we have $4n$. Adding the four corner tiles, we get $4n + 4$.   **(b)** (i) 206 (ii) $2n + 6$ **4-2. (1)** by calculating $2 \cdot 50$ first   **(2)** by writing $70 \cdot 50 = 50 \cdot 70$ **(3)** no, the computations in the parentheses need to be done first   **4-3. (a)** $\square = 3, \triangle = 9$
(b) $\square = 4, \triangle = 2$ **4-4.** to find s, since $\dfrac{6s}{6} = s$

4-5. 1. pounds gained   **2.** $4 + p = 8, p = 4$
4-6. (1) If Bob delivered b papers then Abby delivered $3b$ papers and Connie $3b + 13$ papers. Hence
$$b + 3b + 3b + 13 = 496,$$
$$b = 69$$
$$a = 3b = 207$$
$$c = 3b + 13 = 220.$$

(2) $x + y = 8$
$z + y = 9$
$z + x = 7$

Subtracting the third equation from the second, we get $y - x = 2$. With $y + x = 8$ by adding the equations side by side, we get $2y = 10$; $y = 5$. Hence, $x = 3$ and $z = 4$.
4-7. (a) It is a function from the set of natural numbers to $\{0\}$, because for each natural number there is a unique output in $\{0\}$.   **(b)** It is a function from the set of natural numbers to $\{0,1\}$, because for each natural number there is a unique output in $\{0,1\}$.   **4-8. (4)** They are on the same straight line.   **(5)** Draw a line through the points and extend it. The point $(6, 20)$ is on the line.   **(6)** No, it is not on the line.   **4-9.** $\{5, 10, 15, 20, 25\}$ **4-10. (a)** girls: A, C, D, F, G, I; boys: B, J **(b)** E and H **4-11. (a)** For $x > 10$, y is not a whole number. The range is obtained by substituting $x = 0, 1, 2, 3, \ldots, 10$ in $y = 10 - x$.   **(b)** If x and y are interchanged, we get $x = y + 10$ or $y = x - 10$, which is different from $y = x + 10$.

Answer to Technology Corner

Section 4 3

Answers vary. The graphs are parallel lines because any two equations for different values of b do not have a common solution.

Answer to Preliminary Problem

If the number that a student chooses is x and the answer is a, then

$$2 \cdot \left(\frac{6x + 4}{2} + 5 \right) - 18 = a$$
$$6x - 4 = a$$
$$6x = a + 4$$
$$x = \frac{a + 4}{6}$$

Thus, the teacher adds 4 to each answer and divides the sum by 6.

Chapter 5

Assessment 5-1A

1. (a) $^-2$ **(b)** 5 **(c)** ^-m **(d)** 0 **(e)** m **(f)** $^-a - b$ or $^-(a + b)$ **2. (a)** 2 **(b)** m **(c)** 0 **3. (a)** 5
(b) 10 **(c)** $^-5$ **(d)** $^-5$
4. (a)

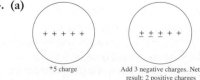
+5 charge   Add 3 negative charges. Net
result: 2 positive charges

(b)

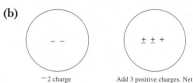
$^-2$ charge   Add 3 positive charges. Net
result: 1 positive charge

(c)

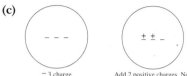
$^-3$ charge   Add 2 positive charges. Net
result: 1 negative charge

(d)

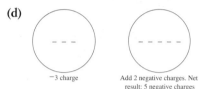
$^-3$ charge   Add 2 negative charges. Net
result: 5 negative charges

5. (a)

(b)

(c)

(d)

6. (a) $3 + {}^-({}^-2) = 5$ **(b)** $^-3 + {}^-2 = {}^-5$
(c) $^-3 + {}^-({}^-2) = {}^-1$ **7. (a)** $3 - ({}^-2) = x$ if, and only if, $3 = {}^-2 + x$. Thus, $x = 5$.   **(b)** $^-3 - 2 = x$ if, and only if, $^-3 = 2 + x$. Therefore, $x = {}^-5$.
(c) $^-3 - ({}^-2) = x$ if, and only if, $^-3 = {}^-2 + x$. Therefore, $x = {}^-1$.   **8. (a)** $^-17 + 10 = {}^-7$ points
(b) $^-10°C + 8°C = {}^-2°C$ **(c)** $5000 \text{ ft} + ({}^-100 \text{ ft}) = 4900 \text{ ft}$ **9. (a)** $({}^-45) + ({}^-55) + ({}^-165) + ({}^-35) + ({}^-100) + 75 + 25 + 400$ **(b)** $\$400$
10. (a)

(b)

11. (a) $^-4 - 2 = {}^-6$; $^-4 - 1 = {}^-5$; $^-4 - 0 = {}^-4$; $^-4 - {}^-1 = {}^-3$ **(b)** $3 - 1 = 2$; $2 - 1 = 1$; $1 - 1 = 0$; $0 - 1 = {}^-1$; $^-1 - 1 = {}^-2$; $^-2 - 1 = {}^-3$ **12. (a)** $^-9$
(b) 3 **(c)** 1 **13. (a) (i)** $55 - 60$ **(ii)** $55 + ({}^-60)$
(iii) $^-5°F$ **(b) (i)** $200 - 220$ **(ii)** $200 + ({}^-220)$
(iii) $\$^-20$ **14. (a)** 10W–40 or 10W–30   **(b)** 5W–30
(c) 10W–40, 5W–30, or 10W–30   **(d)** none   **(e)** 10W–30 or 10W–40   **15. (a)** $1 + 4x$ **(b)** $2x + y$ **16.** The

equation holds if, and only if, $c = 0$ (a and b can be any integers). Justification: It can be shown that $a - (b - c) = a - b + c$ for all integers a, b, and c. Thus, the original equation holds if, and only if, $(a - b) + {}^-c = (a - b) + c$, which in turn holds if, and only if, ${}^-c = c$. This last equation is true if, and only if, $c = 0$ **17. (a)** I **(b)** W
(c) $I - \{0\}$ **(d)** $\varnothing$ **(e)** $\varnothing$ **(f)** I^- **18. (a)** 0
(b) ${}^-101$ **(c)** 1   **(d)** $a - 1$ **(e)** ${}^-4$ **19. (a)** All negative integers   **(b)** All positive integers   **(c)** All integers less than ${}^-1$ **(d)** 2 or ${}^-2$ **20. (a)** 9   **(b)** 2   **(c)** 0 and 2   **(d)** the set of whole numbers   **21. (a)** 0 or 12
(b) ${}^-8$ or 8   **(c)** Every integer satisfies this property.
22. (a) 89   **(b)** 19   **23.** Greatest possible value: $a - (b - c) - d$, or 8; least possible value: $a - b - (c - d)$, or ${}^-6$ **24. (a)** $d = {}^-3$, next terms: ${}^-12, {}^-15$
(b) $d = {}^-y$, next terms: $x - 2y, x - 3y$ **25.** ${}^-1$
26. (a) true   **(b)** true   **(c)** true   **27. (a)** ${}^-4$ **(b)** 3
(c) ${}^-5$ **28. (a)** ${}^-18$ **(b)** ${}^-6$ **(c)** 22   **(d)** ${}^-18$ **(e)** 23

Mathematical Connections 5-1

Communication

1. He could have driven 12 mi in either direction from milepost 68. Therefore, his location could be either at the $68 - 12 = 56$ milepost or at the $68 + 12 = 80$ milepost.
3. One way is to find the difference of the greater absolute value and the lesser absolute value. The sum has the same sign as the integer with the greater absolute value.
5. (a) No; if $x < 0$, then ${}^-x$ is positive.
(b) ${}^-(a - b - c) = {}^-(a + {}^-b + {}^-c)$
$\qquad = {}^-a + {}^-({}^-b) + {}^-({}^-c)$
$\qquad = {}^-a + b + c = b + c - a$
7. $a < b$ if, and only if, there exists a nonzero whole number c such that $a + c = b$. ${}^-8 < {}^-7$ because ${}^-8 + 1 = {}^-7$.

Open-Ended

9. Answers vary; for example, the floors above ground could be numbered as usual $1, 2, 3, 4, \ldots, n$, the zero or ground floor could be called G, and the floors belowground could be called $1B, 2B, 3B, 4B, \ldots, mB$. The system could be modeled on a vertical number line with G replacing 0 and $1B, 2B, 3B, \ldots, mB$ replacing the negative integers.
11. (a) Answers vary. For example, $f(x) = -|x| - 1$.
(b) Answers vary. For example, $f(x) = |x|$.

Cooperative Learning

13. Answers vary.

Questions from the Classroom

15. The algorithm is correct, and the student should be congratulated for finding it. One way to encourage such creative behavior is to name and refer to the procedure after the student who invented it; for example, "David's subtraction method." In fourth grade, the technique can be explained by using a money model. Suppose you have $4 in one checking

account and $80 in another, for a total of $84. You spent $27 by withdrawing $7 from the first account and $20 from the second. The first checking account is overdrawn by $3; that is, the balance is ${}^-\$3$. The balance in the second account is $60. After transferring $3 from the second account to the first, the balance in the first account is $0 and in the second $57; that is, the total balance is $57.   **17.** The picture is supposed to illustrate the fact that an integer and its opposite are mirror images of each other. Because a could be negative, the picture is correct. For example, possible values for a and ${}^-a$ are $a = {}^-1, {}^-a = 1$, and $a = {}^-7, {}^-a = 7$. At this point, the teacher could remind the students that the "$-$" sign in ${}^-a$ *does not* mean that ${}^-a$ is negative. If a is positive, ${}^-a$ is negative, but if a is negative, ${}^-a$ is positive.

Assessment 5-2A

1. $3({}^-1) = {}^-3; 2({}^-1) = {}^-2; 1({}^-1) = {}^-1; 0({}^-1) = 0;$ $({}^-1)({}^-1) = 1$, by continuing the pattern of an arithmetic sequence with fixed difference 1.
2.

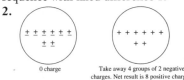

0 charge   Take away 4 groups of 2 negative
                  charges. Net result is 8 positive charges.

3.

$$2({}^-4) = {}^-8$$

4. (a) $({}^-3)({}^-3) = 9$ **(b)** $({}^-5)2 = {}^-10$
5. (a) $4({}^-20)$ or ${}^-80$ **(b)** $({}^-4)({}^-20)$ or 80
(c) $n({}^-20)$ or ${}^-20n$ **(d)** $({}^-n)({}^-20)$ or $20n$
6. (a) 5   **(b)** ${}^-11$ **(c)** Impossible; since $0 \cdot k = {}^-5$ has no integer solution.   **7. (a)** ${}^-10$ **(b)** ${}^-10$ **(c)** Not defined   **(d)** Not defined   **(e)** ${}^-1$ **8. (a)** ${}^-30; {}^-30 \div {}^-6 = 5; {}^-30 \div 5 = {}^-6$ **(b)** $20; 20 \div {}^-5 = {}^-4;$ $20 \div {}^-4 = {}^-5$ **(c)** $0, 0 \div {}^-3 = 0$; division by 0 is not defined.   **(d)** 0, division by 0 is not defined.
9. (a) $(4x) \div 4 = a$ if, and only if, $4x = 4a$ if, and only if, $a = x$. Thus $(4x) \div 4 = x$ **(b)** ${}^-xy \div y = a$ if, and only if, ${}^-xy = ya = ay$ if, and only if, $a = {}^-x$ **10.** All answers are in °C.   **(a)** $32 + ({}^-3)30$ or $32 - 3 \cdot 30 = {}^-58$
(b) $0 + ({}^-4)({}^-25)$ or $4 \cdot 25 = 100$ **(c)** ${}^-20 + ({}^-4)({}^-30) = 100$ **(d)** $25 + 3({}^-20) = 25 - 3 \cdot 20 = {}^-35$ **11.** 108,000 acres   **12. (a)** ${}^-1({}^-5 + {}^-2) = {}^-1({}^-7) = 7; ({}^-1)({}^-5) + ({}^-1)({}^-2) = 5 + 2 = 7$
(b) ${}^-3({}^-3 + 2) = {}^-3({}^-1) = 3; ({}^-3)({}^-3) + ({}^-3)(2) = 9 + {}^-6 = 3$ **13. (a)** ${}^-8$ **(b)** 16
(c) ${}^-1000$ **(d)** 81   **(e)** 1   **(f)** ${}^-1$ **(g)** 12   **(h)** 0
14. (a) 0   **(b)** 0   **(c)** 9   **15.** (b) and (c) are always positive, (a) is always negative. (d) and (e) are neither
16. (b) = (c); (d) = (e)   **17. (a)** Commutative property of multiplication   **(b)** Closure property of multiplication
(c) Associative property of multiplication   **(d)** Distributive property of multiplication over addition   **18. (a)** xy
(b) $2xy$ **(c)** $3x - y$ **(d)** ${}^-x$ **19. (a)** ${}^-2x + 2y$
(b) $x^2 - xy$ **(c)** ${}^-x^2 + xy$ **(d)** ${}^-2x - 2y + 2z$

20. (a) $^-2$ **(b)** 2 **(c)** 0 **(d)** $^-6$ **(e)** $^-36$ **(f)** 6
(g) All integers except 0 **(h)** All integers except 0
21. (a) $^-5$ **(b)** $^-2$ **(c)** No solutions **(d)** $^-2$ **(e)** $^-2$
and 2 **(f)** $^-2$ or 4 **(g)** $^-1$ **(h)** 1 or $^-3$
22. (a) $(50 + 2)(50 - 2) = 50^2 - 2^2 = 2500 - 4 =$
2496 **(b)** $25 - 10,000 = ^-9975$ **(c)** $x^2 - y^2$
23. (a) $8x$ **(b)** $x(y + 1)$ **(c)** $x(x + y)$
(d) $x(3y + 2 - z)$ **(e)** $a[b(c + 1) - 1]$
(f) $(4 + a)(4 - a)$ **(g)** $(2x + 5y)(2x - 5y)$
24. (a) $(a - b)^2 = (a + {}^-b)(a + {}^-b)$
$$= a(a + {}^-b) + {}^-b(a + {}^-b)$$
$$= a^2 + a({}^-b) + ({}^-b)a + ({}^-b)({}^-b)$$
$$= a^2 - 2ab + b^2$$
(b) (i) $98^2 = (100 - 2)^2 = 100^2 - 2(200) + 2^2 =$
$10,000 - 400 + 4 = 9604$ **(ii)** $99^2 = (100 - 1)^2 =$
$100^2 - 2(100) + 1^2 = 10,000 - 200 + 1 = 9801$
(iii) $997^2 = (1000 - 3)^2 = 1000^2 - 2(3000) + 3^2 =$
$1,000,000 - 6000 + 9 = 994,009$ **25. (a)** $8, 11, d = 3$,
nth term is $3n - 13$. **(b)** $^-128, ^-256, r = 2$, nth term is
$^-2^n$. **(c)** $2^7, ^-2^8, r = ^-2$, nth term is $2 \cdot (^-2)^{n-1}$ or
$^-(^-2)^n$. **26.** 13,850 **27. (a)** $^-9, ^-6, ^-1, 6, 15$
(b) $^-2, ^-7, ^-12, ^-17, ^-22$ **(c)** $^-3, 3, ^-9, 15, ^-33$
28. The first term is 7 and the second is $^-4$. **29.** 17 min;
$-108°C$ **30. (a)** If $x \geq 0$ and $y \leq 0$, then $^-|x| \cdot |y| =$
$^-x(^-y) = xy$. Similarly, the statement is true for $x \leq 0$ and
$y \geq 0$. **(b)** True if, and only if, $x = 0$. If $x \neq 0$, $^-x^2$ is
negative and x^2 is positive and hence the expressions cannot
be equal. **(c)** If x and y are positive and $x > y$, the
statement is true.

Mathematical Connections 5-2

Communication

1. No; it is not of the form $(a - b)(a + b)$.
3. (a) $(^-1)a + a = (^-1)a + 1 \cdot a$
$$= (^-1 + 1)a$$
$$= 0 \cdot a$$
$$= 0$$
(b) Part (a) implies that $(^-1) \cdot a$ is the additive inverse of a.
Because ^-a is also the additive inverse of a, it follows that
$(^-1)a = ^-a$.
5. $^-(a + b) = (^-1)(a + b)$ Problem 3(b)
$$= (^-1)a + (^-1)b$$ By distributive property
$$= ^-a + ^-b$$ Problem 3(b)
7. (a) $5x + 3 < ^-20$ if, and only if, $5x < ^-23$. Because
$5(^-5) = ^-25 < ^-23$ and $5(^-4) = ^-20 > ^-23$, the
greatest integer is $^-5$. **(b)** No, there is no least integer.
Because x is an integer, the inequality is equivalent to
$x \leq ^-5$ and there is no such smallest integer.

Open-Ended

9. Answers vary. **11.** Answers vary depending on grade
level and publisher of book selected.

Cooperative Learning

13. (a) Answers vary. **(b)** Answers vary.

Questions from the Classroom

15. The student is correct that a debt of $5 is greater than a
debt of $2. However, what this means is that on a number
line $^-5$ is farther to the left than is $^-2$. The fact that $^-5$ is
farther to the left than $^-2$ on a number line implies that
$^-5 < ^-2$. **17.** The procedure can be justified as follows.
Since for all integers c, $^-c = (^-1)c$, the effect of performing
the opposite of an algebraic expression is the same as multiplying the expression by $^-1$. However, in the expression
$x - (2x - 3)$, the "$-$" is used to denote subtraction, not
simply finding the opposite. If the expression is first rewritten as $x + ^-(2x + ^-3)$, then it is the case that
$^-(2x + ^-3) = ^-1(2x + ^-3)$, or $^-2x + 3$. Now the
expression can be rewritten as $x + ^-2x + 3$, which a
student might obtain from the father's rule. **19.** The
model shows that $^-2(^-3) = 6$ but it does not show that in
general a negative integer times a negative integer is a positive integer. A formal proof showing a negative integer
times a negative integer results in a positive integer for all
integers is desirable.

Review Problems

21. (a) 5 **(b)** $^-7$ **(c)** 0 **23.** 400 lb **25. (a)** $x = 3$
or $^-3$ **(b)** Not possible **(c)** $x \geq 0$ **(d)** $x \leq 0$

Assessment 5-3A

1. (a) true **(b)** true **(c)** true **(d)** true **(e)** true
(f) False; there is no value $c \in I$ such that $30c = 6$.
2. (a) yes **(b)** no **(c)** yes **3. (a)** 2, 3, 4, 6, 11
(b) 2, 3, 6, 9 **(c)** 2, 3, 5, 6, 10 **4. (a)** No, $17|34,000$
and $17 \nmid 15$, so $17 \nmid 34,015$. **(b)** Yes, $17|34,000$ and $17|51$,
so $17|34,051$. **(c)** No, $19|19,000$ and $19 \nmid 31$, so
$19 \nmid 19,031$. **(d)** Yes, 5 is a factor of $2 \cdot 3 \cdot 5 \cdot 7$. **(e)** No,
$5|2 \cdot 3 \cdot 5 \cdot 7$ and $5 \nmid 1$, so $5 \nmid (2 \cdot 3 \cdot 5 \cdot 7) + 1$. **5. (a)** True
by Theorem 5–12 **(b)** True by Theorem 5–13(b)
(c) none **(d)** True by Theorem 5–13(b) **(e)** True by
Theorem 5–12 **6. (a)** False, $2|6$ but $(2 + 5) \nmid (6 + 5)$
(b) True. Because $b|a$, there is c such that $a = bc$. Then
$a^3 = b^3c^3 = (bc^3)b^2$, which means $b^2|a^3$. **(c)** True. Because
$b|a$, there is c such that $a = bc$. Then $^-a = ^-(bc) = b(^-c)$,
which implies that $b|^-a$. Also, $^-a = ^-(bc) = (^-b)c$, which
implies $^-b|^-a$. **7. (a)** $210 = 7 \cdot 30$ **(b)** $19|1900$ and
$19|38$ **(c)** $6|(2 \cdot 3) \cdot 2^2 \cdot 3 \cdot 17^4$ and $6|6 \cdot 2^4 \cdot 3 \cdot 17^4$
(d) $7|4200$ but $7 \nmid 22$ **8. (a)** true **(b)** false **9. (a)** 7
(b) 7 **(c)** 6 **10. (a)** Any digit 0 through 9 **(b)** 1, 4, 7
(c) 1, 3, 5, 7, 9 **(d)** 7 **(e)** 7 **11.** 17 **12.** Each pencil
costs 19¢. **13. (a)** yes **(b)** no **(c)** yes **(d)** no
14. (a) $12,343 + 4546 + 56 = 16,945$; $4 + 1 + 2 = 7$
has a remainder of 7 when divided by 9, as does $1 + 6 +$
$9 + 4 + 5$. **(b)** $987 + 456 + 8765 = 10,208$; $6 + 6 +$
$8 = 20$ has a remainder of 2 when divided by 9, as does

$1 + 0 + 2 + 0 + 8 = 11$. **(c)** $10,034 + 3004 + 400 + 20 = 13,458$; $8 + 7 + 4 + 2 = 21$ has a remainder 3 when divided by 9, as does $1 + 3 + 4 + 5 + 8$.
(d) $1003 - 46 = 957$; $4 - 1 = 3$ has a remainder of 3 when divided by 9, as does $9 + 5 + 7 = 21$.
(e) $345 \cdot 56 = 19,320$. 345 has a remainder of 3 when divided by 9; 56 has a remainder of 2 when divided by 9; $3 \cdot 2 = 6$ has remainder 6 when divided by 9. $1 + 9 + 3 + 2 + 0 = 15$ has a remainder of 6 when divided by 9. **(f)** Answers vary.   **15. (a)** 1, 3, and 7 divide n. 1 divides every number. Also, because $n = 21 \cdot d, d \in I$, then $n = (3 \cdot 7)d = 3(7d) = 7(3d)$, which implies that n is divisible by 3 and by 7. **(b)** 1, 2, 4, and 8 divide n. 1 divides every number. Also, because $16|n$, then $n = 16 \cdot d, d \in I$. Therefore, $n = (2 \cdot 8)d = 2(8d) = 8(2d) = 4(4d)$. Hence, n is divisible by 2, 4, and 8.
16. (a) Yes, if $5|x$ and $5|y$, then $5|(x + y)$. **(b)** Yes, if $5|y$, then $5|(^-y)$. If $5|x$ and $5|(^-y)$, then $5|(x + (^-y))$ or $5|x - y$. **(c)** Yes, if $5|x$, then 5 divides all the multiples of x and in particular $5|xy$. **17.** 6,868,395 is divisible by 15 because it is divisible by both 3 and 5. The last digit is 5 and the sum of the digits is 45, which is divisible by 3.
18. (a) False; $2|4$, but $2 \nmid 1$ and $2 \nmid 3$. **(b)** False (same example as in (a)) **(c)** False; $12|72$, but $12 \nmid 8$ and $12 \nmid 9$.
(d) True **(e)** False; if $a = 5$ and $b = -5$, then $a|b$ and $b|a$, but $a \neq b$.
19. Let $n = a10^4 + b10^3 + c10^2 + d10 + e$.

$$a10^4 = a(10,000) = a(9999 + 1) = a9999 + a$$
$$b10^3 = b(1000) = b(999 + 1) = b999 + b$$
$$c10^2 = c(100) = c(99 + 1) = c99 + c$$
$$d10 = d(10) = d(9 + 1) = d9 + d$$

Thus, $n = (a9999 + b999 + c99 + d9) + (a + b + c + d + e)$. Because $9|9, 9|99, 9|999, 9|9999$, it follows that if $9|(a + b + c + d + e)$ then $9|[(a9999 + b999 + c99 + d9) + (a + b + c + d + e)]$; that is, $9|n$. If, on the other hand, $9 \nmid (a + b + c + d + e)$, it follows that $9 \nmid n$.

Mathematical Connections 5-3

Communication

1. No, any amount of postage must be a multiple of 3 (being the sum of a multiple of 6 and 9, both of which are multiples of 3). **3. (a)** Yes, $4|52,832$, so 4 divides any integer times 52,832. Therefore, 4 divides $52,832 \cdot 324,518$, which is the area. **(b)** Yes, 2 is a factor of 52,834 and 2 is a factor of 324,514, so $2 \cdot 2$ or 4 is a factor of $52,834 \cdot 324,514$, which is the area.   **5. (a)** No. If $16|x$, then $x = 10n = 5 \cdot 2n$ for some integer n. Therefore x is divisible by 5. **(b)** Yes, all odd multiples of 5 are not divisible by 10 but are divisible by 5.
7. 243; yes; Consider any number n of the form $abcabc$. Then we have the following:

$$n = (a10^5) + (b10^4) + (c10^3) + (a10^2)$$
$$+ (b10^1) + c$$
$$= a(10^5 + 10^2) + b(10^4 + 10^1) + c(10^3 + 1)$$

$$= a(100,000 + 100) + b(10,000 + 10) + c1001$$
$$= a(1001 \cdot 100) + b(1001 \cdot 10) + c1001$$
$$= (1001)[a100 + b10 + c1]$$
$$= (7 \cdot 11 \cdot 13)[a100 + b10 + c1]$$

Therefore, if you divide by 1001, the quotient is abc.
9. (a) Let $abcd$ be a number. Subtracting the last digit gives $abc0$. Thus, the result is divisible by 2, 5, and 10. **(b)** If one subtracts cd from $abcd$, one obtains $ab00$. This number is divisible by 2, 4, 5, 10, 20, 25, 50, and 100.
(c) $abcd - (a + b + c + d) = a \cdot 10^3 + b \cdot 10^2 + c \cdot 10 + d - a - b - c - d = a \cdot 999 + b \cdot 99 + c \cdot 9 = 9(a \cdot 111 + b \cdot 11 + c)$. Thus, the result is divisible by 3 and 9.
(d) • Consider a four-digit number in base five, $abcd_{\text{five}}$. Then subtracting the last digit produces $abc0_{\text{five}} = 10_{\text{five}} \cdot abc_{\text{five}}$, which is divisible by 10_{five} or 5 in base ten.
• $abcd_{\text{five}} - cd_{\text{five}} = ab00_{\text{five}}$, which is divisible by 10_{five} and 100_{five} or 5 and 25 in base ten.
• $abcd_{\text{five}} - a - b - c - d = 444_{\text{five}} \cdot a + 44_{\text{five}} \cdot b + 4_{\text{five}} \cdot c = 4_{\text{five}}(111_{\text{five}} \cdot a + 11_{\text{five}} \cdot b + c)$, which is divisible by 2 and 4.

Open-Ended

11. (a) Answers vary; for example, upon inspection of the given numbers we notice that all the numbers are multiples of 3. Since 3 divides each number, then 3 divides the sum of any of the numbers. Because 100 is not divisible by 3, there is no winning combination of the given numbers that will sum to 100. **(b)** Answers vary; for example, since many of these multiples of 3 sum to 99 ($33 + 66, 45 + 51 + 3$, etc.), the company could place at most 1000 cards with the number 1 on the card. This would ensure that there are at most 1000 winners. Other numbers could also be used.

Cooperative Learning

13. Answers vary.

Questions from the Classroom

15. Although these two expressions look quite similar, they are not equal; a/b means "a divided by b," an operation, and it has a numerical answer if $b \neq 0$; $a|b$ means "a divides b," a relation that is either true or false. Note if a/b is an integer, then $b|a$.   **17.** Yes if $a \neq 0$; the student's conclusion is that $a|0$ and this is true because if $a \neq 0$, the equation $a \cdot k = 0$ has a unique solution, $k = 0$.   **19.** It has been shown that any four-digit number n can be written in the form $n = a10^3 + b10^2 + c10 + d = (a999 + b99 + c9) + (a + b + c + d)$. The test for divisibility by some number g will depend on the sum of the digits $a + b + c + d$ if, and only if, $g|(a999 + b99 + c9)$ regardless of the values of a, b, and c. Since the only numbers greater than 1 that divide 9, 99, and 999 are 3 and 9, the test for divisibility by dividing the sum of the digits by the number works only for 3 and 9. A similar argument works for any m-digit number.

21. It is true that a number is divisible by 21 if, and only if, it is divisible by 3 and by 7. However, the general statement is false. For example, 12 is divisible by 4 and by 6 but not by $4 \cdot 6$, or 24. One part of the statement is true, that is, "if a number is divisible by $a \cdot b$, then it is divisible by a and b." The statement "if a number is divisible by a and by b, it is divisible by ab" is true if a and b are relatively prime. To see why this is true, suppose $GCD(a,b) = 1$ and m is an integer such that $a|m$ and $b|m$. Since $a|m$, $m = ka$ for some integer k. Now $b|m$ implies that $b|ka$. Since a and b are relatively prime, it follows from the Fundamental Theorem of Arithmetic and the fact that $b|ka$ that $b|k$ (why?), and therefore $k = jb$, for some integer j. Substituting $k = jb$ in $m = ka$, we obtain $m = jba$, and consequently $ab|m$.

Review Problems

23. (a) $^-2$ **(b)** No such integers exist. **(c)** No such integers exist. **(d)** All integers **(e)** No such integers exist. **(f)** All positive integers   **25. (a)** $f(^-5) = 10 - 3 = 7$ **(b)** $^-10$ **(c)** No, if $^-2x - 3 = 2$, then $^-2x = 5$, which has no solution in integers. **(d)** The set of all odd integers

Assessment 5-4A

1. 30   **2. (a)** prime **(b)** not prime **(c)** not prime **(d)** prime **(e)** prime **(f)** prime **(g)** prime **(h)** not prime
3. (a)

```
        504
       /\
      2  252
         /\
        2  126
           /\
          2  63
             /\
            3  21
               /\
              3  7
```
$504 = 2^3 \cdot 3^2 \cdot 7$

(b)
```
        2475
       /\
      3  825
         /\
        3  275
           /\
          5  55
             /\
            5  11
```
$2475 = 3^2 \cdot 5^2 \cdot 11$

(c) 11,250
```
       11,250
       /\
      2  5625
         /\
        3  1875
           /\
          3  625
             /\
            5  125
               /\
              5  25
                 /\
                5  5
```
$11,250 = 2 \cdot 3^2 \cdot 5^4$

4. (a)
```
        210
       /  \
      2    105
          /  \
        21    5
       /  \
      3    7
```
(b) You could multiply $2 \cdot 3 \cdot 7 \cdot 5$. **5.** 73
6. (a) $2^8 \cdot 3^4 \cdot 5^2 \cdot 7$ **(b)** $2^3 \cdot 5^2 \cdot 7^{20} \cdot 13$ **(c)** 251

(d) $7 \cdot 11 \cdot 13$ **7. (a)** 1 by 48, 2 by 24, 3 by 16, 4 by 12 **(b)** Only one, 1 by 47   **8. (a)** The Fundamental Theorem of Arithmetic states that n can be written as a product of primes in one and only one way. Since $2|n$ and $3|n$ and 2 and 3 are both prime, they must be included in the unique factorization:

$2 \cdot 3 \cdot p_1 \cdot \cdots \cdot p_m = n;$
$(2 \cdot 3)(p_1 \cdot p_2 \cdot \cdots \cdot p_m) = n;$ therefore
$6|n.$

(b) Yes. If $a|n$, there exists an interger c such that $ca = n$. If $b|n$ there exists an integer d such that $db = n$. Therefore $(ca)(db) = n^2 \Rightarrow (dc)(ab) = n^2 \Rightarrow ab|n^2$.   **9. (a)** 2, 3, 4, 6, 9, 12, 18, or 36 **(b)** 1, 2, 4, 7, 14, or 28 **(c)** 1 or 17 **(d)** 1, 2, 3, 4, 6, 8, 9, 12, 16, 18, 24, 36, 48, 72, or 144
10. 90   **11.** 101, 103, 107, 109, 113, 127, 131, 137, 139, 149, 151, 157, 163, 167, 173, 179, 181, 191, 193, 197, 199
12. 3, 5; 5, 7; 11, 13; 17, 19; 29, 31; 41, 43; 59, 61; 71, 73; 101, 103; 107, 109; 137, 139; 149, 151; 179, 181; 191, 193; 197, 199
13. 1, 2, 3, 6, 7, 14, 21   **14.** There are 16 total factors of 1000. The other 15 factors are 1, 2, 4, 5, 8, 10, 20, 25, 40, 50, 100, 125, 200, 250 and 500.   **15.** There are infinitely many composites of the form 1, 11, 111, 1111, 11111, 111111 ... since every third member of this sequence will be divisible by 3 (also every other member of this sequence is divisible by 11).   **16.** Yes, because $(3^2 \cdot 2^4)(3^2 \cdot 2^3) = 3^4 \cdot 2^7$
17. (a) $3 \cdot 5 \cdot 7 \cdot 11 \cdot 13$ is composite because it is divisible by 3, 5, 7, 11, and 13. **(b)** $(3 \cdot 4 \cdot 5 \cdot 6 \cdot 7 \cdot 8) + 2 = 2[(3 \cdot 2 \cdot 5 \cdot 6 \cdot 7 \cdot 8) + 1]$ and so it is composite.
(c) $(3 \cdot 5 \cdot 7 \cdot 11 \cdot 13) + 5 = 5[(3 \cdot 7 \cdot 11 \cdot 13) + 1]$ and so it is composite. **(d)** $10! + 7 = 7[(10 \cdot 9 \cdot 8 \cdot 6 \cdot 5 \cdot 4 \cdot 3 \cdot 2 \cdot 1) + 1]$ and so it is composite.   **18.** $2^3 \cdot 3^2 \cdot 25^3$ is not a prime factorization because 25 is not prime. The prime factorization is $2^3 \cdot 3^2 \cdot 5^6$.   **19. (a)** $2^{35} \cdot 3^{35} \cdot 7^{40}$
(b) $2^{200} \cdot 3^{40} \cdot 5^{200}$ **(c)** $2 \cdot 3^6 \cdot 5^{110}$ **(d)** 2311   **20.** 73

Mathematical Connections 5-4

Communication

1. In any set of three consecutive numbers, there is one number that is divisible by 3 and at least one of the other two numbers is divisible by 2. Therefore, the product will be divisible by 2 and by 3 and so it is divisible by 6.
3. No, they are not both correct. Using 3 and 4 is correct because they have no common divisors. But 2 and 6 have a common divisor of 2. Using this test will ensure only that the number is divisible by 6.   **5.** Let $a = 2 \cdot 3 \cdot 5 \cdot 7$ and $b = 11 \cdot 13 \cdot 17 \cdot 19$. Then each prime p less than or equal to 19 appears in the prime factorization of a or of b but not in both. If p is in the prime factorization of a, then $p|a$ but $p \nmid b$ and hence $p \nmid a + b$. A similar argument holds if p is in the prime factorization of b.   **7.** Suppose n is composite and d is its least positive divisor other than 1. We need to show that d is prime. If not, then some prime p less than d will divide d and hence will divide n, which contradicts the fact that d is the smallest divisor of n greater than 1.

Open-Ended

9. (a) Answers vary. **(i)** 25 **(ii)** 21 **(b)** 13, in the interval 100–199 **(c) (i)** 8 **(ii)** 7 **(d)** Answers vary.
10. (a) (i) $1 + 2 + 3 + 4 + 6 = 16$, so 12 is abundant.
(ii) $1 + 2 + 4 + 7 + 14 = 28$, so 28 is perfect.
(iii) $1 + 5 + 7 = 13$, so 35 is deficient. **(b)** Answers vary; for example, 10 and 14 are deficient, 18 is abundant, and 496 and 8128 are perfect.

Cooperative Learning

11. The students must have had the first 23 prime numbers for the number of tiles; that is, 2, 3, 5, 7, 11, 13, 17, 19, 23, 29, 31, 37, 41, 43, 47, 53, 59, 61, 67, 71, 73, 79, 83 tiles. Therefore, the number of tiles is the sum of the first 23 prime numbers, which is 874 tiles.

Questions from the Classroom

13. Bob has almost the right idea but it needs a little work. It should be pointed out that if a number is not divisible by 2 and 3, then it couldn't be divisible by 6, so there is no need to check 6. Also, if a number is not divisible by 2, then it can't be divisible by 4 or 8, so there is no need to check 4 and 8. Next, if a number is not divisible by 5, then it can't be divisible by 10, so there is no need to check 10. Thus we have cut Bob's list to 2, 3, and 5, which are all primes. The Sieve of Eratosthenes can be used to motivate these concepts. Next we need to explore what happens when the number to be checked is large and show that just checking 2, 3, and 5 is not adequate. For example, just checking for divisibility by 2, 3, and 5 is not enough to check if 169 is prime. We can use the sieve to show that if we are trying to determine if a positive integer n is prime, we only have to check primes p such that $p^2 \le n$. **15.** Only the perfect squares have an odd number of divisors. The perfect squares less than 1000 are $1^2, 2^2, 3^2, \ldots, 31^2$. Therefore, there are 31 perfect squares between 1 and 1000 and hence 31 numbers with an odd number of divisors. Consequently, there are $1000 - 31 = 969$ numbers between 1 and 1000 that have an even number of divisors. **17.** The student is right. In every sequence of six consecutive numbers greater than 3, only the numbers before and after a multiple of 6 can be prime. For example, consider the numbers 17, 18, 19, 20, 21, and 22. 18, 20 and 22 are even. 18 and 21 are multiples of 3. Only 17 and 19, the numbers before and after 18 (a multiple of 6), can be prime. All of the others are multiples of 2 or 3.

Review Problems

19. (a) 2, 3, 6 **(b)** 2, 3, 5, 6, 9, 10 **21.** Yes, among eight people, each would get $422.

Assessment 5-5A

1. (a) $D_{18} = \{1, 2, 3, 6, 9, 18\}$
$D_{10} = \{1, 2, 5, 10\}$
$\text{GCD}(18, 10) = 2$
$M_{18} = \{18, 36, 54, 72, 90, \ldots\}$
$M_{10} = \{10, 20, 30, 40, 50, 60, 70, 80, 90, \ldots\}$
$\text{LCM}(18, 10) = 90$
(b) $D_{24} = \{1, 2, 3, 4, 6, 8, 12, 24\}$
$D_{36} = \{1, 2, 3, 4, 6, 9, 12, 18, 36\}$
$\text{GCD}(24, 36) = 12$
$M_{24} = \{24, 48, 72, 96, 120, 144, 168, \ldots\}$
$M_{36} = \{36, 72, 108, 144, 180, \ldots\}$
$\text{LCM}(24, 36) = 72$
(c) $D_8 = \{1, 2, 4, 8\}$
$D_{24} = \{1, 2, 3, 4, 6, 8, 12, 24\}$
$D_{52} = \{1, 2, 4, 13, 26, 52\}$
$\text{GCD}(8, 24, 52) = 4$
$M_8 = \{8, 16, 24, 32, 40, 48, 56, 64, 72, 80, 88, 96, \ldots\}$
$M_{24} = \{24, 48, 72, 96, 120, 144, 168, 192, 216, 240,$
$\quad 264, 288, 312, \ldots\}$
$M_{52} = \{52, 104, 156, 208, 260, 312, \ldots\}$
$\text{LCM}(8, 24, 52) = 312$
(d) $D_7 = \{1, 7\}, D_9 = \{1, 9\}$
$\text{GCD}(7, 9) = 1$
$\text{LCM}(7, 9) = 7 \cdot 9 = 63$
$M_7 = \{7, 14, 21, 28, 35, 42, 49, 56, 63, \ldots\}$
$M_9 = \{9, 18, 27, 36, 45, 54, 63, \ldots\}$
2. (a) $132 = 2^2 \cdot 3 \cdot 11$
$504 = 2^3 \cdot 3^2 \cdot 7$
$\text{GCD}(132, 504) = 2^2 \cdot 3 = 12$
$\text{LCM}(132, 504) = 2^3 \cdot 3^2 \cdot 7 \cdot 11 = 5544$
(b) $65 = 5 \cdot 13$
$1690 = 2 \cdot 5 \cdot 13^2$
$\text{GCD}(65, 1690) = 5 \cdot 13 = 65$
$\text{LCM}(65, 1690) = 2 \cdot 5 \cdot 13^2 = 1690$
(c) $96 = 2^5 \cdot 3$
$900 = 2^2 \cdot 3^2 \cdot 5^2$
$630 = 2 \cdot 3^2 \cdot 5 \cdot 7$
$\text{GCD}(96, 900, 630) = 2 \cdot 3 = 6$
$\text{LCM}(96, 900, 630) = 2^5 \cdot 3^2 \cdot 5^2 \cdot 7 = 50,400$
(d) $108 = 2^2 \cdot 3^3$
$360 = 2^3 \cdot 3^2 \cdot 5$
$\text{GCD}(108, 360) = 2^2 \cdot 3^2 = 36$
$\text{LCM}(108, 360) = 2^3 \cdot 3^3 \cdot 5 = 1080$
3. (a) $\text{GCD}(2924, 220) = \text{GCD}(220, 64) =$
$\text{GCD}(64, 28) = \text{GCD}(28, 8) = \text{GCD}(8, 4) = \text{GCD}(4, 0)$
$= 4$ **(b)** $\text{GCD}(14595, 10856) = \text{GCD}(10856, 3739)$
$= \text{GCD}(3739, 3378) = \text{GCD}(3378, 361) =$
$\text{GCD}(361, 129) = \text{GCD}(129, 103) =$
$\text{GCD}(103, 26) = \text{GCD}(26, 25) = \text{GCD}(25, 1) = 1$
4. (a) 72 **(b)** 1440 **(c)** 630 **(d)** $9^{100} \cdot 25^{100}$ or $3^{200} \cdot 5^{200}$ or 15^{200} **5. (a)** $220 \cdot 2924/4$, or 160,820
(b) $14,595 \cdot 10,856/1$, or 158,443,320
6. $\text{GCD}(6, 10) = 2, \text{LCM}(6, 10) = 30$

7. (a) $\mathrm{LCM}(15, 40, 60) = 120 \, \mathrm{min} = 2 \, \mathrm{hr}$, so the clocks alarm again together at 8:00 A.M. **(b)** no **8.** 24
9. 15 cookies **10.** 36 min **11.** They should pass the starting point after $\mathrm{LCM}(12, 18, 16) = 144$ min.
12. (a) ab **(b)** $\mathrm{GCD}(a, a) = a$; $\mathrm{LCM}(a, a) = a$
(c) $\mathrm{GCD}(a^2, a) = a$, $\mathrm{LCM}(a^2, a) = a^2$
(d) $\mathrm{GCD}(a, b) = a$; $\mathrm{LCM}(a, b) = b$ **13. (a)** True, if a and b are even, then $\mathrm{GCD}(a, b) \geq 2$. **(b)** True, $\mathrm{GCD}(a, b) = 2$ implies that a and b are even. **(c)** False, the GCD could be a multiple of 2; for example, $\mathrm{GCD}(8, 12) = 4$. **14. (a)** 15 **(b)** 1 **15.** $4 = 2^2$. Since 97,219,988,751 is odd, it has no prime factors of 2. Consequently, 1 is their only common divisor and they are relatively prime. **16.** The 60th caller **17.** $48
18. Two packages of plates, four packages of cups, and three packages of napkins
19. (a)

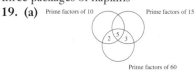

Prime factors of 10   Prime factors of 15
Prime factors of 60

(b)

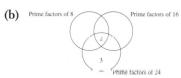

Prime factors of 8   Prime factors of 16
Prime factors of 24

20. $1, 2, 2^2, 2^3, \ldots, 2^{20}$ **21. (a)** $6x^3 y^3 (2x + 3y)$
(b) $6x^2 y^2 z^2 (2x + 3y^2 z + 4x^2 yz^2)$ **22. (a)** Always true; the common divisors of a and b are the same as the common divisors of $|a|$ and b and the same as the common divisors of $|a|$ and $|b|$. **(b)** Always true; same reason as in part (a)
23. (a) $\mathrm{GCD}(10!, 11!) - 10!$ and $\mathrm{LCM}(10!, 11!) = 11!$
(b) $\mathrm{GCD}(10!, 10! + 1) = 1$ and $\mathrm{LCM}(10!, 10! + 1) = 10!(10! + 1)$ **24.** $5^9 = 1,953,125$ and $2^9 = 512$

Mathematical Connections 5-5

Communication

1. No, the set of common multiples is infinite and therefore there could be no greatest common multiple.
3. No; for example, consider $\mathrm{GCD}(2, 4, 10) = 2$. $\mathrm{LCM}(2, 4, 10) = 20$, and the $\mathrm{GCD} \cdot \mathrm{LCM} = 2 \cdot 20 = 40$, while $abc = 2 \cdot 4 \cdot 10 = 80$. **5.** No. Let $a = 2 \cdot 3$, $b = 3 \cdot 5, c = 5 \cdot 7$. Then $\mathrm{GCD}(a, b, c) = 1$, but $\mathrm{GCD}(a, b) = 3$ and $\mathrm{GCD}(b, c) = 5$. **7.** The LCM equals the product of the numbers if, and only if, the numbers have no prime factors in common. Because $\mathrm{GCD}(a, b) \cdot \mathrm{LCM}(a, b) = ab$, $\mathrm{LCM}(a, b) = ab$ if, and only if, $\mathrm{GCD}(a, b) = 1$; that is, a and b have no prime factors in common.
9. • $m = 6$ and $n = 9$; then $\mathrm{GCD}(6, 9) = 3$ and $\mathrm{LCM}(6, 9) = 18$. $\mathrm{GCD}(15, 18) = 3$.
• $m = 7$ and $n = 11$; then $\mathrm{GCD}(7, 11) = 1$ and $\mathrm{LCM}(7, 11) = 77$. $\mathrm{GCD}(18, 77) = 1$.
• $m = 8$ and $n = 16$; then $\mathrm{GCD}(8, 16) = 8$ and $\mathrm{LCM}(8, 16) = 16$. $\mathrm{GCD}(24, 16) = 8$.

11. Answers vary. From the answer to problem 8 it follows that $\mathrm{LCM}(a, b) < ab$ if, and only if, $\mathrm{GCD}(a, b) > 1$; that is, if, and only if, a and b have at least one prime in common.

Cooperative Learning

13. Answers vary.

Questions from the Classroom

15. It does not make much sense to talk about the LCD of numbers since the least common divisor of a set of numbers is always 1. Likewise, it does not make sense to talk about the GCM of a set of numbers since there is no greatest common multiple because the set of common multiples is an infinite set and hence there is no greatest element. **17.** In finding the least common denominator of fractions, one must find the LCM of the denominators. Thus, the LCM of a set of denominators is the least common denominator.
19. We have seen that $\mathrm{GCD}(a, b) = \mathrm{GCD}(a - b, b)$. Applying this k times, we have $\mathrm{GCD}(a, b) = \mathrm{GCD}(a - kb, b)$ for all integers $k \geq 1$. Also $\mathrm{GCD}(a, b) = \mathrm{GCD}(-a, b)$. Hence,

$$\begin{aligned}
\mathrm{GCD}(2132, 534) &= \mathrm{GCD}(2132 - 4 \cdot 534, 534) \\
&= \mathrm{GCD}(^-4, 534) \\
&= \mathrm{GCD}(4, 534) \\
&= 2
\end{aligned}$$

Review Problems

21. (a) 83,151; 83,451; 83,751 **(b)** 86,691 **(c)** 10,396
23. Answers may vary. $30,030 = 2 \cdot 3 \cdot 5 \cdot 7 \cdot 11 \cdot 13$
25. 43

Assessment 5-6A

1. 2:00 P.M. **2. (a)** 3 **(b)** 2 **(c)** 6 **(d)** 8 **(e)** 3
(f) 4 **(g)** Does not exist **(h)** 10 **3. (a)** 2 **(b)** 1
(c) 2 **(d)** 4 **(e)** 1 **(f)** 1 **(g)** 2 **(h)** 4
4. (a)

$\oplus$	1	2	3	4	5	6	7	8	9
1	2	3	4	5	6	7	8	9	1
2	3	4	5	6	7	8	9	1	2
3	4	5	6	7	8	9	1	2	3
4	5	6	7	8	9	1	2	3	4
5	6	7	8	9	1	2	3	4	5
6	7	8	9	1	2	3	4	5	6
7	8	9	1	2	3	4	5	6	7
8	9	1	2	3	4	5	6	7	8
9	1	2	3	4	5	6	7	8	9

(b) $5 \ominus 6 = 8, 2 \ominus 5 = 6$ **(c)** Every subtraction problem can be written as an addition problem, which can always be solved using the 9-hr addition table.

5. (a)

⊗	1	2	3	4	5	6	7	8	9
1	1	2	3	4	5	6	7	8	9
2	2	4	6	8	1	3	5	7	9
3	3	6	9	3	6	9	3	6	9
4	4	8	3	7	2	6	1	5	9
5	5	1	6	2	7	3	8	4	9
6	6	3	9	6	3	9	6	3	9
7	7	5	3	1	8	6	4	2	9
8	8	7	6	5	4	3	2	1	9
9	9	9	9	9	9	9	9	9	9

(b) $3 \oplus 5 = 6$; $4 \oplus 6$ is not defined.　**(c)** No, division by numbers other than 9 is not always possible because not every row (except the row for the identity) contains 1 through 9.　**6. (a)** 3　**(b)** 2　**(c)** 1　**(d)** 1　**(e)** 1　**(f)** 4　**7. (a)** 2, 9, 16, 30　**(b)** 3, 10, 17, 24, 31　**(c)** $366 \equiv 2$ (mod 7); Wednesday　**8. (a)** 4　**(b)** 0　**(c)** 0　**(d)** 7　**9. (a)** $x = 2k$, k is an integer.　**(b)** $x - 1 = 2k$ implies $x = 2k + 1$, where k is an integer.　**(c)** $x - 3 = 5k$ implies $x = 3 + 5k$, where k is an integer.　**10.** Tuesday at 2:00 A.M.　**11.** C

Mathematical Connections 5-6

Communication

1. (a) Let $abcd$ be a four-digit number, for example. Then $abcd = a10^3 + b10^2 + c10 + d = 10(100a + 10b + c) + d$. Note that $10(100a + 10b + c) \equiv 0$ (mod 10); therefore, $abcd \equiv d$ (mod 10).　**(b)** 5　**(c)** Let $abcd$ be a four-digit number, for example. Then $abcd = a10^3 + b10^2 + c10 + d = 100(10a + b) + (c10 + d)$. Note that $100(10a + b) \equiv 0$ (mod 100); therefore, $abcd \equiv cd$ (mod 100).

Open-Ended

3. On a 12-hr clock, 12 is the additive identity, whereas for integers 0 is the additive identity. We define $^-3$ as the solution of the equation $x + 3 = 12$. Because $9 + 3 = 12$, we have $^-3 = 9$. In general, if a is a whole number other than 12 on the 12-hr clock, then $^-a = 12 - a$ and $^-12 = 12$. The last statement is analogous to $^-0 = 0$ in integers. Properties of additive inverse such as $^-(a + b) = ^-a + ^-b$ hold on the clock.

Questions from the Classroom

5. Dan is right in that if $\frac{1}{4}$ is interpreted as $1 \oplus 4$, then $1 \oplus 4 = 4$ and $4 > 3$ but $<$ and $>$ have no meaning in clock arithmetic.　**7.** On a 5-hr clock, the number 5 plays the role of 0 in integers and is the additive identity. Notice that $1 \oplus 5 = 1$, $2 \oplus 5 = 2$, $3 \oplus 5 = 3$, $4 \oplus 5 = 4$, and $5 \oplus 5 = 5$. Because addition on the 5-hr clock is commutative, each of these additions can be reversed.

Chapter Review

1. (a) $^-3$　**(b)** a　**(c)** -1　**(d)** $^-x - y$　**(e)** $x - y$　**(f)** $x + y$　**(g)** 32　**(h)** 32　**2. (a)** $^-7$　**(b)** 8　**(c)** 8　**(d)** 0　**(e)** 8　**(f)** 15　**3. (a)** 3　**(b)** $^-5$　**(c)** Any integer except 0　**(d)** No integer will work.　**(e)** $^-41$　**(f)** Any integer　**4.** $2(^-3) = ^-6; 1(^-3) = ^-3$; $0(^-3) = 0$; if the pattern continued, then $^-1(^-3) = 3$; $^-2(^-3) = 6$.　**5. (a)** $10 - 5 = 5$　**(b)** $1 - (^-2) = 3$　**6. (a)** ^-x　**(b)** $y - x$　**(c)** $3x - 1$　**(d)** $2x^2$　**(e)** 0　**(f)** $^-x^2 - 6x - 9$　**(g)** $4 - x^2$　**7. (a)** ^-2x　**(b)** $x(x + 1)$　**(c)** $(x - 6)(x + 6)$　**(d)** $(9y^2 + 4x^2)(3y - 2x)(3y + 2x)$　**(e)** $5(1 + x)$　**(f)** $(x - y)x$　**8. (a)** False; it is not positive for $x = 0$.　**(b)** False, if one value is positive and one is negative　**(c)** False, let $a = 2, b = ^-5$　**(d)** true　**9. (a)** $1 \div 2 \neq 2 \div 1$　**(b)** $3 - (4 - 5) \neq (3 - 4) - 5$　**(c)** $1 \div 2$ is not an integer.　**(d)** $8 \div (4 - 2) \neq (8 \div 4) - (8 \div 2)$　**10. (a)** $^-10$　**(b)** $^-2^{99}$　**(c)** 2^{89}　**(d)** 0　**(e)** 3 or $^-3$　**(f)** $x \leq 0$; that is, $0, ^-1, ^-2, ^-3, \ldots$　**(g)** $x \geq 4$ or $x \leq ^-4$; that is, $\{ \ldots ^-6, ^-5, ^-4 \} \cup \{4, 5, 6, 7, \ldots \}$　**(h)** $x = 11$ or $^-9$　**11. (a)** $^-1, 1, ^-1, 1, ^-1, 1$　**(b)** 0, 2, 0, 4, 0, 6　**(c)** $^-2, 4, ^-8, 16, ^-32, 64$　**(d)** $^-5, ^-8, ^-11, ^-14, ^-17, ^-20$　**12. (a)** Geometric, ratio $^-1$　**(c)** Geometric, ratio $^-2$　**(d)** Arithmetic, difference $^-3$　**13. (a)** false　**(b)** false　**(c)** true　**(d)** False; 12, for example　**(e)** False; 9, for example　**14. (a)** False; $7|7$ and $7 \nmid 3$, yet $7 | 3 \cdot 7$.　**(b)** False; $3 \nmid (3 + 4)$, but $3|3$ and $3 \nmid 4$.　**(c)** true　**(d)** true　**(e)** False; $4 \nmid 2$ and $4 \nmid 22$, but $4|44$.　**15. (a)** Divisible by 2, 3, 4, 5, 6, 8, 9, 11　**(b)** Divisible by 3, 11　**16.** If 10,007 is prime, $17 \nmid 10,007$. We know $17|17$, so $17 \nmid (10,007 + 17)$ by Theorem 5–13(b).　**17. (a)** 87<u>2</u>4; 87<u>5</u>4; 87<u>8</u>4　**(b)** 41,856; 44,856; 47,856　**(c)** 87,<u>1</u>74; 87,<u>4</u>64; 87,<u>7</u>54　**18. (a)** The student's claim is true. Examples vary.　**(b)** Let n be an integer. Then $n + (n + 1) + (n + 2) + (n + 3) + (n + 4) = 5n + 10 = 5(n + 2)$. Thus, the sum is divisible by 5.　**19. (a)** composite　**(b)** prime　**20.** Check for divisibility by 3 and 8, $24|4152$.　**21.** No, they are the same if the numbers are equal.　**22.** $\text{LCM}(a, b, c) = \text{LCM}(m, c)$, where $m = \text{LCM}(a, b)$, $\text{LCM}(a, b) = \dfrac{ab}{\text{GCD}(a, b)}$, and $\text{LCM}(m, c) = \dfrac{mc}{\text{GCD}(m, c)}$. Each of these GCDs can be found using the Euclidean algorithm.　**23.** The number is not divisible by 2, 3, 5, 7, 11, and 13 because each of these primes divides one product in the sum $2 \cdot 3 \cdot 5 \cdot 7 + 11 \cdot 13$ but not the other. The student checked that $17 \nmid 353$ and because $19^2 = 361 > 353$, no other primes need to be tested.　**24. (a)** 4　**(b)** 73　**25. (a)** $2^4 \cdot 5^3 \cdot 7^4 \cdot 13 \cdot 29$　**(b)** 77,562　**26.** Answers vary; for example, 16. To obtain five divisors, we raise a prime (2) to the $(5 - 1)$ power.　**27.** 1, 2, 3, 4, 6, 8, 9, 12, 16, 18, 24, 36, 48, 72, 144　**28. (a)** $2^2 \cdot 43$　**(b)** $2^5 \cdot 3^2$　**(c)** $2^2 \cdot 5 \cdot 13$　**(d)** $3 \cdot 37$　**29.** The LCM of all positive integers less than or equal to 10 is $2^3 \cdot 3^2 \cdot 5 \cdot 7$, or 2520.　**30.** \$0.31　**31.** 9:30 A.M.

32. We know that the $\text{GCD}(a, b) \cdot \text{LCM}(a, b) = ab$. Because $\text{GCD}(a, b) = 1$, then $\text{LCM}(a, b) = ab$.
33. 5 packages   **34.** 15 min   **35.** 71 lattes. Because $9869 = 71 \cdot 139$ and 71 as well as 139 are prime, she sold 71 lattes at \$1.39 each.   **36. (a)** $2^{10} \cdot 3^{10}$
(b) $(2 \cdot 17)^n = 2^n \cdot 17^n$ **(c)** 97^4 since 97 is prime
(d) $(2^3)^4 \cdot (2 \cdot 3)^3 \cdot (2 \cdot 13)^2$
$\quad = 2^{12} \cdot 2^3 \cdot 3^3 \cdot 2^2 \cdot 13^2$
$\quad = 2^{17} \cdot 3^3 \cdot 13^2$
(e) $2^3 \cdot 3^2(1 + 2 \cdot 3 \cdot 7) = 2^3 \cdot 3^2 \cdot 43$
(f) $2^4 \cdot 5^6(3 \cdot 5 + 1) = 2^8 \cdot 5^6$ **37.** By the division algorithm every prime number greater than 3 can be written as $12q + r$, where $r = 1, 5, 7,$ or 11, since for other values of $r, 12q + r$ is composite as 12 and r will share a common factor, making it not prime.
38. $n = a \cdot 10^2 + b \cdot 10 + c$
$\quad n = a(99 + 1) + b(9 + 1) + c$
$\quad n = 99a + 9b + c + b + a$
Since $9|99a$ and $9|9b$, then $9|[99a + 9b + (a + b + c)]$ if, and only if, $9|(a + b + c)$.   **39.** We first show that among any three consecutive odd integers, there is always one that is divisible by 3. For that purpose, suppose that the first integer in the triplet is not divisible by 3. Then by the division algorithm, that integer can be written in the form $3n + 1$ or $3n + 2$ for some integer n. Then the three consecutive odd integers are $3n + 1, 3n + 3, 3n + 5$ or $3n + 2, 3n + 4, 3n + 6$. In the first triplet, $3n + 3$ is divisible by 3, and in the second, $3n + 6$ is divisible by 3. This implies that if the first odd integer is greater than 3 and not divisible by 3, then the second or the third must be divisible by 3, and hence cannot be prime.   **40.** Friday   **41.** mod 360. It would cover all the area encircling the lighthouse.

Answers to Now Try This

5-1. (a) yes   **(b)** It can be either depending on the values of the numbers.
(c)

$\begin{array}{c} \xleftarrow{\hspace{0.3cm}-2\hspace{0.3cm}} \xrightarrow{\hspace{1cm}8\hspace{1cm}} \\ \xrightarrow{\hspace{0.8cm}6\hspace{0.8cm}} \\ \overset{\longleftrightarrow}{\underset{-4\ -3\ -2\ -1\ 0\ 1\ 2\ 3\ 4\ 5\ 6}{}} \end{array}$

5-2. (a) Since $x \le 0$, $|x| = {}^-x$ and $|x| + x = {}^-x + x = 0$.
(b) Since $x \le 0$, ${}^-|x| + x = {}^-({}^-x) + x = 2x$.
(c) Since $x \ge 0$, ${}^-|x| + x = {}^-x + x = 0$.
5-3. (a) Answers will vary. For example, a mail carrier brings you three letters, one with a check for \$23 and the other two with bills for \$13 and \$12, respectively. Are you richer or poorer and by how much?   **(b)** Answers will vary. For example, a mail carrier brings you one letter with a check for \$18 and takes away a bill for \$37 that was intended for someone else. Are you richer or poorer and by how much?   **5-4. (a)** Yes, because $a - b = a + {}^-b$ and the sum of two integers is an integer.   **(b)** None of the properties holds for integers because

$$a - b \ne b - a \;(\text{if } a \ne b)$$
$$(a - b) - c \ne a - (b - c) \;(\text{if } c \ne 0)$$

There is no single integer i such that for all integers $a - i = a$ and $i - a = a$ (the first equation implies that $i = 0$ but 0 does not satisfy the second equation).
5-5. (a) $101 \cdot 99 = (100 + 1)(100 - 1) = 100^2 - 1^2 = 10{,}000 - 1 = 9999$ **(b)** $22 \cdot 18 = (20 + 2)(20 - 2) = 20^2 - 2^2 = 400 - 4 = 396$ **(c)** $24 \cdot 36 = (30 - 6)(30 + 6) = 30^2 - 6^2 = 900 - 36 = 864$
(d) $998 \cdot 1002 = (1000 - 2)(1000 + 2) = 1000^2 - 2^2 = 1{,}000{,}000 - 4 = 999{,}996$ **5-6.** $a \div 0 = x$ if, and only if, $0 \cdot x = a$ and x is unique. Because $0 \cdot x = 0$ for all integers x, the equation has no solution if $a \ne 0$. If $a = 0$, then for all integers $x, 0 \cdot x = 0$. Because the solution is not unique, $0 \div 0$ is not defined.
5-7.

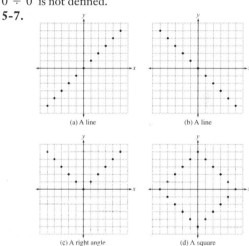

(a) A line   (b) A line

(c) A right angle   (d) A square

5-8. If $5 \nmid a$ and $5 \nmid b$, then $5 \nmid (a + b)$ is not true. For example $5 \nmid 8$ and $5 \nmid 12$ But $5 | (8 + 12)$. On the other hand, $5 \nmid 7$ and $5 \nmid 12$ and $5 \nmid (7 + 12)$.   **5-9.** Yes, it is true. If $3|x$, then 3 divides any number times x and in particular $3|xy$.
5-10. $1 + 2 + 5 + 0 + 6 + 5 = 19$, so we must find numbers x and y such that $9|[19 + (x + y)]$. Any two numbers that sum to 8 or 17 will satisfy this. Therefore, the blanks could be filled with 8 and 9, or 9 and 8, or the pairs $(8, 0)$, $(0, 8), (1, 7), (7, 1), (2, 6), (6, 2), (3, 5), (5, 3), (4, 4)$.
5-11. (a) Answers vary. For example, only square numbers are listed in column 3; 2 is the only even number that will ever be in column 2, and column 2 contains prime numbers. The powers of 2 appear in successive columns.   **(b)** There will never be other entries in column 1 because 1 is the only number with one factor. Other numbers have at least the number itself and 1.   **(c)** $49, 121, 169$ **(d)** 64   **(e)** The square numbers have an odd number of factors. Factors occur in pairs; for example, for 16 we have 1 and 16, 2 and 8, and 4 and 4. When we list the factors, we list only the distinct factors, so 4 is not listed twice, thereby making the number of factors of 16 an odd number. Similar reasoning holds for all square numbers.   **5-12. (a)** When you obtain a whole-number answer, it means the number you started with is divisible by the number you divided by.   **(b)** 2261 is divisible by both 17 and 19 and therefore the choice of color is not determined uniquely.   **5-13. (a)** $1, 2, 3, 6, 9$ **(b)** $1, 2, 3, 4,$

6, 8 **(c)** Only white rods can be used to form one-color trains for prime numbers if two or more rods must be used. **(d)** The number must have at least 8 factors: 1, 2, 3, 5, 6, 10, 15, 30. **5-14. (a)** No, because the multiples of 2 have 2 as a factor **(b)** The multiples of 3: $\{3, 6, 9, 12, 15, \dots\}$ **(c)** The multiples of 5: $\{5, 10, 15, 20, \dots\}$ **(d)** The multiples of 7: $\{7, 14, 21, \dots\}$ **(e)** We have to check only divisibility by 2, 3, 5, and 7. **5-15.** The 1, 2, 3, and 6 rods can all be used to build both the 24 and 30 train. The greatest of these is 6, so $\mathrm{GCD}(24, 30) = 6$. **5-16. (a)** The leftmost area is the factors of 24 that are not factors of 40. The center or intersection area is the factors of both 24 and 40. The rightmost area is the factors of 40 that are not factors of 24. **(b)** 8 **(c)**

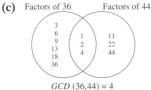

Factors of 36   Factors of 44

$GCD\,(36, 44) = 4$

5-17.

Multiples of 8   Multiples of 12

$LCM\,(8, 12) = 24$

5-18. (a) yes **(b)** Yes, 12 is the identity. **(c)** yes

Answers to Brain Teasers

Section 5-1

$123 - 45 - 67 + 89 = 100$

Section 5-2

Page 280 Answers may vary. $1 = 4^4/4^4; 2 = (4 \cdot 4)/(4 + 4); 3 = 4 - (4/4)^4; 4 = [(4 - 4)/4] + 4;$ $5 = 4 + 4^{(4-4)}; 6 = 4 + [(4 + 4)/4]; 7 = (44/4) - 4;$ $8 = [(4 + 4)/4] \cdot 4; 9 = 4 + 4 + 4/4; 10 = (44 - 4)/4$

Page 285 0 because $(x - x) = 0$

Section 5-3

Page 295 The part of the explanation that is incorrect is the division by $(e - a - d)$, which is equal to 0. Division by 0 is impossible.

Page 300 The number is 381-65-4729.

Section 5-4

Assuming that the ages are whole numbers, we list the decompositions of 2450 into three factors each followed by the sum of the three factors.

1, 1, 2450	2452	1, 2, 1225	1228
1, 5, 490	496	1, 7, 350	358
1, 10, 245	256	1, 14, 175	190
1, 35, 70	106	1, 25, 98	124
1, 49, 50	100	2, 5, 245	252
2, 25, 49	76	2, 7, 175	184
2, 35, 35	72	5, 14, 35	54
5, 10, 49	**64**	5, 5, 98	108
5, 7, 70	82	**7, 7, 50**	**64**
7, 10, 35	52		
7, 14, 25	46		

The only sum of three factors that occurs more than once is indicated in bold. If all sums were different, Natasha would have known which one it is, because she was told that the "right" sum was twice her age. Because she needed more information, we can conclude that her age is 32 and the ages of Jody's friends are 5, 10, and 49 or 7, 7, and 50. That Natasha determined the answer after Jody's reply that she is at least one year younger than the oldest of the three friends eliminates 5, 10, 49 among these ages. If Jody's age had been 48 or less, Natasha would still need more information. Hence, the friend's ages must be 7, 7, and 50 and Jody is 49 years old.

Section 5-5

If n is the width of the rectangle and m is the length of the rectangle, then the number of squares the diagonal crosses is $(n + m) - \mathrm{GCD}(n, m)$ or $(n + m) - 1$.

Section 5-6

There are no primes in this list.

Answer to Laboratory Activity

Section 5-4

Yes, each of the primes on the diagonal can be obtained from the formula. The reason is as follows: Because of the geometrical structure of the spiral, the "distance" from the center square (where 41 is located) to the next square on the diagonal along the spiral is 2 steps. From there to the next one on the diagonal along the spiral, 4 steps, and from there to the next, 6 steps, and so on. In general, from any spot on the diagonal to the next one on the diagonal along the spiral is 2 steps further than it took to reach the previous spot on the diagonal. It can be checked (there is no other known way) that for $0 \le n \le 39$ the formula $n^2 + n + 41$ yields primes. For $n = 0$, we get 41. Whenever $n^2 + n + 41$ is known, the next number resulting from the formula is $(n + 1)^2 + (n + 1) + 41 = (n^2 + n + 41) + 2(n + 1)$. Thus, for $0 \le n \le 38$ whenever $n^2 + n + 41$ yields a prime, the next prime obtained from the formula is $2n + 2$ steps away and hence, as explained earlier, on the diagonal. Notice that the spiral can be continued with only primes on the diagonal up to $n = 39$, that is, until we get $39^2 + 39 + 41$, or 1601. For $n = 40$, we get $1601 + 2(39 + 1)$, or $1681 = 41^2$, which will end up on the diagonal (because it is $2 \cdot 40$ steps away from 1601) but is not a prime.

Answers to Technology Corner

Section 5-1

1. The entries in column A stay 4 while the entries in column B start with 3 and decrease by 1. The sum of columns A and B is entered in column C starting with 7. The entries in column C are the integers in decreasing order starting with 7. The patterns show that the sum of two positive numbers is positive. The sum of a positive and a negative number is positive if the absolute value of the positive number is greater than the absolute value of the negative number. The sum is 0 if both numbers have the same absolute value. The sum is negative otherwise. Similar results can be obtained if column A is changed to $^-4$.
2. (a) The graph should appear as shown below.

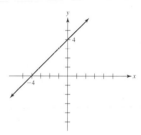

(b) When x is less than $^-4$, the y-values are negative; when $x = {}^-4$, $y = 0$; when x is greater than $^-4$, the y-values are positive.

Section 5-2

The product of two negative numbers is a positive number. The numbers in the C column are perfect squares.

Section 5-5

1. The intersection is the first twelve multiples of 12.
2. You only need to fill down to 47.   **3. (a)** 180
(b) You need to use one of the techniques in the section to find LCM (6, 9, 12, 15).

Answer to Preliminary Problem

The difference-of-squares formula can be used to help answer this problem. We rewrite the given expression as follows to obtain a sum that is familiar from Chapter 1:

$$50^2 - 49^2 + 48^2 - 47^2 + \ldots + 2^2 - 1^2 =$$
$$(50^2 - 49^2) + (48^2 - 47^2) + \ldots + (2^2 - 1^2)$$
$$= (50 - 49)(50 + 49) + (48 - 47)(48 + 47) +$$
$$\ldots + (4 - 3)(4 + 3) + (2 - 1)(2 + 1)$$
$$= 1(50 + 49) + 1(48 + 47) + \ldots$$
$$+ 1(4 + 3) + 1(2 + 1)$$
$$= 50 + 49 + 48 + 47 + 46 + \ldots + 3 + 2 + 1$$

At this point we can use the work done with the Gauss problem in Chapter 1 to find the sum of the integers from 1 to 50. $51(50)/2 = 1275$ and therefore the value of the initial expression is 1275.

Chapter 6

Assessment 6-1A

1. (a) The solution to $8x = 7$ is $\frac{7}{8}$.   **(b)** Jane ate $\frac{7}{8}$ of the pizza.   **(c)** The ratio of boys to girls is 7 to 8.
2. (a) $\frac{1}{6}$ **(b)** $\frac{1}{4}$ **(c)** $\frac{2}{6}$ or $\frac{1}{3}$ **(d)** $\frac{7}{12}$ **3. (a)** $\frac{2}{3}$ **(b)** $\frac{4}{6}$ or $\frac{2}{3}$ **(c)** $\frac{6}{9}$ or $\frac{2}{3}$ **(d)** $\frac{8}{12}$ or $\frac{2}{3}$. The diagram illustrates the Fundamental Law of Fractions.   **4. (a)** no, not equal parts **(b)** yes   **(c)** yes
5. (a)   **(b)**   **(c)**

$\frac{2}{8}$ $\frac{3}{9}$ $\frac{3}{6}$

6. (a) $\frac{9}{24}$ or $\frac{3}{8}$ **(b)** $\frac{12}{24}$ or $\frac{1}{2}$ **(c)** $\frac{4}{24}$ or $\frac{1}{6}$ **(d)** $\frac{8}{24}$ or $\frac{1}{3}$
7. (a) $\frac{4}{18}, \frac{6}{27}, \frac{8}{36}$ **(b)** $\frac{-4}{10}, \frac{2}{-5}, \frac{-10}{25}$ **(c)** $\frac{0}{1}, \frac{0}{2}, \frac{0}{4}$
(d) $\frac{2a}{4}, \frac{3a}{6}, \frac{4a}{8}$ **8. (a)** $\frac{52}{31}$ **(b)** $\frac{3}{5}$ **(c)** $\frac{-5}{7}$
9. (a) undefined   **(b)** undefined   **(c)** 0   **(d)** cannot be simplified   **(e)** cannot be simplified
10. (a) $\frac{a - b}{3}, a \ne {}^-b$ **(b)** $\frac{2x}{9y}, x \ne 0, y \ne 0$
11. (a) equal   **(b)** equal   **12. (a)** not equal   **(b)** not equal
13.   **14.** $\frac{36}{48}$

15.   **16. (a)** $\frac{32}{3}$ **(b)** $^-36$

17. (a) > **(b)** > **(c)** <   **18. (a)** $\dfrac{11}{13}, \dfrac{11}{16}, \dfrac{11}{22}$

(b) $\dfrac{^-1}{5}, \dfrac{^-19}{36}, \dfrac{^-17}{30}$ **19.** The nth term of the sequence is

$\dfrac{n}{n+2}$. Hence, the $(n+1)$st term is $\dfrac{n+1}{n+3}$. We need to

show that $\dfrac{n+1}{n+3} > \dfrac{n}{n+2}$ for $n \geq 1$. By Theorem 6–4 this

is true if, and only if, $(n+1)(n+2) > n(n+3)$, which is

equivalent to $n^2 + 2n + n + 2 > n^2 + 3n$, or $2 > 0$.

Because the last statement is true, the statement

$\dfrac{n+1}{n+3} > \dfrac{n}{n+2}$ is true. Another approach is to use

Theorem 6–6 and notice that each term of the sequence

starting from the second can be obtained by adding the nu-

merators and the denominators of the neighboring terms.

20. Answers may vary. The following are possible

answers: **(a)** $\dfrac{10}{21}, \dfrac{11}{21}$ **(b)** $\dfrac{^-22}{27}, \dfrac{^-23}{27}$ **21.** 456 mi

22. (a) $\dfrac{6}{16}$, or $\dfrac{3}{8}$ of a pound; $\dfrac{6}{32,000}$, or $\dfrac{3}{16,000}$ of a ton

(b) $\dfrac{10}{100}$, or $\dfrac{1}{10}$ **(c)** $\dfrac{15}{60}$ or $\dfrac{1}{4}$ **(d)** $\dfrac{8}{24}$, or $\dfrac{1}{3}$

Mathematical Connections 6-1

Communication

1. Answers vary; for example, $\dfrac{3}{4}$ of a quantity is found by

dividing that quantity by 4 and then taking 3 of the parts.

Hence, $\dfrac{3}{4}$ of 4 is 3. On the other hand, four $\dfrac{3}{4}$ is 12 quarters,

or 3.   **3.** Answers vary; for example, the student is think-

ing of the Fundamental Law of Fractions that says

$\dfrac{a}{b} = \dfrac{a \cdot c}{b \cdot c}$. This only works for multiplication. $\dfrac{2}{3} \neq \dfrac{6}{7}$ because

$2 \cdot 7 \neq 3 \cdot 6$. In general, $\dfrac{a}{b} \neq \dfrac{a+c}{b+c}$ unless $a = b$. An alterna-

tive approach is to apply Theorem 6–6 to the fractions $\dfrac{a}{b}$ and

$\dfrac{c}{c}$.   **5.** The new fraction is equal to $\dfrac{1}{2}$. The principle can be

generalized as follows:

$$\dfrac{a}{b} = \dfrac{ar_1}{br_1} = \dfrac{ar_2}{br_2} = \dfrac{ar_3}{br_3} = \ldots = \dfrac{ar_n}{br_n}$$
$$= \dfrac{a(r_1 + r_2 + r_3 + \ldots + r_n)}{b(r_1 + r_2 + r_3 + \ldots + r_n)} = \dfrac{a}{b}$$

7. (a) Suppose the rational numbers are $\dfrac{2}{16}$ and $\dfrac{1}{4}$. In this

case, $\dfrac{1}{4} > \dfrac{2}{16}$ and Iris is incorrect.   **(b)** Suppose the rational

numbers are $\dfrac{2}{3}$ and $\dfrac{1}{2}$. In this case, $\dfrac{2}{3} > \dfrac{1}{2}$. Shirley is incorrect.

9. Because 36 in. = 1 yd, 1 in. = $\dfrac{1}{36}$ yd , x in. is $\dfrac{1}{36}$ of x yd,

that is $\dfrac{x}{36}$ of yd.

Open-Ended

11. Answers will certainly vary. Some mathematics

educators argue that positive and negative integers are easier

for students to comprehend than are rational numbers. Such

educators also argue that the operations on integers are

easier than are operations on rational numbers.

Cooperative Learning

13. (a) Answers vary; students must work together to

determine the heights and to order the people according

to height. For example, if the heights are 52", 56",

and 58", the fractions could be $\dfrac{52}{60}, \dfrac{56}{60}$, and $\dfrac{58}{60}$.

(b) Answers vary depending on part (a).

Questions from the Classroom

15. The first student's approach is correct. What the second

student has done is to treat the problem as if it had been

$\left(\dfrac{1}{5}\right)\left(\dfrac{5}{3}\right) = \dfrac{1}{3}$, when in reality, the problem is $\dfrac{15}{53}$. One

cancels factors not digits.   **17.** According to the Denseness

Property for Rational Numbers, there is another rational

number between any two given rational numbers. $\dfrac{9999}{10000}$ is

one rational number between $\dfrac{999}{1000}$ and 1.   **19.** The

student was probably thinking that more pieces meant more

pizza. A pizza (or circle) could be cut into 6 pieces, then

each piece could be cut into 2 pieces. This shows the

amount of pizza did not change from these last cuts only the

number of pieces changed.   **21.** The condition implies

that $a \neq b$ and $c \neq d$. Also $b \neq 0$ and $a \neq 0$. If $\dfrac{a}{b} = \dfrac{c}{d}$, if and

only if $ad = bc$. $\dfrac{d}{c-d} = \dfrac{b}{a-b}$ if, and only if, $ad - bd =$

$bc - bd$ or $ad = bc$, which is given. Joe is correct.

23. Daryl is not correct because the whole is not divided

into 3 equal-size pieces.   **25.** Looking at the number

line you will notice that

$$\left\{ x \,\middle|\, -\dfrac{1}{2} < x < \dfrac{1}{2} \right\} \subset \left\{ x \,\middle|\, -\dfrac{3}{2} < x < \dfrac{3}{2} \right\}$$

Thus any x in the first set is also in the second set.

Assessment 6-2A

1. (a) $\dfrac{7}{6}$ or $1\dfrac{1}{6}$ **(b)** $\dfrac{-4}{12}$ or $\dfrac{-1}{3}$ **(c)** $\dfrac{5y - 3x}{xy}$

(d) $\dfrac{-3y + 5x + 14y^2}{2x^2y^2}$ **(e)** $\dfrac{71}{24}$ or $2\dfrac{23}{24}$ **(f)** $\dfrac{-23}{3}$ or $-7\dfrac{2}{3}$

2. (a) $18\dfrac{2}{3}$ **(b)** $-2\dfrac{93}{100}$ **3. (a)** $\dfrac{27}{4}$ **(b)** $\dfrac{-29}{8}$

4. (a) $\dfrac{1}{3}$, high **(b)** $\dfrac{1}{6}$, low **(c)** $\dfrac{3}{4}$, low **(d)** $\dfrac{1}{2}$, high

5. (a) Beavers **(b)** Ducks **(c)** Bears

6.

7. (a) $\dfrac{1}{2}$, high **(b)** 0, low **(c)** $\dfrac{3}{4}$, high **(d)** 1, high

8. (a) 2 **(b)** $\dfrac{3}{4}$ **9. (a)** $\dfrac{1}{4}$ **(b)** 0 **10. (a)** A

(b) H **(c)** T **(d)** H **11.** Approximately, $4 \cdot 3 = 12$.

12. $\dfrac{1}{4}$ **13.** $6\dfrac{7}{12}$ yd **14.** $2\dfrac{5}{6}$ yd **15. (a)** Team 4,

$76\dfrac{11}{16}$ lb **(b)** $3\dfrac{11}{16}$ lb **16. (a)** $\dfrac{1}{2} + \dfrac{3}{4} \in Q$

(b) $\dfrac{1}{2} + \dfrac{3}{4} = \dfrac{3}{4} + \dfrac{1}{2}$ **(c)** $\left(\dfrac{1}{2} + \dfrac{1}{3}\right) + \dfrac{1}{4} = \dfrac{1}{2} + \left(\dfrac{1}{3} + \dfrac{1}{4}\right)$

17. (a) $\dfrac{3}{2}, \dfrac{7}{4}, 2$ **(b)** $\dfrac{6}{7}, \dfrac{7}{8}, \dfrac{8}{9}$, not arithmetic,

$\dfrac{2}{3} - \dfrac{1}{2} \neq \dfrac{3}{4} - \dfrac{2}{3}$ **18. (a)** $\dfrac{n}{4}$ **(b)** $\dfrac{n}{n+1}$

19. $1, \dfrac{7}{6}, \dfrac{8}{6}, \dfrac{9}{6}, \dfrac{10}{6}, \dfrac{11}{6}, 2$ **20. (a) (i)** $\dfrac{3}{4}$ **(ii)** $\dfrac{25}{12}$, or $2\dfrac{1}{12}$

(iii) 0 **(b) (i)** $\dfrac{1}{4}$ **(ii)** $\dfrac{-7}{4}$ **(iii)** $\dfrac{-1}{4}$ **21. (a)** -2

(b) 0 **(c)** $\dfrac{1}{2}$ **(d)** $\dfrac{7}{4}$ **22. (a) (i)** $\dfrac{1}{4} + \dfrac{1}{3 \cdot 4} = \dfrac{1}{4} + \dfrac{1}{12} =$

$\dfrac{16}{48} = \dfrac{1}{3}$ **(ii)** $\dfrac{1}{5} + \dfrac{1}{4 \cdot 5} = \dfrac{1}{5} + \dfrac{1}{20} = \dfrac{25}{100} = \dfrac{1}{4}$

(iii) $\dfrac{1}{6} + \dfrac{1}{5 \cdot 6} = \dfrac{1}{6} + \dfrac{1}{30} = \dfrac{36}{180} = \dfrac{1}{5}$ **(b)** $\dfrac{1}{n} = \dfrac{1}{n+1} +$

$\dfrac{1}{n(n+1)}$ **23. (a)** $\dfrac{5}{6}$ **(b)** $\dfrac{21}{6}$ or $3\dfrac{1}{2}$ **24. (a)** $\dfrac{3x^2 + y^3}{x^2y^2}$

(b) $\dfrac{az - by}{xy^2z}$ **(c)** $\dfrac{2ab - b^2}{a^2 - b^2}$

Mathematical Connections 6-2

Communication

1. Answers vary; for example, $\dfrac{1}{3} + \dfrac{1}{4}$ or $\dfrac{7}{12}$ does not represent the amount received since the fractions did not come from the same size "whole." **3. (a)** No, but it is

easier to reduce if we do choose a least common denominator.

(b) No, for example, $\dfrac{1}{3} + \dfrac{1}{6} = \dfrac{3}{6}$, which is not in simplest form. **5.** Answers vary; for example, the teacher is thinking each rectangle which is divided into 4 equal parts represents a whole divided into fourths, so the picture is $2\dfrac{3}{4}$. Ken is thinking the three rectangles represent the whole and 11 of 12 parts are shaded. **7.** Answers vary; for example,

$3\dfrac{3}{4} + 5\dfrac{1}{3} = \dfrac{15}{4} + \dfrac{16}{3} = \dfrac{45 + 64}{12} = \dfrac{109}{12} = 9\dfrac{1}{12}$ or

$3\dfrac{3}{4} + 5\dfrac{1}{3} = (3 + 5) + \left(\dfrac{3}{4} + \dfrac{1}{3}\right) = 8\dfrac{13}{12} = 9\dfrac{1}{12}$. In the second method, the numbers are smaller to work with.

9. (a) Yes. If a, b, c, and d are integers, $b \neq 0$, $d \neq 0$, then $\dfrac{a}{b} - \dfrac{c}{d} = \dfrac{ad - bc}{bd}$ is a rational number. **(b)** No; for

example, $\dfrac{1}{2} - \dfrac{1}{4} \neq \dfrac{1}{4} - \dfrac{1}{2}$. **(c)** No; for example,

$\dfrac{1}{2} - \left(\dfrac{1}{4} - \dfrac{1}{8}\right) \neq \left(\dfrac{1}{2} - \dfrac{1}{4}\right) - \dfrac{1}{8}$. **(d)** No. If there is

an identity for subtraction, it must be 0, since only for 0 does $\dfrac{a}{b} - 0 = \dfrac{a}{b}$. However, in general, $0 - \dfrac{a}{b} \neq \dfrac{a}{b} - 0$, and hence there is no identity. **(e)** No; since there is no identity, an inverse cannot be defined. **11. (a)** Decreasing because $\dfrac{1}{2} > \dfrac{1}{3} > \dfrac{1}{4} > \dfrac{1}{5} > \dots$, and adding 1 to each term does not change the inequalities. **(b)** Increasing because $\dfrac{1}{2} = 1 - \dfrac{1}{2}, \dfrac{2}{3} = 1 - \dfrac{1}{3}, \dfrac{3}{4} = 1 - \dfrac{1}{4}, \dots, \dfrac{n}{n+1} = 1 - \dfrac{1}{n+1}$. Therefore, in each successive term we subtract a smaller number from 1 and hence we get closer and closer to 1. More formally:

$$\dfrac{1}{n+1} < \dfrac{1}{n}$$
$$\dfrac{-1}{n+1} > \dfrac{-1}{n}$$
$$1 - \dfrac{1}{n+1} > 1 - \dfrac{1}{n}$$

(c) Decreasing because $\dfrac{1}{n+1} < \dfrac{1}{n}$ implies $\dfrac{1}{(n+1)^2} < \dfrac{1}{n^2}$

Open-Ended

13. Answers vary; for example: **(a)** $\dfrac{2}{3}$ and $\dfrac{1}{3}, \dfrac{b-a}{b}$

(b) $\dfrac{2}{7}, \dfrac{4}{7}$, and $\dfrac{1}{7}$ **(c)** $\dfrac{150}{100}, \dfrac{49}{100}$

Cooperative Learning

15. Depending on the people interviewed, students may hear an answer like the following from a teacher. I use fractions in determining total grades for my classes. For example, if a paper is $\frac{1}{2}$ of the grade and a test is another $\frac{1}{3}$ of the grade, I need to know what fractional part of the grade is yet to be determined.

Questions from the Classroom

17. Kendra's picture shows that $\frac{1}{3}$ of the 3-square whole combined with $\frac{3}{4}$ of the 4-square whole gives $\frac{4}{7}$ of a 7-square whole. To do what Kendra is trying to do, she would need to use the same size whole rather than 3 different wholes. When $\frac{1}{3} + \frac{3}{4}$ are added, the same whole must be used.

Review Problems

19. (a) $\frac{2}{3}$ **(b)** $\frac{13}{17}$ **(c)** $\frac{25}{49}$ **(d)** $\frac{a}{1}$, or $a \neq -1$
(e) simplified **(f)** $a + b$ (if $a \neq b$). **21. (a)** February
(b) $\frac{184}{365}$

Assessment 6-3A

1. (a) $\frac{1}{4}, \frac{1}{3}$ and $\frac{1}{12}$ **(b)** $\frac{2}{4} \cdot \frac{3}{5} = \frac{6}{20}$

2. (a) 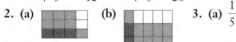 **(b)** **3. (a)** $\frac{1}{5}$

(b) $\frac{b}{a}$ **(c)** $\frac{za}{x^2 y}$ **4. (a)** $10\frac{1}{2}$ **(b)** $8\frac{1}{3}$ **5. (a)** $^-3$
(b) $\frac{3}{10}$ **(c)** $\frac{y}{x}$ **(d)** $\frac{-1}{7}$ **6. (a)** $\frac{21}{12}$ or $\frac{7}{4}$ **(b)** $\frac{6}{4}$ or $\frac{3}{2}$
(c) $\frac{-1}{8}$ **(d)** $\frac{3}{2}$ **7.** Answers vary; for example,
(a) $6 \div 2 \neq 2 \div 6$ **(b)** $(8 \div 4) \div 2 \neq 8 \div (4 \div 2)$
8. (a) 26 **(b)** 29 **(c)** 92 **(d)** 18 **9. (a)** 20
(b) 16 **(c)** 1 **10. (a)** 18 **(b)** 25 **11. (a)** less than 1
(b) less than 1 **(c)** greater than 2 **12. (c)** **13.** 9600
students **14.** $\frac{1}{6}$ **15.** $240 **16.** $225 **17.** 32 marbles
18. (a) $\frac{1}{3^{13}}$ **(b)** 3^{13} **(c)** 5^{11} **(d)** 5^{19} **(e)** $\frac{1}{(^-5)^2}$ or $\frac{1}{5^2}$
(f) a^5 **19. (a)** $\left(\frac{1}{2}\right)^{10}$ **(b)** $\left(\frac{1}{2}\right)^3$ **(c)** $\left(\frac{2}{3}\right)^9$
(d) 1 **20. (a)** False; $2^3 \cdot 2^4 \neq (2 \cdot 2)^{3+4}$. **(b)** False;
$2^3 \cdot 2^2 \neq (2 \cdot 2)^{3 \cdot 2}$ **(c)** False; $2^3 \cdot 2^3 \neq (2 \cdot 2)^{2 \cdot 3}$.
(d) True, since $ab \neq 0$. **(e)** False; $(2 + 3)^2 \neq 2^2 + 3^2$.
(f) False; $(2 + 3)^{-2} \neq \frac{1}{2^2} + \frac{1}{3^2}$. **21. (a)** 5 **(b)** 6 or $^-6$
(c) $^-2$ **(d)** $^-4$ **22. (a)** $x \leq 4$ **(b)** $x \leq 1$ **(c)** $x \geq 2$

(d) $x \geq 1$ **23. (a)** $\left(\frac{1}{2}\right)^3$ **(b)** $\left(\frac{3}{4}\right)^8$ **(c)** $\left(\frac{4}{3}\right)^{10}$
(d) $\left(\frac{4}{5}\right)^{10}$ **24. (a)** 10^{10} **(b)** $10^{10} \cdot \left(\frac{6}{5}\right)^2 =$
$1.44 \cdot 10^{10} = 14.4$ billion

25. (a) and **(b)** $2S = 1 + \frac{1}{2} + \frac{1}{2^2} + \cdots + \frac{1}{2^{63}}$

$2S - S = 1 + \left(\frac{1}{2} + \frac{1}{2^2} + \cdots + \frac{1}{2^{63}}\right)$

$\qquad - \left(\frac{1}{2} + \frac{1}{2^2} + \cdots + \frac{1}{2^{63}}\right) - \frac{1}{2^{64}}$

$\qquad = 1 - \frac{1}{2^{64}}$ Note that $2S = 1 + S - \frac{1}{2^{64}}$.

Hence, $2S - S = 1 + S - \frac{1}{2^{64}} - S = 1 - \frac{1}{2^{64}}$.

(c) $1 - \frac{1}{2^n}$ **26. (a)** $\frac{2}{5}, \frac{6}{5}, \frac{2}{15}$ **(b)** 20
27. (a) $\frac{3}{2}, \frac{3}{4}, \frac{3}{8}, \frac{3}{16}, \frac{3}{32}$ **(b)** The common ratio is $\frac{1}{2}$.
(c) $n \geq 10$ **28. (a)** 32^{50}, since $32^{50} = (2^5)^{50} = 2^{250}$ and
$4^{100} = (2^2)^{100} = 2^{200}$ **(b)** $(^-3)^{-75}$, since
$(^-27)^{-15} = (^-3)^{-45} = \frac{^-1}{3^{45}} < \frac{^-1}{3^{75}}$ **29.** Assume
$x < y$, i.e., $y - x > 0$. Then $\frac{x + y}{2} - x = \frac{x + y - 2x}{2} =$
$y - x > 0$. Hence $\frac{x + y}{2} < y$. Similarly $\frac{x + y}{2} < y$.

Mathematical Connections 6-3

Communication

1. Answers vary; for example, $\frac{1}{2}$ of a number x is equivalent to $\frac{1}{2} \cdot x$ or $\frac{x}{2}$, whereas dividing a number x by $\frac{1}{2}$ is equivalent to $\frac{x}{\frac{1}{2}} = \frac{x}{1} \cdot \frac{2}{1} = \frac{2x}{1} = 2x$. Therefore, they are not the same.

3. Never less than n. $0 < \frac{a}{b} < 1$ implies $0 < 1 < \frac{b}{a}$. The last inequality implies $0 < n < n \cdot \left(\frac{b}{a}\right)$. Therefore $n \div \left(\frac{a}{b}\right) = n \cdot \frac{b}{a} > n$. **5.** The second number must be the reciprocal of the first and must be a positive number less than 1. **7.** Answers vary; for example, Carl is not correct since 0 is a rational number and the inverse of $\frac{0}{1}$ does not exist.

9. Joel is wrong. $2\frac{3}{5} \cdot 3\frac{4}{5} = \left(2 + \frac{3}{5}\right)\left(3 + \frac{4}{5}\right) \neq$
$\left(2 + \frac{4}{5}\right)\left(3 + \frac{3}{5}\right)$ **11. (a)** Ben ran at 6 mph for 3 hr,

that is, $6 \cdot 3$, or 18, mi—the distance between the towns. Noah ran $\frac{3}{4}$ of 18 mi, that is, $\frac{3 \cdot 9}{2} = 13\frac{1}{2}$ mi. His speed was $\frac{\text{distance}}{\text{time}} = \frac{13\frac{1}{2}}{3} = \frac{27}{2 \cdot 3} = 4\frac{1}{2}$ mph.   **(b)** Ben's new speed is 5 mph and $\frac{1}{4}$ of the distance is $\frac{1}{4} \cdot 18 = \frac{9}{2}$, or $4\frac{1}{2}$ miles.

The corresponding time is $\frac{\text{distance}}{\text{speed}} = \frac{\frac{9}{2}}{5} = \frac{9}{10}$ of an hour, or $\frac{9}{10} \cdot 60$ min., or 54 min.

Open-Ended

13. (a) Answers vary; for example, if you have a board $1\frac{3}{4}$ yd long, how many $\frac{1}{2}$ yd lengths can you divide it into?   **(b)** Answers vary.   **(c)** Answers vary.

Cooperative Learning

15. The answers depend on the size of the bricks and the size of the joints. In all likelihood, the measurements will be made in fractions of inches for the size of the joints. The size of the bricks may be done in inches. An alternative is to measure in centimeters. All measurement is approximate and some rounding or estimation may occur.

Questions from the Classroom

17. The student is generalizing the distributive property of multiplication over addition to the distributive property of multiplication over multiplication. The latter does not hold.
19. The student is wrong unless $n = 0$ or $p = m$. The Fundamental Law of Fractions holds only for multiplication.
21. Jillian is right, 3 R2 is not a number, so strictly speaking it can't equal $\frac{17}{5}$. Notice that from the division algorithm $17 = 5 \cdot 3 + 2$, so when dealing with integers the result of the division is given as 3R2. However when we move to rational numbers we can proceed as follows:

$$17 = \frac{5 \cdot 3 + 2}{5}$$
$$= \frac{5 \cdot 3}{5} + \frac{2}{5}$$
$$= 3 + \frac{2}{5}$$
$$= 3\frac{2}{5}.$$

23. Fran probably does not understand the notation for mixed numbers and she thought of the problem as $2\frac{1}{2} \cdot \frac{3}{5}$ is $2 + \frac{1}{2} \cdot \frac{3}{5}$ or $2 + \frac{3}{10}$ or $2\frac{3}{10}$. If the distributive property is used correctly, we have $\left(2 + \frac{1}{2}\right) \cdot \frac{3}{5} = 2 \cdot \frac{3}{5} +$

$\frac{1}{2} \cdot \frac{3}{5} = \frac{6}{5} + \frac{3}{10} = \frac{12}{10} + \frac{3}{10} = \frac{15}{10}$, or $\frac{3}{2}$. We could also convert each number to an improper fraction and multiply as shown: $\frac{5}{2} \cdot \frac{3}{5} = \frac{3}{2}$.

Review Problems

25. 120

Chapter Review

1. (a)   **(b)**

(c)

2. Answers may vary: for example $\frac{10}{12}, \frac{15}{18}, \frac{20}{24}$.   **3. (a)** $\frac{6}{7}$
(b) $\frac{ax}{b}$ **(c)** 0   **(d)** $\frac{5}{9}$ **(e)** b **(f)** $\frac{2}{27}$ **(g)** Cannot be further reduced.   **(h)** cannot be further reduced
4. (a) $=$ **(b)** $>$ **(c)** $>$ **(d)** $<$ **5. (a)** $^-3, \frac{1}{3}$
(b) $^-3\frac{1}{7}, \frac{7}{22}$ **(c)** $\frac{^-5}{6}, \frac{6}{5}$ **(b)** $\frac{3}{4}, \frac{^-4}{3}$ **6.** $^-2\frac{1}{3}, ^-1\frac{7}{8}, 0,$
$\left(\frac{71}{140}\right)^{300}, \frac{69}{140}, \frac{1}{2}, \frac{71}{140}, \left(\frac{74}{73}\right)^{300}$ **7.** Yes. By the definition of multiplication and the commutative and associative laws of multiplication, we can do the following:

$$\frac{4}{5} \cdot \frac{7}{8} \cdot \frac{5}{14} = \frac{4 \cdot 7 \cdot 5}{5 \cdot 8 \cdot 14}$$
$$= \frac{4 \cdot 7 \cdot 5}{8 \cdot 14 \cdot 5}$$
$$= \frac{4}{8} \cdot \frac{7}{14} \cdot \frac{5}{5}$$

8. (a) 24; because $\frac{1}{3}(8 \cdot 9)$ is equal to $\left(\frac{1}{3} \cdot 9\right) \cdot 8 = 3 \cdot 8$
$= 24$.   **(b)** 66; because $36 \cdot 1\frac{5}{6}$ is equal to $36 \cdot \frac{11}{6} = 6 \cdot 11$
$= 66$.   **9.** 17 pieces, $\frac{11}{6}$ yd remain.   **10. (a)** 15   **(b)** 15
(c) 4   **11.** Answers vary; see Section 6.3.   **12.** Answers vary.   **13.** $\frac{76}{100}, \frac{78}{100}$, but answers may vary.
14. $\boxed{5}\,\boxed{0}\,\boxed{4}\,\boxed{7}\,\boxed{9}\,\boxed{2}\,\boxed{\times}\,\boxed{2}\,\boxed{3}\,\boxed{=}$
15. $333\frac{1}{3}$ calories   **16.** 752 times   **17.** $\frac{240}{1000} = \frac{6}{25}$
18. It is not reasonable to say that the university won $\frac{3}{4} + \frac{5}{8}$, or $\frac{11}{8}$, of its basketball games. The correct fraction

cannot be determined without additional information but it is between $\frac{5}{8}$ and $\frac{3}{4}$.   **19.** The numerators of the rational numbers are integers and follow the properties of integers; the same is true of the denominators. Thus, both the numerator and denominator of the answer are integers, and we can apply another property of integers to determine the sign of the answer.   **20.** You should show him that the given fraction could be written as an integer over an integer. In this case, the result is $\frac{8}{9}$.   **21.** 112 bags   **22.** $\frac{4}{15}$

23. $\frac{^-12}{10}$ is greater than $\frac{^-11}{9}$ because $\frac{^-12}{10} - \left(\frac{^-11}{9}\right)$ is a positive number.   **24. (a)** 3   **(b)** 3   **25. (a)** $\frac{5}{4}$, or $1\frac{1}{4}$

(b) $\frac{19}{6}$, or $3\frac{1}{6}$ **(c)** $^-100$ **(d)** $\frac{13}{3}$, or $4\frac{1}{3}$ **26. (a)** $\frac{a^3}{x^7}$

(b) $\frac{y^8}{x^{10}}$ **27. (a)** $\frac{3ax+b}{x^2y^2}$ **(b)** $\frac{15-2y^2}{3xy^2}$

(c) $\frac{a-bx^2y}{x^3y^2z}$ **(d)** $\frac{31}{2^33^3}$, or $\frac{31}{216}$ **28.** Answers vary; for example, the problem is to find how many $\frac{1}{2}$ yd pieces of ribbon there are in $1\frac{3}{4}$ yd. There are 3 pieces of length $\frac{1}{2}$ yd with $\frac{1}{4}$ yd left over. This $\frac{1}{4}$ yd is $\frac{1}{2}$ of a $\frac{1}{2}$ yd piece. Therefore, there are 3 pieces of $\frac{1}{2}$ yd ribbon and 1 piece that is $\frac{1}{2}$ of the $\frac{1}{2}$ yd piece or $3\frac{1}{2}$ of the $\frac{1}{2}$ yd pieces. This "3 pieces and $\frac{1}{4}$ yard left" and "$3\frac{1}{2}$ pieces" are correct answers.

29. $\dfrac{2ab}{ab+bc+ac}$

Answers to Now Try This

6-1. (a) One quarter of 100¢ is 25¢, one quarter of 60 min. is 15 min.   **(b)** Fractions are defined in relation to a whole. In this case, there are 2 wholes. In Figure 6-2(a), $\frac{1}{3}$ of a large circular whole is shaded. In Figure 6-2(b), $\frac{1}{2}$ of a smaller square whole is shaded. To show $\frac{1}{3} > \frac{1}{2}$, the fractions would have to be associated with the same whole, for example, by comparing $\frac{1}{3}$ of the circle with $\frac{1}{2}$ of the circle. It is true that the area shaded in the circle is larger than the area shaded in the square, but this does not show $\frac{1}{3} > \frac{1}{2}$.

6-2. (a)

▭	$\frac{3}{4}$
▭	$\frac{1}{4}$
▭	$\frac{1}{2}$
▭	$\frac{3}{2}$
▭	$\frac{2}{3}$
▭	$\frac{4}{3}$

(b) number line from $-2\frac{7}{4}$ to 2 marked at $-2\frac{7}{4}$, -1, $-\frac{1}{2}$, 0, $\frac{1}{2}$, 1, $\frac{3}{2}$, 2

6-3. (1) $\frac{1}{3}$ **(2)** $\frac{0}{2}, \frac{1}{2}, \frac{2}{2}$ (or $0, \frac{1}{2}, 1$)   **(3)** $\frac{1}{4}$ **6-4.** A Venn diagram depicting the relationship among natural numbers, whole numbers, integers, and rational numbers follows:

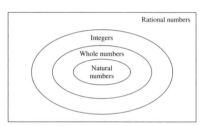

6-5. If n were zero, then bn would be zero and division by zero is undefined.   **6-6.** Theorem 6–3 states that if a, b, and c are integers and $b > 0$, then $\frac{a}{b} > \frac{c}{b}$ if, and only if, $a > c$. The question is to investigate if the theorem is true if $b < 0$. Consider $2 > 1$ and $^-1 < 0$. Now $2/(^-1) < 1/(^-1)$, which contradicts an expanded theorem when $b < 0$.

6-7. $\frac{9}{16}, \frac{5}{8}, \frac{2}{3}$, and $\frac{3}{4}$ **6-8.** If $a > 0$ and $c > b > 0$, then $\frac{a}{b} > \frac{a}{c}$.   **6-9.** Consider two rational numbers $\frac{a}{b}$ and $\frac{c}{d}$, where $\frac{a}{b} < \frac{c}{d}$. By the denseness property of rational numbers (Theorem 6–5) we can find a rational number x, between the two fractions. Since $\frac{a}{b} < x_1$, there is a rational number x_2 between $\frac{a}{b}$ and x_1. We next can find a rational number x_3 between $\frac{a}{b}$ and x_2 and so on. This process can be repeated indefinitely and hence we obtain infinitely many rational numbers $x_1, x_2, x_3, \ldots$ between $\frac{a}{b}$ and $\frac{c}{d}$.

6-10. Because $\frac{a}{b} < \frac{c}{d}$ with $b > 0$ and $d > 0$, Theorem 6–4 implies that $ad < bc$. Now $\frac{a}{b} < \frac{a+c}{b+d}$ if, and only if, $a(b+d) < b(a+c)$, which is equivalent to $ab + ad < ba + bc$ or $ad < bc$. Thus $\frac{a}{b} < \frac{a+c}{b+d}$. Similarly we can show that $\dfrac{a+c}{b+d} < \dfrac{c}{d}$.   **6-11.** $2\dfrac{3}{4} + 5\dfrac{3}{8} = 2 + \dfrac{3}{4} + 5 + \dfrac{3}{8} =$

$7 + \dfrac{6}{8} + \dfrac{3}{8} = 7 + \dfrac{9}{8} = 7 + 1 + \dfrac{1}{8} = 8\dfrac{1}{8}$.   **6-12.** $\dfrac{3}{4}$ is

greater than $\dfrac{1}{2}$; $\dfrac{1}{2} + \dfrac{1}{2} = 1$, so $\dfrac{3}{4} + \dfrac{1}{2} > 1$. $\dfrac{4}{6}$ is less than 1,

so it cannot be the correct answer for $\dfrac{3}{4} + \dfrac{1}{2}$.

6-13. **(a)** Folding, see student page.   **(b)** $\dfrac{1}{4}$ **(c)** $\dfrac{1}{8}$

(d) Shading, see student page.   **(e)** $\dfrac{3}{8}$ **(f)** $\dfrac{1}{4}, \dfrac{1}{8}, \dfrac{1}{16}, \dfrac{9}{16}, \dfrac{5}{16}$

(g) $\dfrac{3}{12}, \dfrac{4}{15}, \dfrac{15}{24}$ **(h)** The product of the fractions is the

product of the numerators over the product of the
denominators.   **6-14.** Answers vary; for example, consider
the following. If Caleb has $10.00, how many chocolate bars
can he buy if **(a)** the price of one bar is $2.00?   **(b)** the

price of one bar is $\$\dfrac{1}{2}$? For (a) the answer is $10 \div 2$ or 5.

For (b) the answer is $10 \div \dfrac{1}{2}$, which is the same as finding

the number of $\dfrac{1}{2}$s in 10. Since there are two halves in 1, in

10 there are 20. Hence Caleb can buy 20 bars.

6-15. $\dfrac{a}{c} \div \dfrac{c}{d} = \dfrac{\frac{a}{b}}{\frac{c}{d}} = \dfrac{\frac{a}{b} \cdot \frac{d}{c}}{\frac{c}{d} \cdot \frac{d}{c}} = \dfrac{\frac{a}{b} \cdot \frac{d}{c}}{1} = \dfrac{a}{b} \cdot \dfrac{d}{c}$

6-16. $\dfrac{a \div c}{b \div d} = \dfrac{\frac{a}{c}}{\frac{b}{d}} = \dfrac{\frac{a}{c} \cdot \frac{b}{d}}{\frac{b}{d} \cdot \frac{b}{d}} = \dfrac{\frac{a}{c} \cdot \frac{b}{d}}{1} = \dfrac{ad}{bc} = \dfrac{a}{b} \div \dfrac{c}{d}$

Answers to Brain Teasers

Section 6-1

At 6:00, the hands on the clock form a straight line, but the
second hand is on 12. After that, the hands form a straight
line approximately every 1 hr, 5 min, and 27 seconds. So
after 6:00 the minute and hour hands form a straight line at:
7:05:27, 8:10:55, 9:16:22, 10:21:49, 11:27:16, 12:32:44,
1:38:11, 2:43:38, 3:49:05, 4:54:33.

Section 6-3

Page 398   Observe that after crossing each bridge, the
prince was left with half the bags he had previously minus
one additional bag of gold. To determine the number he had
prior to crossing the bridge, we can use the inverse
operations; that is, add 1 and multiply by 2. The prince had
one bag left after crossing the fourth bridge. He must have
had two before he gave the guard the extra bag. Finally, he

must have had four bags before he gave the guard at the
fourth bridge any bags. The entire procedure is summarized
in the following table.

Bridge	Bags After Crossing	Bags Before Guard Given Extra	Bags Prior to Crossing
Fourth	1	2	4
Third	4	5	10
Second	10	11	22
First	22	23	46

Page 405   No. The will is impossible because the fraction
of cats to be shared do not add up to the whole units of cats.
$\dfrac{1}{2}x + \dfrac{1}{3}x + \dfrac{1}{9}x = \dfrac{17}{18}x$, but the sum should be $1x$, or $\dfrac{18}{18}x$.

Answers to Laboratory Activity

Section 6-1

(a) Area of $a = \dfrac{1}{4}$ **(b)** Area of $a = 1$

Area of $b = \dfrac{1}{4}$ Area of $b = 1$

Area of $c = \dfrac{1}{16}$ Area of $c = \dfrac{1}{4}$

Area of $d = \dfrac{1}{8}$ Area of $d = \dfrac{1}{2}$

Area of $e = \dfrac{1}{16}$ Area of $e = \dfrac{1}{4}$

Area of $f = \dfrac{1}{8}$ Area of $f = \dfrac{1}{2}$

Area of $g = \dfrac{1}{8}$ Area of $g = \dfrac{1}{2}$

Solution to Preliminary Problem

If x is the amount spent on the horse's keep, then on the
one hand Brandy's loss is $(270 + x) - 540$, or $x - 270$,
and on the other hand it is $\dfrac{1}{2} \cdot 270 + \dfrac{1}{4}x$, or $135 + \dfrac{1}{4}x$.
Thus,

$$135 + \dfrac{1}{4}x = x - 270$$

$$405 = \dfrac{3}{4}x$$

$$x = \dfrac{4}{3} \cdot 405 = 540$$

Brandy's loss is $540 - 270$, or $270.

Chapter 7

Assessment 7-1A

1. (a) $0 \cdot 10^0 + 0 \cdot 10^{-1} + 2 \cdot 10^{-2} + 3 \cdot 10^{-3}$
(b) $2 \cdot 10^2 + 0 \cdot 10^1 + 6 \cdot 10^0 + 0 \cdot 10^{-1} + 6 \cdot 10^{-2}$
(c) $3 \cdot 10^2 + 1 \cdot 10^1 + 2 \cdot 10^0 + 0 \cdot 10^{-1} + 1 \cdot 10^{-2} + 0 \cdot 10^{-3} + 3 \cdot 10^{-4}$ (d) $0 \cdot 10^0 + 0 \cdot 10^{-1} + 0 \cdot 10^{-2} + 0 \cdot 10^{-3} + 1 \cdot 10^{-4} + 3 \cdot 10^{-5} + 2 \cdot 10^{-6}$
2. (a) 4356.78 (b) 4000.608 (c) 40,000.03
(d) 0.2004007 **3.** (a) 536.0076 (b) 3.008
(c) 0.000436 (d) 5,000,000.2 **4.** (a) Thirty-four hundredths (b) Twenty and thirty-four hundredths (c) Two and thirty-four thousandths (d) Thirty-four millionths
5. (a) $\dfrac{436}{1000} = \dfrac{109}{250}$ (b) $\dfrac{2516}{100} = \dfrac{629}{25}$ (c) $\dfrac{^-316,027}{1000}$
(d) $\dfrac{281,902}{10,000} = \dfrac{140,951}{5000}$ (e) $\dfrac{^-43}{10}$ (f) $\dfrac{^-6201}{100}$
6. (a) yes (b) yes (c) yes (d) yes (e) yes (f) yes
(g) no (h) yes (i) no (j) yes **7.** (a) 0.8 (b) 3.05
(c) 0.5 (d) 0.03125 (e) 0.01152 (f) 0.2128 (h) 0.08
(j) 0.4 **8.** Nonterminating; the fraction $\dfrac{7}{60}$ does not terminate when written as a decimal. The denominator has a factor of 3 when in simplest form. **9.** 3.56 **10.** Answers may vary. Some of the numbers that are composed of whole-number powers of 2 and 5 are 1, 5, 10, 25, and 50. These all divide 100 and can be written as a two-digit decimal between 0 and 1, but there are others. **11.** (a) 13.492, 13.49199, 13.4919, 13.49183 (b) $^-1.4053, ^-1.45, ^-1.453, ^-1.493$
12. (a) 0.014 in. (b) 365.24 days **13.** (a)

There are 32 of 100 squares shaded, representing $\dfrac{32}{100}$ of the whole grid, or 0.32 of the grid. **14.** 0.84 **15.** for example, 8.3401 **16.** (a) One method for doing this is to convert both to the form $\dfrac{a}{b}$ where $a, b \in I$, and $b \neq 0$.

Find a rational number between them whose denominator in simplest form is of the form $2^m \cdot 5^n$. Convert this fraction to a terminating decimal. (b) The process described in part (a) can be used to find terminating decimals greater than the least number and less than the original greatest number.
17. (a) If the block is used as 1 unit, then 0.613 can be represented by 6 flats, 1 long, and 3 cubes. (b) It cannot because the set can only be used to represent four non-zero digits. **18.** Three-twenty-two could be interpreted as three hundred twenty-two, but the batting average is 0.322. A batting average cannot be greater than 1.000.
19. A meaning could be as follows: $3 \cdot 6^0 + 1 \cdot 6^{-1} + 4 \cdot 6^{-2} + 5 \cdot 6^{-3}$ in base 10. In base six, 10_{six} would be used instead of 6. **20.** Rhonda, Martha, Kathy, Molly, Emily

Mathematical Connections 7-1

Communication

1. (a) $3^① 2^② 5^③ 6^④$ (b) $0^① 0^② 3^③ 2^④$ **3.** One day cannot be expressed as a terminating decimal because $\dfrac{1}{365}$ cannot be written as a terminating decimal. **5.** *Tithe* means a one-tenth, or 0.1. **7.** Answers vary; for example, a fraction can be written as a terminating decimal if it can be written as a fraction with a denominator that is a power of 10. The denominator can be written as a power of 10 if it contains only factors of 2 and 5. Other factors may appear in the denominator if the fraction is not in simplest form. For example, in $\dfrac{28}{35}$ the denominator of 35 has a factor of 7, but in its simplest form, $\dfrac{4}{5}$, there is no factor of 7.

Open-Ended

9. Answers vary, but many countries will be similar to ours.

Cooperative Learning

11. Answers vary depending on the objects chosen. For example, if a 5×5 flat is used to represent 1, then a 1×5 long would represent 0.1, and a 1×1 unit would represent 0.01.

Questions from the Classroom

13. The student is mistaken. $0.36 = \dfrac{36}{100}; 0.9 = \dfrac{90}{100}$ and $\dfrac{90}{100} > \dfrac{36}{100}$, so $0.9 > 0.36$.

Assessment 7-2A

1. $231.24 **2.** (a)

8.2	1.9	6.4
3.7	5.5	7.3
4.6	9.1	2.8

(b) (i) yes (ii) 19.05
(c) yes; 8.25 **3.** 73.005 **4.** $27.746192, which would amount to $27.75 in Canadian dollars **5.** (a) about $5.80 to heat the house for 1 day (b) 199 hr (rounded)
6. approximately 8.64 liters **7.** 21.3 mph (rounded to nearest tenth mph) **8.** (a) 5.4, 6.3, 7.2 (b) 1.3, 1.5, 1.7
9. 1.12464 **10.** 0.2222 could be written as the sum of a geometric sequence as follows:
$$\dfrac{2}{10} + \left(\dfrac{2}{10}\right)\left(\dfrac{1}{10}\right) + \left(\dfrac{2}{10}\right)\left(\dfrac{1}{10}\right)^2 + \left(\dfrac{2}{10}\right)\left(\dfrac{1}{10}\right)^3 \text{ or as}$$
$0.2 + (0.2)(0.1) + (0.2)(0.1)^2 + (0.2)(0.1)^3$.
11. 0.06, 0.018, 0.0054, 0.00162
12.

![number line with (b) near 0 and (a) near 1, marks at 0, 0.5, 1]

13. No, the bank is over $7.74. **14.** (a) 0.0000000032
(b) 3,200,000,000 (c) 0.42 (d) 620,000
15. (a) $1.27 \cdot 10^7$ m (b) $4.486 \cdot 10^9$ km (c) $5 \cdot 10^7$ cans
16. (a) 0.000000000753 g (b) 298,000 km/ sec

(c) 778,570,000 km   **17. (a)** $4.8 \cdot 10^{28}$ **(b)** $4 \cdot 10^7$
(c) $2 \cdot 10^2$ **18. (a)** 200   **(b)** 200   **(c)** 204   **(d)** 203.7
(e) 203.65   **19.** 19 mpg   **20.** $55 + 5 + 18$, 78 dollars
21. Estimate may vary. Exact answers are the following:
(a) 122.06   **(b)** 57.31   **(c)** 25.40   **(d)** 136.15
22. Answers may vary. 2.0×10^2 and 5.0×10^{-3}
23. $2.3 \cdot 1 = 2.3; 8.7 \cdot 9 = 78.3$ **24.** 49,736.5281
25. Answers vary; for example, $40 \cdot \$8 = \$320$;
$40\left(\dfrac{1}{4}\right) = \$10$; so her salary is $330.   **26. (a)** 12   **(b)** 0.6
(c) $2b$ **(d)** $2b$ **27. (a), (b),** and **(d)** have equal
quotients.   **28. (a)** 181.56   **(b)** 148.551

Mathematical Connections 7-2

Communication

1. Suppose a deposit is recorded with $10.00 over the true amount and a check is recorded with $10.00 more than the true amount. The ultimate balance is correct.   **3.** Answers vary; for example, many of the estimation techniques that work for whole-number division also work for decimal division. The long division algorithm is more efficient when good estimates are used. Also, estimates are important to determine whether an answer obtained by long division is reasonable. Estimation techniques can also be used to place the decimal point in the quotient when decimals are divided.
5. Answers vary; for example, lining up the decimal points acts like using place value.

Open-Ended

7. Answers vary; for example, the calculator could be used to explore the result of placing the decimal point when multiplying by a power of 10 or of placing the decimal point when multiplying two decimals.   **9.** Answers may vary; it tends to support.

Cooperative Learning

11. Follow the flowchart and play the game.

Questions from the Classsroom

13. It is evident what happens when 0.5 is raised to large powers. Because $\left(\dfrac{1}{2}\right)^{10} = \dfrac{1}{1024}$ and $\dfrac{1}{2^{20}} =$
$\left[\left(\dfrac{1}{2}\right)^{10}\right]^2 = \dfrac{1}{1,048,576}$, therefore $\dfrac{1}{2}$ raised to a positive integer gets quickly close to 0. In fact, any number between 0 and 1 when raised to a sufficiently large exponent will get as close to 0 as we wish. At the level of this course, it may be sufficient to use a calculator to see what happens when 0.999 is raised to large powers. Using $\boxed{x^2}$ the key repeatedly, we get: $0.998001, 0.9960058, 0.9920274, 0.9841185, \ldots$, which are approximate values of $0.999^2, 0.999^4, 0.999^8$, $0.999^{16}, \ldots, 0.999^{1024}$. We see that the 10th term in the sequence is less than 0.5 and hence further squaring should quickly result in numbers closer and closer to 0.

Review Problems

15. $14.0479 = 1 \cdot 10^1 + 4 \cdot 10^0 + 0 \cdot 10^{-1} + 4 \cdot 10^{-2} +$
$7 \cdot 10^{-3} + 9 \cdot 10^{-4}$ **17.** yes; for example, $\dfrac{13}{26} = \dfrac{1}{2} = 0.5$.

Assessment 7-3A

1. (a) $0.\overline{4}$ **(b)** $0.\overline{285714}$ **(c)** $0.\overline{27}$ **(d)** $0.0\overline{6}$ **(e)** $0.02\overline{6}$
(f) $0.0\overline{1}$ **(g)** $0.8\overline{3}$ **(h)** $0.0\overline{76923}$ **(i)** $0.04\overline{7619}$
(j) $0.\overline{157894736842105263}$ **2. (a)** $\dfrac{4}{9}$ **(b)** $\dfrac{61}{99}$ **(c)** $\dfrac{461}{330}$
(d) $\dfrac{5}{9}$ **(e)** $\dfrac{-211}{90}$ **(f)** $\dfrac{-2}{90}$ **3.** $0.01\overline{6}$ hr
4. $^-1.454 > {}^-1.45\overline{4} > {}^-1.\overline{454} > {}^-1.4\overline{54} = {}^-1.\overline{45}$
5. Answers may vary. $\dfrac{6}{7}, \dfrac{7}{8}, \dfrac{8}{9}$, or $0.\overline{857142}, 0.875, 0.\overline{8}$
6. 0.01   **7.** The repeating decimal part is determined by the 7 in the denominator of $\dfrac{22}{7}$. The denominator, 7, is not a factor of 10.   **8. (a)** $0.\overline{446355}$; six digits   **(b)** $1.3\overline{5775}$ The answer is a rational number and there are four digits in the repetend.   **9.** Yes. Zeros can be repeated or 9s could be repeated.   **10. (a)** Answers vary; for example, 3.221, 3.2211, 3.22111.   **(b)** Answers vary; for example, 462.2425, 462.2426, 462.2427.   **11.** $0.47\overline{2}$ **12. (a)** Answers vary; for example, 0.751, 0.752, 0.753.   **(b)** Answers vary; for example, 0.334, 0.335, 0.336.   **13. (a)** 8   **(b)** 7
14. (a) (i) $\dfrac{1}{9}$ **(ii)** $\dfrac{1}{99}$ **(iii)** $\dfrac{1}{999}$ **(b)** $\dfrac{1}{9999}$ **(c)** $0.0\overline{1}$
15. (a) $\dfrac{2}{9}$ **(b)** $\dfrac{3}{9}$ or $\dfrac{1}{3}$ **(c)** $\dfrac{5}{9}$ **(d)** $\dfrac{25}{9}$ **(e)** 10
16. (a) $\dfrac{5}{99}$ **(b)** $\dfrac{3}{999}$ or $\dfrac{1}{333}$ **(c)** $\dfrac{322}{99}$ **(d)** $\dfrac{3122}{999}$
17. 0.775   **18. (a)** $\dfrac{3}{10}$ **(b)** $\dfrac{2009}{990}$
19. $1.\overline{0}$; the zero repeats in this representation but adds no value.   **20. (a)** No; for example, $0.\overline{3} + 0.\overline{6} = 1$. The answer could be yes if one considered $1.\overline{0} = 1$ as a repeating decimal.   **(b)** Consider the place values when the sum is found. The repeating blocks in place values to the right of the place value of the rightmost digit of the terminating decimal will appear in the sum.   **(c)** See part **(a)** if 0 is considered a repeating block.   **(d)** Consider the example $0.\overline{20} + 0.0\overline{3}$. The sum is $0.\overline{23}$, which does not terminate. If $0.\overline{23}$ is converted to a reduced fraction, the denominator contains factors other than 2 and 5.

Mathematical Connections 7-3

Communication

1. (a) Mathematically the cost is $66\dfrac{2}{3}¢$ but realistically, the cost is 67¢.   **(b)** See part (a).   **(c)** The cost is rounded up.   **(d)** Most cash registers do not allow repeating decimals, so grocery stores do not use them.   **(e)** See part (d).

3. Answers vary; for example, it is easier to compute addition of fractions when there is a common deno-minator, as in $\frac{1}{7} + \frac{5}{7}$. When nonrepeating decimals are involved, it becomes hard to do additions because of lining up the place value. When denominators are different and terminating decimals are obtained, it is easier to use decimals, as in $\frac{2}{5} + \frac{1}{4} = 0.40 + 0.25 = 0.65$.

Open-Ended

5. **(a)** Any integer n can be written as a decimal by appending .0 to the right.   **(b)** **(i)** $0.\overline{6}$ **(ii)** 1 or $1.\overline{0}$; 1 is the simplest form because it requires fewer symbols.   **(iii)** $1.\overline{06}$
(c) It would have to be adapted to allow multiplication from the left.   **(d)** A repeating decimal can be written as a rational number in the form $\frac{a}{b}$, where $b \neq 0$. The fractions can be multiplied and the product can be converted to a decimal.   **7.** Most would prefer $\frac{7}{3}$ as a solution.

Questions from the Classroom

9. Answers may vary. A calculator is a tool that may aid in the computation of decimals. It can help by furnishing estimates for answers. This has to do with the value of the calculator as a tool. The value of repeating decimals is a different question. One major white-collar crime dealt with rounding off fractions of pennies and depositing that money in a bank account. These fractions could have been the result of using repeating decimals and rounding. Depending upon the context, the result of not using such decimals could lead to serious errors.

Review Problems

11. $22,761.95   **13.** 0.077. The rule says that the placement should be four places or 0.0770. Because 0.077 and 0.0770 are equivalent, the rule still works.

Assessment 7-4A

1. Answers vary. One answer is 0.232233222333...
2. **(a)** $\sqrt{6}$ **(b)** 2 **(c)** 3   **3.** $0.\overline{9}, 0.9\overline{98}, \sqrt{0.98}, 0.\overline{98},$ $0.9\overline{88}, 0.9, 0.\overline{898}$ **4.** **(a)** yes **(b)** no **(c)** no **(d)** yes **(e)** yes **(f)** yes   **5.** **(a)** 15 **(b)** 13 **(c)** impossible **(d)** 25   **6.** **(a)** 2.65 **(b)** 0.11   **7.** **(a)** false; $\sqrt{2} + 0$ **(b)** false; $^-\sqrt{2} + \sqrt{2}$ **(c)** false; $\sqrt{2} \cdot \sqrt{2}$ **(d)** true; $\sqrt{2} - \sqrt{2} = 0$ **8.** Answers vary; for example, $\sqrt{2}, \sqrt{3},$ and $\sqrt{5}$.   **9.** Answers vary. For example, assume the following pattern continues: 0.54544544454444....
10. Answers vary; for example, 0.5155155515555....
11. The answer must be an irrational number. Suppose it were rational; then the difference of it and the other rational number is rational. Then you have a rational number equal to an irrational number—a contradiction. Thus, the sum cannot be a rational number.   **12.** There are

infinitely many positive rational numbers. Add $\sqrt{2}$ to each of those and infinitely many irrational numbers result.
13. **(a)** R **(b)** $\varnothing$ **(c)** Q **(d)** $\varnothing$ **(e)** R **(f)** R
14. **(a)** Q, R **(b)** N, I, Q, R **(c)** R, S **(d)** I, Q, R **(e)** Q, R **15.** **(a)** N, I, Q, R **(b)** Q, R **(c)** R, S **(d)** N, I, Q, R **(e)** None **(f)** Q, R **16.** **(a)** 64 **(b)** none **(c)** $^-64$ **(d)** none **(e)** all real numbers greater than zero **(f)** none   **17.** 6.4 ft   **18.** **(a)** $6\sqrt{5}$ **(b)** $11\sqrt{3}$ **(c)** $6\sqrt{7}$ **19.** **(a)** $^-3\sqrt[3]{2}$ **(b)** $2\sqrt[5]{3}$ **(c)** $5\sqrt[3]{2}$ **(d)** $^-3$ **20.** **(a)** $5, 5\sqrt[3]{2}, 5\sqrt[3]{4}, 10$
(b) Answers may vary. $2, 2\sqrt[4]{\frac{1}{2}}, 2\sqrt[4]{\frac{1}{4}}, 2\sqrt[4]{\frac{1}{8}}, 1$.   **21.** **(a)** 2^{10}
(b) 2^{11} **(c)** 2^{12} **22.** **(a)** 4 **(b)** $\frac{3}{2}$ **(c)** $\frac{-4}{7}$ **(d)** $\frac{5}{6}$
23. **(a)** rational **(b)** rational   **24.** **(a)** Yes, when $a \leq 0$.
(b) **(i)** $x \geq \sqrt{2}$ or $x \leq -\sqrt{2}$ **(ii)** all real numbers x
25. between 18 and 19   **26.** $Guess2 = \dfrac{\frac{n}{Guess1} + Guess1}{2}$
If *Guess1* is accurate, then let $x = Guess1 = Guess2$.

$$x = \frac{\frac{n}{x} + x}{2}$$
$$2x = \frac{n}{x} + x$$
$$2x^2 = n + x^2$$
$$x^2 = n$$
$$x = \sqrt{n}$$

Mathematical Connections 7-4

Communication

1. The mathematician probably meant that there more irrational numbers than rational numbers. Georg Cantor proved this.   **3.** false: $\sqrt{64 + 36} \neq \sqrt{64} + \sqrt{36}$
5. No; $\sqrt{13}$ is an irrational number. So when it is expressed as a decimal, it is nonterminating and nonrepeating.
7. Notice that $\left(\frac{4}{25}\right)^{-1/3} = \left(\frac{25}{4}\right)^{1/3}$ and

$\left(\frac{4}{25}\right)^{-1/4} = \left(\frac{25}{4}\right)^{1/4}$. Because $\left(\frac{25}{4}\right)^{1/4} < \left(\frac{25}{4}\right)^{1/3}$, we

have $\left(\frac{4}{25}\right)^{-1/4} < \left(\frac{4}{25}\right)^{-1/3} = \left(\frac{25}{4}\right)^{1/3}$.

Open-Ended

9. Answers vary; for example:
(a) $\sqrt{1/2}, \pi/6, \sqrt[3]{2/5}, 0.505005000500005..., \sqrt{3/10}$
(b) $0.505005000500005..., 0.505105000500005...,$
$0.505105100500005..., 0.505105100510005...,...$

Cooperative Learning

11. $3.7^{2.4} = 3.7^{\frac{24}{10}} = 3.7^{\frac{12}{5}} = \sqrt[5]{3.7^{12}}$

13. To be a rational number a number must be able to be written in the form $\frac{a}{b}$ with $b \neq 0$ and both a and b must be integers. $\sqrt{2}$ is not an integer.   **15.** The principal square root of 25 is 5 and not $^-5$. While it is true that $(^-5)^2 = 25$, $^-5$ is not the principal square root of 25; the roots don't have to be equal.   **17.** The student is incorrect. One example in carpentry is to find the hypotenuse of a right triangle. Likely the result will be estimated, but if the lengths of the shorter sides are 1 in. and 1 in., the length of the hypotenuse is $\sqrt{2}$ in.   **19.** Answers may vary. Anecdotal evidence certainly supports Brown. It is the case that there is less understanding of decimals than whole numbers.

Review Problems

21. (a) $\frac{418}{25}$ **(b)** $\frac{3}{1000}$ **(c)** $\frac{^-507}{100}$ **(d)** $\frac{123}{1000}$

23. $\frac{3}{12,500}$ **25. (a)** 208,000 **(b)** 0.00038

Assessment 7-5A

1. (a)

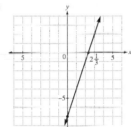

(b)

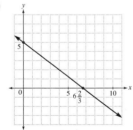

(c)

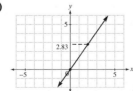

2. (a)

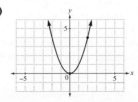

(b)

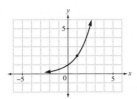

3. You can graph the function but the values for x have to be less than or equal to 0. A sample graph follows:

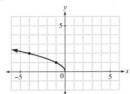

4. $y = ^-x + b$ **5.** The solution is (2.4, 2.2).
6. (a) 18 **(b)** 42 **(c)** $^-2$ **(d)** $\sqrt{2}$ or $^-\sqrt{2}$

(e) 7 or $^-3$ **(f)** $^-5$ or 1 **(g)** 1 or $\frac{^-5}{2}$ **7. (a)** 2856 ft

(b) 4600 ft   **8.** a line passing through the origin with slope π **9. (a)** 0.618 **(b)** It is an approximation to a solution of the equation.   **10.**

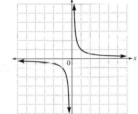

Mathematical Connections 7-5

Communication

1. Answers will vary depending upon the chapter read.

Open-Ended

3. Answers will vary. There are many approximations of π.

Cooperative Learning

5. Answers will vary depending upon the textbook chosen.

Questions from the Classroom

7. The student is incorrect. The equation represents a line that passes through the points with coordinates (0, 0) and $(1, 1 + \sqrt{2})$.

Review Problems

9. 2, $1.\overline{3}$, $1.\overline{7}$, $1.\overline{42}$ **11. (a)** 2.5 **(b)** 2.5   **13.** Answers vary; for example, $1.2\overline{1}$.

Chapter Review

1. (a) $A = 0.02, B = 0.05, C = 0.11$

(b)   **2. (a)** $\dfrac{32012}{1000}$ or $\dfrac{8003}{250}$

(b) $\dfrac{103}{100,000}$ **3.** A fraction in simplest form, $\dfrac{a}{b}$, can be written as a terminating decimal if, and only if, the prime factorization of the denominator contains no primes other than 2 or 5.   **4.** 8 shelves   **5. (a)** $0.\overline{571428}$ **(b)** 0.125   **(c)** $0.\overline{6}$ **(d)** 0.625   **6. (a)** $\dfrac{7}{25}$ **(b)** $\dfrac{^-607}{100}$ **(c)** $\dfrac{1}{3}$ **(d)** $\dfrac{94}{45}$

7. (a) 307.63   **(b)** 307.6   **(c)** 308   **(d)** 300

8. (a) $4.26 \cdot 10^5$ **(b)** $3.24 \cdot 10^{-4}$ **(c)** $2.37 \cdot 10^{-6}$ **(d)** $^-3.25 \cdot 10^{-1}$ **9.** $1.451\overline{9}, 1.45\overline{19}, 1.45\overline{19}, 1.451\overline{9},$ $0.13\overline{401}, ^-0.134, ^-0.13\overline{401}.$ **10. (a)** $5\sqrt{2}$ or $^-5\sqrt{2}$

(b) $\dfrac{1}{\sqrt[4]{4}}, \dfrac{1}{\sqrt[4]{16}}, \dfrac{1}{\sqrt[4]{64}}$ or $\dfrac{^-1}{\sqrt[4]{4}}, \dfrac{1}{\sqrt[4]{16}}, \dfrac{^-1}{\sqrt[4]{64}}$

11. (a) $1.78341156 \cdot 10^6$ **(b)** $3.47 \cdot 10^{-6}$ **(c)** $4.93 \cdot 10^9$ **(d)** $2.94 \cdot 10^{17}$ **(e)** $4.7 \cdot 10^{35}$ **(f)** $1.536 \cdot 10^{-6}$

12. (a) Answers vary; for example: 0.105, 0.104, 0.103, 0.102, 0.101   **(b)** Answers may vary; 0.0005, 0.001, 0.002, 0.004   **(c)** 0.15, 0.175, 0.1875, 0.19375.   **13. (a)** irrational   **(b)** irrational   **(c)** rational   **(d)** rational   **(e)** irrational   **(f)** irrational   **14. (a)** $11\sqrt{2}$ **(b)** $12\sqrt{2}$ **(c)** $6\sqrt{5}$ **(d)** $3\sqrt[3]{6}$ **15. (a)** No; $\sqrt{2} + (^-\sqrt{2})$ is rational.   **(b)** No; see (a).   **(c)** No; $\sqrt{2} \cdot \sqrt{2}$ is rational.   **(d)** No; $\sqrt{2}/\sqrt{2}$ is rational.   **16.** 4.796   **17.** approximately 1.26

18. (a)

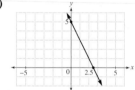

(b)

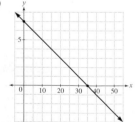

(c)

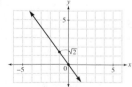

19. (a)

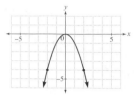

(b)

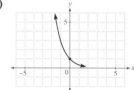

20. $\left(\dfrac{13}{3}, \dfrac{5}{3}\right)$ or $(4.\overline{3}, 1.\overline{6})$ **21. (a)** $\dfrac{^-11}{3}$ **(b)** 36   **(c)** $^-1$ **(d)** 3 or $^-3$ **(e)** $\dfrac{7}{2}$ or $\dfrac{^-2}{3}$ **(f)** $^-2$ or 3   **(g)** $\dfrac{1}{2}$ or 1   **22.** $\dfrac{1}{\pi}$

23. No; cash registers do not handle irrational numbers.

Answers to Now Try This

7-1. (a) $\dfrac{1}{10}$ or 0.1   **(b)** $\dfrac{1}{100}$ or 0.01

(c)

7-2. The cartoon reports the cost of tomatoes is .99¢, which is less than 1¢. This mistake is commonly made in stores when the money amount is less than 1 dollar; for example, an item is marked for .50¢ when it really should be $0.50 or 50¢. Note that the clerk did not raise the price.

7-3. (a) 2.5   **(b)** $\dfrac{3}{8} = \dfrac{3 \cdot 5^3}{2^3 \cdot 5^3} = \dfrac{375}{1000} = 0.375$

(c) $\dfrac{3}{20} = \dfrac{3 \cdot 5}{2^2 \cdot 5 \cdot 5} = \dfrac{15}{2^2 \cdot 5^2} = \dfrac{15}{100} = 0.15$

7-4. (a) $3.6 \cdot 1000 = 3.6 \cdot 10^3 = \left(3 + \dfrac{6}{10}\right)10^3 =$

$3 \cdot 10^3 + \dfrac{6}{10} \cdot 10^3 = 3 \cdot 10^3 + 6 \cdot 10^2 = 3 \cdot 10^3 + 6 \cdot 10^2 +$

$0 \cdot 10^1 + 0 \cdot 1 = 3600$. Thus, we see that multiplication by 1000 results in moving the decimal point three places to the right.   **(b)** In general, multiplication by 10^n, where n is a positive integer, results in moving the decimal point n places to the right.   **7-5.** $1.19/32$ oz is about 0.037/oz, while $1.43/48$ oz is about 0.030/oz, so the 48-oz jar is a better buy.   **7-6.** Answers vary; for example, using the front digits the first estimate is $2 + $0 + $6 + $4 + $5 = $17. Next, we adjust the estimate. Because $0.89 + $0.13 is about $1.00 and $0.75 + $0.05 is $0.80 and $0.80 + $0.39 is about $1.20, the adjustment is $2.20 and the estimate is $19.20.   **7-7. (a)** The answer is rounded to 0.0000014.

(b) The answer is 0.00000136. **7-8. (a)** $\frac{1}{9} = 0.\overline{1}$

(b) (i) $\frac{2}{9} = 2(0.\overline{1}) = 0.\overline{2}$ **(ii)** $\frac{3}{9} = 3(0.\overline{1}) = 0.\overline{3}$

(iii) $\frac{5}{9} = 5(0.\overline{1}) = 0.\overline{5}$ **(iv)** $\frac{8}{9} = 8(0.\overline{1}) = 0.\overline{8}$

7-9. (a) If $r = 1$, the denominator of $\frac{a}{1 - r}$ is 0. **(b)** As n becomes greater, so does r^{n+1} when $r > 1$.

(c) Let $S = a + ar + ar^2 + \ldots + ar^n$
$$\frac{-rS = {}^{-}(ar + ar^2 + \ldots + ar^n + ar^{n+1})}{S - rS = a - ar^{n+1}}$$
$$S = \frac{a(1 - r^{n+1})}{1 - r}$$

7-10. $0.\overline{235} = \frac{235}{999}, 2.3\overline{45} = \frac{129}{55}$ **7-11.** Answers may vary. **(a)** $0.3\overline{55}; 0.36\overline{5}$ **(b)** With each repeating decimal (like 0.36) one can move halfway between the number and the two found to write more repeating decimals—for example, $0.3\overline{55} < 0.36 < 0.36\overline{5}$ and continue the process indefinitely. **7-12. (a)** $\sqrt{13} \approx 3.6056$

(b) $Guess2 = \dfrac{\dfrac{n}{Guess1} + Guess1}{2}$ **7-13. (a)** The approach works because
$$\sqrt{\sqrt{\sqrt{a}}} = \left(\left(a^{1/2}\right)^{1/2}\right)^{1/2} = \left(a^{1/4}\right)^{1/2}$$
$$= a^{1/8}$$
$$= \sqrt[8]{a}$$

(b) For $n = 2^k$, where k is a positive integer. As shown in part (a), by repeatedly applying the square-root function to a we get $\sqrt[2^k]{a}$.

7-14. Lines are symmetric about the line $y = x$.

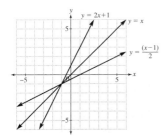

7-15. Lines will be parallel; slopes are the same.

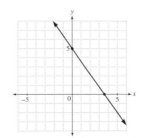

7-16. (a)
$$x^2 + 5x + 6 = 0$$
$$x^2 + 5x + \left(\frac{5}{2}\right)^2 + 6 - \left(\frac{5}{2}\right)^2 = 0$$
$$\left(x + \frac{5}{2}\right)^2 + \frac{-1}{4} = 0$$
$$\left(x + \frac{5}{2}\right)^2 - \left(\frac{1}{2}\right)^2 = 0$$
$$\left(x + \frac{5}{2} - \frac{1}{2}\right)\left(x + \frac{5}{2} + \frac{1}{2}\right) = 0$$
$$(x + 2)(x + 3) = 0$$
$$x = {}^{-}2 \text{ or } {}^{-}3$$

(b) 4 or 3 **(c)** There is no real number solution when $b^2 - 4ac < 0$. We do not take square roots of negative numbers in this course.

Answers to Brain Teasers

Section 7–1

Two base-two decimals are 0.1_{two} and 0.01_{two}. Every base-two decimal represents a fraction whose denominator is a power of 2 and is a terminating decimal in base-ten.

Section 7–4

March 14 is 3/14, the first three digits of π. Many schools also have a special event at 1:59 P.M. to celebrate the first six digits of π (3.14159).

Solution to Technology Corner

Section 7-5

Solution to Preliminary Problem

There are many answers. Among them is the following:
$$\begin{array}{r} .\overline{98765} \\ + \ .\overline{01234} \\ \hline 0.\overline{99999}, \text{ or } 1 \end{array}$$

Chapter 8

Assessment 8-1A

1. (a) $5:21$ **(b)** $21:5$ **(c)** $21:26$ **(d)** Answers vary; for example, minor. **2. (a)** $2:5$. Because the ratio is $2:3$, there are $2x$ boys and $3x$ girls; hence, the ratio of boys to all students is $\dfrac{2x}{2x + 3x} = \dfrac{2}{5}$.   **(b)** $m:(m + n)$ **(c)** $3:2$

3. (a) 30   **(b)** $^-3\dfrac{1}{3}$ **(c)** $23\dfrac{1}{3}$ **(d)** $10\dfrac{1}{2}$ **4.** 36 lb

5. 12 grapefruits for \$1.80   **6.** 270 mi   **7.** 64 pages

8. (a) 42, 56   **(b)** 24 and 32 (or $^-$24 and $^-$32)

9. \$14,909.09; \$29,818.18; \$37,272.73   **10.** \$77 and \$99

11. 135   **12. (a)** $\dfrac{1}{6}$ **(b)** $\dfrac{1}{1}$ **(c)** $\dfrac{7}{12}$

13. $\dfrac{36 \text{ oz}}{12¢} = \dfrac{48 \text{ oz}}{16¢}; \dfrac{12¢}{16¢} = \dfrac{36 \text{ oz}}{48 \text{ oz}}, \dfrac{16¢}{12¢} = \dfrac{48 \text{ oz}}{36 \text{ oz}}$

14. (a) $\dfrac{5}{7}$ **(b)** 6 ft   **15. (a)** 27   **(b)** 20

16. approximately 34 cm   **17.** 312 lb   **18. (a)** $\dfrac{2}{3}$ tsp

mustard seeds, 1 c scallions, $2\dfrac{1}{6}$ c beans   **(b)** $\dfrac{2}{3}$ tsp mustard

seeds, 2 c tomato sauce, $2\dfrac{1}{6}$ c beans   **(c)** $\dfrac{7}{13}$ tsp mustard

seeds, $1\dfrac{8}{13}$ c tomato sauce, $\dfrac{21}{26}$ c scallions   **19.** 15.12 Ω

20. about 74.6 cm   **21.** The ratio between the mass of the gold in the ring and the mass of the ring is 18/24. If x is the number of ounces of pure gold in the ring that weighs 0.4 oz. we have $18/24 = x/0.4$. Hence, $x = (18 \cdot 0.4)/24$, or 0.3 oz. Consequently, the price of the gold in the ring is $0.3 \cdot \$300$, or \$90.   **22. (a)** \$320   **(b)** 8   **23. (a)** $1:2$

(b) Let $\dfrac{a}{b} = \dfrac{c}{d} = \dfrac{e}{f} = r$.

Then, $a = br$
$\qquad c = dr$
$\qquad e = fr$
So, $a + c + e = br + dr + fr$
$\qquad a + c + e = r(b + d + f)$
$\qquad \dfrac{a + c + e}{b + d + f} = r$

Mathematical Connections 8-1

Communication

1. (a) 40/700, or 4/70, or 2/35   **(b)** 525 cm   **(c)** For the first set, $\dfrac{\text{footprint length}}{\text{thighbone length}} = \dfrac{40}{100} = \dfrac{20}{50}$; that is, a 50-cm thighbone would correspond to a 20-cm footprint. Thus, it is not likely that the 50-cm thighbone is from the animal that left the 30-cm footprint. $\left(\text{Notice that } \dfrac{20}{50} \ne \dfrac{30}{50}.\right)$

3. Yes, the proportion can be set up in a variety of ways because the following are equivalent:

$$a/b = c/d, \ a/c = b/d, \ b/a = d/c, \text{ and } c/a = d/b$$

However, you must keep corresponding parts in the correct positions.   **5.** No, the dimensions don't vary proportionately because $4/6 \ne 5/7$ and $4/6 \ne 8/10$. This can be seen by checking cross products or just by reducing all three

fractions to simplest terms—$\dfrac{2}{3}, \dfrac{5}{7}$, and $\dfrac{4}{5}$—and noticing that no two are equal.   **7.** Answers vary; for example, let m be the number of adult men living in the condo and w be the number of adult women. The number of married men is equal to the number of married women, so $\dfrac{2}{3} m = \dfrac{3}{4} w$. The ratio of married people to the total adult population is

$$\dfrac{\dfrac{2m}{3} + \dfrac{3w}{4}}{m + w} = \dfrac{\dfrac{2m}{3} + \dfrac{2m}{3}}{m + w} = \dfrac{\dfrac{4m}{3}}{m + w}$$

Because $\dfrac{2m}{3} = \dfrac{3w}{4}$, then $w = \dfrac{8m}{9}$. Thus,

$$\dfrac{\dfrac{4m}{3}}{m + w} = \dfrac{\dfrac{4m}{3}}{m + \dfrac{8m}{9}} = \dfrac{\dfrac{4m}{3}}{\dfrac{17m}{9}} = \dfrac{12}{17}$$

which is the desired ratio.

Open-Ended

9. Answers vary. For example, ratios and proportions are seen in census reports, economic data about joblessness, and baseball statistics.   **11. (a)** 57.6 lb per sq in.   **(b)** Answers vary. For example, distance (d) = rate (r) · time (t).

Cooperative Learning

13. (a) Answers vary; most measurements will be close.   **(b)** approximately 36 cm   **(c)** 1:1 go to the Internet to observe the art with ratios displayed   **(d)** Answers vary; for example 1:6 is most likely.   **(e)** Answers vary.

Questions from the Classroom

15. Yes, she is correct. Because each ratio is equal to the same number, then they must equal each other and hence form a proportion.   **17.** Al probably thought the height of the tree was 15 ft and set up the proportion in that way. The correct proportion comparing the object to its shadow is

$$\dfrac{5 \text{ ft}}{\dfrac{3}{2} \text{ ft}} = \dfrac{x \text{ ft}}{15 \text{ ft}}, \quad \text{so } x = 50 \text{ ft}.$$

The key is to have the same units for shadows and the same units for heights.

Assessment 8-2A

1. (a) 789%   **(b)** 19,310%   **(c)** $83\dfrac{1}{3}$%   **(d)** 12.5%
(e) 62.5%   **(f)** 80%   **2. (a)** 0.16   **(b)** 0.002
(c) $0.13\overline{6}$ **(d)** $0.00\overline{3}$ **3. (a)** 4   **(b)** 2   **(c)** 25
(d) 200   **(e)** 12.5   **4.** Depends on calculator.

5. (a) 2.04   **(b)** 50%   **(c)** 60   **(d)** 3.43   **6. (a)** $\dfrac{5x}{100}$

(b) $10a$ **7.** 63 boxes   **8.** $16,960
9. $25,500   **10. (a)** Bill sold 221.   **(b)** Joe sold 90%.
(c) Ron started with 265.   **11.** 20%
12. approximately 23% increase   **13.** 100%
14. $22.40   **15.** $336   **16.** 35%   **17.** $3200
18. 1200 employees   **19.** $\dfrac{325}{500}; \dfrac{325}{500} = \dfrac{650}{1000} =$
$\dfrac{65}{100} = 65\%$, while $\dfrac{600}{1000} = \dfrac{60}{100} = 60\%$.   **20. (a)** $76; $76
(b) They cost the same.   **21.** 11.1% or approximately
11%   **22.** $440   **23. (a)** $3.30   **(b)** $24.00
(c) $1.90   **(d)** $24.50   **24. (a)** Answers vary.
(b) (i) about 0.4 sec. between beats   **(ii)** about 0.006 min
between beats   **25.** Apprentice makes $700. Journeyman
makes $1400. Master makes $2100.   **26. (a)** 4%
(b) (i) 44   **(ii)** 8.8%   **27.** $82,644.63   **28. (a)** The
answer is essentially true. Spending at that rate will amount
to $2.57 trillion.   **(b)** Approximately 0.12%   **29.** No;
the respective salaries would be approximately $2000 and
$1696.43.   **30. (a)** 3   **(b)** 0.003%   **31. (a)** $4.50
(b) 50%   **(c)** 100%   **32.** 550 students

Mathematical Connections 8-2

Communication

1. Answers vary; for example, 10% of 850 is 85 and 1%
of 850 is 8.5, so 11% of 850 is 93.5.   **3.** It means that
not only did you meet 100% of your savings goal, you
surpassed it by 25%. If your savings goal was $100, then
you saved $100 plus an extra $25.   **5. (a)** Greater,
because if 25% of $x = 55$, the x must be 4 times greater than
55, or 210.   **(b)** Less, because if 150% of $x = 55$, the x is
only $\dfrac{2}{3}$ of 55, or $36\dfrac{2}{3}$.   **7.** The *whole* in each part is
different, so 50% of the greater quantity is greater than 50%
of the smaller quantity. To be equal, we would have to have
the same size whole to begin with.   **9.** Let x be the amount
invested. The first stock will be worth $(1.15x)0.85$ after 2 yr.
The second stock will be worth $(0.85x)1.15$. Because each of
these equals $(1.15 \cdot 0.85)x$, the investments are equally good.

Open-Ended

11. Answers vary.   **13.** Answers vary.

Questions from the Classroom

15. $3\dfrac{1}{4}\% = 3\% + \dfrac{1}{4}\% = \dfrac{3}{100} + \dfrac{\left(\frac{1}{4}\right)}{100} = 0.03 + 0.0025 =$

0.0325. Knowing that $\dfrac{1}{4} = 0.25$, the student incorrectly

wrote $\dfrac{1}{4}\% = 0.25$.   **17.** Let s denote the amount of

salary. After a $p\%$ increase, the new salary is $s\left(1 + \dfrac{p}{100}\right)$.

When this amount is decreased by $q\%$, the result is

$s\left(1 + \dfrac{p}{100}\right)\left(1 - \dfrac{q}{100}\right)$. Similarly, if the initial salary is first

decreased by $q\%$ and then the new amount is raised by $p\%$,

the final salary is $s\left(1 - \dfrac{q}{100}\right)\left(1 + \dfrac{p}{100}\right)$. Because the two

expressions are equal by the commutative and associative
property of multiplication, the student is right.

Review Problems

19. No, he should have put in 30 fl oz.   **21.** Always, because
$4x \cdot 9 = 3 \cdot 12x = 36x$ or because $\dfrac{12x}{9} = \dfrac{3 \cdot 4x}{3 \cdot 3} = \dfrac{4x}{3}$.

Assessment 8-3A

1.

Interest Rate per Period	Number of Periods	Amount of Interest Paid	Total Amount in Account
3%	4	$125.51	$1125.51
2%	12	$268.24	$1268.24
$\dfrac{10}{12}\%$ or $\dfrac{5}{6}\%$ or $0.8\overline{3}\%$	60	$645.31	$1645.31
$\dfrac{12}{365}\%$	1460	$615.95	$1615.95

2. $3675.00   **3.** $24.45   **4.** $43,059.50   **5.** $64,800
6. $12,905.80   **7.** $1015.20   **8. (iii)** it pays 13.2%
9. approximately $2.53   **10.** approximately 14.03%
11. $3483.81   **12.** The Pay More Bank offers a better
rate.   **13.** an arithmetic sequence with difference of 0 or
a geometric sequence with ratio equal to 1.

Mathematical Connections 8-3

Communication

1. Let a be the original value of the house. Because it
depreciated 10% each year for the first 3 yr, using
compound depreciation the price, after 3 yr is
$a(1 - 0.10)^3$, or $a \cdot 0.9^3$. Because of compound
appreciation, after another 3 yr, the value of the house is
$a(0.9^3) \cdot 1.1^3$, or $a(0.9^3 \cdot 1.1^3)$, which equals approximately
$a \cdot 0.9703$. Because $a \cdot 0.9703 < a$, the value of the house
decreased after 6 yr. The value of the house decreases by
approximately 3%.   **3.** No, the percentages cannot be
added because each time the percent is of a different quantity. After 5 yr, the car would have depreciated 67%.

Open-Ended

5. Answers vary.

Cooperative Learning

7. (a) Answers vary.   **(b)** Answers vary.   **(c)** Answers vary.

Questions from the Classroom

9. No, she is not correct. Men make 25¢ more for every
75¢ that women make, so men make $\dfrac{25}{75} = 33\dfrac{1}{3}\%$ more or
women make 25% less than men.

Chapter Review

1. (a) $17:30$ **(b)** $17:13$ **(c)** $13:17$ **2.** 64 fl oz for $3.60 **3.** No, $18/6 \neq 12/3$. **4. (a)** 16.8 **(b)** 192.5
(c) 1 **5.** 44 grapes and $8\frac{1}{4}$ oranges **6.** 7.5 m
7. The ratio of hydrogen to the total is $1:9$. Therefore, $\frac{1}{9} = \frac{x}{16}$ implies $x = 1\frac{7}{9}$ oz. **8.** 560 fish **9.** No, the ratio depends on how manychips came from each plant.
10. $1:r^2$ **11. (a)** $18:7$ **(b)** $18:25$ **12. (a)** $1:5$
(b) $8:15$ **13.** 9 **14. (a)** 25% **(b)** 192 **(c)** $56.\overline{6}$
(d) 20% **15. (a)** 12.5% **(b)** 7.5% **(c)** 627%
(d) 1.23% **(e)** 150% **16. (a)** 0.60 **(b)** $0.00\overline{6}$ **(c)** 1
17. $9280 **18.** $3.\overline{3}$% **19.** 88.6% **20.** $5750
21. It makes no difference, the discount is always 31.6%.
22. $80 **23.** approximately 31% **24.** Answers vary; for example, if the dress was originally priced at $100, then 60% off would result in a sale price of $ 40. Then the 40% coupon would give a final price of $24. This model could be applied to the actual list price of the dress. **25.** All are mathematically meaningful. **26.** $15,000 **27.** approximately $15,110.69

Answers to Now Try This

8-1. $\dfrac{a}{b} = \dfrac{c}{d}$

$$\frac{a}{b} \cdot \frac{bd}{1} = \frac{c}{d} \cdot \frac{bd}{1}$$

$$\frac{ad}{1} = \frac{cb}{1}$$

$$ad = cb$$

$$ad = bc$$

8-2. (a) $\dfrac{11}{20}$, 55% **(b)** $\dfrac{3}{5}$, 60% **(c)** $\dfrac{13}{25}$, 52% **(d)** $\dfrac{1}{5}$, 20%

8-3. (a) Answers vary; for example, most calculators will convert the decimal form of a number to a percent by moving the decimal point two places to the left. Other calculators actually place a % symbol in the display when the %̲ key is pushed. **(b)** $33.\overline{3}$% **8-4.** Olives—$37\frac{1}{2}$%,

3 slices; Plain—$12\frac{1}{2}$%, 1 slice; Remainder—50%, 4 slices

Answers to Brain Teasers

Section 8-1

Collect data on the ratio of arm length to nose length by measuring the arms and noses of other students in her class. Find the average ratio, then use it to find the expected length of the nose of the Statue of Liberty.

Section 8-2

Let $C =$ amount of crust and $P =$ amount of pie; $x =$ percent of crust reduced. $C = 25\%$ of P, so $\dfrac{C(100 - x)}{100} = \left(\dfrac{20}{100}\right)P$. Hence, $x = 20\%$.

Answer to Technology Corner

Section 8-2

a. Column (d) shows that the 30% mixture is reached before 5 L of lemon juice and 12 L of water are mixed.
b. $\dfrac{5}{5 + x} \cdot 100$ changes the ratio of lemon juice to water to a percentage, where x is the number of liters of water added and $5 + x$ is the number of total liters in the mixture.

Answer to Preliminary Problem

If the first car was worth A dollars and the dealer made 10%, then we have $A(1.1) = \$9999$. Solving for A, we find $A = \$9090$. If the second car was worth B dollars, then we have $B(0.9) = \$9999$. Solving for B, we find $B = \$11,110$. Therefore, the total worth of the two cars is $9090 + $11,110 = $20,200. The dealer received $2(\$9999) = \$19,998$ for the two cars. Therefore, the dealer lost $20,200 - $19,998 = $202.

Chapter 9

Assessment 9-1A

1. (a) No **(b)** Yes **(c)** Yes **(d)** No **2. (a)** $\dfrac{5}{26}$ **(b)** $\dfrac{21}{26}$
(c) $\dfrac{11}{26}$ **3. (a)** $\dfrac{3}{8}$ **(b)** $\dfrac{2}{8}$, or $\dfrac{1}{4}$ **(c)** $\dfrac{4}{8}$, or $\dfrac{1}{2}$ **(d)** $\dfrac{2}{8}$, or $\dfrac{1}{4}$
(e) 0 **(f)** $\dfrac{3}{8}$ **(g)** $\dfrac{1}{8}$ **4. (a)** $\dfrac{26}{52}$, or $\dfrac{1}{2}$ **(b)** $\dfrac{12}{52}$, or $\dfrac{3}{13}$
(c) $\dfrac{28}{52}$, or $\dfrac{7}{13}$ **(d)** $\dfrac{4}{52}$, or $\dfrac{1}{13}$ **(e)** $\dfrac{48}{52}$, or $\dfrac{12}{13}$ **(f)** $\dfrac{22}{52}$, or
$\dfrac{11}{26}$ **(g)** $\dfrac{3}{52}$ **(h)** $\dfrac{30}{52}$, or $\dfrac{15}{26}$ **5. (a)** $\dfrac{4}{12}$, or $\dfrac{1}{3}$
(b) $\dfrac{8}{12}$, or $\dfrac{2}{3}$ **(c)** 0 **(d)** $\dfrac{6}{12}$, or $\dfrac{1}{2}$ **(e)** 1 **6. (a)** $\dfrac{1}{6}$
(b) $\dfrac{4}{6}$, or $\dfrac{2}{3}$ **7.** 70%, $P(\text{No Rain}) = 1 - P(\text{Rain}) =$
$1 - 0.30 = 0.70$ **8. (a)** $\dfrac{18}{38}$, or $\dfrac{9}{19}$ **(b)** $\dfrac{2}{38}$, or $\dfrac{1}{19}$
(c) $\dfrac{26}{38}$, or $\dfrac{13}{19}$ **(d)** $\dfrac{20}{38}$, or $\dfrac{10}{19}$ **9.** 10 times
10. (a) $\dfrac{2}{4}$, or $\dfrac{1}{2}$ **(b)** $\dfrac{3}{4}$ **(c)** $\dfrac{3}{4}$ **11.** $\dfrac{350}{1380}$, or $\dfrac{35}{138}$
12. 0.7 **13. (a)** $\dfrac{45}{80}$, or $\dfrac{9}{16}$ **(b)** $\dfrac{10}{80}$, or $\dfrac{1}{8}$ **(c)** $\dfrac{60}{80}$, or $\dfrac{3}{4}$

(d) $\frac{30}{80}$, or $\frac{3}{8}$ **14. (a)** 22 **(b)** 10 **15.** No, she is not

correct. The probability of drawing a white from box #1 is

$\frac{3}{4}$, or $\frac{6}{8}$. The probability of drawing a white from box #2 is

$\frac{5}{8}$. Because $\frac{6}{8} > \frac{5}{8}$, the probability of drawing a white ball is

greater is box #1. **16.** The probability is $\frac{1}{2}$ because it is a

fair coin. The coin has no memory, so the probability of a
head on the 16th toss is the same as the probability of a head

on any toss regardless of this history. **17. (a)** $\frac{4}{10}$ **(b)** $\frac{6}{10}$,

or $\frac{3}{5}$ **18. (a)** likely **(b)** unlikely **(c)** likely **19.** No;

0 and 1 are not the lead digits of any normal a phone number
area code.

Mathematical Connections 9-1

Communication

1. All probabilities are less than or equal to 1 and greater
than or equal to 0. If events are mutually exclusive, then
$P(A \cup B) = P(A) + P(B)$. Therefore, if A and B are
mutually exclusive, $P(A \cup B) = P(A) + P(B) = 0.8 + 0.9$
$= 1.7$, which is impossible. Therefore, events A and B are not
mutually exclusive. **3.** Joe's conjecture is incorrect. Each

of the numbers 1 through 4 has probability $\frac{1}{4}$ of occurring

because each angle where the arrow is located seems to
measure 90 degrees and the spinner has the same chance of
landing in any of the regions.

Open-Ended

5. Answers vary depending on the book selected.
7. Answers vary; for example, event A is an impossible event
such as rolling a 10 on a single roll of a standard die. Event B
has low probability, such as the chance of rain being 20%.
Event C is around 0.5, so this might be something like
obtaining a head when tossing a fair coin. Event D has a high
probability of happening, but it is not certain (for example,
tossing a number less than 6 on a toss of a standard die).
Event E has probability 1, so it has to happen (for example,
tossing either a head or a tail on the toss of a fair coin). Event
F has probability greater than 1, but this cannot happen, so no
event is possible.

Cooperative Learning

9. (a) Answers vary. **(b)** The person who receives 4 times
the value of the number on the die wins when 1, 2, and 3
are tossed. The person who receives the square of the
number showing wins when a 5 or a 6 is tossed. When a 4 is
tossed, the game is a draw. The person who receives 4 times
the value of the die has a greater chance of winning, and
therefore the game is not fair.

Questions from the Classroom

11. If the four regions corresponding to the colors were determined by equal-sized angles, the events of the spinners landing
on each of the regions would be equally likely, and the student
would be correct. However, since the four angles are not the
same sizes, the events are not equally likely, and the student
is wrong. **13.** For an experiment with sample
space S with equally likely outcomes, the probability of an

event A is given by $P(A) = \frac{n(A)}{n(S)}$. Because an event A must

be a subset of S, the smallest that $n(A)$ could be is 0. This
occurs when $A = \varnothing$. Because $n(S)$ is never negative and
$n(A)$ is never negative, then $P(A)$ can never be negative.

Assessment 9-2A

1. (a)

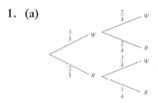

The probability that two balls are of different colors is
$\frac{3}{5} \cdot \frac{2}{4} + \frac{2}{5} \cdot \frac{3}{4} = \frac{3}{5}$. **(b)** $\frac{12}{25}$

2. (a) $\frac{1}{216}$ **(b)** $\frac{1}{120}$ **3.** $\frac{1}{30}$ **4. (a)** $\frac{64}{75}$ **(b)** $\frac{11}{75}$

5. $\frac{1}{16}$ **6. (a)** $\frac{4}{10}$, or $\frac{2}{5}$ **(b)** Yes, now $P(\text{Even}) = \frac{9}{10}$

7. (a) $\frac{1}{320}$ **(b)** $\frac{63}{4000}$ **(c)** 0 **(d)** $\frac{171}{320}$ **8. (a)** The

first spinner. If you choose the first spinner, you win if, and
only if, the spinning combinations are as follows:

Outcome on spinner A	Outcome on spinner B
4	3
6	3 or 5
8	3 or 5

The probability of this happening is
$\frac{1}{3} \cdot \frac{1}{3} + \frac{1}{3} \cdot \frac{2}{3} + \frac{1}{3} \cdot \frac{2}{3} = \frac{5}{9}$.

If you choose spinner B, the probability of winning is only $\frac{4}{9}$.
(b) Answers vary. One choice is as follows:

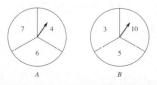

If you choose A, the probability of winning is $\frac{5}{9}$. If you choose B, the probability of winning is only $\frac{4}{9}$.   **9.** $\frac{1}{32}$

10. (a) $\frac{16}{81}$ **(b)** $\frac{8}{27}$ **11. (a)** $\frac{1}{25}$ **(b)** $\frac{8}{25}$ **(c)** $\frac{16}{25}$

12. (a) 100 square units

  (b) $P(\text{Region } A) = \frac{4}{100}$, or $\frac{1}{25}$

  $P(\text{Region } B) = \frac{12}{100}$, or $\frac{3}{25}$

  $P(\text{Region } C) = \frac{20}{100}$, or $\frac{1}{5}$

  $P(\text{Region } D) = \frac{28}{100}$, or $\frac{7}{25}$

  $P(\text{Region } E) = \frac{36}{100}$, or $\frac{9}{25}$

(c) $\frac{1}{625}$ **(d)** $\frac{36}{100}$, or $\frac{9}{25}$ **13.** $\frac{1}{12}$ **14.** $\frac{2}{5}$

15. $\frac{2}{28}$, or $\frac{1}{14}$

16. (a)

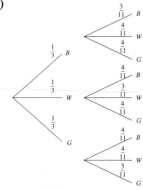

(b) $\frac{1}{3} \cdot \frac{3}{11} + \frac{1}{3} \cdot \frac{3}{11} + \frac{1}{3} \cdot \frac{3}{11}$, or $\frac{3}{11}$ **(c)** $\frac{1}{3} \cdot \frac{3}{11} = \frac{1}{11}$

(d) 1   **17. (a)** $\frac{11}{16}$ **(b)** $\frac{11}{16}$ **18.** 4   **19.** Billie-Bobby-Billie because the probability of winning two in a row is greater this way. Note, it does not say win two out of three.

Mathematical Connections 9-2

Communication

1. If the die is fair, the probability that Jim will roll a 3 on the next roll is $\frac{1}{6}$, which is the same probability of rolling any one of the other numbers. The results of the rolls are independent.   **3.** We could ask José to draw a tree diagram for this experiment. He will then see the sample space is $\{BB, BG, GB, GG\}$ with each outcome having probability $\frac{1}{4}$. Therefore, the probability of having two boys is $\frac{1}{4}$, not $\frac{1}{3}$.

Open–Ended

5. Answers vary depending on how complex the game is. The game could be as simple as player A wins if the sum of the digits is even and player B wins if the sum of the digits is odd.   **7.** Answers vary, but $\frac{3}{5}$ of the area of the board must be labeled in a certain way.

Cooperative Learning

9. If the first player chooses TT and the second HT, then the first player wins the game if, and only if, tails appears on the first and second flips. The probability of this happening is $\frac{1}{4}$. The probabilities of winning the game for other choices are summarized below:

		First Player's Choice			
		HH	*HT*	*TH*	*TT*
Second Player's Choice	*HH*	—	0.50	0.75	0.50
	HT	0.50	—	0.50	0.25
	TH	0.25	0.50	—	0.50
	TT	0.50	0.75	0.50	—

Of the 12 possible games, only 8 result in choices with equally likely probabilities. Therefore, the game is not fair.

Questions from the Classroom

11. The student is wrong. The sample space for this event is not $\{HH, HT, TT\}$, but rather $\{HH, HT, TH, TT\}$. Consequently, the probability of HH is $\frac{1}{4}$.   **13.** The student is not correct. The confusion probably lies in the fact that the student thinks that probabilities are additive. The student does not understand the Multiplication Rule for Probabilities. A tree diagram for the experiment could possibly help. A partial tree diagram is

$\xrightarrow{\frac{1}{6}} 5 \xrightarrow{\frac{1}{6}} 5$

Thus, $P(5, 5) = \frac{1}{6} \cdot \frac{1}{6} = \frac{1}{36}$.   **15.** The sample space for the outcomes of the experiment is $\{BB, BW, WB, WW\}$. The probability of two blacks is $\frac{1}{4}$, not $\frac{1}{2}$.

Review Problems

17. (a) v **(b)** iii **(c)** ii **(d)** i **(e)** iv   **19.** $\frac{1}{3}$

Assessment 9-3A

1. Answers vary; likely this would not work because the outcomes are not equally likely.   **2.** Answers vary; for example, let the digits 1 and 2 represent Diamonds, digits 3 and 4 represent Hearts, digits 5 and 6 represent Spades, and

7 and 8 represent Clubs. If 0 or 9 appear, then disregard these digits. Read the random digits in pairs to simulate two draws.   **3. (a)** Let 1, 2, 3, 4, 5, and 6 represent the numbers on the die and ignore the numbers 0, 7, 8, 9.   **(b)** Number the persons 01, 02, 03, . . . , 18, 19, 20. Go to the random-digit table and mark off groups of two. The three persons chosen are the first three whose numbers appear.   **(c)** Represent red by the numbers 0, 1, 2, 3, 4; green by the numbers 5, 6, 7; yellow by the number 8; and white by the number 9.   **4.** Mark off 30 blocks of three digits in a random-digit table. Disregard 000 or any numbers that are greater than 500. Also disregard duplicates. These are the numbers of the 30 students who will be chosen for the trip.   **5.** To simulate Monday, let the digits 1 through 8 represent rain and 0 and 9 represent no rain. If rain occurred on Monday, repeat the same process for Tuesday. If it did not rain on Monday, let the digits 1 through 7 represent rain and 0, 8, and 9 represent dry. Repeat a similar process for the rest of the week.   **6.** Answers vary; for example, mark off blocks of two digits and let the digits 00, 01, 02, . . . , 13, 14 represent contracting the disease and 15 to 99 represent no disease. Mark off blocks of six digits to represent the three children. If at least one of the numbers is in the range 00 to 14, then this represents a child in the three-child family having strep.   **7.** $\frac{3}{10}$ **8.** 1200 fish

9. (a) 7   **(b)** Answers vary; for example, let the digits 0, 1, 2, 3, 4 represent a victory by team A and the digits 5, 6, 7, 8, 9 represent a victory by team B. Then go to the random-digit table, pick a starting place, and count the number of digits (games) it takes for one of the teams to win. Repeat this experiment a number of times for four games and for seven games and use the definition of probability to compute the experimental probability.   **10.** One way to simulate the problem is to use a random-digit table. Because Carmen's probability of making any basket is 80%, we could use the occurrence of a 0, 1, 2, 3, 4, 5, 6, or 7 to simulate making the basket and the occurrence of an 8 or a 9 to simulate missing the basket. Another way to simulate the problem is to construct a spinner with 80% of the spinner devoted to making a basket and 20% of the spinner devoted to missing the basket. This could be done by constructing the spinner with 80% of the 360 degrees (that is, 288 degrees) devoted to making the basket and 72 degrees devoted to missing the basket.

We can compute the theoretical probability for this experiment by using a *tree diagram*, as shown here. Thus, theoretical probabilities for the number of points scored in 25 attempts can be computed. These theoretical estimates are given in the following table. Compare these results with the experimental probability obtained by simulation.

Number of Points	Expected Number of Times Points Are Scored in 25 Attempts
0	5
1	4
2	16

Mathematical Connections 9-3

Communication

1. Answers will vary. One possibility is to see Kanold, C. "Teaching Probability Theory Modeling Real Problems." *Mathematics Teacher* 85 (April 1994): 232–235.

Open-Ended

3. Answers vary; one could use a random-digit table with blocks of two digits, or a spinner could be designed with 12 sections representing the different months. The spinner could be spun 5 times to represent the birthdays of five people. We could then keep track of how many times at least two people have the same birthday.   **5.** Let the 10 ducks be represented by the digits 0, 1, 2, 3, . . . , 8, 9. Then pick a starting point in the table and mark off 10 digits to simulate which ducks the hunters shoot at. Count how many of the digits 0 through 9 are not in the 10 digits; this represents the ducks that escaped. Do this experiment many times and take the average to determine an answer. See how close your simulation comes to 3.49 ducks.

Cooperative Learning

7. (a) Answers vary.   **(b)** Answers vary.   **(c)** Answers vary.   **(d)** $\left(\frac{1}{2}\right)^{10} = \frac{1}{1024}$ **9.** Pick a starting spot in the table and count the number of digits it takes before all the numbers 1 through 9 are obtained. Ignore 0. Repeat this experiment many times and find the average number of boxes.

Questions from the Classroom

11. Maximilian could be correct, but typically people are not unbiased when it comes to "making up results" to experimental probabilities.

Review Problems

13. (a) $\frac{1}{4}$ **(b)** $\frac{1}{52}$ **(c)** $\frac{48}{52}$, or $\frac{12}{13}$ **(d)** $\frac{3}{4}$ **(e)** $\frac{1}{2}$ **(f)** $\frac{1}{52}$ **(g)** $\frac{16}{52}$, or $\frac{4}{13}$ **(h)** 1   **15.** $\frac{8}{27}$

Assessment 9-4A

1. (a) 12 to 40, or 3 to 10   **(b)** 40 to 12, or 10 to 3

2. 30 to 6, or 5 to 1   **3.** 15 to 1   **4.** $\dfrac{5}{8}$ **5.** $\dfrac{1}{27}$

6. 73 : 27   **7. (a)** $\dfrac{1}{2}$ **(b)** $\dfrac{1}{4}$ **8.** 8.4¢   **9.** $10,000

10. (a) the expected loss is 22¢   **(b)** 20¢   **11.** no

12. 5 : 88   **13.** 5 : 3   **14.** 3 : 5   **15.** $\dfrac{2}{11}$ **16.** 1 : 1

Mathematical Connections 9-4

Communication

1. Odds are determined from probabilities. The *odds in favor* of an event, E, are given by $\dfrac{P(E)}{P(\overline{E})}$. The *odds against* an event are given by $\dfrac{P(\overline{E})}{P(E)}$.

Open-Ended

3. Answers vary; for example, you pay $1.00 to play the game; $2 if you win. If both flips are heads or both are tails, you win. Otherwise, you lose.   **5.** Answers vary. One possibility is a game involving a spinner with five sectors of equal area numbered 1 through 5. You pay $2 for playing the game. If the spinner lands on regions 1 or 2, you win $5. The expected winnings in this game are $5 \cdot \dfrac{2}{5} - 2$, or 0.

Questions from the Classroom

7. Suppose the odds in favor of winning a game are 1 : 2. Even when the outcomes are equally likely, it does not mean that out of every three games the player will win one game. It only means that if a large number of games are played, the ratio between the number of wins and the number of losses is close to $\dfrac{1}{2}$. However, the answer is partially correct because while the ratio of wins to losses is $\dfrac{P(W)}{1 - P(W)} = \dfrac{a}{b}$; $P(Win) = \dfrac{a}{a + b}$.   **9.** The probability is $\dfrac{1}{3}$ that an individual does not vote. Maria expects $48 - \dfrac{1}{3} \cdot 48$, or 32 to vote. Since she is guaranteed 24 votes and 24 > (51%)(32), she could be confident of winning.

Another way to consider the situation is to consider that 32 are expected to vote, BUT perhaps $\dfrac{1}{3}$ of 24, or 8, that she expected to vote for her might not. In that situation, she is guaranteed $24 - 8$, or 16, votes but needs 51%(32), or 17. She is missing 1 vote. Her ground is certainly less safe in this scenario.

Review Problems

11. The blue section must have angle measure 300 degrees; the red has angle measure 60 degrees.

Assessment 9-5A

1. 224   **2.** 10,000   **3.** 180   **4. (a)** true   **(b)** false   **(c)** false   **5.** 40,320   **6.** 15   **7. (a)** 24,360   **(b)** 4060

8. $\dfrac{1}{120}$ **9.** 45   **10.** 1260   **11.** eight people

12. 2,598,960 different 5-card hands (Order within the hand is not important.)   **13.** $\dfrac{1}{25,827,165}$

14. (a) $\dfrac{{}_{12}C_4}{{}_{22}C_4}$, or $\dfrac{9}{133}$ **(b)** $\dfrac{{}_{10}C_4}{{}_{22}C_4}$, or $\dfrac{6}{209}$

(c) $1 - \dfrac{{}_{10}C_4}{{}_{22}C_4}$, or $\dfrac{203}{209}$ **15. (a)** $({}_{20}C_2 \cdot {}_{21}C_4 \cdot {}_4C_2) \div {}_{45}C_8 \approx 0.032$ **(b)** $\dfrac{{}_{25}C_8}{{}_{45}C_8} \approx 0.005$ **(c)** $1 - \dfrac{{}_{25}C_8}{{}_{45}C_8} \approx 0.995$

(d) $\dfrac{{}_{20}C_8}{{}_{45}C_8} \approx 5.84 \cdot 10^{-4}$ **16. (a)** $\dfrac{1}{120}$ **(b)** 0   **(c)** 0

17. (a) 10^4, or 10,000   **(b)** 5040   **(c)** $\dfrac{625}{10,000}$, or $\dfrac{1}{16}$

18. ${}_6C_4$ or 15   **19.** 9! or 362,880   **20.** 56 choices

Mathematical Connections 9-5

Communication

1. Answers vary. For example, the Fundamental Counting Principle (FCP) says that to find the number of ways of making several decisions in a row, multiply the number of choices that can be made for each decision. The FCP can be used to find the number of permutations. A permutation is an arrangement of things in a definite order. A combination is a selection of things in which the order is not important. We could find the number of combinations by using the FCP and then dividing by the number of ways in which the things can be arranged.   **3. (a)** $8! \cdot 3!$. If the family is considered a unit and each of the remaining people also a unit, we have 8 units. There are 8! ways to arrange the 8 units. For each of the 8! ways, the family unit can be arranged in 3! ways and hence the number of seating arrangements is $8! \cdot 3!$.   **(b)** $10! - 8! \cdot 3!$. We need to subtract the answer in part (a) from the number of all possible seating arrangements.

Open-Ended

5. (a) 10^6, or 1,000,000   **(b)** Yes, Montana, Wyoming, Alaska, Delaware, North Dakota, South Dakota, Vermont   **(c)** Answers vary. For example, you would first find the population of California and then experiment with using letters in the license plates. This would help because the choice is for 26 letters in a slot rather than 10 numbers.

Questions from the Classroom

7. The student is confused about choosing four objects none at a time. There is one way to choose no objects. Therefore, we say $_4P_0 = 1$.   **9.** If the principle is used properly, one can compute the number of permutations. That does not imply that they are not needed.

Review Problems

11. $\dfrac{3}{36}$, or $\dfrac{1}{12}$ **13. (a)** no   **(b)** The expected payoff is $\dfrac{1}{38} \cdot \$37$, or approximately 97¢. Therefore, you can expect to lose on the average about 3¢ a game if you play it a large number of times.

Chapter Review

1. (a) {*HHH, HHT, HTH, HTT, TTT, TTH, THH, THT*}
(b) {*HHH, HHT, HTH, THH*}   **(c)** $\dfrac{4}{8}$, or $\dfrac{1}{2}$

2. (a) {Monday, Tuesday, Wednesday, Thursday, Friday, Saturday, Sunday}   **(b)** {Tuesday, Thursday}   **(c)** $\dfrac{2}{7}$

3. There are 800 blue ones, 125 red ones, and 75 that are neither blue nor red.   **4. (a)** approximately 0.497
(b) approximately 0.503   **(c)** about 1.01 to 1

5. (a) $\dfrac{5}{12}$ **(b)** $\dfrac{9}{12}$, or $\dfrac{3}{4}$ **(c)** $\dfrac{5}{12}$ **(d)** $\dfrac{9}{12}$, or $\dfrac{3}{4}$ **(e)** 0

(f) 1   **6. (a)** $\dfrac{13}{52}$, or $\dfrac{1}{4}$ **(b)** $\dfrac{1}{52}$ **(c)** $\dfrac{22}{52}$, or $\dfrac{11}{26}$

(d) $\dfrac{48}{52}$, or $\dfrac{12}{13}$ **7. (a)** $\dfrac{64}{729}$ **(b)** $\dfrac{24}{504}$, or $\dfrac{1}{21}$ **8.** $\dfrac{6}{25}$

9. $\dfrac{14}{80}$, or $\dfrac{7}{40}$ **10.** $\dfrac{7}{45}$ **11.** 4 to 48, or 1 to 12

12. 3 to 3, or 1 to 1   **13.** $\dfrac{3}{8}$ **14.** $0.30   **15.** $33\dfrac{1}{3}$¢

16. The expected value is $1.50. The expected net winnings are ⁻$0.50.   **17.** 900   **18.** 120   **19.** 5040

20. $\dfrac{2}{20}$, or $\dfrac{1}{10}$ **21. (a)** $5 \cdot 4 \cdot 3$, or 60   **(b)** $\dfrac{3}{60}$, or $\dfrac{1}{20}$

(c) $\dfrac{1}{60}$ **22.** $\dfrac{15}{36}$, or $\dfrac{5}{12}$ **23.** $\dfrac{2}{5}$ **24.** $7! = 5040$

25. 0.027   **26.** $\dfrac{63}{80}$ **27.** Answers vary.   **28.** These events are not equally likely. If each probability is computed, we have $P(3H) = \dfrac{1}{8}$, $P(2H) = \dfrac{3}{8}$, $P(1H) = \dfrac{3}{8}$, and $P(0H) = \dfrac{1}{8}$.   **29. (a)** $\dfrac{1}{8}$ **(b)** $\dfrac{1}{4}$ **(c)** $\dfrac{1}{16}$

30. $\dfrac{8}{20}$, or $\dfrac{2}{5}$

Answers to Now Try This

9-1. (a) 1   **(b)** 1   **(c)** Yes, they always sum to 1. 1 is the sum of the probabilities of all the different elements in any sample space.   **9-2. (a)** $\dfrac{1}{10}$ **(b)** $\dfrac{1}{10}$ **(c)** Answers may vary depending on the phone. Possibly $\dfrac{6}{10}$, or $\dfrac{3}{5}$.

(d) Answers may vary depending on the phone. Possibly $\dfrac{2}{10}$, or $\dfrac{1}{5}$.   **(e)** Answers may vary depending on the phone. Possibly $\dfrac{2}{10}$, or $\dfrac{1}{5}$ **(f)** No; the probabilities are not equal.

9-3. 1. (a) $P(\text{both balls are red}) = \dfrac{1}{9}$ **(b)** $P(\text{no ball is red}) = \dfrac{4}{9}$ **(c)** $P(\text{at least one ball is red}) = \dfrac{5}{9}$ **(d)** $P(\text{at most one ball is red}) = \dfrac{8}{9}$ **(e)** $P(\text{both balls are the same color}) = \dfrac{3}{9} - \dfrac{1}{3}$ **2.** The answers are the same as in part 1.

9-4.

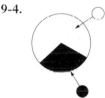

9-5. (a) $P(\text{rain both days}) = (0.3)(0.6) = 0.18$
(b) $P(\text{no rain either day}) = (0.7)(0.4) = 0.28$
(c) $P(\text{rain exactly one day}) = (0.3)(0.4) + (0.7)(0.6) = 0.12 + 0.42 = 0.54$ **(d)** $P(\text{rain at least one day}) = 1 - 0.28 = 0.72$ **(e)** No. If conditions are favorable for rain tonight, they are more likely to be favorable for rain tomorrow.   **9-6. (a)** $\dfrac{2}{3}$ **(b)** $\dfrac{1}{3}$ **(c)** $\dfrac{8}{15}$

9-7. (a) yes   **(b)** With replacement: any game with the same number of white and colored marbles. Without replacement: 3 colored and 6 white or 3 white and 6 colored, 6 colored and 10 white or 6 white and 10 colored marbles.
(c) Without replacement: the white and colored marbles need to be two consecutive triangular numbers, that is, $1 + 2 + 3 + \ldots + n$ and $1 + 2 + 3 + \ldots + n + n + 1$ (this is easier to discover and verify using combinations introduced in Section 9–5).   **9-8. (a)** Answers will vary. Earth is approximately 30% land.   **(b)** 0.7

(c) $500(0.3) = 150$ **9-9. (a)** Answers vary.   **(b)** $\dfrac{3}{8}$

(c) No, simulations will not always result in the same probability as the theoretical probability. However, if the experiment is repeated a great number of times, the simulated probability should approach the theoretical probability.

9-10. The odds in favor of winning are $\dfrac{1}{55,000,000}$.

9-11. (b) There is one way to toss a head and one way of not tossing a head, so the odds in favor are $1:1$.   **(c)** There are 4 ways to draw an ace and 48 ways of not drawing an ace, so the odds in favor are $4:48$, or $1:12$.   **(d)** There are 13 ways of drawing a heart and 39 ways of not drawing a heart, so the odds in favor are $13:39$, or $1:3$.   **9-12.** The mathematical expectation is about \$168,516.84.
9-13. (a) $n(n-1)$, $n(n-1)(n-2)$, $n(n-1)(n-2)(n-3)$
(b) $n(n-1)(n-2)\cdot\ldots\cdot(n-r+1)$ **(c)** $4\cdot3\cdot2$ $=24$ ways to choose a president, vice president, and secretary
9-14. Answers vary.   **(a)** We get an error message because 100! and 98! are too large for the calculator to handle.
(b) $\dfrac{100!}{98!} = \dfrac{98!\cdot99\cdot100}{98!} = 9900$ **9-15.** $\dfrac{1}{9}$

Answers to Brain Teasers

Section 9-2

These dice are truly remarkable in that no matter which die the other player chooses, you can always choose one that will beat it $\dfrac{2}{3}$ of the time. Therefore, the strategy for playing this game is to go second and make your choice of die based on the information in the following table:

First Person's Choice	Second Person's Choice	Probability of Second Choice Winning
A	D	$\dfrac{2}{3}$
B	A	$\dfrac{2}{3}$
C	B	$\dfrac{2}{3}$
D	C	$\dfrac{2}{3}$

Section 9-4

Page 572   In order to have the same number of heads, Al and Betty must both toss 0 heads or they must both toss 1 head. The probability that Al tosses 0 heads is $\dfrac{1}{2}$, and the probability he tosses 1 head is $\dfrac{1}{2}$. For Betty, the probability that she tosses 0 heads is the same as tossing 2 tails, which is $\dfrac{1}{4}$. The probability that she tosses 1 head is equal to $P(HT) + P(TH) = \dfrac{1}{4} + \dfrac{1}{4} = \dfrac{1}{2}$. Next, the probability of Al tossing 0 heads and Betty tossing 0 heads is $\dfrac{1}{2}\cdot\dfrac{1}{4} = \dfrac{1}{8}$. The

probability of Al tossing 1 head and Betty tossing 1 head is $\dfrac{1}{2}\cdot\dfrac{1}{2} = \dfrac{1}{4}$. Therefore, the probability that they toss the same number of heads is $\dfrac{1}{8} + \dfrac{1}{4} = \dfrac{3}{8}$.

Page 575   We first find the probability that all 40 children have different birthdays and then subtract the result from 1. We get

$$1 - \left(\frac{365-1}{365}\right)\left(\frac{365-2}{365}\right)\ldots\left(\frac{365-39}{365}\right) = 0.891$$

Thus, the probability that the friend wins the bet is approximately 0.891, or 89.1%. For 50 people, the probability rises to approximately 0.97.

A different approach to solving the problem is to think about it as an occupancy problem where the names of 40 children are randomly put in 365 boxes, one box for each day of the year. The sample space S is the set of all permutations of 365 possible birthdays (repetition allowed). Thus, S has 365^{40} elements. The probability is

$$\frac{365^{40} - 365\cdot364\cdot\ldots\cdot(365-39)}{365^{40}}$$

This expression can of course be written in the form found in the first approach.

Section 9-5

(a) The probability of a successful flight with two engines is 0.9999.   **(b)** The probability of a successful flight with four engines is 0.99999603.

Answer to Laboratory Activity

Section 9-1

1.–3. Answers vary.

Answers to Technology Corner

Section 9-3

1. (a) Use **randint**$(1, 6)$ to randomly generate a number from 1 to 6. **randint**$(1, 6, 5)$ will randomly generate five numbers from 1 to 6.   **(b)** Use **randint**$(1, 6)$ + **randint**$(1, 6)$ to find the sum of the two randomly generated numbers. Or use the **dice** function to simulate the sum of two dice. Using **dice**$(5, 2)$, for example, gives the sum of the numbers on two dice on five rolls of the dice.   **2. (a)** Combine the integer function with the rand function. **ipart**$(6*\textbf{randint} + 1)$ will randomly generate integers from 1 to 6.
(b) **ipart**$(6*\textbf{randint} + 1)$ + **ipart**$(6*\textbf{randint} + 1)$ will give the sum of two randomly generated numbers.
3. Answers will vary.

Answer to Preliminary Problem

The probability of the second two-digit number matching the first is $\frac{1}{10} \cdot \frac{1}{10}$ or $\frac{1}{100}$. The probability of the two numbers not matching is $1 - \frac{1}{100}$, or $\frac{99}{100}$. For the 10 cases, the probability of being able to draw a segment every time is $\left(\frac{99}{100}\right)^{10}$. Thus, the probability of not being able to draw at least one segment is $1 - \left(\frac{99}{100}\right)^{10}$, or approximately 9.6%.

Chapter 10

Assessment 10-1A

1. **(a)** 225 million **(b)** 375 million **(c)** 550 million
2. **(a)** Answer may vary. An example follows:

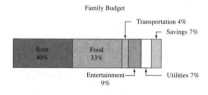

(b) Given the condition (constant rate), the graph is reasonable. However, it is probably unreasonable to expect a constant death rate. **3.** Percentages are approximate.

4. **(a)** 72, 74, 81, 81, 82, 85, 87, 88, 92, 94, 97, 98, 103, 123, 125 **(b)** 72 lb **(c)** 125 lb
5.

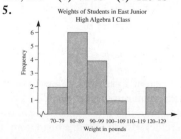

6. Answers will vary. **(a)** A dot plot will have two columns of *x*'s, one for Heads and one for Tails. **(b)** A bar graph will have two bars, one for Heads and one for Tails. The vertical axis will be partitioned to show frequencies of each. **7. (a)** November, 30 cm **(b)** 50 cm
8. **(a)**   Ages of HKM Employees

```
1 | 889
2 | 0111333334566679
3 | 224447
4 | 1115568          3 | 4 represents
5 | 2248             34 years old
6 | 233
```

(b) There are more employees in their 40s. **(c)** 20
(d) 17.5% **9. (a)** approximately 3800 km **(b)** approximately 1900 km
10.

Tosses of Five Coins

11. **(a)** Answers may vary. For example, a histogram might be used to depict real or estimated census data over the 200 year period. **(b)** A histogram could depict the collected numerical data over time.
12. **(a)** 19% of 21 people is approximately 4 people.
14% of 21 people ≈ 3 people.
52% of 21 people ≈ 11 people.
Each of the 5% sectors possibly represents 1 person. There are only 21 people total. **(b)** The answers in this part will depend on the person answering. **13.** Answers will vary. A graph of this type might be suspect. To have a survey where there are more people in the 70s–90s might mean the people are in a home for the aging. Also to have more people in their 90s than 80s reading might make one suspicious. Without more information, one cannot tell.
14. From this particular survey, one would suspect that people in their 70s read the most, **but** the graph only depicts the number of readers of mysteries.
15. **(a)**   Fall Textbook Costs

```
1 | 6
2 | 33
3 | 0357799
4 | 0122589
5 | 00138
6 | 022              2 | 3 represents $23
```

(b) Fall Textbook Costs

Classes	Tally	Frequency
$15–19	|	1
$20–24	||	2
$25–29		0
$30–34	||	2
$35–39	⧄|	5
$40–44	||||	4
$45–49	|||	3
$50–54	||||	4
$55–59	|	1
$60–64	|||	3
		25

(c)

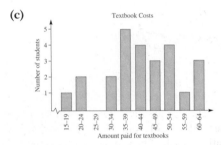

Textbook Costs

16. (a) women **(b)** approximately 1 yr **(c)** about 5.5 yr
17. (a) Asia **(b)** Africa **(c)** It is about 2/3 as large.
(d) Asia and Africa **(e)** 5 : 16 **(f)** approximately
58.6 million mi^2

Mathematical Connections 10-1

Communication

1. (a) Answers vary; for example, in a pictograph, we can
observe change over time and make comparisons between
similar situations. In a circle graph, it is hard to make these
observations. **(b)** Answers vary; for example, circle graphs
are used when comparing parts to a whole. Circle graphs
allow for visual comparisons of fractional parts. Bar graphs
can't handle these comparisons as well. **(c)** Answers vary;
for example, in a stem and leaf plot, we can see the same
shape and information as in a histogram but no data points
are lost in the display as they may be in a histogram,
especially when intervals are used. **3.** Answers vary;
for example, the percents are of a whole and the circle represents the complete whole, or 100%. Because of rounding,
it may happen that the sum of the percents is close to 100%
but not exactly 100% **5. (a)** Answers vary; for example,
a different graph may be more appropriate since we have
continuous data changing over time. A line graph is given
though students see this in the next section.

(b) The data fall into distinct categories and are not continuous, so we use a bar graph.

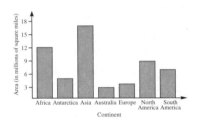

7. Answers vary; a good source of graphs is *U.S.A. Today*.
9. (a)

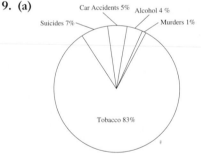

Estimated Death Causes in Canada in 1996

Note: The data here are from the cigarette package which
does not depict *all* Canadian deaths.
(b) Answers may vary. **(c)** Answers may vary.

Cooperative Learning

11. Answers may vary. A circle graph might be a logical
choice with the total graph representing 100% of time and
sectors representing teaching, service, and research. The
emphases in the different sectors may represent the
emphases of the university.

Questions from the Classroom

13. Jackson is correct that a stem and leaf plot does not
have to be constructed with small numbers at the top. An
advantage to constructing one in this way is that one can rotate the graph 90° counterclockwise to obtain a histogram.

Assessment 10-2A

1. (a) approximately $8400 **(b)** $14,000
(c) approximately $7000 **(d)** right after 2 yr
2. (a)

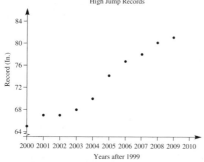

High Jump Records

(b) positive **3. (a)** negative **(b)** approximately 10
(c) about 22 yr old
4. (a), (b) 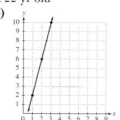 **(c)** $y = 4 - 2x$

5. (a) (i) negative
(ii) Answers will vary.

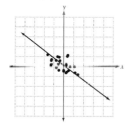

(b) no association **6. (a)** $y = \frac{3}{8}x - \frac{3}{10}$

(b) $y = \frac{1}{2}x + \frac{3}{2}$ **(c)** $y = \frac{3}{4}x + \frac{1}{2}$

7.

Olympic Discus Lengths Tossed

Answers will vary.

8.

Cost per Carat of Diamonds

Using ordered data and drawing a trend line, an estimate for 0.5 carat is about $1863, but it is unwise to estimate outside the given range of data.

9.

Year and Age of Best Actress

A trend line is sketched. The predicted age is about 41.
10. (8D) September; (9D) **11. (a)** -1 **(b)** 47 **(c)** 5
(d) 20 **12.** negative **13.** The data has no apparent association or the data could be constant.

Mathematical Connections 10-2

Communications

1. Answers will vary, but if association is positive, the slope of the trend line is positive; if association is negative, so is the slope of the trend line.

Cooperative Learning

3. Answers will vary depending upon books chosen.

Open-Ended

5. Answers will vary. If equations of trend lines are found with technology, there will likely be many decimal places involved.

Questions from the Classroom

7. Jacquie is incorrect. A trend line is used to predict and might contain none of the original data points. **9.** Arthur is likely incorrect. A trend line will only be flat if the currency is pegged to the dollar at a fixed rate or if there is no association shown.

Review Problems

11. If the miscellaneous category were only 10%, the stacked graph follows:

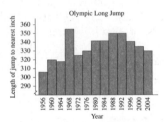

Smith Expenses

| Misc. 10% |
| Gas 13% |
| Utilities 5% |
| Food 20% |
| Rent 32% |
| Taxes 20% |

13. Depictions will vary.

Olympic Long Jump

Assessment 10-3A

1. (a) Mean = 6.625, median = 7.5, mode = 8.
(b) Mean $13.\overline{4}$, median = 12, mode = 12.
(c) Mean ≈ 19.9, median = 18, modes = 18 and 22.
2. Answers will vary; for example, 1, 2, 3, 4, 5, 6, 6.
3. 1500   **4.** $78.\overline{3}$ **5.** Answers will vary.   **(a)** Suppose the scores were 100, 100, 50, 50. The mean is 75.
(b) Suppose the scores are 100, 100, and 50. The mean is $83.\overline{3}$.   **6.** 62.5   **7.** $\dfrac{100m + 50n}{m + n}$ = mean
8. approximately 2.59   **9.** approximately 215 lb
10. (a) $41,275   **(b)** $38,000   **(c)** $38,000
11. (a) $132,000   **(b)** $12,143.75   **(c)** approximately $21,756.60   **(d)** $14,500   **12. (a)** Answers may vary. The median of $38,000 reported with the interquartile range of $14,500 would be one appropriate set of choices.
(b) Based on the answer in (a), we know that at least 50% of the salaries are at least within $14,500 of $38,000.

13. $19\frac{2}{3}$ mpg   **14.** 58 years old   **15.** 91   **16.** 96, 90, and 90   **17. (a)** A—$25, B—$50   **(b)** B   **(c)** $80 at B
(d) Answers vary. There is more variation at Theater B; also there are higher prices at Theater B.   **18.** $s ≈ 7.3$ cm
19. (a) approximately 76.8   **(b)** 76   **(c)** 71   **(d)** 14
(e) approximately 156.8   **(f)** approximately 12.5
(g) approximately 9.04   **20.** 0.68   **21.** 16%

22. They are equal.   **23.** 90   **24.** between 60.5 in. and 70.5 in.   **25. (a)** 65   **(b)** 53   **(c)** 77   **(d)** 65
26. (a) cum/n

0
2.5
2.5
10.3
15.4
30.8
43.6
61.5
74.4
94.9
100

(b)

Cumulative Frequency Curve

(c) Yes; the curves should always slope up to 100%.

Mathematical Connections 10-3

Communication

1. One might use the mode. If you considered the most common age at which one could get a license, that would be the mode. Both the mean and median might be nonwhole numbers which would not be used for a driver's license.
3. The government probably uses the mean of data collected over a period of time.   **5. (a)** for example, 10, 30, 70, and 90   **(b)** Choose four numbers whose mean is 50.   **(c)** The mean of the new numbers is 50.
7. Answers may vary. The mode could be an extreme value in a data set (for example, the least value) and not represent the center at all.   **9.** Answers may vary. A set of data that contains wait time might be appropriate *if* the times were determined during peak people-period events. In that case, the mean with mean absolute deviation and standard deviation would be appropriate for this numerical data.

Open-Ended

11. (a) Answers vary; for example, the more time spent in school, the better the math scores will be.   **(b)** The box plot should be used.

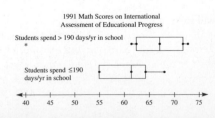

1991 Math Scores on International Assessment of Educational Progress

Students spend > 190 days/yr in school

Students spend ≤190 days/yr in school

(c) Answers vary; for example, the graph shows that overall the more time that is spent in school, the better the math scores. It also shows that the score of 40 for countries that spend more than 190 days is an outlier. It appears that almost 50% of the countries that spent more than 190 days outscored all of the countries that spent 190 days or fewer.

(d)

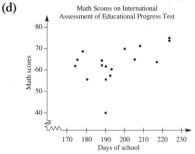

Math Scores on International Assessment of Educational Progress Test

Cooperative Learning

13. Answers will vary. To do this problem, the class will have to be divided in such a way that the choice of newspapers does not overlap. A teacher may want to discuss how one might do a random sample of newspapers.

Questions from the Classroom

15. The median is found by finding the mean of the two middle values. Because it is found by dividing by 2, the digit in the tenths place will always be either a 0 or a 5.   **17.** In a grouped frequency table, the precise value of the raw data are not displayed, and hence it is impossible to conclude from the table which value occurs most often. Consequently, it is impossible to find the exact mode from the information given in a grouped frequency table. In this situation, the mode is usually given as a class interval.   **19.** No, it is not possible to have a standard deviation of $^-5$. By definition, the standard deviation is the prinicipal square root of the variance.

Review Problems

21. Answers vary; for example, the number of men in the workforce has increased by a factor of about 4 over the last 110 years, while the number of women in the workforce has increased by about a factor of 17. As late as 1960, there were more than two times more men in the workforce than there were women. In 1970, this was no longer true, and in 2000 there were only about 15% more men.

23. (a) Everest, approximately 8500 m   **(b)** Aconcagua, Everest, McKinley

Assessment 10-4A

1. Answers may vary in all parts of this question.   **(a)** A question to ask is whether the car is running. If it is not running, then there is no sound and the car would be quieter.   **(b)** One question is how many motorcycles were sold in what years.   **(c)** 11% is not much more than 10%

with the given base for the percentage.   **(d)** Any time there is a percentage involved, one should ask "percentage of what?"   **2.** Answers may vary. One possibility is that the temperature is always 25°C .   **3.** The horizontal axis does not have uniformly sized intervals and both the horizontal axis and the graph are not labeled.   **4.** There were more scores above the mean than below, but the mean was affected more by low scores.   **5.** It could very well be that most of the pickups sold in the last 10 yr were actually sold during the last 2 yr. In such a case, most of the pickups have been on the road for only 2 yr, and therefore the given information might imply, but would not substantiate, that the average life of a pickup is around 10 yr.   **6.** Answers may vary. It is expected that the 892 respondents represent a large number of those surveyed and further that the 892 respondents are representative of the entire population of Americans.   **7.** The three-dimensional drawing distorts the graph. The result of doubling the radius and the height of the can is to increase the volume by a factor of 8.
8. One would need more information; for example, do men drive more than women?   **9. (a)** False; prices vary only by $30.   **(b)** False; the bar has 4 times the area but this is not true of prices.   **(c)** true   **10. (a)** This bar graph could have perhaps 20 accidents at the point where the scale starts. Then 38 in 2008 would appear to be over four times the 24 of 2004, when in fact it is only 58% higher.   **(b)** A bar graph with a vertical scale going to 100 would minimize the effect.   **11.** Answers may vary, but one such would be 5, 5, 5, 5, 5, 5, 100, 100. The mean would be 28.75 and the median 5. Neither is representative.   **12.** Answers may vary.   **(a)** As a teacher at either level, if I were not using calculators/computers, I would question why more than 40% and 50% at the two respective levels were doing something I was not. Not knowing about the number of teachers surveyed and the population allows for misuse at all levels by all types of people.   **(b)** As a principal, I might question how I could help the 57% and 43% of teachers, respectively, who were not using technology. I would need to question whether my teachers fit these survey results.
(c) As a parent, I would need to know how my child's teachers would have answered the survey before reaching any conclusions pro or con on calculator/computer usage.
13. Answers may vary.   **(a)** An immediate reaction might be that I need to address the 61% of middle-school teachers who appear not to be highly qualified.   **(b)** The data could be misused to justify actions if the state's middle-school teachers did not match the survey results.   **14.** Answers may vary.   **(a)** A reaction might be to introduce a bill to raise starting teacher salaries based on these data.
(b) A knowledge of the mean, mean absolute deviation, and standard deviation for a particular state's comparison would help avoid misuse.   **15.** Answers may vary.   **(a)** An advisor might recommend teaching in a nonurban area for the first 5 yr.   **(b)** There are many factors that affect teachers who quit. An informed counselor might seek specific data about the area or region where an advisee plans to teach.

16. A median might be misleading, depending on the number of data points given. Also the mean would not be sufficient. A report of the mean, median, and standard deviation would be the most helpful of all "averages" studied.
17. It is possible. Such a country might be the Netherlands. The country has dikes to help keep out the sea.
18. Answers may vary. A sample size is needed. Then, one way to pick a random sample of adults in the town is to use the telephone book or a voter registration list. These methods will not list all adults in the town, but these are probably the most accessible sets of data. To pick the sample, one might roll a die and consider the number n that appears on it. Then starting at a random point on the adult list, choose every nth person after the start on the list.   **19.** One of the primary reasons that this statement could be true is that the cost of replacing defective chips is less expensive than attempting to correct the problems that caused the chips to be defective in the first place. The claim seems impossible, but it is true.   **20.** The student is incorrect and, in addition, is misusing percentages. She mixes percentages of "effectiveness" with percentages of times taking the drug.
21. Sam is possibly incorrect. Without more information, Sam's conclusion cannot be verified. One needs information about the remainder of the total population and the spread of the disease.

Mathematical Connections 10-4

Review Problems

1. (a) about 74.17   **(b)** 75   **(c)** 65   **(d)** about 237.97
(e) about 15.43   **(f)** 13   **3.** $76.\overline{6}$
5.

Women's Olympic 100 m Swim Times 1960–2004

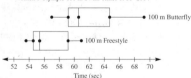

From examining the box plot, we can see that the times on the 100 m butterfly are much greater (relatively speaking) than the times on the 100 m freestyle.

Chapter Review

1. If the average is 2.41 children, then the mean is being used. If the average is 2.5, then the mean or the median might have been used.   **2.** 12   **3. (a)** Mean = 30, median = 30, mode = 10.   **(b)** Mean = 5, median = 5, modes = 3, 5, 6.   **4. (a)** Range = 50, variance ≈ 371.4, standard deviation ≈ 19.3, mean absolute deviation ≈ 17.14, IQR = 40.   **(b)** Range = 8, variance = 5.2, standard deviation = 2.28, mean absolute deviation = 1.8, IQR = 3.
5. (a)

Miss Rider's Class
Masses in Kilograms

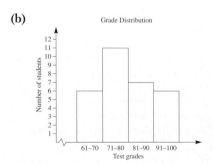

(b) Miss Rider's Class
Masses in Kilograms

3 | 99
4 | 001122223345678999   4 | 0 represents 40 kg

(c) Miss Rider's Class
Masses in Kilograms

Mass	Tally	Frequency
39	||	2
40	||	2
41	||	2
42	||||	4
43	||	2
44	|	1
45	|	1
46	|	1
47	|	1
48	|	1
49	|||	3
		20

(d)

Miss Rider's Class
Masses in Kilograms

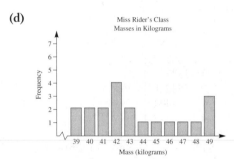

6. (a) Test Grades

Classes	Tally	Frequency
61–70	⫼⫼ |	6
71–80	⫼⫼ ⫼⫼ |	11
81–90	⫼⫼ ||	7
91–100	⫼⫼ |	6
		30

(b)

Grade Distribution

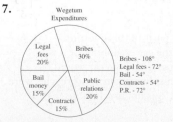

7.

Wegetum Expenditures

Legal fees 20%
Bribes 30%
Bail money 15%
Public relations 20%
Contracts 15%

Bribes - 108°
Legal fees - 72°
Bail - 54°
Contracts - 54°
P.R. - 72°

8. The widths of the bars are not uniform.   **9.** $2840

10.

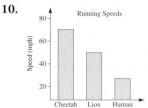

Running Speeds

11. (a)

Life Expectancies
of Males and Females

Females		Males
	67	1446
	68	28
	69	156
	70	0049
	71	02235578
	72	01145
	73	1689
7	74	1478
9310	75	2224
86	76	
88532	77	
9999854332211	78	
98554210	79	
4441	80	
2	83	

7 | 74 | represents
74.7 years old

| 67 | 1 represents
67.1 years old

(b)

Life Expectancies

(c) 75% of the women's data is significantly above that of men; women are expected to live longer.   **12.** Larry was correct because his average was $3.2\overline{6}$, while Marc's was $2.7\overline{3}$.
13. (a) 360   **(b)** none   **(c)** 350   **(d)** $s \approx 108.2$
(e) 320   **(f)** 200   **(g)** about 11,711.11   **(h)** $84.\overline{4}$
14. (a) 67 mph   **(b)** $Q_3 = 74, Q_1 = 64$
(c)

Car Speeds

(d) approximately 50%   **(e)** 30%   **(f)** About 50% of people checked drove between 65 and 75 mph.   **(g)** 64; 74; 67
15. (a) positive   **(b)** 170 lb   **(c)** 67 in.   **(d)** 64 in.
(e) 50 lb   **16. (a)** Collette is more consistent. There is less deviation from the mean in her scoring.   **(b)** Collette: 18–30; Rudy: 10–38   **(c)** Collette scored more than might have been expected.   **17.** 2.5%   **18. (a)** the score for each   **(b)** same as (a)   **(c)** same as (a)   **(d)** 0   **(e)** 0

19. The bars would be approximately the same height in a vertical bar graph.   **20. (a)** 19 lb   **(b)** It is possible, but the probability is near 0.   **(c)** Since the company claims a person can lose up to 6 lbs, any average loss from 0 to 6 is legitimate.   **21.** Answers may vary. The 15% of women versus 85% appears as if there might be bias, but other factors such as available applicants for a position are a significant factor.   **22.** No reportable single number would be meaningful for such a pool.   **23.** Answers may vary. A reasonable way to make the claim is to pull boxes at random after they are packed and weigh the contents. If more than 98% of the boxes' contents have weight within 1 g of 138 g, the claim is reasonable.   **24. (a)** Answers will vary. Perhaps to be with those of comparable age a typical senior might choose the first.   **(b)** the first   **25.** Answers may vary. The validity is suspect, though kids may reflect parents' voting habits.   **26.** Answers may vary.   **(a)** One way would be to leave the television on, even if no one is watching.
(b) They show very popular shows during "ratings sweeps" periods.   **27.** Answers may vary. For example, graphs may show area or volume instead of relative size; another is to select a horizontal baseline that will support the point being made.   **28.** The advertisement is not reasonable. One would expect the snow to be okay, but with reports like those given, there is little information about the snow in the middle of the mountain. If it is cold enough to have 23 in. at the bottom of a hill, one would expect there to be snow all the way to the top. However, if the bottom and the top are shaded, then there could be much variation if a part of the hill were sunny.   **29. (a)** Answers vary; for example, 5.1.
(b) between 83–84 and 84–85   **(c)** A line graph is more appropriate because the data is continuous over time.
(d) Answers vary; for example, you could change the scale on the y-axis to go up by 5s or 10s rather than by 20s.
30. (a) 25   **(b)** 475   **(c)** 16%   **31.** 475   **32. (a)** 525
(b) 600   **(c)** 675

Answers to Now Try This

10-1.   Ages of Presidents at Death

4	69
5	36778
6	003344567778
7	0112347889
8	01358
9	0033

4 | 9 represents
49 years old

10-2. It appears that the fifth-period class did better. We can see that there are more scores grouped toward the bottom, which is where the higher grades are located.
10-3. The implication is that the data are continuous.
10-4. Answers will vary with the number of terms.
(a)

# of Term	Value of Term
1	$^{-}10$
2	$^{-}6$
3	$^{-}2$
4	$+2$

(b)

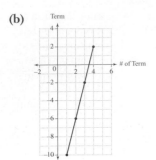

(c) $y = 4x - 14$, where x is a natural number
10-5. (a) A histogram is more appropriate to compare numbers of data that are grouped into numerical intervals. With the line graph there appears to be a frequency for *Times per month* such as 4.5, etc.   **(b)** Connecting the dots is visually okay, but is mathematically meaningless.
(c) (i) A circle graph is appropriate for showing the division of a whole into parts.   **(ii)** A line graph is appropriate for showing how data values change over time.
10-6. (a) Answers vary; for example, all the puppies could weigh 7 lb. Then $(7 \cdot 7)/7 = 7$ lb. Another possibility is 4, 5, 6, 7, 8, 9, 10. The mean for these scores is also 7.   **(b)** If all the scores were equal, then the mean is equal to the high score and also the low score.   **10-7.** Answers may vary. For example, one set of data is 92, 94, 94, 94, 96, in which the mean, median, and mode are all 94.   **10-8. (a)** The average of 2.58 could only be a mean. To be a mode it would have to be a whole number. To be a median it would have to be a whole number plus $\frac{1}{2}$.   **(b)** The average of 2.5 could be a mean or a median. It could not be a mode since the mode for the number of children must be a whole number.   **10-9.** One possible set is 0, 0, 40, 40, 40, 70, 70, 80, 90, 90, 100.   **10-10.** One would not expect any outliers when the lower end of the salary scale was dropped. One would expect all the five-number summary points to increase. There will probably be little noticeable difference in the story the data tell.   **10-11.** Because there is no base for the percentages given, arguments are difficult to make. Assuming the base is the set of all parents, consider the following: **(a)** The argument cannot be made because we do not know even if the other 82% answered this part of the survey.   **(b)** The argument cannot be made. The statement may be the result of considering test/learning to be two equal parts that make 18%. Hence, half of 18% is 9%.   **(c)** The argument cannot be made; there is no cause-effect relationship shown.   **(d)** This argument cannot be made. Possibly it could be made for 11%.   **(e)** The argument cannot be made. The indicated 9% does not tell us about the other 91% of parents.

Answers to Brain Teasers

Section 10-2

Page 629   Answers will vary.

Section 10-3

Page 636   The completed table follows:

	Overall Mean	Mean for Females	Mean for Males	Percent Females
Ramirez	218	230	205	15
Jonsey	221	224	199	88

The percentage of males in Ms. Jonsey's class is 100% − 88%, or 12%. Also the mean for Ms. Jonsey's class is determined as follows:

$$(88\%)(224) + 12\%(\text{Mean for males}) = 221$$

Solving, we find that the mean for males in Ms. Jonsey's class is 199.

For the entry for Mr. Ramirez's class, we know that the percentages of males and females must sum to 100%. If F is the percentage of females, then the percentage of males in that class must be $1.00 - F$. Further, we know that the overall mean is determined as follows:

$$(F)(230) + (1.00 - F)(205) = 218$$

Solving for F, we find that $F = 52\%$.

Finally, we observe that the means for both females and males were higher in Mr. Ramirez's class, but the overall mean is higher in Ms. Jonsey's class.

Page 650   We know the sum of the ages of the first seven guests divided by 7 is 21. Therefore, the sum of the ages is $7 \cdot 21 = 147$. If we add another age of 29 to the sum, we have $147 + 29 = 176$. If we divide by 8, we see the mean is now 22. When another 29-year-old enters the room, the sum of the ages increases to $176 + 29 = 205$ and we now have nine people. If we let the honoree's age be represented by x, then the sum of the ages of the ten people now present is $205 + x$. If we divide this sum by 10, then we would obtain the mean of 27. If we solve the equation for x, we will have the honoree's age.

$$\frac{205 + x}{10} = 27$$
$$205 + x = 270$$
$$x = 65$$

Therefore, the honoree is 65 years old.

Section 10-4

The sum of Ouida's current test scores is 492. We divide by 6 to find that the average (mean) is 82. If she replaces the highest and lowest scores with one score, she will have only five scores. If the final exam score is represented by x, the sum of the five scores is $326 + x$. If we solve the equation, we will have the needed score.

$$\frac{326 + x}{5} = 82$$
$$326 + x = 410$$
$$x = 84$$

Ouida needs to earn 84 points on the final exam to maintain her average.

Answer to Laboratory Activity

Section 10-3

1. (a) 60 **(b)** Answers vary. 57, 59 **(c)** 58, 58, 59, 60
2. The mean in each case is 56 and that is also the number in the middle of the strip.   **3.** The mean would not change as the balance point does not change.   **4.** The median is 56, and the mode is 58.   **5.** There is not a balance around the median or mode.   **6.** Answers vary. If the median or mode are equal to the mean, then there would be balance.

Answers to Technology Corners

Section 10-3

Page 640   Answers will vary. Check out the given web site.

Page 657   The mean is 46. The standard deviation is approximately 8.85. The variance is 78.36. The mean absolute deviation is 7.6.

Answer to the Preliminary Problem

The headline can easily be correct. Suppose that the number of men in the professional nonfaculty category is very high, the number of women in the clerical category is very high, and there are more women than men in the faculty, technical, and skilled craft categories. There are many varied numerical examples that can be created to show this. One example created with a spreadsheet is a bit extreme but does show that the headlines could be correct.

Chapter 11

Assessment 11-1A

1. (a) in 12 ways, if different order implies different name **(b)** 60 ways, if different order implies different name **2. (a)** for example: $\overleftrightarrow{BC}$ and $\overleftrightarrow{DH}$ or $\overleftrightarrow{AE}$ and $\overleftrightarrow{BD}$ **(b)** parallel **(c)** no **(d)** empty set (since $\overleftrightarrow{BD}$ and plane EFG are parallel) **(e)** In the floor plane FHG consider $\overleftrightarrow{FK}$ (not shown) perpendicular to $\overleftrightarrow{FH}$. In plane BDH we have $\overleftrightarrow{FB} \perp \overleftrightarrow{FH}$. Because $\overleftrightarrow{FB}$ is perpendicular to the floor, $\overleftrightarrow{FB} \perp \overleftrightarrow{FK}$ and hence the dihedral angle between the planes measures 90°. **3. (a)** $\varnothing$ **(b)** C **(c)** A **(d)** Answers vary; for example, $\overleftrightarrow{AC}$ and $\overleftrightarrow{BE}$. **(e)** $\overleftrightarrow{AC}$ and $\overleftrightarrow{DE}$ or $\overleftrightarrow{AD}$ and $\overleftrightarrow{CE}$ **(f)** plane BCD or plane BEA **4.** 14 **5.** Answers vary; for example: **(a)** intersecting crossroads, **(b)** angle on a yield sign, **(c)** the top of a coat hanger. **6. (a)** 110° **(b)** 40° **(c)** 20° **(d)** 130° **7. (a) (i)** 41°31′10″ **(ii)** 11°44′ **(b) (i)** 54′ **(ii)** 15°7′48″ **8. (a) (i)** 90° **(ii)** 12°30′ **(iii)** 205° **(b)** 52°30′ **9. (a)** Let $m(\angle AOB) = x$. Then $x + 90° + 3x = 180°$, $x = 22°30'$, or 22.5°. Thus, $m(\angle AOB) = 22°30'$, $m(\angle COD) = 67°30'$, or 67.5°. **(b)** $m(\angle BOC) = x$, $3x - 35 + x = 90$, $x = 31.25°$, $m(\angle BOC) = 31.25°$, $m(\angle AOB) = 58.75°$ **(c)** $3x + 3y = 180$, $x + y = 60$, $m(\angle BOC) = 60°$. If we change the position of $\overrightarrow{OE}$, $3x$ and $3y$ will have different values and hence x and y will have different values, but $3x + 3y$ remains 180° and therefore $x + y = 60°$. **10. (a)** 6 **(b)** 8 **(c)** $2(n - 1)$ **(b)** $n(n - 1)/2$
11. (a)

| Number of Lines | Number of Intersection Points | | | | | |
	0	1	2	3	4	5
2			Not possible	Not possible	Not possible	Not possible
3					Not possible	Not possible
4				Not possible		
5			Not possible	Not possible		
6			Not possible	Not possible	Not possible	

	Men Salaries	Women Salaries	No. of Men	No. of Women
Executive/Managerial	$100,000	$81,000	1	1
Faculty	$70,000	$56,000	20	25
Prof. Nonfaculty	$40,000	$31,600	267	10
Clerical	$25,000	$26,750	50	600
Tech.	$40,000	$32,000	20	60
Skilled Craft	$40,000	$33,600	10	50
		Total	368	746
		Approx. Average Salary	$39,755 (Men)	$27,033 (Women)
		68% of men's average salary	$27,033	

12. Answers vary. **13.** Answers may vary.

Mathematical Connections 11-1

Communication

1. (a) Answers vary. **(b)** The fire is located near the intersection of the two bearing lines. **(c)** Answers vary.

3. $2 + 2 + 3 + 4 + \ldots + n$, or $1 + \dfrac{n(n + 1)}{2}$

5. Yes. Through P in plane α, draw a line k perpendicular to $\overleftrightarrow{AB}$. By definition, since n is perpendicular to α it is perpendicular to every line in α through P. Thus n and k are perpendicular, which implies that planes α and β are perpendicular.

Open-Ended

7. Answers depend on the classroom.

Questions from the Classroom

9. From the Denseness Property, we know that for any two real numbers, there is a real number between them. Because points of a line and the set of real numbers can be put in a one-to-one correspondence, we can always find a point between any two points. Since this is true for any two points, we see that there are an infinite number of points in a segment. **11.** The measure of an angle has nothing to do with the fact that rays cannot be measured. The measure of an angle in degrees is based on constructing a circle with center at the vertex of the angle and dividing the circle into 360 congruent parts. The number of parts in the arc that the angle intercepts is the measure of the given angle in degrees. The number of parts in the intercepted arc is the same regardless of the size of the circle (or protractor). **13.** If two parallel lines are defined as lines that are in the same plane and have no points in common, then two identical lines cannot be parallel, because they have infinitely many points in common. It is possible to define two identical lines as parallel or nonparallel. Some books define it one way; other books the other way. **15.** The carpenter can check if the dihedral angle is 90° by placing a well-constructed box on the floor and moving it as close to the wall as possible. If an entire face of the box touches the wall, then the wall is perpendicular to the floor.

Assessment 11-2A

1. (a) 1, 4, 6, 7, 8 **(b)** 1, 6, 7, 8 **(c)** 6, 7 **(d)** 1, 8 **2.** 8 **3.** a concave polygon **4. (a)** and **(c)** are convex, since no matter which two points in the interior of the figure are chosen, the entire segment connecting the points lies within the figure; **(b)** and **(d)** are concave, since two points within the figure can be chosen such that the segment formed by the two points does not lie entirely within the figure. **5. (d)** and **(e)** are impossible because each angle of an equilateral triangle has measure 60°. **6. (a)** 5 **(b)** 35

(c) 170 **(d)** $\dfrac{n(n - 3)}{2}$ **7. (a)** isosceles and equilateral **(b)** isosceles **(c)** scalene **8.** all squares **9.** Construction. Answers vary; for example, base angles, are congruent, the angle bisector of the angle opposite the base bisects the base and is perpendicular to the base. **10. (a)** A parallelogram. Answers vary; for example, opposite angles and opposite sides are congruent. **(b) i.** isosceles with $AC = BC$ **ii.** a right triangle **iii.** an isosceles right triangle

Mathematical Connections 11-2

Communication

1. (a) Answers vary. An $8\text{-}\frac{1}{2} \times 11$-in. piece of rectangular paper requires only one fold. **(b)** Folding the square along one diagonal and then along the other, 4 congruent angles are formed. Each is $\dfrac{360°}{4}$ or 90°. The paper folding implies that the diagonals bisect each other. Unfolding and then folding the square so that opposite sides fall onto each other shows that the diagonals are congruent.

Cooperative Learning

3. Answers vary. **5. (a)** Reuleaux triangles aren't truly triangles. They are curves of constant width. **(b)** Reuleaux triangles have equal-sized arcs, whereas equilateral triangles have sides of the same length.

Questions from the Classroom

7. A regular polygon is a polygon in which all the angles are congruent and all the sides are congruent. In general neither condition implies the other, and hence neither is sufficient to describe a regular polygon. For example, a rhombus that is not a square has all sides congruent, but all its angles are not congruent. A rectangle that is not a square has all its angles congruent, but not all its sides congruent. Neither a rhombus nor a rectangle is a regular polygon unless they are squares. **9.** Since an angle is a set of points determined by two rays with the same endpoint, to say that two angles are equal implies that the two sets of points determining the angles are equal. The only way this can happen is if the two angles are actually the same angle. To say that two angles are congruent is to say that the angles have the same size or measure. **11.** To be regular, all sides must be congruent but also all angles must be congruent. All angles are not congruent unless the rhombus is a square.

Review Problems

13. (a) 45 **(b)** $n(n - 1)/2$ **15. (a)** False. A ray has only one endpoint. **(b)** True **(c)** False. By definition skew lines cannot be contained in a single plane. **(d)** False. $\overrightarrow{MN}$ has endpoint M and extends in the direction of point N; $\overrightarrow{NM}$ has endpoint N and extends in the direction of point M. **(e)** True **(f)** False. Their intersection is a line.

Assessment 11-3A

1. 6   **2. (a)** 60°   **(b)** 45°   **(c)** 60°   **(d)** 60°
3. (a) Yes. A pair of corresponding angles are 50° each.
(b) Yes. A pair of corresponding angles are 70° each.
(c) Yes. A pair of alternate interior angles are 40° each.
(d) Yes. A pair of corresponding angles are 90° each.
4. 70° and 20°   **5. (a)** 20   **(b)** 150°   **6. (a)** 70°
(b) 70°   **(c)** 65°   **(d)** 45°   **7.** $m(\angle 1) = 40°$,
$m(\angle 2) = 140°$, $m(\angle 3) = 115°$, $m(\angle 4) = 40°$,
$m(\angle 5) = 40°$, $m(\angle 6) = 65°$ **8. (a)** 50°   **(b)** 85°
9. (a) $x = 40°$ **(b)** $x = 18°$ **10. (a)** The marked
angles are congruent corresponding angles. By the Angles
and Parallel Lines property, $k \parallel \ell$.

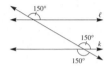

(b) One way is to extend $\overline{AC}$ to intersect k at D. Because x
is the measure of an exterior angle in triangle BCD,
$x = z + w$ (x is a supplement of $\angle BCD$ and so is $z + w$).
But it is given that $x = y + z$. Hence, $z + w = z + y$,
$w = y$. Because w and y are measures of alternate interior
angles, $k \parallel \ell$.

11. (a) 60°   **(h)** 90°   **12.** 360°   **13.** 111°
14. $m(\angle 1) = 60°$, $m(\angle 2) = 30°$, $m(\angle 3) = 110°$
15. 135°   **16. (a)** 100°   **(b)** The measure of the
exterior angle equals the sum of the measures of the remote
interior angles. Since $\angle 1$ and $\angle ACB$ are supplementary,
then $m(\angle 1) + m(\angle ACB) = 180°$. We also know that the
sum of the measures of the angles of a triangle equals 180°;
that is, $m(\angle A) + m(\angle B) + m(\angle C) = 180°$. So
$m(\angle 1) = m(\angle A) + m(\angle B)$.   **17.** 60°, 120°, 60°
18. $x = \frac{180°}{13}$ or $x \approx 13.846°$, $y = \frac{360°}{13}$ or $y \approx 27.692°$,
$z = \frac{720°}{13}$ or $z \approx 53.385°$, $w = \frac{1080°}{13}$ or $w \approx 83.077°$
19. (a) $x = 80°, y = 150°$ **(b)** $x = 45°, y = 60°, z = 75°$

Mathematical Connections 11-3

Communication

1. (a) No. Two or more obtuse angles will produce a sum of
more than 180°.   **(b)** Yes. For example, each angle may
have measure 60°.   **(c)** No. The sum of the measures of the
three angles would be more than 180°.   **(d)** No. It may
have an obtuse or right angle as well.   **3. (a)** Five
triangles will be constructed in which the sum of the angles
of each triangle is 180°. The sum of the measures of the
angles of all the triangles equals 5(180°), from which we
subtract 360° (the sum of all the measures of the angles of
the triangles with vertex P). Thus, $5(180°) - 360° = 540°$.

(b) Here, n triangles are constructed, so we have $n(180°)$
from which we subtract 360° (the sum of all the measures of
the angles of the triangles with vertex P). Thus, we obtain
$n(180°) - 360°$ or $(n - 2)180$, which is the same answer
stated in Section 11-3.   **5. (a)** Fold the isosceles triangle
so that the congruent sides match up. The two angles will
fall on top of each other.   **(b)** Fold the triangle so that the
two congruent angles fall on top of each other. The two
sides will match up.   **(c)** Choose a vertex and fold the triangle at that vertex so that the other two vertices fall on top
of each other. Two angles of the triangle should fall on top
of each other. Repeat at a second vertex.   **(d)** Fold the
trapezoid so that its congruent sides fall on top of each
other and check that the base angles fall on top of each
other as well.   **7.** No. Regular hexagons fit because the
measure of each vertex angle is 120°, three hexagons fit to
form 360°, and the plane can be filled. For a regular
pentagon, the measure of each vertex angle is 108°, so
pentagons cannot be placed together to form 360° (360 is
not divisible by 108) and the plane cannot be filled.
9. $\beta = w - z + y - x$ Draw through A the line m parallel
to $\overline{BC}$, as shown.

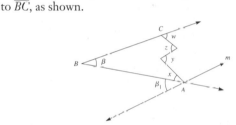

Then $\beta_1 = \beta$ and from problem 3(b), $y = (x + \beta_1) +$
$z - w$, $\beta_1 = w - z + y - x$, $\beta = w - z + y - x$.
11. $ABCD$ is a parallelogram. Let the measures of the congruent angles at A and C be x degrees, and the measures of
the congruent angles at B and D be y degrees. Then,

$$x + y + x + y = 360$$
$$2x + 2y = 360$$
$$x + y = 180$$

Consequently, $m(\angle A) + m(\angle B) = x + y = 180$, and as
shown in problem 8 above lines AD and BC are parallel.
Similarly, $m(\angle A) + m(\angle D) = x + y = 180$, and hence
lines AB and DC are parallel. Thus, $ABCD$ is a
parallelogram.   **13.** $m(\angle E) = 70°, m(\angle D) = 50°$,
$m(\angle F) = 60°$. Because the sides of $\triangle DEF$ are parallel to
corresponding sides of $\triangle BCA$, the quadrilaterals $EBCA$,
$BFCA$, and $BCDA$ are parallelograms. The answers follow
from the fact that opposite angles in a parallelogram are
congruent (see the remark following Example 11-2).
15. (a) Let the congruent angles at P each measure
x degrees and at Q, y degrees each. Then,

$$3x + 3y = 180$$
$$x + y = 60$$

Hence, in $\triangle PQB$, $m(\angle B) = 180 - (x + y) = 120$.
Thus, $m(\angle CBA) = 120°$.

Now in $\triangle PDQ$:

$$m(\angle D) = 180° - (2x + 2y)$$
$$= 180° - 2(x + y)$$
$$= 180° - 2 \cdot 60°$$
$$= 60°$$

Hence, $m(\angle CDA) = 60°$.

(b) $\angle A$ is an exterior in $\triangle PAQ$. Using the notation in part (a) we have

$$m(\angle A) = m(\angle APQ) + m(\angle PQA) = 2x + y$$
$$= x + 60$$

Since $m(\angle P) = 3x$, and it can be any number between 0 and 180°, $x + 60$ varies between 0 and 120°.

Open-Ended

17. Answers vary. The following are some properties of figures in a plane and corresponding properties on a sphere:

Property in a Plane	Property on a Sphere
Two lines in a plane may intersect in one point or may be parallel.	Two great circles (lines on a sphere) always intersect in two points.
Two points determine a unique line.	Two points do not always determine a unique great circle
A line has no length, and any point on the line separates it into two parts.	A great circle has a finite length ($2\pi r$, where r is the radius of the sphere) and is not separated into two parts by a point on the circle.

Cooperative Learning

19. (a) because congruent supplementary angles are formed **(b)** the theorem concerning the sum of the measures of the angles of a triangle **(c)** When A and C are folded, congruent supplementary angles are formed and, hence, each is a right angle. When B is folded along $\overline{BB'}$, the crease $\overline{DE}$ formed is perpendicular to $\overline{BB'}$ (again because congruent supplementary angles are formed); consequently, $\overline{DE} \parallel \overline{GF}$. Thus, D and E are also right angles.
(d) $GF = GB' + B'F = \frac{1}{2}AB' + \left(\frac{1}{2}\right)B'C = \frac{1}{2}(AB' + B'C) = \frac{1}{2}AC$; hence, $GF = \frac{1}{2}AC$.

Questions from the Classroom

21. The only possible figures are acute triangles. To see why, let an n-gon ($n \geq 4$) have the property that each angle is acute. Then all the exterior angles would be obtuse. A quadrilateral would have more than $90 \cdot 4$, or 360°, as the sum of the measures of the exterior angles and for $n > 4$ the polygon would have more than $90 \cdot 4$ or 360° as the sum of the measures of its exterior angles, which is not possible.

Review Problems

23. Answers can vary. **(a)** All angles must be right angles, and all diagonals are the same length. **(b)** All sides are the same length, and all angles are right angles. **(c)** This is impossible because all squares are parallelograms.
25. Answers will vary but should include the following:
• Congruent opposite sides describe a parallelogram.
• All sides congruent describe a rhombus.
• Four right angles describe a rectangle.
• Two pairs of consecutive sides congruent describe a kite.

Assessment 11-4A

1. (a) quadrilateral pyramid **(b)** quadrilateral prism; possibly a trapezoid prism **(c)** pentagonal pyramid
2. (a) A, D, R, W **(b)** $\overline{AR}, \overline{RD}, \overline{AD}, \overline{AW}, \overline{WR}, \overline{WD}$ **(c)** $\triangle ARD, \triangle AWD, \triangle AWR, \triangle WDR$ **(d)** R **(e)** $\overline{DW}$
3. Answers vary; for example, rectangular prism, cylinder, pyramid, and sphere. **4. (a)** 5 **(b)** 4 **(c)** 4
5. (a) true **(b)** false **(c)** true **(d)** false **(e)** false **(f)** false **(g)** false **(h)** true **6.** All are possible.
7. Answers vary. **8. (a)** 6 cubes; 814 faces glued. **(b)** 24 cubes; 392 faces glued. **9. (a)** right hexagonal pyramid **(b)** right square pyramid **(c)** cube **(d)** right square prism **(e)** right hexagonal prism **10. (a)** triangular right prism **(b)** a right square pyramid **11.** Answers vary.
12. (a) iv **(b)** ii **13. (a)** i, ii, and iii **(b)** ii, iii, and iv
14. (a) ![jar illustration] **(b)** ![tilted jar illustration]

15. (a) (2) **(b)** (4)
16. (a) ![Triangle] Triangle **(b)** ![Rectangle] Rectangle **(c)** ![Circle] Circle

17. (a) a rectangle whose opposite sides are diagonals of two opposite faces and the other two sides, edges of the cube **(b)** No. Answers vary. **18.** Answers vary.
19. (a) three pairs of faces determined by three pairs of opposite sides of a hexagonal base **(b)** 120° since $\angle FAB$ is an angle whose measure is the measure of the dihedral angle formed by two adjacent faces. **20. (a)** $10 + 7 - 15 = 2$ **(b)** $9 + 9 - 16 = 2$

Mathematical Connections 11-4

Communication

1. Three. Each pair of parallel faces could be considered bases. **3.** Both could be drawings of a quadrilateral pyramid. In **(a)**, we are directly above the pyramid, and in **(b)**, we are directly below the pyramid. **5. (a)** A rectangle. $\overline{AD}$ is perpendicular to $\overline{AB}$ and to $\overline{AE}$ and therefore perpendicular to the plane $AEHB$. Since $\overline{MN} \parallel \overline{AD}$, $\overline{MN}$ is perpendicular to $AEHB$. Thus $\overline{MN}$ is perpendicular to every line through M

in that plane. So $\overline{MN} \perp \overline{MH}$ and $\angle NMH$ is a right angle. Similarly the other angles of *MHGN* are right angles.
(b) No, the edges of the pyramid are not the same length.
(c) Yes, all the edges are the same length.

Cooperative Learning

7. Answers vary. **9. (a)** parallelogram, rectangle, square, scalene triangle, isosceles triangle, equilateral triangle, pentagon, hexagon, a square and its interior, trapezoid that is not a parallelogram and a parallelogram that is neither a rhombus nor a rectangle **(b)** triangle, quadrilateral, line segment.

Questions from the Classroom

11. There are 2 different nets shown below (assuming that the faces are indistinguishable).

Review Problems

13. $m(\angle BCD) = 60°$ **15. (a)** true **(b)** true **(c)** false (e.g., equilateral triangle)

Chapter Review

1. (a) $\overleftrightarrow{AB}$, $\overleftrightarrow{BC}$ and $\overleftrightarrow{AC}$ **(b)** $\overrightarrow{BA}$ and $\overrightarrow{BC}$ **(c)** $\overline{AB}$
(d) $\overline{AB}$ **2. (a)** Lines PQ and AB **(b)** Planes APQ and BPQ **(c)** $\overleftrightarrow{AQ}$ **(d)** No, $\overleftrightarrow{PQ}$ and $\overleftrightarrow{AB}$ are skew lines, so no single plane contains them. **3.** Answers vary.
4. (a) No. The sum of the measures of two obtuse angles is greater than 180°, which is the sum of the measures of the angles of any triangle. **(b)** No. The sum of the measures of the four angles in a parallelogram must be 360°. If all the angles are acute, the sum would be less than 360°.
5. (a) 132°11′ **(b)** 56°28′ **(c)** 132°21′ **(d)** 0°45′
6. 18°, 36°, 126° **7. (a)** Given any convex *n*-gon, pick any vertex and draw all possible diagonals from this vertex. This will determine $n - 2$ triangles. Because the sum of the measures of the angles in each triangle is 180°, the sum of the measures of the angles in the *n*-gon is $(n - 2)180°$.
(b) 90 sides **8.** Answers vary. **9.** Answers vary, but the possibilities are a point, a segment, a triangle, a triangle and its interior, a quadrilateral, a square, a square and its interior, or the empty set. **10.** $m(\angle 3) = m(\angle 4) = 45°$
11. 35°8′35″ **12. (a)** 60° **(b)** 120° **(c)** 120°
13. (a) Alternate interior angles are congruent.
(b) Corresponding angles are congruent.
(c) $m(\angle B) + m(\angle C) + m(\angle BAC) = m(\angle BAD) + m(\angle DAE) + m(\angle BAC) = 180°$ **14.** Draw any line t intersecting the three lines. $a \parallel b$ implies $\alpha = \beta$. Next $b \parallel c$ implies $\beta = \gamma$. Consequently, $\alpha = \gamma$, which implies $a \parallel b$.

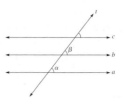

15. Out of context of a geometry course, assuming that a rectangle has four right angles and that a diagonal divides it into two triangles with the same angles, Wafiq's claim is correct. As explained in the problem, the sum of the measures of the interior angles in each right triangle is 180°. Therefore, the sum of the measures of the angles in $\triangle ACD$ and $\triangle BCD$ is $2 \cdot 180°$, or 360°. In this sum all the angles of the original triangle are included as well as the two right angles at D. Hence the sum of the measures of the angles in the original triangle is $360° - 2 \cdot 90°$, or 180°. On the other hand, to prove that a rectangle has four right angles requires the use of the property that if lines are parallel and cut by a transversal then corresponding angles are congruent. (This last property is also required to prove that the sum of the measures of the interior angles in a triangle is 180°.) Thus wafiq's reasoning is incomplete.

16. One approach is to draw through A a line c parallel to a and then prove that it is parallel to b. If $c \parallel a$, then $\alpha_1 = \alpha$, as these are corresponding angles. Since $x = \alpha + \beta$ and also $x = \alpha_1 + \beta_1 = \alpha + \beta_1$, we get $\beta_1 = \beta$, which implies that $c \parallel b$.

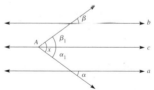

17. (a) $5 \cdot 180° - 360°$, or $3 \cdot 180°$ **(b)** From the sum of the measures of all the interior angles of the *n* triangles we need to subtract the measures of the angles with vertex at P, that is, 360°. We get $n \cdot 180° - 360°$ or $(n - 2) \cdot 180°$.
(c) Answers vary. One way is to connect B with E and A with F. We get two quadrilaterals and a triangle. Thus, the sum of all the measures of the interior angles of the polygon is $2 \cdot 360° + 180°$, or 900°. **18.** $m(\angle 1) = m(\angle 2) = 65°$, $m(\angle 3) = m(\angle 4) = 115°$, $m(\angle 5) = 70°$. **19.** 55°
20. (a) $\angle 3$ and $\angle 4$ are supplements of congruent angles.
(b) $m(\angle 3) > m(\angle 4)$. Reason: $m(\angle 3) = 180° - m(\angle 1)$ and $m(\angle 4) = 180° - m(\angle 2)$. Because $m(\angle 1) < m(\angle 2)$, $^-m(\angle 1) > ^-m(\angle 2)$. After adding 180° to both sides of the last inequality, the result follows. **21. (a)** The measure of one of the dihedral angles formed by planes α and γ is $m(\angle APS) = 90°$ because α and γ intersect in $\overleftrightarrow{PQ}$, $\overleftrightarrow{AP} \perp \overleftrightarrow{PQ}$, and $\overleftrightarrow{SP} \perp \overleftrightarrow{PQ}$ (since the dihedral angle $S - AB - Q$ measures 90°). The measure of one of the

dihedral angles formed by planes β and γ is $m(\angle BPQ) = 90°$ because line PQ is perpendicular to line AB (given).   **(b)** Line AB is perpendicular to the lines PQ and PS in plane γ.

22. 8   **23. (a)** $_{10}C_3 = \dfrac{10 \cdot 9 \cdot 8}{3!} = 120$. The number of triangles equals the number of ways of choosing 3 points out of 10 points where order is not important.

(b) $_nC_3 = \dfrac{n(n-1)(n-2)}{6}$ **24. (a)** $\dfrac{360}{20}$, or 18   **(b)** Does not exist; $25 \nmid 360$.   **(c)** Does not exist; the sum is always 360°.   **(d)** Does not exist; the equation $\dfrac{n(n-3)}{2} = 4860$, or $n(n-3) = 9720$, has no solution in integers because if $n = 100$ then $100 \cdot 97 = 9700$ and if $n = 101$ then $101 \cdot 98 = 9898 > 9700$.   **25. (a)** 188°

(b) $\dfrac{9000°}{53}$ or approximately 169.8°   **26.** Answers vary.

27. (a) Answers vary.   **(b)** Answers vary.   **28.** 48°
29. Answers may vary.   **30.** Yes if the model is scaled down; no if the model is large. In the latter case, perspective will make the segments appear nonparallel.   **31.** No; see definition.   **32.** Answers vary; for example,

Answers to Now Try This

11-1. (a) No, the points must be collinear. We assumed that A, B, and C are different points   **(b) i.** $\overleftrightarrow{AB}$ **ii.** $\overline{AB}$
11-2. (a) No. If skew lines had a point in common there would be a single plane that contains the lines, which would contradict the definition of skew lines.   **(b)** Skew lines cannot be parallel. By definition, parallel lines are in the same planes. Skew lines are not.   **(c)** Let O be the center of the globe. First consider two points A and B on the globe that are the endpoints of a diameter, that is, A, O, and B are collinear. There are infinitely many planes containing A and B and hence the center of the globe, O. Each of these planes intersects the globe in a great circle. Consequently, if A, O, B are collinear, there is no unique "line" through A and B. If A and B are not the endpoints of a diameter, then consider the plane through A, B and O. That plane intersects the sphere in a great circle and there is no other "line" through A and B.   **11-3.** The problem on page xxx gives the number of lines that can be drawn through n points. This problem is a model for the number of handshakes that take place if everyone shakes hands with everyone else. Everyone shakes hands with everyone except himself. Because we don't want to count one hand shake twice, we need to divide by 2. For n people there will be $\dfrac{n(n-1)}{2}$ handshakes. For 20 people there will be $\dfrac{20 \cdot 19}{2} = 190$ handshakes.

11-4. $8.42° = 8° + 0.42(60)' = 8°25.2' = 8°25' + (.2)(60)'' = 8°25'12''$ **11-5. (a)** It is possible for a line intersecting a plane to be perpendicular to only one line in the plane. Drawings will vary.   **(b)** It is not possible for a line intersecting a plane to be perpendicular to two

distinct lines and not be perpendicular to the plane.
(c) Yes. If a line intersects a plane in point P and is perpendicular to two lines in the plane through P, then it is perpendicular to every line in the plane through P.   **(d)** ℓ is perpendicular to α.   **11-6.** Answers vary, e.g., the Greek letter α.   **11-7.** One approach to this problem is to start shading the area surrounding point X. If we stay between the lines, we should be able to decide whether the shaded area is inside or outside the curve. The shaded part of the following figure indicates that point X is located outside the curve.   **11-8.** At the endpoint A of an arbitrary ray k construct $\angle A$ of 120° and an arbitrary point B on the side of this angle. At B construct another 120° angle and point C so that $BC \neq AB$. Then construct a 120° angle at C and point D on the next side. Similarly at D make a 120° and choose E on the next side. Finally at E construct a 120° angle and F as the intersection of a side of $\angle E$ with k. The angle at F is $4 \cdot 180° - 5 \cdot 120°$ or 120°.

(a)   **(b)**

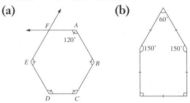

11-9. (1) True. From the definition (Table 11-6) we know that an equilateral triangle has three congruent angles. Therefore it has at least two congruent sides and hence it is isosceles. (2) True. By definition, a regular quadrilateral is a four-sided figure that is both equilateral and equiangular. A square has four sides of equal length and four right angles—so a square is a regular quadrilateral. (3) True. Because a rhombus is a parallelogram and a rectangle is a parallelogram with a right angle, if one angle of a rhombus is a right angle then all of its angles are right. (4) True; follows from part (3). (5) True, because a rectangle is a parallelogram and if one angle of a parallelogram is a right angle all its angles are right angles. (6) True. A rectangle is a trapezoid since it is a quadrilateral with at least one pair of parallel sides. (7) True. A square is a rectangle that is an isosceles trapezoid. Also a square is a kite. Hence there are isosceles trapezoids that are kites. (8) False, as shown in the following figure:

11-10. (a) There are infinitely many. Consider the number of lines of longitude on the globe.   **(b)** A spherical triangle could have as vertices the North Pole, Quito, Ecuador, and Kisumu, Kenya.   **(c)** Consider any triangle with the North Pole as one vertex and the other two vertices on the equator such that the angle formed at the North Pole is a right angle.   **11-11. (a)** The sum of the measures of the interior and exterior angles is $n \cdot 180°$.

Therefore the sum of the measures of the interior angles is $n \cdot 180° - 360°$ or $(n - 2)180°$

(b) $\dfrac{(n - 2) \cdot 180°}{n}$ or $180° - \dfrac{360°}{n}$

11-12. $V = 60$, $E = 90$, $F = 32$ and
$$\begin{aligned} V - E + F &= 60 - 90 + 32 \\ &= 2 \end{aligned}$$

Answer to Brain Teasers

Section 11-3

The problem can be solved by walking around the star, but a simpler approach is to notice that the vertices of the star can be obtained by extending pairs of sides of a regular 12-gon. Choosing any side and a sixth side of the 12-gon we obtain an isosceles triangle whose base is a longest diagonal of the 12-gon and whose base angles can be shown to measure 75° each. Thus the measure of an interior angle of the star is $180° - 2 \cdot 75°$ or 30°.

Section 11-4

See "Making a Better Beer Glass" by A. Hoffer in the May, 1982 issue of *Mathematics Teacher*, pages 378–379, to see the shape of the glass.

Answer to Laboratory Activities

Section 11-1

a. By folding a crease onto itself, we have created two angles that are both supplementary and congruent. They must have measure 90°, making the creases perpendicular. **c.** One obtains a 45° angle by bisecting the 90° angle. The 135° angle is 90° + 45°. The 22°30′ angle is obtained by bisecting the 45° angle. **d.** The smallest angle measures 90°. **e.** The smallest angle measures 45°. **f.** The smallest angle measures $\dfrac{360°}{2^n}$.

Section 11-2

One way to construct the figure is to fix the first square in place then work your way around the circle by overlapping the previous square by 30°.

Section 11-4

One needs to use cubes to build the structure. There is more than one structure that meets the specifications.

Answer to Preliminary Problem

1. There are two types of concave kites as shown below.

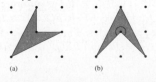

(a)   (b)

There are four of the first kind (left figure) since any four corners of the array can be used as the vertex currently in the bottom left corner. There are four of the second shape, since the marked obtuse angle can open toward any of the 4 edges of the square array. Thus, there are altogether eight possible concave kites (since location matters). **2.** In addition to the eight concave kites, there are two kinds of concave quadrilaterals, as shown here.

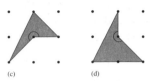

(c)   (d)

There are eight of the kind in (a), since in addition to the four choices for the corner vertex, there are two choices for which side the obtuse angle will open toward. There are also eight of the shape in (b), since there are four choices for the side of two-unit length and two choices for which side the obtuse angle will open toward. Thus, there are altogether $4 + 4 + 8 + 8$, or 24, concave quadrilaterals.

Chapter 12

Assessment 12-1A

1. Otherwise the definition applies to three-dimensional space and thus a sphere. **2.** on a plane perpendicular to $\overline{AB}$ and containing its midpoint. **3. (a)** $m(\angle A) > m(\angle B)$ **(b)** The side of greater length is opposite the angle of greater measure. **4. (a)** construction **(b)** construction **(c)** a scalene right triangle **(d)** Not possible **(e)** construction **(f)** construction **(g)** construction **(h)** construction **(i)** construction **5. (b)** yes, by SSS **(c)** yes, by SSS **(d)** There is no triangle. **(e)** yes, by SSS **(f)** yes, by SAS **(g)** The triangle is not unique. **(h)** yes, by SAS **(i)** yes, by SAS **6.** 22 **7. (a)** yes; SAS **(b)** yes; SSS **(c)** no **8.** No; there could be no right angle if the triangle is equilateral. **9.** yes, by SSS **10.** Fold the circle onto itself using two different fold lines. The intersection of the lines is the center of the circle. **11.** The circumcircles must be concentric. The perpendicular bisectors of the sides of both the large and small triangles intersect at the same point, the circumcenter. **12.** construction **13.** The construction makes $\triangle ABC \cong \triangle DBC$ by SAS, so $AB = DB$; by measuring DB, AB is known. **14. (a)** construction **(b)** construction **(c)** construction **15. (a)** construction **(b)** construction **16. (a)** One method is to start at any point on the circle and with a compass opened to the length of the radius of the circle mark off equally spaced points; there will be six equal arcs. **(b)** construction **(c)** Because $m(\angle AOF) = 360° - 5 \cdot 60°$ or 60°, the isosceles triangle $\triangle ADF$ is equilateral and therefore $AF = r$. So the six triangles are congruent by SSS. **(d)** The length of each side of the hexagon is equal to r, the radius of the circle.

(e) Anywhere on the circle construct a chord $\overline{AB}$ congruent to the radius of the circle, then construct $\overline{BC}$, $\overline{CD}$, $\overline{DE}$, and $\overline{EF}$ all congruent to the radius of the circle; the hexagon $ABCDEF$ is a regular hexagon. **(f)** From part (e), $\triangle ACE$ is an equilateral triangle inscribed in the circle. **17.** Congruent arcs are arcs of kcircles with equal radii whose central angles have the same measure. **18. (a)** 6 **(b)** 24 **(c)** $n(n - 1)(n - 2) \cdot \ldots \cdot 3 \cdot 2 \cdot 1$, or $n!$ **19.** congruent by SAS **20.** The center of the circle is the intersection of the perpendicular bisectors of any two sides of the triangle. Because the triangle is obtuse, its center is exterior to the triangle. The radius of the circle is the distance from the center to any of the vertices of the triangle. **21. (a)** yes **(b)** yes **22.** If the obtuse angle is at vertex A, then $\overline{BE}$ and $\overline{CF}$ will be in the exterior of the triangle. **23.** A circle can be circumscribed about the sign. **24. (a)** A square is formed. **(b)** In the figure shown, triangles ABO, BCO, CDO, and DAO are isosceles right triangles. They are congruent by SAS, so $\overline{AB} \cong \overline{BC} \cong \overline{CD} \cong \overline{DA}$. We can argue that each base angle of the isosceles triangles has measure 45°, so $m(\angle ABC) = m(\angle BCD) = m(\angle CDA) = m(\angle DAB) = 90°$. Thus, $ABCD$ is a square.

Mathematical Connections 12-1

Communication

1. P is not on the circle because it doesn't fit the definition. **3. (a)** $\triangle ABC \cong \triangle ADC$ by SSS. Hence, $\angle BAC \cong \angle DAC$ and $\angle BCM \cong \angle DCM$ by CPCTC. Therefore, $\overleftrightarrow{AC}$ bisects $\angle A$ and $\angle C$. **(b)** The angles formed are right angles. By part (a), $\angle BAM \cong \angle DAM$. Hence, $\triangle ABM \cong \triangle ADM$ by SAS. $\angle BMA \cong \angle DMA$ by CPCTC. Since $\angle BMA$ and $\angle DMA$ are adjacent congruent angles, each must be a right angle. Since vertical angles formed are congruent, all four angles formed by the diagonals are right angles. **(c)** By part (a) and SAS, $\triangle BAM \cong \triangle DAM$. Hence, $\overline{BM} \cong \overline{MD}$; CPCTC. **5.** The result does not follow. **7.** $\overline{CB} \cong \overline{CD}$ since each is 2 units long. $\overline{AB} \cong \overline{AD}$ because each can be viewed as a hypotenuse of two congruent right triangles with legs 1 and 3 units long.

Open-Ended

9. (a) Answers may vary. Some objects are tins of food and floor tiles. **(b)** Answers may vary. Photographs and their enlargements, an original and its projected image in an overhead projector, and a slide and its image are examples. **11.** Answers vary.

Cooperative Learning

13. Answer may vary.

15. The symbol $\cong$ is used only for congruent parts. Because AB and CD designate length of segments and not the segments themselves, it is not true that that $AB \cong CD$. Notice that if segments are congruent, then they are of the same length; hence it is correct to write $AB = CD$. **17.** Perhaps the "best" definition relies on transformational geometry, discussed in Chapter 14. Two figures can be defined to be congruent if, and only if, one figure can be mapped onto the other by successively applying translation, reflection, rotation, or glide reflection. For similarity, add a size transformation. **19.** Because the diagonals in $PTQS$ are perpendicular bisectors of each other, the quadrilateral is a rhombus and hence $\overleftrightarrow{PT} \parallel \overleftrightarrow{SQ}$.

Assessment 12-2A

1. constructions **2. (a)** No; by ASA, the triangle is unique. **(b)** No; by AAS, the triangle is unique. **(c)** Yes, the sides can be of any length. **3. (a)** yes; ASA **(b)** yes; AAS **4.** When the parallel ruler is open at any setting, the distance $BC = BC$; it is given that $AB = DC$ and $AC = BD$, so $\triangle ABC \cong \triangle DCB$ by SSS. Hence, $\angle ABC \cong \angle DCB$ by CPCTC. Because these are alternate interior angles formed by lines $\overleftrightarrow{AB}$ and $\overleftrightarrow{CD}$ with transversal line $\overleftrightarrow{BC}$, then $\overline{AB} \parallel \overline{DC}$. **5. (a)** parallelogram **(b)** none **(c)** none **6.** An isosceles trapezoid is formed and the angles formed with one base as a side are congruent, as are the angles formed on the other base. The congruent angles are sometimes referred to as the base angles of the isosceles trapezoid. **7. (a)** true **(b)** true **(c)** true **(d)** False. A trapeziod may have two consecutive right angles with the other two not right angles. **8. (a)** Answers may vary. **(b)** No. If the quadrilateral has three right angles, then the fourth must also be a right angle. **(c)** No; any parallelogram with a pair of right angles must have right angles as its other pair and hence it must also be a rectangle. **9.** Because $AB = BC$, B is equidistant from A and C. By Theorem 12–5, point B is on the perpendicular bisector of $\overline{AC}$. Similarly D is on the perpendicular bisector of $\overline{AC}$. Because two points determine a unique line, $\overleftrightarrow{BD}$ is the perpendicular bisector of $\overline{AC}$. **10. (a)** rectangle **(b)** isosceles trapezoid **11.** It appears to be a parallelogram whose opposite sides are congruent. Because the opposite sides are congruent, it is a parallelogram. **12.** a rhombus, because all the sides are congruent **13.** Either the arcs or the central angles must have the same measure (radii are the same since the sectors are part of the same circle). An alternative condition is that the segments joining endpoints of each arc are congruent. **14. (a)** Kite. $\overline{DX} \cong \overline{AX}$; $\angle D$ and $\angle A$ are right angles; $\overline{DP} \cong \overline{AQ}$. Thus, $\triangle PDX \cong \triangle QAX$ by SAS. Hence, $\overline{PX} \cong \overline{QX}$ by CPCTC. $\overline{DC} \cong \overline{BA}$ as opposite sides of a rectangle. $\overline{PD} \cong \overline{QA}$. Hence, $\overline{CP} \cong \overline{BQ}$. $\overline{CY} \cong \overline{BY}$, given; and $\angle C$ and $\angle B$ are right angles. Therefore, $\triangle CYP \cong \triangle BYQ$ by SAS so that $\overline{QY} \cong \overline{PY}$ and $PXQY$ is a kite. **(b)** The answer does not change. However, when P and Q are midpoints of $\overline{DC}$ and $\overline{AB}$, respectively, $PXQY$ is a rhombus. **15.** If placed together properly, the matching side becomes a diagonal of a parallelogram. The triangles are congruent, and if placed properly, a set of congruent alternate interior angles is formed, creating one set of parallel sides. Properties of the quadrilateral allow for completing the proof that it is a parallelogram. **16. (a)** Answers may vary. For example, ⊞ **(b)** Yes, divide the rectangle into four congruent rectangles. **17.** A square can circumscribe a circle, and results in a case that is stackable and easily shippable. Also, a circular CD would not be stable when put on its edge. **18.** Make one of the quadrilaterals a square and the other a rectangle. **19.** Construct the first kite by constructing a segment to become a diagonal and then construct two isosceles triangles with the segment as a common base. Construct the second kite starting with a segment not congruent to the first and construct two isosceles triangles with that segment as their common base but the sides congruent to the corresponding sides of the isosceles triangles in the first construction. **20.** Rhombus; use SAS to prove that $\triangle ECF \cong \triangle GBF \cong \triangle EDH \cong \triangle GAH$. **21. (a)** The lengths of one side of each square must be equal. **(b)** The lengths of the sides of two perpendicular sides of the rectangles must be equal. **(c)** Answers may vary; one solution is that two adjacent sides must have equal lengths and the included angle of one must be congruent to the other. **22.** Answers may vary; e.g., the polygons must have the same number of sides with one pair congruent (all regular polygons with the same number of sides are similar, so if they have the same number of sides with one pair congruent, they are congruent).

Mathematical Connections 12-2

Communication

1. The triangle formed by Stan's head, Stan's feet, and the opposite bank is congruent to the triangle formed by Stan's head, Stan's feet, and the spot just obscured by the bill of his cap. These triangles are congruent by ASA since the angle at Stan's feet is 90° in both triangles, Stan's height is the same in both triangles, and the angle formed by the bill of his cap is the same in both triangles. The distance across the river is approximately equal to the distance he paced off, since these distances are corresponding parts of congruent triangles. **3.** Given $\angle BAC$, construct a circle with center A and any radius. Let D and E be the points where the circle intersects the sides of the angle. Next construct the perpendicular bisector of $\overline{DE}$. **5. (a)** The distances are equal. **(b)** The distances from every point on the angle bisector of the angle to the sides of the angle are equal. **(c)** $\triangle APC \cong \triangle APB$ by AAS, hence $\overline{PC} \cong \overline{PB}$. **(d)** If a point P is equidistant from the sides of an angle, then it is on the angle bisector of the angle.

To prove this statement we assume that $PC = PB$ and prove that $\overrightarrow{AP}$ bisects $\angle A$. We have $\triangle APC \cong \triangle APB$ by Hypotenuse-Leg congruency condition. Thus $\angle CAP \cong PAB$ as these are corresponding angles in the congruent triangles.

Open-Ended

7. Answers vary. Possible questions and answers concerning figures with vertices on the nails (or dots of the dot paper). **(a)** How many non-congruent isosceles trapezoids that are not rectangles are there Answer: 4 **(b)** Same as (a) for isosceles trapezoids that are not rectangles. Answer: 6

Questions from the Classroom

9. The student is wrong. $\angle 1 \cong \angle 2$ implies that $\overline{AD}$ and $\overline{BC}$ are parallel, but does not imply that the other two sides are parallel. **11.** The SSS congruency condition assures that triangles are rigid and hence contribute to supporting the structure. **13.** This is true because the distance from A or from A' to the line BC is the distance between the two parallel lines.

Review Problems

15. constructions **17. (a)** yes, SAS **(b)** yes, SSS **(c)** yes, Hypotenuse-Leg

Assessment 12-3A

1. (a) construction **(b)** construction **2.** Constructions. Paper folding is a tactile approach to the problem. The Mira is easy to use when the paper on which the constructions are to be performed may not be altered. Using a compass and straightedge is the classical way to do constructions. The Geometer's Sketchpad computes exact measurement of objects used on the screen. **3. (a)** a right triangle **(b)** The altitude is the extension of the cable from vertex A to the ground. **4. (a)** The perpendicular bisectors of the sides of an acute triangle meet inside the triangle. **(b)** The perpendicular bisectors of the sides of a right triangle meet at the midpoint of the hypotenuse. **(c)** The perpendicular bisectors of the sides of an obtuse triangle meet outside the triangle. **(d)** construction **5. (a)** This point is equidistant from all vertices because it is on all three perpendicular bisectors. Being at the intersection of two of the perpendicular bisectors forces the point to be equidistant from all three vertices. **(b)** same as part (a) **6.** Construction. To find the radius, construct the perpendicular to one of the sides. The perpendicular segment is the radius. **7. (a)** If the rectangle is not a square, it is impossible to construct an inscribed circle. The angle bisectors of a rectangle do not intersect in a single point. **(b)** Possible. The center of the circle is the intersection of the diagonals (which are also the angle bisectors of the vertices) and the radius of the circle is the distance from the center to any of the sides. **(c)** Possible. The intersection of the three longest diagonals is the center of the circle.   **8.** construction   **9.** construction   **10.** 4″   **11. (a)** If the parallelogram is not a rectangle, cut along an altitude, which is in the center of the parallelogram. If the parallelogram is a rectangle, cut along any line through the point where the diagonals meet and such that the line is not a diagonal and is not parallel to any side.   **(b)** Make a copy of the given trapezoid and put it upside down next to $\overline{CD}$ as shown. More precisely, extend $\overline{BC}$ so that $CE = a$ and extend $\overline{AD}$ so that $\overline{DF} = b$. Because $\overline{BE} \parallel \overline{AF}$ and $BE = AF$ (the length of each is $a + b$), $ABEF$ is a parallelogram.

12. The center is where the perpendicular bisectors of the sides meet (the incenter). The radius is the distance from the incenter to a side.   **13.** $x = 3″$ **14.** as close to 26 in. as the jack can close   **15.** trapezoid because $\overline{XT} \parallel \overline{IO}$ **16. (a)** possible   **(b)** There are infinitely many non-congruent rectangles. The endpoints of two segments bisecting each other and congruent to the given diagonal determine a rectangle, but since the segments may intersect at any angle, there are infinitely many such rectangles.   **(c)** Not possible because the sum of the measurements of the angles would be greater than 180°.   **(d)** There is no unique parallelogram because the fourth angle must also be a right angle.   **17. (a)** Construct an equilateral triangle and bisect one of its angles.   **(b)** Bisect a 30° angle, then add 30° and 15° angles or bisect a right angle.   **(c)** Add 60° and 15° angles or 45° and 30° angles.   **18.** construction   **19. (a)** Because the triangles are congruent, the acute angles formed by the hypotenuse and the line are congruent; the corresponding angles are congruent, so the hypotenuses are parallel (the line formed by the top of the ruler is the transversal).   **(b)** construction   **20. (a)** The point is determined by the intersection of the angle bisector of $\angle A$ and the perpendicular bisector of $\overline{BC}$. Because the point is on the angle bisector of $\angle BAD$, it is equidistant from its sides. Because it is on the perpendicular bisector of $\overline{BC}$, it is equidistant from B and C.   **(b)** The point is determined by the intersection of the angle bisector of $\angle A$ and $\angle B$.   **21.** Construct two perpendicular diameters. Their endpoints will determine a square.

Mathematical Connections 12-3

Communication

1. Given $\angle BAC$, place one strip of tape so that an edge of the tape is along $\overline{AB}$ and another strip of the tape so that one of its edges is on $\overline{AC}$, as shown. Two edges of the strips of tape intersect in the interior of the angle at D. Connect A with D. $\overline{AD}$ is the angle bisector. Because the diagonals of a rhombus bisect its angles, this contruction can be justified by showing that $AEDF$ is a rhombus. (E and F are the

points of intersection of the tops of the tape pieces and the opposite sides.) $AEDF$ is a parallelogram. (Why?) It remains to be shown that $\overline{AF} \cong \overline{AE}$. For that purpose, we show that $\triangle FAG \cong \triangle EAH$. We have $\overline{FG} \cong \overline{HE}$ because the two strips of tape have the same width. $\angle A \cong \angle A$, and the angles at H and G are right angles. Thus, the triangles are congruent by AAS.

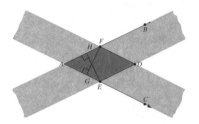

3. Answers vary.

Open-Ended

5. (a) Concentric circles are circles with the same center.   **7.** No, any two great circles on a sphere intersect.

Questions from the Classroom

9. The construction is not valid if the only tools allowed are straightedge and compass. The procedure is sometimes referred to as eyeballing, where we depend on our sight to judge if the ruler touches the circle only at one point.   **11.** $\overline{OD}$ is the radius if, and only if, it is perpendicular to $\overline{AC}$. If $AB \neq BC$, then $\overline{BD}$ is not perpendicular to $\overline{AC}$.

Review Problems

13. construction

Assesment 12-4A

1. 83/70   **2. (a)** yes; AAA   **(b)** Yes; sides are proportional and angles are congruent.   **(c)** no   **(d)** always similar because the ratio of the longer side equals the ratio of the corresponding shorter sides and the angles are congruent.   **3.** Make all dimensions 3 times as long; that is, in part (c) each side would be 3 diagonal units long. One possible solution set is shown here:

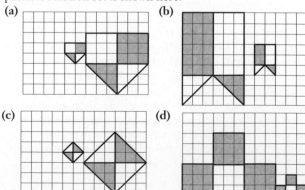

4. Answers may vary.   **(a)** two rectangles, one of which is a square and the other is not   **(b)** two rhombuses with the same length sides but with different angles   **5. (a)** The ratio of the perimeters is the same as the ratio of the sides. **(b)** If a, b, c, d are the sides of one quadrilateral and a_1, b_1, c_1, d_1 the corresponding sides of a similar quadrilateral then

$$\frac{a}{a_1} = \frac{b}{b_1} = \frac{c}{c_1} = \frac{d}{d_1} = r \text{ (the scale factor)}$$

Now, $\dfrac{a + b + c + d}{a_1 + b_1 + c_1 + d_1} = \dfrac{a_1 r + b_1 r + c_1 r + d_1 r}{a_1 + b_1 + c_1 + d_1}$

$$= \frac{(a_1 + b_1 + c_1 + d_1)r}{a_1 + b_1 + c_1 + d_1} = r.$$

Hence the ratio of the perimeters is r. An analogous proof works for any two similar n-gons.
6. (a) (i) $\triangle ABC \sim \triangle DEF$ (by AA)
(ii) $\triangle ABC \sim \triangle EDA$ (by AA)   **(iii)** $\triangle ACD \sim \triangle ABE$ (by AA)   **(iv)** $\triangle ABE \sim \triangle DBC$ by SAS similarity condition since $\dfrac{2}{3} = \dfrac{3}{4.5}$ and the vertical angles at 3 are congruent.   **(b) (i)** $\dfrac{2}{3}$ **(ii)** $\dfrac{1}{2}$ **(iii)** $\dfrac{4}{3}$ **(iv)** $\dfrac{2}{3}$
7. (a) 7   **(b)** $\dfrac{24}{7}$ **(c)** $\dfrac{16}{5}$ **(d)** $\dfrac{36}{11}$ **(e)** $\dfrac{8}{3}$

8. construction
9. Construct as follows:

10. No. The maps are similar and even though the scales may change, the actual distances do not.   **11.** sketch
12. $\triangle ACO \sim \triangle BDO$ by AA. Thus, $\dfrac{a}{b} = \dfrac{d}{c}$, which implies $ac = bd$.   **13.** Sketch one regular hexagon and the other with all angles 120° but not all sides congruent.
14. Answers may vary. For example, ▱
15. Without knowing the lengths of the rulers, one cannot tell. However, 1" corresponds to 2.54 cm so it is possible to have them similar with ratio $\dfrac{1}{2.54}$.   **16.** One would expect them to be, with ratio $\dfrac{4}{9}$.   **17.** $133\dfrac{1}{3}\%$, but most copy machines will not allow this setting   **18.** 15 m   **19.** about 232.6 in., or 19.4 feet   **20. (a)** rectangle   **(b)** rectangle   **(c)** rhombus   **(d)** rectangle   **21.** All sides in each regular octagon are congruent and all interior angles are congruent.
22. They have exactly the same shape but not necessarily the same diameter; i.e., they are similar.

Mathematical Connections 12-4

Communication

1. Any two cubes are similar because all the faces are squares and the dihedral angles between two adjacent faces measure 90°   **3.** Lay the licorice diagonally on the paper so that it spans a number of spaces equal to the number of children. (See the figure.) Cut on the lines. Equidistant parallel lines will divide any transversal into congruent segments.

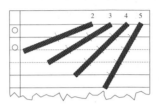

Open-Ended

5. Points lie along the same line because similar triangles are formed.

Questions from the Classroom

7. No, they are not necessarily similar. For example, a rectangle that is not a square and a square have all angles congruent but are not similar.   **9.** The student is incorrect. The corresponding sides are not proportional.

Review Problems

11. Start with the given base and construct the perpendicular bisector of the base. The vertex of the required triangle must be on that perpendicular bisector. Starting at the point where the perpendicular bisector intersects the base, mark on the perpendicular bisector a segment congruent to the given altitude. The endpoint of the segment not on the base is the vertex of the required isosceles triangle.   **13.** Answers may vary. Students may suggest that angles of measure 45° be constructed with the endpoints of the hypotenuse as vertices of the 45° angles and the hypotenuse as one of the sides of the angles. Both angles need to be constructed on the same side of the hypotenuse.   **15.** SASAS or ASASA

Assessment 12-5A

1. The graph of $y = mx + 3$ contains the point $(0, 3)$ and is parallel to the line $y = mx$. Similarly, the graph of $y = mx - 3$ contains the point $(0, ^-3)$ and is parallel to $y = mx$.

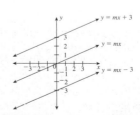

2. (a) 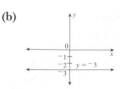 **(b)**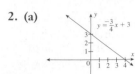

(c)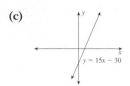

3. x-intercept y-intercept
(a) 4   3
(b) None $^-3$
(c) 2 $^-30$

4. (a) Using $(0, 32)$ and $(100, 212)$, the slope
$(212 - 32)/(100 - 0) = \dfrac{9}{5}$. So $F = \left(\dfrac{9}{5}\right)C + b$. Plug in
the point $(0, 32)$. Thus, $b = 32$, and the equation is
$F = \left(\dfrac{9}{5}\right)C + 32$.   **(b)** $C = \dfrac{5}{9}(F - 32)$

5. (a) $y = \left(\dfrac{1}{3}\right)x$; slope $\dfrac{1}{3}$, y intercept 0   **(b)** $y = {}^-x + 3$;
slope $^-1$, y intercept 3   **(c)** $y = \left(\dfrac{1}{3}\right)x$; slope $\dfrac{1}{3}$, y intercept 0

6. (a) $y = {}^-x - 1$ **(b)** $y = \left(\dfrac{1}{2}\right)x$ **(c)** $y = x - \dfrac{1}{2}$

7. Answers vary.   **8. (a)** $x = {}^-2$; y is any real number.
(b) $x > 0$ and $y < 0$; x and y are real numbers.
9. perimeter $= 12$ units; area $= 8$ sq units   **10. (a)** $x = 3$
(b) $y = 5$ **11. (a)** $\dfrac{1}{3}$ **(b)** 0   **(c)** 1 if $a \neq b$

12. (a) $y = \dfrac{1}{3}x + \dfrac{5}{3}$ **(b)** $y = 2$ **(c)** $y = x$ **13.** Answers
may vary, depending on estimates from the fitted line; for
example: **(a)** From the fitted line, estimate point coordinates of $(50, 8)$ and $(60, 18)$; the slope is $\dfrac{10}{10} = 1$. Use the
point $(50, 8)$ and substitute $T = 50$ and $C = 8$ into $C =$
$1T + b$ (i.e., an equation of the form $y = mx + b$), so
$8 = 1(50) + b$ ⟹ $b = {}^-42$; the equation is then
$C = T - 42$.   **(b)** 48 chirps in 15 sec   **(c)** $N = 4T - 168$
14. (a) $(3, 0)$ or $(^-3, 0)$ **(b)** $x = 3$ or $x = {}^-3$

15. (a) $y = 4$ **(b)** $x = {}^-7$ **16. (a)** $(2, 3)$ and $\left(12\dfrac{1}{2}, 3\right)$
(b) $y = 3$
17. (a)    **(b)** 60

18. (a) (i) The plotted points are shown in the figure.

(ii) Answers may vary. One possible answer is $y = 2x + 6$.
(b) Answers vary.   **19. (a)** unique solution of $(2, 5)$
(b) unique solution of $(1, {}^-5)$ **20.** 4000 gal of gasoline
and 1000 gal of kerosene   **21.** 17 quarters, 10 dimes
22. Answers vary; for example, $(50, 106)$.   **23.** It is negative.

Mathematical Connections 12-5

Communication

1. No (unless $r = 1$), because the slopes between two
successive points are not the same   **3.** Answers will
vary. Explanation 1: If two distinct lines have the same slope
m, then the equations are $y = mx + b$ and $y = mx + c$ for
some real numbers b and c (with $b \neq c$). To show that the
lines are parallel, it is sufficient to show that the lines do not
intersect; that is, that the system of equations has no
solution. Indeed, if we try to solve the equations, we get
$mx + b = mx + c$. Because $b \neq c$, this equation has no solution. Explanation 2: With $y = mx + b$ and $y = mx + c$,
each is a vertical shift of the same line, $y = mx$.

Open-Ended

5. (a) Answers vary.   **(b)** Answers vary.

Questions from the Classroom

7. The slope of a vertical line is undefined because division
by 0 is impossible. When a line is closer and closer to being a
vertical line, the slope is getting bigger and bigger. One way
to see this is to take two points whose y-coordinates differ by
1. The corresponding x-coordinates of the points will be
close to each other and hence their difference will be close to
0. The slope resulting from dividing 1 by a very small number will be very large.   **9.** The student is correct. If α is the
angle that a line makes with the x-axis, then $\tan \alpha = \dfrac{\text{Rise}}{\text{Run}}$
(see figure).

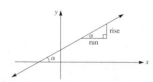

11. A line with undefined slope is perpendicular to the
x-axis. Two lines perpendicular to another line are parallel.

Review Problems

13. The four smaller triangles are congruent. Using the
Midsegment Theorem we find that each side of a smaller
triangle is half as long as a side of $\triangle ABC$. Hence all the
smaller triangles are congruent by SSS. Since $\overline{MN} \parallel \overline{AC}$,
$\angle BMN \cong \angle BAC$ and $\angle BNM \cong BCA$. Thus by AA
$\triangle BMN \sim \triangle ABC$. The other three triangles being
congruent to $\triangle MNP$ are also similar to $\triangle ABC$.   **15.** 8.2

Chapter Review

1. (a) $\triangle ADB \cong \triangle CDB$ by SAS   **(b)** $\triangle GAC \cong \triangle EDB$ by SAS   **(c)** $\triangle ABC \cong \triangle EDC$ by AAS   **(d)** $\triangle BAD \cong \triangle EAC$ by ASA   **(e)** $\triangle ABD \cong \triangle CBD$ by ASA or SAS
(f) $\triangle ABD \cong \triangle CBD$ by SAS   **(g)** $\triangle ABD \cong \triangle CBE$ by SSS; $\triangle ABE \cong \triangle CBD$ by SSS   **(h)** $\triangle ABC \cong \triangle ADC$ by SSS; $\triangle ABE \cong \triangle ADE$ by SSS or SAS; $\triangle EBC \cong \triangle EDC$ by SSS or SAS   **2.** Parallelogram; $\triangle ADE \cong \triangle CBF$ by SAS, so $\angle DEA \cong \angle CFB$; $\angle DEA \cong \angle EAF$ (alternate interior angles between the parallels $\overleftrightarrow{DC}$ so $\overleftrightarrow{AB}$ and the transversal $\overleftrightarrow{AE}$), so $\angle EAF \cong \angle CFB$; and therefore $\overline{AE} \parallel \overline{FC}$. $\overline{EC} \parallel \overline{AF}$ (parallel sides of the square); two pairs of parallel opposite sides implies a parallelogram.   **3.** constructions
4. (a) $x = 8$ cm $y = 5$ cm   **(b)** $x = 6.5$ m

5. construction   **6.** $\dfrac{a}{b} = \dfrac{c}{d}$ **7.** *Hint:* Find the intersection of the perpendicular bisector of $\overline{AB}$ and line ℓ.

8. (a) $\triangle ACB \sim \triangle DEB$ by AA, $x = \dfrac{24}{5}$ in.

(b) (i) $\triangle AED \sim \triangle ACB$ by AA   **(ii)** $y = \dfrac{4'}{3}$ **(iii)** $x = \dfrac{55'}{6}$

9. (a) False: A chord has its endpoints on the circle.
(b) true   **10.** 12 m   **11. (a)** polygons (ii) and (iii)
(b) Any convex regular polygon can be inscribed in a circle.   **12.** $b = 6$ m   **13.** $d = \dfrac{256}{5}$ m   **14.** Cut as shown. Slide shaded triangle to the right.

15. (a) AD is a diameter.   **(b)** $\overline{AB}$ and $\overline{BE}$ are diameters that bisect each other (at the center). Hence $ABDE$ is a rectangle and so   **16.** They are equal unless vertical when the slopes do not exist.   **17.** All midsegments are congruent.
18. (a) Use AA.   **(b)** A midsegment is parallel to a side. Thus, a trapezoid is formed.   **19.** True in some cases and false in others. If the diagonals bisect each other, then the quadrilateral is a square. If not, it is not a square.
20. (a) $y = \dfrac{-4}{3}x - \dfrac{1}{3}$ **(b)** $y = \dfrac{1}{3}x + 1$ **21.** No; if a line connects all three points, the slope between any two must be the same. Slope between $(4, 2)$ and $(0, {}^{-}1) = \dfrac{{}^{-}1 - 2}{0 - 4} = \dfrac{3}{4}$; slope between $(0, {}^{-}1)$ and $(7, {}^{-}5) = \dfrac{{}^{-}5 - {}^{-}1}{7 - 0} = \dfrac{{}^{-}4}{7}$. Slopes are unequal, so there is no line through the points.   **22. (a)** unique solution of $\left(4\dfrac{1}{5}, \dfrac{-3}{5}\right)$
(b) unique solution of $\left(\dfrac{10}{9}, \dfrac{4}{3}\right)$ **(c)** no solution; parallel lines   **23.** Congruent triangles have two corresponding angles that are congruent. This makes them similar by AA.

Answers to Now Try This

12-1.

12-2. There are six possible ways to write the congruence: ABC paired with $A'B'C'$, ACB paired with $A'C'B'$, BAC paired with $B'A'C'$, BCA paired with $B'C'A'$, CAB paired with $C'A'B'$, and CBA paired with $C'B'A'$.   **12-3. (a)** In order to construct a triangle from three lengths of straws, the length of one piece must be less than the sum of the lengths of the other two pieces.   **(b)** The triangle would be equilateral and equiangular.   **(c)** There can be no triangle constructed. The three pieces would lie along a segment.
(d) No. See the answer to part (a).   **12-4.** An example is seen in the following drawing

If the nonincluded angle is a right angle, the triangles are congruent. More generally, if the nonincluded angle is opposite the larger of the two sides, the triangles are congruent.
12-5. Draw two intersecting arcs with the same radius, one with center at A and the other with center at B. The two points of intersection are opposite vertices of a rhombus. The line connecting the points is the perpendicular bisector of the segment.   **12-6.** The answer is no. Consider the following drawing:

12-7. In a parallelogram $ABCD$ draw diagonal $\overline{AC}$. Because opposite sides are parallel (by definition), $\angle BAC \cong \angle ACD$ as well as $\angle BCA \cong \angle DAC$. Hence, $\triangle ABC \cong \triangle CDA$ by ASA since $\overline{AC}$ is a common side. By CPCTC (corresponding parts of congruent triangles), $\overline{AB} \cong \overline{CD}$ and $\overline{BC} \cong \overline{DA}$.
12-8. Yes. If $ABCD$ is a quadrilateral whose diagonals intersect at M, then congruent vertical angles with vertex at M are formed. Because the diagonals bisect each other, it follows from SAS that $\triangle AMD \cong \triangle CMB$. Hence, $\angle MAD \cong \angle MCB$ and therefore $\overline{BC}$ is parallel to $\overline{AD}$. In a similar way, we can show that $\overline{AB}$ is parallel to $\overline{DC}$ and hence $ABCD$ is a parallelogram.   **12-9.** One way to accomplish the construction is to place the hypotenuse of the triangle on the given line and to place the ruler so that one of the legs of the triangle will be on the ruler. Now, keeping the ruler fixed, slide the triangle so that the side of the triangle on the ruler touches the ruler all the time. Slide the triangle on the ruler until the given point is on the hypotenuse. The line containing the hypotenuse is parallel to the given line.   **12-10.** In Figure 12-47 if $PD = PE$ then by the Hypotenuse-Leg congruence condition (Theorem 12-3), $\triangle APD \cong \triangle APE$ and hence $\angle PAD \cong \angle PAE$.
12-11. Construct the diagonals of the square. Use the intersection of the diagonals as the center of the inscribed circle.   **12-12.** The statement is not true for polygons. For example, a rectangle and a square have congruent

angles but do not have to be similar.   **12-13. (a)** $y = 0$ (This is the x-axis.)   **(b)** As m increases, the slope is positive and the graph of the line becomes steeper as m becomes larger.   **(c)** As m decreases, the slope is negative and the graph of the line becomes steeper as the absolute value of m becomes larger.   **12-14. (a)** All the points on a horizontal line have the same y-coordinate. Thus, two points on a horizontal line have the form (x_1, y_1) and (x_2, y_2), where $y_2 = y_1$. The slope is $\dfrac{y_2 - y_1}{x_2 - x_1} = \dfrac{0}{x_2 - x_1} = 0$.   **(b)** For any vertical line, $x_2 = x_1$ and therefore $x_2 - x_1 = 0$. If we attempted to find the slope, we would have to divide by 0, which is impossible. Hence, the slope of a vertical line is not defined.   **12-15. (a)** The graphs are lines that intersect at $x = 4$ and $y = 3$.   **(b)** The lines are parallel and hence the system has no solution. Algebraic approach: Assuming that there is a solution x and y, we multiply the first equation by 2 and add it side-by-side to the second. Therefore, x and y must satisfy $0 \cdot x + 0 \cdot y = 5$. However, no x and y satisfies this equation (a solution would imply $0 = 5$). This contradicts our assumption that the original system has a solution.   **(c)** The lines are parallel. The algebraic approach is similar to part (b).

Answers to Technology Corner

Section 12-1

5. (b) When **Mark Angle** and **Transform** are selected, an arc appears briefly that looks like it is measuring the angle. When **Rotate By Marked Angle** is chosen, another line appears making an angle congruent to $\angle PQR$.

Section 12-2

(a) It is true that if one pair of sides of a quadrilateral is congruent and parallel, then the quadrilateral is a parallelogram.   **(b)** If $\overline{AD} \parallel \overline{BC}$ and $AD = BC$ then $\triangle ABD \cong \triangle CDB$ by SAS ($\angle ADB \cong \angle CBD$ as alternate interior angles created by parallel lines, $\overline{AD} \cong \overline{BC}$ and $\overline{DB} \cong \overline{BD}$). Thus $\angle ABD \cong \angle CDB$, which implies that $\overline{AB} \parallel \overline{CD}$.

Section 12-3

Page 792   1. (b) Answers vary.   **(c)** $m(\angle BOC) = 2m(\angle BPC)$ **2.** $90°$

Section 12-4

All of the ratios are equal. You can conclude that the triangles are similar because the ratios of their sides are equal. The perimeter of the S_1 is 3; the perimeter of S_2 is 4; and the perimeter of S_3 is $\dfrac{16}{3}$. The perimeter of the snow flake curver is infinite. Students may have trouble thinking about the perimeters of the snowflake curves. **(d)** A cube

Answers to Brain Teasers

Section 12-4

Page 801   If x is the height of the fence post, show that $\dfrac{x}{15} + \dfrac{x}{10} = 1$. Hence, the fence post must be 6 ft high regardless of how far apart the flag poles are.

Page 809   When P is located at the midpoint of a side, the length of the path is one-half the perimeter of the triangle. In general, if P is not on a midpoint of a side, the length of the path is equal to the perimeter of the triangular base.

Answer to Laboratory Activity

Section 12-4

The pantograph is designed so that point E must be fixed (nailed to a table or a board), the strips pivoted A, B, C, and D, and $ABCD$ is a parallelogram. Also, points E, D, and F are collinear (it can be shown that if E, D, and F are collinear for a particular position of the pantograph, then the three points will be collinear for every position of the pantograph). Because $\overline{CD}$ is parallel to $\overline{BF}$, we have: $\dfrac{EF}{ED} = \dfrac{BE}{EC}$. Since $\dfrac{BE}{EC}$ is fixed (the same for every position of the pantograph), it follows that $\dfrac{EF}{ED}$ is fixed; it equals some real number r. This implies that if D' and F' are new positions of the pointer and the pencil, then $\dfrac{EF'}{ED'} = \dfrac{EF}{ED} = r$. If the pointer traces the segment $\overline{DD'}$, the pencil will draw the segment $\overline{FF'}$ and $\dfrac{FF'}{DD'} = r$. Because each segment traced is enlarged by the same factor r, when any figure is traced, the pencil will draw a similar figure. This can be explained in a simpler way using the concept of a size transformation in Chapter 14 (the size transformation has center E and scale factor r).

Answer to Preliminary Problem

Section 12-5

We are to explain how two triangles with one 18-in. side and angles of measures 36°, 84°, and 60° could be different sizes. To explain how this is possible, we try to construct two noncongruent triangles with Rosa's specifications. First, we need to construct a triangle with the given side and angles. Then, because the second triangle has the same angles, it must be similar to the first by AAA. Consequently, we try to construct a triangle similar to the first but not congruent to it. In the figure $\overline{AB}$ represents the 18-in. side. We then use a protractor to construct $\angle A$ measuring 60° and $\angle B$ measuring 36°. Hence $m(\angle C) = 84°$. Next, we are to construct a noncongruent triangle similar to $\triangle ABC$ with one side as long as $\overline{AB}$. Because this may seem difficult, we consider a *simpler but related problem*. We drop the condition that the triangle must have a side of length AB and try to construct a triangle similar to $\triangle ABC$. This can be achieved by using one of the existing vertices of $\triangle ABC$ and drawing a side parallel to the opposite side. For example, in the figure $\overline{C_1B_1} \parallel \overline{CB}$ and $\triangle ACB \sim \triangle AC_1B_1$. There are infinitely many such triangle AC_1B_1. We imagine sliding $\overline{BC}$ along $\overrightarrow{AB}$ and $\overrightarrow{AC}$ so that $\overline{B_1C_1} \parallel \overline{BC}$ and until $AC_1 = AB$. Consequently, $\triangle ACB$ and $\triangle AC_1B_1$ are two non-congruent similar triangles such that $AC_1 = AB$, and therefore there are two non-congruent triangles with Rosa's specifications.

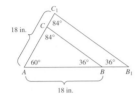

Chapter 13

Assessment 13-1A

1. (a) 1 cm **(b)** 1 cm **(c)** 8 cm **(d)** 0.5 cm **(e)** 0.7 cm
(f) 5 cm **(g)** 7.3 cm **(h)** 5.2 cm **2.** Answers vary.
3. (a) 960 **(b)** 1.5 **(c)** 2880
4. (a) Perimeter is 12 units.

(b) Perimeter is 4 units.

(c) Perimeter is $2 + \sqrt{2}$ units.

5. (a) $2\frac{7}{9}$ yd **(b)** 14,400 **(c)** 100 **(d)** 31 **6.** Answers vary. **7. (a)** 65 mm **(b)** 6.5 cm **8. (a)** centimeters **(b)** centimeters or millimeters **(c)** centimeters or meters **9. (a)** inches **(b)** inches **(c)** feet **10. (a)** 0.35 m, 350 mm **(b)** 163 cm, 1630 mm **(c)** 0.035 m, 3.5 cm **(d)** 0.1 m, 10 cm **(e)** 200 cm, 2000 mm **11. (a)** 10.00 **(b)** 0.770 **(c)** 10.0 **(d)** 15.5 **12.** 6 m, 5218 mm, 245 cm, 700 mm, 91 mm, 8 cm **13.** Answers vary.

14. (a) π **(b)** $\frac{\pi}{2}$ **(c)** yes; $\frac{1}{2}$ **15. (a)** 7 cm **(b)** 10 cm

16. (a) 1 **(b)** 0.262 **(c)** 3000 **(d)** 0.03
17. (a) $AB + BC > AC$ **(b)** $BC + CA > AB$
(c) $AB + CA > BC$ **18. (a)** Can be **(b)** Cannot be

because $10 + 40 = 50$, which contradicts the triangle inequality. **19. (a)** The minimum perimeter is attained when the second longest side is not part of the perimeter.

(b) The maximum perimeter is attained when the longer sides are part of the perimeter; for example,

20. (a) 6 cm **(b)** $3/\pi$ m **21. (a)** 6π cm **(b)** 4 cm
22. The circumference will double. **23. (a)** 3096 km/hr
(b) 1032 m/sec **(c)** approximately Mach 4.04
24. $(50 + 6\pi)$ ft **25.** 49 cm **26. (a)** 0.5 m
(b) 0.05 cm **(c)** 0.005 m or 5 mm

Mathematical Connections 13-1

Communication

1. Answers may vary, but the accomplishment was not trivial.
3. (a) Answers vary; for example, the metric system is a base-ten system and works "almost like" our money system. Conversions are much easier because of the decimal relationship between certain units of measure. **(b)** Metric relationships are easier to remember. Also, because it is a base-ten system, conversions are easier. **(c)** Answers vary; for example, things like original property descriptions and some building materials will probably not change.

Open-Ended

5. (a) Answers vary. For example: 1-1-1 and 2-2-1 work; 2-3-6 and 1-2-3 do not. **(b)** A triangle can be constructed if, and only if, $a + b > c$ and $a + c > b$ and $b + c > a$, where a, b, and c are the lengths of the three sides.

Cooperative Learning

7. (a) Answers vary. **(b)** If the circles are the same size, then regardless of the radius of each circle the amount of wire needed is the same. **(c)** Let the radius of each circle be r cm. Then the number of circles needed for each chain is $\frac{60}{2r}$ (why?). The circumference of each circle is $2\pi r$ and hence the amount of wire, which is the total circumference of all the circles, is $\frac{60}{2r} \cdot 2\pi r$, or 60π cm. Because this number does not depend on the value of r, the amount of wire needed for the chain is the same regardless of the radius of each circle.

9. The height is $3d$, where d is the diameter of a tennis ball. The perimeter of the can is πd, or about $3.14d$, which is greater than $3d$.

Assessment 13-2A

1. Answers vary.   **2. (a)** cm^2, in.^2 **(b)** cm^2 or mm^2, in.^2
(c) m^2 or cm^2, yd^2 or in.^2 **(d)** m^2, yd^2 **3.** Answers vary, for example,   **(a)** 1.6 m^2 **(b)** 2500 cm^2.
4. (a) 0.0588 m^2, $58,800 \text{ mm}^2$ **(b)** 0.000192 m^2, 1.92 cm^2
(c) $15,000 \text{ cm}^2$, $1,500,000 \text{ mm}^2$ **(d)** 0.01 m^2, $10,000 \text{ mm}^2$
(e) 0.0005 m^2, 500 mm^2 **5. (a)** $444\frac{4}{9}$ **(b)** approx. 0.32
(c) 6400 **(d)** $130,680$ **6. (a)** 4900 **(b)** 98 **(c)** 0.98
7. (a) 3 sq. units   **(b)** 3 sq. units   **(c)** 2 sq. units
(d) 5 sq. units   **8. (a)** 20 cm^2 **(b)** 7.5 m^2 **9.** r^2
10. (a) Yes. All squares are similar.   **(b)** $a^2 : b^2$
11. (a) 9 cm^2 **(b)** 96 cm^2 **(c)** 20 cm^2 **(d)** 84 cm^2
12. (a) (i) 1.95 km^2 **(ii)** 195 ha **(b) (i)** 0.63 mi^2
approx.   **(ii)** approx. 403 acres   **(c)** Answers vary; for example, the metric system is easier because you have only to move the decimal point to convert units within the system.   **13. (a)** true   **(b)** Don't know. It could be.
(c) False, the maximum area is 60 cm^2.   **(d)** Don't know, since the height is unknown. The area is less than *or equal to* 60 cm^2.   **14.** 30 cm^2 **15. (a)** 405.11 **(b)** 550
16. (a) $25\pi \text{ cm}^2$ **(b)** $(8/3)\pi \text{ cm}^2$ **17.** 1200 tiles
18. (a) $24\sqrt{3} \text{ cm}^2$ **(b)** $9\sqrt{3} \text{ cm}^2$ **19. (a)** $16\pi \text{ cm}^2$
(b) $r = \dfrac{s}{\sqrt{\pi}}$ **20. (a)** $2\pi \text{ cm}^2$ **(b)** $\left(\dfrac{\pi}{2} + 2\right) \text{ cm}^2$
(c) $2\pi \text{ cm}^2$ **(d)** $(50\pi - 100) \text{ cm}^2$ **21.** $7\pi \text{ m}^2$
22. (a) 48 cm **(b)** 64 cm^2 **23. (a)** The area is quadrupled.   **(b)** $1 : 25$ **24.** $(320 + 64\pi) \text{ m}^2$ **25.** $MA = 32$ units and $AT = 33$ units; area 1056 sq. units
26. (a)

$a(b + c) = ab + ac$

(b)

$(a + b)(c + d) = ac + ad + bc + bd$

27. The area of each triangle is 10 cm^2 because the base of each triangle is $\overline{AB}$ and the height of each triangle is the perpendicular distance between the two lines. Because each triangle has the same base and height, the areas of the triangles are the same.   **28. (a)** 6 cm^2 **(b)** 30 cm^2

Mathematical Connections 13-2

Communication

1. Answers may vary. The triangle may be drawn on graph paper and estimated. Students may look up Heron of Alexandria to find the formula $A = \sqrt{s(s - a)(s - b)(s - c)}$, where s is one-half of the perimeter of the triangle with sides a, b, and c. In this case, the area is

$\sqrt{15(15 - 6)(15 - 11)(15 - 13)}$, or $\sqrt{1080}$, approximately 32.9 in.^2.   **3. (a)** The area of the 10-in. pizza is $25\pi \text{ in}^2$. The area of the 20-in. pizza is $100\pi \text{ in}^2$. Because the area of the 20-in. pizza is 4 times as great, this pizza might cost 4 times as much, or $40. However, this is not the case because other factors are considered rather than just the area of the pizza.   **(b)** If the price is based only on the area of pizza, then the ratio between prices should be $1 : k^2$.   **5.** After the rotations, a rectangle is formed. The area of the rectangle is length times width, which in the case of the parallelogram is the same as the base times the height.

Open-Ended

7. (a) Answers vary depending on the size of the hand.   **(b)** Answers vary; for example, many people will trace their hands on square-centimeter paper and then count the number of squares that are entirely contained in the outline. Next, they count the number of squares that are partially contained in the hand outline and multiply this number by $1/2$. The final estimate is the sum of the two numbers.

Cooperative Learning

9. (a) 5 sq. units   **(b)** 12 units   **(c) (i)** 3   **(ii)** 15   **(iii)** 20
(d) Answers vary depending on chosen shape.   **(e)** Answers vary depending on chosen shape.

Questions from the Classroom

11. No. An angle is a union of two rays. The student probably means the area of the interior of an angle. However, because the area of the interior of an angle is infinite, it does not have a measurable area.   **13.** The area of a garden does not depend on the perimeter of the garden. For example, a garden that is 2 m by 6 m has a perimeter of 16 m and an area of 12 m^2. A garden that is 4 m by 4 m also has a perimeter of 16 m but has an area of 16 m^2.

Review Problems

15. $\left(8 + \dfrac{\pi}{2}\right)$ ft, or approximately 9.57 ft   **17.** $(2\pi - 6)r$, or approximately $0.28r$ is the difference.

Assessment 13-3A

1. (a) $\sqrt{20} \approx 4.5$ **(b)** $\sqrt{17} \approx 4.1$ **2. (a)** 6 **(b)** $5a$
(c) 12 **(b)** $\dfrac{s}{2}\sqrt{3}$ **3. (a)** Make a right triangle with legs of length 2 and 3, and then $h = \sqrt{13}$.   **(b)** Make a right triangle legs of length 1 and 2, and then $h = \sqrt{5}$.
4. Answers may vary.

5. $8 + 4\sqrt{2} \approx 13.66$ units   **6.** $6\sqrt{5} \text{ cm}$, $12\sqrt{3} \text{ cm}$

7. (a) no **(b)** yes **(c)** yes   **8.** $\sqrt{450}$ cm, or $15\sqrt{2}$ cm
9. $\sqrt{125}$ mi, or about 11.2 mi   **10.** $6\sqrt{6}$, or about 14.7 ft
11. $4\sqrt{6}$ m or approx 9.8 m.   **12. (a)** $x = 8, y = 2\sqrt{3}$

(b) $x = 4, y = 2$ **13.** $\dfrac{2.6}{\sqrt{3}}$ m or approx 1.5 m   **14.** 12.5

cm; 15cm   **15.** $12\sqrt{2}$ in. or approximately 16.97 in., so
17.0 in   **16.** $10\sqrt{99}$ ft or approximately 99.5 ft
17. The area of the trapezoid is equal to the sum of the areas of the three triangles. Thus,

$$\frac{1}{2}(a + b)(a + b) = \left(\frac{1}{2}\right)ab + \left(\frac{1}{2}\right)ab + \left(\frac{1}{2}\right)c^2$$

$$\frac{1}{2}(a^2 + 2ab + b^2) = ab + \left(\frac{1}{2}\right)c^2$$

$$\frac{a^2}{2} + ab + \frac{b^2}{2} = ab + \frac{c^2}{2}$$

Subtracting ab from both sides and multiplying both sides by 2, we have $a^2 + b^2 = c^2$. The reader should also verify that the angle formed by the two sides of the length c has measure $90°$.   **18.** The area of the large square is equal to the sum of the area of the small square and the four right triangles. Therefore, $(a + b)^2 = c^2 + 4(ab/2)$. Thus, $a^2 + 2ab + b^2 = c^2 + 2ab$, which in turn implies that $c^2 = a^2 + b^2$. It must also be shown that the inside quadrilateral is really a square.   **19. (a)** 4 **(b)** 5 **(c)** $2\sqrt{13}$

(d) $\dfrac{\sqrt{365}}{4}$, or approx. 4.78   **20.** $\dfrac{c\sqrt{3}}{4}$ **21. (a)** $x = 0$;

$y = 0; y = \dfrac{3}{4}x$ **(b)** $(0, 0)$ **22. (a)** $(16, 0)$ **(b)** $8\sqrt{2}$,

$8\sqrt{2}$, and 16   **(c)** $(8\sqrt{2})^2 + (8\sqrt{2})^2 = 128 + 128 = 256 = 16^2$ **23. (a)** $(x + 3)^2 + (y - 4)^2 = 16$

(b) $(x + 3)^2 + (y + 2)^2$ 2   **24. (a)** $(0, 0), 4$
(b) $(3, 2), 10$ **(c)** $(^-2, 3), \sqrt{5}$ **(d)** $(0, ^-3), 3$

Mathematical Connections 13-3

Communication

1. (a) Let the length of the side of the square be s. Draw the diagonal of the square and make the new square have side lengths equal to the diagonal. Then the area is $(\sqrt{2}s)^2 = 2s^2$. **(b)** Make the lengths of the new square 1/2 the length of the diagonal, or $(\sqrt{2}s/2)$; then the new square is $(\sqrt{2}s/2)^2 = s^2/2$.   **3.** Draw a perpendicular segment from y to the segment representing 200 m. Draw segment $\overline{XY}$. A right triangle is formed with side lengths 100 m and 150 m. Using the Pythagorean theorem, we find that $XY = \sqrt{100^2 + 150^2} = \sqrt{32,500} = 180.3$ m.

Open-Ended

5. Areas vary depending on the triangles drawn, but the Pythagorean theorem works only for the right triangles.

Cooperative Learning

7. Answers vary depending on the triangles that are formed.

Questions from the Classroom

9. This can't be the equation of a circle because of the $^-16$ value. The radius squared is always a positive number.

Review Problems

11. 0.032 km, 322 cm, 3.2 m, 3020 mm.   **13. (a)** 10 cm, 10π cm, 25π cm^2 **(b)** 12 cm, 24π cm, 144π cm^2
(c) $\sqrt{17}$ m, $2\sqrt{17}$ m, $2\pi\sqrt{17}$ m **(d)** 10 cm, 20 cm, 100π cm^2

Assessment 13-4A

1. (a) no **(b)** yes **(c)** no **(b)** yes   **2. (a)** 96 cm^2
(b) 216π cm^2 **(c)** 236 cm^2 **(d)** 64π cm^2 **(e)** 96π cm^2
3. 2.5 L so buy 3 L   **4.** 2688π mm^2 **5.** 162,307,600π km^2
6. 4:9   **7.** $(108\sqrt{21} + 216\sqrt{3})$ m^2 **8.** $10.5\,\pi$ in.^{2}

or approx. 33 in.^{2} **9.** $\left(100 + \dfrac{50}{\pi}\right)$, or 115.92 cm^2

approx.   **10.** $\ell = 11$ cm, $w = 8$ cm, $h = 4$ cm
11. (a) The lateral surface area is multiplied by 3.
(b) The lateral surface area is multiplied by 3. **(c)** The lateral area is multiplied by 9.   **12.** $(100 + 100\sqrt{17})$ cm^2
13. (a) 3 units **(b)** 5 units **(c)** 4 units **(d)** $216°$
14. (a) 1.5π m^2 **(b)** 2.5π in^2 **15.** 10π in.^{2}
16. 1:4   **17.** 4:3   **18. (a)** $\sqrt{\frac{10648}{6}}$ or approx. 42 cm
(b) $\sqrt{5324}$ cm or approximately 72.97 cm
19. $(6400\pi\sqrt{2} + 13,600\pi)$ cm^2

Mathematical Connections 13-4

Communication

1. She would need 4 times as much cardboard. If each measurement is doubled, then the area of each face is increased by a factor of 4; that is, $A_1 = \ell w$ and $A_2 = (2\ell)(2w) = 4\ell w = 4A_1$. Because this is true for all faces, the surface area is multiplied by 4.   **3.** doubling the radius

Open-Ended

5. (a) 2% **(b)** Estimates vary. **(c)** Estimates vary depending on the size of the person. **(d)** Answers vary depending on the size of the person and the desk.

Cooperative Learning

7. (a)

Number of Faces Painted	Number of Pieces
6	0
5	0
4	0
3	8
2	12
1	6
0	1

(b) Number of Faces

Painted	Number of Pieces
6	0
5	0
4	0
3	8
2	24
1	24
0	8

(c) There could never be more than three faces painted. The number of pieces with three faces painted is always 8. The number of pieces with two faces painted is always a multiple of 12. The number of pieces with two faces painted is $12(n-2)$. The number of pieces with one face painted is the number of pieces in the center square of each face. The length of the square is $(n-2)$, so there are $(n-2)^2$ pieces on each of the six faces. Therefore, there are $6(n-2)^2$ pieces. The number of pieces with no faces painted is the number of interior cubes, which is 2 less than the dimensions of the original cube, or $(n-2)^3$.

Questions from the Classroom

9. Abi is correct. She is using the factored form of the formula for surface area of a cone. She should use the distributive property of multiplication over addition to confirm that the two forms are equivalent.

Review Problems

11. (a) 100,000   **(b)** 1.3680   **(c)** 500   **(d)** 2,000,000
(e) 1   **(f)** 10^6 **13.** $20\sqrt{5}$ cm   **15.** Length of side is 25 cm. Diagonal $\overline{BD}$ is 30 cm long.

Assessment 13-5A

1. (a) 8000   **(b)** 0.000675   **(c)** 7   **(d)** $\dfrac{25}{2916}$ or 0.00857 approx.   **(e)** 345.6   **2.** 32.4 L   **3. (a)** $\left(\dfrac{256}{3}\right)\pi$ cm^3
(b) 216 cm^3 **(c)** 15π cm^3 **(d)** $(4000/3)\pi$ cm^3
(e) $(20,000/3)\pi$ ft^3 **4. (a)** 2000, 2, 2000   **(b)** 0.5, 0.5, 500   **(c)** 1500, 1.5, 1500   **(d)** 5000, 5, 5   **(e)** 0.75, 0.75, 750   **(f)** 4.8, 4.8, 4800   **5. (a)** 200.0   **(b)** 0.320
(c) 1.0   **(d)** 5.00   **6.** 8:27   **7.** It is multiplied by 8.
8. (a) 2000, 2, 2   **(b)** 6000, 6, 6   **(c)** 2 dm, 4000, 4
(d) 2.5 dm, 7500, 7.5   **9.** 64 to 1   **10.** 2,500,000 L
11. π mL   **12. (a)** It is multiplied by 8.   **(b)** It is multiplied by 27.   **(c)** It is multiplied by n^3.   **13.** The Great Pyramid has the greater volume. It is approx. 25.6 times greater.   **14.** 1/8 of the cone is filled.   **15.** It is multiplied by 2.197; 119.7% increase.   **16.** The 2 ft $\times$ 2 ft $\times$ 4 ft freezer is a better buy at \$25/ft^3.

17. $66\frac{2}{3}$% occupied by balls, so $33\frac{1}{3}$% occupied by air.

18. $1 - \frac{\pi}{4}$ or about 21.5%   **19.** They are equal.

20. $\dfrac{125}{18\pi}$ or approx. 2.2 cm   **21. (a)** Answers vary; for example, a square base with sides 5 m and a height of 12 m.
(b) Infinitely many. Because $V = 100 = (1/3)a^2h$, where a is a side of the square base, then $300 = a^2h$. This equation has infinitely many solutions.   **22.** 512,000 cm^3
23. (a) metric tons   **(b)** kilograms   **(c)** grams
(d) metric tons   **24. (a)** milligrams   **(b)** kilograms
(c) milligrams   **25. (a)** 15   **(b)** 36   **(c)** 4.320   **(d)** 30
(e) 1.5625   **26. (a)** no   **(b)** possibly   **(c)** yes   **(d)** yes
(e) yes   **27.** 16 kg   **28. (a)** $^-$12°C   **(b)** $^-$1°C
(c) 100°C   **29. (a)** no   **(b)** yes   **(c)** yes   **(d)** not

Mathematical Connections 13-5

Communication

1. (a) Doubling the height will only double the volume. Doubling the radius will multiply the volume by 4. This happens because the value of the radius is squared after it is doubled.   **(b)** yes   **3. (i)** The volume is approx. 144.7 ft^3, the capacity is approx. 1082.4 gal, and the weight of the water is approx. 9043.7 lb.   **(ii)** The volume is 8 m^3, the capacity is 8000 L, and the mass is 8000 kg. The metric problem is much easier to work because the conversions are much easier. They just involve moving the decimal point. The relationships among length, volume, capacity, and mass are much easier than in the English system.

Open-Ended

5. Answers vary but should have volumes close to 24π or 75.4 in.^{3}; for example, a rectangular prism that is 6 in. long, 4 in. wide, and 3.14 in. high.   **7.** Answers vary, but the volume must be 1000 cm^3. Some students will worry about shelf space while others will worry about what shape is easiest to hold.

Questions from the Classroom

9. Consider a cube with side 6 cm. The volume is 216 cm^3 and its surface area is 216 cm^2.   **11.** For each degree change in Celsius, there is a 9/5 degree change in Fahrenheit. When a person's temperature is 2 degrees above normal Celsius, it is (9/5)2 or 3.6, degrees above normal Fahrenheit. Therefore, being 2 degrees above normal Celsius is more serious than being 2 degrees above normal Fahrenheit.   **13.** Suppose the volume of each of the small containers is πr^2h. Then Jamie would receive $2(\pi r^2h)$ for two containers. The volume of the large container is $\pi(2r)^2h = \pi 4r^2h = 4\pi r^2h$. Therefore, the volume of the large container is twice that of the two small containers combined.

Review Problems

15. (a) 35   **(b)** 0.16   **(c)** 400,000   **(d)** 0.00052
17. (a) 2400π cm^2 **(b)** $(6065 + 40\sqrt{5314})$ cm^2

Chapter Review

1. (a) $16\frac{2}{3}$ **(b)** $\frac{947}{1760}$ or approx. 0.538 **(c)** 3960

(d) $9\frac{25}{36}$ or approx. 9.694 **(e)** 5000 **(f)** 1.65 **(g)** 520

(h) 0.125 **2. (a)** Not possible, $p > q + r$ **(b)** Not

possible, $p = q + r$ **3. (a)** $8\frac{1}{2}$ cm^2 **(b)** $6\frac{1}{2}$ cm^2

(c) 7 cm^2 **4.** The pieces of the trapezoid are rearranged
to form a rectangle with width $h/2$ and length $(b_2 + b_1)$.
The area is $A = h/2(b_2 + b_1)$, which is the area of the
initial trapezoid. **5.** Area$(\triangle ABC) <$ Area$(\triangle ABD) =$
Area$(\triangle ABE) <$ Area$(\triangle ABF)$. All the triangles have the
same base, so the ordering is just by height and $\triangle ABD$ and
$\triangle ABE$ have the same height. **6. (a)** $54\sqrt{3}$ cm^2
(b) 36π cm^2 **7. (a)** 12π cm^2 **(b)** $(12 + 4.5\pi)$ cm^2
(c) 24 cm^2 **(d)** 4π cm^2 **(e)** 64.5 cm^2 **(f)** 178.5 m^2
8. 5400 cm^2 **9.** $\sqrt{16,200} = 90\sqrt{2}$ ft or approx. 127.3 ft
10. 3 **11. (a)** yes **(b)** no **12. (a)** S.A. $= 32(2 +$
$\sqrt{13})$ cm^2; $V = 128$ cm^3 **(b)** S.A. $= 96\pi$ cm^2;
$V = 96\pi$ cm^3 **(c)** S.A. $= 100\pi$ m^2; $V = (500\pi)/3$ m^3
(d) S.A. $= 54\pi$ cm^2; $V = 54\pi$ cm^3 **(e)** S.A. $= 304$ m^2;
$V = 320$ m^3 **13.** 65π m^2 **14.** The graph on the right
has 8 times the volume of the figure on the left, rather than
double as it should be **15.** 252 cm^2 **16. (a)** 340 cm
(b) 6000 cm^2 **17.** $2\sqrt{2}$ m^2 **18.** 62 cm **19.** 30 cm
20. (a) $y = 0; y = {}^-5x + 10; y = (5/3)x$
(b) Midpoints are at $(4, 0)$, $(3, 5)$, and $(7, 5)$.
(c) $y = \frac{5}{7}x; y = {}^-x + 8; y = 5x - 20$ **(d)** $\left(\frac{14}{3}, \frac{10}{3}\right)$
(e) The ratio is $2:1$. (*Hint:* Use the distance formula.)
21. (a) Answers vary.

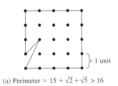

(a) Perimeter $= 15 + \sqrt{2} + \sqrt{5} > 16$

(b) Answers may vary.

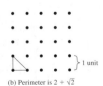

(b) Perimeter is $2 + \sqrt{2}$

(c)

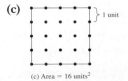

(c) Area $= 16$ units^2

22. $\sqrt{193.25}$ or approximately 13.9 in. **23.** 10π in.

24. $2\sqrt{3}$ in. **25. (a)** $(x - 3)^2 + (y + 4)^2 = 25$
(b) $6, (5, {}^-3)$ **26. (a)** metric tons **(b)** 1 cm^3 **(c)** 1 g
(d) 1 **(e)** 25 **(f)** 2000 **(g)** $51,800$ **(h)** $10,000,000$
(i) $50,000$ **(j)** 5.830 **(k)** $25,000$ **(l)** $75,000$
(m) 52.813 **(n)** 4.8 **27. (a)** 6000 kg **(b)** approx.
1.5565 m **28. (a)** L **(b)** kg **(c)** g **(d)** g **(e)** kg
(f) t **(g)** mL **29. (a)** unlikely **(b)** likely
(c) unlikely **(d)** unlikely **(e)** unlikely **30. (a)** 2000
(b) 1000 **(c)** 3 **(d)** 0.0042 **(e)** 0.0002

Answers to Now Try This

13-1. (a) Answers will vary. **(b)** Answers may vary
depending upon the size of paper clips used. The length is
approximately 8 paper clips. **(c)** Answers will vary.
(d) Answers will vary. **13-2. (a)** decidollar **(b)** centi-
dollar **(c)** dekadollar **(d)** hectodollar **(e)** kilodollar
13-3. (a) 115 cm **(b)** 55 cm **13-4.** π is an irrational
number. The decimal representation never terminates and
it is not a repeating decimal. **13-5. (a)** 12 **(b)** 6 **(c)** 4

13-6. approximately 100 cm^2 **13-7. (a)** $\frac{b}{2} \cdot \frac{h}{2} = \frac{bh}{4}$

(b) The triangle was twice as large as the folded rectangle, so

multiply by 2. Thus, $2 \cdot \frac{bh}{4} = \frac{1}{2}bh$ **13-8.** The areas of all
the triangles have the same measure because the base of each
triangle is the same and all have the same height.

13-9. Let h_1 be the height of $\triangle ABC$ and h_2 be the height of

$\triangle ADC$. Then the area of $\triangle ABC$ is $\frac{1}{2}(AC)h_1$ and the area of

$\triangle ADC$ is $\frac{1}{2}(AC)h_2$. Therefore, the area of quadrilateral

$ABCD$ is $\frac{1}{2}(AC)h_1 + \frac{1}{2}(AC)h_2 = \frac{1}{2}(AC)(h_1 + h_2)$. AC is

the measure of one diagonal and $(h_1 + h_2)$ is the measure
of the other diagonal, so the formula works.

13-10. The new figure is a parallelogram with base $(b_1 + b_2)$
and height h, where b_1 and b_2 are the bases of the original
trapezoid. The area of the parallelogram is $A = h(b_1 + b_2)$.
Because this is twice the area of the original trapezoid, we

divide by 2 to obtain $A = \frac{h(b_1 + b_2)}{2}$, or $A = h/2(b_1 + b_2)$,

which is the formula for the area of the original trapezoid.
13-11. (a) construction. **(b)** 25; parallelogram.
26; The length is approximately equal to $\pi \cdot$ radius, half
of the circumference. The height is approximately equal
to the radius. The area is approximately equal to 27: As the
circle is cut into more and more sectors and put back
together, the shape becomes more and more like a

parallelogram. 28; $A = \frac{1}{2}(Cr) = \frac{1}{2}(2\pi r) \cdot r = \pi r^2$.

13-12. The square on one leg that is labeled 1 could be cut off and placed in the dashed space on the square of the hypotenuse. Then pieces 2, 3, 4, and 5 could be cut off and placed around piece 1 so that the square on the hypotenuse is filled exactly with the five pieces. This shows that the sum of the areas of the squares on the two legs of a right triangle is equal to the area of the square of the hypotenuse.

13-13. Jason ran the hypotenuse of a right triangle with legs each 10 yd long.   **13-14. (a)** You could build the triangle and then measure the angles to see if there was a right angle. If the angle is a right angle, then the triangle is a right triangle. You could measure the three sides and use the converse of the Pythagorean theorem to see if a right triangle is formed.   **(b)** If the three lengths of a right triangle are multiplied by a fixed number, then the resulting lengths determine a right triangle; for example, if the right triangle lengths are 3-4-5, and the fixed number is 5, then 15-20-25 is a right triangle.   **(c)** If the three lengths of a right triangle are multiplied by a fixed number, then the resulting numbers determine a right triangle.   **13-15.** It makes no difference in the distance formula if $(x_1 - x_2)$ and $(y_1 - y_2)$ are used instead of $(x_2 - x_1)$ and $(y_2 - y_1)$, respectively. Because both quantities in the formula are squared, the result is the same whether the difference is positive or negative.   **13-16. (a)** $S.A. = 2 \cdot (5/2) \cdot 8 + 2 \cdot (8 \cdot 11) + 2 \cdot (5/2) \cdot 11 = 271$ in.^{2} **(b)** No, the rectangle would have to be 21 in. by 16 in.   **13-17.** Because we want the surface area of a right prism, we must include the top and bottom, so we need $2B$ (where B is the area of the base, which is the same as the area of the top) in the formula $S.A. = ph + 2B$. From the net, we see that the lateral surface area opens up into a rectangle that has width equal to the height, h. The length of the rectangle is equal to the sum of the lengths of the sides of the base, which is the perimeter of the base. Therefore, the area of the rectangle (lateral surface area of the prism) is $A = \ell w = ph$. Hence, the surface area for any right prism is given by $S.A. = ph + 2B$.

13-18. (a) Both figures have a volume of 9 cubic units.   **(b)** No, the second figure has the greater surface area.   **(c)** The first figure has surface area 34 square units and the second figure has area 36 square units.   **13-19. (a)** Move the decimal point twice as many places as in a linear equation and in the same direction. For example, the area of a square that is 1 m on each side is 1 m^2. 1 m = 10 dm. 1 m^2 = 100 dm^2.   **(b)** Move the decimal point 3 times as many places as in a linear equation and in the same direction. 1 m^3 = 1000 dm^3.   **13-20. (a)** The two figures have bases in the same plane and the figures have the same height. Figure 13-81 shows that if a plane parallel to the base is passed through the figures, then equal areas are obtained. By Cavalieri's Principle, these two figures have equal volumes.   **(b) (i)** By Cavalieri's Principle, the volumes are the same.   **(ii)** By Cavalieri's Principle, the volumes are the same.   **13-21. (a)** We know that the height of each figure is $2r$ because the height of the sphere is $2r$. The volume of the cylinder is $\pi r^2 2r = 2\pi r^3$. The

volume of the cone is $\dfrac{1}{3}\pi r^2 \cdot 2r = \dfrac{2}{3}\pi r^3$. The volume of the sphere is $\dfrac{4}{3}\pi r^3$.   **(b)** Using a common denominator of 3, the three formulas are $\dfrac{6}{3}\pi r^3$, $\dfrac{2}{3}\pi r^3$, and $\dfrac{4}{3}\pi r^3$. The ratio is $6:2:4$, which simplifies to $3:1:2$.   **13-22. (a)** g   **(b)** kg   **(c)** dm^3 **(d)** mL   **(e)** g   **(f)** kL   **(g)** t   **13-23.** Yes, when it is ⁻40°C it is ⁻40°F.

Answers to Brain Teasers

Section 13-1

The tallest person on Earth could walk under the wire. Suppose the two concentric circles represent Earth and the lengthened wire. $\overline{OA}$ and $\overline{OB}$ are the radii of the respective circles and have lengths r and $r + x$. Because the circumference of the Earth plus 20 m equals the circumference of the lengthened wire, we have $2\pi r + 20 = 2\pi(r + x)$. Consequently, $x = 10/\pi$, or approximately 3.18 m.

Section 13-2

1. 64 square units
2. 65 square units
3. Although the pieces look like they should fit together, they do not really fit. To see this, assume the pieces do fit. We then obtain the following figure:

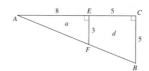

Since $\triangle AEF \sim \triangle ACB$, we have $\dfrac{8}{13} = \dfrac{3}{5}$, which is a contradiction. This implies that pieces like those in the figure cannot fit together in order to form a triangle. In order for the pieces to fit together, the measure of $\overline{EF}$ must be given as $\dfrac{8}{13} = \dfrac{EF}{5}$; hence, $EF = \dfrac{40}{13} = 3\dfrac{1}{13}$. Since $3\dfrac{1}{13}$ is close to 3, the discrepancy is so small that the pieces appear to fit.

Section 13-3

Let s be the length of the side of the square field. Then the area of the field is s^2 and the area covered by the large sprinkler is $\left(\dfrac{1}{2}s\right)^2 \pi$. The area of the field not covered by the sprinkler is

$$s^2 - \left(\frac{1}{2}s\right)^2 \pi = s^2 - \frac{1}{4}s^2\pi$$

The length of the side of each small square is $\frac{1}{3}s$. The

radius of each small circle is $\frac{1}{6}s$. The area of each small field

that is not covered by the small sprinkler is

$$\left(\frac{1}{3}s\right)^2 - \left(\frac{1}{6}s\right)^2 \pi = \frac{1}{9}s^2 - \frac{1}{36}s^2\pi$$

Multiply this by 9 to find the total area of the field that is
not covered:

$$9\left(\frac{1}{9}s^2 - \frac{1}{36}s^2\pi\right) = s^2 - \frac{1}{4}s^2\pi$$

which is the same as the area not covered by the large
sprinkler.

    Therefore, both sprinkler systems cover the same
percentage of the square field and it does not matter which
system is used if the only selection criterion is the amount
of land covered by the system.

    A related problem is to change the number of circles
contained in the square. Will the percentage always be the
same? Changing the shape of the field can also vary the
problem.

Section 13-4

The cone and the flattened region obtained by slitting the
cone along a slant height are shown.

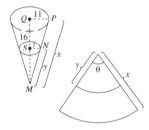

To construct the flattened ring we need to find x, y, and θ.
Because $\triangle MQP \sim \triangle MSN$, we have $\dfrac{16 + MS}{MS} = \dfrac{11}{7}$.
Hence, $MS = 28$ cm. In $\triangle MSN$, we have $28^2 + 7^2 = y^2$,
or $y \doteq 28.86$. In $\triangle PQM$: $x^2 = 11^2 + 44^2$, or $x \doteq 45.35$ cm.
To find θ, we roll the sector with radius y and central angle
θ into the cone whose base is 7 cm and whose slant height
is y. Hence, $2\pi y \cdot \dfrac{\theta}{360} = 2\pi \cdot 7$, or $\theta = \dfrac{7 \cdot 360}{28.86} = 87°19'$.

Answers to Technology Corners

Section 13-2

Geometer's Sketchpad is used to derive the formulas for the
area of a parallelogram, a triangle, and a trapezoid.

Section 13-3

Page 877   Geometer's Sketchpad is used to explore the
Pythagorean theorem.
Page 893   *GSP* can be used to investigate the properties of
45°-45°-90° and 30°-60°-90° triangles.

Section 13-4

GSP can be used to investigate surface area and perimeter
using nets.

Answers to Laboratory Activities

Section 13-1

Page 848   Answers vary but the average should be approximately 3.1. The line should have an equation
approximately $y = 3.1x$. The slope is 3.1.
Page 854 **(a)** perimeter 39 in.
(b) Answers will vary, but cutting might start as follows:

Section 13-2

Answers will vary.

Section 13-4

The peeling of an orange can be used to motivate the
formula for the surface area of a sphere, $S.A. = 4\pi r^2$.

Answer to Preliminary Problem

Because the radius of the 14 in. pizza is 7 in., the area is
49π sq. in. Without the crust, the radius is 6 in. and
the area is 36π sq. in. Therefore, the area of the crust is
$49\pi - 36\pi = 13\pi$ sq. in. and the percent of crust is
$13\pi/49\pi$, or 13/49. The area of the 10 in. pizza is 25π sq.
in. and therefore the area of the crust must be $(13/49)(25\pi)$,
or approximately 6.63π sq. in. The area without the crust is
approximately $25\pi - 6.63\pi$, or 18.37π sq. in., and to find
the radius without the crust we have $\pi r^2 = 18.37\pi$ and
$r = \sqrt{18.37}$, or about 4.29 in. The radius of the 10 in. pizza
is 5 in., so the width of the crust is approximately
$5 - 4.29 = 0.71$. Therefore, the width of the crust in the
10 in. pizza is 0.7 rounded to the nearest tenth of an inch.

Chapter 14

Assessment 14-1A

1. **(a)** 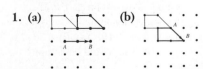   **(b)**

2. (a) Construct as suggested by the following: Trace $\overline{BC}$ and the line containing the slide arrow on the tracing paper and label the trace of B as B' and the trace of C as C'. Mark on the original paper and on the tracing paper the initial point of the arrow by P and the head of the arrow by Q. Slide the tracing paper along the line $\overleftrightarrow{PQ}$ so that P will fall on Q. The segment $\overline{B'C'}$ overlays the image of $\overline{BC}$ under the translation. **(b)** Construct a parallelogram with $\overline{BC}$ and $\overline{B'C'}$ as opposite sides. **3. (a)** $(3, {}^-4)$ **(b)** $(0, 0)$ **(c)** $({}^-3, {}^-13)$ **4. (a)** $(3, {}^-4)$ **(b)** $(0, 0)$ **(c)** $({}^-3, {}^-13)$

5. (a) 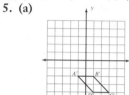 **(b)**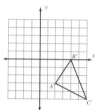

6. 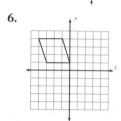 **7.**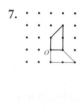

8. (a) $y = 2x - 3$ **(b)** $y = 2x - 7$ **9.** $y = {}^-2x - 1$
10. Choose points P and Q on ℓ. Construct P' on $\overrightarrow{OP}$ such that $OP = OP'$. Similarly find Q'. $\overleftrightarrow{P'Q'}$ is the desired image.
11. Reverse the rotation (to the counterclockwise direction) to locate $\overline{AB}$, that is, the preimage.

12. Answers may vary, but H, I, N, O, S, X, or Z could appear in such rotational words. Examples include SOS. Variations could use M and W in rotational images, for example, MOW. **13. (a)** The image is the line ℓ itself. **(b)** construction **(c)** ℓ and ℓ' are parallel. If P and Q are any points on ℓ and P' and Q' their respective images, then from the definition of a translation $\overline{PP'}$ and $\overline{QQ'}$ are parallel and congruent to $\overline{AB}$. Hence, $\overline{PP'}$ and $\overline{QQ'}$ are parallel and congruent. Thus, $PP'Q'Q$ is a parallelogram and therefore $\ell'\|\ell$. **(d)** The image is $\angle ABC$ itself. **14.** *Hints:* An angle whose measure is $45°$ can be constructed by bisecting a right angle. An angle whose measure is $60°$ can be constructed by first constructing an equilateral triangle. **15. (a)** $({}^-4, 0)$ **(b)** $({}^-2, {}^-4)$ **(c)** $(2, 4)$ **(d)** $({}^-a, {}^-b)$ **16.** *Hint*: Find the images of the vertices. **17. (a)** $\ell' = \ell$ **(b)** $\ell' \perp \ell$
18. (a) First rotate $\triangle ABC$ by angle α to obtain $\triangle A'B'C'$, and then rotate $\triangle A'B'C'$ by angle β to obtain $\triangle A'B'C'$. **(b)** no **(c)** yes, by rotation about O by angle $|\alpha - \beta|$ in the direction of the larger of α and β **19. (a) (i)** $({}^-3, 2)$ **(ii)** $({}^-2, {}^-1)$ **(iii)** $({}^-m, n)$ **(b)** This could be demon-

strated with graph paper. The image of the point with coordinates (a, b) under a half-turn with the origin as center must be on a circle with center at $(0, 0)$ having equation $x^2 + y^2 = a^2 + b^2$ and being the other endpoint of the diameter having (a, b) as one endpoint. The only point satisfying these conditions is $({}^-a, {}^-b)$. **(c)** $(b, {}^-a)$
20. (a) $C(5, 0), ({}^-5, 0), (6, 0)$ or $\left(\dfrac{25}{6}, 0\right)$ **(b)** $B(8, 4)$,

$({}^-2, 4), (3, {}^-4)$ or $\left(\dfrac{-7}{6}, 4\right)$ **(c)** The product of the

slopes of respective diagonals is ${}^-1$. **21. (a)** The image of $A(b, k)$ under the translation from O to C is $B(b + a, k + 0)$ or $B(b + a, k)$ **(b)** Using the distance formula, we get $OA^2 = b^2 + k^2$. Since $OA = OC$, $b^2 + k^2 = a^2$. **(c)** Using part (b), the product of the slopes of $\overline{OB}$ and $\overline{AC}$ is

$$\frac{k}{b + a} \cdot \frac{k}{b - a} = \frac{k^2}{b^2 - a^2} = \frac{k^2}{-k^2} = -1$$

22. (a) construction **(b)** It is the same. **(c)** The images of Q are the same. **(d)** a translation taking N to M (where O is the midpoint of $\overline{MN}$) **(e)** $\overline{OM}$ is a midsegment in $\triangle PP'P''$. Hence, $\overline{OM}\|\overline{PP''}$ and $OM = \dfrac{1}{2}PP''$. Because O and M are the same for all points P in the plane, the vectors from N to M and from P to P'' are parallel and have the same length and direction. **23. (a)** The image of every point of the plane is itself. **(b)** One half-turn "undoes" the other.
24. Four that can be their own images are a circle, a square, a regular octagon, and a regular hexagon. An equilateral triangle cannot be its own image. **25.** The possibilities are points with coordinates $(5, 8), ({}^-1, 4)$, and $(3, 2)$.

Mathematical Connections 14-1

Communication

1. Yes. If the congruent segments are $\overline{AB}$ and $\overline{CD}$, connect A with C and B with D, as shown. The perpendicular bisectors of $\overline{AC}$ and $\overline{BD}$ intersect at O. (If they do not intersect, connect A with D and B with C.) The point O is the center of the required rotation. The angle of rotation is $\angle AOC$ in the direction from A to C. The image of A will be C because the rotation is by $\angle AOC$. To show that under this rotation the image of $\overline{AB}$ is $\overline{CD}$, we need only show that $\angle AOC \cong \angle BOD$. This can be checked experimentally by performing the rotation or by measuring $\angle AOC$ and $\angle BOD$.

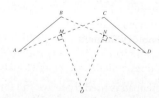

3. (a) A parallelogram with one diagonal; under a half-turn, the image of a line is parallel to the line. Thus,

$\overline{AB} \parallel \overline{CD}$ and $\overline{AC} \parallel \overline{DB}$; therefore, *ABCD* is a parallelogram.

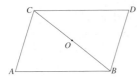

(b) A parallelogram with an extra segment; the image of $\overline{AB}$ is $\overline{FE}$ and thus $\overline{AB} \parallel \overline{EF}$. Consequently, $\overline{BF} \parallel \overline{AE}$ and *ABFE* is a parallelogram.

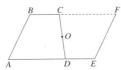

(c) A square; because the image of *A* is *C* and the image of *B* is *D*, the image of *ABCD* is *CDAB*.

Open-Ended

5. Answers vary.

Cooperative Learning

7. (a) The path will look like the one shown in the following figure. Such a path traced by *P* on the circle is called a *cycloid*.

(b) The path is not an arc of a circle. The perpendicular bisectors of all the chords (segments connecting two points on the arc) do not intersect in a single point.   **(c)** The length of $\overline{AB}$ is the circumference of the circle.

Questions from the Classroom

9. The result of a 180° rotation clockwise or counterclockwise is the same, so no direction is needed.

Assessment 14-2A

1. Locate the image of vertices directly across (perpendicular to) ℓ on the geoboard.

(a)   **(b)**

2. Reflecting lines are described for each.   **(a)** all diameters (infinitely many)   **(b)** perpendicular bisector of the segment or the line containing the segment   **(c)** the line containing the ray   **(d)** any line perpendicular to the given line or the line itself   **(e)** perpendicular besectors of the sides and lines containing the diagonals **(f)** perpendicular bisectors of pairs of parallel sides **(g)** none   **(h)** if not equilateral, perpendicular bisector of the side that is not congruent to the other two, if equilateral, see (i)   **(i)** perpendicular bisectors of each side

(j) none   **(k)** perpendicular bisector of parallel sides **(l)** perpendicular bisector of the chord connecting the endpoints of the arc   **(m)** if not a rhombus, the line containing the diagonal determined by vertices of the noncongruent angles; if a rhombus, see (n)   **(n)** the lines containing the diagonals   **(o)** perpendicular bisectors of parallel sides and three diameters determined by vertices on the circumscribed circle   **(p)** There will be *n* reflecting lines in all. If *n* is even, the lines are determined as in part (o). If *n* is odd, the lines are the perpendicular bisectors of the sides.   **3.** the original figure   **4. (a)** The final images are congruent but in different locations, and hence not the same.   **(b)** A translation determined by a slide arrow from *P* to *R* is determined as follows: Let *P* be any point on ℓ and *Q* on *m* such that $\overrightarrow{PQ} \perp \ell$. Point *R* is on $\overrightarrow{PQ}$ such that $PQ = QR$.

5. (a)

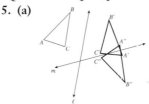

(b) a rotation about *O* by 2α from ℓ to *m* as shown **(c)** a half-turn about *O*   **6.** construction   **7.** The line of reflection is the perpendicular bisector of $\overline{BB'}$.
8. (a) Examples include MOM, WOW, TOOT, and HAH.   **(b)** Examples include BOX, HIKE, CODE, etc. B, C, D, E, H, I, K (depending on construction), O, and X may be used.   **(c)** 1, 8, 11, 88, 101, 111, 181, 808, 818, 888, 1001, 1111, 1881   **9. (a)** The images are the same. **(b)** yes   **10.** None of the images has a reverse orientation, so there are no reflections or glide reflections involved. Thus,

    1 to 2 is a rotation.
    1 to 3 is a rotation.
    1 to 4 is a translation down.
    1 to 5 is a rotation (with an exterior point as the center of rotation).
    1 to 6 is a translation (sides are parallel to 1).
    1 to 7 is a translation (sides are parallel).

11. (a) $A'(3, {}^-4)$, $B'(2, 6)$, $C'({}^-2, {}^-5)$ **(b)** $A'(4, 3)$, $B'({}^-6, 2)$, $C'(5, {}^-2)$ **12. (a) (i)** $(x, {}^-y)$ **(ii)** (y, x) **(b)** $({}^-x, {}^-y)$; the following figure:

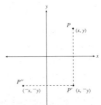

13. (a) $y = x - 3$ **(b)** $y = 0$ **14. (a)** $y = x + 3$ **(b)** $y = 0$ **15. (a)** It is a square.   **(b)** The respective images are $(0, a)$, $(a, 0)$, $(0, {}^-a)$, $({}^-a, 0)$.   **16.** the line through P perpendicular to $\overleftrightarrow{O_1O_2}$ or the perpendicular bisector of QQ', where Q and Q' are the points of intersection other than P of $\overleftrightarrow{O_1O_2}$ with circle O_2 and circle O_1, respectively   **17. (a)** $\overrightarrow{AB}$ **(b)** yes, a translation taking O_1 to O_2 **18. (a)** $x = 0$ or $y = 0$ **(b)** $x = 0$ or $y = 0$ **19.** Let H represent one house, P the pole, and T the other house, and T' (the reflection of T in the road r). The intersection of $\overline{HT'}$ and r determines the point on the road at which the pole should be placed.

Mathematical Connections 14-2

Communication

1. Answers may vary. Find A', the image of A under reflection in $\overline{EH}$ and B', the image of B under reflection in $\overline{GH}$. Mark the intersections of $\overline{A'B'}$ with $\overline{EH}$ and $\overline{GH}$, respectively, by C and D. The player should aim the ball at A toward the point C. The ball will hit D, bounce off, and hit B. To justify the answer, we need to show that the path A–C–D–B is such that $\angle 1 \cong \angle 3$ and $\angle 4 \cong \angle 6$. Notice that $\angle 1 \cong \angle 2$ and $\angle 6 \cong \angle 5$ because the image of an angle under reflection is congruent to the original angle. Also, $\angle 2 \cong \angle 3$ and $\angle 4 \cong \angle 5$, as each pair constitutes vertical angles. Consequently, $\angle 1 \cong \angle 3$ and $\angle 4 \cong \angle 6$.

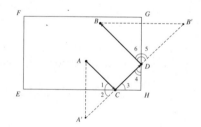

3. The angle of incidence is the same as the angle of reflection. With the mirrors tilted 45°, the object's image reflects to 90° down the tube and then 90° to the eyepiece. The two reflections "counteract" each other, leaving the image upright.

Open-Ended

5. Answers vary.

Cooperative Learning

7. (a) Constructions will vary.   **(b)** The experiment will work for rectangular tables in which the length is twice the width and for any position of B.

Questions from the Classroom

9. Again, having only a segment and its image is not enough to determine the transformation. It requires three noncollinear points. For a further examination of this problem, see *Transformational Geometry*, by Richard Brown (Palo Alto, CA: Dale Seymour Publications, 1989).

Review Problems

11. depending on how they are drawn: 0, 1, and 8
13. A rotation by any angle about the center of the circle will result in the same circle.

Assessment 14-3A

1. (a) Slide the smaller triangle down 3 units (translation). Then complete a size transformation with scale factor 2 using the top-right vertex as the center.   **(b)** Slide the smaller triangle right 5, and up 1. Then complete the size transformation as in (a).

2.

3. (a) translation taking B to B' followed by a size transformation with center B' (and scale factor 2)   **(b)** size transformation with scale factor $1/2$ and center A followed by a half-turn with the midpoint of $\overline{C'B'}$ as center
4. $x = 6, y = 5.2$, scale factor $2/5$ **5.** 12 cm   **6.** scale factor $3, x = 12, y = 10$ **7. (a)** $(12, 18), (^-12, 18)$
(b) same as in (a)   **8.** the size transformation with center O and scale factor $1/r$ **9. (a)** Answers vary. A possible explanation is:

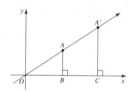

Let $A'(x', y')$ be the image of A under the size transformation. From the definition of a size transformation, we have $\dfrac{OA'}{OA} = r$. Since $\triangle OA'C \cong \triangle OAB$, $\dfrac{OC}{OB} = r$, which implies that $\dfrac{x'}{x} = r$ and hence $x' = rx$. Similarly, $y' = ry$.

(b) Answers vary; the size transformation with center at the origin and scale factor 2 followed by a half-turn in the origin   **(c) (i)** $y = 2x$ **(ii)** $y = 2x$ **(iii)** $y = 2x + \dfrac{1}{3}$
(iv) $y = ^-x - 3$ **10.** Answers vary; a translation taking O_1 to O_2 followed by a size transformation with center at O_2 and scale factor $\dfrac{3}{2}$ **11. (a)** The net result is the identity transformation. Every point is its own image.   **(b)** A scale factor of 0 maps every point in the plane to the center of the

size transformations. Here lines are no longer mapped to parallel lines but to a single point. This transformation is not a size transformation. **12.** The set of images of the set of integers on a number line is the set of points where coordinates are multiples of 3; that is, { ..., ⁻6, ⁻3, 0, 3, 6, 9, ... }. **13.** The area of the image is 9 times as great, or 63 sq. units. **14.** The enlargement of a 2" × 3" photograph to a 4" × 6" photograph can be achieved by a size transformation, but there are many potential centers of the size transformation. One might be the lower left corner of the 2" × 3" photograph.

Mathematical Connections 14-3

Communication

1. (a) It does change. For example, consider the segment whose endpoints are $(0, 0)$ and $(1, 1)$ and that has length $\sqrt{2}$. Under the size transformation with center at $(0, 0)$ and scale factor 2, the image of the segment is a segment whose endpoints are $(0, 0)$ and $(2, 2)$. That segment has length $2\sqrt{2}$. **(b)** It does not change. Under a size transformation, the image of a triangle is a similar triangle and the angles of corresponding angles of two similar triangles are congruent. **(c)** It does not change. Given two parallel lines, draw a transversal that intersects each line. Because the lines are parallel, the corresponding angles are congruent. From (b), the images of the angles will also be congruent and hence the image lines will be parallel. **3. (a)** a single size transformation with center O and scale factor $\frac{1}{2} \cdot \frac{1}{3}$ or $\frac{1}{6}$; let P be any point and P' its image under the first size transformation and P'' the image of P' under the second size transformation. Then $\frac{OP'}{OP} = \frac{1}{2}$ and $\frac{OP''}{OP'} = \frac{1}{3}$. Consequently, $\left(\frac{OP'}{OP}\right) \cdot \left(\frac{OP''}{OP'}\right) = \frac{1}{2} \cdot \frac{1}{3}$, or $\frac{OP''}{OP} = \frac{1}{6}$. Thus, P' can be obtained from P by a size transformation with center O and scale factor $\frac{1}{2} \cdot \frac{1}{3}$ or $\frac{1}{6}$. **(b)** the size transformation with center O and scale factor $r_1 r_2$

Open-Ended

5. Answers vary; for example, a 50% reduction in size when making a photocopy.

Cooperative Learning

7. (a) If p is the perimeter of the figure and p' the perimeter of the image, then $p' = 3p$. **(b)** If A is the area of the figure and A' the area of the image, then $A' = 9A$. **(c)** $p' = rp$ **(d)** $A' = r^2 A$ **(e)** Answers vary.

Questions from the Classroom

9. The student is incorrect. In the original grid, a square might have area 1 sq. unit but in the image grid, that same square has area 4 sq. units. The student is correct that the image is a grid, but it is not the same size.

Review Problems

11. (a) $(4, 3)$ reflects about m to $(4, 1)$; $(4, 1)$ reflects about n to $(2, 1)$. **(b)** $(0, 1) \rightarrow (0, 3) \rightarrow (6, 3)$ **(c)** $(⁻1, 0) \rightarrow (⁻1, 4) \rightarrow (7, 4)$ **(d)** $(0, 0) \rightarrow (0, 4) \rightarrow (6, 4)$

Assessment 14-4A

1. The symbol can be read the same backward, forward, and upside down. **2. (a) (i)** Yes. A line may be drawn through the center of the circle, either horizontally or vertically. A line may also be drawn through any of the sets of arrows. **(ii)** Yes. The figure will match the original figure after rotations of 90°, 180°, or 270° about the center of the circle. **(iii)** Yes. Any figure having 180° rotational symmetry has point symmetry about the turn center. **(b) (i)** Yes. A vertical line through the middle of the bulb is a line of symmetry. **(ii)** No. The figure will not match the original under rotation of less than 360°. **(iii)** No. The figure does not have 180° rotational symmetry. **3.** Answers may vary, but some possibilities are the following: **(a)** See the drawing of exercise 2b **(b)** A regular pentagon (Chrysler symbol) **(c)** The letter N **4.** Reflect the given portions about ℓ.
(a) **(b)**

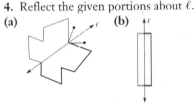

5. (a) (i) four lines of symmetry; the diagonals and horizontal or vertical lines through the center **(ii)** no lines of symmetry
(b) (i) **(ii)** No lines of symmetry

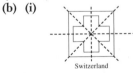

Switzerland

6. (a) 6 **(b)** Yes. The figure has 60° rotational symmetry. **7. (a)** one; vertically through the center **(b)** one; vertically through the center
8. (a) **(b)**

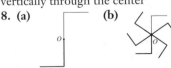

9. (a) seven; three through the "peaks," three through the "valleys," and one perpendicular to the others through the width of the figure **(b)** two; one through the middle "peak" and one through the width **10. (a)** point symmetry; 180° rotational symmetry **(b)** rotational symmetry about the center by 36°; point symmetry **11. (a)** line symmetry

in lines that bisect opposite sides of the rectangle, 180°
rotational symmetry in O the point of intersection of the
diagonals, and point symmetry in center **(b)** line symmetry in the perpendicular bisector of the bases **(c)** line symmetry in either diagonal, 180° rotational symmetry about
the point of intersection of the diagonals, point symmetry
(d) line symmetry in the angle bisector **(e)** four line symmetries: in lines that connect the midpoints of opposite sides
and two lines determined by the diagonals; also, turn
symmetries about the center of the square (intersection of
the diagonals) by 90° or 180° or 270° (also, rotation by 360°,
which is the identity transformation); point symmetry in
center **(f)** six line symmetries, three in perpendicular
bisectors of each pair of opposite sides and three in lines
determined by the three longest diagonals (segments
connecting a pair of opposite vertices); also, six turn symmetries about the center of the hexagon by 60°, 2 · 60°, 3 · 60°,
4 · 60°, 5 · 60°, and 6 · 60°; point symmetry in center
(g) If the circles have equal radii, then there are two
symmetries—one in $\overleftrightarrow{O_1O_2}$ and the other in $\overleftrightarrow{AB}$. If the radii
are different; only one symmetry, determined by the line of
reflection $\overleftrightarrow{O_1O_2}$. **(h)** infinitely many line symmetries in
lines through the center; also, infinitely many rotational
symmetries about the center by any angle **12.** No; a
counterexample follows:

The figure has point symmetry but no line symmetry.
13. (a) The figure has plane symmetry in four ways.
(b) a vertical plane of symmetry

Mathematical Connections 14-4

Communication

1. (a) Yes. The definition of point symmetry is that it is rotational symmetry of 180°. **(b)** No. It may have rotational
symmetry of other than 180°; an equilateral triangle is an
example. **(c)** Yes. A circle is an example **(d)** No.
Consider the letter Z. (Nor is the converse true. Consider a
heart shape.) **(e)** Yes. Point symmetry implies 180°
rotational symmetry.

Open-Ended

3. Answers may vary, but examples include the letter S and
the Chevrolet logo. **5.** Answers vary.

Cooperative Learning

7. Answers vary; for example, consider a rectangle. You
would report that your figure has two lines of symmetry and
a rotational symmetry of 180°.

Questions from the Classroom

9. The student is wrong. One counterexample is a right triangular prism whose bases are scalene triangles. **11.** The
symmetries are reflections in each of two lines that bisect a
pair of vertical angles as well as a half-turn about the point
of intersection. **13.** For the graphs of $y = x^2$ as well as
$y = |x|$, the y-axis is a line of symmetry. To see why, notice
that when a point (a, b) is reflected in the y-axis, its image is
$(^-a, b)$. If (a, b) is on the graph, we show that $(^-a, b)$ is
also on the graph. Thus we need to show that if $b = a^2$,
then $b = (^-a)^2$, which is true since $(^-a)^2 = a^2$. A similar
argument works for $y = |x|$; if $b = |a|$, then $b = |^-a|$, since
$|^-a| = |a|$. When we graph $y = x^3$, we can see that when a
point is on the graph, so is its image under half-turn in the
origin. This can be shown precisely as follows: If (a, b) is on
the graph, then $b = a^3$. We want to show that $(^-a, ^-b)$, the
image of (a, b) under half-turn about the origin, also satisfies the equation $y = x^3$, that is, that $^-b = (^-a)^3$ is a true
statement. But that statement is true if, and only if,
$^-b = ^-a^3$ or $b = a^3$, which is true.

Review Problems

15. Find the images of the vertices.

Assessment 14-5A

1.

2. (a) Perform half-turns about the midpoints of all sides.
(b) Yes. If a polygon tessellates the plane, the sum of the
angles around every vertex must be 360°. Successive 180°
turns of a quadrilateral about the midpoints of its sides will
produce four congruent quadrilaterals around a common
vertex, with each of the quadrilateral's angles being
represented at each vertex. These angles must add up to
360°, as angles of any quadrilateral do. **3.** Experimen-
tation by cutting out shapes and moving them about is one
way to learn about these types of problems.

4. *Hint:* Consider figures like a pentagon formed by
combining a square and an equilateral triangle.
5. (a) construction **(b)** It combines a translation and a
reflection. **(c)** It will tessellate the plane. **6. (a)** The
dual is another tessellation of squares (congruent to those
given). **(b)** a tessellation of isosceles triangles **(c)** The
tessellation of equilateral triangles; it is illustrated in the
statement of the problem.

7.

8. (a) is not semiregular; (b) and (c) are semiregular.

Mathematical Connections 14-5

Communication

1. (a) The image of *ABCD* under a half-turn in *M* (the midpoint of *CD*) is the trapezoid *FEDC*. Because the trapezoids are congruent, *ABFE* is a parallelogram. The area of the parallelogram is $AE \cdot h$ or $(a + b) \cdot h$. The parallelogram is the union of two nonoverlapping congruent trapezoids. The area of each trapezoid is $(a + b)(h/2)$.

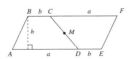

(b) To tessellate the plane with $\triangle ABC$, we find $\triangle A'BC$—the image of $\triangle ABC$ under a half-turn about *M*, the midpoint of $\overline{BC}$. If *N* is the midpoint of $\overline{AB}$, then the image of *N* is *N'*. It can be shown that *N*, *M*, and *N'* are collinear and hence that *ANN'C* is a parallelogram. Because the image of $\overline{NM}$ is $\overline{MN'}$, it follows that $NM = MN'$. Hence, $NM = \frac{1}{2}NN' = \frac{1}{2}AC$. Thus, $NM = \frac{1}{2}AC$. Also $\overline{NM} \| \overline{AC}$.

3. First tessellate the plane with regular hexagons and then divide each hexagon into congruent pentagons by three segments connecting the center of a hexagon to the midpoint of every second side. The segments inside one hexagon should not have points in common with segments inside a neighboring hexagon. Actually, one could draw the *Y*-shape segments in one hexagon and then translate the shape in all directions by a translation, taking the center of one hexagon to the center of a neighboring one.

Open-Ended

5. Answers vary.

Cooperative Learning

7. Answers vary.   **9.** All will tessellate. Arguments vary.

Questions from the Classroom

11. This can be done; she should be asked to explain and give an example. One example might be a right circular cylinder and its "axis" through the centers of the bases.

13. (a) 90°, 180°, 270° rotation about center; point symmetry   **(b)** 180° rotational symmetry, two line symmetries; point symmetry

Chapter Review

1. (a)

(b)

(c)

2. In each part find the images of the vertices.   **3. (a)** 4   **(b)** 1   **(c)** 1   **(d)** none   **(e)** 2   **(f)** 2   **4. (a)** line and rotational   **(b)** line, rotational, and point   **(c)** line   **5. (a)** infinitely many   **(b)** infinitely many   **(c)** 3   **(d)** 12   **6.** This answer depends totally upon how the letters are made, but generally we have the following: *c* has line symmetry; *i* has line symmetry; *o* has line, rotational, and point symmetry; *s* has rotational and point symmetry; *t* has line symmetry; *v* has line symmetry; *w* has line symmetry; *x* has line, rotational, and point symmetry; *z* has rotational and point symmetry.   **7.** $A = A'$, *B* is the midpoint of $\overline{A'B'}$, and *C* is the midpoint of $\overline{A'C'}$.   **8.** in each case, half-turn about *X*   **9. (a)** clockwise rotation by 120° about the center of the hexagon   **(b)** a reflection in the perpendicular bisector of $\overline{BY}$ **10.** reflection in $\overleftrightarrow{SO}$ **11.** Let $\triangle H'O'R'$ be the image of $\triangle HOR$ under a half-turn about *R*. Then $\triangle SER$ is the image of $\triangle H'O'R'$ under a size transformation with center *R* and scale factor $\frac{2}{3}$. Thus, $\triangle SER$ is the image of $\triangle HOR$ under the half-turn about *R* followed by the size transformation described above.   **12.** Rotate $\triangle PIG$ 180° (half-turn) about the midpoint of $\overline{PT}$, then perform a size transformation with scale factor 2 and center $P'(= T)$.   **13. (a)** $A(^-3, 12.91)$, $B(^-8, 0.07)$, $C(1.83, 5)$ **(b)** under the translation $(x, y) \rightarrow (x - 3, y + 5)$ **14.** $(x, y) \rightarrow (x, y)$. It is the identity transformation with every point its own image.   **15. (a)** $A(^-3, ^-2)$, $B(^-2, 3)$, $C(^-5, 5)$ **(b)** $(x, y) \rightarrow (x + 3, y + 2)$ **16. (a) (i)** a translation from *A* to *C*   **(ii)** same as (i)   **(b)** a rotation about *O* by 60° counterclockwise.   **(c) (i)** a size transformation with center *O* and scale factor 6   **(ii)** same as (i)   **17. (a)** Square. A square has four lines of symmetry, a 90° rotational symmetry about its center, which implies that it has a 180° rotational symmetry (i.e., point symmetry) as well as 270° rotational symmetry. A rhombus has only two lines of symmetry and a 180° rotational symmetry (point symmetry). **(b)** Equally symmetrical. An isosceles trapezoid has one line of symmetry (the line through the midpoints of the bases). A parallelogram has one point symmetry (a 180° rotational symmetry about the point formed by the intersection of the diagonals).

18. (a) $y = 2x + 5$ **(b)** $x = {}^-1$ **(c)** $y = {}^-\frac{1}{2}x + \frac{5}{2}$

19. (a) $y = {}^-x + 2$ **(b)** $y = x - 3$ **(c)** $y = x + 3$
(d) $y = {}^-x + 3$ **(e)** $y = {}^-x - 3$ **(f)** $y = {}^-x + 6$
20. (a) $(x, y) \rightarrow (x - 3, y + 5)$ **(b)** the identity translation: $(x, y) \rightarrow (x, y)$ **21.** The measure of each exterior angle of a regular octagon is $\frac{360°}{8}$, or 45°. Hence the measure of each interior angle is $180° - 45°$, or 135°. Because $135 \nmid 360$, a regular octagon does not tessellate the plane. **22.** no **23.** Yes, but unless the shape fits exactly into the object being measured, it may be difficult.

Answers to Now Try This

14-1. (a) One way to do this is to connect A and M. Draw a line parallel to line MN through A; find a point A' such that $MN = AA'$. **(b)** The direction of $\overrightarrow{AA'}$ must be the same as that of the direction of $\overrightarrow{MN}$. **14-2.** Draw a circle with center O and radius OP. Next draw the same size circle with center as vertex of the angle given. Use your compass to measure the arc of the circle cut off by the angle. On circle O, mark the image P' of P. **14-3.** Hint: Construct a perpendicular from P to line m. Find P' so that line m is the perpendicular bisector of $\overline{PP'}$. Points P and P' are the endpoints of a diagonal of a rhombus. The other diagonal lies along line m. **14-4.** Find the line such that the figure folds onto the image. The fold line is the reflecting line. **14-5.** The accompanying figure suggests the reason why the image of $P(a, b)$ is $P'(b, a)$. Notice that the line $y = x$ bisects the right angle in the first quadrant formed by the x- and y-axes. Because $y = x$ is the perpendicular bisector of $\overline{AA'}$, $OA = OA'$. Thus $\triangle OAM \cong \triangle OA'M$. Consequently the marked angles with vertex O are congruent. Now it is easy to show that $\angle BOA \cong \angle B'OA'$ and consequently that $\triangle OBA \cong \triangle OB'A'$, which in turn implies that $OB' = b$ and $A'B' = a$. **14-6.** $MB = MB'$ and $BP = B'P$. Points A, P, and B' make a triangle. Therefore, they are not all on the same line and by the triangle inequality we know that $AP + PB'$ must be longer than AB'. A, M, and B' are collinear. Therefore, $AM + MB < AP + PB$.
14-7. Answers vary, but a square will do. **14-8.** A 180-sided regular polygon has the desired symmetry.
14-9. (a) 120 degrees. 4 axes of rotation **(b)** 180 degrees. 6 axes of rotation **14-10.** Answers will vary.
14-11. Continue to create the shapes and rotate them to form the tessellation.

Answers to Brain Teasers

Section 14-1

It will be as it now is—that is, upside down.

Section 14-2

Translate B towards A in the direction perpendicular to the banks of the river a distance equal to the width of the river. Connect A with the image B'. The point P where $\overline{AB'}$ intersects the far bank is the point where the bridge should be built.

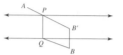

Section 14-4

In the drawing below, $\triangle AHT \cong \triangle AET$ and $\triangle DEN \cong \triangle DBN$. (Why?) Thus, $HT = ET$, $EN = BN$, and $HT + TE + EN + NB = 2(TE + EN)$. The image length HB is twice the length of the body because $TN = AD$.

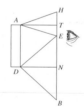

Section 14-5

In figure (a) $\triangle BT'A'$ is the image of $\triangle BTA$ under rotation clockwise by 60° about B. Because $\angle TBT'$ is 60° and $BT' = BT$, $\triangle TBT'$ must be equilateral (the base angles at T and are congruent, their sum is and hence each is 60°). Thus

$$CT + BT + AT = CT + TT' + T'A'$$

Consequently, for each point T in the interior of the triangle the sum of the distances to the three cities is the length of the path $C - T - T' - A'$. The shortest path is along a straight line connecting C to A' (see Figure (b)). The location of the required point T for which the sum of the distances is minimal is on line CA' and can be determined by the fact that $\triangle TBT'$ is equilateral and thus $\angle BTA'$ is 60°. Hence we need only construct through B a line that makes 60° angle with line CA' (see figure (b)). For that purpose, we construct an equilateral triangle DEF with vertex D anywhere on line DA' so that F is on line CA'. Through B we draw a line parallel to line DE (see figure (b)). The point of intersection of that line with line CA' is the required point T.

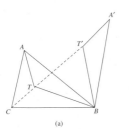
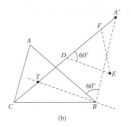

(a)   (b)

Answer for Laboratory Activities

Section 14-5
Answers will depend upon the type of patterns chosen. Each group should tessellate the plane.

Answers to Technology Corner

Section 14-1

Page 938 Answer to GSP Lab 11   **7. (a)** congruent
(b) yes **(c)** no   **8. (a)** The segments are congruent.
(c) yes

Page 943

- The object and its rotation image have the same orientation.
- The object and image are congruent and the object and its image are the same figure.
- An equilateral triangle has 120° rotational symmetry. Other triangles do not have this property.

Section 14-2

Answer to GSP Lab 11
5. (a) The measure of an angle and the measure of its reflected angle are equal.   **(b)** The length of a segment and the length of its reflected segment are equal.   **(c)** A polygon and its reflection are congruent.   **(d)** The orientations of an object and its reflection are opposite.
(e) The reflecting line contains the midpoint of each segment joining a point and its reflection.

Answer to GSP Lab 12
1. (i) It might be called a "slide" reflection because it is a composition of a "slide" and a reflection. Glide is used instead of "slide."   **(j)** They lie on the reflecting line.

Section 14-3

It is always possible to find such a line. The line, x, needed is such that the angle formed by lines m and n in that order is the same as the angle formed by the lines x and p in that order.

Answer to the Preliminary Problem

For a figure to have 80° rotational symmetry, the figure must be such that it can be turned about its turn center to fit onto itself. Consider a regular 360-gon. Each angle with vertex at its "center" has measure 1°, so by turning it 1°, 80 times in succession, it will fit onto itself and have 80° rotational symmetry. Similarly, any regular n-gon whose central angle has as measure a factor of 80 might have 80° rotational symmetry. The factors of 80 are 1, 2, 4, 8, 16, 5, 10, 20, 40, and 80. So as we have seen, a regular 360-gon has 80° rotational symmetry. We start a table of n-gons that work:

Measure of central angle	n-gon (regular)
1°	360-gon
2°	180-gon
4°	90-gon
8°	45-gon
16°	Does not work because the factor of 80 must divide 360 as well.
5°	72-gon
10°	36-gon
20°	18-gon
40°	nonagon
80°	Does not work (see 16°).

But are these the only regular n-gons that work? The answer is no. Consider a regular 720-gon. Its central angle has measure $\frac{1°}{2}$, and rotating it 160 times gives 80° turn symmetry. Similarly, a 1440-gon with $\frac{1°}{4}$ central angle can be rotated 320 times to have 80° symmetry. In this manner, we can see that there are infinitely many regular n-gons with 80° rotational symmetry. Thus, a central angle measure of the form $\frac{k}{n}$ degrees, where $\frac{k}{n}$ is reduced, k is a divisor of 80 and 360, and n is a positive integer, determines the polygon.

INDEX